Coulson & Richardson's

CHEMICAL ENGINEERING

VOLUME 1

Fluid Flow, Heat Transfer and Mass Transfer

Coulson & Richardson's Chemical Engineering

Chemical Engineering, Volume 1, Sixth edition
Fluid Flow, Heat Transfer and Mass Transfer
J. M. Coulson and J. F. Richardson
with J. R. Backhurst and J. H. Harker

Chemical Engineering, Volume 2, Fourth edition
Particle Technology and Separation Processes
J. M. Coulson and J. F. Richardson
with J. R. Backhurst and J. H. Harker

Chemical Engineering, Volume 3, Third edition
Chemical & Biochemical Reactors & Process Control
Edited by J. F. Richardson and D. G. Peacock

Chemical Engineering
Solutions to the Problems in Volume 1
J. R. Backhurst, J. H. Harker and J. F. Richardson

Chemical Engineering
Solutions to the Problems in Volumes 2 and 3
J. R. Backhurst, J. H. Harker and J. F. Richardson

Chemical Engineering, Volume 6, Third edition
Chemical Engineering Design
R. K. Sinnott

Coulson & Richardson's

CHEMICAL ENGINEERING

VOLUME 1

SIXTH EDITION

Fluid Flow, Heat Transfer and Mass Transfer

J. M. COULSON

Late Emeritus Professor of Chemical Engineering
University of Newcastle-upon-Tyne

and

J. F. RICHARDSON

Department of Chemical and Biological Process Engineering
University of Wales, Swansea

WITH

J. R. BACKHURST and J. H. HARKER

Department of Chemical and Process Engineering
University of Newcastle-upon-Tyne

ELSEVIER
BUTTERWORTH
HEINEMANN

AMSTERDAM • BOSTON • HEIDELBERG • LONDON • NEW YORK • OXFORD
PARIS • SAN DIEGO • SAN FRANCISCO • SINGAPORE • SYDNEY • TOKYO

Butterworth-Heinemann is an imprint of Elsevier
Linacre House, Jordan Hill, Oxford OX2 8DP, UK
30 Corporate Drive, Suite 400, Burlington, MA 01803, USA

First published by Pergamon Press 1954
Second edition 1964
Third edition 1977
Fourth edition 1990
Fifth edition 1996
Fifth edition (revised) 1997, 1999
Sixth edition 1999
Reprinted 2000, 2003, 2004, 2005, 2007, 2008

British Library Cataloguing in Publication Data
A catalogue record for this book is available from the British Library

Library of Congress Cataloging-in-Publication Data
A catalog record for this book is available from the Library of Congress

ISBN: 978-0-7506-4444-0

Printed and bound in *Great Britain*

08 09 10 10 9 8 7

Contents

Professor J. M. Coulson

JOHN COULSON, who died on 6 January 1990 at the age of 79, came from a family with close involvement with education. Both he and his twin brother Charles (renowned physicist and mathematician), who predeceased him, became professors. John did his undergraduate studies at Cambridge and then moved to Imperial College where he took the postgraduate course in chemical engineering — the normal way to qualify at that time — and then carried out research on the flow of fluids through packed beds. He then became an Assistant Lecturer at Imperial College and, after war-time service in the Royal Ordnance Factories, returned as Lecturer and was subsequently promoted to a Readership. At Imperial College he initially had to run the final year of the undergraduate course almost single-handed, a very demanding assignment. During this period he collaborated with Sir Frederick (Ned) Warner to write a model design exercise for the I. Chem. E. Home Paper on "The Manufacture of Nitrotoluene". He published research papers on heat transfer and evaporation, on distillation, and on liquid extraction, and co-authored this textbook of Chemical Engineering. He did valiant work for the Institution of Chemical Engineers which awarded him its Davis medal in 1973, and was also a member of the Advisory Board for what was then a new Pergamon journal, *Chemical Engineering Science*.

In 1954 he was appointed to the newly established Chair at Newcastle-upon-Tyne, where Chemical Engineering became a separate Department and independent of Mechanical Engineering of which it was formerly part, and remained there until his retirement in 1975. He took a period of secondment to Heriot Watt University where, following the splitting of the joint Department of Chemical Engineering with Edinburgh, he acted as adviser and *de facto* Head of Department. The Scottish university awarded him an Honorary D.Sc. in 1973.

John's first wife Dora sadly died in 1961; they had two sons, Anthony and Simon. He remarried in 1965 and is survived by Christine.

JFR

Preface to Sixth Edition

It is somewhat sobering to realise that the sixth edition of Volume 1 appears 45 years after the publication of the first edition in 1954. Over the intervening period, there have been considerable advances in both the underlying theory and the practical applications of Chemical Engineering; all of which are reflected in parallel developments in undergraduate courses. In successive editions, we have attempted to adapt the scope and depth of treatment in the text to meet the changes in the needs of both students and practitioners of the subject.

Volume 1 continues to concentrate on the basic processes of Momentum Transfer (as in fluid flow), Heat Transfer, and Mass Transfer, and it is also includes examples of practical applications of these topics in areas of commercial interest such as the pumping of fluids, the design of shell and tube heat exchangers and the operation and performance of cooling towers. In response to the many requests from readers (and the occasional note of encouragement from our reviewers), additional examples and their solutions have now been included in the main text. The principal areas of application, particularly of the theories of Mass Transfer across a phase boundary, form the core material of Volume 2 however, whilst in Volume 6, material presented in other volumes is utilised in the practical design of process plant.

The more important additions and modifications which have been introduced into this sixth edition of Volume 1 are:

Dimensionless Analysis. The idea and advantages of treating length as a vector quantity and of distinguishing between the separate role of mass in representing a quantity of matter as opposed to its inertia are introduced.

Fluid Flow. The treatment of the behaviour of non-Newtonian fluids is extended and the methods used for pumping and metering of such fluids are updated.

Heat Transfer. A more detailed discussion of the problem of unsteady-state heat transfer by conduction where bodies of various shapes are heated or cooled is offered together with a more complete treatment of heat transfer by radiation and a re-orientation of the introduction to the design of shell and tube heat exchangers.

Mass Transfer. The section on mass transfer accompanied by chemical reaction has been considerably expanded and it is hoped that this will provide a good basis for the understanding of the operation of both homogeneous and heterogeneous catalytic reactions.

As ever, we are grateful for a great deal of help in the preparation of this new edition from a number of people. In particular, we should like to thank Dr. D.G. Peacock for the great enthusiasm and dedication he has shown in the production of the Index, a task he has undertaken for us over many years. We would also mention especially Dr. R.P. Chhabra of the Indian Institute of Technology at Kanpur for his contribution on unsteady-state heat transfer by conduction, those commercial organisations which have so generously contributed new figures and diagrams of equipment, our publishers who cope with our perhaps overwhelming number of suggestions and alterations with a never-failing patience and, most of all, our readers who with great kindness, make so many extremely useful and helpful suggestions all of which, are incorporated wherever practicable. With their continued help and support, the signs are that this present work will continue to be of real value as we move into the new Millenium.

Swansea, 1999
Newcastle upon Tyne, 1999.

J.F. RICHARDSON
J.R. BACKHURST
J.H. HARKER

Preface to Fifth Edition

This textbook has been the subject of continual up-dating since it was first published in 1954. There have been numerous revised impressions and the opportunity has been taken on the occasion of each reprinting to make corrections and revisions, many of them in response to readers who have kindly pointed out errors or who have suggested modifications and additions. When the summation of the changes has reached a sufficiently high level, a new edition has been produced. We have now reached this point again and the fifth edition incorporates all the alterations in the 1993 revision of the fourth edition, together with new material, particularly on simultaneous mass transfer and chemical reaction for unsteady-state processes.

There have been changes in publisher too. Since the appearance of the fourth edition in 1990, Pergamon Press has become part of Elsevier Science and now, following a re-organisation in the Reed Elsevier group of companies, the responsibility for publishing the Chemical Engineering series has passed to Butterworth-Heinemann, another Reed Elsevier company.

We are grateful to our readers for their interest and very much hope they will continue to make suggestions for the improvement of the series.

JFR

Preface to Fourth Edition

THE First Edition of Volume 1 was published in 1954 and Volume 2 appeared a year later. In the intervening 35 years or so, there have been far-reaching developments in Chemical Engineering and the whole approach to the subject has undergone a number of fundamental changes. The question therefore arises as to whether it is feasible to update a textbook written to meet the needs of the final year students of an undergraduate course in the 1950s so that it can continue to fulfil a useful purpose in the last decade of the century. Perhaps it would have been better if a new textbook had been written by an entirely new set of authors. Although at one stage this had seemed likely through the sponsorship of the Institution of Chemical Engineers, there is now no sign of any such replacement book appearing in the United Kingdom.

In producing the Fourth Edition, it has been necessary to consider whether to start again with a clean sheet of paper — an impossibly daunting task — or whether to retain the original basic structure with relatively small modifications. In following the latter course, the authors were guided by the results of a questionnaire sent to a wide range of University (and Polytechnic) Departments throughout the English-speaking-world. The clear message which came back was "Do not tamper over-much with the devil we know, in spite of all his faults!"

It was in 1971 that Volume 3 was added to the series, essentially to make good some of the more glaring omissions in the earlier volumes. Volume 3 contains a series of seven specialist chapters written by members of the staff of the Chemical Engineering Department at the University College of Swansea, with Dr D. G. Peacock of the School of Pharmacy, London as a joint editor. In 1977/9, as well as contributing significantly to the new editions of Volumes 1 and 2, two colleagues at the University of Newcastle-upon-Tyne, Dr J. R. Backhurst and The Revd. Dr J. H. Harker, prepared Volumes 4 and 5, the solutions to the problems in Volumes 1 and 2, respectively. The final major development was the publication of Volume 6 on Chemical Engineering Design by Mr R. K. Sinnott in 1983. With the preparation of a Fourth Edition, the opportunity has presented itself for a degree of rationalisation, without introducing major changes to the structure. This has led to the following format:

Volume 1	Fluid Flow, Heat Transfer and Mass Transfer
Volume 2	Particle Technology and Separation Processes
Volume 3	Chemical and Biochemical Reactor Engineering and Control
Volume 4/5	Solutions to the Problems in Volumes 1, 2 and 3
Volume 6	Chemical Engineering Design

The details of this new arrangement are as follows:

1. Volume 1 has acquired an abbreviated treatment of non-Newtonian Flow, formerly in Volume 3.

2. Liquid Mixing appears as a new Chapter in Volume 1, which incorporates the relevant material formerly in Volumes 2 and 3.
3. Separate chapters now appear in Volume 1 on Compressible Flow and on Multiphase Flow, the latter absorbing material previously scattered between Volumes 1 and 2.
4. New chapters are added to Volume 2 to cover four separation processes of increasing importance — Adsorption (from Volume 3), Ion Exchange, Chromatographic Separations and Membrane Separations.
5. Volume 3 is now devoted to various aspects of Reaction Engineering and Control, material which is considerably expanded.
6. Some aspects of design, previously in the earlier volumes, are now transferred to a more appropriate home in Volume 6.

As far as Volume 1 is concerned, the opportunity has been taken to update existing material. The major changes in Fluid Flow include the incorporation of non-Newtonian Flow, an extensive revision of Compressible Flow and the new chapters on Multiphase Flow and Liquid Mixing. Material for this last chapter has been contributed by Dr R P Chhabra of the Indian Institute of Technology at Kanpur. There has also been a substantial revision of the presentation of material on Mass Transfer and Momentum, Heat and Mass Transfer. To the Appendix have been added the Tables of Laplace Transform and Error Functions which were formerly in Volume 3, and throughout this new edition, all the diagrams have been redrawn. Some further problems have been added at the end.

Sadly, John Coulson was not able to contribute as he had done previously and his death in January 1990 leaves us with a gap which is difficult to fill. John Backhurst and John Harker, who made a substantial contribution to the preparation of the Third Edition in 1977, have taken an increased share of the burden of revising the book and contributing new material, and have taken a special responsibility for those sections which originated from John Coulson, in addition to the special task of up-dating the illustrations. Without their continued support and willing co-operation there would have been no Fourth Edition.

Finally, we would all like to thank our many readers who have made such helpful suggestions in the past and have pointed out errors, many of which the authors would never have spotted. It is hoped that readers will continue to act in this way as unseen authors.

June 1990 JFR

NOTE TO REVISED REPRINT 1993

The reprint incorporates corrections and minor amendments, but the opportunity has been taken to effect some re-arrangements and additions, as follows:

Chapter 5 Multiphase Flow

Revision of section on pneumatic conveying.

Chapter 9 Heat Transfer

(i) Re-arrangement of material on plasmas and transfer to particles.
(ii) Inclusion of an Example on lagging.

Chapter 10 Mass Transfer

(i) Re-arrangement at the beginning to give a more logical sequence.
(ii) Revision of final section on practical results.

Chapter 11 Boundary Layer

Addition of section on flow with constant heat flux at surface.

Steam Tables

These have been recast to facilitate interpolation.

Preface to Third Edition

THE introduction of the SI system of units by the United Kingdom and many other countries has itself necessitated the revision of this engineering text. This clear implementation of a single system of units will be welcomed not only by those already in the engineering profession, but even more so by those who are about to join. The system which is based on the c.g.s. and m.k.s. systems using length (**L**), mass (**M**), and time (**T**) as the three basic dimensions, as is the practice in the physical sciences, has the very great advantage that it removes any possible confusion between mass and force which arises in the engineering system from the common use of the term *pound* for both quantities. We have therefore presented the text, problems, and examples in the SI system, but have arranged the tables of physical data in the Appendix to include both SI and other systems wherever possible. This we regard as important because so many of the physical data have been published in c.g.s. units. For similar reasons, engineering units have been retained as an alternative where appropriate.

In addition to the change to the SI system of units, we have taken the opportunity to update and to clarify the text. A new section on the flow of two-phase gas–liquid mixtures has been added to reflect the increased interest in the gas and petroleum industries and in its application to the boiling of liquids in vertical tubes.

The chapter on Mass Transfer, the subject which is so central and specific to chemical engineering, has been considerably extended and modernised. Here we have thought it important in presenting some of the theoretical work to stress its tentative nature and to show that, although some of the theories may often lack a full scientific basis, they provide the basis of a workable technique for solving problems. In the discussion on Fluid Flow reference has been made to American methods, and the emphasis on Flow Measurement has been slanted more to the use of instruments as part of a control system. We have emphasised the importance of pipe-flow networks which represent a substantial cost item in modern large-scale enterprises.

This text covers the physical basis of the three major transfer operations of fluid flow, heat transfer, and mass transfer. We feel that it is necessary to provide a thorough grounding in these operations before introducing techniques which have been developed to give workable solutions in the most convenient manner for practical application. At the same time, we have directed the attention of the reader to such invaluable design codes as TEMA and the British Standards for heat exchanger design and to other manuals for pipe-flow systems.

It is important for designers always to have in their minds the need for reliability and safety: this is likely to follow from an understanding of the basic principles involved, many of which are brought out in the text.

We would like to thank our many friends from several countries who have written with suggestions, and it is our hope that this edition will help in furthering growth and interest in the profession. We should also like to thank a number of industrialists who have made available much useful information for incorporation in this edition; this help is acknowledged at the appropriate point. Our particular thanks are due to Dr B. Waldie for his contribution to the high temperature aspects of heat transfer and to the Kellogg International Corporation and Humphreys and Glasgow Limited for their help. In conclusion, we would like to thank Dr J. R. Backhurst and Dr J. H. Harker for their editorial work and for recalculating the problems in SI units and converting the charts and tables.

Since the publication of the Second Edition of this Volume, Volume 3 of *Chemical Engineering* has been published in order to give a more complete coverage of those areas of chemical engineering which are of importance in both universities and industry in the 1970's.

January 1976

JMC
JFR

Preface to Second Edition

In presenting this second edition, we should like to thank our many friends from various parts of the world who have so kindly made suggestions for clarifying parts of the text and for additions which they have felt to be important. During the last eight years there have been changes in the general approach to chemical engineering in the universities with a shift in emphasis towards the physical mechanisms of transport processes and with a greater interest in unsteady state conditions. We have taken this opportunity to strengthen those sections dealing with the mechanisms of processes, particularly in Chapter 7 on mass transfer and in the chapters on fluid mechanics where we have laid greater emphasis on the use of momentum exchange. Many chemical engineers are primarily concerned with the practical design of plant and we have tried to include a little more material of use in this field in Chapter 6 on heat transfer. An introductory section on dimensional analysis has been added but it has been possible to do no more than outline the possibilities opened up by the use of this technique. Small changes will be found throughout the text and we have tried to meet many readers' requests by adding some more worked examples and a further selection of problems for the student. The selection of material and its arrangement are becoming more difficult and must be to a great extent a matter of personal choice, but we hope that this new edition will provide a sound basis for the study of the fundamentals of the subject and will perhaps be of some value to practising engineers.

J. M. Coulson
J. F. Richardson

Preface to First Edition

THE idea of treating the various processes of the chemical industry as a series of unit operations was first brought out as a basis for a new technology by Walker, Lewis and McAdams in their book in 1923. Before this, the engineering of chemical plants had been regarded as individual to an industry and there was little common ground between one industry and another. Since the early 1920s chemical engineering as a separate subject has been introduced into the universities of both America and England and has expanded considerably in recent years so that there are now a number of university courses in both countries. During the past twenty years the subject matter has been extensively increased by various researches described in a number of technical journals to which frequent reference is made in the present work.

Despite the increased attention given to the subject there are few general books, although there have been a number of specialised books on certain sections such as distillation, heat transfer, etc. It is the purpose of the present work to present to the student an account of the fundamentals of the subject. The physical basis of the mechanisms of many of the chemical engineering operations forms a major feature of chemical engineering technology. Before tackling the individual operations it is important to stress the general mechanisms which are found in so many of the operations. We have therefore divided the subject matter into two volumes, the first of which contains an account of these fundamentals—diffusion, fluid flow and heat transfer. In Volume 2 we shall show how these theoretical foundations are applied in the design of individual units such as distillation columns, filters, crystallisers, evaporators, etc.

Volume 1 is divided into four sections, fluid flow, heat transfer, mass transfer and humidification. Since the chemical engineer must handle fluids of all kinds, including compressible gases at high pressures, we believe that it is a good plan to consider the problem from a thermodynamic aspect and to derive general equations for flow which can be used in a wide range of circumstances. We have paid special attention to showing how the boundary layer is developed over plane surfaces and in pipes, since it is so important in controlling heat and mass transfer. At the same time we have included a chapter on pumping since chemical engineering is an essentially practical subject, and the normal engineering texts do not cover the problem as experienced in the chemical and petroleum industries.

The chapter on heat transfer contains an account of the generally accepted techniques for calculation of film transfer coefficients for a wide range of conditions, and includes a section on the general construction of tubular exchangers which form a major feature of many works. The possibilities of the newer plate type units are indicated.

In section three, the chapter on mass transfer introduces the mechanism of diffusion and this is followed by an account of the common relationships between heat, mass and momentum transfer and the elementary boundary layer theory. The final section includes the practical problem of humidification where both heat and mass transfer are taking place simultaneously.

It will be seen that in all chapters there are sections in small print. In a subject such as this, which ranges from very theoretical and idealised systems to the practical problems with empirical or experimentally determined relations, there is much to be said for omitting the more theoretical features in a first reading, and in fact this is frequently done in the more practical courses. For this reason the more difficult theoretical sections have been put in small print and the whole of Chapter 9 may be omitted by those who are more concerned with the practical utility of the subject.

In many of the derivations we have given the mathematical analysis in more detail than is customary. It is our experience that the mathematical treatment should be given in full and that the student should then apply similar analysis to a variety of problems.

We have introduced into each chapter a number of worked examples which we believe are essential to a proper understanding of the methods of treatment given in the text. It is very desirable for a student to understand a worked example before tackling fresh practical problems himself. Chemical engineering problems require a numerical answer and it is essential to become familiar with the different techniques so that the answer is obtained by systematic methods rather than by intuition.

In preparing this text we have been guided by courses of lectures which we have given over a period of years and have presented an account of the subject with the major emphasis on the theoretical side. With a subject that has grown so rapidly, and which extends from the physical sciences to practical techniques, the choice of material must be a matter of personal selection. It is, however, more important to give the principles than the practice, which is best acquired in the factory. We hope that the text may also prove useful to those in industry who, whilst perhaps successfully employing empirical relationships, feel that they would like to find the extent to which the fundamentals are of help.

We should like to take this opportunity of thanking a number of friends who have helped by their criticism and suggestions, amongst whom we are particularly indebted to Mr F. E. Warner, to Dr M. Guter, to Dr D. J. Rasbash and to Dr L. L. Katan. We are also indebted to a number of companies who have kindly permitted us to use illustrations of their equipment. We have given a number of references to technical journals and we are grateful to the publishers for permission to use illustrations from their works. In particular we would thank the Institution of Chemical Engineers, the American Institute of Chemical Engineers, the American Chemical Society, the Oxford University Press and the McGraw-Hill Book Company.

South Kensington,
London S.W.7
1953

Acknowledgements

THE authors and publishers acknowledge the kind assistance of the following organisations in providing illustrative material.

Figure 6.6*a*, 6.6*b* Bowden Sedema Ltd.
Figures 6.7, 6.6*c* Foxboro Great Britain Ltd, Redhill, Surrey.
Figures 6.17, Kent Meters Ltd.
Figure 6.21 Rotameter Manufacturing Co Ltd.
Figure 6.27 Endress and Hauser.
Figure 6.29 Baird and Tatlock Ltd.
Figures 7.26, 7.27, 7.28 Chemineer Ltd, Derby.
Figure 7.29 Sulzer Ltd, Switzerland.
Figure 8.1 Worthington-Simpson Ltd.
Figure 8.3 ECD Co Ltd.
Figure 8.9 Varley Pumps and Engineering Ltd.
Figure 8.11 Watson-Marlow Ltd.
Figures 8.12, 8.13, 8.14 Mono Pumps Ltd, Manchester.
Figures 8.19, 8.20 Sigmund Pumps Ltd.
Figure 8.29 Crane Packing Ltd, Wolverhampton, W. Midlands.
Figure 8.31 International Combustion Ltd, Derby.
Figure 8.36 Nash Europe, Manchester.
Figures 8.37, 8.38, Reavell and Co Ltd.
Figures 8.45, 8.46 Hick Hargreaves and Co Ltd.
Figure 9.65 W. J. Fraser and Co
Figure 9.86*a* Brown Fintube Co.
Figure 9.86*b* G. A. Harvey and Co Ltd.
Figure 9.89*a* Alfa Laval Ltd. Brentford, Middlesex.
Figure 9.89*b* APV Ltd.
Figure 9.90 Ashmore, Benson, Pease and Co Ltd.
Figures 11.2, 11.3 Professor F. N. M. Brown, University of Notre Dame.
Figure 13.10*a b d* Casella London Ltd, Bedford.
Figure 13.10*c* Protimeter plc, Marlow, UK.
Figures 13.12, 13.13 Visco Ltd, Croydon, Surrey.
Figure 13.14 Davenport Engineering.

CHAPTER 1

Units and Dimensions

1.1. INTRODUCTION

Students of chemical engineering soon discover that the data used are expressed in a great variety of different units, so that quantities must be converted into a common system before proceeding with calculations. Standardisation has been largely achieved with the introduction of the *Système International d'Unités* (SI)[1,2] to be discussed later, which is used throughout all the Volumes of this series of books. This system is now in general use in Europe and is rapidly being adopted throughout the rest of the world, including the USA where the initial inertia is now being overcome. Most of the physical properties determined in the laboratory will originally have been expressed in the cgs system, whereas the dimensions of the full-scale plant, its throughput, design, and operating characteristics appear either in some form of general engineering units or in special units which have their origin in the history of the particular industry. This inconsistency is quite unavoidable and is a reflection of the fact that chemical engineering has in many cases developed as a synthesis of scientific knowledge and practical experience. Familiarity with the various systems of units and an ability to convert from one to another are therefore essential, as it will frequently be necessary to access literature in which the SI system has not been used. In this chapter the main systems of units are discussed, and the importance of understanding dimensions emphasised. It is shown how dimensions can be used to help very considerably in the formulation of relationships between large numbers of parameters.

The magnitude of any physical quantity is expressed as the product of two quantities; one is the magnitude of the unit and the other is the number of those units. Thus the distance between two points may be expressed as 1 m or as 100 cm or as 3.28 ft. The metre, centimetre, and foot are respectively the size of the units, and 1, 100, and 3.28 are the corresponding numbers of units.

Since the physical properties of a system are interconnected by a series of mechanical and physical laws, it is convenient to regard certain quantities as basic and other quantities as derived. The choice of basic dimensions varies from one system to another although it is usual to take length and time as fundamental. These quantities are denoted by **L** and **T**. The dimensions of velocity, which is a rate of increase of distance with time, may be written as $\mathbf{LT}^{-1}$, and those of acceleration, the rate of increase of velocity, are $\mathbf{LT}^{-2}$. An area has dimensions $\mathbf{L}^2$ and a volume has the dimensions $\mathbf{L}^3$.

The volume of a body does not completely define the amount of material which it contains, and therefore it is usual to define a third basic quantity, the amount of matter in the body, that is its mass **M**. Thus the density of the material, its mass per unit volume, has the dimensions $\mathbf{ML}^{-3}$. However, in the British Engineering System (Section 1.2.4) force **F** is used as the third fundamental and mass then becomes a derived dimension.

Physical and mechanical laws provide a further set of relations between dimensions. The most important of these is that the force required to produce a given acceleration of a body is proportional to its mass and, similarly, the acceleration imparted to a body is proportional to the applied force.

Thus force is proportional to the product of mass and acceleration (Newton's law), or:

$$\mathbf{F} = \text{const}\ \mathbf{M}(\mathbf{LT}^{-2}) \tag{1.1}$$

The proportionality constant therefore has the dimensions:

$$\frac{\mathbf{F}}{\mathbf{M}(\mathbf{LT}^{-2})} = \mathbf{FM}^{-1}\mathbf{L}^{-1}\mathbf{T}^{2} \tag{1.2}$$

In any set of *consistent* or *coherent* units the proportionality constant in equation 1.1 is put equal to unity, and unit force is that force which will impart unit acceleration to unit mass. Provided that no other relationship between force and mass is used, the constant may be arbitrarily regarded as dimensionless and the dimensional relationship:

$$\mathbf{F} = \mathbf{MLT}^{-2} \tag{1.3}$$

is obtained.

If, however, some other physical law were to be introduced so that, for instance, the attractive force between two bodies would be proportional to the product of their masses, then this relation between **F** and **M** would no longer hold. It should be noted that mass has essentially two connotations. First, it is a measure of the amount of material and appears in this role when the density of a fluid or solid is considered. Second, it is a measure of the inertia of the material when used, for example, in equations 1.1–1.3. Although mass is taken normally taken as the third fundamental quantity, as already mentioned, in some engineering systems force is used in place of mass which then becomes a derived unit.

1.2. SYSTEMS OF UNITS

Although in scientific work mass is taken as the third fundamental quantity and in engineering force is sometimes used as mentioned above, the fundamental quantities **L**, **M**, **F**, **T** may be used interchangeably. A summary of the various systems of units, and the quantities associated with them, is given in Table 1.1. In the cgs system which has historically been used for scientific work, metric units are employed. From this has been developed the mks system which employs larger units of mass and length (kilogram in place of gram, and metre in place of centimetre); this system has been favoured by electrical engineers because the fundamental and the practical electrical units (volt, ampere and ohm) are then identical. The SI system is essentially based on the mks system of units.

1.2.1. The centimetre–gram–second (cgs) system

In this system the basic units are of length **L**, mass **M**, and time **T** with the nomenclature:

Length:	Dimension **L**:	Unit 1 centimetre	(1 cm)
Mass:	Dimension **M**:	Unit 1 gram	(1 g)
Time:	Dimension **T**:	Unit 1 second	(1 s)

Table 1.1
Units

Quantity	cgs	SI	fps	Dimensions in M, L, T, θ	Engineering system	Dimensions F, L, T, θ	Dimensions in F, M, L, T, θ
Mass	gram	kilogram	pound	$\mathbf{M}$	slug	$\mathbf{FL^{-1}T^2}$	$\mathbf{M}$
Length	centimetre	metre	foot	$\mathbf{L}$	foot	$\mathbf{L}$	$\mathbf{L}$
Time	second	second	second	$\mathbf{T}$	second	$\mathbf{T}$	$\mathbf{T}$
Force	dyne	Newton	poundal	$\mathbf{MLT^{-2}}$	pound force	$\mathbf{F}$	$\mathbf{F}$
Energy	erg (= 10^{-7} joules)	Joule	foot-poundal	$\mathbf{ML^2T^{-2}}$	foot-pound	$\mathbf{FL}$	$\mathbf{FL}$
Pressure	dyne/square centimetre	Newton/sq metre	poundal/square foot	$\mathbf{ML^{-1}T^{-2}}$	pound force/square foot	$\mathbf{FL^{-2}}$	$\mathbf{FL^{-2}}$
Power	erg/second	Watt	foot-poundal/second	$\mathbf{ML^2T^{-3}}$	foot-pound/second	$\mathbf{FLT^{-1}}$	$\mathbf{FLT^{-1}}$
Entropy per unit mass	erg/gram °C	Joule/kilogram K	foot-poundal/pound °C	$\mathbf{L^2T^{-2}\theta^{-1}}$	foot-pound/slug °F	$\mathbf{L^2T^{-2}\theta^{-1}}$	$\mathbf{FM^{-1}L\theta^{-1}}$
Universal gas constant	8.314×10^7 erg/mole °C	8314 J/kmol K	8.94 ft-poundal/lb mol °C	$\mathbf{MN^{-1}L^2T^{-2}\theta^{-1}}$	4.96×10^4 foot-pound/slug mol °F	$\mathbf{MN^{-1}L^2T^{-2}\theta^{-1}}$	$\mathbf{FN^{-1}L\,\theta^{-1}}$

Heat units

Quantity	cgs	SI	British/American engineering system	Dimensions in M, L, T, θ	Dimensions in H, M, L, T, θ
Temperature	degree centigrade	degree Kelvin	degree Fahrenheit	$\boldsymbol{\theta}$	$\boldsymbol{\theta}$
Thermal energy or heat	calorie	joule	British thermal unit (Btu)	$\mathbf{M\theta}$	$\mathbf{H}$
Entropy per unit mass, specific heat	calorie/gram °C	joule/kilogram K	Btu/pound °F	—	$\mathbf{HM^{-1}\theta^{-1}}$
Mechanical equivalent of heat, *J*	4.18×10^7 erg/gram-°C	1 J (heat energy) = 1 J (mechanical energy)	2.50×10^4 foot-poundal/pound °F	$\mathbf{L^2T^{-2}\theta^{-1}}$	$\mathbf{H^{-1}ML^2\,T^{-2}}$
Universal gas constant **R**	1.986 calorie/mole °C	8314 J/kmol K	1.986 Btu/lb-mol °F	$\mathbf{MN^{-1}L^2T^{-2}\theta^{-1}}$	$\mathbf{HN^{-1}\theta^{-1}}$

The unit of force is that force which will give a mass of 1 g an acceleration of 1 cm/s^2 and is known as the dyne:

Force:	Dimension $\mathbf{F} = \mathbf{MLT}^{-2}$:	Unit	1 dyne (1 dyn)
Energy:	Dimensions $\mathbf{ML}^2\mathbf{T}^{-2}$	Unit	1 erg
Power:	Dimensions $\mathbf{ML}^2\mathbf{T}^{-3}$	Unit	1 erg/s

1.2.2. The metre–kilogram–second (mks system and the *Système International d'Unités* (SI)

These systems are in essence modifications of the cgs system but employ larger units. The basic dimensions are again of **L**, **M**, and **T**.

Length:	Dimension **L**:	Unit 1 metre	(1 m)
Mass:	Dimension **M**:	Unit 1 kilogram	(1 kg)
Time:	Dimension **T**:	Unit 1 second	(1 s)

The unit of force, known as the *Newton*, is that force which will give an acceleration of 1 m/s^2 to a mass of one kilogram. Thus 1 N = 1 kg m/s^2 with dimensions $\mathbf{MLT}^{-2}$, and one Newton equals 10^5 dynes. The energy unit, the Newton-metre, is 10^7 ergs and is called the *Joule*; and the power unit, equal to one Joule per second, is known as the *Watt*.

Thus:	Force:	Dimensions $\mathbf{MLT}^{-2}$:	Unit 1 Newton (1 N)	or 1 kg m/s^2
	Energy:	Dimensions $\mathbf{ML}^2\mathbf{T}^{-2}$:	Unit 1 Joule (1 J)	or 1 kg m^2/s^2
	Power:	Dimensions $\mathbf{ML}^2\mathbf{T}^{-3}$:	Unit 1 Watt (1 W)	or 1 kg m^2/s^3

For many purposes, the chosen unit in the SI system will be either too large or too small for practical purposes, and the following prefixes are adopted as standard. Multiples or sub-multiples in powers of 10^3 are preferred and thus, for example, millimetre should always be used in preference to centimetre.

10^{18}	exa	(E)	10^{-1}	deci	(d)
10^{15}	peta	(P)	10^{-2}	centi	(c)
10^{12}	tera	(T)	10^{-3}	milli	(m)
10^{9}	giga	(G)	10^{-6}	micro	(μ)
10^{6}	mega	(M)	10^{-9}	nano	(n)
10^{3}	kilo	(k)	10^{-12}	pico	(p)
10^{2}	hecto	(h)	10^{-15}	femto	(f)
10^{1}	deca	(da)	10^{-18}	alto	(a)

These prefixes should be used with great care and be written immediately adjacent to the unit to be qualified; furthermore only one prefix should be used at a time to precede a given unit. Thus, for example, 10^{-3} metre, which is one millimetre, is written 1 mm. 10^3 kg is written as 1 Mg, not as 1 kkg. This shows immediately that the name *kilogram* is an unsuitable one for the basic unit of mass and a new name may well be given to it in the future.

Some special terms are acceptable, and commonly used in the SI system and, for example, a mass of 10^3 kg (1 Mg) is called a *tonne* (t); and a pressure of 100 kN/m^2 is called a *bar*.

The most important practical difference between the mks and the SI systems lies in the units used for thermal energy (heat), and this topic is discussed in Secton 1.2.7.

A detailed account of the structure and implementation of the SI system is given in a publications of the British Standards Institution[(1)], and of Her Majesty's Stationery Office[(2)].

1.2.3. The foot–pound–second (fps) system

The basic units in this system are:

Length:	Dimension **L**:	Unit 1 foot	(1 ft)
Mass:	Dimension **M**:	Unit 1 pound	(1 lb)
Time:	Dimension **T**:	Unit 1 second	(1 s)

The unit of force gives that which a mass of 1 lb an acceleration of 1 ft/s^2 is known as the poundal (pdl).

The unit of energy (or work) is the foot-poundal, and the unit of power is the foot-poundal per second.

Thus:	Force	Dimensions $\mathbf{MLT}^{-2}$	Unit	1 poundal (1 pdl)
	Energy	Dimensions $\mathbf{ML}^2\mathbf{T}^{-2}$	Unit	1 ft-poundal
	Power	Dimensions $\mathbf{ML}^2\mathbf{T}^{-3}$	Unit	1 foot-poundal/s

1.2.4. The British engineering system

In an alternative form of the fps system (*Engineering system*) the units of length (ft) and time (s) are unchanged, but the third fundamental is a unit of force (**F**) instead of mass and is known as the pound force (lb_f). This is defined as the force which gives a mass of 1 lb an acceleration of 32.1740 ft/s^2, the "standard" value of the acceleration due to gravity. It is therefore a fixed quantity and must not be confused with the pound weight which is the force exerted by the earth's gravitational field on a mass of one pound and which varies from place to place as g varies. It will be noted therefore that the pound force and the pound weight have the same value only when g is 32.1740 ft^2/s.

The unit of mass in this system is known as the slug, and is the mass which is given an acceleration of 1 ft/s^2 by a one pound force:

$$1 \text{ slug} = 1 \text{ lb}_f \text{ ft}^{-1}\text{s}^2$$

Misunderstanding often arises from the fact that the pound which is the unit of mass in the fps system has the same name as the unit of force in the engineering system. To avoid confusion the pound mass should be written as lb or even lb_m and the unit of force always as lb_f.

It will be noted that:

$$1 \text{ slug} = 32.1740 \text{ lb mass} \quad \text{and} \quad 1 \text{ lb}_f = 32.1740 \text{ pdl}$$

To summarise:

The basic units are:

Length	Dimension **L**	Unit 1 foot (1 ft)
Force	Dimension **F**	Unit 1 pound-force (1 lb_f)
Time	Dimension **T**	Unit 1 second (1 s)

The derived units are:

Mass	Dimensions $\mathbf{FL^{-1}T^{-2}}$	Unit 1 slug (= 32.1740 pounds)
Energy	Dimensions **FL**	Unit 1 foot pound-force (1 ft lb_f)
Power	Dimensions $\mathbf{FLT^{-1}}$	Unit 1 foot-pound force/s (1 ft-lb_f/s)
		Note: 1 horsepower is defined as 550 ft-lb_f/s.

1.2.5. Non-coherent system employing pound mass and pound force simultaneously

Two units which have never been popular in the last two systems of units (Sections 1.2.3 and 1.2.4) are the poundal (for force) and the slug (for mass). As a result, many writers, particularly in America, use both the pound mass and pound force as basic units in the same equation because they are the units in common use. This is an essentially incoherent system and requires great care in its use. In this system a proportionality factor between force and mass is defined as g_c given by:

Force (in pounds force) =(mass in pounds) (acceleration in ft/s^2)/g_c

Thus in terms of dimensions: $$\mathbf{F} = (\mathbf{M})(\mathbf{LT^{-2}})/g_c \tag{1.4}$$

From equation 1.4, it is seen that g_c has the dimensions $\mathbf{F^{-1}MLT^{-2}}$ or, putting $\mathbf{F} = \mathbf{MLT^{-2}}$, it is seen to be dimensionless. Thus:

$$g_c = 32.1740 \text{ lb}_f/(\text{lb}_m \text{ft s}^{-2})$$

or: $$g_c = \frac{32.1740 \text{ ft s}^{-2}}{1 \text{ ft s}^{-2}} = 32.1740$$

i.e. g_c is a dimensionless quantity whose numerical value corresponds to the acceleration due to gravity expressed in the appropriate units.

(It should be noted that a force in the cgs system is sometimes expressed as a *gram force* and in the mks system as *kilogram force*, although this is not good practice. It should also be noted that the gram force = 980.665 dyne and the kilogram force = 9.80665 N)

1.2.6. Derived units

The three fundamental units of the SI and of the cgs systems are length, mass, and time. It has been shown that force can be regarded as having the dimensions of $\mathbf{MLT^{-2}}$, and the dimensions of many other parameters may be worked out in terms of the basic **MLT** system.

For example:

energy is given by the product of force and distance with dimensions $\mathbf{ML^2T^{-2}}$, and pressure is the force per unit area with dimensions $\mathbf{ML^{-1}T^{-2}}$.

viscosity is defined as the shear stress per unit velocity gradient with dimensions $(\mathbf{MLT}^{-2}/\mathbf{L}^2)/(\mathbf{LT}^{-1}/\mathbf{L}) = \mathbf{ML}^{-1}\mathbf{T}^{-1}$.

and kinematic viscosity is the viscosity divided by the density with dimensions $\mathbf{ML}^{-1}\mathbf{T}^{-1}/\mathbf{ML}^{-3} = \mathbf{L}^2\mathbf{T}^{-1}$.

The units, dimensions, and normal form of expression for these quantities in the SI system are:

Quantity	Unit	Dimensions	Units in kg, m, s
Force	Newton	$\mathbf{MLT}^{-2}$	1 kg m/s^2
Energy or work	Joule	$\mathbf{ML}^2\mathbf{T}^{-2}$	1 kg m^2/s^2 (= 1 N m = 1 J)
Power	Watt	$\mathbf{ML}^2\mathbf{T}^{-3}$	1 kg m^2/s^3 (= 1 J/s)
Pressure	Pascal	$\mathbf{ML}^{-1}\mathbf{T}^{-2}$	1 kg/m s^2 (= 1 N/m^2)
Viscosity	Pascal-second	$\mathbf{ML}^{-1}\mathbf{T}^{-1}$	1 kg/m s (= 1 N s/m^2)
Frequency	Hertz	$\mathbf{T}^{-1}$	1 s^{-1}

1.2.7. Thermal (heat) units

Heat is a form of energy and therefore its dimensions are $\mathbf{ML}^2\mathbf{T}^{-2}$. In many cases, however, no account is taken of interconversion of heat and "mechanical" energy (for example, kinetic, potential and kinetic energy), and heat can treated as a quantity which is conserved. It may then be regarded as having its own independent dimension **H** which can be used as an additional fundamental. It will be seen in Section 1.4 on dimensional analysis that increasing the number of fundamentals by one leads to an additional relation and consequently to one less dimensionless group.

Wherever heat is involved temperature also fulfils an important role: firstly because the heat content of a body is a function of its temperature and, secondly, because temperature difference or temperature gradient determines the rate at which heat is transferred. Temperature has the dimension θ which is independent of **M**,**L** and **T**, provided that no resort is made to the kinetic theory of gases in which temperature is shown to be directly proportional to the square of the velocity of the molecules.

It is not *incorrect* to express heat and temperature in terms of the **M**,**L**,**T** dimensions, although it is unhelpful in that it prevents the maximum of information being extracted from the process of dimensional analysis and reduces the insight that it affords into the physical nature of the process under consideration.

Dimensionally, the relation between **H**, **M** and θ can be expressed in the form:

$$\mathbf{H} \propto \mathbf{M}\theta = C_p\mathbf{M}\theta \qquad (1.5)$$

where C_p the specific heat capacity has dimensions $\mathbf{H}\,\mathbf{M}^{-1}\theta^{-1}$.

Equation 1.5 is similar in nature to the relationship between force mass and accelaration given by equation 1.1 with one important exception. The proportionality constant in equation 1.1 is not a function of the material concerned and it has been possible arbitrarily to put it equal to unity. The constant in equation 1.5, the specific heat capacity C_p, differs from one material to another.

In the SI system, the unit of heat is taken as the same as that of mechanical energy and is therefore the Joule. For water at 298 K (the datum used for many definitions), the specific heat capacity C_p is 4186.8 J/kg K.

Prior to the now almost universal adoption of the SI system of units, the unit of heat was defined as the quantity of heat required to raise the temperature of unit mass of water by one degree. This heat quantity is designated the calorie in the cgs system and the kilocalorie in the mks system, and in both cases temperature is expressed in degrees Celsius (Centigrade). As the specific heat capacity is a function of temperature, it has been necessary to set a datum temperature which is chosen as 298 K or 25°C.

In the British systems of units, the pound, but never the slug, is taken as the unit of mass and temperature may be expressed either in degrees Centigrade or in degrees Fahrenheit. The units of heat are then, respectively, the *pound-calorie* and the *British thermal unit* (Btu). Where the Btu is too small for a given application, the *therm* (= 10^5 Btu) is normally used.

Thus the following definitions of heat quantities therefore apply:

System	*Mass unit*	*Temperature scale (degrees)*	*Unit of Heat*
cgs	gram	Celsius	calorie
mks	kilogram	Celsius	kilocalorie
fps	pound	Celsius	pound calorie or Centigrade heat unit (CHU)
fps	pound	Fahrenheit	British thermal unit (Btu)
			1 CHU = 1.8 Btu

In all of these systems, by definition, the specific heat capacity of water is unity. It may be noted that, by comparing the definitions used in the SI and the mks systems, the kilocalorie is equivalent to 4186.8 J/kg K. This quantity has often been referred to as the *mechanical equivalent of heat* J.

1.2.8. Molar units

When working with ideal gases and systems in which a chemical reaction is taking place, it is usual to work in terms of *molar* units rather than mass. The *mole* (*mol*) is defined in the SI system as the quantity of material which contains as many entities (atoms, molecules or formula units) as there are in 12 g of carbon 12. It is more convenient, however, to work in terms of the *kilomole* (*kmol*) which relates to 12 kg of carbon 12, and the kilomole is used exclusively in this book. The number of molar units is denoted by dimensional symbol **N**. The number of kilomoles of a substance **A** is obtained by dividing its mass in kilograms (**M**) by its *molecular weight* M_A. M_A thus has the dimensions $\mathbf{MN}^{-1}$. The Royal Society recommends the use of the term *relative molecular mass* in place of *molecular weight*, but *molecular weight* is normally used here because of its general adoption in the processing industries.

1.2.9. Electrical units

Electrical current (I) has been chosen as the basic SI unit in terms of which all other electrical quantities are defined. Unit current, the *ampere* (A, or amp), is defined in terms of the force exerted between two parallel conductors in which a current of 1 amp is flowing. Since the unit of power, the watt, is the product of current and potential difference, the *volt* (V) is defined as watts per amp and therefore has dimensions of $\mathbf{ML^2T^{-3}I^{-1}}$. From Ohm's law the unit of resistance, the *ohm*, is given by the ratio volts/amps and therefore has dimensions of $\mathbf{ML^2T^{-3}I^{-2}}$. A similar procedure may be followed for the evaluation of the dimensions of other electrical units.

1.3. CONVERSION OF UNITS

Conversion of units from one system to another is simply carried out if the quantities are expressed in terms of the fundamental units of mass, length, time, temperature. Typical conversion factors for the British and metric systems are:

Mass	$1\ \text{lb} = \left(\frac{1}{32.2}\right)\ \text{slug} = 453.6\ \text{g} = 0.4536\ \text{kg}$
Length	$1\ \text{ft} = 30.48\ \text{cm} = 0.3048\ \text{m}$
Time	$1\ \text{s} = \left(\frac{1}{3600}\right)\ \text{h}$
Temperature difference	$1^\circ\text{F} = \left(\frac{1}{1.8}\right)\ ^\circ\text{C} = \left(\frac{1}{1.8}\right)\ \text{K (or deg.K)}$
Force	$1\ \text{pound force} = 32.2\ \text{poundal} = 4.44 \times 10^5\ \text{dyne} = 4.44\ \text{N}$

Other conversions are now illustrated.

Example 1.1

Convert 1 poise to British Engineering units and SI units.

Solution

$$1\ \text{Poise} = 1\ \text{g/cm s} = \frac{1\ \text{g}}{1\ \text{cm} \times 1\ \text{s}}$$

$$= \frac{(1/453.6)\ \text{lb}}{(1/30.48)\ \text{ft} \times 1\ \text{s}}$$

$$= 0.0672\ \text{lb/ft s}$$

6.7197×10⁻⁴ lb/ft.s
1cP

$$= \underline{\underline{242\ \text{lb/ft h}}}$$

$$1\ \text{Poise} = 1\ \text{g/cm s} = \frac{1\ \text{g}}{1\ \text{cm} \times 1\ \text{s}}$$

$$= \frac{(1/1000)\ \text{kg}}{(1/100)\ \text{m} \times 1\ \text{s}}$$

$$= 0.1\ \text{kg/m s}$$

$$= \underline{\underline{0.1\ \text{N s/m}^2}}\ [(\text{kg m/s}^2)\text{s/m}^2]$$

Example 1.2

Convert 1 kW to h.p.

Solution

$$1\ \text{kW} = 10^3\ \text{W} = 10^3\ \text{J/s}$$

$$= 10^3 \times \left(\frac{1\ \text{kg} \times 1\ \text{m}^2}{1\ \text{s}^3}\right)$$

$$= \frac{10^3 \times (1/0.4536)\ \text{lb} \times (1/0.3048)^2\ \text{ft}^2}{1\ \text{s}^3}$$

$$= 23{,}730\ \text{lb ft}^2/\text{s}^3$$

$$= \left(\frac{23{,}730}{32.2}\right) = 737\ \text{slug ft}^2/\text{s}^3$$

$$= 737\ \text{lb}_\text{f}\ \text{ft/s}$$

$$= \left(\frac{737}{550}\right) = \underline{\underline{1.34\ \text{h.p.}}}$$

or: $$1\ \text{h.p.} = 0.746\ \text{kW}.$$

Conversion factors to SI units from other units are given in Table 1.2 which is based on a publication by MULLIN[3].

Table 1.2. Conversion factors for some common SI units[4]
(An asterisk * denotes an exact relationship.)

Length	*1 in.	: 25.4 mm
	*1 ft	: 0.304,8 m
	*1 yd	: 0.914,4 m
	1 mile	: 1.609,3 km
	*1 Å (angstrom)	: 10^{-10} m
Time	*1 min	: 60 s
	*1 h	: 3.6 ks
	*1 day	: 86.4 ks
	1 year	: 31.5 Ms
Area	*1 in.^{2}	: 645.16 mm^2
	1 ft^2	: 0.092,903 m^2
	1 yd^2	: 0.836,13 m^2
	1 acre	: 4046.9 m^2
	1 mile^2	: 2.590 km^2
Volume	1 in.^{3}	: 16.387 cm^3
	1 ft^3	: 0.028,32 m^3
	1 yd^3	: 0.764,53 m^3
	1 UK gal	: 4546.1 cm^3
	1 US gal	: 3785.4 cm^3
Mass	1 oz	: 28.352 g
	*1 lb	: 0.453,592,37 kg
	1 cwt	: 50.802,3 kg
	1 ton	: 1016.06 kg
Force	1 pdl	: 0.138,26 N
	1 lbf	: 4.448,2 N
	1 kgf	: 9.806,7 N
	1 tonf	: 9.964,0 kN
	*1 dyn	: 10^{-5} N

(Continued on facing page)

Table 1.2. (*continued*)

Temperature difference	*1 deg F (deg R)	: $\frac{5}{9}$ deg C (deg K)
Energy (work, heat)	1 ft lbf	: 1.355,8 J
	1 ft pdl	: 0.042,14 J
	*1 cal (international table)	: 4.186,8 J
	1 erg	: 10^{-7} J
	1 Btu	: 1.055,06 kJ
	1 hp h	: 2.684,5 MJ
	*1 kW h	: 3.6 MJ
	1 therm	: 105.51 MJ
	1 thermie	: 4.185,5 MJ
Calorific value (volumetric)	1 Btu/ft^3	: 37.259 kJ/m^3
Velocity	1 ft/s	: 0.304,8 m/s
	1 mile/h	: 0.447,04 m/s
Volumetric flow	1 ft^3/s	: 0.028,316 m^3/s
	1 ft^3/h	: 7.865,8 cm^3/s
	1 UK gal/h	: 1.262,8 cm^3/s
	1 US gal/h	: 1.051,5 cm^3/s
Mass flow	1 lb/h	: 0.126,00 g/s
	1 ton/h	: 0.282,24 kg/s
Mass per unit area	1 lb/in.^{2}	: 703.07 kg/m^2
	1 lb/ft^2	: 4.882,4 kg/m^2
	1 ton/sq mile	: 392.30 kg/km^2
Density	1 lb/in^3	: 27.680 g/cm^3
	1 lb/ft^3	: 16.019 kg/m^3
	1 lb/UK gal	: 99.776 kg/m^3
	1 lb/US gal	: 119.83 kg/m^3
Pressure	1 lbf/in.^{2}	: 6.894,8 kN/m^2
	1 tonf/in.^{2}	: 15.444 MN/m^2
	1 lbf/ft^2	: 47.880 N/m^2
	*1 standard atm	: 101.325 kN/m^2
	*1 atm (1 kgf/cm^2)	: 98.066,5 kN/m^2
	*1 bar	: 10^5 N/m^2
	1 ft water	: 2.989,1 kN/m^2
	1 in. water	: 249.09 N/m^2
	1 in. Hg	: 3.386,4 kN/m^2
	1 mm Hg (1 torr)	: 133.32 N/m^2
Power (heat flow)	1 hp (British)	: 745.70 W
	1 hp (metric)	: 735.50 W
	1 erg/s	: 10^{-7} W
	1 ft lbf/s	: 1.355,8 W
	1 Btu/h	: 0.293,07 W
	1 ton of refrigeration	: 3516.9 W
Moment of inertia	1 lb ft^2	: 0.042,140 kg m^2
Momentum	1 lb ft/s	: 0.138,26 kg m/s
Angular momentum	1 lb ft^2/s	: 0.042,140 kg m^2/s
Viscosity, dynamic	*1 P (Poise)	: 0.1 N s/m^2
	1 lb/ft h	: 0.413,38 mN s/m^2
	1 lb/ft s	: 1.488,2 Ns/m^2
Viscosity, kinematic	*1 S (Stokes)	: 10^{-4} m^2/s
	1 ft^2/h	: 0.258,06 cm^2/s

(*continued overleaf*)

Table 1.2. (*continued*)

Surface energy	1 erg/cm^2	: 10^{-3} J/m^2
(surface tension)	(1 dyn/cm)	: (10^{-3} N/m)
Mass flux density	1 lb/h ft^2	: 1.356,2 g/s m^2
Heat flux density	1 Btu/h ft^2	: 3.154,6 W/m^2
	*1 kcal/h m^2	: 1.163 W/m^2
Heat transfer		
coefficient	1 Btu/h ft^2°F	: 5.678,3 W/m^{2}K
Specific enthalpy		
(latent heat, etc.)	*1 Btu/lb	: 2.326 kJ/kg
Specific heat capacity	*1 Btu/lb °F	: 4.186,8 kJ/kgK
Thermal	1 Btu/h ft °F	: 1.730,7 W/mK
conductivity	1 kcal/h m °C	: 1.163 W/mK

1.4. DIMENSIONAL ANALYSIS

Dimensional analysis depends upon the fundamental principle that any equation or relation between variables must be *dimensionally consistent*; that is, each term in the relationship must have the same dimensions. Thus, in the simple application of the principle, an equation may consist of a number of terms, each representing, and therefore having, the dimensions of length. It is not permissible to add, say, lengths and velocities in an algebraic equation because they are quantities of different characters. The corollary of this principle is that if the whole equation is divided through by any one of the terms, each remaining term in the equation must be dimensionless. The use of these *dimensionless groups*, or *dimensionless numbers* as they are called, is of considerable value in developing relationships in chemical engineering.

The requirement of dimensional consistency places a number of constraints on the form of the functional relation between variables in a problem and forms the basis of the technique of *dimensional analysis* which enables the variables in a problem to be grouped into the form of dimensionless groups. Since the dimensions of the physical quantities may be expressed in terms of a number of fundamentals, usually mass, length, and time, and sometimes temperature and thermal energy, the requirement of dimensional consistency must be satisfied in respect of each of the fundamentals. Dimensional analysis gives no information about the form of the functions, nor does it provide any means of evaluating numerical proportionality constants.

The study of problems in fluid dynamics and in heat transfer is made difficult by the many parameters which appear to affect them. In most instances further study shows that the variables may be grouped together in dimensionless groups, thus reducing the effective number of variables. It is rarely possible, and certainly time consuming, to try to vary these many variables separately, and the method of dimensional analysis in providing a smaller number of independent groups is most helpful to the investigated.

The application of the principles of dimensional analysis may best be understood by considering an example.

It is found, as a result of experiment, that the pressure difference (ΔP) between two ends of a pipe in which a fluid is flowing is a function of the pipe diameter d, the pipe length l, the fluid velocity u, the fluid density ρ, and the fluid viscosity μ.

The relationship between these variables may be written as:

$$\Delta P = \mathrm{f}_1(d, l, u, \rho, \mu) \tag{1.6}$$

The form of the function is unknown, though since any function can be expanded as a power series, the function may be regarded as the sum of a number of terms each consisting of products of powers of the variables. The simplest form of relation will be where the function consists simply of a single term, or:

$$\Delta P = \text{const}\ d^{n_1} l^{n_2} u^{n_3} \rho^{n_4} \mu^{n_5} \tag{1.7}$$

The requirement of dimensional consistency is that the combined term on the right-hand side will have the same dimensions as that on the left; that is, it must have the dimensions of pressure.

Each of the variables in equation 1.7 may be expressed in terms of mass, length, and time. Thus, dimensionally:

$$\Delta P \equiv \mathbf{ML^{-1}T^{-2}} \qquad u \equiv \mathbf{LT^{-1}}$$

$$d \equiv \mathbf{L} \qquad \rho \equiv \mathbf{ML^{-3}}$$

$$l \equiv \mathbf{L} \qquad \mu \equiv \mathbf{ML^{-1}T^{-1}}$$

and:

$$\mathbf{ML^{-1}T^{-2}} \equiv \mathbf{L}^{n_1}\mathbf{L}^{n_2}(\mathbf{LT^{-1}})^{n_3}(\mathbf{ML^{-3}})^{n_4}(\mathbf{ML^{-1}T^{-1}})^{n_5}$$

The conditions of dimensional consistency must be met for each of the fundamentals of **M**, **L**, and **T** and the indices of each of these variables may be equated. Thus:

$$\mathbf{M} \qquad 1 = n_4 + n_5$$

$$\mathbf{L} \qquad -1 = n_1 + n_2 + n_3 - 3n_4 - n_5$$

$$\mathbf{T} \qquad -2 = -n_3 - n_5$$

Thus three equations and five unknowns result and the equations may be solved in terms of any two unknowns. Solving in terms of n_2 and n_5:

$$n_4 = 1 - n_5 \text{ (from the equation in } \mathbf{M}\text{)}$$

$$n_3 = 2 - n_5 \text{ (from the equation in } \mathbf{T}\text{)}$$

Substituting in the equation for **L**:

$$-1 = n_1 + n_2 + (2 - n_5) - 3(1 - n_5) - n_5$$

or:

$$0 = n_1 + n_2 + n_5$$

and:

$$n_1 = -n_2 - n_5$$

Thus, substituting into equation 1.7:

$$\Delta P = \text{const}\ d^{-n_2-n_5} l^{n_2} u^{2-n_5} \rho^{1-n_5} \mu^{n_5}$$

or:

$$\frac{\Delta P}{\rho u^2} = \text{const}\left(\frac{l}{d}\right)^{n_2}\left(\frac{\mu}{du\rho}\right)^{n_5} \tag{1.8}$$

Since n_2 and n_5 are arbitrary constants, this equation can only be satisfied if each of the terms $\Delta P/\rho u^2$, l/d, and $\mu/du\rho$ is dimensionless. Evaluating the dimensions of each group shows that this is, in fact, the case.

The group $ud\rho/\mu$, known as the *Reynolds number*, is one which frequently arises in the study of fluid flow and affords a criterion by which the type of flow in a given geometry may be characterised. Equation 1.8 involves the reciprocal of the Reynolds number, although this may be rewritten as:

$$\frac{\Delta P}{\rho u^2} = \text{const}\left(\frac{l}{d}\right)^{n_2}\left(\frac{ud\rho}{\mu}\right)^{-n_5} \tag{1.9}$$

The right-hand side of equation 1.9 is a typical term in the function for $\Delta P/\rho u^2$. More generally:

$$\frac{\Delta P}{\rho u^2} = \text{f}_2\left(\frac{l}{d}, \frac{ud\rho}{\mu}\right) \tag{1.10}$$

Comparing equations 1.6 and 1.10, it is seen that a relationship between six variables has been reduced to a relationship between three dimensionless groups. In subsequent sections of this chapter, this statement will be generalised to show that the number of dimensionless groups is *normally* the number of variables less the number of fundamentals (but see the note in Section 1.5).

A number of important points emerge from a consideration of the preceding example:

1 If the index of a particular variable is found to be zero, this indicates that this variable is of no significance in the problem.
2 If two of the fundamentals always appear in the same combination, such as **L** and **T** always occuring as powers of $\mathbf{LT}^{-1}$, for example, then the same equation for the indices will be obtained for both **L** and **T** and the number of effective fundamentals is thus reduced by one.
3 The form of the final solution will depend upon the method of solution of the simultaneous equations. If the equations had been solved, say, in terms of n_3 and n_4 instead of n_2 and n_5, the resulting dimensionless groups would have been different, although these new groups would simply have been products of powers of the original groups. Any number of fresh groups can be formed in this way.

Clearly, the maximum degree of simplification of the problem is achieved by using the greatest possible number of fundamentals since each yields a simultaneous equation of its own. In certain problems, force may be used as a fundamental in addition to mass, length, and time, provided that at no stage in the problem is force defined in terms of mass and acceleration. In heat transfer problems, temperature is usually an additional fundamental, and heat can also be used as a fundamental provided it is not defined in terms of mass and temperature and provided that the equivalence of mechanical and thermal energy is not utilised. Considerable experience is needed in the proper use of dimensional analysis, and its application in a number of areas of fluid flow and heat transfer is seen in the relevant chapters of this Volume.

The choice of physical variables to be included in the dimensional analysis must be based on an understanding of the nature of the phenomenon being studied although, on occasions there may be some doubt as to whether a particular quantity is relevant or not.

If a variable is included which does not exert a significant influence on the problem, the value of the dimensionless group in which it appears will have little effect on the final numerical solution of the problem, and therefore the exponent of that group must approach zero. This presupposes that the dimensionless groups are so constituted that the variable in question appears in only one of them. On the other hand if an important variable is omitted, it may be found that there is no unique relationship between the dimensionless groups.

Chemical engineering analysis requires the formulation of relationships which will apply over a wide range of size of the individual items of a plant. This problem of scale up is vital and it is much helped by dimensional analysis.

Since linear size is included among the variables, the influence of scale, which may be regarded as the influence of linear size without change of shape or other variables, has been introduced. Thus in viscous flow past an object, a change in linear dimension **L** will alter the Reynolds number and therefore the flow pattern around the solid, though if the change in scale is accompanied by a change in any other variable in such a way that the Reynolds number remains unchanged, then the flow pattern around the solid will not be altered. This ability to change scale and still maintain a design relationship is one of the many attractions of dimensional analysis.

It should be noted that it is permissible to take a function only of a dimensionless quantity. It is easy to appreciate this argument when account is taken of the fact that any function may be expanded as a power series, each term of which must have the same dimensions, and the requirement of dimensional consistency can be met only if these terms and the function are dimensionless. Where this principle appears to have been invalidated, it is generally because the equation includes a further term, such as an integration constant, which will restore the requirements of dimensional consistency. For example, $\int_{x_o}^{x} \frac{dx}{x} = \ln x - \ln x_o$, and if x is not dimensionless, it appears at first sight that the principle has been infringed. Combining the two logarithmic terms, however, yields $\ln\left(\frac{x}{x_o}\right)$, and $\frac{x}{x_o}$ is clearly dimensionless. In the case of the indefinte integral, $\ln x_o$ would, in effect, have been the integration constant.

1.5. BUCKINGHAM'S Π THEOREM

The need for dimensional consistency imposes a restraint in respect of each of the fundamentals involved in the dimensions of the variables. This is apparent from the previous discussion in which a series of simultaneous equations was solved, one equation for each of the fundamentals. A generalisation of this statement is provided in Buckingham's Π theorem[(4)] which states that the number of dimensionless groups is equal to the number of variables minus the number of fundamental dimensions. In mathematical terms, this can be expressed as follows:

If there are n variables, $Q_1, Q_2, \ldots, Q_n$, the functional relationship between them may be written as:

$$f_3(Q_1, Q_2, \ldots, Q_n) = 0 \tag{1.11}$$

If there are m fundamental dimensions, there will be $(n - m)$ dimensionless groups $(\Pi_1, \Pi_2, \ldots, \Pi_{n-m})$ and the functional relationship between them may be written as:

$$f_4(\Pi_1, \Pi_2, \ldots, \Pi_{n-m}) = 0 \tag{1.12}$$

The groups Π_1, Π_2, and so on must be independent of one another, and no one group should be capable of being formed by multiplying together powers of the other groups.

By making use of this theorem it is possible to obtain the dimensionless groups more simply than by solving the simultaneous equations for the indices. Furthermore, the functional relationship can often be obtained in a form which is of more immediate use.

The method involves choosing m of the original variables to form what is called a *recurring set*. Any set m of the variables may be chosen with the following two provisions:

(1) Each of the fundamentals must appear in at least one of the m variables.
(2) It must not be possible to form a dimensionless group from some or all of the variables within the recurring set. If it were so possible, this dimensionless group would, of course, be one of the Π terms. Thus, the number of dimensionless groups is increased by one for each of the independent groups that can be so formed.

The procedure is then to take each of the remaining $(n - m)$ variables on its own and to form it into a dimensionless group by combining it with one or more members of the recurring set. In this way the $(n - m)$ Π groups are formed, the only variables appearing in more than one group being those that constitute the recurring set. Thus, if it is desired to obtain an explicit functional relation for one particular variable, that variable should not be included in the recurring set.

In some cases, the number of dimensionless groups will be greater than predicted by the Π theorem. For instance, if two of the fundamentals always occur in the same combination, length and time always as $\mathbf{LT}^{-1}$, for example, they will constitute a single fundamental instead of two fundamentals. By referring back to the method of equating indices, it is seen that each of the two fundamentals gives the same equation, and therefore only a single constraint is placed on the relationship by considering the two variables. Thus, although m is normally the number of fundamentals, it is more strictly defined as the *maximum number of variables from which a dimensionless group cannot be formed.*

The procedure is more readily understood by consideration of the illustration given previously. The relationship between the variables affecting the pressure drop for flow of fluid in a pipe may be written as:

$$f_5(\Delta P, d, l, \rho, \mu, u) = 0 \tag{1.13}$$

Equation 1.13 includes six variables, and three fundamental quantities (mass, length, and time) are involved. Thus:

$$\text{Number of groups} = (6 - 3) = 3$$

The recurring set must contain three variables that cannot themselves be formed into a dimensionless group. This imposes the following two restrictions:

(1) Both l and d cannot be chosen as they can be formed into the dimensionless group l/d.
(2) ΔP, ρ and u cannot be used since $\Delta P/\rho u^2$ is dimensionless.

Outside these constraints, any three variables can be chosen. It should be remembered, however, that the variables forming the recurring set are liable to appear in all the dimensionless groups. As this problem deals with the effect of conditions on the pressure difference ΔP, it is convenient if ΔP appears in only one group, and therefore it is preferable not to include it in the recurring set.

If the variables d, u, ρ are chosen as the recurring set, this fulfils all the above conditions. Dimensionally:

$$d \equiv \mathbf{L}$$

$$u \equiv \mathbf{LT}^{-1}$$

$$\rho \equiv \mathbf{ML}^{-3}$$

Each of the dimensions **M**, **L**, **T** may then be obtained explicitly in terms of the variables d, u, ρ, to give:

$$\mathbf{L} \equiv d$$

$$\mathbf{M} \equiv \rho d^3$$

$$\mathbf{T} \equiv du^{-1}$$

The three dimensionless groups are thus obtained by taking each of the remaining variables ΔP, l, and μ in turn.

ΔP has dimensions $\mathbf{ML}^{-1}\mathbf{T}^{-2}$, and $\Delta P\mathbf{M}^{-1}\mathbf{LT}^{2}$ is therefore dimensionless.

Group Π_1 is, therefore, $\Delta P(\rho d^3)^{-1}(d)(du^{-1})^2 = \dfrac{\Delta P}{\rho u^2}$

l has dimensions **L**, and $l\mathbf{L}^{-1}$ is therefore dimensionless.

Group Π_2 is therefore: $l(d^{-1}) = \dfrac{l}{d}$

μ has dimensions $\mathbf{ML}^{-1}\mathbf{T}^{-1}$, and $\mu\mathbf{M}^{-1}\mathbf{LT}$ is therefore dimensionless.

Group Π_3 is, therefore: $\mu(\rho d^3)^{-1}(d)(du^{-1}) = \dfrac{\mu}{du\rho}$.

Thus: $$f_6\left(\frac{\Delta P}{\rho u^2}, \frac{l}{d}, \frac{\mu}{ud\rho}\right) = 0 \quad \text{or} \quad \frac{\Delta P}{\rho u^2} = f_7\left(\frac{l}{d}, \frac{ud\rho}{\mu}\right)$$

$\mu/ud\rho$ is arbitrarily inverted because the Reynolds number is usually expressed in the form $ud\rho/\mu$.

Some of the important dimensionless groups used in Chemical Engineering are listed in Table 1.3.

Example 1.3

A glass particle settles under the action of gravity in a liquid. Obtain a dimensionless grouping of the variables involved. The falling velocity is found to be proportional to the square of the particle diameter when the other variables are constant. What will be the effect of doubling the viscosity of the liquid?

Solution

It may be expected that the variables expected to influence the terminal velocity of a glass particle settling in a liquid, u_0, are:

Table 1.3. Some important dimensionless groups

Symbol	Name of group	In terms of other groups	Definition	Application
Ar	Archimedes	Ga	$\dfrac{\rho(\rho_s - \rho)gd^3}{\mu^2}$	Gravitational settling of particle in fluid
Db	Deborah		$\dfrac{t_P}{t_F}$	Flow of viscoelastic fluid
Eu	Euler		$\dfrac{P}{\rho u^2}$	Pressure and momentum in fluid
Fo	Fourier		$\dfrac{D_H t}{l^2}, \dfrac{Dt}{l^2}$	Unsteady state heat transfer/mass transfer
Fr	Froude		$\dfrac{u^2}{gl}$	Fluid flow with free surface
Ga	Galileo	Ar	$\dfrac{\rho(\rho_s - \rho)gd^3}{\mu^2}$	Gravitational settling of particle in fluid
Gr	Grashof		$\dfrac{l^3\rho^2\beta g\Delta T}{\mu^2}$	Heat transfer by natural convection
Gz	Graetz		$\dfrac{GC_p}{kl}$	Heat transfer to fluid in tube
He	Hedström		$\dfrac{R_Y\rho d^2}{\mu_p^2}$	Flow of fluid exhibiting yield stress
Le	Lewis	$Sc \cdot Pr^{-1}$	$\dfrac{k}{C_p\rho D} = \dfrac{D_H}{D}$	Simultaneous heat and mass transfer
Ma	Mach		$\dfrac{u}{u_w}$	Gas flow at high velocity
Nu	Nusselt		$\dfrac{hl}{k}$	Heat transfer in fluid
Pe	Peclet	$Re \cdot Pr$	$\dfrac{ul}{D_H}$	Fluid flow and heat transfer
		$Re \cdot Sc$	$\dfrac{ul}{D}$	Fluid flow and mass transfer
Pr	Prandtl		$\dfrac{C_p\mu}{k}$	Heat transfer in flowing fluid
Re	Reynolds		$\dfrac{ul\rho}{\mu}$	Fluid flow involving viscous and inertial forces
Sc	Schmidt		$\dfrac{\mu}{\rho D}$	Mass transfer in flowing fluid
Sh	Sherwood		$\dfrac{h_D l}{D}$	Mass transfer in fluid
St	Stanton	$Nu \cdot Pr^{-1} \cdot Re^{-1}$	$\dfrac{h}{C_p\rho u}$	Heat transfer in flowing fluid
We	Weber		$\dfrac{\rho u^2 l}{\sigma}$	Fluid flow with interfacial forces
ϕ	Friction factor		$\dfrac{R}{\rho u^2}$	Fluid drag at surface
$\mathbf{N}_p$	Power number		$\dfrac{\mathbf{P}}{\rho N^3 d^5}$	Power consumption for mixers

particle diameter d; particle density, ρ_s; liquid density, ρ; liquid viscosity, μ and the acceleration due to gravity, g.

Particle density ρ_s is important because it determines the gravitational (accelerating) force on the particle. However when immersed in a liquid the particle receives an upthrust which is proportional to the liquid density ρ. The effective density of the particles $(\rho_s - \rho)$ is therefore used in this analysis. Then:

$$u_0 = f(d, (\rho_s - \rho), \rho, \mu, g)$$

The dimensions of each variable are:

$$u_0 = \mathbf{LT}^{-1}, \quad d = \mathbf{L}, \quad \rho_s - \rho = \mathbf{ML}^{-3}, \quad \rho = \mathbf{ML}^{-3},$$

$$\mu = \mathbf{ML}^{-1}\mathbf{T}^{-1} \quad \text{and} \quad g = \mathbf{LT}^{-2}.$$

With six variables and three fundamental dimensions, $(6 - 3) = 3$ dimensionless groups are expected. Choosing d, ρ and μ as the recurring set:

$$d \equiv \mathbf{L} \qquad \mathbf{L} = d$$
$$\rho \equiv \mathbf{ML}^{-3} \qquad \mathbf{M} = \rho\mathbf{L}^3 = \rho d^3$$
$$\mu \equiv \mathbf{ML}^{-1}\mathbf{T}^{-1} \qquad \mathbf{T} = \mathbf{M}/\mu L = \rho d^3/(\mu d) = \rho d^2/\mu$$

Thus:

dimensionless group 1: $$u_0\mathbf{TL}^{-1} = u_0\rho d^2/(\mu d) = u_0\rho d/\mu$$

dimensionless group 2: $$(\rho_s - \rho)\mathbf{L}^3\mathbf{M}^{-1} = \rho_s d^3/(\rho d^3) = (\rho_s - \rho)/\rho$$

dimensionless group 3: $$g\mathbf{T}^2\mathbf{L}^{-1} = g\rho^2 d^4/(\mu^2 d) = g\rho^2 d^3/\mu^2$$

and: $$(u_0\rho d/\mu) \propto ((\rho_s - \rho)/\rho)(g\rho^2 d^3/\mu^2)$$

or: $$(u_0\rho d/\mu) = \kappa((\rho_s - \rho)/\rho)^{n_1}(g\rho^2 d^3/\mu^2)^{n_2}$$

when $u_0 \propto d^2$, when $(3n_2 - 1) = 2$ and $n_2 = 1$.

Thus: $$(u_0\rho d/\mu) = \kappa((\rho_s - \rho)/\rho)^{n_1}(g\rho^2 d^3/\mu^2)$$

or: $$u_0 = \kappa((\rho_s - \rho)/\rho)^{n_1}(d^2\rho g/\mu)$$

and: $$u_0 \propto (1/\mu)$$

In this case, doubling the viscosity of the liquid will *halve the terminal velocity of the particle*, suggesting that the flow is in the Stokes' law regime.

Example 1.4

A drop of liquid spreads over a horizontal surface. Obtain dimensionless groups of the variables which will influence the rate at which the liquid spreads.

Solution

The rate at which a drop spreads, say u_R m/s, will be influenced by:

$$\text{viscosity of the liquid, } \mu\text{–dimensions} = \mathbf{ML}^{-1}\mathbf{T}^{-1}$$

$$\text{volume of the drop, } V\text{–dimensions} = \mathbf{L}^3$$

$$\text{density of the liquid, } \rho\text{–dimensions} = \mathbf{ML}^{-3}$$

$$\text{acceleration due to gravity, } g\text{–dimensions} = \mathbf{LT}^{-2}.$$

and possibly, surface tension of the liquid, σ–dimensions $= \mathbf{MT}^{-2}$.

Noting the dimensions of u_R as $\mathbf{LT}^{-1}$, there are six variables and hence $(6 - 3) = 3$ dimensionless groups. Taking V, ρ and g as the recurring set:

$$V \equiv \mathbf{L}^3 \text{ and } \mathbf{L} = V^{0.33}$$

$$\rho \equiv \mathbf{ML}^{-3} \text{ and } \mathbf{M} = \rho \mathbf{L}^3 = \rho V$$

$$g \equiv \mathbf{LT}^{-2} \text{ and } \mathbf{T}^2 = \mathbf{L}/g \text{ or } \mathbf{T} = V^{0.16}/g^{0.5}$$

Thus:

dimensionless group 1: $u_R\mathbf{TL}^{-1} = u_R V^{0.16}/(V^{0.33}g^{0.5}) = u_R/(V^{0.33}g)^{0.5}$

dimensionless group 2: $\mu\mathbf{LTM}^{-1} = \mu V^{0.33}(V^{0.16}/g^{0.5})(\rho V)^{-1} = \mu/(g^{0.5}V^{0.5})$

dimensionless group 3: $\sigma\mathbf{T}^2\mathbf{M}^{-1} = \sigma(V^{0.33}/g)/(\rho V) = \sigma/(g\rho V^{0.67})$

and: $u_R/(V^{0.33}g)^{0.5} \propto (\mu/(g^{0.5}\rho V^{0.5}))/(\sigma/(g\rho V^{0.67}))$

or: $\underline{\underline{u_R^2/V^{0.33}g = \kappa(\mu^2/g\rho^2 V)^{n_1}(\sigma/g\rho V^{0.67})^{n_2}}}$

1.6. REDEFINITION OF THE LENGTH AND MASS DIMENSIONS

1.6.1. Vector and scalar quantities

It is important to recognise the differences between *scalar* quantities which have a magnitude but no direction, and vector quantities which have both magnitude and direction. Most length terms are vectors in the Cartesian system and may have components in the X, Y and Z directions which may be expressed as $\mathbf{L_X}$, $\mathbf{L_Y}$ and $\mathbf{L_Z}$. There must be dimensional consistency in all equations and relationships between physical quantities, and there is therefore the possibility of using all three length dimensions as fundamentals in dimensional analysis. This means that the number of dimensionless groups which are formed will be less.

Combinations of length dimensions in areas, such as $\mathbf{L_X L_Y}$, and velocities, accelerations and forces are all *vector* quantities. On the other hand, mass, volume and heat are all scalar quantities with no directional significance. The power of dimensional analysis is thus increased as a result of the larger number of fundamentals which are available for use. Furthermore, by expressing the length dimension as a vector quantity, it is possible to obviate the difficulty of two quite different quantities having the same dimensions. For example, the units of *work* or *energy* may be obtained by multiplying a force in the X-direction (say) by a distance also in the X-direction. The dimensions of energy are therefore:

$$(\mathbf{ML_X T}^{-2})(\mathbf{L_X}) = \mathbf{ML_X^2 T}^{-2}$$

It should be noted in this respect that a *torque* is obtained as a product of a force in the X-direction and an arm of length $\mathbf{L_Y}$, say, in a direction at right-angles to the Y-direction. Thus, the dimensions of torque are $\mathbf{ML_X L_Y T}^{-2}$, which distinguish it from energy.

Another benefit arising from the use of vector lengths is the ability to differentiate between the dimensions of frequency and angular velocity, both of which are $\mathbf{T}^{-1}$ if length is treated as a scalar quantity. Although an angle is dimensionless in the sense that it can be defined by the ratio of two lengths, its dimensions become $\mathbf{L_X/L_Y}$ if these two lengths are treated as vectors. Thus angular velocity then has the dimensions $\mathbf{L_X L_Y^{-1} T}^{-1}$ compared with $\mathbf{T}^{-1}$ for frequency.

Of particular interest in fluid flow is the distinction between shear stress and pressure (or pressure difference), both of which are defined as force per unit area. For steady-state flow of a fluid in a pipe, the forces attributable to the pressure difference and the shear stress must balance. The pressure difference acts in the axial X-direction, say, and the area A on which it acts lies in the Y–Z plane and its dimensions can therefore be expressed as $\mathbf{L_Y L_Z}$. On the other hand, the shear stress R which is also exerted in the X-direction acts on the curved surface of the walls whose area S has the dimensions $\mathbf{L_X L_R}$ where $\mathbf{L_R}$ is length in the radial direction. Because there is axial symetry, $\mathbf{L_R}$ can be expressed as $\mathbf{L_Y^{1/2} L_Z^{1/2}}$ and the dimensions of S are then $\mathbf{L_X L_Y^{1/2} L_Z^{1/2}}$.

The force F acting on the fluid in the X (axial)-direction has dimensions $\mathbf{ML_X T^{-2}}$, and hence:

$$\Delta P = F/A \text{ has dimensions } \mathbf{ML_X T^{-2}/L_Y L_Z} = \mathbf{ML_X L_Y^{-1} L_Z^{-1} T^{-2}}$$

and $$R = F/S \text{ has dimensions } \mathbf{ML_X T^{-2}/L_X L_Y^{1/2} L_Z^{1/2}} = \mathbf{ML_Y^{-1/2} L_Z^{-1/2} T^{-2}}$$

giving dimensions of $\Delta P/R$ as $\mathbf{L_X L_Y^{-1/2} L_Z^{-1/2}}$ or $\mathbf{L_X L_R^{-1}}$ (which would have been dimensionless had lengths not been treated as vectors).

For a pipe of radius r and length l, the dimensions of r/l are $\mathbf{L_X^{-1} L_R}$ and hence $(\Delta P/R)(r/l)$ is a dimensionless quantity. The role of the ratio r/l would not have been established had the lengths not been treated as vectors. It is seen in Chapter 3 that this conclusion is consistent with the results obtained there by taking a force balance on the fluid.

1.6.2 Quantity mass and inertia mass

The term mass **M** is used to denote two distinct and different properties:

1 The quantity of matter $\mathbf{M}_\mu$, and
2 The inertial property of the matter $\mathbf{M_i}$.

These two quantities are proportional to one another and may be numerically equal, although they are essentially different in kind and are therefore not identical. The distinction is particularly useful when considering the energy of a body or of a fluid.

Because inertial mass is involved in mechanical energy, the dimensions of all energy terms are $\mathbf{M_i L^2 T^{-2}}$. Inertial mass, however, is not involved in thermal energy (heat) and therefore specific heat capacity C_p has the dimensions $\mathbf{M_i L^2 T^2}/\mathbf{M}_\mu \theta = \mathbf{M_i M_\mu^{-1} L^2 T^{-2}} \theta^{-1}$ or $\mathbf{HM}_\mu^{-1} \theta^{-1}$ according to whether energy is expressed in, joules or kilocalories, for example.

In practical terms, this can lead to the possibility of using both mass dimensions as fundamentals, thereby achieving similar advantages to those arising from taking length as a vector quantity. This subject is discussed in more detail by HUNTLEY[(5)].

WARNING

Dimensional analysis is a very powerful tool in the analysis of problems involving a large number of variables. However, there are many pitfalls for the unwary, and the technique should never be used without a thorough understanding of the underlying basic principles of the physical problem which is being analysed.

1.7. FURTHER READING

ASTARITA, G.: *Chem. Eng. Sci* **52** (1997) 4681. Dimensional analysis, scaling and orders of magnitude.
BLACKMAN, D. R.: *SI Units in Engineering* (Macmillan, 1969).
BRIDGMAN, P. W.: *Dimensional Analysis* (Yale University Press, 1931).
FOCKEN, C. M.: *Dimensional Methods and their Applications* (Edward Arnold, London, 1953).
HEWITT, G. F.: *Proc 11th International Heat Transfer Conference* Kyongju, Korea (1998). 1.
HUNTLEY, H. E.: *Dimensional Analysis* (Macdonald and Co. (Publishers) Ltd, London, 1952).
IPSEN, D. C.: *Units, Dimensions, and Dimensionless Numbers* (McGraw-Hill, 1960).
JOHNSTONE, R. E. and THRING, M. W.: *Pilot Plants, Models and Scale-up in Chemical Engineering* (McGraw-Hill, 1957).
KLINKENBERG, A. and MOOY, H. H.: *Chem. Eng. Prog.* **44** (1948) 17. Dimensionless groups in fluid friction, heat and material transfer.
MASSEY, B. S.: *Units, Dimensional Analysis and Physical Similarity* (VAN Nostrand Reinhold, 1971).
MULLIN, J. W.: *The Chemical Engineer* (London) No. 211 (Sept. 1967) 176. SI units in chemical engineering.
MULLIN, J. W.: *The Chemical Engineer* (London) No. 254 (1971) 352. Recent developments in the change-over to the International System of Units (SI).
British Standards Institution Publication PD 5686 (1967). *The Use of SI Units.*
Quantities, Units and Symbols. (The Symbols Committee of the Royal Society, 1971).
SI. The International System of Units (HMSO, 1970).

1.8 REFERENCES

1. British Standards Institution Publication PD 5686 (1967). *The Use of SI Units.*
2. *SI. The International System of Units* (HMSO, 1970).
3. MULLIN, J. W. *The Chemical Engineer* (London) No 211 (Sept 1967) 176, SI units in chemical engineering.
4. BUCKINGHAM, E.: *Phys. Rev. Ser.*, 2, **4** (1914) 345. On physically similar systems: illustrations of the use of dimensional equations.
5. HUNTLEY, H.E.: *Dimensional Analysis* (Madonald and Co (Publishers) Ltd, London,(1952).

1.9. NOMENCLATURE

		Units in SI system	Dimensions in **M, N, L, T, θ, H, I**
C_p	Specific heat capacity at constant pressure	J/kg K	$\mathbf{L^2T^{-2}\theta^{-1}}$ ($\mathbf{HM^{-1}\theta^{-1}}$)
D	Diffusion coefficient, molecular diffusivity	m²/s	$\mathbf{L^2T^{-1}}$
D_H	Thermal diffusivity $k/C_p\rho$	m²/s	$\mathbf{L^2T^{-1}}$
d	Diameter	m	**L**
f	A function	—	—
G	Mass rate of flow	kg/s	$\mathbf{MT^{-1}}$
g	Acceleration due to gravity	m/s²	$\mathbf{LT^{-2}}$
g_c	Numerical constant equal to standard value of "g"	—	—
h	Heat transfer coefficient	W/m²K	$\mathbf{MT^{-3}\theta^{-1}}$
h_D	Mass transfer coefficient	m/s	$\mathbf{LT^{-1}}$
I	Electric current	A	**I**
J	Mechanical equivalent of heat	—	—
k	Thermal conductivity	W/m K	$\mathbf{MLT^{-3}\theta^{-1}}$
l	Characteristic length or length of pipe	m	**L**
M_A	Molecular weight (relative molecular mass) of **A**	kg/kmol	$\mathbf{MN^{-1}}$
m	Number of fundamental dimensions	—	—
N	Rotational speed	s^{-1}	$\mathbf{T^{-1}}$

		Units in SI system	Dimensions in $\mathbf{M, N, L, T, \theta, H, I}$
n	Number of variables	—	—
$\mathbf{P}$	Power	W	$\mathbf{ML^2T^{-3}}$
P	Pressure	N/m^2	$\mathbf{ML^{-1}T^{-2}}$
ΔP	Pressure difference	N/m^2	$\mathbf{ML^{-1}T^{-2}}$
Q	Physical quantity	—	—
R	Shear stress	N/m^2	$\mathbf{ML^{-1}T^{-2}}$
R_y	Yield stress	N/m^2	$\mathbf{ML^{-1}T^{-2}}$
$\mathbf{R}$	Universal gas constant	8314 J/kmol K	$\mathbf{MN^{-1}L^2T^{-2}\theta^{-1}}$ $(\mathbf{HN^{-1}\theta^{-1}})$
r	Pipe radius	m	$\mathbf{L}$
ΔT	Temperature difference	K	$\boldsymbol{\theta}$
t	Time	s	$\mathbf{T}$
t_F	Characteristic time for fluid	s	$\mathbf{T}$
t_P	Characteristic time for process	s	$\mathbf{T}$
u	Velocity	m/s	$\mathbf{LT^{-1}}$
u_w	Velocity of a pressure wave	m/s	$\mathbf{LT^{-1}}$
V	Potential difference	V	$\mathbf{ML^2T^{-3}I^{-1}}$
β	Coefficient of cubical expansion of fluid	K^{-1}	$\boldsymbol{\theta}^{-1}$
Π	A dimensionless group	—	—
μ	Viscosity	$N\ s/m^2$	$\mathbf{ML^{-1}T^{-1}}$
μ_p	Plastic viscosity	$N\ s/m^2$	$\mathbf{ML^{-1}T^{-1}}$
ρ	Density of fluid	kg/m^3	$\mathbf{ML^{-3}}$
ρ_s	Density of solid	kg/m^3	$\mathbf{ML^{-3}}$
σ	Surface or interfacial tension	N/m	$\mathbf{MT^{-2}}$
Ω	Electrical resistance	Ohm	$\mathbf{ML^2T^{-3}I^{-2}}$

Dimensions

$\mathbf{H}$	Heat		
$\mathbf{I}$	Electric current	Amp	I
$\mathbf{L}$	Length		
$\mathbf{L_X\ L_Y\ L_Z}$	Length vectors in X-Y-Z directions		
$\mathbf{M}$	Mass		
$\mathbf{M_i}$	Inertial mass		
$\mathbf{M_\mu}$	Quantity mass		
$\mathbf{N}$	Moles		
$\mathbf{T}$	Time		
θ	Temperature		

PART 1

Fluid Flow

CHAPTER 2

Flow of Fluids — Energy and Momentum Relationships

2.1. INTRODUCTION

Chemical engineers are interested in many aspects of the problems involved in the flow of fluids. In the first place, in common with many other engineers, they are concerned with the transport of fluids from one location to another through pipes or open ducts, which requires the determination of the pressure drops in the system, and hence of the power required for pumping, selection of the most suitable type of pump, and measurement of the flow rates. In many cases, the fluid contains solid particles in suspension and it is necessary to determine the effect of these particles on the flow characteristics of the fluid or, alternatively, the drag force exerted by the fluid on the particles. In some cases, such as filtration, the particles are in the form of a fairly stable bed and the fluid has to pass through the tortuous channels formed by the pore spaces. In other cases the shape of the boundary surfaces must be so arranged that a particular flow pattern is obtained: for example, when solids are maintained in suspension in a liquid by means of agitation, the desired effect can be obtained with the minimum expenditure of energy as the most suitable flow pattern is produced in the fluid. Further, in those processes where heat transfer or mass transfer to a flowing fluid occurs, the nature of the flow may have a profound effect on the transfer coefficient for the process.

It is necessary to be able to calculate the energy and momentum of a fluid at various positions in a flow system. It will be seen that energy occurs in a number of forms and that some of these are influenced by the motion of the fluid. In the first part of this chapter the thermodynamic properties of fluids will be discussed. It will then be seen how the thermodynamic relations are modified if the fluid is in motion. In later chapters, the effects of frictional forces will be considered, and the principal methods of measuring flow will be described.

2.2. INTERNAL ENERGY

When a fluid flows from one location to another, energy will, in general, be converted from one form to another. The energy which is attributable to the physical state of the fluid is known as internal energy; it is arbitrarily taken as zero at some reference state, such as the absolute zero of temperature or the melting point of ice at atmospheric pressure. A change in the physical state of a fluid will, in general, cause an alteration in the internal energy. An elementary reversible change results from an infinitesimal change in one of the intensive factors acting on the system; the change proceeds at an infinitesimal rate and a small change in the intensive factor in the opposite direction would have caused the process to take place in the reverse direction. Truly reversible changes never occur in practice but they provide a useful standard with which actual processes can be compared. In an irreversible process, changes are caused by a finite difference in the intensive factor and take place at a finite rate. In general the process will be accompanied by the conversion of electrical or mechanical energy into heat, or by the reduction of the temperature difference between different parts of the system.

For a stationary material the change in the internal energy is equal to the difference between the net amount of heat added to the system and the net amount of work done by the system on its surroundings. For an infinitesimal change:

$$\mathrm{d}U = \delta q - \delta W \tag{2.1}$$

where $\mathrm{d}U$ is the small change in the internal energy, δq the small amount of heat added, and δW the net amount of work done on the surroundings.

In this expression consistent units must be used. In the SI system each of the terms in equation 2.1 is expressed in Joules per kilogram (J/kg). In other systems either heat units (e.g. cal/g) or mechanical energy units (e.g. erg/g) may be used. $\mathrm{d}U$ is a small change in the internal energy which is a property of the system; it is therefore a perfect differential. On the other hand, δq and δW are small quantities of heat and work; they are not properties of the system and their values depend on the manner in which the change is effected; they are, therefore, not perfect differentials. For a reversible process, however, both δq and δW can be expressed in terms of properties of the system. For convenience, reference will be made to systems of unit mass and the effects on the surroundings will be disregarded.

A property called entropy is defined by the relation:

$$\mathrm{d}S = \frac{\delta q}{T} \tag{2.2}$$

where $\mathrm{d}S$ is the small change in entropy resulting from the addition of a small quantity of heat δq, at a temperature T, under reversible conditions. From the definition of the thermodynamic scale of temperature, $\oint \delta q/T = 0$ for a reversible cyclic process, and the net change in the entropy is also zero. Thus, for a particular condition of the system, the entropy has a definite value and must be a property of the system; $\mathrm{d}S$ is, therefore, a perfect differential.

For an irreversible process:

$$\frac{\delta q}{T} < \mathrm{d}S = \frac{\delta q}{T} + \frac{\delta F}{T} \quad \text{(say)} \tag{2.3}$$

δF is then a measure of the degree of irreversibility of the process. It represents the amount of mechanical energy converted into heat or the conversion of heat energy at one temperature to heat energy at another temperature. For a finite process:

$$\int_{S_1}^{S_2} T\,\mathrm{d}S = \Sigma\delta q + \Sigma\delta F = q + F \quad \text{(say)} \tag{2.4}$$

When a process is isentropic, $q = -F$; a reversible process is isentropic when $q = 0$, that is a reversible adiabatic process is isentropic.

The increase in the entropy of an irreversible process may be illustrated in the following manner. Considering the spontaneous transfer of a quantity of heat δq from one part of a system at a temperature T_1 to another part at a temperature T_2, then the net change in the entropy of the system as a whole is then:

$$\mathrm{d}S = \frac{\delta q}{T_2} - \frac{\delta q}{T_1}$$

T_1 must be greater than T_2 and dS is therefore positive. If the process had been carried out reversibly, there would have been an infinitesimal difference between T_1 and T_2 and the change in entropy would have been zero.

The change in the internal energy may be expressed in terms of properties of the system itself. For a reversible process:

$$\delta q = T\,\mathrm{d}S \quad \text{(from equation 2.2)} \quad \text{and} \quad \delta W = P\,\mathrm{d}v$$

if the only work done is that resulting from a change in volume, dv.

Thus, from equation 2.1:

$$\mathrm{d}U = T\,\mathrm{d}S - P\,\mathrm{d}v \tag{2.5}$$

Since this relation is in terms of properties of the system, it must also apply to a system in motion and to irreversible changes where the only work done is the result of change of volume.

Thus, in an irreversible process, for a stationary system:

from equations 2.1 and 2.2: $\mathrm{d}U = \delta q - \delta W = T\,\mathrm{d}S - P\,\mathrm{d}v$

and from equation 2.3: $\delta q + \delta F = T\,\mathrm{d}S$

$$\therefore \quad \delta W = P\,\mathrm{d}v - \delta F \tag{2.6}$$

that is, the useful work performed by the system is less than $P\,\mathrm{d}v$ by an amount δF, which represents the amount of mechanical energy converted into heat energy.

The relation between the internal energy and the temperature of a fluid will now be considered. In a system consisting of unit mass of material and where the only work done is that resulting from volume change, the change in internal energy after a reversible change is given by:

$$\mathrm{d}U = \delta q - P\,\mathrm{d}v \quad \text{(from equation 2.1)}$$

If there is no volume change:

$$\mathrm{d}U = \delta q = C_v\,\mathrm{d}T \tag{2.7}$$

where C_v is the specific heat at constant volume.

As this relation is in terms of properties of the system, it must be applicable to all changes at constant volume.

In an irreversible process:

$$\mathrm{d}U = \delta q - (P\,\mathrm{d}v - \delta F) \quad \text{(from equations 2.1 and 2.6)} \tag{2.8}$$

$$= \delta q + \delta F \quad \text{(under conditions of constant volume)}$$

This quantity δF thus represents the mechanical energy which has been converted into heat and which is therefore available for increasing the temperature.

Thus: $$\delta q + \delta F = C_v \, dT = dU. \tag{2.9}$$

For changes that take place under conditions of constant pressure, it is more satisfactory to consider variations in the enthalpy H. The enthalpy is defined by the relation:

$$H = U + Pv. \tag{2.10}$$

Thus: $$dH = dU + P\,dv + v\,dP$$

$$= \delta q - P\,dv + \delta F + P\,dv + v\,dP \quad \text{(from equation 2.8)}$$

for an irreversible process: (For a reversible process $\delta F = 0$)

$\therefore$ $$dH = \delta q + \delta F + v\,dP \tag{2.11}$$

$$= \delta q + \delta F \quad \text{(at constant pressure)}$$

$$= C_p \, dT \tag{2.12}$$

where C_p is the specific heat at constant pressure.

No assumptions have been made concerning the properties of the system and, therefore, the following relations apply to all fluids.

From equation 2.7: $$\left(\frac{\partial U}{\partial T}\right)_v = C_v \tag{2.13}$$

From equation 2.12: $$\left(\frac{\partial H}{\partial T}\right)_P = C_p \tag{2.14}$$

2.3. TYPES OF FLUID

Fluids may be classified in two different ways; either according to their behaviour under the action of externally applied pressure, or according to the effects produced by the action of a shear stress.

If the volume of an element of fluid is independent of its pressure and temperature, the fluid is said to be incompressible; if its volume changes it is said to be compressible. No real fluid is completely incompressible though liquids may generally be regarded as such when their flow is considered. Gases have a very much higher compressibility than liquids, and appreciable changes in volume may occur if the pressure or temperature is altered. However, if the percentage change in the pressure or in the absolute temperature is small, for practical purposes a gas may also be regarded as incompressible. Thus, in practice, volume changes are likely to be important only when the pressure or temperature of a gas changes by a large proportion. The relation between pressure, temperature, and volume of a real gas is generally complex though, except at very high pressures the behaviour of gases approximates to that of the ideal gas for which the volume of a given mass is inversely proportional to the pressure and directly proportional to the absolute temperature. At high pressures and when pressure changes are large, however, there may be appreciable deviations from this law and an approximate equation of state must then be used.

The behaviour of a fluid under the action of a shear stress is important in that it determines the way in which it will flow. The most important physical property affecting the stress distribution within the fluid is its viscosity. For a gas, the viscosity is low and even at high rates of shear, the viscous stresses are small. Under such conditions the gas approximates in its behaviour to an *inviscid* fluid. In many problems involving the flow of a gas or a liquid, the viscous stresses are important and give rise to appreciable velocity gradients within the fluid, and dissipation of energy occurs as a result of the frictional forces set up. In gases and in most pure liquids the ratio of the shear stress to the rate of shear is constant and equal to the viscosity of the fluid. These fluids are said to be *Newtonian* in their behaviour. However, in some liquids, particularly those containing a second phase in suspension, the ratio is not constant and the apparent viscosity of the fluid is a function of the rate of shear. The fluid is then said to be *non-Newtonian* and to exhibit rheological properties. The importance of the viscosity of the fluid in determining velocity profiles and friction losses is discussed in Chapter 3.

The effect of pressure on the properties of an incompressible fluid, an ideal gas, and a non-ideal gas is now considered.

2.3.1. The incompressible fluid (liquid)

By definition, v is independent of P, so that $(\partial v/\partial P)_T = 0$. The internal energy will be a function of temperature but not a function of pressure.

2.3.2. The ideal gas

An *ideal gas* is defined as a gas whose properties obey the law:

$$PV = n\mathbf{R}T \tag{2.15}$$

where V is the volume occupied by n molar units of the gas, $\mathbf{R}$ the universal gas constant, and T the absolute temperature. Here n is expressed in kmol when using the SI system.

This law is closely obeyed by real gases under conditions where the actual volume of the molecules is small compared with the total volume, and where the molecules exert only a very small attractive force on one another. These conditions are met at very low pressures when the distance apart of the individual molecules is large. The value of $\mathbf{R}$ is then the same for all gases and in SI units has the value of 8314 J/kmol K.

When the only external force on a gas is the fluid pressure, the equation of state is:

$$\mathrm{f}(P, V, T, n) = 0$$

Any property may be expressed in terms of any three other properties. Considering the dependence of the internal energy on temperature and volume, then:

$$U = \mathrm{f}(T, V, n)$$

For unit mass of gas:

$$U = \mathrm{f}(T, v)$$

and:

$$Pv = \frac{\mathbf{R}T}{M} \tag{2.16}$$

where M is the molecular weight of the gas and v is the volume per unit mass.

Thus:
$$\mathrm{d}U = \left(\frac{\partial U}{\partial T}\right)_v \mathrm{d}T + \left(\frac{\partial U}{\partial v}\right)_T \mathrm{d}v \tag{2.17}$$

and:
$$T\,\mathrm{d}S = \mathrm{d}U + P\,\mathrm{d}v \quad \text{(from equation 2.5)}$$

$\therefore$
$$T\,\mathrm{d}S = \left(\frac{\partial U}{\partial T}\right)_v \mathrm{d}T + \left[P + \left(\frac{\partial U}{\partial v}\right)_T\right] \mathrm{d}v$$

and:
$$\mathrm{d}S = \left(\frac{\partial U}{\partial T}\right)_v \frac{\mathrm{d}T}{T} + \frac{1}{T}\left[P + \left(\frac{\partial U}{\partial v}\right)_T\right] \mathrm{d}v \tag{2.18}$$

Thus:
$$\left(\frac{\partial S}{\partial T}\right)_v = \frac{1}{T}\left(\frac{\partial U}{\partial T}\right)_v \tag{2.19}$$

and:
$$\left(\frac{\partial S}{\partial v}\right)_T = \frac{1}{T}\left[P + \left(\frac{\partial U}{\partial v}\right)_T\right] \tag{2.20}$$

Then differentiating equation 2.19 by v and equation 2.20 by T and equating:

$$\frac{1}{T}\frac{\partial^2 U}{\partial T \partial v} = \frac{1}{T}\left[\left(\frac{\partial P}{\partial T}\right)_v + \frac{\partial^2 U}{\partial v \partial T}\right] - \frac{1}{T^2}\left[P + \left(\frac{\partial U}{\partial v}\right)_T\right]$$

or
$$\left(\frac{\partial U}{\partial v}\right)_T = T\left(\frac{\partial P}{\partial T}\right)_v - P \tag{2.21}$$

This relation applies to any fluid. For the particular case of an ideal gas, since $Pv = \mathbf{R}T/M$ (equation 2.16):

$$T\left(\frac{\partial P}{\partial T}\right)_v = T\frac{\mathbf{R}}{Mv} = P$$

so that:
$$\left(\frac{\partial U}{\partial v}\right)_T = 0 \tag{2.22}$$

and:
$$\left(\frac{\partial U}{\partial P}\right)_T = \left(\frac{\partial U}{\partial v}\right)_T \left(\frac{\partial v}{\partial P}\right)_T = 0 \tag{2.23}$$

Thus the internal energy of an ideal gas is a function of temperature only. The variation of internal energy and enthalpy with temperature will now be calculated.

$$\mathrm{d}U = \left(\frac{\partial U}{\partial T}\right)_v \mathrm{d}T + \left(\frac{\partial U}{\partial v}\right)_T \mathrm{d}v \quad \text{(equation 2.17)}$$

$$= C_v\,\mathrm{d}T \quad \text{(from equations 2.13 and 2.22)} \tag{2.24}$$

Thus for an ideal gas under all conditions:

$$\frac{dU}{dT} = C_v \tag{2.25}$$

In general, this relation applies only to changes at constant volume. For the particular case of the ideal gas, however, it applies under all circumstances.

Again, since $H = f(T, P)$:

$$\begin{aligned} dH &= \left(\frac{\partial H}{\partial T}\right)_P dT + \left(\frac{\partial H}{\partial P}\right)_T dP \\ &= C_p dT + \left(\frac{\partial U}{\partial P}\right)_T dP + \left(\frac{\partial (Pv)}{\partial P}\right)_T dP \quad \text{(from equations 2.12 and 2.10)} \\ &= C_p dT \end{aligned}$$

since $(\partial U/\partial P)_T = 0$ and $[\partial(Pv)/\partial P]_T = 0$ for an ideal gas.

Thus under all conditions for an ideal gas:

$$\frac{dH}{dT} = C_p \tag{2.26}$$

$$\therefore \quad C_p - C_v = \frac{dH}{dT} - \frac{dU}{dT} = \frac{d(Pv)}{dT} = \frac{\mathbf{R}}{M} \tag{2.27}$$

Isothermal processes

In fluid flow it is important to know how the volume of a gas will vary as the pressure changes. Two important idealised conditions which are rarely obtained in practice are changes at constant temperature and changes at constant entropy. Although not actually reached, these conditions are approached in many flow problems.

For an *isothermal* change in an ideal gas, the product of pressure and volume is a constant. For unit mass of gas:

$$Pv = \frac{\mathbf{R}T}{M} = \text{constant} \qquad \text{(equation 2.16)}$$

Isentropic processes

For an *isentropic* process the enthalpy may be expressed as a function of the pressure and volume:

$$\begin{aligned} H &= f(P, v) \\ dH &= \left(\frac{\partial H}{\partial P}\right)_v dP + \left(\frac{\partial H}{\partial v}\right)_P dv \\ &= \left(\frac{\partial H}{\partial T}\right)_v \left(\frac{\partial T}{\partial P}\right)_v dP + \left(\frac{\partial H}{\partial T}\right)_P \left(\frac{\partial T}{\partial v}\right)_P dv \end{aligned}$$

Since: $$\left(\frac{\partial H}{\partial T}\right)_P = C_p \qquad \text{(equation 2.14)}$$

and:
$$\left(\frac{\partial H}{\partial T}\right)_v = \left(\frac{\partial U}{\partial T}\right)_v + \left(\frac{\partial (Pv)}{\partial T}\right)_v$$

$$= C_v + v\left(\frac{\partial P}{\partial T}\right)_v \quad \text{(from equation 2.13)}$$

Further:
$$dH = dU + P\,dv + v\,dP \quad \text{(from equation 2.10)}$$

$$= T\,dS - P\,dv + P\,dv + v\,dP \quad \text{(from equation 2.5)}$$

$$= T\,dS + v\,dP \tag{2.28}$$

$$= v\,dP \quad \text{(for an isentropic process)} \tag{2.29}$$

Thus, for an isentropic process:

$$v\,dP = \left[C_v + v\left(\frac{\partial P}{\partial T}\right)_v\right]\left(\frac{\partial T}{\partial P}\right)_v dP + C_p\left(\frac{\partial T}{\partial v}\right)_P dv$$

or:
$$\left(\frac{\partial T}{\partial P}\right)_v dP + \frac{C_p}{C_v}\left(\frac{\partial T}{\partial v}\right)_P dv = 0$$

From the equation of state for an ideal gas (equation 2.15):

$$\left(\frac{\partial T}{\partial P}\right)_v = \frac{T}{P} \quad \text{and} \quad \left(\frac{\partial T}{\partial v}\right)_P = \frac{T}{v}$$

$$\therefore \quad \left(\frac{dP}{P}\right) + \gamma\left(\frac{dv}{v}\right) = 0$$

where $\gamma = C_p/C_v$.

Integration gives:

$$\ln P + \gamma \ln v = \text{constant}$$

or:
$$Pv^\gamma = \text{constant} \tag{2.30}$$

This relation holds only approximately, even for an ideal gas, since γ has been taken as a constant in the integration. It does, however, vary somewhat with pressure.

2.3.3. The non-ideal gas

For a non-ideal gas, equation 2.15 is modified by including a compressibility factor Z which is a function of both temperature and pressure:

$$PV = Zn\mathbf{R}T \tag{2.31}$$

At very low pressures, deviations from the ideal gas law are caused mainly by the attractive forces between the molecules and the compressibility factor has a value less than unity. At higher pressures, deviations are caused mainly by the fact that the volume of the molecules themselves, which can be regarded as incompressible, becomes significant compared with the total volume of the gas.

Many equations have been given to denote the approximate relation between the properties of a non-ideal gas. Of these the simplest, and probably the most commonly used, is van der Waals' equation:

$$\left(P + a\frac{n^2}{V^2}\right)(V - nb) = n\mathbf{R}T \tag{2.32}$$

where b is a quantity which is a function of the incompressible volume of the molecules themselves, and a/V^2 is a function of the attractive forces between the molecules. Values of a and b can be expressed in terms of the critical pressure P_c and the critical temperature T_c as $a = \dfrac{27\mathbf{R}^2T_c^2}{64P_c}$ and $b = \dfrac{\mathbf{R}T_c}{8P_c}$. It is seen that as P approaches zero and V approaches infinity, this equation reduces to the equation of state for the ideal gas.

A chart which correlates experimental $P - V - T$ data for all gases is included as Figure 2.1 and this is known as the generalised compressibility-factor chart.[1] Use is made of *reduced* coordinates where the *reduced temperature* T_R, the *reduced pressure* P_R, and the *reduced volume* V_R are defined as the ratio of the actual temperature, pressure, and volume of the gas to the corresponding values of these properties at the critical state. It is found that, at a given value of T_R and P_R, nearly all gases have the same molar volume, compressibility factor, and other thermodynamic properties. This empirical relationship applies to within about 2 per cent for most gases; the most important exception to the rule is ammonia.

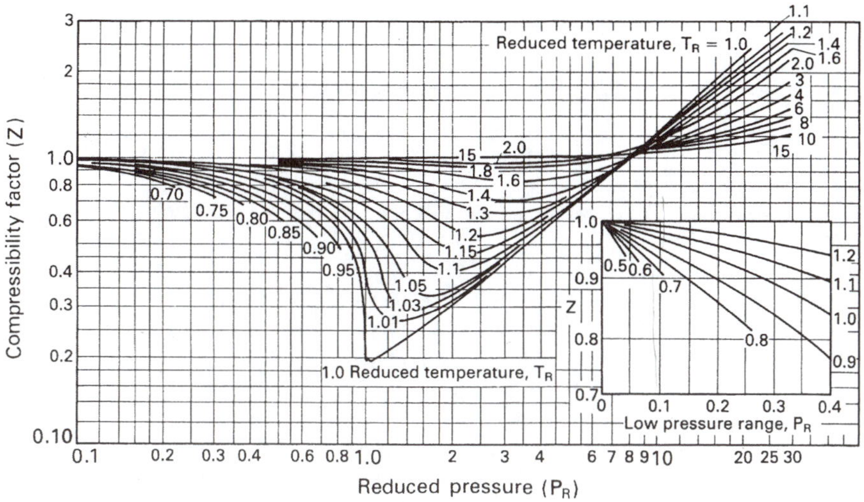

Figure 2.1. Compressibility factors of gases and vapours

The generalised compressibility-factor chart is not to be regarded as a substitute for experimental $P - V - T$ data. If accurate data are available, as they are for some of the more common gases, they should be used.

It will be noted from Figure 2.1 that Z approaches unity for all temperatures as the pressure approaches zero. This serves to confirm the statement made previously that all gases approach ideality as the pressure is reduced to zero. For most gases the critical pressure is 3 MN/m^2 or greater. Thus at atmospheric pressure (101.3 kN/m^2), P_R is 0.033 or less. At this pressure, for any temperature above the critical temperature ($T_R = 1$), it will be seen that Z deviates from unity by no more than 1 per cent. Thus at atmospheric pressure for temperatures greater than the critical temperature, the assumption that the ideal gas law is valid usually leads to errors of less than 1 per cent. It should also be noted that for reduced temperatures between 3 and 10 the compressibility factor is nearly unity for reduced pressures up to a value of 6. For very high temperatures the isotherms approach a horizontal line at $Z = 1$ for all pressures. Thus all gases tend towards ideality as the temperature approaches infinity.

Example 2.1

It is required to store 1 kmol of methane at 320 K and 60 MN/m^2. Using the following methods, estimate the volume of the vessel which must be provided:

(a) ideal gas law;
(b) van der Waals' equation;
(c) generalised compressibility-factor chart;

Solution

For 1 kmol of methane,
(a) $PV = 1 \times \mathbf{R}T$, where $\mathbf{R} = 8314$ J/kmol K.

In this case: $$P = 60 \times 10^6 \text{ N/m}^2;\ T = 320 \text{ K}$$

$$\therefore \quad V = 8314 \times \frac{320}{(60 \times 10^6)} = \underline{\underline{0.0443 \text{ m}^3}}$$

(b) In van der Waals' equation (2.32), the constants may be taken as:

$$a = \frac{27\mathbf{R}^2T_c^2}{64P_c}; \quad b = \frac{\mathbf{R}T_c}{8P_c}$$

where the critical temperature $T_c = 191$ K and the critical pressure $P_c = 4.64 \times 10^6$ N/m^2 for methane as shown in the Appendix tables.

$$\therefore \quad a = \frac{27 \times 8314^2 \times 191^2}{(64 \times 4.64 \times 10^6)} = 229{,}300 \text{ (N/m}^2\text{)(m}^3\text{)}^2/\text{(kmol)}^2$$

and: $$b = 8314 \times \frac{191}{(8 \times 4.64 \times 10^6)} = 0.0427 \text{ m}^3/\text{kmol}$$

Thus in equation 2.32:

$$\left(60 \times 10^6 + 229{,}300 \times \frac{1}{V^2}\right)\left(V - (1 \times 0.0427)\right) = 1 \times 8314 \times 320$$

or: $$V^3 - 0.0427V^2 + 0.000382 = 0.0445$$

Solving by trial and error: $$V = \underline{\underline{0.066 \text{ m}^3}}$$

(c) $$T_r = \frac{T}{T_c} = \frac{320}{191} = 1.68$$

$$P_r = \frac{P}{P_c} = \frac{60 \times 10^3}{4640} = 12.93$$

Thus from Figure 2.1, $Z = 1.33$

and: $$V = \frac{Zn\mathbf{R}T}{P} \quad \text{(from equation 2.31)}$$

$$= \frac{1.33 \times 1.0 \times 8314 \times 320}{(60 \times 10^6)} = \underline{\underline{0.0589 \text{ m}^3}}$$

Example 2.2

Obtain expressions for the variation of:

(a) internal energy with change of volume,
(b) internal energy with change of pressure, and
(c) enthalpy with change of pressure,

all at constant temperature, for a gas whose equation of state is given by van der Waals' Law.

Solution

van der Waals' equation (2.32) may be written for n kmol of gas as:

$$[P + (a/V^2)](V - b) = n\mathbf{R}T \quad \text{(equation 2.32)}$$

or: $$P = [n\mathbf{R}T/(V - b)] - (aV^2)$$

(a) Internal energy and temperature are related by:

$$\left(\frac{\partial U}{\partial V}\right)_T = T\left(\frac{\partial P}{\partial T}\right)_V - P \quad \text{(equation 2.21)}$$

From van der Waals' equation:

$$\left(\frac{\partial P}{\partial T}\right)_V = \frac{n\mathbf{R}}{(V - b)} \quad \text{and} \quad T\left(\frac{\partial P}{\partial T}\right)_V = n\mathbf{R}T/(V - b)$$

Hence: $$\underline{\underline{\left(\frac{\partial U}{\partial V}\right)_T = \frac{n\mathbf{R}T}{(V - b)} - P = \frac{a}{V^2}}}$$

(For an ideal gas: $b = 0$ and $(\partial U/\partial V)_T = (n\mathbf{R}T/V) - P = 0$)

(b) $$\left(\frac{\partial U}{\partial P}\right)_T = \left(\frac{\partial U}{\partial V}\right)_T \left(\frac{\partial V}{\partial P}\right)_T \quad \text{and} \quad (\partial V/\partial P) = 1/(\partial P/\partial V)$$

Hence: $$\frac{\partial P}{\partial V} = \frac{-n\mathbf{R}T}{(V - b)^2} + \frac{2a}{V^3} = \frac{2a(V - b)^2 - n\mathbf{R}TV^3}{V^3(V - b)^2}$$

$$\left(\frac{\partial U}{\partial V}\right)_T = \frac{n\mathbf{R}T}{(V-b)} - P$$

and thus:

$$\left(\frac{\partial U}{\partial P}\right)_T = \left(\frac{n\mathbf{R}T}{(V-b)} - P\right)\left(\frac{V^3(V-b)^2}{2a(V-b)^2 - n\mathbf{R}TV^3}\right)$$

$$= \frac{[n\mathbf{R}T - P(V-b)][V^3(V-b)^2]}{[2a(V-b)^2 - n\mathbf{R}TV^3]}$$

(For an ideal gas, $a = b = 0$ and $(\partial U/\partial P)_T = 0$)

(c) Since H is a function of T and P:

$$\mathrm{d}H = \left(\frac{\partial H}{\partial T}\right)_P \mathrm{d}T + \left(\frac{\partial H}{\partial P}\right)_T \mathrm{d}P$$

$$= C_p\,\mathrm{d}T + \left(\frac{\partial U}{\partial P}\right)_T \mathrm{d}P + \left(\frac{\partial (PV)}{\partial P}\right)_T \mathrm{d}P$$

For a constant temperature process:

$$C_p\,\mathrm{d}T = 0 \quad \text{and} \quad \frac{\mathrm{d}H}{\mathrm{d}P} = \left(\frac{\partial U}{\partial P}\right)_T + \left(\frac{\partial (PV)}{\partial P}\right)_T$$

Thus:

$$\frac{\partial (PV)}{\partial T} = \frac{\partial}{\partial T}[n\mathbf{R}T + bP - (a/V) + (ab/V^2)] = b$$

and:

$$\frac{\mathrm{d}H}{\mathrm{d}P} = \frac{[n\mathbf{R}T - P(V-b)][V^3(V-b)^2] - b}{2a(V-b)^2 - n\mathbf{R}TV^3}$$

Joule–Thomson effect

It has already been shown that the change of internal energy of unit mass of fluid with volume at constant temperature is given by the relation:

$$\left(\frac{\partial U}{\partial v}\right)_T = T\left(\frac{\partial P}{\partial T}\right)_v - P \qquad \text{(equation 2.21)}$$

For a non-ideal gas:

$$T\left(\frac{\partial P}{\partial T}\right)_v \neq P$$

and therefore $(\partial U/\partial v)_T$ and $(\partial U/\partial P)_T$ are not equal to zero.

Thus the internal energy of the non-ideal gas is a function of pressure as well as temperature. As the gas is expanded, the molecules are separated from each other against the action of the attractive forces between them. Energy is therefore stored in the gas; this is released when the gas is compressed and the molecules are allowed to approach one another again.

A characteristic of the non-ideal gas is that it has a finite *Joule–Thomson effect*. This relates to the amount of heat which must be added during an expansion of a gas from a pressure P_1 to a pressure P_2 in order to maintain isothermal conditions. Imagine a gas flowing from a cylinder, fitted with a piston at a pressure P_1 to a second cylinder at a pressure P_2 (Figure 2.2).

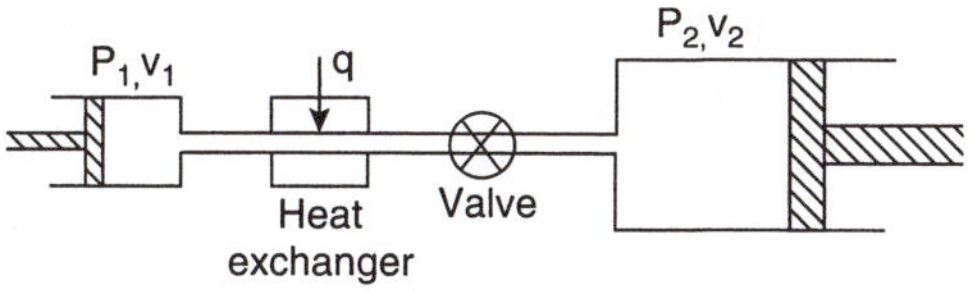

Figure 2.2. Joule-Thomson effect

The net work done by unit mass of gas on the surroundings in expanding from P_1 to P_2 is given by:

$$W = P_2v_2 - P_1v_1 \tag{2.33}$$

A quantity of heat (q, say) is added during the expansion so as to maintain isothermal conditions. The change in the internal energy is therefore given by:

$$\Delta U = q - W \quad \text{(from equation 2.1)}$$

$$\therefore \quad q = \Delta U - P_1v_1 + P_2v_2 \tag{2.34}$$

For an ideal gas, under isothermal conditions, $\Delta U = 0$ and $P_2v_2 = P_1v_1$. Thus $q = 0$ and the ideal gas is said to have a zero Joule-Thomson effect. A non-ideal gas has a Joule-Thomson effect which may be either positive or negative.

2.4. THE FLUID IN MOTION

When a fluid flows through a duct or over a surface, the velocity over a plane at right angles to the stream is not normally uniform. The variation of velocity can be shown by the use of streamlines which are lines so drawn that the velocity vector is always tangential to them. The flowrate between any two streamlines is always the same. Constant velocity over a cross-section is shown by equidistant streamlines and an increase in velocity by closer spacing of the streamlines. There are two principal types of flow which are discussed in detail later, namely streamline and turbulent flow. In streamline flow, movement across streamlines occurs solely as the result of diffusion on a molecular scale and the flowrate is steady. In turbulent flow the presence of circulating current results in transference of fluid on a larger scale, and cyclic fluctuations occur in the flowrate, though the time-average rate remains constant.

A group of streamlines can be taken together to form a streamtube, and thus the whole area for flow can be regarded as being composed of bundles of streamtubes.

Figures 2.3, 2.4, and 2.5 show the flow patterns in a straight tube, through a constriction and past an immersed object. In the first case, the streamlines are all parallel to one another, whereas in the other two cases the streamlines approach one another as the passage becomes constricted, indicating that the velocity is increasing.

2.4.1. Continuity

Considering the flow of a fluid through a streamtube, as shown in Figure 2.6, then equating the mass rates of flow at sections 1 and 2:

$$dG = \rho_1 \dot{u}_1 \, dA_1 = \rho_2 \dot{u}_2 \, dA_2 \tag{2.35}$$

where ρ_1, ρ_2 are the densities; $\dot{u}_1$, $\dot{u}_2$ the velocities in the streamtube; and dA_1, dA_2 the flow areas at sections 1 and 2 respectively.

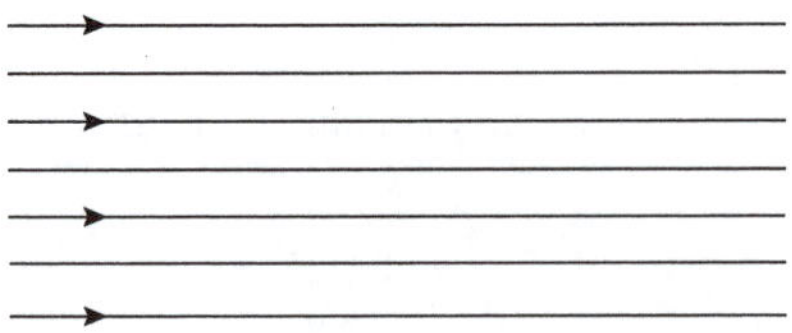

Figure 2.3. Streamlines in a straight tube

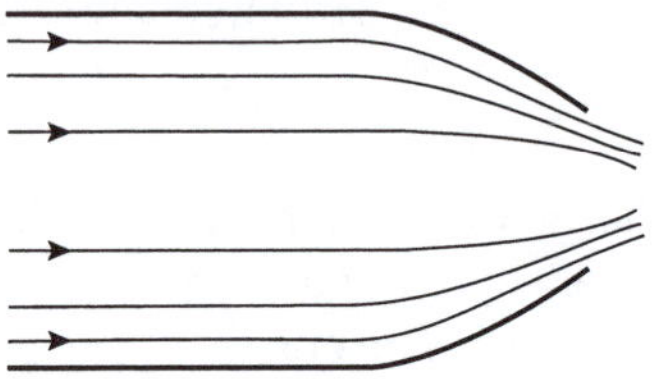

Figure 2.4. Streamlines in a constriction

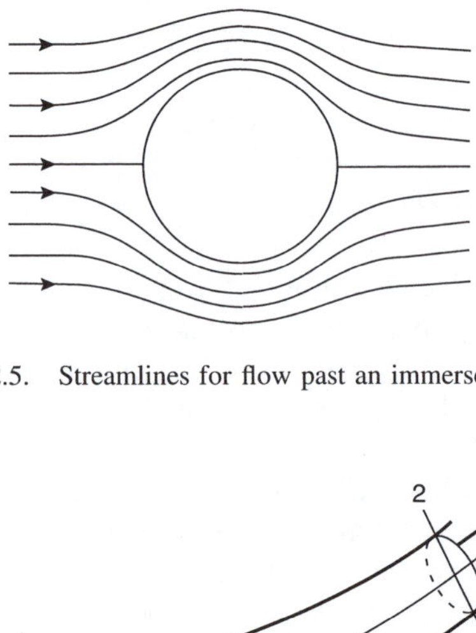

Figure 2.5. Streamlines for flow past an immersed object

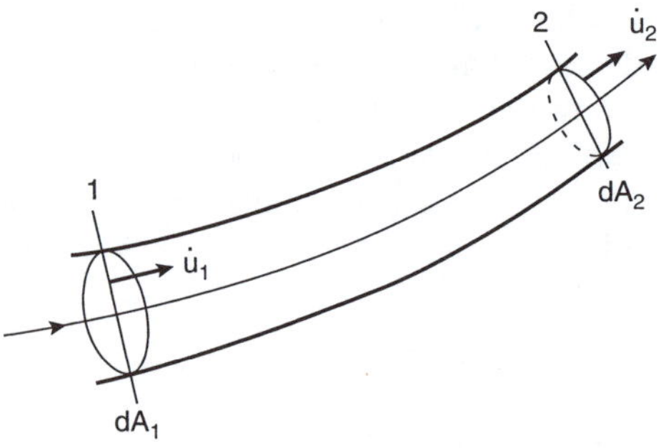

Figure 2.6. Flow through a streamtube

On integration:

$$G = \int \rho_1 \dot{u}_1 \, dA_1 = \int \rho_2 \dot{u}_2 \, dA_2 = \rho_1 u_1 A_1 = \rho_2 u_2 A_2 \tag{2.36}$$

where u_1, u_2 are the average velocities (defined by the previous equations) at the two sections. In many problems, the mass flowrate per unit area G' is the important quantity.

$$G' = \frac{G}{A} = \rho u \tag{2.37}$$

For an incompressible fluid, such as a liquid or a gas where the pressure changes are small:

$$u_1 A_1 = u_2 A_2 \tag{2.38}$$

It is seen that it is important to be able to determine the velocity profile so that the flowrate can be calculated, and this is done in Chapter 3. For streamline flow in a pipe the mean velocity is 0.5 times the maximum stream velocity which occurs at the axis. For turbulent flow, the profile is flatter and the ratio of the mean velocity to the maximum velocity is about 0.82.

2.4.2. Momentum changes in a fluid

As a fluid flows through a duct its momentum and pressure may change. The magnitude of the changes can be considered by applying the momentum equation (force equals rate of change of momentum) to the fluid in a streamtube and then integrating over the cross-section of the duct. The effect of frictional forces will be neglected at first and the relations thus obtained will strictly apply only to an inviscid (frictionless) fluid. Considering an element of length dl of a streamtube of cross-sectional area dA, increasing to $dA + (d(dA)/dl)dl$, as shown in Figure 2.7, then the upstream pressure $= P$ and force attributable to upstream pressure $= P\,dA$.

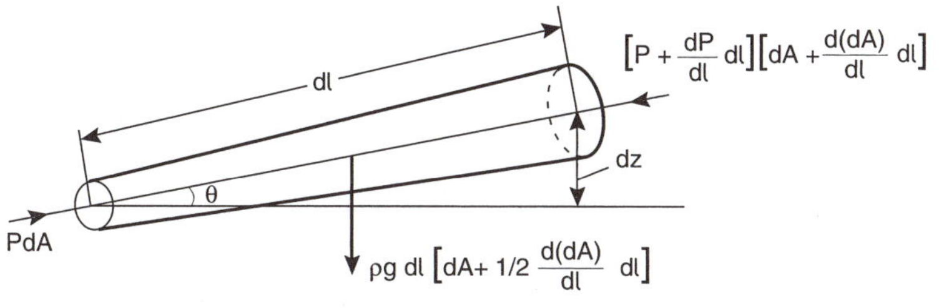

Figure 2.7. Forces on fluid in a streamtube

The downstream pressure $= P + \left(\dfrac{dP}{dl}\right) dl$

This pressure acts over an area $dA + (d(dA)/dl)dl$ and gives rise to a total force of $-\{P + (dP/dl)dl\}\{dA + [d(dA)/dl]dl\}$.

In addition, the mean pressure of $P + \frac{1}{2}(dP/dl)dl$ acting on the sides of the streamtube will give rise to a force having a component $[P + \frac{1}{2}(dP/dl)dl][d(dA)/dl]dl$ along the streamtube. Thus, the net force along the streamtube due to the pressure gradient is:

$$P\,\mathrm{d}A - \left(P + \frac{\mathrm{d}P}{\mathrm{d}l}\mathrm{d}l\right)\left(\mathrm{d}A + \frac{\mathrm{d}(\mathrm{d}A)}{\mathrm{d}l}\mathrm{d}l\right) + \left(P + \frac{1}{2}\frac{\mathrm{d}P}{\mathrm{d}l}\mathrm{d}l\right)\frac{\mathrm{d}(\mathrm{d}A)}{\mathrm{d}l}\mathrm{d}l \approx -\frac{\mathrm{d}P}{\mathrm{d}l}\mathrm{d}l\,\mathrm{d}A$$

The other force acting is the weight of the fluid

$$= \rho g\,\mathrm{d}l\left(\mathrm{d}A + \frac{1}{2}\frac{\mathrm{d}(\mathrm{d}A)}{\mathrm{d}l}\mathrm{d}l\right)$$

The component of this force along the streamtube is

$$-\rho g\,\mathrm{d}l\left(\mathrm{d}A + \frac{1}{2}\frac{\mathrm{d}(\mathrm{d}A)}{\mathrm{d}l}\mathrm{d}l\right)\sin\theta$$

Neglecting second order terms and noting that $\sin\theta = \mathrm{d}z/\mathrm{d}l$:

$$\text{Total force on fluid} = -\frac{\mathrm{d}P}{\mathrm{d}l}\mathrm{d}l\mathrm{d}A - \frac{\mathrm{d}z}{\mathrm{d}l}\rho g\,\mathrm{d}l\,\mathrm{d}A \tag{2.39}$$

The rate of change of momentum of the fluid along the streamtube

$$= (\rho\dot{u}\,\mathrm{d}A)\left[\left(\dot{u} + \frac{\mathrm{d}\dot{u}}{\mathrm{d}l}\mathrm{d}l\right) - \dot{u}\right]$$

$$= \rho\dot{u}\frac{\mathrm{d}\dot{u}}{\mathrm{d}l}\mathrm{d}l\,\mathrm{d}A \tag{2.40}$$

Equating equations 2.39 and 2.40:

$$\rho\dot{u}\,\mathrm{d}l\,\mathrm{d}A\frac{\mathrm{d}\dot{u}}{\mathrm{d}l} = -\mathrm{d}l\,\mathrm{d}A\frac{\mathrm{d}P}{\mathrm{d}l} - \rho g\,\mathrm{d}l\,\mathrm{d}A\frac{\mathrm{d}z}{\mathrm{d}l}$$

$$\therefore \qquad \dot{u}\,\mathrm{d}\dot{u} + \frac{\mathrm{d}P}{\rho} + g\,\mathrm{d}z = 0 \tag{2.41}$$

On integration:

$$\frac{\dot{u}^2}{2} + \int\frac{\mathrm{d}P}{\rho} + gz = \text{constant} \tag{2.42}$$

For the simple case of the incompressible fluid, ρ is independent of pressure, and:

$$\frac{\dot{u}^2}{2} + \frac{P}{\rho} + gz = \text{constant} \tag{2.43}$$

Equation 2.43 is known as Bernoulli's equation, which relates the pressure at a point in the fluid to its position and velocity. Each term in equation 2.43 represents energy per unit mass of fluid. Thus, if all the fluid is moving with a velocity u, the total energy per unit mass ψ is given by:

$$\psi = \frac{u^2}{2} + \frac{P}{\rho} + gz \tag{2.44}$$

Dividing equation 2.44 by g:

$$\frac{u^2}{2g} + \frac{P}{\rho g} + z = \text{constant} \tag{2.45}$$

In equation 2.45 each term represents energy per unit weight of fluid and has the dimensions of length and can be regarded as representing a contribution to the total fluid head.

Thus: $\dfrac{u^2}{2g}$ is the velocity head

$\dfrac{P}{\rho g}$ is the pressure head

and: z is the potential head

Equation 2.42 can also be obtained from consideration of the energy changes in the fluid.

Example 2.3

Water leaves the 25 mm diameter nozzle of a fire hose at a velocity of 25 m/s. What will be the reaction force at the nozzle which the fireman will need to counterbalance?

Solution

$$\begin{aligned}
\text{Mass rate of discharge of water, } G &= \rho u A \\
&= 1000 \times 25 \times \frac{\pi}{4}(0.025)^2 \\
&= 12.27 \text{ kg/s} \\
\text{Momentum of fluid per second} &= Gu \\
&= 12.27 \times 25 \\
&= 307 \text{ N} \\
\text{Reaction force} = \text{Rate of change of momentum} &= \underline{\underline{307 \text{ N}}}
\end{aligned}$$

Example 2.4

Water is flowing at 5 m/s in a 50 mm diameter pipe which incorporates a 90° bend, as shown in Figure 2.8. What is the additional force to which a retaining bracket will be subjected, as a result of the momentum changes in the liquid, if it is arranged symmetrically in the pipe bend?

Solution

Momentum per second of approaching liquid in Y-direction

$$\begin{aligned}
&= \rho u^2 A \\
&= 1000 \times 25 \times \frac{\pi}{4}(0.050)^2 \\
&= 49.1 \text{ N}
\end{aligned}$$

The pipe bracket must therefore exert a reaction force of −49.1 N in the Y-direction, that is in the direction in which the fluid is accelerating. Similarly, the force in the X-direction = 49.1 N

$$\begin{aligned}
\text{The resultant force in direction of arm of bracket} &= 49.1 \cos 45^\circ + 49.1 \sin 45^\circ \\
&= 49.1\left(\frac{1}{\sqrt{2}} + \frac{1}{\sqrt{2}}\right) \\
&= \underline{\underline{69.4 \text{ N}}}
\end{aligned}$$

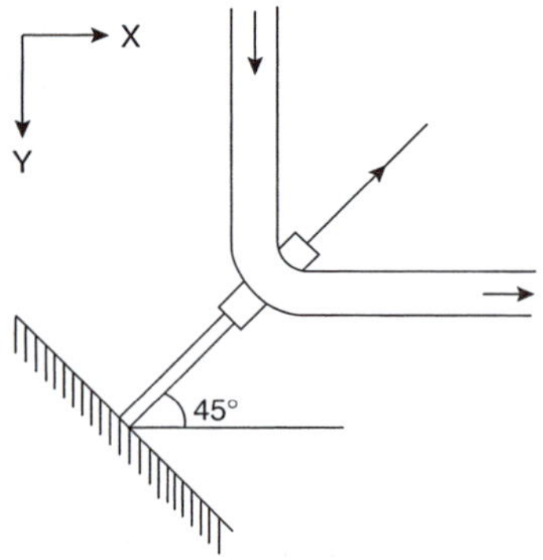

Figure 2.8. Force on support for pipe bend

Water hammer

If the flowrate of a liquid in a pipeline is suddenly reduced, such as by rapid closure of a valve for example, its rate of change of momentum can be sufficiently high for very large forces to be set up which may cause damage to the installation. In a pipeline carrying water, the resulting pressure wave may have a velocity as high as 1200 m/s. The behaviour of the pipe network will be influenced by a large number of factors, including the density and the bulk modulus of elasticity of the liquid, Young's modulus for the material of the pipe, and the design and layout of the installation. The phenomenon is complex[(2)] and reference should be made to one of the specialised texts, such as those by PARMAKIAN[(3)] and SHARP[(4)] for a detailed analysis. The situation can arise with the flow of any liquid, but it is usually referred to as *water hammer* on account of the characteristic sound arising from water distribution systems.

2.4.3. Energy of a fluid in motion

The total energy of a fluid in motion is made up of a number of components. For unit mass of fluid and neglecting changes in magnetic and electrical energy, the magnitutes of the various forms of energy are as follows.

Internal energy U

This has already been discussed in Section 2.2.

Pressure energy

This represents the work which must be done in order to introduce the fluid, without change in volume, into the system. It is therefore given by the product Pv, where P is the pressure of the system and v is the volume of unit mass of fluid.

Potential energy

The potential energy of the fluid, due to its position in the earth's gravitational field, is equal to the work which must be done on it in order to raise it to that position from some arbitrarily chosen datum level at which the potential energy is taken as zero. Thus, if the fluid is situated at a height z above the datum level, the potential energy is zg, where g is the acceleration due to gravity which is taken as constant unless otherwise stated.

Kinetic energy

The fluid possesses kinetic energy by virtue of its motion with reference to some arbitrarily fixed body, normally taken as the earth. If the fluid is moving with a velocity $\dot{u}$, the kinetic energy is $\dot{u}^2/2$.

The total energy of unit mass of fluid is, therefore:

$$U + Pv + gz + \frac{\dot{u}^2}{2} \tag{2.46}$$

If the fluid flows from section 1 to section 2 (where the values of the various quantities are denoted by suffixes 1 and 2 respectively) and q is the net heat absorbed from the surroundings and W_s is the net work done by the fluid on the surroundings, other than that done by the fluid in entering or leaving the section under consideration, then:

$$U_2 + P_2v_2 + gz_2 + \frac{\dot{u}_2^2}{2} = U_1 + P_1v_1 + gz_1 + \frac{\dot{u}_1^2}{2} + q - W_s \tag{2.47}$$

$$\Delta U + \Delta(Pv) + g\Delta z + \Delta\frac{\dot{u}^2}{2} = q - W_s \tag{2.48}$$

where Δ denotes a finite change in the quantities.

Thus:
$$\Delta H + g\Delta z + \frac{\Delta\dot{u}^2}{2} = q - W_s \tag{2.49}$$

It should be noted that the shaft work W_s is related to the total work W by the relation:

$$W = W_s + \Delta(Pv) \tag{2.50}$$

For a small change in the system:

$$dH + g\,dz + \dot{u}\,d\dot{u} = \delta q - \delta W_s \tag{2.51}$$

For many purposes it is convenient to eliminate H by using equation 2.11:

$$dH = \delta q + \delta F + v\,dP \qquad \text{(equation 2.11)}$$

Here δF represents the amount of mechanical energy irreversibly converted into heat.

Thus:
$$\dot{u}\,d\dot{u} + g\,dz + v\,dP + \delta W_s + \delta F = 0 \tag{2.52}$$

When no work is done by the fluid on the surroundings and when friction can be neglected, it will be noted that equation 2.52 is identical to equation 2.41 derived from consideration of a momentum balance, since:

$$v = \frac{1}{\rho}$$

Integrating this equation for flow from section 1 to section 2 and summing the terms δW_s and δF:

$$\Delta\frac{u^2}{2} + g\Delta z + \int_{P_1}^{P_2} v\,\mathrm{d}P + W_s + F = 0 \tag{2.53}$$

Equations 2.41 to 2.53 are quite general and apply therefore to any type of fluid.

With incompressible fluids the energy F is either lost to the surroundings or causes a very small rise in temperature. If the fluid is compressible, however, the rise in temperature may result in an increase in the pressure energy and part of it may be available for doing useful work.

If the fluid is flowing through a channel or pipe, a frictional drag arises in the region of the boundaries and gives rise to a velocity distribution across any section perpendicular to the direction of flow. For the unidirectional flow of fluid, the mean velocity of flow has been defined by equation 2.36 as the ratio of the volumetric flowrate to the cross-sectional area of the channel. When equation 2.52 is applied over the whole cross-section, therefore, allowance must be made for the fact that the mean square velocity is not equal to the square of the mean velocity, and a correction factor α must therefore be introduced into the kinetic energy term. Thus, considering the fluid over the whole cross-section, for small changes:

$$\frac{u\,\mathrm{d}u}{\alpha} + g\,\mathrm{d}z + v\,\mathrm{d}P + \delta W_s + \delta F = 0 \tag{2.54}$$

and for finite changes:

$$\Delta\left(\frac{u^2}{2\alpha}\right) + g\Delta z + \int_{P_1}^{P_2} v\,\mathrm{d}P + W_s + F = 0 \tag{2.55}$$

Before equation 2.55 may be applied to any particular flow problem, the term $\int_{P_1}^{P_2} v\,\mathrm{d}P$ must be evaluated.

Equation 2.50 becomes:

$$\Delta\left(\frac{u^2}{2\alpha}\right) + g\Delta z + \Delta H = q - W_s \tag{2.56}$$

For flow in a pipe of circular cross-section α will be shown to be exactly 0.5 for streamline flow and to approximate to unity for turbulent flow.

For turbulent flow, and where no external work is done, equation 2.54 becomes:

$$u\,\mathrm{d}u + g\,\mathrm{d}z + v\,\mathrm{d}P = 0 \tag{2.57}$$

if frictional effects can be neglected.

For horizontal flow, or where the effects of change of height may be neglected, as normally with gases, equation 2.57 simplifies to:

$$u\,\mathrm{d}u + v\,\mathrm{d}P = 0 \tag{2.58}$$

2.4.4. Pressure and fluid head

In equation 2.54 each term represents energy per unit mass of fluid. If the equation is multiplied throughout by density ρ, each term has the dimensions of pressure and represents energy per unit volume of fluid:

$$\rho\frac{u\,\mathrm{d}u}{\alpha} + \rho g\,\mathrm{d}z + \mathrm{d}P + \rho\delta W_s + \rho\delta F = 0 \tag{2.59}$$

If equation 2.54 is divided throughout by g, each term has the dimensions of length, and, as already noted, may be regarded as a component of the total head of the fluid and represents energy per unit weight:

$$\frac{1}{g}\frac{u\,\mathrm{d}u}{\alpha} + \mathrm{d}z + v\frac{\mathrm{d}P}{g} + \frac{\delta W_s}{g} + \frac{\delta F}{g} = 0 \tag{2.60}$$

For an incompressible fluid flowing in a horizontal pipe of constant cross-section, in the absence of work being done by the fluid on the surroundings, the pressure change due to frictional effects is given by:

$$v\frac{\mathrm{d}P_f}{g} + \frac{\delta F}{g} = 0$$

or:

$$-\,\mathrm{d}P_f = \frac{\delta F}{v} = \rho g\,\mathrm{d}h_f \tag{2.61}$$

where $\mathrm{d}h_f$ is the loss in head corresponding to a change in pressure due to friction of $\mathrm{d}P_f$.

2.4.5. Constant flow per unit area

When the flow rate of the fluid per unit area G' is constant, equation 2.37 can be written:

$$\frac{G}{A} = G' = \frac{u_1}{v_1} = \frac{u_2}{v_2} = \frac{u}{v} \tag{2.62}$$

or:

$$G' = u_1\rho_1 = u_2\rho_2 = u\rho \tag{2.63}$$

Equation 2.58 is the momentum balance for horizontal turbulent flow:

$$u\,\mathrm{d}u + v\,\mathrm{d}P = 0 \qquad \text{(equation 2.58)}$$

or:

$$u\frac{\mathrm{d}u}{v} + \mathrm{d}P = 0$$

Because u/v is constant, on integration this gives:

$$\frac{u_1(u_2 - u_1)}{v_1} + P_2 - P_1 = 0$$

or:

$$\frac{u_1^2}{v_1} + P_1 = \frac{u_2^2}{v_2} + P_2 \tag{2.64}$$

2.4.6. Separation

It may be noted that the energy and mass balance equations assume that the fluid is continuous. This is so in the case of a liquid, provided that the pressure does not fall to such a low value that boiling, or the evolution of dissolved gases, takes place. For water at normal temperatures the pressure should not be allowed to fall below the equivalent of a head of 1.2 m of liquid. With gases, there is no lower limit to the pressures at which the fluid remains continuous, but the various equations which are derived need modification if the pressures are so low that the linear dimensions of the channels become comparable with the mean free path of the molecules, that is when the so-called *molecular flow* sets in.

2.5. PRESSURE–VOLUME RELATIONSHIPS

2.5.1. Incompressible fluids

For incompressible fluids v is independent of pressure so that

$$\int_{P_1}^{P_2} v\,\mathrm{d}P = (P_2 - P_1)v \tag{2.65}$$

Therefore equation 2.55 becomes:

$$\frac{u_1^2}{2\alpha_1} + gz_1 + P_1 v = \frac{u_2^2}{2\alpha_2} + gz_2 + P_2 v + W_s + F \tag{2.66}$$

or:

$$\Delta\frac{u^2}{2\alpha} + g\Delta z + v\Delta P + W_s + F = 0 \tag{2.67}$$

In a frictionless system in which the fluid does not work on the surroundings and α_1 and α_2 are taken as unity (turbulent flow), then:

$$\frac{u_1^2}{2} + gz_1 + P_1 v = \frac{u_2^2}{2} + gz_2 + P_2 v \tag{2.68}$$

Example 2.5

Water flows from a tap at a pressure of 250 kN/m^2 above atmospheric. What is the velocity of the jet if frictional effects are neglected?

Solution

From equation 2.68:

$$0.5(u_2^2 - u_1^2) = g(z_1 - z_2) + \frac{(P_1 - P_2)}{\rho}$$

Using suffix 1 to denote conditions in the pipe and suffix 2 to denote conditions in the jet and neglecting the velocity of approach in the pipe:

$$0.5(u_2^2 - 0) = 9.81 \times 0 + \frac{250 \times 10^3}{1000}$$

$$\underline{\underline{u_2 = 22.4 \text{ m/s}}}$$

2.5.2. Compressible fluids

For a gas, the appropriate relation between specific volume and pressure must be used although, for small changes in pressure or temperature, little error is introduced by using a mean value of the specific volume.

The term $\int_{P_1}^{P_2} v\,dP$ will now be evaluated for the ideal gas under various conditions. In most cases the results so obtained may be applied to the non-ideal gas without introducing an error greater than is involved in estimating the other factors concerned in the process. The only common exception to this occurs in the flow of gases at very high pressures and for the flow of steam, when it is necessary to employ one of the approximate equations for the state of a non-ideal gas, in place of the equation for the ideal gas. Alternatively, equation 2.56 may be used and work expressed in terms of changes in enthalpy. For a gas, the potential energy term is usually small compared with the other energy terms.

The relation between the pressure and the volume of an ideal gas depends on the rate of transfer of heat to the surroundings and the degree of irreversibility of the process. The following conditions will be considered.

(a) an isothermal process;
(b) an isentropic process;
(c) a reversible process which is neither isothermal nor adiabatic;
(d) an irreversible process which is not isothermal.

Isothermal process

For an isothermal process, $Pv = \mathbf{R}T/M = P_1v_1$, where the subscript 1 denotes the initial values and M is the molecular weight.

Thus
$$\int_{P_1}^{P_2} v\,dP = P_1v_1\int_{P_1}^{P_2}\frac{1}{P}dP = P_1v_1\ln\frac{P_2}{P_1} \tag{2.69}$$

Isentropic process

From equation 2.30, for an isentropic process:

$$Pv^\gamma = P_1v_1^\gamma = \text{constant}$$

$$\int_{P_1}^{P_2} v\,dP = \int_{P_1}^{P_2}\left(\frac{P_1v_1^\gamma}{P}\right)^{1/\gamma} dP$$

$$= P_1^{1/\gamma}v_1\int_{P_1}^{P_2} P^{-1/\gamma}\,dP$$

$$= P_1^{1/\gamma}v_1\frac{1}{1-(1/\gamma)}(P_2^{1-(1/\gamma)} - P_1^{1-(1/\gamma)})$$

$$= \frac{\gamma}{\gamma-1}P_1v_1\left[\left(\frac{P_2}{P_1}\right)^{(\gamma-1)/\gamma} - 1\right] \tag{2.70}$$

$$= \frac{\gamma}{\gamma-1}\left[P_1\left(\frac{P_2}{P_1}\right)^{(\gamma-1)/\gamma}\left(\frac{P_2}{P_1}\right)^{1/\gamma} v_2 - P_1v_1\right]$$

$$= \frac{\gamma}{\gamma-1}(P_2v_2 - P_1v_1) \tag{2.71}$$

Further, from equations 2.29 and 2.26, taking C_p as constant:

$$\int_{P_1}^{P_2} v\,dP = \int_{H_1}^{H_2} dH = C_p \Delta T \tag{2.72}$$

The above relations apply for an ideal gas to a reversible adiabatic process which, as already shown, is isentropic.

Reversible process— neither isothermal nor adiabatic

In general the conditions under which a change in state of a gas takes place are neither isothermal nor adiabatic and the relation between pressure and volume is approximately of the form Pv^k = constant for a reversible process, where k is a numerical quantity whose value depends on the heat transfer between the gas and its surroundings. k usually lies between 1 and γ though it may, under certain circumstances, lie outside these limits; it will have the same value for a reversible compression as for a reversible expansion under similar conditions. Under these conditions therefore, equation 2.70 becomes:

$$\int_{P_1}^{P_2} v\,dP = \frac{k}{k-1} P_1 v_1 \left[\left(\frac{P_2}{P_1} \right)^{(k-1)/k} - 1 \right] \tag{2.73}$$

Irreversible process

For an irreversible process it may not be possible to express the relation between pressure and volume as a continuous mathematical function though, by choosing a suitable value for the constant k, an equation of the form Pv^k = constant may be used over a limited range of conditions. Equation 2.73 may then be used for the evaluation of $\int_{P_1}^{P_2} v\,dP$. It may be noted that, for an irreversible process, k will have different values for compression and expansion under otherwise similar conditions. Thus, for the irreversible adiabatic compression of a gas, k will be greater than γ, and for the corresponding expansion k will be less than γ. This means that more energy has to be put into an irreversible compression than will be received back when the gas expands to its original condition.

2.6. ROTATIONAL OR VORTEX MOTION IN A FLUID

In many chemical engineering applications a liquid undergoes rotational motion, such as for example, in a centrifugal pump, in a stirred vessel, in the basket of a centrifuge or in a cyclone-type separator. In the first instance, the effects of friction may be disregarded and consideration will be given to how the forces acting on the liquid determine the pressure distribution. If the liquid may be considered to be rotating about a vertical axis, it will then be subjected to vertical forces due to gravity and centrifugal forces in a horizontal plane. The total force on the liquid and the pressure distribution is then obtained by summing the two components. The vertical pressure gradient attributed to the force of gravity is given by:

$$\frac{\partial P}{\partial z} = -\rho g = -\frac{g}{v} \tag{2.74}$$

The centrifugal force acts in a horizontal plane and the resulting pressure gradient may be obtained by taking a force balance on a small element of liquid as shown in Figure 2.9.

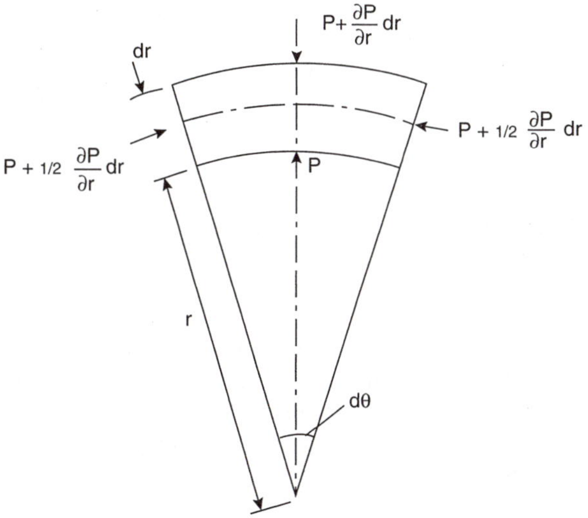

Figure 2.9. Forces acting on element of fluid in a vortex

At radius r, the pressure is P.

At radius $r + \mathrm{d}r$, the pressure is $P + (\partial P/\partial r)\mathrm{d}r$.

For small values of dr, the pressure on the "cut" faces may be taken as $P + \frac{1}{2}(\partial P/\partial r)\mathrm{d}r$.

Then, a force balance in the radial direction on an element of inner radius r, outer radius $r + \mathrm{d}r$, depth dz and subtending a small angle dθ at the centre gives:

$$\left(P + \frac{\partial P}{\partial r}\mathrm{d}r\right)(r + \mathrm{d}r)\mathrm{d}\theta\,\mathrm{d}z - Pr\,\mathrm{d}\theta\,\mathrm{d}z - 2\left(P + \frac{1}{2}\frac{\partial P}{\partial r}\mathrm{d}r\right)\mathrm{d}r\,\mathrm{d}z\,\sin\left(\frac{\mathrm{d}\theta}{2}\right)$$
$$- r\,\mathrm{d}\theta\,\mathrm{d}r\,\mathrm{d}z\,\rho r\omega^2 = 0$$

Simplifying and neglecting small quantities of second order and putting $\sin(\mathrm{d}\theta/2)$ equal to $\mathrm{d}\theta/2$ for a small angle:

$$\frac{\partial P}{\partial r} = \rho\omega^2 r = \frac{\rho u_t^2}{r} \tag{2.75}$$

where u_t is the tangential component of the liquid velocity at radius r.

Now:

$$\mathrm{d}P = \frac{\partial P}{\partial z}\mathrm{d}z + \frac{\partial P}{\partial r}\mathrm{d}r$$

Substituting for $\partial P/\partial z$ and $\partial P/\partial r$ from equations 2.74 and 2.75:

$$\mathrm{d}P = (-\rho g)\mathrm{d}z + (r\rho\omega^2)\,\mathrm{d}r \tag{2.76}$$

Equation 2.76 may be integrated provided that the relation between ω and r is specified. Two important cases are considered:

(a) The *forced vortex* in which ω is constant and independent of r, and
(b) The *free vortex* in which the energy per unit mass of the liquid is constant.

2.6.1. The forced vortex

In a forced vortex the angular velocity of the liquid is maintained constant by mechanical means, such as by an agitator rotating in the liquid or by rotation in the basket of a centrifuge, and:

$$\frac{u_t}{r} = \omega = \text{constant} \tag{2.77}$$

Thus, on integration of equation 2.76, for a constant value of ω:

$$P = -\rho g z + \frac{\rho\omega^2 r^2}{2} + \text{constant}$$

If the z-coordinate is z_a at the point on the axis of rotation which coincides with the free surface of the liquid (or the extension of the free surface), then the corresponding pressure P_0 must be that of the atmosphere in contact with the liquid.

That is, when $r = 0$, $z = z_a$ and $P = P_0$, as shown in Figure 2.10.
Then, on evaluation of the constant:

$$P - P_0 = \frac{\rho\omega^2 r^2}{2} - \rho g(z - z_a) \tag{2.78}$$

For any constant pressure P, equation 2.78 is the equation of a parabola, and therefore all surfaces of constant pressure are paraboloids of revolution. The free surface of the liquid is everywhere at the pressure P_0 of the surrounding atmosphere and therefore is itself a paraboloid of revolution. Putting $P = P_0$ in equation 2.78 for the free surface ($r = r_0, z = z_0$):

$$(z_0 - z_a) = \frac{\omega^2}{2g} r_0^2 \tag{2.79}$$

Differentiating equation 2.79:

$$\frac{dz_0}{dr_0} = \frac{r_0\omega^2}{g} \tag{2.80}$$

Thus the greater the speed of rotation ω, the steeper is the slope. If $r_0\omega^2 \gg g$, $dz_0/dr_0 \to \infty$ and the surface is nearly vertical, and if $r_0\omega^2 \ll g$, $dz_0/dr_0 \to 0$ and the surface is almost horizontal.

The total energy ψ per unit mass of fluid is given by equation 2.44:

$$\psi = \frac{u_t^2}{2} + \frac{P}{\rho} + gz \qquad \text{(equation 2.44)}$$

where u_t denotes the tangential velocity of the liquid.

Substituting $u_t = \omega r$ and for P/ρ from equation 2.78:

$$\psi = \frac{\omega^2 r^2}{2} + \left(\frac{P_0}{\rho} + \frac{\omega^2 r^2}{2} - g(z - z_a)\right) + gz$$

$$= \omega^2 r^2 + \frac{P_0}{\rho} + g z_a \tag{2.81}$$

Thus, the energy per unit mass increases with radius r and is independent of depth z. In the absence of an agitator or mechanical means of rotation energy transfer will take place to equalise ψ between all elements of fluid. Thus the *forced vortex* tends to decay into a *free vortex* (where energy per unit mass is independent of radius).

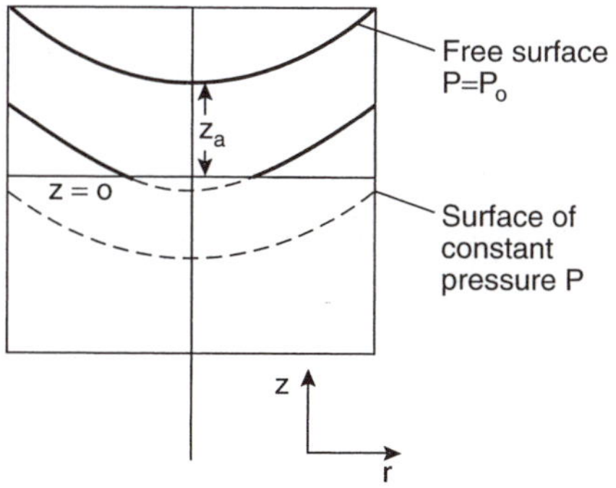

Figure 2.10. Forced vortex

Application of the forced vortex—the centrifuge

Some of the important cases of forced vortexes are:

(a) The movement of liquid within the impeller of a centrifugal pump when there is no flow as, for example, when the outlet valve is closed.
(b) The rotation of liquid within the confines of a stirrer in an agitated tank.
(c) The rotation of liquid in the basket of a centrifuge. This application will now be considered. The operation of centrifuges is considered in detail in Volume 2, Chapter 9.

If liquid is contained in a cylindrical basket which is rotated about a vertical axis, the surfaces of constant pressure, including the free surface are paraboloids of revolution. Thus, in general, the pressure at the walls of the basket is not constant, but varies with height. However, at normal operating speeds the centrifugal force will greatly exceed the gravitational force, and the inner surface of the liquid will be approximately vertical and the wall pressure will be nearly constant. At high operating speeds, where the gravitational force is relatively small, the functioning of the centrifuge is independent of the orientation of the axis of rotation. If mixtures of liquids or suspensions are to be separated in a centrifuge it is necessary to calculate the pressure at the walls arising from the rotation of the basket.

From equation 2.75:

$$\frac{\partial P}{\partial r} = \rho\omega^2 r \qquad \text{(equation 2.75)}$$

If it is assumed that there is no slip between the liquid and the basket, ω is constant and a forced vortex is created.

For a basket of radius R and with the radius of the inner surface of the liquid equal to r_0, the pressure P_R at the walls of the centrifuge is given by integration of equation 2.75

for a given value of z:

$$P_R - P_0 = \frac{\rho\omega^2}{2}(R^2 - r_0^2) \tag{2.82}$$

that is the pressure difference across the liquid at any horizontal level is

$$\frac{\rho\omega^2}{2}(R^2 - r_0^2). \tag{2.82a}$$

Example 2.6

Water is contained in the basket of a centrifuge of 0.5 m internal diameter, rotating at 50 revolutions per second. If the inner radius of the liquid is 0.15 m, what is the pressure at the walls of the basket?

Solution

Angular speed of rotation $= (2\pi \times 50) = 314$ rad/s

The wall pressure is given by equation 2.82 as:

$$\frac{(1000 \times 314^2)}{2}(0.25^2 - 0.15^2)$$

$$= \underline{\underline{1.97 \times 10^6 \text{ N/m}^2}}$$

2.6.2. The free vortex

In a free vortex the energy per unit mass of fluid is constant, and thus a free vortex is inherently stable. The variation of pressure with radius is obtained by differentiating equation 2.44 with respect to radius at constant depth z to give:

$$u_t \frac{\partial u_t}{\partial r} + \frac{1}{\rho}\frac{\partial P}{\partial r} = 0 \tag{2.83}$$

But: $$\frac{\partial P}{\partial r} = \frac{\rho u_t^2}{r} \quad \text{(equation 2.75)}$$

$\therefore$ $$\frac{\partial u_t}{\partial r} + \frac{u_t}{r} = 0$$

and: $$u_t r = \text{constant} = \kappa \tag{2.84}$$

Hence the angular momentum of the liquid is everywhere constant.

Thus: $$\frac{\partial P}{\partial r} = \frac{\rho\kappa^2}{r^3} \tag{2.85}$$

Substituting from equations 2.74 and 2.85 into equation 2.76 and integrating:

$$P - P_\infty = (z_\infty - z)\rho g - \frac{\rho\kappa^2}{2r^2} \tag{2.86}$$

where P_∞ and z_∞ are the values of P and z at $r = \infty$.

Putting $P = P_\infty$ = atmospheric pressure:

$$z = z_\infty - \frac{\kappa^2}{2r^2 g} \tag{2.87}$$

Substituting into equation 2.44 gives:

$$\psi = \frac{P_\infty}{\rho} + g z_\infty \tag{2.88}$$

ψ is constant by definition and equal to the value at $r = \infty$ where $u = 0$.

A free vortex (Figure 2.11) exists:

(a) outside the impeller of a centrifugal pump;
(b) outside the region of the agitator in a stirred tank;
(c) in a cyclone separator or hydrocyclone;
(d) in the flow of liquid into a drain, as in a sink or bath;
(e) in liquid flowing round a bend in a pipe.

In all of these cases the free vortex may be modified by the frictional effect exerted by the external walls.

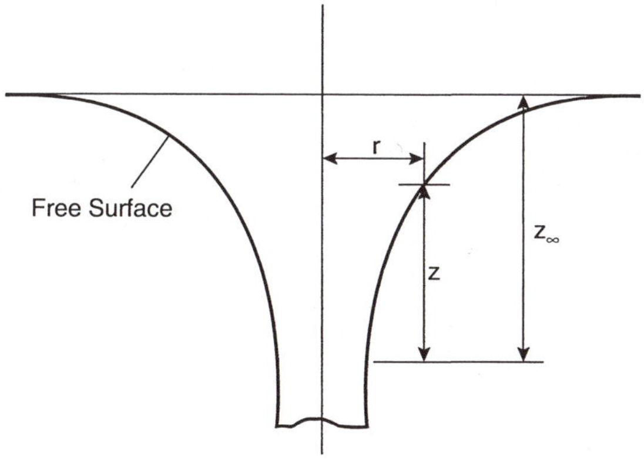

Figure 2.11. Free vortex

2.7. FURTHER READING

DODGE, B. F.: *Chemical Engineering Thermodynamics* (McGraw-Hill, New York, 1944).
DE NEVERS, N.: *Fluid Mechanics for Chemical Engineers*, 2nd edn (McGraw-Hill, New York, 1970).
DOUGLAS, J. F.: *Solution of Problems in Fluid Mechanics* (Pitman, London, 1971).
MASSEY, B. S.: *Mechanics of Fluids,* 6th edn (Chapman and Hall, London, 1989).
MILNE-THOMSON, L. M.: *Theoretical Hydromechanics* (Macmillan, London, 1968).
SCHLICHTING, H.: *Boundary Layer Theory,* 5th edn (McGraw-Hill, New York, 1968).
SMITH, J. M. and VAN NESS, H. C.: *Introduction to Chemical Engineering Thermodynamics,* 5th edn (McGraw-Hill, New York, 1995).

2.8. REFERENCES

1. HOUGEN, O. A. and WATSON, K. M.: *Chemical Process Principles* (Wiley, New York, 1964).
2. CLARKE, D.: *The Chemical Engineer (London)* No.452 (1988) 34. Waterhammer: 1.
3. PARMAKIAN, J.: *Waterhammer Analysis* (Prentice-Hall, Englewood Cliffs, 1955).
4. SHARP, B. B.: *Water Hammer: Problems and Solutions* (Edward Arnold, London, 1981).

2.9. NOMENCLATURE

		Units in SI system	Dimensions in M, N, L, T, θ
A	Area perpendicular to direction of flow	m^2	$\mathbf{L}^2$
a	Coefficient in van der Waals' equation	$(kN/m^2)(m^3)^2/(kmol)^2$	$\mathbf{MN^{-2}L^5T^{-2}}$
b	Coefficient in van der Waals' equation	$m^3/kmol$	$\mathbf{N^{-1}L^3}$
C_p	Specific heat at constant pressure per unit mass	J/kg K	$\mathbf{L^2T^{-2}}\boldsymbol{\theta}^{-1}$
C_v	Specific heat at constant volume per unit mass	J/kg K	$\mathbf{L^2T^{-2}}\boldsymbol{\theta}^{-1}$
F	Energy per unit mass degraded because of irreversibility of process	J/kg	$\mathbf{L^2T^{-2}}$
G	Mass rate of flow	kg/s	$\mathbf{MT^{-1}}$
G'	Mass rate of flow per unit area	kg/m^2s	$\mathbf{ML^{-2}T^{-1}}$
g	Acceleration due to gravity	$9.81\ m/s^2$	$\mathbf{LT^{-2}}$
H	Enthalpy per unit mass	J/kg	$\mathbf{L^2T^{-2}}$
h_f	Head lost due to friction	m	$\mathbf{L}$
k	Numerical constant used as index for compression	—	—
l	Length of streamtube	m	$\mathbf{L}$
M	Molecular weight	kg/kmol	$\mathbf{MN^{-1}}$
n	Number of molar units of fluid	kmol	$\mathbf{N}$
P	Pressure	N/m^2	$\mathbf{ML^{-1}T^{-2}}$
P_B	Pressure at wall of centrifuge basket	N/m^2	$\mathbf{ML^{-1}T^{-2}}$
P_R	Reduced pressure	—	—
q	Net heat flow into system	J/kg	$\mathbf{L^2T^{-2}}$
R	Radius of centrifuge basket	m	$\mathbf{L}$
$\mathbf{R}$	Universal gas constant	(8314)J/kmol K	$\mathbf{MN^{-1}L^2T^{-2}}\boldsymbol{\theta}^{-1}$
r	Radius	m	$\mathbf{L}$
S	Entropy per unit mass	J/kg K	$\mathbf{L^2T^{-2}}\boldsymbol{\theta}^{-1}$
T	Absolute temperature	K	$\boldsymbol{\theta}$
T_R	Reduced temperature	—	—
t	Time	s	$\mathbf{T}$
U	Internal energy per unit mass	J/kg	$\mathbf{L^2T^{-2}}$
u	Mean velocity	m/s	$\mathbf{LT^{-1}}$
u_t	Tangential velocity	m/s	$\mathbf{LT^{-1}}$
$\dot{u}$	Velocity in streamtube	m/s	$\mathbf{LT^{-1}}$
V	Volume of fluid	m^3	$\mathbf{L^3}$
V_R	Reduced volume	—	—
v	Volume per unit mass of fluid	m^3/kg	$\mathbf{M^{-1}L^3}$
W	Net work per unit mass done by system on surroundings	J/kg	$\mathbf{L^2T^{-2}}$
W_s	Shaft work per unit mass	J/kg	$\mathbf{L^2T^{-2}}$
Z	Compressibility factor for non-ideal gas	—	—
z	Distance in vertical direction	m	$\mathbf{L}$
z_a	Value of z_0 at $r_0 = 0$	m	$\mathbf{L}$
α	Constant in expression for kinetic energy of fluid	—	—
γ	Ratio of specific heats C_p/C_v	—	—
θ	Angle	—	—

		Units in SI system	Dimensions in $\mathbf{M, N, L, T}, \theta$
ρ	Density of fluid	kg/m^3	$\mathbf{ML}^{-3}$
ψ	Total mechanical energy per unit mass of fluid	J/kg	$\mathbf{L}^2\mathbf{T}^{-2}$
ω	Angular velocity of rotation	rad/s	$\mathbf{T}^{-1}$

Suffix

c	Value at critical condition
0	Value at free surface
∞	Value at $r = \infty$

CHAPTER 3

Flow of liquids in Pipes and Open Channels

3.1. INTRODUCTION

In the processing industries it is often necessary to pump fluids over long distances, and there may be a substantial drop in pressure in both the pipeline and in individual units. Intermediate products are often pumped from one factory site to another, and raw materials such as natural gas and petroleum products may be pumped very long distances to domestic or industrial consumers. It is necessary, therefore, to consider the problems concerned with calculating the power requirements for pumping, with designing the most suitable flow system, with estimating the most economical sizes of pipes, with measuring the rate of flow, and frequently with controlling this flow at a steady rate. Fluid flow may take place at high pressures, when process streams are fed to a reactor, for instance, or at low pressures when, for example, vapour leaves the top of a vacuum distillation column.

Fluids may be conveniently categorised in a number of different ways. First, the response of the fluid to change of pressure needs to be considered. In general, liquids may be regarded as incompressible in the sense that their densities are substantially independent of the pressure to which they are subjected, and volume changes are insufficient to affect their flow behaviour in most applications of practical interest. On the other hand, gases are highly compressible, and their isothermal densities are approximately directly proportional to the pressure–exactly so when their behaviour follows the *ideal gas law*. Again, compressibility is of only minor importance if the pressure changes by only a small proportion of the total pressure; it is then satisfactory to regard the gas as an incompressible fluid whose properties may be taken as those at the mean pressure. When pressure ratios differ markedly from unity, the effects of compressibility may give rise to fundamental changes in flow behaviour.

All gases and most liquids of simple molecular structure exhibit what is termed *Newtonian* behaviour, and their viscosities are independent of the way in which they are flowing. Temperature may, however, exert a strong influence on viscosity which, for highly viscous liquids, will show a rapid decrease as the temperature is increased. Gases, show the reverse tendency, however, with viscosity rising with increasing temperature, and also with increase of pressure.

Liquids of complex structure, such a polymer solutions and melts, and pseudo-homogeneous suspensions of fine particles, will generally exhibit *non-Newtonian* behaviour, with their apparent viscosities depending on the rate at which they are sheared, and the time for which they have been subjected to shear. They may also exhibit significant elastic properties–similar to those normally associated with solids. The flow behaviour of such fluids is therefore very much more complicated than that of Newtonian fluids.

The fluids discussed so far consist essentially of a single phase and the composition does not vary from place to place within the flow field. Many of the fluids encountered in processing operations consist of more than one phase, however, and the flow behaviour depends on how the phases are distributed. Important cases which will be considered include the flow of gas–liquid mixtures, where the flow pattern will be influenced by the properties of the two phases, their relative proportions and the flow velocity, and by the geometry of the flow passages. Liquids, both Newtonian and non-Newtonian, are frequently used for the transport of particulate solids both in pipelines and in open channels, and it is important to be able to design such systems effectively so that they will operate both reliably and economically. Gases are also used for the transportation of suspended solids in pipelines and, in this case, there is the added complication that the transporting fluid is compressible and the flow velocity will increase along the length of the pipeline.

The treatment of fluid flow in this Volume is structured as follows.

Chapter 3 Flow of Newtonian and non-Newtonian Liquids
Chapter 4 Flow of Compressible Fluids (Gases)
Chapter 5 Flow of Multiphase Systems (gas–liquid, liquid–solids, gas–solids)
Chapter 6 Flow Measurement
Chapter 7 Mixing of Liquids
Chapter 8 Pumping of Liquids and Gases

3.2. THE NATURE OF FLUID FLOW

When a fluid flows through a tube or over a surface, the pattern of the flow varies with the velocity, the physical properties of the fluid, and the geometry of the surface. This problem was first examined by REYNOLDS[1] in 1883 using an apparatus shown in Figure 3.1. A glass tube with a flared entrance was immersed in a glass tank fed with water and, by means of the valve, the rate of flow from the tank through the glass tube was controlled. By introducing a fine filament of coloured water from a small reservoir centrally into the flared entrance of the glass tube, the nature of the flow was observed. At low rates of flow the coloured filament remained at the axis of the tube indicating that the flow was in the form of parallel streams which did not interact with each other. Such flow is called *laminar* or *streamline* and is characterised by the absence of bulk movement at right angles to the main stream direction, though a small amount of radial dispersion will

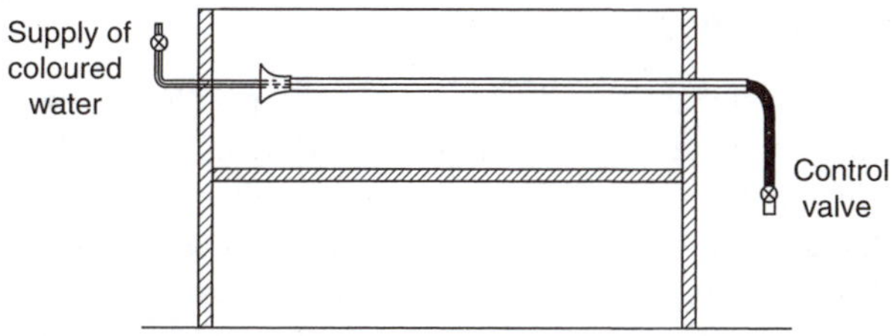

Figure 3.1. Reynolds' method of for tracing flow patterns

occur as a result of diffusion. As the flowrate was increased, oscillations appeared in the coloured filament which broke up into eddies causing dispersion across the tube section. This type of flow, known as *turbulent flow*, is characterised by the rapid movement of fluid as eddies in random directions across the tube. The general pattern is as shown in Figure 3.2. These experiments clearly showed the nature of the transition from streamline to turbulent flow. Below the critical velocity, oscillations in the flow were unstable and any disturbance quickly disappeared. At higher velocities, however, the oscillations were stable and increased in amplitude, causing a high degree of radial mixing. It was found, however, that even when the main flow was turbulent there was a region near the wall (the *laminar sub-layer*) in which streamline flow persisted.

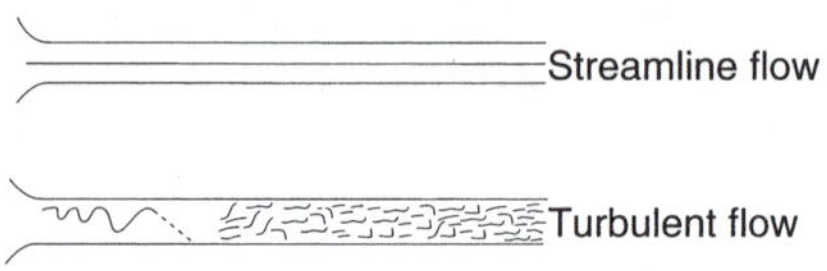

Figure 3.2. Break-up of laminar thread in Reynolds' experiment

In the present discussion only the problem of steady flow will be considered in which the time average velocity in the main stream direction X is constant and equal to u_x. In laminar flow, the instantaneous velocity at any point then has a steady value of u_x and does not fluctuate. In turbulent flow the instantaneous velocity at a point will vary about the mean value of u_x. It is convenient to consider the components of the eddy velocities in two directions—one along the main stream direction X and the other at right angles to the stream flow Y. Since the net flow in the X-direction is steady, the instantaneous velocity u_i may be imagined as being made up of a steady velocity u_x and a fluctuating velocity u_{Ex}, so that:

$$u_i = u_x + u_{Ex} \tag{3.1}$$

Since the average value of the main stream velocity is u_x, the average value of u_{Ex}, is zero, although the fluctuating component may at any instant amount to a significant proportion of the stream velocity. The fluctuating velocity in the Y-direction also varies but, again, this must have an average value of zero since there is no net flow at right angles to the stream flow. Turbulent flow is of great importance in fluids processing because it causes rapid mixing of the fluid elements and is therefore responsible for promoting high rates of heat and mass transfer.

3.2.1. Flow over a surface

When a fluid flows over a surface the elements in contact with the surface will be brought to rest and the adjacent layers retarded by the viscous drag of the fluid. Thus the velocity in the neighbourhood of the surface will change with distance at right angles to the stream flow. It is important to realise that this change in velocity originates at the walls or surface. If a fluid flowing with uniform velocity approaches a plane surface, as shown in Figure 3.3, a velocity gradient is set up at right angles to the surface because of the viscous forces acting within the fluid. The fluid in contact with the surface must be brought to rest as otherwise there would be an infinite velocity gradient at the wall, and a corresponding infinite stress. If u_x is the velocity in the X-direction at distance y from the surface, u_x will increase from zero at the surface ($y = 0$) and will gradually approach the stream velocity u_s at some distance from the surface. Thus, if the values of u_x are measured, the velocity profile will be as shown in Figure 3.3. The velocity distributions are shown for three different distances downstream, and it is seen that in each case there is a rapid change in velocity near the wall and that the thickness of the layer in which the fluid is retarded becomes greater with distance in the direction of flow. The line AB divides the stream into two sections; in the lower part the velocity is increasing with distance from the surface, whilst in the upper portion the velocity is approximately equal to u_s. This line indicates the limits of the zone of retarded fluid which was termed the *boundary layer* by PRANDTL.[2] As shown in Chapter 11, the main stream velocity is approached asymptotically, and therefore the boundary layer strictly has no precise outer limit. However, it is convenient to define the boundary layer thickness such that the velocity at its outer edge equals 99 per cent of the stream velocity. Other definitions are given later. Thus, by making certain assumptions concerning the velocity profile, it is shown in Chapter 11 that the boundary layer thickness δ at a distance x from the leading edge of a surface is dependent on the Reynolds number.

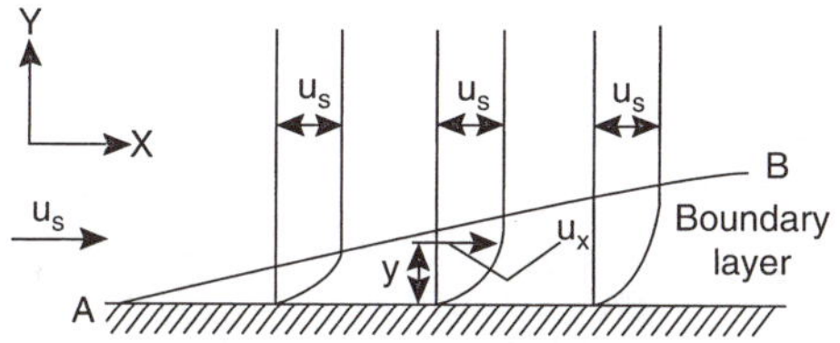

Figure 3.3. Development of boundary layer

Near the leading edge of the surface, the flow in the boundary layer is laminar, and then at a critical distance eddies start to form giving a turbulent boundary layer. In the turbulent layer there is a thin region near the surface where the flow remains laminar, and this is known as the *laminar sub-layer*. The change from laminar to turbulent flow in the boundary layer occurs at different distances downstream depending on the roughness of the surface and the physical properties of the fluid. This is discussed at length in Chapter 11.

3.2.2. Flow in a pipe

When a fluid flowing at a uniform velocity enters a pipe, the layers of fluid adjacent to the walls are slowed down as they are on a plane surface and a boundary layer forms at the entrance. This builds up in thickness as the fluid passes into the pipe. At some distance downstream from the entrance, the boundary layer thickness equals the pipe radius, after which conditions remain constant and *fully developed flow* exists. If the flow in the boundary layers is streamline where they meet, laminar flow exists in the pipe. If the transition has already taken place before they meet, turbulent flow will persist in the region of fully developed flow. The region before the boundary layers join is known as the *entry length* and this is discussed in greater detail in Chapter 11.

3.3. NEWTONIAN FLUIDS

3.3.1. Shearing characteristics of a Newtonian fluid

As a fluid is deformed because of flow and applied external forces, frictional effects are exhibited by the motion of molecules relative to each other. The effects are encountered in all fluids and are due to their *viscosities*. Considering a thin layer of fluid between two parallel planes, distance y apart as shown in Figure 3.4 with the lower plane fixed and a shearing force F applied to the other, since fluids deform continuously under shear, the upper plane moves at a steady velocity u_x relative to the fixed lower plane. When conditions are steady, the force F is balanced by an internal force in the fluid due to its viscosity and the shear force per unit area is proportional to the velocity gradient in the fluid, or:

$$\frac{F}{A} = R_y \propto \frac{u_x}{y} \propto \frac{\mathrm{d}u_x}{\mathrm{d}y} \tag{3.2}$$

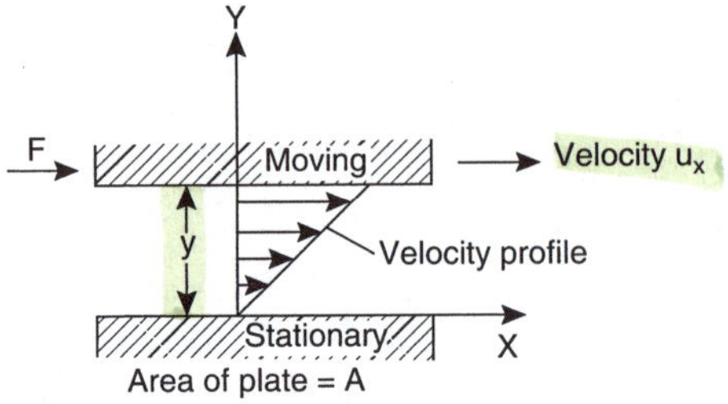

Figure 3.4. Shear stress and velocity gradient in a fluid

R is the shear stress in the fluid and $\mathrm{d}u_x/\mathrm{d}y$ is the velocity gradient or the rate of shear. It may be noted that R corresponds to τ used by many authors to denote shear stress; similarly, shear rate may be denoted by either $\mathrm{d}u_x/\mathrm{d}y$ or $\dot{\gamma}$. The proportionality sign may be replaced by the introduction of the proportionality factor μ, which is the coefficient of viscosity, to give:

$$R_y = \pm\mu \frac{\mathrm{d}u_x}{\mathrm{d}y} \tag{3.3}$$

A *Newtonian* fluid is one in which, provided that the temperature and pressure remain constant, the shear rate increases linearly with shear stress over a wide range of shear rates. As the shear stress tends to retard the fluid near the centre of the pipe and accelerate the slow moving fluid towards the walls, at any radius within the pipe it is acting simultaneously in a negative direction on the fast moving fluid and in the positive direction on the slow moving fluid. In strict terms equation 3.3 should be written with the incorporation of modulus signs to give:

$$\mu = \frac{|R_y|}{|\mathrm{d}u_x/\mathrm{d}y|} \tag{3.4}$$

The viscosity strongly influences the shear stresses and hence the pressure drop for the flow. Viscosities for liquids are generally two orders of magnitude greater than for gases at atmospheric pressure. For example, at 294 K, $\mu_{\text{water}} = 1.0 \times 10^{-3}$ N s/m^2 and $\mu_{\text{air}} = 1.8 \times 10^{-5}$ N s/m^2. Thus for a given shear rate, the shear stresses are considerably greater for liquids. It may be noted that with increase in temperature, the viscosity of a liquid decreases and that of a gas increases. At high pressures, especially near the critical point, the viscosity of a gas increases with increase in pressure.

3.3.2. Pressure drop for flow of Newtonian liquids through a pipe

Experimental work by REYNOLDS,[1] NIKURADSE,[3] STANTON and PANNELL,[4] MOODY,[5] and others on the drop in pressure for flow through a pipe is most conveniently represented by plotting the head loss per unit length against the average velocity through the pipe. In this way, the curve shown in Figure 3.5 is obtained in which $i = h_f/l$ is known as the hydraulic gradient. At low velocities the plot is linear showing that i is directly proportional to the velocity, but at higher velocities the pressure drop increases more rapidly. If logarithmic axes are used, as in Figure 3.6, the results fall into three sections. Over the region of low velocity the line PB has a slope of unity although beyond this region, over section BC, there is instability with poorly defined data. At higher velocities, the line CQ has a slope of about 1.8. If QC is produced, it cuts PB at the point A, corresponding in Reynolds' earlier experiments to the change from laminar to turbulent

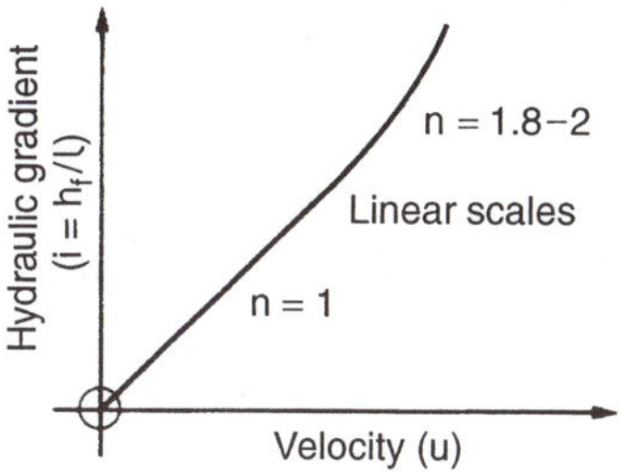

Figure 3.5. Hydraulic gradient versus velocity

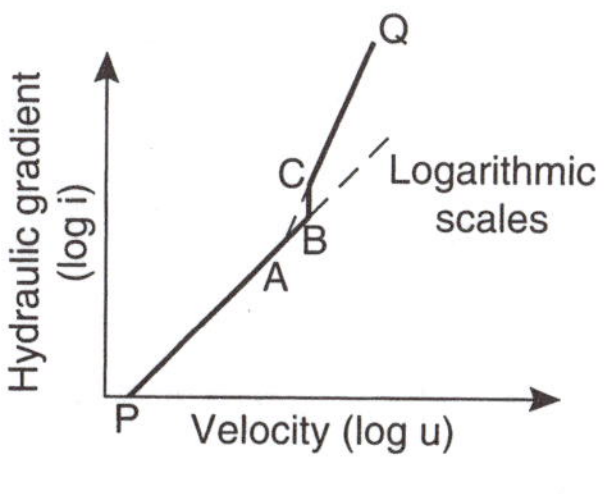

Figure 3.6. Critical velocities

flow and representing the critical velocity. Thus for streamline flow the pressure gradient is directly proportional to the velocity, and for turbulent flow the pressure gradient is proportional to the velocity raised to the power of approximately 1.8.

The velocity corresponding to point A is taken as the lower critical velocity and that corresponding to B as the higher critical velocity. Experiments with pipes of various sizes showed that the critical velocity was inversely proportional to the diameter, and that it was less at higher temperatures where the viscosity was lower. This led Reynolds to develop a criterion based on the velocity of the fluid, the diameter of the tube, and the viscosity and density of the fluid. The dimensionless group $du\rho/\mu$ is termed the *Reynolds number* (*Re*) and this is of vital importance in the study of fluid flow. It has been found that for values of *Re* less than about 2000 the flow is usually laminar and for values above 4000 the flow is usually turbulent. The precise velocity at which the transition occurs depends on the geometry and on the pipe roughness. It is important to realise that there is no such thing as stable transitional flow.

If a turbulent fluid passes into a pipe so that the Reynolds number there is less than 2000, the flow pattern will change and the fluid will become streamline at some distance from the point of entry. On the other hand, if the fluid is initially streamline ($Re < 2000$), the diameter of the pipe can be gradually increased so that the Reynolds number exceeds 2000 and yet streamline flow will persist in the absence of any disturbance. Unstable streamline flow has been obtained in this manner at Reynolds numbers as high as 40,000. The initiation of turbulence requires a small force at right angles to the flow to promote the formation of eddies.

The property of the fluid which appears in the Reynolds number is the kinematic viscosity μ/ρ. The kinematic viscosity of water at 294 K and atmospheric pressure is 10^{-6} m^2/s compared with 15.5×10^{-6} m^2/s for air. Thus, gases typically have higher kinematic viscosities than liquids at atmospheric pressure.

Shear stress in fluid

It is now convenient to relate the pressure drop due to fluid friction $-\Delta P_f$ to the shear stress R_0, at the walls of a pipe. If R_y is the shear stress at a distance y from the wall of the pipe, the corresponding value at the wall R_0 is given by:

$$R_0 = -\mu \left(\frac{du_x}{dy}\right)_{y=0} \qquad \text{(from equation 3.4)}$$

In this equation the negative sign is introduced in order to maintain a consistency of sign convention when shear stress is related to momentum transfer as in Chapter 11. Since $(du_x/dy)_{y=0}$ must be positive (velocity increases towards the pipe centre), R_0 is negative. It is therefore more convenient to work in terms of R, the shear stress exerted by the fluid on the surface $(= -R_0)$ when calculating friction data.

If a fluid is flowing through a length l of pipe, and of radius r (diameter d) over, which the change in pressure due to friction is ΔP_f, then a force balance on the fluid in the direction of flow in the pipe gives:

$$-\Delta P_f \pi r^2 = 2\pi r l(-R_0) = 2\pi r l R$$

or:

$$R = -R_0 = -\Delta P_f \frac{r}{2l} \qquad (3.5)$$

or:
$$-R_0 = R = -\Delta P_f \frac{r}{2l} = -\Delta P_f \frac{d}{4l} \tag{3.6}$$

If a force balance is taken over the central core of fluid of radius s:

$$-\Delta P_f \pi s^2 = 2\pi s l(-R_y)$$

or:
$$-R_y = -\Delta P_f \frac{s}{2l} \tag{3.7}$$

Thus from equations 3.5 and 3.7:

$$\frac{R_y}{-R_0} = \frac{-R_y}{-R_0} = \frac{s}{r} = 1 - \frac{y}{r} \tag{3.8}$$

Thus the shear stress increases linearly from zero at the centre of the pipe to a maximum at the walls, and:

$$\frac{|R_y|}{R} = 1 - \frac{y}{r} \tag{3.9}$$

It may be noted that at the pipe walls, the shear stress acting on the walls R (positive) is equal and opposite to the shear stress acting on the fluid in contact with the walls R_0 (negative)

Thus:
$$R = -R_0 \tag{3.10}$$

Resistance to flow in pipes

STANTON and PANNELL[4] measured the drop in pressure due to friction for a number of fluids flowing in pipes of various diameters and surface roughnesses. The results were expressed by using the concept of a friction factor, defined as the dimensionless group $R/\rho u^2$, which is plotted as a function of the Reynolds number, where here $R(= -R_0)$ represents the resistance to flow per unit area of pipe surface. For a given pipe surface a single curve was found to express the results for all fluids, pipe diameters, and velocities. As with the results of Reynolds the curve was in three parts, as shown in Figure 3.7. At low values of Reynolds number ($Re < 2000$), $R/\rho u^2$ was independent of the surface roughness, but at high values ($Re > 2500$), $R/\rho u^2$ varied with the surface roughness. At very high Reynolds numbers the friction factor became independent of Re and a function of the surface roughness only. Over the transition region of Re, from 2000 to 2500, $R/\rho u^2$ increased very rapidly, showing the great increase in friction as soon as turbulent motion commenced. This general relationship is one of the most widely used in all problems associated with fluid motion, heat transfer, and mass transfer. MOODY[5] worked in terms of a friction factor (here denoted by f') equal to $8R/\rho u^2$ and expressed this factor as a function of the two dimensionless terms Re and e/d where e is a length representing the magnitude of the surface roughness. This relationship may be obtained from dimensional analysis.

Thus if R is a function of u, d, ρ, μ, e, the analysis gives:

$$\frac{R}{\rho u^2} = \text{function of } \frac{ud\rho}{\mu} \text{ and } \frac{e}{d}$$

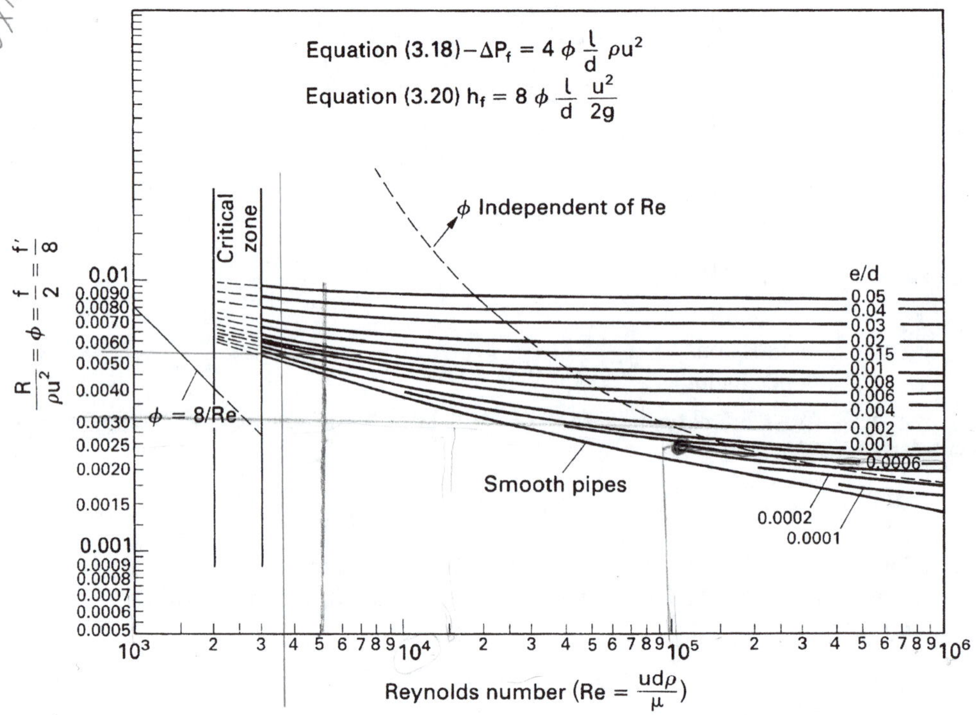

Figure 3.7. Pipe friction chart ϕ versus Re (also see fold-out in the Appendix)

Thus a single curve will correlate the friction factor with the Reynolds group for all pipes with the same degree of roughness of e/d. This curve is of very great importance since it not only determines the pressure loss in the flow but can often be related to heat transfer or mass transfer, as shown in Chapter 12. Such a series of curves for varying values of e/d is shown in Figure 3.7 where the values of ϕ and the Fanning friction f and of the Moody factor f' are related to the Reynolds group. Four separate regions may be distinguished:

Region 1 ($Re < 2000$) corresponds to streamline motion and a single curve represents all the data, irrespective of the roughness of the pipe surface. The equation of the curve is $R/\rho u^2 = 8/Re$.

Region 2 ($2000 < Re < 3000$) is a transition region between streamline and turbulent flow conditions. Reproducible values of pressure drop cannot be obtained in this region, but the value of $R/\rho u^2$ is considerably higher than that in the streamline region. If an unstable form of streamline flow does persist at a value of Re greater than 2000, the frictional force will correspond to that given by the curve $R/\rho u^2 = 8/Re$, extrapolated to values of Re greater than 2000.

Region 3 ($Re > 3000$) corresponds to turbulent motion of the fluid and $R/\rho u^2$ is a function of both Re and e/d, with rough pipes giving high values of $R/\rho u^2$. For smooth pipes there is a lower limit below which $R/\rho u^2$ does not fall for any particular value of Re.

Region 4 corresponds to rough pipes at high values of *Re*. In this region the friction factor becomes independent of *Re* and depends only on (e/d) as follows:

$$\frac{e}{d} = 0.05 \quad Re > 1 \times 10^5 \quad \phi = \frac{R}{\rho u^2} = 0.0087$$

$$\frac{e}{d} = 0.0075 \quad Re > 1 \times 10^5 \quad \phi = \frac{R}{\rho u^2} = 0.0042$$

$$\frac{e}{d} = 0.001 \quad Re > 1 \times 10^6 \quad \phi = \frac{R}{\rho u^2} = 0.0024$$

A number of expressions have been proposed for calculating $R/\rho u^2 (= \phi)$ in terms of the Reynolds number including the following:

Smooth pipes: $2.5 \times 10^3 < Re < 10^5$ $\quad \phi = 0.0396 Re^{-0.25}$ (3.11)

Smooth pipes: $2.5 \times 10^3 < Re < 10^7$ $\quad \phi^{-0.5} = 2.5 \ln(Re\phi^{0.5}) + 0.3$ (see equation 12.77) (3.12)

Rough pipes: $\quad \phi^{-0.5} = -2.5 \ln\left(0.27\frac{e}{d} + 0.885 Re^{-1}\phi^{-0.5}\right)$ (3.13)

Rough pipes: $\frac{e}{d} Re\phi^{0.5} \gg 3.3$ $\quad \phi^{-0.5} = 3.2 - 2.5 \ln\frac{e}{d}$ (3.14)

Equation 3.11 is due to BLASIUS(6) and the others are derived from considerations of velocity profile. In addition to the Moody friction factor $f' = 8R/\rho u^2$, the Fanning or Darcy friction factor $f = 2R/\rho u^2$ is often used. It is extremely important therefore to be clear about the exact definition of the friction factor when using this term in calculating head losses due to friction.

Calculation of pressure drop for liquid flowing in a pipe

For the flow of a fluid in a pipe of length l and diameter d, the total frictional force at the walls is the product of the shear stress R and the surface area of the pipe $(R\pi d l)$. This frictional force results in a change in pressure ΔP_f so that for a horizontal pipe:

$$R\,\pi\,d\,l = -\Delta P_f \pi \frac{d^2}{4} \tag{3.15}$$

or:

$$-\Delta P_f = 4R\frac{l}{d} = 4\frac{R}{\rho u^2}\frac{l}{d}\rho u^2 = 4\phi\frac{l}{d}\rho u^2 \tag{3.16}$$

and:

$$\phi = \frac{R}{\rho u^2} = \frac{-\Delta P d}{4l\rho u^2} \tag{3.17}$$

The head lost due to friction is then:

$$h_f = \frac{-\Delta P_f}{\rho g} = 4\frac{R}{\rho u^2}\frac{l}{d}\frac{u^2}{g} \tag{3.18}$$

The energy dissipated per unit mass F is then given by equation 3.19:

$$F = \frac{-\Delta P_f}{\rho} = 4\frac{R}{\rho u^2}\frac{l}{d}u^2 = 4\phi\frac{l}{d}u^2 \tag{3.19}$$

To calculate $-\Delta P_f$ it is therefore necessary to evaluate e/d and obtain the corresponding value of $\phi = R/\rho u^2$ from a knowledge of the value of Re. This value of ϕ is then used in equation 3.16 to give $-\Delta P_f$ or the head loss due to friction h_f as:

$$h_f = \frac{-\Delta P_f}{\rho g} = 4\phi\frac{l}{d}\frac{u^2}{g} = 8\phi\frac{l}{d}\frac{u^2}{2g} \tag{3.20}$$

With the friction factors used by Moody and Fanning, f' and f respectively, the head loss due to friction is obtained from the following equations:

Moody: $$h_f = f'\frac{l}{d}\frac{u^2}{2g} \tag{3.21}$$

Fanning: $$h_f = 4f\frac{l}{d}\frac{u^2}{2g} \tag{3.22}$$

The energy dissipated per unit mass due to the irreversibility of the process is given by $F = -\Delta P_f/\rho = 4\phi(l/d)u^2$ (equation 3.19).

If it is necessary to calculate the flow in a pipe where the pressure drop is specified, the velocity u is required but the Reynolds number is unknown, and this approach cannot be used to give $R/\rho u^2$ directly. One alternative here is to estimate the value of $R/\rho u^2$ and calculate the velocity and hence the corresponding value of Re. The value of $R/\rho u^2$ is then determined and, if different from the assumed value, a further trial becomes necessary.

An alternative approach to this problem is to use a friction group formed by combining ϕ and Re as follows:

$$\phi Re^2 = \frac{R}{\rho u^2}\left(\frac{\rho u d}{\mu}\right)^2 = \frac{Rd^2\rho}{\mu^2} = \frac{-\Delta P_f d^3\rho}{4l\mu^2} \tag{3.23}$$

If Re is plotted as a function of ϕRe^2 and e/d as shown in Figure 3.8, the group $(-\Delta P_f d^3\rho/4l\mu^2) = \phi Re^2$ may be evaluated directly as it is independent of velocity. Hence Re may be found from the graph and the required velocity $\left(u = \dfrac{Re\mu}{\rho d}\right)$ obtained. Similarly, if the diameter of pipe is required to transport fluid at a mass rate of flow G with a given fall in pressure, the following group, which is independent of d, may be used:

$$\left(\frac{R}{\rho u^2}\right)\left(\frac{ud\rho}{\mu}\right)^{-1} = \phi Re^{-1} = \frac{-\Delta P_f\mu}{4\rho^2u^3l} \tag{3.24}$$

Effect of roughness of pipe surfaces

The estimation of the roughness of the surface of the pipe often presents considerable difficulty. The use of an incorrect value is not usually serious, however, even for turbulent

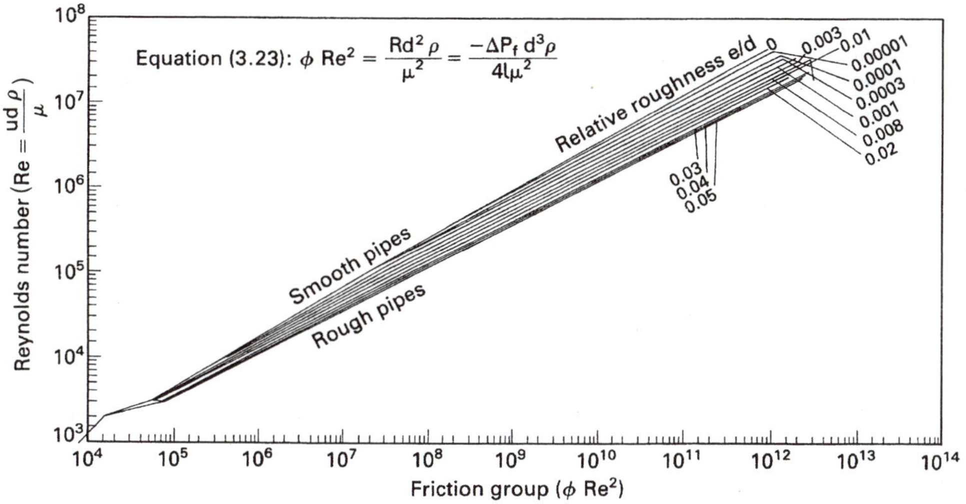

Figure 3.8. Pipe friction chart ϕRe^2 versus Re for various values of e/d (also see fold-out in the Appendix)

flow at low Reynolds numbers because the pressure drop is not critically dependent on the roughness in this region. However, at high values of Reynolds number, the effect of pipe roughness is considerable, as may be seen from the plot of $R/\rho u^2$ against the Reynolds number shown in Figure 3.7. The values of the absolute roughness have been measured for a number of materials and typical data are given in Table 3.1. Where the value for the pipe surface in question is not given, it is necessary to estimate an approximate value based on the available data. Where pipes have become corroded, the value of the roughness is commonly increased, up to tenfold.

Values of roughness applicable to materials used in the construction of open channels are also included in Table 3.1.

Table 3.1. Values of absolute roughness e

	(ft)	(mm)
Drawn tubing	0.000005	0.0015
Commercial steel and wrought-iron	0.00015	0.046
Asphalted cast-iron	0.0004	0.12
Galvanised iron	0.0005	0.15
Cast-iron	0.00085	0.26
Wood stave	0.0006–0.003	0.18–0.9
Concrete	0.001–0.01	0.3–3.0
Riveted steel	0.003–0.03	0.9–9.0

With practical installations it must be remembered that the frictional losses cannot be estimated with very great accuracy because the roughness will change with use and the pumping unit must therefore always have ample excess capacity.

Example 3.1

Ninety-eight per cent sulphuric acid is pumped at 4.5 tonne/h (1.25 kg/s) through a 25 mm diameter pipe, 30 m long, to a reservoir 12 m higher than the feed point. Calculate the pressure drop in the pipeline.

$$\text{Viscosity of acid} = 25 \text{ mN s/m}^2 \text{ or } 25 \times 10^{-3} \text{ N s/m}^2$$

$$\text{Density of acid} = 1840 \text{ kg/m}^3$$

Solution

Reynolds number:

$$Re = \frac{ud\rho}{\mu} = \frac{4G}{\pi\mu d}$$

$$= \frac{4 \times 1.25}{\pi \times 25 \times 10^{-3} \times 25 \times 10^{-3}}$$

$$= 2545$$

For a mild steel pipe, suitable for conveying the acid, the roughness e will be between 0.05 and 0.5 mm (0.00005 and 0.0005 m).

The relative roughness is thus: $\frac{e}{d} = 0.002 \text{ to } 0.02$

From Figure 3.7: $\frac{R}{\rho u^2} = 0.006$ over this range of $\frac{e}{d}$

and the velocity is:

$$u = \frac{G}{\rho A} = \frac{1.25}{1840 \times (\pi/4)(0.025)^2}$$

$$= 1.38 \text{ m/s}$$

The kinetic energy attributable to this velocity will be dissipated when the liquid enters the reservoir. The pressure drop may now be calculated from the energy balance equation and equation 3.19. For turbulent flow of an incompressible fluid:

$$\Delta\frac{u^2}{2} + g\Delta z + v(P_2 - P_1) + 4\frac{R}{\rho u^2}\frac{l}{d}u^2 = 0 \quad \text{(from equation 2.67)}$$

$$\therefore \quad -\Delta P = (P_1 - P_2) = \rho\left\{\left[0.5 + 4\frac{R}{\rho u^2}\frac{l}{d}\right]u^2 + g\Delta z\right\}$$

$$= 1840\left\{\left[0.5 + 4 \times 0.006\frac{30}{0.025}\right]1.38^2 + 9.81 \times 12\right\}$$

$$= 3.19 \times 10^5 \text{ N/m}^2$$

or: $\underline{\underline{320 \text{ kN/m}^2}}$

Example 3.2

Water flows in a 50 mm pipe, 100 m long, whose roughness e is equal to 0.013 mm. If the pressure drop across this length of pipe is not to exceed 50 kN/m^2, what is the maximum allowable water velocity? The density and viscosity of water may be taken as 1000 kg/m^3 and 1.0 mN s/m^2 respectively.

Solution

From equation 3.23:

$$\phi Re^2 = \frac{R}{\rho u^2}Re^2 = -\frac{\Delta P_f d^3 \rho}{4l\mu^2} = \phi Re^2$$

$$-\frac{\Delta P_f d^3 \rho}{4l\mu^2} = \frac{50{,}000(0.05)^3 1000}{4 \times 100(1 \times 10^{-3})^2}$$

$$= 1.56 \times 10^7$$

and:
$$\frac{e}{d} = \frac{0.013}{50} = 0.00026$$

From Figure 3.8, for $\phi Re^2 = 1.56 \times 10^7$ and $(e/d) = 0.00026$, then:

$$Re = \frac{\rho u d}{\mu} = 7.9 \times 10^4$$

Hence:
$$u = \frac{7.9 \times 10^4 (1 \times 10^{-3})}{1000 \times 0.05}$$

$$= \underline{\underline{1.6 \text{ m/s}}}$$

Example 3.3

A cylindrical tank, 5 m in diameter, discharges water through a horizontal mild steel pipe, 100 m long and 225 mm in diameter, connected to the base. What is the time taken for the water level in the tank to drop from 3 m to 0.3 m above the bottom? The viscosity of water may be taken as 1 mN s/m^2.

Solution

If at time t the liquid level is D m above the bottom of the tank, then designating point 1 as the liquid level and point 2 as the pipe outlet, and applying the energy balance equation (2.67) for turbulent flow, then:

$$\Delta\frac{u^2}{2} + g\Delta z + v(P_2 - P_1) + F = 0$$

$$P_2 = P_1 = 101.3 \text{ kN/m}^2$$

$$\frac{u_1}{u_2} = \left(\frac{0.225}{5}\right)^2 = 0.0020, \quad \text{and hence } u_1 \text{ may be neglected}$$

and:
$$\Delta z = -D$$

Thus:
$$\frac{u_2^2}{2} - Dg + 4\frac{R}{\rho u^2}\frac{l}{d}u_2^2 = 0$$

or:
$$u_2^2 - 19.62D + 8\frac{R}{\rho u^2}(444)u_2^2 = 0$$

from which:
$$u_2 = \frac{4.43\sqrt{D}}{\sqrt{1 + 3552(R/\rho u^2)}}$$

As the level of liquid in the tank changes from D to $(D + \mathrm{d}D)$, the quantity of fluid discharged $= (\pi/4)5^2(-D) = -19.63\,\mathrm{d}D$ m^3.

The time taken for the level to change by an amount dD is given by:

$$\mathrm{d}t = \frac{-19.63\,\mathrm{d}D}{(\pi/4)0.225^2 \times 4.43\sqrt{D}/\sqrt{[1 + 3552(R/\rho u^2)]}}$$

$$= -111.5\sqrt{\left[1 + 3552\frac{R}{\rho u^2}\right]}D^{-0.5}\mathrm{d}D$$

and the total time:
$$t = -\int_3^{0.3} 111.5\sqrt{\left[1 + 3552\frac{R}{\rho u^2}\right]}D^{-0.5}\mathrm{d}D$$

Assuming that $R/\rho u^2$ is constant over the range of flow rates considered, then:

$$t = 264\sqrt{\left[1 + 3552\frac{R}{\rho u^2}\right]} \text{ s}$$

If it is assumed that the kinetic energy of the liquid is small compared with the frictional losses, then an approximate value of $R/\rho u^2$ may be calculated.

Pressure drop along the pipe $= D\rho g = \dfrac{4Rl}{d}$:

$$\frac{R}{\rho u^2}Re^2 = \frac{Rd^2\rho}{\mu^2} = \frac{Dg\rho^2 d^3}{4l\mu^2}$$

$$= \frac{(D \times 9.81 \times 1000^2 \times 0.225^3)}{(4 \times 100 \times 0.001^2)}$$

$$= 2.79 \times 10^8 D$$

As D varies from 3 m to 0.3 m, $(R/\rho u^2)Re^2$ varies from 8.38×10^8 to 0.838×10^8. It is of interest to consider whether the difference in roughness of a new or old pipe has a significant effect at this stage.

For a mild steel pipe:

new: $e = 0.00005$ m, $\dfrac{e}{d} = 0.00022$, $Re = 7.0$ to 2.2×10^5 (from Figure 3.7)

old: $e = 0.0005$ m, $\dfrac{e}{d} = 0.0022$, $Re = 6.0$ to 2.2×10^5 (from Figure 3.7)

For a new pipe $R/\rho u^2$ therefore varies from 0.0019 to 0.0020 and for an old pipe $R/\rho u^2 = 0.0029$ (from Figure 3.7).

Taking a value of 0.002 for a new pipe, and assuming that this is constant, then:

$$t = 264\sqrt{(1 + 3552 \times 0.002)} = 264\sqrt{8.1} = 750 \text{ s}$$

The pressure drop due to friction is approximately $(7.1/8.1) = 0.88$ or 88 per cent of the total pressure drop, that is the drop due to friction plus the change in kinetic energy.

Thus: $\dfrac{R}{\rho u^2}Re^2$ varies from about 7.4×10^8 to 0.74×10^8

Re varies from about 6.2×10^5 to 1.9×10^5

and: $\dfrac{R}{\rho u^2}$ varies from about 0.0019 to 0.0020

which is sufficiently close to the assumed value of $R/\rho u^2 = 0.002$.

The time taken for the level to fall is therefore about 750 s or 12.5 min.

Example 3.4

Two storage tanks, A and B, containing a petroleum product, discharge through pipes each 0.3 m in diameter and 1.5 km long to a junction at D, as shown in Figure 3.9. From D the liquid is passed through a 0.5 m diameter pipe to a third storage tank C, 0.75 km away. The surface of the liquid in A is initially 10 m above that in C and the liquid level in B is 6 m higher than that in A. Calculate the initial rate of discharge of liquid into tank C assuming the pipes are of mild steel. The density and viscosity of the liquid are 870 kg/m^3 and 0.7 mN s/m^2 respectively.

Solution

Because the pipes are long, the kinetic energy of the fluid and minor losses at the entry to the pipes may be neglected.

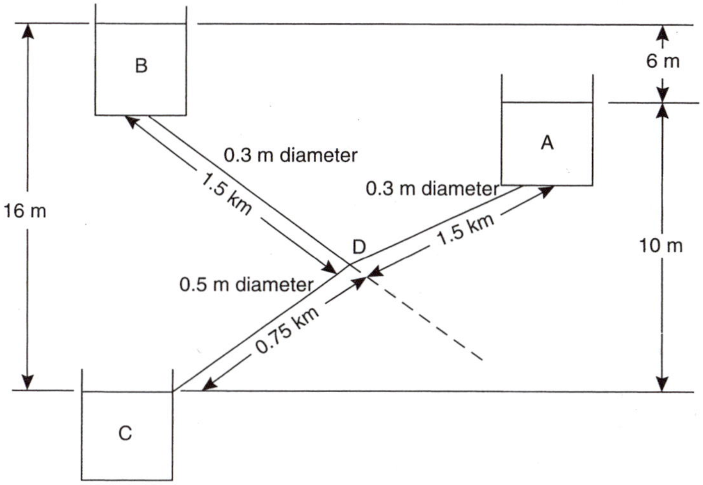

Figure 3.9. Tank layout for Example 3.4

It may be assumed, as a first approximation, that $R/\rho u^2$ is the same in each pipe and that the velocities in pipes AD, BD, and DC are u_1, u_2, and u_3 respectively, if the pressure at D is taken as P_D and point D is z_d m above the datum for the calculation of potential energy, the liquid level in C.

Then applying the energy balance equation between D and the liquid level in each of the tanks gives:

A-D $$(z_d - 10)g + vP_D + 4\frac{R}{\rho u^2}\left(\frac{1500}{0.3}\right)u_1^2 = 0 \qquad (1)$$

B-D $$(z_d - 16)g + vP_D + 4\frac{R}{\rho u^2}\left(\frac{1500}{0.3}\right)u_2^2 = 0 \qquad (2)$$

D-C $$-z_d g - vP_D + 4\frac{R}{\rho u^2}\left(\frac{750}{0.5}\right)u_3^2 = 0 \qquad (3)$$

From equations 1 and 2: $$6g + 20{,}000\frac{R}{\rho u^2}(u_1^2 - u_2^2) = 0 \qquad (4)$$

and from equations 2 and 3: $$-16g + 20{,}000\frac{R}{\rho u^2}(u_2^2 + 0.30u_3^2) = 0 \qquad (5)$$

Taking the roughness of mild steel pipe e as 0.00005 m, e/d varies from 0.0001 to 0.00017.

As a first approximation, $R/\rho u^2$ may be taken as 0.002 in each pipe, and substituting this value in equations 4 and 5 then:

$$58.9 + 40(u_1^2 - u_2^2) = 0 \qquad (6)$$

$$-156.0 + 40(u_2^2 + 0.30u_3^2) = 0 \qquad (7)$$

The flowrate in DC is equal to the sum of the flowrates in AD and BD.

$\therefore$ $$\frac{\pi}{4}0.3^2u_1 + \frac{\pi}{4}0.3^2u_2 = \frac{\pi}{4}0.5^2u_3$$

or: $$u_1 + u_2 = 2.78u_3 \qquad (8)$$

From equation 6: $$u_1^2 = u_2^2 - 1.47 \qquad (9)$$

From equations 7, 8, and 9:

$$-156.0 + 40\left\{u_2^2 + 0.3 \times \left(\frac{1}{2.78}\right)^2 [u_2^2 + u_2^2 - 1.47 + 2u_2\sqrt{(u_2^2 - 1.47)}]\right\} = 0$$

$\therefore$ $$u_2\sqrt{(u_2^2 - 1.47)} = 50.7 - 13.8u_2^2$$

or: $$u_2^4 - 7.38u_2^2 + 13.57 = 0$$

$\therefore$ $$u_2^2 = 0.5[7.38 \pm \sqrt{(54.46 - 54.28)}] = 3.90 \text{ or } 3.48$$

and: $$u_2 = 1.975 \text{ or } 1.87 \text{ m/s}$$

Substituting in equation 9: $$u_1 = 1.56 \text{ or } 1.42 \text{ m/s}$$

Substituting in equation 8: $$u_3 = 1.30 \text{ or } 1.18 \text{ m/s}$$

When these values of u_1, u_2, and u_3 are substituted in equation 7, the lower set of values satisfies the equation and the higher set, introduced as false roots during squaring, does not.

Thus: $$u_1 = 1.42 \text{ m/s}$$

$$u_2 = 1.87 \text{ m/s}$$

and: $$u_3 = 1.18 \text{ m/s}$$

The assumed value of 0.002 for $R/\rho u^2$ must now be checked:

For pipe *AD*: $$Re = \frac{0.3 \times 1.42 \times 870}{0.7 \times 10^{-3}} = 5.3 \times 10^5$$

For pipe *BD*: $$Re = \frac{0.3 \times 1.87 \times 870}{0.7 \times 10^{-3}} = 6.9 \times 10^5$$

For pipe *DC*: $$Re = \frac{0.5 \times 1.18 \times 870}{0.7 \times 10^{-3}} = 7.3 \times 10^5$$

For $e/d = 0.0001$ to 0.00017 and this range of *Re*, $R/\rho u^2$ varies from 0.0019 to 0.0017. The assumed value of 0.002 is therefore sufficiently close.

Thus, the volumetric flowrate is:

$$\frac{\pi}{4}(0.5^2 \times 1.18) = \underline{\underline{0.23 \text{ m}^3/\text{s}}}$$

3.3.3. Reynolds number and shear stress

For a fluid flowing through a pipe the momentum per unit cross-sectional area is given by ρu^2. This quantity, which is proportional to the inertia force per unit area, is the force required to counterbalance the momentum flux.

The ratio u/d represents the velocity gradient in the fluid, and thus the group $(\mu u/d)$ is proportional to the shear stress in the fluid, so that $(\rho u^2)/(\mu u/d) = (du\rho/\mu) = Re$ is proportional to the ratio of the inertia forces to the viscous forces. This is an important physical interpretation of the Reynolds number.

In turbulent flow with high values of *Re*, the inertia forces become predominant and the viscous shear stress becomes correspondingly less important.

In steady streamline flow the direction and velocity of flow at any point remain constant and the shear stress R_y at a point where the velocity gradient at right angles to the direction of flow is $\mathrm{d}u_x/\mathrm{d}y$ and is given, for a Newtonian fluid, by the relation:

$$R_y = -\mu \frac{\mathrm{d}u_x}{\mathrm{d}y} = -\frac{\mu}{\rho}\frac{\mathrm{d}(\rho u_x)}{\mathrm{d}y} \tag{3.25}$$

which gives the relation between shear stress and momentum per unit volume (ρu_x) (equation 3.3). The negative sign in equation 3.25 indicates that the shear stress on the fluid exerts a retarding force on the faster-moving fluid.

In turbulent motion, the presence of circulating or eddy currents brings about a much-increased exchange of momentum in all three directions of the stream flow, and these eddies are responsible for the random fluctuations in velocity u_E. The high rate of transfer in turbulent flow is accompanied by a much higher shear stress for a given velocity gradient.

Thus:

$$R_y = -\left(\frac{\mu}{\rho} + E\right)\frac{\mathrm{d}(\rho u_x)}{\mathrm{d}y} \tag{3.26}$$

where E is known as the *eddy kinematic viscosity* of the fluid, which will depend upon the degree of turbulence in the fluid, is not a physical property of the fluid and varies with position.

In streamline flow, E is very small and approaches zero, so that μ/ρ determines the shear stress. In turbulent flow, E is negligible at the wall and increases very rapidly with distance from the wall. LAUFER(7), using very small hot-wire anemometers, measured the velocity fluctuations and gave a valuable account of the structure of turbulent flow. In the operations of mass, heat, and momentum transfer, the transfer has to be effected through the laminar layer near the wall, and it is here that the greatest resistance to transfer lies.

The Reynolds group will often be used where a moving fluid is concerned. Thus the drag produced as a fluid flows past a particle is related to the Reynolds number in which the diameter of the particle is used in place of the diameter of the pipe. Under these conditions the transition from streamline to turbulent flow occurs at a very much lower value. Again for the flow of fluid through a bed composed of granular particles, a mean dimension of the particles is often used, and the velocity is usually calculated by dividing the flowrate by the total area of the bed. In this case there is no sharp transition from streamline to turbulent flow because the sizes of the individual flow passages vary.

If the surface over which the fluid is flowing contains a series of relatively large projections, turbulence may arise at a very low Reynolds number. Under these conditions, the frictional force will be increased but so will the coefficients for heat transfer and mass transfer, and therefore turbulence is often purposely induced by this method.

3.3.4. Velocity distributions and volumetric flowrates for streamline flow

The velocity over the cross-section of a fluid flowing in a pipe is not uniform. Whilst this distribution in velocity over a diameter can be calculated for streamline flow this is not possible in the same basic manner for turbulent flow.

The pressure drop due to friction and the velocity distribution resulting from the shear stresses within a fluid in streamline Newtonian flow are considered for three cases: (a) the flow through a pipe of circular cross-section, (b) the flow between two parallel plates, and (c) the flow through an annulus. The velocity at any distance from the boundary surfaces are calculated and the mean velocity of the fluid are related to the pressure gradient in the system. For flow through a circular pipe, the kinetic energy of the fluid may be calculated in terms of the mean velocity of flow.

Pipe of circular cross-section

A horizontal pipe with a concentric element marked *ABCD* is shown in Figure 3.10. Since the flow is steady, the net force on this element must be zero. The forces acting are the normal pressures over the ends and shear forces over the curved sides.

$$\text{The force over } AB = P\pi s^2$$

$$\text{The force over } CD = -(P + \Delta P)\pi s^2$$

$$\text{and the force over curved surface} = 2\pi s l R_y$$

$$\text{where the shear stress } R_y = \mu \frac{du_x}{ds} \left(= -\mu \frac{du_x}{dy}\right) \tag{3.27}$$

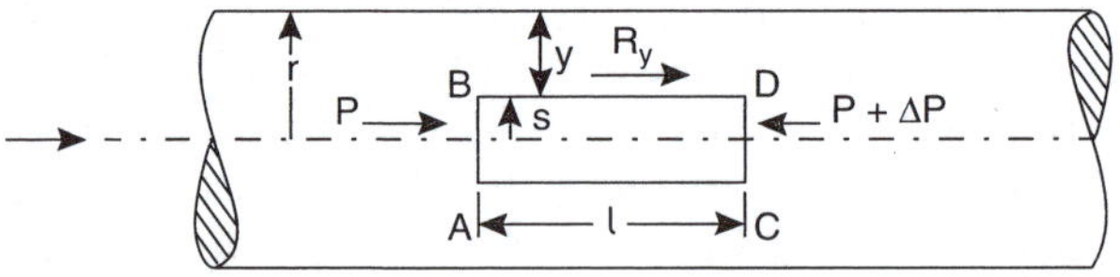

Figure 3.10. Flow through pipe

Taking a force balance:

$$P\pi s^2 - (P + \Delta P)\pi s^2 + 2\pi s l\, \mu \frac{du_x}{ds} = 0$$

or:

$$\left(\frac{-\Delta P}{l}\right) s + 2\mu \frac{du_x}{ds} = 0 \tag{3.28}$$

From equation 3.27:

$$\frac{du_x}{dy} = -\frac{du_x}{ds}$$

and hence in equation 3.28:

$$\frac{du_x}{dy} = \left(\frac{-\Delta P}{l}\right) \frac{s}{2\mu}$$

and the shear rate at the wall is given by:

$$\left(\frac{du_x}{dy}\right)_{y=0} = \left(\frac{-\Delta P}{l}\right) \frac{r}{2\mu} = \left(\frac{-\Delta P}{l}\right) \frac{d}{4\mu} \tag{3.29}$$

The velocity at any distance s from the axis of the pipe may now be found by integrating equation 3.28 to give:

$$u_x = \frac{1}{2\mu}\left(\frac{\Delta P}{l}\right)\frac{s^2}{2} + \text{constant}$$

At the walls of the pipe, that is where $s = r$, the velocity u_x must be zero in order to satisfy the condition of zero wall slip. Substituting the value $u_x = 0$, when $s = r$, then:

$$\text{constant} = \frac{1}{2\mu}\left(\frac{-\Delta P}{l}\right)\frac{r^2}{2}$$

and:

$$u_x = \frac{1}{4\mu}\left(\frac{-\Delta P}{l}\right)(r^2 - s^2) \tag{3.30}$$

Thus the velocity over the cross-section varies in a parabolic manner with the distance from the axis of the pipe. The velocity of flow is seen to be a maximum when $s = 0$, that is at the pipe axis.

Thus the maximum velocity, at the pipe axis, is given by u_{CL} where:

$$u_{\max} = u_{CL} = \frac{1}{4\mu}\left(\frac{-\Delta P}{l}\right)r^2 = \left(\frac{-\Delta P}{l}\right)\frac{d^2}{16\mu} \tag{3.31}$$

Hence:

$$\frac{u_x}{u_{CL}} = 1 - \frac{s^2}{r^2} \tag{3.32}$$

or:

$$= 1 - \frac{4s^2}{d^2} \tag{3.33}$$

The velocity is thus seen to vary in a parabolic manner over the cross-section of the pipe, and this agrees well with experimental measurements.

Volumetric rate of flow and average velocity

If the velocity is taken as constant over an annulus of radii s and $(s + \mathrm{d}s)$, the volumetric rate of flow $\mathrm{d}Q$ through the annulus is given by:

$$\mathrm{d}Q = 2\pi s\,\mathrm{d}s\,u_x$$

$$= 2\pi u_{CL} s\left(1 - \frac{s^2}{r^2}\right)\mathrm{d}s \tag{3.34}$$

The total flow over the cross-section is then given by integrating equation 3.34:

$$Q = 2\pi u_{CL}\int_0^r s\left(1 - \frac{s^2}{r^2}\right)\mathrm{d}s$$

$$= 2\pi u_{CL}\left[\frac{s^2}{2} - \frac{s^4}{4r^2}\right]_0^r$$

$$= \frac{\pi}{2}r^2 u_{CL} = \frac{\pi}{8}d^2 u_{CL} \tag{3.35}$$

Thus the average velocity u is given by:

$$u = \frac{Q}{(\pi d^2/4)}$$

On substitution from equation 3.35 into equation 3.31:

$$u = \left(\frac{-\Delta P}{l}\right)\frac{r^2}{8\mu} = \left(\frac{-\Delta P}{l}\right)\frac{d^2}{32\mu} = \frac{u_{CL}}{2} = \frac{u_{\max}}{2} \tag{3.36}$$

This relation was derived by HAGEN[8] in 1839 and independently by POISEUILLE[9] in 1840.

From equations 3.16 and 3.36:

$$32\mu u\frac{l}{d^2} = -\Delta P = 4\frac{R}{\rho u^2}\frac{l}{d}(\rho u^2)$$

and:

$$\frac{R}{\rho u^2} = \frac{8\mu}{ud\rho} = 8Re^{-1} \tag{3.37}$$

as shown in Figure 3.7.

From equations 3.32 and 3.35:

$$\frac{u_x}{u} = \frac{2(d^2 - 4s^2)}{d^2} = 2\left(1 - \frac{s^2}{r^2}\right) \tag{3.38}$$

Equation 3.38 is plotted in Figure 3.11 which shows the shape of the velocity profile for streamline flow.

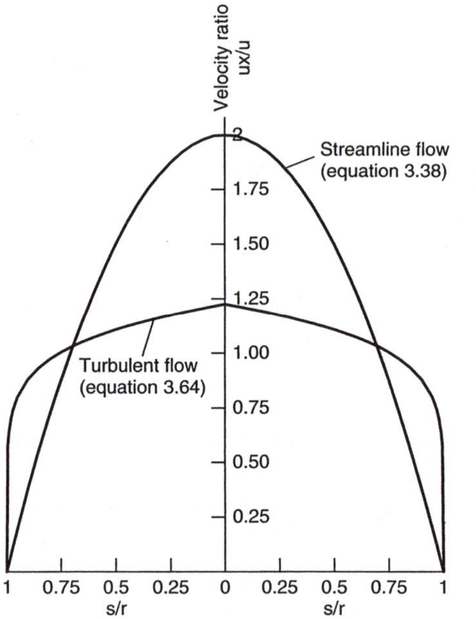

Figure 3.11. Shape of velocity profiles for streamline and turbulent flow

The velocity gradient at any point is obtained by differentiating equation 3.38 with respect to s, or:

$$\frac{\mathrm{d}u_x}{\mathrm{d}y} = -\frac{\mathrm{d}u_x}{\mathrm{d}s} = \frac{4su}{r^2}$$

$$\text{At the wall, } s = r \text{ and } \left(\frac{\mathrm{d}u_x}{\mathrm{d}y}\right)_{y=0} = \frac{4u}{r} = \frac{8u}{d} \tag{3.39}$$

$$\text{Thus, the shear stress at the wall } R = \frac{8\mu u}{d} \tag{3.40}$$

Kinetic energy of fluid

In order to obtain the kinetic energy term for use in the energy balance equation, it is necessary to obtain the average kinetic energy per unit mass in terms of the mean velocity.

The kinetic energy of the fluid flowing per unit time in the annulus between s and $(s + \mathrm{d}s)$ is given by:

$$(\rho 2\pi s\,\mathrm{d}s\,u_x)\frac{u_x^2}{2} = \pi\rho s u_x^3\,\mathrm{d}s$$

$$= \pi\rho u^3 8s\left[1 - \frac{s^2}{r^2}\right]^3 \mathrm{d}s \quad \text{(from equation 3.38)}$$

The total kinetic energy per unit time of the fluid flowing in the pipe is then:

$$= -4\pi\rho u^3 r^2 \int_{s=0}^{s=r}\left(1 - \frac{s^2}{r^2}\right)^3 d\left(1 - \frac{s^2}{r^2}\right)$$

Integrating gives:

$$= -\frac{4\pi\rho u^3 r^2}{4}\left[\left(1 - \frac{s^2}{r^2}\right)^4\right]_{s=0}^{s=r}$$

$$= \rho\pi r^2 u^3$$

Hence the kinetic energy per unit mass is given by:

$$\frac{\rho\pi r^2 u^3}{\rho\pi r^2 u} = u^2 \tag{3.41}$$

Since in the energy balance equation, the kinetic energy per unit mass is expressed as $u^2/2\alpha$, hence $\alpha = 0.5$ for the streamline flow of a fluid in a round pipe.

Flow between two parallel plates

Considering the flow of fluid between two plates of unit width, a distance d_a apart, as shown in Figure 3.12, then for the equilibrium of an element $ABCD$, a force balance may be set up in a similar manner to that used for flow through pipes to give:

$$P\,2s - (P + \Delta P)2s + 2l\mu\frac{\mathrm{d}u_x}{\mathrm{d}s} = 0$$

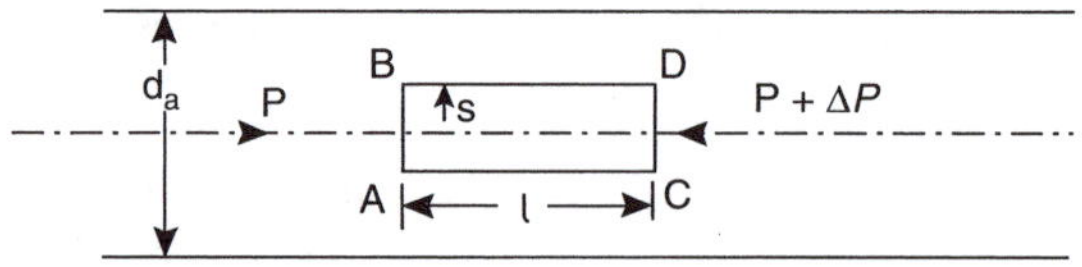

Figure 3.12. Streamline flow between parallel plates

or:
$$\frac{-\Delta P}{l}s + \mu\frac{du_x}{ds} = 0 \tag{3.42}$$

and:
$$u_x = -\frac{1}{\mu}\left(\frac{-\Delta P}{l}\right)\frac{s^2}{2} + \text{constant}$$

When $s = d_a/2$, $u_x = 0$, and:

$$\text{constant} = \frac{d_a^2}{8}\frac{1}{\mu}\left(\frac{-\Delta P}{l}\right)$$

or:
$$u_x = \frac{1}{2\mu}\left(\frac{-\Delta P}{l}\right)\left(\frac{d_a^2}{4} - s^2\right) \tag{3.43}$$

The total rate of flow of fluid between the plates is obtained by calculating the flow through two laminae of thickness ds and situated at a distance s from the centre plane and then integrating. The flow through the laminae is given by:

$$dQ' = \frac{1}{2\mu}\left(\frac{-\Delta P}{l}\right)\left(\frac{d_a^2}{4} - s^2\right)2\,ds$$

On integrating between the limits of s from 0 to $d_a/2$,

the total rate of flow:
$$Q' = \frac{1}{\mu}\left(\frac{-\Delta P}{l}\right)\left(\frac{d_a^3}{8} - \frac{d_a^3}{24}\right)$$
$$= \left(\frac{-\Delta P}{l}\right)\frac{d_a^3}{12\mu} \tag{3.44}$$

The average velocity of the fluid is: $u = \dfrac{Q'}{d_a}$

$$= \left(\frac{-\Delta P}{l}\right)\frac{d_a^2}{12\mu} \tag{3.45}$$

The maximum velocity occurs at the centre plane, and this is obtained by putting $s = 0$ in equation 3.43 to give:

$$\text{Maximum velocity} = u_{max} = \left(\frac{-\Delta P}{l}\right)\frac{d_a^2}{8\mu} = 1.5u \tag{3.46}$$

It has been assumed that the width of the plates is large compared with the distance between them so that the flow may be considered as unidirectional.

Flow through an annulus

The velocity distribution and the mean velocity of a fluid flowing through an annulus of outer radius r and inner radius r_i is more complex. If, as shown in Figure 3.13, the pressure changes, by an amount ΔP as a result of friction in a length l of annulus, the resulting force may be equated to the shearing force acting on the fluid. For the flow of the fluid situated at a distance not greater than s from the centre line of the annuli, the shear force acting on this fluid consists of two parts; one is the drag on its outer surface which may be expressed in terms of the viscosity of the fluid and the velocity gradient at that radius, and the other is the drag occurring at the inner boundary of the annulus, which cannot be estimated at present and will be denoted by the symbol λ for unit length of pipe.

Then:
$$\left(\frac{-\Delta P}{l}\right) l\pi(s^2 - r_i^2) = \mu 2\pi s\, l\left(\frac{du_x}{ds}\right) + \lambda\, l \tag{3.47}$$

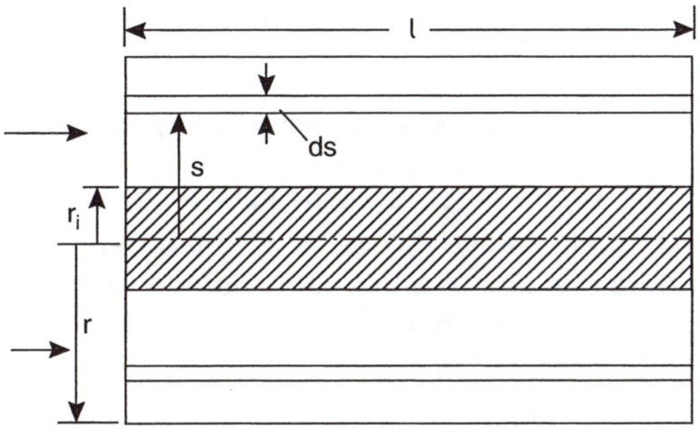

Figure 3.13. Streamline flow through annulus

where u_x is the velocity of the fluid at radius s.

Thus:
$$du_x = \frac{s^2 - r_i^2}{2\mu s}\frac{\Delta P}{l}ds - \lambda\frac{ds}{2\pi\mu s} \tag{3.48}$$

On integration:
$$u_x = \frac{1}{2\mu}\frac{\Delta P}{l}\left(\frac{s^2}{2} - r_i^2\ln s\right) - \frac{\lambda}{2\pi\mu}\ln s + u_c \tag{3.49}$$

where u_c is an integration constant with dimensions of velocity.

Substituting the boundary conditions $s = r_i$, $u_x = 0$, and $s = r$, $u_x = 0$ in equation 3.49 and solving for λ and u_c gives:

$$\lambda = \pi\frac{\Delta P}{l}\left(\frac{r^2 - r_i^2}{2\ln(r/r_i)} - r_i^2\right) \tag{3.50}$$

and:
$$u_c = \frac{1}{2\mu}\left(\frac{-\Delta P}{l}\right)\left(\frac{r^2}{2} - \frac{r^2 - r_i^2}{2\ln(r/r_i)}\ln r\right) \tag{3.51}$$

Substituting these values of λ and u_c into equation 3.49, and simplifying:

$$u_x = \frac{1}{4\mu}\left(\frac{-\Delta P}{l}\right)\left(r^2 - s^2 + \frac{r^2 - r_i^2}{\ln(r/r_i)}\ln\frac{s}{r}\right) \tag{3.52}$$

The rate of flow of fluid through a small annulus of inner radius s and outer radius $(s + ds)$, is given by:

$$dQ = 2\pi s \, ds \, u_x$$

$$= \frac{\pi}{2\mu}\left(\frac{-\Delta P}{l}\right)\left(r^2 s - s^3 + \frac{r^2 - r_i^2}{\ln(r/r_i)} s \ln\frac{s}{r}\right) ds \tag{3.53}$$

Integrating between the limits $s = r_i$ and $s = r$:

$$Q = \frac{-\Delta P \pi}{8\mu l}\left(r^2 + r_i^2 - \frac{r^2 - r_i^2}{\ln(r/r_i)}\right)(r^2 - r_i^2) \tag{3.54}$$

The average velocity is given by: $$u = \frac{Q}{\pi(r^2 - r_i^2)}$$

$$= \frac{-\Delta P}{8\mu l}\left(r^2 + r_i^2 - \frac{r^2 - r_i^2}{\ln(r/r_i)}\right) \tag{3.55}$$

3.3.5. The transition from laminar to turbulent flow in a pipe

Laminar flow ceases to be stable when a small perturbation or disturbance in the flow tends to increase in magnitude rather than decay. For flow in a pipe of circular cross-section, the critical condition occurs at a Reynolds number of about 2100. Thus although laminar flow can take place at much higher values of Reynolds number, that flow is no longer stable and a small disturbance to the flow will lead to the growth of the disturbance and the onset of turbulence. Similarly, if turbulence is artificially promoted at a Reynolds number of less than 2100 the flow will ultimately revert to a laminar condition in the absence of any further disturbance.

In connection with the transition, RYAN and JOHNSON[10] have proposed a *stability parameter* Z. If the critical value Z_c of that parameter is exceeded at any point on the cross-section of the pipe, then turbulence will ensue. Based on a concept of a balance between energy supply to a perturbation and energy dissipation, it was proposed that Z could be defined as:

$$Z = \frac{r\rho u_x}{R}\frac{du_x}{dy} \tag{3.56}$$

Z will be zero at the pipe wall ($u_x = 0$) and at the axis ($\partial u_x/\partial y = 0$), and it will reach a maximum at some intermediate position in the cross-section. From equation 3.6, the wall shear stress R for laminar flow may be expressed in terms of the pressure gradient along the pipe $(-\Delta P/l)$:

$$R = \frac{-\Delta P d}{4l} = \frac{-\Delta P r}{2l} \qquad \text{(equation 3.6)}$$

Thus: $$Z = \frac{2l\rho u_x}{-\Delta P}\frac{\partial u_x}{\partial y}$$

The velocity distribution over the pipe cross-section is given by:

$$u_x = \frac{-\Delta P}{4\mu l}(r^2 - s^2)$$

and:
$$\frac{\mathrm{d}u_x}{\mathrm{d}y} = -\frac{\mathrm{d}u_x}{\mathrm{d}s} = \frac{-\Delta P s}{2\mu l}$$

Thus:
$$Z = \frac{2l\rho}{-\Delta P}\left[\frac{-\Delta P}{4\mu l}(r^2 - s^2)\right]\frac{-\Delta P s}{2\mu l}$$
$$= \frac{-\Delta P \rho}{4\mu^2 l}(r^2 s - s^3)$$

The value of s at which Z is a maximum is obtained by differentiating Z with respect to s and equating the derivative to zero.

$$\frac{\mathrm{d}Z}{\mathrm{d}s} = \frac{-\Delta P \rho}{4\mu^2 l}(r^2 - 3s^2)$$

and, when $\frac{\mathrm{d}Z}{\mathrm{d}s} = 0$:
$$s = \frac{r}{\sqrt{3}}$$

Thus, the position in the cross-section where laminar flow will first break down is where $s/r = 1/\sqrt{3}$

and:
$$Z_{\max} = \frac{-\Delta P}{6\sqrt{3}}\frac{\rho r^3}{\mu^2 l} \tag{3.57}$$

Substituting for $(-\Delta P/l)$ from equation 3.36:

$$u = \frac{-\Delta P r^2}{8\mu l} \qquad \text{(equation 3.36)}$$

$$Z_{\max} = \frac{4}{3\sqrt{3}}\frac{u\rho r}{\mu}$$
$$= \frac{2}{3\sqrt{3}}\frac{ud\rho}{\mu} = 0.384\, Re_{\text{crit}} \tag{3.58}$$

Taking $Re_{\text{crit}} = 2100$, then:
$$Z_{\max} = 808$$

3.3.6. Velocity distributions and volumetric flowrates for turbulent flow

No exact mathematical analysis of the conditions within a turbulent fluid has yet been developed, though a number of semi-theoretical expressions for the shear stress at the walls of a pipe of circular cross-section have been suggested, including that proposed by BLASIUS.[6]

The shear stresses within the fluid are responsible for the frictional force at the walls and the velocity distribution over the cross-section. A given assumption for the shear stress at the walls therefore implies some particular velocity distribution. It will be shown in Chapter 11 that the velocity at any point in the cross-section will be proportional to the one-seventh power of the distance from the walls if the shear stress is given by the Blasius equation (equation 3.11). This may be expressed as:

$$\frac{u_x}{u_{CL}} = \left(\frac{y}{r}\right)^{1/7} \tag{3.59}$$

where u_x is the velocity at a distance y from the walls, u_{CL} the velocity at the axis of the pipe, and r the radius of the pipe.

This equation is sometimes referred to as the *Prandtl one-seventh power law.*

Pipe of circular cross-section

Mean velocity

In a thin annulus of inner radius s and outer radius $s + ds$, the velocity u_x may be taken as constant (see Figure 3.10) and:

$$\therefore \quad dQ = 2\pi s\, ds\, u_x$$

$$= -2\pi(r - y)dy\, u_x (\text{since } s + y = r) \tag{3.60}$$

$$= -2\pi(r - y)dy\, u_{CL}\left(\frac{y}{r}\right)^{1/7} \tag{3.61}$$

Thus:

$$Q = \int_{y=r}^{y=0} -2\pi r^2 u_{CL}\left(1 - \frac{y}{r}\right)\left(\frac{y}{r}\right)^{1/7} d\left(\frac{y}{r}\right)$$

$$= -2\pi r^2 u_{CL}\left[\frac{7}{8}\left(\frac{y}{r}\right)^{8/7} - \frac{7}{15}\left(\frac{y}{r}\right)^{15/7}\right]_{y=r}^{y=0}$$

$$= 2\pi r^2 u_{CL}\left[\tfrac{7}{8} - \tfrac{7}{15}\right]$$

$$= \tfrac{49}{60}\pi r^2 u_{CL} = 0.817\pi r^2 u_{CL} \tag{3.62}$$

The mean velocity of flow is then: $$u = \frac{Q}{\pi r^2} = \frac{49}{60}u_{CL} = 0.817u_{CL} \tag{3.63}$$

This relation holds provided that the one-seventh power law may be assumed to apply over the whole of the cross-section of the pipe. This is strictly the case only at high Reynolds numbers when the thickness of the laminar sub-layer is small. By combining equations 3.59 and 3.63, the velocity profile is given by:

$$\frac{u_x}{u} = \frac{1}{0.817}\left(\frac{y}{r}\right)^{1/7}$$

$$= 1.22\left(1 - \frac{s}{r}\right)^{1/7} \tag{3.64}$$

Equation 3.64 is plotted in Figure 3.11, from which it may be noted that the velocity profile is very much flatter than for streamline flow.

The variation of $(u/u_{\max})$ with Reynolds number is shown in Figure 3.14, from which the sharp change at a Reynolds number between 2000 and 3000 may be noted.

Kinetic energy

Since: $$dQ = -2\pi(r - y)dy\, u_x \quad \text{(equation 3.60)}$$

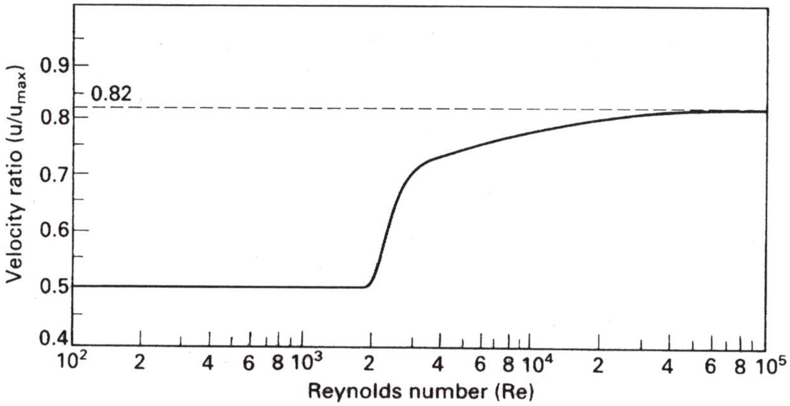

Figure 3.14. Variation of (u/u_{max}) with Reynolds number in a pipe

the kinetic energy per unit time of the fluid passing through the annulus is given by:

$$= -2\pi(r - y)\rho \, dy \, u_x \frac{u_x^2}{2}$$

$$= -\pi u_{CL}^3 \rho r^2 \left(1 - \frac{y}{r}\right)\left(\frac{y}{r}\right)^{3/7} d\left(\frac{y}{r}\right) \tag{3.65}$$

Integration of equation 3.65 gives the total kinetic energy per unit time as:

$$= \pi u_{CL}^3 \rho r^2 \left[\tfrac{7}{10} - \tfrac{7}{17}\right]$$

$$= \tfrac{49}{170}\pi r^2 \rho u_{CL}^3 \tag{3.66}$$

The mean kinetic energy per unit mass of fluid from equation 3.63 is:

$$= \frac{\left(\frac{49}{170}\pi r^2 \rho u_{CL}^3\right)}{\left(\frac{49}{60}\pi r^2 \rho u_{CL}\right)}$$

$$= \tfrac{6}{17} u_{CL}^2$$

$$= \tfrac{6}{17}\left(\tfrac{60}{49} u\right)^2 \quad \text{(from equation 3.60)}$$

$$= 0.53u^2 \approx \frac{u^2}{2}$$

$$= \frac{u^2}{2\alpha} \quad \text{(from the definition of } \alpha) \tag{3.67}$$

Thus for turbulent flow at high Reynolds numbers, where the thickness of the laminar sub-layer may be neglected, $\alpha \approx 1$.

When the thickness of the laminar sub-layer may not be neglected, α will be slightly less than 1.

Example 3.5

On the assumption that the velocity profile in a fluid in turbulent flow is given by the Prandtl one-seventh power law, calculate the radius at which the flow between it and the centre is equal to that between it and the wall, for a pipe 100 mm in diameter.

Solution

The Prandtl one-seventh power law gives the velocity at a distance y from the wall, u_x, as:

$$u_x = u_{CL}(y/r)^{1/7} \qquad \text{(equation 3.59)}$$

where u_{CL} is the velocity at the centre line of the pipe, and r is the radius of the pipe.

The total flow is then:

$$Q = 2\pi r^2 u_{CL} \int_0^1 \left[\left(\frac{y}{r}\right)^{1/7} - \left(\frac{y}{r}\right)^{8/7}\right] \mathrm{d}\left(\frac{y}{r}\right)$$

$$= 2\pi r^2 u_{CL} \left[\frac{7}{8}\left(\frac{y}{r}\right)^{8/7} - \frac{7}{15}\left(\frac{y}{r}\right)^{15/7}\right]_0^1$$

$$= \frac{49}{60}\pi r^2 u_{CL} \qquad \text{(equation 3.62)}$$

When the flow in the central core is equal to the flow in the surrounding annulus, then taking $a = y/r$, the flow in the central core is:

$$Q_c = 2\pi r^2 u_{CL} \left[\frac{7}{8}a^{8/7} - \frac{7}{15}a^{15/7}\right]_0^a$$

$$= \frac{\pi r^2}{60} u_{CL}(105a^{8/7} - 56a^{15/7})$$

Since: flow in the core = 0.5(flow in the whole pipe)

then: $$0.5 \times \left(\frac{49}{60}\right)\pi r^2 u_{CL} = \left(\frac{\pi r^2}{60}\right) u_{CL}(105a^{8/7} - 56a^{15/7})$$

or: $$105a^{8/7} - 56a^{15/7} = 24.5$$

Solving by trial and error:

$$a = y/r = 0.33$$

and: $$y = (0.33 \times 50) = \underline{\underline{16.5 \text{ mm}}}$$

Non-circular ducts

For turbulent flow in a duct of non-circular cross-section, the hydraulic mean diameter may be used in place of the pipe diameter and the formulae for circular pipes may then be applied without introducing a large error. This approach is entirely empirical.

The *hydraulic mean diameter* d_m is defined as four times the cross-sectional area divided by the wetted perimeter. For a circular pipe, for example, the hydraulic mean diameter is:

$$d_m = \frac{4(\pi/4)d^2}{\pi d} = d \qquad (3.68)$$

For an annulus of outer radius r and inner radius r_i:

$$d_m = \frac{4\pi(r^2 - r_i^2)}{2\pi(r + r_i)} = 2(r - r_i) = d - d_i \qquad (3.69)$$

and for a duct of rectangular cross-section d_a by d_b:

$$d_m = \frac{4d_a d_b}{2(d_a + d_b)}$$

$$= \frac{2d_a d_b}{d_a + d_b} \tag{3.70}$$

The method is not entirely satisfactory for streamline flow, and exact expressions relating the pressure drop to the velocity may be obtained only for ducts of certain shapes.

3.3.7. Flow through curved pipes

If a pipe is not straight, the velocity distribution over the section is altered and the direction of flow of the fluid is continuously changing. The frictional losses are therefore somewhat greater than for a straight pipe of the same length. If the radius of the pipe divided by the radius of the bend is less than about 0.002, the effects of the curvature are negligible, however.

WHITE[11] found that stable streamline flow persists at higher values of the Reynolds number in coiled pipes. Thus, for example, when the ratio of the diameter of the pipe to the diameter of the coil is 1 to 15, the transition occurs at a Reynolds number of about 8000.

3.3.8. Miscellaneous friction losses

Friction losses occurring as a result of a sudden enlargement or contraction in the cross-section of the pipe, and the resistance of various standard pipe fittings, are now considered.

Sudden enlargement

If the diameter of the pipe suddenly increases, as shown in Figure 3.15, the effective area available for flow gradually increases from that of the smaller pipe to that of the larger one, and the velocity of flow progressively decreases. Thus fluid with a relatively high velocity will be injected into relatively slow moving fluid; turbulence will be set up and much of the excess kinetic energy will be converted into heat and therefore wasted. If the change of cross-section is gradual, the kinetic energy may be recovered as pressure energy.

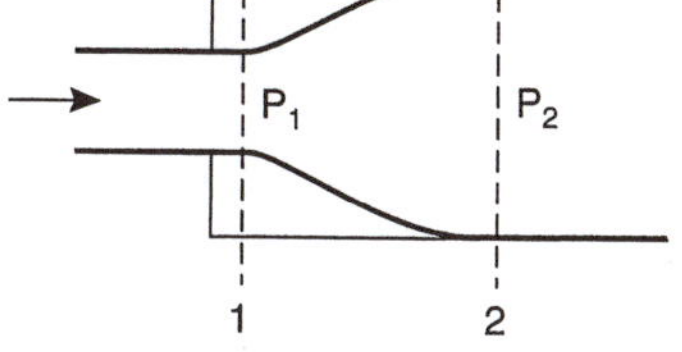

Figure 3.15. Sudden enlargement in a pipe

For the fluid flowing as shown in Figure 3.15, from section 1 (the pressure just inside the enlargement is found to be equal to that at the end of the smaller pipe) to section 2, the net force = the rate of change of momentum, or:

$$P_1A_2 - P_2A_2 = \rho_2 A_2 u_2(u_2 - u_1)$$

and:
$$(P_1 - P_2)v_2 = u_2^2 - u_1u_2 \tag{3.71}$$

For an incompressible fluid:
$$(P_1 - P_2)v_2 = -\int_{P_1}^{P_2} v\,dP \tag{3.72}$$

Applying the energy equation between the two sections:

$$\frac{u_1^2}{2\alpha_1} = \frac{u_2^2}{2\alpha_2} + \int_1^2 v\,dP + F \tag{3.73}$$

or:
$$F = \frac{u_1^2}{2\alpha_1} - \frac{u_2^2}{2\alpha_2} + u_2^2 - u_1u_2 \tag{3.74}$$

For fully turbulent flow: $\alpha_1 = \alpha_2 = 1$

and:
$$F = \frac{(u_1 - u_2)^2}{2} \tag{3.75}$$

The change in pressure $-\Delta P_f$ is therefore given by:

$$-\Delta P_f = \rho\frac{(u_1 - u_2)^2}{2} \tag{3.76}$$

The loss of head h_f is given by:

$$h_f = \frac{(u_1 - u_2)^2}{2g} \tag{3.77}$$

Substituting from the relation $u_1A_1 = u_2A_2$ into equation 3.76:

$$-\Delta P_f = \frac{\rho u_1^2}{2}\left[1 - \left(\frac{A_1}{A_2}\right)\right]^2 \tag{3.78}$$

The loss can be substantially reduced if a tapering enlarging section is used. For a circular pipe, the optimum angle of taper is about 7° and for a rectangular duct it is about 11°.

If A_2 is very large compared with A_1, as, for instance, at a pipe exit, then:

$$-\Delta P_f = \frac{\rho u_1^2}{2} \tag{3.79}$$

The use of a small angle enlarging section is a feature of the venturi meter, as discussed in Chapter 6.

Example 3.6

Water flows at 7.2 m^3/h through a sudden enlargement from a 40 mm to a 50 mm diameter pipe. What is the loss in head?

Solution

$$\text{Velocity in 50 mm pipe} = \frac{(7.2/3600)}{(\pi/4)(50 \times 10^{-3})^2} = 1.02 \text{ m/s}$$

$$\text{Velocity in 40 mm pipe} = \frac{(7.2/3600)}{(\pi/4)(40 \times 10^{-3})^2} = 1.59 \text{ m/s}$$

The head lost is given by equation 3.77 as:

$$h_f = \frac{(u_1 - u_2)^2}{2g}$$

$$= \frac{(1.59 - 1.02)^2}{(2 \times 9.81)} = 0.0165 \text{ m water}$$

or: $\underline{\underline{16.5 \text{ mm water}}}$

Sudden contraction

As shown in Figure 3.16, the effective area for flow gradually decreases as a sudden contraction is approached and then continues to decrease, for a short distance, to what is known as the *vena contracta*. After the vena contracta the flow area gradually approaches that of the smaller pipe. As the fluid moves towards the vena contracta it is accelerated and pressure energy is converted into kinetic energy; this process does not give rise to eddy formation and losses are very small. Beyond the vena contracta, however, the velocity falls as the flow area increases and conditions are equivalent to those for a sudden enlargement. The expression for the loss at a sudden enlargement can therefore be applied for the fluid flowing from the vena contracta to some section a small distance downstream, where the whole of the cross-section of the pipe is available for flow.

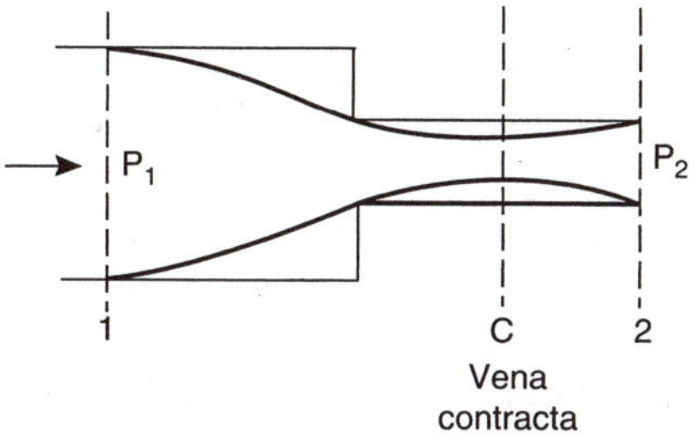

Figure 3.16. Sudden contraction in a pipe

Applying equation 3.75 between sections C and 2, as shown in Figure 3.16 the frictional loss per unit mass of fluid is then given by:

$$F = \frac{(u_c - u_2)^2}{2}$$

$$= \frac{u_2^2}{2}\left[\frac{u_c}{u_2} - 1\right]^2 \tag{3.80}$$

Denoting the ratio of the area at section C to that at section 2 by a coefficient of contraction C_c:

$$F = \frac{u_2^2}{2}\left[\frac{1}{C_c} - 1\right]^2 \tag{3.81}$$

Thus the change in pressure ΔP_f is $-(\rho u^2/2)[(1/C_c) - 1]^2$ and the head lost is:

$$\frac{u_2^2}{2g}\left[\frac{1}{C_c} - 1\right]^2$$

C_c varies from about 0.6 to 1.0 as the ratio of the pipe diameters varies from 0 to 1. For a common value of C_c of 0.67:

$$F = \frac{u_2^2}{8} \tag{3.82}$$

It may be noted that the maximum possible frictional loss which can occur at a change in cross-section is equal to the entire kinetic energy of the fluid.

Pipe fittings

Most pipes are fabricated from steel with or without small alloying ingredients, and they are welded or drawn to give a seamless pipe. Tubes with diameters of 6 to 50 mm are frequently made from non-ferrous metals such as copper, brass, or aluminium, and these are very widely used in heat exchangers. For special purposes there is a very wide range of materials including industrial glass, many varieties of plastics, rubber, stoneware, and ceramic materials. The normal metal piping is supplied in standard lengths of about 6 m and these are joined to give longer lengths as required. Such jointing is by screw flanging or welding, and small diameter copper or brass tubes are often brazed or soldered or jointed by compression fittings.

A very large range of pipe fittings is available to enable branching and changes in size to be incorporated into industrial pipe layouts. For the control of the flow of a fluid, valves of various designs are used, the most important being gate, globe, and needle valves. In addition, check valves are supplied for relieving the pressure in pipelines, and reducing valves are available for controlling the pressure on the downstream side of the valve. Gate and globe valves are supplied in all sizes and may be controlled by motor units operated from an automatic control system. Hand wheels are usually fitted for emergency use.

In general, gate valves give coarse control, globe valves give finer and needle valves give the finest control of the rate of flow. Diaphragm valves are also widely used for the handling of corrosive fluids since the diaphragm may be made of corrosion resistant materials.

Some representative figures are given in Table 3.2 for the friction losses in various pipe fittings for turbulent flow of fluid, and are expressed in terms of the equivalent length of straight pipe with the same resistance, and as the number of velocity heads $(u^2/2g)$ lost. Considerable variation occurs according to the exact construction of the fittings.

Typical pipe-fittings are shown in Figures 3.17 and 3.18 and details of other valve types are given in Volume 6.

Table 3.2. Friction losses in pipe fittings

	Number of pipe diameters	Number of velocity heads ($u^2/2g$)
45° elbows (a)*	15	0.3
90° elbows (standard radius) (b)	30–40	0.6–0.8
90° square elbows (c)	60	1.2
Entry from leg of T-piece (d)	60	1.2
Entry into leg of T-piece (d)	90	1.8
Unions and couplings (e)	Very small	Very small
Globe valves fully open	60–300	1.2–6.0
Gate valves: fully open	7	0.15
$\frac{3}{4}$ open	40	1
$\frac{1}{2}$ open	200	4
$\frac{1}{4}$ open	800	16

*See Figure 3.17.

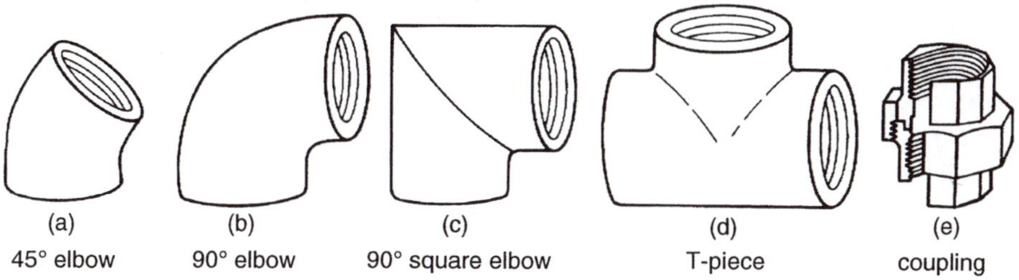

Figure 3.17. Standard pipe fittings

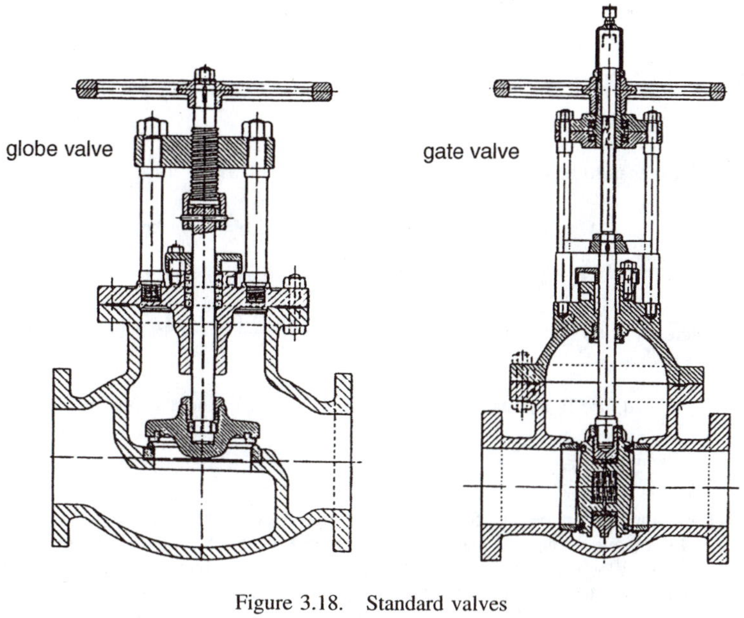

Figure 3.18. Standard valves

Example 3.7

2.27 m^3/h water at 320 K is pumped in a 40 mm i.d. pipe through a distance of 150 m in a horizontal direction and then up through a vertical height of 10 m. In the pipe there is a control valve for which the friction loss may be taken as equivalent to 200 pipe diameters and also other pipe fittings equivalent to 60 pipe diameters. Also in the line is a heat exchanger across which there is a loss in head of 1.5 m of water. If the main pipe has a roughness of 0.2 mm, what power must be supplied to the pump if it is 60 per cent efficient?

Solution

Relative roughness: $$\frac{e}{d} = \left(\frac{0.2}{40}\right) = 0.005$$

Viscosity at 320 K: $$\mu = 0.65 \text{ mN s/m}^2 \text{ or } 0.65 \times 10^{-3} \text{ N s/m}^2$$

$$\text{Flowrate} = 2.27 \text{ m}^3\text{/h} = 6.3 \times 10^{-4} \text{ m}^3\text{/s}$$

$$\text{Area for flow} = \frac{\pi}{4}(40 \times 10^{-3})^2 = 1.26 \times 10^{-3} \text{ m}^2$$

Thus: $$\text{Velocity} = \frac{6.3 \times 10^{-4}}{1.26 \times 10^{-3}} = 0.50 \text{ m/s}$$

and: $$Re = \frac{(40 \times 10^{-3} \times 0.50 \times 1000)}{(0.65 \times 10^{-3})} = 30{,}770$$

giving: $$\frac{R}{\rho u^2} = 0.004 \quad \text{(from Figure 3.7)}$$

$$\text{Equivalent length of pipe} = 150 + 10 + (260 \times 40 \times 10^{-3}) = 170.4 \text{ m}$$

$$h_f = 4\frac{R}{\rho u^2}\frac{l}{d}\frac{u^2}{g}$$

$$= 4 \times 0.004 \left(\frac{170.4}{40 \times 10^{-3}}\right)\left(\frac{0.5^2}{9.81}\right)$$

$$= 1.74 \text{ m}$$

$$\text{Total head to be developed} = (1.74 + 1.5 + 10) = 13.24 \text{ m}$$

$$\text{Mass throughput} = (6.3 \times 10^{-4} \times 1000) = 0.63 \text{ kg/s}$$

$\therefore$ $$\text{Power required} = (0.63 \times 13.24 \times 9.81) = 81.8 \text{ W}$$

Since the pump efficiency is 60 per cent, the power required $= \left(\frac{81.8}{0.60}\right) = 136.4$ W or 0.136 kW

The kinetic energy head, $u^2/2g$ amounts to $0.5^2/(2 \times 9.81) = 0.013$ m, and this may be neglected.

$$\frac{P}{\rho} = \Sigma F + \frac{v^2}{2} + g(z_2 - z_1)$$

Example 3.8

Water in a tank flows through an outlet 25 m below the water level into a 0.15 m diameter horizontal pipe 30 m long, with a 90° elbow at the end leading to a vertical pipe of the same diameter 15 m long. This is connected to a second 90° elbow which leads to a horizontal pipe of the same diameter, 60 m long, containing a fully open globe valve and discharging to atmosphere 10 m below the level of the water in the tank. Taking $e/d = 0.01$ and the viscosity of water as 1 mN s/m^2, what is the initial rate of discharge?

Solution

From equation 3.20, the head lost due to friction is given by:

$$h_f = 4\phi \frac{l}{d}\frac{u^2}{g} \text{ m water}$$

The total head loss is:
$$h = \frac{u^2}{2g} + h_f + \text{losses in fittings}$$

From Table 3.2., the losses in the fittings are:

$$= \frac{2 \times 0.8u^2}{2g} \text{ (for the elbows)} + \frac{5.0u^2}{2g} \text{ (for the valve)}$$

$$= \frac{6.6u^2}{2g} \text{ m water}$$

Taking ϕ as 0.0045, then:

$$10 = \frac{(6.6 + 1)u^2}{2g} + 4 \times 0.0045 \left[\frac{30 + 15 + 60}{0.15}\right]\frac{u^2}{g}$$

$$= \frac{(3.8 + 12.6)u^2}{g}$$

from which
$$u^2 = 5.98 \text{ m}^2/\text{s}^2$$

and:
$$u = 2.45 \text{ m/s}$$

The assumed value of ϕ may now be checked.

$$Re = \frac{du\rho}{\mu} = \frac{(0.15 \times 2.45 \times 1000)}{(1 \times 10^{-3})} = 3.67 \times 10^5$$

For $Re = 3.67 \times 10^5$ and $e/d = 0.01$, $\phi = 0.0045$ (from Figure 3.7) which agrees with the assumed value. Thus the initial rate of discharge $= 2.45 \times (\pi/4)0.15^2 = 0.043$ m^3/s or $(0.043 \times 1000) = \underline{\underline{43 \text{ kg/s}}}$.

3.3.9. Flow over banks of tubes

The frictional loss for a fluid flowing parallel to the axes of the tubes may be calculated in the normal manner by considering the hydraulic mean diameter of the system, although this applies strictly to turbulent flow only.

For flow at right angles to the axes of the tubes, the cross-sectional area is continually changing, and the problem may be treated as one involving a series of sudden enlargements and sudden contractions. Thus the friction loss would be expected to be directly proportional to the number of banks of pipes j in the direction of flow and to the kinetic energy of the fluid. The pressure drop $-\Delta P_f$ may be written as:

$$-\Delta P_f = \frac{C_f j \rho u_t^2}{6} \tag{3.83}$$

where C_f is a coefficient dependent on the arrangement of the tubes and the Reynolds number. The values of C_f given in Chapter 9 (Tables 9.3 and 9.4) are based on the velocity u_t of flow at the narrowest cross-section.

3.3.10. Flow with a free surface

If a liquid is flowing with a free surface exposed to the surroundings, the pressure at the liquid surface will everywhere be constant and equal to atmospheric pressure. Flow will take place therefore only as a result of the action of the gravitational force, and the surface level will necessarily fall in the direction of flow.

Two cases are considered. The first, the laminar flow of a thin film down an inclined surface, is important in the heat transfer from a condensing vapour where the main resistance to transfer lies in the condensate film, as discussed in Chapter 9 (Section 9.6.1). The second is the flow in open channels which are frequently used for transporting liquids down a slope on an industrial site.

Laminar flow down an inclined surface

In any liquid flowing down a surface, a velocity profile is established with the velocity increasing from zero at the surface itself to a maximum where it is in contact with the surrounding atmosphere. The velocity distribution may be obtained in a manner similar to that used in connection with pipe flow, but noting that the driving force is that due to gravity rather than a pressure gradient.

For the flow of liquid of depth s down a plane surface of width w inclined at an angle θ to the horizontal, as shown in Figure 3.19, a force balance in the X-direction (parallel to the surface) may be written. In an element of length dx the gravitational force acting on that part of the liquid which is at a distance greater than y from the surface is given by:

$$(s - y)w\,dx\,\rho g \sin\theta$$

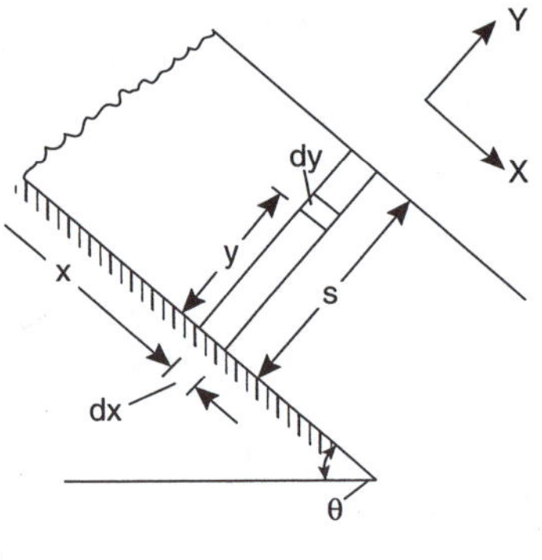

Figure 3.19. Flow of liquid over a surface

If the drag force of the atmosphere is negligible, the retarding force for laminar flow is attributable to the viscous drag in the liquid at the distance y from the surface given by:

$$\mu \frac{du_x}{dy} w\,dx$$

where u_x is the velocity of the fluid at that position.

Thus, at equilibrium:

$$(s - y)w\,\mathrm{d}x\,\rho g \sin\theta = \mu \frac{\mathrm{d}u_x}{\mathrm{d}y} w\,\mathrm{d}x \tag{3.84}$$

Since there will normally be no slip between the liquid and the surface, then $u_x = 0$ when $y = 0$, and:

$$\int_0^{u_x} \mathrm{d}u_x = \frac{\rho g \sin\theta}{\mu} \int_0^y (s - y)\,\mathrm{d}y$$

and:

$$u_x = \frac{\rho g \sin\theta}{\mu}(sy - \tfrac{1}{2}y^2) \tag{3.85}$$

The mass rate of flow G of liquid down the surface is now calculated.

$$\begin{aligned} G &= \int_0^s \left(\frac{\rho g \sin\theta}{\mu} w(sy - \tfrac{1}{2}y^2)\right) \rho\,\mathrm{d}y \\ &= \frac{\rho^2 g \sin\theta}{\mu} w \left(\frac{s^3}{2} - \frac{s^3}{6}\right) \\ &= \frac{\rho^2 g \sin\theta\, w s^3}{3\mu} \end{aligned} \tag{3.86}$$

The mean velocity of the fluid is then:

$$u = \frac{\rho g \sin\theta\, s^2}{3\mu} \tag{3.87}$$

For a vertical surface: $\sin\theta = 1$ and $u = \dfrac{\rho g s^2}{3\mu}$

The maximum velocity, which occurs at the free surface is given by:

$$u_s = \frac{\rho g \sin\theta\, s^2}{2\mu} \tag{3.88}$$

and this is 1.5 times the mean velocity of the liquid.

Flow in open channels

For flow in an open channel, only turbulent flow is considered because streamline flow occurs in practice only when the liquid is flowing as a thin layer, as discussed in the previous section. The transition from streamline to turbulent flow occurs over the range of Reynolds numbers, $u\rho d_m/\mu = 4000 - 11{,}000$, where d_m is the hydraulic mean diameter discussed earlier under *Flow in non-circular ducts*.

Three different types of turbulent flow may be obtained in open channels. They are *tranquil flow*, *rapid flow*, and *critical flow*. In tranquil flow the velocity is less than that at which some disturbance, such as a surge wave, will be transmitted, and the flow is influenced by conditions at both the upstream and the downstream end of the channel. In rapid flow, the velocity of the fluid is greater than the velocity of a surge wave and the conditions at the downstream end do not influence the flow. Critical flow occurs when the velocity is exactly equal to the velocity of a surge wave.

Uniform flow

For a channel of constant cross-section and slope, the flow is said to be uniform when the depth of the liquid D is constant throughout the length of the channel. For these conditions, as shown in Figure 3.20, for a length l of channel:

$$\text{The accelerating force acting on liquid} = lA\rho g \sin\theta$$

$$\text{The force resisting the motion} = R_m pl$$

where R_m is the mean value of the shear stress at the solid surface of the channel, p the wetted perimeter and A the cross-sectional area of the flowing liquid.

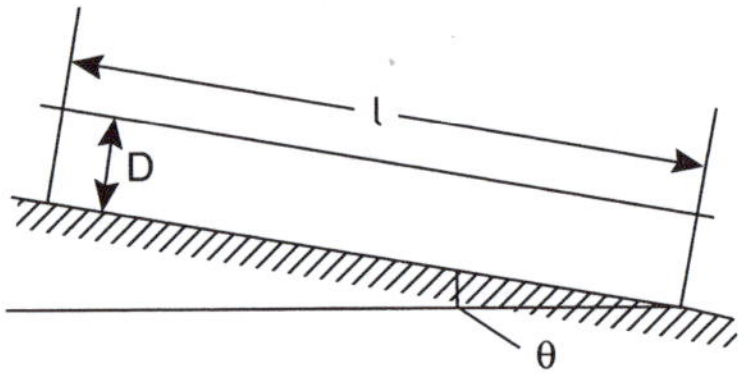

Figure 3.20. Uniform flow in open channel

For uniform motion:

$$lA\rho g \sin\theta = R_m pl$$

$$R_m = \frac{A}{p}\rho g \sin\theta$$

$$= \tfrac{1}{4} d_m \rho g \sin\theta \tag{3.89}$$

where d_m is the hydraulic mean diameter $= 4\ A/p$ (see equations 3.68–3.70).

Dividing both sides of the equation by ρu^2, where u is the mean velocity in the channel gives:

$$\frac{R_m}{\rho u^2} = \frac{d_m}{4}\frac{g \sin\theta}{u^2} \tag{3.90}$$

or:

$$u^2 = \frac{d_m}{4(R_m/\rho u^2)} g \sin\theta \tag{3.91}$$

For turbulent flow, $R_m/\rho u^2$ is almost independent of velocity although it is a function of the surface roughness of the channel. Thus the resistance force is proportional to the square of the velocity. $R_m/\rho u^2$ is found experimentally to be proportional to the one-third power of the relative roughness of the channel surface and may be conveniently written as:

$$\frac{R_m}{\rho u^2} = \frac{1}{16}\left(\frac{e}{d_m}\right)^{1/3} \tag{3.92}$$

for the values of the roughness e given in Table 3.3.

Thus:
$$u^2 = 4d_m\left(\frac{d_m}{e}\right)^{1/3} g\sin\theta \tag{3.93}$$
$$= 4d_m^{4/3}e^{-1/3}g\sin\theta \tag{3.94}$$

Table 3.3. Values of the roughness e, for use in equation 3.92

	ft	mm
Planed wood or finished concrete	0.00015	0.046
Unplaned wood	0.00024	0.073
Unfinished concrete	0.00037	0.11
Cast iron	0.00056	0.17
Brick	0.00082	0.25
Riveted steel	0.0017	0.51
Corrugated metal	0.0055	1.68
Rubble	0.012	3.66

The volumetric rate of flow Q is then given by:
$$Q = uA$$
$$= 2Ad_m^{2/3}e^{-1/6}\sqrt{g\sin\theta} \tag{3.95}$$

The loss of energy due to friction for unit mass of fluid flowing isothermally through a length l of channel is equal to its loss of potential energy because the other forms of energy remain unchanged.

Thus:
$$F = gl\sin\theta = 4\frac{R_m}{\rho u^2}\frac{l}{d_m}u^2 \tag{3.96}$$

An empirical equation for the calculation of the velocity of flow in an open channel is the *Chezy equation*, which may be expressed as:
$$u = C\sqrt{\tfrac{1}{4}d_m\sin\theta} \tag{3.97}$$

where the value of the coefficient C is a function of the units of the other quantities in the equation. This expression takes no account of the effect of surface roughness on the velocity of flow.

The velocity of the liquid varies over the cross-section and is usually a maximum at a depth of between 0.05 D and 0.25 D below the surface, at the centre line of the channel. The velocity distribution may be measured by means of a pitot tube as described in Chapter 6.

The shape and the proportions of the channel may be chosen for any given flow rate so that the perimeter, and hence the cost of the channel, is a minimum.

From equation 3.95:
$$Q = 2Ad_m^{2/3}e^{-1/6}\sqrt{g\sin\theta}$$

Assuming that slope and the roughness of the channel are fixed then, for a given flowrate Q, from equation 3.95:
$$Ad_m^{2/3} = \text{constant}$$

or:
$$A^{5/3}p^{-2/3} = \text{constant} \tag{3.98}$$

The perimeter is therefore a minimum when the cross-section for flow is a minimum.

For a rectangular channel, of depth D and width B:

$$A = DB = D(p - 2D)$$

$\therefore$
$$D^{5/3}(p - 2D)^{5/3}p^{-2/3} = \text{constant}$$

and:
$$Dp^{3/5} - 2D^2p^{-2/5} = \text{constant}$$

p is a minimum when $dp/dD = 0$. Differentiating with respect to D and putting dp/dD equal to 0:

$$p^{3/5} - p^{-2/5}4D = 0$$

$\therefore$
$$4D = p = 2D + B$$

or:
$$B = 2D \tag{3.99}$$

Thus the most economical rectangular section is one where the width is equal to twice the depth of liquid flowing. The most economical proportions for other shapes may be determined in a similar manner.

Specific energy of liquid

For a liquid flowing in a channel inclined at an angle θ to the horizontal as shown in Figure 3.21, the various energies associated with unit mass of fluid at a depth h below the liquid surface (measured at right angles to the bottom of the channel) are:

$$\text{Internal energy:} = U$$

$$\text{Pressure energy:} = (P_a + h\rho g \sec\theta)v = P_a v + hg\sec\theta$$

where P_a is atmospheric pressure.

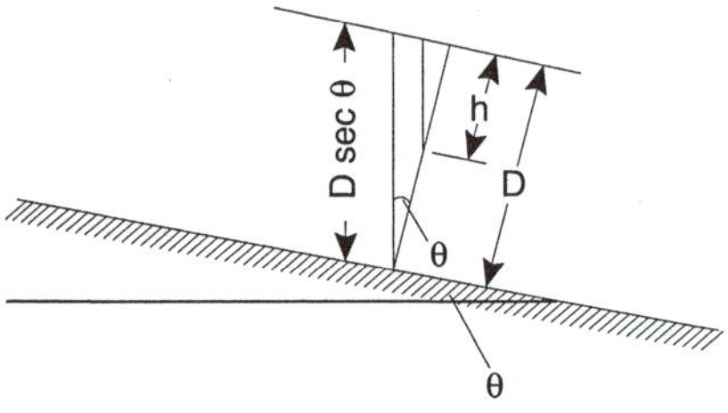

Figure 3.21. Energy of fluid in open channel

$$\text{Potential energy:} = zg + (D - h)g\sec\theta$$

where z is the height of the bottom of the channel above the datum level at which the potential energy is reckoned as zero.

$$\text{Kinetic energy} = \frac{u^2}{2}$$

The total energy per unit mass is, therefore:

$$U + P_a v + zg + Dg\sec\theta + \frac{u^2}{2} \quad (3.100)$$

It may be seen that at any cross-section the total energy is independent of the depth h below the liquid surface. As the depth is increased, the pressure energy increases at the same rate as the potential energy decreases. If the fluid flows through a length $\mathrm{d}l$ of channel, the net change in energy per unit mass is given by:

$$\delta q - \delta W_s = \mathrm{d}U + g\mathrm{d}z + g\sec\theta\,\mathrm{d}D + u\,\mathrm{d}u \quad (3.101)$$

For an irreversible process:

$$\mathrm{d}U = T\,\mathrm{d}S - P\,\mathrm{d}v \quad \text{(equation 2.5)}$$

$$= \delta q + \delta F \quad \text{(equation 2.9)}$$

assuming the fluid is incompressible, and $\mathrm{d}v = 0$.

If no work is done on the surroundings, $\delta W_s = 0$ and:

$$g\,\mathrm{d}z + g\sec\theta\,\mathrm{d}D + u\,\mathrm{d}u + \delta F = 0 \quad (3.102)$$

For a fluid at a constant temperature, the first three terms in equation 3.100 are independent of the flow conditions within the channel. On the other hand, the last two terms are functions of the velocity and the depth of liquid. The *specific energy* of the fluid is defined by:

$$J = Dg\sec\theta + \frac{u^2}{2} \quad (3.103)$$

For a horizontal channel, rectangular in section:

$$J = Dg + \frac{u^2}{2} = Dg + \frac{Q^2}{2B^2D^2}$$

The specific energy will vary with the velocity of the liquid and will be a minimum for some critical value of D; for a given rate of flow Q the minimum will occur when $\mathrm{d}J/\mathrm{d}D = 0$.

Thus:

$$g + \frac{(-2Q^2)}{2B^2D^3} = 0$$

$$\frac{u^2}{D} = g$$

and:

$$u = \sqrt{gD} \quad (3.104)$$

This value of u is known as the *critical velocity*.

The corresponding values of D the *critical depth* and J are given by:

$$D = \left(\frac{Q^2}{B^2g}\right)^{1/3} \quad (3.105)$$

and:

$$J = Dg + \frac{Dg}{2} = \frac{3Dg}{2} \quad (3.106)$$

Similarly it may be shown that, at the critical conditions, the flowrate is a maximum for a given value of the specific energy J. At the critical velocity, (u^2/gD) is equal to unity. This dimensionless group is known as the *Froude number* Fr. For velocities greater than the critical velocity Fr is greater than unity, and vice versa. It may be shown that the velocity with which a small disturbance is transmitted through a liquid in an open channel is equal to the critical velocity, and hence the Froude number is the criterion by which the type of flow, tranquil or rapid, is determined. Tranquil flow occurs when Fr is less than unity and rapid flow when Fr is greater than unity.

Velocity of transmission of a wave

For a liquid which is flowing with a velocity u in a rectangular channel of width B, the depth of liquid is initially D_1. As a result of a change in conditions at the downstream end of the channel, the level there suddenly increases to some value D_2. A wave therefore tends to move upstream against the motion of the oncoming fluid. For two sections, 1 and 2, one on each side of the wave at any instant, as shown in Figure 3.22, the rate of accumulation of fluid between the two sections is given by:

$$u_1 D_1 B - u_2 D_2 B$$

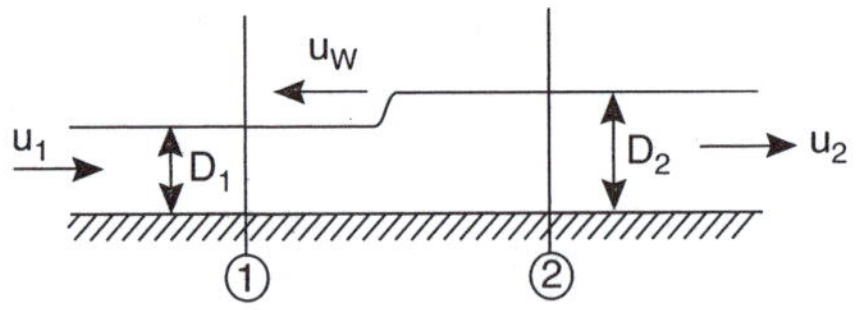

Figure 3.22. Transmission of a wave

This accumulation of fluid results from the propagation of the wave and is therefore equal to $u_w B(D_2 - D_1)$.

Thus: $$u_1 D_1 - u_2 D_2 = u_w(D_2 - D_1)$$

or: $$u_2 = \frac{u_w(D_1 - D_2) + u_1 D_1}{D_2} \tag{3.107}$$

The velocity of the fluid is changed from u_1 to u_2 by the passage of the wave. The rate of travel of the wave relative to the upstream liquid is $(u_1 + u_w)$ and therefore the mass of fluid whose velocity is changed in unit time is:

$$(u_1 + u_w)BD_1\rho \tag{3.108}$$

The force acting on the fluid at any section where the liquid depth is D is:

$$\int_0^D (h\rho g)B\,\mathrm{d}h = \tfrac{1}{2}B\rho g D^2 \tag{3.109}$$

where h is any depth below the liquid surface.

The net force acting on the fluid between sections 1 and 2 is:

$$\tfrac{1}{2}B\rho g(D_1^2 - D_2^2) \quad \text{in the direction of flow}$$

Thus, neglecting the frictional drag of the walls of the channel between sections 1 and 2, the net force can be equated to the rate of increase of momentum and thus:

$$\tfrac{1}{2}B\rho g(D_1^2 - D_2^2) = (u_1 + u_w)BD_1\rho(u_2 - u_1)$$

$$\therefore \quad (u_1 + u_w)D_1\left\{\frac{1}{D_2}[u_w(D_1 - D_2) + u_1 D_1] - u_1\right\} = \tfrac{1}{2}g(D_1^2 - D_2^2)$$

and:

$$(u_1 + u_w)^2 = \frac{D_2}{D_1}(D_1 + D_2)\frac{g}{2}$$

$$= \frac{gD_2}{2}\left(1 + \frac{D_2}{D_1}\right) \tag{3.110}$$

where $u_1 + u_w$ is the velocity of the wave relative to the oncoming fluid. For a very small wave, $D_1 \rightarrow D_2$ and:

$$u_1 + u_w = \sqrt{(gD_2)} \tag{3.111}$$

It is thus seen that the velocity of an elementary wave is equal to the critical velocity, at which the specific energy of the fluid is a minimum for a given flowrate. The criterion for critical conditions is therefore that the Froude number, (u^2/gD), must be equal to unity.

Hydraulic jump

If a liquid enters a channel under a gate, it will flow at a high velocity through and just beyond the gate and the depth will be correspondingly low. This is an unstable condition, and at some point the depth of the liquid may suddenly increase and the velocity fall. This change is known as the *hydraulic jump*, and it is accompanied by a reduction of the specific energy of the liquid as the flow changes from rapid to tranquil, any excess energy being dissipated as a result of turbulence.

If a liquid is flowing in a rectangular channel in which a hydraulic jump occurs between sections 1 and 2, as shown in Figure 3.23, then the conditions after the jump can be determined by equating the net force acting on the liquid between the sections to the rate of change of momentum, if the frictional forces at the walls of the channel may be neglected.

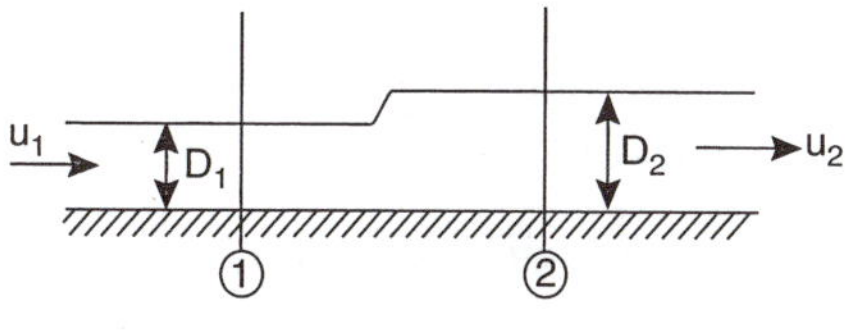

Figure 3.23. Hydraulic jump

The net force acting on the fluid is given by:

$$\tfrac{1}{2}B\rho g(D_1^2 - D_2^2)$$

The rate of change of momentum of fluid is given by:

$$= u_1 B D_1 \rho (u_2 - u_1)$$

or:
$$\tfrac{1}{2} g (D_1^2 - D_2^2) = u_1 D_1 (u_2 - u_1)$$

The volumetric rate of flow of the fluid is the same at sections 1 and 2.

Thus:
$$Q = u_1 B D_1 = u_2 B D_2$$

or:
$$\tfrac{1}{2} g (D_1^2 - D_2^2) = u_1^2 D_1 \left[\left(\frac{D_1}{D_2} \right) - 1 \right]$$

If $D_1 \neq D_2$, then:
$$\tfrac{1}{2} g (D_1 + D_2) = \frac{u_1^2 D_1}{D_2}$$

and:
$$D_2^2 + D_1 D_2 - \frac{2 u_1^2 D_1}{g} = 0 \tag{3.112}$$

or:
$$D_2 = \tfrac{1}{2} \left(-D_1 \overset{+}{[\neq]} \sqrt{D_1^2 + \frac{8 u_1^2 D_1}{g}} \right) \tag{3.113}$$

This expression gives D_2 as a function of the conditions at the upstream side of the hydraulic jump. The corresponding velocity u_2 is obtained by substituting in the equation:

$$u_1 D_1 = u_2 D_2$$

Corresponding values of D_1 and D_2 are referred to as *conjugate depths*.

The minimum depth at which a hydraulic jump can occur is found by putting $D_1 = D_2 = D$ in equation 3.112 giving:

$$2D^2 = 2D \frac{u_1^2}{g}$$

or:
$$D = \frac{u_1^2}{g} \tag{3.114}$$

This value of D corresponds to the *critical depth* for flow in a channel.

Thus a hydraulic jump can occur provided that the depth of the liquid is less than the critical depth. After the jump the depth will be greater than the critical depth, the flow having changed from rapid to tranquil.

The energy dissipated in the hydraulic jump is now calculated. For a small change in the flow of a fluid in an open channel:

$$u\,\mathrm{d}u + g\,\mathrm{d}z + g \sec\theta\,\mathrm{d}D + \delta F = 0 \qquad \text{(equation 3.102)}$$

Then, for a horizontal channel:

$$F = g(D_1 - D_2) + \frac{(u_1^2 - u_2^2)}{2}$$

From equation 3.112:
$$u_1^2 = \frac{g D_2 (D_1 + D_2)}{2 D_1}$$

Similarly: $$u_2^2 = \frac{gD_1(D_1 + D_2)}{2D_2}$$

Thus: $$F = g(D_1 - D_2) + \frac{1}{4}g(D_1 + D_2)\left(\frac{D_2}{D_1} - \frac{D_1}{D_2}\right)$$

$$= \tfrac{1}{4}g(D_1 - D_2)\left[4 + (D_1 + D_2)\frac{-(D_1 + D_2)}{D_1 D_2}\right]$$

$$= \frac{(D_2 - D_1)^3 g}{4D_1 D_2} \quad (3.115)$$

The hydraulic jump may be compared with the shock wave for the flow of a compressible fluid, discussed in Chapter 4.

Example 3.9

A hydraulic jump occurs during the flow of a liquid discharging from a tank into an open channel under a gate so that the liquid is initially travelling at a velocity of 1.5 m/s with a depth of 75 mm. Calculate the corresponding velocity and the liquid depth after the jump.

Solution

The depth of fluid in the channel after the jump is given by:

$$D_2 = 0.5\{-D_1 + \surd[D_1^2 + (8u_1^2 D_1/g)]\} \quad \text{(equation 3.113)}$$

where D_1 and u_1 are the depth and velocity of the fluid before the jump.

Thus: $D_1 = 0.075$ m and $u_1 = 1.5$ m/s

and hence: $$D_2 = 0.5\{-0.075 + \surd[0.075^2 + (8 \times 1.5^2 \times 0.075/9.81)]\}$$

$$= 0.152 \text{ m} = \underline{\underline{152 \text{ mm}}}$$

If the channel is of uniform cross-sectional area, then:

$$u_1 D_1 = u_2 D_2$$

and: $$u_2 = u_1 D_1/D_2$$

$$= (1.5 \times 0.075)/0.152) = \underline{\underline{0.74 \text{ m/s}}}$$

3.4. NON-NEWTONIAN FLUIDS

In the previous sections of this chapter, the calculation of frictional losses associated with the flow of simple *Newtonian* fluids has been discussed. A Newtonian fluid at a given temperature and pressure has a constant viscosity μ which does not depend on the shear rate and, for streamline (laminar) flow, is equal to the ratio of the shear stress (R_y) to the shear rate (du_x/dy) as shown in equation 3.4, or:

$$\mu = \frac{|R_y|}{|du_x/dy|} \quad (3.116)$$

The modulus sign is used because shear stresses within a fluid act in both the positive and negative senses. Gases and simple low molecular weight liquids are all Newtonian, and viscosity may be treated as constant in any flow problem unless there are significant variations of temperature or pressure.

Many fluids, including some that are encountered very widely both industrially and domestically, exhibit non-Newtonian behaviour and their apparent viscosities may depend on the rate at which they are sheared and on their previous shear history. At any position and time in the fluid, the apparent viscosity μ_a which is defined as the ratio of the shear stress to the shear rate at that point is given by:

$$\mu_a = \frac{|R_y|}{|\mathrm{d}u_x/\mathrm{d}y|} \tag{3.117}$$

When the apparent viscosity is a function of the shear rate, the behaviour is said to be *shear-dependent*; when it is a function of the duration of shearing at a particular rate, it is referred to as *time-dependent*. Any shear-dependent fluid must to some extent be time-dependent because, if the shear rate is suddenly changed, the apparent viscosity does not alter instantaneously, but gradually moves towards its new value. In many cases, however, the time-scale for the flow process may be sufficiently long for the effects of time-dependence to be negligible.

The apparent viscosity of a fluid may either decrease or increase as the shear rate is raised. The more common effect is for the apparent viscosity to fall as the shear rate is raised; such behaviour is referred to as *shear-thinning*. Most paints are shear-thinning: in the can and when loaded on to the brush, the paint is subject only to low rates of shear and has a high apparent viscosity. When applied to the surface the paint is sheared by the brush, its apparent viscosity becomes less, and it flows readily to give an even film. However, when the brushing ceases, it recovers its high apparent viscosity and does not drain from the surface. This non-Newtonian behaviour is an important characteristic of a good paint. It is frequently necessary to build-in non-Newtonian characteristics to give a product the desired properties. Paints will usually exhibit appreciable time-dependent behaviour. Thus, as the paint is stirred its apparent viscosity will decrease progressively until it reaches an asymptotic equilibrium value characteristic of that particular rate of shear.

Some materials have the characteristics of both solids and liquids. For instance, tooth paste behaves as a solid in the tube, but when the tube is squeezed the paste flows as a plug. The essential characteristic of such a material is that it will not flow until a certain critical shear stress, known as the *yield stress* is exceeded. Thus, it behaves as a solid at low shear stresses and as a fluid at high shear stress. It is a further example of a shear-thinning fluid, with an infinite apparent viscosity at stress values below the yield value, and a falling finite value as the stress is progressively increased beyond this point.

A further important property which may be shown by a non-Newtonian fluid is *elasticity*–which causes the fluid to try to regain its former condition as soon as the stress is removed. Again, the material is showing some of the characteristics of both a solid and a liquid. An ideal (Newtonian) liquid is one in which the stress is proportional to the *rate* of shear (or rate of strain). On the other hand, for an ideal solid (obeying Hooke's Law) the stress is proportional to the strain. A fluid showing elastic behaviour is termed *viscoelastic* or *elastoviscous*.

The branch of science which is concerned with the flow of both simple (Newtonian) and complex (non-Newtonian) fluids is known as *rheology*. The flow characteristics are represented by a *rheogram*, which is a plot of shear stress against rate of shear, and normally consists of a collection of experimentally determined points through which a curve may be drawn. If an equation can be fitted to the curve, it facilitates calculation of the behaviour of the fluid. It must be borne in mind, however, that such equations are approximations to the actual behaviour of the fluid and should not be used outside the range of conditions (particularly shear rates) for which they were determined.

An understanding of non-Newtonian behaviour is important to the chemical engineer from two points of view. Frequently, non-Newtonian properties are desirable in that they can confer desirable properties on the material which are essential if it is to fulfil the purpose for which it is required. The example of paint has already been given. Toothpaste should not flow out of the tube until it is squeezed and should stay in place on the brush until it is applied to the teeth. The texture of foodstuffs is largely attributable to rheology.

Second, it is necessary to take account of non-Newtonian behaviour in the design of process plant and pipelines. Heat and mass transfer coefficients are considerably affected by the behaviour of the fluid, and special attention must be devoted to the selection of appropriate mixing equipment and pumps.

In this section, some of the important aspects of non-Newtonian behaviour will be quantified, and some of the simpler approximate equations of state will be discussed. An attempt has been made to standardise nomenclature in the British Standard, BS 5168[12].

Shear stress is denoted by R in order to be consistent with other parts of the book; τ is frequently used elsewhere to denote shear stress. R without suffix denotes the shear stress acting on a surface in the direction of flow and $R_0(= -R)$ denotes the shear stress exerted by the surface on the fluid. R_s denotes the positive value in the fluid at a radius s and R_y the positive value at a distance y from a surface. Strain is defined as the ratio $\mathrm{d}x/\mathrm{d}y$, where $\mathrm{d}x$ is the shear displacement of two elements a distance $\mathrm{d}y$ apart and is often denoted by γ. The rate of strain or rate of shear is $(\mathrm{d}x/\mathrm{d}t)/\mathrm{d}y$ or $\mathrm{d}u_x/\mathrm{d}y$ and is denoted by $\dot{\gamma}$.

Thus equation 3.117 may be written:

$$\mu_a = \frac{\tau}{\dot{\gamma}} \tag{3.118}$$

3.4.1. Steady-state shear-dependent behaviour

In this section, consideration will be given to the equilibrium relationships between shear stress and shear rate for fluids exhibiting non-Newtonian behaviour. Whenever the shear stress or the shear rate is altered, the fluid will gradually move towards its new equilibrium state and, for the present, the period of "adjustment" between the two equilibrium states will be ignored.

The fluid may be either shear-thinning or, less often, shear-thickening, and in either case the shear stress and the *apparent viscosity* μ_a are functions of shear rate, or:

$$|R_y| = f_1\left(\left|\frac{\mathrm{d}u_x}{\mathrm{d}y}\right|\right) \tag{3.119}$$

and:
$$\mu_a = |R_y| \bigg/ \left|\frac{du_x}{dy}\right| = f_2\left(\left|\frac{du_x}{dy}\right|\right) \tag{3.120}$$

Here y is distance measured from a boundary surface.

Typical forms of curve of shear stress versus shear rate are shown in Figure 3.24 for a shear-thinning (or *pseudoplastic*) fluid, a *Bingham-plastic*, a Newtonian fluid and a shear-thickening (or *dilatant*) fluid. For the particular case chosen the apparent viscosity is the same for all four at the shear rate where they intersect. At lower shear rates the shear-thinning fluid is more viscous than the Newtonian fluid, and at higher shear rates it is less viscous. For the shear-thickening fluid, the situation is reversed. The corresponding curves showing the variation of apparent viscosity are given in Figure 3.25. Because the rates of shear of interest can cover several orders of magnitude, it is convenient to replot the curves using log-log coordinates, as shown in Figures 3.26 and 3.27.

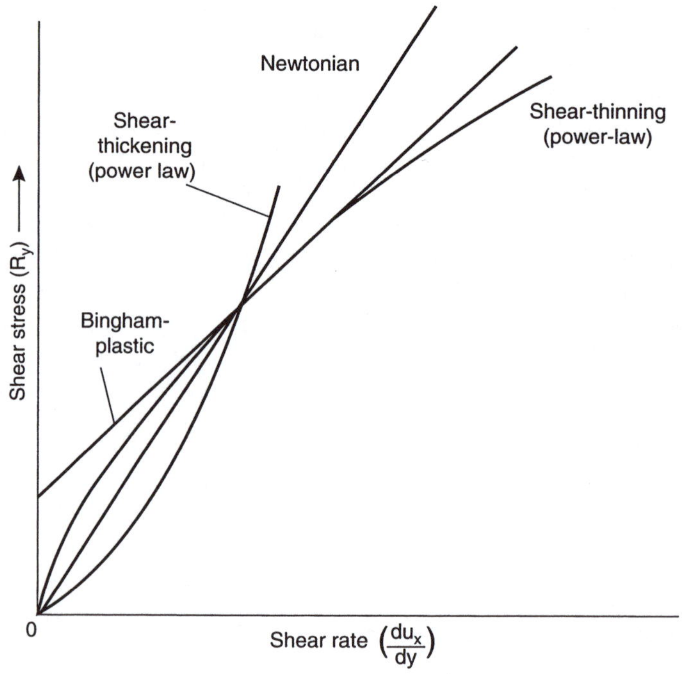

Figure 3.24. Shear stress–shear rate behaviour of Newtonian and non-Newtonian fluids plotted using linear coordinates

The relation between shear stress and shear rate for the Newtonian fluid is defined by a single parameter μ, the viscosity of the fluid. No single parameter model will describe non-Newtonian behaviour and models involving two or even more parameters only approximate to the characteristics of real fluids, and can be used only over a limited range of shear rates.

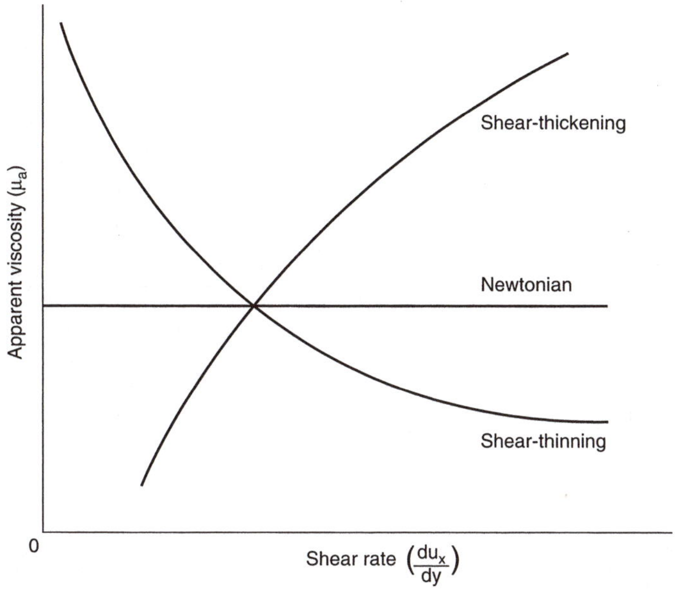

Figure 3.25. Shear rate dependence of apparent viscosity for Newtonian and non-Newtonian fluids plotted on linear co-ordinates

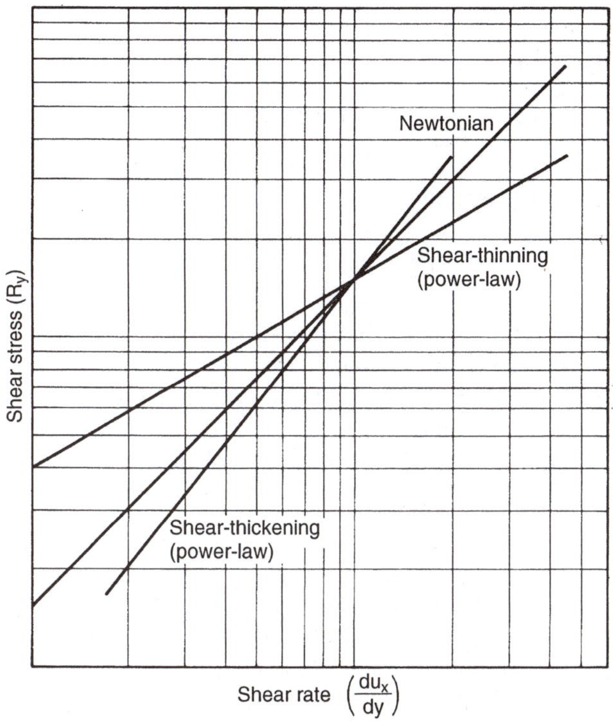

Figure 3.26. The relation between shear-stress and shear-rate using logarithmic axes

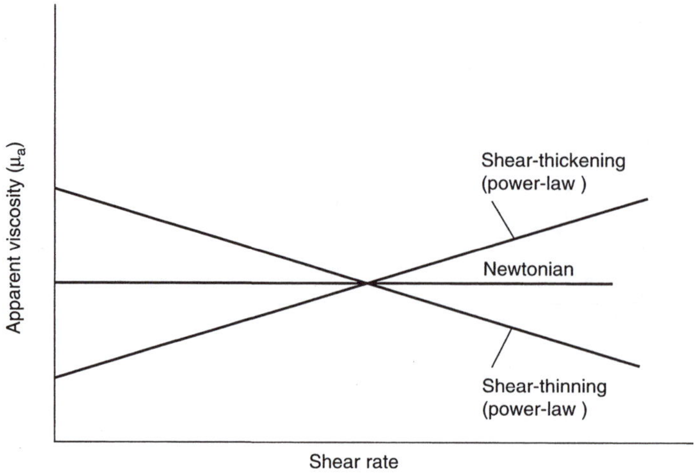

Figure 3.27. Plot of apparent viscosity–shear rate relation using logarithmic axes

A useful two-parameter model is the *power-law* model, or Ostwald–de Waele law to identify its first proponents. The relation between shear stress and shear rate is given by:

$$|R_y| = k\left(\left|\frac{du_x}{dy}\right|\right)^n \tag{3.121}$$

where n is known as the *power-law index* and k is known as the *consistency* coefficient. This equation may be written:

$$|R_y| = k\left(\left|\frac{du_x}{dy}\right|\right)^{n-1}\left|\frac{du_x}{dy}\right| \tag{3.122}$$

It is therefore seen that the apparent viscosity μ_a is given by:

$$\mu_a = k\left(\left|\frac{du_x}{dy}\right|\right)^{n-1} \tag{3.123}$$

From equation 3.123: when $n > 1$, μ_a increases with increase of shear rate, and shear-thickening behaviour is described;
when $n = 1$, μ_a is constant and equal to the Newtonian viscosity μ of the fluid;
when $n < 1$, μ_a decreases with increase of shear rate, and the behaviour is that of a shear-thinning fluid.

Thus, by selecting an appropriate value of n, both shear-thinning and shear-thickening behaviour can be represented, with $n = 1$ representing Newtonian behaviour which essentially marks the transition from shear-thinning to shear-thickening characteristics.

It will be noted that the dimensions of k are $\mathbf{ML^{-1}T}^{n-2}$, that is they are dependent on the value of n. Values of k for fluids with different n values cannot therefore be compared. *Numerically*, k is the value of the apparent viscosity (or shear stress) at unit shear rate and this numerical value will depend on the units used; for example the value of k at a shear rate of 1 s^{-1} will be different from that at a shear rate of 1 h^{-1}.

In Figure 3.28, the shear stress is shown as a function of shear rate for a typical shear-thinning fluid, using logarithmic coordinates. Over the shear rate range (*ca* 10^0 to 10^3 s^{-1}), the fluid behaviour is described by the power-law equation with an index n of 0.6, that is the line CD has a slope of 0.6. If the power-law were followed at all shear rates, the extrapolated line $C'CDD'$ would be applicable. Figure 3.29 shows the corresponding values of apparent viscosity and the line $C'CDD'$ has a slope of $n - 1 = -0.4$. It is seen that it extrapolates to $\mu_a = \infty$ at zero shear rate and to $\mu_a = 0$ at infinite shear rate.

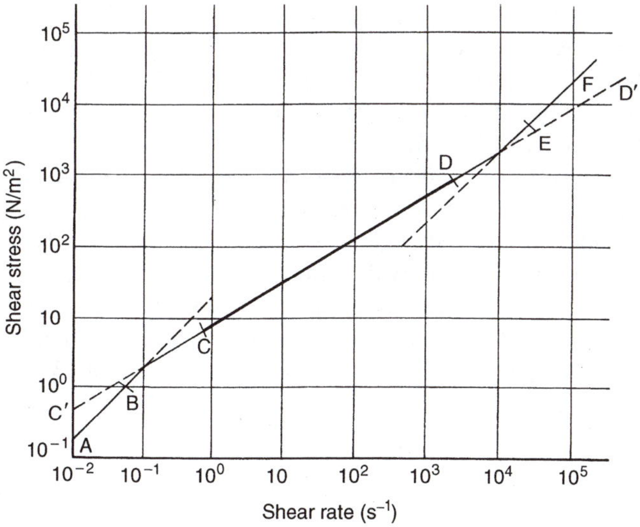

Figure 3.28. Behaviour of a typical shear-thinning fluid plotted logarithmically for several orders of shear rate

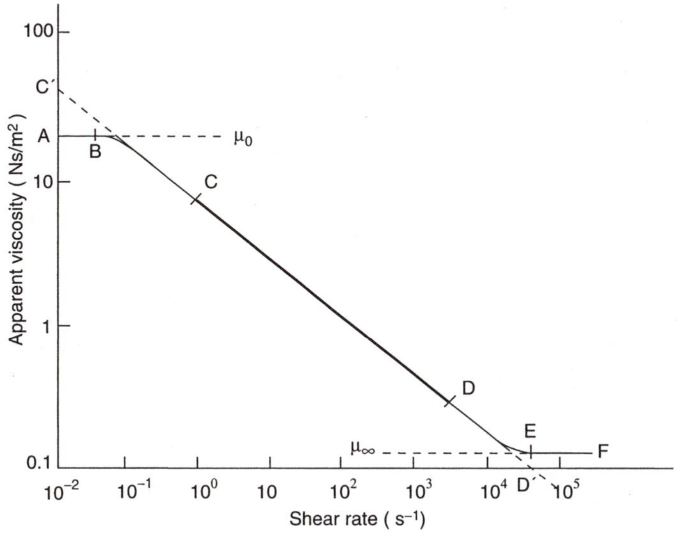

Figure 3.29. Apparent viscosity corresponding to the data of Figure 3.28

The viscosities of most real shear-thinning fluids approach constant values both at *very low shear rates* and at *very high shear rates*; that is, they tend to show Newtonian properties at the extremes of shear rates. The limiting viscosity at low shear rates μ_0 is referred to as the lower-Newtonian (or zero-shear μ_0) viscosity (see lines *AB* in Figures 3.28 and 3.29), and that at high shear rates μ_∞ is the upper-Newtonian (or infinite-shear) viscosity (see lines *EF* in Figures 3.28 and 3.29).

By reference to Figure 3.29, it is seen that the power law model can, at best therefore, represent the behaviour of a real fluid over only a limited range of shear rates. Effectively, it represents what is happening in the region where apparent viscosity is changing most rapidly with shear rate. However, over any limited range of shear rates, it is possible to approximate to the curve in Figure 3.29 by a straight line, whose slope $(n - 1)$ will determine the "best" value of the power-law index n which can be used over that range.

In practice, shear rates ranging from 10^{-6} s^{-1} to 10^6 s^{-1}, or 12 orders of magnitudes, are encountered and over that wide a range the behaviour depicted in Figure 3.29 is of greater importance. Frequently, however, a much lower range of shear rates occurs in a particular application and rheological measurements can then be restricted to this narrower shear-rate range and a relatively simple model, such as the power-law, may be used. It is always convenient to use the simplest model which adequately describes the rheology over the range of interest. As an example, the shear-rate ranges which apply in the production and subsequent use of foodstuffs are given in Table 3.4

Table 3.4. Typical shear-rate ranges for food products and processes

Situation	Shear rate range, s^{-1}	Application
Sedmentation of particles in a suspending fluid	10^{-3} to 10^{-6}	Spices in salad dressing
Levelling due to surface tension	0.01 to 0.1	Frosting
Draining under gravity	0.1 to 10	Vats, small food containers
Extrusion	1 to 1000	Snack foods, cereals, pasta
Calendering	10 to 100	Dough sheeting
Chewing and swallowing	10 to 100	All foods
Dip coating	10 to 100	Confectionery
Mixing and stirring	10 to 1000	Numerous
Pumping/Pipeflow	1 to 1000	Numerous
Spraying and brushing	10^3 to 10^4	Spray Drying

*Courtesy Neil Alderman, AEA Technology.

A convenient form of 3-parameter equation which extrapolates to a constant limiting apparent viscosity (μ_0 or μ_∞) as the shear rate approaches both zero and infinity has been proposed by CROSS[13]:

$$\frac{\mu_0 - \mu_\infty}{\mu_a - \mu_\infty} = 1 + \beta\dot{\gamma}^{2/3} \tag{3.124}$$

where $\dot{\gamma}$ is the shear rate and μ_0, μ_∞ and β must be determined for each fluid. For some materials a slight modification of the exponent of $\dot{\gamma}(2/3)$ leads to improved agreement between predicted and experimental results.

This equation is based on the assumption that pseudoplastic (shear-thinning) behaviour is associated with the formation and rupture of structural linkages. It is based on an experimental study of a wide range of fluids–including aqueous suspensions of flocculated inorganic particles, aqueous polymer solutions and non-aqueous suspensions and solutions–over a wide range of shear rates ($\dot{\gamma}$) (~ 10 to 10^4 s^{-1}).

For a shear-thickening fluid the same arguments can be applied, with the apparent viscosity rising from zero at zero shear rate to infinity at infinite shear rate, on application of the power law model. However, shear-thickening is generally observed over very much narrower ranges of shear rate and it is difficult to generalise on the type of curve which will be obtained in practice.

In order to overcome the shortcomings of the power-law model, several alternative forms of equation between shear rate and shear stress have been proposed. These are all more complex involving three or more parameters. Reference should be made to specialist works on non-Newtonian flow[14–17] for details of these *Constitutive Equations*.

Some fluids exhibit a *yield stress*. When subjected to stresses below the yield stress they do not flow and effectively can be regarded as fluids of infinite viscosities, or alternatively as solids. When the yield stress is exceeded they flow as fluids. Such behaviour cannot be described by a power-law model.

The simplest type of behaviour for a fluid exhibiting a yield stress is known as *Bingham-plastic*. The shear rate is directly proportional to the amount by which the stress exceeds the yield stress.

Thus:
$$|R_y| - R_Y = \mu_p \left|\frac{du_x}{dy}\right| \qquad (|R_y| \geqslant R_Y) \tag{3.125}$$
$$\frac{du_x}{dy} = 0 \qquad (|R_y| \leqslant R_Y)$$

μ_p is known as the *plastic viscosity*.

The apparent viscosity μ_a is given, by definition, as:

$$\mu_a = |R_y| \Big/ \left|\frac{du_x}{dy}\right| \qquad \text{(equation 3.117)}$$
$$= \mu_p + R_Y \Big/ \left|\frac{du_x}{dy}\right| \qquad (|R_y| \geqslant R_Y) \tag{3.126}$$

Thus, the apparent viscosity falls from infinity at zero shear rate ($|R_y \leqslant R_Y$) to μ_p at infinite shear rate, i.e. the fluid shows shear-thinning characteristics.

Because it is very difficult to measure the flow characteristics of a material at very low shear rates, behaviour at zero shear rate can often only be assessed by extrapolation of experimental data obtained over a limited range of shear rates. This extrapolation can be difficult, if not impossible. From Example 3.10 in Section 3.4.7, it can be seen that it is sometimes possible to approximate the behaviour of a fluid *over the range of shear rates for which experimental results are available*, either by a power-law or by a Bingham-plastic equation.

Some materials give more complex behaviour and the plot of shear stress against shear rate approximates to a curve, rather than to a straight line with an intercept R_Y on the shear stress axis. An equation of the following form may then be used:

$$|R_y| - R_Y = \mu'_p \left(\left| \frac{\mathrm{d}u_x}{\mathrm{d}y} \right| \right)^m \tag{3.127}$$

Thus, equation 3.127, which includes three parameters, is effectively a combination of equations 3.121 and 3.125. It is sometimes called the *generalised Bingham equation* or *Herschel–Bulkley equation*, and the fluids are sometimes referred to as having *false body*. Figures 3.30 and 3.31 show shear stress and apparent viscosity, respectively, for Bingham plastic and false body fluids, using linear coordinates.

In many flow geometries, the shear stress will vary over the cross-section and there will then be regions where the shear stress exceeds the yield stress and the fluid will then be sheared. In other regions, the shear stress will be less than the yield value and the fluid will flow there as an unsheared plug. Care must be taken therefore in the application of equations 3.125 and 3.127. Thus, for the case of flow in a circular tube which will be considered later, the shear stress varies linearly from a maximum value at the wall, to zero at the centre-line (see equation 3.9). If the shear stress at the wall is less than the yield stress, no flow will occur. If it is greater than the yield stress, shear will take place in the region between the wall and the point where the shear stress equals the yield stress (Figure 3.32). Inside this region the fluid will flow as an unsheared plug. As the pressure difference over the tube is increased, the wall shear stress will increase in direct proportion and the unsheared plug will be of smaller radius. For any fluid subjected to a finite pressure gradient there must always be some region, however small, near the pipe axis in which shear does not take place.

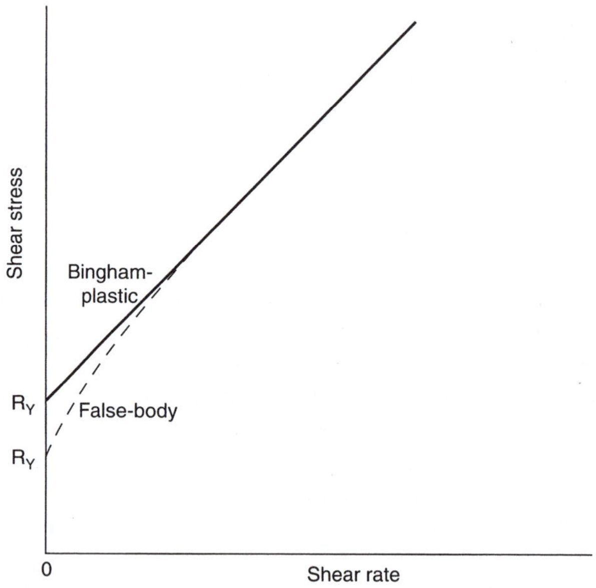

Figure 3.30. Shear stress-shear rate data for Bingham-plastic and false-body fluids using linear scale axes

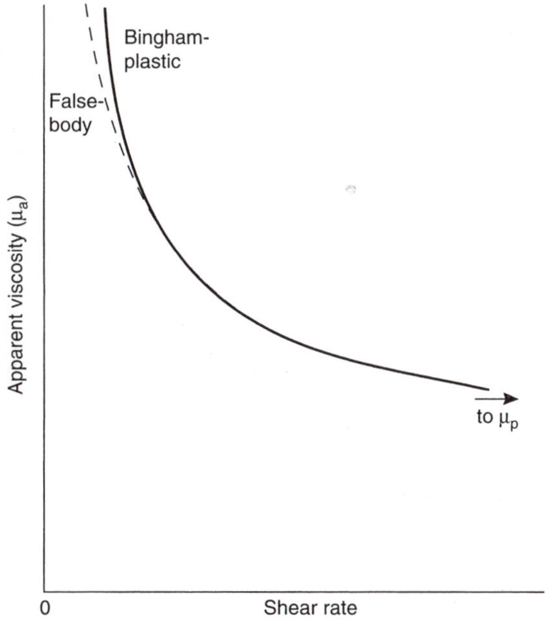

Figure 3.31. Apparent viscosity for Bingham-plastic and false-body fluids using linear axes

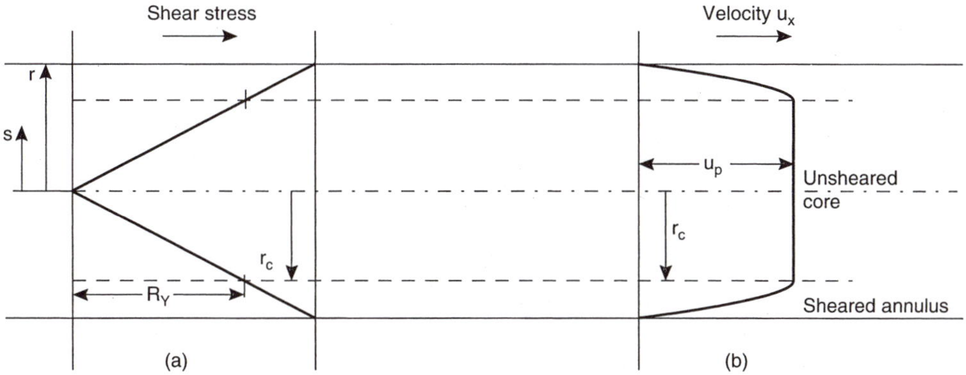

Figure 3.32. (*a*) Shear stress distribution in pipe (*b*) Velocity profile for Bingham plastic fluid in pipe

3.4.2 Time-dependent behaviour

In a *time-dependent fluid*, the shear rate depends upon the time for which it has been subjected to a given shear stress. Conversely, if the shear rate is kept constant, the shear stress will change with time. However, with all *time-dependent* fluids an equilibrium condition is reached if the imposed condition (e.g. shear rate or shear stress) is maintained constant. Some fluids respond so quickly to changes that the effect of time dependence can be neglected. Others may have a much longer constant, and in changing flow situations will never be in the equilibrium state.

In general, for shear-thinning pseudoplastic fluids the apparent viscosity will gradually decrease with time if there is a step increase in its rate of shear. This phenomenon is known as *thixotropy*. Similarly, with a shear-thickening fluid the apparent viscosity increases under these circumstances and the fluid exhibits *rheopexy* or *negative-thixotropy*.

The effect of increasing and then decreasing the rate of shear of a thixotropic fluid is shown in Figure 3.33 in which the shear stress is plotted against shear rate in an experiment; the shear rate is steadily increased from zero and subsequently decreased again. It is seen that a hysteresis loop is formed, with the shear stress always lagging behind its equilibrium value. If the shear rate is changed rapidly the hysteresis loop will have a large area. At low rates of change the area will be small and will eventually become zero as the two curves coincide when sufficient time is allowed for equilibrium to be reached at each point on the curve. In Figure 3.34, the effect on the apparent viscosity is seen of a step increase in the shear rate; it gradually decreases from the initial to the final equilibrium value. This is a picture of what happens to a material which does not suffer any irreversible changes as a result of shearing. Some materials, particular gels, suffer structural breakdown when subjected to a shear field, but this effect is not considered here.

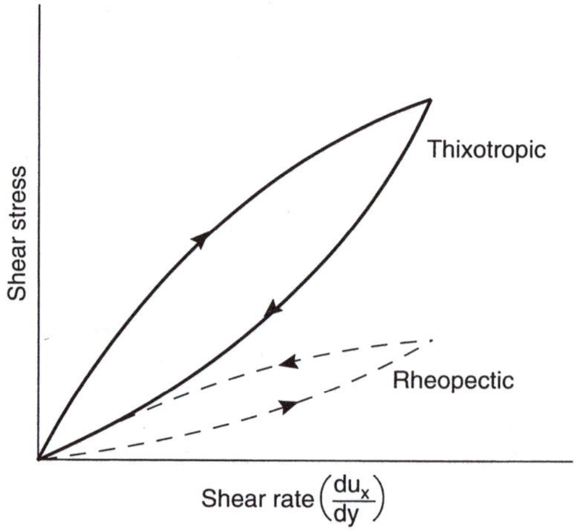

Figure 3.33. Thixotropic and rheopectic behaviour

The behaviour of a *rheopectic* fluid is the reverse of that of a thixotropic fluid and is illustrated by the broken lines in Figures 3.33 and 3.34.

3.4.3. Viscoelastic behaviour

A true fluid flows when it is subjected to a shear field and motion ceases as soon as the stress is removed. In contrast, an ideal solid which has been subjected to a stress recovers its original state as soon as the stress is removed. The two extremes of behaviour are therefore represented by:

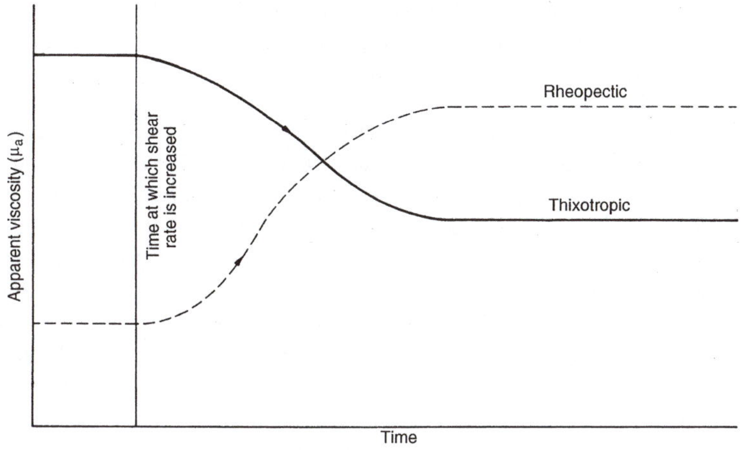

Figure 3.34. Effect of sudden change of shear rate on apparent viscosity of time-dependent fluid

(a) The ideal Newtonian fluid in which:

$$|R_y| = \mu \left|\frac{du_x}{dy}\right| \qquad \text{(equation 3.4)}$$

$$(\text{or } \tau = \mu\dot{\gamma})$$

where $du_x/dy(=\dot{\gamma})$ is the rate of shear.

(b) The ideal elastic solid which obeys Hooke's law in which the relation between shear distortion and stress is:

$$|R_y| = \mathbf{G}\left|\frac{dx}{dy}\right| \qquad (3.128)$$

$$(\text{or } \tau = \mathbf{G}\gamma)$$

where **G** is Young's modulus, and the shear $dx/dy(=\gamma)$ is the ratio of the shear displacement of two elements to their distance apart.

Many materials of practical interest (such as polymer solutions and melts, foodstuffs, and biological fluids) exhibit *viscoelastic* characteristics; they have some ability to store and recover shear energy and therefore show some of the properties of both a solid and a liquid. Thus a solid may be subject to *creep* and a fluid may exhibit *elastic* properties. Several phenomena ascribed to fluid elasticity including die swell, rod climbing (Weissenberg effect), the tubeless siphon, bouncing of a sphere, and the development of secondary flow patterns at low Reynolds numbers, have recently been illustrated in an excellent photographic study[18]. Two common and easily observable examples of viscoelastic behaviour in a liquid are:

(a) The liquid in a cylindrical vessel is given a swirling motion by means of a stirrer. When the stirring is stopped, the fluid gradually comes to rest and, if viscoelastic, may then start to rotate in the opposite direction, that is to unwind.

(b) A viscoelastic fluid, on emerging from a tube or from a die, may form a jet which is of larger diameter than the aperture. The phenomenon, referred to above as "die-swell", results from the sudden removal of a constraining force on the fluid.

Viscoelastic fluids are thus capable of exerting *normal stresses*. Because most materials, under appropriate circumstances, show simultaneously *solid-like* and *fluid-like* behaviours in varying proportions, the notion of an ideal elastic solid or of a purely viscous fluid represents the commonly encountered limiting condition. For instance, the viscosity of ice and the elasticity of water may both pass unnoticed! The response of a material may also depend upon the type of deformation to which it is subjected. A material may behave like a highly elastic solid in one flow situation, and like a viscous fluid in another.

Generally, viscoelastic effects are of particular importance in unsteady state flow and where there are rapid changes in the pressure to which the fluid is subjected. They can give rise to very complex behaviour and mechanical analogues have sometimes been found useful for calculating behaviour at least qualitatively. In this approach the fluid properties are represented by a "dashpot", a piston in a cylinder with a small outlet in which the flowrate is linearly related to the pressure difference across it. The elastic properties are represented by a spring. By combination of such elements in a variety of ways it is possible to simulate the behaviour of very complex materials. A single dashpot and spring in series is known as the Maxwell model; and the dashpot and spring in parallel is the Voigt model, as shown in Figure 3.35.

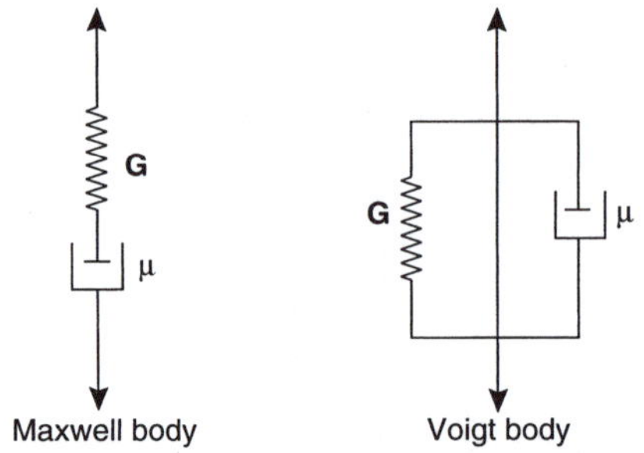

Figure 3.35. Mechanical analogues for viscoelastic fluids

Because of the assumption that linear relations exist between shear stress and shear rate (equation 3.4) and between distortion and stress (equation 3.128), both of these models, namely the Maxwell and Voigt models, and all other such models involving combinations of springs and dashpots, are restricted to small strains and small strain rates. Accordingly, the equations describing these models are known as line viscoelastic equations. Several theoretical and semi-theoretical approaches are available to account for non-linear viscoelastic effects, and reference should be made to specialist works[14–16] for further details.

For process design calculations, there are two basic matters of concern:

(1) to characterise the viscoelastic behaviour of a substance, and
(2) to ascertain whether viscoelastic effects are significant in a given flow situation.

Depending upon the particular application and the type of deformation to which an element is likely to be subjected, there are several ways of characterising viscoelastic behaviour of a material, not all of which are mutually exclusive; in fact, in some cases, the information deduced using different experimental techniques may be interconnected. In general, viscoelastic effects are of greater significance in flow domains remote from boundary surfaces. If an infinitesimal element of viscoelastic fluid flowing in the x-direction is subjected to a steady-state shear stress which arises from the velocity gradient in the y-direction it will, unlike an inelastic fluid, be subject to normal shear stresses I_x, I_y and I_z acting on the faces, A, C and B, respectively, as shown in Figure 3.36. In this case, the fluid has zero velocity components in the y- and z-directions and there is no velocity gradient in either the x- or the z-direction. Of great importance in characterising viscoelastic behaviour are the so-called normal shear-stress differences $N_1(= I_x - I_y)$ and $N_2(= I_y - I_z)$ which are more easily measured than the normal shear stresses themselves. Generally, N_1 is considerably greater than N_2, and the ratio N_1/R_c, where R_c is the shear stress acting on face C, is an indication of the degree of viscoelasticity of the fluid. The higher the value of N_1/R_c, the greater is the elasticity, the ratio being zero for an inelastic fluid.

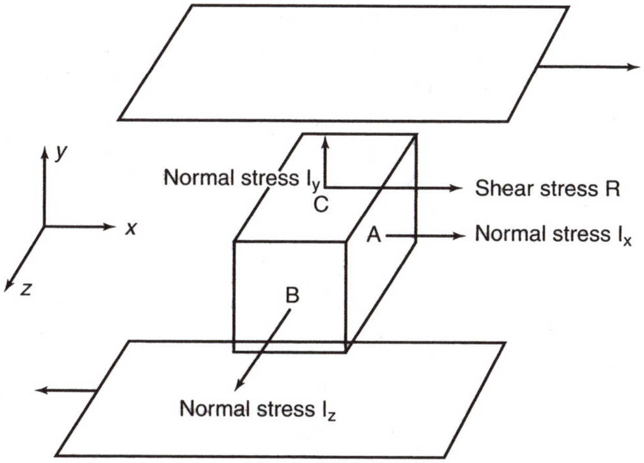

Figure 3.36. Shear stress and normal stresses on element of fluid

The other commonly encountered type of deformation is stretching, or extensional flow. This type of flow occurs in the film-blowing process, and in flow through dense porous media where it arises from the successive convergent–divergent nature of the flow passages. Viscoelastic fluids in extensional flow behave in a significantly different manner from purely viscous fluids. It is well established that the so-called *Trouton ratio*, defined as the ratio of the extensional viscocity (tension/rate of stretch) to the shear viscocity is 3 for a Newtonian fluid. For many viscoelastic fluids the Trouton ratio is again 3, but values as high as 1000 have sometimes been observed. Again, the higher the value of the Trouton ratio, the more elastic is the fluid behaviour. In general, extensional viscocities are high when the aspect ratios of the molecules are large.

Many other techniques of measuring viscoelastic parameters, such as transient shear, creep and sinusoidally-varying shear, are available. A good description, together with the merits and demerits of each of these techniques, is available in WHORLOW[19].

3.4.4. Characterisation of non-Newtonian fluids

The characterisation of non-Newtonian fluids is a major area of science in itself and the equipment required for measurement of all the relevant properties, including the normal stresses of viscoelastic fluids, is extremely complex. The range of shear rates of interest is very great covering many orders of magnitude, and it is frequently better to use a number of instruments each covering a relatively narrow range, rather than to try to carry out all the measurements with a single instrument. Some instruments are designed to operate at a series of constant shear rates and the resulting stresses are measured. Others, constant stress instruments, are more suitable for measurements at low shear rate conditions, particularly with materials that have yield stresses.

Even the measurement of the steady-state characteristics of shear-dependent fluids is more complex than the determination of viscosities for Newtonian fluids. In simple geometries, such as capillary tubes, the shear stress and shear rate vary over the cross-section and consequently, at a given operating condition, the apparent viscosity will vary with location. Rheological measurements are therefore usually made with instruments in which the sample to be sheared is subjected to the same rate of shear throughout its whole mass. This condition is achieved in concentric cylinder geometry (Figure 3.37) where the fluid is sheared in the annular space between a fixed and a rotating cylinder; if the gap is small compared with the diameters of the cylinders, the shear rate is approximately

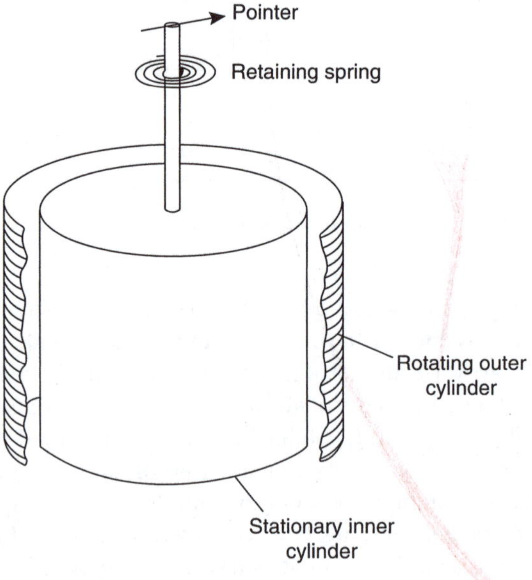

Figure 3.37. Partial section of a concentric-cylinder viscometer

constant. Alternatively, a cone and plate geometry (Figure 3.38) gives a constant shear rate, provided that the angle θ between the cone and the plate is sufficiently small that $\sin\theta \approx \theta$.

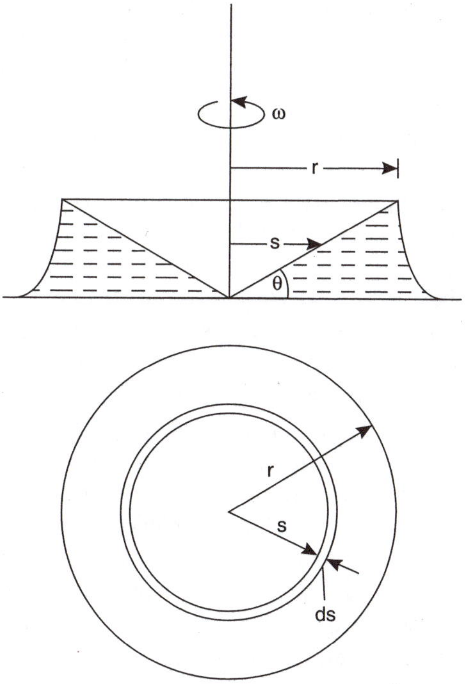

Figure 3.38. Cone and plate viscometer

In making rheological measurements, and in applying the results to a particular flow geometry, it should be noted that many non-Newtonian fluids exhibit the phenomenon of *wall slip* as a result of which the fluid layers in contact with a surface have a finite velocity relative to that surface. This may occur when the fluid consists of a solution of macromolecules or of a melt, of a polymer of high molecular weight, or of a suspension of very fine particles. If the size of the individual molecules or particles is large compared with the roughness dimensions of the surface, the particles or macromolecules may *ride over* the tops of the small protuberance. The difficulty can usually be overcome by artificial roughening of the surface.

If it is known that a particular form of relation, such as the *power-law* model, is applicable, it is not necessary to maintain a constant shear rate. Thus, for instance, a capillary tube viscometer can be used for determination of the values of the two parameters in the model. In this case it is usually possible to allow for the effects of wall-slip by making measurements with tubes covering a range of bores and extrapolating the results to a tube of infinite diameter. Details of the method are given by FAROOQI and RICHARDSON[20].

3.4.5 Dimensionless characterisation of viscoelastic flows

The enormous practical value obtained from the consideration of viscous flows in terms of Reynolds numbers has led to the consideration of the possibility of using analogous dimensionless groups to characterise elastic behaviour. The Reynolds group represents the ratio of inertial to viscous forces, and it might be expected that a ratio involving elastic and inertial forces would be relevant. Attempts to achieve useful correlations have not been very successful, perhaps most frequently because of the complexity of natural situations and of real materials. One simple parameter that proves to be of value is the ratio of a *characteristic time of deformation* to a *natural time constant of the fluid*. The precise definition for these times is somewhat arbitrary, but it is evident that, for processes involving very slow deformation of the fluid elements, it is possible for the elastic forces to be released by the normal processes of relaxation as fast as they build up. Examples of the flow of rigid (apparently infinitely viscous) material over long periods of time include the thickening of the lower parts of mediaeval glass windows, and the plastic flow and deformation that lead to the folded strata of geological structures. In operations that are carried out rapidly, the extent of viscous flow will be minimal and the deformation will be followed by rapid recovery when the stress is removed. To obtain some idea of the possible regions in which such an analysis can provide guidance, consideration may be given be the flow of a 1 per cent solution of polyacrylamide in water for which the relaxation time (in Maxwell model terms) is of the order of 10^{-2} s. If the fluid were flowing through a packed bed, it would be subject to alternating accelerations and decelerations as it flowed through the interstices of the bed. With a particle size of the order of 25 mm diameter and a superficial flowrate of 0.25 m/s, the elastic properties would not be expected to influence the flow significantly. However, in a free jet discharge with a velocity of the order of 30 m/s some evidence of elastic behaviour would be expected to be evident near to the point of discharge.

The *Deborah number Db* has been defined by METZNER, WHITE and DENN[21] as:

$$Db = \frac{\text{Characteristic process time}}{\text{Characteristic fluid time}}$$

In the above example of the packed bed, *Db* would be ~ 10; for the jet and the nozzle *Db* would be $\sim 10^{-1}$. The smaller the value of *Db*, the more likely is elasticity to be of practical significance.

Unfortunately, this group *Db* depends on the assignment of a single characteristic time to the fluid (perhaps a *relaxation time*). While this has led to some success, it appears to be inadequate for many viscoelastic materials which show different relaxation behaviour under differing conditions.

For further information on viscoelastic behaviour, reference should again be made to specialist sources[14–16].

3.4.6 Relation between rheology and structure of material

An understanding of the contribution of the relevant physical and chemical properties of the system to rheological behaviour is an area which has made little progress until recent years, but a clearer picture is now emerging of how the various types of rheological behaviour arise.

Most non-Newtonian fluids are either two-phase systems, or single phase systems in which large molecules are in solution in a liquid which itself may be Newtonian or non-Newtonian, or are in the form of a melt.

Flocculated suspensions of fine particles (coal, china clay, pigments, etc.) are usually shear-thinning. The flocs are weak mechanically and tend to break up in a shear field and the individual particles are capable of regrouping to give structures offering a lower resistance to shear. When the shear field is removed, the original structure is regained. The particles are generally of the order of 1 μm in size and the flocs (loose agglomerates occluding liquid) may be 10-100 μm in size. The specific surface of the particles is very high, and surface chemistry plays an important role in determining the structure of the flocs. The rheology of many of these suspensions can be approximated over a reasonable range of flowrates by a power-law equation ($0.1 < n < 0.6$). There is evidence that some have a finite yield stress and conform to the Bingham-plastic model. Frequently however, the rheology can equally well be described by a power-law or Bingham-plastic form of equation, as seen in Example 3.10.

Non-flocculated suspensions can exist at very much higher concentrations and, at all but the highest volumetric concentrations, are often Newtonian. When such suspensions are sheared, some *dilation* occurs as a result of particles trying to "climb over each other". If the amount of liquid present is then insufficient fully to fill the void spaces, particle-particle solid friction can come into play and the resistance to shear increases. This is just one way in which shear-thickening can occur.

Many polymers form shear-thinning solutions in water. The molecules are generally long and tend to be aligned and to straighten out in a shear field, and thus to offer less resistance to flow. Such solutions are sometimes viscoelastic and this effect may be attributable to a tendency of the molecules to recover their previous configuration once the stress is removed. Molten polymers are usually viscoelastic.

3.4.7 Streamline Flow in pipes and channels of regular geometry

As in the case of Newtonian fluids, one of the most important practical problems involving non-Newtonian fluids is the calculation of the pressure drop for flow in pipelines. The flow is much more likely to be streamline, or laminar, because non-Newtonian fluids usually have very much higher apparent viscosities than most simple Newtonian fluids. Furthermore, the difference in behaviour is much greater for laminar flow where viscosity plays such an important role than for turbulent flow. Attention will initially be focused on laminar-flow, with particular reference to the flow of power-law and Bingham-plastic fluids.

In order to predict the transition point from stable streamline to stable turbulent flow, it is necessary to define a modified Reynolds number, though it is not clear that the same sharp transition in flow regime always occurs. Particular attention will be paid to flow in pipes of circular cross-section, but the methods are applicable to other geometries (annuli, between flat plates, and so on) as in the case of Newtonian fluids, and the methods described earlier for flow between plates, through an annulus or down a surface can be adapted to take account of non-Newtonian characteristics of the fluid.

Power-law fluids

The distribution of shear stress over the cross-section of a pipe is determined by a force balance and is independent of the nature of the fluid or the type of flow.

From equation 3.8 and Figure 3.32*a* it is seen that the shear stress $|R_s|$ at a radius s in a pipe of radius r is given by:

$$\frac{|R_s|}{R} = \frac{s}{r} \tag{3.129}$$

that is, the shear stress varies linearly from the centre of the pipe ($s = 0$) to the wall ($s = r$).

When the fluid behaviour can be described by a power-law, the apparent viscosity for a shear-thinning fluid will be a minimum at the wall where the shear stress is a maximum, and will rise to a theoretical value of infinity at the pipe axis where the shear stress is zero. On the other hand, for a shear-thickening fluid the apparent viscosity will fall to zero at the pipe axis. It is apparent, therefore, that there will be some error in applying the power-law near the pipe axis since all real fluids have a limiting viscosity μ_0 at zero shear stress. The procedure is exactly analogous to that used for the Newtonian fluid, except that the power-law relation is used to relate shear stress to shear rate, as opposed to the simple Newtonian equation.

For a power-law fluid, equation 3.28 becomes:

$$\left(\frac{-\Delta P}{l}\right) s - 2k\left(-\frac{\mathrm{d}u_x}{\mathrm{d}s}\right)^n = 0$$

$$\therefore \quad \mathrm{d}u_x = -\left(\frac{-\Delta P}{2kl}\right)^{1/n} s^{1/n}\,\mathrm{d}s$$

and:

$$u_x = -\left(\frac{-\Delta P}{2kl}\right)^{1/n} \frac{n}{n+1} s^{(n+1)/n} + \text{constant}$$

At the pipe wall, $s = r$ and, for the no-slip condition, $u_x = 0$.

So:

$$\text{constant} = \left(\frac{-\Delta P}{2kl}\right)^{1/n} \frac{n}{n+1} r^{(n+1)/n}$$

$$\therefore \quad u_x = \left(\frac{-\Delta P}{2kl}\right)^{1/n} \frac{n}{n+1} \left(r^{(n+1)/n} - s^{(n+1)/n}\right) \tag{3.130}$$

The velocity at the centre line, u_{CL}, is then given by putting $s = 0$:

$$u_{CL} = \left(\frac{-\Delta P}{2kl}\right)^{1/n} \frac{n}{n+1} r^{(n+1)/n} \tag{3.131}$$

Dividing equation 3.130 by equation 3.131 gives:

$$\frac{u_x}{u_{CL}} = 1 - \left(\frac{s}{r}\right)^{(n+1)/n} \tag{3.132}$$

The mean velocity of flow u is given by:

$$u = \frac{1}{\pi r^2} \int_0^r u_x 2\pi s\,\mathrm{d}s$$

$$= \frac{2}{r^2} u_{CL} r^2 \int_0^1 \left(1 - \frac{s}{r}\right)^{(n+1)/n} \frac{s}{r} d\left(\frac{s}{r}\right)$$

$$= 2u_{CL} \left[\frac{1}{2}\left(\frac{s}{r}\right)^2 - \frac{n}{3n+1}\left(\frac{s}{r}\right)^{(3n+1)/n}\right]_0^1$$

$$= 2u_{CL} \left(\frac{1}{2} - \frac{n}{3n+1}\right)$$

or:

$$\frac{u}{u_{CL}} = \frac{n+1}{3n+1} \tag{3.133}$$

Combining equations 3.132 and 3.133, gives the velocity profile in terms of the mean velocity u in place of the centre-line velocity u_{CL} as:

$$\frac{u_x}{u} = \frac{3n+1}{n+1}\left[1 - \left(\frac{s}{r}\right)^{(n+1)/n}\right] \tag{3.134}$$

Substituting into equation 3.131:

$$u = \left(\frac{-\Delta P}{2kl}\right)^{1/n} \frac{n}{3n+1} r^{(n+1)/n} \tag{3.135}$$

Working in terms of pipe diameter:

$$u = \left(\frac{-\Delta P}{4kl}\right)^{1/n} \frac{n}{6n+2} d^{(n+1)/n} \tag{3.136}$$

For a Newtonian fluid $n = 1$,

and:

$$u = \frac{-\Delta P d^2}{32kl} \qquad \text{(equation 3.36)}$$

This is identical to equation 3.36, bearing in mind that $k = \mu$ for a Newtonian fluid.

The shear rate (velocity gradient) at the tube wall is obtained by differentiating equation 3.134 with respect to s, and then putting $s = r$.

$$\frac{1}{u}\frac{du_x}{ds} = \frac{3n+1}{n+1}\left[-\frac{n+1}{n}\left(\frac{s}{r}\right)^{(n+1)/n}\frac{1}{s}\right]$$

If y is distance from the wall, $y + s = r$ and:

$$\left(-\frac{du_x}{ds}\right)_{s=r} = \left(\frac{du_x}{dy}\right)_{y=0} = \frac{3n+1}{n}\frac{u}{r} = \frac{6n+2}{n}\frac{u}{d} \tag{3.137}$$

For a Newtonian fluid, equation 3.137 gives a wall shear rate of $8u/d$ (corresponding to equation 3.39) and a shear stress of $8\mu u/d$ (corresponding to equation 3.40).

For a Newtonian fluid, the data for pressure drop may be represented on a pipe friction chart as a friction factor $\phi = (R/\rho u^2)$ expressed as a function of Reynolds number $Re = (udp/\mu)$. The friction factor is independent of the rheological properties of the fluid, but the Reynolds number involves the viscosity which, for a non-Newtonian fluid, is dependent on shear rate. METZNER and REED[22] defined a Reynolds number Re_{MR} for a power-law fluid in such a way that it is related to the friction factor for streamline flow in exactly the same way as for a Newtonian fluid.

Thus, from equation 3.39:

$$\phi = \frac{R}{\rho u^2} = 8Re_{MR}^{-1} \tag{3.138}$$

For the flow of a power-law fluid in a pipe of length l, the pressure drop $-\Delta P$ is obtained from equation 3.136 as:

$$-\Delta P = \left(\frac{6n+2}{n}\right)^n 4klu^n d^{-(n+1)}$$

Thus:

$$\phi = \frac{R}{\rho u^2} = \frac{-\Delta P d}{4l\rho u^2} \quad \text{(from equation 3.17)}$$

$$= \left(\frac{6n+2}{n}\right)^n \frac{ku^{n-2}d^{-n}}{\rho} \tag{3.139}$$

Substituting into equation 3.138:

$$Re_{MR} = 8\left(\frac{n}{6n+2}\right)^n \frac{\rho u^{2-n} d^n}{k} \tag{3.140}$$

As indicated in Section 3.7.9, this definition of Re_{MR} may be used to determine the limit of stable streamline flow. The transition value $(Re_{MR})_c$ is approximately the same as for a Newtonian fluid, but there is some evidence that, for moderately shear-thinning fluids, streamline flow may persist to somewhat higher values. Putting $n = 1$ in equation 3.140 leads to the standard definition of the Reynolds number.

The effect of power-law index on the velocity profile is seen by plotting equation 3.134 for various values of n, as shown in Figure 3.39.

Compared with the parabolic profile for a Newtonian fluid ($n = 1$), the profile is flatter for a shear-thinning fluid ($n < 1$) and sharper for a shear-thickening fluid ($n > 1$). The ratio of the centre line (u_{CL}) to mean (u) velocity, calculated from equation 3.133, is:

n	2.0	1.5	1.0	0.8	0.6	0.4	0.2	0.1
u_{CL}/u	2.33	2.2	2	1.89	1.75	1.57	1.33	1.18

Bingham-plastic fluids

For the flow of a Bingham-plastic fluid, the cross-section may be considered in two parts, as shown in Figure 3.32:

(1) The central unsheared core in which the fluid is all travelling at the centre-line velocity.
(2) The annular region separating the core from the pipe wall, over which the whole of the velocity change occurs.

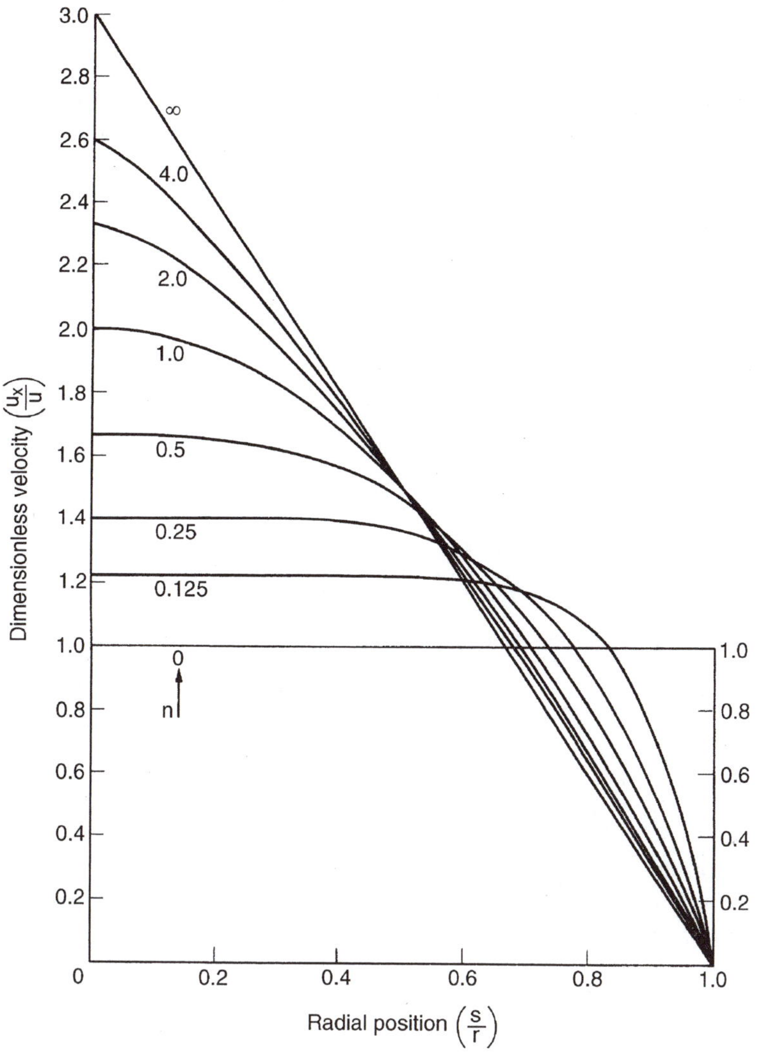

Figure 3.39. Fully-developed laminar velocity profiles for power-law fluids in a pipe (from equation 3.134)

The boundary between the two regions and the radius r_c of the core is determined by the position in the cross-section at which the shear stress is exactly equal to the yield stress R_Y of the fluid. Since the shear stress is linearly related to the radial position:

$$\frac{R_Y}{R} = \frac{r_c}{r} \qquad \text{(equation 3.8)}$$

where R is the shear stress at the wall (radius r).

In a given pipe, R is determined solely by the pressure drop $-\Delta P$ and is completely independent of the rheology of the fluid and, from equation 3.6:

$$R = -\Delta P \frac{r}{2l} \qquad (3.141)$$

and:
$$\frac{r_c}{2l} = \frac{R_Y}{-\Delta P} \tag{3.142}$$

Thus, for a given fluid (R_Y constant), the critical radius r_c is determined entirely by the pressure drop and becomes progressively larger as the pressure drop is reduced. Flow ceases when the shear stress at the wall R falls to a value equal to the yield stress R_Y.

Flow in the annular region ($s > r_c$)

In this region, the relation between the shear stress R_s and the velocity gradient $\mathrm{d}u_x/\mathrm{d}s$ is given by:

$$|R_s| - R_Y = \mu_p \left(-\frac{\mathrm{d}u_x}{\mathrm{d}s}\right)$$

Writing: $R_s = R(s/r)$ and rearranging:

$$\mathrm{d}u_x = \frac{1}{\mu_p}\left(R_Y - \frac{s}{r}R\right)\mathrm{d}s$$

Integrating:
$$u_x = \frac{1}{\mu_p}\left\{R_Y s - R\frac{s^2}{2r}\right\} + \text{constant.}$$

For the no-slip condition at the wall, $u_x = 0$ at $s = r$.

$$\therefore \qquad \text{constant} = -\frac{1}{\mu_p}\left\{R_Y r - R\frac{r^2}{2r}\right\}$$

Thus:
$$u_x = \frac{1}{\mu_p}\left\{\frac{R}{2r}(r^2 - s^2) - R_Y(r - s)\right\} \tag{3.143}$$

This equation for the velocity profile, reduces to the parabolic form for a Newtonian fluid, when $R_Y = 0$, and applies in the region $r > s > r_c$.

The volumetric flowrate Q_A in the annular region is given by:

$$Q_A = \int_{s=2l(R_Y/-\Delta P)}^{s=r} \left\{\frac{1}{\mu_p}\left[\frac{R}{2r}(r^2 - s^2) - R_Y(r - s)\right]\right\} 2\pi s\,\mathrm{d}s$$

On integration:
$$Q_A = \frac{-\Delta P \pi r^4}{8\mu_p l}\left\{1 - \frac{4}{3}X - 2X^2 + 4X^3 - \frac{5}{3}X^4\right\} \tag{3.144}$$

where $X = \dfrac{2R_Y l}{-\Delta P r} = \dfrac{R_Y}{R} = \dfrac{\text{Yield stress}}{\text{Wall shear stress}}$.

Flow in the centre plug

The velocity u_p of the centre plug is obtained by putting:

$$s = r_c = 2l\frac{R_Y}{-\Delta P}$$

in equation 3.143.

Thus: $$u_p = \frac{1}{\mu_p}\left\{\frac{-\Delta P r}{2l}\frac{1}{2r}\left[r^2 - 4l^2\left(\frac{R_Y}{-\Delta P}\right)^2\right] - R_Y\left(r - 2l\frac{R_Y}{-\Delta P}\right)\right\}$$

This simplifies to:

$$u_p = \frac{-\Delta P r^2}{8\mu_p l}\{2 - 4X + 2X^2\} \tag{3.145}$$

The volumetric flowrate in the plug Q_p is then given by:

$$\begin{aligned} Q_p &= u_p \pi r_c^2 \\ &= \frac{-\Delta P r^2}{8\mu_p l}(2 - 4X + 2X^2)\pi\frac{4l^2R_Y^2}{(-\Delta P)^2} \\ &= \frac{-\Delta P \pi r^4}{8\mu_p l}\{2X^2 - 4X^3 + 2X^4\} \end{aligned} \tag{3.146}$$

Total flow through the pipe

The total flowrate Q through the pipe is given by:

$$\begin{aligned} Q &= Q_A + Q_p \\ &= \frac{-\Delta P \pi r^4}{8\mu_p l}\left\{1 - \frac{4}{3}X + \frac{1}{3}X^4\right\} \end{aligned} \tag{3.147}$$

Thus, the mean velocity of flow u is given by:

$$\begin{aligned} u &= \frac{Q}{\pi r^2} = \frac{-\Delta P r^2}{8\mu_p l}\left\{1 - \frac{4}{3}X + \frac{1}{3}X^4\right\} \\ &= \frac{-\Delta P d^2}{32\mu_p l}\left\{1 - \frac{4}{3}X + \frac{1}{3}X^4\right\} \end{aligned} \tag{3.148}$$

For a Newtonian fluid both R_Y and X are zero and the mean velocity is:

$$u = \frac{-\Delta P d^2}{32\mu_p l} \qquad \text{(cf. equation 3.35)}$$

Equation 3.148 is sometimes known as *Buckingham's equation.*

Example 3.10

The rheological properties of a particular suspension may be approximated reasonably well by either a *power-law* or a *Bingham-plastic* model over the shear rate range of 10 to 50 s^{-1}. If the consistency coefficient k is 10 N s^n/m^{-2} and the flow behaviour index n is 0.2 in the power law model, what will be the approximate values of the yield stress and of the plastic viscosity in the Bingham-plastic model?

What will be the pressure drop, when the suspension is flowing under laminar conditions in a pipe 200 m long and 40 mm diameter, when the centre line velocity is 1 m/s, according to the power-law model? Calculate the centre-line velocity for this pressure drop for the Bingham-plastic model.

Solution

Using the *power-law model* (equation 3.121):

$$|R_y| = k\left(\left|\frac{du_x}{dy}\right|\right)^n = 10\left(\left|\frac{du_x}{dy}\right|\right)^{0.2}$$

When: $\left|\frac{du_x}{dy}\right| = 10\ s^{-1} : |R_y| = 10 \times 10^{0.2} = 15.85\ N/m^2$

$\left|\frac{du_x}{dy}\right| = 50\ s^{-1} : |R_y| = 10 \times 50^{0.2} = 21.87\ N/m^2$

Using the *Bingham-plastic model* (equation 3.125):

$$|R_y| = R_Y + \mu_p\left|\frac{du_x}{dy}\right|$$

When: $\left|\frac{du_x}{dy}\right| = 10\ s^{-1} : \quad 15.85 = R_Y + 10\mu_p$

$\left|\frac{du_x}{dy}\right| = 50\ s^{-1} : \quad \underline{21.87 = R_Y + 50\mu_p.}$

Subtracting: $6.02 = 40\mu_p.$

Thus: $\mu_p = 0.150\ N\ s/m^2$

and: $R_Y = 14.35\ N/m^2$

Thus, the Bingham-plastic equation is:

$$|R_y| = 14.35 + 0.150\left|\frac{du_x}{dy}\right|$$

For the *power-law fluid*:

Equation 3.131 gives: $$u_{CL} = \left(\frac{-\Delta P}{2kl}\right)^{1/n}\frac{n}{n+1}r^{(n+1)/n}$$

Rearranging: $$-\Delta P = 2klu_{CL}^n\left(\frac{n+1}{n}\right)^n r^{-(n+1)}.$$

The numerical values in SI units are:

$$u_{CL} = 1\ m/s,\quad l = 200\ m,\quad r = 0.02\ m,\quad k = 10\ Ns^n m^{-2},\quad n = 0.2$$

and: $-\Delta P = \underline{\underline{626{,}000\ N/m^2}}$

For a Bingham-plastic fluid:

The centre line velocity is given by equation 3.145:

$$u_p = \frac{-\Delta P r^2}{8\mu_p l}(2 - 4X + 2X^2)$$

where: $$X = \frac{l}{r}\frac{2R_Y}{(-\Delta P)}$$

$$= \frac{200}{0.02} \times \frac{2 \times 14.35}{626{,}000} = 0.458$$

$\therefore \quad 2 - 4X + 2X^2 = 0.589$

$$u_p = \frac{626{,}000 \times (0.02)^2}{8 \times 0.150 \times 200} \times 0.589$$

$$= \underline{\underline{0.61 \text{ m/s}}}$$

Example 3.11

A Newtonian liquid of viscosity 0.1 N s/m^2 is flowing through a pipe of 25 mm diameter and 20 m in length, and the pressure drop is 10^5 N/m^2. As a result of a process change a small quantity of polymer is added to the liquid and this causes the liquid to exhibit non-Newtonian characteristics; its rheology is described adequately by the *power-law* model and the *flow index* is 0.33. The apparent viscosity of the modified fluid is equal to the viscosity of the original liquid at a shear rate of 1000 s^{-1}.

If the pressure difference over the pipe is unaltered, what will be the ratio of the volumetric flowrates of the two liquids?

Solution

For a power-law fluid: $$|R_y| = k\left(\left|\frac{du_x}{dy}\right|\right)^n = \left[k\left(\left|\frac{du_x}{dy}\right|\right)^{n-1}\right]\frac{du_x}{dy}$$

Apparent viscosity $$\mu_a = k\left(\left|\frac{du_x}{dy}\right|\right)^{-0.67}$$

When: $\left|\dfrac{du_x}{dy}\right| = 1000 \text{ s}^{-1}$ $$\mu_a = k(1000)^{-0.67} = \frac{k}{100} = 0.1 \text{ kg/m s}$$

$\therefore$ $$k = 10 \text{ Ns}^{0.33}/\text{m}^2$$

The rheological equation is $$|R_y| = 10\left(\left|\frac{du_x}{dy}\right|\right)^{0.33}$$

From equation 3.136:

For a power-law fluid:

$$u = \left(\frac{-\Delta P}{4kL}\right)^{1/n}\frac{n}{2(3n+1)}d^{(n+1)/n}.$$

For the polymer solution ($n = 0.33$):

$$u_2 = \left(\frac{10^5}{4 \times 10 \times 20}\right)^3 \frac{1}{12} \cdot 0.025^4$$

$$= 0.0636 \text{ m/s}$$

For the original Newtonian fluid ($n = 1$):

$$u_1 = \left(\frac{10^5}{4 \times 0.1 \times 20}\right) \cdot \left(\frac{1}{8}\right) \cdot (0.025^2)$$

$$= 0.977 \text{ m/s}$$

and: $$\frac{u_2}{u_1} = \left(\frac{0.0636}{0.977}\right) \underline{\underline{0.065}}$$

Example 3.12

Two liquids of equal densities, one Newtonian and the other a non-Newtonian '*power-law*' fluid, flow at equal volumetric rates per unit width down two wide vertical surfaces. The non-Newtonian fluid has a *power-law* index of 0.5 and it has the same apparent viscosity as the Newtonian fluid when its shear rate is 0.01 s^{-1}. Show that, for equal surface velocities of the two fluids, the film thickness to the non-Newtonian fluid is 1.132 times that of the Newtonian fluid.

Solution

For a *power-law* fluid:

$$|R_y| = k\left|\frac{du_x}{dy}\right|^n \qquad \text{(equation 3.121)}$$

$$= k\left|\frac{du_x}{dy}\right|^{n-1}\left|\frac{du_x}{dy}\right| \qquad \text{(equation 3.122)}$$

and from equation 3.123, the apparent viscosity, $\mu_a = k|du_x/dy|^{n-1}$
For a Newtonian fluid:

$$R_y = \mu|du_x/dy| \qquad \text{(equation 3.3)}$$

For $n = 0.5$ and $|du_x/dy| = 0.01\ s^{-1}$, then, using SI units:

$$\mu_a = \mu = k|du_x/dy|^{n-1}$$

$$\mu = k(0.01)^{-0.5} = 10k \text{ and } k = 0.1\ \mu.$$

The equation of state of the power-law fluid is therefore in SI units:

$$R_y = 0.1\mu|du_x/dy|^{0.5}$$

For a fluid film of thickness s flowing down a vertical surface of length l and width w, a force balance on the fluid at a distance greater than y from the surface (fluid depth $s - y$) gives:

$$(s - y)wl\rho g = R_y wl = k(du_x/dy)^n wl$$

or:

$$du_x/dy = (\rho g/k)^{1/n}(s - y)^{1/n}$$

and:

$$u_x = (\rho g/k)^{1/n}(s - y)^{(n+1)/n}[-n/(n + 1)] + \text{constant}$$

At the surface, $y = 0$, $u_x = 0$ and the constant $= (\rho g/k)^{1/n}s^{(n+1)/n}[n/(n + 1)]$

and:

$$u_x = \left(\frac{\rho g}{k}\right)^{1/n}\frac{n}{n+1}\left[s^{\frac{n+1}{n}} - (s - y)^{\frac{n+1}{n}}\right]$$

At the free surface where $y = s$:

$$u_s = \left(\frac{\rho g}{k}\right)^{1/n}\frac{n}{n+1}s^{\frac{n+1}{n}} \qquad \text{(i)}$$

The volumetric flowrate, Q, is given by:

$$Q = \int_0^s w\,dy\left(\frac{\rho g}{k}\right)^{1/n}\frac{n}{n+1}\left[s^{\frac{n+1}{n}} - [s - y]^{\frac{n+1}{n}}\right]$$

$$= w\left(\frac{\rho g}{k}\right)^{1/n}\left(\frac{n}{2n+1}\right)s^{\frac{2n+1}{n}} \qquad \text{(ii)}$$

For the non-Newtonian fluid, $k = 0.1\mu$, $n = 0.5$ and equation (ii) becomes:

$$Q = w(\rho g/(0.1\mu))^2 \times 0.25\ s^4$$

$$= 25w\left(\frac{\rho g}{\mu}\right)^2 s^4 \qquad \text{(iii)}$$

For the Newtonian fluid, $n = 1$ and $k = \mu$ and substituting in equation (ii):

$$Q = w\left(\frac{\rho g}{\mu}\right)\left(\frac{s_N^3}{3}\right) \quad \text{(iv)}$$

where s_N is the thickness of the Newtonian fluid.

For equal flowrates, then from equations (iii) and (iv):

$$25w(\rho g/\mu)^2 s^4 = 0.33w(\rho g/\mu)s_N^3$$

or:

$$s_N^3 = 75(\rho g/\mu)s^4$$

For equal surface velocities, the term $(\rho g/k)$ may be substituted from equation (iv) into equation (i) to give,

for the non-Newtonian fluid:

$$u_s = (\rho g/0.1\mu)^2 0.33s^3$$

$$= 100(s_N^3/75s^4)^2 0.33s^3$$

$$= 0.00592s_N^6/s^5$$

for the Newtonian fluid:

$$u_s = (\rho g/\mu)0.5s_N^2$$

$$= (s_N^3/75s^4)0.5s_N^2$$

$$= 0.0067s_N^5/s^4$$

and hence:

$$s_N/s = (0.0067/0.00592) = \underline{\underline{1.132}}$$

General equations for pipeline flow

Fluids whose behaviour can be approximated by the power-law or Bingham-plastic equation are essentially special cases, and frequently the rheology may be very much more complex so that it may not be possible to fit simple algebraic equations to the flow curves. It is therefore desirable to adopt a more general approach for time-independent fluids in fully-developed flow which is now introduced. For a more detailed treatment and for examples of its application, reference should be made to more specialist sources.[14–17]

If the shear stress is a function of the shear rate, it is possible to invert the relation to give the shear rate, $\dot{\gamma} = -du_x/ds$, as a function of the shear stress, where the negative sign is included here because velocity decreases from the pipe centre outwards.

Thus:

$$\dot{\gamma} = -\frac{du_x}{ds} = f(R_s) = f\left(R\frac{s}{r}\right) \quad (3.149)$$

where R_s is the positive value of the shear stress at radius s.

The advantage of using the relation between shear rate and shear stress in this form is that R_s, unlike $\dot{\gamma}$, is known at all values of s.

On integration, and noting that $u_x = 0$ at $s = r$:

$$u_x = \int_s^r f\left(R\frac{s}{r}\right) ds \quad (3.150)$$

The total volumetric flowrate Q through the pipe is found by integrating over the whole cross-section to give:

$$Q = \int_0^r 2\pi s u_x \, ds = \pi \int_0^{r^2} u_x \, d(s^2)$$

$$= \pi \left[s^2 u_x - \int s^2 \, du_x \right]_{s=0}^{s=r}$$

$$= \pi \int_0^r s^2 f\left(R\frac{s}{r}\right) ds \quad (\text{since } u_x = 0 \text{ at } s = r) \tag{3.151}$$

Substituting $s = r(R_s/R)$:

$$Q = \pi \int_0^R \left(r\frac{R_s}{R}\right)^2 f(R_s)\frac{r}{R} \, dR_s$$

or:

$$\frac{Q}{\pi r^3} = \frac{1}{R^3} \int_0^R R_s^2 f(R_s) \, dR_s. \tag{3.152}$$

Multiplying equation 3.152 by R^3 and differentiating with respect to R gives:

$$\frac{d}{dR}\left(R^3 \frac{Q}{\pi r^2}\right) = \frac{d}{dR} \int_0^R Rs^2 f(R_s) dR_s.$$

Use of parameters n' and k'

Noting that R is the upper limit of the integral, and using the Leibnitz rule:

$$R^3 \frac{d(Q/\pi r^3)}{dR} + 3R^2 \frac{Q}{\pi r^3} = R^2 f(R)$$

Dividing by R^2 and putting $R = -\Delta P \frac{r}{2l}$ gives:

$$-\Delta P \frac{r}{2l} \frac{d(Q/\pi r^3)}{d\left(-\Delta P \frac{r}{2l}\right)} + \frac{3Q}{\pi r^3} = f(R) = \left(-\frac{du_x}{ds}\right)_{s=r}$$

$$\therefore \quad \frac{4Q}{\pi r^3} \frac{d[\ln(Q/\pi r^3)]}{4 d\ln\left[-\Delta P \frac{r}{2l}\right]} + \frac{3Q}{\pi r^3} = \left(-\frac{du_x}{ds}\right)_{s=r} \tag{3.153}$$

Writing:

$$n' = \frac{d\left[\ln\left(-\Delta P \frac{r}{2l}\right)\right]}{d[\ln(Q/\pi r^3)]} \tag{3.154}$$

where n' is the slope of the log/log plot of $-\Delta P(r/2l)$ against $Q/\pi r^3$ gives:

$$\frac{4Q}{\pi r^3}\left(\frac{1}{4n'} + \frac{3}{4}\right) = \frac{4Q}{\pi r^3}\left(\frac{3n'+1}{4n'}\right) = \left(-\frac{du_x}{ds}\right)_{s=r} \tag{3.155}$$

Equation 3.155 is frequently referred to as the RABINOWITSCH(23)–MOONEY(24) relation. Now the rheological data for a fluid may be represented as a curve of $R = -\Delta P(r/2l)$ plotted against $4Q/\pi r^3 \left(= \frac{8u}{d}\right)$ using logarithmic coordinates, and the slope n' may be measured over any small range, irrespective of whether or not an equation may be fitted to the curve. This provides a basis for utilising practical rheological data for *any shear-dependent fluid*.

From the definition of n':

$$R = -\Delta P \frac{r}{2l} = \text{constant}\left(\frac{Q}{\pi r^3}\right)^{n'} = k'\left(\frac{4Q}{\pi r^3}\right)^{n'} \tag{3.156}$$

Thus n' and k' are parameters which can be measured for any fluid, and the method may be applied in a wide range of rheological properties.

For a *power-law fluid*, the slope n' (equation 3.154) has a constant value n and hence:

$$n' = n \tag{3.157}$$

Thus, using equation 3.135 to substitute for $-\Delta P$ in equation 3.156:

$$k' = k\left(\frac{3n+1}{4n}\right)^n \tag{3.158}$$

Generalised Reynolds Number

The Metzner and Reed Reynolds number Re_{MR} may be expressed in terms of n' and k'. From equation 3.140, derived for a power-law fluid:

$$Re_{MR} = 8\left(\frac{n}{6n+2}\right)^n \frac{\rho u^{2-n} d^n}{k}$$

Putting $n = n'$ and $k = k'\left(\dfrac{4n'}{3n'+1}\right)^{n'}$, then:

$$Re_{MR} = 8\left(\frac{n'}{6n'+2}\right)^{n'}\left(\frac{3n'+1}{4n'}\right)^{n'}\frac{\rho u^{2-n'} d^{n'}}{k'}$$

$$= 8^{1-n'}\frac{\rho u^{2-n'} d^{n'}}{k'} \tag{3.159}$$

Working in terms of the apparent viscosity μ_w at the wall shear rate, by definition:

$$\mu_w = \frac{R}{\left(-\dfrac{\mathrm{d}u_x}{\mathrm{d}s}\right)_{s=r}} \tag{3.160}$$

From equation 3.156:

$$R = k'\left(\frac{4Q}{\pi r^3}\right)^{n'} = k'\left(\frac{8u}{d}\right)^{n'} \tag{3.161}$$

and from equation 3.155:

$$\left(-\frac{\mathrm{d}u_x}{\mathrm{d}s}\right)_{s=r} = \frac{4Q}{\pi r^3}\frac{3n'+1}{4n'} = \frac{8u}{d}\frac{3n'+1}{4n'} \tag{3.162}$$

Substituting from equations 3.161 and 3.162 into equation 3.160:

$$\mu_w = \frac{k'\left(\dfrac{8u}{d}\right)^{n'}}{\dfrac{8u}{d}\cdot\dfrac{3n'+1}{4n'}} = k'\left(\frac{8u}{d}\right)^{n'-1}\frac{4n'}{3n'+1} \tag{3.163}$$

Then substituting for k' in equation 3.159:

$$Re_{MR} = \frac{8^{1-n'}\rho u^{2-n'} d^{n'}}{\mu_w \left(\frac{8u}{d}\right)^{1-n'} \cdot \frac{3n'+1}{4n'}}$$

or:

$$Re_{MR} = \frac{4n'}{3n'+1}\frac{\rho u d}{\mu_w} \tag{3.164}$$

Velocity–pressure gradient relationships for fluids of specified rheology

Equation 3.152 provides a method of determining the relationship between pressure gradient and mean velocity of flow in a pipe for fluids whose rheological properties may be expressed in the form of an explicit relation for shear rate as a function of shear stress.

$$\frac{Q}{\pi r^3} = \frac{1}{R^3}\int_0^R R_s^2 \mathrm{f}(R_s)\mathrm{d}R_s \qquad \text{(equation 3.152)}$$

For a *power-law fluid*, from equation 3.121:

$$\mathrm{f}(R_s) = \dot{\gamma} = \left(\frac{R_s}{k}\right)^{\frac{1}{n}} \tag{3.165}$$

$$\begin{aligned}
\frac{Q}{\pi r^3} &= \frac{1}{R^3}\int_0^R R_s^2 \left(\frac{R_s}{k}\right)^{\frac{1}{n}} \mathrm{d}R_s \\
&= \frac{1}{R^3 k^{\frac{1}{n}}}\int_0^R R_s^{\frac{2n+1}{n}} \mathrm{d}R_s \\
&= \frac{1}{R^3 k^{\frac{1}{n}}} \cdot \frac{n}{3n+1} R^{\frac{3n+1}{n}} \\
&= \frac{1}{k^{\frac{1}{n}}} \frac{n}{3n+1} R^{\frac{1}{n}}
\end{aligned} \tag{3.166}$$

Writing the mean velocity as $u = \dfrac{Q}{\pi r^2}$ and $R = -\Delta P \cdot \dfrac{r}{2l}$, then

$$\begin{aligned}
u &= \frac{r}{k^{\frac{1}{n}}} \frac{n}{3n+1}\left(\frac{-\Delta P r}{2l}\right)^{\frac{1}{n}} \\
&= \left(\frac{-\Delta P d}{4kl}\right)^{\frac{1}{n}} \frac{n}{6n+2} d
\end{aligned} \tag{3.167}$$

For a *Bingham-plastic fluid*, from equation 3.125:

$$\mathrm{f}(R_s) = \dot{\gamma} = \frac{R_s - R_Y}{\mu_p} \quad (R_s \geq R_Y) \tag{3.168a}$$

$$f(R_s) = \dot{\gamma} = 0 \quad (R_s \leq R_Y) \tag{3.168b}$$

$$\therefore \quad \frac{Q}{\pi r^3} = \frac{1}{R^3}\int_{R_Y}^{R} R_s^2 \frac{1}{\mu_p}(R_s - R_Y)\mathrm{d}R_s \quad \left(\text{noting that } \int_0^{R_s} = 0\right)$$

$$= \frac{1}{\mu_p R^3}\int_{R_Y}^{R}(R_s^3 - R_s^2 R_Y)\mathrm{d}R_s$$

$$= \frac{1}{\mu_p R^3}\left[\frac{R_s^4}{4} - \frac{R_s^3 R_Y}{3}\right]_{R_Y}^{R}$$

$$= \frac{1}{\mu_p R^3}\left(\frac{R^4}{4} - \frac{R^3 R_Y}{3} - \frac{R_Y^4}{4} + \frac{R_Y^4}{3}\right)$$

$$= \frac{1}{4\mu_p} R\left(1 - \frac{4}{3}\frac{R_Y}{R} + \frac{1}{3}\left(\frac{R_Y}{R}\right)^4\right) \tag{3.169}$$

Again, noting that $u = \dfrac{Q}{\pi r^2}$ and $R = -\Delta P \cdot \dfrac{r}{2l}$ and putting $\dfrac{R_Y}{R} = X$

$$u = \frac{-\Delta P d^2}{32\mu_p l}\left(1 - \frac{4}{3}X + \frac{1}{3}X^4\right) \tag{3.170}$$

The above procedure may be followed using any other equation of state, provided that $\dot{\gamma}$ can be expressed as an explicit function of shear stress R_s.

It may be noted that equations 3.167 and 3.170 are identical to equations 3.136 and 3.148 derived earlier. Although these derivations are simpler to carry out, the method does not allow the velocity profile in the pipe to be obtained.

For a *Herschel–Bulkley fluid*, from equation 3.127:

$$f(R_s) = \dot{\gamma} = \left(\frac{R_s - R_Y}{\mu_p}\right)^{\frac{1}{m}} \quad (R_s \geq R_Y) \tag{3.171a}$$

$$f(R_s) = \dot{\gamma} = 0 \quad (R_s \leq R_Y) \tag{3.171b}$$

$$\therefore \quad \frac{Q}{\pi r^3} = \frac{1}{R^3}\int_{R_Y}^{R} R_s^2\left(\frac{R_s - R_Y}{\mu_p}\right)^{\frac{1}{m}} \mathrm{d}R_s$$

Putting $R_s - R_Y = R'$:

$$\frac{Q}{\pi r^3} = \frac{1}{R^3 \mu_p^{\frac{1}{m}}}\int_0^{R-R_Y}(R' + R_Y)^2 R'^{\frac{1}{m}}\mathrm{d}R'$$

$$= \frac{1}{R^3 \mu_p^{\frac{1}{m}}}\int_0^{R-R_Y}\left(R'^{\frac{2m+1}{m}} + 2R_Y R'^{\frac{m+1}{m}} + R_Y^2 R'^{\frac{1}{m}}\right)\mathrm{d}R'$$

$$= \frac{1}{R^3 \mu_p^{\frac{1}{m}}}\left[\frac{m}{3m+1}R'^{\frac{3m+1}{m}} + R_Y\frac{2m}{2m+1}R'^{\frac{2m+1}{m}} + R_Y^2\frac{m}{m+1}R'^{\frac{m+1}{m}}\right]_0^{R-R_Y}$$

giving:
$$\frac{Q}{\pi r^3} = \frac{1}{R^3 \mu_p^{\frac{1}{m}}} \left\{ \frac{m}{3m+1}(R - R_Y)^{\frac{3m+1}{m}} + \frac{2m}{2m+1} R_Y (R - R_Y)^{\frac{2m+1}{m}} + \frac{m}{m+1} R_Y^2 (R - R_Y)^{\frac{m+1}{m}} \right\} \tag{3.172}$$

$$\therefore \quad \frac{Q}{\pi r^2} = u = \frac{r}{R^3 \mu_p^{\frac{1}{m}}} (R - R_Y)^{\frac{m+1}{m}} \left\{ \frac{m}{3m+1}(R - R_Y)^2 + \frac{2m}{2m+1} R_Y (R - R_Y) + \frac{m}{m+1} R_Y^2 \right\} \tag{3.173}$$

Equation 3.172 reduces to equation 3.166 for a *power-law fluid* ($R_Y = 0$) and to equation 3.169 for a *Bingham-plastic fluid* ($m = 1$)

Expressing equation 3.173 in terms of pressure gradient $\left(R = -\frac{\Delta P d}{4l}\right)$ and $X = \frac{R_Y}{R}$:

$$u = \frac{1}{2}\left(-\frac{\Delta P}{4\mu_p l}\right)^{\frac{1}{m}} d^{\frac{m+1}{m}} (1 - X)^{\frac{m+1}{m}} \left\{ \frac{m}{3m+1}(1 - X)^2 + \frac{2m}{2m+1} X(1 - X) + \frac{m}{m+1} X^2 \right\} \tag{3.174}$$

3.4.8. Turbulent flow

Surprising though it may be, there is no completely reliable method of predicting pressure drop for turbulent flow of non-Newtonian fluids in pipes. There is strong evidence that fluids showing similar flow characteristics in laminar flow do not necessarily behave in the same way in turbulent flow. Thus, for instance, a flocculated suspension and a polymer solution following the power-law model and having similar values of n and k may give different results in turbulent flow. HEYWOOD and CHENG[25] have illustrated the difficulties in a paper in which they have, unsuccessfully, attempted to calculate pressure drops for the turbulent flow of sewage sludge, using the various equations given in the literature.

As indicated earlier, non-Newtonian characteristics have a much stronger influence on flow in the streamline flow region where viscous effects dominate than in turbulent flow where inertial forces are of prime importance. Furthermore, there is substantial evidence to the effect that, for shear-thinning fluids, the standard friction chart tends to over-predict pressure drop if the Metzner and Reed Reynolds number Re_{MR} is used. Furthermore, laminar flow can persist for slightly higher Reynolds numbers than for Newtonian fluids. Overall, therefore, there is a factor of safety involved in treating the fluid as Newtonian when flow is expected to be turbulent.

HARTNETT and KOSTIC[26] have recently examined the published correlations for turbulent flow of shear-thinning "power-law" fluids in pipes and in non-circular ducts, and have concluded that, for smooth pipes, DODGE and METZNER'S[27] modification of equation 3.11 (to which it reduces for Newtonian fluids) is the most satisfactory.

DODGE and METZNER[27] carried out experimental work using pipes of nominal diameters 1/2 in. (12.7 mm), 1 in. (25.4 mm) and 2 in. (50.8 mm), using polymer gels and solids–liquid suspensions, at Metzner and Reed Reynolds numbers (Re_{MR}) up to 36,000. The flow charcteristics of the fluids corresponded approximately to the *power-law* relation, with n values ranging from 0.3 to 1, though only two of the fluids conformed to the power-law relation over the whole range of Reynolds numbers. Figure 3.40 is a reproduction of Dodge and Metzner's graph of friction factor against Reynolds number (Re_{MR}) with friction factor expressed as $\phi(= R/\rho u^2)$, in order to conform with the standard used elsewhere in this chapter, instead of the Fanning friction factor $f(= 2\phi)$ used by the authors. For turbulent flow, the friction factor at a given value of Re_{MR} becomes progressively less as the degree of shear-thinning increases (n decreasing). It will be noted that the experimental results and extrapolated values are separately designated. Extrapolated values should never be used for non-Newtonian fluids; certainly not the extrapolated values for shear-thickening fluids which should be ignored, as virtually no experimental data are available in the literature for n greater than 1.

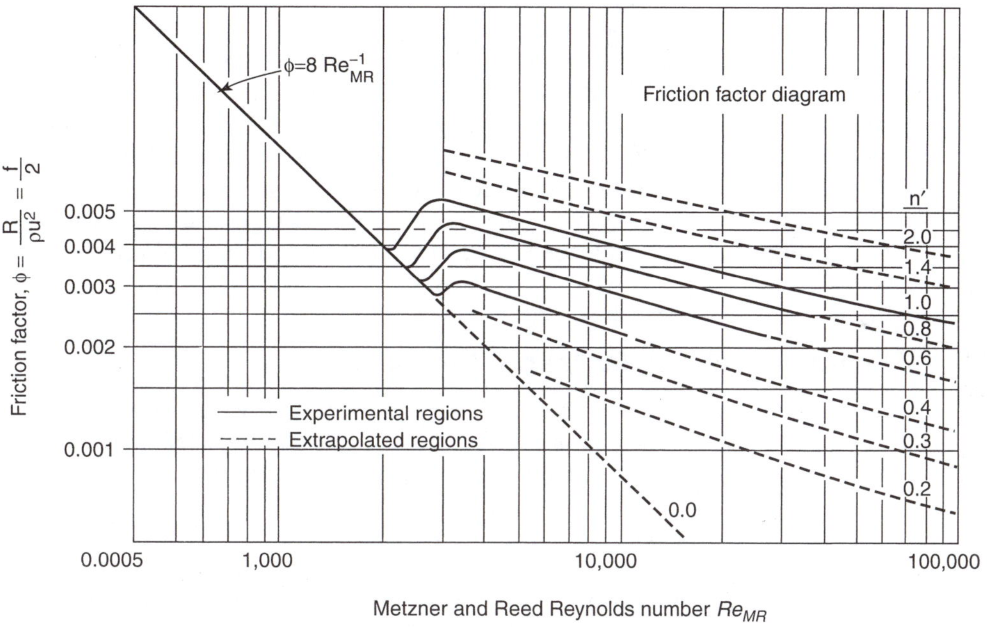

Figure 3.40. Metzner and Reed correlation of friction factor and Reynolds number

YOO[28] has proposed a simple modification to the Blasius equation for turbulent flow in a pipe, which gives values of the friction factor accurate to within about ±10 per cent. The friction factor is expressed in terms of the METZNER and REED[22] generalised Reynolds number Re_{MR} and the power-law index n.

$$\frac{f}{2} = \phi = \frac{R}{\rho u^2} = 0.0396 n^{0.675} Re_{MR}^{-0.25} \tag{3.175}$$

Equation 3.175 reduces to the simple Blasius relation (equation 3.10) for a Newtonian fluid ($n = 1$).

Rearranging equation 3.175 gives:

$$\frac{R}{\rho u^2} n^{-0.675} = 0.0396 Re_{MR}^{-0.25} \tag{3.176}$$

Thus, the pipe friction chart for a Newtonian fluid (Figure 3.3) may be used for shear-thinning power-law fluids if Re_{MR} is used in place of Re. In the turbulent region, the ordinate is equal to $(R/\rho u^2)n^{-0.675}$. For the streamline region the ordinate remains simply $R/\rho u^2$, because Re_{MR} has been defined so that it shall be so (see equation 3.140). More recently, IRVINE[29] has proposed an improved form of the modified Blasius equation which predicts the friction factor for inelastic shear-thinning polymer-solutions to within ± 7 per cent.

$$\phi = \frac{R}{\rho u^2} = \left[\frac{2^{3-2n}}{7^{7n}} \left(\frac{4}{3n+1} \right)^{n(3n+2)} Re_{MR}^{-1} \right]^{1/(3n+1)} \tag{3.177}$$

Both equations 3.176 and 3.177 are easy to use as they give $R/\rho u^2$ as explicit functions of Re_{MR} and n. Equation 3.175 should be used with caution, particularly if the fluid exhibits any elastic properties, bearing in mind the conclusions of HEYWOOD and CHENG[25].

BOWEN[30] gives a generalised scale-up method method by which the pressure drop in an industrial pipeline may be esimated from the results of laboratory experiments. It involves making measurements on the small-scale using a sample of the fluid. The method obviates the difficulty that erroneous results may be obtained by predicting the pressure drop for flow in the turbulent regime using parameters measured under laminar flow conditions. Reference may also be made to CHHABRA and RICHARDSON[17] for details of the procedure.

3.4.9. The transition from laminar to turbulent flow

The critical value of the Reynolds number (Re_{MR}) for the transition from laminar to turbulent flow may be calculated from the RYAN and JOHNSON[10] stability parameter, defined earlier by equation 3.56. For a *power-law* fluid, this becomes:

$$(Re_{MR})_c = \frac{6464}{(3n+1)^2}(2+n)^{\frac{2+n}{1+n}} \tag{3.178}$$

From equation 3.178, the critical Reynolds number for a Newtonian fluid is 2100. As n decreases, $(Re_{MR})_c$ rises to a maximum of 2400, and then falls to 1600 for $n = 0.1$. Most of the experimental evidence suggests, however, that the transition occurs at a value close to 2000.

3.5. FURTHER READING

ASTARITA, G. and MARRUCCI, G.: *Principles of Non-Newtonian Fluid Mechanics* (McGraw-Hill, New York, 1974).

BIRD, R. B., ARMSTRONG, R. C. and HASSAGER, O.: *Dynamics of Polymeric Liquids Vol. 1, Fluid Mechanics* (Wiley, New York 1977).

BRASCH, D. J. and WHYMAN, D.: *Problems in Fluid Flow* (Edward Arnold, London, 1986).

CHHABRA, R. P. and RICHARDSON J. F.: *Non-Newtonian Flow in the Process Industries* (Butterworth-Heinemann, 1999)
CHURCHILL, S. W.: *Viscous Flows: The Practical Use of Theory in Fluid Flow* (Butterworth, London, 1988).
CRANE: Technical Paper No. 410. *Flow Through Valves, Fittings and Pipes*. 14th printing (Crane, New York, 1974).
DUNCAN, W. J., THOM, A. S. and YOUNG, A. D.: *Mechanics of Fluids*, 2nd edn (Edward Arnold, London, 1972).
FRANCIS, J. R. D.: *A Textbook of Fluid Mechanics*, 3rd edn (Edward Arnold, London, 1971).
HOLLAND, F. A.: *Fluid Flow* (Edward Arnold, London, 1986).
HUTTON, J. F., PEARSON, J. R. A. and WALTERS, K. (eds): *Theoretical Rheology* (Applied Science Publishers, London, 1975).
KAY, J. M. and NEDDERMAN, R. M.: *An Introduction to Fluid Mechanics and Heat Transfer*, 3rd edn (Cambridge University Press, Cambridge, 1974).
MASSEY, B. S.: *Mechanics of Fluids*, 5th edn (Van Nostrand/Reinhold, London, 1987).
PATERSON, A. R.: *A First Course in Fluid Mechanics* (Cambridge University Press, Cambridge 1985).
STREETER, V. L. and WYLIE, E. B.: *Fluid Mechanics*, 8th edn (McGraw-Hill, New York, 1985).
ZAHORSKI, S.: *Mechanics of Viscoelastic Fluids* (PNN, Polish Scientific Publishers and Martinus Nijheff, 1981).

3.6. REFERENCES

1. REYNOLDS, O.: *Papers on Mechanical and Physical Subjects* **2** (1881–1901) 51. An experimental investigation of the circumstances which determine whether the motion of water shall be direct or sinuous and the law of resistance in parallel channels. 535. On the dynamical theory of incompressible viscous fluids and the determination of the criterion.
2. PRANDTL, L.: *Z. Ver. deut. Ing.* **77** (1933) 105. Neuere Ergebnisse der Turbulenzforschung.
3. NIKURADSE, J.: *Forsch. Ver. deut. Ing.* **361** (1933). Strömungsgesetze in rauhen Röhren.
4. STANTON, T. and PANNELL, J.: *Phil. Trans. R. Soc.* **214** (1914) 199. Similarity of motion in relation to the surface friction of fluids.
5. MOODY, L.F.: *Trans. Am. Soc. Mech. Engrs.* **66** (1944) 671. Friction factors for pipe flow.
6. BLASIUS, H.: *Forsch. Ver. deut. Ing.* **131** (1913). Das Ähnlichkeitsgesetz bei Reibungsvorgängen in Flüssigkeiten.
7. LAUFER, J.: United States National Advisory Committee for Aeronautics. Report No. 1174 (1955). The structure of turbulence in fully developed pipe flow.
8. HAGEN, G.: *Ann Phys. (Pogg. Ann.)* **46** (1839) 423. Ueber die Bewegung des Wassers in engen zylindrischen Röhren.
9. POISEUILLE, J.: *Inst. de France Acad. des Sci. Mémoires présentés par divers savantes* **9** (1846) 433. Recherches experimentales sur le mouvement des liquides dans les tubes de très petit diamètre.
10. RYAN, N. W. and JOHNSON, M. A.: *A.I.Ch.E.Jl.* **5** (1959) 433. Transition from laminar to turbulent flow in pipes.
11. WHITE, C. M.: *Proc. Roy. Soc.* A, **123** (1929) 645. Streamline flow through curved pipes.
12. BS 5168: 1975: British Standard 5168 (British Standards Institution, London). Glossary of Rheological Terms.
13. CROSS, M. M.: *J. Colloid Sci.* **20** (1965) 417. Rheology of non-Newtonian fluids: a new flow equation for pseudoplastic systems.
14. ASTARITA, G. and MARRUCCI, G.: *Principles of Non-Newtonian Fluid Mechanics* (McGraw-Hill, New York, 1974).
15. SKELLAND, A.H.P.: *Non-Newtonian Flow and Heat Transfer* (Wiley, 1967).
16. HUTTON, J. F., PEARSON, J. R. A. and WALTERS, K. (eds): *Theoretical Rheology* (Applied Science Publishers, London, 1975).
17. CHHABRA, R. P. and RICHARDSON J. F.: *Non-Newtonian Flow in the Process industries* (Butterworth-Heinemann, 1999).
18. BOGER D. V. and WALTERS K.: *Rheological Phenomena in Focus* (Elsevier, Amsterdam, 1993).
19. WHORLOW, R. H.: *Rheological techniques* 2nd edn. (Wiley, 1992).
20. FAROOQI, S. I. and RICHARDSON, J. F.: *Trans. I. Chem. E.* **58** (1980) 116. Rheological behaviour of kaolin suspensions in water and water-glycerol mixtures.
21. METZNER, A. B., WHITE, J. L. and DENN, H. M.: *A.I. Ch. E. Jl.* **12** (1966). 836. Constitutive equations for viscoelastic fluids for short deformation periods and for rapidly changing flows.
22. METZNER, A. B. and REED, J. C.: *A. I. Ch. E. Jl.* **1** (1955) 434. Flow of non-Newtonian fluids — correlation of the laminar, transition and turbulent flow regions.
23. RABINOWITSCH, B.: *Z. Phys. Chem.* **A145** (1929). Über die Viskosität and Elastizität van Solen.
24. MOONEY, M. J.: Rheology **2** (1931) 210. Explicit formulas for slip and fluidity.

25. HEYWOOD, N. I. and CHENG, D. C.-H.: *Trans Inst. Measurement and Control* **6** (1984) 33. Comparison of methods for predicting head loss in turbulent pipe flow of non-Newtonian fluids.
26. HARTNETT, J. P. and KOSTIC, M.: *Int. Comm. Heat & Mass Transfer* **17** (1990) 59. Turbulent friction factor correlations for power law fluids in circular and non-circular channels.
27. DODGE, D. W. and METZNER, A. B.: *A. I. Ch. E. Jl.* **5** (1959) 189. [See also correction: *A. I. Ch. E. Jl.* **5** (1962) 143] Turbulent flow of non-Newtonian systems.
28. YOO, S. S.: Ph.D. Thesis, University of Illinois, Chicago (1974), Heat transfer and friction factors for non-Newtonian fluids in circular tubes.
29. IRVINE, T. F.: *Chem. Eng. Comm.* **65** (1988) 39. A generalized Blasius equation for power law fluids.
30. BOWEN, R. L.: *Chemical Eng.* (Albany) **68** (24 July, 1961) 143, Designing turbulent-flow systems.

3.7. NOMENCLATURE

		Units in SI system	Dimensions in **M, L, T, θ**
A	Area perpendicular to direction of flow	m^2	$\mathbf{L}^2$
B	Width of rectangular channel or notch	m	**L**
C_c	Coefficient of contraction	—	—
C_f	Coefficient for flow over a bank of tubes	—	—
D	Depth of liquid in channel	m	**L**
d	Diameter of pipe	m	**L**
d_a	Dimension of rectangular duct, or distance apart of parallel plates	m	**L**
d_b	Dimension of rectangular duct	m	**L**
d_m	Hydraulic mean diameter ($= 4\,A/p$)	m	**L**
E	Eddy kinematic viscosity	m^2/s	$\mathbf{L}^2\mathbf{T}^{-1}$
e	Surface roughness	m	**L**
F	Energy per unit mass degraded because of irreversibility of process	J/kg	$\mathbf{L}^2\mathbf{T}^{-2}$
f	Fanning friction factor ($= 2\,R/\rho u^2$)	—	—
f'	Moody friction factor ($8R/\rho u^2$)	—	—
G	Mass rate of flow	kg/s	$\mathbf{MT}^{-1}$
$\mathbf{G}$	Young's modulus	N/m^2	$\mathbf{ML}^{-1}\mathbf{T}^{-2}$
g	Acceleration due to gravity	m/s^2	$\mathbf{LT}^{-2}$
h	Depth below surface measured perpendicular to bottom of channel or notch	m	**L**
h_f	Head lost due to friction	m	**L**
I_x, I_y, I_z	Normal stresses in x, y, z directions (surfaces A, C, B in Figure 3.36)	N/m^2	$\mathbf{ML}^{-1}\mathbf{T}^{-2}$
i	Hydraulic gradient (h_f/l)	—	—
J	Specific energy of fluid in open channel	J/kg	$\mathbf{L}^2\mathbf{T}^{-2}$
j	Number of banks of pipes in direction of flow	—	—
k	Consistency coefficient in power-law equation	N s^n/m^2	$\mathbf{ML}^{-1}\mathbf{T}^{n-2}$
k'	Coefficient defined by equation 3.93	N s$^{n'}$/m^2	$\mathbf{ML}^{-1}\mathbf{T}^{n'-2}$
L	Characteristic linear dimension	m	**L**
l	Length of pipe or channel	m	**L**
m	Index in equation 3.127	—	—
N_1	Normal stress difference $I_x - I_y$	N/m^2	$\mathbf{ML}^{-1}\mathbf{T}^{-2}$
N_2	Normal stress difference $I_y - I_z$	N/m^2	$\mathbf{ML}^{-1}\mathbf{T}^{-2}$
n	Index in power-law equation	—	—
n'	Slope defined by equation 3.154	—	—
P	Pressure	N/m^2	$\mathbf{ML}^{-1}\mathbf{T}^{-2}$
P_f	Pressure due to friction	N/m^2	$\mathbf{ML}^{-1}\mathbf{T}^{-2}$
ΔP	Pressure difference or change	N/m^2	$\mathbf{ML}^{-1}\mathbf{T}^{-2}$
p	Wetted perimeter	m	**L**
Q	Volumetric rate of flow	m^3/s	$\mathbf{L}^3\mathbf{T}^{-1}$
Q_A	Volumetric flowrate in sheared annulus	m^3/s	$\mathbf{L}^3\mathbf{T}^{-1}$

		Units in SI system	Dimensions in **M, L, T, θ**
Q_p	Volumetric flowrate in unsheared plug	m^3/s	$\mathbf{L^3T^{-1}}$
q	Net heat flow into system	J/kg	$\mathbf{L^2T^{-2}}$
R	Shear stress on surface	N/m^2	$\mathbf{ML^{-1}T^{-2}}$
R_c	Shear stress on surface C of element in Figure 3.36	N/m^2	$\mathbf{ML^{-1}T^{-2}}$
R_m	Mean value of shear stress at surface	N/m^2	$\mathbf{ML^{-1}T^{-2}}$
R_S	Shear stress at radius s	N/m^2	$\mathbf{ML^{-1}T^{-2}}$
R_Y	Yield stress	N/m^2	$\mathbf{ML^{-1}T^{-2}}$
R_y	Shear stress at some point in fluid distance y from surface	N/m^2	$\mathbf{ML^{-1}T^{-2}}$
R_0	Shear stress $(-R)$ in fluid at boundary surface $(y = 0)$	N/m^2	$\mathbf{ML^{-1}T^{-2}}$
r	Radius of pipe, or outer pipe in case of annulus, or radius of cone	m	**L**
r_c	Radius of unsheared plug	m	**L**
r_i	Radius of inner pipe of annulus	m	**L**
S	Entropy per unit mass	J/kg K	$\mathbf{L^2T^{-2}\theta^{-1}}$
s	Distance from axis of pipe or from centre-plane or of rotation or thickness of liquid film	m	**L**
t	Time	s	**T**
T	Temperature	K	**θ**
U	Internal energy per unit mass	J/kg	$\mathbf{L^2T^{-2}}$
u	Mean velocity	m/s	$\mathbf{LT^{-1}}$
u_{CL}	Velocity in pipe at centre line	m/s	$\mathbf{LT^{-1}}$
u_{Ex}	Fluctuating velocity component	m/s	$\mathbf{LT^{-1}}$
u_i	Instantaneous value of velocity	m/s	$\mathbf{LT^{-1}}$
u_p	Velocity of unsheared plug	m/s	$\mathbf{LT^{-1}}$
u_s	Velocity at free surface	m/s	$\mathbf{LT^{-1}}$
u_{max}	Maximum velocity	m/s	$\mathbf{LT^{-1}}$
u_t	Velocity at narrowest cross-section of bank of tubes	m/s	$\mathbf{LT^{-1}}$
u_x	Velocity in X-direction at distance y from surface	m/s	$\mathbf{LT^{-1}}$
u_w	Velocity of propagation wave	m/s	$\mathbf{LT^{-1}}$
v	Volume per unit mass of fluid	m^3/kg	$\mathbf{M^{-1}L^3}$
W_s	Shaft work per unit mass	J/kg	$\mathbf{L^2T^{-2}}$
w	Width of surface	m	**L**
X	$(2l/r)(R_Y/-\Delta P)$	—	—
x	Distance in X-direction or in direction of motion or parallel to surface	m	**L**
y	Distance in Y-direction or perpendicular distance from surface	m	**L**
Z	Stability criterion defined in equation 3.56	—	—
z	Distance in vertical direction	m	**L**
α	Constant in expression for kinetic energy of fluid	—	—
β	Constant in equation 3.124	$s^{2/3}$	$\mathbf{T^{2/3}}$
γ	Strain	—	—
$\dot{\gamma}$	Rate of shear or of strain	s^{-1}	$\mathbf{T^{-1}}$
δ	Boundary layer thickness	m	**L**
λ	Shear force acting on unit length of inner surface of annulus	N/m	$\mathbf{MT^{-2}}$
μ	Viscosity of fluid	$N\ s/m^2$	$\mathbf{ML^{-1}T^{-1}}$
μ_a	Apparent viscosity defined by equation 3.117	$N\ s/m^2$	$\mathbf{ML^{-1}T^{-1}}$
μ_p	Plastic viscosity	$N\ s/m^2$	$\mathbf{ML^{-1}T^{-1}}$
μ_0	Apparent viscosity $(\dot{\gamma} \to 0)$	$N\ s/m^2$	$\mathbf{ML^{-1}T^{-1}}$
μ_∞	Apparent viscosity $(\dot{\gamma} \to \infty)$	$N\ s/m^2$	$\mathbf{ML^{-1}T^{-1}}$
ϕ	Friction factor $(= R/\rho u^2)$	—	—

ρ	Density of fluid	kg/m^3	$\mathbf{ML^{-3}}$
θ	Angle between cone and plate in viscometers or angle	—	—
τ	Shear stress (= R)	N/m^2	$\mathbf{ML^{-1}T^{-2}}$
ω	Angular speed of rotation	s^{-1}	$\mathbf{T^{-1}}$
Re	Reynolds number with respect to pipe diameter	—	—
Re_{MR}	Generalised (Metzner and Reed) Reynolds number	—	—
$Re_{(MR)c}$	Critical value of Re_{MR} at laminar-turbulent transition	—	—
Re_x	Reynolds number with respect to distance x	—	—

CHAPTER 4

Flow of Compressible Fluids

4.1. INTRODUCTION

Although all fluids are to some degree compressible, compressibility is sufficiently great to affect flow under normal conditions only for a gas. Furthermore, if the pressure of the gas does not change by more than about 20 per cent, it is usually satisfactory to treat the gas as an incompressible fluid with a density equal to that at the mean pressure.

When compressibility is taken into account, the equations of flow become very much more complex than they are for an incompressible fluid, even if the simplest possible equation of state (the *ideal gas law*) is used to describe their behaviour. In this chapter, attention is confined to consideration of the flow of ideal gases. The physical property of a gas which varies, but which is constant for an incompressible fluid, is the density ρ or specific volume v $(= 1/\rho)$. Density is a function of both temperature and pressure and it is necessary therefore to take account of the effects of both of these variables. The relation between the pressure and the density will be affected by the heat transfer to the gas as it passes through the system. Isothermal conditions can be maintained only if there is very good heat transfer to the surroundings and normally exist only at low flowrates in small equipment. At the opposite extreme, in large installations with high flowrates, conditions are much more nearly adiabatic. It should be noted that, except for isothermal flow, the relation between pressure and density is influenced by the way in which the change is caused (for example, the degree of reversibility).

In this chapter consideration is given to the flow of gases through orifices and nozzles, and to flow in pipelines. It is found that, in all these cases, the flow may reach a limiting maximum value which is independent of the downstream pressure; this is a phenomenon which does not arise with incompressible fluids.

4.2. FLOW OF GAS THROUGH A NOZZLE OR ORIFICE

This is one of the simplest applications of the flow of a compressible fluid and it can be used to illustrate many of the features of the process. In practical terms, it is highly relevant to the design of relief valves or bursting discs which are often incorporated into pressurised systems in order to protect the equipment and personnel from dangers which may arise if the equipment is subjected to pressures in excess of design values. In many cases it is necessary to vent gases evolved in a chemical reaction.

For this purpose, the gas flowrate at an aperture through which it discharges from a vessel maintained at a constant pressure P_1 to surroundings at a pressure P_2 (Figure 4.1) is considered.

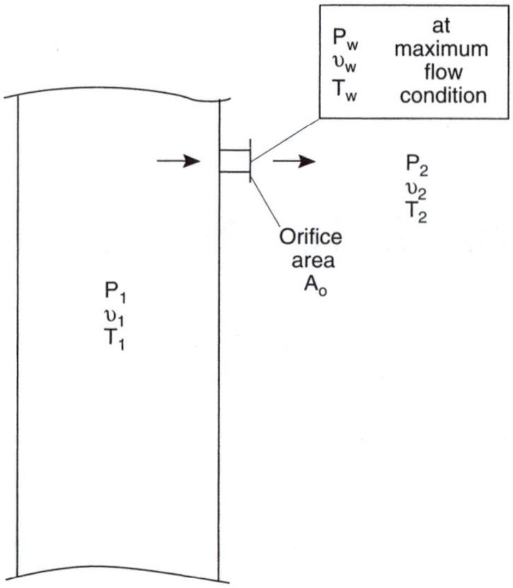

Figure 4.1. Discharge through an orifice

The energy balance, equation 2.54, is as follows:

$$\frac{1}{\alpha} u \, \mathrm{d}u + g \, \mathrm{d}z + v \, \mathrm{d}P + \delta W_s + \delta F = 0 \tag{4.1}$$

If the flow is horizontal ($\mathrm{d}z = 0$), the velocity distribution is assumed to be flat ($\alpha = 1$) and friction neglected ($\delta F = 0$) then, as no external work is performed:

$$u \, \mathrm{d}u + v \, \mathrm{d}P = 0 \tag{4.2}$$

If the velocity in the vessel at which the gas approaches the outlet is negligible ($u_1 = 0$), integration of equation 4.2 gives:

$$\frac{u_2^2}{2} = -\int_{P_1}^{P_2} v \, \mathrm{d}P \tag{4.3}$$

The value of the integral depends on the pressure–volume relation.

4.2.1. Isothermal flow

For the isothermal flow of an ideal gas, then from equation 2.69:

$$-\int_{P_1}^{P_2} v \, \mathrm{d}P = -P_1 v_1 \ln \frac{P_2}{P_1} = P_1 v_1 \ln \frac{P_1}{P_2} \tag{4.4}$$

Substituting into equation 4.3, gives the discharge velocity u_2:

$$\frac{u_2^2}{2} = P_1 v_1 \ln \frac{P_1}{P_2}$$

or:
$$u_2 = \sqrt{2P_1 v_1 \ln \frac{P_1}{P_2}} \tag{4.5}$$

The mass rate of flow is given by:

$$G = u_2 A_2 \rho_2 = \frac{u_2 A_2}{v_2}$$
$$= \frac{A_2}{v_2}\sqrt{2P_1 v_1 \ln \frac{P_1}{P_2}} \tag{4.6}$$

It is shown in Chapter 6 that the minimum flow area A_2 tends to be somewhat smaller than the area A_0 of the aperture because the gas leaves with a small radial inwards velocity component. Furthermore, there will be some reduction of discharge rate because of the frictional effects which have been neglected. Grouping these factors together by means of a coefficient of discharge C_D, where $C_D < 1$, gives:

$$G = \frac{C_D A_0}{v_2}\sqrt{2P_1 v_1 \ln \frac{P_1}{P_2}} \tag{4.7}$$

Substituting for isothermal flow:

$$v_2 = v_1 \frac{P_1}{P_2}$$

$$G = C_D A_0 \frac{P_2}{P_1 v_1}\sqrt{2P_1 v_1 \ln \frac{P_1}{P_2}}$$

$$= C_D A_0 P_2 \sqrt{\frac{2}{P_1 v_1} \ln \frac{P_1}{P_2}} = C_D A_0 P_2 \sqrt{\frac{2M}{\mathbf{R}T} \ln \frac{P_1}{P_2}} \tag{4.8}$$

where $\mathbf{R}$ is the universal gas constant, and
M is the molecular weight of the gas.

Maximum flow conditions

Equation 4.8 gives $G = 0$ for $P_2 = P_1$, and for $P_2 = 0$. As it is a continuous function, it must yield a maximum value of G at some intermediate pressure P_w, where $0 < P_w < P_1$.

Differentiating both sides with respect to P_2:

$$\frac{dG}{dP_2} = C_D A_0 \sqrt{\frac{2}{P_1 v_1}} \frac{d}{dP_2}\left\{P_2 \left[\ln\left(\frac{P_1}{P_2}\right)\right]^{1/2}\right\}$$

$$= C_D A_0 \sqrt{\frac{2}{P_1 v_1}} \left\{\left[\ln\left(\frac{P_1}{P_2}\right)\right]^{1/2} + P_2 \frac{1}{2}\left[\ln\left(\frac{P_1}{P_2}\right)\right]^{-1/2} \left(\frac{P_2}{P_1}\right)\left(\frac{P_1}{-P_2^2}\right)\right\}$$

For a maximum value of G, $dG/dP_2 = 0$, and:

Then:
$$\ln \frac{P_1}{P_2} = \tfrac{1}{2} \tag{4.9}$$

or: $$\frac{P_1}{P_2} = 1.65 \tag{4.10}$$

and the critical pressure ratio, $$\frac{P_2}{P_1} = w_c = 0.607 \tag{4.11}$$

Substituting into equation 4.8 to give the maximum value of $G(G_{max})$:

$$G_{max} = C_D A_0 P_2 \sqrt{\frac{2}{P_1 v_1} \cdot \frac{1}{2}}$$

$$= C_D A_0 \sqrt{\frac{P_2}{v_2}}$$

$$= C_D A_0 \rho_2 \sqrt{P_2 v_2} \left(\rho_2 = \frac{1}{v_2}\right) \tag{4.12}$$

Thus the velocity, $$u_w = \sqrt{P_2 v_2} = \sqrt{P_0 v_0} = \sqrt{\frac{\mathbf{R}T}{M}} \tag{4.13}$$

It is shown later in Section 4.3 that this corresponds to the velocity at which a small pressure wave will be propagated under isothermal conditions, sometimes referred to as the "isothermal sonic velocity", though the heat transfer rate will not generally be sufficient to maintain truly isothermal conditions.

Writing equation 4.12 in terms of the upstream conditions (P_1, v_1):

$$G_{max} = C_D A_0 \frac{1}{v_1} \frac{P_2}{P_1} \sqrt{P_1 v_1}$$

or, on substitution from equation 4.11:

$$G_{max} = 0.607 C_D A_0 \sqrt{\frac{P_1}{v_1}} \tag{4.14}$$

At any given temperature, $P_1 v_1 = P_0 v_0 =$ constant, where v_0 is the value of v at some reference pressure P_0.

Then: $$G_{max} = 0.607 C_D A_0 P_1 \sqrt{\frac{1}{P_0 v_0}} = 0.607 C_D A_0 P_1 \sqrt{\frac{M}{\mathbf{R}T}} \tag{4.15}$$

Thus, G_{max} is linearly related to P_1.

It will be seen when the pressure ratio P_2/P_1 is less than the critical value ($w_c = 0.607$) the flow rate becomes independent of the downstream pressure P_2. The fluid at the orifice is then flowing at the velocity of a small pressure wave and the velocity of the pressure wave *relative to the orifice* is zero. That is the upstream fluid cannot be influenced by the pressure in the downstream reservoir. Thus, the pressure falls to the critical value at the orifice, and further expansion to the downstream pressure takes place in the reservoir with the generation of a *shock wave*, as discussed in Section 4.6.

4.2.2. Non-isothermal flow

If the flow is non-isothermal, it may be possible to represent the relationship between pressure and volume by an equation of the form:

$$Pv^k = \text{constant} \tag{4.16}$$

Then from equation 2.73:

$$-\int_{P_1}^{P_2} v\,dP = \frac{k}{k-1}P_1v_1\left[1-\left(\frac{P_2}{P_1}\right)^{(k-1)/k}\right] \tag{4.17}$$

The discharge velocity u_2 is then given by:

$$\frac{u_2^2}{2} = \frac{k}{k-1}P_1v_1\left[1-\left(\frac{P_2}{P_1}\right)^{(k-1)/k}\right]$$

i.e.:

$$u_2 = \sqrt{\frac{2k}{k-1}P_1v_1\left[1-\left(\frac{P_2}{P_1}\right)^{(k-1)/k}\right]} \tag{4.18}$$

Allowing for a discharge coefficient, the mass rate of flow is given by:

$$G = \frac{C_DA_0}{v_2}\sqrt{\frac{2k}{k-1}P_1v_1\left[1-\left(\frac{P_2}{P_1}\right)\right]^{(k-1)/k}} \tag{4.19}$$

$$= \frac{C_DA_0}{v_1}\left(\frac{P_2}{P_1}\right)^{1/k}\sqrt{\frac{2k}{k-1}P_1v_1\left[1-\left(\frac{P_2}{P_1}\right)^{(k-1)/k}\right]} \tag{4.20}$$

Maximum flow conditions

This equation also gives $G = 0$ for $P_2 = 0$ and for $P_2 = P_1$.

Differentiating both sides of equation 4.20 with respect to P_2/P_1:

$$\frac{dG}{d(P_2/P_1)} = \frac{C_DA_0}{v_1}\sqrt{\frac{2k}{k-1}P_1v_1}\left\{\left(\frac{P_2}{P_1}\right)^{1/k}\frac{1}{2}\left[1-\left(\frac{P_2}{P_1}\right)^{(k-1)/k}\right]^{-1/2}\right.$$
$$\left.\times\left(-\frac{k-1}{k}\right)\left(\frac{P_2}{P_1}\right)^{-1/k} + \left[1-\left(\frac{P_2}{P_1}\right)^{(k-1)/k}\right]^{1/2}\frac{1}{k}\left(\frac{P_2}{P_1}\right)^{(1/k)-1}\right\}$$

Putting $dG/d(P_2/P_1) = 0$ for the maximum value of $G(G_{max})$:

$$1-\left(\frac{P_2}{P_1}\right)^{(k-1)/k} = \frac{1}{2}(k-1)\left(\frac{P_2}{P_1}\right)\left(\frac{P_2}{P_1}\right)^{-1/k} \tag{4.21}$$

Substituting into equation 4.19:

$$G_{max} = \frac{C_DA_0}{v_2}\sqrt{\frac{2k}{k-1}P_1v_1\frac{1}{2}(k-1)\frac{P_2}{P_1}\left(\frac{P_2}{P_1}\right)^{-1/k}}$$

$$= C_DA_0\rho_2\sqrt{kP_2v_2} \tag{4.22}$$

Hence: $$u_w = \sqrt{kP_2 v_2} \tag{4.23}$$

The velocity $u_w = \sqrt{kP_2 v_2}$ is shown to be the velocity of a small pressure wave if the pressure-volume relation is given by Pv^k = constant. If the expansion approximates to a reversible adiabatic (isentropic) process $k \approx \gamma$, the ratio of the specific heats of the gases, as indicated in equation 2.30.

Equation 4.22 then becomes:

$$G_{max} = C_D A_0 \rho_2 \sqrt{\gamma P_2 v_2} \tag{4.24}$$

and: $$u_w = \sqrt{\gamma P_2 v_2} = \sqrt{\frac{\gamma \mathbf{R} T_2}{M}} \tag{4.25}$$

where $u_w = \sqrt{\gamma P_2 v_2}$ is the velocity of propagation of a small pressure wave under isentropic conditions.

As for isothermal conditions, when maximum flow is occurring, the velocity of a small pressure wave, *relative to the orifice*, is zero, and the fluid at the orifice is not influenced by the pressure further downstream.

Substituting the critical pressure ratio $w_c = P_2/P_1$ in equation 4.21:

$$1 + w_c^{(k-1)/k} = \frac{k-1}{2} w_c^{(k-1)/k}$$

giving: $$w_c = \left[\frac{2}{k+1}\right]^{k/(k-1)} \tag{4.26}$$

For isentropic conditions:

$$w_c = \left[\frac{2}{\gamma+1}\right]^{\gamma/(\gamma-1)} \tag{4.26a}$$

For a diatomic gas at approximately atmospheric pressure, $\gamma = 1.4$, and $w_c = 0.53$. From equation 4.22:

$$\begin{aligned} G_{max} &= C_D A_0 \sqrt{\frac{kP_2}{v_2}} \\ &= C_D A_0 \sqrt{\frac{k}{v_1}\left(\frac{P_2}{P_1}\right)^{(k+1)/k} P_2} \\ &= C_D A_0 \sqrt{\frac{kP_1}{v_1}\left(\frac{P_2}{P_1}\right)^{(k+1)/k}} \end{aligned} \tag{4.27}$$

Substituting from equation 4.26:

$$G_{max} = C_D A_0 \sqrt{\frac{kP_1}{v_1}\left[\frac{2}{k+1}\right]^{(k+1)/(k-1)}} \tag{4.28}$$

For isentropic conditions, $k = \gamma$, and:

$$G_{max} = C_D A_0 \sqrt{\frac{\gamma P_1}{v_1}\left[\frac{2}{\gamma+1}\right]^{(\gamma+1)/(\gamma-1)}} \tag{4.28a}$$

For a given upstream temperature T_1, $P_1 v_1 = P_0 v_0 =$ constant where v_0 is the value of v at some reference pressure P_0 and temperature T_1.
Thus:

$$G_{max} = C_D A_0 P_1 \sqrt{\frac{k}{P_0 v_0}\left[\frac{2}{k+1}\right]^{(k+1)/(k-1)}} = C_D A_0 P_1 \sqrt{\frac{M}{\mathbf{R}T_1} k \left[\frac{2}{k+1}\right]^{(k+1)/(k-1)}} \tag{4.29}$$

For isentropic conditions $k = \gamma$,
and:

$$G_{max} = C_D A_0 P_1 \sqrt{\frac{\gamma}{P_0 v_0}\left[\frac{2}{\gamma+1}\right]^{(\gamma+1)/(\gamma-1)}} = C_D A_0 P_1 \sqrt{\frac{M}{\mathbf{R}T_1} \gamma \left[\frac{2}{\gamma+1}\right]^{(\gamma+1)/(\gamma-1)}} \tag{4.30}$$

For a diatomic gas, $\gamma \approx 1.4$ for pressures near atmospheric

and: $$G_{max} = 0.685 C_D A_0 P_1 \sqrt{\frac{1}{P_0 v_0}} = 0.685 C_D A_0 P_1 \sqrt{\frac{M}{\mathbf{R}T_1}} \tag{4.31}$$

It may be noted that equations 4.29, 4.30 and 4.31 give a linear relation between G_{max} and P_1. Comparison with equation 4.15 shows that the maximum flowrate G_{max} is $(0.685/0.607) = 1.13$ times greater than for isothermal flow for a diatomic gas.

In Figure 4.2, the mass flowrate is plotted as a function of cylinder pressure for discharge through an orifice to an atmosphere at a constant downstream pressure P_2 — for

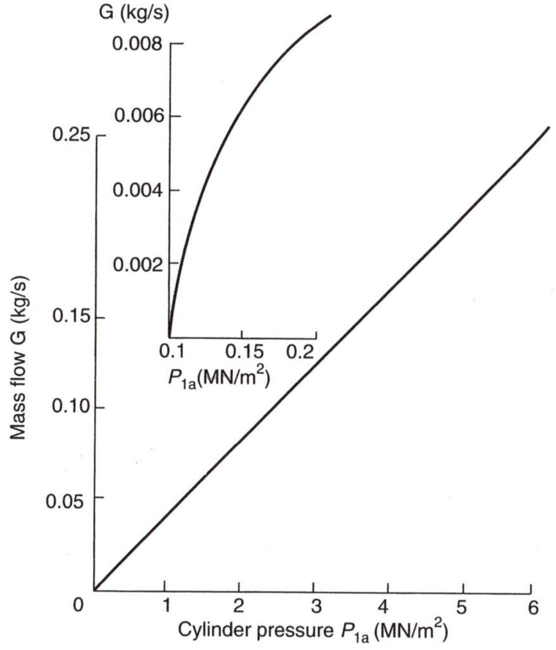

Figure 4.2. Mass rate of discharge of gas as function of cylinder pressure for constant downstream pressure

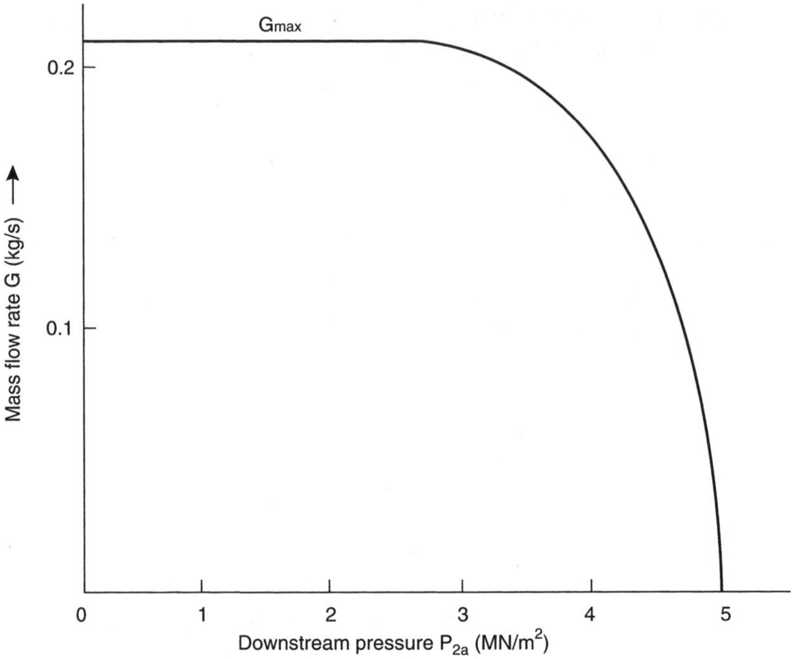

Figure 4.3. Mass rate of discharge of gas as function of downstream pressure for constant cylinder pressure

the conditions given in Example 4.1. In Figure 4.3, the cylinder pressure P_1 is maintained constant and the effect of the downstream pressure P_2 on the flowrate is shown.

Example 4.1

A cylinder contains air at a pressure of 6.0 MN/m^2 and discharges to atmosphere through a valve which may be taken as equivalent to a sharp-edged orifice of 6 mm diameter (coefficient of discharge = 0.6). (i) Plot the rate of discharge of air from the cylinder against cylinder pressure. Assume that the expansion at the valve is approximately isentropic and that the temperature of the air in the cylinder remains unchanged at 273 K. (ii) For a constant pressure of 5 MN/m^2 in the cylinder, plot the discharge rate as a function of the pressure P_2 of the surroundings.

Solution

The critical pressure ratio for discharge through the valve

$$= \left[\frac{2}{\gamma + 1}\right]^{[\gamma/(\gamma-1)]} \qquad \text{(equation 4.26)}$$

where γ varies from 1.40 to 1.54 over the pressure range 0.1 to 6.0 MN/m^2. Critical conditions occur at a relatively low cylinder pressure. Taking $\gamma = 1.40$, then:

$$\text{critical ratio} = \left(\frac{2}{2.40}\right)^{1.4/0.4} = 0.53$$

(i) Sonic velocity will occur until the pressure in the cylinder falls to $(101.3/0.53) = 191.1$ kN/m^2.

The rate of discharge for cylinder pressures greater than 191.1 kN/m^2 is given by equation 4.30:

$$G_{max} = C_D A_0 P_1 \sqrt{\frac{\gamma M}{\mathbf{R}T_1}\left(\frac{2}{\gamma+1}\right)^{(\gamma+1)/(\gamma-1)}}$$

Taking a mean value for γ of 1.47, then:

$$\gamma\left(\frac{2}{\gamma+1}\right)^{(\gamma+1)/(\gamma-1)} = 0.485.$$

$$A_0 = \frac{\pi}{4} \times (0.006)^2 = 2.83 \times 10^{-5}\ \text{m}^2$$

$$\frac{1}{P_1 v_1} = \frac{M}{\mathbf{R}T_1} = \frac{29}{8314 \times 273} = 1.28 \times 10^{-5}\ \text{s}^2/\text{m}^2$$

$$P_1 v_1 = \frac{\mathbf{R}T_1}{M} = 78{,}100\ \text{s}^2/\text{m}^2$$

$$G_{max} = 0.6 \times (2.83 \times 10^{-5}) \times P_1\sqrt{(1.28 \times 10^{-5}) \times 0.485}$$

$$= 4.23 \times 10^{-8} P_1\ \text{kg/s} \quad (P_1 \text{ in N/m}^2)$$

$$= 4.23 \times 10^{-2} P_{1a}\ \text{kg/s} \quad (P_{1a} \text{ in MN/m}^2)$$

For cylinder pressures below 191.1 kN/m^2, the mass flowrate is given by equation 4.20. Putting $k = \gamma$ for isentropic conditions:

$$G = \frac{C_D A_0}{v_1}\left(\frac{P_2}{P_1}\right)^{1/\gamma}\sqrt{\frac{2\gamma}{\gamma-1}P_1 v_1\left[1-\left(\frac{P_2}{P_1}\right)^{(\gamma-1)/\gamma}\right]}$$

Using a value of 1.40 for γ in this low pressure region, and writing:

$$\frac{1}{v_1} = \frac{1}{P_1 v_1} \cdot P_1$$

$$G = \frac{0.6 \times (2.83 \times 10^{-5})}{78{,}100} P_1 \left(\frac{P_2}{P_1}\right)^{0.714}\sqrt{7 \times 78{,}100\left[1-\left(\frac{P_2}{P_1}\right)^{0.286}\right]}$$

$$= 1.608 \times 10^{-7} P_1 \left(\frac{P_2}{P_1}\right)^{0.714}\sqrt{\left[1-\left(\frac{P_2}{P_1}\right)^{0.286}\right]}$$

For discharge to atmospheric pressure, $P_2 = 101{,}300$ N/m^2, and:

$$G = 6.030 \times 10^{-4} P_1^{0.286}\sqrt{1 - 27P_1^{-0.286}}\ \text{kg/s} \quad (P_1 \text{ in N/m}^2)$$

Putting pressure P_{1a} in MN/m^2, then:

$$G = 0.0314 P_{1a}^{0.286}\sqrt{1 - 0.519 P_{1a}^{-0.286}}\ \text{kg/s} \quad (P_{1a} \text{ in MN/m}^2)$$

The discharge rate is plotted in Figure 4.2.

(ii)

G vs P_{1a} data

(i) Above $P_{1a} = 0.19$ MN/m^2		(ii) Below $P_{1a} = 0.19$ MN/m^2	
P_{1a} (MN/m^2)	G (kg/s)	P_{1a} (MN/m^2)	G (kg/s)
0.19	0.0080		
0.2	0.0084	0.10	0
0.5	0.021	0.125	0.0042
1.0	0.042	0.15	0.0060
2.0	0.084	0.17	0.0070
3.0	0.126	0.19	0.0079
4.0	0.168		
5.0	0.210		
6.0	0.253		

Note: The slight mismatch in the two columns for $P_{1a} = 0.19$ MN/m^2 is attributable to the use of an average value of 1.47 for γ in column 2 and 1.40 for γ in column 4.

From this table, for a constant upstream pressure of 5 MN/m^2, the mass flowrate G remains at 0.210 kg/s for all pressures P_2 below $5 \times 0.53 = 2.65$ MN/m^2.

For higher values of P_2, the flowrate is obtained by substitution of the constant value of P_1 (5 MN/m^2) in equation 4.20.

or:

$$G = (1.608 \times 10^{-7}) \times (82.40) \times P_2^{0.714}\sqrt{1 - 0.01214P_2^{0.286}} \quad (P_2 \text{ in N/m}^2)$$

$$= 0.2548P_{2a}^{0.714}\sqrt{1 - 0.631P_{2a}^{0.286}} \text{ kg/s} \quad (P_2 \text{ in MN/m}^2)$$

This relationship is plotted in Figure 4.3 and values of G as a function of P_{2a} are:

P_{2a} MN/m^2	G kg/s
<2.65	0.210
3.0	0.206
3.5	0.194
4.0	0.171
4.5	0.123
4.9	0.061
4.95	0.044
5	0

4.3. VELOCITY OF PROPAGATION OF A PRESSURE WAVE

When the pressure at some point in a fluid is changed, the new condition takes a finite time to be transmitted to some other point in the fluid because the state of each intervening element of fluid has to be changed. The velocity of propagation is a function of the bulk modulus of elasticity ε, where ε is defined by the relation:

$$\varepsilon = \frac{\text{increase of stress within the fluid}}{\text{resulting volumetric strain}} = \frac{dP}{-(dv/v)} = -v\frac{dP}{dv} \tag{4.32}$$

If a pressure wave is transmitted at a velocity u_w over a distance dl in a fluid of cross-sectional area A, from section A to section B, as shown in Figure 4.4, it may be

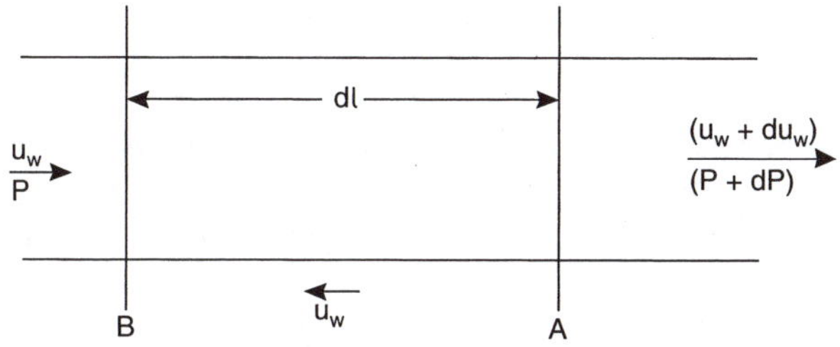

Figure 4.4. Propagation of pressure wave

brought to rest by causing the fluid to flow at a velocity u_w in the opposite direction. The pressure and specific volume are at B P and v, and at A $(P + \mathrm{d}P)$ and $(v + \mathrm{d}v)$, respectively. As a result of the change in pressure, the velocity of the fluid changes from u_w at B to $(u_w + \mathrm{d}u_w)$ at A and its mass rate of flow is:

$$G = \frac{u_w A}{v} = \frac{(u_w + \mathrm{d}u_w)A}{v + \mathrm{d}v}$$

The net force acting on the fluid between sections A and B is equal to the rate of change of momentum of the fluid, or:

$$PA - (P + \mathrm{d}P)A = G\,\mathrm{d}u_w$$

and:

$$-A\,\mathrm{d}P = G\frac{G}{A}\,\mathrm{d}v$$

$$-\frac{\mathrm{d}P}{\mathrm{d}v} = \frac{G^2}{A^2}$$

$$= \frac{\varepsilon}{v} \quad \text{(from equation 4.32)}$$

$\therefore$

$$u_w = \sqrt{\varepsilon v} \tag{4.33}$$

For an ideal gas, ε may be calculated from the equation of state. Under isothermal conditions:

$$Pv = \text{constant.}$$

$\therefore$

$$-\frac{\mathrm{d}P}{\mathrm{d}v} = \frac{P}{v}$$

$\therefore$

$$\varepsilon = P \tag{4.34}$$

and:

$$u_w = \sqrt{Pv} = \sqrt{\frac{\mathbf{R}T}{M}} \tag{4.35}$$

Under isentropic conditions:

$$Pv^\gamma = \text{constant}$$

$$-\frac{\mathrm{d}P}{\mathrm{d}v} = \frac{\gamma P}{v}$$

$\therefore$

$$\varepsilon = \gamma P \tag{4.36}$$

and:
$$u_w = \sqrt{\gamma P v} = \sqrt{\gamma \frac{\mathbf{R}T}{M}} \tag{4.37}$$

This value of u_w corresponds closely to the velocity of sound in the fluid. That is, for normal conditions of transmission of a small pressure wave, the process is almost isentropic. When the relation between pressure and volume is $Pv^k =$ constant, then:

$$u_w = \sqrt{kPv} = \sqrt{k \frac{\mathbf{R}T}{M}} \tag{4.38}$$

4.4. CONVERGING–DIVERGING NOZZLES FOR GAS FLOW

Converging-diverging nozzles, as shown in Figure 4.5, sometimes known as Laval nozzles, are used for the expansion of gases where the pressure drop is large. If the nozzle is carefully designed so that the contours closely follow the lines of flow, the resulting expansion of the gas is almost reversible. Because the flow rate is large for high-pressure differentials, there is little time for heat transfer to take place between the gas and surroundings and the expansion is effectively isentropic. In the analysis of the nozzle, the change in flow is examined for various pressure differentials across the nozzle.

The specific volume v_2 at a downstream pressure P_2, is given by:

$$v_2 = v_1 \left(\frac{P_1}{P_2}\right)^{1/\gamma} = v_1 \left(\frac{P_2}{P_1}\right)^{-1/\gamma} \tag{4.39}$$

If gas flows under turbulent conditions from a reservoir at a pressure P_1, through a horizontal nozzle, the velocity of flow u_2, at the pressure P_2 is given by:

$$\frac{u_2^2}{2} + \int_1^2 v\,dP = 0 \quad \text{(from equation 2.42)}$$

Thus:
$$u_2^2 = \frac{2\gamma}{\gamma - 1} P_1 v_1 \left[1 - \left(\frac{P_2}{P_1}\right)^{(\gamma-1)/\gamma}\right] \tag{4.40}$$

Since:
$$A_2 = \frac{G v_2}{u_2} \quad \text{(from equation 2.36)} \tag{4.41}$$

the required cross-sectional area for flow when the pressure has fallen to P_2 may be found.

4.4.1. Maximum flow and critical pressure ratio

In the flow of a gas through a nozzle, the pressure falls from its initial value P_1 to a value P_2 at some point along the nozzle; at first the velocity rises more rapidly than the specific volume and therefore the area required for flow decreases. For low values of the pressure ratio P_2/P_1, however, the velocity changes much less rapidly than the specific volume so that the area for flow must increase again. The effective area for flow presented by the nozzle must therefore pass through a minimum. It is shown that this occurs if the pressure ratio P_2/P_1 is less than the critical pressure ratio (usually approximately 0.5) and that the velocity at the throat is then equal to the velocity of sound. For expansion

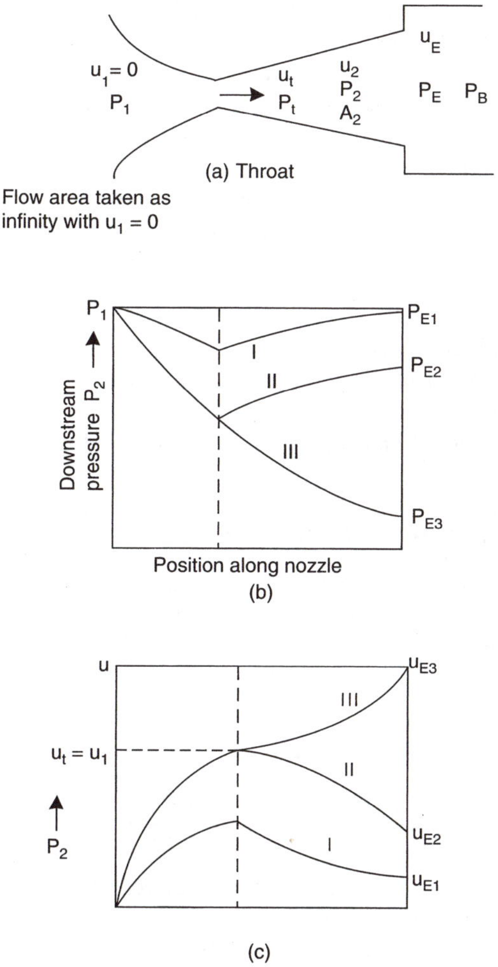

Figure 4.5. Flow through converging-diverging nozzles

to pressures below the critical value the flowing stream must be diverging. Thus in a converging nozzle the velocity of the gas stream will never exceed the sonic velocity though in a converging-diverging nozzle supersonic velocities may be obtained in the diverging section. The flow changes which occur as the back pressure P_B is steadily reduced are shown in Figure 4.5. The pressure at the exit of the nozzle is denoted by P_E, which is not necessarily the same as the back pressure P_B. In the following three cases, the exit and back pressures correspond and the flow is approximately isentropic. The effect of a mismatch in the two pressures is considered later.

Case I. Back-pressure P_B quite high. Curves I show how pressure and velocity change along the nozzle. The pressure falls to a minimum at the throat and then rises to a value $P_{E1} = P_B$. The velocity increases to maximum at the throat (less than sonic velocity) and then decreases to a value of u_{E1} at the exit of the nozzle. This situation corresponds to conditions in a venturi operating entirely at subsonic velocities.

Case II. Back-pressure reduced (curves II). The pressure falls to the critical value at the throat where the velocity is sonic. The pressure then rises to $P_{E2} = P_B$ at the exit. The velocity rises to the sonic value at the throat and then falls to u_{E2} at the outlet.

Case III. Back-pressure low, with pressure less than critical value at the exit. The pressure falls to the critical value at the throat and continues to fall to give an exit pressure $P_{E3} = P_B$. The velocity increases to sonic at the throat and continues to increase to supersonic in the diverging cone to a value u_{E3}.

With a converging-diverging nozzle, the velocity increases beyond the sonic velocity only if the velocity at the throat is sonic and the pressure at the outlet is lower than the throat pressure. For a converging nozzle the rate of flow is independent of the downstream pressure, provided the critical pressure ratio is reached and the throat velocity is sonic.

4.4.2. The pressure and area for flow

As indicated in Section 4.4.1, the area required at any point depends upon the ratio of the downstream to the upstream pressure P_2/P_1 and it is helpful to establish the minimum value of $A_2 \cdot A_2$ may be expressed in terms of P_2 and $w[= (P_2/P_1)]$ using equations 4.39, 4.40 and 4.41.

Thus:

$$A_2^2 = G^2 \frac{\gamma - 1}{2\gamma} \frac{v_1^2 (P_2/P_1)^{-2/\gamma}}{P_1 v_1 [1 - (P_2/P_1)^{(\gamma-1)/\gamma}]}$$

$$= \frac{G^2 v_1 (\gamma - 1)}{2P_1 \gamma} \frac{w^{-2/\gamma}}{1 - w^{(\gamma-1)/\gamma}} \tag{4.42}$$

For a given rate of flow G, A_2 decreases from an effectively infinite value at pressure P_1 at the inlet to a minimum value given by:

$$\frac{dA_2^2}{dw} = 0$$

or, when:

$$(1 - w^{(\gamma-1)/\gamma}) \frac{-2}{\gamma} w^{-1-2/\gamma} - w^{-2/\gamma} \frac{1-\gamma}{\gamma} w^{-1/\gamma} = 0$$

or:

$$w = \left(\frac{2}{\gamma + 1}\right)^{\gamma/(\gamma-1)} \tag{4.43}$$

The value of w given by equation 4.43 is the critical pressure ratio w_c given by equation 4.26*a*. Thus the velocity at the throat is equal to the sonic velocity. Alternatively, equation 4.42 may be put in terms of the flowrate (G/A_2) as:

$$\left(\frac{G}{A_2}\right)^2 = \frac{2\gamma}{\gamma - 1} \left(\frac{P_2}{P_1}\right)^{2/\gamma} \frac{P_1}{v_1} \left[1 - \left(\frac{P_2}{P_1}\right)^{(\gamma-1)/\gamma}\right] \tag{4.44}$$

and the flowrate G/A_2 may then be shown to have a maximum value of $G_{max}/A_2 = \sqrt{\gamma P_2/v_2}$.

A nozzle is correctly designed for any outlet pressure between P_1 and P_{E1} in Figure 4.5. Under these conditions the velocity will not exceed the sonic velocity at any point, and the flowrate will be independent of the exit pressure $P_E = P_B$. It is also correctly designed for supersonic flow in the diverging cone for an exit pressure of P_{E3}.

It has been shown above that when the pressure in the diverging section is greater than the throat pressure, subsonic flow occurs. Conversely, if the pressure in the diverging section is less than the throat pressure the flow will be supersonic beyond the throat. Thus at a given point in the diverging cone where the area is equal to A_2 the pressure may have one of two values for isentropic flow.

As an example, when γ is 1.4:

$$v_2 = v_1 w^{-0.71} \tag{4.45}$$

and:

$$u_2^2 = 7P_1 v_1 (1 - w^{0.29}) \tag{4.46}$$

Thus:

$$A_2^2 = \frac{v_1 w^{-1.42}}{7P_1(1 - w^{0.29})} G^2 \tag{4.47}$$

In Figure 4.6 values of v_2/v_1, $u_2/\sqrt{P_1 v_1}$, and $(A_2/G)\sqrt{P_1 v_1}$ which are proportional to v_2, u_2, and A_2 respectively are plotted as abscissae against P_2/P_1. It is seen that the area A_2 decreases to a minimum and then increases again. At the minimum cross-section the velocity is equal to the sonic velocity and P_2/P_1 is the critical ratio. $u_w/\sqrt{P_1 v_1}$ is also plotted and, by comparing with the curve for $u_2/\sqrt{P_1 v_1}$, it is easy to compare the corresponding values of u_2 and u_w. The flow is seen to be sub-sonic for $P_2/P_1 > 0.53$ and supersonic for $P_2/P_1 < 0.53$.

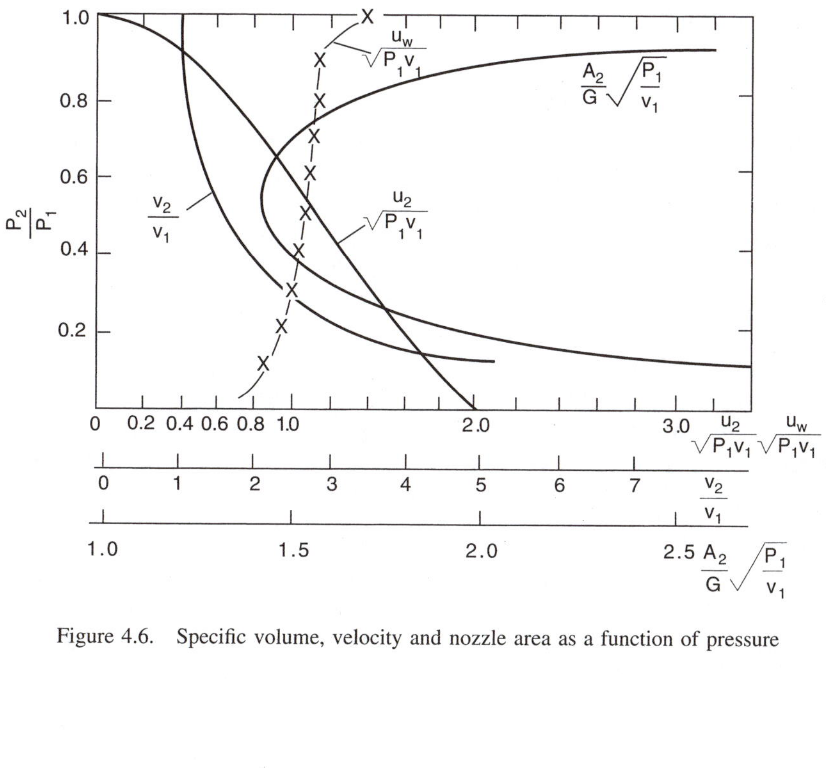

Figure 4.6. Specific volume, velocity and nozzle area as a function of pressure

4.4.3. Effect of back-pressure on flow in nozzle

It is of interest to study how the flow in the nozzle varies with the back-pressure P_B.

(1) $P_B > P_{E2}$. The flow is subsonic throughout the nozzle and the rate of flow G is determined by the value of the back pressure P_B. Under these conditions $P_E = P_B$. It is shown as P_{E1} in Figure 4.5.
(2) $P_B = P_{E2}$. The flow is subsonic throughout the nozzle, except at the throat where the velocity is sonic. Again $P_E = P_B$.
(3) $P_{E3} < P_B < P_{E2}$. The flow in the converging section and at the throat is exactly as for Case 2. For this range of exit pressures the flow is not isentropic through the whole nozzle and the area at the exit section is greater than the design value for a back pressure P_B. One of two things may occur. Either the flow will be supersonic but not occupy the whole area of the nozzle, or a shock wave will occur at some intermediate point in the diverging cone, giving rise to an increase in pressure and a change from supersonic to subsonic flow.
(4) $P_B = P_{E3}$. The flow in the converging section and at the throat is again the same as for Cases 2, 3, and 4. The flow in the diverging section is supersonic throughout and the pressure of P_B is reached smoothly. P_{E3} is therefore the design pressure for supersonic flow at the exit.
(5) $P_B < P_{E3}$. In this case the flow throughout the nozzle is exactly as for Case 4. The pressure at the exit will again be P_{E3} and the pressure will fall beyond the end of the nozzle from P_{E3} to P_B by continued expansion of the gas.

It may be noted that the flowrate through the nozzle is a function of back pressure only for Case 1.

A practical example of flow through a converging-diverging nozzle is given in Example 4.4 after a discussion of the generation of shock waves.

4.5. FLOW IN A PIPE

Compressibility of a gas flowing in a pipe can have significant effect on the relation between flowrate and the pressures at the two ends. Changes in fluid density can arise as a result of changes in either temperature or pressure, or in both, and the flow will be affected by the rate of heat transfer between the pipe and the surroundings. Two limiting cases of particular interest are for isothermal and adiabatic conditions.

Unlike the orifice or nozzle, the pipeline maintains the area of flow constant and equal to its cross-sectional area. There is no possibility therefore of the gas expanding laterally. Supersonic flow conditions can be reached in pipeline installations in a manner similar to that encountered in flow through a nozzle, but *not within the pipe itself* unless the gas enters the pipe at a supersonic velocity. If a pipe connects two reservoirs and the upstream reservoir is maintained at constant pressure P_1, the following pattern will occur as the pressure P_2 in the downstream reservoir is reduced.

(1) Starting with $P_2 = P_1$ there is, of course, no flow and $G = 0$.
(2) Reduction of P_2 initially results in an increase in G. G increases until the gas velocity at the outlet of the pipe just reaches the velocity of propagation of a pressure wave ("sonic velocity"). This value of P_2 will be denoted by P_w.

(3) Further reduction of P_2 has no effect on the flow in the pipeline. The pressure distribution along the length of the pipe remains unaltered and the pressure at its outlet remains at P_w. The gas, on leaving the pipe, expands laterally and its pressure falls to the reservoir pressure P_2.

In considering the flow in a pipe, the differential form of the general energy balance equation 2.54 are used, and the friction term δF will be written in terms of the energy dissipated per unit mass of fluid for flow through a length dl of pipe. In the first instance, isothermal flow of an ideal gas is considered and the flowrate is expressed as a function of upstream and downstream pressures. Non-isothermal and adiabatic flow are discussed later.

4.5.1. Energy balance for flow of ideal gas

In Chapter 2, the general energy equation for the flow of a fluid through a pipe has been expressed in the form:

$$\frac{u\,\mathrm{d}u}{\alpha} + g\,\mathrm{d}z + v\,\mathrm{d}P + \delta W_s + \delta F = 0 \quad \text{(equation 2.54)}$$

For a fluid flowing through a length dl of pipe of constant cross-sectional area A:

$$W_s = 0$$

$$\delta F = 4\left(\frac{R}{\rho u^2}\right)u^2\frac{\mathrm{d}l}{d} \qquad \text{(from equation 3.19)}$$

$$\therefore \quad \frac{u\,\mathrm{d}u}{\alpha} + g\,\mathrm{d}z + v\,\mathrm{d}P + 4\frac{R}{\rho u^2}u^2\frac{\mathrm{d}l}{d} = 0 \qquad (4.48)$$

This equation cannot be integrated directly because the velocity u increases as the pressure falls and is, therefore, a function of l (Figure 4.7). It is, therefore, convenient to work in terms of the mass flow G which remains constant throughout the length of pipe.

The velocity, $$u = \frac{Gv}{A} \qquad \text{(from equation 2.36)}$$

and hence: $$\frac{1}{\alpha}\left(\frac{G}{A}\right)^2 v\,\mathrm{d}v + g\,\mathrm{d}z + v\,\mathrm{d}P + 4\left(\frac{R}{\rho u^2}\right)\left(\frac{G}{A}\right)^2 v^2\frac{\mathrm{d}l}{d} = 0 \qquad (4.49)$$

For turbulent flow, which is usual for a gas, $\alpha = 1$.

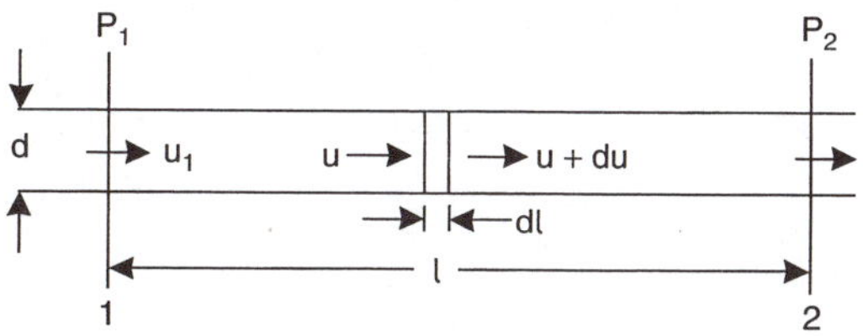

Figure 4.7. Flow of compressible fluid through pipe

Then, for flow in a horizontal pipe ($dz = 0$):

$$\left(\frac{G}{A}\right)^2 v\,dv + v\,dP + 4\left(\frac{R}{\rho u^2}\right)\left(\frac{G}{A}\right)^2 v^2\frac{dl}{d} = 0 \tag{4.50}$$

Dividing by v^2:

$$\left(\frac{G}{A}\right)^2 \frac{dv}{v} + \frac{dP}{v} + 4\left(\frac{R}{\rho u^2}\right)\left(\frac{G}{A}\right)^2 \frac{dl}{d} = 0 \tag{4.51}$$

Now the friction factor $R/\rho u^2$ is a function of the Reynolds number Re and the relative roughness e/d of the pipe surface which will normally be constant along a given pipe.

The Reynolds number is given by:

$$Re = \frac{ud\rho}{\mu} = \frac{Gd}{A\mu} = \frac{4G}{\pi d\mu} \tag{4.52}$$

Since G is constant over the length of the pipe, Re varies only as a result of changes in the viscosity μ. Although μ is a function of temperature, and to some extent of pressure, it is not likely to vary widely over the length of the pipe. Furthermore, the friction factor $R/\rho u^2$ is only a weak function of Reynolds number when Re is high, and little error will therefore arise from regarding it as constant.

Thus, integrating equation 4.51 over a length l of pipe:

$$\left(\frac{G}{A}\right)^2 \ln\frac{v_2}{v_1} + \int_{P_1}^{P_2}\frac{dP}{v} + 4\left(\frac{R}{\rho u^2}\right)\frac{l}{d}\left(\frac{G}{A}\right)^2 = 0 \tag{4.53}$$

The integral will depend on the P-v relationship during the expansion of the gas in the pipe, and several cases are now considered.

4.5.2. Isothermal flow of an ideal gas in a horizontal pipe

For isothermal changes in an ideal gas:

$$\int_{P_1}^{P_2}\frac{dP}{v} = \frac{1}{P_1 v_1}\int_{P_1}^{P_2} P\,dP = \frac{P_2^2 - P_1^2}{2P_1 v_1} \tag{4.54}$$

and, therefore, substituting in equation 4.53:

$$\left(\frac{G}{A}\right)^2 \ln\frac{P_1}{P_2} + \frac{P_2^2 - P_1^2}{2P_1 v_1} + 4\left(\frac{R}{\rho u^2}\right)\frac{l}{d}\left(\frac{G}{A}\right)^2 = 0 \tag{4.55}$$

Since v_m, the specific volume at the mean pressure in the pipe, is given by:

$$\frac{P_1 + P_2}{2} v_m = P_1 v_1$$

and:

$$\left(\frac{G}{A}\right)^2 \ln\frac{P_1}{P_2} + (P_2 - P_1)\frac{1}{v_m} + 4\left(\frac{R}{\rho u^2}\right)\frac{l}{d}\left(\frac{G}{A}\right)^2 = 0 \tag{4.56}$$

If the pressure drop in the pipe is a small proportion of the inlet pressure, the first term is negligible and the fluid may be treated as an incompressible fluid at the mean pressure in the pipe.

It is sometimes convenient to substitute $\mathbf{R}T/M$ for $P_1 v_1$ in equation 4.55 to give:

$$\left(\frac{G}{A}\right)^2 \ln\frac{P_1}{P_2} + \frac{P_2^2 - P_1^2}{2\mathbf{R}T/M} + 4\left(\frac{R}{\rho u^2}\right)\frac{l}{d}\left(\frac{G}{A}\right)^2 = 0 \tag{4.57}$$

Equations 4.55 and 4.57 are the most convenient for the calculation of gas flowrate as a function of P_1 and P_2 under isothermal conditions. Some additional refinement can be added if a compressibility factor is introduced as defined by the relation $Pv = Z\mathbf{R}T/M$, for conditions where there are significant deviations from the *ideal gas law* (equation 2.15).

Maximum flow conditions

Equation 4.55 expresses G as a continuous function of P_2, the pressure at the downstream end of the pipe for a given upstream pressure P_1. If there is no pressure change over the pipe then, $P_2 = P_1$, and substitution of P_1 for P_2 in equation 4.55 gives $G = 0$, as would be expected. Furthermore, substituting $P_2 = 0$ also gives $G = 0$. Thus, for some intermediate value of P_2 ($= P_w$, say), where $0 < P_w < P_1$, the flowrate G must be a maximum.

Multiplying equation 4.55 by $(A/G)^2$:

$$-\ln\left(\frac{P_2}{P_1}\right) + \left(\frac{A}{G}\right)^2 \frac{(P_2^2 - P_1^2)}{2P_1 v_1} + 4\left(\frac{R}{\rho u^2}\right)\frac{l}{d} = 0 \tag{4.58}$$

Differentiating with respect to P_2, for a constant value of P_1:

$$-\frac{P_1}{P_2}\frac{1}{P_1} + \left(\frac{A}{G}\right)^2 \frac{2P_2}{2P_1 v_1} + \frac{A^2}{2P_1 v_1}(P_2^2 - P_1^2)\frac{-2}{G^3}\frac{dG}{dP_2} = 0$$

The rate of flow is a maximum when $dG/dP_2 = 0$. Denoting conditions at the downstream end of the pipe by suffix w, when the flow is a maximum:

$$\frac{1}{P_w} = \left(\frac{A}{G}\right)^2 \frac{P_w}{P_1 v_1}$$

or:

$$\left(\frac{G}{A}\right)^2 = \frac{P_w^2}{P_1 v_1} = \frac{P_w^2}{P_w v_w} = \frac{P_w}{v_w} \tag{4.59}$$

or:

$$u_w = \sqrt{P_w v_w} \tag{4.60}$$

It has been shown in equation 4.35 that this velocity v_w is equal to the velocity of transmission of a small pressure wave in the fluid at the pressure P_w if heat could be transferred sufficiently rapidly to maintain isothermal conditions. If the pressure at the downstream end of the pipe were P_w, the fluid there would then be moving with the velocity of a pressure wave, and therefore a wave could not be transmitted through the fluid in the opposite direction because its velocity relative to the pipe would be zero. If, at the downstream end, the pipe were connected to a reservoir in which the pressure was reduced below P_w, the flow conditions within the pipe would be unaffected and the pressure at the exit of the pipe would remain at the value P_w as shown in Figure 4.8. The drop in pressure from P_w to P_2 would then take place by virtue of lateral expansion of the gas beyond the end of the pipe. If the pressure P_2 in the reservoir at the downstream end were gradually reduced from P_1, the rate of flow would increase until the pressure reached P_w and then remain constant at this maximum value as the pressure was further reduced.

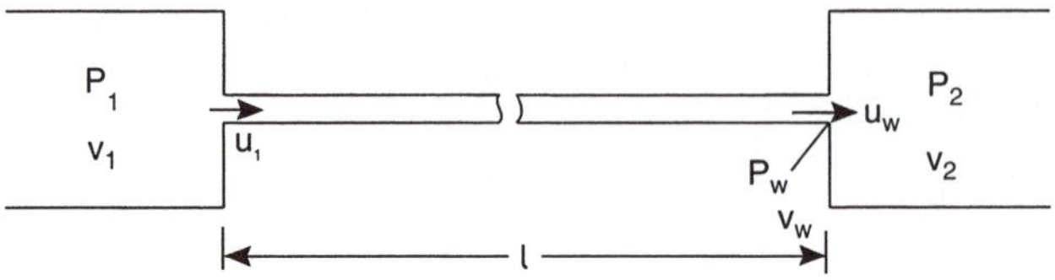

Figure 4.8. Maximum flow conditions

Thus, with compressible flow there is a maximum mass flowrate G_w which can be attained by the gas for a given upstream pressure P_1, and further reduction in pressure in the downstream reservoir below P_w will not give any further increase.

The maximum flowrate G_w is given by equation 4.59:

$$G_w = A\sqrt{\frac{P_w}{v_w}}$$

Substituting $v_w = v_1(P_1/P_w)$:

$$G_w = AP_w\sqrt{\frac{1}{P_1 v_1}} \tag{4.61}$$

P_w is given by substitution in equation 4.58,

that is:

$$\ln\left(\frac{P_1}{P_w}\right) + \frac{v_w}{P_w}\frac{1}{P_w v_w}\frac{P_w^2 - P_1^2}{2} + 4\left(\frac{R}{\rho u^2}\right)\frac{l}{d} = 0$$

or:

$$\ln\left(\frac{P_1}{P_w}\right)^2 + 1 - \left(\frac{P_1}{P_w}\right)^2 + 8\left(\frac{R}{\rho u^2}\right)\frac{l}{d} = 0 \tag{4.62}$$

or:

$$8\left(\frac{R}{\rho u^2}\right)\frac{l}{d} = \left(\frac{1}{w_c}\right)^2 - \ln\left(\frac{1}{w_c}\right)^2 - 1 \tag{4.63}$$

where $w_c = P_w/P_1$ (the critical value of the pressure ratio $w = (P_2/P_1)$.

Equations 4.62 and 4.63 are very important in that they give:

(a) The maximum value of the pressure ratio $P_1/P_2(= P_1/P_w)$ for which the whole of the expansion of the gas can take place in the pipe.

If $P_1/P_2 > P_1/P_w$, the gas expands from P_1 to P_w in the pipe, and from P_w to P_2 in the downstream reservoir.

Equations 4.55, 4.56 and 4.57 may be used to calculate what is happening in the pipe only provided $P_1/P_2 \not> P_1/P_w$ as given by equation 4.62.

(b) The minimum value of $8(R/\rho u^2)(l/d)$ for which, for any pressure ratio P_1/P_2, the fall in gas pressure will take place entirely within the pipe.

In Table 4.1, the relation between P_1/P_w and $8(R/\rho u^2)(l/d)$ is given for a series of numerical values of w_c.

It is seen that P_1/P_w increases with $8(R/\rho u^2)(l/d)$; in other words, in general, the longer the pipe (assuming $R/\rho u^2 \approx$ constant) the greater is the ratio of upstream to downstream pressure P_1/P_2 which can be accommodated in the pipe itself.

For an assumed average pipe friction factor $(R/\rho u^2)$ of 0.0015, this has been expressed as an l/d ratio. For a 25-mm diameter pipeline, the corresponding pipelengths are given in metres. In addition, for an upstream pressure P_1 of 100 bar, the *average* pressure gradient in the 25-mm pipe is given in the last column. At first sight it might seem strange that limiting conditions should be reached in such very short pipes for values of $P_1/P_w \approx 1$, but it will be seen that the *average pressure gradients* then become very high.

In Figure 4.9 values of P_1/P_w and w_c are plotted against $8(R/\rho u^2)(l/d)$. This curve gives the limiting value of P_1/P_2 for which the whole of the expansion of the gas can take place within the pipe.

It is seen in Table 4.1, for instance, that for a 25-mm pipeline and an assumed value of 0.0015 for $R/\rho u^2$ that for a pipe of length 14 m, the ratio of the pressure at the pipe inlet to that at the outlet cannot exceed 3.16 ($w_c = 0.316$). If $P_1/P_2 > 3.16$, the gas expands to the pressure $P_w(= 0.316P_1)$ inside the pipe and then it expands down to the pressure P_2 *within the downstream reservoir.*

Flow with fixed upstream pressure and variable downstream pressure

As an example, the flow of air at 293 K in a pipe of 25 mm diameter and length 14 m is considered, using the value of 0.0015 for $R/\rho u^2$ employed in the calculation of the figures in Table 4.1; $R/\rho u^2$ will, of course, show some variation with Reynolds number, but this effect will be neglected in the following calculation. The variation in flowrate G is examined, for a given upstream pressure of 10 MN/m^2, as a function of downstream pressure P_2. As the critical value of P_1/P_2 for this case is 3.16 (see Table 4.1), the maximum flowrate will occur at all values of P_2 less than $10/3.16 = 3.16$ MN/m^2. For values of P_2 greater than 3.16 MN/m^2, equation 4.57 applies:

$$\left(\frac{G}{A}\right)^2 \ln\frac{P_1}{P_2} + \frac{P_2^2 - P_1^2}{2\mathbf{R}T/M} + 4\left(\frac{R}{\rho u^2}\right)\frac{l}{d}\left(\frac{G}{A}\right)^2 = 0$$

Multiplying through by $(A/G)^2$ and inserting the numerical values with pressures P_1 and P_2 expressed in MN/m^2 (P_{1a} and P_{2a}):

$$\ln\frac{P_{1a}}{P_{2a}} - 5.95\times 10^6(P_{1a}^2 - P_{2a}^2)\left(\frac{A}{G}\right)^2 + 3.35 = 0$$

For $P_{1a} = 10$ MN/m^2:

$$2.306 - \ln P_{2a} - 5.95\times 10^6(100 - P_{2a}^2)\left(\frac{A}{G}\right)^2 + 3.35 = 0$$

Table 4.1. Limiting pressure ratios for pipe flow

					For $\frac{R}{\rho u^2} = 0.0015$	For $d = 25$ mm	For $P_1 = 100$ bar	
$\frac{P_1}{P_w}$	$\frac{P_w}{P_1}$	$\left(\frac{1}{w_c}\right)^2$	$\ln\left(\frac{1}{w_c}\right)^2$	$8\frac{R}{\rho u^2}\frac{l}{d}$	$\frac{l}{d}$	l(m)	$P_1 - P_w$ (bar)	Average $\frac{P_1 - P_w}{l}$ (bar/m)
1	1	1	0	0	0	0	0	—
1.1	0.9091	1.21	0.1906	0.0194	1.62	0.0404	9.1	225
1.78	0.562	3.16	1.151	1.01	84.2	2.11	43.8	20.8
3.16	0.316	10	2.302	6.70	558	14.0	68.4	4.89
5.62	0.178	31.6	3.453	27.1	2258	56.5	82.2	1.45
10	0.1	100	4.605	94.4	7867	197	90	0.457
17.8	0.0562	316	5.756	309.2	25,770	644	94.4	0.146
31.6	0.0316	1000	6.901	992	82,700	2066	96.8	0.046
56.2	0.0178	3160	8.058	3151	263,000	6575	98.2	0.015
100	0.01	10,000	9.210	9990	832,500	20,800	99	0.005

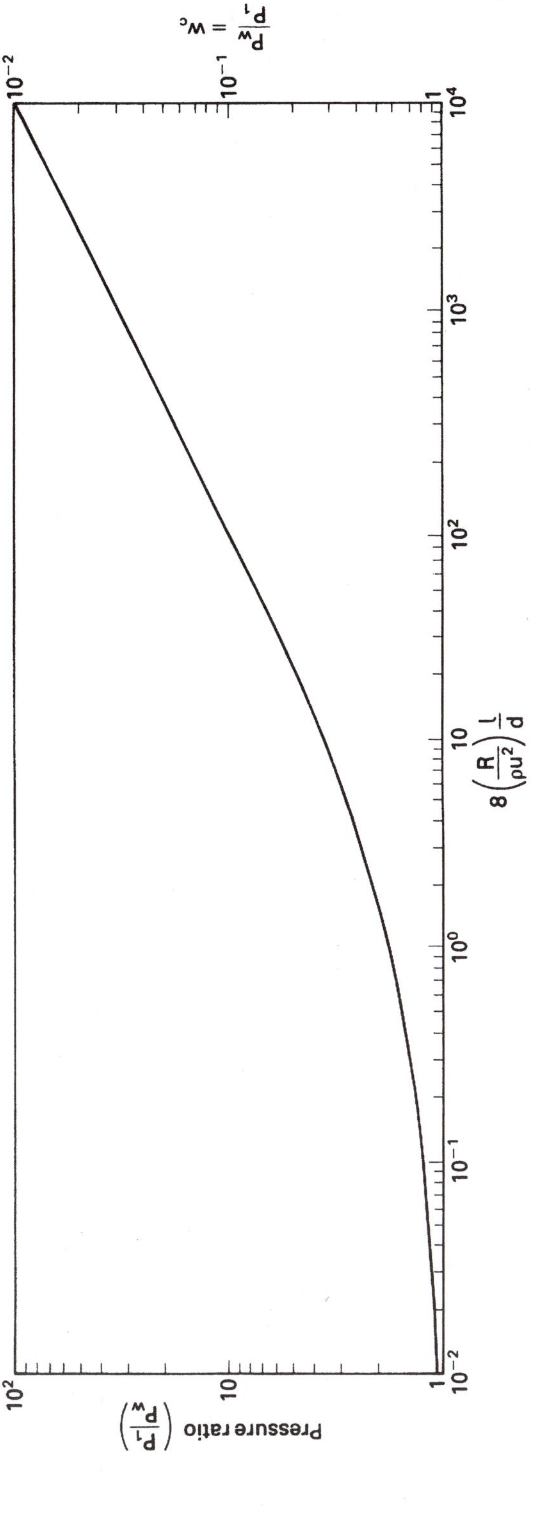

Figure 4.9. Critical pressure ratio P_1/P_w (maximum value of P_1/P_2 which can occur in a pipe) as function of $8(R/\rho u^2)(l/d)$

giving:
$$\left(\frac{G}{A}\right)^2 = \frac{5.95 \times 10^6(100 - P_{2a}^2)}{5.656 - \ln P_{2a}} \quad (\text{kg}^2/\text{m}^4\ \text{s}^2)$$

Values of G calculated from this relation are given in Table 4.2 and plotted in Figure 4.10.

Table 4.2. Mass flowrate as function of downstream pressure P_{2a} for constant upstream pressure of 10 MN/m^2

Pressure P_{2a} MN/m^2	$\frac{G}{A} \times 10^{-4}$ kg/m^{2}s	G kg/s	$\frac{G^*}{A} \times 10^{-4}$ kg/m^{2}s
10	0	0	0
9.5	0.41	2.0	0.42
9	0.57	2.8	0.58
8	0.78	3.8	0.80
7	0.90	4.4	0.95
6	0.99	4.9	1.07
5	1.05	5.2	1.15
4	1.08	5.3	1.22
<3.16	1.09	5.4	1.26

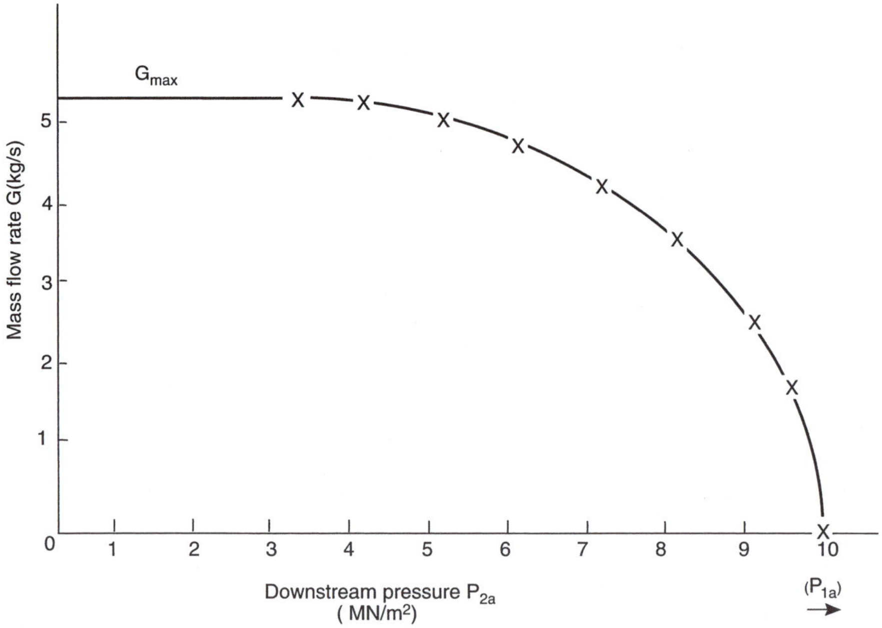

Figure 4.10. Mass flowrate as function of downstream pressure for isothermal flow of air in pipe ($l = 14$ m, $d = 25$ mm)

In the last column of Table 4.2, values are given for G^*/A calculated by ignoring the effect of changes in the kinetic energy of the gas, that is by neglecting the log term in equation 4.57. It will be noted that the effects are small ranging from about 2 per cent for $P_{2a} = 9$ MN/m^2 to about 13 per cent for $P_{2a} = 4$ MN/m^2.

Values of G_{max} for maximum flow conditions can also be calculated from equation 4.61:

$$G_{max} = AP_w\sqrt{\frac{1}{P_1 v_1}}$$

Substitution of the numerical values gives:

$G_{max} = 5.4$ kg/s, which is in agreement with the value in Table 4.2.

For $G/A = 10^4$ kg/m^2 s, the Reynolds number $Re \approx \dfrac{(10^4 \times 25 \times 10^{-3})}{10^{-5}} \approx 2.5 \times 10^7$

The chosen value of 0.0015 for $R/\rho u^2$ is reasonable for a smooth pipe over most of the range of Reynolds numbers encountered in this exercise.

Example 4.2

Over a 30 m length of a 150 mm vacuum line carrying air at 295 K, the pressure falls from 0.4 kN/m^2 to 0.13 kN/m^2. If the relative roughness e/d is 0.003 what is the approximate flowrate? It may be noted that flow of gases in vacuum systems is discussed fully by GRIFFITHS[1].

Solution

$$\left(\frac{G}{A}\right)^2 \ln\frac{P_1}{P_2} + \frac{P_2^2 - P_1^2}{2P_1 v_1} + 4\left(\frac{R}{\rho u^2}\right)\frac{l}{d}\left(\frac{G}{A}\right)^2 = 0 \qquad \text{(equation 4.55)}$$

$$\left(\frac{G}{A}\right)^2 \ln\frac{0.4}{0.13} + \frac{(0.13^2 - 0.40^2) \times 10^6}{2 \times 8.46 \times 10^4} + 4 \times 0.004 \times \frac{30}{0.150} \times \left(\frac{G}{A}\right)^2 = 0$$

$$1.124\left(\frac{G}{A}\right)^2 - 0.846 + 3.2\left(\frac{G}{A}\right)^2 = 0$$

$$\frac{G}{A} = 0.44 \text{ kg/m}^2\text{ s}$$

The viscosity of air at 295 K $= 1.8 \times 10^{-5}$ N s/m^2

$$\text{Reynolds number, } Re = \frac{0.44 \times 0.15}{1.8 \times 10^{-5}} = 3670 \text{ (from equation 4.52)}$$

$$\text{From Figure 3.7, } \frac{R}{\rho u^2} = 0.005$$

Inserting new values:

$$1.124\left(\frac{G}{A}\right)^2 - 0.846 + 4.0\left(\frac{G}{A}\right)^2 = 0$$

and:

$$\frac{G}{A} = \underline{\underline{0.41 \text{ kg/m}^2\text{ s}}}$$

A check is made that this flowrate is possible.

$$8\frac{R}{\rho u^2}\frac{l}{d} = 8.0$$

From Figure 4.9: $$\frac{P_1}{P_w} = 3.5$$

Actual value: $$\frac{P_1}{P_2} = \left(\frac{0.4}{0.13}\right) = 3.1$$

Thus, the flow condition just falls within the possible range.

Example 4.3

A flow of 50 m^3/s methane, measured at 288 K and 101.3 kN/m^2, has to be delivered along a 0.6 m diameter line, 3.0 km long with a relative roughness of 0.0001, linking a compressor and a processing unit. The methane is to be discharged at the plant at 288 K and 170 kN/m^2 and it leaves the compressor at 297 K. What pressure must be developed at the compressor in order to achieve this flowrate?

Solution

Taking mean temperature of 293 K,

$$P_1 v_1 = \frac{\mathbf{R}T}{M} = \frac{8314 \times 293}{16} = 1.5225 \times 10^5 \text{ m}^2/\text{s}^2$$

At 288 K and 101.3 kN/m^2, $v = \dfrac{1.5225 \times 10^5}{1.013 \times 10^5} \times \dfrac{288}{293} = 1.477 \text{ m}^3/\text{kg}$

Mass flowrate of methane, $G = \dfrac{50}{1.497} = 33.85$ kg/s

Cross-sectional area of pipe $A = \dfrac{\pi}{4} \times (0.6)^2 = 0.283 \text{ m}^2$

$$\frac{G}{A} = \frac{33.85}{0.283} = 119.6 \text{ kg/m}^2\text{ s}$$

$$\left(\frac{G}{A}\right)^2 = 1.431 \times 10^4 (\text{kg/m}^2\text{ s})^2.$$

Viscosity of methane at 293 K $\approx 0.01 \times 10^{-3}$ N s/m^2

$$\text{Reynolds number } Re = \frac{119.6 \times 0.6}{10^{-5}} = 7.18 \times 10^6$$

$$\text{For } \frac{e}{d} = 0.0001, \quad \frac{R}{\rho u^2} = 0.0015 \quad \text{(from Figure 3.7)}$$

The upstream pressure is calculated using equation 4.55:

$$\left(\frac{G}{A}\right)^2 \ln\frac{P_1}{P_2} + \frac{P_2^2 - P_1^2}{2P_1 v_1} + 4\left(\frac{R}{\rho u^2}\right)\frac{l}{d}\left(\frac{G}{A}\right)^2 = 0$$

Substituting: $1.431 \times 10^4 \ln \dfrac{P_1}{1.7 \times 10^5} - \dfrac{P_1^2 - (1.7 \times 10^5)^2}{2 \times 1.5225 \times 10^5} + 4 \times 0.0015 \times \dfrac{3000}{0.6} \times 1.431 \times 10^4 = 0$

Dividing by 1.431×10^4 gives:

$$\ln P_1 - 12.04 - 2.29 \times 10^{-10} P_1^2 + 6.63 + 30.00 = 0$$

$$2.29 \times 10^{-10} P_1^2 - \ln P_1 = 24.59$$

A trial and error solution gives: $$\underline{\underline{P_1 = 4.05 \times 10^5 \text{ N/m}^2}}$$

It is necessary to check that this degree of expansion is possible *within* pipe

$$\frac{P_1}{P_2} = 2.38$$

By reference to Figure 4.9:

$$8\frac{R}{\rho u^2}\frac{l}{d} = 60 \quad \text{and} \quad \frac{P_1}{P_w} = 8.1$$

Thus, the pressure ratio is within the possible range.

Heat flow required to maintain isothermal conditions

As the pressure in a pipe falls, the kinetic energy of the fluid increases at the expense of the internal energy and the temperature tends to fall. The maintenance of isothermal conditions therefore depends on the transfer of an adequate amount of heat from the surroundings. For a small change in the system, the energy balance is given in Chapter 2 as:

$$\delta q - \delta W_s = \mathrm{d}H + g\,\mathrm{d}z + u\,\mathrm{d}u \quad \text{(from equation 2.51)}$$

For a horizontal pipe, $\mathrm{d}z = 0$, and for isothermal expansion of an ideal gas $\mathrm{d}H = 0$. Thus if the system does no work on the surroundings:

$$\delta q = u\,\mathrm{d}u \tag{4.64}$$

and the required transfer of heat (in mechanical energy units) per unit mass is $\Delta u^2/2$. Thus the amount of heat required is equivalent to the increase in the kinetic energy of the fluid. If the mass rate of flow is G, the total heat to be transferred per unit time is $G\Delta u^2/2$.

In cases where the change in the kinetic energy is small, the required flow of heat is correspondingly small, and conditions are almost adiabatic.

4.5.3. Non-isothermal flow of an ideal gas in a horizontal pipe

In general, where an ideal gas expands or is compressed, the relation between the pressure P and the specific volume v can be represented approximately by:

$$Pv^k = \text{a constant} = P_1 v_1^k$$

where k will depend on the heat transfer to the surroundings.

Evaluation of the integral gives:

$$\int_{P_1}^{P_2} \frac{\mathrm{d}P}{v} = \frac{k}{k+1}\frac{P_1}{v_1}\left[\left(\frac{P_2}{P_1}\right)^{(k+1)/k} - 1\right] \tag{4.65}$$

Inserting this value in equation 4.53:

$$\left(\frac{G}{A}\right)^2 \frac{1}{k}\ln\left(\frac{P_1}{P_2}\right) + \frac{k}{k+1}\frac{P_1}{v_1}\left[\left(\frac{P_2}{P_1}\right)^{(k+1)/k} - 1\right] + 4\left(\frac{R}{\rho u^2}\right)\frac{l}{d}\left(\frac{G}{A}\right)^2 = 0 \tag{4.66}$$

For a given upstream pressure P_1 the maximum flow rate occurs when $u_2 = \sqrt{kP_2 v_2}$, the velocity of transmission of a pressure wave under these conditions (equation 4.38).

Flow under adiabatic conditions is considered in detail in the next section, although an approximate result may be obtained by putting k equal to γ in equation 4.66; this is only approximate because equating k to γ implies reversibility.

4.5.4. Adiabatic flow of an ideal gas in a horizontal pipe

The conditions existing during the adiabatic flow in a pipe may be calculated using the approximate expression Pv^k = a constant to give the relation between the pressure and the specific volume of the fluid. In general, however, the value of the index k may not be known for an irreversible adiabatic process. An alternative approach to the problem is therefore desirable.[2,3]

For a fluid flowing under turbulent conditions in a pipe, $\delta W_s = 0$ and:

$$\delta q = \mathrm{d}H + g\,\mathrm{d}z + u\,\mathrm{d}u \qquad \text{(from equation 2.51)}$$

In an adiabatic process, $\delta q = 0$, and the equation may then be written for the flow in a pipe of constant cross-sectional area A to give:

$$\left(\frac{G}{A}\right)^2 v\,\mathrm{d}v + g\,\mathrm{d}z + \mathrm{d}H = 0 \tag{4.67}$$

Now:

$$\mathrm{d}H = \mathrm{d}U + \mathrm{d}(Pv)$$

$$= C_v\,\mathrm{d}T + \mathrm{d}(Pv) \quad \text{for an ideal gas (from equation 2.24)}$$

Further:

$$C_p\,\mathrm{d}T = C_v\,\mathrm{d}T + \mathrm{d}(Pv) \quad \text{for an ideal gas (from equation 2.26)}$$

$\therefore$

$$\mathrm{d}T = \frac{\mathrm{d}(Pv)}{C_p - C_v} \tag{4.68}$$

so that:

$$\mathrm{d}H = \mathrm{d}(Pv)\left(\frac{C_v}{C_p - C_v} + 1\right) \tag{4.69}$$

$$= \frac{\gamma}{\gamma - 1}\,\mathrm{d}(Pv) \tag{4.70}$$

Substituting this value of $\mathrm{d}H$ in equation 4.67 and writing $g\,\mathrm{d}z = 0$ for a horizontal pipe:

$$\left(\frac{G}{A}\right)^2 v\,\mathrm{d}v + \frac{\gamma}{\gamma - 1}\,\mathrm{d}(Pv) = 0 \tag{4.71}$$

Integrating, a relation between P and v for adiabatic flow in a horizontal pipe is obtained:

$$\frac{1}{2}\left(\frac{G}{A}\right)^2 v^2 + \frac{\gamma}{\gamma - 1}Pv = \frac{1}{2}\left(\frac{G}{A}\right)^2 v_1^2 + \frac{\gamma}{\gamma - 1}P_1 v_1 = \text{constant, } K\text{(say)} \tag{4.72}$$

From equation 4.72:

$$P = \frac{\gamma - 1}{\gamma}\left[\frac{K}{v} - \frac{1}{2}\left(\frac{G}{A}\right)^2 v\right] \tag{4.73}$$

$$\mathrm{d}P = \frac{\gamma - 1}{\gamma}\left[-\frac{K}{v^2} - \frac{1}{2}\left(\frac{G}{A}\right)^2\right]\mathrm{d}v$$

$$\frac{\mathrm{d}P}{v} = \frac{\gamma - 1}{\gamma}\left[-\frac{K}{v^3} - \frac{1}{2}\left(\frac{G}{A}\right)^2\frac{1}{v}\right]\mathrm{d}v \tag{4.74}$$

$$\int_{P_1}^{P_2}\frac{\mathrm{d}P}{v} = \frac{\gamma - 1}{\gamma}\left[\frac{K}{2}\left(\frac{1}{v_2^2} - \frac{1}{v_1^2}\right) - \frac{1}{2}\left(\frac{G}{A}\right)^2\ln\frac{v_2}{v_1}\right] \tag{4.75}$$

Substituting for K from equation 4.72:

$$\int_{P_1}^{P_2}\frac{\mathrm{d}P}{v} = \frac{\gamma - 1}{\gamma}\left[\left(\frac{G}{A}\right)^2\frac{v_1^2}{4}\left(\frac{1}{v_2^2} - \frac{1}{v_1^2}\right) + \frac{\gamma P_1 v_1}{2(\gamma - 1)}\left(\frac{1}{v_2^2} - \frac{1}{v_1^2}\right) - \frac{1}{2}\left(\frac{G}{A}\right)^2\ln\frac{v_2}{v_1}\right]$$

$$= \frac{\gamma - 1}{4\gamma}\left(\frac{G}{A}\right)^2\left(\frac{v_1^2}{v_2^2} - 1 - 2\ln\frac{v_2}{v_1}\right) + \frac{P_1 v_1}{2}\left(\frac{1}{v_2^2} - \frac{1}{v_1^2}\right) \tag{4.76}$$

Inserting the value of $\int_{P_1}^{P_2}\mathrm{d}P/v$ from equation 4.76 into equation 4.53:

$$\left(\frac{G}{A}\right)^2\ln\frac{v_2}{v_1} + \frac{\gamma - 1}{4\gamma}\left(\frac{G}{A}\right)^2\left(\frac{v_1^2}{v_2^2} - 1 - 2\ln\frac{v_2}{v_1}\right)$$
$$+ \frac{P_1 v_1}{2}\left(\frac{1}{v_2^2} - \frac{1}{v_1^2}\right) + 4\left(\frac{R}{\rho u^2}\right)\frac{l}{d}\left(\frac{G}{A}\right)^2 = 0$$

Simplifying:

$$8\left(\frac{R}{\rho u^2}\right)\frac{l}{d} = \left[\frac{\gamma - 1}{2\gamma} + \frac{P_1}{v_1}\left(\frac{A}{G}\right)^2\right]\left[1 - \left(\frac{v_1}{v_2}\right)^2\right] - \frac{\gamma + 1}{\gamma}\ln\frac{v_2}{v_1} \tag{4.77}$$

This expression enables v_2, the specific volume at the downstream end of the pipe, to be calculated for the fluid flowing at a mass rate G from an upstream pressure P_1.

Alternatively, the mass rate of flow G may be calculated in terms of the specific volume of the fluid at the two pressures P_1 and P_2.

The pressure P_2 at the downstream end of the pipe is obtained by substituting the value of v_2 in equation 4.72.

For constant upstream conditions, the maximum flow through the pipe is found by differentiating with respect to v_2 and putting $(\mathrm{d}G/\mathrm{d}v_2)$ equal to zero. The maximum flow is thus shown to occur when the velocity at the downstream end of the pipe is the sonic velocity $\sqrt{\gamma P_2 v_2}$ (equation 4.37).

The rate of flow of gas under adiabatic conditions is never more than 20 per cent greater than that obtained for the same pressure difference with isothermal conditions. For pipes of length at least 1000 diameters, the difference does not exceed about 5 per cent. In practice the rate of flow may be limited, not by the conditions in the pipe itself, but by the development of sonic velocity at some valve or other constriction in the pipe. Care should, therefore, be taken in the selection of fittings for pipes conveying gases at high velocities.

Analysis of conditions for maximum flow

It will now be shown from purely thermodynamic considerations that for, adiabatic conditions, supersonic flow cannot develop in a pipe of constant cross-sectional area because the fluid is in a condition of maximum entropy when flowing at the sonic velocity. The condition of the gas at any point in the pipe where the pressure is P is given by the equations:

$$Pv = \frac{1}{M}\mathbf{R}T \qquad \text{(equation 2.16)}$$

and:

$$\frac{\gamma}{\gamma - 1}Pv + \frac{1}{2}\left(\frac{G}{A}\right)^2 v^2 = K \qquad \text{(equation 4.72)}$$

It may be noted that if the changes in the kinetic energy of the fluid are small, the process is almost isothermal.

Eliminating v, an expression for T is obtained as:

$$\frac{\gamma}{\gamma - 1}\frac{\mathbf{R}T}{M} + \frac{1}{2}\left(\frac{G}{A}\right)^2 \frac{\mathbf{R}^2 T^2}{P^2 M^2} = K \qquad (4.78)$$

The corresponding value of the entropy is now obtained:

$$\mathrm{d}H = T\,\mathrm{d}S + v\,\mathrm{d}P = C_p\,\mathrm{d}T \quad \text{for an ideal gas (from equations 2.28 and 2.26)}$$

$$\therefore \qquad \mathrm{d}S = C_p\frac{\mathrm{d}T}{T} - \frac{\mathbf{R}}{MP}\,\mathrm{d}P$$

$$\therefore \qquad S = C_p \ln\frac{T}{T_0} - \frac{\mathbf{R}}{M}\ln\frac{P}{P_0} \quad \text{(if } C_p \text{ is constant)} \qquad (4.79)$$

where T_0, P_0 represents the condition of the gas at which the entropy is arbitrarily taken as zero.

The temperature or enthalpy of the gas may then be plotted to a base of entropy to give a *Fanno line*.[4] This line shows the condition of the fluid as it flows along the pipe. If the velocity at entrance is subsonic (the normal condition), then the enthalpy will decrease along the pipe and the velocity will increase until sonic velocity is reached. If the flow is supersonic at the entrance, the velocity will decrease along the duct until it becomes sonic. The entropy has a maximum value corresponding to sonic velocity as shown in Figure 4.11. (Mach number $Ma < 1$ represents sub-sonic conditions; $Ma > 1$ supersonic.)

Fanno lines are also useful in presenting conditions in nozzles, turbines, and other units where supersonic flow arises.[5]

For small changes in pressure and entropy, the kinetic energy of the gas increases only very slowly, and therefore the temperature remains almost constant. As the pressure is further reduced, the kinetic energy changes become important and the rate of fall of temperature increases and eventually $\mathrm{d}T/\mathrm{d}S$ becomes infinite. Any further reduction of the pressure would cause a decrease in the entropy of the fluid and is, therefore, impossible.

The condition of maximum entropy occurs when $\mathrm{d}S/\mathrm{d}T = 0$, where:

$$\frac{\mathrm{d}S}{\mathrm{d}T} = \frac{C_p}{T} - \frac{\mathbf{R}}{MP}\frac{\mathrm{d}P}{\mathrm{d}T} \qquad \text{(from equation 4.79)}$$

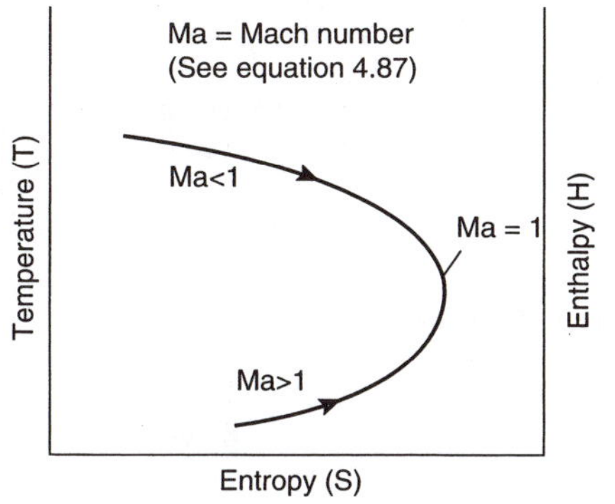

Figure 4.11. The Fanno line

The entropy is, therefore, a maximum when:

$$\frac{dP}{dT} = \frac{MPC_P}{\mathbf{R}T} \tag{4.80}$$

For an ideal gas:

$$C_p - C_v = \frac{\mathbf{R}}{M} \qquad \text{(equation 2.27)}$$

Substituting in equation 4.80:

$$\frac{dP}{dT} = \frac{PC_v}{(C_p - C_v)T} = \frac{P}{T}\frac{\gamma}{\gamma - 1} \tag{4.81}$$

The general value of dP/dT may be obtained by differentiating equation 4.78 with respect to T, or:

$$\frac{\mathbf{R}}{M} + \frac{\gamma - 1}{2\gamma}\left(\frac{G}{A}\right)^2 \frac{\mathbf{R}^2}{M^2}\left(\frac{P^2 2T - T^2 2P(dP/dT)}{P^4}\right) = 0$$

$$\therefore \qquad 1 + \frac{\gamma - 1}{\gamma}\left(\frac{G}{A}\right)^2 \frac{\mathbf{R}}{M}\left(\frac{T}{P^2} - \frac{T^2}{P^3}\frac{dP}{dT}\right) = 0$$

$$\therefore \qquad \frac{dP}{dT} = \frac{P}{T} + \frac{\gamma}{\gamma - 1}\left(\frac{A}{G}\right)^2 \frac{M}{\mathbf{R}}\frac{P^3}{T^2} \tag{4.82}$$

The maximum value of the entropy occurs when the values of dP/dT given by equations 4.81 and 4.82 are the same, that is when:

$$\frac{\gamma}{\gamma - 1}\frac{P}{T} = \frac{P}{T} - \frac{\gamma}{\gamma - 1}\left(\frac{A}{G}\right)^2 \frac{M}{\mathbf{R}}\frac{P^3}{T^2}$$

and

$$\left(\frac{G}{A}\right)^2 = \frac{\gamma}{\gamma - 1}\frac{M}{\mathbf{R}}\frac{P^2}{T}(\gamma - 1)$$

$$= \gamma\frac{P^2}{T}\frac{T}{Pv} - \gamma\frac{P}{v}$$

i.e., when: $$u = \sqrt{\gamma P v} = u_w \qquad \text{(from equation 4.37)}$$

which has already been shown to be the velocity of propagation of a pressure wave. This represents the limiting velocity which can be reached in a pipe of constant cross-sectional area.

4.5.5. Flow of non-ideal gases

Methods have been given for the calculation of the pressure drop for the flow of an incompressible fluid and for a compressible fluid which behaves as an ideal gas. If the fluid is compressible and deviations from the ideal gas law are appreciable, one of the approximate equations of state, such as van der Waals' equation, may be used in place of the law $PV = n\mathbf{R}T$ to give the relation between temperature, pressure, and volume. Alternatively, if the enthalpy of the gas is known over a range of temperature and pressure, the energy balance, equation 2.56, which involves a term representing the change in the enthalpy, may be employed:

$$\Delta\frac{u^2}{2\alpha} + g\Delta z + \Delta H = q - W_s \qquad \text{(equation 2.56)}$$

This method of approach is useful in considering the flow of steam at high pressures.

4.6. SHOCK WAVES

It has been seen in deriving equations 4.33 to 4.38 that for a small disturbance the velocity of propagation of the pressure wave is equal to the velocity of sound. If the changes are much larger and the process is not isentropic, the wave developed is known as a shock wave, and the velocity may be much greater than the velocity of sound. Material and momentum balances must be maintained and the appropriate equation of state for the fluid must be followed. Furthermore, any change which takes place must be associated with an increase, never a decrease, in entropy. For an ideal gas in a uniform pipe under adiabatic conditions a material balance gives:

$$\frac{u_1}{v_1} = \frac{u_2}{v_2} \qquad \text{(equation 2.62)}$$

Momentum balance: $$\frac{u_1^2}{v_1} + P_1 = \frac{u_2^2}{v_2} + P_2 \qquad \text{(equation 2.64)}$$

Equation of state: $$\frac{u_1^2}{2} + \frac{\gamma}{\gamma - 1}P_1 v_1 = \frac{u_2^2}{2} + \frac{\gamma}{\gamma - 1}P_2 v_2 \qquad \text{(from equation 4.72)}$$

Substituting from equation 2.62 into 2.64:

$$v_1 = \frac{u_1^2 - u_1 u_2}{P_2 - P_1} \qquad (4.83)$$

$$v_2 = \frac{u_2^2 - u_1 u_2}{P_1 - P_2} \qquad (4.84)$$

Then, from equation 4.72:

$$\frac{u_1^2 - u_2^2}{2} + \frac{\gamma}{\gamma - 1}\frac{P_1}{P_2 - P_1}u_1(u_1 - u_2) - \frac{\gamma}{\gamma - 1}\frac{P_1}{P_1 - P_2}u_2(u_2 - u_1) = 0$$

that is: $(u_1 - u_2) = 0$, representing no change in conditions, or:

$$\frac{u_1 + u_2}{2} + \frac{\gamma}{\gamma - 1}\frac{1}{P_2 - P_1}(u_1P_1 - u_2P_2) = 0$$

Hence:

$$\frac{u_2}{u_1} = \frac{(\gamma - 1)(P_2/P_1) + (\gamma + 1)}{(\gamma + 1)(P_2/P_1) + (\gamma - 1)} \tag{4.85}$$

and:

$$\frac{P_2}{P_1} = \frac{(\gamma + 1) - (\gamma - 1)(u_2/u_1)}{(\gamma + 1)(u_2/u_1) - (\gamma - 1)} \tag{4.86}$$

Equation 4.86 gives the pressure changes associated with a sudden change of velocity. In order to understand the nature of a possible velocity change, it is convenient to work in terms of Mach numbers. The Mach number (*Ma*) is defined as the ratio of the velocity at a point to the corresponding velocity of sound where:

$$Ma_1 = \frac{u_1}{\sqrt{\gamma P_1 v_1}} \tag{4.87}$$

and:

$$Ma_2 = \frac{u_2}{\sqrt{\gamma P_2 v_2}} \tag{4.88}$$

From equations 4.83, 4.84, 4.87, and 4.88:

$$v_1 = \frac{u_1^2}{Ma_1^2}\frac{1}{\gamma P_1} = \frac{u_1(u_1 - u_2)}{P_2 - P_1} \tag{4.89}$$

$$v_2 = \frac{u_2^2}{Ma_2^2}\frac{1}{\gamma P_2} = \frac{u_2(u_2 - u_1)}{P_1 - P_2} \tag{4.90}$$

Giving:

$$\frac{Ma_2^2}{Ma_1^2} = \frac{u_2}{u_1}\frac{P_1}{P_2} \tag{4.91}$$

From equation 4.89:

$$\frac{1}{\gamma Ma_1^2} = \frac{1 - (u_2/u_1)}{(P_2/P_1) - 1} \tag{4.92}$$

$$= \frac{2}{(\gamma + 1)(P_2/P_1) + (\gamma - 1)} \quad \text{(from equation 4.85)} \tag{4.93}$$

Thus:

$$\frac{P_2}{P_1} = \frac{2\gamma Ma_1^2 - (\gamma - 1)}{\gamma + 1} \tag{4.94}$$

and:

$$\frac{u_2}{u_1} = \frac{(\gamma - 1)Ma_1^2 + 2}{Ma_1^2(\gamma + 1)} \tag{4.95}$$

From equation 4.91:

$$\frac{Ma_2^2}{Ma_1^2} = \frac{(\gamma - 1)Ma_1^2 + 2}{Ma_1^2(\gamma + 1)} \frac{(\gamma + 1)}{2\gamma Ma_1^2 - (\gamma - 1)}$$

or:

$$Ma_2^2 = \frac{(\gamma - 1)Ma_1^2 + 2}{2\gamma Ma_1^2 - (\gamma - 1)} \tag{4.96}$$

For a sudden change or normal shock wave to occur, the entropy change per unit mass of fluid must be positive.

From equation 4.79, the change in entropy is given by:

$$\begin{aligned} S_2 - S_1 &= C_p \ln \frac{T_2}{T_1} - \frac{\mathbf{R}}{M} \ln \frac{P_2}{P_1} \\ &= C_p \ln \frac{P_2}{P_1} + C_p \ln \frac{v_2}{v_1} - \frac{\mathbf{R}}{M} \ln \frac{P_2}{P_1} \\ &= C_v \ln \frac{P_2}{P_1} + C_p \ln \frac{u_2}{u_1} \quad \text{(from equations 2.62 and 2.27)} \\ &= C_v \ln \frac{2\gamma Ma_1^2 - (\gamma - 1)}{\gamma + 1} - C_p \ln \frac{Ma_1^2(\gamma + 1)}{(\gamma - 1)Ma_1^2 + 2} \end{aligned} \tag{4.97}$$

$S_2 - S_1$ is positive when $Ma_1 > 1$. Thus a normal shock wave can occur only when the flow is supersonic. From equation 4.96, if $Ma_1 > 1$, then $Ma_2 < 1$, and therefore the flow necessarily changes from supersonic to subsonic. If $Ma_1 = 1$, $Ma_2 = 1$ also, from equation 4.96, and no change therefore takes place. It should be noted that there is no change in the energy of the fluid as it passes through a shock wave, though the entropy increases and therefore the change is irreversible.

For flow in a pipe of constant cross-sectional area, therefore, a shock wave can develop only if the gas enters at supersonic velocity. It cannot occur spontaneously.

Example 4.4

A reaction vessel in a building is protected by means of a bursting disc and the gases are vented to atmosphere through a stack pipe having a cross-sectional area of 0.07 m^2. The ruptured disc has a flow area of 4000 mm^2 and the gases expand to the full area of the stack pipe in a divergent section. If the gas in the vessel is at a pressure of 10 MN/m^2 and a temperature of 500 K, calculate: (a) the initial rate of discharge of gas, (b) the pressure and Mach number immediately upstream of the shock wave, and (c) the pressure of the gas immediately downstream of the shock wave.

Assume that isentropic conditions exist on either side of the shock wave and that the gas has a mean molecular weight of 40 kg/kmol, a ratio of specific heats of 1.4, and obeys the ideal gas law.

Solution

The pressure ratio w_c at the throat is given by equation 4.43:

$$w_c = \frac{P_c}{P_1} = \left[\frac{2}{\gamma + 1}\right]^{\gamma/(\gamma-1)} = \left(\frac{2}{2.4}\right)^{1.4/0.4} = 0.53$$

Thus, the throat pressure = (10 × 0.53) = 5.3 MN/m^2.

Specific volume of gas in reactor:

$$v_1 = \left(\frac{22.4}{40}\right)\left(\frac{500}{273}\right)\left(\frac{101.3}{10{,}000}\right) = 0.0103\ \text{m}^3/\text{kg}$$

Specific volume of gas at the throat $= (0.0103)(1/0.53)^{1/1.4} = (0.0103 \times 1.575) = 0.0162\ \text{m}^3/\text{kg}$

$$\therefore \text{ velocity at the throat} = \text{sonic velocity}$$

$$= \sqrt{\gamma P v} \quad \text{(from equation 4.37)}$$

$$= \sqrt{1.4 \times 5.3 \times 10^6 \times 0.0162} = 347\ \text{m/s}$$

Initial rate of discharge of gas; $G = uA/v$ (at throat)

$$= \frac{347 \times 4000 \times 10^{-6}}{0.0162}$$

$$= \underline{\underline{85.7\ \text{kg/s}}}$$

The gas continues to expand isentropically and the pressure ratio w is related to the flow area by equation 4.47. If the cross-sectional area of the exit to the divergent section is such that $w^{-1} = (10{,}000/101.3) = 98.7$, the pressure here will be atmospheric and the expansion will be entirely isentropic. The duct area, however, has nearly twice this value, and the flow is *over-expanded,* atmospheric pressure being reached within the divergent section. In order to satisfy the boundary conditions, a shock wave occurs further along the divergent section across which the pressure increases. The gas then expands isentropically to atmospheric pressure.

If the shock wave occurs when the flow area is A, then the flow conditions at this point can be calculated by solution of the equations for:

(1) *The isentropic expansion from conditions at the vent.* The pressure ratio w (pressure/pressure in the reactor) is given by equation 4.47 as:

$$w^{-1.42}(1 - w^{0.29}) = \left(\frac{A}{G}\right)^2\left(\frac{7P_1}{v_1}\right)$$

$$= \left(\frac{A}{85.7}\right)^2 \frac{(7 \times 10{,}000 \times 10^3)}{0.0103} = 9.25 \times 10^5\ A^2 \qquad (1)$$

The pressure at this point is $10 \times 10^6 w$ N/m^2.

Specific volume of gas at this point is given by equation 4.45 as:

$$v = v_1 w^{-0.71} = 0.0103 w^{-0.71}$$

The velocity is given by equation 4.46 as:

$$u^2 = 7P_1 v_1 (1 - w^{0.29})$$

$$= (7 \times 10 \times 10^6 \times 0.0103)(1 - w^{0.29})$$

$$\therefore \quad u = 0.849 \times 10^3 (1 - w^{0.29})^{0.5}\ \text{m/s}$$

Velocity of sound at a pressure of $10 \times 10^6 w$ N/m^2

$$= \sqrt{1.4 \times 10 \times 10^6 w \times 0.0103 w^{-0.71}}$$

$$= 380 w^{0.145}\ \text{m/s}$$

$$\text{Mach number} = \frac{0.849 \times 10^3 (1 - w^{0.29})^{0.5}}{380 w^{0.145}} = 2.23(w^{-0.29} - 1)^{0.5} \qquad (2)$$

(2) *The non-isentropic compression across the shock wave.* The velocity downstream from the shock wave (suffix s) is given by equation 4.95 as:

$$u_s = u_1 \frac{(\gamma - 1)Ma_1^2 + 2}{Ma_1^2(\gamma + 1)}$$

$$= \frac{0.849 \times 10^3(1 - w^{0.29})^{0.5}[0.4 \times 4.97(w^{-0.29} - 1) + 2]}{4.97(w^{-0.29} - 1) \times 2.4}$$

$$= 141(1 - w^{0.29})^{-0.5} \text{ m/s} \tag{3}$$

The pressure downstream from the shock wave P_s is given by equation 4.94:

$$\frac{P_s}{10} \times 10^6 w = \left[2\gamma Ma_1^2 - \frac{\gamma - 1}{\gamma + 1}\right] \tag{4}$$

Substituting from equation 2:

$$P_s = 56.3w(w^{-0.29} - 1) \times 10^6 \text{ N/m}^2$$

(3) *The isentropic expansion of the gas to atmospheric pressure.* The gas now expands isentropically from P_s to P_a (= 101.3 kN/m^2) and the flow area increases from A to the full bore of 0.07 m^2. Denoting conditions at the outlet by suffix a, then from equation 4.46:

$$u_a^2 - u_s^2 = 7P_s v_s \left[1 - \left(\frac{P_a}{P_s}\right)^{0.25}\right] \tag{5}$$

$$\frac{u_a}{v_a} = \frac{85.7}{0.07} = 1224 \text{ kg/m}^2\text{s} \tag{6}$$

$$\frac{u_s}{v_s} = \frac{85.7}{A} \text{ kg/m}^2\text{s} \tag{7}$$

$$\frac{v_a}{v_s} = \left(\frac{P_a}{P_s}\right)^{-0.71} \tag{8}$$

Equations 1, 3 to 8, involving seven unknowns, may be solved by trial and error to give $w = 0.0057$. Thus the pressure upstream from the shock wave is:

$$(10 \times 10^6 \times 0.0057) = 0.057 \times 10^6 \text{ N/m}^2$$

or: $\underline{\underline{57 \text{ kN/m}^2}}$

The Mach Number, from equation (2): $= \underline{\underline{4.15}}$

The pressure downstream from shock wave P_s, from equation (4),

$$= \underline{\underline{1165 \text{ kN/m}^2}}$$

4.7. FURTHER READING

BECKER, E. (Trans. E. L. CHU): *Gas Dynamics* (Academic Press, New York and London, 1968).

DODGE, B. F.: *Chemical Engineering Thermodynamics* (McGraw-Hill, New York, 1944).

LEE, J. F. and SEARS, F. W.: *Thermodynamics*, 2nd edn (Addison-Wesley, 1962).

LIEPMANN, H. W. and ROSHKO, A.: *Elements of Gasdynamics* (Wiley, New York, 1957).

MAYHEW, Y. R. and ROGERS, G. F. C.: *Thermodynamics and Transport Properties of Fluids*, 2nd edn (Blackwell, Oxford, 1971).

SHAPIRO, A. H.: *The Dynamics and Thermodynamics of Compressible Fluid Flow*, Vols. I and II (Ronald, New York, 1953 and 1954).

4.8. REFERENCES

1. GRIFFITHS, H.: *Trans. Inst. Chem. Eng.* **23** (1945) 113. Some problems of vacuum technique from a chemical engineering standpoint.
2. LAPPLE, C. E.: *Trans. Am. Inst. Chem. Eng.* **39** (1948) 385. Isothermal and adiabatic flow of compressible fluids.
3. WOOLLATT, E.: *Trans. Inst. Chem. Eng.* **24** (1946) 17. Some aspects of chemical engineering thermodynamics with particular reference to the development and use of the steady flow energy balance equations.
4. STODOLA, A. and LOWENSTEIN, L. C.: *Steam and Gas Turbines* (McGraw-Hill, New York, 1945).
5. LEE, J. F. and SEARS, F. W.: *Thermodynamics*, 2nd edn (Addison-Wesley, Reading, Mass., 1962).

4.9. NOMENCLATURE

		Units in SI system	Dimensions in **M, L, T, θ**
A	Cross-sectional area of flow	m^2	$\mathbf{L}^2$
A_0	Area of orifice	m^2	$\mathbf{L}^2$
C_D	Coefficient of discharge	—	—
C_p	Specific heat capacity at constant pressure	J/kg K	$\mathbf{L}^2\mathbf{T}^{-2}\theta^{-1}$
C_v	Specific heat capacity at constant volume	J/kg K	$\mathbf{L}^2\mathbf{T}^{-2}\theta^{-1}$
d	Pipe diameter	m	$\mathbf{L}$
e	Pipe roughness	m	$\mathbf{L}$
F	Energy dissipated per unit mass	J/kg	$\mathbf{L}^2\mathbf{T}^{-2}$
G	Mass flowrate	kg/s	$\mathbf{MT}^{-1}$
G_{max}	Mass flowrate under conditions of maximum flow	kg/s	$\mathbf{MT}^{-1}$
G^*	Value of G calculated ignoring changes in kinetic energy	kg/s	$\mathbf{MT}^{-1}$
g	Acceleration due to gravity	m/s^2	$\mathbf{LT}^{-2}$
H	Enthalpy per unit mass	J/kg	$\mathbf{L}^2\mathbf{T}^{-2}$
K	Energy per unit mass	J/kg	$\mathbf{L}^2\mathbf{T}^{-2}$
k	Gas expansion index	—	—
l	Pipe length	m	$\mathbf{L}$
M	Molecular weight	kg/kmol	$\mathbf{MN}^{-1}$
n	Number of moles	k mol	$\mathbf{N}$
P	Pressure	N/m^2	$\mathbf{ML}^{-1}\mathbf{T}^{-2}$
P_B	Back-pressure at nozzle	N/m^2	$\mathbf{ML}^{-1}\mathbf{T}^{-2}$
P_E	Exit pressure of gas	N/m^2	$\mathbf{ML}^{-1}\mathbf{T}^{-2}$
P_w	Downstream pressure P_2 at maximum flow condition	N/m^2	$\mathbf{ML}^{-1}\mathbf{T}^{-2}$
q	Heat added per unit mass	J/kg	$\mathbf{L}^2\mathbf{T}^{-2}$
R	Shear stress at pipe wall	N/m^2	$\mathbf{ML}^{-1}\mathbf{T}^{-2}$
$\mathbf{R}$	Universal gas constant	8314 J/kmol K	$\mathbf{MN}^{-1}\mathbf{L}^2\mathbf{T}^{-2}\theta^{-1}$
S	Entropy per unit mass	J/kg K	$\mathbf{L}^2\mathbf{T}^{-2}\theta^{-1}$
T	Temperature (absolute)	K	θ
U	Internal energy per unit mass	J/kg	$\mathbf{L}^2\mathbf{T}^{-2}\theta^{-1}$
u	Velocity	m/s	$\mathbf{LT}^{-1}$
u_E	Velocity at exit of nozzle	m/s	$\mathbf{LT}^{-1}$
u_w	Velocity of pressure wave	m/s	$\mathbf{LT}^{-1}$
v	Specific volume (= ρ^{-1})	m^3/kg	$\mathbf{M}^{-1}\mathbf{L}^3$
W_s	Shaft work per unit mass	J/kg	$\mathbf{L}^2\mathbf{T}^{-2}$
w	Pressure ratio P_2/P_1	—	—
w_c	Pressure ratio P_w/P_1	—	—
z	Vertical height	m	$\mathbf{L}$
α	Kinetic energy correction factor	—	—
γ	Specific heat ratio (C_p/C_v)	—	—
ε	Bulk modulus of elasticity	N/m^2	$\mathbf{ML}^{-1}\mathbf{T}^{-2}$
μ	Viscosity	Ns/m^2, kg/m s	$\mathbf{ML}^{-1}\mathbf{T}^{-1}$
ρ	Density	kg/m^3	$\mathbf{ML}^{-3}$

		Units in SI system	Dimensions in **M**, **L**, **T**, θ
Ma	Mach number	—	—
Re	Reynolds number	—	—

Suffixes

0	Reference condition
1	Upstream condition
2	Downstream condition
w	Maximum flow condition

Superscript

*	Value obtained neglecting kinetic energy changes

CHAPTER 5

Flow of Multiphase Mixtures

5.1. INTRODUCTION

The flow problems considered in previous chapters are concerned with homogeneous fluids, either single phases or suspensions of fine particles whose settling velocities are sufficiently low for the solids to be completely suspended in the fluid. Consideration is now given to the far more complex problem of the flow of multiphase systems in which the composition of the mixture may vary over the cross-section of the pipe or channel; furthermore, the components may be moving at different velocities to give rise to the phenomenon of "slip" between the phases.

Multiphase flow is important in many areas of chemical and process engineering and the behaviour of the material will depend on the properties of the components, the flowrates and the geometry of the system. In general, the complexity of the flow is so great that design methods depend very much on an analysis of the behaviour of such systems in practice and, only to a limited extent, on theoretical predictions. Some of the more important systems to be considered are:

Mixtures of liquids with gas or vapour.
Liquids mixed with solid particles ("hydraulic transport").
Gases carrying solid particles wholly or partly in suspension ("pneumatic transport").
Multiphase systems containing solids, liquids and gases.

Mixed materials may be transported horizontally, vertically, or at an inclination to the horizontal in pipes and, in the case of liquid-solid mixtures, in open channels. Although there is some degree of common behaviour between the various systems, the range of physical properties is so great that each different type of system must be considered separately. Liquids may have densities up to three orders of magnitude greater than gases but they do not exhibit any significant compressibility. Liquids themselves can range from simple Newtonian liquids such as water, to non-Newtonian fluids with very high apparent viscosities. These very large variations in density and viscosity are responsible for the large differences in behaviour of solid-gas and solid-liquid mixtures which must, in practice, be considered separately. For, all multiphase flow systems, however, it is important to understand the nature of the interactions between the phases and how these influence the *flow patterns* — the ways in which the phases are distributed over the cross-section of the pipe or duct. In design it is necessary to be able to predict *pressure drop* which, usually, depends not only on the flow pattern, but also on the relative velocity of the phases; this *slip velocity* will influence the *hold-up*, the fraction of the pipe volume which is occupied by a particular phase. It is important to note that, in the flow of a two-component mixture, the hold-up (or *in situ* concentration) of a component will differ from that in the mixture discharged at the end of the pipe because, as a result of *slip* of the phases relative to one another, their residence times in the pipeline will not be the same. Special attention is therefore focused on three aspects of the flow of these complex mixtures.

(1) The flow patterns.
(2) The hold-up of the individual phases and their relative velocities.
(3) The relationship between pressure gradient in a pipe and the flowrates and physical properties of the phases.

The difference in density between the phases is important in determining flow pattern. In gas-solid and gas-liquid mixtures, the gas will always be the lighter phase, and in liquid-solid systems it will be usual for the liquid to be less dense than the solid. In vertical upward flow, therefore, there will be a tendency for the lighter phase to rise more quickly than the denser phase giving rise to a slip velocity. For a liquid-solid or gas-solid system this *slip* velocity will be close to the terminal falling velocity of the particles. In a liquid-gas system, the slip velocity will depend on the flow pattern in a complex way. In all cases, there will be a net upwards force resulting in a transference of energy from the faster to the slower moving phase, and a vertically downwards gravitational force will be balanced by a vertically upwards drag force. There will be axial symmetry of flow.

In horizontal flow, the flow pattern will inevitably be more complex because the gravitational force will act perpendicular to the pipe axis, the direction of flow, and will cause the denser component to flow preferentially nearer the bottom of the pipe. Energy transfer between the phases will again occur as a result of the difference in velocity, but the net force will be horizontal and the suspension mechanism of the particles, or the dispersion of the fluid will be a more complex process. In this case, the flow will not be symmetrical about the pipe axis.

In practice, many other considerations will affect the design of an installation. For example, wherever solid particles are present, there is the possibility of *blockage* of the pipe and it is therefore important to operate under conditions where the probability of this occurring is minimised. Solids may be *abrasive* and cause undue wear if the velocities are too high or changes in direction of flow are too sudden. Choice of suitable materials of construction and operating conditions is therefore important. In pneumatic transport, *electrostatic charging* may take place and cause considerable increase in pressure gradient.

5.2. TWO-PHASE GAS (VAPOUR)–LIQUID FLOW

5.2.1. Introduction

Some of the important features of the flow of two-phase mixtures composed of a liquid together with a gas or vapour are discussed in this section. There are many applications in the chemical and process industries, ranging from the flow of mixtures of oil and gas from well heads to flow of vapour-liquid mixtures in boilers and evaporators.

Because of the presence of the two phases, there are considerable complications in describing and quantifying the nature of the flow compared with conditions with a single phase. The lack of knowledge of the velocities at a point in the individual phases makes it impossible to give any real picture of the velocity distribution. In most cases the gas phase, which may be flowing with a much greater velocity than the liquid, continuously accelerates the liquid thus involving a transfer of energy. Either phase may be in *streamline* or in *turbulent* flow, though the most important case is that in which both phases are turbulent. The criterion for streamline or turbulent flow of a phase is whether the Reynolds number for its flow at the same rate on its own is less or greater than 1000–2000. This distinction is to some extent arbitrary in that injection of a gas into a liquid initially in streamline flow may result in turbulence developing.

If there is no heat transfer to the flowing mixture, the mass rate of flow of each phase will remain substantially constant, though the volumetric flowrates (and velocities) will increase progressively as the gas expands with falling pressure. In a boiler or evaporator, there will be a progressive vaporisation of the liquid leading to a decreased mass flowrate of liquid and corresponding increase for the vapour, with the total mass rate of flow remaining constant. The volumetric flowrate will increase very rapidly as a result of the combined effects of falling pressure and increasing vapour/liquid ratio.

A gas-liquid mixture will have a lower density than the liquid alone. Therefore, if in a U-tube one limb contains liquid and the other a liquid-gas mixture, the equilibrium height in the second limb will be higher than in the first. If two-phase mixture is discharged at a height less than the equilibrium height, a continuous flow of liquid will take place from the first to the second limb, provided that a continuous feed of liquid and gas is maintained. This principle is used in the design of the air lift pump described in Chapter 8.

Consideration will now be given to the various flow regimes which may exist and how they may be represented on a "Flow Pattern Map"; to the calculation and prediction of hold-up of the two phases during flow; and to the calculation of pressure gradients for gas-liquid flow in pipes. In addition, when gas-liquid mixtures flow at high velocities serious erosion problems can arise and it is necessary for the designer to restrict flow velocities to avoid serious damage to equipment.

A more detailed treatment of the subject is given by GOVIER and AZIZ[(1)], by CHISHOLM[(2)] and by HEWITT[(3)].

5.2.2. Flow regimes and flow patterns

Horizontal flow

The flow pattern is complex and is influenced by the diameter of the pipe, the physical properties of the fluids and their flowrates. In general, as the velocities are increased and as the gas-liquid ratio increases, changes will take place from "bubble flow" through to "mist flow" as shown in Figure 5.1[(1–7)]; the principal characteristics are described in Table 5.1. At high liquid-gas ratios, the liquid forms the continuous phase and at low values it forms the disperse phase. In the intervening region, there is generally some instability; and sometimes several flow regimes are lumped together. In plug flow and slug flow, the gas is flowing faster than the liquid and liquid from a slug tends to become detached, to move as a relatively slow moving film along the surface of the pipe and then to be reaccelerated when the next liquid slug catches it up. This process can account for a significant proportion of the total energy losses. Particularly in short pipelines, the flow develops an oscillating pattern arising largely from discontinuities associated with the expulsion of successive liquid slugs.

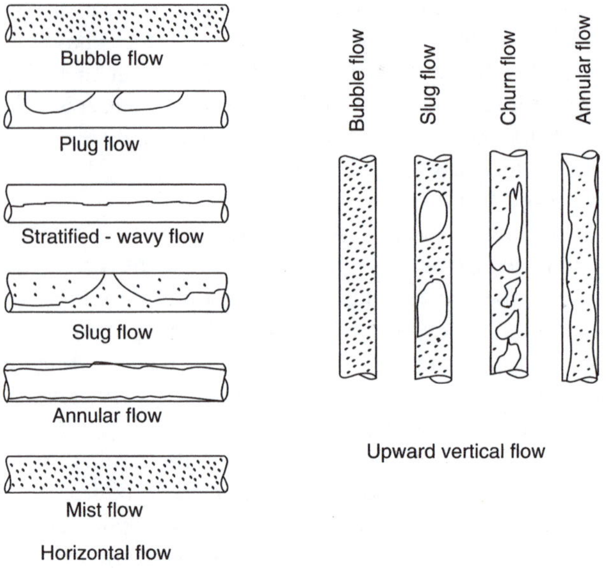

Figure 5.1. Flow patterns in two-phase flow

Table 5.1. Flow regimes in horizontal two-phase flow

Regime	Description	Typical velocities (m/s) Liquid	Vapour
1. Bubble flow[a]	Bubbles of gas dispersed throughout the liquid	1.5–5	0.3–3
2. Plug flow[a]	Plugs of gas in liquid phase	0.6	<1.0
3. Stratified flow	Layer of liquid with a layer of gas above	<0.15	0.6–3
4. Wavy flow	As stratified but with a wavy interface due to higher velocities	<0.3	>5
5. Slug flow[a]	Slug of gas in liquid phase	Occurs over a wide range of velocities	
6. Annular flow[b]	Liquid film on inside walls with gas in centre		>6
7. Mist flow[b]	Liquid droplets dispersed in gas		>60

[a] Frequently grouped together as *intermittent* flow
[b] Sometimes grouped as *annular/mist* flow

The regions over which the different types of flow can occur are conveniently shown on a "Flow Pattern Map" in which a function of the gas flowrate is plotted against a function of the liquid flowrate and boundary lines are drawn to delineate the various regions. It should be borne in mind that the distinction between any two flow patterns is not clear-cut and that these divisions are only approximate as each flow regime tends to merge in with its neighbours; in any case, the whole classification is based on highly subjective observations. Several workers have produced their own maps[4–8].

Most of the data used for compiling such maps have been obtained for the flow of water and air at near atmospheric temperature and pressure, and scaling factors have been introduced to extend their applicability to other systems. However, bearing in mind the diffuse nature of the boundaries between the regimes and the relatively minor effect of changes in physical properties, such a refinement does not appear to be justified. The flow pattern map for horizontal flow illustrated in Figure 5.2 which has been prepared by CHHABRA and RICHARDSON[9] is based on those previously presented by MANDHANE *et al.*[8] and WEISMAN *et al.*[7] The axes of this diagram are superficial liquid velocity u_L and superficial gas velocity u_G (in each case the volumetric flowrate of the phase divided by the total cross-sectional area of the pipe).

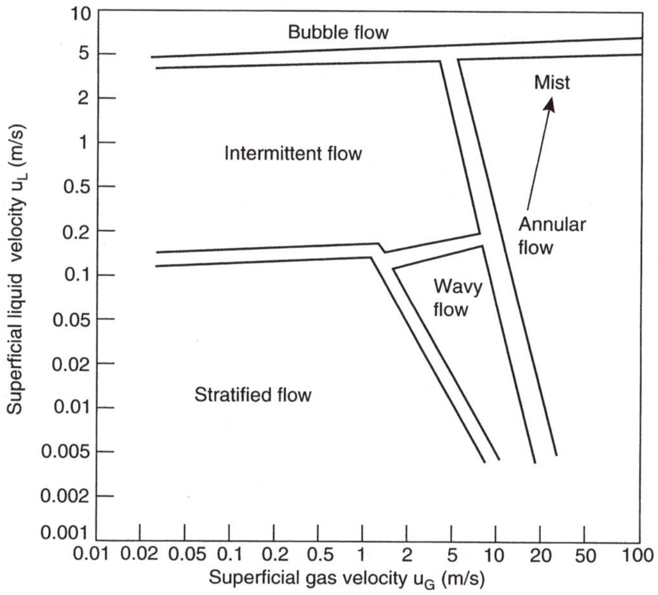

Figure 5.2. Flow pattern map

Slug flow should be avoided when it is necessary to obviate unsteady conditions, and it is desirable to design so that annular flow still persists at loadings down to 50 per cent of the normal flow rates. Even though in many applications both phases should be turbulent, excessive gas velocity will lead to a high pressure drop, particularly in small pipes.

Although most of the data relate to flow in pipes of small diameters (<42 mm), results of experiments carried out in a 205 mm pipe fit well on the diagram. The flow pattern map, shown in Figure 5.2, also gives a good representation of results obtained for the flow of mixtures of gas and shear-thinning non-Newtonian liquids, including very highly shear-thinning suspensions (power law index $n \approx 0.1$) and viscoelastic polymer solutions.

Vertical flow

In vertical flow, axial symmetry exists and flow patterns tend to be somewhat more stable. However, with slug flow in particular, oscillations in the flow can occur as a result of sudden changes in pressure as liquid slugs are discharged from the end of the pipe.

The principal flow patterns are shown in Figure 5.1. In general, the flow pattern map (Figure 5.2) is also applicable to vertical flow. Further reference to flow of gas–liquid mixtures in vertical pipes is made in Section 8.4.1 with reference to the operation of the air-lift pump.

5.2.3. Hold-up

Because the gas always flows at a velocity greater than that of the liquid, the *in situ* volumetric fraction of liquid at any point in a pipeline will be greater than the input volume fraction of liquid; furthermore it will progressively change along the length of the pipe as a result of expansion of the gas.

There have been several experimental studies of two-phase flow in which the hold-up has been measured, either directly or indirectly. The direct method of measurement involves suddenly isolating a section of the pipe by means of quick-acting valves and then determining the quantity of liquid trapped.(10,11) Such methods are cumbersome and are subject to errors arising from the fact that the valves cannot operate instantaneously. Typical of the indirect methods is that in which the pipe cross-section is scanned by γ-rays and the hold-up is determined from the extent of their attenuation.(12,13,14)

LOCKHART and MARTINELLI(15) expressed hold-up in terms of a parameter X, characteristic of the relative flowrates of liquid and gas, defined as:

$$X = \sqrt{\frac{-\Delta P_L}{-\Delta P_G}} \tag{5.1}$$

where $-\Delta P_L$ and $-\Delta P_G$ are the frictional pressure drops which would arise from the flow of the respective phases on their own at the same rates. Their correlation is reproduced in Figure 5.3. As a result of more recent work it is now generally accepted that the correlation overpredicts values of liquid hold-up. Thus FAROOQI and RICHARDSON(16), the results of whose work are also shown in Figure 5.3, have given the following expression for liquid hold-up ϵ_L for co-current flow of air and Newtonian liquids in horizontal pipes:

$$\left.\begin{aligned} \epsilon_L &= 0.186 + 0.0191X && 1 < X < 5 \\ \epsilon_L &= 0.143X^{0.42} && 5 < X < 50 \\ \epsilon_L &= \frac{1}{0.97 + 19/X} && 50 < X < 500 \end{aligned}\right\} \tag{5.2}$$

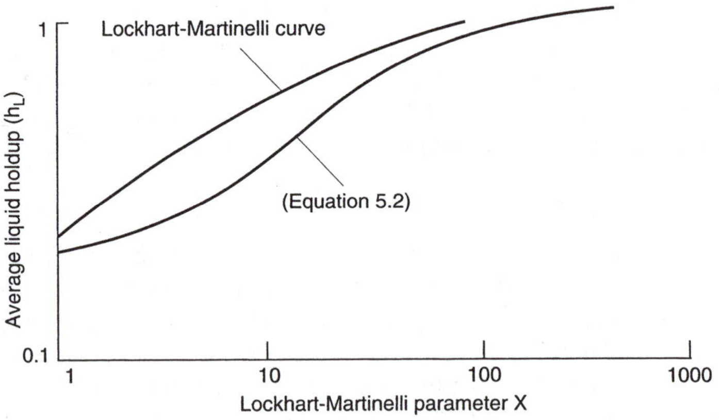

Figure 5.3. Correlation for average liquid hold-up, ϵ_L

It should be noted that, for turbulent flow of each phase, pressure drop is approximately proportional to the square of velocity and X is then equal to the ratio of the superficial velocities of the liquid and gas.

Equation 5.2 is found to hold well for non-Newtonian shear-thinning suspensions as well, provided that the liquid flow is turbulent. However, for laminar flow of the liquid, equation 5.2 considerably overpredicts the liquid hold-up ϵ_L. The extent of overprediction increases as the degree of shear-thinning increases and as the liquid Reynolds number becomes progressively less. A modified parameter X' has therefore been defined[16,17] for a power-law fluid (Chapter 3) in such a way that it reduces to X both at the superficial velocity u_L equal to the transitional velocity $(u_L)_c$ from streamline to turbulent flow and when the liquid exhibits Newtonian properties. The parameter X' is defined by the relation

$$X' = X\left(\frac{u_L}{(u_L)_c}\right)^{1-n} \tag{5.3}$$

where n is the *power-law* index. It will be seen that the correction factor becomes progressively more important as n deviates from unity and as the velocity deviates from the critical velocity. Equation 5.2 may then be applied provided X' is used in place of X in the equation.

Thus, in summary, liquid hold-up can be calculated using equation 5.2 for:

Newtonian fluids in laminar or turbulent flow.
Non-Newtonian fluids in *turbulent flow only.*

Equation 5.2, with the modified parameter X' used in place of X, may be used for laminar flow of shear-thinning fluids whose behaviour can be described by the *power-law* model.

A knowledge of hold-up is particularly important for vertical flow since the hydrostatic pressure gradient, which is frequently the major component of the total pressure gradient, is directly proportional to liquid hold-up. However, in slug flow, the situation is complicated by the fact that any liquid which is in the form of an annular film surrounding the gas slug does not contribute to the hydrostatic pressure[14].

5.2.4. Pressure, momentum, and energy relations

Methods for determining the drop in pressure start with a physical model of the two-phase system, and the analysis is developed as an extension of that used for single-phase flow. In the *separated flow* model the phases are first considered to flow separately; and their combined effect is then examined.

The total pressure gradient in a horizontal pipe, $(-\mathrm{d}P_{TPF}/\mathrm{d}l)$, consists of two components which represent the frictional and the acceleration pressure gradients respectively, or:

$$\frac{-\mathrm{d}P_{TPF}}{\mathrm{d}l} = \frac{-\mathrm{d}P_f}{\mathrm{d}l} + \frac{-\mathrm{d}P_a}{\mathrm{d}l} \tag{5.4}$$

A momentum balance for the flow of a two-phase fluid through a horizontal pipe and an energy balance may be written in an expanded form of that applicable to single-phase fluid flow. These equations for two-phase flow cannot be used in practice since the individual phase velocities and local densities are not known. Some simplification is possible if it is assumed that the two phases flow separately in the channel occupying fixed fractions of the total area, but even with this assumption of separated flow regimes, progress is difficult. It is important to note that, as in the case of single-phase flow of a compressible fluid, it is no longer possible to relate the shear stress to the pressure drop in a simple form since the pressure drop now covers both frictional and acceleration losses. The shear at the wall is proportional to the total rate of momentum transfer, arising from friction and acceleration, so that the total drop in pressure $-\Delta P_{TPF}$ is given by:

$$-\Delta P_{TPF} = (-\Delta P_f) + (-\Delta P_a) \tag{5.5}$$

The pressure drop due to acceleration is important in two-phase flow because the gas is normally flowing much faster than the liquid, and therefore as it expands the liquid phase will accelerate with consequent transfer of energy. For flow in a vertical direction, an additional term $-\Delta P_{\text{gravity}}$ must be added to the right hand side of equation 5.5 to account for the hydrostatic pressure attributable to the liquid in the pipe, and this may be calculated approximately provided that the liquid hold-up is known.

Analytical solutions for the equations of motion are not possible because of the difficulty of specifying the flow pattern and of defining the precise nature of the interaction between the phases. Rapid fluctuations in flow frequently occur and these cannot readily be taken into account. For these reasons, it is necessary for design purposes to use correlations which have been obtained using experimental data. Great care should be taken, however, if these are used outside the limits used in the experimental work.

Practical methods for evaluating pressure drop

Probably the most widely used method for estimating the drop in pressure due to friction is that proposed by LOCKHART and MARTINELLI(15) and later modified by CHISHOLM(18). This is based on the physical model of separated flow in which each phase is considered separately and then a combined effect formulated. The two-phase pressure drop due to friction $-\Delta P_{TPF}$ is taken as the pressure drop $-\Delta P_L$ or $-\Delta P_G$ that would arise for either phase flowing alone in the pipe at the stated rate, multiplied by some factor Φ_L^2 or Φ_G^2. This factor is presented as a function of the ratio of the individual single-phase pressure drops and:

$$\frac{-\Delta P_{TPF}}{-\Delta P_G} = \Phi_G^2 \tag{5.6}$$

$$\frac{-\Delta P_{TPF}}{-\Delta P_L} = \Phi_L^2 \tag{5.7}$$

The relation between Φ_G and Φ_L and X (defined by equation 5.1) is shown in Figure 5.4, where it is seen that separate curves are given according to the nature of the flow of the two phases. This relation was developed from studies on the flow in small tubes of up to 25 mm diameter with water, oils, and hydrocarbons using air at a pressure of up to 400 kN/m^2. For mass flowrates per unit area of L' and G' for the liquid and gas, respectively, Reynolds numbers $Re_L(L'd/\mu_L)$ and $Re_G(G'd/\mu_G)$ may be used as criteria for defining the flow regime; values less than 1000 to 2000, however, do not necessarily imply that the fluid is in truly laminar flow. Later experimental work showed that the total pressure has an influence and data presented by GRIFFITH(19) may be consulted where

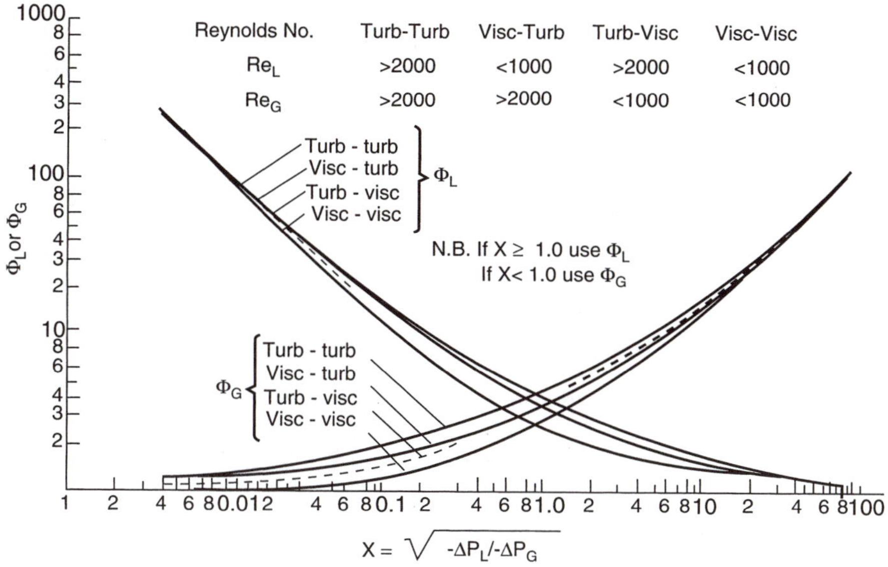

Figure 5.4. Relation between Φ and *X* for two-phase flow

pressures are in excess of 3 MN/m^2. CHISHOLM[18] has developed a relation between Φ_L and X which he puts in the form:

$$\Phi_L^2 = 1 + \frac{c}{X} + \frac{1}{X^2} \tag{5.8}$$

where c has a value of 20 for turbulent/turbulent flow, 10 for turbulent liquid/streamline gas, 12 for streamline liquid/turbulent gas, and 5 for streamline/streamline flow. If the densities of the fluids are significantly different from those of water and air at atmospheric temperature and pressure, the values of c are somewhat modified.

CHENOWETH and MARTIN[20,21] have presented an alternative method for calculating the drop in pressure, which is empirical and based on experiments with pipes of 75 mm and pressures up to 0.7 MN/m^2. They have plotted the volume fraction of the inlet stream that is liquid as abscissa against the ratio of the two-phase pressure drop to that for liquid flowing at the same volumetric rate as the mixture. An alternative technique has been described by BAROCZY[22]. If heat transfer gives rise to evaporation then reference should be made to work by DUKLER *et al*[23].

An illustration of the method of calculation of two-phase pressure drop is included here as Example 5.1.

Critical flow

For the flow of a compressible fluid, conditions of sonic velocity may be reached, thus limiting the maximum flowrate for a given upstream pressure. This situation can also occur with two-phase flow, and such critical velocities may sometimes be reached with a drop in pressure of only 30 per cent of the inlet pressure.

Example 5.1

Steam and water flow through a 75 mm i.d. pipe at flowrates of 0.05 and 1.5 kg/s respectively. If the mean temperature and pressure are 330 K and 120 kN/m^2, what is the pressure drop per unit length of pipe assuming adiabatic conditions?

Solution

$$\text{Cross-sectional area for flow} = \frac{\pi(0.075)^2}{4} = 0.00442 \text{ m}^2$$

$$\text{Flow of water} = \frac{1.5}{1000} = 0.0015 \text{ m}^3/\text{s}$$

$$\text{Water velocity} = \frac{0.0015}{0.00442} = 0.339 \text{ m/s}$$

Density of steam at 330 K and 120 kN/m^2

$$= \left(\frac{18}{22.4}\right)\left(\frac{273}{330}\right)\left(\frac{120}{101.3}\right) = 0.788 \text{ kg/m}^3$$

$$\text{Flow of steam} = \frac{0.05}{0.788} = 0.0635 \text{ m}^3/\text{s}$$

$$\text{Steam velocity} = \frac{0.0635}{0.00442} = 14.37 \text{ m/s}$$

Viscosities at 330 K and 120 kN/m^2:

$$\text{steam} = 0.0113 \times 10^{-3} \text{ N s/m}^2; \quad \text{water} = 0.52 \times 10^{-3} \text{ N s/m}^2$$

Therefore:
$$Re_L = \frac{0.075 \times 0.339 \times 1000}{0.52 \times 10^{-3}} = 4.89 \times 10^4$$

$$Re_G = \frac{0.075 \times 14.37 \times 0.788}{0.0113 \times 10^{-3}} = 7.52 \times 10^4$$

That is, both the gas and liquid are in turbulent flow.

From the friction chart (Figure 3.7), assuming $e/d = 0.00015$:

$$\left(\frac{R}{\rho u^2}\right)_L = 0.0025 \quad \text{and} \quad \left(\frac{R}{\rho u^2}\right)_G = 0.0022$$

$\therefore$ From equation 3.18:

$$-\Delta P_L = 4\left(\frac{R}{\rho u^2}\right)_L \frac{l}{d}\rho u^2 = 4 \times 0.0025\left(\frac{1}{0.075}\right)(1000 \times 0.339^2) = 15.32 \text{ (N/m}^2\text{)/m}$$

$$-\Delta P_G = 4 \times 0.0022\left(\frac{1}{0.075}\right)(0.778 \times 14.37^2) = 18.85 \text{ (N/m}^2\text{)/m}$$

$\therefore$
$$\frac{-\Delta P_L}{-\Delta P_G} = \frac{15.32}{18.85} = 0.812$$

and:
$$X^2 = 0.812 \quad \text{and} \quad X = 0.901$$

From Figure 5.4, for turbulent-turbulent flow,

$$\Phi_L = 4.35 \quad \text{and} \quad \Phi_G = 3.95$$

Therefore:
$$\frac{-\Delta P_{TPF}}{-\Delta P_G} = 3.95^2 = 15.60$$

and:
$$-\Delta P_{TPF} = 15.60 \times 18.85 = 294 \text{ (N/m}^2\text{)/m}$$

or:
$$\underline{\underline{-\Delta P_{TPF} = 0.29 \text{ (kN/m}^2\text{)/m}}}$$

Non-Newtonian flow

When a liquid exhibits non-Newtonian characteristics, the above procedures for Newtonian fluids are valid provided that the liquid flow is turbulent.

For streamline flow of non-Newtonian liquids, the situation is completely different and the behaviour of two-phase mixtures in which the liquid is a shear-thinning fluid is now examined.

The injection of air into a shear-thinning liquid in laminar flow may result in a substantial reduction in the pressure drop[24] and values of the *drag ratio* ($\Phi_L^2 = -\Delta P_{TPF}/-\Delta P_L$) may be substantially below unity. For a constant flowrate of liquid, the drag ratio gradually falls from unity as the gas rate is increased. At a critical air flowrate, the drag ratio passes through a minimum $(\Phi_L^2)_{\min}$ and then increases, reaching values in excess of unity at high gas flowrates. This effect has been observed with shear-thinning solutions of polymers and with flocculated suspensions of fine kaolin and anthracite coal, and is confined to conditions where the liquid flow would be laminar in the absence of air.

A typical graph of drag ratio as a function of superficial air velocity is shown in Figure 5.5 in which each curve refers to a constant superficial liquid velocity. The liquids in question exhibited power law rheology and the corresponding values of the Metzner and Reed Reynolds numbers Re_{MR} based on the superficial liquid velocity u_L (see Chapter 3) are given. The following characteristics of the curves may be noted:

(1) For a given liquid, the value of the minimum drag ratio $(\Phi_L^2)_{\min}$ decreases as the superficial liquid velocity is decreased.
(2) The superficial air velocity required to give the minimum drag ratio increases as the liquid velocity decreases.

For a more highly shear-thinning liquid, the minimum drag ratio becomes even smaller although more air must be added to achieve the condition.

If, for a given liquid, drag ratio is plotted (Figure 5.6) against total mixture velocity (superficial air velocity + superficial liquid velocity) as opposed to superficial gas velocity, it is found that the minima all occur at the same mixture velocity, irrespective of the liquid flowrate. For liquids of different rheological properties the minimum occurs at the same Reynolds number Re_{MR} (based on mixture as opposed to superficial liquid velocity) — about 2000 which corresponds to the limit of laminar flow. This suggests that the extent of drag reduction increases progressively until the liquid flow ceases to be laminar. Thus, at low flowrates more air can be injected before the liquid flow becomes turbulent, as indicated previously.

At first sight, it seems anomalous that, on increasing the total volumetric flowrate by injection of air, the pressure drop can actually be reduced. Furthermore, the magnitude of the effect can be very large with values of drag ratio as low as 0.2, i.e. the pressure drop can be reduced by a *factor of 5* by air injection. How this can happen can be illustrated by means of a highly simplified model. If the air and liquid flow as a series of separate plugs, as shown in Figure 5.7, the total pressure drop can then be taken as the sum of the pressure drops across the liquid and gas slugs; the pressure drop across the gas slugs however will be negligible compared with that contributed by the liquid.

For a 'power-law' fluid in laminar flow at a velocity u_L in a pipe of length l, the pressure drop $-\Delta P_L$ will be given by:

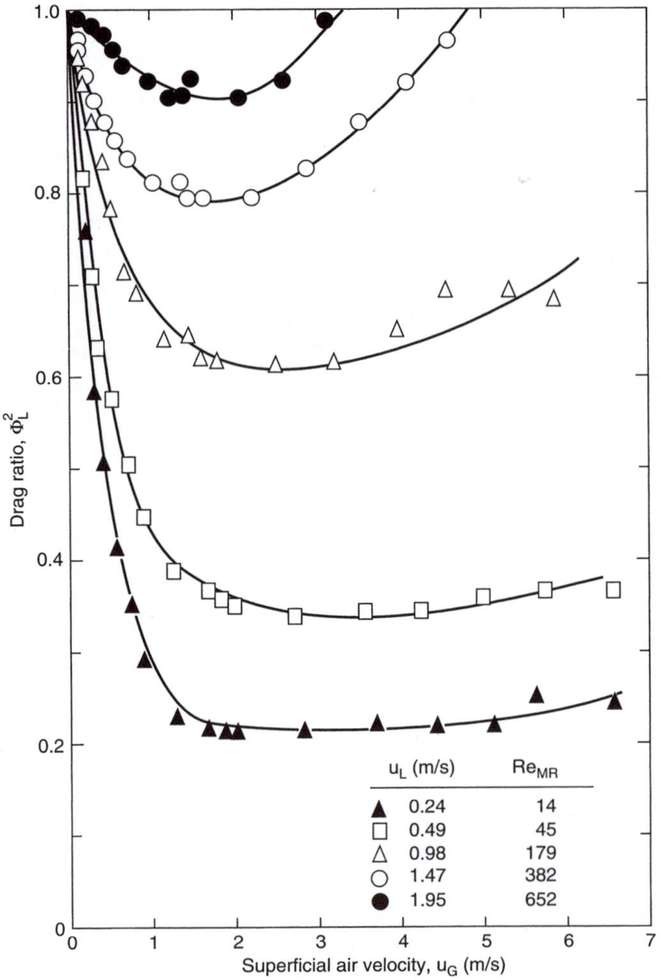

Figure 5.5. Drag ratio as function of superficial gas velocity (liquid velocity as parameter)

$$-\Delta P_L = K u_L^n l \tag{5.9}$$

If air is injected so that the mixture velocity is increased to bu_L, then the total length of liquid slugs in the pipe will be $(1/b)l$. Then, neglecting the pressure drop across the air slugs, the two-phase pressure drop $-\Delta P_{TP}$ will be given by:

$$-\Delta P_{TPF} = K(bu_L)^n \left(\frac{1}{b} l\right)$$

$$= K b^{n-1} u^n l \tag{5.10}$$

Thus, dividing equation 5.10 by equation 5.9:

$$\Phi_L^2 = \frac{-\Delta P_{TPF}}{-\Delta P_L} = b^{n-1} \tag{5.11}$$

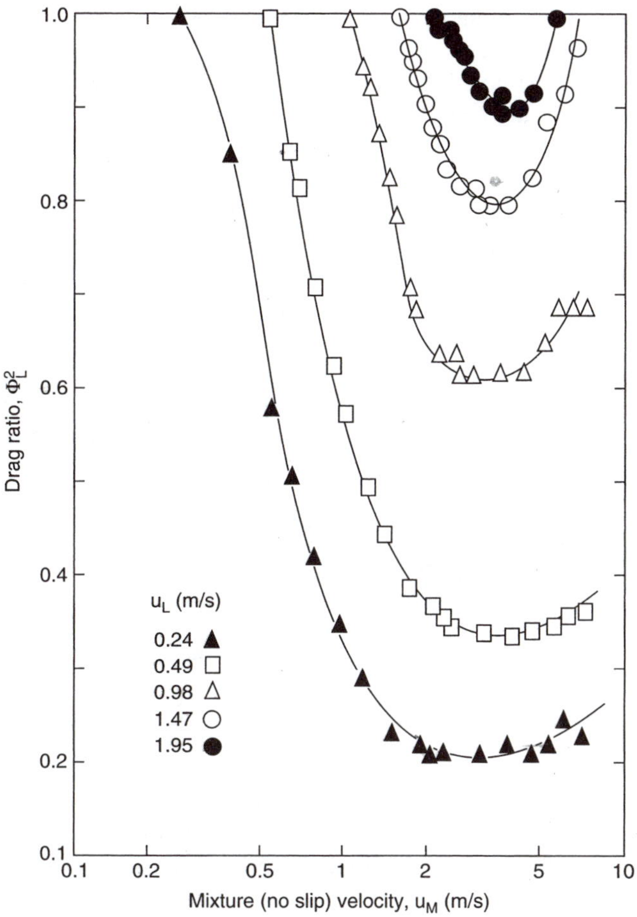

Figure 5.6. Drag ratio as function of mixture velocity

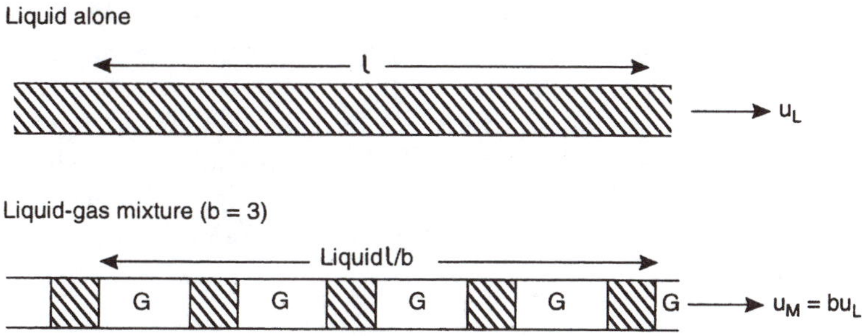

Figure 5.7. Air-liquid flow in form of separate discrete plugs

Because $n < 1$ for a shear thinning fluid, Φ_L^2 will be less than unity and a reduction in pressure drop occurs. The lower the value of n and the larger the value of b, the greater the effect will be. It will be noted that the effects of expansion of the air as the pressure falls have not been taken into account.

It has been found experimentally that equation 5.11 does apply for small rates of injection of air. At higher injection rates the model is not realistic and the pressure drop is reduced by a smaller amount than predicted by the equation.

It is seen that for a fluid with a power-law index of 0.2 (a typical value for a flocculated kaolin suspension) and a flowrate of air equal to that of liquid ($b = 2$), $\Phi_L^2 = 0.57$, that is there is a pressure drop reduction of 43 per cent. For $n = 0.1$ and $b = 5$, $\Phi_L^2 = 0.23$, corresponding to a 77 per cent reduction.

It will be noted that for a Newtonian fluid ($n = 1$) equation 5.11 gives $\Phi_L^2 = 1$ for all values of b. In other words, the pressure drop will be unaffected by air injection provided that the liquid flow remains laminar. In practice, because of losses not taken into account in the simplified model, the pressure drop for a Newtonian fluid always increases with air injection.

Furthermore, it can be seen that for turbulent flow when $n \to 2$, air injection would result in substantial increases in pressure drop.

Air injection can be used, in practice, in two ways:

(1) To reduce the pressure drop, and hence the upstream pressure in a pipeline, for a given flowrate of shear-thinning liquid.
(2) To increase the flowrate of liquid for any given pipeline pressure drop.

It may be noted that energy will be required for compressing the air to the injection pressure which must exceed the upstream pressure in the pipeline. The conditions under which power-saving is achieved have been examined by DZIUBINSKI[25], who has shown that the relative efficiency of the liquid pump and the air compressor are critically important factors.

The pressure drop for a fluid exhibiting a yield stress, such as a *Bingham plastic* material, can be similarly reduced by air injection.

In a practical situation, air injection can be beneficial in that, when a pipeline is shut down, it may be easier to start up again if the line is not completely full of a slurry. On the other hand, if the pipeline follows an undulating topography difficulties can arise if air collects at the high spots.

Air injection may sometimes be an alternative to deflocculation. In general, deflocculated suspensions flow more readily but they tend to give much more highly consolidated sediments which can be difficult to resuspend on starting up following a shutdown. Furthermore, deflocculants are expensive and may adversely affect the suitability of the solids for subsequent use.

5.2.5. Erosion

The flow of two-phase systems often causes erosion, and many empirical relationships have been suggested to define exactly when the effects are likely to be serious. Since high velocities may be desirable to avoid the instability associated with slug flow, there is a danger that any increase in throughput above the normal operating condition will lead to a situation where erosion may become a serious possibility.

An indication of the velocity at which erosion becomes significant may be obtained from:

$$\rho_M u_M^2 = 15{,}000$$

where ρ_M is the mean density of the two-phase mixture (kg/m^3) and u_M the mean velocity of the two-phase mixture (m/s). Here:

$$\rho_M = [L' + G'] \Big/ \left[\frac{L'}{\rho_L} + \frac{G'}{\rho_G}\right]$$

and:

$$u_M = u_L + u_G$$

where u_L and u_G are the superficial velocities of the liquid and gas respectively.

It is apparent that some compromise may be essential between avoiding a slug-flow condition and velocities which are likely to cause erosion.

5.3. FLOW OF SOLIDS–LIQUID MIXTURES

5.3.1. Introduction

Hydraulic transport is the general name frequently given to the transportation of solid particles in liquids — the term *hydraulic* relates to the fact that most of the earlier applications of the technique involved water as the carrier fluid. Today there are many industrial plants, particularly in the mining industries, where particles are transported in a variety of liquids. Transport may be in vertical or horizontal pipes and, as is usually the case in long pipelines, it may follow the undulations of the land over which the pipeline is constructed. The diameter and length of the pipeline and its inclination, the properties of the solids and of the liquid, and the flowrates all influence the nature of the flow and the pressure gradient. Design methods are, in general, not very reliable, particularly for the transportation of coarse particles in horizontal pipelines and calculated values of pressure drop and energy requirements should be treated with caution. In practice, it may be more important to ensure that the system operates reliably, and without the risk of blockage and without excessive erosion, than to achieve optimum conditions in relation to power requirements.

The most important variables which must be considered in estimating power consumption and pressure drop are:

(1) The pipeline — length, diameter, inclination to the horizontal, necessity for bends, valves, etc.

(2) The liquid — its physical properties, including density, viscosity and rheology, and its corrosive nature, if any.

(3) The solids — their particle size and size distribution, shape and density — all factors which affect their behaviour in a fluid.

(4) The concentration of particles and the flowrates of both solids and liquid.

Suspensions are conveniently divided into two broad classes — *fine suspensions* in which the particles are reasonably uniformly distributed in the liquid; and *coarse suspensions* in which particles tend to travel predominantly in the bottom part of a horizontal pipe at a lower velocity than the liquid, and to have a significantly lower velocity than the liquid in a vertical pipe. This is obviously not a sharp classification and, furthermore, is influenced by the flowrate and concentration of solids. However, it provides a useful initial basis on which to consider the behaviour of solid-liquid mixtures.

5.3.2. Homogeneous non-settling suspensions

Fine suspensions are reasonably homogeneous and segregation of solid and liquid phases does not occur to any significant extent during flow. The settling velocities of the particles are low in comparison with the liquid velocity and the turbulent eddies within the fluid are responsible for the suspension of the particles. In practice, turbulent flow will always be used, except when the liquid has a very high viscosity or exhibits non-Newtonian characteristics. The particles may be individually dispersed in the liquid or they may be present as flocs.

In disperse, or deflocculated, suspensions, the particles generally all have like charges and therefore repel each other. Such conditions exist when the pH range and the concentration of ions in the liquid is appropriate; this subject is discussed in Volume 2. Disperse suspensions tend to exhibit Newtonian behaviour. High fractional volumetric concentrations (0.4–0.5) are achievable and pressure drops for pipeline flow are comparatively low and correspond closely with those calculated for homogeneous fluids of the same density as the suspension. The particles do not have a very large effect on the viscosity of the liquid except at very high concentrations, when non-Newtonian shear-thickening characteristics may be encountered as a result of the build up of a "structure". There are two possible reasons for not transporting particles in the deflocculated state. First, they tend to form dense coherent sediments and, if the flow in the pipe is interrupted, there may be difficulties in restarting operation. Secondly, the cost of the chemicals which need to be added to maintain the particles in the dispersed state may be considerable.

In most large-scale pipelines the suspensions of fine particles are usually transported in the flocculated state. Flocs consist of large numbers of particles bound loosely together with liquid occluded in the free space between them. They therefore tend to behave as relatively large particles of density intermediate between that of the liquid and the solid. Because some of the liquid is immobilised within the flocs, the maximum solids concentration obtainable is less than for deflocculated suspensions. The flocs are fragile and can break down and deform in the shear fields which exist within a flowing fluid. They therefore tend to exhibit non-Newtonian behaviour. At all but the highest concentrations they are *shear-thinning* and frequently exhibit a *yield stress* (Chapter 3). Their behaviour, over limited ranges of shear rates, can usually be reasonably well-described by the *power law* or *Bingham plastic* models (see equations 3.120 and 3.122). Thixotropic and viscoelastic effects are usually negligible under the conditions existing in hydraulic transport lines.

Because concentrated flocculated suspensions generally have high apparent viscosities at the shear rates existing in pipelines, they are frequently transported under laminar flow conditions. Pressure drops are then readily calculated from their rheology, as described in Chapter 3. When the flow is turbulent, the pressure drop is difficult to predict accurately and will generally be somewhat *less* than that calculated assuming Newtonian behaviour. As the Reynolds number becomes greater, the effects of non-Newtonian behaviour become progressively less. There is thus a safety margin for design purposes if the suspensions are treated as Newtonian fluids when in turbulent flow.

Since, by definition, the settling velocity of the particles is low in a fine suspension, its behaviour is not dependent on its direction of flow and, if allowance is made for the hydrostatic pressure, pressure gradients are similar in horizontal and vertical pipelines.

In a series of experiments on the flow of flocculated kaolin suspensions in laboratory and industrial scale pipelines[26,27,28], measurements of pressure drop were made as a function of flowrate. Results were obtained using a laboratory capillary-tube viscometer, and pipelines of 42 mm and 205 mm diameter arranged in a recirculating loop. The rheology of all of the suspensions was described by the *power-law* model with a power law index less than unity, that is they were all shear-thinning. The behaviour in the laminar region can be described by the equation:

$$|R_y| = k\left|\frac{\mathrm{d}u_x}{\mathrm{d}y}\right|^n \quad \text{(see equation 3.121)} \tag{5.12}$$

where R_y is the shear stress at a distance y from the wall,
u_x is the velocity at that position,
n is the *flow index*, and
k is the *consistency*.

Values of n and k for the suspensions used are given in Table 5.2. Experimental results are shown in Figure 5.8 as wall shear stress R as a function of wall shear rate $(\mathrm{d}u_x/\mathrm{d}y)_{y=0}$ using logarithmic coordinates.

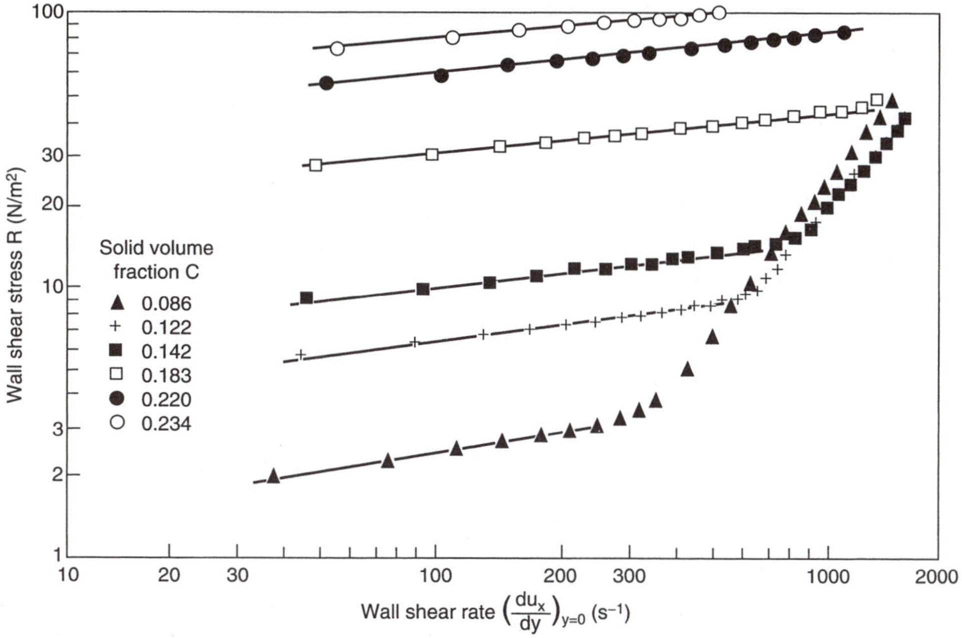

Figure 5.8. Rheograms for flocculated kaolin suspensions

Table 5.2. Power-law parameters for flocculated kaolin suspensions

Solids volume fraction C	Flow index n	Consistency k (Ns^n/m^2)
0.086	0.23	0.89
0.122	0.18	2.83
0.142	0.16	4.83
0.183	0.15	15.3
0.220	0.14	32.4
0.234	0.13	45.3

It is shown in Chapter 3 (equation 3.137) that:

$$\left[\frac{du_x}{dy}\right]_{y=0} = \frac{6n+2}{n}\frac{u}{d} \tag{5.13}$$

Figure 5.8 shows clearly the transition point from laminar to turbulent flow for each of the suspensions when flowing in the 42 mm diameter pipe.

5.3.3. Coarse solids

The flow behaviour of suspensions of coarse particles is completely different in horizontal and vertical pipes. In horizontal flow, the concentration of particles increases towards the bottom of the pipe, the degree of non-uniformity increasing as the velocity of flow is decreased. In vertical transport, however, axial symmetry is maintained with the solids evenly distributed over the cross-section. The two cases are therefore considered separately.

5.3.4. Coarse solids in horizontal flow

Only with fine solids are the particles uniformly distributed over the cross-section of a horizontal pipe. For coarse particles, the following principal types of flow are observed as the velocity is decreased:

(a) Heterogeneous suspension with all the particles suspended but with a significant concentration gradient vertically.
(b) Heterogeneous suspension in the upper part of the bed but a sliding bed moving along the bottom of the pipe.
(c) A similar pattern to (b), but with the bed composed of moving layers at the top and a stationary deposit at the bottom.
(d) Transport as a bed with the lower layers stationary and a few particles moving over the surface of the bed in intermittent suspension.

In addition, it is possible to obtain what is known as *dense phase flow*[(29)] with the particles filling the whole bore of the pipe and sliding with little relative movement between the particles.

In all cases where the two phases are moving with different velocities, it is important to differentiate between the concentration of particles in the pipe (their holdup ϵ_S) and the volume fraction of particles (C) in the discharge. The implications of this will now be considered, together with possible means of experimentally determining the holdup.

Hold-up and slip velocity

In any two-phase flow system in which the two phases are flowing at different velocities, the in-line concentration of a component will differ from that in the stream which leaves the end of the pipe. The in-line concentration of the component with the *lower* velocity, will be greater than its concentration in the exit stream because it will have a longer residence time. It is important to understand what is happening within the pipe because the relative velocity between the phases results in energy transfer from the faster to the slower moving component. Thus, in hydraulic transport the liquid will transfer energy to the solid particles at a rate which is a function of the *slip velocity*. The solid particles will, in turn, be losing energy as a result of impact with the walls and frictional effects. In the steady state, the rate of gain and of loss of energy by the particles will be equal.

It is not possible to calculate the in-line concentrations and slip velocity from purely *external* measurements on the pipe, i.e. a knowledge of the rates at which the two components are delivered from the end of the pipe provides no evidence for what is happening within the pipe. It is thus necessary to measure *one* or more of the following variables:

The absolute linear velocity of the particles u'_S
The absolute linear velocity of the liquid u'_L
The slip velocity $u_R = u'_L - u'_S$
The hold-up of the solids ϵ_S
The hold-up of the liquid $\epsilon_L = 1 - \epsilon_S$

In a recent study of the transport of solids by liquid in a 38 mm diameter pipe,[30] the following variables were measured:

u mixture velocity (by electromagnetic flowmeter)
u'_L linear velocity of liquid (by salt injection method)
ϵ_S hold-up of solid particles in the pipe (by γ-ray absorption method).

In the salt injection method[31] a pulse of salt solution is injected into the line and the time is measured for it to travel between two electrode pairs situated a known distance apart, downstream from the injection point.

The γ-ray absorption method of determining in-line concentration (hold-up) of particles depends on the different degree to which the solid and the liquid attenuate γ-rays; details of the method are given in the literature[13,14].

All the other important parameters of the systems can be determined from a series of material balances as follows:

The superficial velocity of the liquid: $$u_L = u'_L(1 - \epsilon_S) \tag{5.14}$$

The superficial velocity of the solids: $$u_S = u - u_L$$
$$= u - u'_L(1 - \epsilon_S) \tag{5.15}$$

where the superficial velocity of a component is defined as the velocity it would have at the same volumetric flowrate if it occupied the total cross-section of the pipe.

The absolute velocity of the solids: $u'_S = \frac{u_S}{\epsilon_S}$

$$= u'_L - \frac{1}{\epsilon_S}(u'_L - u) \tag{5.16}$$

The slip, or relative, velocity: $u_R = u'_L - u'_S$

$$= \frac{1}{\epsilon_S}(u'_L - u) \tag{5.17}$$

The fractional volumetric concentration of solids C in the mixture issuing from the end of the pipe can then be obtained simply as the ratio of the superficial velocity of the solids (u_S) to the mixture flowrate.

$$u = (u_S + u_L)$$

or:

$$C = \frac{u_S}{u} = \frac{u'_S \epsilon_S}{u} = 1 - \frac{u'_L}{u}(1 - \epsilon_S)$$

or:

$$\frac{u_R}{u} = \frac{\epsilon_S - C}{(1 - \epsilon_S)\epsilon_S} \tag{5.18}$$

The consistency of the data can be checked by comparing values calculated using equation 5.18 with measured values for samples collected at the outlet of the pipe.

When an industrial pipeline is to be designed, there will be no *a priori* way of knowing what the in-line concentration of solids or the slip velocity will be. In general, the rate at which solids are to be transported will be specified and it will be necessary to predict the pressure gradient as a function of the properties of the solid particles, the pipe dimensions and the flow velocity. The main considerations will be to select a pipeline diameter, such that the liquid velocity and concentrations of solids in the discharged mixture will give acceptable pressure drops and power requirements and will not lead to conditions where the pipeline is likely to block.

It is found that the major factor which determines the behaviour of the solid particles is their terminal falling velocity in the liquid. This property gives a convenient way of taking account of particle size, shape and density.

In the experimental study which has just been referred to, it was found that the slip velocity was of the same order as the terminal falling velocity of the particles. Although there is no theoretical basis, its assumption does provide a useful working guide and enables all the internal parameters, including holdup, to be calculated for a given mixture velocity u and delivered concentration C using equations 5.14–5.18. It is of interest to note that, in pneumatic conveying (discussed in Section 5.4), slip velocity is again found to approximate to terminal falling velocity.

Some of the experimental results of different workers and methods of correlating results are now described.

Predictive methods for pressure drop

A typical curve for the conveying of solids in water is shown in Figure 5.9 which refers to the transport of *ca.* 200 μm sand in a small pipeline of 25 mm diameter.[32] It shows

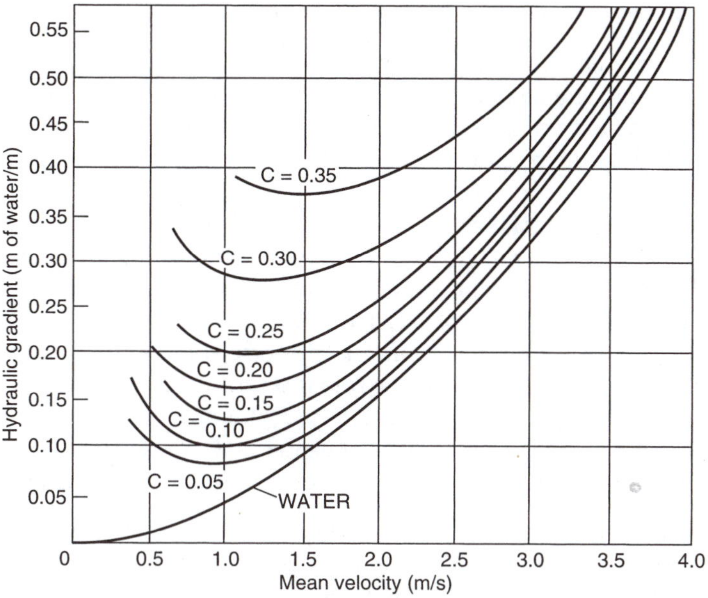

Figure 5.9. Hydraulic gradient-velocity curves for 200 μm sand in 25 mm diameter hydraulic conveying line

hydraulic gradient i (head lost per unit length of pipe) as a function of mean velocity, with delivered concentration (C) as parameter. Each curve shows a minimum, which corresponds approximately to the transition between flow with a bed and suspended flow. The economic operating condition is frequently close to the critical velocity. It will be noted that as concentration is increased, the minimum occurs at a progressively high velocity. The contribution of the solids to hydraulic gradient is seen to be greatest at low velocities. Great care should be exercised in operating at velocities below the critical values as the system is then unstable and blockage can easily occur; it is a region in which the pressure drop increases as the velocity is reduced. The prediction of conditions under which blockage is liable to occur is a complex area and the subject has been discussed in depth by HISAMITSU, ISE and TAKEISHI(33).

There were several studies of hydraulic transport in the 1950s, sparked off particularly by an interest in the economic possibilities of transportation of coal and other minerals over long distances. NEWITT *et al.*,(32) working with solids of a range of particle sizes (up to 5 μm) and densities (1180–4600 kg/m^3) in a 25 mm diameter pipe, suggested separate correlations for flow with a bed deposit and for conditions where the particles were predominantly in heterogeneous suspension.

For flow where a bed deposit tends to form:

$$\frac{i - i_w}{Ci_w} = 66\frac{gd}{u^2}(s-1) = 66\left[\frac{u^2}{gd(s-1)}\right]^{-1} \tag{5.19}$$

and for heterogeneous suspensions of particles of terminal falling velocity u_0:

$$\frac{i - i_w}{Ci_w} = 1100\frac{gd}{u^2}(s-1)\frac{u_0}{u} = 1100\frac{u_0}{u}\left[\frac{u^2}{gd(s-1)}\right]^{-1} \tag{5.20}$$

The following features of equations 5.19 and 5.20 should be noted.

(a) They both represent pressure gradient in the form of the quotient of a dimensionless excess hydraulic gradient $(i - i_w)/i_w$ and the delivered concentration (C). This implies that the excess pressure gradient is linearly related to concentration. More recent work casts doubt on the validity of this assumption, particularly for flow in suspension.[34]

(b) The excess pressure gradient is seen to be inversely proportional to a modified Froude number $u^2/[gd(s-1)]$ in which s is the ratio of the densities of the solids and the liquid. The pipe diameter d has been included to make the right hand side dimensionless but its effect was not studied.

(c) Particle characteristics do not feature in the correlation for flow with a bed. This is because the additional hydraulic gradient is calculated using a force balance in which the contribution of the particles to pressure drop is attributed to solid-solid friction at the walls of the pipe. This equation does not include the coefficient of friction, the importance of which will be referred to in a later section.

In a comprehensive study carried out at roughly the same time by DURAND[35,36,37] the effect of pipe diameter was examined using pipes of large diameter (40–560 mm) and a range of particle sizes d_p. The experimental data were correlated by:

$$\frac{i - i_w}{Ci_w} = 121\left\{\frac{gd}{u^2}(s-1)\frac{u_0}{[gd_p(s-1)]^{1/2}}\right\}^{1.5} \tag{5.21}$$

Reference to Volume 2 (Chapter 3) shows that at low particle Reynolds numbers (Stokes' Law region), $u_0 \propto d_p^2$ and that at high Reynolds number $u_0 \propto d_p^{1/2}$ — at intermediate Reynolds numbers the relation between u_0 and d_p is complex, but over a limited range $u_0 \propto d_p^m$ where $\frac{1}{2} < m < 2$. It will be seen, therefore, that the influence of particle diameter is greatest for small particles (low Reynolds numbers) and becomes progressively less as particle size increases, becoming independent of size at high Reynolds numbers — as in equation 5.19 which refers to flow with a moving bed. DURAND and CONDOLIOS[38] found in their experiments using large diameter pipes that particle size when in excess of 20 mm did not affect the pressure difference. JAMES and BROAD[39] also used pipelines ranging from 102 to 207 mm diameter for the transportation of coarse particles under conditions which tended to give rise to the formation of a bed deposit. Their results suggest that the coefficient in equation 5.19 is dependent on pipe diameter and that the constant value of 66 should be replaced by $(60 + 0.24d)$ where d is in mm. Most of their results related to conditions of heterogeneous suspension.

Equations 5.21 and 5.20 give results which are reasonably consistent, and they both give $(i - i_w)/Ci_w$ proportional to u^{-3}.

The term $u_0/[gd_p(s-1)]^{1/2}$ is shown in Volume 2 (Chapter 3) to be proportional to the reciprocal square root of the *drag coefficient* (C_D) for a particle settling at its terminal falling velocity.

Substituting $\frac{4}{3}(s-1)(gd_p/u_0^2) = C_D$ into equation 5.21 gives:

$$\frac{i - i_w}{Ci_w} = 150\left\{\frac{gd}{u^2}(s-1)\frac{1}{\sqrt{C_D}}\right\}^{1.5} \tag{5.22}$$

If the concentration term C is transferred to the right-hand side of equation 5.22 (taking account of the fact that $(i - i_w)$ is not necessarily linearly related to C), it may be written as:

$$\frac{i - i_w}{i_w} = \mathrm{f}\left[\frac{u^2\sqrt{C_D}}{gd(s-1)C}\right] \tag{5.23}$$

In Volume 2, the drag coefficient $C'_D (= C_D/2)$ is used in the calculation of the behaviour of single particles. However, C_D is used in this Chapter to facilitate comparison with the results of other workers in the field of Hydraulic Transport.

In Figure 5.10,[40] results of a number of workers[33,34,39,40,41,42,43,44] covering a wide range of experimental conditions, are plotted as:

$$\frac{i - i_w}{i_w} vs \frac{u^2\sqrt{C_D}}{gd(s-1)C}$$

The scatter of the results is considerable but this arrangement of the groups seems to be the most satisfactory. The best line through all the points is given by:

$$\frac{i - i_w}{i_w} = 30\left[\frac{u^2\sqrt{C_D}}{gd(s-1)C}\right]^{-1} \tag{5.24}$$

ZANDI and GOVATOS[45] suggest that the transition from flow with bed formation to flow as a heterogeneous suspension occurs at the condition where:

$$x = \frac{u^2\sqrt{C_D}}{Cgd(s-1)} = 40 \tag{5.25}$$

For $x < 40$, corresponding to flow with a bed, the best value for the slope of the line in Figure 5.10 is about -1. For higher values of x, corresponding to heterogeneous suspension, there is some evidence for a slope nearer -1.5, in line with equations 5.21 and 5.22 of Durand. However, in this region values of $(i - i_w)/i_w$ are so low that the experimental errors will be very great, for example at $x = 300$, $(i - i_w)/i_w \approx 0.1$ and friction losses differ from those for water alone by only about 10 per cent. What the figure does show is that the excess pressure gradient due to the solids becomes very large only under conditions where bed formation tends to occur ($x < 40$). As this is also the region of greatest practical interest, further consideration will be confined to this region.

It will be noted that in this region ($x < 40$) the experimental data show an approximately fourfold spread of ordinate at any given value of x.

Even taking into account the different experimental conditions of the various workers and errors in their measurements, and the generally unstable nature of solid-liquid flow, equation 5.24 is completely inadequate as a design equation.

An alternative approach to the representation of results for solid-liquid flow is to use the *two-layer model* which will be described in the following section. It will be seen that the coefficient of friction between the particles and the wall of the pipe is an important parameter in the model. It is suggested that its complete absence in equation 5.24 may be an important reason for the extent of the scatter. Unfortunately, it is a quantity which has been measured in only a very few investigations. It is interesting to note that the form of equation 5.19 was obtained by NEWITT *et al.*[32] using a force balance similar to that

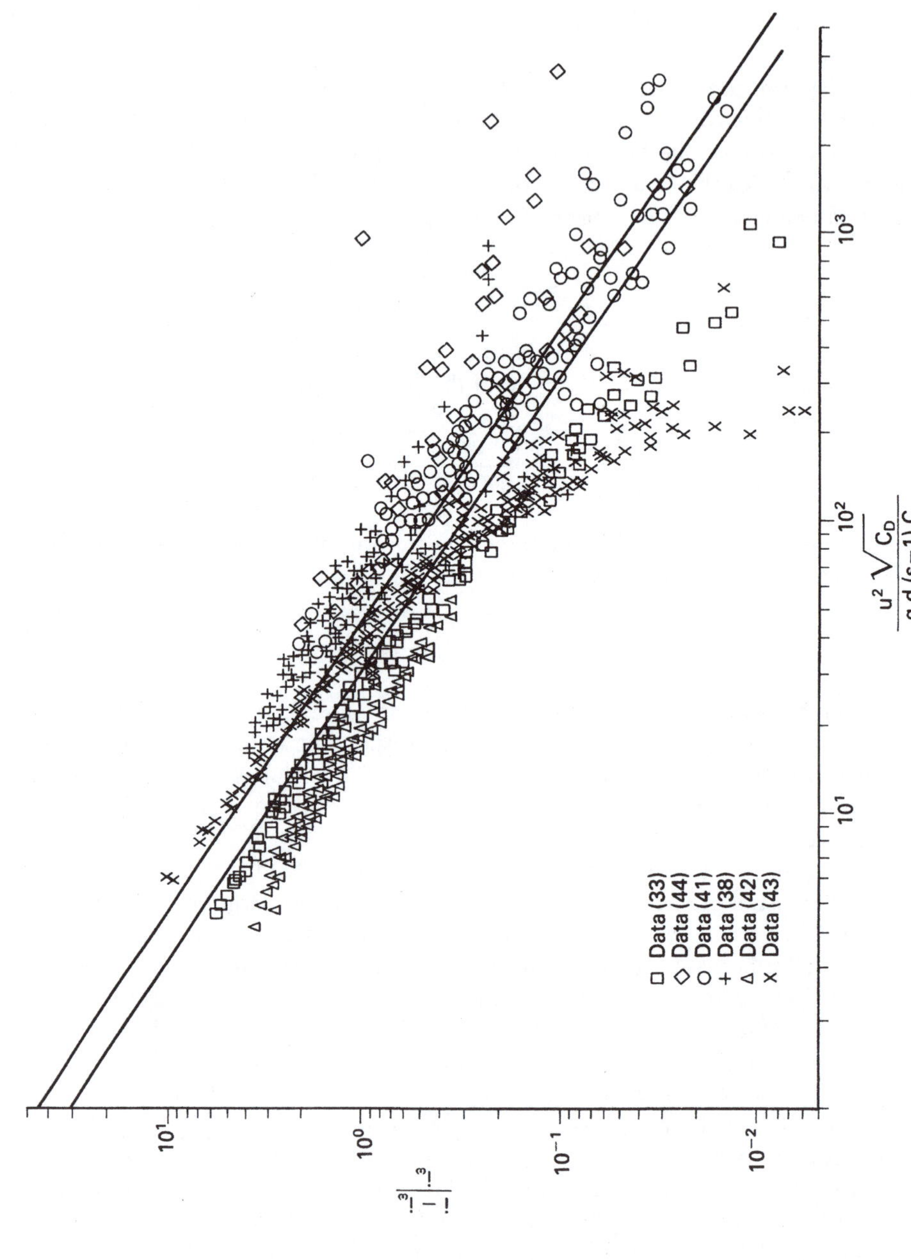

Figure 5.10. Excess hydraulic gradient due to solids as function of modified Froude number. Comparison of results of different workers

employed in the two-layer model and it is implicit in their analysis that the coefficient (66 in equation 5.19) will be a function of the coefficient of friction between the particles at the wall of the pipe.

ROCO and SHOOK[46,47] do not accept that the suspended solids and the bed constitute two identifiable regions in the pipe and have developed a model for the prediction of concentrations and velocities as continuous functions of position in the cross-section. The prediction of pressure drop from their theory involves complex numerical calculations and there is some difficulty in assessing the range of conditions over which it can be used because of the paucity of experimental data available. Some of the earlier experiments were carried out by SHOOK and DANIEL[48] who used a γ-ray absorption system for measuring in line concentration profiles. An extensive programme of experimental work is continuing at the Saskatchewan Research Council with a view to providing much needed data on large diameter pipelines (250 mm).

The two-layer model

It is seen that the models put forward do not adequately explain the behaviour of hydraulic conveying systems and they poorly correlate the results of different workers.

A two-layer model was first proposed by WILSON[49] in 1976 to describe the flow of solid-liquid mixtures in pipes under conditions where part of the solids are present in a moving bed and part are in suspended flow in the upper part of the pipe. He carried out a steady-state force balance on the two layers in order to calculate the pressure drop in the pipeline. The original model has been modified considerably by Wilson himself and by others[50,51]. However, it has not been used widely for design purposes, partly because of its complexity and the difficulty of obtaining convergent solutions to the iterative calculations which are necessary. Furthermore, it is necessary to be able to predict both the proportion of the solids which are present in each layer and the interfacial shear at the upper surface of the bed. The former is dependent on the experimental measurement of solids distribution and the generation of a reliable method of predicting it in terms of system properties and flowrates. The latter, the interfacial shear stress, cannot readily be measured and is calculated on the basis that the upper liquid layer is moving over the surface of a bed whose effective roughness is a function of the size of the particles.

The force balance is as follows (Figure 5.11):

For the bed: $$F_B = -\Delta P A_B = (\mu_F \Sigma F_N + R_B S_B - R_i S_i)l \tag{5.26}$$

For the upper layer: $$F_L = -\Delta P A_L = (R_i S_i + R_L S_L)l \tag{5.27}$$

where $-\Delta P$ is the pressure drop over the pipe of length l,
A_B is the cross-section occupied by bed,
A_L is the cross-section occupied by the upper layer,
R_B is the shear stress at the wall in the liquid in the bed,
R_i is the interfacial shear stress,
R_L is the shear stress at the wall in the liquid above the bed,
S_B is the perimeter of the cross-section in contact with the bed,

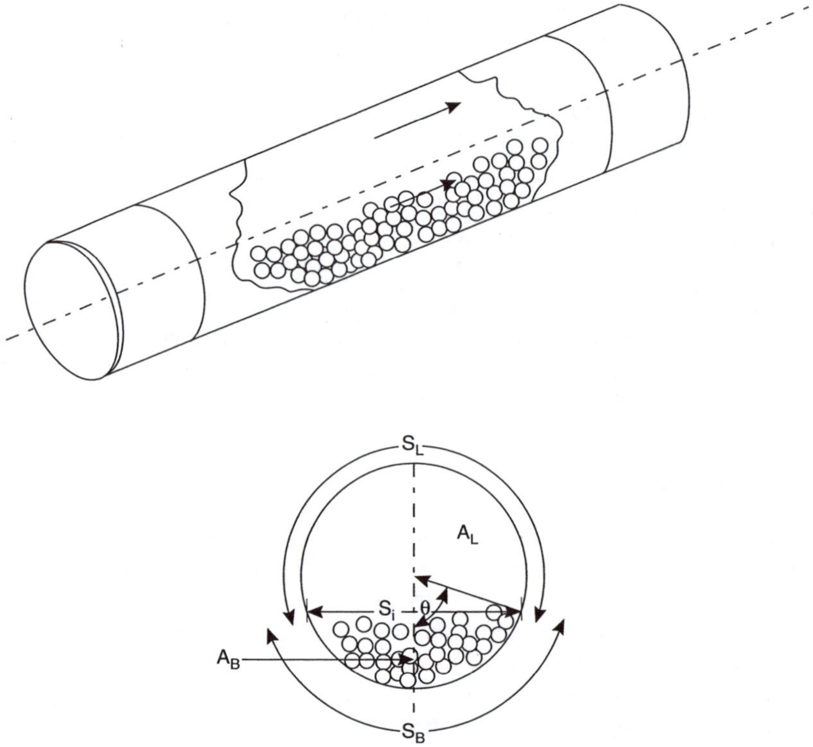

Figure 5.11. Solids distribution over pipe section

S_i is the length of the chord between the two layers, and
S_L is the perimeter of the cross-section in contact with the upper layer.

The term $\mu_F \Sigma F_N$ is the product of the coefficient of friction μ_F and the sum ΣF_N of the normal forces per unit length between the particles and the wall.

Adding equations 5.26 and 5.27 and dividing by A, the total cross-section area of the pipe, then:

$$\frac{-\Delta P}{l} = \frac{R_L S_L + \mu_F \Sigma F_N + R_B S_B}{A} \tag{5.28}$$

It may be noted that the term $R_i S_i$ cancels out, the interfacial drag forces on the two layers being equal and opposite.

The two perimeters S_L and S_B are a function of the depth of bed and the angle subtended by the bed surface to the centre of the pipe. In order to calculate their values, it is necessary to know the in-line concentration of solids in the pipe and the voidage of the bed.

The shear stress R_L at the pipe wall in the upper portion of the pipe may be calculated on the assumption that the liquid above the bed is flowing through a non-circular duct, bounded at the top by the wall of the pipe and at the bottom by the upper surface of the bed. The hydraulic mean diameter may then be used in the calculation of wall shear stress. However, this does not take account of the fact that the bottom boundary, the top surface of the bed, is not stationary, and will have a greater effective roughness than the pipe wall. The effects of these two assumptions do, however, operate in opposite directions and therefore partially cancel each other out. The shear stress R_B due to the flow of liquid in the bed, relative to the pipe wall, is calculated on the assumption that it is unaltered by the presence of the particles, since their size is likely to be considerably greater than the thickness of the laminar sub-layer for turbulent flow in a pipe.

Before the pressure drop can be calculated from equation 5.28, it is still necessary to know the in-line concentration of solids in the pipe for a given flowrate and discharge concentration. Furthermore, it is necessary to know how the liquid and solid are distributed between the two layers.

The *two-layer model* is being progressively updated as fresh experimental results and correlations become available. The most satisfactory starting-point for anyone wishing to use the model to calculate pressure gradients for flow of solids–liquid mixtures in a pipeline is the text of SHOOK and ROCO[52] which includes a worked example. However, there are many pitfalls to be avoided in this area, and there is no substitute for practical experience gained by working in the field.

In a recent study of the transport of coarse solids in a horizontal pipeline of 38 mm diameter,[30] measurements were made of pressure drop, as a function not only of mixture velocity (determined by an electromagnetic flowmeter) but also of in-line concentration of solids and liquid velocity. The solids concentration was determined using a *γ-ray absorption* technique, which depends on the difference in the attenuation of γ-rays by solid and liquid. The liquid velocity was determined by a *salt injection* method,[31] in which a pulse of salt solution was injected into the flowing mixture, and the time taken for the pulse to travel between two electrode pairs a fixed distance apart was measured. It was then possible, using equation 5.17, to calculate the relative velocity of the liquid to the solids. This relative velocity was found to increase with particle size and to be of the same order as the terminal falling velocity of the particles in the liquid.

The values of R_L, S_L, R_B and S_B were then calculated on the following assumptions:

(1) That the relative velocity u_R was equal to the terminal falling velocity of the particles. This assumption does not necessarily apply to large pipes.
(2) That the particles all travelled in a bed of porosity equal to 0.5. It had previously been shown that, as soon as bed formation occurred, the major part of the contribution of the solids was attributable to those present within the bed.
(3) That the particles and liquid in *the bed* were travelling at the same velocity, i.e. there was no slip within the bed.

In addition, off-line experiments were carried out to measure the coefficient of solid friction μ_F between the particles and the surface of the pipe. The term $\mu_F \Sigma F_N$ was then calculated on the assumption that ΣF_N was equal to the net weight (actual minus buoyancy) of the particles in the pipeline. This assumption was reasonable provided that the bed did not occupy more than about 30 per cent of the pipe cross-section.

A sensitivity analysis showed that the calculated value of pressure gradient was very sensitive to the value of the coefficient of friction μ_F, but was relatively insensitive to the slip velocity u_R and to the bed voidage.

Some examples of calculated pressure gradients are shown in Figure 5.12 in which solids concentration is a parameter. Experimental points are given, with separate designations for each concentration band. The difference between experimental and predicted value does not generally exceed about 15 per cent.

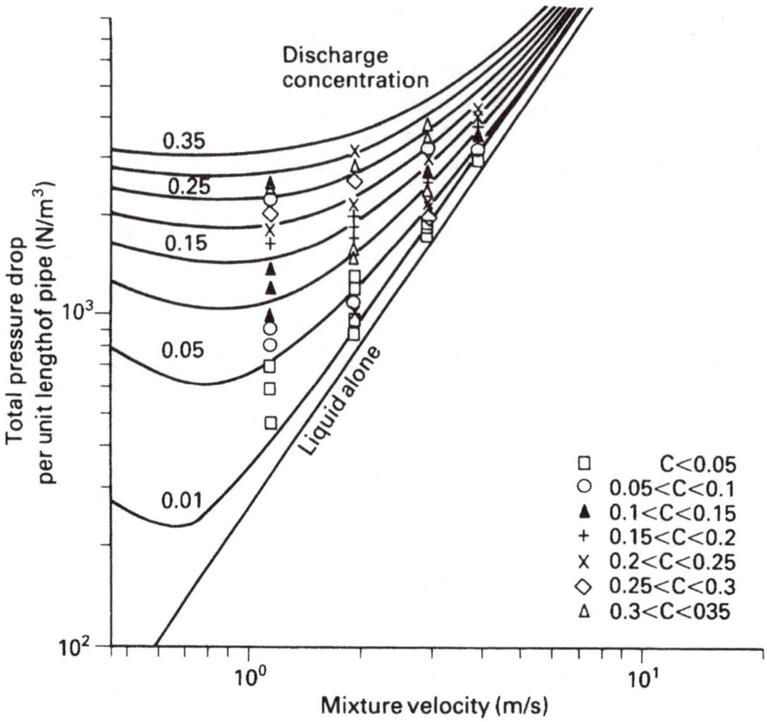

Figure 5.12. Comparison of predicted and experimental results for simplified "two-layer" model (3.5 mm gravel in 38 mm pipe[30])

The sensitivity of the pressure drop to the coefficient of solids-surface friction μ_F may well account for the wide scatter in the results shown earlier in Figure 5.10. Unfortunately this quantity has been measured by only very few investigators. It must be emphasised that in the design of any hydraulic transport system it is extremely important to have a knowledge of the coefficient of friction.

Transport of coarse particles in heavy and shear-thinning media

There have been several studies involving the use of media consisting of fine dense particles suspended in water for transporting coarse particles. The fine suspension behaves as a homogeneous fluid of increased density, but its viscosity is not sufficiently altered to have a significant effect on the pressure drop during turbulent flow, the normal condition for hydraulic transport. The cost of the dense particles may, however, be appreciable and their complete separation from the coarse particles may be difficult.

More recently attention has been focused on the use of suspensions of fines at very high concentrations; the desired increase in density can then be achieved using lighter particles. Furthermore, if the particles are flocculated, the suspension will generally exhibit shear-thinning rheology conforming approximately to the power-law or the Bingham-plastic model (see Chapter 3). Contamination of the product can frequently be eliminated by using fines of the same material as the coarse particles. For example, coarse particles of coal can be transported in suspensions of fine coal with beneficial results.

Concentrated (20–30 per cent by volume) flocculated suspensions of coal and of china clay are highly shear-thinning (power-law index $0.1 < n < 0.2$) and, at the shear rates encountered in pipe line flow, have high apparent viscosities, as a result of which the flow is frequently laminar. In these circumstances, the suspension has a very high apparent viscosity in the low shear region in the core of the flow, and therefore suspension of coarse particles is facilitated. At the same time, the apparent viscosity in the high-shear region near the pipe wall is much lower and pressure drops are not excessive. Thus, in effect, a shear-thinning suspension is a tailor-made fluid for hydraulic transport giving a highly desirable variation of apparent viscosity over the cross-section of the pipe.

Solutions of some polymeric materials also have similar rheological properties and behave in the same way as suspensions in pipe flow. However, they do in time break down and they do not give any appreciable increase in buoyancy as their densities differ little from that of water.

One of the earlier studies was carried out in 1971 by CHARLES and CHARLES[53] who investigated the feasibility of transporting coarse materials in heavy media (sand in flocculated suspensions of clay in water). They concluded that the power requirement for the transport of one million tonnes/year of solids in a 200 mm diameter pipe could be reduced by a factor of about 6 by using heavy media in place of water. KENCHINGTON[54], who also studied the transport of coarse particles in china clay suspensions, concluded that a significant proportion of the coarse particles may be suspended in the dense medium without a noticeable increase in pressure drop. DUCKWORTH *et al.*[55–57] showed that it is possible to transport coal with a maximum particle size of 19 mm in a 250 mm diameter pipeline under laminar conditions using a fine coal slurry as the carrier fluid. Work has also been carried out on characterising suspensions of coarse particles in fines by means of a form of rotary viscometer, a "shearometer", and designing pipelines on the basis of the resulting measurements[58].

CHHABRA and RICHARDSON[59] transported coarse particles in a 42 mm pipe in a wide range of fluids, including Newtonian liquids of high viscosity, shear-thinning polymer solutions and shear-thinning suspensions of fine flocculated particles. They correlated their own and Kenchington's experimental results by means of linear regression to give:

$$\frac{i - i_w}{i_w}\frac{\phi}{C} = 0.20\left[\frac{u^2}{gd(s-1)}\right]^{-1} \tag{5.29}$$

A somewhat improved correlation is obtained by non-linear regression as:

$$\frac{i - i_w}{i_w}\frac{\phi}{C} = 0.28\left[\frac{u^2}{gd(s-1)}\right]^{-1.25} \tag{5.30}$$

ϕ is the friction factor, $R/\rho u^2$, for the flow of the liquid alone at the same superficial velocity u. Because most of the work was carried out in the laminar-flow region (high apparent viscosity of the medium), ϕ is very sensitive to the flowrate. This contrasts with the work described in the previous section using water. The flow was then turbulent and the friction factor did not vary significantly over the range of velocities studied; in effect, the value of ϕ was implicitly incorporated in the coefficient in the equations. It is of interest to note that equation 5.29 is consistent with equation 5.19 for heterogeneous suspension in water for a value of ϕ equal to 0.003, corresponding to a Reynolds number of about 22,000 for flow in a smooth pipe.

Transport of particles of low density

Hydraulic transport has been extensively used in the minerals industry over a long period of time. A characteristic of mineral particles is that they have higher densities than the transporting fluid and always travel at a lower velocity than the liquid. Recently, developments in the continuous processing of foodstuffs[60,61] have led to the use of shear-thinning non-Newtonian fluids (usually starch-based) to transport food particles through heat exchangers for the purpose of sterilisation. Laminar flow conditions usually prevail, but the particles are readily maintained in suspension because they are almost neutrally-buoyant in the carrier fluid which usually has a very high apparent viscosity, and in many cases a yield stress. As the particles tend to migrate away from the walls of the heat exchanger to the core region where the velocity of the fluid is greatest, their velocities are frequently greater than the mean velocity of the fluid, and the slip velocity is then negative. The residence time of the particles will then be lower than that calculated from the mean velocity of flow, and it is important that this difference be taken into account in order to avoid under-sterilisation of the foodstuff.

5.3.5. Coarse solids in vertical flow

DURAND[62] has also studied vertical transport of sand and gravel of particle size ranging between 0.18 mm and 4.57 mm in a 150 mm diameter pipe, and WORSTER and DENNY[63] conveyed coal and gravel in vertical pipes of diameters 75, 100, and 150 mm. They concluded that the pressure drop for the slurry was the same as for the water alone, if due allowance was made for the static head attributable to the solids in the pipe.

NEWITT *et al.*[64] conveyed particles of densities ranging from 1190 to 4560 kg/m^3, of sizes 0.10 to 3.8 mm, in a 25 mm diameter pipe 12.8 m tall and in a 50 mm pipe 6.7 m tall. The particles used had a thirtyfold variation in terminal falling velocity. It was found that the larger particles had little effect on the frictional losses, provided the static head due to the solids was calculated on the assumption that the particles had a velocity relative to the liquid equal to their terminal falling velocities. Furthermore, it was shown photographically that at high velocities these particles travel in a central core, and thus the frictional forces at the wall will be unaffected by their presence. Very fine particles of sand give suspensions which behave as homogeneous fluids, and the hydraulic gradient due to friction is the same as for horizontal flow. When the settling velocity could not be neglected in comparison with the liquid velocity, the hydraulic gradient was found to be given by:

$$\frac{i - i_w}{Ci_w} = 0.0037 \left(\frac{gd}{u^2}\right)^{1/2} \frac{d}{d_p} u_0^2 \tag{5.31}$$

For the transport of coarse particles, the relative velocity between the liquid and solids is an important factor determining the hold-up, and hence the in-line concentration of solids. CLOETE *et al.*[65] who conveyed sand and glass ballotini particles through vertical pipes of 12.5 and 19 mm diameter at velocities up to 3 m/s, obtained values of the in-line concentration by means of a γ-ray technique similar to that discussed earlier. The vertical pressure gradient in the pipe was measured and the component due to the hydrostatic pressure of the mixture was subtracted in order to obtain the frictional component. It was found that the frictional component was similar to that for water alone for velocities up to 0.7 m/s, but tended to increase at higher velocities.

In the absence of a direct measurement of the in-line concentration of solids ϵ_S, it is necessary to make an estimate of its value in order that the hydrostatic pressure gradient in the pipe may be calculated. This can be done for a given mixture velocity u and delivered concentration C, provided that the velocity of the particles relative to the liquid u_R is known.

Following the same argument as used for horizontal flow (equations 5.14–5.18):

For unit area of cross-section:

Mixture flowrate = Solids flowrate + Liquid flowrate

$$u = u'_S\epsilon_S + u'_L(1 - \epsilon_S) \tag{5.32}$$

where u'_S and u'_L are the linear velocities of solid and liquid, respectively

and
$$u_R = u'_L - u'_S. \tag{5.33}$$

Considering the flow of solids:

$$uC = u'_S\epsilon_S \tag{5.34}$$

From equations 5.32, 5.33 and 5.34, the relation between ϵ_S and C is given by:

$$C = \epsilon_S\left[1 - \frac{u_R}{u}(1 - \epsilon_S)\right] \tag{5.35}$$

For the transport of a dilute suspension of solids, u_R will approximate to the free-falling velocity u_0 of the particles in the liquid. For concentrated suspensions, a correction must be applied to take account of the effect of neighbouring particles. This subject is considered in detail in Volume 2 (Chapter 5) from which it will be seen that the simplest form of correction takes the form:

$$\frac{u_R}{u_0} = (1 - \epsilon_S)^{m-1} \tag{5.36}$$

where m ranges from about 4.8 for fine particles to 2.4 for coarse particles. Equations 5.35 and 5.36 are then solved simultaneously to give the value of ϵ_S, from which the contribution of the solids to the hydrostatic pressure may be calculated.

KOPKO *et al.*(66) made measurements of the pressure drop for the flow of suspensions of iron shot (0.0734-0.131 mm diameter) in a vertical pipe 61 mm diameter and 4.3 m tall.

The frictional pressure drop was obtained by subtracting the hydrostatic component from the total measured pressure drop and it was found that it constituted only a very small proportion of the total measured. More recently, AL-SALIHI(67) has confirmed that, if allowance is made for the hydrostatic component using the procedure outlined above, the frictional pressure drop is largely unaltered by the presence of the solids.

In summary, therefore, it is recommended that the frictional pressure drop for vertical flow be calculated as follows:

For *"non-settling" suspensions*: The standard equation for a single phase fluid is used with the physical properties of the suspension in place of those of the liquid.

For *suspensions of coarse particles*: The value calculated for the carrier fluid flowing alone at the mixture velocity is used.

Example 5.2

Sand with a mean particle diameter of 0.2 mm is to be conveyed in water flowing at 0.5 kg/s in a 25 mm internal diameter horizontal pipe 100 m long. Assuming fully suspended flow, what is the maximum amount of sand which may be transported in this way if the head developed by the pump is limited to 300 kN/m^2?

The terminal falling velocity of the sand particles in water may be taken as 0.0239 m/s. This value may be confirmed using the method given in Volume 2.

Solution

Assuming the mean velocity of the suspension is equal to the water velocity, that is, neglecting slip, then:

$$u_m = 0.5/(1000\pi 0.025^2/4) = 1.02 \text{ m/s}$$

For water alone, flowing at 1.02 m/s:

$$Re = (0.025 \times 1.02 \times 1000)/0.001 = 25{,}500$$

Assuming $e/d = 0.008$, then, from Figure 3.7:

$$\phi = 0.0046 \quad f = 0.0092$$

From, equation 3.20, the head loss is:

$$h_f = (8 \times 0.0046)(100/0.025)(1.02^2/(2 \times 981)) = 7.8 \text{ m water}$$

and the hydraulic gradient is:

$$i_w = (7.8/100) = 0.078 \text{ m water/m.}$$

A pressure drop of 300 kN/m^2 is equivalent to:

$$(300 \times 1000)/(1000 \times 9.81) = 30.6 \text{ m water}$$

and hence:

$$i = (30.6/100) = 0.306 \text{ m water/m.}$$

Substituting in equation 5.20:

$$(0.306 - 0.078)/0.078C = 1100(8.81 \times 0.025/1.08^2)(0.0239/1.02)(2.6 - 1)$$

from which:

$$C = 0.30$$

[Equation 5.21 may also be used, in which case:

$$(0.306 - 0.078)/0.078C = 121\{[8.81 \times 0.025(2.6 - 1)0.0239]/[1.02^2(9.81 \times 0.0002(2.6 - 1)^{0.5}]\}^{1.5}$$

from which:

$$C = 0.36$$

which is a very similar result.]

If G kg/s is the mass flow of sand, then:

$$\text{volumetric flow of sand} = (G/2600) = 0.000385G \text{ m}^3/\text{s}$$

$$\text{volumetric flow of water} = (0.5/1000) = 0.0005 \text{ m}^3/\text{s}$$

and : $$0.000385G/(0.000385G + 0.0005) = 0.30$$

from which: $$\underline{\underline{G = 0.56 \text{ kg/s}}} = \underline{\underline{2 \text{ tonne/h}}}$$

5.4. FLOW OF GAS–SOLIDS MIXTURES

5.4.1. General considerations

Pneumatic conveying involves the transport of particulate materials by air or other gases. It is generally suitable for the transport of particles in the size range 20 μm to 50 mm. Finer particles cause problems arising from their tendency to adhere together and to the walls of the pipe and ancillary equipment. Sticky and moist powders are the worst of all. Large particles may need excessively high velocities in order to maintain them in suspension or to lift them from the bottom of the pipe in horizontal systems. Pneumatic conveying lines may be horizontal, vertical or inclined and may incorporate bends, valves and other fittings, all of which may exert a considerable influence on the behaviour of the whole installation. Whereas it is possible to make reasonable predictions of pressure gradients in long straight pipes, the effects of such fittings can be ascertained reliably only from the results of practical studies under conditions close to those that will be experienced in the final installation. The successful operation of a pneumatic conveyor may well depend much more on the need to achieve reliable operation, by removing the risks of blockage and of damage by erosion, than on achieving conditions which optimise the performance of the straight sections of the pipeline. It is important to keep changes in direction of flow as gradual as possible, to use suitable materials of construction (polyurethane lining is frequently employed) and to use velocities of flow sufficiently high to keep the particles moving, but not so high as to cause serious erosion. Whenever possible, it is desirable to carry out pilot scale tests with a sample of the solids.

Two characteristics of the conveying fluid result in considerable differences in the behaviour of pneumatic and hydraulic conveying systems: fluid density and compressibility. In hydraulic conveying the densities of the solids and fluid are of the same order of magnitude, with the solids usually having a somewhat higher density than the liquid. Practical flow velocities are commonly in the range of 1 to 5 m/s. In pneumatic transport, the solids may have a density two to three orders of magnitude greater than the gas and velocities will be considerably greater—up to 20–30 m/s. In a mixture with a volume fraction of solids, of say 0.05, the mass ratio of solids to fluid will be about 0.15 to 0.20 for hydraulic conveying compared with about 50 for pneumatic conveying. Compressibility of the gas is important because it expands as it flows down the pipe and if, for example, the ratio of the pressures at the upstream and downstream ends is 4, the velocity will increase by this factor and it would be necessary to double the diameter of the pipe to maintain the same linear gas velocity as at the beginning of the pipe. In practice, "stepped" pipelines are commonly used with progressively larger pipes used towards the downstream end. Horizontal pipelines up to 500 m long are in common use and a few 2000 m lines now exist; vertical lifts usually do not exceed about 50 m.

There are three basic modes of transport which are employed. The first, and most common, is termed *dilute phase* or *lean phase* transport in which the volume fraction of solids in this suspension does not exceed about 0.05 and a high proportion of the particles spend most of their time in suspension. The second is transport which takes place largely in the form of a *moving bed* in which the solids volume fraction may be as high as 0.6; this is relevant only for horizontal or slightly inclined pipelines. The third form is *dense phase* transport in which fairly close packed "slugs" of particles, with volume fractions of up to 0.5, alternate with slugs of gas and are propelled along the pipe. Because the velocities ($\sim$3 m/s) are very much lower than in dilute phase transport, there is less attrition of particles and less pipe wear. The gas usage is reduced but the risk of blockage is serious in horizontal pipelines.

Considerably more work has been carried out on horizontal as opposed to vertical pneumatic conveying. A useful review of relevant work and of correlations for the calculation of pressure drops has been given by KLINZING *et al.*[(68)]. Some consideration will now be given to horizontal conveying, with particular reference to dilute phase flow, and this is followed by a brief analysis of vertical flow.

5.4.2. Horizontal transport

Flow patterns

In a horizontal pipeline the distribution of the solids over the cross-section becomes progressively less uniform as the velocity is reduced. The following flow patterns which are commonly encountered in sequence at decreasing gas velocities have been observed in pipelines of small diameter.

1. *Uniform suspended flow*

   The particles are evenly distributed over the cross-section over the whole length of pipe.

2. *Non-uniform suspended flow*

   The flow is similar to that described above but there is a tendency for particles to flow preferentially in the lower portion of the pipe. If there is an appreciable size distribution, the larger particles are found predominantly at the bottom.

3. *Slug flow*

   As the particles enter the conveying line, they tend to settle out before they are fully accelerated. They form dunes which are then swept bodily downstream giving an uneven longitudinal distribution of particles along the pipeline.

4. *Dune flow*

   The particles settle out as in slug flow but the dunes remain stationary with particles being conveyed above the dunes and also being swept from one dune to the next.

5. *Moving bed*

   Particles settle out near the feed point and form a continuous bed on the bottom of the pipe. The bed develops gradually throughout the length of the pipe and moves slowly forward. There is a velocity gradient in the vertical direction in the bed, and conveying continues in suspended form above the bed.

6. *Stationary bed*

The behaviour is similar to that of a moving bed, except that there is virtually no movement of the bed particles. The bed can build up until it occupies about three-quarters of the cross-section. Further reduction in velocity quickly gives rise to a complete blockage.

7. *Plug flow*

Following slug flow, the particles, instead of forming stationary dunes, gradually build up over the cross-section until they eventually cause a blockage. This type of flow is less common than dune flow.

Suspension mechanisms

The mechanism of suspension is related to the type of flow pattern obtained. Suspended types of flow are usually attributable to dispersion of the particles by the action of the turbulent eddies in the fluid. In turbulent flow, the vertical component of the eddy velocity will lie between one-seventh and one-fifth of the forward velocity of the fluid and, if this is more than the terminal falling velocity of the particles, they will tend to be supported in the fluid. In practice it is found that this mechanism is not as effective as might be thought because there is a tendency for the particles to damp out the eddy currents.

If the particles tend to form a bed, they will be affected by the lateral dispersive forces described by BAGNOLD[69,70]. A fluid in passing through a loose bed of particles exerts a dilating action on the system. This gives rise to a dispersion of the particles in a direction at right angles to the flow of fluid.

If a particle presents a face inclined at an angle to the direction of motion of the fluid, it may be subjected to an upward lift due to the *aerofoil* effect. In Figure 5.13 a flat plate is shown at an angle to a stream of fluid flowing horizontally. The fluid pressure acts normally at the surface and thus produces forces with vertical and horizontal components as shown.

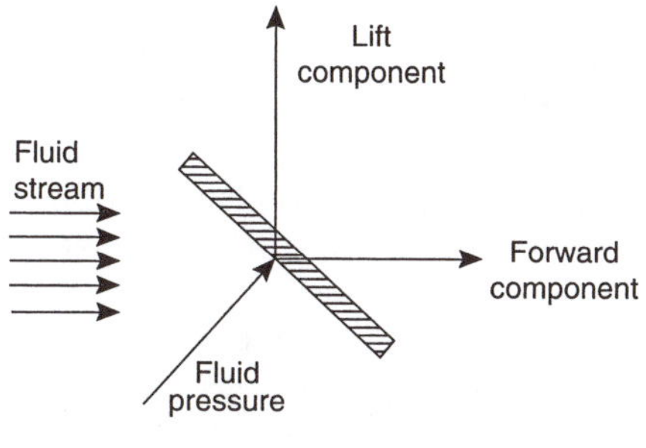

Figure 5.13. Aerofoil effect

If the particle rotates in the fluid, it will be subjected to a lift on the *Magnus Principle*. Figure 5.14 shows a section through a cylinder rotating in a fluid stream. At its upper edge the cylinder and the fluid are both moving in the same direction, but at its lower edge they are moving in opposite directions. The fluid above the cylinder is therefore accelerated, and that below the cylinder is retarded. Thus the pressure is greater below the cylinder and an upward force is exerted.

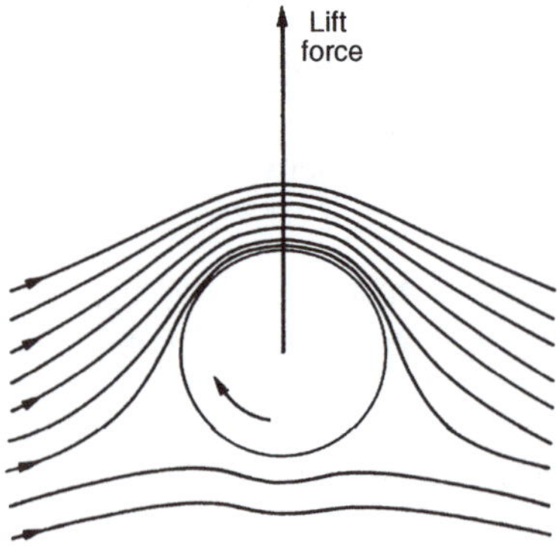

Figure 5.14. Magnus effect

The above processes, other than the action of the dispersive forces, can result in an upward or downward displacement of an individual particle, there being approximately equal chances of the net force acting upwards or downwards. However, because the gravitational force gives rise to a tendency for the concentration of particles to be greater at the bottom of the pipe, the overall effect is to lift the particles.

The principal characteristic of a particle which determines the dominant suspension mechanism is its terminal falling velocity. Particles with low falling velocities will be readily suspended by the action of the eddies, whereas the dispersive forces will be most important with particles of high falling velocities. In a particular case, of course, the fluid velocity will also be an important factor, full suspension of a given particle occurring more readily at high velocities.

Energy requirements for dilute phase conveying

The energy required for conveying can conveniently be considered in two parts: that required for the flow of the air alone, and the additional energy necessitated by the presence of the particles. It should be noted, however, that the fluid friction will itself be somewhat modified for the following reasons: the total cross-sectional area will not be available for the flow of fluid; the pattern of turbulence will be affected by the solids; and the pressure distribution through the pipeline will be different, and hence the gas density at a given point will be affected by the solids.

The presence of the solids is responsible for an increased pressure gradient for a number of reasons. If the particles are introduced from a hopper, they will have a lower forward velocity than the fluid and therefore have to be accelerated. Because the relative velocity is greatest near the feed point and progressively falls as the particles are accelerated, their velocity will initially increase rapidly and, as the particles approach their limiting velocities, the acceleration will become very small. The pressure drop due to acceleration is therefore greatest near the feed point. Similarly, when solids are transported round a bend, they are retarded and the pressure gradient in the line following the bend is increased as a result of the need to accelerate the particles again. In pneumatic conveying, the air is expanding continuously along the line and therefore the solid velocity is also increasing. Secondly, work must be done against the action of the earth's gravitational field because the particles must be lifted from the bottom of the pipe each time they drop. Finally, particles will collide with one another and with the walls of the pipe, and therefore their velocities will fall and they will need to be accelerated again. Collisions between particles will be less frequent and result in less energy loss than impacts with the wall, because the relative velocity is much lower in the former case.

The transference of energy from the gas to the particles arises from the existence of a relative velocity. The particles will always be travelling at a lower velocity than the gas. The loss of energy by a particle will generally occur on collision and thus be a discontinuous process. The acceleration of the particle will be a gradual process occurring after each collision, the rate of transfer of energy falling off as the particle approaches the gas velocity.

The accelerating force exerted by the fluid on the particle will be a function of the properties of the gas, the shape and size of the particle, and the relative velocity. It will also depend on the dispersion of the particles over the cross-section and the shielding of individual particles. The process is complex and therefore it is not possible to develop a precise analytical treatment, but it is obviously important to know the velocity of the particles.

Determination of solid velocities

The determination of the velocity of individual particles can be carried out in a number of ways. First, the particles in a given section of pipe can be isolated using two rapidly acting shutters or valves, and the quantity of particles trapped measured by removing the intervening section of pipe. This method was used by SEGLER(71), CLARK *et al.*(72), and MITLIN(73). It is, however, cumbersome, very time-consuming, and dependent upon extremely good synchronisation of the shutters. Another method is to take two photographs of the particles in the pipeline at short time-intervals and to measure the distance the particles have travelled in the time. This method is restricted to very low concentrations of relatively large particles, and does not permit the measurement of more than a few particles.(74)

A third method consists of measuring the time taken for a "tagged" particle (e.g. radioactive or magnetic) to travel between two points(75). The method gives results applicable only to an isolated particle which may not be representative of the bulk of the particles. These techniques can readily be used in experimental equipment but are not practicable for industrial plant.

Cross-correlation methods

Because of the general oscillations in the flow condition in pneumatic conveying, it is possible to use sensing probes to measure the high frequency fluctuations at two separate locations a specified distance apart and then to determine the mean time interval between a given pattern appearing at the two locations. KLINZING and MATHUR(76) used a dielectric measuring device for this purpose. Such methods are readily applied to both laboratory and large scale plant.

MCLEMAN(77) used a method which was really a pre-cursor of the cross-correlation technique. He injected a pulse of air into the conveying line as a result of which there was a very short period during which the walls at any particular point were not subject to bombardment by particles, and the noise level was substantially reduced. By placing two transducers in contact with the wall of the pipe at a known distance apart and connecting each to a thyratron, it was possible to arrange for the first to start a frequency counter and for the second to stop the counter. A very accurate method of timing the air pulse was thus provided. It was found that the pulse retained its identity over a long distance, and this suggested that the velocities of all the particles tended to be the same. The method enabled extremely rapid and accurate measurements of the solids velocity to be obtained. Over a 16 m distance the error was less than 1 per cent.

The importance of obtaining accurate measurements of solid velocity is associated with the fact that the drag exerted by the fluid on the particle is approximately proportional to the square of the relative velocity. As the solid velocity frequently approaches the air velocity, the necessity for very accurate values is apparent.

Pressure drops and solid velocities for dilute phase flow

When solid particles are introduced into an air stream a large amount of energy is required to accelerate the particles, and the acceleration period occupies a considerable length of pipe. In order to obtain values of pressure gradients and solid velocities under conditions approaching equilibrium, measurements must be made at a considerable distance from the feed point. Much of the earlier experimental work suffered from the facts that conveying lines were much too short and that the pressure gradients were appreciably influenced by the acceleration of the particles. A typical curve obtained by CLARK *et al.*[72] for the pressure distribution along a 25 mm diameter conveying line, is shown in Figure 5.15. It may be noted that the pressure gradient gradually diminishes from a very high value in the neighbourhood of the feed point to an approximately constant value at distances greater than about 2 m. It has been checked experimentally that the solid velocity is increasing

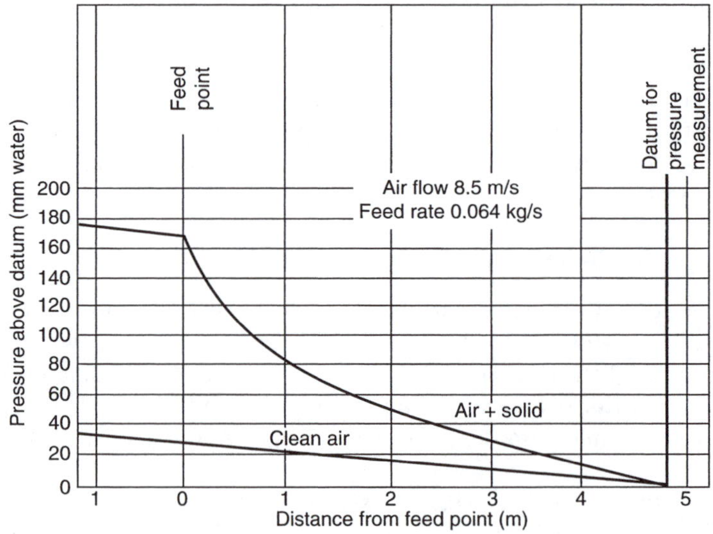

Figure 5.15. Pressure in horizontal 25 mm conveying line for transport of cress seed

in the region of decreasing pressure gradient and that the length of the acceleration period increases with the mass of the particles, as might be expected. When the pressure gradient is approximately constant, so is the solid velocity. A further factor which makes it necessary to obtain measurements over a long section of pipe is that the pressure gradient does in many cases exhibit a wave form.[77] This appears to be associated with the tendency for dune formation to occur within the pipe, and thus the measured value of the pressure gradient may be influenced by the exact location of the pressure tappings. It is therefore concluded that measurements should be made over a length of at least 15 m of pipe, and remote from the solids feed point.

If the pressure drop is plotted against air flow rate as in Figure 5.16, it is seen that at a given feed rate the curve always passes through a minimum. At air velocities above the minimum of the curve, the solids are in suspended flow, but at lower velocities particles are deposited on the bottom of the pipe and there is a serious risk of blockage occurring.

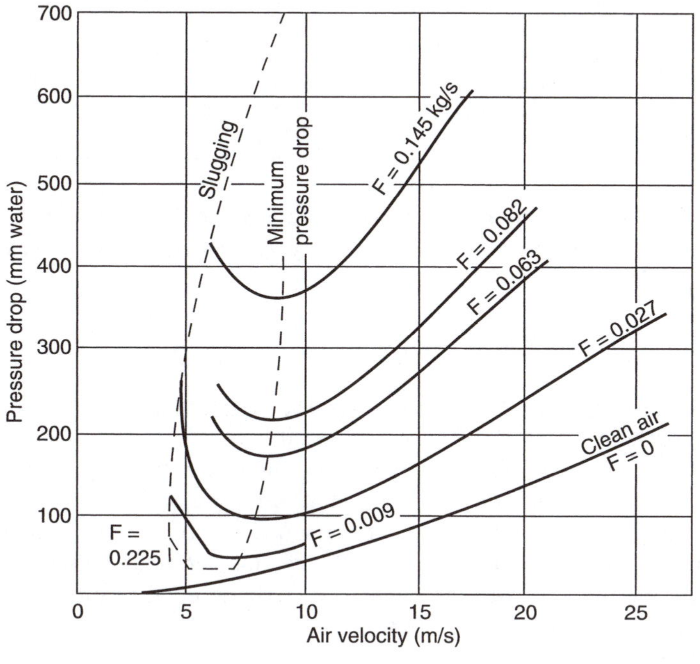

Figure 5.16. Effect of air rate on pressure drop for transport of cress seed in 4.8 m horizontal length of 25 mm pipe

Accurate measurements of solid velocities and pressure gradients have been made by McLeman[77] using a continuously operating system in which the solids were separated from the discharged air in a cyclone separator and introduced again to the feed hopper at the high pressure end of the system by means of a specially constructed rotary valve. A 25 mm diameter pipeline was used with two straight lengths of about 35 m joined by a semicircular bend. Experiments were carried out with air velocities up to 35 m/s. The solids used and their properties are listed in Table 5.3.

Table 5.3. Physical properties of the solids used for pneumatic conveying[77]

Material	Shape	Particle size (mm) Range	Particle size (mm) Mean	Density (kg/m^3)	Free-falling velocity u_0 (m/s)
Coal A	Rounded	1.5–Dust	0.75	1400	2.80
Coal B	Rounded	1.3–Dust	0.63		2.44
Coal C	Rounded	1.0–Dust	0.50		2.13
Coal D	Rounded	2.0–Dust	1.0		3.26
Coal E	Rounded	4.0–Dust	2.0		3.72
Perspex A	Angular	2.0–1.0	1.5	1185	3.73
Perspex B	Angular	5.0–2.5	3.8		5.00
Perspex C	Spherical	1.0–0.5	0.75		2.35
Polystyrene	Spherical	0.4–0.3	0.36	1080	1.62
Lead	Spherical	1.0–0.15	0.30	11,080	8.17
Brass	Porous, feathery filings	0.6–0.2	0.40	8440	4.08
Aluminium	Rounded	0.4–0.1	0.23	2835	3.02
Rape seed	Spherical	2.0–1.8	1.91	1080	5.91
Radish seed	Spherical	2.8–2.3	2.5	1065	6.48
Sand	Nearly spherical	1.5–1.0	1.3	2610	4.66
Managanese dioxide	Rounded	1.0–0.25	0.75	4000	5.27

The velocities of solid particles u_s (m/s) are represented in terms of the air velocity u_G (m/s), the free-falling velocity of the particles u_0 (m/s), and the density of the solid particles ρ_S (kg/m^3) by the equation:

$$u_G - u_S = \frac{u_0}{0.468 + 7.25\sqrt{u_0/\rho_S}} \tag{5.37}$$

Over the range studied, the slip velocity u_R $(= u_G - u_S)$ is close to the terminal falling velocity of the particles in air $(0.67 < u_R/u_0 < 1.03)$.

Deviations from equation 5.37 are noted only at high loadings with fine solids of wide size distribution. Experimental results are plotted in Figure 5.17.

The additional pressure drop due to the presence of solids in the pipeline $(-\Delta P_x)$ could be expressed in terms of the solid velocity, the terminal falling velocity of the particles and the feed rate of solids F (kg/s). The experimental results for a 25 mm pipe are correlated to within ±10 per cent by:

$$\frac{-\Delta P_x}{-\Delta P_{air}} \frac{u_S^2}{F} = \frac{2805}{u_0} \tag{5.38}$$

In Figure 5.18 $(-\Delta P_x/-\Delta P_{air})u_S^2/F$ is plotted against $1/u_0$ for a 21 m length of pipe. $-\Delta P_{air}$ is the pressure drop for the flow of air alone at the pressure existing within the pipe.

The effect of pipe diameter on the pressure drop in a conveying system is seen by examining the results of SEGLER[71] who conveyed wheat grain in pipes up to 400 mm diameter. Taking a value of the solids velocity for wheat from the work of GASTERSTÄDT[78], $[(-\Delta P_x/-\Delta P_{air})u_S^2 u_0]/F$ is plotted against pipe diameter in Figure 5.19. The constant in equation 5.38 decreases with pipe diameter and is

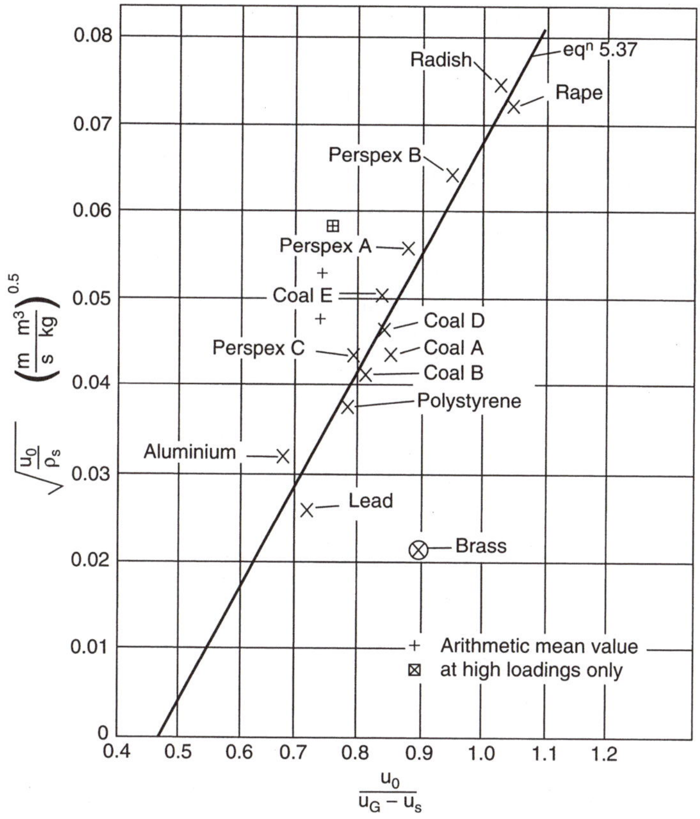

Figure 5.17. Slip velocities $u_G - u_S$ for various materials

proportional to $d^{-0.7}$. Equation 5.38 may thus be written (with d in m) as:

$$\frac{-\Delta P_x}{-\Delta P_{\text{air}}} \frac{u_S^2}{F} = \frac{210}{u_0 d^{0.7}} \tag{5.39}$$

The above correlations all apply in straight lengths of pipe with the solids suspended in the gas stream. Under conditions of slug or dune flow the pressure gradients will be considerably greater. It is generally found that the economic velocity corresponds approximately with the minimum of the curves in Figure 5.16. If there are bends in the conveying line, the energy requirement for conveying will be increased because the particles are retarded and must be re-accelerated following the bend. Bends should be eliminated wherever possible because, in addition, the wear tends to be very high.

Electrostatic charging

Equations 5.37, 5.38 and 5.39 for solid velocity and pressure drop are applicable only in the absence of electrostatic charging of the particles. Many materials, including sand, become charged during transport and cause the deposition of a charged layer on the surface of the pipe. The charge remains on the earthed pipeline for long periods but can

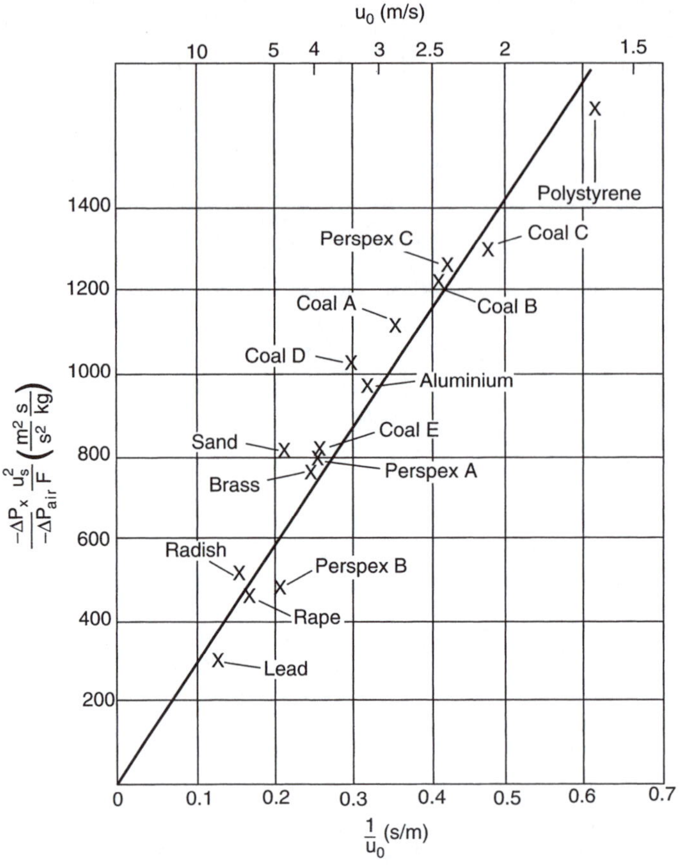

Figure 5.18. Pressure drop for flow of solids in pneumatic conveying

be removed by conveying certain materials, including coal and perspex, through the pipe. The charging is thought to be associated with attrition of the particles and therefore to be a relatively slow process, while the discharging is entirely an electrical phenomenon and therefore more rapid. The effect of electrostatic charging is to increase the frequency of collisions between the particles and the wall. As a result, the velocity of the solids is substantially reduced and the excess pressure difference $(-\Delta P_x)$ may be increased by as much as tenfold. The results included in Figure 5.18 are those obtained before an appreciable charge has built up. Increasing the humidity of the air will substantially reduce the electrostatic effects.

Dense phase conveying

There are several attractive features of dense phase conveying which warrant its use in some circumstances. In general, pipe wear is less and separation of the product is easier. KONRAD(79) has carried out a comprehensive review of the current state of the art. He points out that the technique is not yet well established and that there are considerable design difficulties, not least a satisfactory procedure for predicting pressure drop. The risk

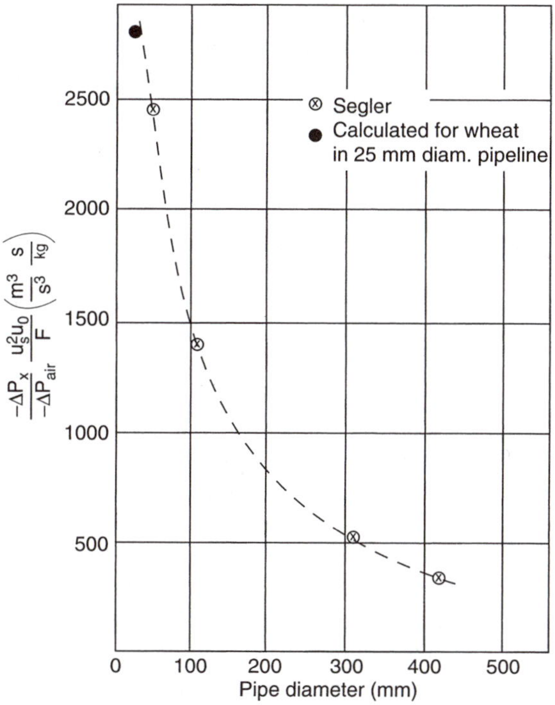

Figure 5.19. Effect of pipe diameter on pressure drop for transport of wheat

of blocking the pipeline if the operating conditions are not correctly chosen is a deterrent to the widescale application of dense phase conveying.

5.4.3. Vertical transport

Vertical pneumatic conveying has received considerably less study than horizontal conveying and interest is usually confined to transport over much smaller distances. It is used for transferring particulate materials between plant units at different levels and for unloading into storage vessels. Another important application is in the recycling of solids in fluidised bed reactor systems (See Volume 2, Chapter 6). As in hydraulic transport, vertical upwards flows are inherently simpler than horizontal flows in that gravitational forces are directly balanced by forces due to the pressure gradient in the pipeline and the drag forces of the fluid on the particles. Furthermore, the flow is axisymmetric and there is no tendency for particles to settle out on to the pipe walls.

Detailed consideration of the interaction between particles and fluids is given in Volume 2 to which reference should be made. Briefly, however, if a particle is introduced into a fluid stream flowing vertically upwards it will be transported by the fluid provided that the fluid velocity exceeds the terminal falling velocity u_0 of the particle; the relative or *slip* velocity will be approximately u_0. As the concentration of particles increases this slip velocity will become progressively less and, for a *slug* of fairly close packed particles, will approximate to the *minimum fluidising velocity* of the particles. (See Volume 2, Chapter 6.)

The evidence from practical studies of vertical transport[75] is that the contribution of the solids to the pressure gradient in the pipeline is attributable predominantly to the weight of the particles. Because the density of the particles ρ_S is much greater than that of the gas, this additional pressure gradient can be written:

$$-\frac{dP}{dl} = (1 - \epsilon_G)\rho_S g \tag{5.40}$$

where ϵ_G is the voidage (gas holdup) of the mixture in the pipeline.

The conveying of fine particles in vertical pipes of diameters 25 mm, 50 mm, and 75 mm has been studied by BOOTHROYD[75]. He measured the pressure gradient in the pipeline, and found that the frictional pressure drop was less than that for air alone in the 25 mm pipe, but was greater in the larger pipes. This effect was attributed to the fact that the extent to which the fluid turbulence was affected by the presence of the particles was markedly influenced by pipe size.

KLINZING *et al.*[68] report the results of experiments on dense-phase plug-flow of a cohesive coal. Pressure drops were measured over plugs up to 0.6 m in length and were found to be linearly related to plug length. There was little wall friction and pressure drops were almost entirely attributable to the gravitational contribution, given by equation 5.40.

LI and KWAUK[80] made a theoretical and practical study of dense phase vertical flow in a tapered rectangular transport line designed to maintain an approximately constant gas velocity. The voidage of the bed was measured at three points along its height, employing a capacitance void detector, and the normal wall shear stress arising from interparticle contact pressure was measured with a probe incorporating a diaphragm and strain gauge. Typically, pressure gradient along the line was about 1.5 times the gravitational component calculated using equation 5.40.

SINCLAIR and JACKSON[81] have presented a theoretical relation between pressure gradient and the flowrates of gas and solids over the whole range of possible conditions for both cocurrent and countercurrent flow. It predicts marked segregation of gas and particles in the radial direction.

MILLER and GIDASPOW[82] have studied transport of 75 μm catalyst particles in a 75 mm diameter vertical pipe. The flow was characterised by a dilute rising core and a dense annular region at the walls which tended to move downwards. The fractional volumetric concentration of the solids was from 0.007 to 0.04 in the core and up to 0.25 in the annular region.

5.4.4. Practical applications

Although pressure drop and energy requirements are important considerations in the design of pneumatic conveying installations, the over-riding factor must be that the plant will operate safely and without giving rise to serious operating and maintenance problems. There is always a considerable reluctance to operate at velocities below that corresponding to the minimum pressure drop shown in Figure 5.16 for horizontal transport because this constitutes an unstable operating region, with the pressure drop increasing as the velocity falls, and the risk of blockage of the line is considerable. As a rule of thumb, it is undesirable to use gas velocities less than about twice the terminal falling velocity of the particles — a figure that is equally valid for vertical transport. Excessive velocities should be avoided, however, not only because of the consequent high costs of power, but also because of the increased effects of abrasion from the solids when moving at high velocity. The effects of abrasion are particularly marked at bends and fittings and changes of flow direction should be as gradual as possible. With combustible solids there are risks of dust explosions and it may be necessary to convey in an inert gas, particularly where there is likely to be a buildup of electrostatic charges.

The solids to gas ratio should be kept as high as possible to avoid the costs of pumping unnecessarily large amounts of gas and of separating the excess gas from the solids at the pipe outlet. However, because fluctuating flow conditions frequently occur it is desirable to avoid high concentrations because of the risk of blockage.

At the downstream end of the pipeline it is necessary to disengage the solids from the gas and this is most usually carried out in cyclone separators (See Volume 2, Chapter 8). However, there may be a carry-over of fine particles which must be eliminated before the gas is vented, and gas filters or electrostatic precipitators may be used for this purpose (See again Volume 2, Chapter 8). At the upstream end, the particles must be introduced using some form of positive feeder, such as a rotary valve or a blow tank.

The design of pneumatic conveying installations presents a number of problems associated with the behaviour of the particles to be transported—particularly their tendency to agglomerate and to electrostatic charging. These factors are very difficult to define, depending as they do on the exact form of solids and the humidity of the air used for conveying. Reliable design methods have therefore not been developed for any but the simplest of problems. It should be borne in mind that variations in excess pressure drops up to 2–10-fold may arise from some of these ill-definable effects. This has led WYPICH and ARNOLD[83] to suggest that it is desirable wherever possible to use a test rig to determine the conveying characteristics of the particles, using standardised procedures, before finalising the design.

Example 5.3

Sand of particle size 1.25 mm and density 2600 kg/m^3 is to be transported in air at the rate of 1 kg/s through a horizontal pipe 200 m long. Estimate the pipe diameter, the pressure drop in the pipeline and the air flow required.

Solution

Assuming a solids:gas mass ratio of 5, then:

$$\text{mass flow of air} = (1/5) = 0.20 \text{ kg/s}$$

and, assuming an air density of 1.0 kg/m^3:

$$\text{volumetric flow of air} = (1.0 \times 0.20) = \underline{\underline{0.20 \text{ m}^3/\text{s}}}$$

In order to avoid an excessive pressure drop, an air velocity of 30 m/s is acceptable.

Ignoring the volume accupied by the sand—about 0.2% of that occupied by the air—the cross-sectional area of the pipe required is $(0.20/30) = 0.0067$ m^2

and thus:

$$\text{pipe diameter} = (4 \times 0.0067/\pi)^{0.5} = 0.092 \text{ m or } 92 \text{ mm}$$

The nearest standard size is $\underline{\underline{100 \text{ mm}}}$

For sand of particle size 1.25 mm and density 2600 kg/m^3, the free-falling velocity is given in Table 5.3 as:

$$u_0 = 4.7 \text{ m/s}$$

In equation 5.37:

$$(u_G - u_S) = 4.7/[0.468 + 7.25(4.7/2600)^{0.5}] = 6.05 \text{ m/s}$$

The cross-sectional area of a 100 mm ID. pipe $= (\pi \times 0.100^2/4) = 0.00786$ m^2

and hence:

$$\text{air velocity, } u_G = (0.20/0.00786) = 25.5 \text{ m/s}$$

and:

$$\text{solids velocity, } u_S = (25.5 - 6.05) = 19.4 \text{ m/s}$$

Taking the viscosity and the density of air as 1.7×10^{-5} Ns/m^2 and 1.0 kg/m^3 respectively, then, for the air alone:

$$Re = (0.100 \times 25.5 \times 1.0/(1.7 \times 10^{-5})) = 150{,}000$$

and from Figure 3.7:

$$\text{friction factor, } \phi = 0.004$$

Assuming isothermal conditions and incompressible flow, then, in equation 3.18:

$$-\Delta P_{\text{air}} = [4 \times 0.004(200/0.100] \times 1.0 \times 25.5^2)/2$$

$$= 10404 \text{ N/m}^2 \text{ or } 10.4 \text{ kN/m}^2$$

and in equation 5.38:

$$(-\Delta P_x/ - \Delta P_{\text{air}})(u_S^2/F) = (2805/u_0)$$

or

$$-\Delta P_x = (2805 \Delta P_{\text{air}} F)/(u_0 u_S)^2$$

$$= (2805 \times 10.4 \times 1.0)/(4.7 \times 19.4^2)$$

$$= \underline{\underline{16.5 \text{ kN/m}^2}}$$

The total pressure drop $= (-\Delta P_a) + (-\Delta P_x) = 10.4 + 16.5 = \underline{\underline{26.8 \text{ kN/m}^2}}$

5.5. FURTHER READING

BERGLES, A. E., COLLIER, J. G., DELHAYE, J. M., HEWITT, G. F. and MAYINGER, F.: *Two-Phase Flow and Heat Transfer in the Power and Process Industries* (Hemisphere Publishing Corporation and McGraw-Hill, New York, 1981).

BROWN, N. P. and HEYWOOD, N. I. (eds): *Slurry Handling. Design of Solid-Liquid Systems* (Elsevier, Amsterdam, 1991).

CHHABRA, R. P.: In *Civil Engineering Practice*, Volume 2, Cheremisinoff, P. N., Cheremisinoff, N. P. and Cheng, S. L. eds (Technomic Pub. Press PA, 1988). Hydraulic transport of solids in horizontal pipes.

CHHABRA, R. P. and RICHARDSON, J. F.: In *Encyclopedia of Fluid Mechanics*, Volume 3, Gas-Liquid Flow Cheremisinoff, N. P. eds (Gulf Publishing Co. 1986). Co-current horizontal and vertical upwards flow of gas and non-Newtonian liquid.

GOVIER, G. W. and AZIZ, K.: *The Flow of Complex Mixtures in Pipes* (Krieger, Florida, 1982).

HEWITT, G. F.: *Int. J. Multiphase Flow*, **9** (1983) 715–749. Two-phase flow studies in the United Kingdom.

KLINZING, G. E.: *Gas-Solid Transport* (McGraw-Hill, New York, 1981).

MILLS, D.: *Pneumatic Conveying Design Guide* (Butterworth, London, 1990).

SHOOK, C. A.: In *Handbook of Fluids in Motion* by Cheremisinoff, N. P. and Gupta, R. (eds) (Ann Arbor, New York, 1983), pp. 929–943: Pipeline flow of coarse particle slurries.

SHOOK, C. A. and ROCO, M. C.: *Slurry Flow. Principles and Practice* (Butterworth-Heinemann, Oxford, 1991).

WILLIAMS, O. A.: *Pneumatic and Hydraulic Conveying of Solids* (Dekker, New York, 1983).

ZANDI, I. (ed): In *Advances in Solid-Liquid Flow in Pipes and its Applications*. Hydraulic Transport of Bulky Materials (Pergamon, Oxford, 1971).

ZENZ, F. A. and OTHMER, D. F.: *Fluidization and Fluid-Particle Systems* (Reinhold, New York, 1960).

5.6. REFERENCES

1. GOVIER, G. W. and AZIZ, K.: *The Flow of Complex Mixtures in Pipes* (Krieger, Florida, 1982).
2. CHISHOLM, D.: *Two-phase Flow in Pipelines and Heat Exchangers* (George Godwin, London, 1983).
3. HEWITT, G. F.: In *Handbook of Multiphase Systems*, G. Hetsroni ed. (McGraw-Hill, New York, 1982), 2–25.
4. BAKER, O.: *Oil and Gas Jl.* **53** (26 July 1954) 185. Simultaneous flow of gas and oil; **56** (10 Nov. 1958) 156. Multiphase flow in pipelines.
5. HOOGENDOORN, C. J.: *Chem. Eng. Sci.* **9** (1959) 205. Gas-liquid flow in horizontal pipes.
6. GRIFFITH, P. and WALLIS, G. B.: *Trans. Am. Soc. Mech. Eng., J. Heat Transfer* **83** (1961) 307. Two-phase slug flow.
7. WEISMAN, J., DUNCAN, D., GIBSON, J. and CRAWFORD, T.: *Intl. Jl. Multiphase Flow* **5** (1979) 437–462. Effects of fluid properties and pipe diameter on two phase flow patterns in horizontal lines.
8. MANDHANE, J. M., GREGORY, G. A. and AZIZ, K.: *Intl. Jl. Multiphase Flow* **1** (1974) 537–553. A flow pattern map for gas-liquid flow in horizontal pipes.
9. CHHABRA, R. P. and RICHARDSON, J. F.: *Can. J. Chem. Eng.* **62** (1984) 449. Prediction of flow patterns for the cocurrent flow of gas and non-Newtonian liquid in horizontal pipes.
10. HEWITT, G. F., KING, I. and LOVEGROVE, P. C.: *Brit. Chem. Eng.* **8** (1963) 311–318. Holdup and pressure drop measurements in the two phase annular flow of air-water mixtures.
11. CHEN, J. J. J. and SPEDDING, P. L.: *Intl. Jl. Multiphase Flow* **9** (1983) 147–159. An analysis of holdup in horizontal two phase gas-liquid flow.
12. PETRICK, P. and SWANSON, B. S.: *Rev. Sci. Instru.* **29** (1958) 1079–1085. Radiation attenuation method of measuring density of a two phase fluid.
13. HEYWOOD, N. I. and RICHARDSON, J. F.: *Chem. Eng. Sci.* **34** (1979) 17–30. Slug flow of air-water mixtures in a horizontal pipe: determination of liquid holdup by γ-ray absorption.
14. KHATIB, Z. and RICHARDSON, J. F.: *Chem. Eng. Res. Des.* **62** (1984) 139. Vertical co-current flow of air and shear thinning suspensions of kaolin.
15. LOCKHART, R. W. and MARTINELLI, R. C.: *Chem. Eng. Prog.* **45** (1949) 39. Proposed correlation of data for isothermal two-phase, two-component flow in pipes.
16. FAROOQI, S. I. and RICHARDSON, J. F.: *Trans. Inst. Chem. Eng.* **60** (1982) 292–305. Horizontal flow of air and liquid (Newtonian and non-Newtonian) in a smooth pipe: Part I: Correlation for average liquid holdup.
17. CHHABRA, R. P., FAROOQI, S. I. and RICHARDSON, J. F.: *Chem. Eng. Res. Des.* **62** (1984) 22–32. Isothermal two phase flow of air and aqueous polymer solutions in a smooth horizontal pipe.
18. CHISHOLM, D.: *Intl. Jl. Heat and Mass Transfer* **10** (1967) 1767. A theoretical basis for the Lockhart-Martinelli correlation for two-phase flow.
19. GRIFFITH, P.: In *Handbook of Heat Transfer*, edited by W. M. Rohsenow and J. P. Hartnett (McGraw-Hill, New York, 1973) Section 14. Two-phase flow.
20. CHENOWETH, J. M. and MARTIN, M. W.: *Pet. Ref.* **34** (No. 10) (1955) 151. Turbulent two-phase flow.
21. CHENOWETH, J. M. and MARTIN, M. W.: *Trans. Am. Soc. Mech. Eng.*, Paper 55 PET-9 (1955). A pressure drop correlation for turbulent two-phase flow of gas-liquid mixtures in horizontal pipes.
22. BAROCZY, C. J.: *Chem. Eng. Prog. Symp. Ser.* No. 64, **62** (1966) 232. A systematic correlation for two-phase pressure drop.
23. DUKLER, A. E. MOYE WICKS III, and CLEVELAND, R. G.: *A. I. Ch. E. Jl.* **10** (1964) 38. Frictional pressure drop in two-phase flow. A comparison of existing correlations for pressure loss and holdup.
24. FAROOQI, S. I. and RICHARDSON, J. F.: *Trans. Inst. Chem. Eng.* **60** (1982) 323. Horizontal flow of air and liquid (Newtonian and non-Newtonian) in a smooth pipe: Part II: Average pressure drop.
25. DZIUBINSKI, M. and RICHARDSON, J. F.: *J. Pipelines* **5** (1985) 107. Two-phase flow of gas and non-Newtonian liquids in horizontal pipes — superficial velocity for maximum power saving.
26. HEYWOOD, N. I. and RICHARDSON, J. F.: *J. Rheology* **22** (1978) 599. Rheological behaviour of flocculated and dispersed aqueous kaolin suspensions in pipe flow.
27. FAROOQI, S. I. and RICHARDSON, J. F.: *Trans. Inst. Chem. Eng.* **58** (1980) 116. Rheological behaviour of kaolin suspensions in water and water-glycerol mixtures.
28. CHHABRA, R. P., FAROOQI, S. I., RICHARDSON, J. F. and WARDLE, A. P.: *Chem. Eng. Res. Des.* **61** (1983) 56. Cocurrent flow of air and china clay suspensions in large diameter pipes.
29. STREAT, M.: *Hydrotransport* 10 (BHRA, Fluid Engineering, Innsbruck, Austria), (Oct. 1986). Paper B1. Dense phase flow of solid-water mixtures in pipelines: A state of the art review.

30. PIRIE, R. L.: Ph.D. Thesis, University of Wales (1990). Transport of coarse particles in water and shear-thinning suspensions in horizontal pipes.
31. PIRIE, R. L., DAVIES, T., KHAN, A. R. and RICHARDSON, J. F.: 2nd International Conf. on Flow Measurement—BHRA, London (1988) Paper F3, 187. Measurement of liquid velocity in multiphase flow by salt injection method.
32. NEWITT, D. M., RICHARDSON, J. F., ABBOTT, M., and TURTLE, R. B.: *Trans. Inst. Chem. Eng.* **33** (1955) 93. Hydraulic conveying of solids in horizontal pipes.
33. HISAMITSU, N., ISE, T. and TAKEISHI, Y.: *Hydrotransport-7* (BHRA Fluid Engineering, Sendai, Japan) (Nov. 1980) B4 71. Blockage of slurry pipeline.
34. CHHABRA, R. P. and RICHARDSON, J. F.: *Chem. Eng. Res. Des.* **61** (1983) 313. Hydraulic transport of coarse gravel particles in a smooth horizontal pipe.
35. DURAND, R.: *Houille Blanche* **6** (1951) 384. Transport hydraulique des materiaux solides en conduite.
36. DURAND, R.: *Houille Blanche* **6** (1951) 609. Transport hydraulique de graviers et galets en conduite.
37. DURAND, R.: Proc. of the Minnesota International Hydraulic Convention (1953) 89. Basic relationships of transportation of solids in pipes—experimental research.
38. DURAND, R. and CONDOLIOS, E.: Proceedings of a Colloquium on the Hydraulic Transport of Coal, National Coal Board, London (1952), Paper IV. The hydraulic transportation of coal and solid materials in pipes.
39. JAMES, J. G. and BROAD, B. A.: Transport and Road Research Laboratory, TRRL Supplementary Report 635 (1980): Conveyance of coarse particle solids by hydraulic pipeline: Trials with limestone aggregates in 102, 156 and 207 mm diameter pipes.
40. KHAN, A. R., PIRIE, R. L. and RICHARDSON, J. F.: *Chem. Eng. Sci.* **42** (1987) 767. Hydraulic transport of solids in horizontal pipelines—predictive methods for pressure gradient.
41. ABBOTT, M.: Ph.D. Thesis, University of London (1955). The Hydraulic Conveying of Solids in Pipe Lines.
42. TURTLE, R. B.: Ph.D. Thesis, University of London (1952). The Hydraulic conveying of Granular Material.
43. HAAS, D. B., GILLIES, R., SMALL, M. and HUSBAND, W. H. W.: Saskatchewan Research Council publication No. E-835-1-C80 (March 1980). Study of the hydraulic properties of coarse particles of metallurgical coal when transported in slurry form through pipelines of various diameters.
44. GAESSLER, H.: Experimentelle und theoretische Untersuchungen über die Strömungsvorgänge beim Transport von Feststoffen in Flüssigkeiten durch horizontale Rohrleitungen, Dissertation Technische Hochschule, Karlsruhe, Germany. (1966).
45. ZANDI, I. and GOVATOS, G.: *J. Hydr. Div. Amer. Soc. Civ. Engrs.*, **93** (1967) 145. Heterogeneous flow of solids in pipelines.
46. ROCO, M. and SHOOK, C. A.: *Can. J. Chem. Eng.* **61** (1983) 494–503. Modelling of slurry flow: The effect of particle size.
47. ROCO, M. and SHOOK, C. A.: *J. Pipelines*, **4** (1984) 3–13. A model for turbulent slurry flow.
48. SHOOK, C. A. and DANIEL, S. M.: *Can. J. Chem. Eng.* **43** (1965) 56. Flow of suspensions of solids in pipelines. Part I. Flow with a stable stationary deposit.
49. WILSON, K. C.: *Hydrotransport* **4** (BHRA Fluid Engineering, Banff, Alberta, Canada) (May 1976) A1.1. A unified physically-based analysis of solid-liquid pipeline flow.
50. WILSON, K. C. and BROWN, N. P.: *Can. J. Chem. Eng.* **60** (1982) 83. Analysis of fluid friction in dense-phase pipeline flow.
51. WILSON, K. C. and PUGH, F. J.: *Can. Jl. Chem. Eng.* **66** (1988) 721. Dispersive-force modelling of turbulent suspensions in heterogeneous slurry flow.
52. SHOOK, C. A. and ROCO, M. C.: *Slurry Flow. Principles and Practice* (Butterworth-Heinemann, Oxford, 1991).
53. CHARLES, M. E. and CHARLES, R. A.: Paper A12, *Advances in Solid-Liquid Flow in Pipes and its Application*, Zandi, I. (ed.). Pergamon Press (1971). The use of heavy media in the pipeline transport of particulate solids.
54. KENCHINGTON, J. M.: Hydrotransport 5, (BHRA Fluid Engineering, Hanover, Germany) (May 1978) Paper D7. Prediction of pressure gradient in dense phase conveying.
55. DUCKWORTH, R. A., PULLUM, L. and LOCKYEAR, C. F.: *J. Pipelines* **3** (1983) 251. The hydraulic transport of coarse coal at high concentration.
56. DUCKWORTH, R. A., PULLUM, L., ADDIE, G. R. and LOCKYEAR, C. F.: Hydrotransport 10 (BHRA Fluid Engineering, Innsbruck, Austria) (October 1986). Paper C2. The pipeline transport of coarse materials in a non-Newtonian carrier fluid.
57. DUCKWORTH, R. A., ADDIE, G. R. and MAFFETT, J.: 11th Int. Conf. on Slurry Technology—The Second Decade, organised by STA Washington DC, Hilton Head SC (March 1986) 187. Minewaste disposal by pipeline using a fine slurry carrier.
58. TATSIS, A., JACOBS, B. E. A., OSBORNE, B. and ASTLE, R. D., 13th Int. Conf. on Slurry Technology, organised by STA Washington DC, Hilton Head SC (March 1988). A comparative study of pipe flow prediction for high concentration slurries containing coarse particles.

59. CHHABRA, R. P. and RICHARDSON, J. F.: *Chem. Eng. Res. Des.* **63** (1985) 390. Hydraulic transport of coarse particles in viscous Newtonian and non-Newtonian media in a horizontal pipe.
60. LAREO, C., FRYER, P. J. and BARIGOU, M.: *Trans.I.Chem.E.* **75** Part C (1997) 73. The fluid mechanics of solid–liquid food flows.
61. TUCKER, G. and HEYDON, C.: *Trans.I.Chem.E.* **76** Part C (1998) 208. Food particle residence time measurement for the design of commercial tubular heat exchangers suitable for processing suspensions of solids in liquids.
62. DURAND, R.: *Houille Blanche* **8** (1953) 124. Écoulements de mixture en conduites verticales — influence de la densité des matériaux sur les caractéristiques de refoulement en conduite horizontale.
63. WORSTER, R. C. and DENNY, D. F.: *Proc. Inst. Mech. Eng.* **169** (1955) 563. The hydraulic transport of solid material in pipes.
64. NEWITT, D. M., RICHARDSON, J. F., and GLIDDON, B. J.: *Trans. Inst. Chem. Eng.* **39** (1961) 93. Hydraulic conveying of solids in vertical pipes.
65. CLOETE, F. L. D., MILLER, A. I., and STREAT, M.: *Trans. Inst. Chem. Eng.* **45** (1967) T392. Dense phase flow of solid-water mixtures through vertical pipes.
66. KOPKO, R. J., BARTON, P. and MCCORMICK, R. H.: *Ind. Eng. Chem. (Proc. Des. and Dev.)* **14** (1975) 264. Hydrodynamics of vertical liquid-solids transport.
67. AL-SALIHI, L.: Ph.D. Thesis, University of Wales (1989). Hydraulic transport of coarse particles in vertical pipelines.
68. KLINZING, G. E., ROHATGI, N. D., ZALTASH, A. and MYLER, C. A.: *Powder Technol.* **51** (1987) 135. Pneumatic transport — a review (Generalized phase diagram approach to pneumatic transport).
69. BAGNOLD, R. A.: *Proc. Inst. Civ. Eng.* (iii) **4** (1955) 174. Some flume experiments on large grains but little denser than the transporting fluid, and their implications.
70. BAGNOLD, R. A.: *Phil. Trans.* **249** (1957) 235. The flow of cohesionless grains in fluids.
71. SEGLER, G.: *Z. Ver. deut. Ing.* **79** (1935) 558. Untersuchungen an Körnergebläsen und Grundlagen für ihre Berechnung. (Pneumatic Grain Conveying (1951), National Institute of Agricultural Engineering.)
72. CLARK, R. H., CHARLES, D. E., RICHARDSON, J. F., and NEWITT, D. M.: *Trans. Inst. Chem. Eng.* **30** (1952) 209. Pneumatic conveying. Part I. The pressure drop during horizontal conveyance.
73. MITLIN, L.: University of London, Ph.D. thesis (1954). A study of pneumatic conveying with special reference to solid velocity and pressure drop during transport.
74. JONES, C. and HERMGES, G.: *Brit. J. Appl. Phys.* **3** (1952) 283. The measurement of velocities for solid-fluid flow in a pipe.
75. BOOTHROYD, R. G.: *Trans. Inst. Chem. Eng.* **44** (1966) T306. Pressure drop in duct flow of gaseous suspensions of fine particles.
76. KLINZING, G. E. and MATHUR, M. P.: *Can. J. Chem. Eng.* **59** (1981) 590. The dense and extrusion flow regime in gas-solid transport.
77. RICHARDSON, J. F. and MCLEMAN, M.: *Trans. Inst. Chem. Eng.* **38** (1960) 257. Pneumatic conveying. Part II. Solids velocities and pressure gradients in a one-inch horizontal pipe.
78. GASTERSTÄDT, J.: *Forsch. Arb. Geb. Ing. Wes.* No. 265 (1924) 1–76. Die experimentelle Untersuchung des pneumatischen Fördervorganges.
79. KONRAD, K.: *Powder Technol.* **49** (1986) 1. Dense-phase pneumatic conveying: a review.
80. LI, H. and KWAUK, M.: *Chem. Eng. Sci.* **44** (1989) 249, 261. Vertical pneumatic moving bed transport. I. Analysis of flow dynamics. II Experimental findings.
81. SINCLAIR, J. L. and JACKSON, R.: *A. I. Ch. E. Jl.* **35** (1989) 1473. Gas-particle flow in a vertical pipe with particle — particle interactions.
82. MILLER, A. and GIDASPOW, D.: *A. I. Ch. E. Jl.* **38** (1992) 1801. Dense, vertical gas-solid flow in a pipe.
83. WYPYCH, P. W. and ARNOLD, P. C.: *Powder Technol.* **50** (1987) 281. On improving scale-up procedures for pneumatic conveying design.

5.7. NOMENCLATURE

		Units in SI system	Dimensions in **M**, **L**, **T**
A	Total cross-sectional area of pipe	m^2	$\mathbf{L}^2$
A_B	Cross-sectional area of bed deposit in pipe	m^2	$\mathbf{L}^2$
A_L	Cross-sectional area of pipe over which suspended flow occurs	m^2	$\mathbf{L}^2$
b	Ratio of volumetric flowrates of mixture and liquid	—	—
C	Fractional volumetric concentration of solids in *flowing* mixture	—	—

C_D	Drag coefficient for particle settling at its terminal falling velocity $\left[= \frac{4}{3}(s' - 1)\frac{d_p g}{u_0^2}\right]$	—	—
c	Constant in equation 5.8	—	—
d	Pipe diameter	m	**L**
d_p	Particle diameter	m	**L**
e	Roughness of pipe wall	m	**L**
F_B	Force on bed layer in length l of pipe	N	$\mathbf{MLT^{-2}}$
F_L	Force on material in suspended flow in length l of pipe	N	$\mathbf{MLT^{-2}}$
ΣF_N	Sum of normal forces between particles and wall in length l of pipe	N	$\mathbf{MLT^{-2}}$
G	Mass flowrate	kg/s	$\mathbf{MT^{-1}}$
G'	Mass flowrate per unit area for gas	kg/m^{2}s	$\mathbf{ML^{-2}T^{-1}}$
g	Acceleration due to gravity	m/s^2	$\mathbf{LT^{-2}}$
h_f	Friction head	L	m
i	Hydraulic gradient for two-phase flow	—	—
i_w	Hydraulic gradient for liquid (water) in two-phase flow	—	—
K	Proportionality constant in equation 5.9	—	—
k	Consistency coefficient for power-law fluid	N s^n/m^2	$\mathbf{ML^{-1}T^{n-2}}$
L'	Mass flowrate per unit area for liquid	kg/m^{2}s	$\mathbf{ML^{-2}T^{-1}}$
l	Pipe length	m	**L**
M	Mass rate of feed of solids	kg/s	$\mathbf{MT^{-1}}$
m	Exponent in equation 5.36	—	—
n	Flow index for power-law fluid	—	—
P	Pressure	N/m^2	$\mathbf{ML^{-1}T^{-2}}$
$-\Delta P$	Pressure drop	N/m^2	$\mathbf{ML^{-1}T^{-2}}$
$-\Delta P_a$	Pressure drop associated with acceleration of fluid	N/m^2	$\mathbf{ML^{-1}T^{-2}}$
$-\Delta P_{air}$	Pressure drop for flow of air alone in pipe	N/m^2	$\mathbf{ML^{-1}T^{-2}}$
$-\Delta P_f$	Pressure drop associated with friction	N/m^2	$\mathbf{ML^{-1}T^{-2}}$
$-\Delta P_G$	Pressure drop for gas flowing *alone* at same superficial velocity as in two-phase flow	N/m^2	$\mathbf{ML^{-1}T^{-2}}$
$-\Delta P_{gravity}$	Pressure drop associated with hydrostatic effects	N/m^2	$\mathbf{ML^{-1}T^{-2}}$
$-\Delta P_L$	Pressure drop for liquid flowing *alone* at same superficial velocity as in two-phase flow	N/m^2	$\mathbf{ML^{-1}T^{-2}}$
$-\Delta P_{TPF}$	Pressure drop for two-phase flow over length l of pipe	N/m^2	$\mathbf{ML^{-1}T^{-2}}$
$-\Delta P_x$	Excess pressure drop due to solids in pneumatic conveying	N/m^2	$\mathbf{ML^{-1}T^{-2}}$
R	Shear stress at pipe wall	N/m^2	$\mathbf{ML^{-1}T^{-2}}$
R_B	Shear stress at pipe wall in fluid within bed	N/m^2	$\mathbf{ML^{-1}T^{-2}}$
R_i	Interfacial shear stress at surface of bed	N/m^2	$\mathbf{ML^{-1}T^{-2}}$
R_L	Shear stress at pipe wall in fluid above bed	N/m^2	$\mathbf{ML^{-1}T^{-2}}$
R_y	Shear stress at distance y from wall	N/m^2	$\mathbf{ML^{-1}T^{-2}}$
S_B	Perimeter of pipe in contact with bed	m	**L**
S_i	Length of chord at top surface of bed	m	**L**
S_L	Perimeter of pipe in contact with suspended flow	m	**L**
s	Ratio of density of solid to density of liquid (ρ_s/ρ)	—	—
u	Mean velocity in pipe or mixture velocity	m/s	$\mathbf{LT^{-1}}$
u_G	Superficial velocity of gas	m/s	$\mathbf{LT^{-1}}$
u_L	Superficial velocity of liquid	m/s	$\mathbf{LT^{-1}}$
u_m	Mixture velocity	m/s	$\mathbf{LT^{-1}}$
$(u_L)_c$	Critical value of u_L for laminar-turbulent transition	m/s	$\mathbf{LT^{-1}}$
u_R	Relative velocity of phases; slip velocity	m/s	$\mathbf{LT^{-1}}$
u_S	Superficial velocity of solids	m/s	$\mathbf{LT^{-1}}$
u_X	Velocity in x-direction at distance y from surface	m/s	$\mathbf{LT^{-1}}$
u_0	Terminal falling velocity of particle in fluid	m/s	$\mathbf{LT^{-1}}$
u'_L	Actual linear velocity of liquid	m/s	$\mathbf{LT^{-1}}$
u'_S	Actual linear velocity of solids	m/s	$\mathbf{LT^{-1}}$
X	Lockhart and Martinelli parameter $\sqrt{-\Delta P_L / -\Delta P_G}$	—	—
X'	Modified parameter X, defined by equation 5.3	—	—

y	Distance perpendicular to surface	m	**L**
z	Vertical height	m	**L**
Re	Reynolds number	—	—
Re_G	Reynolds number for gas flowing alone	—	—
Re_L	Reynolds number for liquid flowing alone	—	—
Re_{MR}	Metzner and Reed Reynolds number for power-law fluid (see Chapter 3)	—	—
Re'_0	Particle Reynolds number for terminal falling conditions	—	—
ϵ_G	Average hold-up for gas in two-phase flow	—	—
ϵ_L	Average hold-up for liquid in two-phase flow	—	—
ϵ_S	Average hold-up for solids in two-phase flow	—	—
μ_F	Coefficient of friction between solids and pipe wall	—	—
μ_G	Viscosity of gas	Ns/m^2	$\mathbf{ML^{-1}T^{-1}}$
μ_L	Viscosity of liquid	Ns/m^2	$\mathbf{ML^{-1}T^{-1}}$
ρ	Density	kg/m^3	$\mathbf{ML^{-3}}$
ρ_G	Density of gas	kg/m^3	$\mathbf{ML^{-3}}$
ρ_L	Density of liquid	kg/m^3	$\mathbf{ML^{-3}}$
ρ_M	Density of mixture	kg/m^3	$\mathbf{ML^{-3}}$
ρ_S	Density of solid	kg/m^3	$\mathbf{ML^{-3}}$
Φ_G^2	Drag ratio for two-phase flow $(-\Delta P_{TPF}/-\Delta P_G)$	—	—
Φ_L^2	Drag ratio for two-phase flow $(-\Delta P_{TPF}/-\Delta P_L)$	—	—

CHAPTER 6

Flow and Pressure Measurement

6.1. INTRODUCTION

The most important parameters measured to provide information on the operating conditions in a plant are flowrates, pressures, and temperatures. The instruments used may give either an instantaneous reading or, in the case of flow, may be arranged to give a cumulative flow over any given period. In either case, the instrument may be required to give a signal to some control unit which will then govern one or more parameters on the plant. It should be noted that on industrial plants it is usually more important to have information on the change in the value of a given parameter than to use meters that give particular absolute accuracy. To maintain the value of a parameter at a desired value a control loop is used.

A simple control system, or *loop*, is illustrated in Figure 6.1. The temperature T_0 of the water at Y is measured by means of a thermocouple, the output of which is fed to a controller mechanism. The latter can be divided into two sections (normally housed in the same unit). In the first (the *comparator*), the *measured value* (T_0) is compared with the desired value (T_d) to produce an *error (e)*, where:

$$e = T_d - T_0 \tag{6.1}$$

The second section of the mechanism (the *controller*) produces an output which is a function of the magnitude of e. This is fed to a control valve in the steam line, so that the valve closes when T_0 increases and vice versa. The system as shown may be used to counteract fluctuations in temperature due to extraneous causes such as variations in water flowrate or upstream temperature—termed *load* changes. It may also be employed to change the water temperature at Y to a new value by adjustment of the desired value.

It is very important to note that in this loop system the parameter T_0, which must be kept constant, is measured, though all subsequent action is concerned with the magnitude of the error and not with the actual value of T_0. This simple loop will frequently be complicated by there being several parameters to control, which may necessitate considerable instrumental analysis and the control action will involve operation of several control valves.

This represents a simple form of control for a single variable, though in a modern plant many parameters are controlled at the same time from various measuring instruments, and the variables on a plant such as a distillation unit are frequently linked together, thus increasing the complexity of control that is required.

On industrial plants, the instruments are therefore required not only to act as indicators but also to provide some link which can be used to help in the control of the plant. In this chapter, pressure measurement is briefly described and methods of measurement of flowrate are largely confined to those which depend on the application of the energy

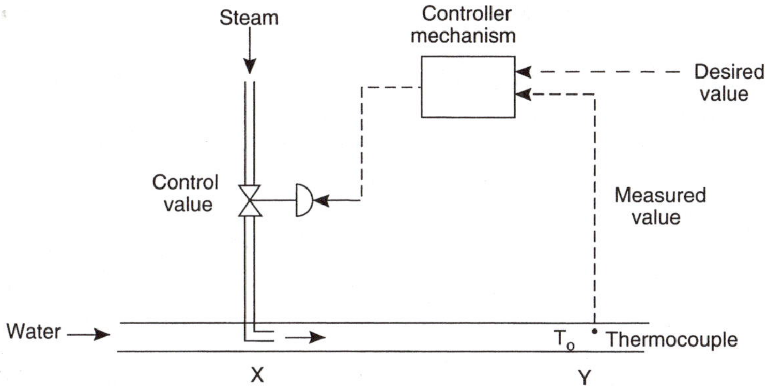

Figure 6.1. Simple feedback control system

balance equation (2.55). For further detail, reference should be made to Volume 3, Chapter 6 (Measurement) and Chapter 7 (Process Control) and to WIGHTMAN[1].

6.2. FLUID PRESSURE

In a stationary fluid the pressure is exerted equally in all directions and is referred to as the *static pressure*. In a moving fluid, the static pressure is exerted on any plane parallel to the direction of motion. The pressure exerted on a plane at right angles to the direction of flow is greater than the static pressure because the surface has, in addition, to exert sufficient force to bring the fluid to rest. This additional pressure is proportional to the kinetic energy of the fluid; it cannot be measured independently of the static pressure.

6.2.1. Static pressure

The energy balance equation can be applied between any two sections in a continuous fluid. If the fluid is not moving, the kinetic energy and the frictional loss are both zero, and therefore:

$$v\,dP + g\,dz = 0 \quad \text{(from equation 2.57)}$$

For an incompressible fluid:

$$v(P_2 - P_1) + g(z_2 - z_1) = 0$$

or:

$$(P_2 - P_1) = -\rho g(z_2 - z_1) \tag{6.2}$$

Thus the pressure difference can be expressed in terms of the height of a vertical column of fluid.

If the fluid is compressible and behaves as an ideal gas, for isothermal conditions:

$$P_1 v_1 \ln \frac{P_2}{P_1} + g(z_2 - z_1) = 0 \quad \text{(from equation 2.69)}$$

$$\therefore \quad \frac{P_2}{P_1} = \exp \frac{-gM}{\mathbf{R}T}(z_2 - z_1) \quad \text{(from equation 2.16)} \tag{6.3}$$

This expression enables the pressure distribution within an ideal gas to be calculated for isothermal conditions.

When the static pressure in a moving fluid is to be determined, the measuring surface must be parallel to the direction of flow so that no kinetic energy is converted into pressure energy at the surface. If the fluid is flowing in a circular pipe the measuring surface must be perpendicular to the radial direction at any point. The pressure connection, which is known as a *piezometer tube*, should terminate flush with the wall of the pipe so that the flow is not disturbed: the pressure is then measured near the walls where the velocity is a minimum and the reading would be subject only to a small error if the surface were not quite parallel to the direction of flow. A piezometer tube of narrow diameter is used for accurate measurements.

The static pressure should always be measured at a distance of not less than 50 diameters from bends or other obstructions, so that the flow lines are almost parallel to the walls of the tube. If there are likely to be large cross-currents or eddies, a piezometer ring should be used. This consists of four pressure tappings equally spaced at 90° intervals round the circumference of the tube; they are joined by a circular tube which is connected to the pressure measuring device. By this means, false readings due to irregular flow are avoided. If the pressure on one side of the tube is relatively high, the pressure on the opposite side is generally correspondingly low; with the piezometer ring a mean value is obtained. The cross-section of the piezometer tubes and ring should be small to prevent any appreciable circulation of the fluid.

6.2.2. Pressure measuring devices

(a) *The simple manometer*, shown in Figure 6.2*a*, consists of a transparent U-tube containing the fluid **A** of density ρ whose pressure is to be measured and an immiscible fluid **B** of higher density ρ_m. The limbs are connected to the two points between which the pressure difference $(P_2 - P_1)$ is required; the connecting leads should be completely full of fluid **A**. If P_2 is greater than P_1, the interface between the two liquids in limb 2 will be depressed a distance h_m (say) below that in limb 1. The pressure at the level $a - a$ must be the same in each of the limbs and, therefore:

$$P_2 + z_m \rho g = P_1 + (z_m - h_m)\rho g + h_m \rho_m g$$

and:

$$\Delta P = P_2 - P_1 = h_m(\rho_m - \rho)g \tag{6.4}$$

If fluid **A** is a gas, the density ρ will normally be small compared with the density of the manometer fluid ρ_m so that:

$$\Delta P = h_m \rho_m g \tag{6.5}$$

(b) In order to avoid the inconvenience of having to read two limbs, the *well-type manometer* shown in Figure 6.2*b* can be used. If A_w and A_c are the cross-sectional areas of the well and the column and h_m is the increase in the level of the column and h_w the decrease in the level of the well, then:

$$P_2 = P_1 + \rho g(h_m + h_w)$$

or:

$$P_2 - P_1 = \rho g(h_m + h_w)$$

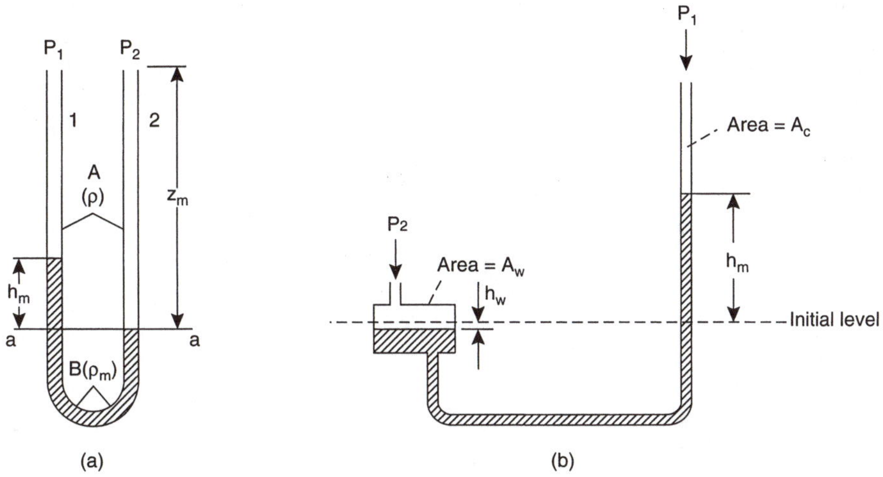

Figure 6.2. (*a*) The simple manometer (*b*) The well-type manometer

The quantity of liquid expelled from the well is equal to the quantity pushed into the column so that:

$$A_w h_w = A_c h_m$$

and:

$$h_w = \frac{A_c}{A_w} h_m$$

Substituting:

$$P_2 - P_1 = \rho g h_m \left(1 + \frac{A_c}{A_w}\right) \tag{6.6}$$

If the well is large in comparison to the column then:

$$P_2 - P_1 = \rho g h_m \tag{6.7}$$

(c) The *inclined manometer* shown in Figure 6.3 enables the sensitivity of the manometers described previously to be increased by measuring the length of the column of liquid.

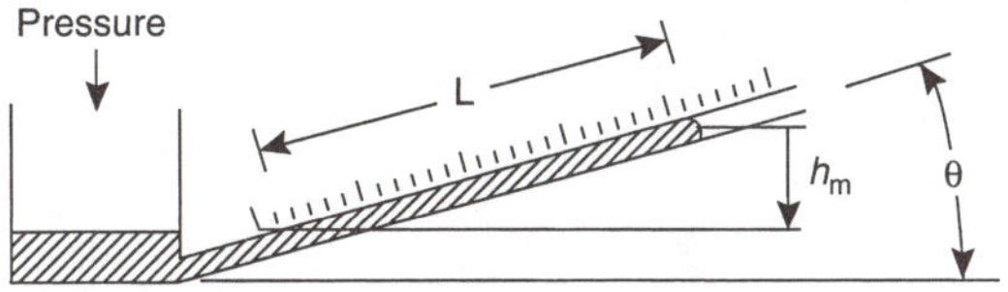

Figure 6.3. An inclined manometer

If θ is the angle of inclination of the manometer (typically about 10–20°) and L is the movement of the column of liquid along the limb, then:

$$L = \frac{h_m}{\sin\theta} \tag{6.8}$$

and if $\theta = 10^\circ$, the manometer reading L is increased by about 5.7 times compared with the reading h_m which would have been obtained from a simple manometer.

(d) The *inverted manometer* (Figure 6.4) is used for measuring pressure differences in liquids. The space above the liquid in the manometer is filled with air which can be admitted or expelled through the tap A in order to adjust the level of the liquid in the manometer.

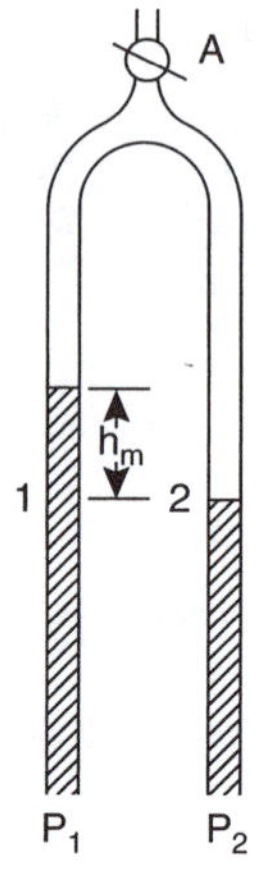

Figure 6.4. Inverted manometer

(e) *The two-liquid manometer*. Small differences in pressure in gases are often measured with a manometer of the form shown in Figure 6.5. The reservoir at the top of each limb is of a sufficiently large cross-section for the liquid level to remain approximately the

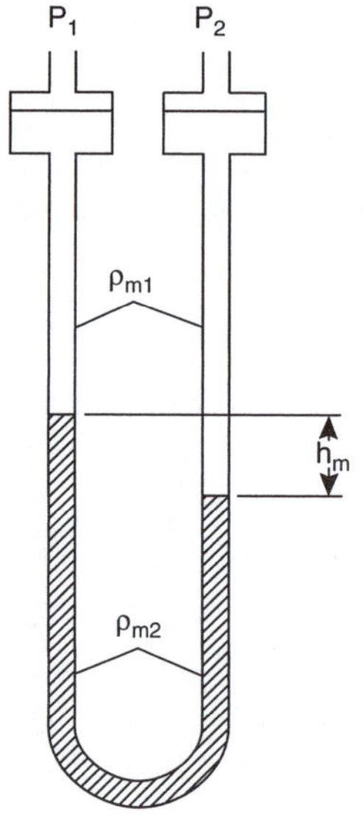

Figure 6.5. Two-liquid manometer

same on each side of the manometer. The difference in pressure is then given by:

$$\Delta P = (P_2 - P_1) = h_m(\rho_{m1} - \rho_{m2})g \tag{6.9}$$

where ρ_{m1} and ρ_{m2} are the densities of the two manometer liquids. The sensitivity of the instrument is very high if the densities of the two liquids are nearly the same. To obtain accurate readings it is necessary to choose liquids which give sharp interfaces: paraffin oil and industrial alcohol are commonly used. According to OWER and PANKHURST[(2)], benzyl alcohol (specific gravity 1.048) and calcium chloride solutions give the most satisfactory results. The difference in density can be varied by altering the concentration of the calcium chloride solution.

(f) *The Bourdon gauge* (Figure 6.6). The pressure to be measured is applied to a curved tube, oval in cross-section, and the deflection of the end of the tube is communicated through a system of levers to a recording needle. This gauge is widely used for steam and compressed gases, and frequently forms the indicating element on flow controllers. The simple form of the gauge is illustrated in Figures 6.6*a* and 6.6*b*. Figure 6.6*c* shows a Bourdon sensing element in the form of a helix; this element has a very much greater sensitivity and is suitable for very high pressures.

It may be noted that the pressure measuring devices (a) to (e) all measure a pressure *difference* $\Delta P(= P_2 - P_1)$. In the case of the Bourdon gauge (f), the pressure indicated is the difference between that communicated by the system to the tube and the external (ambient) pressure, and this is usually referred to as the *gauge* pressure. It is then necessary to add on the ambient pressure in order to obtain the (absolute) pressure. Even the mercury barometer measures, not atmospheric pressure, but the difference between atmospheric pressure and the vapour pressure of mercury which, of course, is negligible. Gauge pressures are not, however, used in the SI System of units.

6.2.3. Pressure signal transmission — the differential pressure cell

The meters described so far provide a measurement usually in the form of a pressure differential, though in most modern plants these readings must be transmitted to a central control facility where they form the basis for either recording or automatic control. The quantity being measured is converted into a signal using a device consisting of a sensing element and a conversion or control element. The sensor may respond to movement, heat, light, magnetic or chemical effects and these physical quantities may be converted to change in an electrical parameter such as voltage, resistance, capacitance or inductance. The transmission is most conveniently effected by pneumatic or electrical methods, and one typical arrangement known as a differential pressure (d.p.) cell system, shown in Figure 6.7, which operates on the *force balance* system illustrated in Figure 6.8.

In Figure 6.8 increase in the pressure on the "in" bellows causes the force-bar to turn clockwise, thus reducing the separation in the flapper — nozzle. This nozzle is fed with air, and the increase in back pressure from the nozzle in turn increases the force developed by the negative feedback bellows producing a counterclockwise movement which restores the flapper-nozzle separation. Balance is achieved when the feedback pressure is proportional to the applied pressure.

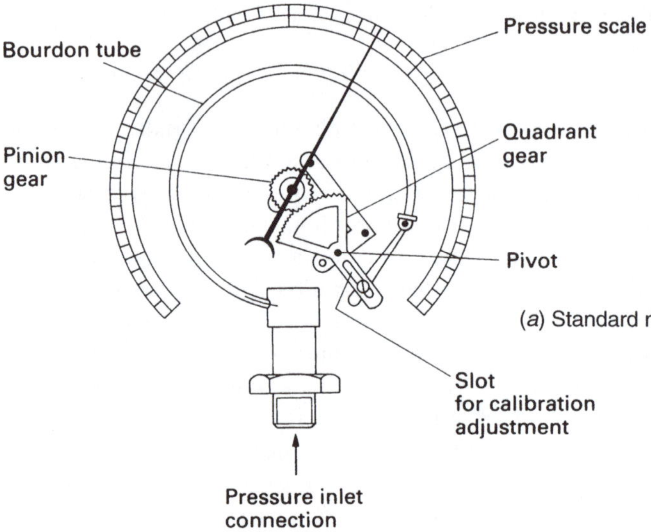

(*a*) Standard mechanism

(*b*) Single coil

Figure 6.6. Bourdon gauge

(*c*) Multiple coil

Figure 6.6. (*continued*)

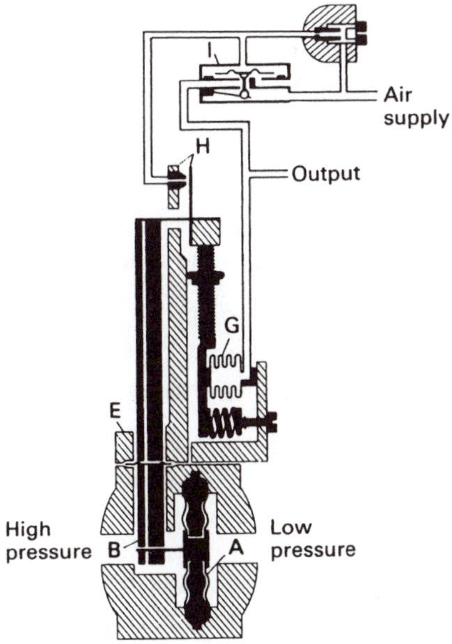

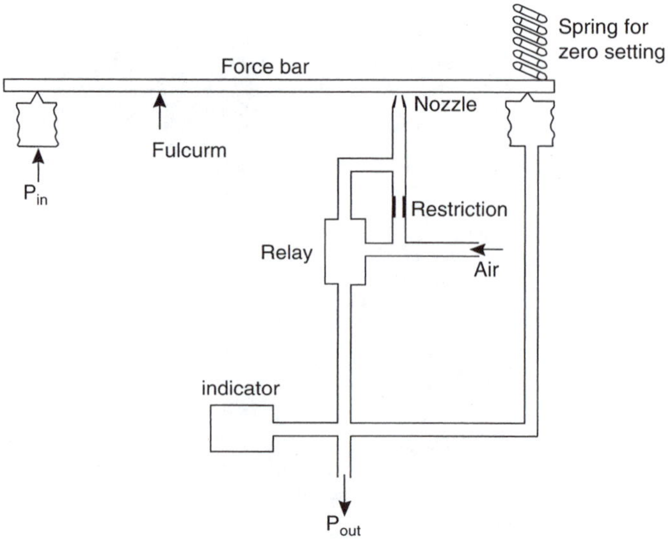

Figure 6.8. Force balance system

In this way, the signal pressure is made to respond to the input pressure and, by adjusting the distance of the bellows from the fulcrum, any range of force may be made to give an output pressure in the range 120–200 kN/m^2. It is important to note that since the flapper–nozzle unit has a high *gain* (large pressure change for small change in separation), the actual movements are very small, usually within the range 0.004–0.025 mm.

An instrument in which this feedback arrangement is used is shown in Figure 6.7. The differential pressure from the meter is applied to the two sides of the diaphragm A giving a force on the bar B. The closure E in the force bar B acts as the fulcrum, so that the flapper-nozzle separation at H responds to the initial difference in pressure. This gives the change in air pressure or signal which can be transmitted through a considerable distance. A unit of this kind is thus an essential addition to all pressure meters from which a signal in a central control room is required.

6.2.4. Intelligent electronic pressure transmitters

Intelligent pressure transmitters have two major components: (1) a sensor module which comprises the process connections and sensor assembly, and (2) a two-compartment electronics housing with a terminal block and an electronics module that contains signal conditioning circuits and a microprocessor. Figure 6.9 illustrates how the primary output signal is processed and compensated for errors caused in pressure-sensor temperature. An internal sensor measures the temperature of the pressure sensor. This measurement is fed into the microprocessor where the primary measurement signal is appropriately corrected. This temperature measurement is also transmitted to receivers over the communications network.

The solid state sensor consists of a Wheatstone Bridge circuit shown in Figure 6.9 which is diffused into a silicon chip, thereby becoming a part of the atomic structure of the

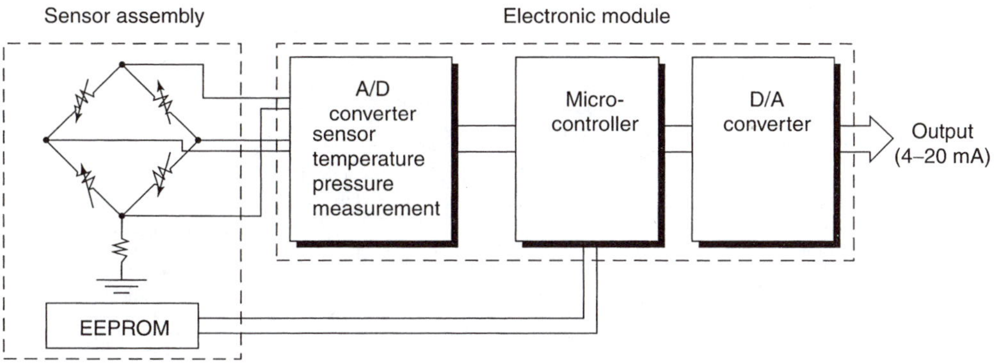

Figure 6.9. The intelligent transmitter

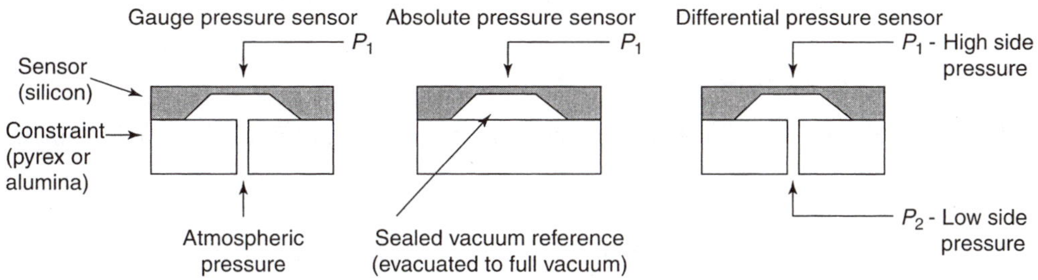

Figure 6.10. Types of pressure sensors

silicon. As pressure is applied to the diaphragm (Figure 6.10), strain is created in the bridge resistors. Piezo-resistive effects created by this strain change resistances in the legs of the bridge, producing a voltage proportional to pressure. Output from the bridge is typically in the range of 75 to 150 mV at full scale pressure for a bridge excitation of 1.0 mA.

The micro-machined silicon sensor is fabricated in three basic types of pressure sensors. The three types which are shown in Figure 6.10, are:

(a) *Gauge pressure* with the sensor referenced to atmosphere.
(b) *Absolute pressure* with the sensor referenced to a full vacuum.
(c) *Differential pressure* (dp) where the sensor measures the difference between two pressures P_1 and P_2.

Because of the wide range of the sensors, only four different sensor units are needed to cover the entire range of dp spans from 10 kN/m^2 to 20 MN/m^2 (0.5 in water to 3000 lb/in^2).

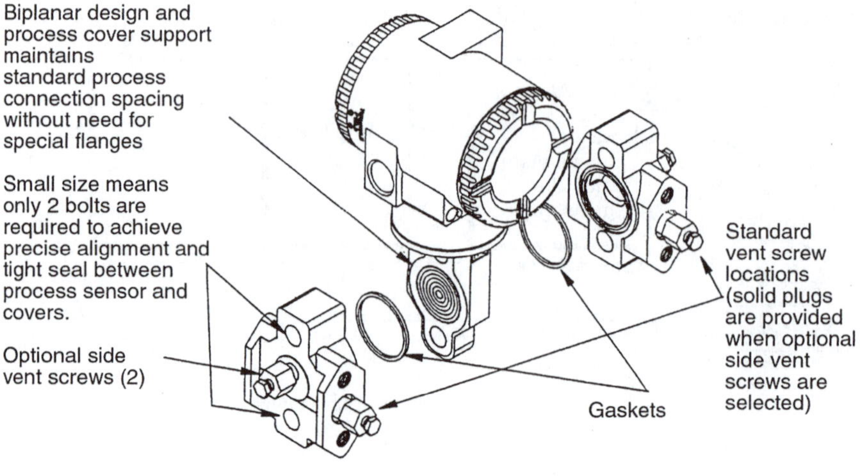
Biplanar design and process cover support maintains standard process connection spacing without need for special flanges
Small size means only 2 bolts are required to achieve precise alignment and tight seal between process sensor and covers.
Optional side vent screws (2)
Gaskets
Standard vent screw locations (solid plugs are provided when optional side vent screws are selected)

Figure 6.11. Intelligent differential-pressure cell with transmitter

6.2.5. Impact pressure

The pressure exerted on a plane at right angles to the direction of flow of the fluid consists of two components:

(a) static pressure;
(b) the additional pressure required to bring the fluid to rest at the point.

Consider a fluid flowing between two sections, 1 and 2 (Figure 6.12), which are sufficiently close for friction losses to be negligible between the two sections; they are a sufficient distance apart, however, for the presence of a small surface at right angles to the direction of flow at section 2 to have negligible effect on the pressure at section 1. These conditions are normally met if the distance between the sections is one pipe diameter.

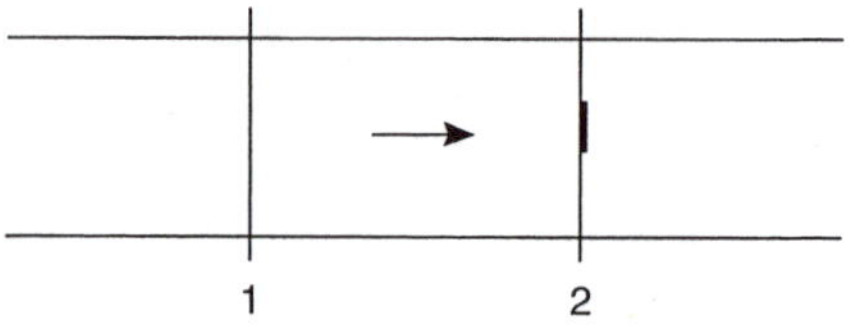
1
2

Figure 6.12. Impact pressure

Considering a small filament of liquid which is brought to rest at section 2, and applying the energy balance equation between the two sections, since $g\Delta z$, W_s, and F are all zero:

$$\frac{\dot{u}_1^2}{2} = \frac{\dot{u}_2^2}{2} + \int_{P_1}^{P_2} v\,\mathrm{d}P \qquad \text{(from equation 2.53)}$$

where $\dot{u}_1$ and $\dot{u}_2$ are velocities at 1 and 2.

If the fluid is incompressible or if the change in the density of the fluid between the sections is negligible, then (since $\dot{u}_2 = 0$):

$$\dot{u}_1^2 = 2v(P_2 - P_1) = 2h_i g$$

or:

$$\dot{u}_1 = \sqrt{2v(P_2 - P_1)} = \sqrt{2gh_i} \tag{6.10}$$

where h_i is the difference between the impact pressure head at section 2 and the static pressure head at section 1.

Little error is introduced if this expression is applied to the flow of a compressible fluid provided that the velocity is not greater than about 60 m/s. When the velocity is high, the equation of state must be used to give the relation between the pressure and the volume of the gas. For non-isothermal flow, Pv^k = a constant,

and:

$$\int_{P_1}^{P_2} v\,dP = \frac{k}{k-1} P_1 v_1 \left[\left(\frac{P_2}{P_1}\right)^{(k-1)/k} - 1\right] \quad \text{(equation 2.73)}$$

so that:

$$\frac{\dot{u}_1^2}{2} = \frac{k}{k-1} P_1 v_1 \left[\left(\frac{P_2}{P_1}\right)^{(k-1)/k} - 1\right]$$

and:

$$P_2 = P_1 \left(1 + \frac{\dot{u}_1^2}{2} \frac{k-1}{k} \frac{1}{P_1 v_1}\right)^{k/(k-1)} \tag{6.11}$$

For isothermal flow:

$$\frac{\dot{u}_1^2}{2} = P_1 v_1 \ln \frac{P_2}{P_1} \quad \text{(from equation 2.69)}$$

$$\therefore \quad P_2 = P_1 \exp \frac{\dot{u}_1^2}{2P_1 v_1}$$

or:

$$P_2 = P_1 \exp \frac{\dot{u}_1^2 M}{2\mathbf{R}T} \tag{6.12}$$

Equations 6.11 and 6.12 can be used for the calculation of the fluid velocity and the impact pressure in terms of the static pressure a short distance upstream. The two sections are chosen so that they are sufficiently close together for frictional losses to be negligible. Thus P_1 will be approximately equal to the static pressure at both sections and the equations give the relation between the static and impact pressure—and the velocity—at any point in the fluid.

6.3. MEASUREMENT OF FLUID FLOW

The most important class of flowmeter is that in which the fluid is either accelerated or retarded at the measuring section and the change in the kinetic energy is measured by recording the pressure difference produced.

This class includes:

The pitot tube, in which a small element of fluid is brought to rest at an orifice situated at right angles to the direction of flow. The flowrate is then obtained from the difference between the impact and the static pressure. With this instrument the velocity measured is that of a small filament of fluid.

The orifice meter, in which the fluid is accelerated at a sudden constriction (the orifice) and the pressure developed is then measured. This is a relatively cheap and reliable instrument though the overall pressure drop is high because most of the kinetic energy of the fluid at the orifice is wasted.

The venturi meter, in which the fluid is gradually accelerated to a throat and gradually retarded as the flow channel is expanded to the pipe size. A high proportion of the kinetic energy is thus recovered but the instrument is expensive and bulky.

The nozzle, in which the fluid is gradually accelerated up to the throat of the instrument but expansion to pipe diameter is sudden as with an orifice. This instrument is again expensive because of the accuracy required over the inlet section.

The notch or weir, in which the fluid flows over the weir so that its kinetic energy is measured by determining the head of the fluid flowing above the weir. This instrument is used in open-channel flow and extensively in tray towers(3) where the height of the weir is adjusted to provide the necessary liquid depth for a given flow.

Each of these devices will now be considered in more detail together with some less common and special purpose meters.

6.3.1. The pitot tube

The pitot tube is used to measure the difference between the impact and static pressures in a fluid. It normally consists of two concentric tubes arranged parallel to the direction of flow; the impact pressure is measured on the open end of the inner tube. The end of the outer concentric tube is sealed and a series of orifices on the curved surface give an accurate indication of the static pressure. The position of these orifices must be carefully chosen because there are two disturbances which may cause an incorrect reading of the static pressure. These are due to:

(1) the head of the instrument;
(2) the portion of the stem which is at right angles to the direction of flow of the fluid.

These two disturbances cause errors in opposite directions, and the static pressure should therefore be measured at the point where the effects are equal and opposite.

If the head and stem are situated at a distance of 14 diameters from each other as on the standard instrument,(4) the two disturbances are equal and opposite at a section 6 diameters from the head and 8 from the stem. This is, therefore, the position at which the static pressure orifices should be located. If the distance between the head and the stem is too great, the instrument will be unwieldy; if it is too short, the magnitude of each of the disturbances will be relatively great, and a small error in the location of the static pressure orifices will appreciably affect the reading.

The two standard instruments are shown in Figure 6.13; the one with the rounded nose is preferred, since this is less subject to damage.

For Reynolds numbers of 500–300,000, based on the external diameter of the pitot tube, an error of not more than 1 per cent is obtained with this instrument. A Reynolds number of 500 with the standard 7.94 mm pitot tube corresponds to a water velocity of 0.070 m/s or an air velocity of 0.91 m/s. Sinusoidal fluctuations in the flowrate up to

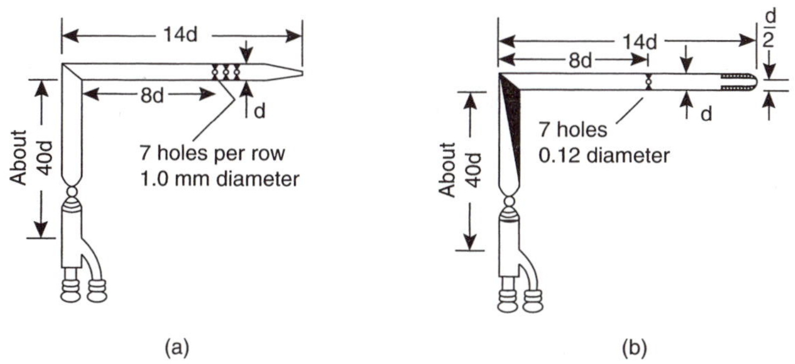

Figure 6.13. Pitot tubes

20 per cent do not affect the accuracy by more than 1 per cent, and calibration of the instrument is not necessary.

A very small pressure difference is obtained for low rates of flow of gases, and the lower limit of velocity that can be measured is usually set by the minimum difference in pressure that can be measured. This limitation is serious, and various methods have been adopted for increasing the reading of the instrument although they involve the need for calibration. Correct alignment of the instrument with respect to the direction of flow is important; this is attained when the differential reading is a maximum.

For the flow not to be appreciably disturbed, the diameter of the instrument must not exceed about one-fiftieth of the diameter of the pipe; the standard instrument (diameter 7.94 mm) should therefore not be used in pipes of less than 0.4 m diameter. An accurate measurement of the impact pressure can be obtained using a tube of very small diameter with its open end at right angles to the direction of flow; hypodermic tubing is convenient for this purpose. The static pressure is measured using a single piezometer tube or a piezometer ring upstream at a distance equal approximately to the diameter of the pipe: measurement should be made at least 50 diameters from any bend or obstruction.

The pitot tube measures the velocity of only a filament of fluid, and hence it can be used for exploring the velocity distribution across the pipe section. If, however, it is desired to measure the total flow of fluid through the pipe, the velocity must be measured at various distances from the walls and the results integrated. The total flowrate can be calculated from a single reading only if the velocity distribution across the section is already known.

Although a single pitot tube measures the velocity at only one point in a pipe or duct, instruments such as the *averaging pitot tube* or *Annubar*, which employ multiple sampling points over the cross-section, provide information on the complete velocity profile which may then be integrated to give the volumetric flowrate. An instrument of this type has the advantage that it gives rise to a lower pressure drop than most other flow measuring devices, such as the orifice meter described in Section 6.3.2.

6.3.2. Measurement by flow through a constriction

In measuring devices where the fluid is accelerated by causing it to flow through a constriction, the kinetic energy is thereby increased and the pressure energy therefore

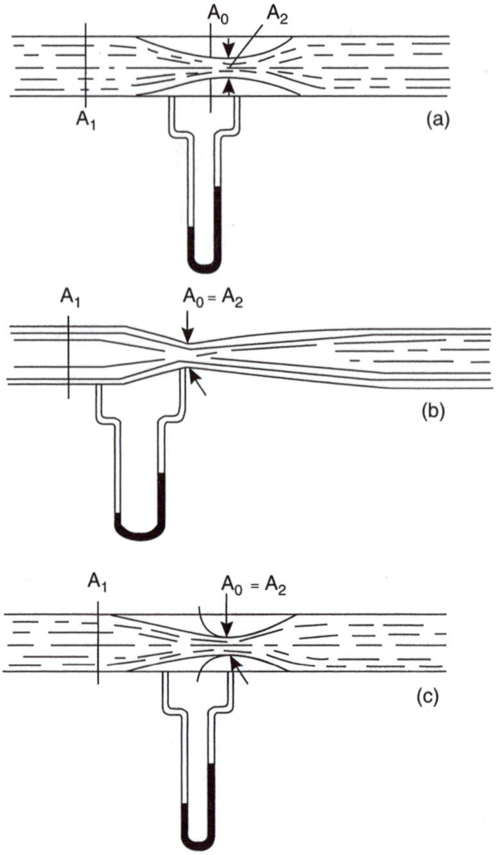

Figure 6.14. (*a*) Orifice meter (*b*) Venturi meter (*c*) Nozzle

decreases. The flowrate is obtained by measuring the pressure difference between the inlet of the meter and a point of reduced pressure, as shown in Figure 6.14 where the orifice meter, the nozzle and the venturi meter are illustrated. If the pressure is measured a short distance upstream where the flow is undisturbed (section 1) and at the position where the area of flow is a minimum (section 2), application of the energy and material balance equations gives:

$$\frac{u_2^2}{2\alpha_2} - \frac{u_1^2}{2\alpha_1} + g(z_2 - z_1) + \int_{P_1}^{P_2} v\,\mathrm{d}P + W_s + F = 0 \quad \text{(from equation 2.55)}$$

and the mass flow, $G = \dfrac{u_1 A_1}{v_1} = \dfrac{u_2 A_2}{v_2}$ (from equation 2.37)

If the frictional losses are neglected, and the fluid does no work on the surroundings, that is W_s and F are both zero, then:

$$\frac{u_2^2}{2\alpha_2} - \frac{u_1^2}{2\alpha_1} = g(z_1 - z_2) - \int_{P_1}^{P_2} v\,\mathrm{d}P \tag{6.13}$$

Inserting the value of u_1 in terms of u_2 in equation 6.13 enables u_2 and G to be obtained.

For an incompressible fluid:

$$\int_{P_1}^{P_2} v\,dP = v(P_2 - P_1) \qquad \text{(equation 2.65)}$$

and:

$$u_1 = u_2 \frac{A_2}{A_1}$$

Substituting these values in equation 6.13:

$$\frac{u_2^2}{2\alpha_2}\left(1 - \frac{\alpha_2}{\alpha_1}\frac{A_2^2}{A_1^2}\right) = g(z_1 - z_2) + v(P_1 - P_2)$$

Thus:

$$u_2^2 = \frac{2\alpha_2[g(z_1 - z_2) + v(P_1 - P_2)]}{1 - \dfrac{\alpha_2}{\alpha_1}\left(\dfrac{A_2}{A_1}\right)^2} \qquad (6.14)$$

For a horizontal meter $z_1 = z_2$, and:

$$u_2 = \sqrt{\frac{2\alpha_2 v(P_1 - P_2)}{1 - \dfrac{\alpha_2}{\alpha_1}\left(\dfrac{A_2}{A_1}\right)^2}}$$

and:

$$G = \frac{u_2 A_2}{v_2} = \frac{A_2}{v}\sqrt{\frac{2\alpha_2 v(P_1 - P_2)}{1 - \dfrac{\alpha_2}{\alpha_1}\left(\dfrac{A_2}{A_1}\right)^2}} \qquad (6.15)$$

For an ideal gas in isothermal flow:

$$\int_{P_1}^{P_2} v\,dP = P_1 v_1 \ln \frac{P_2}{P_1}. \qquad \text{(equation 2.69)}$$

and:

$$u_1 = u_2 \frac{A_2 v_1}{A_1 v_2}$$

And again neglecting terms in z, from equation 6.13:

$$\frac{u_2^2}{2\alpha_2}\left[1 - \frac{\alpha_2}{\alpha_1}\left(\frac{v_1 A_2}{v_2 A_1}\right)^2\right] = P_1 v_1 \ln \frac{P_1}{P_2}$$

and:

$$u_2^2 = \frac{2\alpha_2 P_1 v_1 \ln \dfrac{P_1}{P_2}}{1 - \dfrac{\alpha_2}{\alpha_1}\left(\dfrac{v_1 A_2}{v_2 A_1}\right)^2} \qquad (6.16)$$

and the mass flow G is again $u_2 A_2 / v_2$.

For an ideal gas in non-isothermal flow. If the pressure and volume are related by Pv^k = constant, then a similar analysis gives:

$$\int_{P_1}^{P_2} v\,dP = P_1 v_1 \frac{k}{k-1}\left[\left(\frac{P_2}{P_1}\right)^{(k-1)/k} - 1\right] \qquad \text{(equation 2.73)}$$

and, hence:
$$\frac{u_2^2}{2\alpha_2}\left[1 - \frac{\alpha_2}{\alpha_1}\left(\frac{v_1 A_2}{v_2 A_1}\right)^2\right] = -P_1 v_1 \frac{k}{k-1}\left[\left(\frac{P_2}{P_1}\right)^{(k-1)/k} - 1\right]$$

or:
$$u_2^2 = \frac{2\alpha_2 P_1 v_1 \dfrac{k}{k-1}\left[1 - \left(\dfrac{P_2}{P_1}\right)^{(k-1)/k}\right]}{1 - \dfrac{\alpha_2}{\alpha_1}\left(\dfrac{v_1 A_2}{v_2 A_1}\right)^2} \qquad (6.17)$$

and the mass flow G is again $u_2 A_2 / v_2$.

It should be noted that equations 6.16 and 6.17 apply provided that P_2/P_1 is greater than the critical pressure ratio w_c. This subject is discussed in Chapter 4, where it is shown that when $P_2/P_1 < w_c$, the flowrate becomes independent of the downstream pressure P_2 and conditions of *maximum flow* occur.

6.3.3. The orifice meter

The most important factors influencing the reading of an orifice meter (Figure 6.15) are the size of the orifice and the diameter of the pipe in which it is fitted, though a number of other factors do affect the reading to some extent. Thus the exact position and the method of fixing the pressure tappings are important because the area of flow, and hence the velocity, gradually changes in the region of the orifice. The meter should be located not less than 50 pipe diameters from any pipe fittings. Details of the exact shape of the orifice, the orifice thickness, and other details are given in BS1042[(4)] and the details must be followed if a standard orifice is to be used without calibration — otherwise the meter must be calibrated. It should be noted that the standard applies only for pipes of at least 150 mm diameter.

A simple instrument can be made by inserting a drilled orifice plate between two pipe flanges and arranging suitable pressure connections. The orifice must be drilled with sharp edges and is best made from stainless steel which resists corrosion and abrasion. The size of the orifice should be chosen to give a convenient pressure drop. Although the flowrate is proportional to the square root of the pressure drop, it is difficult to cover a wide range in flow with any given size of orifice. Unlike the pitot tube, the orifice meter gives the average flowrate from a single reading.

The most serious disadvantage of the meter is that most of the pressure drop is not recoverable, that is it is inefficient. The velocity of the fluid is increased at the throat without much loss of energy. The fluid is subsequently retarded as it mixes with the relatively slow-moving fluid downstream from the orifice. A high degree of turbulence is set up and most of the excess kinetic energy is dissipated as heat. Usually only about 5 or 10 per cent of the excess kinetic energy can be recovered as pressure energy. The pressure drop over the orifice meter is therefore high, and this may preclude it from being used in a particular instance.

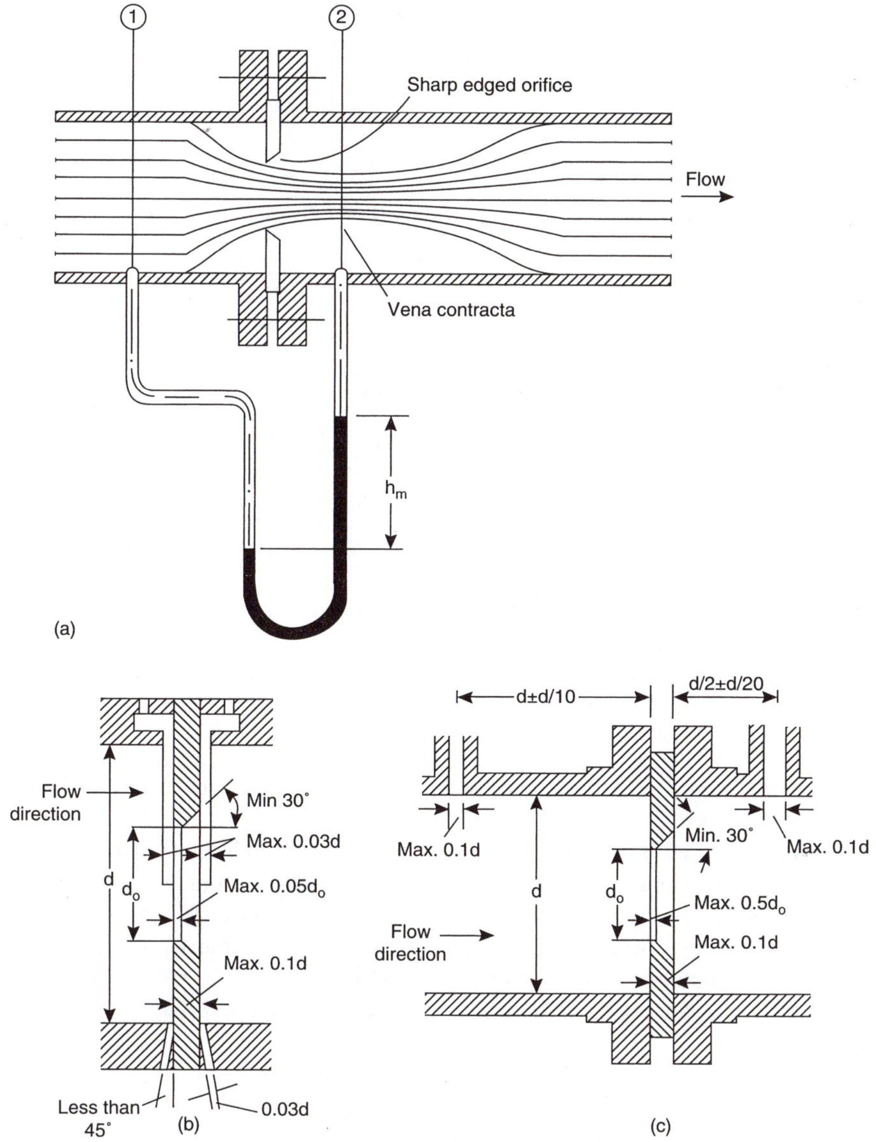

Figure 6.15. (*a*) General arrangement (*b*) Orifice plate with corner tappings. Upper half shows construction with piezometer ring. Lower half shows construction with tappings into pipe flange (*c*) Orifice plate with d and $d/2$ tappings. Nipples must finish flush with wall of pipe without burrs

The area of flow decreases from A_1 at section 1 to A_0 at the orifice and then to A_2 at the *vena contracta* (Figure 6.14). The area at the vena contracta can be conveniently related to the area of the orifice by the coefficient of contraction C_c, defined by the relation:

$$C_c = \frac{A_2}{A_0}$$

Inserting the value $A_2 = C_c A_0$ in equation 6.15, then for an *incompressible fluid* in a horizontal meter:

$$G = \frac{C_c A_0}{v} \sqrt{\frac{2\alpha_2 v(P_1 - P_2)}{1 - \frac{\alpha_2}{\alpha_1}\left(C_c \frac{A_0}{A_1}\right)^2}} \tag{6.18}$$

Using a coefficient of discharge C_D to take account of the frictional losses in the meter and of the parameters C_c, α_1, and α_2:

$$G = \frac{C_D A_0}{v} \sqrt{\frac{2v(P_1 - P_2)}{1 - \left(\frac{A_0}{A_1}\right)^2}} \tag{6.19}$$

For a meter in which the area of the orifice is small compared with that of the pipe:

$$\sqrt{1 - \left(\frac{A_0}{A_1}\right)^2} \to 1$$

and:

$$G = \frac{C_D A_0}{v} \sqrt{2v(P_1 - P_2)}$$

$$= C_D A_0 \sqrt{2\rho(P_1 - P_2)} \tag{6.20}$$

$$= C_D A_0 \rho \sqrt{2gh_0} \tag{6.21}$$

where h_0 is the difference in head across the orifice expressed in terms of the fluid in question.

This gives a simple working equation for evaluating G though the coefficient C_D is not a simple function and depends on the values of the Reynolds number in the orifice and the form of the pressure tappings. A value of 0.61 may be taken for the standard meter for Reynolds numbers in excess of 10^4, though the value changes noticeably at lower values of Reynolds number as shown in Figure 6.16.

For the isothermal flow of an ideal gas, from equation 6.16 and using C_D as above:

$$G = \frac{C_D A_0}{v_2} \sqrt{\frac{2P_1 v_1 \ln\left(\frac{P_1}{P_2}\right)}{1 - \left(\frac{v_1 A_0}{v_2 A_1}\right)^2}} \tag{6.22}$$

For a meter in which the area of the orifice is small compared with that of the pipe:

$$G = \frac{C_D A_0}{v_2} \sqrt{2P_1 v_1 \ln\left(\frac{P_1}{P_2}\right)} \tag{6.23}$$

$$= C_D A_0 \sqrt{2\frac{P_2}{v_2} \ln \frac{P_1}{P_2}}$$

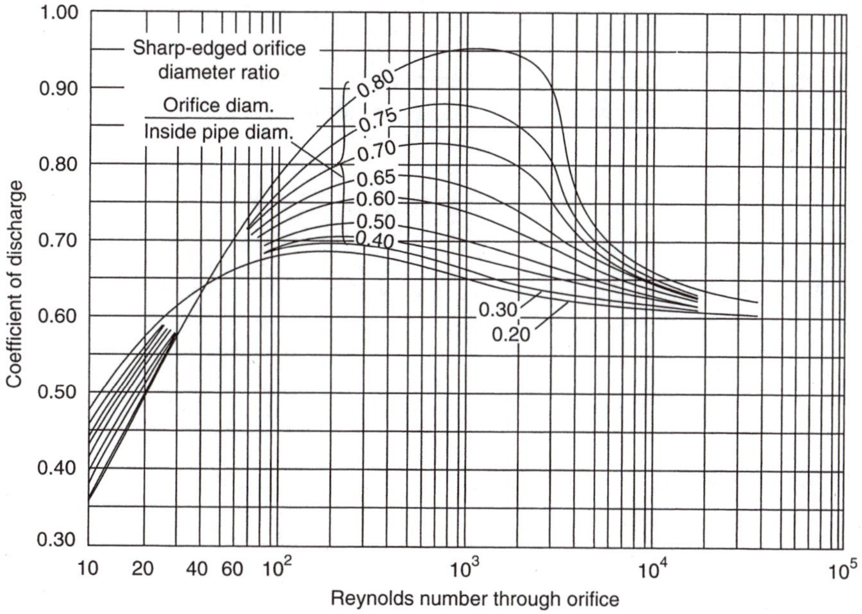

Figure 6.16. Coefficient for orifice meter

$$= C_D A_0 \sqrt{2\frac{P_2}{v_2}\ln\left(1-\frac{\Delta P}{P_2}\right)} \quad (\text{where}\,\Delta P = P_2 - P_1)$$

$$= C_D A_0 \sqrt{2\left(\frac{-\Delta P}{v_2}\right)} \quad \text{if } \Delta P \text{ is small compared with } P_2 \tag{6.24}$$

For non-isothermal flow of an ideal gas, from equation 6.17:

$$G = \frac{C_D A_0}{v_2}\sqrt{\frac{2P_1 v_1 \left[\dfrac{k}{k-1}\right]\left[1-\left(\dfrac{P_2}{P_1}\right)^{(k-1)/k}\right]}{1-\left(\dfrac{v_1 A_0}{v_2 A_1}\right)^2}} \tag{6.25}$$

For a horizontal orifice in which (A_0/A_1) is small:

$$G = \frac{C_D A_0}{v_2}\sqrt{2P_1 v_1 \frac{k}{k-1}\left[1-\left(\frac{P_2}{P_1}\right)^{(k-1)/k}\right]} \tag{6.26}$$

$$= \frac{C_D A_0}{v_2}\sqrt{2P_1 v_1 \frac{k}{k-1}\left[1-\left(1+\frac{\Delta P}{P_1}\right)^{(k-1)/k}\right]}$$

$$= C_D A_0 \sqrt{2\left(\frac{-\Delta P}{v_2}\right)} \quad \text{if } \Delta P \text{ is small compared with } P_2 \tag{6.27}$$

Equations 6.22, 6.23, 6.25 and 6.26 hold provided that P_2/P_1 is greater than the critical pressure ratio w_c.

For $P_2/P_1 < w_c$, the flowrate through the orifice is maintained at the maximum rate. From equation 4.15, the flowrate is given by equation 6.28 and *not equation* 6.23 for isothermal conditions:

$$G_{max} = 0.607 C_D A_0 \sqrt{\frac{P_1}{v_1}} \tag{6.28}$$

Similarly, when the pressure-volume relation is Pv^k = constant, equation 4.30 replaces equation 6.26:

$$G_{max} = C_D A_0 \sqrt{\frac{kP_1}{v_1}\left(\frac{2}{k+1}\right)^{(k+1)/(k-1)}} \tag{6.29}$$

For isentropic conditions $k = \gamma$ and:

$$G_{max} = C_D A_0 \sqrt{\frac{\gamma P_1}{v_1}\left(\frac{2}{\gamma+1}\right)^{(\gamma+1)/(\gamma-1)}} \tag{6.30}$$

For a diatomic gas, $\gamma = 1.4$ and:

$$G_{max} = 0.685 C_D A_0 \sqrt{\frac{P_1}{v_1}} \tag{6.31}$$

For the flow of steam, a highly non-ideal gas, it is necessary to apply a correction to the calculated flowrate, the magnitude of which depends on whether the steam is saturated, wet or superheated. Correction charts are given by LYLE[5] who also quotes a useful approximation[6] — that a steam meter registers 1 per cent low for every 2 per cent of liquid water in the steam, and 1 per cent high for every 8 per cent of superheat.

Example 6.1

Water flows through an orifice of 25 mm diameter situated in a 75 mm diameter pipe, at a rate of 300 cm^3/s. What will be the difference in level on a water manometer connected across the meter? The viscosity of water is 1 mN s/m^2.

Solution

$$\text{Area of orifice} = \frac{\pi}{4} \times 25^2 = 491 \text{ mm}^2 \text{ or } 4.91 \times 10^{-4} \text{ m}^2$$

$$\text{Flow of water} = 300 \text{ cm}^3/\text{s or } 3.0 \times 10^{-4} \text{ m}^3/\text{s}$$

$$\therefore \text{ Velocity of water through the orifice} = \frac{3.0 \times 10^{-4}}{4.91 \times 10^{-4}} = 0.61 \text{ m/s}$$

$$Re \text{ at the orifice} = \frac{25 \times 10^{-3} \times 0.61 \times 1000}{1 \times 10^{-3}} = 15250$$

From Figure 6.16, the corresponding value of $C_D = 0.61$ (diameter ratio = 0.33):

$$\left[1 - \left(\frac{A_0}{A_1}\right)^2\right]^{0.5} = \left[1 - \left(\frac{25^2}{75^2}\right)^2\right]^{0.5} = 0.994 \approx 1$$

Equation 6.21 may therefore be applied:

$$G = 3.0 \times 10^{-4} \times 10^3 = 0.30 \text{ kg/s.}$$

$\therefore$ $$0.30 = (0.61 \times 4.91 \times 10^{-4} \times 10^3)\sqrt{(2 \times 9.81 \times h_0)}.$$

Hence: $$\sqrt{h_0} = 0.226$$

and: $$h_0 = 0.051 \text{ m of water}$$

$$= \underline{\underline{51 \text{ mm of water}}}$$

Example 6.2

Sulphuric acid of density 1300 kg/m^3 is flowing through a pipe of 50 mm, internal diameter. A thin-lipped orifice, 10 mm in diameter is fitted in the pipe and the differential pressure shown on a mercury manometer is 0.1 m. Assuming that the leads to the manometer are filled with the acid, calculate (a) the mass flow rate of acid and (b) the approximate drop in pressure caused by the orifice in kN/m^2. The coefficient of discharge of the orifice may be taken as 0.61, the density of mercury as 13,550 kg/m^3 and the density of the water as 1000 kg/m^3.

Solution

(a) The mass flow-rate G is given by:

$$G = [(C_D A_0)/v]\sqrt{[(2v(P_1 - P_2))/(1 - (A_0/A_1)^2]} \qquad \text{(equation 6.19)}$$

where the area of the orifice is small compared with the area of the pipe; that is $\sqrt{[1 - (A_0/A_1)^2]}$ approximates to 1.0 and:

$$G = C_D A_0 \rho \sqrt{(2gh_0)} \qquad \text{(equation 6.21)}$$

where h_0 is the difference in head across the orifice expressed in terms of the fluid.
The area of the orifice, is given by:

$$A_0 = (\pi/4)(10 \times 10^{-3})^2 = 7.85 \times 10^{-5} \text{m}^2$$

and the area of the pipe by:

$$A_1 = (\pi/4)(50 \times 10^{-3})^2 = 196.3 \times 10^{-5} \text{ m}^2$$

Thus:

$$[1 - (A_0/A_1)^2] = [1 - (7.85/196.3)^2] = 0.999$$

The differential pressure is given by:

$$h_0 = 0.1 \text{ m mercury}$$

$$= 0.1(13550 - 1300)/1300 = 0.94 \text{ m sulphuric acid}$$

and, substituting in equation 6.21 gives the mass flowrate as:

$$G = (0.61 \times 7.85 \times 10^{-15})(1300)\sqrt{(2 \times 9.81 \times 0.94)}$$

$$= \underline{\underline{0.268 \text{ kg/s}}}$$

(b) the drop in pressure is then:

$$-\Delta P = \rho g h_0$$

$$= (1300 \times 9.81 \times 0.94) = 11988 \text{ N/m}^2 \text{ or } \underline{\underline{12 \text{ kN/m}^2}}$$

6.3.4. The nozzle

The nozzle is similar to the orifice meter other than that it has a converging tube in place of the orifice plate, as shown in Figures 6.14*c* and 6.17. The velocity of the fluid is gradually increased and the contours are so designed that almost frictionless flow takes place in the converging portion; the outlet corresponds to the vena contracta on the orifice meter. The nozzle has a constant high coefficient of discharge (*ca.* 0.99) over a wide range of conditions because the coefficient of contraction is unity, though because the simple nozzle is not fitted with a diverging cone, the head lost is very nearly the same as with an orifice. Although much more costly than the orifice meter, it is extensively used for metering steam. When the ratio of the pressure at the nozzle exit to the upstream pressure is less than the critical pressure ratio w_c, the flowrate is independent of the downstream pressure and can be calculated from the upstream pressure alone.

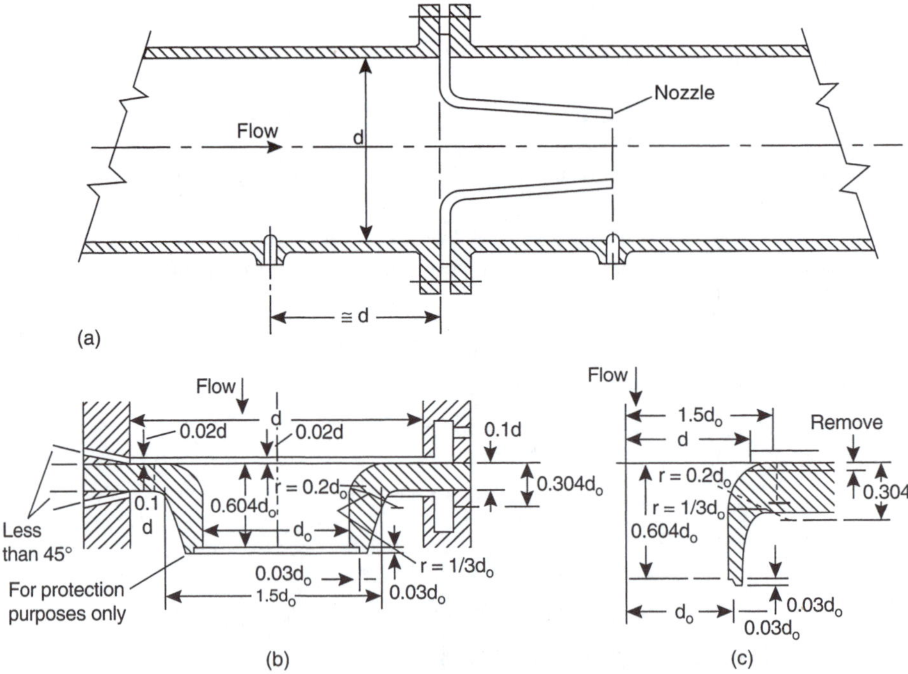

Figure 6.17. (*a*) General arrangement (*b*) Standard nozzle (A_0/A_1) is less than 0.45. Left half shows construction for corner tappings. Right half shows construction for piezometer ring (*c*) Standard nozzle where A_0/A_1 is greater than 0.45

6.3.5. The venturi meter

In this meter, illustrated in Figures 6.14*b* and 6.18, the fluid is accelerated by its passage through a converging cone of angle 15–20°. The pressure difference between the upstream end of the cone (section 1) and the throat (section 2) is measured and provides the signal for the rate of flow. The fluid is then retarded in a cone of smaller angle (5–7°) in which a large proportion of the kinetic energy is converted back to pressure energy. Because of the gradual reduction in the area of flow there is no vena contracta and the flow area is a minimum at the throat so that the coefficient of contraction is unity. The attraction of this meter lies in its high energy recovery so that it may be used where only a small pressure head is available, though its construction[4] is expensive. The flow relationship is given by a similar equation to that for the orifice.

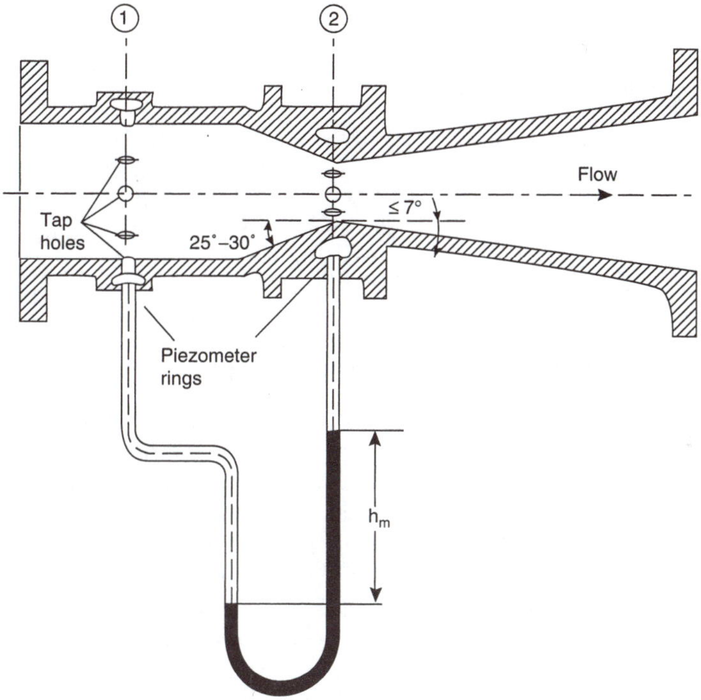

Figure 6.18. The venturi meter

For an incompressible fluid in horizontal meter:

$$G = \frac{C_D A_2}{v}\sqrt{\frac{2v(P_1 - P_2)}{1 - (A_2/A_1)^2}} \qquad \text{(equation 6.19)}$$

$$= C_D \rho \frac{A_1 A_2}{\sqrt{A_1^2 - A_2^2}}\sqrt{2v(P_1 - P_2)} \qquad (6.32)$$

$$= C_D \rho C' \sqrt{2gh_v} \qquad (6.33)$$

where C' is a constant for the meter and h_v is the loss in head over the converging cone expressed as height of fluid. The coefficient C_D is high, varying from 0.98 to 0.99. The meter is equally suitable for compressible and incompressible fluids.

Example 6.3

The rate of flow of water in a 150 mm diameter pipe is measured with a venturi meter with a 50 mm diameter throat. When the pressure drop over the converging section is 121 mm of water, the flowrate is 2.91 kg/s. What is the coefficient for the converging cone of the meter at this flowrate?

Solution

From equation 6.32, the mass rate of flow,

$$G = \frac{C_D \rho A_1 A_2 \sqrt{(2gh_v)}}{\sqrt{(A_1^2 - A_2^2)}}$$

The coefficient for the meter is therefore given by:

$$2.91 = C_D \times 1000 \left(\frac{\pi}{4}\right)^2 (150 \times 10^{-3})^2 (50 \times 10^{-3})^2$$

$$\times \frac{\sqrt{(2 \times 9.81 \times 121 \times 10^{-3})}}{(\pi/4)\sqrt{[(150 \times 10^{-3})^4 - (50 \times 10^{-3})^4]}}$$

and: $C_D = \underline{\underline{0.985}}$

6.3.6. Pressure recovery in orifice-type meters

In the orifice meter little of the excess kinetic energy is converted back into pressure energy and there is a substantial drop in pressure over the instrument. This is seen in Figure 6.19 where the pressure is shown in relation to the position along the tube. There is a steep drop in pressure of about 10 per cent as the vena contracta is reached and the subsequent pressure recovery is poor, amounting to about 50 per cent of the loss. The *Dall Tube* (Figure 6.20) has been developed to provide a low head loss over the meter while still giving a reasonably high value of pressure head over the orifice and thus offering great advantages. A short length of parallel lead-in tube is followed by the converging upstream cone and then a diverging downstream cone. This recovery cone is formed by a liner which fits into the meter, the *throat* being formed by a circular slot located between the two cones. One pressure connection is taken to the throat through the annular chamber shown, and the second tapping is on the upstream side. The flow leaves the throat as a diverging jet which follows the walls of the downstream cone so that eddy losses are almost eliminated. These instruments are made for pipe sizes greater than 150 mm and the pressure loss is only 2–8 per cent of the differential head. They are cheaper than a venturi, much shorter in length, and correspondingly lighter.

It should be noted that all these restricted flow type meters are primarily intended for pipe sizes greater than 50 mm, and when used on smaller tubes they must be individually calibrated.

Figure 6.19 shows the pressure changes over the four instruments considered. If the upstream pressure is P_1, the throat pressure P_2, and the final recovery pressure P_3, then

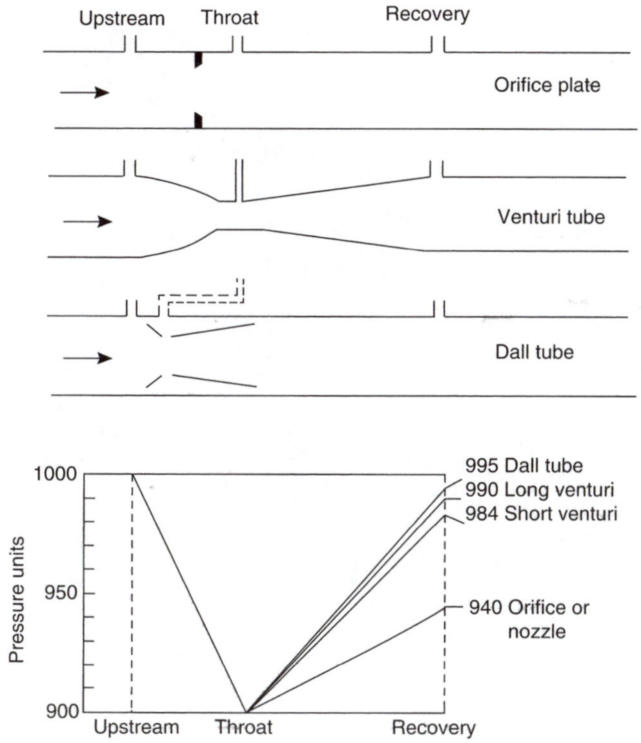

Figure 6.19. Pressure distribution using orifice plate, venturi meter, and Dall tube. Pressure falls by 10% from upper pressure tapping to throat in each case

the pressure recovery is conveniently expressed as $(P_1 - P_3)/(P_1 - P_2)$, and for these meters the values are given in Table 6.1.

Table 6.1. Typical meter pressure recovery

Meter type	Value of $(P_1 - P_3)/(P_1 - P_2)$ (%)
Orifice	40–50
Nozzle	40–50
Venturi	80–90
Dall	92–98

Because of its simplicity the orifice meter is commonly used for process measurements, and this instrument is suitable for providing a signal of the pressure to some comparator as indicated in Figure 6.1.

6.3.7. Variable area meters — rotameters

In the meters so far described the area of the constriction or orifice is constant and the drop in pressure is dependent on the rate of flow. In the variable area meter, the drop in pressure is constant and the flowrate is a function of the area of the constriction.

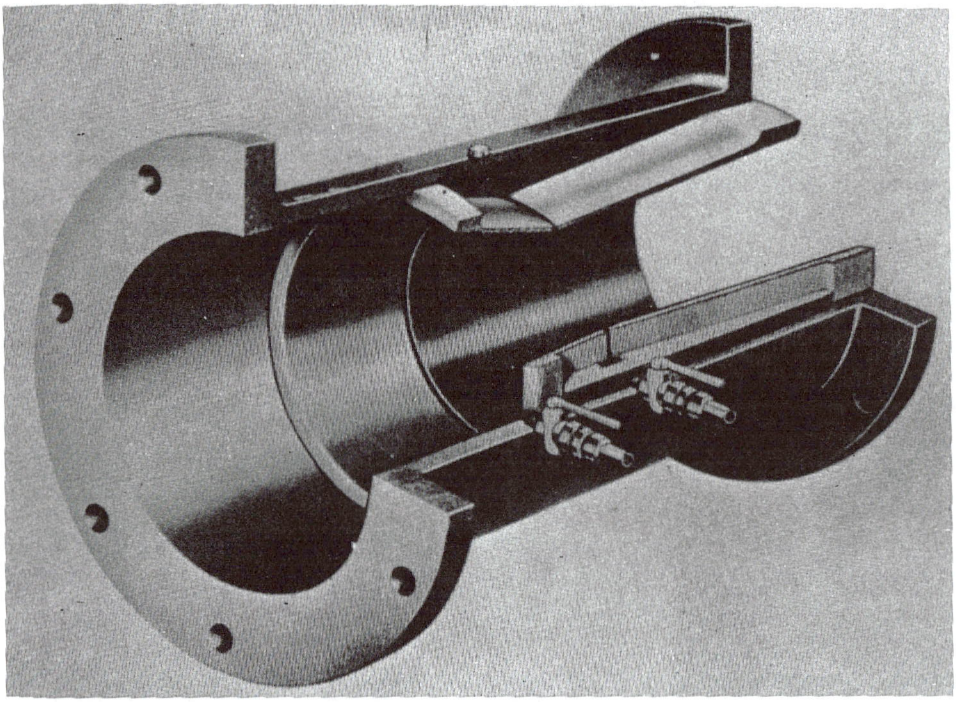

Figure 6.20. Dall tube

A typical meter of this kind, which is commonly known as a *rotameter* (Figure 6.21), consists of a tapered tube with the smallest diameter at the bottom. The tube contains a freely moving float which rests on a stop at the base of the tube. When the fluid is flowing the float rises until its weight is balanced by the upthrust of the fluid, its position then indicating the rate of flow. The pressure difference across the float is equal to its weight divided by its maximum cross-sectional area in a horizontal plane. The area for flow is the annulus formed between the float and the wall of the tube.

This meter may thus be considered as an orifice meter with a variable aperture, and the formulae already derived are therefore applicable with only minor changes. Both in the orifice-type meter and in the rotameter the pressure drop arises from the conversion of pressure energy to kinetic energy and from frictional losses which are accounted for in the coefficient of discharge. The pressure difference over the float $-\Delta P$, is given by:

$$-\Delta P = \frac{V_f(\rho_f - \rho)g}{A_f} \tag{6.34}$$

where V_f is the volume of the float, ρ_f the density of the material of the float, and A_f is the maximum cross-sectional area of the float in a horizontal plane.

If the area of the annulus between the float and tube is A_2 and the cross-sectional area of the tube is A_1, then from equation 6.19:

$$G = C_D A_2 \sqrt{\frac{2\rho(-\Delta P)}{1 - (A_2/A_1)^2}} \tag{6.35}$$

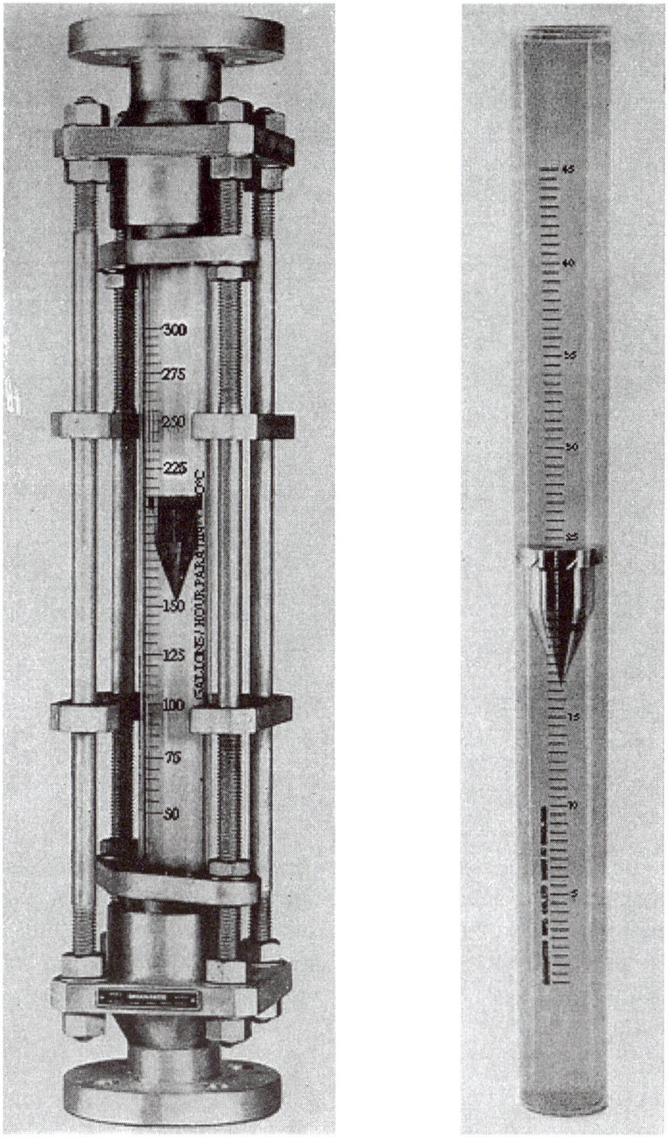

Figure 6.21. Variable area flowmeters

Substituting for $-\Delta P$ from equation 6.34:

$$G = C_D A_2 \sqrt{\frac{2gV_f(\rho_f - \rho)\rho}{A_f[1 - (A_2/A_1)^2]}} \tag{6.36}$$

The coefficient C_D depends on the shape of the float and the Reynolds number (based on the velocity in the annulus and the mean hydraulic diameter of the annulus) for the flow through the annular space of area A_2. In general, floats which give the most nearly constant coefficient are of such a shape that they set up eddy currents and give low values of C_D. The variation in C_D largely arises from differences in viscous drag of fluid on the float, and if turbulence is artificially increased, the drag force rises quickly to a limiting but high value. As seen in Figure 6.22, float A does not promote turbulence and the coefficient rises slowly to a high value of 0.98. Float C promotes turbulence and C_D rises quickly but only to a low value of 0.60.

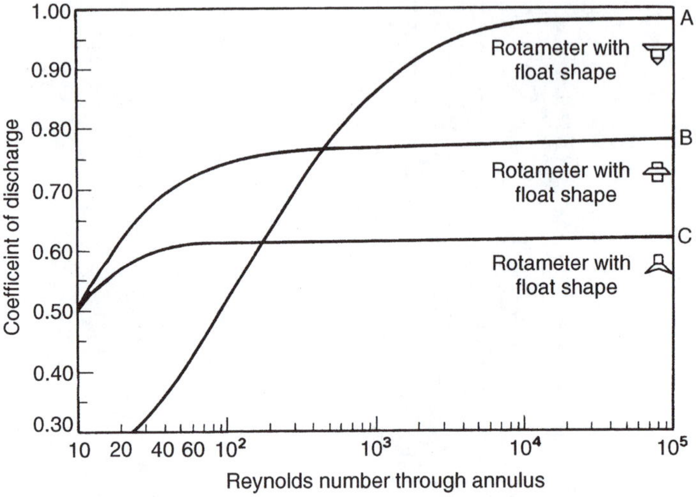

Figure 6.22. Coefficients for rotameters

The constant coefficient for float C arises from turbulence promotion, and for this reason the coefficient is also substantially independent of the fluid viscosity. The meter can be made relatively insensitive to changes in the density of the fluid by selection of the density of the float, ρ_f. Thus the flowrate for a given meter will be independent of ρ when $\mathrm{d}G/\mathrm{d}\rho = 0$.
From equation 6.36:

$$\frac{\mathrm{d}G}{\mathrm{d}\rho} = \frac{C_D A_2}{2}\sqrt{\frac{2gV_f}{A_f[1-(A_2/A_1)^2]}}\left\{(\rho_f-\rho)^{1/2}\tfrac{1}{2}\rho^{-1/2} - \rho^{1/2}\tfrac{1}{2}(\rho_f-\rho)^{-1/2}\right\} \tag{6.37}$$

When:
$$\frac{\mathrm{d}G}{\mathrm{d}\rho} = 0,$$

$$\rho_f = 2\rho \tag{6.38}$$

Thus if the density of the float is twice that of the fluid, then the position of the float for a given flow is independent of the fluid density.

The range of a meter can be increased by the use of floats of different densities, a given float covering a flowrate range of about 10:1. For high pressure work the glass tube is replaced by a metal tube. When a metal tube is used or when the liquid is very dark or dirty an external indicator is required.

Example 6.4

A rotameter tube is 0.3 m long with an internal diameter of 25 mm at the top and 20 mm at the bottom. The diameter of the float is 20 mm, its density is 4800 kg/m^3 and its volume is 6.0 cm^3. If the coefficient of discharge is 0.7, what is the flowrate of water (density 1000 kg/m^3) when the float is halfway up the tube?

Solution

From equation 6.36:

$$G = C_D A_2 \sqrt{\frac{2gV_f(\rho_f - \rho)\rho}{A_f[1 - (A_2/A_1)^2]}}$$

Cross-sectional area at top of tube $= \frac{\pi}{4} 25^2 = 491 \text{ mm}^2 \text{ or } 4.91 \times 10^{-4} \text{ m}^2$

Cross-sectional area at bottom of tube $= \frac{\pi}{4} 20^2 = 314 \text{ mm}^2 \text{ or } 3.14 \times 10^{-4} \text{ m}^2$

Area of float: $A_f = 3.14 \times 10^{-4} \text{ m}^2$

Volume of float: $V_f = 6.0 \times 10^{-6} \text{ m}^3$

When the float is halfway up the tube, the area at the height of the float A_1 is given by:

$$A_1 = \frac{\pi}{4} 22.5^2 \times 10^{-4}$$

or:

$$A_1 = 3.98 \times 10^{-4} \text{ m}^2$$

The area of the annulus A_2 is given by:

$$A_2 = A_1 - A_f = 0.84 \times 10^{-4} \text{ m}^2$$

$$A_2/A_1 = 0.211 \quad \text{and} \quad [1 - (A_2/A_1)^2] = 0.955$$

Substituting into equation 6.36:

$$G = (0.70 \times 0.84 \times 10^{-4}) \sqrt{\left\{ \frac{[2 \times 9.81 \times 6.0 \times 10^{-6}(4800 - 1000)1000]}{(3.14 \times 10^{-4} \times 0.955)} \right\}}$$

$$= \underline{\underline{0.072 \text{ kg/s}}}$$

6.3.8. The notch or weir

The flow of a liquid presenting a free surface can be measured by means of a weir. The pressure energy of the liquid is converted into kinetic energy as it flows over the weir which may or may not cover the full width of the stream, and a calming screen may be fitted before the weir. Then the height of the weir crest gives a measure of the rate of flow. The velocity with which the liquid leaves depends on its initial depth below the surface. For unit mass of liquid, initially at a depth h below the free surface, discharging through a notch, the energy balance (equation 2.67) gives:

$$\Delta \frac{u^2}{2} + v\Delta P = 0 \tag{6.39}$$

for a frictionless flow under turbulent conditions. If the velocity upstream from the notch is small and the flow upstream is assumed to be unaffected by the change of area at the notch:

$$\frac{u_2^2}{2} = -v\Delta P = gh$$

and:

$$u_2 = \sqrt{2gh} \tag{6.40}$$

Rectangular notch

For a rectangular notch (Figure 6.23) the rate of discharge of fluid at a depth h through an element of cross-section of depth $\mathrm{d}h$ will be given by:

$$\mathrm{d}Q = C_D B\,\mathrm{d}h\sqrt{2gh} \tag{6.41}$$

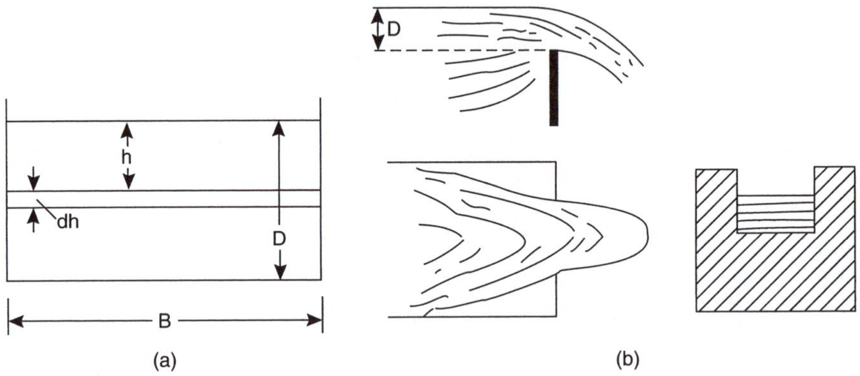

Figure 6.23. Rectangular notch

where C_D is the coefficient of discharge (usually about 0.6) and B is the width of the notch.

Thus, the total rate of flow,

$$Q = C_D B\sqrt{2g}(\tfrac{2}{3})D^{1.5} \tag{6.42}$$

where D is the total depth of liquid above the bottom of the notch.

The empirical Francis formula:

$$Q = 1.84(B - 0.1nD)D^{1.5} \tag{6.43}$$

gives the flowrate in m^3/s if the dimensions of the notch are expressed in metres with $C_D = 0.62$; $n = 0$ if the notch is the full width of the channel; $n = 1$ if the notch is narrower than the channel but is arranged with one edge coincident with the edge of the channel; $n = 2$ if the notch is narrower than the channel and is situated symmetrically (see Figure 6.23*b*): n is known as the number of end contractions.

Example 6.5

Water flows in an open channel across a weir which occupies the full width of the channel. The length of the weir is 0.5 m and the height of water over the weir is 100 mm. What is the volumetric flowrate of water?

Solution

Use is made of the Francis formula:

$$Q = 1.84(L - 0.1nD)D^{1.5} \text{ (m}^3\text{/s)} \qquad \text{(equation 6.43)}$$

where L is the length of the weir (m), n the number of end contractions (in this case $n = 0$), and D the height of liquid above the weir (m).

Thus:

$$Q = 1.84(0.5)0.100^{1.5}$$
$$= \underline{\underline{0.030 \text{ m}^3\text{/s}}}$$

Example 6.6

An organic liquid flows across a distillation tray and over a weir at the rate of 15 kg/s. The weir is 2 m long and the liquid density is 650 kg/m^3. What is the height of liquid flowing over the weir?

Solution

Use is made of the Francis formula (equation 6.43), where, as in the previous example, $n = 0$. In the context of this example the height of liquid flowing over the weir is usually designated h_{ow} and the volumetric liquid flow by Q. Rearrangement of equation 6.43 gives:

$$h_{ow} = 0.666\left(\frac{Q}{L_w}\right)^{0.67} \text{ (m)}$$

where Q is the liquid flowrate (m^3/s) and L_w the weir length (m).
In this case:

$$Q = \left(\frac{15}{650}\right) = 0.0230 \text{ m}^3\text{/s} \quad \text{and} \quad L_w = 2.0 \text{ m}$$

$$\therefore \quad h_{ow} = 0.666\left(\frac{0.0230}{2}\right)^{0.67} = 0.033 \text{ m}$$
$$= \underline{\underline{33.0 \text{ mm}}}$$

Triangular notch

For a triangular notch of angle 2θ, the flow $\mathrm{d}Q$ through the thin element of cross-section (Figure 6.24) is given by:

$$\mathrm{d}Q = C_D(D - h)2\tan\theta \mathrm{d}h\sqrt{2gh} \tag{6.44}$$

The total rate of flow,

$$Q = 2C_D\tan\theta\sqrt{2g}\int_0^D (Dh^{1/2} - h^{3/2})\mathrm{d}h$$
$$= 2C_D\tan\theta\sqrt{2g}(\tfrac{2}{3}D^{5/2} - \tfrac{2}{5}D^{5/2})$$
$$= \tfrac{8}{15}C_D\tan\theta\sqrt{2g}D^{5/2} \tag{6.45}$$

For a 90° notch for which $C_D = 0.6$, and using SI units:

$$Q = 1.42D^{2.5} \tag{6.46}$$

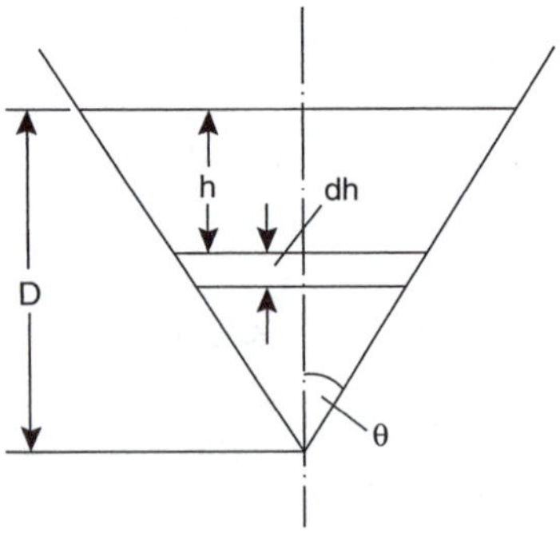

Figure 6.24. Triangular notch

Thus for a rectangular notch the rate of discharge is proportional to the liquid depth raised to a power of 1.5, and for a triangular notch to a power of 2.5. A triangular notch will, therefore, handle a wider range of flowrates. In general, the index of D is a function of the contours of the walls of the notch, and any desired relation between flowrate and depth can be obtained by a suitable choice of contour. It is sometimes convenient to employ a notch in which the rate of discharge is directly proportional to the depth of liquid above the bottom of the notch. It can be shown that the notch must have curved walls giving a large width to the bottom of the notch and a comparatively small width towards the top.

While open channels are not used frequently for the flow of material other than cooling water between plant units, a weir is frequently installed for controlling the flow within the unit itself, for instance in a distillation column or reactor.

6.3.9. Other methods of measuring flowrates

The meters which have been described so far depend for their operation on the conversion of some of the kinetic energy of the fluid into pressure energy or vice versa. They form by far the largest class of flowmeters. Other meters are used for special purposes, and brief reference will now be made to a few of these.

For further details reference should be made to Volume 3, which gives a list of the characteristics of flow meters in common use, together with an account of the principles on which they operate and their operational range.

Hot-wire anemometer

If a heated wire is immersed in a fluid, the rate of loss of heat will be a function of the flowrate. In the hot-wire anemometer a fine wire whose electrical resistance has a high temperature coefficient is heated electrically. Under equilibrium conditions the rate of loss of heat is then proportional to $I^2\Omega$, where Ω is the resistance of the wire and I is the current flowing.

Either the current or the resistance (and hence the temperature) of the wire is maintained constant. The following is an example of a method in which the resistance is maintained constant. The wire is incorporated as one of the resistances of a Wheatstone network

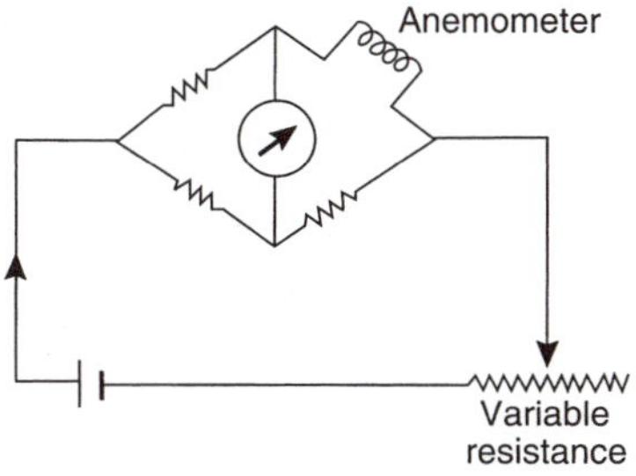

Figure 6.25. Typical circuit for a hot-wire anemometer

(Figure 6.25) in which the other three resistances have low temperature coefficients. The circuit is balanced when the wire is immersed in the stationary fluid but, when the fluid is set in motion, the rate of loss of heat increases and the temperature of the wire falls. Its resistance therefore changes and the bridge is thrown out of balance. The balance can be restored by increasing the current so that the temperature and resistance of the wire are brought back to their original values; the other three resistances will be unaffected by the change in current because of their low temperature coefficients. The current flowing in the wire is then measured using either an ammeter or a voltmeter. The rate of loss of heat is found to be approximately proportional to $\sqrt{u\rho + b'}$, so that

$$a'\sqrt{u\rho + b'} = I^2\Omega \quad \text{under equilibrium conditions} \tag{6.47}$$

where u is the velocity of the fluid (m/s), ρ the density (kg/m^3), and a' and b' are constants for a given meter.

Since the resistance of the wire is maintained constant:

$$u\rho = \frac{I^4\Omega^2}{a'^2} - b' = a''I^4 - b' \tag{6.48}$$

where $a'' = \Omega^2/a'^2$ remains constant, i.e. the mass rate of flow per unit area is a function of the fourth power of the current, which can be accurately measured.

The hot-wire anemometer is very accurate even for very low rates of flow. It is one of the most convenient instruments for the measurement of the flow of gases at low velocities; accurate readings are obtained for velocities down to about 0.03 m/s. If the ammeter has a high natural frequency, pulsating flows can be measured. Platinum wire is commonly used.

The magnetic flowmeter

The magnetic flowmeter is a volumetric metering device based on a discovery by Faraday that electrical currents are induced in a conductor that is moved through a magnetic field. Faraday's Law states that the voltage induced in a conductor is proportional to the width of the conductor, the intensity of the magnetic field, and the velocity of the conductor. Thus the output of the flowmeter, a millivoltage, is linearly related to the velocity of the conductor, which is the flowing process fluid. The principle of operation and meter itself is shown in Figure 6.26.

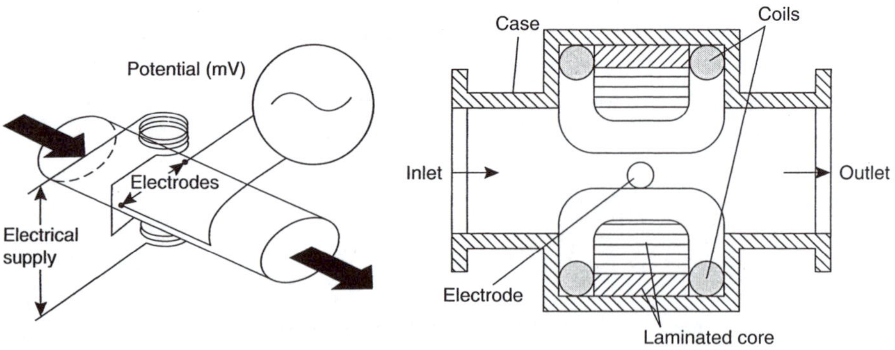

Figure 6.26. The magnetic meter

The induced voltage is in a plane that is mutually perpendicular to both the motion of the conductor and the magnetic field. The process fluid passes through the magnetic field induced by electromagnetic coils or permanent magnets built around the short length of pipe called the metering section. Because of the magnetic field used for the induction, the metering section must be of non-magnetic material. Further, if it is metal, it must have a lining that serves as an electrical insulator from the flowing liquid.

These meters have been very useful for handling liquid metal coolants where the conductivity is high though they have now been developed for liquids such as tap water which have a poor conductivity. Designs are now available for all sizes of tubes, and the instrument is particularly suitable for pharmaceutical and biological purposes where contamination of the fluid must be avoided.

This meter has a number of special features which should be noted. Thus it gives rise to a negligible drop in pressure for the fluid being metered; fluids containing a high percentage of solids can be metered and the flow can be in either direction. The liquid must have at least a small degree of electrical conductivity. Further details are given by GINESI(8).

Vortex-shedding flowmeters

When a bluff body is interspersed in a fluid stream, the flow is split into two parts. The boundary layer (see Chapter 11) which forms over the surface of the obstruction develops instabilities and vortices are formed and then shed successively from alternate sides of the body, giving rise to what is known as a *von Karman vortex street*. This process sets up regular pressure variations downstream from the obstruction whose frequency is proportional to the fluid velocity, as shown by STROUHAL(9). Vortex flowmeters are very versatile and can be used with almost any fluid — gases, liquids and multi-phase fluids. The operation of the vortex meter, illustrated in Figure 6.27, is described in more detail in Volume 3, by GINESI(8) and in a publication by a commercial manufacturer, ENDRESS and HAUSER.(10)

The time-of-flight ultrasonic flowmeter

A high-frequency pressure wave is transmitted at an acute angle to the walls of the pipe and impinges on a receiver on the other side of the pipe. The elapsed time between

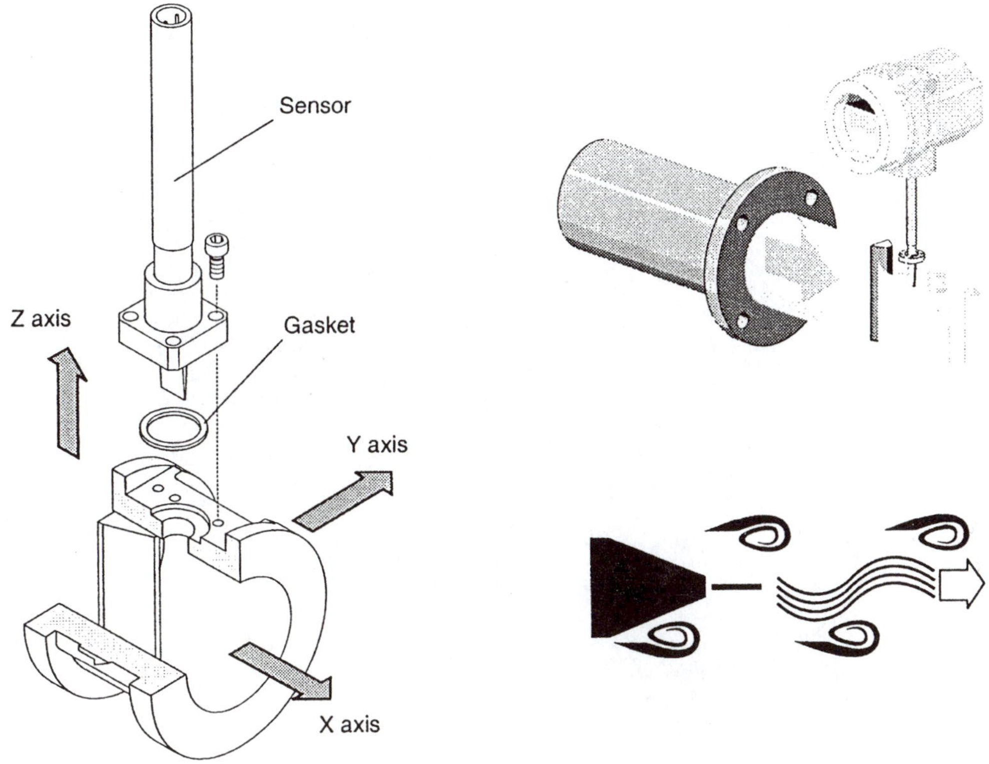

Figure 6.27. Vortex flow measuring system

transmission and reception is a function of the velocity and velocity profile in the fluid, and on the velocity of sound in the fluid and the angle at which the wave is transmitted. These meters which are suitable for the measurement of flowrates of clean liquids only, are described in detail by LYNNWORTH[11].

The Doppler ultrasonic flowmeter

The Doppler meter may be used wherever small particulate solids, bubbles or droplets are dispersed in the fluid and are moving at essentially the same velocity as the fluid stream which is to be metered. A continuous ultrasonic wave is transmitted, again at an acute angle to the wall of the duct, and the shift in frequency between the transmitted and scattered waves is measured. This method of measurement of flowrate is frequently used for slurries and dispersions which present considerable difficulties when other methods are used.

The Coriolis meter

The Coriolis meter (Figure 6.28) contains a sensor consisting of one or more tubes which are vibrated at their resonant frequency by electromagnetic drivers, and their harmonic vibrations impart an angular motion to the fluid as it passes through the tubes which,

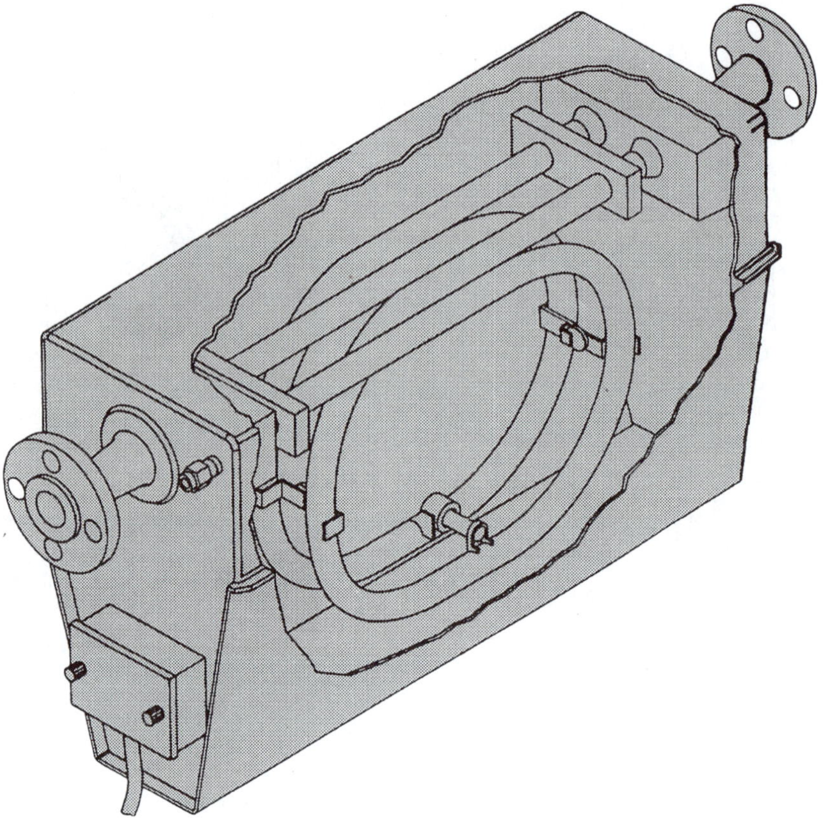

Figure 6.28. The Coriolis Meter[8]

in turn, exert Coriolis forces on the tube walls. The magnitude of the Coriolis forces is proportional to the product of the velocity and density of the fluid. A secondary movement of the tubes occurs which is proportional to the mass flowrate and this then becomes superimposed on the primary vibration. A sensor then detects and measures the magnitude of this secondary oscillation. Details of the characteristics of the meters are given by GINESI[8] and by PLACHE[12].

Quantity meters

The meters which have been described so far give an indication of the rate of flow of fluid; the total amount passing in a given time must be obtained by integration. Orifice meters are frequently fitted with integrating devices. A number of instruments is available, however, for measuring directly the total quantity of fluid which has passed. An average rate of flow can then be obtained by dividing the quantity by the time of passage.

Gas meters

A simple quantity meter which is used for the measurement of the flow of gas in an accessible duct is the anemometer (Figure 6.29). A shaft carrying radial vanes or cups

Figure 6.29. Vane anemometer

supported on low friction bearings is caused to rotate by the passage of the gas; the relative velocity between the gas stream and the surface of the vanes is low because the frictional resistance of the shaft is small. The number of revolutions of the spindle is counted automatically, using a gear train connected to a series of dials. The meter must be calibrated and should be checked at frequent intervals because the friction of the bearings will not necessarily remain constant. The anemometer is useful for gas velocities above about 0.15 m/s.

Quantity meters, suitable for the measurement of the flow of gas through a pipe, include the standard wet and dry meters. In the wet meter, the gas fills a rotating segment and an equal volume of gas is expelled from another segment (Figure 6.30). The dry gas meter employs a pair of bellows. Gas enters one of the bellows and automatically expels gas from the other; the number of cycles is counted and recorded on a series of dials. Both of these are positive displacement meters and therefore do not need frequent calibration. Gas meters usually appear very bulky for the quantities they are measuring. This is because the linear velocity of a gas in a pipe is normally very high compared with that of a liquid, and the large volume is needed so that the speed of the moving parts can be reduced and wear minimised.

Liquid meters

In the oscillating-piston meter, the principle of which is illustrated in Figure 6.31, the flow of the liquid results in the positive displacement of a rotating element, the cumulative flow being obtained by gearing to a counter.

The body, in the form of a cylindrical chamber, is fitted with a radial partition and a central hub. The circular piston has a slot gap in the circumference to fit over the partition

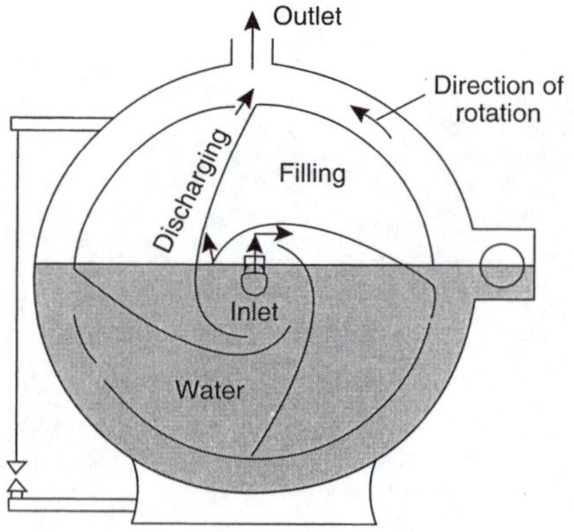

Figure 6.30. Wet gas meter

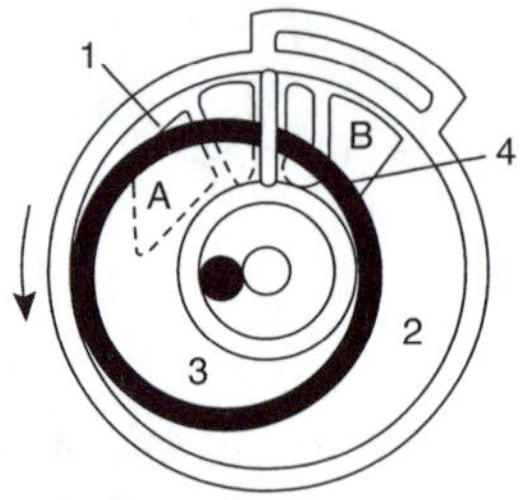

(1)

Spaces 1 and 3 are receiving fluid from the inlet port A, and spaces 2 and 4 are discharging through the outlet port B

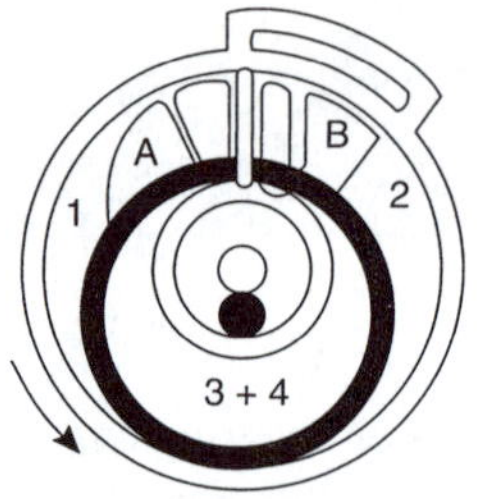

(2)

The piston has advanced and space 1, in connection with the inlet port, has enlarged, and space 2, in connection with the outlet port, has decreased while spaces 3 and 4, which have combined, are about to move into position to discharge through the outlet port

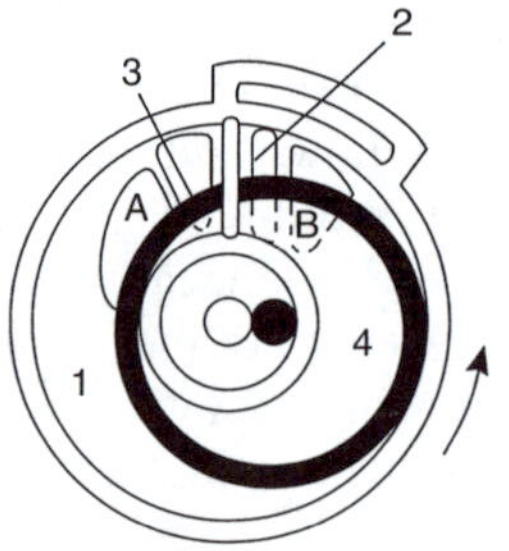

(3)

Space 1 is still admitting fluid from the inlet port and space 3 is just opening up again to the inlet port, while spaces 2 and 4 are discharging through the outlet port

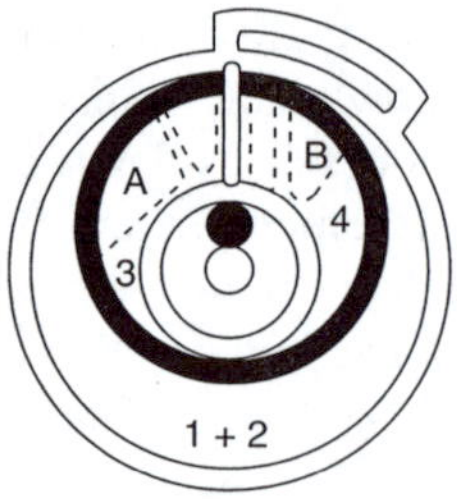

(4)

Fluid is being received into space 3 and discharged from space 4, while spaces 1 and 2 have combined and are about to begin discharging as the piston moves forward again to occupy position as shown in (1)

Figure 6.31. The oscillating-piston meter

and a peg on its upper face to control movement around the central hub. A rolling seal is thus formed between the piston and hub and between the piston and main chamber. The chamber is thus split into four spaces as shown in Figure 6.31. The fluid enters the bottom of the chamber through one port and leaves by the other port, the piston forming a movable division between the inlet and outlet.

In a cycle of operation the liquid enters port A and fills the spaces 1 and 3, thus forcing the piston to oscillate counterclockwise opening spaces 2 and 4 to port B. Because of the partition, the piston moves downwards so that space 3 is cut off from port A and becomes space 4. Further movement allows the exit port to be uncovered, and the measured volume between hub and piston is then discharged. The outer space 1 increases until the piston moves upwards over the partition and space 1 becomes space 2 when a second metered volume is discharged by the filling of the inner space 3. Meters of this type will handle flows of between about 0.005 and 15 litres/s.

Turbine flow meters are composed of some form of rotary device such as a helical rotor, Pelton wheel or a vane mounted in the flow stream. The fluid passing the rotor causes the rotor to turn at an angular velocity which is proportional to the flow velocity and hence the volumetric flowrate through the meter. The rotary motion of the rotor is sensed by some form of pick-up device that produces an electrical pulse output. The frequency of this signal is proportional to the flowrate and the total count of pulses is proportional to the total volume of liquid passed through the meter.

Turbine flow meters range in size from 5 to 600 mm in diameter, are suitable for temperatures of between 20 and 750 K at pressures of up to 300 bar. A normal range of flows falls between 0.02 litre/s and 2000 litre/s (2m^3/s) and a diagram showing a section of an axial turbine flow meter is shown in Figure 6.32.

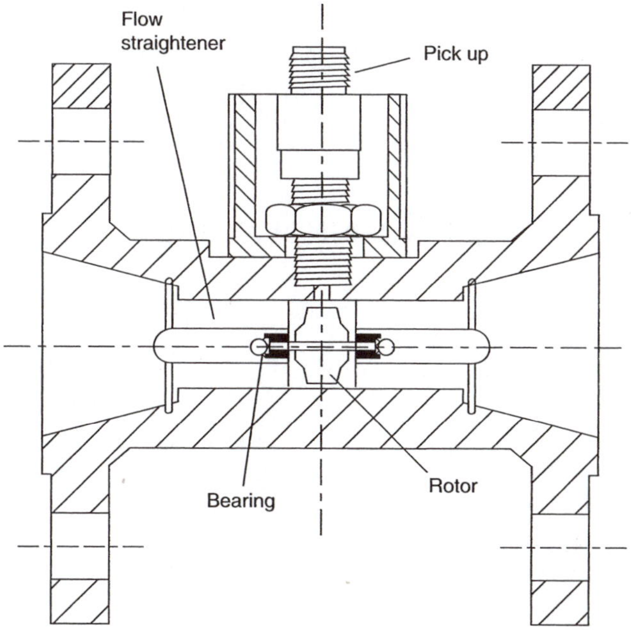

Figure 6.32. Section through an axial turbine flow meter

Further details of these and other measuring equipment are given by OWER and PANKHURST[2] and LINFORD[7] in the *Process Industries and Controls Handbook*[13] and in the *Instrument Manual*[14].

6.4. FURTHER READING

CHEREMISINOFF, N. P. and CHEREMISINOFF, P. N.: *Flow Measurement for Engineers and Scientists* (Marcel Dekker, New York, 1988).

MILLER, J. T.: *The Revised Course on Industrial Instrument Technology* (United Trade Press, London, 1973).

MILLER, J. T. (ed.): *Instrument Manual*, 4th edn (United Trade Press, London, 1971).

MILLER, R. N.: *Flow Measurement Engineering Handbook* (McGraw-Hill, New York, 1983).

MOORE, R. L.: *Basic Instrumentation Lecture Notes and Study Guide*, Vol. 1 — Measurement Fundamentals, Vol. 2 — Process Analyzers and Recorders. Instrument Society of America (Prentice Hall, Englewood Cliffs, 1982).

SCOTT, R. W. W. (ed.): *Developments in Flow Measurement* (Applied Science, London, 1982).

SYDENHAM, P. H.: *Transducers in Measurement and Control* (Adam Hilger, London, 1984).

WIGHTMAN, E. J.: *Instrumentation in Process Control* (Butterworths, London, 1972).

6.5. REFERENCES

1. WIGHTMAN, E. J.: *Instrumentation in Process Control* (Butterworths, London 1972).
2. OWER, E. and PANKHURST, R. C.: *Measurement of Air Flow*, 4th edn (Pergamon, Oxford 1966).
3. SMITH, B. D.: *Design of Equilibrium Stage Processes* (McGraw-Hill, New York, 1963).
4. BS 1042: 1943: British Standard 1042 (British Standards Institution, London): Code for flow measurement.
5. LYLE, O.: *The Efficient Use of Fuel*, 2nd edn, 2nd impression (HMSO, London, 1964). 237–241.
6. LYLE, O.: *The Efficient Use of Steam* (HMSO, London, 1963) §231.
7. LINFORD, A.: *Flow Measurement and Meters* (Spon, London, 1949).
8. GINESI, D.: *Chem Engg* (Albony), (May 1991) 147. A raft of flow meters on tap.
9. STROUHAL, F.: *Ann. Phys. Chem.* **5** (1878), 216. Über eine besonderer Art der Tonerregung.
10. ENDRESS and HAUSER plc: *Technical Information TI 031D/06/e* (1997). Vortex flow measuring system, prowirl 70.
11. LYNNWORTH, L. C.: in *Physical Acoustics*, **14** (eds. MASON, W. P. and THURSTON, R. N.), Chapter 5, Ultrasonic flowmeters (Academic Press, New York, 1979).
12. PLACHE, K. O.: *Mech. Eng.* (March, 1979), 36. Coriolis/gyroscope flow meter.
13. CONSIDINE, D. M.: *Process Industries and Controls Handbook* (McGraw-Hill, New York, 1957).
14. MILLER, J. T. (ed.): *Instrument Manual*, 4th edn (United Trade Press, London, 1971).

6.6. NOMENCLATURE

		Units in SI system	Dimensions in **M, N, L, T, θ, I**
A	Area perpendicular to direction of flow	m^2	$\mathbf{L}^2$
A_c	Cross-sectional area of column	m^2	$\mathbf{L}^2$
A_f	Area of rotameter float	m^2	$\mathbf{L}^2$
A_o	Area of orifice	m^2	$\mathbf{L}^2$
A_w	Cross-sectional area of well	m^2	$\mathbf{L}^2$
B	Width of rectangular channel or notch	m	**L**
C_c	Coefficient of contraction	—	—
C_D	Coefficient of discharge	—	—
C'	Constant for venturi meter	m^2	$\mathbf{L}^2$
D	Depth of liquid or above bottom of notch	m	**L**
d	Diameter of pipe or pitot tube	m	**L**
d_o	Diameter of orifice	m	**L**
e	Error		
F	Energy dissipated per unit mass of fluid	J/kg	$\mathbf{L}^2\mathbf{T}^{-2}$

		Units in SI system	Dimensions in **M, N, L, T, θ, I**
G	Mass rate of flow	kg/s	$\mathbf{MT^{-1}}$
g	Acceleration due to gravity	m/s^2	$\mathbf{LT^{-2}}$
h_i	Difference between impact and static heads on pitot tube	m	**L**
h_m	Reading on manometer	m	**L**
h_v	Fall in head over converging cone of venturi meter	m	**L**
h_w	Decrease in level in well	m	**L**
h_0	Fall in head over orifice meter	m	**L**
I	Electric current	A	**I**
k	Numerical constant used as index for compression	—	—
L	Length of weir or of inclined tube of manometer	m	**L**
M	Molecular weight	kg/kmol	$\mathbf{MN^{-1}}$
n	Number of end contractions	—	—
P	Pressure	N/m^2	$\mathbf{ML^{-1}T^{-2}}$
ΔP	Pressure difference	N/m^2	$\mathbf{ML^{-1}T^{-2}}$
Q	Volumetric rate of flow	m^3/s	$\mathbf{L^3T^{-1}}$
$\mathbf{R}$	Universal gas constant	(8314) J/kmol K	$\mathbf{MN^{-1}L^2T^{-2}}\boldsymbol{\theta}^{-1}$
T	Absolute temperature	K	$\boldsymbol{\theta}$
u	Mean velocity	m/s	$\mathbf{LT^{-1}}$
V_f	Volume of rotameter float	m^3	$\mathbf{L^3}$
v	Volume per unit mass of fluid	m^3/kg	$\mathbf{M^{-1}L^3}$
W_s	Shaft work per unit mass	J/kg	$\mathbf{L^2T^{-2}}$
w	Pressure ratio P_2/P_1	—	—
w_c	Critical pressure ratio	—	—
z	Distance in vertical direction	m	**L**
z_m	Vertical distance between level of manometer liquid and axis of venturi meter	m	**L**
α	Constant in expression for kinetic energy of fluid	—	—
γ	Ratio of specific heats at constant pressure and volume	—	—
Ω	Electrical resistance	ohm	$\mathbf{ML^2T^{-3}I^{-2}}$
ρ	Density of fluid	kg/m^3	$\mathbf{ML^{-3}}$
ρ_f	Density of rotameter float	kg/m^3	$\mathbf{ML^{-3}}$
ρ_m	Density of manometer fluid	kg/m^3	$\mathbf{ML^{-3}}$
θ	Half angle of triangular notch or angle made by inclined manometer tube with horizontal	—	—

CHAPTER 7

Liquid Mixing

7.1. INTRODUCTION—TYPES OF MIXING

Mixing is one of the most common operations carried out in the chemical, processing and allied industries. The term "mixing" is applied to the processes used to reduce the degree of non-uniformity, or gradient of a property in a system such as concentration, viscosity, temperature and so on. Mixing is achieved by moving material from one region to another. It may be of interest simply as a means of achieving a desired degree of homogeneity but it may also be used to promote heat and mass transfer, often where a system is undergoing a chemical reaction.

At the outset it is useful to consider some common examples of problems encountered in industrial mixing operations, since this will not only reveal the ubiquitous nature of the process, but will also provide an appreciation of some of the associated difficulties. Several attempts have been made to classify mixing problems and, for example, REAVELL[1] used as a criterion for mixing of powders, the flowability of the final product. HARNBY *et al.*[2] base their classification on the phases present, that is liquid-liquid, liquid-solid and so on. This is probably the most useful description of mixing as it allows the adoption of a unified approach to the problems encountered in a range of industries. This approach is now followed here.

7.1.1. Single-phase liquid mixing

In many instances, two or more miscible liquids must be mixed to give a product of a desired specification, such as, for example, in the blending of petroleum products of different viscosities. This is the simplest type of mixing as it involves neither heat nor mass transfer, nor indeed a chemical reaction. Even such simple operations can however pose problems when the two liquids have vastly different viscosities. Another example is the use of mechanical agitation to enhance the rates of heat and mass transfer between the wall of a vessel, or a coil, and the liquid. Additional complications arise in the case of highly viscous Newtonian and non-Newtonian liquids.

7.1.2. Mixing of immiscible liquids

When two immiscible liquids are stirred together, one phase becomes dispersed as tiny droplets in the second liquid which forms a continuous phase. Liquid-liquid extraction, a process using successive mixing and settling stages (Volume 2, Chapter 13) is one important example of this type of mixing. The liquids are brought into contact with a solvent that will selectively dissolve one of the components present in the mixture. Vigorous agitation causes one phase to disperse in the other and, if the droplet size is small, a high interfacial area is created for interphase mass transfer. When the agitation is stopped, phase separation takes place, but care must be taken to ensure that the droplets are not so small that a diffuse layer appears in the region of the interface; this can remain in a semi-stable state over a long period of time and prevent effective separation from occurring. Another important example of dispersion of two immiscible liquids is the production of stable emulsions, such as those encountered in food, brewing and pharmaceutical applications. Because the droplets are very small, the resulting emulsion is usually stable over considerable lengths of time.

7.1.3. Gas–liquid mixing

Numerous processing operations involving chemical reactions, such as aerobic fermentation, wastewater treatment, oxidation of hydrocarbons, and so on, require good contacting between a gas and a liquid. The purpose of mixing here is to produce a high interfacial area by dispersing the gas phase in the form of bubbles into the liquid. Generally, gas-liquid mixtures or dispersions are unstable and separate rapidly if agitation is stopped, provided that a foam is not formed. In some cases a stable foam is needed, and this can be formed by injecting gas into a liquid which is rapidly agitated, often in the presence of a surface-active agent.

7.1.4. Liquid–solids mixing

Mechanical agitation may be used to suspend particles in a liquid in order to promote mass transfer or a chemical reaction. The liquids involved in such applications are usually of low viscosity, and the particles will settle out when agitation ceases.

At the other extreme, in the formation of composite materials, especially filled polymers, fine particles must be dispersed into a highly viscous Newtonian or non-Newtonian liquid. The incorporation of carbon black powder into rubber is one such operation. Because of the large surface areas involved, surface phenomena play an important role in such applications.

7.1.5. Gas–liquid–solids mixing

In some applications such as catalytic hydrogenation of vegetable oils, slurry reactors, froth flotation, evaporative crystallisation, and so on, the success and efficiency of the process is directly influenced by the extent of mixing between the three phases. Despite its great industrial importance, this topic has received only limited attention.

7.1.6. Solids–solids mixing

Mixing together of particulate solids, sometimes referred to as blending, is a very complex process in that it is very dependent, not only on the character of the particles — density, size, size distribution, shape and surface properties — but also on the differences of these properties in the components. Mixing of sand, cement and aggregate to form concrete and of the ingredients in gunpowder preparation are longstanding examples of the mixing of solids.

Other industrial sectors employing solids mixing include food, drugs, and the glass industries, for example. All these applications involve only physical contacting, although in recent years, there has been a recognition of the industrial importance of solid-solid reactions, and solid-solid heat exchangers[3]. Unlike liquid mixing, research on solids mixing has not only been limited but also has been carried out only relatively recently. The problems involved in the blending of solids are discussed in Volume 2, Chapter 1.

7.1.7. Miscellaneous mixing applications

Mixing equipment may be designed not only to achieve a predetermined level of homogeneity, but also to improve heat transfer. For example, the rotational speed of an impeller in a mixing vessel is selected so as to achieve a required rate of heat transfer, and the agitation may then be more than sufficient for the mixing duty. Excessive or overmixing should be avoided as it is not only wasteful of energy but may be detrimental to product quality. For example, in biological applications, excessively high impeller speeds or energy input may give rise to shear rates which damage the micro-organisms present. In a similar way, where the desirable rheological properties of some polymer solutions may be attributable to structured long-chain molecules, excessive impeller speeds or agitation over prolonged periods, may damage the structure of the polymer molecules thereby altering their properties. It is therefore important to appreciate that overmixing may often be undesirable because it may result in both excessive energy consumption and impaired product quality. From the examples given of its application, it is abundantly clear that mixing cuts across the boundaries between industries, and indeed it may be required to mix virtually anything with anything else — be it a gas or a liquid or a solid; it is clearly not possible to consider the whole range of mixing problems here. Instead attention will be given primarily to *batch liquid mixing*, and reference will be made to the literature, where appropriate, in this field. This choice has been made largely because it is liquid mixing which is most commonly encountered in the processing industries. In addition, an extensive literature, mainly dealing with experimental studies, now exists which provides the basis for the design and selection of mixing equipment. It also affords some insight into the nature of the mixing process itself.

In mixing, there are two types of problems to be considered — how to design and select mixing equipment for a given duty, and how to assess whether a mixer is suitable for a particular application. In both cases, the following aspects of the mixing process should be understood:

(i) Mechanisms of mixing.
(ii) Scale-up or similarity criteria.
(iii) Power consumption.
(iv) Flow patterns.
(v) Rate of mixing and mixing time.
(vi) The range of mixing equipment available and its selection.

Each of these factors is now considered.

7.2. MIXING MECHANISMS

If mixing is to be carried out in order to produce a uniform mixture, it is necessary to understand how liquids move and approach this condition. In liquid mixing devices, it is necessary that two requirements are fulfilled. Firstly, there must be bulk or convective flow so that there are no dead (stagnant) zones. Secondly, there must be a zone of intensive or high-shear mixing in which the inhomogeneities are broken down. Both these processes are energy-consuming and ultimately the mechanical energy is dissipated as heat; the proportion of energy attributable to each varies from one application to another. Depending upon the fluid properties, primarily viscosity, the flow in mixing vessels may be laminar or turbulent, with a substantial transition zone in between the two, and frequently both flow types will occur simultaneously in different parts of the vessel. Laminar and turbulent flow arise from different mechanisms, and it is convenient to consider them separately.

7.2.1. Laminar mixing

Laminar flow is usually associated with high viscosity liquids (in excess of 10 N s/m^2) which may be either Newtonian or non-Newtonian. The inertial forces therefore tend to die out quickly, and the impeller of the mixer must cover a significant proportion of the cross-section of the vessel to impart sufficient bulk motion. Because the velocity gradients close to the rotating impeller are high, the fluid elements in that region deform and stretch. They repeatedly elongate and become thinner each time the fluid elements pass through the high shear zone. Figure 7.1 shows the sequence for a fluid element undergoing such a process.

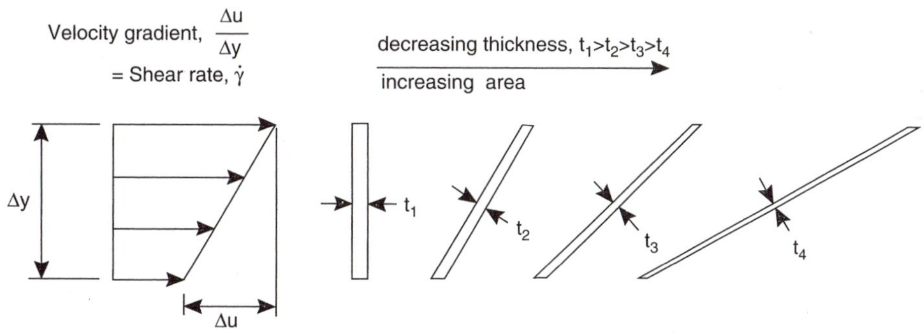

Figure 7.1. Schematic representation of the thinning of fluid elements due to laminar shear flow

In addition, extensional or elongational flow usually occurs simultaneously. As shown in Figure 7.2, this is a result of the convergence of the streamlines and consequential increased velocity in the direction of flow. As the volume remains constant, there must, be a thinning or flattening of the fluid elements, as shown in Figure 7.2. Both of these mechanisms (shear and elongation), give rise to stresses in the liquid which then effect a reduction in droplet size and an increase in interfacial area, by which means the desired degree of homogeneity is obtained.

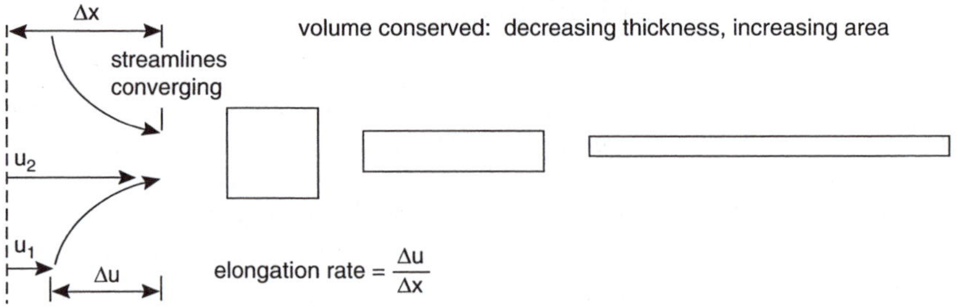

Figure 7.2. Schematic representation of the thinning of fluid elements due to extensional flow

In addition, molecular diffusion is always tending to reduce inhomogeneities but its effect is not significant until the fluid elements have been reduced in size sufficiently for their specific areas to become large. It must be recognized, however, that the ultimate homogenisation of miscible liquids, can be only brought about by molecular diffusion. In the case of liquids of high viscosity, this is a slow process.

In laminar flow, a similar mixing process occurs when the liquid is sheared between two rotating cylinders. During each revolution, the thickness of the fluid element is reduced, and molecular diffusion takes over when the elements are sufficiently thin. This type of mixing is shown schematically in Figure 7.3 in which the tracer is pictured as being introduced perpendicular to the direction of motion.

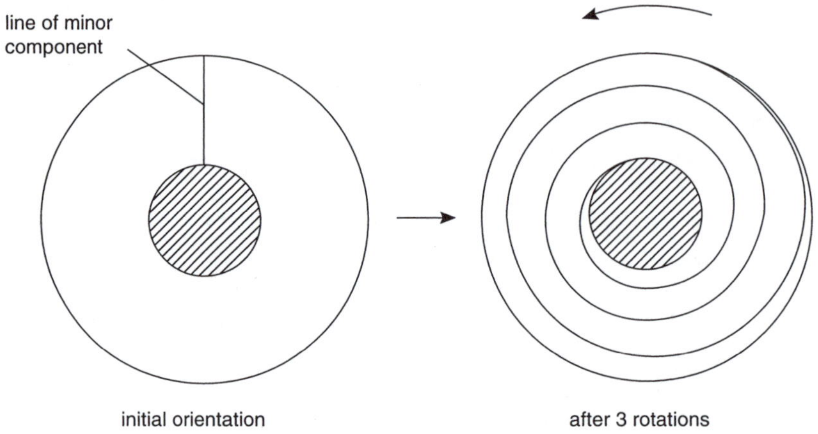

Figure 7.3. Laminar shear mixing in a coaxial cylinder arrangement

Finally, mixing can be induced by physically splicing the fluid into smaller units and re-distributing them. In-line mixers rely primarily on this mechanism, which is shown in Figure 7.4.

Thus, mixing in liquids is achieved by several mechanisms which gradually reduce the size or scale of the fluid elements and then redistribute them in the bulk. If there are initially differences in concentration of a soluble material, uniformity is gradually achieved, and molecular diffusion becomes progressively more important as the element

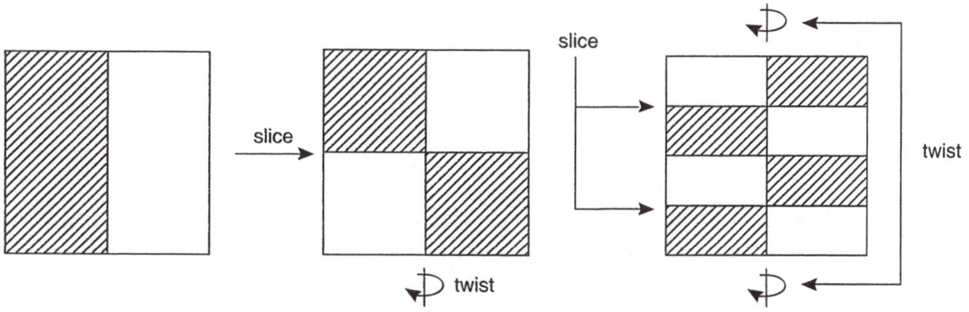

Figure 7.4. Schematic representation of mixing by cutting and folding of fluid elements

size is reduced. OTTINO[4] has recently illustrated the various stages by means of a series of coloured photographs.

7.2.2. Turbulent mixing

For low viscosity liquids (less than 10 mN s/m^2), the bulk flow pattern in mixing vessels with rotating impellers is turbulent. The inertia imparted to the liquid by the rotating impeller is sufficient to cause the liquid to circulate throughout the vessel and return to the impeller. Turbulent eddy diffusion takes place throughout the vessel but is a maximum in the vicinity of the impeller. Eddy diffusion is inherently much faster than molecular diffusion and, consequently, turbulent mixing occurs much more rapidly than laminar mixing. Ultimately homogenisation at the molecular level depends on molecular diffusion, which takes place more rapidly in low viscosity liquids. Mixing is most rapid in the region of the impeller because of the high shear rates due to the presence of trailing vortices, generated by disc-turbine impellers, and associated Reynolds stresses (Chapter 12); furthermore, a high proportion of the energy is dissipated here.

Turbulent flow is inherently complex, and calculation of the flow fields prevailing in a mixing vessel is not amenable to rigorous theoretical treatment. If the Reynolds numbers of the main flow is sufficiently high, some insight into the mixing process can be gained by using the theory of local isotropic turbulence[5]. As is discussed in Chapter 12, turbulent flow may be considered as a spectrum of velocity fluctuations and eddies of different sizes superimposed on an overall time-averaged mean flow. In a mixing vessel, it is reasonable to suppose that the large primary eddies, of a size corresponding approximately to the impeller diameter, would give rise to large velocity fluctuations but would have a low frequency. Such eddies are anisotropic and account for the bulk of the kinetic energy present in the system. Interaction between these primary eddies and slow moving streams produces smaller eddies of higher frequency which undergo further disintegration until finally, dissipate their energy as heat.

During the course of this disintegration process, kinetic energy is transferred from fast moving large eddies to the smaller eddies. The description given here is a gross oversimplification, but it does give a qualitative representation of the salient features of turbulent mixing. This whole process is similar to that for the turbulent flow of a fluid close to a boundary surface described in Chapter 11. Although some quantitative results for the scale size of eddies, have been obtained[5] and some workers[6,7] have reported experimental measurements on the structure of turbulence in mixing vessels, it is not at all clear how these data can be used for design purposes.

7.3. SCALE-UP OF STIRRED VESSELS

One of the problems confronting the designers of mixing equipment is that of deducing the most satisfactory arrangement for a large unit from experiments with small units. In order to achieve the same kind of flow pattern in two units, geometrical, kinematic, and dynamic similarity and identical boundary conditions must be maintained. This problem of scale-up has been discussed by a number of workers including HARNBY *et al.*[2], RUSHTON *et al.*[8], KRAMERS *et al.*[9], SKELLAND[10] and OLDSHUE[11]. It has been found convenient to relate the power used by the agitator to the geometrical and mechanical arrangement of the mixer, and thus to obtain a direct indication of the change in power arising from alteration of any of the factors relating to the mixer. A typical mixer arrangement is shown in Figure 7.5.

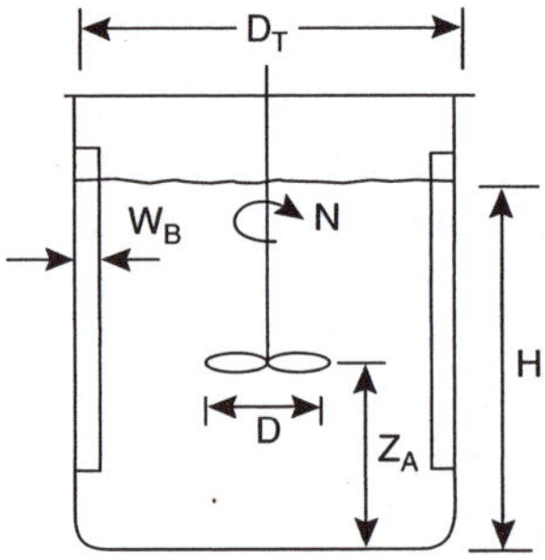

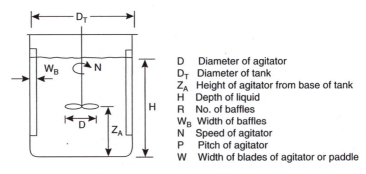

Figure 7.5. Typical configuration and dimensions of an agitated vessel

For similarity in two mixing systems, it is important to achieve geometric kinematic and dynamic similarity.

Geometric similarity prevails between two systems of different sizes if all counterpart length dimensions have a constant ratio. Thus the following ratios must be the same in two systems:

$$\frac{D_T}{D};\ \frac{Z_A}{D};\ \frac{W_B}{D};\ \frac{W}{D};\ \frac{H}{D} \quad \text{and so on.}$$

Kinematic similarity exists in two geometrically similar units when the velocities at corresponding points have a constant ratio. Also, the paths of fluid motion (flow patterns) must be alike.

Dynamic similarity occurs in two geometrically similar units of different sizes if all corresponding forces at counterpart locations have a constant ratio. It is necessary here to distinguish between the various types of force: inertial, gravitational, viscous, surface tension and other forms, such as normal stresses in the case of viscoelastic non-Newtonian liquids. Some or all of these forms may be significant in a mixing vessel. Considering corresponding positions in systems 1 and 2 which refer to the laboratory and large scale, respectively, when the different types of force occurring are F_a, F_b, F_c and so on dynamic similarity requires that:

$$\frac{F_{a1}}{F_{a2}} = \frac{F_{b1}}{F_{b2}} = \frac{F_{c1}}{F_{c2}} = \cdots\cdots \text{ a constant} \tag{7.1}$$

or:

$$\frac{F_{a1}}{F_{b1}} = \frac{F_{a2}}{F_{b2}};\ \frac{F_{a1}}{F_{c1}} = \frac{F_{a2}}{F_{c2}} \text{ etc.} \tag{7.2}$$

Kinematic and dynamic similarities both require geometrical similarity, so the corresponding positions "1" and "2" can be identified in the two systems. Some of the various types of forces that may arise during mixing or agitation will now be formulated.

Inertial force is associated with the reluctance of a body to change its state of rest or motion. Considering a mass m of liquid flowing with a linear velocity u through an area A during the time interval $\mathrm{d}t$; then $\mathrm{d}m = \rho uA\,\mathrm{d}t$ where ρ is the fluid density. The inertial force F_i = (mass × acceleration) or:

$$\mathrm{d}F_i = (\rho uA\,\mathrm{d}t)\left(\frac{\mathrm{d}u}{\mathrm{d}t}\right) = \rho uA\,\mathrm{d}u$$

$\therefore$

$$F_i = \int_0^{F_i} \mathrm{d}F_i = \int_0^u \rho uA\,\mathrm{d}u = \rho A\frac{u^2}{2} \tag{7.3}$$

The area for flow is however, A = (constant) L^2, where L is the characteristic linear dimension of the system. In mixing applications, L is usually chosen as the impeller diameter D, and, likewise, the representative velocity u is taken to be the velocity at the tip of impeller (πDN), where N is revolutions per unit time. Therefore, the expression for inertial force may be written as:

$$F_i \propto \rho D^4N^2 \tag{7.4}$$

where the constants have been omitted.

The rate change in u due to F_i, $\mathrm{d}u/\mathrm{d}t$, may be countered by the rate of change in u due to *viscous forces* F_v, which for a Newtonian fluid is given by:

$$F_v = \mu A'\left(\frac{\mathrm{d}u}{\mathrm{d}y}\right) \tag{7.5}$$

where $\mathrm{d}u/\mathrm{d}y$, the velocity gradient, may be taken to be proportional to u/L, and A' is proportional to L^2. The viscous force is then given by:

$$F_v \propto \mu uL \tag{7.6}$$

which, for an agitated system, becomes μD^2N. In a similar way, it is relatively simple to show that the *force due to gravity*, F_g, is given by:

$$F_g = (\text{mass of fluid} \times \text{acceleration due to gravity})$$

or:

$$F_g \propto \rho D^3g \tag{7.7}$$

and finally the *surface tension* force, $F_s \propto \sigma D$.

Taking F_a, F_b, and so on, in equation 7.1 to represent F_i, F_v, F_g, F_s respectively for systems 1 and 2, dynamic similarity of the two systems requires that:

$$\left(\frac{F_i}{F_v}\right)_1 = \left(\frac{F_i}{F_v}\right)_2 \tag{7.8}$$

which upon substitution of the respective expressions for F_i and F_v leads to:

$$\left(\frac{\rho D^2 N}{\mu}\right)_1 = \left(\frac{\rho D^2 N}{\mu}\right)_2 \tag{7.9}$$

This is, of course, the *Reynolds number*, which determines the nature of the flow, as shown in Chapter 3. In similar fashion, the constancy of the ratios between other forces results in Froude and Weber numbers as follows:

$$\left(\frac{F_i}{F_g}\right)_1 = \left(\frac{F_i}{F_g}\right)_2 \rightarrow \left(\frac{DN^2}{g}\right)_1 = \left(\frac{DN^2}{g}\right)_2 \quad \text{(Froude number)} \tag{7.10}$$

$$\left(\frac{F_i}{F_s}\right)_1 = \left(\frac{F_i}{F_s}\right)_2 \rightarrow \left(\frac{D^3 N^2 \rho}{\sigma}\right)_1 = \left(\frac{D^3 N^2 \rho}{\sigma}\right)_2 \quad \text{(Weber number)} \tag{7.11}$$

Thus, the ratios of the various forces occurring in mixing vessels can be expressed as the above dimensionless groups which, in turn, serve as similarity parameters for scale-up of mixing equipment. It can be shown that the existence of geometric and dynamic similarities also ensures kinematic similarity.

In the case of non-viscoelastic, time-independent, non-Newtonian fluids, the fluid viscosity μ, occurring in Reynolds numbers, must be replaced by an apparent viscosity μ_a evaluated at an appropriate value of shear rate. METZNER and co-workers[12,13] have developed a procedure for calculating an average value of shear rate in a mixing vessel as a function of impeller/vessel geometry and speed of rotation; this, in turn, allows the value of the apparent viscosity to be obtained either directly from the rheological data, or by the use of a rheological model. Though this scheme has proved satisfactory for time-independent non-Newtonian fluids, further complications arise in the scale-up of mixing equipment for viscoelastic liquids (ULBRECHT[14], ASTARITA[15] and SKELLAND[16]).

7.4. POWER CONSUMPTION IN STIRRED VESSELS

From a practical point of view, power consumption is perhaps the most important parameter in the design of stirred vessels. Because of the very different flow patterns and mixing mechanisms involved, it is convenient to consider power consumption in low and high viscosity systems separately.

7.4.1. Low viscosity systems

Typical equipment for low viscosity liquids consists of a vertical cylindrical tank, with a height to diameter ratio of 1.5 to 2, fitted with an agitator. For low viscosity liquids, high-speed propellers of diameter about one-third that of the vessel are suitable, running at 10-25 Hz. Although work on single-phase mixing of low viscosity liquids is of limited value in industrial applications it does, however, serve as a useful starting point for the subsequent treatment of high viscosity liquids.

Considering a stirred vessel in which a Newtonian liquid of viscosity μ, and density ρ is agitated by an impeller of diameter D rotating at a speed N; the tank diameter is D_T, and the other dimensions are as shown in Figure 7.5, then, the functional dependence of the power input to the liquid $\mathbf{P}$ on the independent variables (μ, ρ, N, D, D_T, g, other geometric dimensions) may be expressed as:

$$\mathbf{P} = \mathrm{f}(\mu, \rho, N, g, D, D_T, \text{ other dimensions}) \tag{7.12}$$

In equation 7.12, $\mathbf{P}$ is the impeller power, that is, the energy per unit time dissipated within the liquid. Clearly, the electrical power required to drive the motor will be greater than $\mathbf{P}$ on account of transmission losses in the gear box, motor, bearings, and so on.

It is readily acknowledged that the functional relationship in equation 7.12 cannot be established from first principles. However, by using dimensional analysis, the number of variables can be reduced to give:

$$\frac{\mathbf{P}}{\rho N^3 D^5} = \mathrm{f}\left(\frac{\rho N D^2}{\mu}, \frac{N^2 D}{g}, \frac{D_T}{D}, \frac{W}{D}, \frac{H}{D}, \dots\right) \tag{7.13}$$

where the dimensionless group on the left-hand side is called the *Power number* (N_p); ($\rho N D^2/\mu$) is the *Reynolds number (Re)* and ($N^2 D/g$) is the *Froude number (Fr)*. Other dimensionless length ratios, such as (D_T/D), (W/D) and so on, relate to the specific impeller/vessel arrangement. For geometrically similar systems, these ratios must be equal, and the functional relationship between the Power number and the other dimensionless groups reduces to:

$$N_p = \mathrm{f}(Re, Fr) \tag{7.14}$$

The simplest form of the function in equation 7.14 is a power law, giving:

$$N_p = K' Re^b Fr^c \tag{7.15}$$

where the values of K', b and c must be determined from experimental measurements, and are dependent upon impeller/vessel configuration and on the flow regime, that is laminar, transition or turbulent, prevailing in the mixing vessel. There are several ways of measuring the power input to the impeller, including the mounting of Prony brakes, using a dynamometer, or a simple calculation from the electrical measurements. Detailed descriptions, along with the advantages and disadvantages of different experimental techniques, have been given by OLDSHUE[(11)], and HOLLAND and CHAPMAN[(17)]. It is appropriate to note that the uncertainty regarding the actual losses in gear box, bearings, and so on makes the estimation of $\mathbf{P}$ from electrical measurements rather less accurate than would be desirable.

In equation 7.15, the Froude number is usually important only in situations where gross vortexing occurs, and can be neglected if the value of Reynolds number is less than about 300. Thus, in a plot of Power number (N_p) against Reynolds number (Re) with Froude number as parameter, all data fall on a single line for values of $Re < 300$ confirming that in this region Fr has no significant effect on N_p. This behaviour is clearly seen in Figure 7.6 where the data for a propeller, as reported by RUSHTON *et al.*[(8)], are plotted. Such a plot is known as a *power curve*.

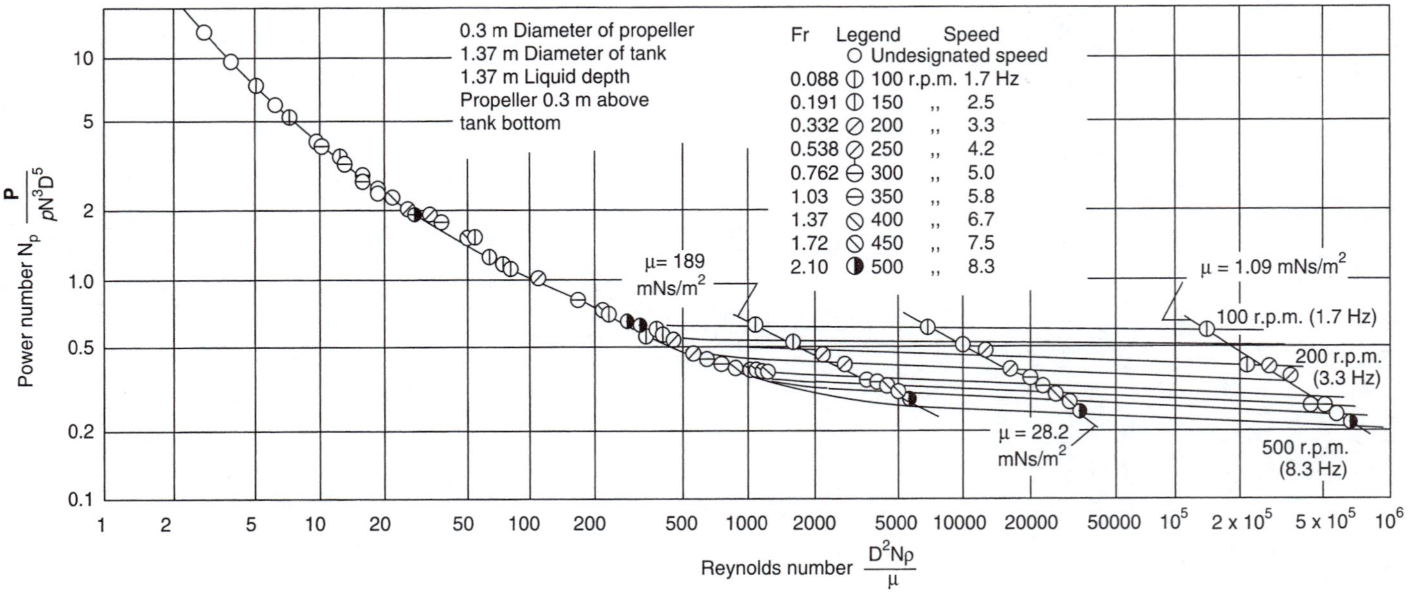

Figure 7.6. Power number as a function of Reynolds number for a propeller mixer

Thus at $Re < 300$:

$$N_p = K' Re^b \tag{7.16}$$

For values of $Re < 10$, b is found to be -1 (see Figure 7.6).
Therefore, power $\mathbf{P}$ is then given by:

$$\mathbf{P} = K' \mu N^2 D^3 \tag{7.17}$$

The value of K' depends on the type of impeller/vessel arrangement, and whether the tank is fitted with baffles. For marine-type three-bladed propellers with pitch equal to diameter, K' has been found to have a value of about 41.

For higher values of Re, the Froude number seems to exert some influence on the value of N_p, and separate lines are drawn for various speeds in Figure 7.6. It will be noted that, on this graph, lines of constant speed of rotation relate to constant values of the Froude number because the diameter of the impeller is constant and equal to 0.3 m. Thus, $Fr = 0.0305N^2$. The Reynolds number was varied by using liquids of different viscosities, as well as different rotational speeds, and the slanting lines in Figure 7.6 represent conditions of constant viscosity. In this region, the effect of Froude number may be minimised, or indeed eliminated, by the use of baffles or installing the impeller off-centre. This point is discussed in more detail in Section 7.5.

Example 7.1

On the assumption that the power required for mixing in a stirred tank is a function of the variables given in equation 7.12, obtain the dimensionless groups which are important in calculating power requirements for geometrically similar arrangements.

Solution

The variables in this problem, together with their dimensions, are as follows:

$\mathbf{P}$	$\mathbf{ML^2T^{-3}}$
μ	$\mathbf{ML^{-1}T^{-1}}$
ρ	$\mathbf{ML^{-3}}$
N	$\mathbf{T^{-1}}$
g	$\mathbf{LT^{-2}}$
D	$\mathbf{L}$
D_T	$\mathbf{L}$

This list includes seven variables and there are three fundamentals (**M, L, T**). By Buckingham's Π theorem, there will be $7 - 3 = 4$ dimensionless groups.

Choosing as the recurring set ρ, N and D, then these three variables themselves cannot be grouped together to give a dimensionless number. **M, L, T** can now be expressed in terms of combinations of ρ, N, D.

$$\mathbf{L} \equiv D$$

$$\mathbf{T} \equiv N^{-1}$$

$$\mathbf{M} \equiv \rho D^3.$$

Dimensionless group 1 : $\mathbf{PM^{-1}L^{-2}T^3} = \mathbf{P}(\rho D^3)^{-1}(D)^{-2}(N^{-1})^3 = \mathbf{P}\rho^{-1}D^{-5}N^{-3}$

Dimensionless group 2 : $\mu\mathbf{M^{-1}LT} = \mu(\rho D^3)^{-1}DN^{-1} = \mu\rho^{-1}D^{-2}N^{-1}$

Dimensionless group 3 : $g\mathbf{L^{-1}T^2} = gD^{-1}(N^{-1})^2 = gD^{-1}N^{-2}$

Dimensionless group 4 : $D_T\mathbf{L}^{-1} \qquad = D_T(D^{-1}) = D_T D^{-1}$

Thus, $\mathbf{P}\rho^{-1}D^{-5}N^{-3} \qquad = \mathrm{f}(\mu\rho^{-1}D^{-2}N^{-1}, gD^{-1}N^{-2}, D_T D^{-1})$

Re-arranging:

$$\frac{\mathbf{P}}{\rho N^3 D^5} = \mathrm{f}\left(\frac{\rho N D^2}{\mu}, \frac{N^2 D}{g}, \frac{D_T}{D} \dots\right)$$

which corresponds with equation 7.13.

Example 7.2

A solution of sodium hydroxide of density 1650 kg/m^3 and viscosity 50 mN s/m^2 is agitated by a propeller mixer of 0.5 m diameter in a tank of 2.28 m diameter, and the liquid depth is 2.28 m. The propeller is situated 0.5 m above the bottom of the tank. What is the power which the propeller must impart to the liquid for a rotational speed of 2 Hz?

Solution

In this problem the geometrical arrangement corresponds with the configuration for which the curves in Figure 7.6 are applicable.

$$\frac{D_T}{D} = \left(\frac{2.28}{0.5}\right) = 4.56; \quad \frac{H}{D} = \left(\frac{2.28}{0.5}\right) = 4.56; \quad \frac{Z_A}{D} = \left(\frac{0.5}{0.5}\right) = 1$$

$$Re = \frac{D^2 N\rho}{\mu} = \frac{(0.5^2 \times 2 \times 1650)}{50 \times 10^{-3}} = 16{,}500$$

$$Fr = \frac{N^2 D}{g} = \frac{(2^2 \times 0.5)}{9.81} = 0.20$$

From Figure 7.6:

$$N_p = \frac{\mathbf{P}}{\rho N^3 D^5} = 0.5$$

and $\mathbf{P} = (0.5 \times 1650 \times (2)^3 \times (0.5)^5)$

$= \underline{\underline{206 \text{ W}}}$

Example 7.3

A reaction is to be carried out in an agitated vessel. Pilot scale tests have been carried out under fully turbulent conditions in a tank 0.6 m in diameter, fitted with baffles and provided with a flat-bladed turbine, and it has been found that satisfactory mixing is obtained at a rotor speed of 4 Hz when the power consumption is 0.15 kW and the Reynolds number 160,000. What should be the rotor speed in order to achieve the same degree of mixing if the linear scale of the equipment if increased by a factor of 6 and what will be the Reynolds number and the power consumption?

Solution

The correlation of power consumption and Reynolds number is given by:

$$\frac{\mathbf{P}}{\rho N^3 D^5} = \mathrm{f}\left(\frac{\rho N D^2}{\mu}\right), \left(\frac{N^2 D}{g}\right) \qquad \text{equation (7.13)}$$

in the turbulent regime, at high values of Re, $\mathbf{P}$ is independent of Re and Fr and:

$$\mathbf{P} = k' N^3 D^5$$

For the *pilot unit*; taking the impeller diameter as $(D_T/3)$, then:

$$0.15 = k'4^3(0.6/3)^5$$

and: $$k' = 7.32$$

and: $$\mathbf{P} = 7.32N^3D^5 \quad \text{(i)}$$

In essence, two criteria may be used to scale-up; constant tip speed and constant power input per unit volume. These are now considered in turn.

Constant impeller tip speed

The tip speed is given by $u_t = \pi ND$

For the pilot unit: $$u_{t_1} = \pi 4(0.6/3) = 2.57 \text{ m/s}$$

For the full-scale unit: $$u_{t_2} = 2.57 = \pi N_2 D_2 \text{ m/s}$$

or: $$2.57 = \pi N_2(6 \times 0.6/3)$$

and: $$N_2 = \underline{\underline{0.66 \text{ Hz}}}$$

In equation (i):

$$\begin{aligned}\mathbf{P}_2 &= 7.32{N_2}^3{D_2}^5\\ &= 7.35 \times 0.66^3(6 \times 0.6/3)^5 = \underline{\underline{5.25 \text{ kW}}}\end{aligned}$$

For thermal similarity, that is the same temperature in both systems:

$$\mu_1 = \mu_2 \text{ and } \rho_1 = \rho_2$$

and: $$Re_2/Re_1 = (N_2{D_2}^2)/(N_1{D_1}^2)$$

Hence: $$Re_2/160{,}000 = [0.66(6 \times 0.6/3)^2]/[4 \times (0.6 \times 3)^2]$$

and: $$Re_2 = \underline{\underline{950{,}000}}$$

Constant power input per unit volume

Assuming the depth of liquid is equal to the tank diameter, then the volume of the pilot scale unit is $[(\pi/4)0.6^2 \times 0.6] = 0.170 \text{ m}^3$ and the power input per unit volume is $(0.157/0.170) = 0.884 \text{ kW/m}^3$

The volume of the full-scale unit is given by:

$$V_2 = (\pi/4)(6 \times 0.6)^2(6 \times 0.6) = 36.6 \text{ m}^3$$

and hence the power requirement is:

$$\mathbf{P} = (0.884 \times 36.6) = \underline{\underline{32.4 \text{ kW}}}$$

From equation (i):

$$32.4 = 7.32{N_2}^3(6 \times 0.6/3)^5$$

and: $$\underline{\underline{N_2 = 1.21 \text{ Hz}}}$$

The Reynolds number is then given by:

$$\begin{aligned}Re &= 160{,}000[(6 \times 0.6/3)^2 \times 1.21]/[(0.6/3)^2 \times 4]\\ &= \underline{\underline{1{,}740{,}000}}\end{aligned}$$

The choice of scale-up technique depends on the particular system. As a general guide, constant tip speed is used where suspended solids are involved, where heat is transferred to a coil or jacket, and for miscible liquids. Constant power per unit volume is used with immiscible liquids, emulsions, pastes and gas-liquid systems. Constant tip speed seems more appropriate in this case, and hence the rotor speed should be $\underline{\underline{0.66 \text{ Hz}}}$. The power consumption will then be $\underline{\underline{5.25 \text{ kW}}}$ giving a Reynolds number of $\underline{\underline{950{,}000}}$.

Figure 7.7, also taken from the work of RUSHTON *et al.*[8], shows N_p *vs* Re data for a 150 mm diameter turbine with six flat blades. In addition, this figure also shows the effect of introducing baffles in the tank. BISSELL *et al.*[18] have studied the effect of different types of baffles and their configuration on power consumption.

Evidently, for most conditions of practical interest, the Froude number is not a significant variable and the Power number is a unique function of Reynolds number for a fixed impeller/vessel configuration and surroundings. The vast amount of work[2] reported on the mixing of low viscosity liquids suggests that, for a given geometrical design and configuration of impeller and vessel, all single phase experimental data fall on a single power curve. Typically there are three discernible zones in a power curve.

At low values of the Reynolds number, less than about 10, a laminar or viscous zone exists and the slope of the power curve on logarithmic coordinates is -1, which is typical of most viscous flows. This region, which is characterised by slow mixing at both macro- and micro-levels, is where the majority of the highly viscous (Newtonian as well as non-Newtonian) liquids are processed.

At very high values of Reynolds number, greater than about 10^4, the flow is fully turbulent and inertia dominated, resulting in rapid mixing. In this region the Power number is virtually constant and independent of Reynolds number, as shown in Figures 7.6 and 7.7, but depends upon the impeller/vessel configuration. Often gas-liquid, solid-liquid and liquid-liquid contacting operations are carried out in this region. Though the mixing itself is quite rapid, the limiting factor controlling the process may be mass transfer.

Between the laminar and turbulent zones, there exists a transition region in which the viscous and inertial forces are of comparable magnitudes. No simple mathematical relationship exists between N_p and Re in this flow region and, at a given value of Re, N_p must be obtained from the appropriate power curve.

For purposes of scale-up, it is generally most satisfactory in the laminar region to maintain a constant speed for the tip of the impeller, and mixing time will generally increase with scale. The most satisfactory basis for scale-up in the turbulent region is to maintain a constant power input per unit volume.

Power curves for many different impeller geometries, baffle arrangements, and so on are to be found in the literature[10,11,17,19–21], but it must always be remembered that though the power curve is applicable to any single phase Newtonian liquid, at any impeller speed, the curve will be valid for only *one system geometry*.

Sufficient information is available on low viscosity systems for the estimation of the power requirements for a given duty under most conditions of practical interest.

7.4.2. High viscosity systems

As noted previously, mixing in highly viscous liquids is slow both at the molecular scale, on account of the low values of diffusivity, as well as at the macroscopic scale, due to poor bulk flow. Whereas in low viscosity liquids momentum can be transferred from a rotating impeller through a relatively large body of fluid, in highly viscous liquids only

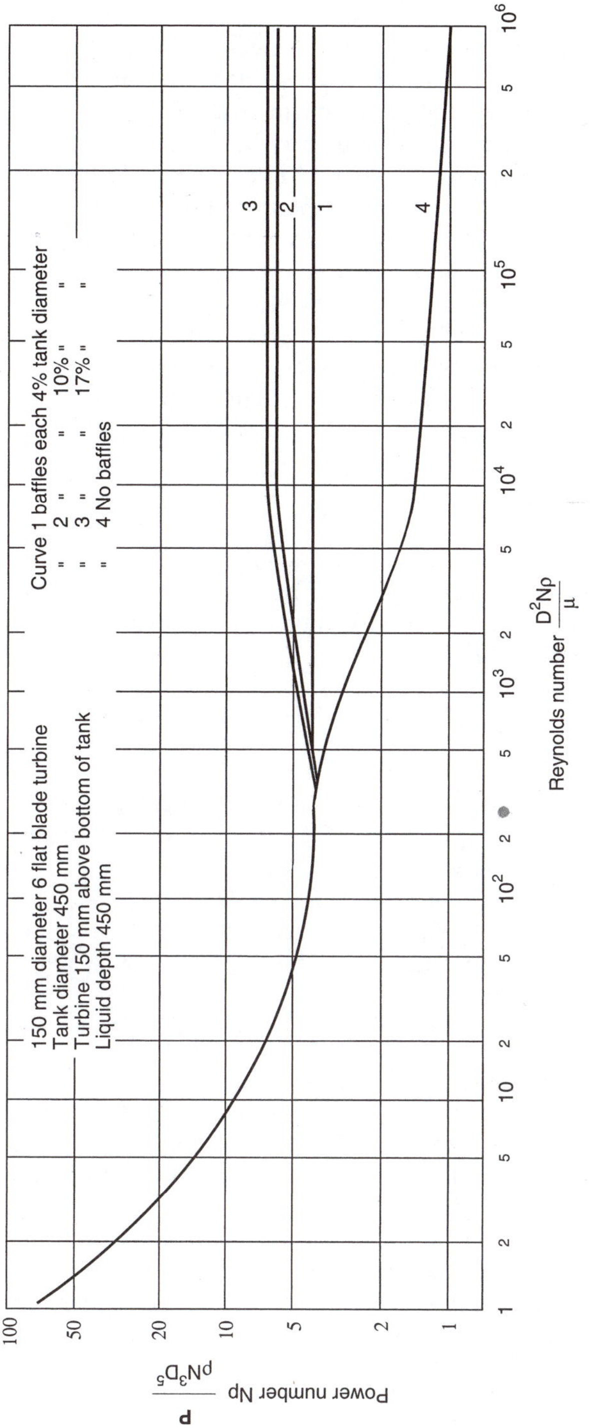

Figure 7.7. Power number as a function of Reynolds number for a turbine mixer

the fluid in the immediate vicinity of the impeller is influenced by the agitator and the flow is normally laminar.

For the mixing of highly viscous and non-Newtonian fluids it is usually necessary to use specially designed impellers involving close clearances with the vessel walls; these are discussed in a later section. High-speed stirring with small impellers merely wastefully dissipates energy at the central portion of the vessel. The power curve approach is usually applicable and the proportionality constant K' in equation 7.17 is a function of the type of rotating member, and the geometrical configuration of the system. Most of the highly viscous fluids of interest in the processing industries exhibit non-Newtonian behaviour, though highly viscous Newtonian fluids include glycerol and many lubricating oils.

A simple relationship has been shown to exist, however, between much of the data on power consumption with time-independent non-Newtonian liquids and Newtonian liquids in the laminar region. This link, which was first established by METZNER and OTTO[12] for pseudoplastic liquids, depends on the fact that there appears to be an average angular shear rate $\dot{\gamma}_{ang}$ for a mixer which characterises power consumption, and which is directly proportional to the rotational speed of impeller:

$$\dot{\gamma}_{ang} = k_s N \tag{7.18}$$

where k_s is a function of the type of impeller and the vessel configuration. If the apparent viscosity corresponding to the average shear rate defined above is used in the equation for a Newtonian liquid, the power consumption for laminar conditions is satisfactorily predicted for the non-Newtonian fluid.

The validity of the linear relationship given in equation 7.18 was subsequently confirmed by METZNER and TAYLOR[13]. The experimental evaluation of k_s for a given geometry is as follows:

(i) The Power number (N_p) is determined for a particular value of N.
(ii) The corresponding value of Re is obtained from the appropriate power curve as if the liquid were Newtonian.
(iii) The equivalent viscosity is computed from the value of Re.
(iv) The value of the corresponding shear rate is obtained, either directly from a flow curve obtained by independent experiment, or by use of an appropriate fluid model such as the power-law model.
(v) The value of k_s is calculated for a particular impeller configuration.

This procedure can be repeated for different values of N. A compilation of the experimental values of k_s for a variety of impellers, turbine, propeller, paddle, anchor, and so on, has been given by SKELLAND[16], and an examination of Table 7.1 suggests that for pseudo-plastic liquids, k_s lies approximately in the range of 10–13 for most configurations of practical interest.[22,23] SKELLAND[16] has also correlated much of the data on the agitation of purely viscous non-Newtonian fluids, and this is shown in Figure 7.8.

The prediction of power consumption for agitation of a given non-Newtonian fluid in a particular mixer, at a desired impeller speed, may be evaluated by the following procedure.

(i) Estimate the average shear rate from equation 7.18.
(ii) Evaluate the corresponding apparent viscosity, either from a flow curve, or by means of the appropriate flow model.

Table 7.1. Values of k_s for various types of impellers and key to Figure 7.8.[16]

Curve	Impeller	Baffles	D (m)	D_T/D	N(Hz)	k_s ($n < 1$)
A-A	Single turbine with 6 flat blades	4, $W_B/D_T = 0.1$	0.051–0.20	1.3–5.5	0.05–1.5	11.5 ± 1.5
A-A$_1$	Single turbine with 6 flat blades	None.	0.051–0.20	1.3–5.5	0.18–0.54	11.5 ± 1.4
B-B	Two turbines, each with 6 flat blades and $D_T/2$ apart	4, $W_B/D_T = 0.1$	—	3.5	0.14–0.72	11.5 ± 1.4
B-B$_1$	Two turbines, each with 6 flat blades and $D_T/2$ apart	4, $W_B/D_T = 0.1$ or none	—	1.023–1.18	0.14–0.72	11.5 ± 1.4
C-C	Fan turbine with 6 blades at 45°	4, $W_B/D_T = 0.1$ or none	0.10–0.20	1.33–3.0	0.21–0.26	13 ± 2
C-C$_1$	Fan turbine with 6 blades at 45°	4, $W_B/D_T = 0.1$ or none	0.10–0.30	1.33–3.0	1.0–1.42	13 ± 2
D-D	Square-pitch marine propellers with 3 blades (downthrusting)	None, (i) shaft vertical at vessel axis, (ii) shaft 10° from vertical, displaced $r/3$ from centre	0.13	2.2–4.8	0.16–0.40	10 ± 0.9
D-D$_1$	Same as for D-D but upthrusting	None, (i) shaft vertical at vessel axis, (ii) shaft 10° from vertical, displaced $r/3$ from centre	0.13	2.2–4.8	0.16–0.40	10 ± 0.9
D-D$_2$	Same as for D-D	None, position (ii)	0.30	1.9–2.0	0.16–0.40	10 ± 0.9
D-D$_3$	Same as for D-D	None, position (i)	0.30	1.9–2.0	0.16–0.40	10 ± 0.9
E-E	Square-pitch marine propeller with 3 blades	4, $W_B/D_T = 0.1$	0.15	1.67	0.16-0.60	10
F-F	Double-pitch marine propeller with 3 blades (downthrusting)	None, position (ii)	—	1.4–3.0	0.16–0.40	10 ± 0.9
F-F$_1$	Double-pitch marine propeller with 3 blades (downthrusting)	None, position (i)	—	1.4–3.0	0.16–0.40	10 ± 0.9
G-G	Square-pitch marine propeller with 4 blades	4, $W_B/D_T = 0.1$	0.12	2.13	0.05–0.61	10
G-G$_1$	Square-pitch marine propeller with 4 blades	4, $W_B/D_T = 0.1$	0.12	2.13	1.28–1.68	—
H-H	2-bladed paddle	4, $W_B/D_T = 0.1$	0.09–0.13	2–3	0.16–1.68	10
—	Anchor	None	0.28	1.02	0.34–1.0	11 ± 5
—	Cone impellers	0 or 4, $W_B/D_T = 0.08$	0.10–0.15	1.92–2.88	0.34–1.0	11 ± 5

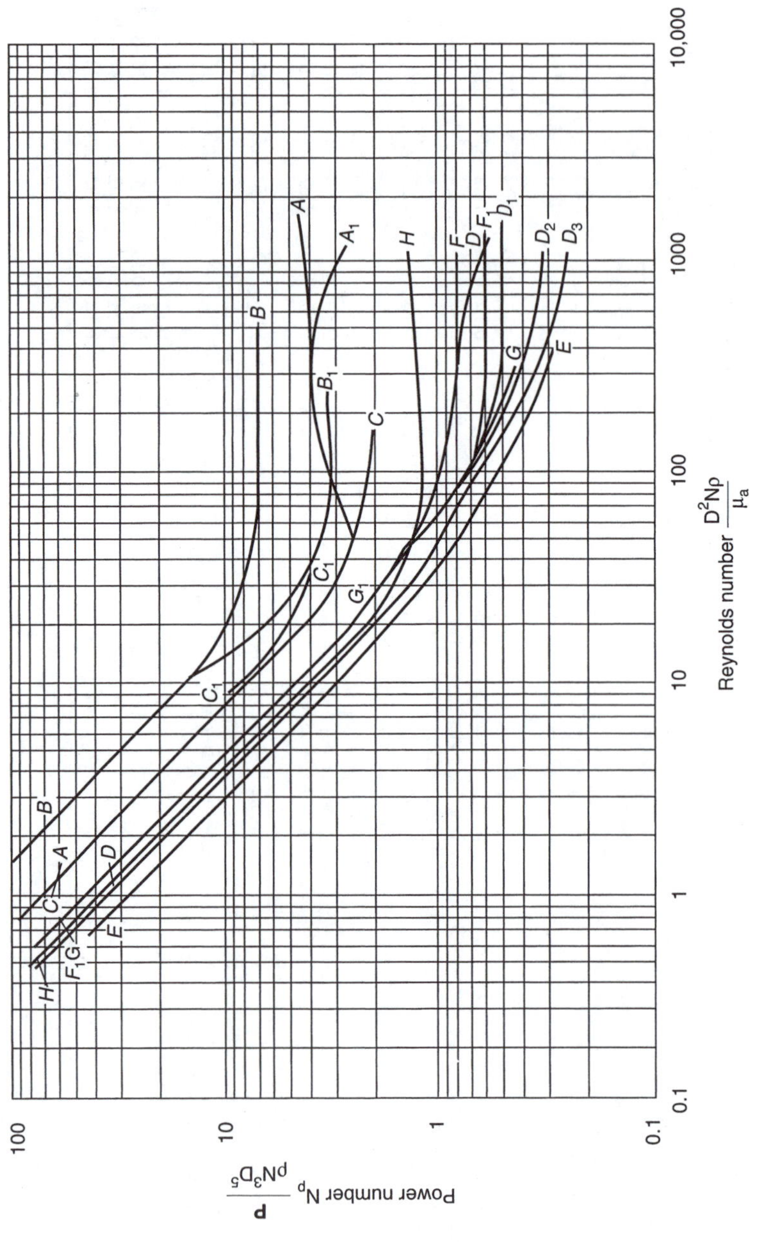

Figure 7.8. Power curve for pseudoplastic fluids agitated by different types of impeller

(iii) Estimate the value of the Reynolds number as $(\rho N D^2/\mu_a)$ and then the value of the Power number, and hence **P**, from the appropriate curve in Figure 7.8.

Although this approach of METZNER and OTTO[12], has gained wide acceptance, it has come under some criticism. For instance, SKELLAND[10] and MITSUISHI and HIRAI[24] have argued that this approach does not always yield a unique power curve for a wide range of the flow behaviour index, n. Despite this, it is safe to conclude that this method predicts power consumption with an accuracy of within about 25–30 per cent. Furthermore, GODFREY[25] has asserted that the constant k_s is independent of equipment size, and thus there are few scale-up problems. It is not yet established, however, how strongly the value of k_s depends upon the rheology. For example, CALDERBANK and MOO-YOUNG[26] and BECKNER and SMITH[27] have related k_s to the impeller/vessel configuration and rheological properties (n in the case of power law liquids); the dependence on n, however, is quite weak.

Data for power consumption of Bingham plastic fluids have been reported and correlated by NAGATA *et al.*[28] and of dilatant fluids by NAGATA *et al.*[28] and METZNER *et al.*[29]. EDWARDS *et al.*[30] have dealt with the mixing of time-dependent thixotropic materials.

Very little is known about the effect of the viscoelasticity of a fluid on power consumption, but early work[31,32] seems to suggest that it is negligible under creeping flow conditions. More recent work by OLIVER *et al.*[33] and by PRUD'HOMME and SHAQFEH[34] has indicated that the power consumption for Rushton-type turbine impellers, illustrated in Figure 7.20, may increase or decrease depending upon the value of Reynolds number and the magnitude of elastic effects or the Deborah number. At higher Reynolds numbers, it appears that the elasticity suppresses secondary flows and this results in a reduction in power consumption in comparison with a purely viscous liquid.[35] Many useful review articles have been published on this subject[2,14,16,36], and theoretical developments relating to the mixing of high viscosity materials have been dealt with by IRVING and SAXTON[37].

Finally, i should be noted that the calculation of the power requirement requires a knowledge of the impeller speed which is necessary to blend the contents of a tank in a given time, or of the impeller speed required to achieve a given mass transfer rate in a gas-liquid system. A full understanding of the mass transfer/mixing mechanism is not yet available, and therefore the selection of the optimum operating speed remains primarily a matter of experience. Before concluding this section, it is appropriate to indicate typical power consumptions in kW/m^3 of liquid for various duties, and these are shown in Table 7.2.

Table 7.2. Typical power consumptions

Duty	Power (kW/m^3)
Low power	
Suspending light solids, blending of low viscosity liquids	0.2
Moderate power	
Gas dispersion, liquid-liquid contacting, some heat transfer, etc.	0.6
High power	
Suspending heavy solids, emulsification, gas dispersion, etc.	2
Very high power	
Blending pastes, doughs	4

7.5. FLOW PATTERNS IN STIRRED TANKS

A qualitative picture of the flow field created by an impeller in a mixing vessel in a single-phase liquid is useful in establishing whether there are stagnant or dead regions in the vessel, and whether or not particles are likely to be suspended. In addition, the efficiency of mixing equipment, as well as product quality, are influenced by the flow patterns prevailing in the vessel.

The flow patterns for single phase, Newtonian and non-Newtonian liquids in tanks agitated by various types of impeller have been reported in the literature.[13,27,38,39] The experimental techniques which have been employed include the introduction of tracer liquids, neutrally buoyant particles or hydrogen bubbles, and measurement of local velocities by means of Pitot tubes, laser-doppler anemometers, and so on. The salient features of the flow patterns encountered with propellers and disc turbines are shown in Figures 7.9 and 7.10.

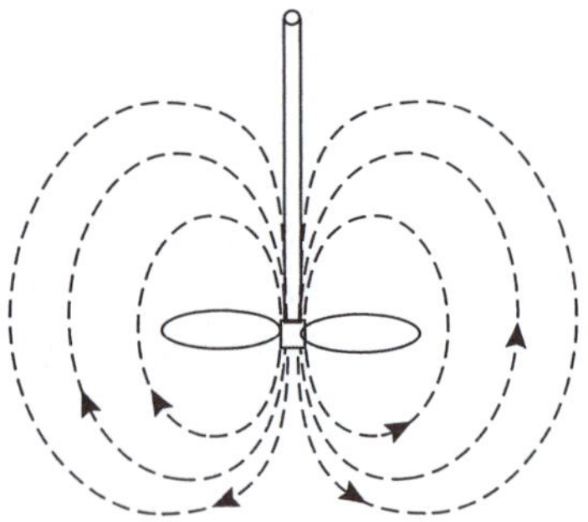

Figure 7.9. Flow pattern from propeller mixer

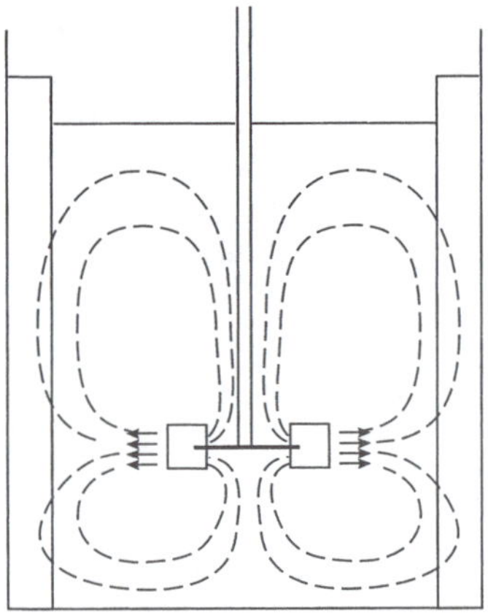

Figure 7.10. Radial flow pattern from disc turbine

Basically, the propeller creates an axial flow through the impeller, which may be upwards or downwards depending upon the direction of rotation. The velocities at any point are three-dimensional, and unsteady, but circulation patterns such as those shown in Figure 7.9 are useful in avoiding dead zones.

If a propeller is mounted centrally, and there are no baffles in the tank, there is a tendency for the lighter fluid to be drawn in to form a vortex, and for the degree of agitation to be reduced. To improve the rate of mixing, and to minimise vortex formation, baffles are usually fitted in the tank, and the resulting flow pattern is as shown in Figure 7.11. The power requirements are, however, considerably increased by the incorporation of baffles. Another way of minimising vortex formation is to mount the agitator, off-centre; where the resulting flow pattern is depicted in Figure 7.12.

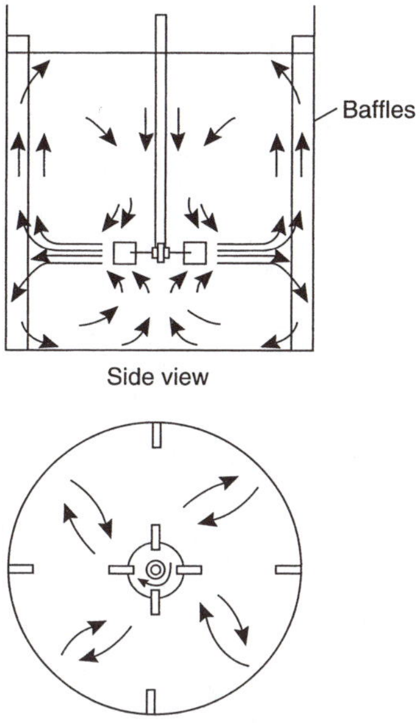

Figure 7.11. Flow pattern in vessel with vertical baffles

The flat-bladed turbine impeller produces a strong radial flow outwards from the impeller (Figure 7.10), thereby creating circulating zones in the top and bottom of the tank. The flow pattern can be altered by changing the impeller geometry and, for example, if the turbine blades are angled to the vertical, a stronger axial flow component is produced. This can be useful in applications where it is necessary to suspend solids. A flat paddle produces a flow field with large tangential components of velocity, and this does not promote good mixing. Propellers, turbines and paddles are the principal types of impellers used for low viscosity systems operating in the transition and turbulent regimes.

A brief mention will be made of flow visualisation studies for some of the agitators used for high viscosity liquids; these include anchors, helical ribbons and screws. Detailed flow pattern studies[40] suggest that both gate and anchor agitators promote fluid motion close to the vessel wall but leave the region in the vicinity of the shaft relatively stagnant. In addition, there is modest top to bottom turnover and thus vertical concentration gradients may exist, but these may be minimised by means of a helical ribbon or a screw added

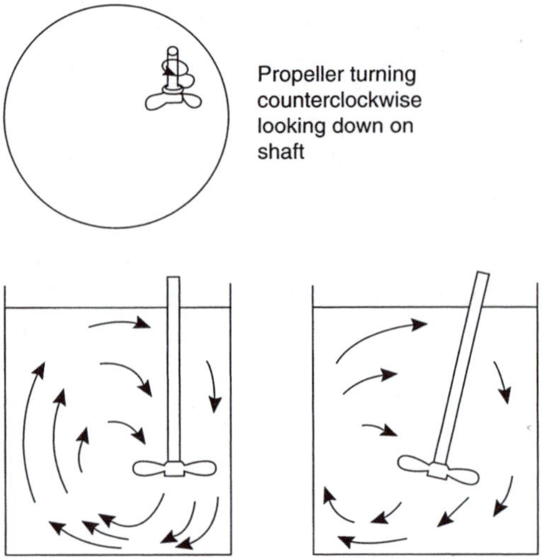

Figure 7.12. Flow pattern with agitator offset from centre

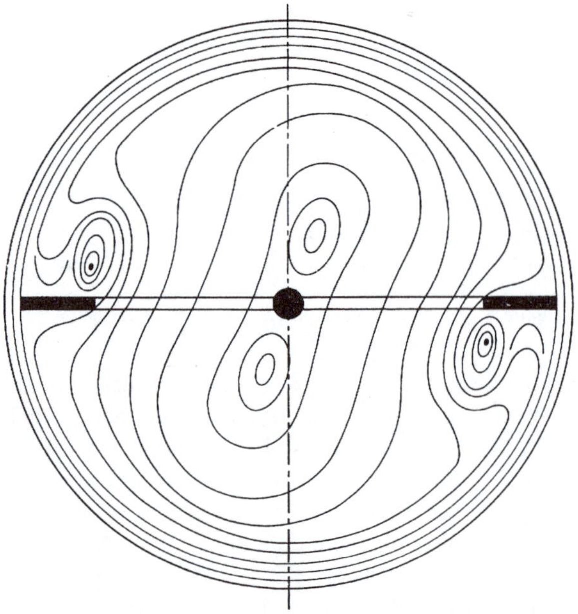

Figure 7.13. Streamlines in a tank with a gate agitator, drawn relative to the arm of the stirrer

to the shaft. Such combined impellers would have a ribbon pumping upwards near the wall with the screw twisted in the opposite sense, pumping the fluid downwards near the shaft region. Typical flow patterns for a gate-type anchor are shown in Figure 7.13.

In these systems the flow patterns change[27] as the stirring speed is increased and the average shear rate cannot be completely described by a linear relationship. Furthermore, any rotational motion induced within a vessel will tend to produce a secondary flow in the vertical direction; the fluid in contact with the bottom of the tank is stationary, while that at higher levels is rotating and will experience centrifugal forces. Consequently, there exist unbalanced pressure forces within the fluid which lead to the formation of a toroidal vortex. This secondary system may be single-celled or double-celled, as shown in Figures 7.14 and 7.15, depending upon the viscosity and type of fluid.

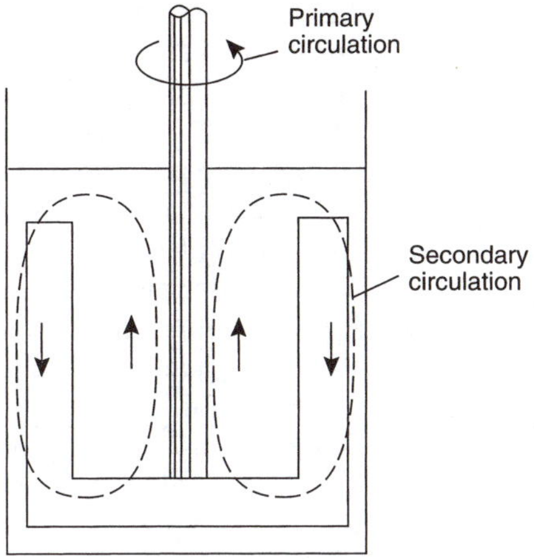

Figure 7.14. Single-celled secondary flow with an anchor agitator

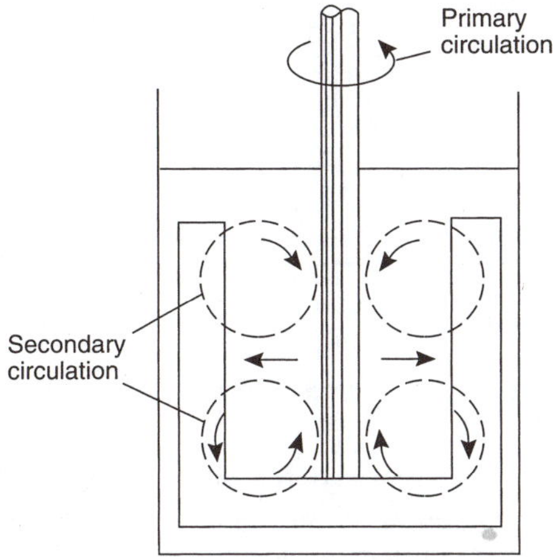

Figure 7.15. Double-celled secondary flow

Clearly, the flow pattern established in a mixing vessel depends critically upon the vessel/impeller configuration and on the physical properties of the liquid (particularly viscosity). In selecting the appropriate combination of equipment, it must be ensured that the resulting flow pattern is suitable for the required application.

7.6. RATE AND TIME FOR MIXING

Before considering rate of mixing and mixing time, it is necessary to have some means of assessing the quality of a mixture, which is the product of a mixing operation. Because of the wide scope and range of mixing problems, several criteria have been developed to assess mixture quality, none of which is universally applicable. One intuitive and convenient, but perhaps unscientific, criterion is whether the product or mixture meets the required specification. Other ways of judging the quality of mixing have been described by HARNBY *et al.*[2]. Whatever the criteria used, mixing time is defined as the time required to produce a mixture or a product of predetermined quality, and the rate of mixing is the rate at which the mixing progresses towards the final state. For a single-phase liquid in a stirred tank to which a volume of tracer material is added, the mixing time is measured from the instant the tracer is added to the time when the contents of the vessel have reached the required degree of uniformity or mixedness. If the tracer is completely miscible and has the same viscosity and density as the liquid in the tank, the tracer concentration may be measured as a function of time at any point in the vessel by means of a suitable detector, such as a colour meter, or by electrical conductivity. For a given amount of tracer, the equilibrium concentration C_∞ may be calculated; this value will be approached asymptotically at any point as shown in Figure 7.16.

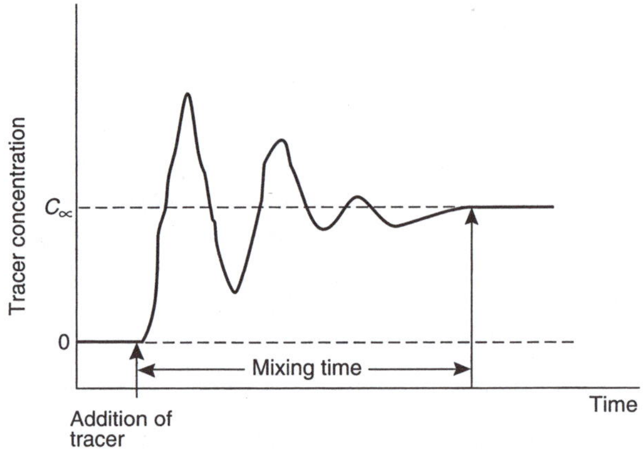

Figure 7.16. Mixing-time measurement curve

The mixing time will be that required for the mixture composition to come within a specified deviation from the equilibrium value and this will be dependent upon the way in which the tracer is added and the location of the detector. It may therefore be desirable to record the tracer concentration at several locations, and to define the variance of concentration σ^2 about the equilibrium value as:

$$\sigma^2 = \frac{1}{\mathbf{n}-1}\sum_{i=0}^{i=\mathbf{n}}(C_i - C_\infty)^2 \tag{7.19}$$

where C_i is the tracer concentration at time t recorded by the ith detector. A typical variance curve is shown in Figure 7.17.

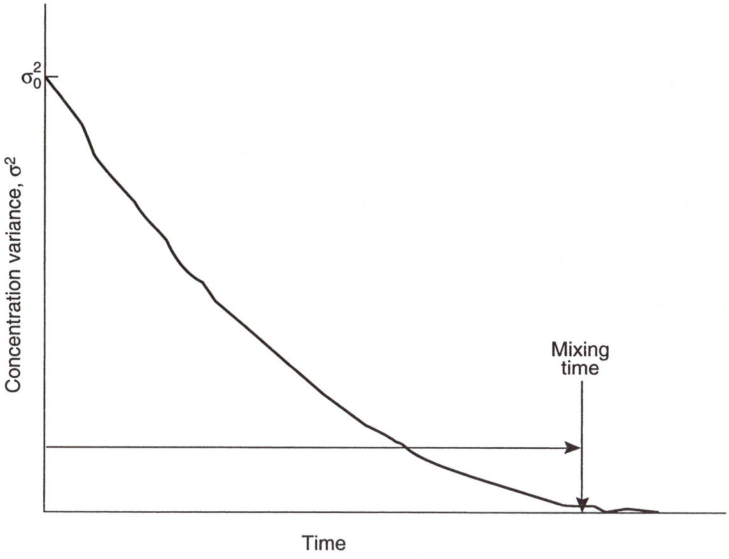

Figure 7.17. Reduction in variance of concentration with time

Several experimental techniques may be used, such as acid/base titration, electrical conductivity measurement, temperature measurement, or measurement of optical properties such as refractive index, light absorption, and so on. In each case, it is necessary to specify the manner of tracer addition, the position and number of recording stations, the sample volume of the detection system, and the criteria used in locating the end-point. Each of these factors will influence the measured value of mixing time, and therefore care must be exercised in comparing results from different investigations.

For a given experiment and configuration, the mixing time will depend upon the process and operating variables as follows:

$$t_m = \mathrm{f}(\rho, \mu, N, D, g, \text{geometrical dimensions of the system}) \tag{7.20}$$

Using dimensional analysis, the functional relationship may be rearranged as:

$$Nt_m = \theta_m = \mathrm{f}\left(\frac{\rho ND^2}{\mu}, \frac{DN^2}{g}, \text{geometrical dimensions as ratios}\right) \tag{7.21}$$

For geometrically similar systems, and assuming that the Froude number DN^2/g is not important:

$$\theta_m = \mathrm{f}\left(\frac{\rho N D^2}{\mu}\right) = \mathrm{f}(Re) \tag{7.22}$$

The available experimental data[2] seem to suggest that the dimensionless mixing time (θ_m) is independent of Reynolds number for both laminar and turbulent zones, in each of which it attains a constant value, and changes from the one value to the other in the transition region. $\theta_m - Re$ behaviour is schematically shown in Figure 7.18, and some typical mixing data obtained by NORWOOD and METZNER[41] for turbine impellers in baffled vessels using an acid/base titration technique are shown in Figure 7.19. It is widely reported that θ_m is quite sensitive to impeller/vessel geometry for low viscosity liquids. BOURNE and BUTLER[42] have made some interesting observations which appear to suggest that the rates of mixing and, hence mixing time, are not very sensitive to the fluid properties for both Newtonian and non-Newtonian materials. On the other hand, GODLESKI and SMITH[43] have reported mixing times for pseudoplastic non-Newtonian fluids fifty times greater than those expected from the corresponding Newtonian behaviour, and this emphasises the care which must be exercised in applying any generalised conclusions to a particular system.

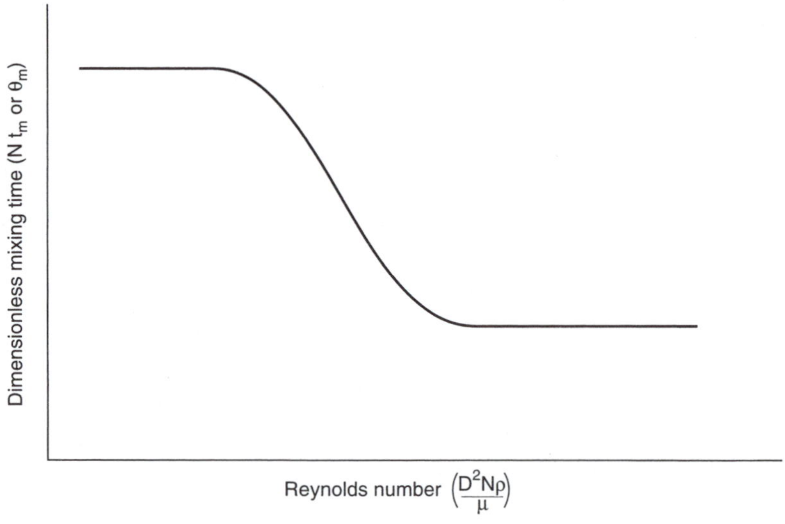

Figure 7.18. Typical mixing-time behaviour

Although very little is known about the influence of fluid elasticity on rates of mixing and mixing times, the limited work appears to suggest that the role of elasticity strongly depends on the impeller geometry. For instance, CHAVAN *et al.*[44] reported a decrease in mixing rate with increasing levels of viscoelasticity for helical ribbons, whereas HALL and GODFREY[45] found the mixing rates to increase with elasticity for sigma blades. Similar conflicting conclusions regarding the influence of viscoelasticity can be drawn

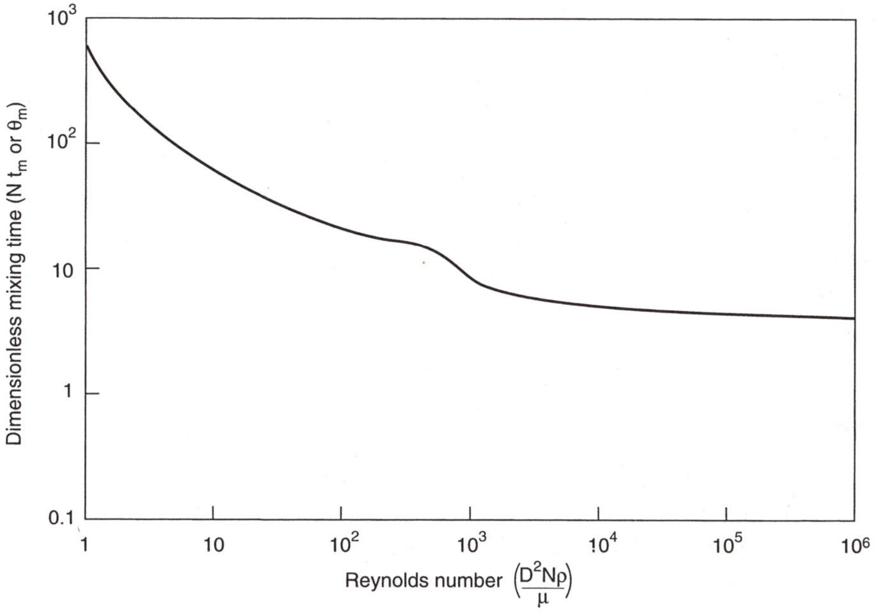

Figure 7.19. Typical mixing time data for turbines

from the work of CARREAU *et al.*[39] who found that the direction of impeller rotation had an appreciable effect on mixing rates and times.

Considerable confusion thus seems to exist in the literature regarding the influence of rheological properties on mixing rates and times. Exhaustive reviews are available on this subject[16,36].

7.7. MIXING EQUIPMENT

The wide range of mixing equipment available commercially reflects the enormous variety of mixing duties encountered in the processing industries. It is reasonable to expect therefore that no single item of mixing equipment will be able to carry out such a range of duties effectively. This has led to the development of a number of distinct types of mixer over the years. Very little has been done, however, by way of standardisation of equipment, and no design codes are available. The choice of a mixer type and its design is therefore primarily governed by experience. In the following sections, the main mechanical features of commonly used types of equipment together with their range of applications are described qualitatively. Detailed description of design and selection of various types of mixers have been presented by OLDSHUE[11].

7.7.1. Mechanical agitation

This is perhaps the most commonly used method of mixing liquids, and essentially there are three elements in such devices.

Vessels

These are often vertically mounted cylindrical tanks, up to 10 m in diameter, which typically are filled to a depth equal to about one diameter, although in some gas-liquid contacting systems tall vessels are used and the liquid depth is up to about three tank diameters; multiple impellers fitted on a single shaft are then frequently used. The base of the tanks may be flat, dished, or conical, or specially contoured, depending upon factors such as ease of emptying, or the need to suspend solids, etc., and so on.

For the batch mixing of viscous pastes and doughs using ribbon impellers and Z-blade mixers, the tanks may be mounted horizontally. In such units, the working volume of pastes and doughs is often relatively small, and the mixing blades are massive in construction.

Baffles

To prevent gross vortexing, which is detrimental to mixing, particularly in low viscosity systems, baffles are often fitted to the walls of the vessel. These take the form of thin strips about one-tenth of the tank diameter in width, and typically four equi-spaced baffles may be used. In some cases, the baffles are mounted flush with the wall, although occasionally a small clearance is left between the wall and the baffle to facilitate fluid motion in the wall region. Baffles are, however, generally not required for high viscosity liquids because the viscous shear is then sufficiently great to damp out the rotary motion. Sometimes, the problem of vortexing is circumvented by mounting impellers off-centre.

Impellers

Figure 7.20 shows some of the impellers which are frequently used. Propellers, turbines, paddles, anchors, helical ribbons and screws are usually mounted on a central vertical shaft in a cylindrical tank, and they are selected for a particular duty largely on the basis of liquid viscosity. By and large, it is necessary to move from a propeller to a turbine and then, in order, to a paddle, to an anchor and then to a helical ribbon and finally to a screw as the viscosity of the fluids to be mixed increases. In so doing the speed of agitation or rotation decreases.

Propellers, turbines and paddles are generally used with relatively low viscosity systems and operate at high rotational speeds. A typical velocity for the tip of the blades of a turbine is of the order of 3 m/s, with a propeller being a little faster and the paddle a little slower. These are classed as remote-clearance impellers, having diameters in the range (0.13–0.67) × (tank diameter). Furthermore, minor variations within each type are possible. For instance, Figure 7.20*b* shows a six-flat bladed Rushton turbine, whereas possible variations are shown in Figure 7.21. Hence it is possible to have retreating-blade turbines, angled-blade turbines, four- to twenty-bladed turbines, and so on. For dispersion of gases in liquid, turbines are usually employed.

Propellers are frequently of the three-bladed marine type and are used for in-tank blending operations with low viscosity liquids, and may be arranged as angled side-entry units, as shown in Figure 7.22. REAVELL[(1)] has shown that the fitting of a cruciform baffle at the bottom of the vessel enables much better dispersion to be obtained, as shown in

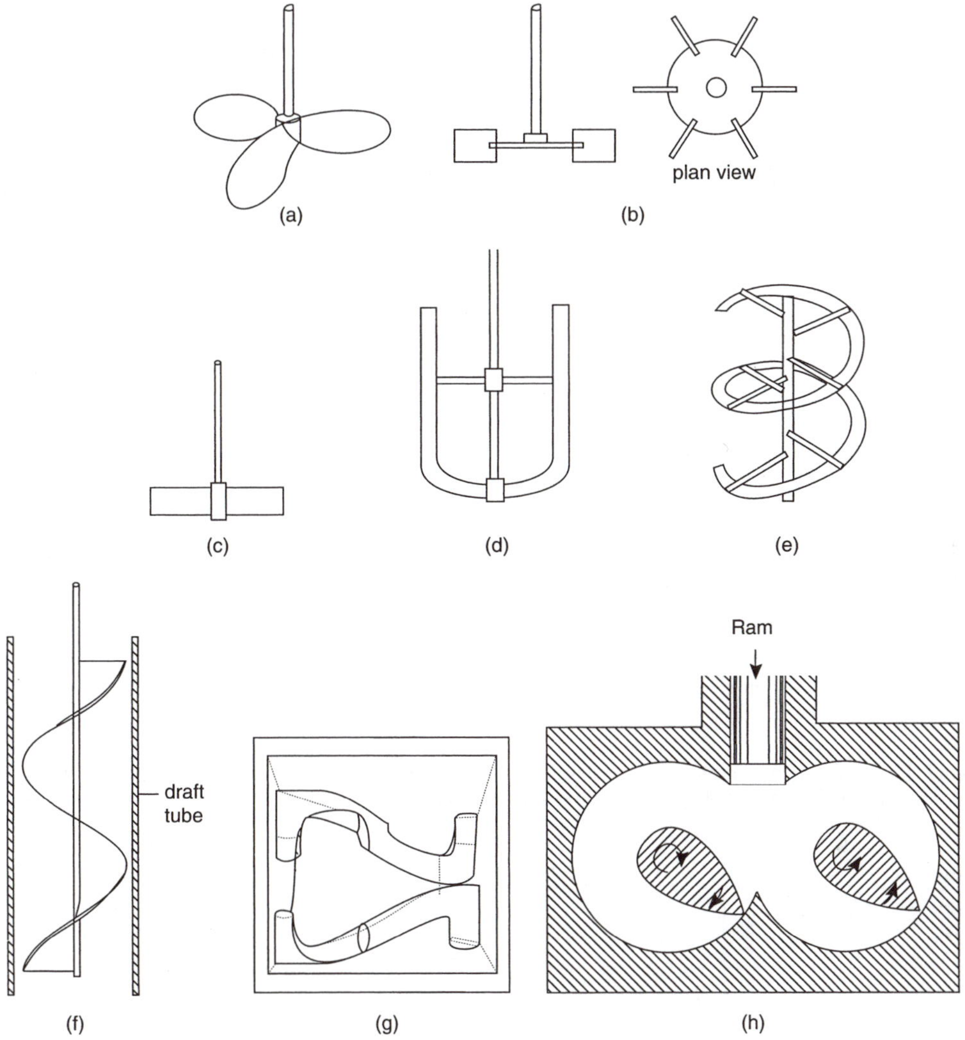

Figure 7.20. Commonly used impellers (*a*) Three-bladed propeller (*b*) Six-bladed disc turbine (Rushton turbine) (*c*) Simple paddle (*d*) Anchor impeller (*e*) Helical ribbon (*f*) Helical screw with draft tube (*g*) Z-blade mixer (*h*) Banbury mixer

Figure 7.23. For large vessels, and when the liquid depth is large compared with the tank diameter, it is a common practice to mount more than one impeller on the same shaft. With this arrangement the unsupported length of the propeller shaft should not exceed about 2 m. In the case of large vessels, there is some advantage to be gained by using side- or bottom-entry impellers to avoid the large length of unsupported shaft, though a good gland or mechanical seal is needed for such installations or alternatively, a foot bearing is employed. Despite a considerable amount of practical experience, foot bearings can be troublesome owing to the difficulties of lubrication, especially when handling corrosive liquids.

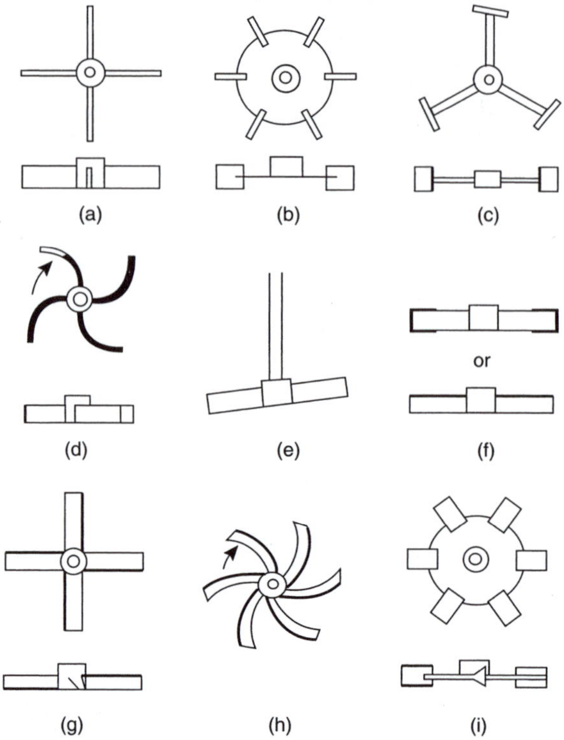

Figure 7.21. Variation in turbine impeller designs (*a*) Flat blade (*b*) Disc flat blade (*c*) Pitched vane (*d*) Curved blade (*e*) Tilted blade (*f*) Shrouded (*g*) Pitched blade (*h*) Pitched curved blade (*i*) Arrowhead

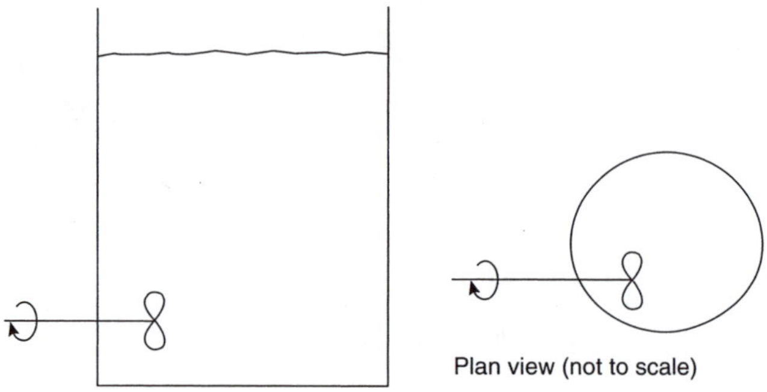

Figure 7.22. Side entering propeller

In comparing propellers and turbines, the following features may be noted:

Propellers

(a) are self-cleaning in operation,
(b) can be used at a wide range of speeds,
(c) give an excellent shearing effect at high speeds,

Figure 7.23. Flow pattern in vessel with cruciform baffle

(d) do not damage dispersed particles at low speeds,
(e) are reasonably economical in power, provided the pitch is adjusted according to the speed,
(f) by offset mounting, vortex formation is avoided,
(g) if horizontally mounted, a stuffing box is required in the liquid, and they are not effective in viscous liquids.

Shrouded turbines

(a) are excellent for providing circulation,
(b) are normally mounted on a vertical shaft with the stuffing box above the liquid,
(c) are effective in fluids of high viscosity,
(d) are easily fouled or plugged by solid particles,
(e) are expensive to fabricate,
(f) are restricted to a narrow range of speeds, and
(g) do not damage dispersed particles at economical speeds.

Open impellers

(a) are less easily plugged than the shrouded type,
(b) are less expensive, and
(c) give a less well-controlled flow pattern.

Anchors, helical ribbons and screws, are generally used for high viscosity liquids. The anchor and ribbon are arranged with a close clearance at the vessel wall, whereas the helical screw has a smaller diameter and is often used inside a draft tube to promote fluid motion throughout the vessel. Helical ribbons or interrupted ribbons are often used in horizontally mounted cylindrical vessels.

Kneaders, Z- and sigma-blade, and Banbury mixers as shown in Figure 7.20, are generally used for the mixing of high-viscosity liquids, pastes, rubbers, doughs, and so on. The tanks are usually mounted horizontally and two impellers are often used. The impellers are massive and the clearances between blades, as well as between the wall and blade, are very small thereby ensuring that the entire mass of liquid is sheared.

7.7.2. Portable mixers

For a wide range of applications, a portable mixer which can be clamped on the top or side of the vessel is often used. This is commonly fitted with two propeller blades so that the bottom rotor forces the liquid upwards and the top rotor forces the liquid downwards. The power supplied is up to about 2 kW, though the size of the motor becomes too great at higher powers. To avoid excessive strain on the armature, some form of flexible coupling should be fitted between the motor and the main propeller shaft. Units of this kind are usually driven at a fairly high rate (15 Hz), and a reduction gear can be fitted to the unit fairly easily for low-speed operations although this increases the mass of the unit.

7.7.3. Extruders

Mixing duties in the plastics industry are often carried out in either single or twin screw extruders. The feed to such units usually contains the base polymer in either granular or powder form, together with additives such as stabilisers, pigments, plasticisers, and so on. During processing in the extruder the polymer is melted and the additives mixed. The extrudate is delivered at high pressure and at a controlled rate from the extruder for shaping by means of either a die or a mould. Considerable progress has been made in the design of extruders in recent years, particularly by the application of finite element methods. One of the problems is that a considerable amount of heat is generated and the fluid properties may change by several orders of magnitude as a result of temperature changes. It is therefore always essential to solve the coupled equations of flow and heat transfer.

In the typical single-screw shown in Figure 7.24, the shearing which occurs in the helical channel between the barrel and the screw is not intense, and therefore this device does not give good mixing. Twin screw extruders, as shown in Figure 7.25, may be co- or counter-rotatory, and here there are regions where the rotors are in close proximity thereby generating extremely high shear stresses. Clearly, twin-screw units can yield a product of better mixture quality than a single-screw machine. Detailed accounts of the design and performance of extruders are available in literature[46–48].

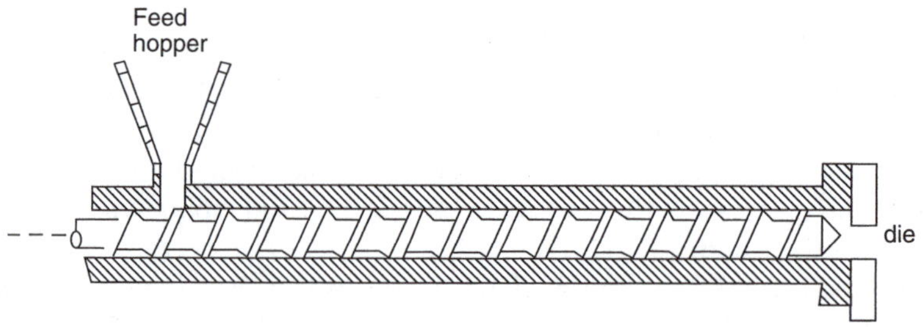

Figure 7.24. Single-screw extruder

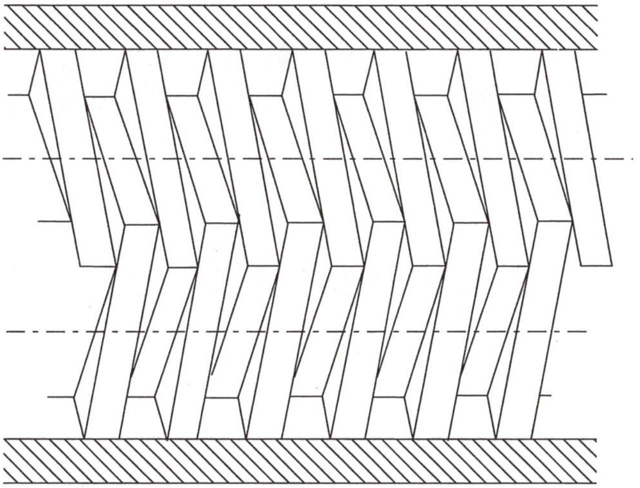

Figure 7.25. Co-rotating twin-screw extruder

7.7.4. Static mixers

All the mixers described so far have been of the dynamic type in the sense that moving blades are used to impart motion to the fluid and produce the mixing effect. In static mixers,[(2)] sometimes called "in-line" or "motionless" mixers, the fluids to be mixed are pumped through a pipe containing a series of specially shaped stationary blades. Static mixers can be used with liquids of a wide range of viscosities in either the laminar or turbulent regimes, but their special features are perhaps best appreciated in relation to laminar flow mixing. The flow patterns within the mixer are complex, though a numerical simulation of the flow has been carried out by LANG *et al.*[(49)]

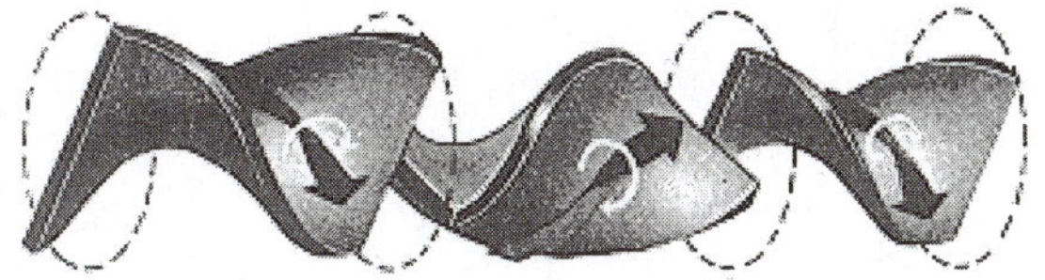

Figure 7.26. Twisted-blade type of static mixer elements

Figure 7.26 shows a particular type of static mixer in which a series of stationary helical blades mounted in a circular pipe is used to divide and twist the flowing streams. In laminar flow, (Figure 7.27) the material divides at the leading edge of each of these elements and follows the channels created by the element shape. At each succeeding element the two channels are further divided, and mixing proceeds by a distributive process similar to the cutting and folding mechanism shown in Figure 7.4. In principle, if each element divided the streams neatly into two, feeding two dissimilar streams to the mixer would give a striated material in which the thickness of each striation would be of the order $D_T/2^{\mathbf{n}}$ where D_T is the diameter of the tube and **n** is the number of

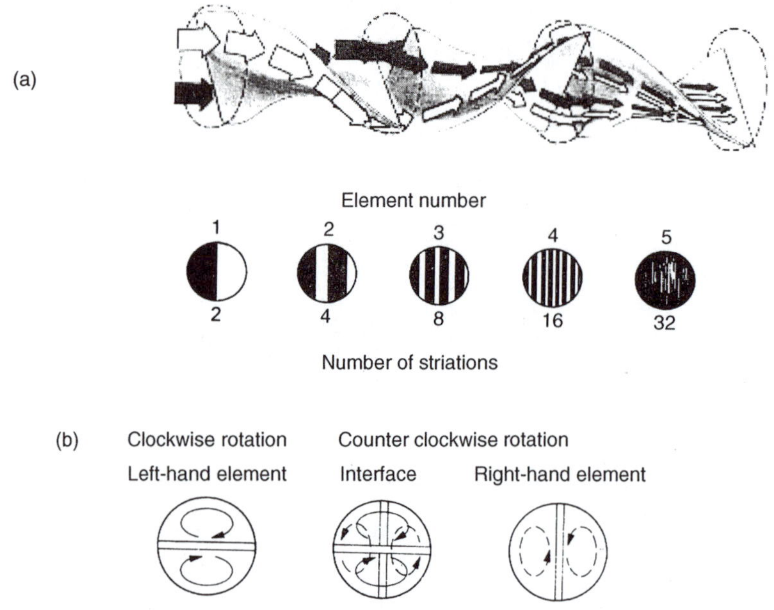

Figure 7.27. Twisted-blade type of static mixer operating in the laminar flow regime (*a*) Distributive mixing mechanism showing, in principle, the reduction in striation thickness produced (*b*) Radial mixing contribution from laminar shear mechanism

elements. However, the helical elements shown in Figure 7.26 also induce further mixing by a laminar shear mechanism (see Figures 7.1 and 7.3). This, combined with the twisting action of the element, helps to promote radial mixing which is important in eliminating any radial gradients of composition, velocity and possibly temperature that might exist in the material. Figure 7.28 shows how these mixing mechanisms together produce after only 10–12 elements, a well-blended material. Figures 7.26 to 7.28 all refer to static mixers made by CHEMINEER Ltd, Derby, U.K.

Figure 7.29 shows a Sulzer type SMX static mixer where the mixing element consists of a lattice of intermeshing and interconnecting bars contained in a pipe 80 mm diameter. It is recommended for viscous materials in laminar flow. The mixer shown is used in food processing, for example mixing fresh cheese with whipped cream.

Quantitatively, a variety of methods[50] has been proposed to describe the degree or quality of mixing produced in a static mixer. One of these measures of mixing quality is the relative standard deviation s/s_0, where s is the standard deviation in composition of a set of samples taken at a certain stage of the mixing operation, and s_0 is the standard deviation for a similar set of samples taken at the mixer inlet. Figure 7.28 shows schematically how the relative standard deviation falls as the number **n** of elements through which the material has passed increases, a perfectly mixed product having a zero relative standard deviation. One of the problems in using relative standard deviation or a similar index as a measure of mixing is that this depends on sample size which therefore needs to be taken into account in any assessment.

One of the most important considerations in choosing or comparing static mixers is the power consumed by the mixer to achieve a desired mixture quality. The pressure

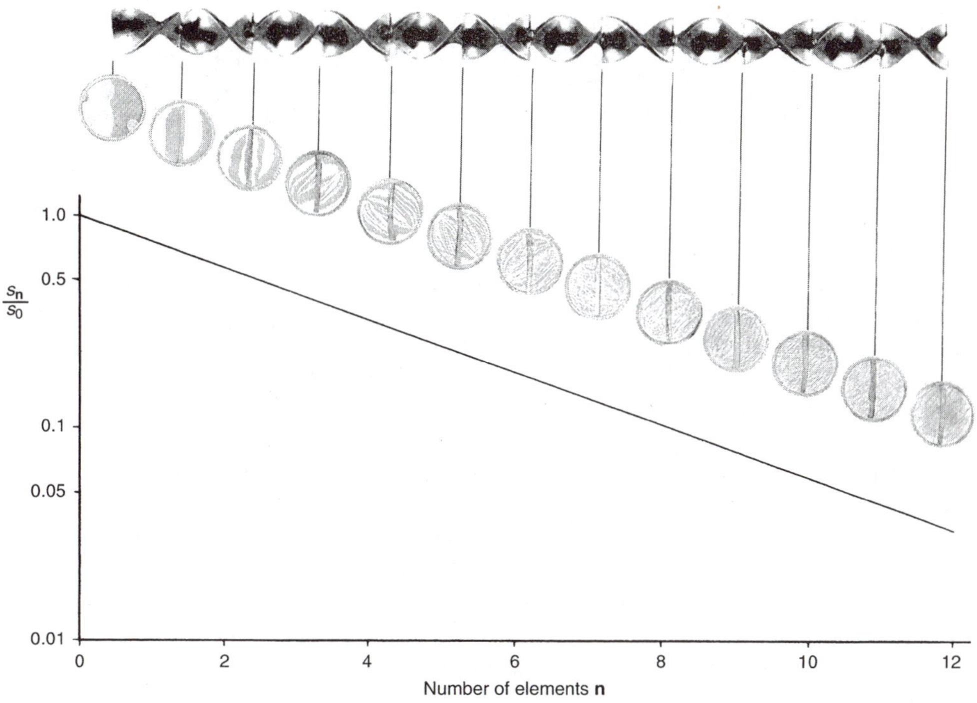

Figure 7.28. Static mixer in laminar flow: reduction in relative standard deviation of samples indicating improvement in mixture quality with increasing number **n** of elements traversed

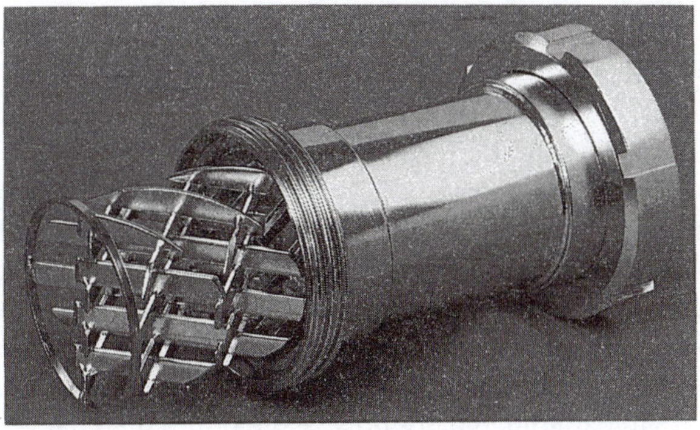

Figure 7.29. Static mixer for viscous materials

drop characteristics of a mixer are most conveniently described as the ratio of mixer pressure drop to empty pipe pressure drop for the same flowrate and diameter. Different static mixer designs can be compared[50] on a basis of mixing quality, pressure-drop ratio, initial cost and convenience of installation.

Static mixers for viscous fluids are widely used in processes producing polymers, fibres and plastics of all kinds where the materials tend to be viscous, hot and often at high pressures. However, static mixers have also achieved widespread use for low viscosity fluid mixing for blending, liquid-liquid, and even gas-liquid dispersions. In some cases the designs used for high viscosity liquids have also proved effective in the turbulent mixing regime for low viscosity fluids. In other cases manufacturers have developed special designs for turbulent flow mixing, and a wide variety of static mixer devices is now available[2].

7.7.5. Other types of mixer

Only a selection of commercially available mixing equipment has been described here. Indeed, the devices described all exist in a variety of configurations. In addition, there are many items of equipment based on altogether different principles; typical examples include jet mixers, in-line dynamic mixers, mills, valve homogenisers, ultrasonic homogenisers, etc. These, as well as many other types, have been discussed by HARNBY *et al.*[2], OLDSHUE[11], and NAGATA[19].

Frequently, it is convenient to mix the contents of a tank without the necessity for introducing an agitator and this may present difficulties in the construction of the vessel. It may then be possible to use an external circulating pump (usually centrifugal). If it is desirable not to make special connections to the vessel, it may be possible to connect the suction side of the pump to the bottom outlet of the tank by means of a T-piece and to discharge the liquid into the tank through its open top or through an existing entry point. In such a system, dispersion is effected in the high-shear region in the pump, and the liquid in the tank is maintained in a state of continuous circulation.

Such an arrangement may well be suitable when it is necessary to prevent fine particles from settling out at the bottom of the tank.

7.8. MIXING IN CONTINUOUS SYSTEMS

The mixing problems considered so far have related to batch systems in which two materials are mixed together and uniformity is maintained by continued operation of the agitator.

Frequently, stirred tanks are used with a continuous flow of material in on one side of the tank and with a continuous outflow from the other. A particular application is the use of the tank as a *continuous stirred-tank reactor* (CSTR). Inevitably, there will be a very wide range of residence times for elements of fluid in the tank. Even if the mixing is so rapid that the contents of the tank are always virtually uniform in composition, some elements of fluid will almost immediately flow to the outlet point and others will continue circulating in the tank for a very long period before leaving. The *mean residence time* of fluid in the tank is given by:

$$t_r = \frac{V}{Q} \tag{7.23}$$

where V is the volume of the contents of the tank (assumed constant), and Q is the volumetric throughput.

In a completely mixed system, the composition of the outlet stream will be equal to the composition in the tank.

The variation of time for which fluid elements remain with the tank is expressed as a *residence time distribution* and this can be calculated from a simple material balance if mixing is complete. For incomplete mixing, the calculation presents difficulties.

The problem is of great significance in the design of reactors because a varying residence time will, in general, lead to different degrees of chemical conversion of various fluid elements, and this is discussed in some detail in Volume 3, Chapter 1.

7.9. FURTHER READING

HARNBY, N., EDWARDS, M. F., and NIENOW, A. W. (eds): *Mixing in the Process Industries* 2nd edn (Butterworth-Heinemann, Oxford, 1992).

HOLLAND, F. A. and CHAPMAN, F. S.: *Liquid Mixing and Processing in Stirred Tanks* (Reinhold, New York, 1966).

METZNER, A. B.: In *Handbook of Fluid Dynamics*, section 7, V. L. Streeter (ed.) (McGraw-Hill, New York, 1961).

NAGATA, S.: *Mixing: Principles and Applications* (Wiley, London, 1975).

OLDSHUE, J. Y.: *Fluid Mixing Technology* (McGraw-Hill, New York, 1983).

TATTERSON, G. B.: *Fluid Mixing and Gas Dispersion in Agitated Tanks* (McGraw-Hill, New York, 1991).

UHL, V. W. and GRAY, J. B. (eds): *Mixing: Theory and Practice,* Vol. 1 (Academic Press, London, 1966); Vol. 2 (1967); Vol. 3 (1986).

7.10. REFERENCES

1. REAVELL, B. N.: *Trans. Inst. Chem. Eng.* **29** (1951) 301. Practical aspects of liquid mixing and agitation.
2. HARNBY, N., EDWARDS, M. F., and NIENOW, A. W. (eds): *Mixing in the Process Industries*, 2nd edn (Butterworth-Heinemann, London, 1992).
3. LEVENSPIEL, O.: *Engineering Flow and Heat Exchange* (Plenum, London, 1985).
4. OTTINO, J. M.: *Sci. Amer.* **260** (1989) 56. The mixing of fluids.
5. LEVICH, V. G.: *Physico-chemical Hydrodynamics* (Prentice-Hall, London, 1962).
6. MOLEN, VAN DER, K. and MAANEN, VAN, H. R. E.: *Chem. Eng. Sci.* **33** (1978) 1161. Laser-Doppler measurements of the turbulent flow in stirred vessels to establish scaling rules.
7. RAO, M. A. and BRODKEY, R. S.: *Chem. Eng. Sci.* **27** (1972) 137. Continuous flow stirred tank turbulence parameters in the impeller stream.
8. RUSHTON, J. H., COSTICH, E. W., and EVERETT, H. J.: *Chem. Eng. Prog.* **46** (1950) 395, 467. Power characteristics of mixing impellers. Parts I and II.
9. KRAMERS, H., BAARS, G. M., and KNOLL, W. H.: *Chem. Eng. Sci.* **2** (1953) 35. A comparative study on the rate of mixing in stirred tanks.
10. SKELLAND, A. H. P.: *Non-Newtonian Flow and Heat Transfer* (Wiley, New York, 1967).
11. OLDSHUE, J. Y.: *Fluid Mixing Technology* (McGraw-Hill, New York, 1983).
12. METZNER, A. B. and OTTO, R. E.: *A.I.Ch.E. Jl.* **3** (1957) 3. Agitation of non-Newtonian fluids.
13. METZNER, A. B. and TAYLOR, J. S.: *A.I.Ch.E. Jl.* **6** (1960) 109. Flow patterns in agitated vessels.
14. ULBRECHT, J.: *The Chemical Engineer* (London) No. 286 (1974) 347. Mixing of viscoelastic fluids by mechanical agitation.
15. ASTARITA, G.: *J. Non-Newt. Fluid Mech.* **4** (1979) 285. Scaleup problems arising with non-Newtonian fluids.
16. SKELLAND, A. H. P.: Mixing and agitation of non-Newtonian fluids. In *Handbook of Fluids in Motion*, Cheremisinoff, N. P. and Gupta, R. (eds) (Ann Arbor, New York, 1983).
17. HOLLAND, F. A. and CHAPMAN, F. S.: *Liquid Mixing and Processing in Stirred Tanks* (Reinhold, New York, 1966).
18. BISSELL, E. S., HESSE, H. C., EVERETT, H. J., and RUSHTON, J. H.: *Chem. Eng. Prog.* **43** (1947) 649. Design and utilization of internal fittings for mixing vessels.
19. NAGATA, S.: *Mixing: Principles and Applications* (Wiley, London, 1975).
20. UHL, V. W. and GRAY, J. B. (eds): *Mixing: Theory and Practice*, Vol. 1 (Academic Press, London, 1966).
21. ŠTERBAČEK Z. and TAUSK, P.: *Mixing in the Chemical Industry*. Translated by Mayer, K. and Bourne, J. R. (Pergamon, Oxford, 1965).
22. HOOGENDOORN, C. J. and DEN HARTOG, A. P.: *Chem. Eng. Sci.* **22** (1967) 1689. Model studies on mixers in the viscous flow region.
23. ULLRICH, H. and SCHREIBER, H.: *Chem. Ing. Tech.* **39** (1967) 218. Rühren in zähen Flüssigkeiten.

24. MITSUISHI, N. and HIRAI, N. J.: *J. Chem. Eng. Japan* **2** (1969) 217. Power requirements in the agitation of non-Newtonian fluids.
25. GODFREY, J. C.: Mixing of high viscosity liquids. In: *Mixing in the Processing Industries*: Harnby, N., Edwards, M. F., and Nienow, A. W. (eds) (Butterworths, London, 1985) pp. 185–201.
26. CALDERBANK, P. H. and MOO-YOUNG, M. B.: *Trans. Inst. Chem. Eng.* **37** (1959) 26. The prediction of power consumption in the agitation of non-Newtonian fluids.
27. BECKNER, J. L. and SMITH, J. M.: *Trans. Inst. Chem. Eng.* **44** (1966) T224. Anchor-agitated systems: Power input with Newtonian and pseudo-plastic fluids.
28. NAGATA, S., NISHIKAWA, M., TADA, H., HIRABAYASHI, H., and GOTOH, S.: *J. Chem. Eng. Japan* **3** (1970) 237. Power consumption of mixing impellers in Bingham plastic fluids.
29. METZNER, A. B., FEEHS, R. H., ROMOS, H. L., OTTO, R. E., and TUTHILL, J. P.: *A.I.Ch.E. Jl.* **7** (1961) 3. Agitation of viscous Newtonian and non-Newtonian fluids.
30. EDWARDS, M. F., GODFREY, J. C., and KASHANI, M. M.: *J. Non-Newt. Fluid Mech.* **1** (1976) 309. Power requirements for the mixing of thixotropic fluids.
31. KELKAR, J. V., MASHELKAR, R. A., and ULBRECHT, J.: *Trans. Inst. Chem. Eng.* **50** (1972) 343. On the rotational viscoelastic flows around simple bodies and agitators.
32. RIEGER, F. and NOVAK, V.: *Trans. Inst. Chem. Eng.* **52** (1974) 285. Power consumption for agitating viscoelastic liquids in the viscous regime.
33. OLIVER, D. R., NIENOW, A. W., MITSON, R. J., and TERRY, K.: *Chem. Eng. Res. Des.* **62** (1984) 123. Power consumption in the mixing of Boger fluids.
34. PRUD'HOMME, R. K. and SHAQFEH, E.: *A.I.Ch.E. Jl.* **30** (1984) 485. Effect of elasticity on mixing torque requirements for Rushton turbine impellers.
35. KALE, D. D., MASHELKAR, R. A., and ULBRECHT, J.: *Chem. Ing. Tech.* **46** (1974) 69. High speed agitation of non-Newtonian fluids: Influence of elasticity and fluid inertia.
36. CHAVAN, V. V. and MASHELKAR, R. A.: *Adv. Transport Proc.* **1** (1980) 210. Mixing of viscous Newtonian and non-Newtonian fluids.
37. IRVING, H. F. and SAXTON, R. L.: In *Mixing Theory and Practice*, Uhl, V.W. and Gray, J. B. (eds) Vol. 2, pp. 169–224, (Academic Press, 1967).
38. GRAY, J. B.: In *Mixing Theory and Practice*, Uhl, V.W. and Gray, J. B. (eds) Vol. 1, pp. 179–278. (Academic Press, 1966).
39. CARREAU, P. J., PATTERSON, I., and YAP, C. Y.: *Can J. Chem. Eng.* **54** (1976) 135. Mixing of viscoelastic fluids with helical ribbon agitators I. Mixing time and flow pattern.
40. PETERS, D. C. and SMITH, J. M.: *Trans. Inst. Chem. Eng.* **45** (1967) 360. Fluid flow in the region of anchor agitator blades.
41. NORWOOD, K. W. and METZNER, A. B.: *A.I.Ch.E. Jl.* **6** (1960) 432. Flow patterns and mixing rates in agitated vessels.
42. BOURNE, J. R. and BUTLER, H.: *Trans. Inst. Chem. Eng.* **47** (1969) T11. On analysis of the flow produced by helical ribbon impellers.
43. GODLESKI, E. S. and SMITH, J. C.: *A.I.Ch.E. Jl.* **8** (1962) 617. Power requirements and blend times in the agitation of pseudoplastic fluids.
44. CHAVAN, V. V., ARUMUGAM, M. and ULBRECHT, J.: *A.I.Ch.E. Jl.* **21** (1975) 613. On the influence of liquid elasticity on mixing in a vessel agitated by a combined ribbon-screw impeller.
45. HALL, K. R. and GODFREY, J. C.: *Trans. Inst. Chem. Eng.* **46** (1968) 205. The mixing rates of highly viscous Newtonian and non-Newtonian fluids in a laboratory sigma blade mixer.
46. MCKELVEY, J. M.: *Polymer Processing* (Wiley, New York, 1962).
47. JANSSEN, L. P. B. M.: *Twin Screw Extrusion* (Elsevier, Amsterdam, 1978).
48. SCHENKEL, G.: *Plastics Extrusion Technology* (Cliffe Books, London, 1966).
49. LANG, E., DRTINA, P., STREIFF, F. and FLEISCHLI, M.: *Int. Jl. Heat Mass Transfer* **38** (1995) 2239. Numerical simulation of the fluid flow and mixing process in a static mixer.
50. HEYWOOD, N. I., VINEY, L. J., and STEWART, I. W.: *Inst. Chem. Eng.* Symposium Series No. 89, Fluid Mixing II (1984) 147. Mixing efficiencies and energy requirements of various motionless mixer designs for laminar mixing applications.

7.11. NOMENCLATURE

		Units in SI system	Dimensions in **M, L, T**
A	Area of flow	m^2	$\mathbf{L}^2$
A'	Area for shear	m^2	$\mathbf{L}^2$
C	Concentration	—	—

		Units in SI system	Dimensions in **M, L, T**
D	Impeller diameter	m	**L**
D_T	Tank *or* tube diameter	m	**L**
F	Force	N	$\mathbf{MLT^{-2}}$
Fr	Froude number (N^2D/g)	—	—
g	Acceleration due to gravity	m/s^2	$\mathbf{LT^{-2}}$
H	Depth of liquid	m	**L**
k_s	Function of impeller speed (equation 7.18)	—	—
L	Length	m	**L**
m	Mass of liquid	kg	**M**
N	Speed of rotation (revs/unit time)	Hz	$\mathbf{T^{-1}}$
N_p	Power number $(\mathbf{P}/\rho N^3D^5)$	—	—
n	Power law index	—	—
n	Number of elements	—	—
P	Power	W	$\mathbf{ML^2T^{-3}}$
P	Pitch of agitator	m	**L**
Q	Volumetric throughput	m^3/s	$\mathbf{L^3T^{-1}}$
R	Number of baffles	—	—
r	Radius	m	**L**
s, s_0	Standard deviations	—	—
t	Time	s	**T**
t_m	Mixing time	s	**T**
t_r	Residence time	s	**T**
u	Velocity	m/s	$\mathbf{LT^{-1}}$
V	Volume of tank	m^3	$\mathbf{L^3}$
W	Blade width	m	**L**
W_B	Width of baffles	m	**L**
y	Distance	m	**L**
Z_A	Height of agitator from base of tank	m	**L**
$\dot{\gamma}_{ang}$	Shear rate (angular)	Hz	$\mathbf{T^{-1}}$
μ	Viscosity	$N\,s/m^2$	$\mathbf{ML^{-1}T^{-1}}$
μ_a	Apparent viscosity	$N\,s/m^2$	$\mathbf{ML^{-1}T^{-1}}$
ρ	Density	kg/m^3	$\mathbf{ML^{-3}}$
σ	Surface tension	N/m	$\mathbf{MT^{-2}}$
σ^2	Variance	—	—
θ_m	Dimensionless mixing time Nt_m	—	—
Re	Reynolds number $(\rho ND^2/\mu)$	—	—

CHAPTER 8

Pumping of Fluids

8.1. INTRODUCTION

For the pumping of liquids or gases from one vessel to another or through long pipes, some form of mechanical pump is usually employed. The energy required by the pump will depend on the height through which the fluid is raised, the pressure required at delivery point, the length and diameter of the pipe, the rate of flow, together with the physical properties of the fluid, particularly its viscosity and density. The pumping of liquids such as sulphuric acid or petroleum products from bulk store to process buildings, or the pumping of fluids round reaction units and through heat exchangers, are typical illustrations of the use of pumps in the process industries. On the one hand, it may be necessary to inject reactants or catalyst into a reactor at a low, but accurately controlled rate, and on the other to pump cooling water to a power station or refinery at a very high rate. The fluid may be a gas or liquid of low viscosity, or it may be a highly viscous liquid, possibly with non-Newtonian characteristics. It may be clean, or it may contain suspended particles and be very corrosive. All these factors influence the choice of pump.

Because of the wide variety of requirements, many different types are in use including centrifugal, piston, gear, screw, and peristaltic pumps, though in the chemical and petroleum industries the centrifugal type is by far the most important. The main features considered in this chapter are an understanding of the criteria for pump selection, the determination of size and power requirements, and the positioning of pumps in relation to pipe systems. For greater detail, reference may be made to specialist publications included in Section 8.7.

Pump design and construction is a specialist field, and manufacturers should always be consulted before a final selection is made. In general, pumps used for circulating gases work at higher speeds than those used for liquids, and lighter valves are used. Moreover, the clearances between moving parts are smaller on gas pumps because of the much lower viscosity of gases, giving rise to an increased tendency for leakage to occur. When a pump is used to provide a vacuum, it is even more important to guard against leakage.

The work done by the pump is found by setting up an energy balance equation. If W_s is the shaft work done by unit mass of fluid on the surroundings, then $-W_s$ is the shaft work done on the fluid by the pump.

From equation 2.55: $$-W_s = \Delta\frac{u^2}{2\alpha} + g\Delta z + \int_{P_1}^{P_2} v\,dP + F \quad (8.1)$$

and from equation 2.56: $$-W_s = \Delta\frac{u^2}{2\alpha} + g\Delta z + \Delta H - q \quad (8.2)$$

In any practical system, the pump efficiency must be taken into account, and more energy must be supplied by the motor driving the pump than is given by $-W_s$. If liquids are considered to be incompressible, there is no change in specific volume from the inlet to the delivery side of the pump. The physical properties of gases are, however, considerably influenced by the pressure, and the work done in raising the pressure of a gas is influenced by the rate of heat flow between the gas and the surroundings. Thus, if the process is carried out adiabatically, all the energy added to the system appears in the gas and its temperature rises. If an ideal gas is compressed and then cooled to its initial temperature, its enthalpy will be unchanged and the whole of the energy supplied by the compressor is dissipated to the surroundings. If, however, the compressed gas is allowed to expand it will absorb heat and is therefore capable of doing work at the expense of heat energy from the surroundings.

8.2. PUMPING EQUIPMENT FOR LIQUIDS

As already indicated, the liquids used in the chemical industries differ considerably in physical and chemical properties, and it has been necessary to develop a wide variety of pumping equipment. The two main forms are the positive displacement type and centrifugal pumps. In the former, the volume of liquid delivered is directly related to the displacement of the piston and therefore increases directly with speed and is not appreciably influenced by the pressure. This group includes the reciprocating piston pump and the rotary gear pump, both of which are commonly used for delivery against high pressures and where nearly constant delivery rates are required. In the centrifugal type a high kinetic energy is imparted to the liquid, which is then converted as efficiently as possible into pressure energy. For some applications, such as the handling of liquids which are particularly corrosive or contain abrasive solids in suspension, compressed air is used as the motive force instead of a mechanical pump. An illustration of the use of this form of equipment is the transfer of the contents of a reaction mixture from one vessel to another.

The following factors influence the choice of pump for a particular operation.

(1) The quantity of liquid to be handled. This primarily affects the size of the pump and determines whether it is desirable to use a number of pumps in parallel.
(2) The head against which the liquid is to be pumped. This will be determined by the difference in pressure, the vertical height of the downstream and upstream reservoirs and by the frictional losses which occur in the delivery line. The suitability of a centrifugal pump and the number of stages required will largely be determined by this factor.
(3) The nature of the liquid to be pumped. For a given throughput, the viscosity largely determines the friction losses and hence the power required. The corrosive nature will determine the material of construction both for the pump and the packing. With suspensions, the clearances in the pump must be large compared with the size of the particles.
(4) The nature of the power supply. If the pump is to be driven by an electric motor or internal combustion engine, a high-speed centrifugal or rotary pump will be

preferred as it can be coupled directly to the motor. Simple reciprocating pumps can be connected to steam or gas engines.

(5) If the pump is used only intermittently, corrosion problems are more likely than with continuous working.

The cost and mechanical efficiency of the pump must always be considered, and it may be advantageous to select a cheap pump and pay higher replacement or maintenance costs rather than to install a very expensive pump of high efficiency.

8.2.1. Reciprocating pump

The piston pump

The piston pump consists of a cylinder with a reciprocating piston connected to a rod which passes through a gland at the end of the cylinder as indicated in Figure 8.1. The liquid enters from the suction line through a suction valve and is discharged through a delivery valve. These pumps may be single-acting, with the liquid admitted only to the portion of the cylinder in front of the piston or double-acting, in which case the feed is admitted to both sides of the piston. The majority of pumps are of the single-acting type typically giving a low flowrate of say 0.02 m^3/s at a high pressure of up to 100 MN/m^2.[1]

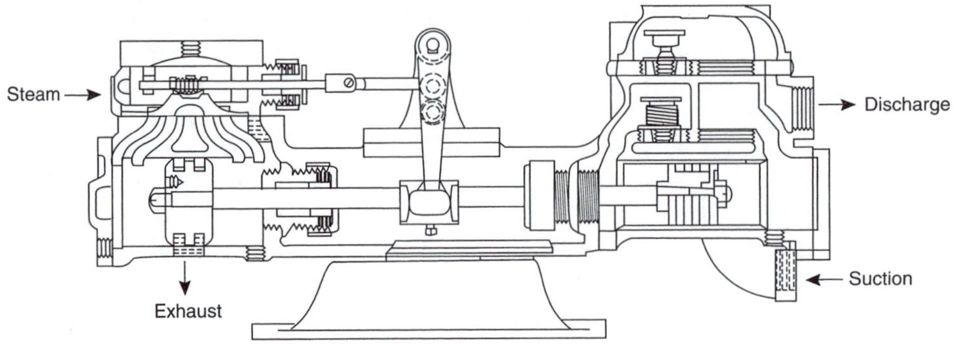

Figure 8.1. A typical steam-driven piston pump

The velocity of the piston varies in an approximately sinusoidal manner and the volumetric rate of discharge of the liquid shows corresponding fluctuations. In a single-cylinder pump the delivery will rise from zero as the piston begins to move forward to a maximum when the piston is fully accelerated at approximately the mid point of its stroke; the delivery will then gradually fall off to zero. If the pump is single-acting there will be an interval during the return stroke when the cylinder will fill with liquid and the delivery will remain zero. On the other hand, in a double-acting pump the delivery will be similar in the forward and return strokes. In many cases, however, the cross-sectional area of the piston rod may be significant compared with that of the piston and the volume delivered during the return stroke will therefore be less than that during the forward stroke. A more even delivery is obtained if several cylinders are suitably compounded. If two double-acting cylinders are used there will be a lag between the deliveries of the two cylinders, and the total delivery will then be the sum of the deliveries from the individual cylinders. Typical curves of delivery rate for a single-cylinder (simplex) pump are shown in Figure 8.2*a*. The delivery from a two-cylinder (duplex) pump in which both the cylinders are double-acting is shown in Figure 8.2*b*; the broken lines indicate the deliveries from the individual cylinders and the unbroken line indicates the total delivery. It will be seen that the delivery is much smoother than that obtained with the simplex pump, the minimum delivery being equal to the maximum obtained from a single cylinder.

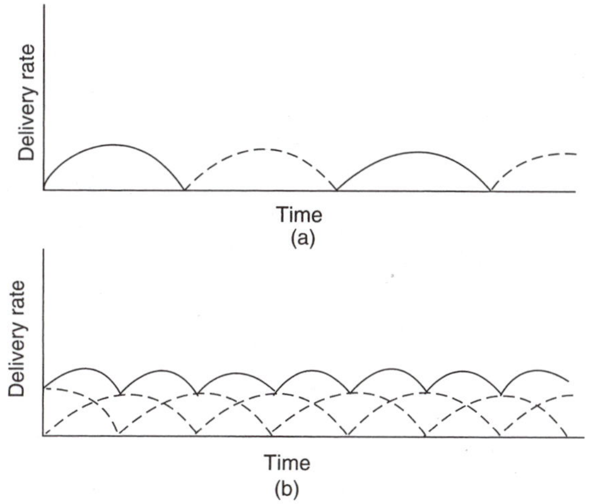

Figure 8.2. Delivery from (*a*) simplex and (*b*) duplex pumps

The theoretical delivery of a piston pump is equal to the total swept volume of the cylinders. The actual delivery may be less than the theoretical value because of leakage past the piston and the valves or because of inertia of the valves. In some cases, however, the actual discharge is greater than theoretical value because the momentum of the liquid in the delivery line and sluggishness in the operation of the delivery valve may result in continued delivery during a portion of the suction stroke. The volumetric efficiency, which is defined as the ratio of the actual discharge to the swept volume, is normally greater than 90 per cent.

The size of the suction and delivery valves is determined by the throughput of the pump. Where the rate of flow is high, two or more valves may be used in parallel.

The piston pump can be directly driven by steam, in which case the piston rod is common to both the pump and the steam engine. Alternatively, an electric motor or an internal combustion engine may supply the motive power through a crankshaft; because the load is very uneven, a heavy flywheel should then be fitted and a regulator in the steam supply may often provide a convenient form of speed control.

The pressure at the delivery of the pump is made up of the following components:

(1) The static pressure at the delivery point.
(2) The pressure required to overcome the frictional losses in the delivery pipe.
(3) The pressure for the acceleration of the fluid at the commencement of the delivery stroke.

The liquid in the delivery line is accelerated and retarded in phase with the motion of the piston, and therefore the whole of the liquid must be accelerated at the commencement of the delivery stroke and retarded at the end of it. Every time the fluid is accelerated, work has to be done on it and therefore in a long delivery line the expenditure of energy is very large since the excess kinetic energy of the fluid is not usefully recovered during the suction stroke. Due to the momentum of the fluid, the pressure at the pump may fall sufficiently low for separation to occur. The pump is then said to *knock*. The flow in the delivery line can be evened out and the energy at the beginning of each stroke reduced, by the introduction of an air vessel at the pump discharge. This consists of a sealed vessel which contains air at the top and liquid at the bottom. When the delivery stroke commences, liquid is pumped into the air vessel and the air is compressed. When the discharge from the pump decreases towards the end of the stroke, the pressure in the air vessel is sufficiently high for some of the liquid to be expelled into the delivery line. If the air vessel is sufficiently large and is fitted close to the pump, the velocity of the liquid in the delivery line can be maintained approximately constant. The frictional losses are also reduced by the incorporation of an air vessel because the friction loss under turbulent conditions is approximately proportional to the linear velocity in the pipe raised to the power 1.8; i.e. the reduced friction losses during the period of minimum discharge do not compensate for the greatly increased losses when the pump is delivering at maximum rate (see Section 8.6). Further, the maximum stresses set up in the pump are reduced by the use of an air vessel.

Air vessels are also incorporated in the suction line for a similar reason. Here they may be of even greater importance because the pressure drop along the suction line is necessarily limited to rather less than one atmosphere if the suction tank is at atmospheric pressure. The flowrate may be limited if part of the pressure drop available must be utilised in accelerating the fluid in the suction line; the air vessel should therefore be sufficiently large for the flowrate to be maintained approximately constant.

The plunger or ram pump

This pump is the same in principle as the piston type but differs in that the gland is at one end of the cylinder making its replacement easier than with the standard piston type. The sealing of piston and ram pumps has been much improved but, because of the nature of the fluids frequently used, care in selecting and maintaining the seal is very important. The piston or ram pump may be used for injections of small quantities of inhibitors to polymerisation units or of corrosion inhibitors to high pressure systems, and also for boiler feed water applications.

The diaphragm pump

The diaphragm pump has been developed for handling corrosive liquids and those containing suspensions of abrasive solids. It is in two sections separated by a diaphragm of rubber, leather, or plastics material. In one section a plunger or piston operates in a cylinder in which a non-corrosive fluid is displaced. The movement of the fluid is transmitted by means of the flexible diaphragm to the liquid to be pumped. The only moving parts of the pump that are in contact with the liquid are the valves, and these can be specially designed to handle the material. In some cases the movement of the diaphragm is produced by direct mechanical action, or the diaphragm may be air actuated as shown in Figure 8.3, in which case a particularly simple and inexpensive pump results, capable of operating up to 0.2 MN/m^2.

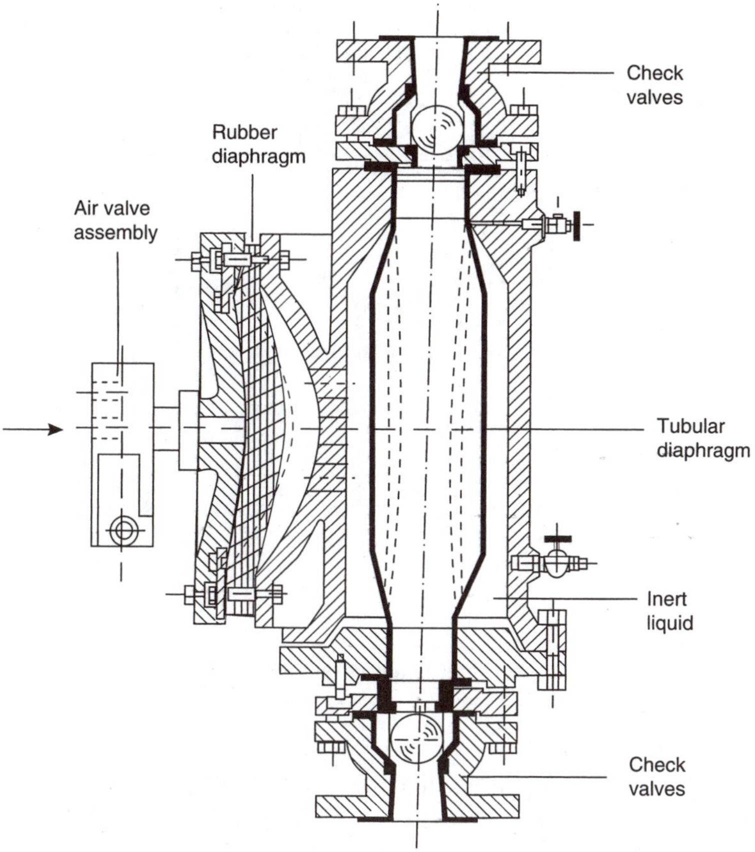

Figure 8.3. Diaphragm pump

When pumping non-Newtonian fluids, difficulties are sometimes experienced in initiating the flow of pseudoplastic materials. Positive displacement pumps can overcome the problem and the diaphragm pump in particular is useful in dealing with agglomerates in suspension. Care must always to be taken that the safe working pressure for the pump is not exceeded. This can be achieved conveniently by using a hydraulic drive for a diaphragm pump, equipped with a pressure limiting relief valve which ensures that no damage is done to the system, as shown in Figure 8.4.

By virtue of their construction, diaphragm pumps cannot be used for high pressure applications. In the Mars pump, there is no need for a diaphragm as the working fluid (oil), of lower density than the liquid to be pumped, forms an interface with it in a vertical chamber. The pump, which is used extensively for concentrated slurries, is really a development of the old Ferrari's acid pump which was designed for corrosive liquids.

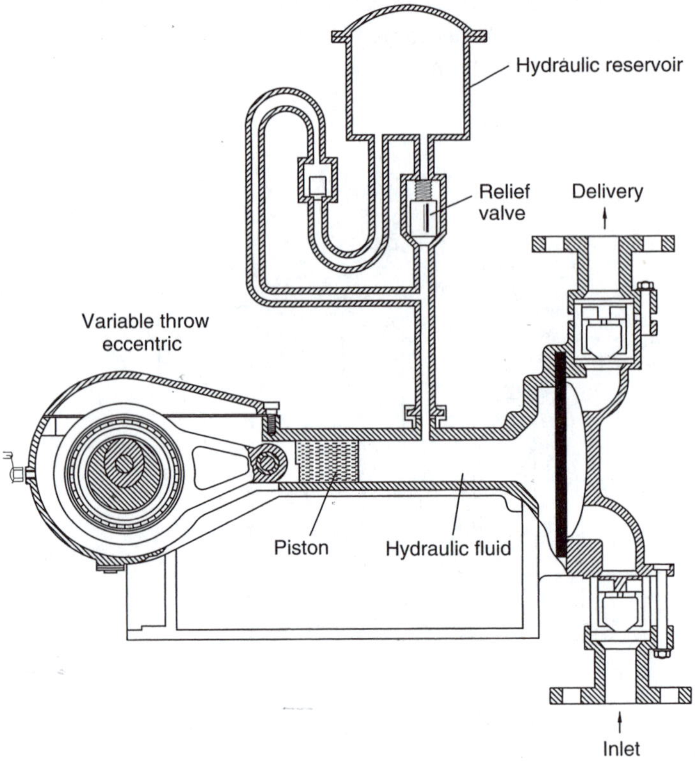

Figure 8.4. A hydraulic drive to protect a positive displacement pump

The metering pump

Metering pumps are positive displacement pumps driven by constant speed electric motors. They are used where a constant and accurately controlled rate of delivery of a liquid is required, and they will maintain this constant rate irrespective of changes in the pressure against which they operate. The pumps are usually of the plunger type for low throughput and high-pressure applications; for large volumes and lower pressures a diaphragm is used. In either case, the rate of delivery is controlled by adjusting the stroke of the piston element, and this can be done whilst the pump is in operation. A single-motor driver may operate several individual pumps and in this way give control of the actual flows and of the flow ratio of several streams at the same time. The output may be controlled from zero to maximum flowrate, either manually on the pump or remotely. These pumps may be used for the dosing of works effluents and water supplies, and the feeding of reactants, catalysts, or inhibitors to reactors at controlled rates, and although a simple method for controlling flowrate is provided, high precision standards of construction are required.

Example 8.1

A single-acting reciprocating pump has a cylinder diameter of 110 mm and a stroke of 230 mm. The suction line is 6 m long and 50 mm in diameter and the level of the water in the suction tank is 3 m below the cylinder of the pump. What is the maximum speed at which the pump can run without an air vessel if separation is not to occur in the suction line? The piston undergoes approximately simple harmonic motion. Atmospheric pressure is equivalent to a head of 10.36 m of water and separation occurs at an absolute pressure corresponding to a head of 1.20 m of water.

Solution

The tendency for separation to occur will be greatest at:

(a) the inlet to the cylinder because here the static pressure is a minimum and the head required to accelerate the fluid in the suction line is a maximum;
(b) the commencement of the suction stroke because the acceleration of the piston is then a maximum.

If the maximum permissible speed of the pump is N Hz:
Angular velocity of the driving mechanism $= 2\pi N$ radians/s
Acceleration of piston $= 0.5 \times 0.230(2\pi N)^2 \cos(2\pi N t)$ m/s^2
Maximum acceleration (when $t = 0$) $= 4.54N^2$ m/s^2
Maximum acceleration of the liquid in the suction pipe

$$= \left(\frac{0.110}{0.05}\right)^2 \times 4.54N^2 = 21.97N^2 \text{ m/s}^2$$

Accelerating force acting on the liquid

$$= 21.97N^2 \frac{\pi}{4}(0.050)^2 \times (6 \times 1000) \text{ N}$$

Pressure drop in suction line due to acceleration $= 21.97N^2 \times 6 \times 1000$ N/m^2

$$= 1.32 \times 10^5 N^2 \text{ N/m}^2$$

or:
$$\frac{(1.32 \times 10^5 N^2)}{(1000 \times 9.81)} = 13.44N^2 \text{ m water}$$

Pressure head at cylinder when separation is about to occur,

$$1.20 = (10.36 - 3.0 - 13.44N^2) \text{ m water}$$

$\therefore$
$$\underline{\underline{N = 0.675 \text{ Hz}}}$$

8.2.2. Positive-displacement rotary pumps

The gear pump and the lobe pump

Gear and lobe pumps operate on the principle of using mechanical means to transfer small elements or "packages" of fluid from the low pressure (inlet) side to the high pressure (delivery) side. There is a wide range of designs available for achieving this end. The general characteristics of the pumps are similar to those of reciprocating piston pumps, but the delivery is more even because the fluid stream is broken down into so much smaller elements. The pumps are capable of delivering to a high pressure, and the pumping rate is approximately proportional to the speed of the pump and is not greatly influenced by the pressure against which it is delivering. Again, it is necessary to provide a pressure relief system to ensure that the safe operating pressure is not exceeded. Recent developments in the use of pumps of this type have been described by HARVEST(2).

One of the commonest forms of the pump is the *gear pump* in which one of the gear wheels is driven and the other turns as the teeth engage; two versions are illustrated in

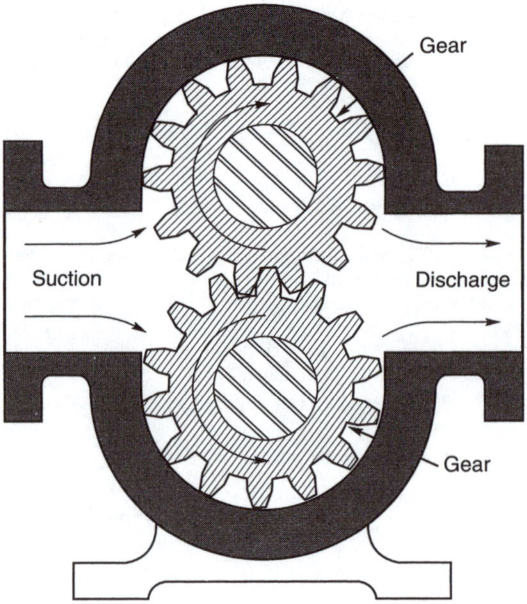

Figure 8.5. Gear pump

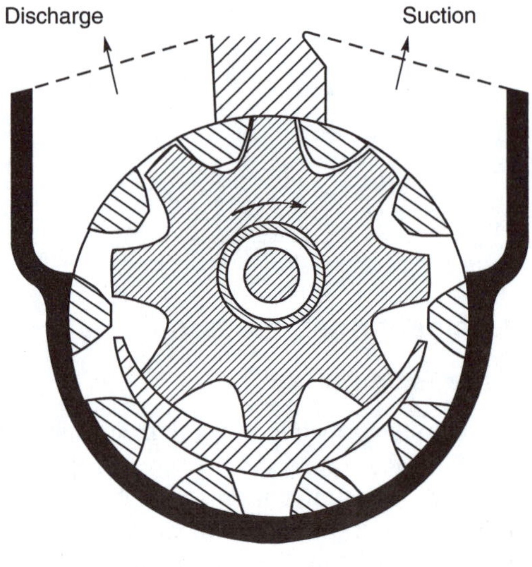

Figure 8.6. Internal gear pumb

Figures 8.5 and 8.6. The liquid is carried round in the spaces between consecutive gear teeth and the outer casing of the pump, and the seal between the high and low pressure sides of the pump is formed as the gears come into mesh and the elements of fluid are squeezed out. Gear pumps are extensively used for both high-viscosity Newtonian liquids and non-Newtonian fluids. The *lobe-pump* (Figures 8.7 and 8.8) is similar, but the gear

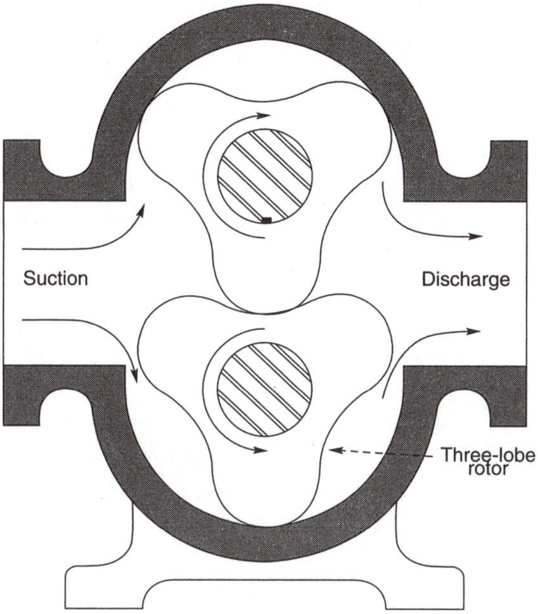

Figure 8.7. Lobe pumb

Figure 8.8. Lobe pump

teeth are replaced by two or three lobes and both axles are driven; it is therefore possible to maintain a small clearance between the lobes, and wear is reduced.

The Cam Pump

A rotating cam is mounted eccentrically in a cylindrical casing and a very small clearance is maintained between the outer edge of the cam and the casing. As the cam rotates it expels liquid from the space ahead of it and sucks in liquid behind it. The delivery and suction sides of the pump are separated by a sliding valve which rides on the cam. The characteristics again are similar to those of the gear pump.

The Vane Pump

The rotor of the vane pump is mounted off centre in a cylindrical casing (Figure 8.9). It carries rectangular vanes in a series of slots arranged at intervals round the curved surface of the rotor. The vanes are thrown outwards by centrifugal action and the fluid is carried in the spaces bounded by adjacent vanes, the rotor, and the casing. Most of the wear is on the vanes and these can readily be replaced.

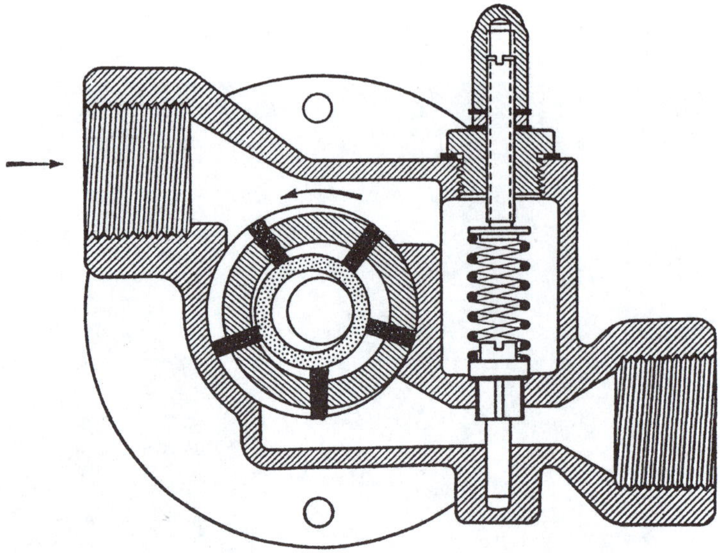

Figure 8.9. Vane pump

The flexible vane pump

The pumps described above will not handle liquids containing solid particles in suspension, and the flexible vane pumps has been developed to overcome this disadvantage. In this case, the rotor (Figure 8.10) is an integral elasomer moulding of a hub with flexible vanes which rotates in a cylindrical casing containing a crescent-shaped block, as in the case of the *internal gear pump*.

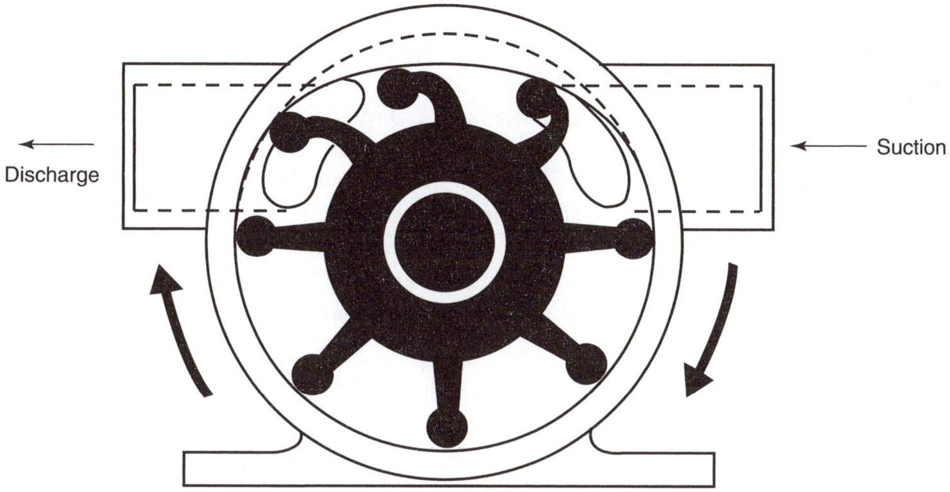

Figure 8.10. Flexible vane pump

The flow inducer or peristaltic pump

This is a special form of pump in which a length of silicone rubber or other elastic tubing, typically of 3 to 25 mm diameter, is compressed in stages by means of a rotor as shown in Figure 8.11. The tubing is fitted to a curved track mounted concentrically with a rotor

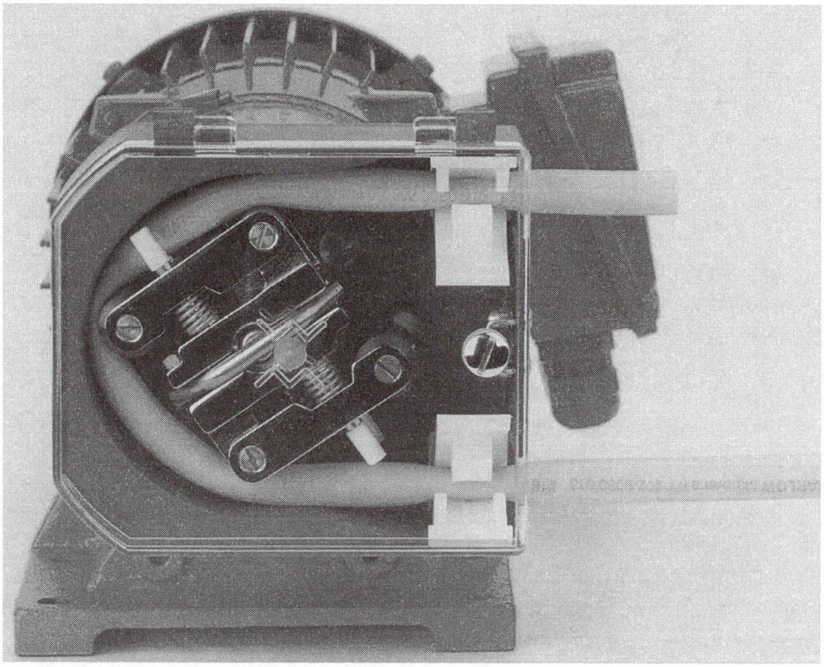

Figure 8.11. Flow inducer

carrying three rollers. As the rollers rotate, they flatten the tube against the track at the points of contact. These "flats" move the fluid by positive displacement, and the flow can be precisely controlled by the speed of the motor.

These pumps have been particularly useful for biological fluids where all forms of contact must be avoided. They are being increasingly used and are suitable for pumping emulsions, creams, and similar fluids in laboratories and small plants where the freedom from glands, avoidance of aeration, and corrosion resistance are valuable, if not essential. Recent developments[1] have produced thick-wall, reinforced moulded tubes which give a pumping performance of up to 0.02 m^3/s at 1 MN/m^2 and these pumps have been further discussed by GADSDEN[3]. The control is such that these pumps may conveniently be used as metering pumps for dosage processes.

The Mono pump

Another example of a positive acting rotary pump is the single screw-extruder pump typified by the Mono pump, illustrated in Figure 8.12, in which a specially shaped helical metal rotor revolves eccentrically within a double-helix, resilient rubber stator of twice the pitch length of the metal rotor. A continuous forming cavity is created as the rotor turns — the cavity progressing towards the discharge, advancing in front of a continuously forming seal line and thus carrying the pumped material with it as shown in Figure 8.13.

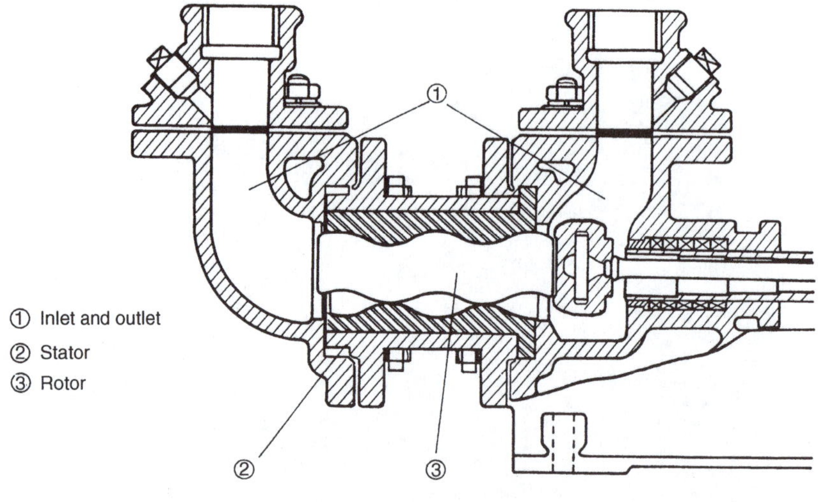

Figure 8.12. Mono pump

The Mono pump gives a uniform flow and is quiet in operation. It will pump against high pressures; the higher the required pressure, the longer are the stator and the rotor and the greater the number of turns. The pump can handle corrosive and gritty liquids and is extensively used for feeding slurries to filter presses. It must never be run dry. The Mono Merlin Wide Throut pump is used for highly viscous liquids (Figure 8.14).

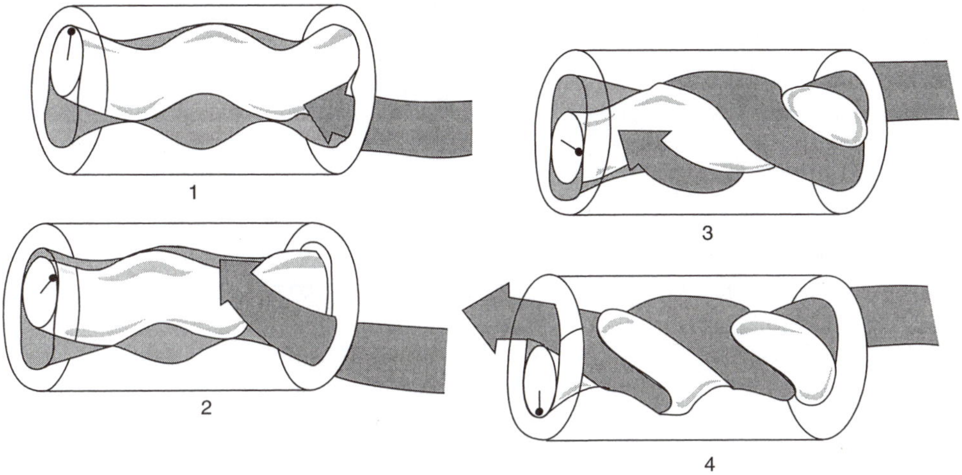

Figure 8.13. The mode of operation of a Mono pump

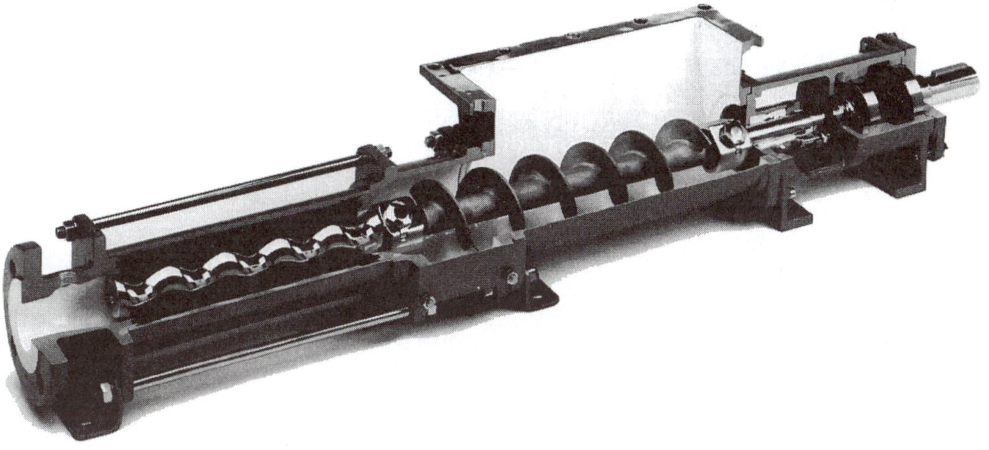

Figure 8.14. Mono Merlin Wide-Throat pump

Screw pumps

A most important class of pump for dealing with highly viscous material is represented by the screw extruder used in the polymer industry. Extruders find their main application in the manufacture of simple and complex sections (rods, tubes, beadings, curtain rails, rainwater gutterings and a multitude of other shapes). However, the shape of section produced in a given material is dependent only on the profile of the hole through which the fluid is pushed just before it cools and solidifies. The screw pump is of more general application and will be considered first.

The principle is shown in Figure 8.15. The fluid is sheared in the channel between the screw and the wall of the barrel. The mechanism that generates the pressure can be

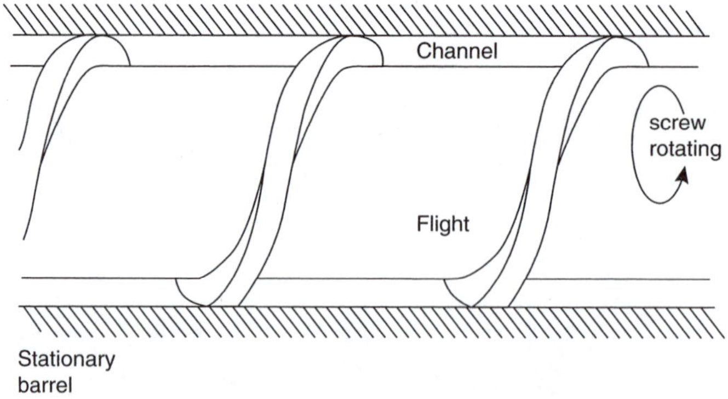

Figure 8.15. Section of a screw pump

visualised in terms of a model consisting of an open channel covered by a moving plane surface (Figure 8.16). This representation of a screw pump takes as the frame of reference a stationary screw with rotating barrel. The planar simplification is not unreasonable, provided that the depth of the screw channel is small with respect to the barrel diameter. It should also be recognised that the distribution of centrifugal forces would be different according to whether the rotating member is the wall or the screw: this distinction would have to be drawn if a detailed force balance were to be attempted, but in any event the centrifugal (inertial) forces are generally far smaller than the viscous forces.

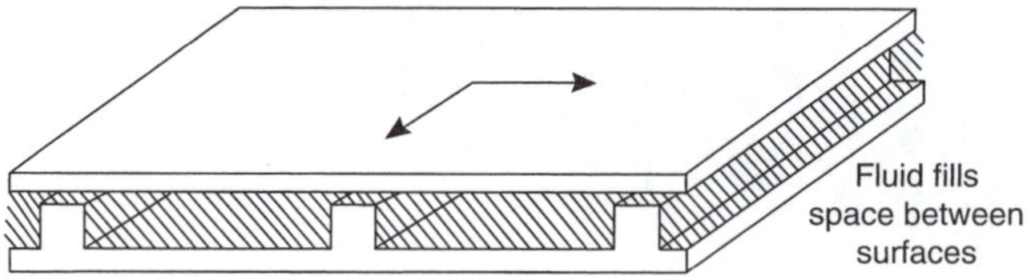

Figure 8.16. Planar model of part of a screw pump

If the upper plate moved along in the direction of the channel, then a velocity profile would be established that would approximate to the linear velocity gradient that exists between planar walls moving parallel to each other. If it moved at right angles to the channel axis, however, a circulation would be developed in the gap, as drawn in Figure 8.17. In fact, the relative movement of the barrel wall is somewhere in between, and is determined by the pitch of the screw. The fluid path in a screw pump is therefore of a complicated helical form within the channel section. The nature of the velocity components along the channel depends on the pressure generated and the amount of resistance at the discharge end. If there is no resistance, the velocity distribution in the channel direction will be the *Couette simple shear* profile shown in Figure 8.18*a*. With a totally closed discharge end the net flow would be zero, but the velocity components at the walls would not be affected. As a result, the flow field necessarily would be of the form shown in Figure 8.18*b*.

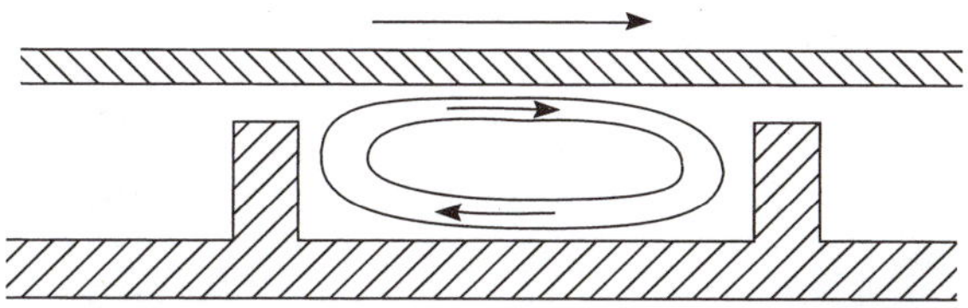

Figure 8.17. Fluid displacement resulting from movement of plane surface

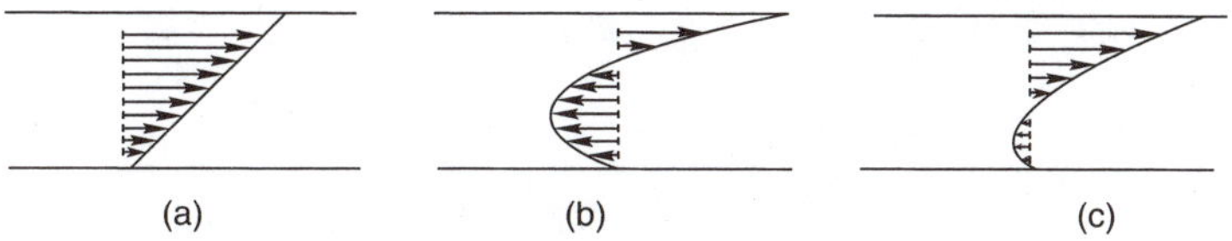

Figure 8.18. Velocity profile produced between screw pump surfaces (*a*) with no restriction on fluid flow (*b*) with no net flow (total restriction) (*c*) with a partially restricted discharge

Viscous forces within the fluid will always prevent a completely unhindered discharge, but in extrusion practice an additional die head resistance is used to generate backflow and mixing, so that a more uniform product is obtained. The flow profile along the channel is then of some intermediate form, such as that shown in Figure 8.18*c*.

It must be emphasised that flow in a screw pump is produced as a result of viscous forces. Pressures achieved with low viscosity materials are negligible. The screw pump is *not* therefore a modification of the Archimedes screw used in antiquity to raise water — that was essentially a positive displacement device using a deep cut helix mounted at an angle to the horizontal, and not running full. If a detailed analysis of the flow in a screw pump is to be carried out, then it is also necessary to consider the small but finite leakage flow that can occur between the flight and the wall. With the large pressure generation in a polymer extruder, commonly 100 bar (10^7 N/m^2), the flow through this gap, which is typically about 2 per cent of the barrel internal diameter, can be significant. The pressure drop over a single pitch length may be of the order of 10 bar (10^6 N/m^2), and this will force fluid through the gap. Once in this region the viscous fluid is subject to a high rate of shear (the rotation speed of the screw is often about 2 Hz), and an appreciable part of the total viscous heat generation occurs in this region of an extruder.

8.2.3. The centrifugal pump

The centrifugal pump is by far the most widely used type in the chemical and petroleum industries. It will pump liquids with very wide-ranging properties and suspensions with a high solids content including, for example, cement slurries, and may be constructed from a very wide range of corrosion resistant materials. The whole pump casing may be constructed from plastics such as polypropylene or it may be fitted with a corrosion-resistant lining. Because it operates at high speed, it may be directly coupled to an electric motor and it will give a high flowrate for its size.

In this type of pump (Figure 8.19), the fluid is fed to the centre of a rotating impeller and is thrown outward by centrifugal action. As a result of the high speed of rotation the

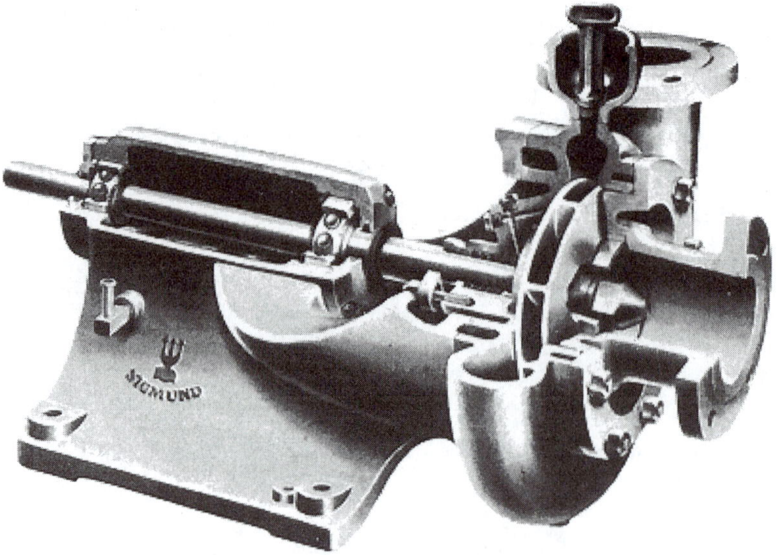

Figure 8.19. Section of centrifugal pump

liquid acquires a high kinetic energy and the pressure difference between the suction and delivery sides arises from the interconversion of kinetic and pressure energy.

The impeller (Figure 8.20) consists of a series of curved vanes so shaped that the flow within the pump is as smooth as possible. The greater the number of vanes on the impeller, the greater is the control over the direction of motion of the liquid and hence the smaller are the losses due to turbulence and circulation between the vanes. In the open impeller, the vanes are fixed to a central hub, whereas in the closed type the vanes are held between

(a) (b) (c)

Figure 8.20. Types of impeller (*a*) for pumping suspensions (*b*) standard closed impeller (*c*) double impeller

two supporting plates and leakage across the impeller is reduced. As will be seen later, the angle of the tips of the blades very largely determines the operating characteristics of the pump.

The liquid enters the casing of the pump, normally in an axial direction, and is picked up by the vanes of the impeller. In the simple type of centrifugal pump, the liquid discharges into a volute, a chamber of gradually increasing cross-section with a tangential outlet. A volute type of pump is shown in Figure 8.21*a*. In the turbine pump (Figure 8.21*b*) the liquid flows from the moving vanes of the impeller through a series of fixed vanes forming a diffusion ring. This gives a more gradual change in direction to the fluid and more efficient conversion of kinetic energy into pressure energy than is obtained with the volute type. The angle of the leading edge of the fixed vanes should be such that the fluid is received without shock. The liquid flows along the surface of the impeller vane with a certain velocity whilst the tip of the vane is moving relative to the casing of the pump. The direction of motion of the liquid relative to the pump casing — and the required angle of the fixed vanes — is found by compounding these two velocities. In Figure 8.22, u_v is the velocity of the liquid relative to the vane and u_t is the tangential velocity of the tip of the vane; compounding these two velocities gives the resultant velocity u_2 of the liquid.

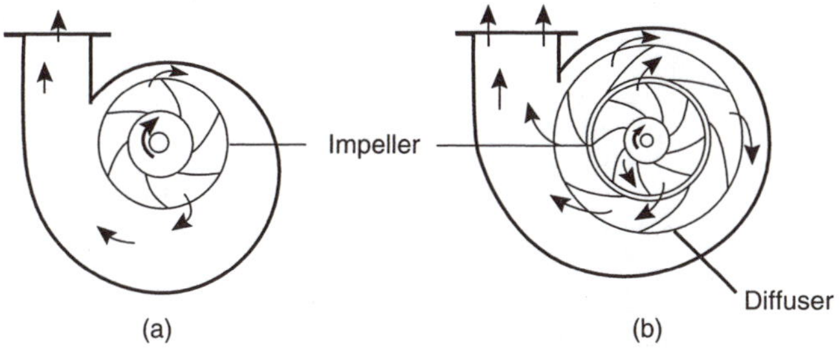

Figure 8.21. Radial flow pumps (*a*) with volute (*b*) with diffuser vanes

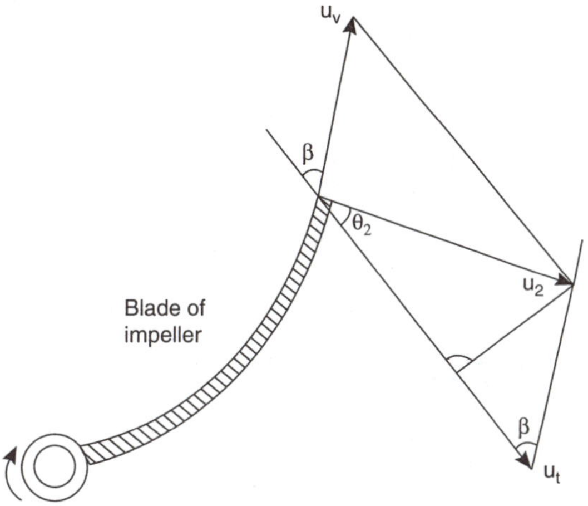

Figure 8.22. Velocity diagram

It is apparent, therefore, that the required vane angle in the diffuser is dependent on the throughput, the speed of rotation, and the angle of the impeller blades. The pump will therefore operate at maximum efficiency only over a narrow range of conditions.

Virtual head of a centrifugal pump

The maximum pressure is developed when the whole of the excess kinetic energy of the fluid is converted into pressure energy. As indicated below, the head is proportional to the square of the radius and to the speed, and is of the order of 60 m for a single-stage centrifugal pump; for higher pressures, multistage pumps must be used. The liquid which is rotating at a distance of between r and $r + \mathrm{d}r$ from the centre of the pump (Figure 8.23) has a mass $\mathrm{d}M$ given by $2\pi r\,\mathrm{d}rb\rho$, where ρ is the density of the fluid and b is the width of the element of fluid.

If the fluid is travelling with a velocity u and at an angle θ to the tangential direction, the angular momentum of this mass of fluid

$$= \mathrm{d}M(ur\cos\theta)$$

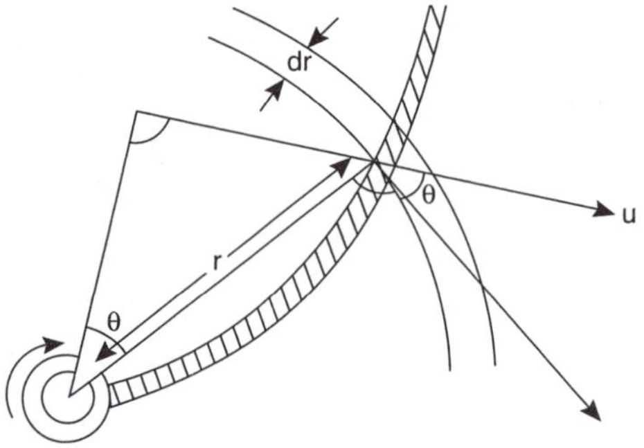

Figure 8.23. Virtual head

The torque acting on the fluid $\mathrm{d}\tau$ is equal to the rate of change of angular momentum with time, as it goes through the pumps or:

$$\mathrm{d}\tau = \mathrm{d}M\frac{\partial}{\partial t}(ur\cos\theta)$$

$$= 2\pi rb\rho\,\mathrm{d}r\frac{\partial}{\partial t}(ur\cos\theta) \qquad (8.3)$$

The volumetric rate of flow of liquid through the pump:

$$Q = 2\pi rb\frac{\partial r}{\partial t} \qquad (8.4)$$

or:

$$\mathrm{d}\tau = Q\rho\,\mathrm{d}(ur\cos\theta) \qquad (8.5)$$

The total torque acting on the liquid in the pump is therefore obtained by integrating $\mathrm{d}\tau$ between the limits denoted by suffix 1 and suffix 2, where suffix 1 refers to the conditions at the inlet to the pump and suffix 2 refers to the conditions at the discharge.

Thus:
$$\tau = Q\rho(u_2 r_2 \cos\theta_2 - u_1 r_1 \cos\theta_1) \tag{8.6}$$

The power **P** developed by the pump is equal to the product of the torque and the angular velocity ω:

$$\therefore \quad \mathbf{P} = Q\rho\omega(u_2 r_2 \cos\theta_2 - u_1 r_1 \cos\theta_1) \tag{8.7}$$

The power can also be expressed as the product Ghg, where G is the mass rate of flow of liquid through the pump, g is the acceleration due to gravity, and h is termed the virtual head developed by the pump.

Thus:
$$Ghg = Q\rho\omega(u_2 r_2 \cos\theta_2 - u_1 r_1 \cos\theta_1)$$

and:
$$h = \frac{\omega(u_2 r_2 \cos\theta_2 - u_1 r_1 \cos\theta_1)}{g} \tag{8.8}$$

Since u_1 will be approximately zero, the virtual head:

$$h = \frac{\omega u_2 r_2 \cos\theta_2}{g} \tag{8.9}$$

where g, ω and r_2 are known in any given instance, and u_2 and θ_2 are to be expressed in terms of known quantities.

From the geometry of Figure 8.22:

$$u_v \sin\beta = u_2 \sin\theta_2 \tag{8.10}$$

and:
$$u_t = u_v \cos\beta + u_2 \cos\theta_2 \tag{8.11}$$

(where β is the angle between the tip of the blade of the impeller and the tangent to the direction of its motion. If the blade curves backwards, β lies between 0 and $\pi/2$ and if it curves forwards, β lies between $\pi/2$ and π).

The volumetric rate of flow through the pump Q is equal to the product of the area available for flow at the outlet of the impeller and the radial component of the velocity, or

$$Q = 2\pi r_2 b u_2 \sin\theta_2$$
$$= 2\pi r_2 b u_v \sin\beta \quad \text{(from equation 8.10)} \tag{8.12}$$

$$\therefore \quad u_v = \frac{Q}{2\pi r_2 b \sin\beta} \tag{8.13}$$

Thus:
$$h = \frac{\omega r_2(u_t - u_v \cos\beta)}{g} \quad \text{(from equations 8.9 and 8.11)}$$

$$= \frac{\omega}{g} r_2 \left(r_2\omega - \frac{Q}{2\pi r_2 b \tan\beta} \right) \quad (\text{since } u_t = r_2\omega)$$

$$= \frac{r_2^2\omega^2}{g} - \frac{Q\omega}{2\pi b g \tan\beta} \tag{8.14}$$

The virtual head developed by the pump is therefore independent of the density of the fluid, and the pressure will thus be directly proportional to the density. For this reason, a centrifugal pump needs priming. If the pump is initially full of air, the pressure developed is reduced by a factor equal to the ratio of the density of air to that of the liquid, and is insufficient to drive the liquid through the delivery pipe.

For a given speed of rotation, there is a linear relation between the head developed and the rate of flow. If the tips of the blades of the impeller are inclined backwards, β is less than $\pi/2$, $\tan\beta$ is positive, and therefore the head decreases as the throughput increases. If β is greater than $\pi/2$ (i.e. the tips of the blades are inclined forwards), the head increases as the delivery increases. The angle of the blade tips therefore profoundly affects the performance and characteristics of the pump. For radial blades the head should be independent of the throughput.

Specific speed

If θ_2 remains approximately constant, u_v, u_t, and u_2 will all be directly proportional to one another, and since $u_t = r_2\omega$, these velocities are proportional to r_2; thus u_v will vary as $r_2\omega$.

The output from a pump is a function of its linear dimensions, the shape, number, and arrangement of the impellers, the speed of rotation, and the head against which it is operating. From equation 8.12, for a radial pump with $\beta = \pi/2$ and $\sin\beta = 1$:

$$Q \propto 2\pi r_2 b u_v$$

$$Q \propto 2\pi r_2 b r_2 \omega$$

$$Q \propto r_2^2 b\omega \tag{8.15}$$

When $\tan\beta = \tan\pi/2 = \infty$, then from equation 8.14:

$$h = \frac{r_2^2\omega^2}{g} \tag{8.16}$$

$$gh = r_2^2\omega^2$$

or:
$$(gh)^{3/4} = r_2^{3/2}\omega^{3/2} \tag{8.17}$$

For a series of geometrically similar pumps, b is proportional to the radius r_2, and thus, from equation 8.15:

$$Q \propto r_2^3\omega \tag{8.18}$$

or:
$$Q^{1/2} \propto r_2^{3/2}\omega^{1/2} \tag{8.19}$$

Eliminating r_2 between equations 8.17 and 8.19:

$$\frac{Q^{1/2}}{(gh)^{3/4}} = \frac{\omega^{1/2}}{\omega^{3/2}}$$

or:
$$\frac{\omega Q^{1/2}}{(gh)^{3/4}} = \text{constant} = N_s \text{ for geometrically similar pumps} \tag{8.20}$$

Criteria for similarity

The dimensionless quantity $\omega Q^{1/2}/(gh)^{3/4}$ is a characteristic for a particular type of centrifugal pump, and, noting that the angular velocity is proportional to the speed N, this group may be rewritten as:

$$N_s = \frac{NQ^{1/2}}{(gh)^{3/4}} \tag{8.21}$$

and is constant for geometrically similar pumps. N_s is defined as the specific speed and is frequently used to classify types of centrifugal pumps. Specific speed may be defined as the speed of the pump which will produce unit flow Q against unit head h under conditions of maximum efficiency.

Equation 8.21 is dimensionless but specific speed is frequently quoted in the form:

$$N_s = \frac{NQ^{1/2}}{h^{3/4}} \tag{8.22}$$

where the impeller speed N is in rpm, the volumetric flowrate Q in US gpm and the total head developed is in ft. In this form, specific speed has dimensions of $(\mathbf{LT}^{-2})^{3/4}$ and for centrifugal pumps has values between 400 and 10,000 depending on the type of impeller. The use of specific speed in pump selection is discussed more fully in Volume 6.

Operating characteristics

The operating characteristics of a pump are conveniently shown by plotting the head h, power **P**, and efficiency η against the flow Q as shown in Figure 8.24. It is important to note that the efficiency reaches a maximum and then falls, whilst the head at first falls slowly with Q but eventually falls off rapidly. The optimum conditions for operation are shown as the duty point, i.e. the point where the head curve cuts the ordinate through the point of maximum efficiency.

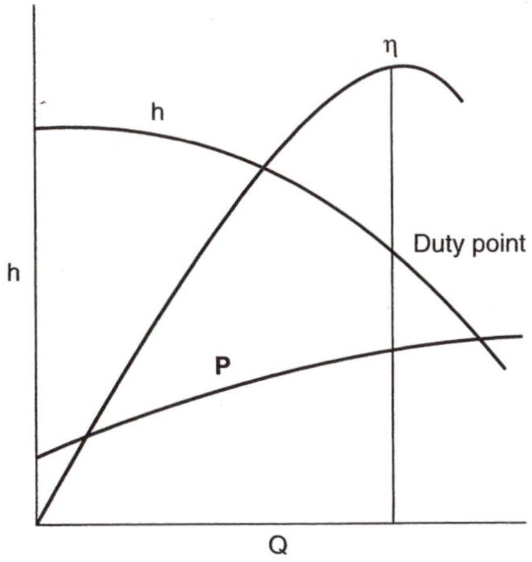

Figure 8.24. Radial flow pump characteristics

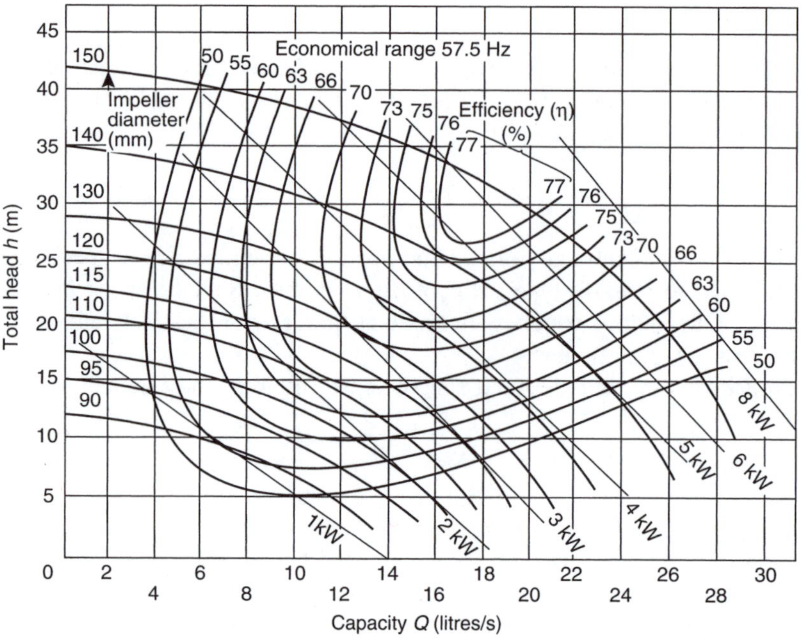

Figure 8.25. Characteristic curves for centrifugal pump

A set of curves for h, η, and $\mathbf{P}$ as a function of Q are shown in Figure 8.25 from which it is seen that when the pump is operating near optimum conditions, its efficiency remains reasonably constant over a wide range of flowrates. A more general indication of the variation of efficiency with specific speed is shown in Figure 8.26 for different types of centrifugal pumps. The power developed by a pump is proportional to $Qgh\rho$:

i.e.: $$\mathbf{P} \propto r_2^2 \omega r_2^2 \omega^2 \rho$$

or: $$\mathbf{P} \propto r_2^4 b \omega^3 \rho \qquad (8.23)$$

so that $Q \propto \omega$; $h \propto \omega^2$; $\mathbf{P} \propto \omega^3$, from equations 8.15, 8.16 and 8.23.

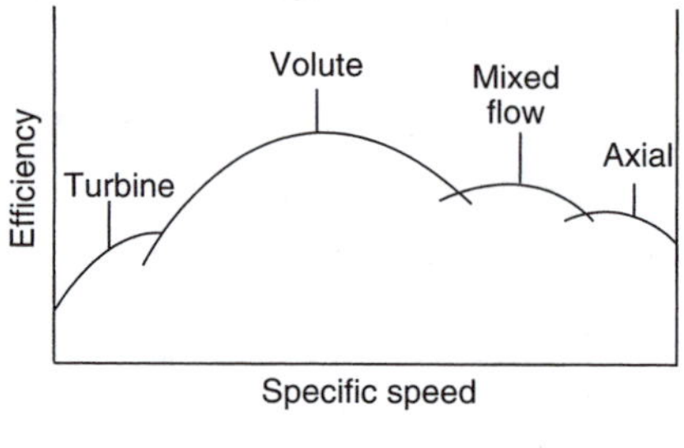

Figure 8.26. Specific speed and efficiency

Cavitation

In designing any installation in which a centrifugal pump is used, careful attention must be paid to check the minimum pressure which will arise at any point. If this pressure is less than the vapour pressure at the pumping temperature, vaporisation will occur and the pump may not be capable of developing the required suction head. Moreover, if the liquid contains gases, these may come out of solution giving rise to pockets of gas. This phenomenon is known as *cavitation* and may result in mechanical damage to the pump as the bubbles collapse. The tendency for cavitation to occur is accentuated by any sudden changes in the magnitude or direction of the velocity of the liquid in the pump. The onset of cavitation is accompanied by a marked increase in noise and vibration as the vapour bubbles collapse, and also a loss of head.

Suction head

Pumps may be arranged so that the inlet is under a suction head or the pump may be fed from a tank. These two systems alter the duty point curves as shown in Figure 8.27. In developing such curves, the normal range of liquid velocities is 1.5 to 3 m/s, but lower values are used for pump suction lines. With the arrangement shown in Figure 8.27*a*, there can be problems in priming the pump and it may be necessary to use a self-priming centrifugal pump.

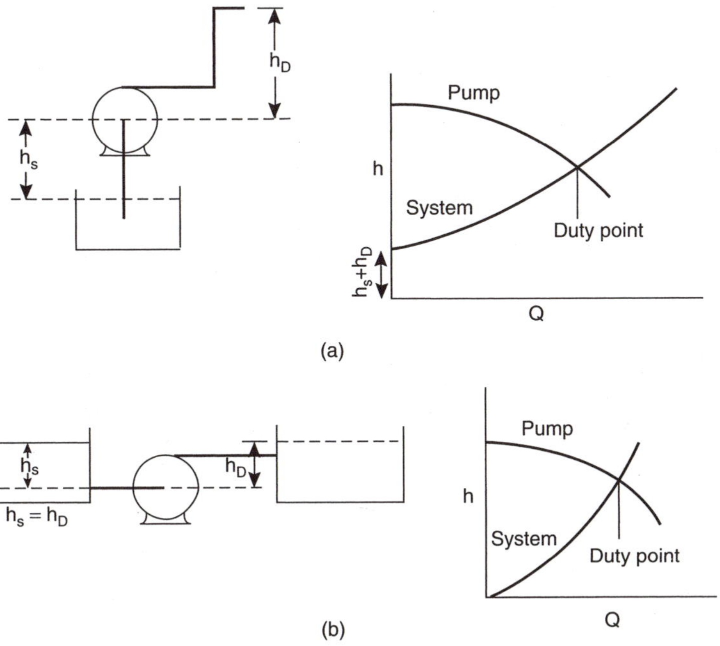

Figure 8.27. Effect of suction head: (*a*) systems with suction lift and friction; (*b*) systems with friction losses only

For any pump, the manufacturers specify the minimum value of the *net positive suction head* (NPSH) which must exist at the suction point of the pump. The NPSH is the amount by which the pressure at the suction point of the pump, expressed as a head of the liquid to be pumped, must exceed the vapour pressure of the liquid. For any installation this must be calculated, taking into account the absolute pressure of the liquid, the level of the pump, and the velocity and friction heads in the suction line. The NPSH must allow for the fall in pressure occasioned by the further acceleration of the liquid as it flows on to the impeller and for irregularities in the flow pattern in the pump. If the required value of NPSH is not obtained, partial vaporisation or liberation of dissolved gas is liable to occur, with the result that both suction head and delivery head may be reduced. The loss of suction head is the more important because it may cause the pump to be starved of liquid.

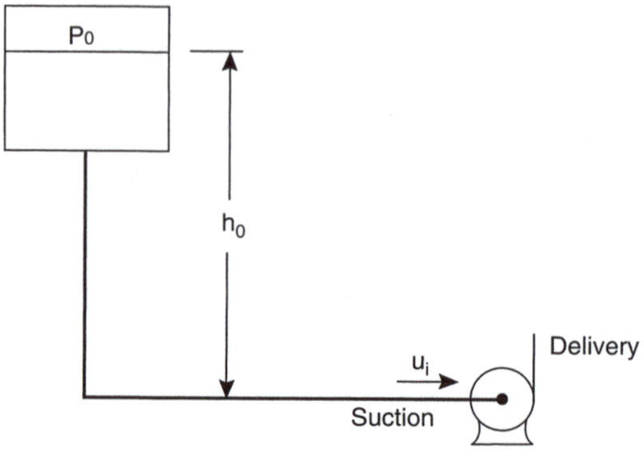

Figure 8.28. Suction system of centrifugal pump

In the system shown in Figure 8.28, the pump is taking liquid from a reservoir at an absolute pressure P_0 in which the liquid level is at a height h_0 above the suction point of the pump. Then, if the liquid velocity in the reservoir is negligible, the *absolute* pressure head h_i at the suction point of the pump is obtained by applying the energy or momentum balance:

$$h_i = \frac{P_0}{\rho g} + h_0 - \frac{u_i^2}{2g} - h_f \tag{8.24}$$

where h_f is the head lost in friction, and u_i is the velocity at the inlet of the pump. If the vapour pressure of the liquid is P_v, the NPSH Z is given by the difference between the *total* head at the suction inlet and the head corresponding to the vapour pressure P_v of the liquid at the pump inlet.

$$Z = \left(h_i + \frac{u_i^2}{2g}\right) - \frac{P_v}{\rho g} \tag{8.25}$$

$$= \frac{P_0}{\rho g} - \frac{P_v}{\rho g} + h_0 - h_f \tag{8.26}$$

In equation 8.26, it is implicitly assumed that the kinetic head of the inlet liquid is available for conversion into pressure head. If this is not so, $u_i^2/2g$ must be deducted from the NPSH.

If cavitation and loss of suction head does occur, it can sometimes be cured by increasing the pressure in the system, either by alteration of the layout to provide a greater hydrostatic pressure or a reduced pressure drop in the suction line. Sometimes, slightly closing the valve on the pump delivery or reducing the pump speed by a small amount may be effective. Generally, a small fast-running pump will require a larger NPSH than a larger slow-running pump.

The efficiency of a centrifugal pump and the head which it is capable of developing are dependent upon achieving a good seal between the rotating shaft and the casing of the pump and inefficient operation is frequently due to a problem with that seal or with the gland. When the pump is fitted with the usual type of packed gland, maintenance costs are often very high, especially when organic liquids of low viscosity are being pumped. A considerable reduction in expenditure on maintenance can be effected at the price of a small increase in initial cost by fitting the pump with a mechanical seal, in which the sealing action is achieved as a result of contact between two opposing faces, one stationary and the other rotating. In Figure 8.29, a mechanical seal is shown in position in a centrifugal pump. The stationary seat *A* is held in position by means of the clamping plate *D*. The seal is made with the rotating face on a carbon ring *B*. The seal between

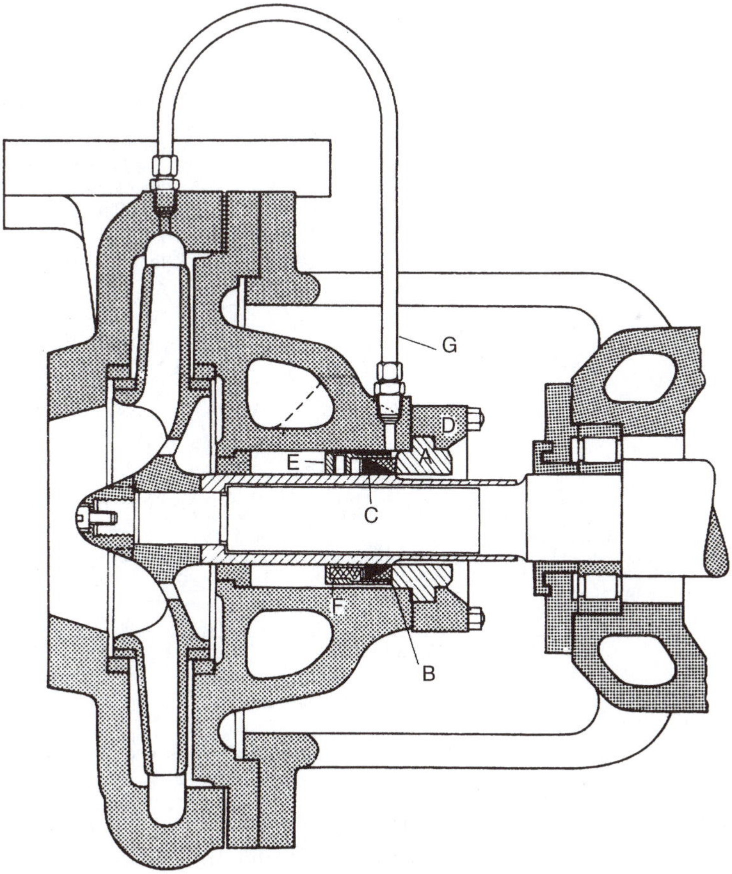

Figure 8.29. Mechanical seal for centrifugal pump

the rotating face and the shaft is made by means of the wedge ring C, usually made of polytetrafluoroethylene (PTFE). The drive is through the retainer E secured to the shaft, usually by Allen screws. Compression between the fixed and rotating faces is provided by the spiral springs F.

It is advantageous to ensure that the seal is fed with liquid which removes any heat generated at the face. In the illustration this is provided by the connection G.

Centrifugal pumps must be fitted with good bearings since there is a tendency for an axial thrust to be produced if the suction is taken only on one side of the impeller. This thrust can be balanced by feeding back a small quantity of the high-pressure liquid to a specially designed thrust bearing. By this method, the risk of air leaking into the pump at the gland — and reducing the pressure developed — is minimised. The glandless centrifugal pump, which is used extensively for corrosive liquids, works on a similar principle, and the use of pumps without glands has been increasing both in the nuclear power industry and in the chemical process industry. Such units are totally enclosed and lubrication is provided by the fluid handled. Figure 8.30 gives the performance characteristics at a particular speed of one group of pumps of this type, and similar data are available for the selection of other types of centrifugal pumps.

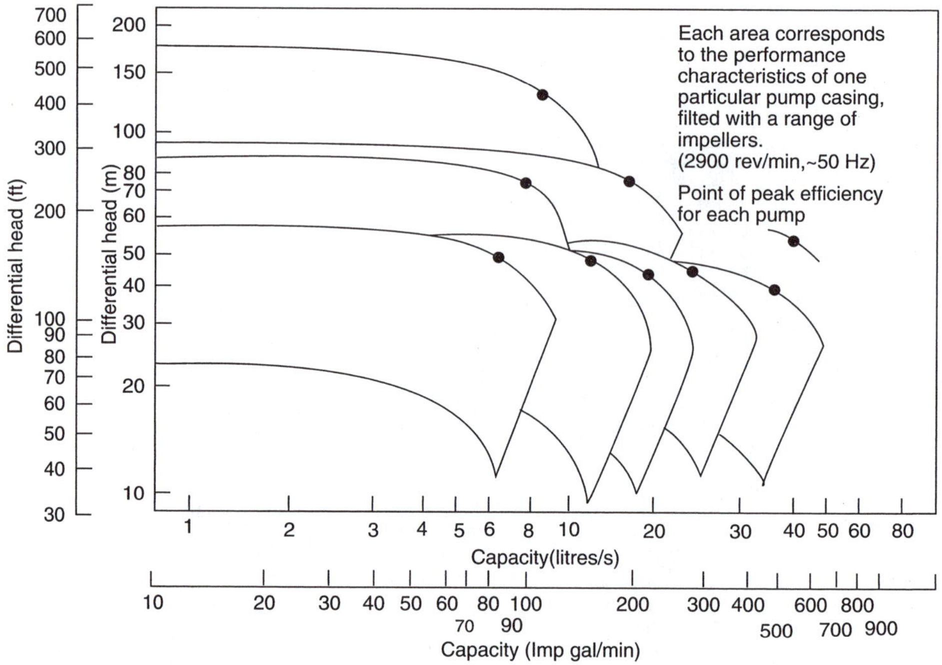

Figure 8.30. Performance characteristics of a family of glandless centrifugal pumps

Centrifugal pumps are made in a wide range of materials, and in many cases the impeller and the casing are covered with resistant material. Thus stainless steel, nickel, rubber, polypropylene, stoneware, and carbon are all used. When the pump is used with suspensions, the ports and spaces between the vanes must be made sufficiently large to eliminate the risk of blockage. This does mean, however, that the efficiency of the pump is reduced. The *Vacseal pump*, developed by the International Combustion Company for pumping slurries, will handle suspensions containing up to 50 per cent by volume of solids. The whole impeller may be rubber-covered and has three small vanes, as shown in Figure 8.31. The back plate of the impeller has a second set of vanes of larger diameter. The pressure at the gland is thereby reduced below atmospheric pressure and below the pressure in the suction line; there is, therefore, no risk of the gritty suspension leaking into the gland and bearings. If leakage does occur, air will enter the pump. As mentioned previously, this may reduce the pressure which the pump can deliver, but this is preferable to damaging the bearings by allowing them to become contaminated with grit. This is another example of the necessity for tolerating rather low efficiencies in pumps that handle difficult materials. The pumping of slurries has been considered by STEELE and ODROWAZ-PIENIAZEK[4] who also discuss the many aspects of selection and efficient operation of centrifugal pumps in these demanding circumstances.

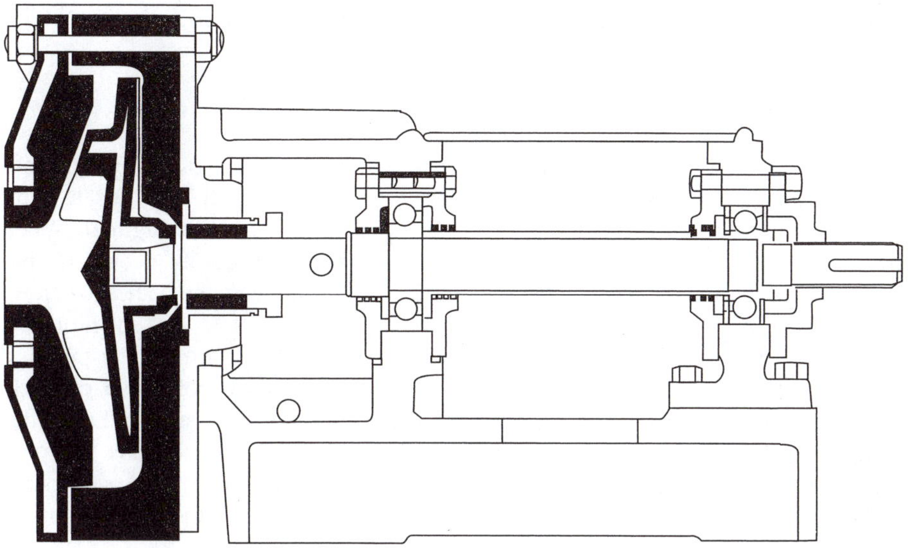

Figure 8.31. Sectioned arrangement of a 38 mm V-type Vacseal pump with moulded rubber impeller

The advantages and disadvantages of the centrifugal pump

The main advantages are:

(1) It is simple in construction and can, therefore, be made in a wide range of materials.
(2) There is a complete absence of valves.
(3) It operates at high speed (up to 100 Hz) and, therefore, can be coupled directly to an electric motor. In general, the higher the speed the smaller the pump and motor for a given duty.
(4) It gives a steady delivery.
(5) Maintenance costs are lower than for any other type of pump.

(6) No damage is done to the pump if the delivery line becomes blocked, provided it is not run in this condition for a prolonged period.
(7) It is much smaller than other pumps of equal capacity. It can, therefore, be made into a sealed unit with the driving motor, and immersed in the suction tank.
(8) Liquids containing high proportions of suspended solids are readily handled.

The main disadvantages are:

(1) The single-stage pump will not develop a high pressure. Multistage pumps will develop greater heads but they are very much more expensive and cannot readily be made in corrosion-resistant material because of their greater complexity. It is generally better to use very high speeds in order to reduce the number of stages required.
(2) It operates at a high efficiency over only a limited range of conditions: this applies especially to turbine pumps.
(3) It is not usually self-priming.
(4) If a non-return valve is not incorporated in the delivery or suction line, the liquid will run back into the suction tank as soon as the pump stops.
(5) Very viscous liquids cannot be handled efficiently.

Pumping of non-Newtonian fluids

The development of the required pressure at the outlet to a centrifugal pump, depends upon the efficient conversion of kinetic energy into pressure energy. For a pump of this type the distribution of shear within the pump will vary with throughput. When the discharge value is completely closed, the highest degree of shearing occurs in the gap between the rotor and the shell, at B in Figure 8.32. Between the vanes of the rotor (region A) there will be some circulation as shown in Figure 8.33, but in the discharge line C the fluid will be essentially static. When fluid is flowing through the pump, there will still be differences between these shear rates, but they will not be so great. If the fluid has pseudoplastic properties, then the effective viscosity will vary in these different regions, being less at B than at A and C. Under steady-state conditions the pressure developed in the pump may be sufficient to establish a uniform flow. However, there may be problems on startup, when the very high apparent viscosities of the fluid might lead to

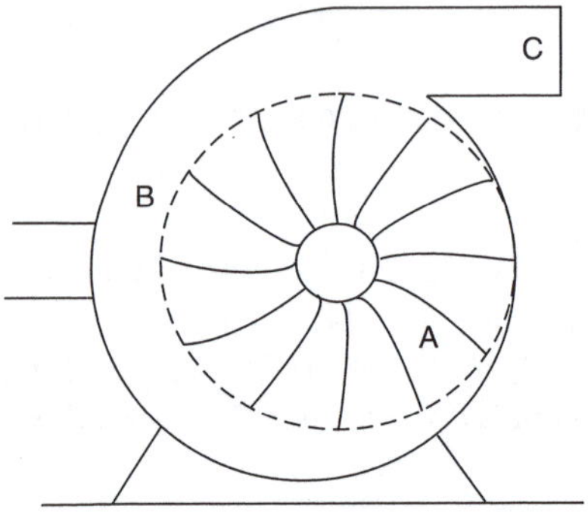

Figure 8.32. Zones of differing shear in a centrifugal pump

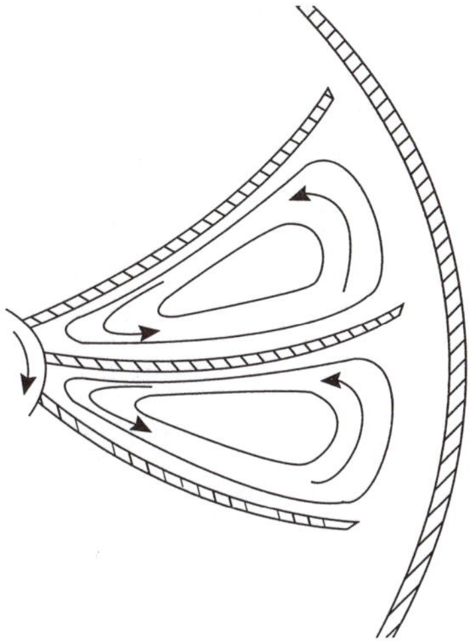

Figure 8.33. Circulation within a centrifugal pump impeller

overloading of the pump motor. The apparent viscosity of the liquid in the delivery line will also be at its maximum value, and the pump may take an inordinately long time to establish the required flowrate. Many pseudoplastic materials are damaged and degraded by prolonged shearing, and such a pump would then be unsuitable. Generally, positive-displacement rotary pumps are more satisfactory with shear-thinning fluids. The question of pump selection for all types of fluids has been subjected to a systems approach by DAVIDSON[(5)].

Example 8.2

A centrifugal pump is required to circulate a liquid of density 800 kg/m^3 and viscosity 0.5×10^{-3} Ns/m^2 from the reboiler of a distillation column through a vaporisor at the rate of 0.004 m^3/s, and to introduce the superheated vapour above the vapour space in the reboiler which contains a 0.07 m depth of liquid. If smooth-bore 25 mm diameter pipe is to be used, the pressure of vapour in the reboiler is 1 kN/m^2 and the Net Positive Suction Head required by the pump is 2 m of liquid, what is the minimum height required between the liquid level in the reboiler and the pump?

Solution

$$\text{Volumetric flowrate of liquid} = 400 \times 10^{-6}\ \text{m}^3/\text{s}$$

$$\text{Cross-sectional area of the pipe} = (\pi/4)(0.025)^2 = 0.00049\ \text{m}^2$$

and hence:

$$\text{velocity in the pipe, } u = (400 \times 10^{-6}/0.00049) = 0.816\ \text{m/s}$$

The Reynolds number is then:

$$\begin{aligned} Re &= du\rho/\mu \\ &= (0.025 \times 0.816 \times 800)/(0.5 \times 10^{-3}) = 32{,}700 \end{aligned}$$

From Figure 3.7, the friction factor for a smooth pipe is:

$$\phi = R/\rho u^2 = 0.0028$$

and the head loss due to friction is given by:

$$h_f = 4\phi(l/d)(u^2/g) \qquad \text{(equation 3.20)}$$

and: $$h_f/l = (4 \times 0.0028)(1/0.025)(0.816^2/(9.81)) = 0.0304 \text{ m/m of pipe}$$

As the liquid is pumped at its boiling point, $(P_0 - P_v)/\rho g = 0$ and equation 8.26 becomes:

$$Z = (h_0 - h_f)$$

and $$h_0 = (Z + h_f) = 2.0 + 0.0304l \text{ m}$$

It should be noted that a slightly additional height will be required if the kinetic energy at the pump inlet cannot be utilised.

Thus the height between the liquid level in the reboiler and the pump, h_0, depends on the length of pipe between the reboiler and the pump. If this is say 10 m, the minimum value of h_0 is 2.3 m.

8.3. PUMPING EQUIPMENT FOR GASES

Essentially the same types of mechanical equipment are used for handling gases and liquids, though the details of the construction are different in the two cases. Again, there are two basic types, *positive displacement*, and *centrifugal*, in which kinetic energy is converted into pressure energy. Over the normal range of operating pressures, the density of a gas is considerably less than that of a liquid with the result that higher speeds of operation can be employed and lighter valves fitted to the delivery and suction lines. Because of the lower viscosity of a gas there is a greater tendency for leakage to occur, and therefore gas compressors are designed with smaller clearances between the moving parts. Further differences in construction are necessitated by the decrease in volume of gas as it is compressed, and this must be allowed for in the design. Since a large proportion of the energy of compression appears as heat in the gas, there will normally be a considerable increase in temperature which may limit the operation of the compressor unless suitable cooling can be effected either within the stages of the compressor or in interstage coolers.

The principal types of compressors for gases will now be described.

8.3.1 Fans and rotary compressors

Fans are used for the supply of gases at relatively low pressures (<3.5 kN/m^2), often at very high flowrates. They may be of the *axial flow* type in which the curved blades directly impart an axial motion to the gas, or of the *centrifugal* type. Centrifugal fans, which operate on the same principle as centrifugal pumps for liquids, depend upon the conversion of the kinetic energy of the gas into pressure energy and are capable of developing somewhat higher pressures than the axial type.

Rotary blowers are of the *positive displacement* type, and a typical lobe-type of machine is shown in Figure 8.34. The rotors are driven in opposite directions so that, as each passes the inlet, it takes in gas which is compressed between the impeller and the casing before being expelled. Machines of this type are capable of developing pressure differentials of up to 100 kN/m^2; they are made in a wide range of sizes, with maximum throughputs of up to 20,000–30,000 m^3/h.

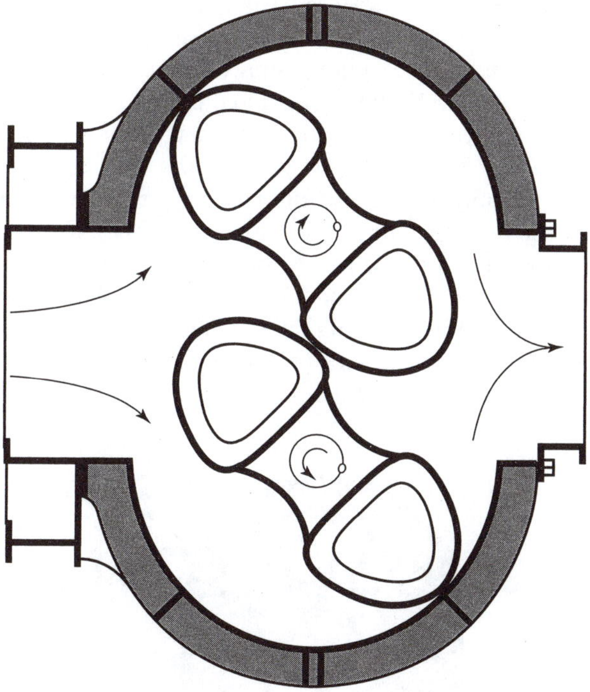

Figure 8.34. Two lobe-compressor

For higher pressures, rotary compressors of the *sliding vane* type will give delivery pressures up to 1 MN/m^2. In a compressor of this type, as illustrated in Figure 8.35, the compression ratio is achieved by eccentric mounting of the rotor which is slotted to take sliding vanes which sub-divide the crescent-shaped space between the rotor and

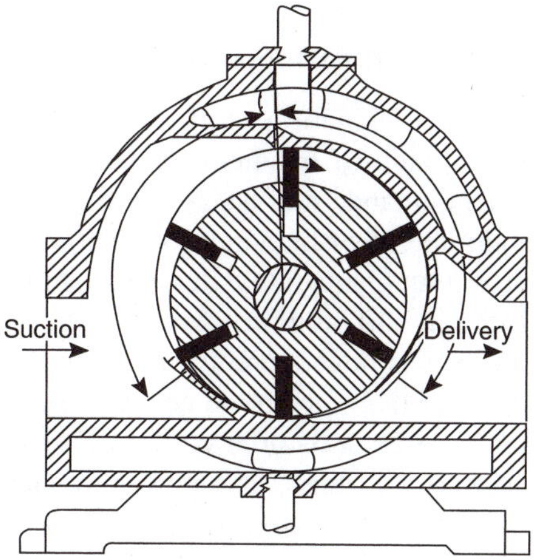

Figure 8.35. The sliding vane rotary compressor and vacuum pump

the casing. On rotation, the vanes are thrown out, trapping pockets of gas which are compressed during the rotation and are discharged, as shown, at the delivery port. Sliding vane compressors are also used as vacuum pumps (see Section 8.5).

Liquid-ring pumps, such as the Nash Hytor pump illustrated in Figure 8.36, are positive-displacement pumps with a specially shaped casing, and a liquid seal which rotates with the impeller. The liquid leaves and re-enters the impeller cells and acts as a piston. The liquid is supplied at a pressure equal to the discharge pressure, and is drawn in automatically to compensate for that discharged from the ports. The energy of compression is converted into heat which is absorbed by the liquid, giving rise to a nearly isothermal process. Downstream, the liquid is separated from the gas, cooled if necessary, and recirculated with make-up liquid. The shaft and the impeller are the only moving mechanical parts and there is no sliding contact, so no lubricants are required, and the gas under compression does not become contaminated.

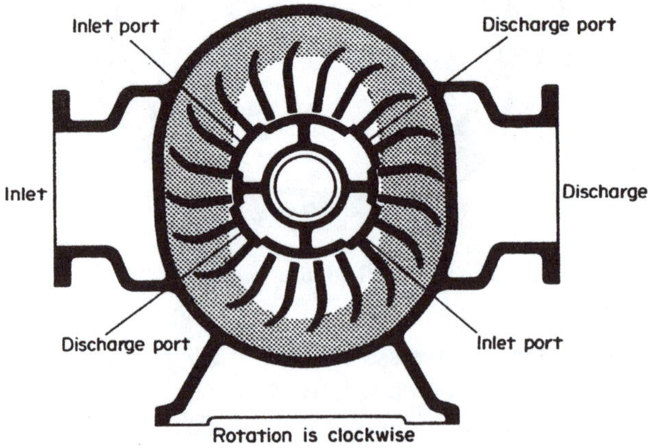

Figure 8.36. Nash Hytor liquid ring pump

8.3.2. Centrifugal and turbocompressors

These depend on the conversion of kinetic energy into pressure energy. Fans are used for low pressures, and can be made to handle very large quantities of gases. For the higher pressure ratios now in demand, multistage centrifugal compressors are mainly used, particularly for the requirements of high capacity chemical plants. Thus in catalytic reforming, petrochemical separation plants (ethylene manufacture), ammonia plants with a production rate of 12 kg/s (45 tonne/h), and for the very large capacity needed for natural gas fields, this type of compressor is now supreme. These units now give flowrates up to 140 m^3/s and pressures up to 5.6 MN/m^2 with the newest range going to 40 MN/m^2. It is important to accept that the very large units offer considerable savings over multiple units and that their reliability is remarkably high. The power required is also very high; thus a typical compressor operating on the process gas stream in the catalytic production of ethylene from naphtha will take 10 MW for a 6.5 kg/s (36 tonne/h) plant. A centrifugal compressor is illustrated in Figure 8.37.

Figure 8.37. A turbocompressor

8.3.3. The reciprocating piston compressor

This type of compressor is the only one capable of developing very high pressures, such as the pressure of 35 MN/m^2 required in the production of polyethylene. Compressors may be either single-stage, or multiple-stage where very high pressures are required. A single-stage two-cylinder unit is illustrated in Figure 8.37. The cylinders are fitted with jackets through which cooling water is circulated, and interstage coolers are provided on multi-stage compressors which may consist of anything from 2 to 12 stages. Cooling is essential to avoid the effects of excessively high temperatures on the mechanical operation of the compressor, and in order to reduce the power requirements. The calculation of the power required for compression, and how this is affected by *clearance volume*, is considered in Section 8.3.4. With recent developments in rotary compressors, the use of piston-type compressors is generally restricted to applications where very high pressures are required.

8.3.4. Power required for the compression of gases

If during the compression of unit mass of gas, its volume changes by an amount $\mathrm{d}v$ at a pressure P, the net work done on the gas, $-\delta W$, for a reversible change is given by:

$$-\delta W = -P\,\mathrm{d}v$$

$$= v\,\mathrm{d}P - \mathrm{d}(Pv) \qquad (8.27)$$

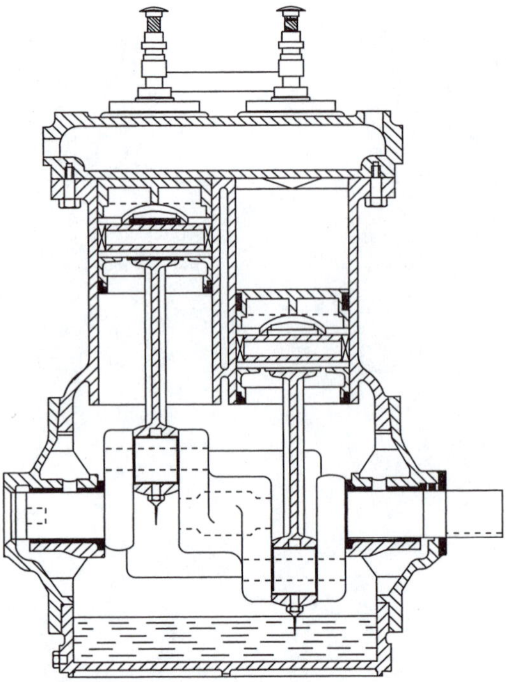

Figure 8.38. Reciprocating compressor

and for an irreversible change by:

$$-\delta W = -P\,dv + \delta F \qquad \text{(from equation 2.6)}$$

$$= v\,dP - d(Pv) + \delta F \qquad (8.28)$$

The work done in a reversible compression will be considered first because this refers to the ideal condition for which the work of compression is a minimum; a reversible compression would have to be carried out at an infinitesimal rate and therefore is not relevant in practice. The actual work done will be greater than that calculated, not only because of irreversibility, but also because of frictional loss and leakage in the compressor. These two factors are difficult to separate and will therefore be allowed for in the overall efficiency of the machine.

The total work of compression from a pressure P_1 to a pressure P_2 is found by integrating equation 8.27. For an ideal gas undergoing an isothermal compression:

$$-\int_{v_1}^{v_2} P\,dv = -W = P_1 v_1 \ln \frac{P_2}{P_1} \qquad \text{(from equation 2.69)}$$

For the isentropic compression of an ideal gas (since $-P\,dv = v\,dP - d(Pv)$):

$$-\int_{v_1}^{v_2} P\,dv = -W = \left[\frac{\gamma}{\gamma - 1}(P_2 v_2 - P_1 v_1)\right] - (P_2 v_2 - P_1 v_1) \qquad \text{(from equation 2.71)}$$

$$= \frac{1}{\gamma - 1}(P_2 v_2 - P_1 v_1) \qquad (8.29)$$

Thus: $$\int_{P_1}^{P_2} v\,\mathrm{d}P = -\gamma \int_{v_1}^{v_2} P\,\mathrm{d}v \quad \text{(comparing equations 2.71 and 8.29)}$$

for isentropic conditions and then:

$$-W = \frac{1}{\gamma - 1} P_1 v_1 \left[\left(\frac{P_2}{P_1} \right)^{(\gamma-1)/\gamma} - 1 \right] \tag{8.30}$$

Under these conditions the whole of the energy of compression appears as heat in the gas.

For unit mass of an ideal gas undergoing an isentropic compression:

$$P_1 v_1^{\gamma} = P_2 v_2^{\gamma} \quad \text{(from equation 2.30)}$$

and: $$\frac{P_1 v_1}{T_1} = \frac{P_2 v_2}{T_2} \quad \text{(from equation 2.16)}$$

Eliminating v_1 and v_2 gives the ratio of the outlet temperature T_2 to the inlet temperatures T_1:

$$\frac{T_2}{T_1} = \left(\frac{P_2}{P_1} \right)^{(\gamma-1)/\gamma} \tag{8.31}$$

In practice, there will be irreversibilities (inefficiencies) associated with the compression and the additional energy needed will appear as heat, giving rise to an outlet temperature higher than T_2 as given by equation 8.31.

For the isentropic compression of a mass m of gas:

$$-Wm = \frac{1}{\gamma - 1} P_1 V_1 \left[\left(\frac{P_2}{P_1} \right)^{(\gamma-1)/\gamma} - 1 \right] \tag{8.32}$$

where V_1 is the volume of a mass m of gas at a pressure P_1.

If the conditions are intermediate between isothermal and isentropic, k must be used in place of γ, where $\gamma > k > 1$.

If the gas deviates appreciably from the ideal gas laws over the range of conditions considered, the work of compression is most conveniently calculated from the change in the thermodynamic properties of the gas.

Thus: $$\mathrm{d}U = T\,\mathrm{d}S - P\,\mathrm{d}v \quad \text{(from equation 2.5)}$$

$$\therefore \quad -\delta W = -P\,\mathrm{d}v = \mathrm{d}U - T\,\mathrm{d}S \quad \text{(for a reversible process)}$$

Under isothermal conditions: $$-W = \Delta U - T\Delta S \tag{8.33}$$

Under isentropic conditions: $$-W = \Delta U \tag{8.34}$$

These equations give the work done during a simple compression of gas in a cylinder but do not take account of the work done either during the admission of the gas prior to compression or during the expulsion of the compressed gas.

If, after the compression of a volume V_1 of gas at a pressure P_1 to a pressure P_2, the whole of the gas is expelled at constant pressure P_2 and a fresh charge of gas is admitted at a pressure P_1, then the cycle can be followed in Figure 8.39, where the pressure P is plotted as ordinate against the volume V as abscissa.

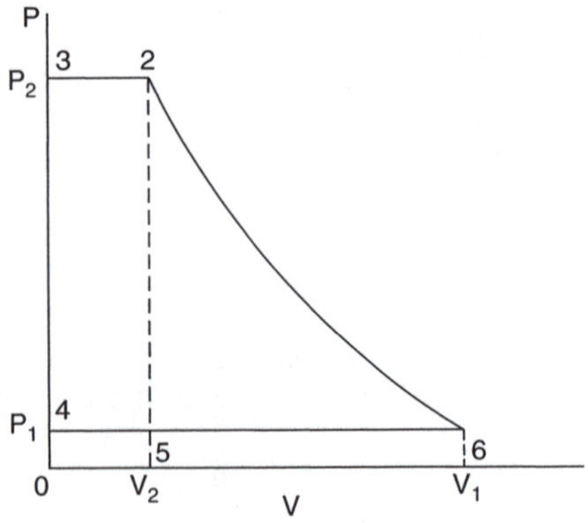

Figure 8.39. Single-stage compression cycle — no clearance

Point 1 represents the initial condition of the gas (P_1 and V_1).

Line 1–2 represents the compression of gas to pressure P_2, volume V_2.

Line 2–3 represents the expulsion of the gas at a constant pressure P_2.

Line 3–4 represents a sudden reduction in the pressure in the cylinder from P_2 to P_1. As the whole of the gas has been expelled, this can be regarded as taking place instantaneously.

Line 4–1 represents the suction stroke of the piston, during which a volume V_1 of gas is admitted at constant pressure, P_1.

It will be noted that the mass of gas in the cylinder varies during the cycle. The work done by the compressor during each phase of the cycle is as follows:

Compression	$-\int_{V_1}^{V_2} P\,dV$	(Area 1–2–5–6)
Expulsion	P_2V_2	(Area 2–3–0–5)
Suction	$-P_1V_1$	– (Area 4–0–6–1)

The total work done per cycle

$$= -\int_{V_1}^{V_2} P\,dV + P_2V_2 - P_1V_1 \quad \text{(Area 1–2–3–4)}$$

$$= \int_{P_1}^{P_2} V\,dP \tag{8.35}$$

The work of compression for an ideal gas per cycle under isothermal conditions:

$$= P_1V_1 \ln \frac{P_2}{P_1} \tag{8.36}$$

Under isentropic conditions, the work of compression

$$= P_1V_1 \frac{\gamma}{\gamma - 1}\left[\left(\frac{P_2}{P_1}\right)^{(\gamma-1)/\gamma} - 1\right] \tag{8.37}$$

Again, working in terms of the thermodynamic properties of the gas:

$$v\,dP = d(Pv) - P\,dv$$

$$= d(Pv) + dU - T\,dS \qquad \text{(from equation 2.5)}$$

$$= dH - T\,dS \tag{8.38}$$

For an isothermal process: $-mW = m(\Delta H - T\Delta S)$ (8.39)

For an isentropic process: $-mW = m\Delta H$ (8.40)

where m is the mass of gas compressed per cycle.

Clearance volume

In practice, it is not possible to expel the whole of the gas from the cylinder at the end of the compression; the volume remaining in the cylinder after the forward stroke of the piston is termed the *clearance volume*. The volume displaced by the piston is termed the *swept volume*, and therefore the total volume of the cylinder is made up of the clearance volume plus the swept volume. The *clearance c* is defined as the ratio of the clearance volume to the swept volume.

A typical cycle for a compressor with a finite clearance volume can be followed by reference to Figure 8.40. A volume V_1 of gas at a pressure P_1 is admitted to the cylinder; its condition is represented by point 1.

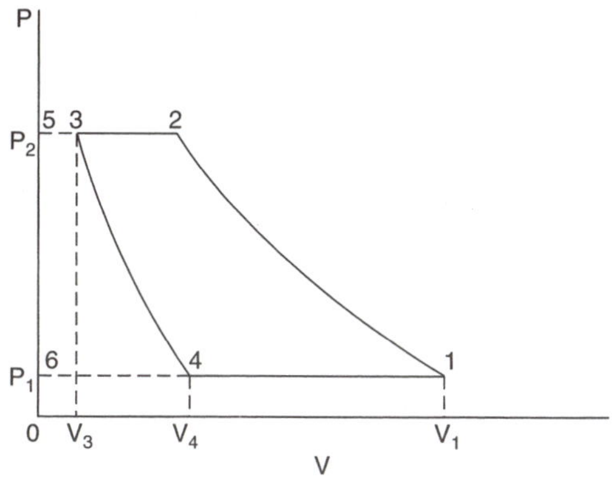

Figure 8.40. Single-stage compression cycle—with clearance

Line 1–2 represents the compression of the gas to a pressure P_2 and volume V_2.

Line 2–3 represents the expulsion of gas at constant pressure P_2, so that the volume remaining in the cylinder is V_3.

Line 3–4 represents an expansion of this residual gas to the lower pressure P_1 and volume V_4 during the return stroke.

Line 4–1 represents the introduction of fresh gas into the cylinder at constant pressure P_1. The work done on the gas during each stage of the cycle is as follows.

Compression	$-\int_{V_1}^{V_2} P\,dV$
Expulsion	$P_2(V_2 - V_3)$
Expansion	$-\int_{V_3}^{V_4} P\,dV$
Suction	$-P_1(V_1 - V_4)$

The total work done during the cycle is equal to the sum of these four components. It is represented by the area 1–2–3–4, which is equal to area 1–2–5–6 less area 4–3–5–6. If the compression and expansion are taken as isentropic, the work done per cycle is therefore:

$$= P_1V_1\frac{\gamma}{\gamma-1}\left[\left(\frac{P_2}{P_1}\right)^{(\gamma-1)/\gamma} - 1\right] - P_1V_4\frac{\gamma}{\gamma-1}\left[\left(\frac{P_2}{P_1}\right)^{(\gamma-1)/\gamma} - 1\right]$$

$$= P_1(V_1 - V_4)\frac{\gamma}{\gamma-1}\left[\left(\frac{P_2}{P_1}\right)^{(\gamma-1)/\gamma} - 1\right] \quad (8.41)$$

Thus, theoretically, the clearance volume does not affect the work done per unit mass of gas, since $V_1 - V_4$ is the volume admitted per cycle. It does, however, influence the quantity of gas admitted and therefore the work done per cycle. In practice, however, compression and expansion are not reversible, and losses arise from the compression and expansion of the clearance gases. This effect is particularly serious at high compression ratios.

The value of V_4 is not known explicitly, but can be calculated in terms of V_3, the clearance volume.

For isentropic conditions:

$$V_4 = V_3\left(\frac{P_2}{P_1}\right)^{1/\gamma}$$

and:

$$V_1 - V_4 = (V_1 - V_3) + V_3 - V_3\left(\frac{P_2}{P_1}\right)^{1/\gamma}$$

$$= (V_1 - V_3)\left[1 + \frac{V_3}{V_1 - V_3} - \frac{V_3}{V_1 - V_3}\left(\frac{P_2}{P_1}\right)^{1/\gamma}\right]$$

Now $(V_1 - V_3)$ is the swept volume, V_s, say; and $V_3/(V_1 - V_3)$ is the clearance c.

Thus:

$$V_1 - V_4 = V_s\left[1 + c - c\left(\frac{P_2}{P_1}\right)^{1/\gamma}\right] \quad (8.42)$$

The total work done on the fluid per cycle is therefore:

$$P_1V_s\frac{\gamma}{\gamma-1}\left[\left(\frac{P_2}{P_1}\right)^{(\gamma-1)/\gamma} - 1\right]\left[1 + c - c\left(\frac{P_2}{P_1}\right)^{1/\gamma}\right] \quad (8.43)$$

The factor $[1 + c - c(P_2/P_1)^{1/\gamma}]$ is called the theoretical volumetric efficiency and is a measure of the effect of the clearance on an isentropic compression. The actual volumetric efficiency will be affected, in addition, by the inertia of the valves and leakage past the piston.

The gas is frequently cooled during compression so that the work done per cycle is less than that given by equation 8.43, and γ is replaced by some smaller quantity k. The greater the rate of heat removal, the less is the work done. The isothermal compression is usually taken as the condition for least work of compression, but clearly the energy consumption can be reduced below this value if the gas is artificially cooled below its initial temperature as it is compressed. This is not a practicable possibility because of the large amount of energy required to refrigerate the cooling fluid. It can be seen that the theoretical volumetric efficiency decreases as the rate of heat removal is increased since γ is replaced by the smaller quantity k.

In practice the cylinders are usually water-cooled. The work of compression is thereby reduced though the effect is usually small. The reduction in temperature does, however, improve the mechanical operation of the compressor and makes lubrication easier.

Multistage compressors

If the required pressure ratio P_2/P_1 is large, it is not practicable to carry out the whole of the compression in a single cylinder because of the high temperatures which would be set up and the adverse effects of clearance volume on the efficiency. Further, lubrication would be difficult due to carbonisation of the oil, and there would be a risk of causing oil mist explosions in the cylinders when gases containing oxygen were being compressed. The mechanical construction also would be difficult because the single cylinder would have to be strong enough to withstand the final pressure and yet large enough to hold the gas at the initial pressure P_1. In the multistage compressor, the gas passes through a number of cylinders of gradually decreasing volume and can be cooled between the stages. The maximum pressure ratio normally obtained in a single cylinder is 10 but values above 6 are unusual.

The operation of the multistage compressor can conveniently be followed again on a pressure–volume diagram (Figure 8.41). The effect of clearance volume will be neglected at first. The area 1–2–3–4 represents the work done in compressing isentropically from P_1 to P_2 in a single stage. The area 1–2–5–4 represents the necessary work for an isothermal compression. Now consider a multistage isentropic compression in which the intermediate pressures are P_{i1}, P_{i2}, etc. The gas will be assumed to be cooled to its initial temperature in an interstage cooler before it enters each cylinder.

Line 1–2 represents the suction stroke of the first stage where a volume V_1 of gas is admitted at a pressure P_1.

Line 2–6 represents an isentropic compression to a pressure P_{i1}.

Line 6–7 represents the delivery of the gas from the first stage at a constant pressure P_{i1}.

Line 7–8 represents the suction stroke of the second stage. The volume of the gas has been reduced in the interstage cooler to V_{i1}; that which would have been obtained as a result of an isothermal compression to P_{i1}.

Line 8–9 represents an isentropic compression in the second stage from a pressure P_{i1} to a pressure P_{i2}.

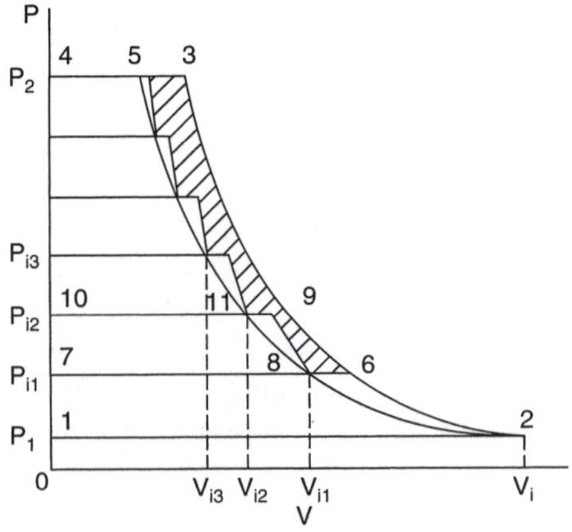

Figure 8.41. Multistage compression cycle with interstage cooling

Line 9–10 represents the delivery stroke of the second stage.

Line 10–11 represents the suction stroke of the third stage. Point 11 again lies on the line 2–5, representing an isothermal compression.

It is seen that the overall work done on the gas is intermediate between that for a single stage isothermal compression and that for an isentropic compression. The net saving in energy is shown as the shaded area in Figure 8.34.

The total work done per cycle W'

$$= P_1 V_1 \frac{\gamma}{\gamma - 1}\left[\left(\frac{P_{i1}}{P_1}\right)^{(\gamma-1)/\gamma} - 1\right] + P_{i1} V_{i1} \frac{\gamma}{\gamma - 1}\left[\left(\frac{P_{i2}}{P_{i1}}\right)^{(\gamma-1)/\gamma} - 1\right] + \cdots$$

for an isentropic compression.

For perfect interstage cooling:

$$P_1 V_1 = P_{i1} V_{i1} = P_{i2} V_{i2} = \cdots$$

$$\therefore \qquad W' = P_1 V_1 \frac{\gamma}{\gamma - 1}\left[\left(\frac{P_{i1}}{P_1}\right)^{(\gamma-1)/\gamma} + \left(\frac{P_{i2}}{P_{i1}}\right)^{(\gamma-1)/\gamma} + \cdots - n\right]$$

where n is the number of stages.

It is now required to find how the total work per cycle W' is affected by the choice of the intermediate pressures P_{i1}, P_{i2}, etc. The work will be a minimum when

$$\frac{\partial W'}{\partial P_{i1}} = \frac{\partial W'}{\partial P_{i2}} = \frac{\partial W'}{\partial P_{i3}} = \cdots = 0.$$

When:
$$\frac{\partial W'}{\partial P_{i1}} = 0$$

$$P_1 V_1 \frac{\gamma}{\gamma - 1}\left[\frac{\gamma - 1}{\gamma}\left(\frac{P_{i1}}{P_1}\right)^{(\gamma-1)/\gamma} P_{i1}^{-1} + \frac{1 - \gamma}{\gamma}\left(\frac{P_{i2}}{P_{i1}}\right)^{(\gamma-1)/\gamma} P_{i1}^{-1}\right] = 0$$

i.e.:
$$\frac{P_{i1}}{P_1} = \frac{P_{i2}}{P_{i1}} \tag{8.44}$$

The same procedure is then adopted for obtaining the optimum value of P_{i2} and hence:

$$\frac{P_{i2}}{P_{i1}} = \frac{P_{i3}}{P_{i2}} \tag{8.45}$$

Thus the intermediate pressures should be arranged so that the compression ratio is the same in each cylinder; and equal work is then done in each cylinder.

The minimum work of compression in a compressor of n stages is therefore:

$$P_1V_1\frac{\gamma}{\gamma-1}\left[n\left(\frac{P_2}{P_1}\right)^{(\gamma-1)/n\gamma} - n\right] = nP_1V_1\frac{\gamma}{\gamma-1}\left[\left(\frac{P_2}{P_1}\right)^{(\gamma-1)/n\gamma} - 1\right] \tag{8.46}$$

The effect of clearance volume may now be taken into account. If the clearances in the successive cylinders are $c_1, c_2, c_3, \ldots$, the theoretical volumetric efficiency of the first cylinder

$$= 1 + c_1 - c_1\left(\frac{P_{i1}}{P_1}\right)^{1/\gamma} \quad \text{(from equation 8.42)}$$

Assuming that the same compression ratio is used in each cylinder, then the theoretical volumetric efficiency of the first stage is:

$$1 + c_1 - c_1\left(\frac{P_2}{P_1}\right)^{1/n\gamma}$$

If the swept volumes of the cylinders are $V_{s1}, V_{s2}, \ldots$, the volume of gas admitted to the first cylinder

$$= V_{s1}\left[1 + c_1 - c_1\left(\frac{P_2}{P_1}\right)^{1/n\gamma}\right] \tag{8.47}$$

The same mass of gas passes through each of the cylinders and, therefore, if the inter-stage coolers are assumed perfectly efficient, the ratio of the volumes of gas admitted to successive cylinders is $(P_1/P_2)^{1/n}$. The volume of gas admitted to the second cylinder is then:

$$V_{s2}\left[1 + c_2 - c_2\left(\frac{P_2}{P_1}\right)^{1/n\gamma}\right] = V_{s1}\left[1 + c_1 - c_1\left(\frac{P_2}{P_1}\right)^{1/n\gamma}\right]\left(\frac{P_1}{P_2}\right)^{1/n}$$

$$\therefore \quad \frac{V_{s1}}{V_{s2}} = \frac{1 + c_2 - c_2(P_2/P_1)^{1/n\gamma}}{1 + c_1 - c_1(P_2/P_1)^{1/n\gamma}}\left(\frac{P_2}{P_1}\right)^{1/n} \tag{8.48}$$

In this manner the swept volume of each cylinder can be calculated in terms of V_{s1} and $c_1, c_2, \ldots$, and the cylinder dimensions determined.

When the gas does not behave as an ideal gas, the change in its condition can be followed on a temperature-entropy or an enthalpy-entropy diagram. The intermediate pressures $P_{i1}, P_{i2}, \ldots$, are then selected so that the enthalpy change (ΔH) is the same in each cylinder.

Several opposing factors will influence the number of stages selected for a given compression. The larger the number of cylinders the greater is the mechanical complexity. Against this must be balanced the higher theoretical efficiency, the smaller mechanical strains set up in the cylinders and the moving parts, and the greater ease of lubrication at the lower temperatures that are experienced. Compressors with as many as nine stages are used for very high pressures.

Compressor efficiencies

The efficiency quoted for a compressor is usually either an isothermal efficiency or an isentropic efficiency. The isothermal efficiency is the ratio of the work required for an ideal isothermal compression to the energy actually expended in the compressor. The isentropic efficiency is defined in a corresponding manner on the assumption that the whole compression is carried out in a single cylinder. Since the energy expended in an isentropic compression is greater than that for an isothermal compression, the isentropic efficiency is always the greater of the two. Clearly the efficiencies will depend on the heat transfer between the gas undergoing compression and the surroundings and on how closely the process approaches a reversible compression.

The efficiency of the compression will also be affected by a number of other factors which are all connected with the mechanical construction of the compressor. Thus the efficiency will be reduced as a result of leakage past the piston and the valves and because of throttling of the gas at the valves. Further, the mechanical friction of the machine will lower the efficiency and the overall efficiency will be affected by the efficiency of the driving motor and transmission.

Example 8.3

A single-acting air compressor supplies 0.1 m^3/s of air measured at, 273 K and 101.3 kN/m^2 which is compressed to 380 kN/m^2 from 101.3 kN/m^2. If the suction temperature is 289 K, the stroke is 0.25 m, and the speed is 4.0 Hz, what is the cylinder diameter? Assuming the cylinder clearance is 4 per cent and compression and re-expansion are isentropic ($\gamma = 1.4$), what are the theoretical power requirements for the compression?

Solution

$$\text{Volume per stroke} = \left(\frac{0.1}{4.0}\right)\left(\frac{289}{273}\right) = 0.0264\ \text{m}^3$$

$$\text{Compression ratio} = \left(\frac{380}{101.3}\right) = 3.75$$

The swept volume is given by equation 8.42:

$$0.0264 = V_s[1 + 0.04 - 0.04(3.75)^{1/1.4}]$$

$$\therefore \quad V_s = \frac{0.0264}{1.04 - 0.04 \times 2.7} = 0.0283\ \text{m}^3$$

$$\text{Thus cross-sectional area of cylinder} = \left(\frac{0.0283}{0.25}\right) = 0.113\ \text{m}^2$$

$$\text{and diameter} = \left(\frac{0.113}{(\pi/4)}\right)^{0.5} = \underline{\underline{0.38\ \text{m}}}$$

From equation 8.41, work of compression per cycle

$$= 101{,}300 \times 0.0264 \left[\frac{1.4}{1.4 - 1.0}\right] [(3.75)^{0.4/1.4} - 1]$$

$$= 9360(1.457 - 1) = 4278 \text{ J}$$

Theoretical power requirements

$$= (4278 \times 4) = 17{,}110 \text{ W} \quad \text{or} \quad \underline{\underline{17.1 \text{ kW}}}$$

Example 8.4

Air at 290 K is compressed from 101.3 kN/m^2 to 2065 kN/m^2 in a two-stage compressor operating with a mechanical efficiency of 85 per cent. The relation between pressure and volume during the compression stroke and expansion of the clearance gas is $PV^{1.25}$ = constant. The compression ratio in each of the two cylinders is the same, and the interstage cooler may be assumed 100 per cent efficient. If the clearances in the two cylinders are 4 per cent and 5 per cent respectively, calculate:

(a) the work of compression per kg of air compressed;
(b) the isothermal efficiency;
(c) the isentropic efficiency ($\gamma = 1.4$), and
(d) the ratio of the swept volumes in the two cylinders.

Solution

$$\text{Overall compression ratio} = \left(\frac{2065}{101.3}\right) = 20.4$$

$$\text{Specific volume of air at 290 K} = \left(\frac{22.4}{28.8}\right)\left(\frac{290}{273}\right) = 0.826 \text{ m}^3\text{/kg}$$

From equation 8.46, work of compression

$$= 101{,}300 \times 0.826 \times 2 \left[\frac{1.25}{1.25 - 1}\right] [(20.4)^{0.25/2.5} - 1]$$

$$= 836{,}700(1.351 - 1)$$

$$= 293{,}700 \text{ J/kg} = 293.7 \text{ kJ/kg}$$

Energy supplied to the compressor, that is the work of compression

$$= \left(\frac{293.7}{0.85}\right) = \underline{\underline{345.5 \text{ kJ/kg}}}$$

From equation 8.36, the work done in isothermal compression of 1 kg of gas

$$= (101{,}300 \times 0.826 \ln 20.4)$$

$$= (83{,}700 \times 3.015) = 252{,}300 \text{ J/kg} = 252.3 \text{ kJ/kg}$$

Isothermal efficiency $= 100 \times \left(\frac{252.3}{345.5}\right) = \underline{\underline{73 \text{ per cent}}}$

From equation 8.37, work done in isentropic compression of 1 kg of gas

$$= 101{,}300 \times 0.826 \frac{1.4}{0.4} [(20.4)^{0.4/1.4} - 1]$$

$$= 292{,}900(2.36 - 1) = 398{,}300 \text{ J/kg} = 398.3 \text{ kJ/kg}$$

Isentropic efficiency $= 100 \times \left(\frac{398.3}{345.5}\right) = \underline{\underline{115 \text{ per cent}}}$

From equation 8.47, volume swept out in first cylinder in compression of 1 kg of gas is given by:

$$0.826 = V_{s_1}[1 + 0.04 - 0.04(20.4)^{1/(2\times 1.25)}]$$

$$= V_{s_1} \times 0.906$$

$$\therefore \quad V_{s_1} = 0.912 \text{ m}^3/\text{kg}$$

Similarly, the swept volume of the second cylinder is given by:

$$0.826\left(\frac{1}{20.4}\right)^{0.5} = V_{s_2}[1 + 0.05 - 0.05(20.4)^{1/(2\times 1.25)}]$$

$$0.183 = 0.883V_{s_2}$$

$$\therefore \quad V_{s_2} = 0.207 \text{ m}^3/\text{kg}$$

and:

$$\frac{V_{s_1}}{V_{s_2}} = \underline{\underline{4.41}}$$

8.4. THE USE OF COMPRESSED AIR FOR PUMPING

Compressed gas is sometimes used for transferring liquid from one position to another in a chemical plant, but more particularly for emptying vessels. It is frequently more convenient to apply pressure by means of compressed gas rather than to install a pump, particularly when the liquid is corrosive or contains solids in suspension. Furthermore, to an increasing extent, chemicals are being delivered in tankers and are discharged by the application of gas under pressure. For instance, phthalic anhydride is now distributed in heated tankers which are discharged into storage vessels by connecting them to a supply of compressed nitrogen or carbon dioxide.

Several devices have been developed to eliminate the necessity for manual operation of valves, and the automatic acid elevator is an example of equipment incorporating such a device. However, such equipment is becoming less important now that it is possible to construct centrifugal pumps in a wide range of corrosion-resistant materials. The *air-lift pump* makes more efficient use of the compressed air and is used for pumping corrosive liquids. Although it is not used extensively in the chemical industry, it is used for pumping oil from wells, and the principle is incorporated into a number of items of chemical engineering equipment, including the climbing film evaporator.

8.4.1. The air-lift pump

In the air-lift pump (Figure 8.42) a high efficiency is obtained by allowing compressed air to expand to atmospheric pressure in contact with the liquid. It can be regarded simply as a U-tube in a state of dynamic equilibrium. One limb containing only liquid is relatively short and this is connected to the feed tank, whilst air is injected near the bottom of the longer limb which therefore contains a mixture of liquid and air that has a lower density. If the air is introduced sufficiently rapidly, liquid will flow from the short to the long limb and be discharged into the delivery tank. The rate of flow will depend on the difference in density and will, therefore, rise as the air rate is increased, but will reach a maximum because the frictional resistance increases with the volumetric rate of flow.

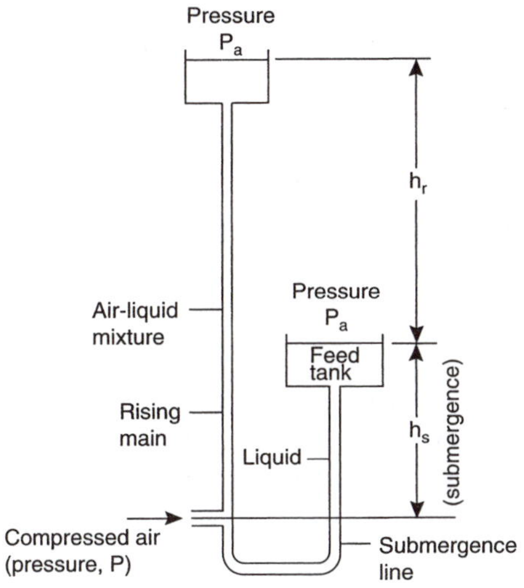

Figure 8.42. Air-lift pump

The liquid feed line is known as the *submergence limb* and the line carrying the aerated mixture as the *rising main*. The ratio of the submergence (h_s) to the total height of rising main above the air injection point ($h_r + h_s$) is known as the *submergence ratio*, $[1 + (h_r/h_s)]^{-1}$.

If a mass G_L of liquid is raised through a net height h_r by a mass G_A of air in unit time, the net rate of energy transfer to the liquid is $G_L g h_r$. If the pressure of the entering air is P, the work done by the air in expanding isothermally to atmospheric pressure P_a is given by:

$$P_a v_a G_A \ln \frac{P}{P_a} \tag{8.49}$$

where v_a is the specific volume of air at atmospheric pressure. The expansion will be essentially isothermal because of the intimate contact between the liquid and the air.

The efficiency of the pump η is therefore given by:

$$\eta = \frac{G_L g h_r}{G_A P_a v_a \ln(P/P_a)} \tag{8.50}$$

The mass of air required to pump unit mass of liquid is, therefore, given by:

$$\frac{G_A}{G_L} = \frac{g h_r}{\eta P_a v_a \ln(P/P_a)} \tag{8.51}$$

If all losses in the operation of the pump were neglected, the pressure at the point of introduction of the compressed air would be equal to atmospheric pressure together with the pressure due to the column of liquid of height h_s, the vertical distance between the liquid level in the suction tank, and the air inlet point. Therefore:

$$P_a = h_a \rho g \quad \text{(say)} \tag{8.52}$$

and:

$$P = (h_a + h_s)\rho g \tag{8.53}$$

where ρ is the density of the liquid.

Thus from equation 8.51 the mass of air required to pump unit mass of liquid (G_A/G_L) would be equal to:

$$\frac{G_A}{G_L} = \frac{h_r g}{P_a v_a \ln[(h_s + h_a)/h_a]} \tag{8.54}$$

This is the minimum air requirement for the pump if all losses are neglected. It will be seen that (G_A/G_L) decreases as h_s increases; if h_s is zero, G_A/G_L is infinite and therefore the pump will not work. A high submergence h_s is therefore desirable. This can be a considerable practical disadvantage in that it may be necessary to provide a deep "pit" to give the demand submergence.

There are a number of important applications of the air-lift pump in the process industries due to its simplicity. It is particularly useful for handling radioactive materials as there are no mechanical parts in contact with the fluid, and the pump will operate virtually indefinitely without the need for maintenance which can prove very difficult when handling radioactive liquids.

Example 8.5

An air-lift pump is used for raising 7.5×10^{-4} m^3/s of a liquid of density 1200 kg/m^3 to a height of 20 m. Air is available at a pressure of 450 kN/m^2. Assuming isentropic compression of the air, what is the power requirement of the pump its efficiency is 30 per cent? ($\gamma = 1.4$). Take the volume of 1 kmol of an ideal gas at 273 K and 101.3 kN/m^2 as 22.4 m^3.

Solution

Mass flow of liquid $= (7.5 \times 10^{-4} \times 1200) = 0.9$ kg/s

Work per unit time done by the pump $= (0.9 \times 9.81 \times 20) = 176.6$ J/s $= 176.6$ W

Actual work of expansion of air per unit time $= (176.6/0.30) = 588.6$ W

Taking the molecular weight of air as 28.9 kg/kmol,
the specific volume of air at 101.3 kN/m^2 and 273 K, $v_a = (22.4/28.9) = 0.775$ m^3/kg

and in equation 8.49: $588.6 = [(101.3 \times 10^3 \times 0.775 G_A)(\ln(450/101.3))]$

from which: $G_A = 0.0050$ kg/s

and: volume flowrate of air, $Q = (0.0050 \times 0.775) = 0.0039$ m^3/s

From equation 8.37:

$$\text{Power for compression} = [101.3 \times 10^3 \times 0.0039][1.4/(1.4 - 1)][(450/101.3)^{(1.4-1)/1.4} - 1]$$

$$= (395.1 \times 3.5 \times 0.53) = 733 \text{ W}$$

and: Power required $= (733/1000) = \underline{\underline{0.733 \text{ kW}}}$

Example 8.6

An air-lift pump raises 0.01 m^3/s of water from a well 100 m deep through a 100 mm diameter pipe. The level of water is 40 m below the surface. The air flow is 0.1 m^3/s of free air compressed to 800 kN/m^2. Calculate the efficiency of the pump and the mean velocity of the mixture in the pipe.

Solution

Mass flow of water = (0.01 × 1000) = 10 kg/s

Work done per unit time = (10 × 40 × 981) = 3924 W

Volumetric flowrate of air used = 0.1 m^3/s

The energy needed to compress 0.1 m^3/s of air is given by:

$$P_1 V_1 [\gamma/(\gamma - 1)][(P_2/P_1)^{(\gamma-1)/\gamma} - 1] \quad \text{(equation 8.37)}$$

$$= (101{,}300 \times 0.1 \times 1.4)[(800/101.3)^{0.286} - 1] = 28{,}750 \text{ J}$$

The power required for this compression is 28,500 J/s = 28,750 W
and hence the efficiency = (3924 × 100)/28,750 = 13.7 per cent

The mean velocity depends upon the pressure of air in the pipe. If the air pressure at the bottom of the well is, say, 60 m of water or (60 × 1000 × 9.81)/1000 = 588.6 kN/m^2, and the pressure at the surface is atmospheric, the mean pressure is (101.3 + 588.6)/2 = 345 kN/m^2.

The specific volume v of the air at this pressure and at 273 K is then given by:

$$v = \frac{\mathbf{R}T}{MP} = \frac{8314 \times 273}{29 \times 345{,}000} = 0.227 \text{ m}^3/\text{kg}$$

The specific volume v of air at 273 K and 101.3 kN/m^2 is given by:

$$v = \frac{(8314 \times 273)}{(29 \times 101{,}300)} = 0.772 \text{ m}^3/\text{kg}$$

and hence the mass flowrate of the air is:

$$(0.10/0.772) = 0.13 \text{ kg/s}$$

Mean volumetric flowrate of air = (0.13 × 0.227) = 0.0295 m^3/s

Volumetric flowrate of water = 0.01 m^3/s

Total volumetric flowrate = 0.0395 m^3/s

Area of pipe = $(\pi/4)0.1^2 = 0.00785$ m^2

and hence the mean velocity of the mixture = (0.0395/0.00785) = 5.03 m/s

Flow of a vertical column of aerated liquid

The behaviour of a rising column of aerated liquid is important in the way it affects the operation of the air-lift pump. In Chapter 5, the flow of gas–liquid mixtures in pipes has been discussed for conditions where the flowrates of the two fluids are controlled independently. In the air-lift pump, the flowrate of liquid is not normally controlled, but is governed by the air rate and other parameters of the system. For gas–liquid flow, at low degrees of aeration the gas is distributed in the form of discrete bubbles which undergo a degree of coalescence. When the proportion of air is increased, '*slug*' flow occurs and the mixture tends to separate into alternate slugs of gas and liquid. The gas slug, however, only occupies the central portion of the cross-section, and is surrounded by liquid which flows backwards relative to the core of gas, though not necessarily relative to the walls of the pipe. The liquid slug normally has gas bubbles in suspension. At somewhat higher velocities, and with higher proportions of gas, a *churn* flow appears, in which the slugs are broken up and lose their individual identity. At very high ratios of air to water and at high velocities, an annular type of flow is obtained with air passing through a central core in the pipe and dragging upwards a film of liquid at the walls. The boundaries between the various types of flow are not precisely defined and the region of flow may change from the bottom to the top of a tall pipe as a result of the expansion of the gas. In air-lift pumping, *slug* flow and *churn* flow are the types most commonly encountered. Furthermore, although the flow pattern near the air-injection point (at the *foot piece*) will be strongly influenced by the nature of the injector, the flow soon assumes a pattern which is independent of that in the immediate vicinity of the injection point.

Thus, if the gas is injected in the form of small dispersed bubbles in order to reduce the *slip velocity*, coalescence rapidly occurs to give large bubbles and slugs.

The distribution of air and liquid in the pipe, and the proportion of the cross-section occupied by liquid (the *holdup* of the liquid) have an important bearing on the flow of the two fluids, firstly because the hydrostatic pressure is affected by the liquid holdup, and secondly because the nature of the flow affects the frictional pressure drop. Furthermore, the velocity of the air relative to the liquid is also dependent upon its pattern of distribution.

There are several methods whereby the holdup of liquid in a flowing mixture can be determined. One involves a measurement of the attenuation of a beam of γ-rays passed through the mixture; in another, the fluid in a test section is suddenly isolated by rapidly operating valves and the volume of liquid retained in the pipe is measured. The former method gives point values and is therefore useful for determining the distribution throughout the length of the pipe, but the second enables an average value to be obtained from a single reading. Results obtained by the isolating method[6] for the liquid holdup in a 25 mm. diameter pipe are shown graphically in Figure. 8.43 which applies to the regime of slug flow, the limits of which are indicated approximately on the diagram.

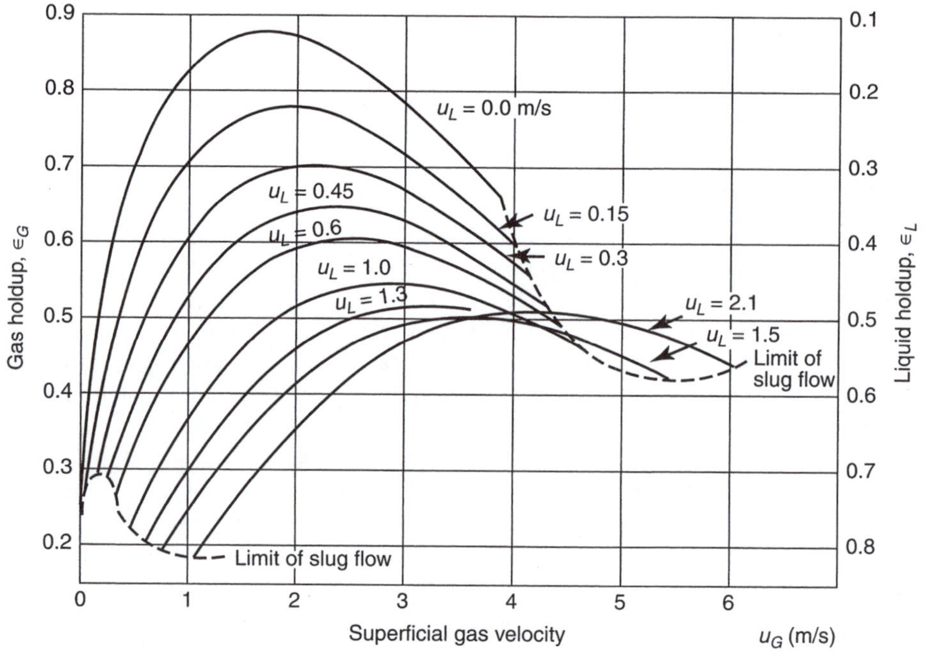

Figure 8.43. Liquid holdup ϵ_L and gas holdup ϵ_G ($= 1 - \epsilon_L$) for slug flow in a 25 mm diameter pipe as a function of superficial gas velocity (u_G) and superficial liquid velocity (u_L)

Values of the holdup may be used to estimate the frictional losses in the pipe, since the overall difference in fluid head Δh of the liquid is equal to the friction head less the hold-up of liquid per unit area;

$$\Delta h = \Delta h_F - \frac{V_L}{A} \tag{8.55}$$

$$= \Delta h_F - l\epsilon_L \tag{8.56}$$

where Δh is the difference in head over a length l of pipe,
Δh_F is the corresponding frictional head,
V_L is the volume of liquid in the pipe of cross-section A, and
ϵ_L is the mean value of the liquid holdup.

In the usual case h and h_F are falling in the direction of flow and Δh and Δh_F are therefore negative. Values of frictional pressure drop, $-\Delta P_{TPF}$ may conveniently be correlated in terms of the pressure drop $-\Delta P_L$ for liquid flowing alone at the same volumetric rate. Experimental results obtained for plug flow in a 25 mm. diameter pipe are given as follows by RICHARDSON and HIGSON[6]:

$$\Phi_G^2 = \frac{-\Delta P_{TPF}}{-\Delta P_L} = \epsilon_L^{-\frac{4}{3}} (\epsilon_L > 0{\cdot}3) \tag{8.57}$$

At high flowrates when churn flow sets in, the pressure drop undergoes a sudden increase and the coefficient in equation 8.57 increases:

$$\Phi_G^2 = \frac{-\Delta P_{TPF}}{-\Delta P_L} = 2.25\epsilon_L^{-\frac{4}{3}} (\epsilon_L < 0{\cdot}3) \tag{8.58}$$

Expressions of this type have also been obtained by several other workers including ARMAND[7], SCHMIDT, BEHRINGER and SCHURIG[8], GOVIER *et al.*[9,10], ISBIN *et al.*[11] and MOORE and WILDE[12].

Equations 8.57 and 8.58 are satisfactory except at low liquid rates when the frictional pressure drop is a very small proportion of the total pressure drop. Frictional effects can then even be negative, because the liquid may then flow downwards at the walls, with the gas passing upwards in slugs.

Operation of the air-lift Pump

In an experimental study of a small air-lift pump[6], (25 mm. diameter and 13.8 m overall height) the results were expressed by plotting the efficiency of the pump, defined as the useful work done on the water divided by the energy required for isothermal compression of the air, to a basis of energy input in the air. In each case, the curve was found to rise sharply to a maximum and then to fall off more gradually. Typical results are shown in Figure 8.37.

The effects of the addition of surface active agent were investigated because the distribution of air might be affected and the slip velocity reduced. As a result of reducing the surface tension from 0.07 to 0.045 N/m, the maximum efficiency of the pump was increased from 49 to 66 per cent.

A characteristic of the pump was the appearance of a cyclic pattern in the rate of discharge. The resulting fluctuations were responsible for a wastage of energy, not only

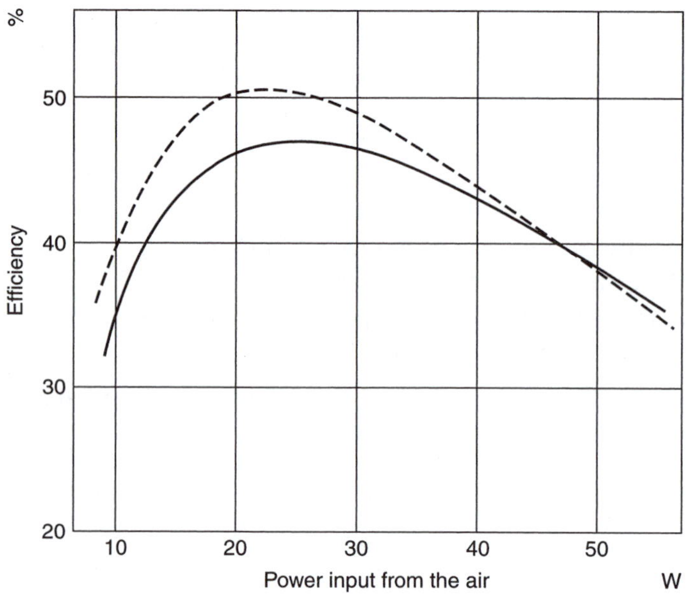

Figure 8.44. Efficiency of air-lift pump as function of energy input from air, showing effect of throttling water inlet. Submergence = 50 per cent

because of the need continually to re-accelerate the fluid above its equilibrium velocity, but also because frictional losses were thereby increased. The fluctuations were reduced, and the efficiency improved, by a limited throttling of the liquid supply, and by reducing the air capacity on the downstream side of the air control valve. There was a limit, of course, to the improvement in efficiency which would be obtained by throttling the liquid inlet, because frictional losses would be excessive in small throttles. In Figure. 8.44, the continuous curve relates to conditions where the inlet was not throttled (diameter 33 mm), and the broken curve refers to conditions where the water inlet was throttled to 15 mm diameter.

The effect of resistances and capacities in the air line and water feed has been studied theoretically and it has been shown that oscillations will tend to be stable if there is a large air capacity downstream from the control valve, and the inclusion of a resistance in the water supply line should dampen them.

Oscillations occur because the liquid column requires more than the equilibrium quantity of air to produce the initial acceleration. It therefore becomes over-accelerated and excess of liquid enters the limb with the result that it is retarded and then subsequently has to be accelerated again. During the period of retardation some liquid may actually run backwards; this can readily be prevented, however, by incorporating a non-return valve in the submergence limb.

8.5 VACUUM PUMPS

Many chemical plants, particularly distillation columns, operate at low pressures, and pumps are required to create and maintain the vacuum. Vacuum pumps take in gas at a very low pressure and normally discharge at atmospheric pressure, and pressure ratios are therefore high. Rotary pumps are generally preferred to piston pumps.

Sliding vane and liquid ring (including the Nash Hytor) types of pumps, already described in Section 8.3.1, are frequently used as vacuum pumps and are capable of achieving pressures down to about 1.3 N/m^2. A pump suitable for vacuum operation is illustrated in Figure 8.45. Care must be taken to avoid cavitation which can occur if the vapour fraction in the rotating cells exceeds about 50 per cent. When water is used as the sealing liquid, its vapour pressure limits the lower limit of pressure which can be achieved to about 3.3 kN/m^2, and a liquid with a lower vapour pressure must be used if lower pressures are required. Lower pressures may also be obtained by using the liquid-ring pump in series with an additional device, such as the steam-jet ejector, described below.

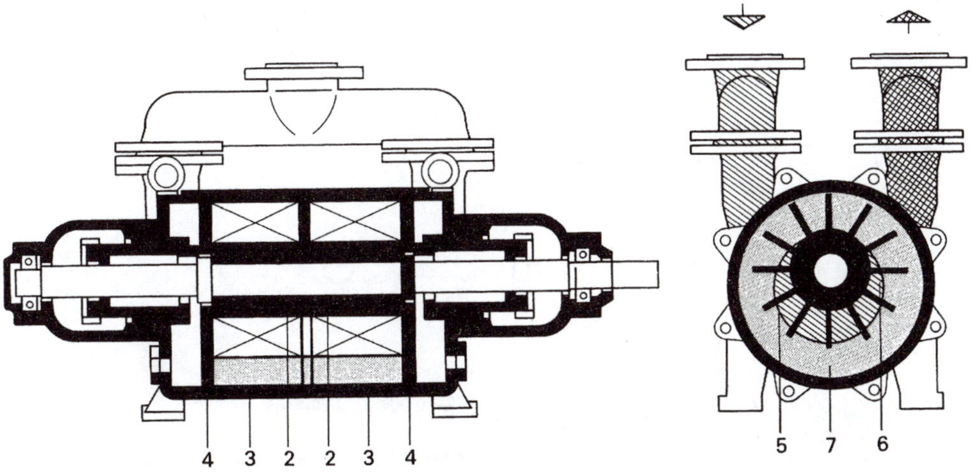

Figure 8.45. Single-stage, single-action liquid ring gas pump: 1, shaft 2, impeller 3, casing 4, guide plate 5, suction port 6, discharge port 7, liquid ring

The *steam-jet ejector*, illustrated in Figure 8.46, is very commonly used in the process industries since it has no moving parts and will handle large volumes of vapour at low pressures. The operation of the unit is shown in Figure 8.47. The steam is fed at constant pressure P_1 and expands nearly isentropically along AC; mixing of the steam and vapour which is drawn in is represented by CE and vapour compression continues to the throat F and to the exit at G. The steam required increases with the compression ratio so that a single-stage unit will operate down to about 17 kN/m^2 corresponding to a compression ratio of 6:1. For lower final pressures multistage units are used, and Figure 8.48 shows the relationship between the number of stages, steam pressure, and vacuum required. If cooling water is applied between the stages an improved performance will be obtained.

As a guide, a single-stage unit gives a vacuum to 13.5 kN/m^2, a double stage from 3.4 to 13.5 kN/m^2, and a three-stage unit from 0.67 to 2.7 kN/m^2.

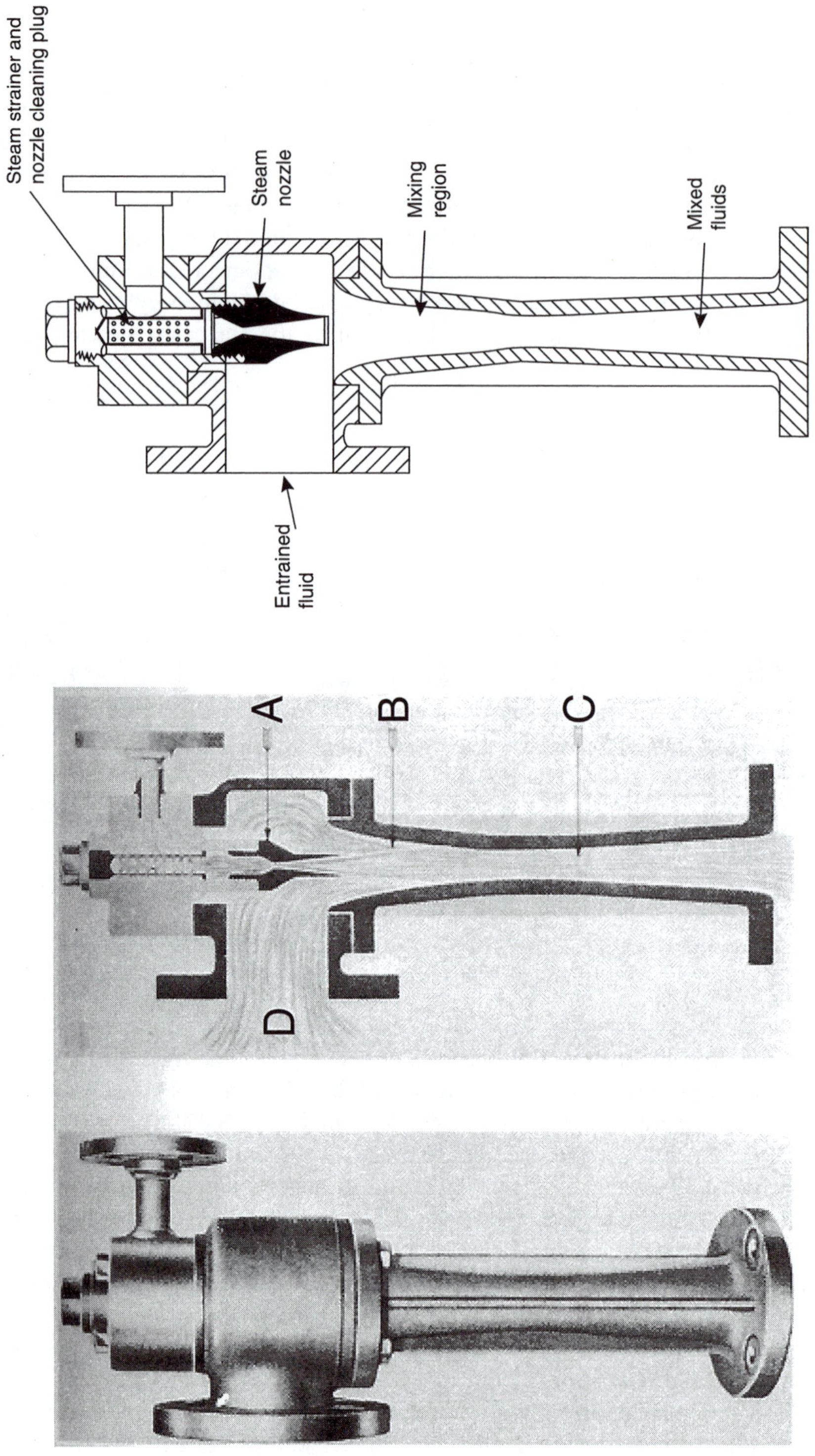

Figure 8.46. Steam jet ejector A: Steam nozzle B: Mixing region C: Mixed fluids D: Entrained fluid

PUMPING OF FLUIDS

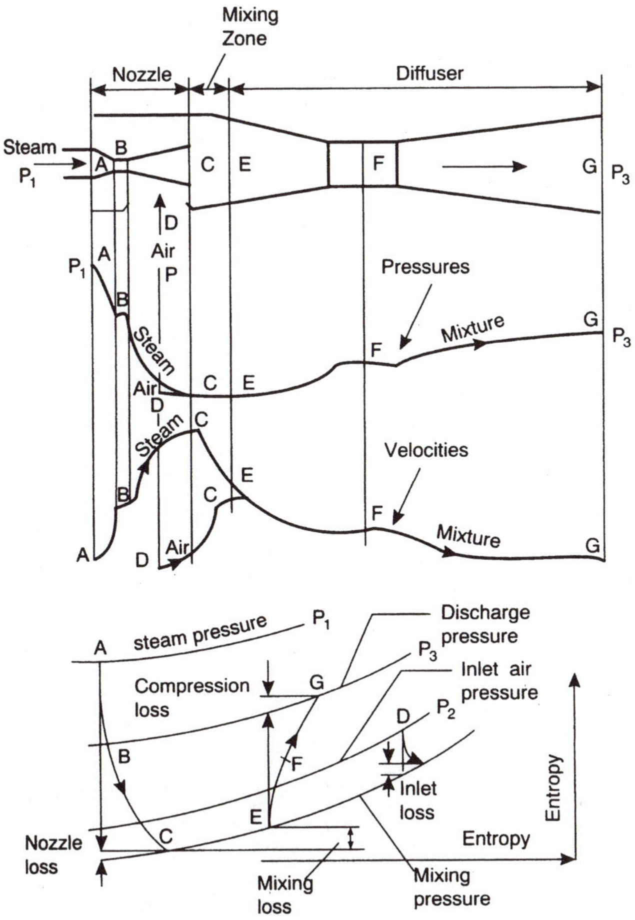

Figure 8.47. Ejector flow phenomena

For very low pressures, a diffusion pump is used with a rotary pump as the first stage. The principle of operation is that the gas diffuses into a stream of oil or mercury and is driven out of the pump by molecular bombardment.

8.6. POWER REQUIREMENTS FOR PUMPING THROUGH PIPELINES

A fluid will flow of its own accord so long as its energy per unit mass decreases in the direction of flow. It will flow in the opposite direction only if a pump is used to supply energy, and to increase the pressure at the upstream end of the system.

The energy balance is given by (equation 2.55):

$$\Delta\frac{u^2}{2\alpha} + g\Delta z + \int_{P_1}^{P_2} v\,\mathrm{d}P + W_s + F = 0 \qquad \text{(equation 2.55)}$$

$$\frac{\Delta u^2}{2} + g\Delta z + v(P_2 - P_1) + 4\left(\frac{R}{\rho v^2}\right)\left(\frac{l}{d}\right)v^2 = 0$$

$$P_1 - P_2 = \rho\left[\frac{\Delta u^2}{2} + g\Delta z + 4\left(\frac{R}{\rho v^2}\right)\left(\frac{l}{d}\right)v^2\right]$$

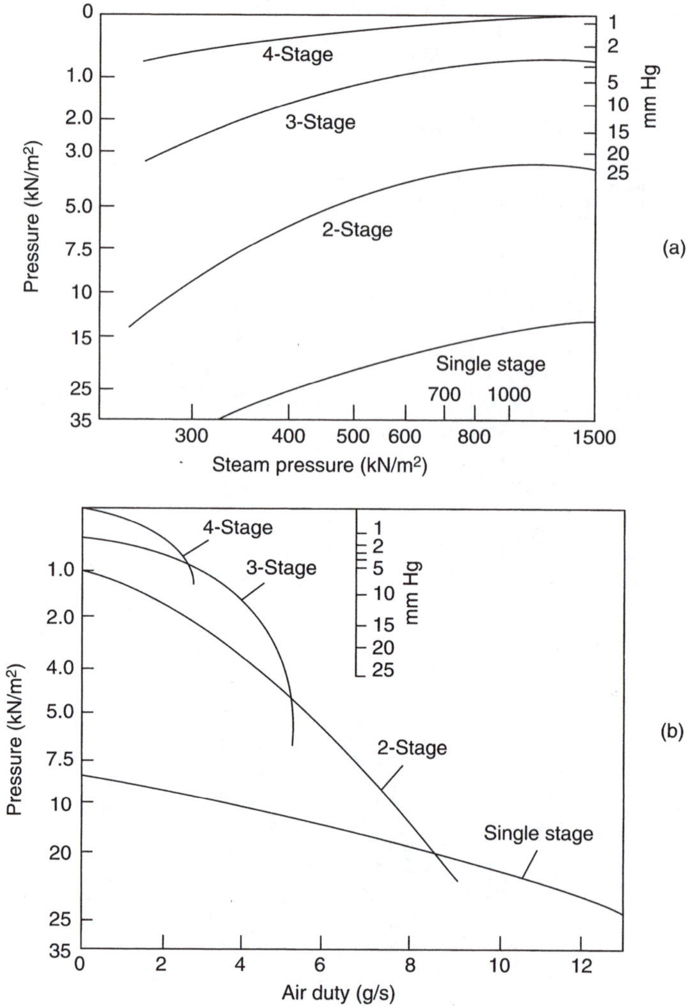

Figure 8.48. (*a*) Diagram showing most suitable number of ejector stages for varying vacua and steam pressures (*b*) Comparative performance curves of ejectors

The work done on unit mass of fluid is $-W_s$, and the total rate at which energy must be transferred to the fluid is $-GW_s$, for a mass rate of flow G.
The power requirement **P** is therefore given by:

$$\mathbf{P} = -GW_s = G\left(\Delta\frac{u^2}{2\alpha} + g\Delta z + \int_{P_1}^{P_2} v\,\mathrm{d}P + F\right) \tag{8.59}$$

8.6.1. Liquids

The term F in equation 8.59 may be calculated directly for the flow of liquid through a uniform pipe. If a liquid is pumped through a height Δz from one open tank to another and none of the kinetic energy is recoverable as pressure energy, the fluid pressure is the same at both ends of the system and $\int_{P_1}^{P_2} v\,\mathrm{d}P$ is zero. The power requirement, is, then:

$$\mathbf{P} = G\left(\frac{u^2}{2\alpha} + g\Delta z + F\right) = Ghg \tag{8.60}$$

Taking into account the pump efficiency η, the overall power requirement is obtained from the product of the mass flowrate G, the total head h and the acceleration due to gravity g as:

$$\mathbf{P} = \frac{1}{\eta}Ghg \tag{8.61}$$

The application of this equation is best illustrated by the following examples.

Example 8.7

2.16 m^3/h (600×10^{-6} m^3/s) water at 320 K is pumped through a 40 mm i.d. pipe, through a length of 150 m in a horizontal direction, and up through a vertical height of 10 m. In the pipe there are a control valve, equivalent to 200 pipe diameters, and other pipe fittings equivalent to 60 pipe diameters. Also in the line is a heat exchanger across which the head lost is 2 m water. Assuming the main pipe has a roughness of 0.0002 m, what power must be supplied to the pump if it is 60 per cent efficient?

Solution

$$\text{Area for flow} = (0.040)^2 \frac{\pi}{4} = 0.0012\ \text{m}^2$$

$$\text{Flow of water} = 600 \times 10^{-6}\ \text{m}^3/\text{s}$$

$$\therefore \quad \text{velocity} = \frac{(600 \times 10^{-6})}{0.0012} = 0.50\ \text{m/s}$$

$$\text{At 320 K, } \mu = 0.65\ \text{mN s/m}^2 = 0.65 \times 10^{-3}\ \text{N s/m}^2$$

$$\rho = 1000\ \text{kg/m}^3$$

$$\therefore \quad Re = \frac{(0.040 \times 0.50 \times 1000)}{(0.65 \times 10^{-3})} = 30{,}780$$

$$\therefore \quad \phi = \frac{R}{\rho u^2} = 0.004 \text{ for a relative roughness of } \frac{e}{d} = \left(\frac{0.0002}{0.040}\right) = 0.005 \quad \text{(from Figure 3.7)}$$

$$\text{Equivalent length of pipe} = 150 + 10 + (260 \times 0.040) = 170.4\ \text{m}$$

$$\therefore \quad h_f = 4\phi \frac{l}{d}\frac{u^2}{g} \quad \text{(from equation 3.20)}$$

$$= 4 \times 0.004 \left(\frac{170.4}{0.040}\right)\left(\frac{0.50^2}{9.81}\right)$$

$$= 1.74\ \text{m}$$

$$\therefore \quad \text{Total head to be developed} = (1.74 + 10 + 2) = 13.74\ \text{m}$$

$$\text{Mass flow of water} = (600 \times 10^{-6} \times 1000) = 0.60\ \text{kg/s}$$

$$\therefore \quad \text{Power required} = (0.60 \times 13.74 \times 9.81) = 80.9\ \text{W}$$

$$\therefore \quad \text{Power to be supplied} = 80.9 \times \left(\frac{100}{60}\right) = \underline{\underline{135\ \text{W}}}$$

In this solution the kinetic energy head, $u^2/2g$, has been neglected as this represents only $[0.5^2/(2 \times 9.81)] = 0.013$ m, or 0.1 per cent of the total head.

Example 8.8

It is required to pump cooling water from a storage pond to a condenser in a process plant situated 10 m above the level of the pond. 200 m of 74.2 mm i.d. pipe is available and the pump has the characteristics given below. The head lost in the condenser is equivalent to 16 velocity heads based on the flow in the 74.2 mm pipe.

If the friction factor $\phi = 0.003$, estimate the rate of flow and the power to be supplied to the pump assuming an efficiency of 50 per cent.

Discharge (m^3/s)	0.0028	0.0039	0.0050	0.0056	0.0059
Head developed (m)	23.2	21.3	18.9	15.2	11.0

Solution

The head to be developed, $h = 10 + 4\phi \dfrac{l}{d}\dfrac{u^2}{g} + 8\dfrac{u^2}{g}$ (from equation 3.20)

$$= 10 + 4 \times 0.003 \left(\frac{200}{0.0742}\right)\frac{u^2}{g} + \frac{8u^2}{g}$$

$$= 10 + 4.12\,u^2 \text{ m water}$$

Discharge, $Q = (\pi d^2/4)u = 0.00434u \text{ m}^3/\text{s}$

$$\therefore \quad u = \frac{Q}{0.00434} = 231.3Q \text{ m/s}$$

$$\therefore \quad h = 10 + 4.12(231.3Q)^2$$

$$= 10 + 2.205 \times 10^5 Q^2 \text{ m water}$$

Values of Q and h are plotted in Figure 8.49 and the discharge at the point of intersection between the pump characteristic equation and the line of the above equation is 0.0054 m^3/s.

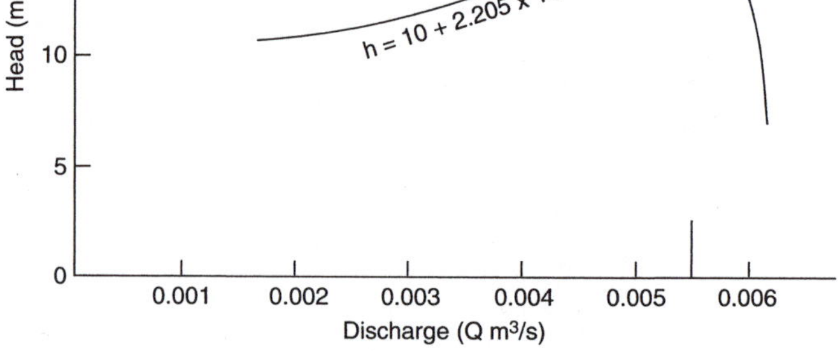

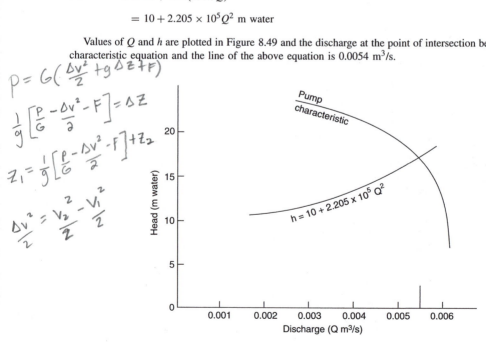

Figure 8.49. Data for Example 8.8

$$\text{The head developed is thus, } h = 10 + (2.205 \times 10^5 \times 0.0054^2)$$

$$= 16.43, \text{ say } 16 \text{ m water}$$

$$\text{Power required} = \frac{(0.0054 \times 1000 \times 16.43 \times 9.81)}{0.50}$$

$$= 1741 \text{ W or } \underline{\underline{1.74 \text{ kW}}}$$

Selection of pipe diameter

If a fluid is to be pumped between two points, the diameter of the pipeline should be chosen so that the overall cost of operation is a minimum. The smaller the diameter, the lower is the initial cost of the line but the greater is the cost of pumping; an economic balance must, therefore, be achieved.

The initial cost of a pipeline and the depreciation and maintenance costs will be approximately proportional to the diameter raised to a power of between 1.0 and 1.5. The power for pumping an incompressible fluid at a given rate G is made up of two parts:

(1) that necessitated by the difference in static pressure and vertical height at the two ends of the system (this is independent of the diameter of the pipe);
(2) that attributable to the kinetic energy of the fluid and the work done against friction. If the kinetic energy is small, this is equal to:

$$G\left[4\phi\frac{l}{d}u^2\right] \tag{8.62}$$

(from equations 3.19 and 8.60)

which is proportional to $d^{-4.5 \text{ to } -5}$ for turbulent flow, since $u \propto d^{-2}$ and $\phi \propto u^{-0.25 \text{ to } 0}$, according to the roughness of the pipe.

The power requirement can, therefore, be calculated as a function of d and the cost obtained. The total cost per annum is then plotted against the diameter of pipe and the optimum conditions are found either graphically or by differentiation as shown in Example 8.9.

Example 8.9

The relation between cost per unit length C of a pipeline installation and its diameter d is given by:

$$C = a + bd$$

where a and b are independent of pipe size. Annual charges are a fraction β of the capital cost. Obtain an expression for the optimum pipe diameter on a minimum cost basis for a fluid of density ρ and viscosity μ flowing at a mass rate of G. Assume that the fluid is in turbulent flow and that the Blasius equation is applicable, that is the friction factor is proportional to the Reynolds number to the power of minus one quarter. Indicate clearly how the optimum diameter depends on flowrate and fluid properties.

Solution

The total annual cost of a pipeline consists of a capital charge plus the running costs. The chief element of the running cost will be the power required to overcome the head loss which is given by equation 3.20:

$$h_f = 4\phi\frac{l}{d}\frac{u^2}{g}$$

If $\phi = R/\rho u^2 = 0.04/Re^{0.25}$, the head loss per unit length l is given by equation 3.20 as:

$$\frac{h_f}{l} = 4\frac{0.04}{Re^{0.25}}\frac{1}{d}\frac{u^2}{g}$$

$$= 0.16\frac{u^2}{gd}\left(\frac{\mu}{\rho u d}\right)^{0.25}$$

$$= 0.16\frac{u^{1.75}\mu^{0.25}}{\rho^{0.25}gd^{1.25}}$$

The velocity, $u = G/\rho A = G/\rho(\pi/4)d^2 = 1.27G/\rho d^2$

$$\therefore \quad \frac{h_f}{l} = \frac{0.16(1.27G/\rho d^2)^{1.75}\mu^{0.25}}{\rho^{0.25}gd^{1.25}}$$

$$= \frac{0.244G^{1.75}\mu^{0.25}}{\rho^2 gd^{4.75}}$$

The power required for pumping if the pump efficiency is η is:

$$\mathbf{P} = Gg\frac{0.244G^{1.75}\mu^{0.25}}{\rho^2 gd^{4.75}}\frac{1}{\eta}$$

If $\eta = 0.5$ (say) $\mathbf{P} = 0.488G^{2.75}\mu^{0.25}/(\rho^2 d^{4.75})$ (Watt)
If c = power cost/W, the cost of pumping is:

$$\frac{0.488cG^{2.75}\mu^{0.25}}{\rho^2 d^{4.75}}$$

The total annual cost is then $C = (\beta a + \beta b d) + \dfrac{\gamma G^{2.75}\mu^{0.25}}{\rho^2 d^{4.75}}$

where $\gamma = 0.488c$.
Differentiating the total cost with respect to the diameter gives:

$$\frac{dC}{dd} = \beta b - \frac{4.75\gamma G^{2.75}\mu^{0.25}}{\rho^2 d^{5.75}}$$

For minimum cost, $dC/dd = 0$ and:

$$d^{5.75} = \frac{4.75\gamma G^{2.75}\mu^{0.25}}{\rho^2 \beta b}$$

or:

$$d = \frac{KG^{0.48}\mu^{0.043}}{\rho^{0.35}}$$

where:

$$K = \left(\frac{4.75\gamma}{\beta b}\right)^{0.174} = \left(\frac{2.32c}{\beta b}\right)^{0.174}$$

Effect of fluctuations in flowrate on power for pumping

The importance of maintaining the flowrate in a pipeline constant may be seen by considering the effect of a sinusoidal variation in flowrate. This corresponds approximately to the discharge conditions in a piston pump during the forward movement of the piston. The flowrate $\dot{Q}$ is given as a function of time t by the relation:

$$\dot{Q} = Q_m \sin \omega t \tag{8.63}$$

The discharge rate rises from zero at $t = 0$ to Q_m when $\omega t = \pi/2$, i.e. $t = \pi/2\omega$.

The mean discharge rate Q is then given by:

$$Q = \frac{1}{\pi/2\omega} \int_0^{\pi/2\omega} Q_m \sin \omega t \, dt$$

$$= \frac{Q_m \cdot 2\omega}{\pi} \left[-\frac{1}{\omega} \cos \omega t \right]_0^{\pi/2\omega}$$

$$= \frac{2}{\pi} Q_m \tag{8.64}$$

The power required for pumping will be given by the product of the volumetric flowrate and the pressure difference between the pump outlet and the discharge end of the pipeline. Taking note of the fluctuating nature of the flow, it is necessary to consider the energy transferred to the fluid over a small time interval and to integrate over the cycle to obtain the mean value of the power.

Thus, the power requirement, $$\mathbf{P} = \frac{1}{\pi/2\omega} \int_0^{\pi/2\omega} Q_m \sin \omega t(-\Delta P) \, dt \tag{8.65}$$

The form of the integral will depend on the relation between $-\Delta P$ and $\dot{Q}$.

Streamline flow

If it is assumed that $-\Delta P = K\dot{Q} = KQ_m \sin \omega t$, then:

$$\mathbf{P} = \frac{2\omega}{\pi} K Q_m^2 \int_0^{\pi/2\omega} \sin^2 \omega t \, dt$$

$$= \frac{2\omega}{\pi} K \left(Q^2 \frac{\pi^2}{4} \right) \frac{1}{2} \int_0^{\pi/2\omega} (1 - \cos 2\omega t) \, dt$$

$$= \frac{KQ^2 \pi \omega}{4} \left[t - \frac{1}{2\omega} \sin 2\omega t \right]_0^{\pi/2\omega}$$

$$= \frac{KQ^2 \pi^2}{8}$$

For steady flow:

$$\mathbf{P} = Q(-\Delta P) = KQ^2 \tag{8.66}$$

Thus power is increased by a factor of $\pi^2/8$ (= 1.23) as a result of the sinusoidal fluctuations.

Turbulent flow

As a first approximation it may be assumed that pressure drop is proportional to the square of flowrate or:

$$-\Delta P = K^1 Q^2 = K^1 Q_m^2 \sin^2 \omega t \tag{8.67}$$

Then:

$$\mathbf{P} = \frac{2\omega}{\pi} K^1 Q_m^3 \int_0^{\pi/2\omega} \sin^3 \omega t \, dt$$

$$= \frac{2\omega}{\pi} K^1 \left(Q^3 \frac{\pi^3}{8} \right) \frac{1}{4} \int_0^{\pi/2\omega} (3 \sin \omega t - \sin 3\omega t) \, dt$$

$$= \frac{K^1 Q^3 \omega \pi^2}{16} \left[-\frac{3}{\omega} \cos \omega t + \frac{1}{3\omega} \cos 3\omega t \right]_0^{\pi/2\omega}$$

$$= \frac{K^1 Q^3 \omega \pi^2}{16} \left(\frac{3}{\omega} - \frac{1}{3\omega} \right)$$

$$= \frac{K^1 Q^3 \pi^2}{6} \tag{8.68}$$

For steady flow:

$$\mathbf{P} = Q(-\Delta P) = K^1 Q^3 \tag{8.69}$$

Thus the power is increased by a factor of $\pi^2/6$ (= 1.64) as a result of fluctuations, compared with 1.23 for streamline flow.

In addition to the increased power to overcome friction, it will be necessary to supply the energy required to accelerate the liquid at the beginning of each cycle.

8.6.2 Gases

If a gas is pumped under turbulent flow conditions from a reservoir at a pressure P_1 to a second reservoir at a higher pressure P_2 through a uniform pipe of cross-sectional area A by means of a pump situated at the upstream end, the power required is:

$$\mathbf{P} = G \left(\frac{u^2}{2} + \int_{P_1}^{P_2} v \, dP + F \right) \tag{8.70}$$

(from equation 8.59)

neglecting pressure changes arising from change in vertical height.

In order to maintain the gas flow, the pump must raise the pressure at the upstream end of the pipe to some value P_3, which is greater than P_2. The required value of P_3 will depend somewhat on the conditions of flow in the pipe. Thus for isothermal conditions, P_3 may be calculated from equation 4.55, since the downstream pressure P_2 and the mass rate of flow G are known. For non-isothermal conditions, the appropriate equation, such as 4.66 or 4.77, must be used.

The power requirement is then that for compression of the gas from pressure P_1 to P_3 and for imparting the necessary kinetic energy to it. Under normal conditions, however, the kinetic energy term is negligible. Thus for an isothermal efficiency of compression η, the power required is:

$$\mathbf{P} = \frac{1}{\eta} G P_1 v_1 \ln \frac{P_3}{P_1} \tag{8.71}$$

Example 8.10

Hydrogen is pumped from a reservoir at 2 MN/m^2 through a clean horizontal mild steel pipe 50 mm in diameter and 500 m long. The downstream pressure is also 2 MN/m^2 and the pressure of the gas is raised to 2.5 MN/m^2 by a pump at the upstream end of the pipe. The conditions of flow are isothermal and the temperature of the gas is 295 K. What is the flowrate and what is the effective rate of working of the pump if it operates with an efficiency of 60 per cent?

Viscosity of hydrogen = 0.009 mN s/m^2 at 295 K.

Solution

Viscosity of hydrogen = 0.009 mN s/m^2 or 9×10^{-6} N s/m^2

Density of hydrogen at the mean pressure of 2.25 MN/m^2

$$= \left(\frac{2}{22.4}\right)\left(\frac{2250}{101.3}\right)\left(\frac{273}{295}\right) = 1.83 \text{ kg/m}^3$$

Firstly, an approximate value of G is obtained by neglecting the kinetic energy of the fluid. Taking P_1 and P_2 as the pressures at the upstream and downstream ends of the pipe, 2.5×10^6 and 2.0×10^6 N/m^2, then in equation 4.56:

$$P_1 - P_2 = 4\phi\frac{l}{d}\rho_m u_m^2$$

or:

$$0.5 \times 10^6 = 4\phi\left(\frac{500}{0.050}\right) \times 1.83u_m^2 \text{ N/m}^2$$

$$\therefore \quad \phi u_m^2 = \frac{R}{\rho u^2}u_m^2 = 6.83$$

$$\therefore \quad \phi Re^2 = \frac{R}{\rho u^2}Re^2 = \frac{6.83 \times 0.050^2 \times 1.83^2}{(9 \times 10^{-6})^2}$$

$$= 7.02 \times 10^8$$

Taking the roughness of the pipe surface, e as 0.00005 m:

$$\frac{e}{d} = 0.001 \text{ and } Re = 5.7 \times 10^5 \text{ from Figure 3.8}$$

$$\therefore \quad \frac{4G}{\pi\mu d} = 5.7 \times 10^5$$

$$\therefore \quad G = (5.7 \times 10^5 \times \frac{\pi}{4} \times 9 \times 10^{-6} \times 0.050)$$

$$= 0.201 \text{ kg/s}$$

From Figure 3.7, $(\phi = R/\rho u^2) = 0.0024$

Taking the kinetic energy of the fluid into account, equation 4.56 may be used:

$$\left(\frac{G}{A}\right)^2 \ln\frac{P_1}{P_2} + (P_2 - P_1)\rho_m + 4\phi\frac{l}{d}\left(\frac{G}{A}\right)^2 = 0$$

Using the value of (ϕ) obtained by neglecting the kinetic energy:

$$\left(\frac{G}{A}\right)^2 \ln\frac{2.5}{2.0} - (0.5 \times 10^6 \times 1.83) + \left[4 \times 0.0024 \times \left(\frac{500}{0.05}\right)\right]\left(\frac{G}{A}\right)^2 = 0$$

$$0.223\left(\frac{G}{A}\right)^2 - 915{,}000 + 96.0\left(\frac{G}{A}\right)^2 = 0$$

$$\therefore \quad \frac{G}{A} = 97.5$$

$$\therefore \quad G = \left(97.5 \times \frac{\pi}{4} \times 0.050^2\right)$$

$$= 0.200 \text{ kg/s}$$

Thus, as is commonly the case, when the pressure drop is a relatively small proportion of the total pressure, the change in kinetic energy is negligible compared with the frictional losses. This would not be true had the pressure drop been much greater.

The power requirements for the pump may now be calculated from equation 8.71.

$$\text{Power} = \frac{GP_m v_m \ln(P_1/P_2)}{\eta}$$

$$= \frac{(0.200 \times 2.25 \times 10^6 \times (1/1.83) \times 0.223)}{0.60}$$

$$= 9.14 \times 10^4 \text{ W or } \underline{\underline{91.4 \text{ kW}}}$$

8.7 FURTHER READING

ANDERSON, J. D. Jr.: *Modern Compressible Flow.* (McGraw-Hill, New York, 1982).

ENGINEERING EQUIPMENT USERS' ASSOCIATION: *Vacuum Producing Equipment*, EEUA Handbook No. 11 (Constable, London, 1961).

ENGINEERING EQUIPMENT USERS' ASSOCIATION: *Electrically Driven Glandless Pumps*, EEUA Handbook No. 26 (Constable, London, 1968).

ENGINEERING EQUIPMENT USERS' ASSOCIATION: *Guide to the Selection of Rotodynamic Pumps*, EEUA Handbook No. 30 (Constable, London, 1972).

GREENE, W. (ed.): *The Chemical Engineering Guide to Compressors* (McGraw-Hill, New York, 1985).

HOLLAND, F. A. and CHAPMAN, F. S.: *Pumping of Liquids* (Reinhold, New York, 1966).

M. W. KELLOGG CO.: *Design of Piping Systems*, 2nd edn (Wiley, Chichester, 1964).

LAZARKIENICZ, S. and TROSKOLANSKI, A. T.: *Impeller Pumps* (Pergamon Press, Oxford, 1965).

MCNAUGHTON, K. (ed.): *The Chemical Engineering Guide to Pumps* (McGraw-Hill, New York, 1985).

PERRY, R. H. and GREEN, D. W (eds): *Perry's Chemical Engineers Handbook.* 7th edn (McGraw-Hill, New York, 1998).

STREETER, V. L. and WYLIE, E. B.: *Fluid Mechanics* (McGraw-Hill, New York, 1985).

TURTON, R. K.: *An Introductory Guide to Pumps and Pumping Systems* (Mechanical Engineering Publications, London, 1993)

WARRING, R. H.: *Seals and Sealing Handbook* (Trade and Technical Press, 1983).

WARRING, R. H.: *Pumps Selection, Systems and Applications* (Trade and Technical Press, 1979).

8.8 REFERENCES

1. MARSHALL, P.: *The Chemical Engineer (London)* No. 418 (1985) 52. Positive displacement pumps—a brief survey.
2. HARVEST, J.: *The Chemical Engineer (London)* No. 403 (1984) 28. Recent developments in gear pumps.
3. GADSDEN, C.: *The Chemical Engineer (London)* No. 404 (1984) 42. Squeezing the best from peristaltic pumps.
4. STEELE, K. and ODROWAZ-PIENIAZEK, S.: *The Chem. Engr.* No. 422, (1986) 34 and No. 423, (1986) 30. Advances in slurry pumps—Parts 1 and 2.
5. DAVIDSON, J. (ed.): *Process Pump Selection—a Systems Approach* (I. Mech. E., Bury St Edmunds, Suffolk, 1986).
6. RICHARDSON, J. F. and HIGSON, D. J. *Trans. Inst. Chem. Eng.* **40** (1962) 169. A study of the energy losses associated with the operation of an air-lift pump.
7. ARMAND, A. A.: USAEC translation (AEC-TR-4490 page 19). *Hydrodynamics and Heat Transfer during Boiling in High Pressure Boilers.* U.S.S.R. Acad. of Sci. (Moscow, 1995).

8. SCHMIDT, E., BEHRINGER, P. and SCHURIG, W.: *Forsch. a. d. Gebiete d. Ing Ausgabe B. Forsch* 365 (1934). Wasserumlauf in Dampfkesseln.
9. GOVIER, G. W., RADFORD, B. A. and DUNN, J. S. C.: *Canad. J. Chem. Eng.* **35** (1957) 58. The upwards vertical flow of air-water mixtures I. Effect of air and water rates on flow pattern, hold-up and pressure drop.
10. GOVIER, G. W. and SHORT, W. L. *Canad. J. Chem. Eng.* **36** (1958) 195. The upward vertical flow of air-water mixtures II. Effect of tubing diameter on flow pattern, hold-up and pressure drop.
11. ISBIN, H. S., SHER, N. C. and EDDY, K. C.: *A.I.Ch.E.Jl* **3** (1957) 136. Void fractions in two-phase steam—water flow.
12. MOORE, T. V. and WILDE, H. D.: *Trans. Amer. Inst. Min. Met. Eng.: (Pet. Div.)* **92** (1931) 296. Experimental measurement of slippage in flow through vertical pipes.

8.9. NOMENCLATURE

		Units in SI system	Dimensions in **M, L, T, θ**
A	Cross-sectional area	m^2	$\mathbf{L^2}$
b	Width of pump impeller	m	**L**
c	Clearance in cylinder, i.e. ratio clearance volume/swept volume	—	—
d	Diameter	m	**L**
F	Energy degraded due to irreversibility per unit mass of fluid	J/kg	$\mathbf{L^2T^{-2}}$
G	Mass flow rate	kg/s	$\mathbf{MT^{-1}}$
G_A	Mass flowrate of air	kg/s	$\mathbf{MT^{-1}}$
G_L	Mass flowrate of liquid	kg/s	$\mathbf{MT^{-1}}$
g	Acceleration due to gravity	m/s^2	$\mathbf{LT^{-2}}$
H	Enthalpy per unit mass	J/kg	$\mathbf{L^2T^{-2}}$
h	Head	m	**L**
h_a	Heat equivalent to atmospheric pressure	m	**L**
h_f	Friction head	m	**L**
h_i	Head at pump inlet	m	**L**
h_0	Height of liquid above pump inlet	m	**L**
h_r	Height through which liquid is raised	m	**L**
h_s	Submergence of air lift pump	m	**L**
K	Coefficient	$(N/m^2)/(m^3/s)$	$\mathbf{ML^{-4}T^{-1}}$
K^1	Coefficient	$(N/m^2)/(m^3/s)^2$	$\mathbf{ML^{-7}}$
l	Length	m	**L**
N	Revolutions per unit time	Hz	$\mathbf{T^{-1}}$
N_s	Specific speed of a pump	—	—
n	Number of stages of compression	—	—
P	Pressure	N/m^2	$\mathbf{ML^{-1}T^{-2}}$
ΔP	Pressure difference	N/m^2	$\mathbf{ML^{-1}T^{-2}}$
P_a	Atmospheric pressure	N/m^2	$\mathbf{ML^{-1}T^{-2}}$
P_i	Intermediate pressure	N/m^2	$\mathbf{ML^{-1}T^{-2}}$
P_v	Vapour pressure of liquid	N/m^2	$\mathbf{ML^{-1}T^{-2}}$
P_0	Pressure at pump suction tank	N/m^2	$\mathbf{ML^{-1}T^{-2}}$
$\mathbf{P}$	Power	W	$\mathbf{ML^2T^{-3}}$
Q	Volumetric rate of flow	m^3/s	$\mathbf{L^3T^{-1}}$
Q_m	Maximum rate of flow	m^3/s	$\mathbf{L^3T^{-1}}$
$\dot{Q}$	Rate of flow at time t	m^3/s	$\mathbf{L^3T^{-1}}$
q	Heat added from surrounding per unit mass of fluid	J/kg	$\mathbf{L^2T^{-2}}$
R	Shear stress at pipe wall	N/m^2	$\mathbf{ML^{-1}T^{-2}}$
r	Radius	m	**L**
S	Entropy per unit mass	J/kg K	$\mathbf{L^2T^{-2}\theta^{-1}}$
T	Temperature	K	$\boldsymbol{\theta}$
t	Time	s	**T**
U	Internal energy per unit mass	J/kg	$\mathbf{L^2T^{-2}}$
u	Velocity of fluid	m/s	$\mathbf{LT^{-1}}$

		Units in SI system	Dimensions in **M, L, T, θ**
u_G	Superficial gas velocity	m/s	$\mathbf{LT^{-1}}$
u_L	Superficial liquid velocity	m/s	$\mathbf{LT^{-1}}$
u_i	Velocity at pump inlet	m/s	$\mathbf{LT^{-1}}$
u_t	Tangential velocity of tip of vane of compressor or pump	m/s	$\mathbf{LT^{-1}}$
u_v	Velocity of fluid relative to vane of compressor	m/s	$\mathbf{LT^{-1}}$
u_1	Velocity at inlet to centrifugal pump	m/s	$\mathbf{LT^{-1}}$
u_2	Velocity at outlet to centrifugal pump	m/s	$\mathbf{LT^{-1}}$
V	Volume	m^3	$\mathbf{L^3}$
V_s	Swept volume	m^3	$\mathbf{L^3}$
v	Specific volume	m^3/kg	$\mathbf{M^{-1}L^3}$
v_a	Specific volume at atmospheric pressure	m^3/kg	$\mathbf{M^{-1}L^3}$
W	Net work done per unit mass	J/kg	$\mathbf{L^2T^{-2}}$
W_s	Shaft work per unit mass	J/kg	$\mathbf{L^2T^{-2}}$
W'	Work of compressor per cycle	J	$\mathbf{ML^2T^{-2}}$
Z	Net positive suction head	m	**L**
z	Height	m	**L**
α	Correction factor for kinetic energy of fluid	—	—
β	Angle between tangential direction and blade of impeller at its tip	—	—
Δ	Finite difference in quantity	—	—
η	Efficiency	—	—
ϵ_G	Mean gas holdup (fraction)	—	—
ϵ_L	Mean liquid holdup (fraction)	—	—
γ	Ratio of specific heat at constant pressure to specific heat at constant volume	—	—
ω	Angular velocity	rad/s	$\mathbf{T^{-1}}$
Π	Dimensionless group	—	—
Φ_G^2	Ratio $-\Delta P_{TPF}/-\Delta P_L$		
ϕ	Friction factor $(R/\rho u^2)$	—	—
ρ	Density	kg/m^3	$\mathbf{ML^{-3}}$
θ	Angle between tangential direction and direction of motion of fluid	—	—
τ	Torque	Nm	$\mathbf{ML^2T^{-2}}$

Suffixes

G	Gas
L	Liquid
TPF	Two-phase friction

PART 2

Heat Transfer

CHAPTER 9

Heat Transfer

9.1. INTRODUCTION

In the majority of chemical processes heat is either given out or absorbed, and fluids must often be either heated or cooled in a wide range of plant, such as furnaces, evaporators, distillation units, dryers, and reaction vessels where one of the major problems is that of transferring heat at the desired rate. In addition, it may be necessary to prevent the loss of heat from a hot vessel or pipe system. The control of the flow of heat at the desired rate forms one of the most important areas of chemical engineering. Provided that a temperature difference exists between two parts of a system, heat transfer will take place in one or more of three different ways.

Conduction. In a solid, the flow of heat by conduction is the result of the transfer of vibrational energy from one molecule to another, and in fluids it occurs in addition as a result of the transfer of kinetic energy. Heat transfer by conduction may also arise from the movement of free electrons, a process which is particularly important with metals and accounts for their high thermal conductivities.

Convection. Heat transfer by convection arises from the mixing of elements of fluid. If this mixing occurs as a result of density differences as, for example, when a pool of liquid is heated from below, the process is known as *natural convection.* If the mixing results from eddy movement in the fluid, for example when a fluid flows through a pipe heated on the outside, it is called *forced convection.* It is important to note that convection requires mixing of fluid elements, and is not governed by temperature difference alone as is the case in conduction and radiation.

Radiation. All materials radiate thermal energy in the form of electromagnetic waves. When this radiation falls on a second body it may be partially reflected, transmitted, or absorbed. It is only the fraction that is absorbed that appears as heat in the body.

9.2. BASIC CONSIDERATIONS

9.2.1. Individual and overall coefficients of heat transfer

In many of the applications of heat transfer in process plants, one or more of the mechanisms of heat transfer may be involved. In the majority of heat exchangers heat passes through a series of different intervening layers before reaching the second fluid (Figure 9.1). These layers may be of different thicknesses and of different thermal conductivities. The problem of transferring heat to crude oil in the primary furnace before it enters the first distillation column may be considered as an example. The heat from the flames passes by radiation and convection to the pipes in the furnace, by conduction through the

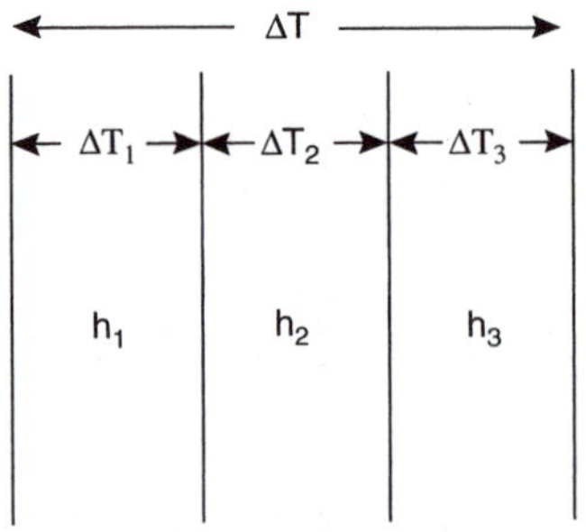

Figure 9.1. Heat transfer through a composite wall

pipe walls, and by forced convection from the inside of the pipe to the oil. Here all three modes of transfer are involved. After prolonged usage, solid deposits may form on both the inner and outer walls of the pipes, and these will then contribute additional resistance to the transfer of heat. The simplest form of equation which represents this heat transfer operation may be written as:

$$Q = UA\Delta T \tag{9.1}$$

where Q is the heat transferred per unit time, A the area available for the flow of heat, ΔT the difference in temperature between the flame and the boiling oil, and U is known as the overall heat transfer coefficient (W/m^2 K in SI units).

At first sight, equation 9.1 implies that the relationship between Q and ΔT is linear. Whereas this is approximately so over limited ranges of temperature difference for which U is nearly constant, in practice U may well be influenced both by the temperature difference and by the absolute value of the temperatures.

If it is required to know the area needed for the transfer of heat at a specified rate, the temperature difference ΔT, and the value of the overall heat-transfer coefficient must be known. Thus the calculation of the value of U is a key requirement in any design problem in which heating or cooling is involved. A large part of the study of heat transfer is therefore devoted to the evaluation of this coefficient.

The value of the coefficient will depend on the mechanism by which heat is transferred, on the fluid dynamics of both the heated and the cooled fluids, on the properties of the materials through which the heat must pass, and on the geometry of the fluid paths. In solids, heat is normally transferred by conduction; some materials such as metals have a high thermal conductivity, whilst others such as ceramics have a low conductivity. Transparent solids like glass also transmit radiant energy particularly in the visible part of the spectrum.

Liquids also transmit heat readily by conduction, though circulating currents are frequently set up and the resulting convective transfer may be considerably greater than the transfer by conduction. Many liquids also transmit radiant energy. Gases are poor conductors of heat and circulating currents are difficult to suppress; convection is therefore much more important than conduction in a gas. Radiant energy is transmitted with only limited absorption in gases and, of course, without any absorption *in vacuo*. Radiation is the only mode of heat transfer which does not require the presence of an intervening medium.

If the heat is being transmitted through a number of media in series, the overall heat transfer coefficient may be broken down into individual coefficients h each relating to a single medium. This is as shown in Figure 9.1. It is assumed that there is good contact between each pair of elements so that the temperature is the same on the two sides of each junction.

If heat is being transferred through three media, each of area A, and individual coefficients for each of the media are h_1, h_2, and h_3, and the corresponding temperature changes are ΔT_1, ΔT_2, and ΔT_3 then, provided that there is no accumulation of heat in the media, the heat transfer rate Q will be the same through each. Three equations, analogous to equation 9.1 can therefore be written:

$$\left.\begin{aligned} Q &= h_1 A \Delta T_1 \\ Q &= h_2 A \Delta T_2 \\ Q &= h_3 A \Delta T_3 \end{aligned}\right\} \tag{9.2}$$

Rearranging:

$$\Delta T_1 = \frac{Q}{A}\frac{1}{h_1}$$

$$\Delta T_2 = \frac{Q}{A}\frac{1}{h_2}$$

$$\Delta T_3 = \frac{Q}{A}\frac{1}{h_3}$$

Adding:

$$\Delta T_1 + \Delta T_2 + \Delta T_3 = \frac{Q}{A}\left(\frac{1}{h_1} + \frac{1}{h_2} + \frac{1}{h_3}\right) \tag{9.3}$$

Noting that $(\Delta T_1 + \Delta T_2 + \Delta T_3)$ = total temperature difference ΔT:

then:

$$\Delta T = \frac{Q}{A}\left(\frac{1}{h_1} + \frac{1}{h_2} + \frac{1}{h_3}\right) \tag{9.4}$$

From equation 9.1:

$$\Delta T = \frac{Q}{A}\frac{1}{U} \tag{9.5}$$

Comparing equations 9.4 and 9.5:

$$\frac{1}{U} = \frac{1}{h_1} + \frac{1}{h_2} + \frac{1}{h_3} \tag{9.6}$$

The reciprocals of the heat transfer coefficients are resistances, and equation 9.6 therefore illustrates that the resistances are additive.

In some cases, particularly for the radial flow of heat through a thick pipe wall or cylinder, the area for heat transfer is a function of position. Thus the area for transfer applicable to each of the three media could differ and may be A_1, A_2 and A_3. Equation 9.3 then becomes:

$$\Delta T_1 + \Delta T_2 + \Delta T_3 = Q\left(\frac{1}{h_1 A_1} + \frac{1}{h_2 A_2} + \frac{1}{h_3 A_3}\right) \tag{9.7}$$

Equation 9.7 must then be written in terms of one of the area terms A_1, A_2, and A_3, or sometimes in terms of a mean area. Since Q and ΔT must be independent of the particular area considered, the value of U will vary according to which area is used as the basis. Thus equation 9.7 may be written, for example:

$$Q = U_1 A_1 \Delta T \quad \text{or} \quad \Delta T = \frac{Q}{U_1 A_1}$$

h = W/m²·K

This will then give U_1 as:

$$\frac{1}{U_1} = \frac{1}{h_1} + \frac{A_1}{A_2}\left(\frac{1}{h_2}\right) + \frac{A_1}{A_3}\left(\frac{1}{h_3}\right) \tag{9.8}$$

In this analysis it is assumed that the heat flowing per unit time through each of the media is the same.

Now that the overall coefficient U has been broken down into its component parts, each of the individual coefficients h_1, h_2, and h_3 must be evaluated. This can be done from a knowledge of the nature of the heat transfer process in each of the media. A study will therefore be made of how these individual coefficients can be calculated for conduction, convection, and radiation.

h = heat transfer coeff.

9.2.2. Mean temperature difference

Where heat is being transferred from one fluid to a second fluid through the wall of a vessel and the temperature is the same throughout the bulk of each of the fluids, there is no difficulty in specifying the overall temperature difference ΔT. Frequently, however, each fluid is flowing through a heat exchanger such as a pipe or a series of pipes in parallel, and its temperature changes as it flows, and consequently the temperature difference is continuously changing. If the two fluids are flowing in the same direction *(co-current flow)*, the temperatures of the two streams progressively approach one another as shown in Figure 9.2. In these circumstances the outlet temperature of the heating fluid must always be higher than that of the cooling fluid. If the fluids are flowing in opposite directions (*countercurrent flow*), the temperature difference will show less variation throughout the heat exchanger as shown in Figure 9.3. In this case it is possible for the cooling liquid to leave at a higher temperature than the heating liquid, and one of the great advantages of

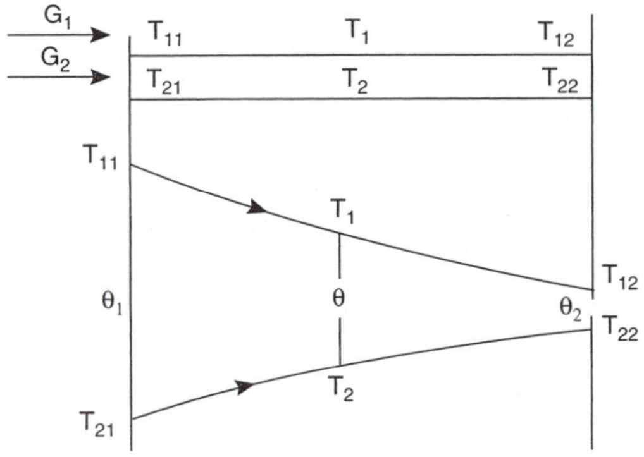

Figure 9.2. Mean temperature difference for co-current flow

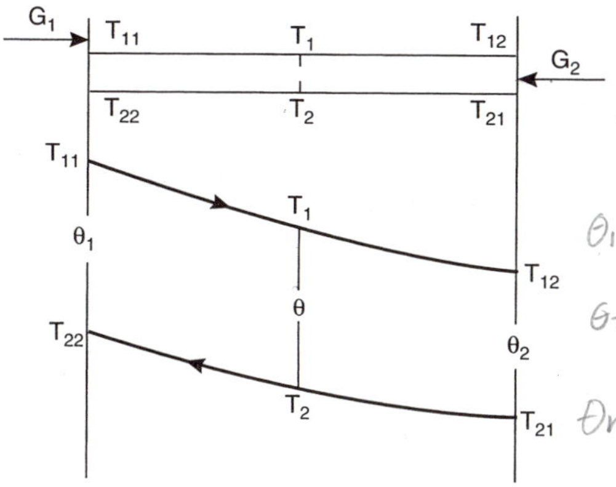

Assume
T @ K_1 constant
$\theta_1 = (T_2 - T_1)$
$\theta_2 = (T_2 - T_1)$
$\theta_m = \dfrac{\theta_1 - \theta_2}{\ln\left(\frac{\theta_1}{\theta_2}\right)}$

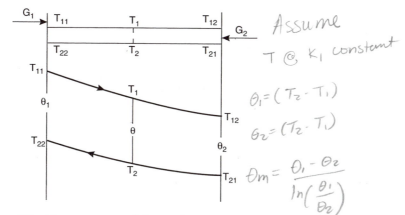

Figure 9.3. Mean temperature difference for countercurrent flow

countercurrent flow is that it is possible to extract a higher proportion of the heat content of the heating fluid. The calculation of the appropriate value of the temperature difference for co-current and for countercurrent flow is now considered. It is assumed that the overall heat transfer coefficient U remains constant throughout the heat exchanger.

It is necessary to find the average value of the temperature difference θ_m to be used in the general equation:

$$Q = UA\theta_m \qquad \text{(equation 9.1)}$$

Figure 9.3 shows the temperature conditions for the fluids flowing in opposite directions, a condition known as countercurrent flow.

The outside stream specific heat C_{p1} and mass flow rate G_1 falls in temperature from T_{11} to T_{12}.

The inside stream specific heat C_{p2} and mass flow rate G_2 rises in temperature from T_{21} to T_{22}.

Over a small element of area dA where the temperatures of the streams are T_1 and T_2. The temperature difference:

$$\theta = T_1 - T_2$$

$$\therefore \quad d\theta = dT_1 - dT_2$$

Heat given out by the hot stream $= dQ = -G_1C_{p1}\,dT_1$

Heat taken up by the cold stream $= dQ = G_2C_{p2}\,dT_2$

$$\therefore \quad d\theta = -\frac{dQ}{G_1C_{p1}} - \frac{dQ}{G_2C_{p2}} = -dQ\left(\frac{G_1C_{p1} + G_2C_{p2}}{G_1C_{p1} \times G_2C_{p2}}\right) = -\psi\,dQ \quad \text{(say)}$$

$$\therefore \quad \theta_1 - \theta_2 = \psi Q$$

Over this element:

$$U\,dA\theta = dQ$$

$$\therefore \quad U\,dA\theta = -\frac{d\theta}{\psi}$$

If U may be taken as constant:

$$-\psi U \int_0^A \mathrm{d}A = \int_{\theta_1}^{\theta_2} \frac{\mathrm{d}\theta}{\theta}$$

$$\therefore \quad -\psi UA = -\ln\frac{\theta_1}{\theta_2}$$

From the definition of θ_m, $Q = UA\theta_m$.

$$\therefore \quad \theta_1 - \theta_2 = \psi Q = \psi UA\theta_m = \ln\frac{\theta_1}{\theta_2}(\theta_m)$$

and:

$$\theta_m = \frac{\theta_1 - \theta_2}{\ln(\theta_1/\theta_2)} \tag{9.9}$$

where θ_m is known as the *logarithmic mean temperature difference*.

UNDERWOOD[1] proposed the following approximation for the logarithmic mean temperature difference:

$$(\theta_m)^{1/3} = \tfrac{1}{2}(\theta_1^{1/3} + \theta_2^{1/3}) \tag{9.10}$$

and, for example, when $\theta_1 = 1$ K and $\theta_2 = 100$ K, θ_m is 22.4 K compared with a logarithmic mean of 21.5 K. When $\theta_1 = 10$ K and $\theta_2 = 100$ K, both the approximation and the logarithmic mean values coincide at 39 K.

If the two fluids flow in the same direction on each side of a tube, co-current flow is taking place and the general shape of the temperature profile along the tube is as shown in Figure 9.2. A similar analysis will show that this gives the same expression for θ_m, the logarithmic mean temperature difference. For the same terminal temperatures it is important to note that the value of θ_m for countercurrent flow is appreciably greater than the value for co-current flow. This is seen from the temperature profiles, where with co-current flow the cold fluid cannot be heated to a higher temperature than the exit temperature of the hot fluid as illustrated in Example 9.1.

Example 9.1

A heat exchanger is required to cool 20 kg/s of water from 360 K to 340 K by means of 25 kg/s water entering at 300 K. If the overall coefficient of heat transfer is constant at 2 kW/m^{2}K, calculate the surface area required in (a) a countercurrent concentric tube exchanger, and (b) a co-current flow concentric tube exchanger.

Solution

Heat load: $Q = 20 \times 4.18(360 - 340) = 1672$ kW

The cooling water outlet temperature is given by:

$$1672 = 25 \times 4.18(\theta_2 - 300) \quad \text{or} \quad \theta_2 = 316 \text{ K}$$

(a) *Counterflow*

In equation 9.9:

$$\theta_m = \frac{44 - 40}{\ln(44/40)} = 41.9 \text{ K}$$

Heat transfer area:

$$A = \frac{Q}{U\theta_m}$$

$$= \frac{1672}{2 \times 41.9}$$

$$= \underline{\underline{19.95\ \text{m}^2}}$$

(b) *Co-current flow*

In equation 9.9:

$$\theta_m = \frac{60 - 24}{\ln(60/24)} = 39.3\ \text{K}$$

Heat transfer area:

$$A = \frac{1672}{2 \times 39.3}$$

$$= \underline{\underline{21.27\ \text{m}^2}}$$

It may be noted that using Underwood's approximation (equation 9.10), the calculated values for the mean temperature driving forces are 41.9 K and 39.3 K for counter- and co-current flow respectively, which agree exactly with the logarithmic mean values.

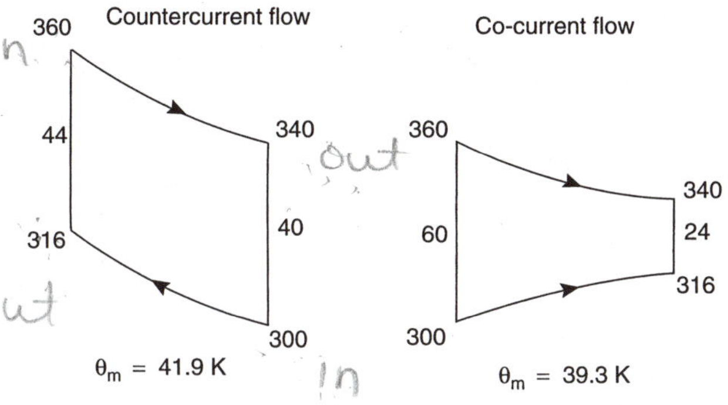

Figure 9.4. Data for Example 9.1

9.3. HEAT TRANSFER BY CONDUCTION

9.3.1. Conduction through a plane wall

This important mechanism of heat transfer is now considered in more detail for the flow of heat through a plane wall of thickness x as shown in Figure 9.5.

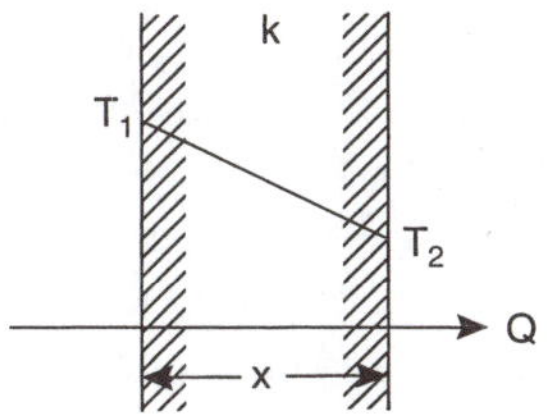

Figure 9.5. Conduction of heat through a plane wall

The rate of heat flow Q over the area A and a small distance $\mathrm{d}x$ may be written as:

$$Q = -kA\left(\frac{\mathrm{d}T}{\mathrm{d}x}\right) \tag{9.11}$$

which is often known as *Fourier's equation*, where the negative sign indicates that the temperature gradient is in the opposite direction to the flow of heat and k is the thermal conductivity of the material. Integrating for a wall of thickness x with boundary temperatures T_1 and T_2, as shown in Figure 9.5:

$$Q = \frac{kA(T_1 - T_2)}{x} \tag{9.12}$$

Thermal conductivity is a function of temperature and experimental data may often be expressed by a linear relationship of the form:

$$k = k_0(1 + k'T) \tag{9.13}$$

where k is the thermal conductivity at the temperature T and k_0 and k' are constants. Combining equations 9.11 and 9.13:

$$-k\,\mathrm{d}T = -k_0(1 + k'T)\mathrm{d}T = \frac{Q\,\mathrm{d}x}{A}$$

Integrating between the temperature limits T_1 and T_2,

$$-\int_{T_1}^{T_2} k\,\mathrm{d}T = (T_1 - T_2)k_0\left\{1 + k'\left(\frac{T_1 + T_2}{2}\right)\right\} = Q\int_{x_1}^{x_2}\frac{\mathrm{d}x}{A} \tag{9.14}$$

Where k is a linear function of T, the following equation may therefore be used:

$$k_a(T_1 - T_2) = Q\int_{x_1}^{x_2}\frac{\mathrm{d}x}{A} \tag{9.15}$$

where k_a is the arithmetic mean of k_1 and k_2 at T_1 and T_2 respectively or the thermal conductivity at the arithmetic mean of T_1 and T_2.

Where k is a non-linear function of T, some mean value, k_m will apply, where:

$$k_m = \frac{1}{T_2 - T_1}\int_{T_1}^{T_2} k\,\mathrm{d}T \tag{9.16}$$

From Table 9.1 it will be seen that metals have very high thermal conductivities, non-metallic solids lower values, non-metallic liquids low values, and gases very low values. It is important to note that amongst metals, stainless steel has a low value, that water has a very high value for liquids (due to partial ionisation), and that hydrogen has a high value for gases (due to the high mobility of the molecules). With gases, k decreases with increase in molecular mass and increases with the temperature. In addition, for gases the dimensionless *Prandtl group* $C_p\mu/k$, which is approximately constant (where C_p is the specific heat at constant pressure and μ is the viscosity), can be used to evaluate k at high temperatures where it is difficult to determine a value experimentally because of the formation of convection currents. k does not vary significantly with pressure, except where this is reduced to a value so low that the mean free path of the molecules becomes

Table 9.1. Thermal conductivities of selected materials

	Temp (K)	k (Btu/h ft^2 °F/ft)	k (W/m^{2}K)		Temp (K)	k (Btu/h ft^2 °F/ft)	k (W/m^{2}K)
Solids—Metals				*Liquids*			
Aluminium	573	133	230	Acetic acid 50%	293	0.20	0.35
Cadmium	291	54	94	Acetone	303	0.10	0.17
Copper	373	218	377	Aniline	273–293	0.1	0.17
Iron (wrought)	291	35	61	Benzene	303	0.09	0.16
Iron (cast)	326	27.6	48	Calcium chloride			
				brine 30%	303	0.32	0.55
Lead	373	19	33	Ethyl alcohol 80%	293	0.137	0.24
Nickel	373	33	57	Glycerol 60%	293	0.22	0.38
Silver	373	238	412	Glycerol 40%	293	0.26	0.45
Steel 1% C	291	26	45	n-Heptane	303	0.08	0.14
Tantalum	291	32	55	Mercury	301	4.83	8.36
				Sulphuric acid 90%	303	0.21	0.36
Admiralty metal	303	65	113	Sulphuric acid 60%	303	0.25	0.43
Bronze	—	109	189	Water	303	0.356	0.62
Stainless Steel	293	9.2	16	Water	333	0.381	0.66
Solids—Non-metals				*Gases*			
Asbestos sheet	323	0.096	0.17	Hydrogen	273	0.10	0.17
Asbestos	273	0.09	0.16	Carbon dioxide	273	0.0085	0.015
Asbestos	373	0.11	0.19	Air	273	0.014	0.024
Asbestos	473	0.12	0.21	Air	373	0.018	0.031
Bricks (alumina)	703	1.8	3.1	Methane	273	0.017	0.029
Bricks (building)	293	0.4	0.69	Water vapour	373	0.0145	0.025
Magnesite	473	2.2	3.8	Nitrogen	273	0.0138	0.024
Cotton wool	303	0.029	0.050	Ethylene	273	0.0097	0.017
Glass	303	0.63	1.09	Oxygen	273	0.0141	0.024
Mica	323	0.25	0.43	Ethane	273	0.0106	0.018
Rubber (hard)	273	0.087	0.15				
Sawdust	293	0.03	0.052				
Cork	303	0.025	0.043				
Glass wool	—	0.024	0.041				
85% Magnesia	—	0.04	0.070				
Graphite	273	87	151				

comparable with the dimensions of the vessel; further reduction of pressure then causes k to decrease.

Typical values for Prandtl numbers are as follows:

Air	0.71	n-Butanol	50
Oxygen	0.63	Light oil	600
Ammonia (gas)	1.38	Glycerol	1000
Water	5–10	Polymer melts	10,000
		Mercury	0.02

The low conductivity of heat insulating materials, such as cork, glass wool, and so on, is largely accounted for by their high proportion of air space. The flow of heat through such materials is governed mainly by the resistance of the air spaces, which should be sufficiently small for convection currents to be suppressed.

It is convenient to rearrange equation 9.12 to give:

$$Q = \frac{(T_1 - T_2)A}{(x/k)} \tag{9.17}$$

where x/k is known as the *thermal resistance* and k/x is the *transfer coefficient*.

Example 9.2.

Estimate the heat loss per square metre of surface through a brick wall 0.5 m thick when the inner surface is at 400 K and the outside surface is at 300 K. The thermal conductivity of the brick may be taken as 0.7 W/mK.

Solution

From equation 9.12:

$$Q = \frac{0.7 \times 1 \times (400 - 300)}{0.5}$$

$$= \underline{\underline{140 \text{ W/m}^2}}$$

9.3.2. Thermal resistances in series

It has been noted earlier that thermal resistances may be added together for the case of heat transfer through a complete section formed from different media in series.

Figure 9.6 shows a composite wall made up of three materials with thermal conductivities k_1, k_2, and k_3, with thicknesses as shown and with the temperatures T_1, T_2, T_3, and T_4 at the faces. Applying equation 9.12 to each section in turn, and noting that the same quantity of heat Q must pass through each area A:

$$T_1 - T_2 = \frac{x_1}{k_1 A}Q, \; T_2 - T_3 = \frac{x_2}{k_2 A}Q \quad \text{and} \quad T_3 - T_4 = \frac{x_3}{k_3 A}Q$$

On addition:

$$(T_1 - T_4) = \left(\frac{x_1}{k_1 A} + \frac{x_2}{k_2 A} + \frac{x_3}{k_3 A}\right) Q \tag{9.18}$$

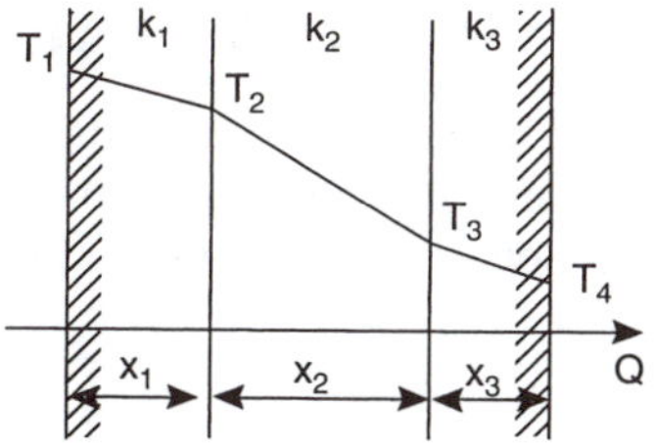

Figure 9.6. Conduction of heat through a composite wall

or:
$$Q = \frac{T_1 - T_4}{\Sigma(x_1/k_1 A)}$$
$$= \frac{\text{Total driving force}}{\text{Total (thermal resistance/area)}} \qquad (9.19)$$

Example 9.3

A furnace is constructed with 0.20 m of firebrick, 0.10 m of insulating brick, and 0.20 m of building brick. The inside temperature is 1200 K and the outside temperature is 330 K. If the thermal conductivities are as shown in Figure 9.7, estimate the heat loss per unit area and the temperature at the junction of the firebrick and the insulating brick.

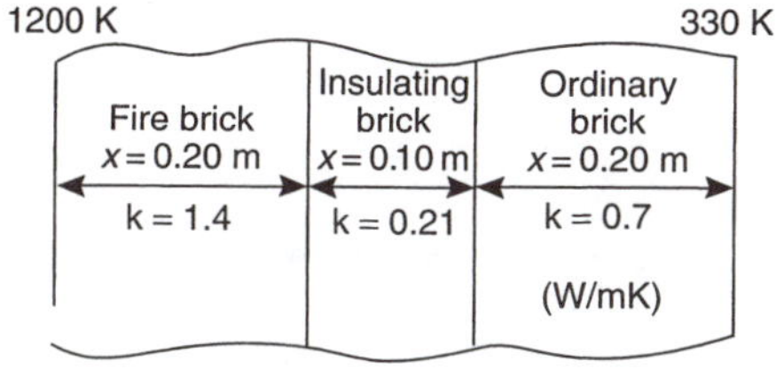

Figure 9.7. Data for Example 9.3

Solution

From equation 9.19:
$$Q = (1200 - 330)\Big/\left[\left(\frac{0.20}{1.4 \times 1}\right) + \left(\frac{0.10}{0.21 \times 1}\right) + \left(\frac{0.20}{0.7 \times 1}\right)\right]$$
$$= \frac{870}{(0.143 + 0.476 + 0.286)} = \frac{870}{0.905}$$
$$= \underline{\underline{961\ \text{W/m}^2}}$$

The ratio (Temperature drop over firebrick)/(Total temperature drop) = (0.143/0.905)

$$\therefore \qquad \text{Temperature drop over firebrick} = \left(\frac{870 \times 0.143}{0.905}\right) = 137 \text{ deg K}$$

Hence the temperature at the firebrick-insulating brick interface = (1200 − 137) = $\underline{\underline{1063\ \text{K}}}$

9.3.3. Conduction through a thick-walled tube

The conditions for heat flow through a thick-walled tube when the temperatures on the inside and outside are held constant are shown in Figure 9.8. Here the area for heat flow is proportional to the radius and hence the temperature gradient is inversely proportional to the radius.

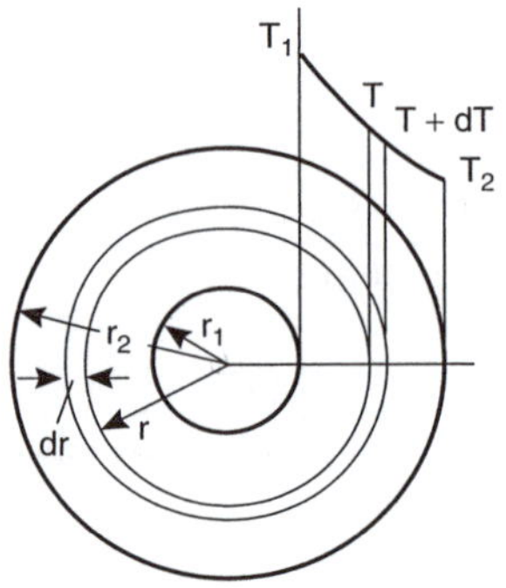

Figure 9.8. Conduction through thick-walled tube or spherical shell

The heat flow at any radius r is given by:

$$Q = -k2\pi rl \frac{dT}{dr} \tag{9.20}$$

where l is the length of tube.

Integrating between the limits r_1 and r_2:

$$Q \int_{r_1}^{r_2} \frac{dr}{r} = -2\pi lk \int_{T_1}^{T_2} dT$$

or:

$$Q = \frac{2\pi lk(T_1 - T_2)}{\ln(r_2/r_1)} \tag{9.21}$$

This equation may be put into the form of equation 9.12 to give:

$$Q = \frac{k(2\pi r_m l)(T_1 - T_2)}{r_2 - r_1} \tag{9.22}$$

where $r_m = (r_2 - r_1)/\ln(r_2/r_1)$, is known as the *logarithmic mean radius*. For thin-walled tubes the arithmetic mean radius r_a may be used, giving:

$$Q = \frac{k(2\pi r_a l)(T_1 - T_2)}{r_2 - r_1} \tag{9.23}$$

9.3.4. Conduction through a spherical shell and to a particle

For heat conduction through a spherical shell, the heat flow at radius r is given by:

$$Q = -k4\pi r^2 \frac{dT}{dr} \tag{9.24}$$

Integrating between the limits r_1 and r_2:

$$Q \int_{r_1}^{r_2} \frac{dr}{r^2} = -4\pi k \int_{T_1}^{T_2} dT$$

$$\therefore \quad Q = \frac{4\pi k(T_1 - T_2)}{(1/r_1) - (1/r_2)} \qquad (9.25)$$

An important application of heat transfer to a sphere is that of conduction through a stationary fluid surrounding a spherical particle or droplet of radius r as encountered for example in fluidised beds, rotary kilns, spray dryers and plasma devices. If the temperature difference $T_1 - T_2$ is spread over a very large distance so that $r_2 = \infty$ and T_1 is the temperature of the surface of the drop, then:

$$\frac{Qr}{(4\pi r^2)(T_1 - T_2)k} = 1$$

or:

$$\frac{hd}{k} = Nu' = 2 \qquad (9.26)$$

where $Q/4\pi r^2(T_1 - T_2) = h$ is the heat transfer coefficient, d is the diameter of the particle or droplet and hd/k is a dimensionless group known as the *Nusselt number* (Nu') for the particle. The more general use of the Nusselt number, with particular reference to heat transfer by convection, is discussed in Section 9.4. This value of 2 for the Nusselt number is the theoretical minimum for heat transfer through a *continuous* medium. It is greater if the temperature difference is applied over a finite distance, when equation 9.25 must be used. When there is relative motion between the particle and the fluid the heat transfer rate will be further increased, as discussed in Section 9.4.6.

In this approach, heat transfer to a spherical particle by conduction through the surrounding fluid has been the prime consideration. In many practical situations the flow of heat from the surface to the internal parts of the particle is of importance. For example, if the particle is a poor conductor then the rate at which the particulate material reaches some desired average temperature may be limited by conduction inside the particle rather than by conduction to the outside surface of the particle. This problem involves unsteady state transfer of heat which is considered in Section 9.3.5.

Equations may be developed to predict the rate of change of diameter d of evaporating droplets. If the latent heat of vaporisation is provided by heat conducted through a hotter stagnant gas to the droplet surface, and heat transfer is the rate controlling step, it is shown by SPALDING[(2)] that d^2 decreases linearly with time. A closely related and important practical problem is the prediction of the residence time required in a combustion chamber to ensure virtually complete burning of the oil droplets. Complete combustion is desirable to obtain maximum utilisation of energy and to minimise pollution of the atmosphere by partially burned oil droplets. Here a droplet is surrounded by a flame and heat conducted back from the flame to the droplet surface provides the heat to vaporise the oil and sustain the surrounding flame. Again d^2 decreases approximately linearly with time though the derivation of the equation is more complex due to mass transfer effects, steep temperature gradients[(3)] and circulation in the drop[(4)].

9.3.5. Unsteady state conduction

Basic considerations

In the problems which have been considered so far, it has been assumed that the conditions at any point in the system remain constant with respect to time. The case of heat transfer by conduction in a medium in which the temperature is changing with time is now considered. This problem is of importance in the calculation of the temperature distribution in a body which is being heated or cooled. If, in an element of dimensions dx by dy by dz (Figure 9.9), the temperature at the point (x, y, z) is θ and at the point $(x + \mathrm{d}x, y + \mathrm{d}y, z + \mathrm{d}z)$ is $(\theta + \mathrm{d}\theta)$, then assuming that the thermal conductivity k is constant and that no heat is generated in the medium, the rate of conduction of heat through the element is:

$$= -k\,\mathrm{d}y\,\mathrm{d}z\left(\frac{\partial \theta}{\partial x}\right)_{yz} \quad \text{in the } x\text{-direction}$$

$$= -k\,\mathrm{d}z\,\mathrm{d}x\left(\frac{\partial \theta}{\partial y}\right)_{zx} \quad \text{in the } y\text{-direction}$$

$$= -k\,\mathrm{d}x\,\mathrm{d}y\left(\frac{\partial \theta}{\partial z}\right)_{xy} \quad \text{in the } z\text{-direction}$$

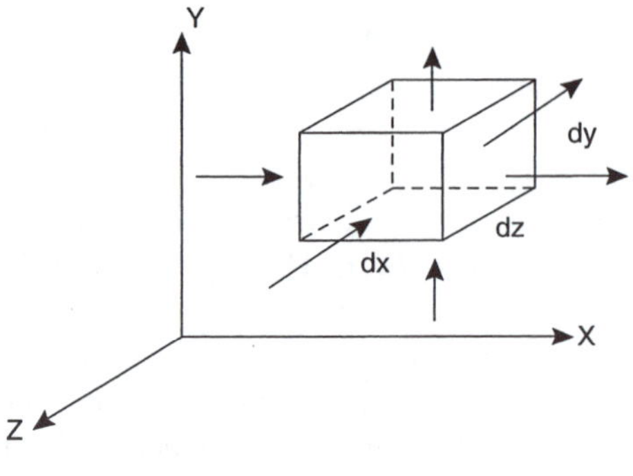

Figure 9.9. Element for heat conduction

The rate of change of heat content of the element is equal to *minus* the rate of increase of heat flow from (x, y, z) to $(x + \mathrm{d}x, y + \mathrm{d}y, z + \mathrm{d}z)$. Thus the rate of change of the heat content of the element is:

$$= k\,\mathrm{d}y\,\mathrm{d}z\left(\frac{\partial^2 \theta}{\partial x^2}\right)_{yz}\mathrm{d}x + k\,\mathrm{d}z\,\mathrm{d}x\left(\frac{\partial^2 \theta}{\partial y^2}\right)_{zx}\mathrm{d}y + k\,\mathrm{d}x\,\mathrm{d}y\left(\frac{\partial^2 \theta}{\partial z^2}\right)_{xy}\mathrm{d}z$$

$$= k\,\mathrm{d}x\,\mathrm{d}y\,\mathrm{d}z\left[\left(\frac{\partial^2 \theta}{\partial x^2}\right)_{yz} + \left(\frac{\partial^2 \theta}{\partial y^2}\right)_{zx} + \left(\frac{\partial^2 \theta}{\partial z^2}\right)_{zy}\right] \qquad (9.27)$$

The rate of increase of heat content is also equal, however, to the product of the heat capacity of the element and the rate of rise of temperature.

Thus: $$k\,\mathrm{d}x\,\mathrm{d}y\,\mathrm{d}z\left[\left(\frac{\partial^2\theta}{\partial x^2}\right)_{yz}+\left(\frac{\partial^2\theta}{\partial y^2}\right)_{zx}+\left(\right.\right.$$

or: $$\frac{\partial\theta}{\partial t}=\frac{k}{C_p\rho}\left[\left(\frac{\partial^2\theta}{\partial x^2}\right)_{yz}+\left(\frac{\partial}{\partial}\right.\right.$$

$$=D_H\left[\left(\frac{\partial^2\theta}{\partial x^2}\right)_{yz}+\left(\frac{\partial^2\theta}{\partial y^2}\right.\right.$$

where $D_H = k/C_p\rho$ is known as the *thermal diff*

This partial differential equation is most conve

transform of temperature with respect to time. As

the problem of the unidirectional flow of heat in a continuous medium will be considered. The basic differential equation for the X-direction is:

$$\frac{\partial\theta}{\partial t}=D_H\frac{\partial^2\theta}{\partial x^2} \tag{9.29}$$

This equation cannot be integrated directly since the temperature θ is expressed as a function of two independent variables, distance x and time t. The method of solution involves transforming the equation so that the Laplace transform of θ with respect to time is used in place of θ. The equation then involves only the Laplace transform $\overline{\theta}$ and the distance x. The Laplace transform of θ is defined by the relation:

$$\overline{\theta}=\int_0^\infty \theta e^{-pt}\,\mathrm{d}t \tag{9.30}$$

where p is a parameter.

Thus $\overline{\theta}$ is obtained by operating on θ with respect to t with x constant.

Then: $$\frac{\partial^2\overline{\theta}}{\partial x^2}=\overline{\frac{\partial^2\theta}{\partial x^2}} \tag{9.31}$$

and: $$\overline{\frac{\partial\theta}{\partial t}}=\int_0^\infty \frac{\partial\theta}{\partial t}e^{-pt}\,\mathrm{d}t$$

$$=\left[\theta e^{-pt}\right]_0^\infty + p\int_0^\infty e^{-pt}\theta\,\mathrm{d}t$$

$$=-\theta_{t=0}+p\overline{\theta} \tag{9.32}$$

Then, taking the Laplace transforms of each side of equation 9.29:

$$\overline{\frac{\partial\theta}{\partial t}}=D_H\overline{\frac{\partial^2\theta}{\partial x^2}}$$

or: $$p\overline{\theta}-\theta_{t=0}=D_H\frac{\partial^2\overline{\theta}}{\partial x^2} \quad \text{(from equations 9.31 and 9.32)}$$

and: $$\frac{\partial^2\overline{\theta}}{\partial x^2}-\frac{p}{D_H}\overline{\theta}=-\frac{\theta_{t=0}}{D_H}$$

...verywhere is constant initially, $\theta_{t=0}$ is a constant and the equation ...s a normal second-order differential equation since p is not a function

$$\overline{\theta} = B_1 e^{\sqrt{(p/D_H)}x} + B_2 e^{-\sqrt{(p/D_H)}x} + \theta_{t=0} p^{-1} \tag{9.33}$$

...herefore:

$$\frac{\partial \overline{\theta}}{\partial x} = B_1 \sqrt{\frac{p}{D_H}} e^{\sqrt{(p/D_H)}x} - B_2 \sqrt{\frac{p}{D_H}} e^{-\sqrt{(p/D_H)}x} \tag{9.34}$$

The temperature θ, corresponding to the transform $\overline{\theta}$, may now be found by reference to tables of the Laplace transform. It is first necessary, however, to evaluate the constants B_1 and B_2 using the boundary conditions for the particular problem since these constants will in general involve the parameter p which was introduced in the transformation.

Considering the particular problem of the unidirectional flow of heat through a body with plane parallel faces a distance l apart, the heat flow is normal to these faces and the temperature of the body is initially constant throughout. The temperature scale will be so chosen that this uniform initial temperature is zero. At time, $t = 0$, one face (at $x = 0$) will be brought into contact with a source at a constant temperature θ' and the other face (at $x = l$) will be assumed to be perfectly insulated thermally.

The boundary conditions are therefore:

$$t = 0, \qquad \theta = 0$$

$$t > 0, \qquad \theta = \theta' \quad \text{when } x = 0$$

$$t > 0, \qquad \frac{\partial \theta}{\partial x} = 0 \quad \text{when } x = l$$

Thus:

$$\overline{\theta}_{x=0} = \int_0^\infty \theta' e^{-pt}\, dt = \frac{\theta'}{p}$$

and:

$$\left(\frac{\partial \overline{\theta}}{\partial x}\right)_{x=l} = 0$$

Substitution of these boundary conditions in equations 9.33 and 9.34 gives:

$$B_1 + B_2 = \frac{\theta'}{p}$$

and:

$$B_1 e^{\sqrt{(p/D_H)}l} - B_2 e^{-\sqrt{(p/D_H)}l} = 0 \tag{9.35}$$

Hence:

$$B_1 = \frac{(\theta'/p) e^{-\sqrt{(p/D_H)}l}}{e^{\sqrt{(p/D_H)}l} + e^{-\sqrt{(p/D_H)}l}}$$

and:

$$B_2 = \frac{(\theta'/p) e^{\sqrt{(p/D_H)}l}}{e^{\sqrt{(p/D_H)}l} + e^{-\sqrt{(p/D_H)}l}}$$

Then:

$$\bar{\theta} = \frac{e^{(l-x)\sqrt{(p/D_H)}} + e^{-(l-x)\sqrt{(p/D_H)}}}{e^{\sqrt{(p/D_H)}l} + e^{-\sqrt{(p/D_H)}l}} \frac{\theta'}{p}$$

$$= \frac{\theta'}{p}(e^{(l-x)\sqrt{(p/D_H)}} + e^{-(l-x)\sqrt{(p/D_H)}})(1 + e^{-2\sqrt{(p/D_H)}l})^{-1}(e^{-\sqrt{(p/D_H)}l})$$

$$= \frac{\theta'}{p}(e^{-x\sqrt{(p/D_H)}} + e^{-(2l-x)\sqrt{(p/D_H)}})(1 - e^{-2l\sqrt{(p/D_H)}} + \cdots$$

$$+ (-1)^N e^{-2Nl\sqrt{(p/D_H)}} + \cdots)$$

$$= \sum_{N=0}^{N=\infty} \frac{\theta'}{p}(-1)^N (e^{-(2lN+x)\sqrt{(p/D_H)}} + e^{-\{2(N+1)l-x\}\sqrt{(p/D_H)}}) \tag{9.36}$$

The temperature θ is then obtained from the tables of inverse Laplace transforms in the Appendix (Table 12, No 83) and is given by:

$$\theta = \sum_{N=0}^{N=\infty} (-1)^N \theta' \left(\operatorname{erfc} \frac{2lN + x}{2\sqrt{D_H t}} + \operatorname{erfc} \frac{2(N+1)l - x}{2\sqrt{D_H t}} \right) \tag{9.37}$$

where:
$$\operatorname{erfc} x = \frac{2}{\sqrt{\pi}} \int_x^{\infty} e^{-\xi^2} \, d\xi$$

Values of erfc x $(= 1 - \operatorname{erf} x)$ are given in the Appendix (Table 13) and in specialist sources.[5]

Equation 9.37 may be written in the form:

$$\frac{\theta}{\theta'} = \sum_{N=0}^{N=\infty} (-1)^N \left\{ \operatorname{erfc} \left[Fo_l^{-1/2} \left(N + \frac{1}{2}\frac{x}{l} \right) \right] + \operatorname{erfc} \left[Fo_l^{-1/2} \left((N+1) - \frac{1}{2}\frac{x}{l} \right) \right] \right\} \tag{9.38}$$

where $Fo_l = (D_H t/l^2)$ and is known as the *Fourier number*.

Thus:
$$\frac{\theta}{\theta'} = f\left(Fo_l, \frac{x}{l} \right) \tag{9.39}$$

The numerical solution to this problem is then obtained by inserting the appropriate values for the physical properties of the system and using as many terms in the series as are necessary for the degree of accuracy required. In most cases, the above series converge quite rapidly.

This method of solution of problems of unsteady flow is particularly useful because it is applicable when there are discontinuities in the physical properties of the material.[6] The boundary conditions, however, become a little more complicated, but the problem is intrinsically no more difficult.

A general method of estimating the temperature distribution in a body of any shape consists of replacing the heat flow problem by the analogous electrical situation and measuring the electrical potentials at various points. The heat capacity per unit volume $C_p\rho$ is represented by an electrical capacitance, and the thermal conductivity k by an electrical conductivity. This method can be used to take account of variations in the thermal properties over the body.

Example 9.4

Calculate the time taken for the distant face of a brick wall, of thermal diffusivity $D_H = 0.0043$ cm^2/s and thickness $l = 0.45$ m, to rise from 295 to 375 K, if the whole wall is initially at a constant temperature of 295 K and the near face is suddenly raised to 900 K and maintained at this temperature. Assume that all the flow of heat is perpendicular to the faces of the wall and that the distant face is perfectly insulated.

Solution

The temperature at any distance x from the near face at time t is given by:

$$\theta = \sum_{N=0}^{N=\infty} (-1)^N \theta' \left\{ \text{erfc}\left[\frac{2lN + x}{2\sqrt{D_H t}}\right] + \text{erfc}\left[\frac{2(N+1)l - x}{2\sqrt{D_H t}}\right]\right\} \quad \text{(equation 9.37)}$$

The temperature at the distant face is therefore given by:

$$\theta = \sum_{N=0}^{N=\infty} (-1)^N \theta' 2\ \text{erfc}\left[\frac{(2N+1)l}{2\sqrt{D_H t}}\right]$$

Choosing the temperature scale so that the initial temperature is everywhere zero, then:

$$\frac{\theta}{2\theta'} = \frac{375 - 295}{2(900 - 295)} = 0.066$$

$$D_H = 4.2 \times 10^{-7}\ \text{m}^2/\text{s} \quad \therefore \quad \sqrt{D_H} = 6.5 \times 10^{-4}$$

$$l = 0.45\ \text{m}$$

Thus:

$$0.066 = \sum_{N=0}^{N=\infty} (-1)^N\ \text{erfc}\left[\frac{0.45(2N+1)}{2 \times 6.5 \times 10^{-4} t^{0.5}}\right]$$

$$= \sum_{N=0}^{N=\infty} (-1)^N\ \text{erfc}\left[\frac{346(2N+1)}{t^{0.5}}\right]$$

$$= \text{erfc}(346t^{-0.5}) - \text{erfc}(1038t^{-0.5}) + \text{erfc}(1730t^{-0.5}) - \cdots$$

An approximate solution is obtained by taking the first term only, to give:

$$346t^{-0.5} = 1.30$$

from which

$$t = 70\,840\ \text{s}$$

$$= \underline{\underline{70.8\ \text{ks}}} \text{ or } \underline{\underline{19.7\ \text{h}}}$$

Schmidt's method

Numerical methods have been developed by replacing the differential equation by a finite difference equation. Thus in a problem of unidirectional flow of heat:

$$\frac{\partial \theta}{\partial t} \approx \frac{\theta_{x(t+\Delta t)} - \theta_{x(t-\Delta t)}}{2\Delta t} \approx \frac{\theta_{x(t+\Delta t)} - \theta_{xt}}{\Delta t}$$

$$\frac{\partial^2\theta}{\partial x^2} \approx \frac{\left(\dfrac{\theta_{(x+\Delta x)t} - \theta_{xt}}{\Delta x} - \dfrac{\theta_{xt} - \theta_{(x-\Delta x)t}}{\Delta x}\right)}{\Delta x}$$

$$= \frac{\theta_{(x+\Delta x)t} + \theta_{(x-\Delta x)t} - 2\theta_{xt}}{(\Delta x)^2}$$

where θ_{xt} is the value of θ at time t and distance x from the surface, and the other values of θ are at intervals Δx and Δt as shown in Figure 9.10.

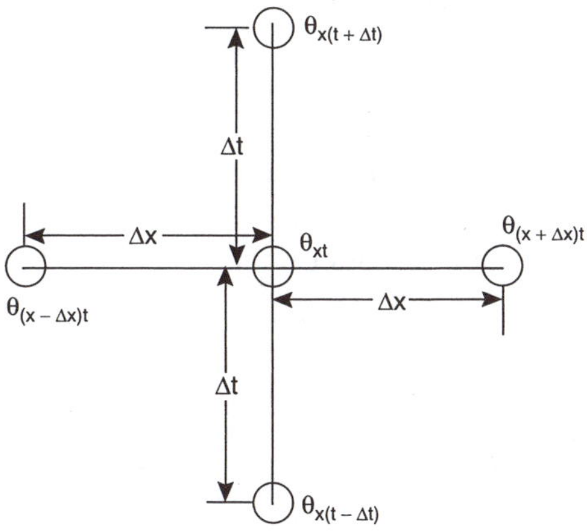

Figure 9.10. Variation of temperature with time and distance

Substituting these values in equation 9.29:

$$\theta_{x(t+\Delta t)} - \theta_{x(t-\Delta t)} = D_H \frac{2\Delta t}{(\Delta x)^2}(\theta_{(x+\Delta x)t} + \theta_{(x-\Delta x)t} - 2\theta_{xt}) \tag{9.40}$$

and:

$$\theta_{x(t+\Delta t)} - \theta_{xt} = D_H \frac{\Delta t}{(\Delta x)^2}(\theta_{(x+\Delta x)t} + \theta_{(x-\Delta x)t} - 2\theta_{xt}) \tag{9.41}$$

Thus, if the temperature distribution at time t, is known, the corresponding distribution at time $t + \Delta t$ can be calculated by the application of equation 9.41 over the whole extent of the body in question. The intervals Δx and Δt are so chosen that the required degree of accuracy is obtained.

A graphical method of procedure has been proposed by SCHMIDT[7]. If the temperature distribution at time t is represented by the curve shown in Figure 9.11 and the points representing the temperatures at $x - \Delta x$ and $x + \Delta x$ are joined by a straight line, then the distance θ_a is given by:

$$\theta_a = \frac{\theta_{(x+\Delta x)t} + \theta_{(x-\Delta x)t}}{2} - \theta_{xt}$$

$$= \frac{(\Delta x)^2}{2D_H\Delta t}(\theta_{x(t+\Delta t)} - \theta_{xt}) \quad \text{(from equation 9.41)} \tag{9.42}$$

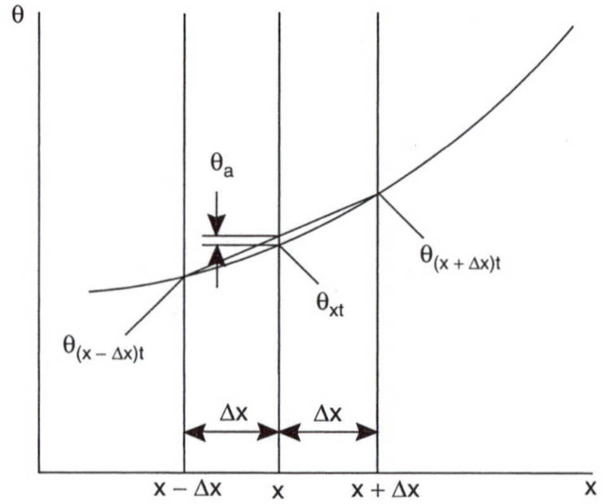

Figure 9.11. Schmidt's method

Thus, θ_a represents the change in θ_{xt} after a time interval Δt, such that:

$$\Delta t = \frac{(\Delta x)^2}{2D_H} \tag{9.43}$$

If this simple construction is carried out over the whole of the body, the temperature distribution after time Δt is obtained. The temperature distribution after an interval $2\Delta t$ is then obtained by repeating this procedure.

The most general method of tackling the problem is the use of the *finite-element* technique[8] to determine the temperature distribution at any time by using the finite difference equation in the form of equation 9.40.

Example 9.5

Solve Example 9.4 using Schmidt's method.

Solution

The development of the temperature profile is shown in Figure 9.12. At time $t = 0$ the temperature is constant at 295 K throughout and the temperature of the hot face is raised to 900 K. The problem will be solved by taking relatively large intervals for Δx.

Choosing $\Delta x = 50$ mm, the construction shown in Figure 9.12 is carried out starting at the hot face.

Points corresponding to temperature after a time interval Δt are marked 1, after a time interval $2\Delta t$ by 2, and so on. Because the second face is perfectly insulated, the temperature gradient must be zero at this point. Thus, in obtaining temperatures at $x = 450$ mm it is assumed that the temperature at $x = 500$ mm will be the same as at $x = 400$ mm, that is, horizontal lines are drawn on the diagram. It is seen that the temperature is less than 375 K after time $23\Delta t$ and greater than 375 K after time $25\Delta t$.

Thus: $t \approx 24\Delta t$

From equation 9.43: $\Delta t = 5.0^2/(2 \times 0.0042) = 2976$ s

Thus time required $= 24 \times 2976 = 71400$ s

or: $\underline{\underline{71.4 \text{ ks}}} = \underline{\underline{19.8 \text{ h}}}$

This value is quite close to that obtained by calculation, even using the coarse increments in Δx.

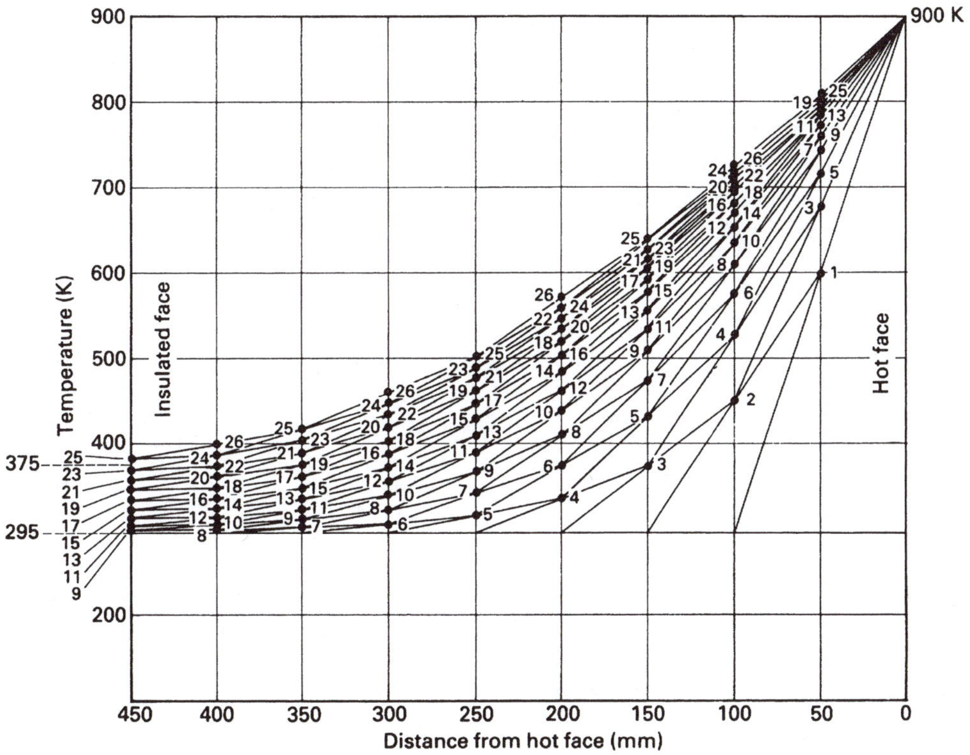

Figure 9.12. Development of temperature profile

Heating and cooling of solids and particles

The exact mathematical solution of problems involving unsteady thermal conduction may be very difficult, and sometimes impossible, especially where bodies of irregular shapes are concerned, and other methods are therefore required.

When a body of characteristic linear dimension L, initially at a uniform temperature θ_0, is exposed suddenly to surroundings at a temperature θ', the temperature distribution at any time t is found from dimensional analysis to be:

$$\frac{\theta' - \theta}{\theta' - \theta_0} = \mathrm{f}\left(\frac{hL}{k_p}, D_H\frac{t}{L^2}, \frac{x}{L}\right) \tag{9.44}$$

where D_H is the thermal diffusivity $(k_p/C_p\rho)$ of the solid, x is distance within the solid body and h is the heat transfer coefficient in the fluid at the surface of the body.

Analytical solutions of equation 9.44 in the form of infinite series are available for some simple regular shapes of particles, such as rectangular slabs, long cylinders and spheres, for conditions where there is heat transfer by conduction or convection to or from the surrounding fluid. These solutions tend to be quite complex, even for simple shapes. The heat transfer process may be characterised by the value of the *Biot number Bi* where:

$$Bi = \frac{hL}{k_p} = \frac{L/k_p}{1/h} \tag{9.45}$$

where h is the external heat transfer coefficient,

L is a characteristic dimension, such as radius in the case of a sphere or long cylinder, or half the thickness in the case of a slab, and

k_p is the thermal conductivity of the particle.

The Biot number is essentially the ratio of the resistance to heat transfer within the particle to that within the external fluid. At first sight, it appears to be similar in form to the Nusselt Number Nu' where:

$$Nu' = \frac{hd}{k} = \frac{2hr_o}{k} \tag{9.46}$$

However, the Nusselt number refers to a single fluid phase, whereas the Biot number is related to the properties of both the fluid and the solid phases.

Three cases are now considered:

(1) Very large Biot numbers, $Bi \to \infty$
(2) Very low Biot numbers, $Bi \to 0$
(3) Intermediate values of the Biot number.

(1) *Bi very large*. The resistance to heat transfer in the fluid is then low compared with that in the solid with the temperature of the surface of the particle being approximately equal to the bulk temperature of the fluid, and the heat transfer rate is independent of the Biot number. Equation 9.44 then simplifies to:

$$\frac{\theta' - \theta}{\theta' - \theta_0} = \text{f}\left(D_H \frac{t}{L^2}, \frac{x}{L}\right) = \text{f}\left(Fo_L, \frac{x}{L}\right) \tag{9.47}$$

where $Fo_L \left(= D_H \frac{t}{L^2}\right)$ is known as the *Fourier* number, using L in this case to denote the characteristic length, and x is distance from the centre of the particle. Curves connecting these groups have been plotted by a number of workers for bodies of various shapes, although the method is limited to those shapes which have been studied experimentally.

In Figure 9.13, taken from CARSLAW and JAEGER[5], the value of $(\theta' - \theta_c)/(\theta' - \theta_0)$ is plotted to give the temperature θ_c at the centre of bodies of various shapes, initially at a uniform temperature θ_0, at a time t after the surfaces have been suddenly altered to and maintained at a constant temperature θ'.

In this case (x/L) is constant at 0 and the results are shown as a function of the particular value of the Fourier number Fo_L $(D_H t/L^2)$.

(2) *Bi very small*. (say, <0.1). Here the main resistance to heat transfer lies within the fluid; this occurs when the thermal conductivity of the particle in very high and/or when the particle is very small. Under these conditions, the temperature within the particle is uniform and a "lumped capacity" analysis may be performed. Thus, if a solid body of volume V and initial temperature θ_0 is suddenly immersed in a volume of fluid large enough for its temperature θ to remain effectively constant, the rate of heat transfer from the body may be expressed as:

$$-\rho C_p V \frac{d\theta}{dt} = hA_e(\theta - \theta') \tag{9.48}$$

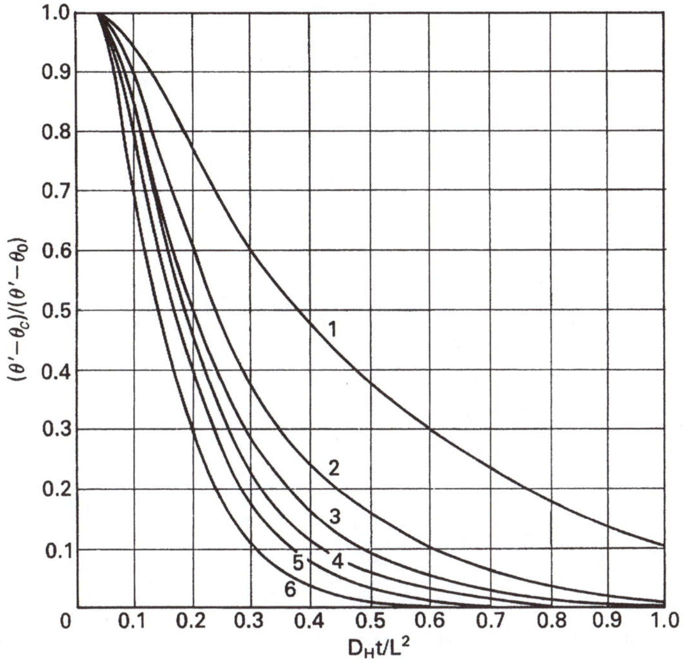

Figure 9.13. Cooling curve for bodies of various shapes: 1, slab ($2L =$ thickness); 2, square bar ($2L =$ side); 3, long cylinder ($L =$ radius); 4, cube ($2L =$ length of side); 5, cylinder ($L =$ radius, length $= 2L$); 6, sphere ($L =$ radius)

where A_e is the external surface area of the solid body.

Then:
$$\int_{\theta_0}^{\theta} \frac{d\theta}{\theta - \theta'} = -\int_0^t \frac{hA_e}{\rho C_v V} dt$$

i.e.:
$$\frac{\theta - \theta'}{\theta_0 - \theta'} = e^{-t/\tau} \tag{9.49}$$

where $\tau = \dfrac{\rho C_p V}{hA_e}$ is known as the *response time constant.*

It will be noted that the relevant characteristic dimension in the Biot number is defined as the ratio of the volume to the external surface area of the particle (V/A_e), and the higher the value of V/A_e, then the slower will be the response time. With the characteristic dimension defined in this way, this analysis is valid for particles of any shape at values of the Biot number less than 0.1

Example 9.6

A 25 mm diameter copper sphere and a 25 mm copper cube are both heated in a furnace to 650 °C (923 K). They are then annealed in air at 95 °C (368 K). If the external heat transfer coefficient h is 75 W/m^{2}K in both cases, what is temperature of the sphere and of the cube at the end of 5 minutes?

The physical properties at the mean temperature for copper are:

$$\rho = 8950 \text{ kg/m}^3 \quad C_p = 0.38 \text{ kJ/kg K} \quad k_p = 385 \text{ W/mK}$$

Solution

$$V/A_e \text{ for the sphere} \quad = \frac{\frac{\pi}{6}d^3}{\pi d^2} = \frac{d}{6} = \frac{25 \times 10^{-3}}{6} = 4.17 \times 10^{-3} \text{ m}$$

$$V/A_e \text{ for the cube} \quad = \frac{l^3}{6l^2} = \frac{l}{6} = \frac{25 \times 10^{-3}}{6} = 4.17 \times 10^{-3} \text{ m}$$

$$\therefore \quad Bi = \frac{h(V/A_e)}{k} = \frac{75 \times 25 \times 10^{-3}}{385 \times 6} = 8.1 \times 10^{-4} \ll 0.1$$

The use of a lumped capacity method is therefore justified.

$$\tau = \frac{\rho C_p V}{h A_e} = \frac{8950 \times 380}{75} \times \frac{25 \times 10^{-3}}{6} = 189 \text{ s}$$

Then using equation 9.49:

$$\frac{\theta - 368}{923 - 368} = \exp\left(-\frac{5 \times 60}{189}\right)$$

and:
$$\theta = 368 + 0.2045(923 - 368) = 481 \text{ K} = 208\,^\circ\text{C}$$

Since the sphere and the cube have the same value of V/A_e, after 5 minutes they will both attain a temperature of 208°C.

(3) *Intermediate values of Bi*. In this case the resistances to heat transfer within the solid body and the fluid are of comparable magnitude. Neither will the temperature within the solid be uniform (case 1), nor will the surface temperature be equal to that in the bulk of the fluid (case 2).

Analytical solutions in the form of infinite series can be obtained for some regular shapes (thin plates, spheres and long cylinders (length $\gg$ radius)), and numerical solutions using *finite element methods*[(8)] have been obtained for bodies of other shapes, both regular and irregular. Some of the results have been presented by HEISLER[(9)] in the form of charts, examples of which are shown in Figures 9.14–9.16 for thin slabs, long cylinders and spheres, respectively. It may be noted that in this case the characteristic length L is the half-thickness of the slab and the external radius r_o of the cylinder and sphere.

Figures 9.14–9.16 enable the temperature θ_c at the centre of the solid (centre-plane, centre-line or centre-point) to be obtained as a function of the Fourier number, and hence of time, with the reciprocal of the Biot number (Bi^{-1}) as parameter.

Temperatures at off-centre locations within the solid body can then be obtained from a further series of charts given by Heisler (Figures 9.17–9.19) which link the desired temperature to the centre-temperature as a function of Biot number, with location within the particle as parameter (that is the distance x from the centre plane in the slab or radius in the cylinder or sphere). Additional charts are given by Heisler for the quantity of heat transferred from the particle in a given time in terms of the initial heat content of the particle.

Figures 9.17–9.19 clearly show that, as the Biot number approaches zero, the temperature becomes uniform within the solid, and the lumped capacity method may be used for calculating the unsteady-state heating of the particles, as discussed in section (2). The charts are applicable for Fourier numbers greater than about 0.2.

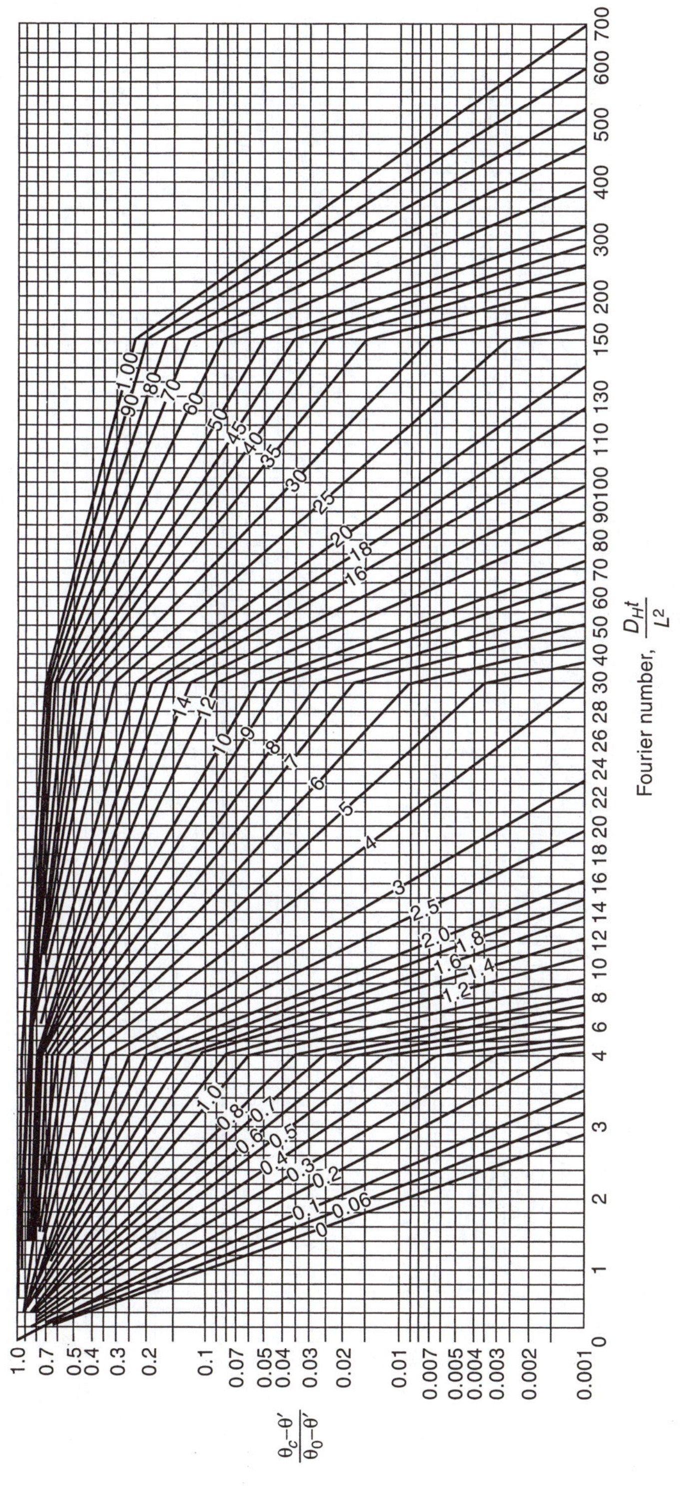

Figure 9.14. Mid-plane temperature for an infinite plate of thickness $2L$, for various values of parameters $k_p/hL(= Bi^{-1})$

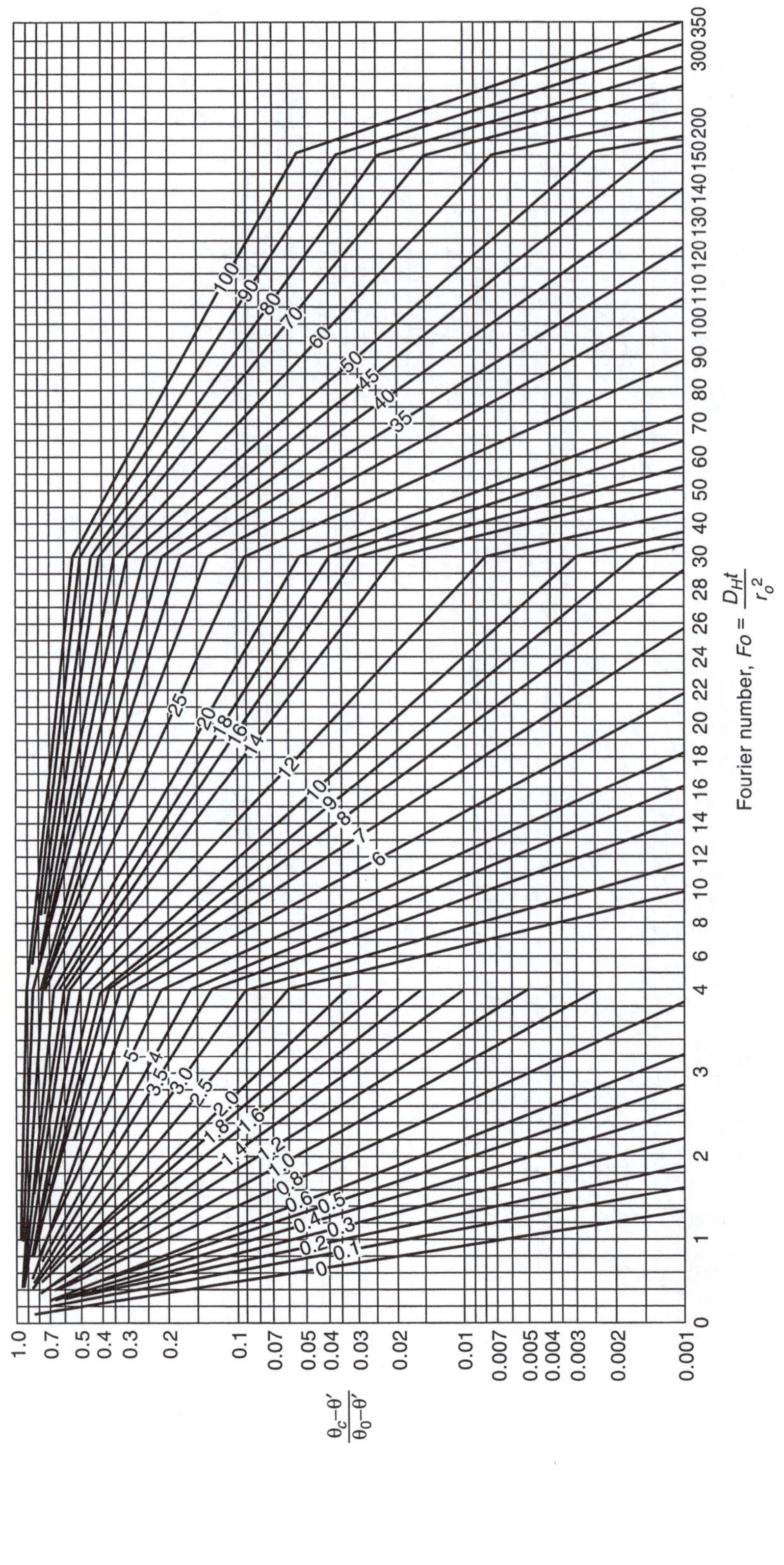

Figure 9.15. Axis temperature for an infinite cyclinder of radius r_o, for various of parameters $k_p/hr_o (= Bi^{-1})$

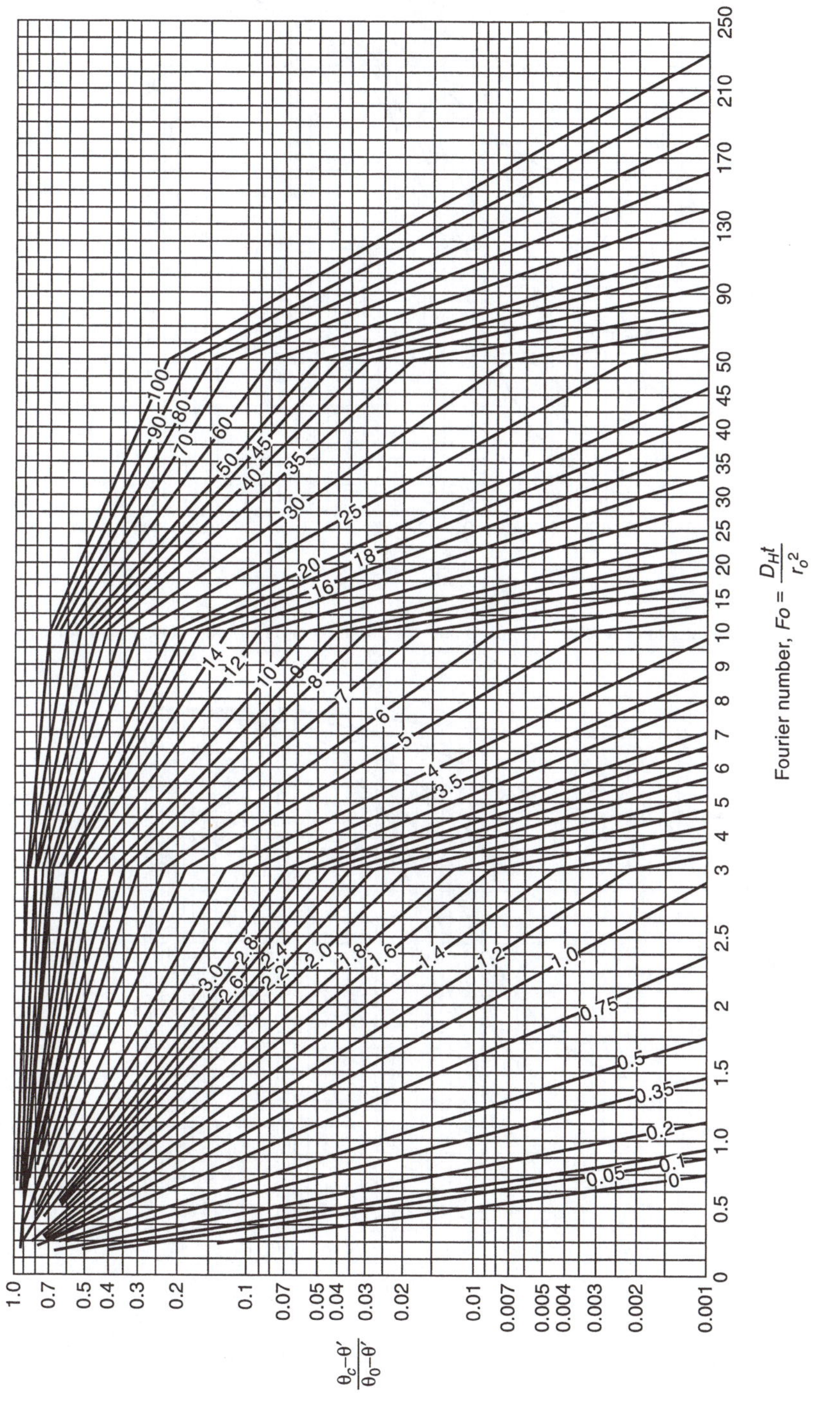

Figure 9.16. Centre-temperature for a sphere of radius r_o, for various values of parameters $\frac{k_p}{hr_o}$ $(= Bi^{-1})$

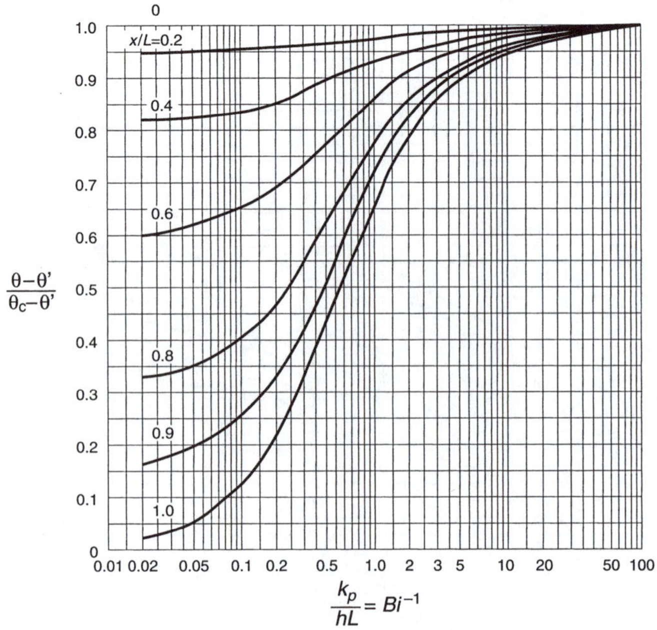

Figure 9.17. Temperature as a function of mid-plane temperature in an infinite plate of thickness $2L$

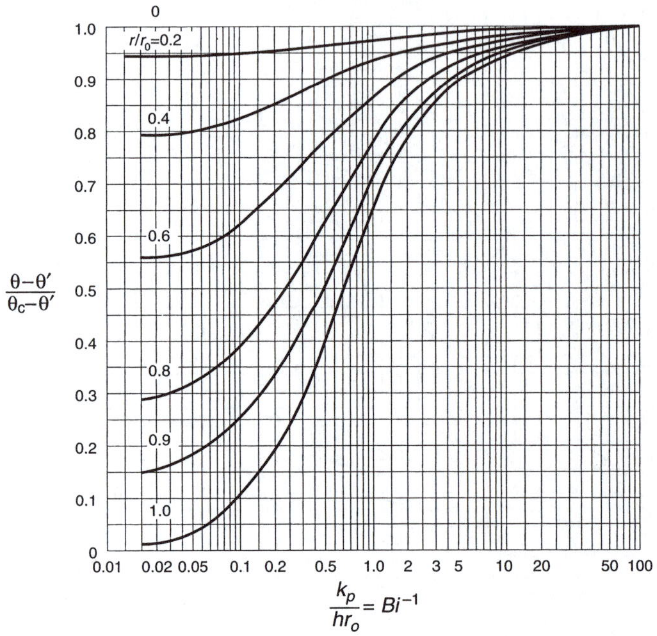

Figure 9.18. Temperature as a function of axis temperature in an infinite cylinder of radius r_o

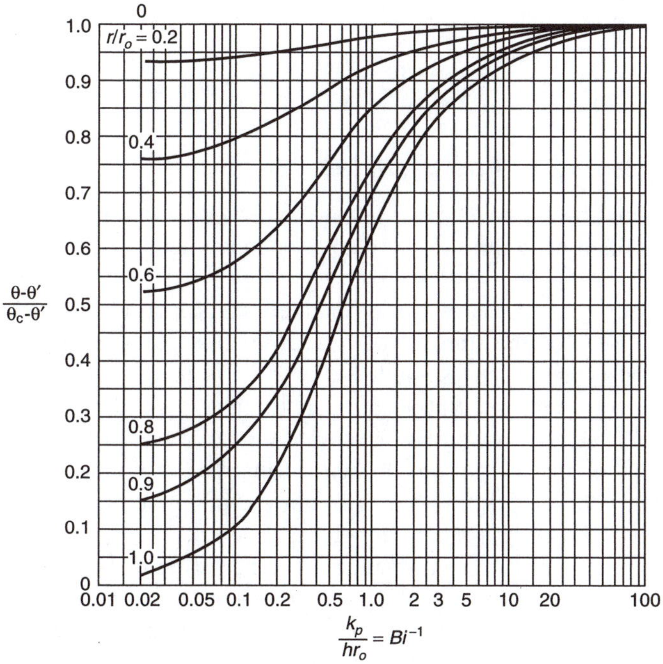

Figure 9.19. Temperature as a function of centre-temperature for a sphere of radius r_o

Example 9.7

A large thermoplastic sheet, 10 mm thick, at an initial temperature of 20 °C (293 K), is to be heated in an oven in order to enable it to be moulded. The oven temperature is maintained at 100 °C (373 K), the maximum temperature to which the plastic may be subjected, and it is necessary to ensure that the temperature throughout the whole of the sheet reaches a minimum of 80 °C (353 K). Calculate the minimum length of time for which the sheet must be heated.

Thermal conductivity k_p of the plastic $= 2.5$ W/m^{2}K

Thermal diffusivity of the surrounding fluid $D_H = 2 \times 10^{-7}$ m^2/s

External heat transfer coefficient h $= 100$ W/m^{2}K

Solution

Throughout the heating process, the temperature within the sheet will be a minimum at the centre-plane ($x = 0$) and therefore the required time is that for the centre to reach 80 °C (353 K).

For this process, the Biot number $Bi = \dfrac{hL}{k_p} = \dfrac{100 \times 5 \times 10^{-3}}{2.5} = 0.2$ and $Bi^{-1} = 5$

(since L, the half-thickness of the plate is 5 mm)

The limiting value of $\dfrac{\theta' - \theta_c}{\theta' - \theta_0} = \dfrac{373 - 353}{373 - 293} = 0.25$

From Figure 9.17, the Fourier number $\dfrac{D_H t}{L^2} \approx 7.7$

Thus: $$t = \frac{7.7 \times (5 \times 10^{-3})^2}{2 \times 10^{-7}} = 960 \text{ s or } \underline{\underline{16 \text{ minutes}}}$$

Heating and melting of fine particles

There are many situations in which particles are heated or cooled by a surrounding gas and these may be classified according to the degree of movement of the particle as follows:
i) Static beds
Although most beds of particles involve relatively large particle diameters, such as in pebble bed units used for the transfer of heat from flue gases to the incoming air for example, smaller particles, such as sand, are used in beds and, again, these are mainly used for heat recovery. One such application is the heating and cooling of buildings in hotter climes where the cool nocturnal air is used to cool a bed of particles which is then used to cool the incoming air during the heat of the day as it enters a building. In this way, an almost constant temperature may be achieved in a given enclosed environment in spite of the widely fluctuating ambient condition. A similar system has been used in less tropical areas where it is necessary to maintain a constant temperature in an environment in which heat is generated, such as a telephone exchange, for example. Such systems have the merit of very low capital and modest operating costs and, in most cases, the resistance to heat transfer by conduction within the solids is not dissimilar to the resistance in the gas film surrounding the particles.
ii) Partial movement of particles
The most obvious example of a process in which particles undergo only limited movement is the fluidised bed which is discussed in some detail in Volume 2. Applications here involve, not only heating and cooling, but also drying as in the case of grain dryers for example, and on occasions, chemical reaction as, for example, with fluidised-bed combustion. In such cases, conditions in the bed may, to all intents and purposes, be regarded as steady-state, with unsteady-state conduction taking place only in the entering 'process stream' which, by and large, is only a small proportion of the total bed mass in the bed.
iii) Falling particles
Particles fall by gravity through either static or moving gas streams in rotary dryers, for example, but they also fall through heating or cooling gases in specially designed columns. Examples here include the cooling of sand after it has been dried — again recovering heat in the process — salt cooling and also the spray drying of materials such as detergents which are sprayed as a concentrated solution of the material at the top of the tower and emerge as a dry powder. A similar situation occurs in fertiliser production where solid particles or granules are obtained from a spray of the molten material by counter-flow against a cooling gas stream. Convection to such materials is discussed in Section 9.4.6

One important problem involving unsteady state conduction of heat to particles is in the melting of powders in plasma spraying[(10)] where Biot numbers can range from 0.005 to 5. In this case, there is initially a very high relative velocity between the fluid and the powder. The plasmas referred to here are partially ionised gases with temperatures of around 10,000 K formed by electric discharges such as arcs. There is an increasing industrial use of the technique of plasma spraying in which powders are injected into a high-velocity plasma jet so that they are both melted and projected at velocities of several hundred metres per second onto a surface. The molten particles with diameters typically of the order 10–100 μm impinge to form an integral layer on the surface. Applications include the building up of worn shafts of pumps, for example, and the deposition of erosion-resistant ceramic layers on centrifugal pump impellers and other equipment prone to erosion damage. When a powder particle first enters the plasma jet, the relative velocity may be hundreds of metres per second and heat transfer to the particle is enhanced by convection, as discussed in Section 9.4.6. Often, and more particularly for smaller particles, the particle is quickly accelerated to essentially the same velocity as the plasma jet[(2)] and conduction becomes the main mechanism of heat transfer from plasma to particle. From a design point of view, neglecting the convective contribution will ease calculations and give a more conservative and safer estimate of the size of the largest particle which can be melted before it strikes the surface. In the absence of complications due to non-continuum conditions discussed later, the value of $Nu' = hd/k$ is therefore often taken as 2, as in equation 9.26.

One complication which arises in the application of this equation to powder heating in high temperature plasmas lies in the dependence of k, the thermal conductivity of the gas or plasma surrounding the particle, on temperature. For example, the temperature of the particle surface may be 1000 K, whilst that of the plasma away from the particle may be about 10,000 K or even higher. The thermal conductivity of argon increases by a factor of about 20 over this range of temperature and that of nitrogen gas passes through a pronounced peak at about 7100 K due to dissociation–recombination effects. Thus, the temperature at which the thermal conductivity k is evaluated will have a pronounced effect on the value of the external heat transfer coefficient. A mean value of k would seem appropriate where:

$$(k)_{\text{mean}} = \frac{1}{T_2 - T_1} \int_{T_1}^{T_2} k \, \mathrm{d}T \qquad \text{(equation 9.16)}$$

Some workers have correlated experimental data in terms of k at the arithmetic mean temperature, and some at the temperature of the bulk plasma. Experimental validation of the true effective thermal conductivity is difficult because of the high temperatures, small particle sizes and variations in velocity and temperature in plasma jets.

In view of the high temperatures involved in plasma devices and the dependence of radiation heat transfer on T^4, as discussed in Section 9.5, it is surprising at first sight that conduction is more significant than radiation in heating particles in plasma spraying. The explanation lies in the small values of d and relatively high values of k for the gas, both of which contribute to high values of h for any given value of Nu'. Also the emissivities of most gases are, as seen later in Section 9.5, rather low.

In situations where the surrounding fluid behaves as a non-continuum fluid, for example at very high temperatures and/or at low pressures, it is possible for Nu' to be less than 2. A gas begins to exhibit non-continuum behaviour when the mean free path between collisions of gas molecules or atoms with each other is greater than about 1/100 of the characteristic size of the surface considered. The molecules or atoms are then sufficiently far apart on average for the gas to begin to lose the character of a homogeneous or continuum fluid which is normally assumed in the majority of heat transfer or fluid dynamics problems. For example, with a particle of diameter 25 μm as encountered in, for example, oil-burner sprays, pulverised coal flames, and in plasma spraying in air at room temperature and atmospheric pressure, the mean free path of gas molecules is about 0.06 μm and the air then behaves as a continuum fluid. If, however, the temperature were say 1800 K, as in a flame, then the mean free path would be about 0.33 μm, which is greater than 1/100 of the particle diameter. Non-continuum effects, leading to values of Nu' lower than 2 would then be likely according to theory[11,12]. The exact value of Nu' depends on the surface accommodation coefficient. This is a difficult parameter to measure for the examples considered here, and hence experimental confirmation of the theory is difficult. At the still higher temperatures that exist in thermal plasma devices, non-continuum effects should be more pronounced and there is limited evidence that values of Nu' below 1 are obtained[10]. In general, non-continuum effects, leading in particular to values of Nu' less than 2, would be more likely at high temperatures, low pressures, and small particle sizes. Thus, there is an interest in these effects in the aerospace industry when considering, for example, the behaviour of small particles present in rocket engine exhausts.

9.3.6. Conduction with internal heat source

If an electric current flows through a wire, the heat generated internally will result in a temperature distribution between the central axis and the surface of the wire. This type of problem will also arise in chemical or nuclear reactors where heat is generated internally. It is necessary to determine the temperature distribution in such a system and the maximum temperature which will occur.

If the temperature at the surface of the wire is T_o and the rate of heat generation per unit volume is Q_G, then considering unit length of a cylindrical element of radius r, the heat generated must be transmitted in an outward direction by conduction so that:

$$-k2\pi r\frac{\mathrm{d}T}{\mathrm{d}r} = \pi r^2 Q_G$$

Hence:

$$\frac{\mathrm{d}T}{\mathrm{d}r} = -\frac{Q_G r}{2k} \tag{9.50}$$

Integrating:

$$T = -\frac{Q_G r^2}{4k} + C$$

$T = T_o$ when $r = r_o$ the radius of wire and hence:

$$T = T_o + Q_G\frac{r_o^2 - r^2}{4k}$$

or:

$$T - T_o = \frac{Q_G r_o^2}{4k}\left(1 - \frac{r^2}{r_o^2}\right) \tag{9.51}$$

This gives a parabolic distribution of temperature and the maximum temperature will occur at the axis of the wire where $(T - T_o) = Q_G r_o^2/4k$. The arithmetic mean temperature difference, $(T - T_o)_{\text{av}} = Q_G r_o^2/8k$.

Since $Q_G \pi r_o^2$ is the rate of heat release per unit length of the wire then, putting T_1 as the temperature at the centre:

$$T_1 - T_o = \frac{\text{rate of heat release per unit length}}{4\pi k} \tag{9.52}$$

Example 9.8

A fuel channel in a natural uranium reactor is 5 m long and has a heat release of 0.25 MW. If the thermal conductivity of the uranium is 33 W/mK, what is the temperature difference between the surface and the centre of the uranium element, assuming that the heat release is uniform along the rod?

Solution

$$\text{Heat release rate} = 0.25 \times 10^6 \text{ W}$$

$$= \frac{0.25 \times 10^6}{5} = 5 \times 10^4 \text{ W/m}$$

Thus, from equation 9.52:

$$T_1 - T_0 = \frac{5 \times 10^4}{4\pi \times 33}$$

$$= \underline{\underline{121 \text{ deg K}}}$$

It should be noted that the temperature difference is independent of the diameter of the fuel rod for a cylindrical geometry, and that the heat released per unit volume has been considered as being uniform.

In practice the assumption of the uniform heat release per unit length of the rod is not valid since the neutron flux, and hence the heat generation rate varies along its length. In the simplest case where the neutron flux may be taken as zero at the ends of the fuel element, the heat flux may be represented by a sinusoidal function, and the conditions become as shown in Figure 9.20.

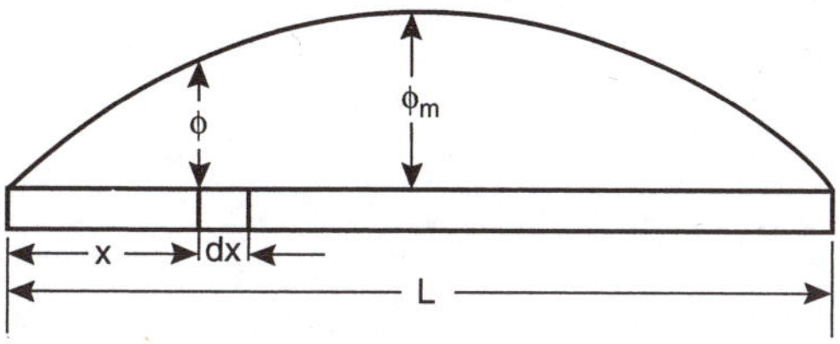

Figure 9.20. Variation of neutron flux along a length of fuel rod

Since the heat generated is proportional to the neutron flux, the heat dQ developed per unit time in a differential element of the fuel rod of length dx may be written as:

$$dQ = C \sin\left(\frac{\pi x}{L}\right) dx$$

The total heat generated by the rod Q is then given by:

$$Q = C\int_0^L \sin\left(\frac{\pi x}{L}\right)\mathrm{d}x = \frac{2CL}{\pi}$$

Thus, $C = \pi Q/2L$. The heat release per unit length at any point is then given by:

$$\frac{\mathrm{d}Q}{\mathrm{d}x} = \frac{\pi Q}{2L}\sin\left(\frac{\pi x}{L}\right)$$

Substituting into equation 9.52 gives:

$$T_1 - T_o = \frac{\left(\dfrac{\pi Q}{2L}\right)\sin\left(\dfrac{\pi x}{L}\right)}{4\pi k} \tag{9.53}$$

It may be noted that when $x = 0$ or $x = L$, then $T_1 - T_o$ is zero as would be expected since the neutron flux was taken as zero at these positions.

9.4. HEAT TRANSFER BY CONVECTION

9.4.1. Natural and forced convection

Heat transfer by convection occurs as a result of the movement of fluid on a macroscopic scale in the form of eddies or circulating currents. If the currents arise from the heat transfer process itself, *natural convection* occurs, such as in the heating of a vessel containing liquid by means of a heat source situated beneath it. The liquid at the bottom of the vessel becomes heated and expands and rises because its density has become less than that of the remaining liquid. Cold liquid of higher density takes its place and a circulating current is thus set up.

In *forced convection*, circulating currents are produced by an external agency such as an agitator in a reaction vessel or as a result of turbulent flow in a pipe. In general, the magnitude of the circulation in forced convection is greater, and higher rates of heat transfer are obtained than in natural convection.

In most cases where convective heat transfer is taking place from a surface to a fluid, the circulating currents die out in the immediate vicinity of the surface and a film of fluid, free of turbulence, covers the surface. In this film, heat transfer is by thermal conduction and, as the thermal conductivity of most fluids is low, the main resistance to transfer lies there. Thus an increase in the velocity of the fluid over the surface gives rise to improved heat transfer mainly because the thickness of the film is reduced. As a guide, the film coefficient increases as (fluid velocity)^n, where $0.6 < n < 0.8$, depending upon the geometry.

If the resistance to transfer is regarded as lying within the film covering the surface, the rate of heat transfer Q is given by equation 9.11 as:

$$Q = kA\frac{(T_1 - T_2)}{x}$$

The effective thickness x is not generally known and therefore the equation is usually rewritten in the form:

$$Q = hA(T_1 - T_2) \tag{9.54}$$

where h is the heat transfer coefficient for the film and $(1/h)$ is the thermal resistance.

9.4.2. Application of dimensional analysis to convection

So many factors influence the value of h that it is almost impossible to determine their individual effects by direct experimental methods. By arranging the variables in a series of dimensionless groups, however, the problem is made more manageable in that the number of groups is significantly less than the number of parameters. It is found that the heat transfer rate per unit area q is dependent on those physical properties which affect flow pattern (viscosity μ and density ρ), the thermal properties of the fluid (the specific heat capacity C_p and the thermal conductivity k) a linear dimension of the surface l, the velocity of flow u of the fluid over the surface, the temperature difference ΔT and a factor determining the natural circulation effect caused by the expansion of the fluid on heating (the product of the coefficient of cubical expansion β and the acceleration due to gravity g). Writing this as a functional relationship:

$$q = \phi[u, l, \rho, \mu, C_p, \Delta T, \beta g, k] \tag{9.55}$$

Noting the dimensions of the variables in terms of length **L**, mass **M**, time **T**, temperature $\boldsymbol{\theta}$, heat **H**:

q	Heat transferred/unit area and unit time	$\mathbf{HL^{-2}T^{-1}}$
u	Velocity	$\mathbf{LT^{-1}}$
l	Linear dimension	$\mathbf{L}$
μ	Viscosity	$\mathbf{ML^{-1}T^{-1}}$
ρ	Density	$\mathbf{ML^{-3}}$
k	Thermal conductivity	$\mathbf{HT^{-1}L^{-1}}\boldsymbol{\theta}^{-1}$
C_p	Specific heat capacity at constant pressure	$\mathbf{HM^{-1}}\boldsymbol{\theta}^{-1}$
ΔT	Temperature difference	$\boldsymbol{\theta}$
(βg)	The product of the coefficient of thermal expansion and the acceleration due to gravity	$\mathbf{LT^{-2}}\boldsymbol{\theta}^{-1}$

It may be noted that both temperature and heat are taken as fundamental units as heat is not expressed here in terms of **M**, **L**, **T**.

With nine parameters and five dimensions, equation 9.55 may be rearranged in four dimensionless groups.

Using the Π-theorem for solution of the equation, and taking as the recurring set: $l, \rho, \mu, \Delta T, k$

The non-recurring variables are: $q, u, (\beta g), C_p$

Then:

$$l \equiv \mathbf{L} \qquad \mathbf{L} = l$$

$$\rho \equiv \mathbf{ML^{-3}} \qquad \mathbf{M} = \rho\mathbf{L}^3 = \rho l^3$$

$$\mu \equiv \mathbf{ML^{-1}T^{-1}} \qquad \mathbf{T} = \mathbf{ML}^{-1}\mu^{-1} = \rho l^3 l^{-1}\mu^{-1} = \rho l^2\mu^{-1}$$

$$\Delta T \equiv \boldsymbol{\theta} \qquad \boldsymbol{\theta} = \Delta T$$

$$k \equiv \mathbf{HL^{-1}T^{-1}}\boldsymbol{\theta}^{-1} \qquad \mathbf{H} = k\mathbf{LT}\boldsymbol{\theta} = kl\rho l^2\mu^{-1}\Delta T = kl^3\rho\mu^{-1}\Delta T$$

The Π groups are then:

$$\Pi_1 = q\mathbf{H}^{-1}\mathbf{L}^2\mathbf{T} = qk^{-1}l^{-3}\rho^{-1}\mu\Delta T^{-1}l^2\rho l^2\mu^{-1} = qk^{-1}l\Delta T^{-1}$$
$$\Pi_2 = u\mathbf{L}^{-1}\mathbf{T} = ul^{-1}\rho l^2\mu^{-1} = u\rho l\mu^{-1}$$
$$\Pi_3 = C_p\mathbf{H}^{-1}\mathbf{M}\theta = C_pk^{-1}l^{-3}\rho^{-1}\mu\Delta T^{-1}\rho l^3\Delta T = C_pk^{-1}\mu$$
$$\Pi_4 = \beta g\mathbf{L}^{-1}\mathbf{T}^2\theta = \beta gl^{-1}\rho^2l^4\mu^{-2}\Delta T = \beta g\Delta T\rho^2\mu^{-2}l^3$$

The relation in equation 9.55 becomes:

$$\frac{ql}{k\Delta T} = \frac{hl}{k} = \phi\left[\left(\frac{lu\rho}{\mu}\right)\left(\frac{C_p\mu}{k}\right)\left(\frac{\beta g\Delta Tl^3\rho^2}{\mu^2}\right)\right] \tag{9.56}$$

or:

$$Nu = \phi[Re, Pr, Gr]$$

This general equation involves the use of four dimensionless groups, although it may frequently be simplified for design purposes. In equation 9.56:

hl/k	is known as the *Nusselt* group Nu (already referred to in equation 9.46),
$lu\rho/\mu$	the *Reynolds* group Re,
$C_p\mu/k$	the *Prandtl* group Pr, and
$\beta g\Delta Tl^3\rho^2/\mu^2$	the *Grashof* group Gr

It is convenient to define other dimensionless groups which are also used in the analysis of heat transfer. These are:

$lu\rho C_p/k$	the *Peclet* group, $Pe = RePr$,
GC_p/kl	the *Graetz* group Gz, and
$h/C_p\rho u$	the *Stanton* group, $St = Nu/(RePr)$

It may be noted that many of these dimensionless groups are ratios. For example, the Nusselt group $h/(k/l)$ is the ratio of the actual heat transfer to that by conduction over a thickness l, whilst the Prandtl group, $(\mu/\rho)/(k/C_p\rho)$ is the ratio of the kinematic viscosity to the thermal diffusivity.

For conditions in which only natural convection occurs, the velocity is dependent on the buoyancy effects alone, represented by the Grashof number, and the Reynolds group may be omitted. Again, when forced convection occurs the effects of natural convection are usually negligible and the Grashof number may be omitted. Thus:

for natural convection: $$Nu = \mathrm{f}(Gr, Pr) \tag{9.57}$$

and for forced convection: $$Nu = \mathrm{f}(Re, Pr) \tag{9.58}$$

For most gases over a wide range of temperature and pressure, $C_p\mu/k$ is constant and the Prandtl group may often be omitted, simplifying the design equations for the calculation of film coefficients with gases.

9.4.3. Forced convection in tubes

Turbulent flow

The results of a number of workers who have used a variety of gases such as air, carbon dioxide, and steam and of others who have used liquids such as water, acetone, kerosene, and benzene have been correlated by DITTUS and BOELTER[13] who used mixed units for their variables. On converting their relations using consistent (SI, for example) units, they become:

for heating of fluids:

$$Nu = 0.0241 Re^{0.8} Pr^{0.4} \tag{9.59}$$

and for cooling of fluids:

$$Nu = 0.0264 Re^{0.8} Pr^{0.3} \tag{9.60}$$

In these equations all of the physical properties are taken at the mean bulk temperature of the fluid $(T_i + T_o)/2$, where T_i and T_o are the inlet and outlet temperatures. The difference in the value of the index for heating and cooling occurs because in the former case the film temperature will be greater than the bulk temperature and in the latter case less. Conditions in the film, particularly the viscosity of the fluid, exert an important effect on the heat transfer process.

Subsequently MCADAMS[14] has re-examined the available experimental data and has concluded that an exponent of 0.4 for the Prandtl number is the most appropriate one for both heating and cooling. He also has slightly modified the coefficient to 0.023 (corresponding to Colburn's value, given below in equation 9.64) and gives the following equation, which applies for $Re > 2100$ and for fluids of viscosities not exceeding 2 mN s/m^2:

$$Nu = 0.023 Re^{0.8} Pr^{0.4} \tag{9.61}$$

WINTERTON[15] has looked into the origins of the "Dittus and Boelter" equation and has found that there is considerable confusion in the literature concerning the origin of equation 9.61 which is generally referred to as the Dittus–Boelter equation in the literature on heat transfer.

An alternative equation which is in many ways more convenient has been proposed by COLBURN[16] and includes the Stanton number ($St = h/C_p \rho u$) instead of the Nusselt number. This equation takes the form:

$$j_H = St Pr^{0.67} = 0.023 Re^{-0.2} \tag{9.62}$$

where j_H is known as the *j-factor for heat transfer*.
It may be noted that:

$$\frac{h}{C_p \rho u} = \left(\frac{hd}{k}\right)\left(\frac{\mu}{ud\rho}\right)\left(\frac{k}{C_p \mu}\right)$$

or:

$$St = Nu Re^{-1} Pr^{-1} \tag{9.63}$$

Thus, multiplying equation 9.62 by $Re Pr^{0.33}$:

$$Nu = 0.023 Re^{0.8} Pr^{0.33} \tag{9.64}$$

which is a form of equations 9.59 and 9.60.

Again, the physical properties are taken at the bulk temperature, except for the viscosity in the Reynolds group which is evaluated at the mean film temperature taken as $(T_{\text{surface}} + T_{\text{bulk fluid}})/2$.

Writing a heat balance for the flow through a tube of diameter d and length l with a rise in temperature for the fluid from T_i to T_o:

$$h\pi dl\Delta T = \frac{\pi d^2}{4} C_p u\rho(T_o - T_i)$$

or:

$$St = \frac{h}{C_p \rho u} = \frac{d(T_o - T_i)}{4l\Delta T} \tag{9.65}$$

where ΔT is the mean temperature difference between the bulk fluid and the walls.

With *very viscous liquids* there is a marked difference at any position between the viscosity of the fluid adjacent to the surface and the value at the axis or at the bulk temperature of the fluid. SIEDER and TATE[17] examined the experimental data available and suggested that a term $\left(\frac{\mu}{\mu_s}\right)^{0.14}$ be included to account for the viscosity variation and the fact that this will have opposite effects in heating and cooling. (μ is the viscosity at the bulk temperature and μ_s the viscosity at the wall or surface). They give a logarithmic plot, but do not propose a correlating equation. However, MCADAMS[14] gives the following equation, based on Sieder and Tate's work:

$$Nu = 0.027Re^{0.8}Pr^{0.33}\left(\frac{\mu}{\mu_s}\right)^{0.14} \tag{9.66}$$

This equation may also be written in the form of the Colburn equation (9.62).

When these equations are applied to *heating or cooling of gases* for which the Prandtl group usually has a value of about 0.74, substitution of $Pr = 0.74$ in equation 9.64 gives:

$$Nu = 0.020Re^{0.8} \tag{9.67}$$

Water is very frequently used as the cooling medium and the effect of the variation of physical properties with temperature may be included in equation 9.64 to give a simplified equation which is useful for design purposes (Section 9.9.4).

There is a very big difference in the values of h for water and air for the same linear velocity. This is shown in Figures 9.21–9.23 and Table 9.2, all of which are based on the work of FISHENDEN and SAUNDERS[18].

The effect of length to diameter ratio (l/d) on the value of the heat transfer coefficient may be seen in Figure 9.24. It is important at low Reynolds numbers but ceases to be significant at a Reynolds number of about 10^4.

It is also important to note that the film coefficient varies with the distance from the entrance to the tube. This is especially important at low (l/d) ratios and an average value is given approximately by:

$$\frac{h_{\text{average}}}{h_\infty} = 1 + \left(\frac{d}{l}\right)^{0.7} \tag{9.68}$$

where h_∞ is the limiting value for a very long tube.

The roughness of the surface of the inside of the pipe can have an important bearing on rates of heat transfer to the fluid, although COPE[19], using degrees of artificial roughness

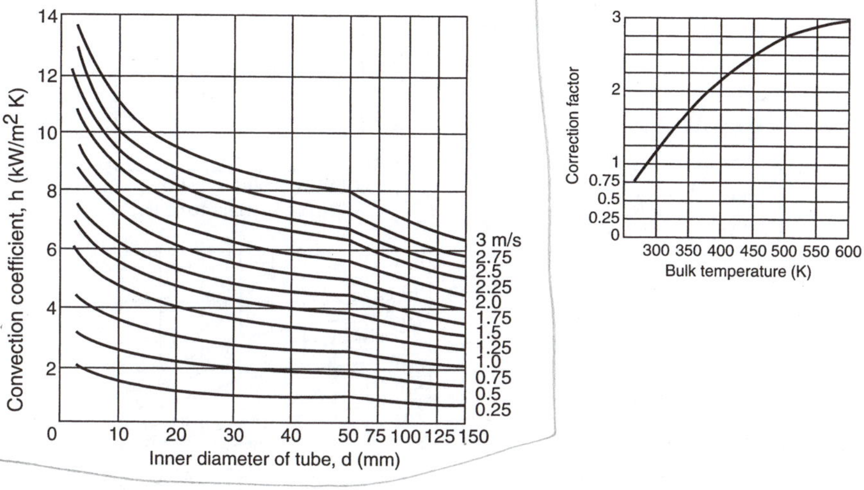

Figure 9.21. Film coefficients of convection for flow of water through a tube at 289 K

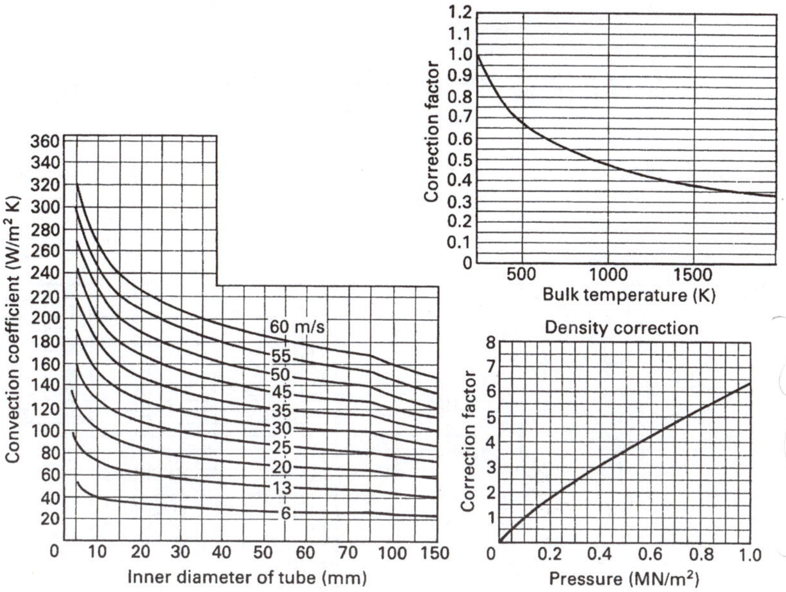

Figure 9.22. Film coefficients of convection for flow of air through a tube at various velocities (289 K, 101.3 kN/m^2)

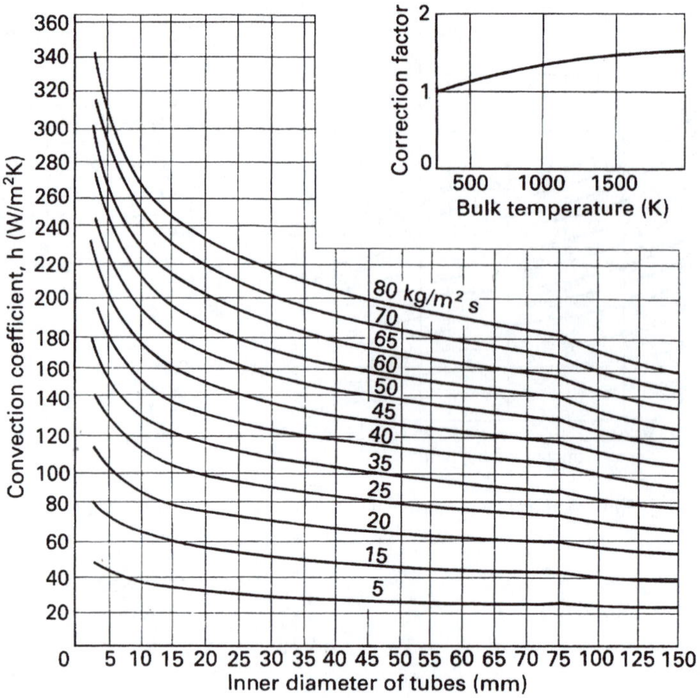

Figure 9.23. Film coefficients of convection for flow of air through a tube for various mass velocities (289 K, 101.3 kN/m^2)

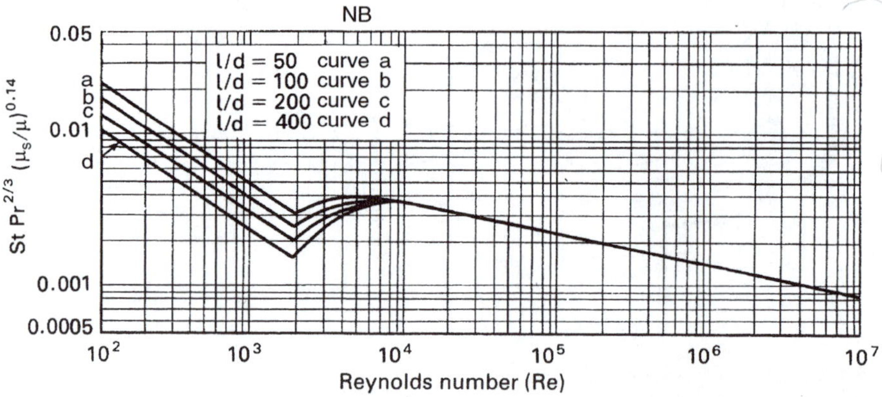

Figure 9.24. Effect of length:diameter ratio on heat transfer coefficient

ranging from 0.022 to 0.14 of the pipe diameter, found that, although the friction loss was some six times greater than for smooth tubes, the heat transfer was only 100–120 per cent higher. It was concluded that, for the same pressure drop, greater heat transfer was obtained from a smooth rather than a rough tube. The effect of a given scale deposit is usually less serious for gases than water because of the higher thermal resistance of

Table 9.2. Film coefficients for air and water (289 K and 101.3 kN/m^2)

Inside diameter of tube		Velocity		Mass velocity		Film coefficient of heat transfer h	
(mm)	(in)	(m/s)	(ft/s)	(kg/m^2 s)	(lb/ft^2 h)	(W/m^2 K) [Ref.18]	(Btu/h ft^2 °F) [Ref.18]
Air							
25	1.0	5	16.4	6.11	4530	31.2	5.5
		10	32.8	12.2	9050	50.0	8.8
		20	65.6	24.5	18,100	84.0	14.8
		40	131	48.9	36,200	146	25.7
		60	197	73.4	54,300	211	37.2
50	2.0	5	16.4	6.11	4530	23.8	4.2
		10	32.8	12.2	9050	44.9	7.9
		20	65.6	24.5	18,100	77.8	13.7
		40	131	48.9	36,200	127	22.4
		60	197	73.4	54,300	181	31.9
75	3.0	5	16.4	6.11	4530	21.6	3.8
		10	32.8	12.2	9050	39.7	7.0
		20	65.6	24.5	18,100	71.0	12.5
		40	131	48.9	36,200	119	21.0
		60	197	73.4	54,300	169	29.8
Water							
25	1.0	0.5	1.64	488	361,000	2160	380
		1.0	3.28	975	722,000	3750	660
		1.5	4.92	1460	1,080,000	5250	925
		2.0	6.55	1950	1,440,000	6520	1150
		2.5	8.18	2440	1,810,000	7780	1370
50	2.0	0.5	1.64	488	361,000	1870	330
		1.0	3.28	975	722,000	3270	575
		1.5	4.92	1460	1,080,000	4540	800
		2.0	6.55	1950	1,440,000	5590	985
		2.5	8.18	2440	1,810,000	6700	1180
75	3.0	0.5	1.64	488	361,000	1760	310
		1.0	3.28	975	722,000	3070	540
		1.5	4.92	1460	1,080,000	4200	740
		2.0	6.55	1950	1,440,000	5220	920
		2.5	8.18	2440	1,810,000	6220	1100

the gas film, although layers of dust or of materials which sublime may seriously reduce heat transfer between gas and solid by as much as 40 per cent.

Streamline flow

Although heat transfer to a fluid in streamline flow takes place solely by conduction, it is convenient to consider it here so that the results may be compared with those for turbulent flow.

In Chapter 3 it has been seen that, for streamline flow through a tube, the velocity distribution across a diameter is parabolic, as shown in Figure 9.25. If a liquid enters a section heated on the outside, the fluid near the wall will be at a higher temperature than that in the centre and its viscosity will be lower. The velocity of the fluid near the wall will therefore be greater in the heated section, and correspondingly less at the centre. The velocity distribution will therefore be altered, as shown. If the fluid enters a section where it is cooled, the same reasoning will show that the distribution in velocity will be altered

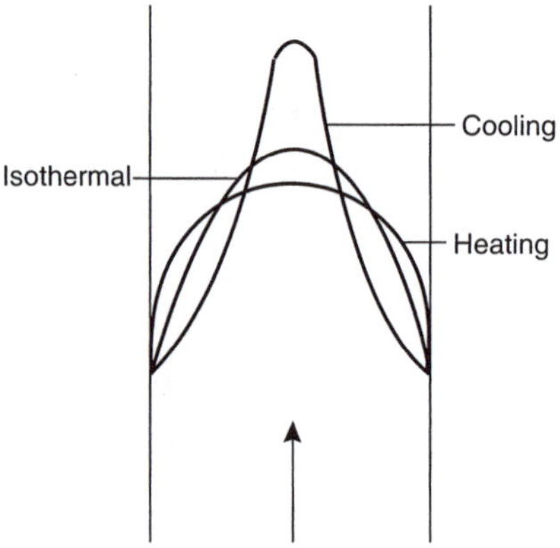

Figure 9.25. Effect of heat transfer on the velocity distribution for a liquid

to that shown. With a gas the conditions are reversed, because of the increase of viscosity with temperature. The heat transfer problem is therefore complex.

For the common problem of heat transfer between a fluid and a tube wall, the boundary layers are limited in thickness to the radius of the pipe and, furthermore, the effective area for heat flow decreases with distance from the surface. The problem can conveniently be divided into two parts. Firstly, heat transfer in the entry length in which the boundary layers are developing, and, secondly, heat transfer under conditions of fully developed flow. Boundary layer flow is discussed in Chapter 11.

For the region of fully developed flow in a pipe of length L, diameter d and radius r, the rate of flow of heat Q through a cylindrical surface in the fluid at a distance y from the wall is given by:

$$Q = -k2\pi L(r - y)\frac{d\theta}{dy} \tag{9.69}$$

Close to the wall, the fluid velocity is low and a negligible amount of heat is carried along the pipe by the flowing fluid in this region and Q is independent of y.

Thus: $$\frac{d\theta}{dy} = -\frac{Q}{k2\pi L}(r - y)^{-1} \quad \text{and} \quad \left(\frac{d\theta}{dy}\right)_{y=0} = -\frac{Q}{2\pi kLr}$$

$$\frac{d^2\theta}{dy^2} = -\frac{Q}{k2\pi L}(r - y)^{-2} \quad \text{and} \quad \left(\frac{d^2\theta}{dy^2}\right)_{y=0} = -\frac{Q}{2\pi kLr^2}$$

Thus: $$\left(\frac{d^2\theta}{dy^2}\right)_{y=0} = r^{-1}\left(\frac{d\theta}{dy}\right)_{y=0} \tag{9.70}$$

Assuming that the temperature of the walls remains constant at the datum temperature and that the temperature at any distance y from the walls is given by a polynomial, then:

$$\theta = a_0 y + b_0 y^2 + c_0 y^3 \tag{9.71}$$

Thus: $$\frac{d\theta}{dy} = a_0 + 2b_0 y + 3c_0 y^2 \quad \text{and} \quad \left(\frac{d\theta}{dy}\right)_{y=0} = a_0$$

$$\frac{d^2\theta}{dy^2} = 2b_0 + 6c_0 y \quad \text{and} \quad \left(\frac{d^2\theta}{dy^2}\right)_{y=0} = 2b_0$$

Thus: $$2b_0 = \frac{a_0}{r} \quad \text{(from equation 9.65)}$$

and: $$b_0 = \frac{a_0}{2r}$$

If the temperature of the fluid at the axis of the pipe is θ_s and the temperature gradient at the axis, from symmetry, is zero, then:

$$0 = a_0 + 2r\left(\frac{a_0}{2r}\right) + 3c_0 r^2$$

giving: $$c_0 = -\frac{2a_0}{3r^2}$$

and: $$\theta_s = a_0 r + r^2\left(\frac{a_0}{2r}\right) + r^3\left(\frac{-2a_0}{3r^2}\right)$$

$$= \frac{5}{6}a_0 r$$

$\therefore$ $$a_0 = \frac{6}{5}\frac{\theta_s}{r}$$

$$b_0 = \frac{3}{5}\frac{\theta_s}{r^2}$$

and: $$c_0 = -\frac{4}{5}\frac{\theta_s}{r^3}$$

Thus: $$\frac{\theta}{\theta_s} = \frac{6}{5}\frac{y}{r} + \frac{3}{5}\left(\frac{y}{r}\right)^2 - \frac{4}{5}\left(\frac{y}{r}\right)^3 \tag{9.72}$$

Thus the rate of heat transfer per unit area at the wall:

$$q = -k\left(\frac{d\theta}{dy}\right)_{y=0}$$

$$= -\frac{6}{5}\frac{k\theta_s}{r} \tag{9.73}$$

In general, the temperature θ_s at the axis is not known, and the heat transfer coefficient is related to the temperature difference between the walls and the bulk fluid. The bulk temperature of the fluid is defined as the ratio of the heat content to the heat capacity of the fluid flowing at any section. Thus the bulk temperature θ_B is given by:

$$\theta_B = \frac{\displaystyle\int_0^r C_p\rho\theta u_x 2\pi(r-y)\,dy}{\displaystyle\int_0^r C_p\rho u_x 2\pi(r-y)\,dy}$$

$$= \frac{\int_0^r \theta u_x(r-y)\,dy}{\int_0^r u_x(r-y)\,dy} \tag{9.74}$$

From Poiseuille's law (equation 3.30):

$$u_x = \frac{-\Delta P}{4\mu L}[r^2 - (r-y)^2] = \frac{-\Delta P}{4\mu L}(2ry - y^2)$$

Hence:

$$u_s = \frac{-\Delta P}{4\mu L}r^2 \tag{9.75}$$

where u_s is the velocity at the pipe axis,

and:

$$\frac{u_x}{u_s} = \frac{2y}{r} - \left(\frac{y}{r}\right)^2 \tag{9.76}$$

Thus:

$$\int_0^r u_x(r-y)\,dy = r^2 u_s \int_0^1 \left[2\frac{y}{r} - \left(\frac{y}{r}\right)^2\right]\left(1 - \frac{y}{r}\right) d\left(\frac{y}{r}\right)$$

$$= r^2 u_s \int_0^1 \left[2\left(\frac{y}{r}\right) - 3\left(\frac{y}{r}\right)^2 + \left(\frac{y}{r}\right)^3\right] d\left(\frac{y}{r}\right)$$

$$= \frac{1}{4}r^2 u_s \tag{9.77}$$

Since:

$$\frac{\theta}{\theta_s} = \frac{6}{5}\frac{y}{r} + \frac{3}{5}\left(\frac{y}{r}\right)^2 - \frac{4}{5}\left(\frac{y}{r}\right)^3 \qquad \text{(equation 9.72)}$$

$$\int_0^r \theta u_x(r-y)\,dy = r^2 u_s\theta_s \int_0^1 \left[\frac{6}{5}\frac{y}{r} + \frac{3}{5}\left(\frac{y}{r}\right)^2 - \frac{4}{5}\left(\frac{y}{r}\right)^3\right]\left[2\left(\frac{y}{r}\right) - 3\left(\frac{y}{r}\right)^2 + \left(\frac{y}{r}\right)^3\right] d\left(\frac{y}{r}\right)$$

$$= r^2 u_s\theta_s \int_0^1 \left[\frac{12}{5}\left(\frac{y}{r}\right)^2 - \frac{12}{5}\left(\frac{y}{r}\right)^3 - \frac{11}{5}\left(\frac{y}{r}\right)^4 + 3\left(\frac{y}{r}\right)^5 - \frac{4}{5}\left(\frac{y}{r}\right)^6\right] d\left(\frac{y}{r}\right)$$

$$= r^2 u_s\theta_s \left(\frac{4}{5} - \frac{3}{5} - \frac{11}{25} + \frac{1}{2} - \frac{4}{35}\right)$$

$$= \frac{51}{350} r^2 u_s \theta_s \tag{9.78}$$

Substituting from equations 9.77 and 9.78 in equation 9.74:

$$\theta_B = \frac{\frac{51}{350} r^2 u_s \theta_s}{\frac{1}{4} r^2 u_s}$$

$$= \frac{102}{175}\theta_s = 0.583\theta_s \tag{9.79}$$

The heat transfer coefficient h is then given by:

$$h = -\frac{q}{\theta_B}$$

where q is the rate of heat transfer per unit area of tube.

Thus, from equations 9.73 and 9.79:

$$h = \frac{6k\theta_s/5r}{0.583\theta_s} = \frac{2.06k}{r} = 4.1\frac{k}{d}$$

and:

$$Nu = \frac{hd}{k} = 4.1 \tag{9.80}$$

This expression is applicable only to the region of fully developed flow. The heat transfer coefficient for the inlet length can be calculated approximately, using the expressions given in Chapter 11 for the development of the boundary layers for the flow over a plane surface. It should be borne in mind that it has been assumed throughout that the physical properties of the fluid are not appreciably dependent on temperature and therefore the expressions will not be expected to hold accurately if the temperature differences are large and if the properties vary widely with temperature.

For values of $(RePr\, d/l)$ greater than 12, the following empirical equation is applicable:

$$Nu = 1.62\left(RePr\frac{d}{l}\right)^{1/3} = 1.75\left(\frac{GC_p}{kl}\right)^{1/3} \tag{9.81}$$

where $G = (\pi d^2/4)\rho u$, i.e. the mass rate of flow.

The product $RePr$ is termed the Peclet number Pe.

Thus:

$$Pe = \frac{ud\rho}{\mu}\frac{C_p\mu}{k} = \frac{C_p\rho u d}{k} \tag{9.82}$$

Equation 9.81 may then be written:

$$Nu = 1.62\left(Pe\frac{d}{l}\right)^{1/3} \tag{9.83}$$

In this equation the temperature difference is taken as the arithmetic mean of the terminal values, that is:

$$\frac{(T_w - T_1) + (T_w - T_2)}{2}$$

where T_w is the temperature of the tube wall which is taken as constant.

If the liquid is heated almost to the wall temperature T_w (that is when GC_p/kl is very small) then, on equating the heat gained by the liquid to that transferred from the pipe:

$$GC_p(T_2 - T_1) = \pi d l h \frac{T_2 - T_1}{2}$$

or:

$$h = \frac{2}{\pi}\frac{GC_p}{dl} \tag{9.84}$$

For values of $(RePr\, d/l)$ less than about 17, the Nusselt group becomes approximately constant at 4.1; the value given in equation 9.80.

Experimental values of h for *viscous oils* are greater than those given by equation 9.81 for heating and less for cooling. This is due to the large variation of viscosity with temperature and the correction introduced for turbulent flow may also be used here, giving:

$$Nu\left(\frac{\mu_s}{\mu}\right)^{0.14} = 1.86\left(RePr\frac{d}{l}\right)^{1/3} = 2.01\left(\frac{GC_p}{kl}\right)^{1/3} \tag{9.85}$$

or: $$Nu\left(\frac{\mu_s}{\mu}\right)^{0.14} = 1.86\left(Pe\frac{d}{l}\right)^{1/3} \qquad (9.86)$$

When $(GC_p/kl) < 10$, the outlet temperature closely approaches that of the wall and equation 9.84 applies. These equations have been obtained with tubes about 10 mm to 40 mm in diameter, and the length of unheated tube preceding the heated section is important. The equations are not entirely consistent since for very small values of ΔT the constants in equations 9.81 and 9.85 would be expected to be the same. It is important to note, when using these equations for design purposes, that the error may be as much as ±25 per cent for turbulent flow and greater for streamline conditions.

With laminar flow there is a marked influence of tube length and the curves shown in Figure 9.24 show the parameter l/d from 50 to 400.

Whenever possible, streamline conditions of flow are avoided in heat exchangers because of the very low heat transfer coefficients which are obtained. With very viscous liquids, however, turbulent conditions can be produced only if a very high pressure drop across the plant is permissible. In the processing industries, streamline flow in heat exchangers is most commonly experienced with heavy oils and brines at low temperatures. Since the viscosity of these materials is critically dependent on temperature, the equations would not be expected to apply with a high degree of accuracy.

9.4.4. Forced convection outside tubes

Flow across single cylinders

If a fluid passes at right angles across a single tube, the distribution of velocity around the tube will not be uniform. In the same way the rate of heat flow around a hot pipe across which air is passed is not uniform but is a maximum at the front and rear, and a minimum at the sides, where the rate is only some 40 per cent of the maximum. The general picture is shown in Figure 9.26 but for design purposes reference is made to the average value.

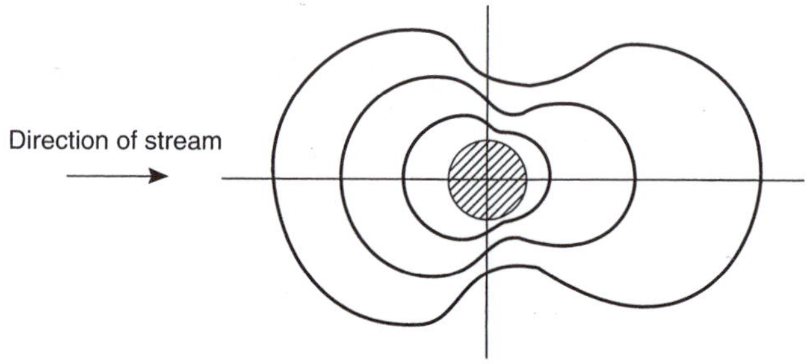

Figure 9.26. Distribution of the film heat transfer coefficient round a cylinder with flow normal to the axis for three different values of *Re*

A number of workers, including, REIHER[20], HILPERT[21], GRIFFITHS and AWBERY[22], have studied the flow of a hot gas past a single cylinder, varying from a thin wire to a tube of 150 mm diameter. Temperatures up to 1073 K and air velocities up to 30 m/s have been used with Reynolds numbers ($d_o u\rho/\mu$) from 1000 to 100,000 (where d_o is the cylinder diameter, or the outside tube diameter). The data obtained may be expressed by:

$$Nu = 0.26Re^{0.6}Pr^{0.3} \tag{9.87}$$

Taking Pr as 0.74 for gases, this reduces to

$$Nu = 0.24Re^{0.6} \tag{9.88}$$

DAVIS[23] has also worked with water, paraffin, and light oils and obtained similar results. For very low values of Re (from 0.2 to 200) with liquids the data are better represented by the equation:

$$Nu = 0.86Re^{0.43}Pr^{0.3} \tag{9.89}$$

In each case the physical properties of the fluid are measured at the mean film temperature T_f, taken as the average of the surface temperature T_w and the mean fluid temperature T_m; where $T_m = (T_1 + T_2)/2$. mean bulk temp.

Flow at right angles to tube bundles

One of the great difficulties with this geometry is that the area for flow is continually changing. Moreover the degree of turbulence is considerably less for banks of tubes in line, as at (*a*), than for staggered tubes, as at (*b*) in Figure 9.27. With the small bundles which are common in the processing industries, the selection of the true mean area for flow is further complicated by the change in number of tubes in the rows.

The results of a number of workers for heat transfer to and from gases flowing across tube banks may be expressed by the equation:

$$Nu = 0.33C_h Re_{\max}^{0.6} Pr^{0.3} \tag{9.90}$$

where C_h depends on the geometrical arrangement of the tubes, as shown in Table 9.3. GRIMISON[24] proposed this form of expression to correlate the data of HUGE[25] and PIERSON[26] who worked with small electrically heated tubes in rows of ten deep. Other workers have used similar equations. Some correction factors have been given by PIERSON[26] for bundles with less than ten rows although there are insufficient reported data from commercial exchangers to fix these values with accuracy. Thus for five rows a factor of 0.92 and for eight rows 0.97 is suggested.

These equations are based on the maximum velocity through the bundle. Thus for an in-line arrangement as is shown in Figure 9.27*a*, $G'_{\max} = G'Y/(Y - d_o)$, where Y is the pitch of the pipes at right-angles to direction of flow; it is more convenient here to use the mass flowrate per unit area G' in place of velocity. For staggered arrangements the maximum velocity may be based on the distance between the tubes in a horizontal line or on the diagonal of the tube bundle, whichever is the less.

It has been suggested that, for in-line arrangements, the constant in equation 9.90 should be reduced to 0.26, but there is insufficient evidence from commercial exchangers to confirm this.

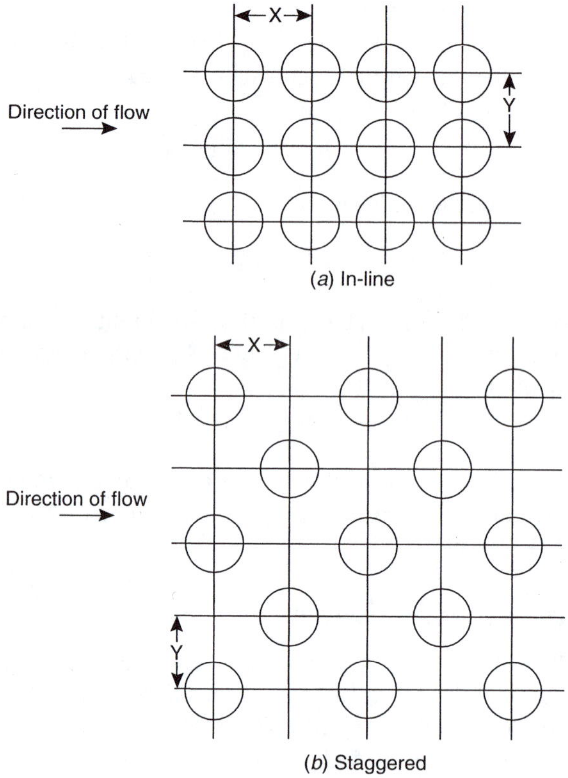

Figure 9.27. Arrangements of tubes in heat exchangers

Table 9.3.[18] Values of C_h and C_f

	$X = 1.25d_o$				$X = 1.5d_o$			
	In-line		Staggered		In-line		Staggered	
Re_{max}	C_h	C_f	C_h	C_f	C_h	C_f	C_h	C_f
				$Y = 1.25d_o$				
2000	1.06	1.68	1.21	2.52	1.06	1.74	1.16	2.58
20,000	1.00	1.44	1.06	1.56	1.00	1.56	1.05	1.74
40,000	1.00	1.20	1.03	1.26	1.00	1.32	1.02	1.50
				$Y = 1.5d_o$				
2000	0.95	0.79	1.17	1.80	0.95	0.97	1.15	1.80
20,000	0.96	0.84	1.04	1.10	0.96	0.96	1.02	1.16
40,000	0.96	0.74	0.99	0.88	0.96	0.85	0.98	0.96

With liquids the same equation may be used, although for *Re* less than 2000, there is insufficient published work to justify an equation. McAdams,[27] however, has given a curve for h for a bundle with staggered tubes ten rows deep.

An alternative approach has been suggested by Kern[28] who worked in terms of the hydraulic mean diameter d_e for flow *parallel* to the tubes:

i.e.:
$$d_e = 4 \times \frac{\text{Free area for flow}}{\text{Wetted perimeter}}$$
$$= 4\left(\frac{Y^2 - (\pi d_o^2/4)}{\pi d_o}\right)$$

for a square pitch as shown in Figure 9.28. The maximum cross-flow area A_s is then given by:

$$A_s = \frac{d_s l_B C'}{Y}$$

where C' is the clearance, l_B the baffle spacing, and d_s the internal diameter of the shell.

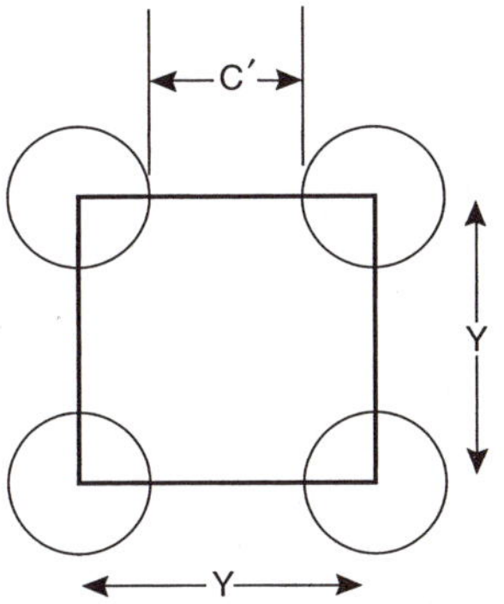

Figure 9.28. Clearance and pitch for tube layouts

The mass rate of flow per unit area G'_s is then given as rate of flow divided by A_s, and the film coefficient is obtained from a Nusselt type expression of the form:

$$\frac{h_o d_e}{k} = 0.36\left(\frac{d_e G'_s}{\mu}\right)^{0.55}\left(\frac{C_p \mu}{k}\right)^{1/3}\left(\frac{\mu}{\mu_s}\right)^{0.14} \tag{9.91}$$

There are insufficient published data to assess the relative merits of equations 9.90 and 9.91.

For 19 mm tubes on 25 mm square pitch:

$$d_e = 4\frac{[25^2 - (\pi/4)19^2]}{\pi \times 19}$$
$$= 22.8 \text{ mm or } 0.023 \text{ m}$$

Example 9.9

14.4 tonne/h (4.0 kg/s) of nitrobenzene is to be cooled from 400 to 315 K by heating a stream of benzene from 305 to 345 K.

Two tubular heat exchangers are available each with a 0.44 m i.d. shell fitted with 166 tubes, 19.0 mm o.d. and 15.0 mm i.d., each 5.0 m long. The tubes are arranged in two passes on 25 mm square pitch with a baffle spacing of 150 mm. There are two passes on the shell side and operation is to be countercurrent. With benzene passing through the tubes, the anticipated film coefficient on the tube side is 1000 W/m^{2}K.

Assuming true cross-flow prevails in the shell, what value of scale resistance could be allowed if these units were used?

For nitrobenzene: $C_p = 2380$ J/kg K, $k = 0.15$ W/m K, $\mu = 0.70$ mN s/m^2

Solution

(i) Tube side coefficient.

$$h_i = 1000 \text{ W/m}^2 \text{ K based on inside area}$$

or:
$$\frac{1000 \times 15.0}{19.0} = 790 \text{ W/m}^2 \text{ K based on outside area}$$

(ii) Shell side coefficient.

$$\text{Area for flow} = \text{shell diameter} \times \text{baffle spacing} \times \text{clearance/pitch}$$
$$= \frac{0.44 \times 0.150 \times 0.006}{0.025} = 0.0158 \text{ m}^2$$

Hence:
$$G'_s = \frac{4.0}{0.0158} = 253.2 \text{ kg/m}^2\text{s}$$

Taking $\mu/\mu_s = 1$ in equation 9.91:

$$h_o = 0.36\frac{k}{d_e}\left(\frac{d_e G'_s}{\mu}\right)^{0.55}\left(\frac{C_p\mu}{k}\right)^{0.33}$$

The hydraulic mean diameter,

$$d_e = 4\left[\left(25^2 - \frac{\pi \times 19.0^2}{4}\right) \Big/ (\pi \times 19.0)\right] = 22.8 \text{ mm} \quad \text{or} \quad 0.023 \text{ m}$$

and here:
$$h_o = \left(\frac{0.15}{0.023}\right) 0.36 \left(\frac{0.023 \times 253.2}{0.70 \times 10^{-3}}\right)^{0.55}\left(\frac{2380 \times 0.70 \times 10^{-3}}{0.15}\right)^{0.33}$$
$$= 2.35 \times 143 \times 2.23 = 750 \text{ W/m}^2 \text{ K}$$

(iii) Overall coefficient.

The logarithmic mean temperature difference is given by:

$$\Delta T_m = \frac{(400 - 345) - (315 - 305)}{\ln(400 - 345)/(315 - 305)}$$
$$= 26.4 \text{ deg K}$$

The corrected mean temperature difference is then $\Delta T_m \times F = 26.4 \times 0.8 = 21.1$ deg K

(Details of the correction factor for ΔT_m are given in Section 9.9.3)

Heat load:
$$Q = 4.0 \times 2380(400 - 315) = 8.09 \times 10^5 \text{ W}$$

The surface area of each tube $= 0.0598$ m^2/m

Thus:
$$U_o = \frac{Q}{A_o \Delta T_m F} = \frac{8.09 \times 10^5}{2 \times 166 \times 5.0 \times 0.0598 \times 21.1}$$
$$= 386.2 \text{ W/m}^2 \text{ K}$$

(iv) Scale resistance.

If scale resistance is R_d, then:

$$R_d = \frac{1}{386.2} - \frac{1}{750} - \frac{1}{1000} = \underline{\underline{0.00026 \text{ m}^2 \text{ K/W}}}$$

This is a rather low value, though the heat exchangers would probably be used for this duty.

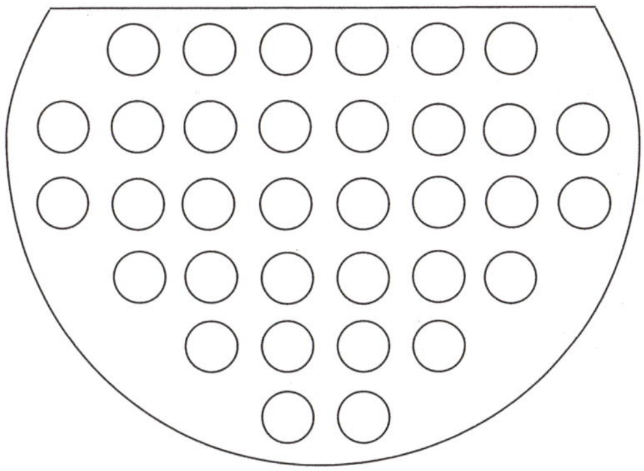

Figure 9.29. Baffle for heat exchanger

As discussed in Section 9.9 it is common practice to fit *baffles* across the tube bundle in order to increase the velocity over the tubes. The commonest form of baffle is shown in Figure 9.29 where it is seen that the cut-away section is about 25 per cent of the total area. With such an arrangement, the flow pattern becomes more complex and the extent of leakage between the tubes and the baffle, and between the baffle and the inside of the shell of the exchanger, complicates the problem, as discussed further in Section 9.9.6. Reference may also be made in Volume 6 and to the work of SHORT(29), DONOHUE(30), and TINKER(31). The various methods are all concerned with developing a method of calculating the true area of flow and of assessing the probable influence of leaks. When using baffles, the value of h_o, as found from equation 9.89, is commonly multiplied by 0.6 to allow for leakage although more accurate approaches have been developed as discussed in Section 9.9.6.

The *drop in pressure* for the flow of a fluid across a tube bundle may be important because of the small pressure head available and because by good design it is possible to get a better heat transfer for the same drop in pressure. $-\Delta P_f$ depends on the velocity u_t through the minimum area of flow and in Chapter 3 an equation proposed by GRIMISON(24) is given as:

$$-\Delta P_f = \frac{C_f j \rho u_t^2}{6} \qquad \text{(equation 3.83)}$$

Table 9.4.(18) Ratio of heat transfer to friction for tube bundles ($Re_{\max} = 20{,}000$)

	$X = 1.25d_o$			$X = 1.5d_o$		
	C_h	C_f	C_h/C_f	C_h	C_f	C_h/C_f
In-line						
$Y = 1.25d_o$	1	1.44	0.69	1	1.56	0.64
$Y = 1.5d_o$	0.96	0.84	1.14	0.96	0.96	1.0
Staggered						
$Y = 1.25d_o$	1.06	1.56	0.68	1.05	1.74	0.60
$Y = 1.5d_o$	1.04	1.10	0.95	1.02	1.16	0.88

where C_f depends on the geometry of the tube layout and j is the number of rows of tubes. It is found that the ratio of C_h, the heat transfer factor in equation 9.90, to C_f

is greater for the in-line arrangement but that the actual heat transfer is greater for the staggered arrangement, as shown in Table 9.4.

The drop in pressure $-\Delta P_f$ over the tube bundles of a heat exchanger is also given by:

$$-\Delta P_f = \frac{f' G_s'^2 (n+1) d_v}{2\rho d_e} \tag{9.92}$$

where f' is the friction factor given in Figure 9.30, G_s' the mass velocity through bundle, n the number of baffles in the unit, d_v the inside shell diameter, ρ the density of fluid, d_e the equivalent diameter, and $-\Delta P_f$ the drop in pressure.

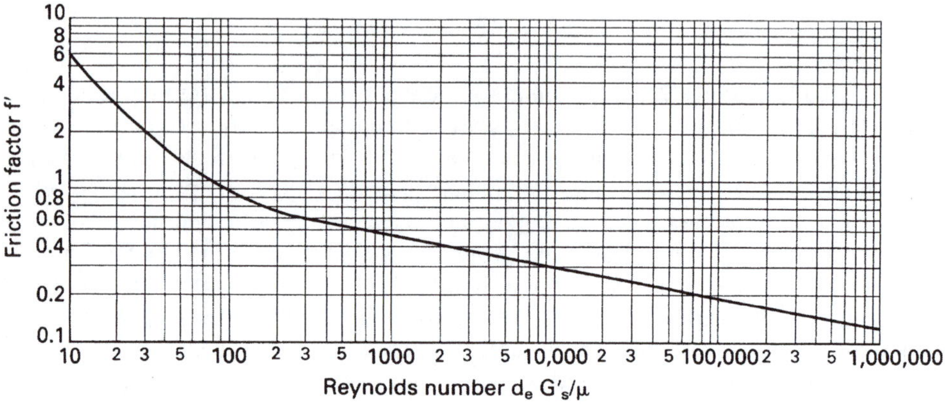

Figure 9.30. Friction factor for flow over tube bundles

Example 9.10

54 tonne/h (15 kg/s) of benzene is cooled by passing the stream through the shell side of a tubular heat exchanger, 1 m i.d., fitted with 5 m tubes, 19 mm o.d. arranged on a 25 mm square pitch with 6 mm clearance. If the baffle spacing is 0.25 m (19 baffles), what will be the pressure drop over the tube bundle? ($\mu = 0.5$ mN s/m^2).

Solution

Cross-flow area: $$A_s = \frac{1.0 \times 0.25 \times 0.006}{0.025} = 0.06 \text{ m}^2$$

Mass flow: $$G_s' = \frac{15}{0.06} = 250 \text{ kg/m}^2 \text{ s}$$

Equivalent diameter: $$d_e = \frac{4[0.025^2 - (\pi/4)0.019^2]}{\pi \times 0.019} = 0.0229 \text{ m}$$

Reynolds number through the tube bundle $$= \frac{250 \times 0.0229}{0.5 \times 10^{-3}} = 11450$$

From Figure 9.29: $f' = 0.280$

Density of benzene $= 881$ kg/m^3

From equation 9.92:

$$-\Delta P_f = \frac{0.280 \times 250^2 \times 20 \times 1.0}{2 \times 881 \times 0.0229} = \underline{\underline{8674 \text{ N/m}^2}}$$

or:

$$\frac{8674}{881 \times 9.81} = \underline{\underline{1.00 \text{ m of benzene}}}$$

9.4.5. Flow in non-circular sections

Rectangular ducts

For the heat transfer for fluids flowing in non-circular ducts, such as rectangular ventilating ducts, the equations developed for turbulent flow inside a circular pipe may be used if an equivalent diameter, such as the hydraulic mean diameter d_e discussed previously, is used in place of d.

The data for heating and cooling water in turbulent flow in rectangular ducts are reasonably well correlated by the use of equation 9.59 in the form:

$$\frac{hd_e}{k} = 0.023 \left(\frac{d_e G'}{\mu}\right)^{0.8} \left(\frac{C_p \mu}{k}\right)^{0.4} \tag{9.93}$$

Whilst the experimental data of COPE and BAILEY[32] are somewhat low, the data of WASHINGTON and MARKS[33] for heating air in ducts are well represented by this equation.

Annular sections between concentric tubes

Concentric tube heat exchangers are widely used because of their simplicity of construction and the ease with which additions may be made to increase the area. They also give turbulent conditions at low volumetric flowrates.

In presenting equations for the film coefficient in the annulus, one of the difficulties is in selecting the best equivalent diameter to use. When considering the film on the outside of the inner tube, DAVIS[34] has proposed the equation:

$$\frac{hd_1}{k} = 0.031 \left(\frac{d_1 G'}{\mu}\right)^{0.8} \left(\frac{C_p \mu}{k}\right)^{0.33} \left(\frac{\mu}{\mu_s}\right)^{0.14} \left(\frac{d_2}{d_1}\right)^{0.15} \tag{9.94}$$

where d_1 and d_2 are the outer diameter of the inner tube, and the inner diameter of the outer tube, respectively.

CARPENTER *et al.*[35] suggest using the hydraulic mean diameter $d_e = (d_2 - d_1)$ in the Sieder and Tate equation (9.66) and recommend the equation:

$$\frac{hd_e}{k} \left(\frac{\mu_s}{\mu}\right)^{0.14} = 0.027 \left(\frac{d_e G'}{\mu}\right)^{0.8} \left(\frac{C_p \mu}{k}\right)^{0.33} \tag{9.95}$$

Their data, which were obtained using a small annulus, are somewhat below those given by equation 9.95 for values of $d_e G'/\mu$ less than 10,000, although this may be because the flow was not fully turbulent: with an index on the Reynolds group of 0.9, the equation fitted the data much better. There is little to choose between these two equations, but they both give rather high values for h.

For the viscous region, Carpenter's results are reasonably well correlated by the equation:

$$\frac{hd_e}{k}\left(\frac{\mu_s}{\mu}\right)^{0.14} = 2.01\left(\frac{GC_p}{kl}\right)^{0.33} \tag{9.96}$$

$$= 1.86\left[\left(\frac{d_eG'}{\mu}\right)\left(\frac{C_p\mu}{k}\right)\left(\frac{d_1+d_2}{l}\right)\right]^{1/3} \tag{9.97}$$

Equations 9.96 and 9.97 are the same as equations 9.85 and 9.86, with d_e replacing d.

These results have all been obtained with small units and mainly with water as the fluid in the annulus.

Flow over flat plates

For the turbulent flow of a fluid over a flat plate the Colburn type of equation may be used with a different constant:

$$j_h = 0.037Re_x^{-0.2} \tag{9.98}$$

where the physical properties are taken as for equation 9.64 and the characteristic dimension in the Reynolds group is the actual distance x along the plate. This equation therefore gives a point value for j_h.

9.4.6. Convection to spherical particles

In Section 9.3.4, consideration is given to the problem of heat transfer by conduction through a surrounding fluid to spherical particles or droplets. Relative motion between the fluid and particle or droplet causes an increase in heat transfer, much of which may be due to convection. Many investigators have correlated their data in the form:

$$Nu' = 2 + \beta''Re'^n Pr^m \tag{9.99}$$

where values of β'', a numerical constant, and exponents n and m are found by experiment. In this equation, $Nu' = hd/k$ and $Re' = du\rho/\mu$, the Reynolds number for the particle, u is the relative velocity between particle and fluid, and d is the particle diameter. As the relative velocity approaches zero, Re' tends to zero and the equation reduces to $Nu' = 2$ for pure conduction.

ROWE *et al.*[(36)], having analysed a large number of previous studies in this area and provided further experimental data, have concluded that for particle Reynolds numbers in the range 20–2000, equation 9.99 may be written as:

$$Nu' = 2.0 + \beta''Re'^{0.5}Pr^{0.33} \tag{9.100}$$

where β'' lies between 0.4 and 0.8 and has a value of 0.69 for air and 0.79 for water. In some practical situations the relative velocity between particle and fluid may change due to particle acceleration or deceleration, and the value of Nu' can then be time-dependent.

For mass transfer, which is considered in more detail in Chapter 10, an analogous relation (equation 10.233) applies, with the Sherwood number replacing the Nusselt number and the Schmidt number replacing the Prandtl number.

9.4.7. Natural convection

If a beaker containing water rests on a hot plate, the water at the bottom of the beaker becomes hotter than that at the top. Since the density of the hot water is lower than that of the cold, the water in the bottom rises and heat is transferred by natural convection. In the same way air in contact with a hot plate will be heated by natural convection currents, the air near the surface being hotter and of lower density than that some distance away. In both of these cases there is no external agency providing forced convection currents, and the transfer of heat occurs at a correspondingly lower rate since the natural convection currents move rather slowly.

For these processes which depend on buoyancy effects, the rate of heat transfer might be expected to follow a relation of the form:

$$Nu = f(Gr, Pr) \qquad \text{(equation 9.57)}$$

Measurements by SCHMIDT[37] of the upward air velocity near a 300 mm vertical plate show that the velocity rises rapidly to a maximum at a distance of about 2 mm from the plate and then falls rapidly. However, the temperature evens out at about 10 mm from the plate. Temperature measurements around horizontal cylinders have been made by RAY[38].

Natural convection from horizontal surfaces to air, nitrogen, hydrogen, and carbon dioxide, and to liquids (including water, aniline, carbon tetrachloride, glycerol) has been studied by several workers, including DAVIS[39], ACKERMANN[40], FISHENDEN and SAUNDERS[18] and SAUNDERS[41]. Most of the results are for thin wires and tubes up to about 50 mm diameter; the temperature differences used are up to about 1100 deg K with gases and about 85 deg K with liquids. The general form of the results is shown in Figure 9.31, where log Nu is plotted against log ($Pr\ Gr$) for streamline conditions. The curve can be represented by a relation of the form:

$$Nu = C'(Gr\ Pr)^n \qquad (9.101)$$

Numerical values of C' and n, determined experimentally for various geometries, are given in Table 9.5[42]. Values of coefficients may then be predicted using the equation:

$$\frac{hl}{k} = C'\left(\frac{\beta g \Delta T l^3 \rho^2}{\mu^2}\frac{C_p\mu}{k}\right)^n \quad \text{or} \quad h = C'\left(\frac{\Delta T}{l}\right)^n k\left(\frac{\beta g \rho^2 C_p}{\mu k}\right)^n \qquad (9.102)$$

Table 9.5. Values of C', C'' and n for use in equations 9.102 and 9.105[42]

Geometry	*GrPr*	C'	n	C'' (SI units) (for air at 294 K)
Vertical surfaces	$< 10^4$	1.36	0.20	
(l = vertical dimension < 1 m)	$10^4 - 10^9$	0.59	0.25	1.37
	$> 10^9$	0.13	0.33	1.24
Horizontal cylinders				
(l = diameter < 0.2 m)	$10^{-5} - 10^{-3}$	0.71	0.04	
	$10^{-3} - 1.0$	1.09	0.10	
	$1.0 - 10^4$	1.09	0.20	
	$10^4 - 10^9$	0.53	0.25	1.32
	$> 10^9$	0.13	0.33	1.24
Horizontal flat surfaces				
(facing upwards)	$10^5 - 2 \times 10^7$	0.54	0.25	1.86
(facing upwards)	$2 \times 10^7 - 3 \times 10^{10}$	0.14	0.33	
(facing downwards)	$3 \times 10^5 - 3 \times 10^{10}$	0.27	0.25	0.88

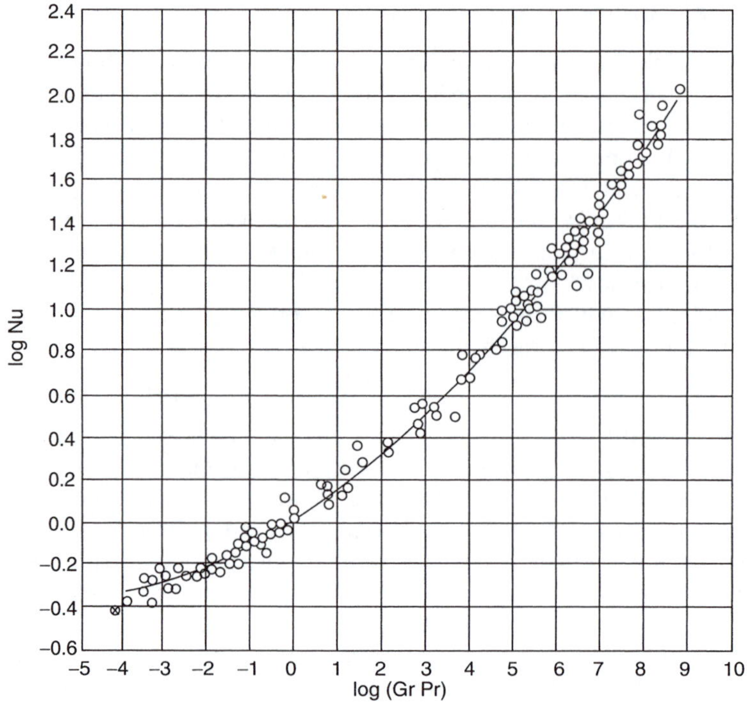

Figure 9.31. Natural convection from horizontal tubes

where the physical properties are at the mean of the surface and bulk temperatures and, for gases, the coefficient of cubical expansion β is taken as $1/T$, where T is the absolute temperature.

For vertical plates and cylinders, KATO *et al.*[(43)] have proposed the following equations for situations where $1 < Pr < 40$:

For $Gr > 10^9$: $$Nu = 0.138Gr^{0.36}(Pr^{0.175} - 0.55) \tag{9.103}$$

and for $Gr < 10^9$: $$Nu = 0.683Gr^{0.25}Pr^{0.25}\left(\frac{Pr}{0.861 + Pr}\right)^{0.25} \tag{9.104}$$

Natural convection to air

Simplified dimensional equations have been derived for air, water and organic liquids by grouping the fluid properties into a single factor in a rearrangement of equation 9.102 to give:

$$h = C''(\Delta T)^n l^{3n-1} \text{ (W/m}^2\text{K)} \tag{9.105}$$

Values of C'' (in SI units) are also given in Table 9.5 for air at 294 K. Typical values for water and organic liquids are 127 and 59 respectively.

Example 9.11

Estimate the heat transfer coefficient for natural convection from a horizontal pipe 0.15 m diameter, with a surface temperature of 400 K to air at 294 K

Solution

Over a wide range of temperature, $k^4(\beta g\rho^2 C_p/\mu k) = 36.0$

For air at a mean temperature of $0.5(400 + 294) = 347$ K, $k = 0.0310$ W/m K (Table 6, Appendix A1)

Thus:

$$\frac{\beta g\rho^2 C_p}{\mu k} = \frac{36.0}{0.0310^4} = 3.9 \times 10^7$$

From Equation 9.102:

$$GrPr = 3.9 \times 10^7 (400 - 294) \times 0.15^3$$
$$= 1.39 \times 10^7$$

From Table 9.5:

$$n = 0.25 \quad \text{and} \quad C'' = 1.32$$

Thus, in Equation 9.104:

$$h = 1.32(400 - 294)^{0.25}(0.15)^{(3\times 0.25)-1}$$
$$= 1.32 \times 106^{0.25} \times 0.15^{-0.25}$$
$$= \underline{\underline{6.81 \text{ W/m}^2 \text{ K}}}$$

Fluids between two surfaces

For the transfer of heat from a hot surface across a thin layer of fluid to a parallel cold surface:

$$\frac{Q}{Q_k} = \frac{h\Delta T}{(k/x)\Delta T} = \frac{hx}{k} = Nu \tag{9.106}$$

where Q_k is the rate at which heat would be transferred by pure thermal conduction between the layers, a distance x apart, and Q is the actual rate.

For $(Gr\ Pr) = 10^3$, the heat transferred is approximately equal to that due to conduction alone, though for $10^4 < Gr\ Pr < 10^6$, the heat transferred is given by:

$$\frac{Q}{Q_k} = 0.15(Gr\ Pr)^{0.25} \tag{9.107}$$

which is noted in Figure 9.32. In this equation the characteristic dimension to be used for the Grashof group is x, the distance between the planes, and the heat transfer is independent of surface area, provided that the linear dimensions of the surfaces are large compared with x. For higher values of $(Gr\ Pr)$, Q/Q_k is proportional to $(Gr\ Pr)^{1/3}$, showing that the heat transferred is not entirely by convection and is not influenced by the distance x between the surfaces.

A similar form of analysis has been given by KRAUSSOLD[(44)] for air between two concentric cylinders. It is important to note from this general analysis that a single layer of air will not be a good insulator because convection currents set in before it becomes 25 mm thick. The good insulating properties of porous materials are attributable to the

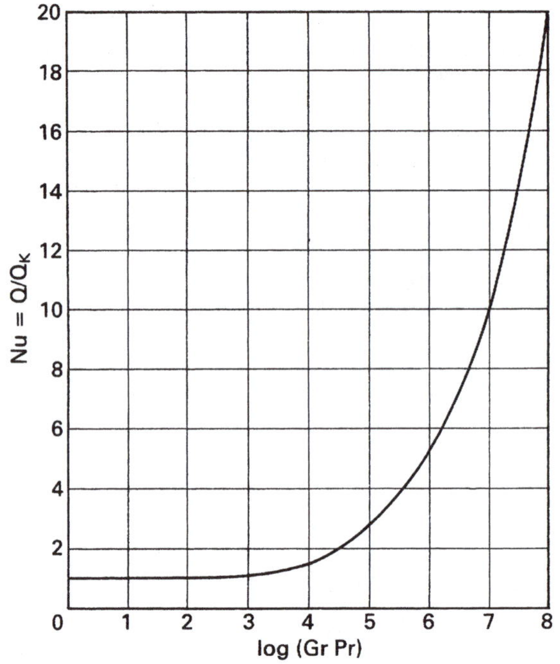

Figure 9.32. Natural convection between surfaces

fact that they offer a series of very thin layers of air in which convection currents are not present.

9.5. HEAT TRANSFER BY RADIATION

9.5.1. Introduction

It has been seen that heat transfer by conduction takes place through either a solid or a stationary fluid and heat transfer by convection takes place as a result of either forced or natural movement of a hot fluid. The third mechanism of heat transfer, radiation, can take place without either a solid or a fluid being present, that is through a vacuum, although many fluids are transparent to radiation, and it is generally assumed that the emission of thermal radiation is by "waves" of wavelengths in the range 0.1–100 μm which travel in straight lines. This means that direct radiation transfer, which is the result of an interchange between various radiating bodies or surfaces, will take place only if a straight line can be drawn between the two surfaces; a situation which is often expressed in terms of one surface "seeing" another. Having said this, it should be noted that opaque surfaces sometimes cast shadows which inhibit radiation exchange and that indirect transfer by radiation can take place as a result of partial reflection from other surfaces. Although all bodies at temperatures in excess of absolute zero radiate energy in all directions, radiation is of especial importance from bodies at high temperatures such as those encountered in furnaces, boilers and high temperature reactors, where in addition to radiation from hot surfaces, radiation from reacting flame gases may also be a consideration.

9.5.2. Radiation from a black body

In thermal radiation, a so-called *black body* absorbs all the radiation falling upon it, regardless of wavelength and direction, and, for a given temperature and wavelength, no surface can emit more energy than a black body. The radiation emitted by a black body, whilst a function of wavelength and temperature, is regarded as *diffuse*, that is, it is independent of direction. In general, most rough surfaces and indeed most engineering materials may be regarded as being diffuse. A black body, because it is a perfect emitter or absorber, provides a standard against which the radiation properties of real surfaces may be compared.

If the *emissive power E* of a radiation source–that is the energy emitted per unit area per unit time–is expressed in terms of the radiation of a single wavelength λ, then this is known as the *monochromatic* or *spectral emissive power* E_λ, defined as that rate at which radiation of a particular wavelength λ is emitted per unit surface area, per unit wavelength in all directions. For a black body at temperature T, the spectral emissive power of a wavelength λ is given by *Planck's Distribution Law*:

$$E_{\lambda,b} = C_1/[\lambda^5(\exp(C_2/\lambda \mathrm{T}) - 1)] \tag{9.108}$$

where, in SI units, $E_{\lambda,b}$ is in W/m^3 and $C_1 = 3.742 \times 10^{-16}$ Wm^2 and $C_2 = 1.439 \times 10^{-2}$ mK are the respective radiation constants. Equation 9.108 permits the evaluation of the emissive power from a black body for a given wavelength and absolute temperature and values obtained from the equation are plotted in Figure 9.33 which is based on the work of INCROPERA and DE WITT(45). It may be noted that, at a given wavelength, the radiation from a black body increases with temperature and that, in general, short wavelengths are associated with high temperature sources.

Example 9.12

What is the temperature of a surface coated with carbon black if the emissive power at a wavelength of 1.0×10^{-6} m is 1.0×10^9 W/m^3? How would this be affected by a +2 per cent error in the emissive power measurement?

Solution

From equation 9.108 $\quad \exp(C_2/\lambda T) = [C_1/E_{\lambda,b}\lambda^5) + 1]$

or: $\quad \exp(1.439 \times 10^{-2}/(1.0 \times 10^{-6}T)) = [3.742 \times 10^{-16}/(1 \times 10^9 \times (1.0 \times 10^{-6})^5)] + 1$

$$= 3.742 \times 10^5$$

Thus: $\quad (1.439 \times 10^4)T = \ln(3.742 \times 10^5) = 12.83$

and: $\quad T = (1.439 \times 10^4)/12.83 = \underline{\underline{1121\ \mathrm{K}}}$

With an error of +2 per cent, the correct value is given by:

$$E_{\lambda,b} = (100 - 2)(1 \times 10^9)/100 = 9.8 \times 10^8\ \mathrm{W/m^3}$$

In equation 9.108:

$$9.8 \times 10^8 = (3.742 \times 10^{-16})/[(1 \times 10^{-6})^5(\exp(1.439 \times 10^{-2}/(1.0 \times 10^{-6}T)) - 1)]$$

and:

$$\underline{\underline{T = 1120\ \mathrm{K}}}$$

Thus, the error in the calculated temperature of the surface is only 1 °K.

The wavelength at which maximum emission takes place is related to the absolute temperature by *Wein's Displacement Law*, which states that the wavelength for maximum emission varies inversely with the absolute temperature of the source, or:

$$\lambda_{max}T = \text{constant}, C_3 (= 2.898 \times 10^{-3} \text{ mK in SI units}) \tag{9.109}$$

Thus, combining equations 9.108 and 9.109:

$$E_{\lambda max,b} = C_1 / \left[(C_3/T)^5[\exp(C_2/C_3) - 1]\right]$$

or:

$$E_{\lambda max,b} = C_4 T^5 \tag{9.110}$$

where, in SI units, the fourth radiation constant, $C_4 = 12.86 \times 10^{-6}$ W/m^3 K^5. Values of the maximum emissive power are shown by the broken line in Figure 9.33.

An interesting feature of Figure 9.33 is that it illustrates the well-known *greenhouse effect* which depends on the ability of glass to transmit radiation from a hot source over only a limited range of wavelengths. This warms the air in the greenhouse, though to a much lower temperature than that of the external source, the sun; a temperature at which the wavelength will be much longer, as seen from Figure 9.33, and one at which the glass will not transmit radiation. In this way, radiation outwards from within a greenhouse is considerably reduced and the air within the enclosure retains its heat. Much the same

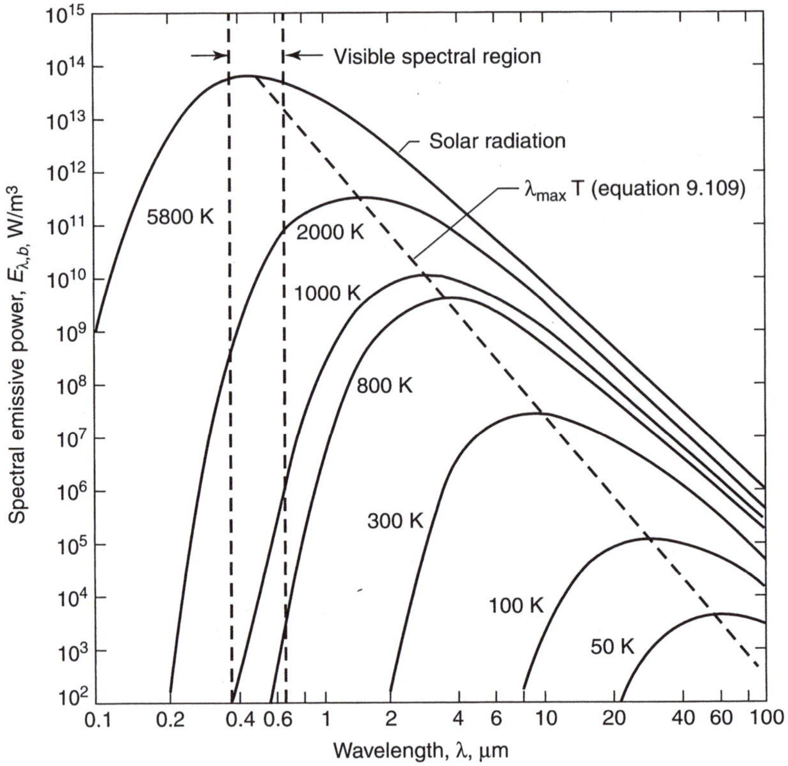

Figure 9.33. Spectral black-body emissive power[45]

phenomenon occurs in the gases above the earth's surface which will transmit incoming radiation from the sun at a given wavelength though not radiation from the earth which, because it is at a lower temperature, emits at a longer wavelength. The passage of radiation through gases and indeed through glass is one example of a situation where the transmissivity **t**, discussed in Section 9.5.3, is not zero.

The *total emissive power E* is defined as the rate at which radiation energy is emitted per unit time per unit area of surface over all wavelengths and in all directions. This may be determined by making a summation of all the radiation at all wavelengths, that is by determining the area corresponding to a particular temperature under the Planck distribution curve, Figure 9.33. In this way, from equation 9.108, the total emissive power is given by:

$$E_b = \int_0^\infty C_1\, d\lambda/[\lambda^5(\exp(C_2/\lambda T) - 1)] \tag{9.111}$$

which is known as the *Stefan–Boltzmann Law*. This may be integrated for any constant value of T to give:

$$E_b = \sigma T^4 \tag{9.112}$$

where, in SI units, the Stefan–Boltzmann constant $\sigma = 5.67 \times 10^{-8}$ W/m^2 K^4.

Example 9.13

Electrically-heated carbide elements, 10 mm in diameter and 0.5 m long, radiating essentially as black bodies, are to be used in the construction of a heater in which thermal radiation from the surroundings is negligible. If the surface temperature of the carbide is limited to 1750 K, how many elements are required to provide a radiated thermal output of 500 kW?

Solution

From equation 9.112, the total emissive power is given by:

$$E_b = \sigma T^4 = (5.67 \times 10^{-8} \times 1750^4) = 5.32 \times 10^5 \text{ W/m}^2$$

The area of one element $= \pi(10/1000)0.5 = 1.571 \times 10^{-2}$ m^2

and: Power dissipated by one element $= (5.32 \times 10^5 \times 1.571 \times 10^{-2}) = 8.367 \times 10^3$ W

Thus: Number of elements required $= (500 \times 1000)/(8.357 \times 10^3) = 59.8$ say $\underline{\underline{60}}$

9.5.3. Radiation from real surfaces

The *emissivity* of a material is defined as the ratio of the radiation per unit area emitted from a "real" or from a grey surface (one for which the emissitivity is independent of wavelength) to that emitted by a black body at the same temperature. Emissivities of "real" materials are always less than unity and they depend on the type, condition and roughness of the material, and possibly on the wavelength and direction of the emitted radiation as well. For perfectly diffuse surfaces emissivities are independent of direction, but for real surfaces there are variations with direction and the average value is known as the *hemispherical emissivity*. For a particular wavelength λ this is given by:

$$\mathbf{e}_\lambda = E_\lambda/E_b \tag{9.113}$$

and, similarly, the total hemispherical emissivity, an average over all wavelengths, is given by:

$$\mathbf{e} = E/E_b \tag{9.114}$$

Equation 9.114 leads to *Kirchoff's Law* which states that the absorptivity, or fraction of incident radiation absorbed, and the emissivity of a surface are equal. If two bodies **A** and **B** of areas A_1 and A_2 are in a large enclosure from which no energy is lost, then the energy absorbed by **A** from the enclosure is $A_1\mathbf{a}_1 I$ where I is the rate at which energy is falling on unit area of **A** and $\mathbf{a}_1$ is the absorptivity. The energy given out by **A** is E_1A_1 and, at equilibrium, these two quantities will be equal or:

$$IA_1\mathbf{a}_1 = A_1E_1$$

and, for **B**: $$IA_2\mathbf{a}_2 = A_2E_2$$

Thus: $$E_1/\mathbf{a}_1 = E_2/\mathbf{a}_2 = E/\mathbf{a} \text{ for any other body.}$$

Since $E/\mathbf{a} = E_b/\mathbf{a}_b$, then, from equation 9.114:

$$\mathbf{e} = E/E_b = \mathbf{a}/\mathbf{a}_b$$

and, as by definition, $\mathbf{a}_b = 1$, the emissivity of any body is equal to its absorptivity, or:

$$\mathbf{e} = \mathbf{a} \tag{9.115}$$

For most industrial, non-metallic surfaces and for non-polished metals, **e** is usually about 0.9. although values as low as 0.03 are more usual for highly polished metals such

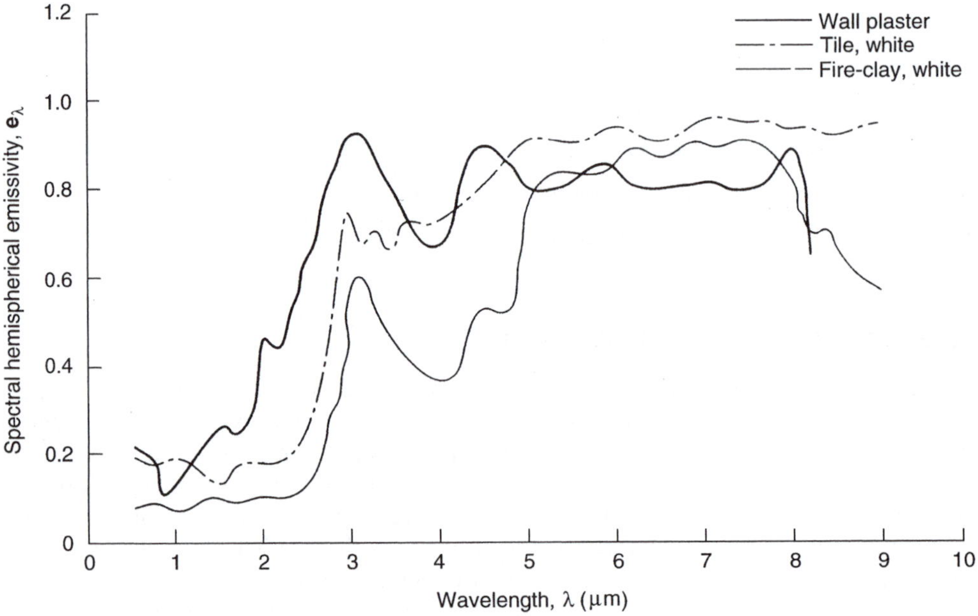

Figure 9.34. Spectral emissivity of non-conductors as a function of wavelength[45]

as copper or aluminium. As explained later, a small cavity in a body acts essentially as a black body with an effective emissivity of unity. The variation of emissivity with wavelength is illustrated in Figure 9.34[45] and typical values are given in Table 9.6 which is based on the work of HOTTEL and SAROFIM[46]. More complete data are available in Appendix A1, Table 10. If equation 9.113 is written as:

$$E_\lambda = \mathbf{e}_\lambda E_{\lambda,b} \tag{9.116}$$

then the spectral emissive power of a grey surface may be obtained from the spectral emissivity, $\mathbf{e}_\lambda$ and the spectral emissive power of a black body $E_{\lambda,b}$ given by equation 9.108. As shown in Figure 9.35, for a temperature of 2000 K for example, the emission curve for a real material may have a complex shape because of the variation of emissivity with wavelength, If, however, the ordinate of the black body curve $E_{\lambda,b}$ at a particular wavelength is multiplied by the spectral emissivity of the source at that wavelength, the ordinates on the curve for the real surface are obtained, and the total emissive power of the real surface is obtained by integrating E_λ over all possible wavelengths to give:

$$E = \int_0^\infty E_\lambda \, d\lambda = \int_0^\infty \mathbf{e}_\lambda E_{\lambda,b} \, d\lambda \tag{9.117}$$

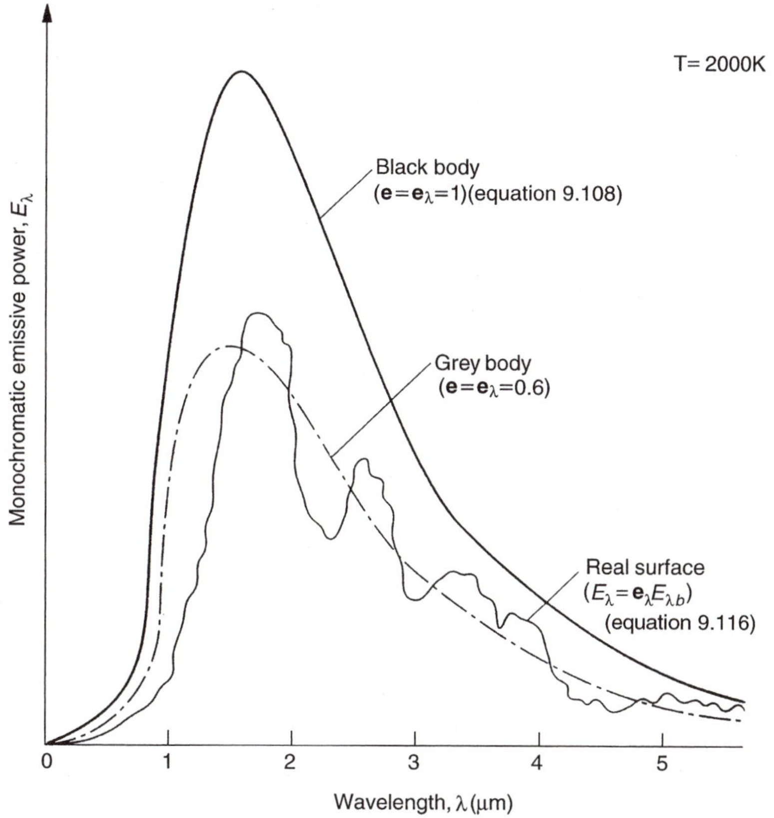

Figure 9.35. Comparison of black body, grey body and real surface radiation at 2000 K.[45]

This integration may be carried out numerically or graphically, though this approach, which has been considered in some detail by INCROPERA and DE WITT[45], can be difficult, especially where the spectral distribution of radiation arrives at a surface of complex structure. The amount of calculation involved cannot often be justified in practical situations and it is more usual to use a mean spectral emissivity for the surface which is assumed to be constant over a range of wavelengths. Where the spectral emissivity does not vary with wavelength then the surface is known as a *grey body*, and, for a diffuse grey body, from equations 9.112 and 9.114:

$$E = \mathbf{e}E_b = \mathbf{e}\sigma T^4 \tag{9.118}$$

In this way, the emissive power of a grey body is a constant proportion of the power emitted by the black body, resulting in the curve shown in Figure 9.35 where, for example, $\mathbf{e} = 0.6$. The assumption that the surface behaves as a grey body is valid for most engineering calculations if the value of emissivity is taken as that for the dominant temperature of the radiation.

From equation 9.117, it is seen that the rate of heat transfer by radiation from a hot body at temperature T_1 to a cooler one at temperature T_2 is then given by:

$$q = Q/A = \mathbf{e}\sigma(T_1^4 - T_2^4) = \mathbf{e}\sigma(T_1 - T_2)(T_1^3 + T_1^2T_2 + T_1T_2^2 + T_2^3)$$

The quantity $q/(T_1 - T_2)$ is a heat transfer coefficient as used in convective heat transfer, and here it may be designated h_r, the heat transfer coefficient for radiation heat transfer where:

$$h_r = q/(T_1 - T_2) = \frac{\mathbf{e}\sigma(T_1^4 - T_2^4)}{T_1 - T_2} = \mathbf{e}\sigma(T_1^3 + T_1^2T_2 + T_1T_2^2 + T_3^3) \tag{9.119}$$

It may be noted that if $(T_1 - T_2)$ is very small, that is T_1 and T_2 are virtually equal, then:

$$h_r = 4\mathbf{e}\sigma T^3$$

Example 9.14

What is the emissivity of a grey surface, 10 m^2 in area, which radiates 1000 kW at 1500 K? What would be the effect of increasing the temperature to 1600 K?

Solution

The emissive power $E = (1000 \times 1000)/10 = 100{,}000\ \text{W/m}^2$

From equation 9.118: $e = E/\sigma T^4$

$$= 100{,}000/(5.67 \times 10^{-8} \times 1500^4) = \underline{\underline{0.348}}$$

At 1600 K: $E = \mathbf{e}\sigma T^4$

$$= (0.348 \times 5.67 \times 10^{-8} \times 1600^4) = \underline{\underline{1295\ \text{kW}}}$$

an increase of 29.5 per cent for a 100 deg K increase in temperature.

Table 9.6. Typical emissivity values[46]

Surface		T K	Emissivity **e**
(A) Metals and metallic oxides			
Aluminium	Polished plate	296	0.040
	Rough plate	299	0.055
Brass	Polished	311–589	0.096
Copper	Polished	390	0.023
	Plate, oxidised	498	0.78
Gold	Highly polished	500–900	0.018–0.35
Iron and steel	Polished iron	700–1300	0.144–0.377
	Cast iron, newly turned	295	0.435
	Smooth sheet iron	1172–1311	0.55–0.60
	Sheet steel, oxidised	295	0.657
	Iron	373	0.736
	Steel plate, rough	311–644	0.94–0.97
Lead	Pure, unoxidised	400–500	0.057–0.075
	Grey, oxidised	297	0.281
Mercury		273–373	0.09–0.12
Molybdenum	Filament	1000–2866	0.096–0.292
Monel	Metal oxidised	472–872	0.41–0.46
Nickel	Polished	500–600	0.07–0.087
	Wire	460–1280	0.096–0.186
	Plate, oxidised	472–872	0.37–0.48
Nickel alloys	Chromonickel	325–1308	0.64–0.76
	Nickelin, grey oxidised	294	0.262
Platinum	Pure, polished plate	500–900	0.054–0.104
	Strip	1200–1900	0.12–0.17
	Filament	300–1600	0.036–0.192
	Wire	500–1600	0.073–0.182
Silver	Polished	310–644	0.0221–0.0312
Tantalum	Filament	1600–3272	0.194–0.31
Tin	Bright tinned iron sheet	298	0.043–0.064
Tungsten	Filament	3588	0.39
Zinc	Pure, polished	500–600	0.045–0.053
	Galvanised sheet	297	0.276
(B) Refractories, building materials, paints etc.			
Asbestos	Board	297	0.96
Brick	Red, rough	294	0.93
	Silica, unglazed	1275	0.80
Carbon	Filament	1311–1677	0.526
	Candle soot	372–544	0.952
	Lampblack	311–644	0.945
Enamel	White fused on iron	292	0.897
Glass	Smooth	295	0.937
Paints, lacquers,	Snow-white enamel	296	0.906
	Black, shiny lacquer	298	0.875
	Black matt shellac	350–420	0.91
Plaster,	Lime, rough	283–361	0.91
Porcellain,	Glazed	295	0.924
Refractory materials	Poor radiators	872–1272	0.65–0.75
	Good radiators	872–1272	0.80–0.90
Rubber	Hard, glossy plate	296	0.945
	Soft, grey, rough	298	0.859
Water		273–373	0.95–0.963

In a real situation, radiation incident upon a surface may be absorbed, reflected and transmitted and the properties, *absorptivity, reflectivity* and *transmissivity* may be used to describe this behaviour. In theory, these three properties will vary with the direction and wavelength of the incident radiation, although, with diffuse surfaces, directional variations may be ignored and mean, *hemispherical* properties used.

If the *absorptivity* **a**, the fraction of the incident radiation absorbed by the body is defined on a spectral basis, then:

$$\mathbf{a}_\lambda = I_{\lambda,\text{abs}}/I_\lambda \tag{9.120}$$

and the total absorptivity, the mean over all wavelengths, is defined as:

$$\mathbf{a} = I_{\text{abs}}/I \tag{9.121}$$

Since a black body absorbs all incident radiation then for a black body:

$$\mathbf{a}_\lambda = \mathbf{a} = 1$$

The absorptivity of a grey body is therefore less than unity.

In a similar way, the *reflectivity*, **r**, the fraction of incident radiation which is reflected from the surface, is defined as:

$$\mathbf{r} = I_{\text{ref}}/I \tag{9.122}$$

and the *transmissivity*, **t**, the fraction of incident radiation which is transmitted through the body, as:

$$\mathbf{t} = I_{\text{trans}}/I \tag{9.123}$$

Since, as shown in Figure 9.36, all the incident radiation is absorbed, reflected or transmitted, then:

$$I_{\text{abs}} + I_{\text{ref}} + I_{\text{trans}} = I$$

or:

$$\mathbf{a} + \mathbf{r} + \mathbf{t} = 1$$

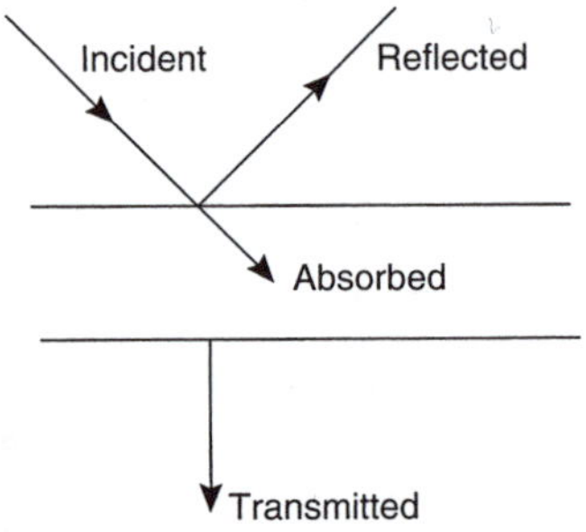

Figure 9.36. Reflection, absorption, and transmission of radiation

Since most solids are opaque to thermal radiation, $\mathbf{t} = 0$ and therefore:

$$\mathbf{a} + \mathbf{r} = 1 \tag{9.124}$$

Kirchoff's Law, discussed previously, states that, at any wavelength, the emissivity and the absorptivity are equal. If this is extended to total properties, then, at a given temperature:

$$\mathbf{e} = \mathbf{a} \qquad \text{(equation 9.115)}$$

For a grey body, the emissivity and the absorptivity are, by definition, independent of temperature and hence equation 9.115 may be applied more generally showing that, where one radiation property (**a**, **r** or **e**) is specified for an opaque body, the other two may be obtained from equations 9.115 and 9.124. Kirchoff's Law explains why a cavity with a small aperture approximates to a black body in that radiation entering is subjected to repeated internal absorption and reflection so that only a negligible amount of the incident radiation escapes through the aperture. In this way, $\mathbf{a} = \mathbf{e} = 1$ and, at T K, the emissive power of the aperture is σT^4.

9.5.4. Radiation transfer between black surfaces

Since radiation arriving at a black surface is completely absorbed, no problems arise from multiple reflections. Radiation is emitted from a diffuse surface in all directions and therefore only a proportion of the radiation leaving a surface arrives at any other given surface. This proportion depends on the relative geometry of the surfaces and this may be taken into account by the *view factor, shape factor* or *configuration F*, which is normally written as F_{ij} for radiation arriving at surface j from surface i. In this way, F_{ij}, which is, of course, completely independent of the surface temperature, is the fraction of radiation leaving i which is directly intercepted by j.

If radiant heat transfer is taking place between two black surfaces, 1 and 2, then:

$$\text{radiation emitted by surface } 1 = A_1 E_{b1}$$

where A_1 and E_{b1} are the area and black body emissive power of surface 1, respectively. The fraction of this radiation which arrives at and is totally absorbed by surface 2 is F_{12} and the heat transferred is then:

$$Q_{1\to 2} = A_1 F_{12} E_{b1}$$

Similarly, the radiation leaving surface 2 which arrives at 1 is given by:

$$Q_{2\to 1} = A_2 F_{21} E_{b2}$$

and the net radiation transfer between the two surfaces is $Q_{12} = (Q_{1\to 2} - Q_{2\to 1})$ or:

$$\begin{aligned} Q_{12} &= A_1 F_{12} E_{b1} - A_2 F_{21} E_{b2} \\ &= \sigma A_1 F_{12} T_1^4 - \sigma A_2 F_{21} T_2^4 \end{aligned} \qquad (9.125)$$

When the two surfaces are at the same temperature, $T_1 = T_2$, $Q_{12} = 0$

and thus: $$Q_{12} = 0 = \sigma T_1^4 (A_1 F_{12} - A_2 F_{21})$$

Since the temperature T_1 can have any value so that, in general $T_1 \neq 0$, then:

$$A_1 F_{12} = A_2 F_{21} \qquad (9.126)$$

Equation 9.126, known as the *reciprocity relationship* or *reciprocal rule*, then leads to the equation:

$$Q_{12} = \sigma A_1 F_{12}(T_1^4 - T_2^4) = \sigma A_2 F_{12}(T_1^4 - T_2^4) \qquad (9.127)$$

The product of an area and an appropriate view factor is known as the *exchange area* which, in SI units, is expressed in m^2. In this way, $A_1 F_{12}$ is known as exchange area 1–2.

Example 9.15

Calculate the view factor, F_{21} and the net radiation transfer between two black surfaces, a rectangle 2 m by 1 m (area A_1) at 1500 K and a disc 1 m in diameter (area A_2) at 750 K, if the view factor, $F_{12} = 0.25$.

Solution

$$A_1 = (2 \times 1) = 2 \text{ m}^2$$

$$A_2 = (\pi \times 1^2)/4 = 0.785 \text{ m}^2$$

From equation 9.126: $A_1 F_{12} = A_2 F_{21}$

or: $(2 \times 0.25) = 0.785 F_{21}$

and: $F_{21} = \underline{\underline{0.637}}$

In equation 9.127: $Q_{12} = \sigma A_1 F_{12}(T_1^4 - T_2^4)$

$$Q_{12} = (5.67 \times 10^{-8} \times 2 \times 0.25)(1500^4 - 750^4)$$

$$= 5.38 \times 10^5 \text{ W or } \underline{\underline{538 \text{ kW}}}$$

View factors, the values of which determine heat transfer rates, are dependent on the geometrical configuration of each particular system. As a simple example, radiation may be considered between elemental areas dA_1 and dA_2 of two irregular-shaped flat bodies, well separated by a distance L between their mid-points as shown in Figure 9.37. If α_1 and α_2 are the angles between the imaginary line joining the mid-points and the normals, the rate of heat transfer is then given by:

$$Q_{12} = \sigma(T_1^4 - T_2^4)\int^{A_1}\int^{A_2}(\cos\alpha_1 \cos\alpha_2 \, dA_1 \, dA_2)/\pi L^2 \qquad (9.128)$$

Equation 9.128 may be extended to much larger surfaces by subdividing these into a series of smaller elements, each of exchange area $A_i F_{ij}$, and summing the exchange areas between each pair of elements to give:

$$A_i F_{ij} = A_j F_{ji} = \int_{Ai}\int_{Aj}(\cos\alpha_i \cos\alpha_j \, dA_i \, dA_j)/\pi L^2 \qquad (9.129)$$

In this procedure, the value of the integrand can be determined numerically for every pair of elements and the double integral, approximately the sum of these values, then becomes:

$$\int_{Aj}\int_{Ai}(\cos\alpha_i \cos\alpha_j \, dA_i \, dA_j)/\pi L^2 = \sum_{Ai}\sum_{Aj}(\cos\alpha_i \cos\alpha_j \, dA_i \, dA_j)/\pi L^2 \qquad (9.130)$$

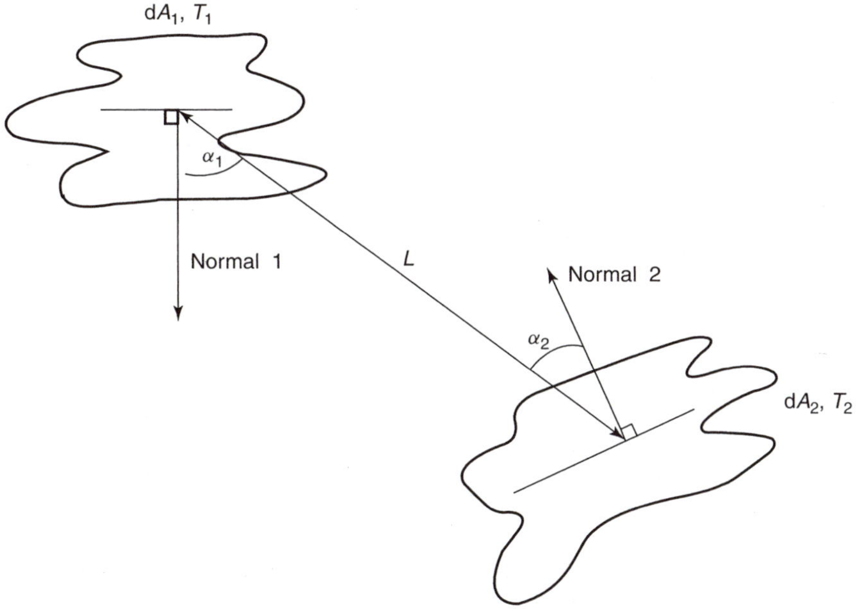

Figure 9.37. Determination of view factor

The amount of calculation involved here can be very considerable and use of a computer is usually required. A simpler approach is to make use of the many expressions, graphs and tables available in the heat transfer literature. Typical data, presented by INCROPERA and DE WITT(45) and by HOWELL(47), are shown in Figures 9.38–9.40, where it will be seen that in many cases, the values of the view factors approach unity. This means that nearly all the radiation leaving one surface arrives at the second surface as, for example, when a sphere is contained within a second larger sphere. Wherever a view factor approaches zero, only a negligible part of one surface can be seen by the other surface.

It is important to note here that if an element does not radiate directly to any part of its own surface, the shape factor with respect to itself, F_{11}, F_{12} and so on, is zero. This applies to any convex surface for which, therefore, $F_{11} = 0$.

Example 9.16

What are the view factors, F_{12} and F_{21}, for (a) a vertical plate, 3 m high by 4 m long, positioned at right angles to one edge of a second, horizontal plate, 4 m wide and 6 m long, and (b) a 1 m diameter sphere positioned within a 2 m diameter sphere?

Solution

(a) Using the nomenclature in Figure 9.40 iii:

$$Y/X = (6/4) = 1.5 \text{ and } Z/X = (3/4) = 0.75$$

From the figure: $F_{12} = \underline{\underline{0.12}}$

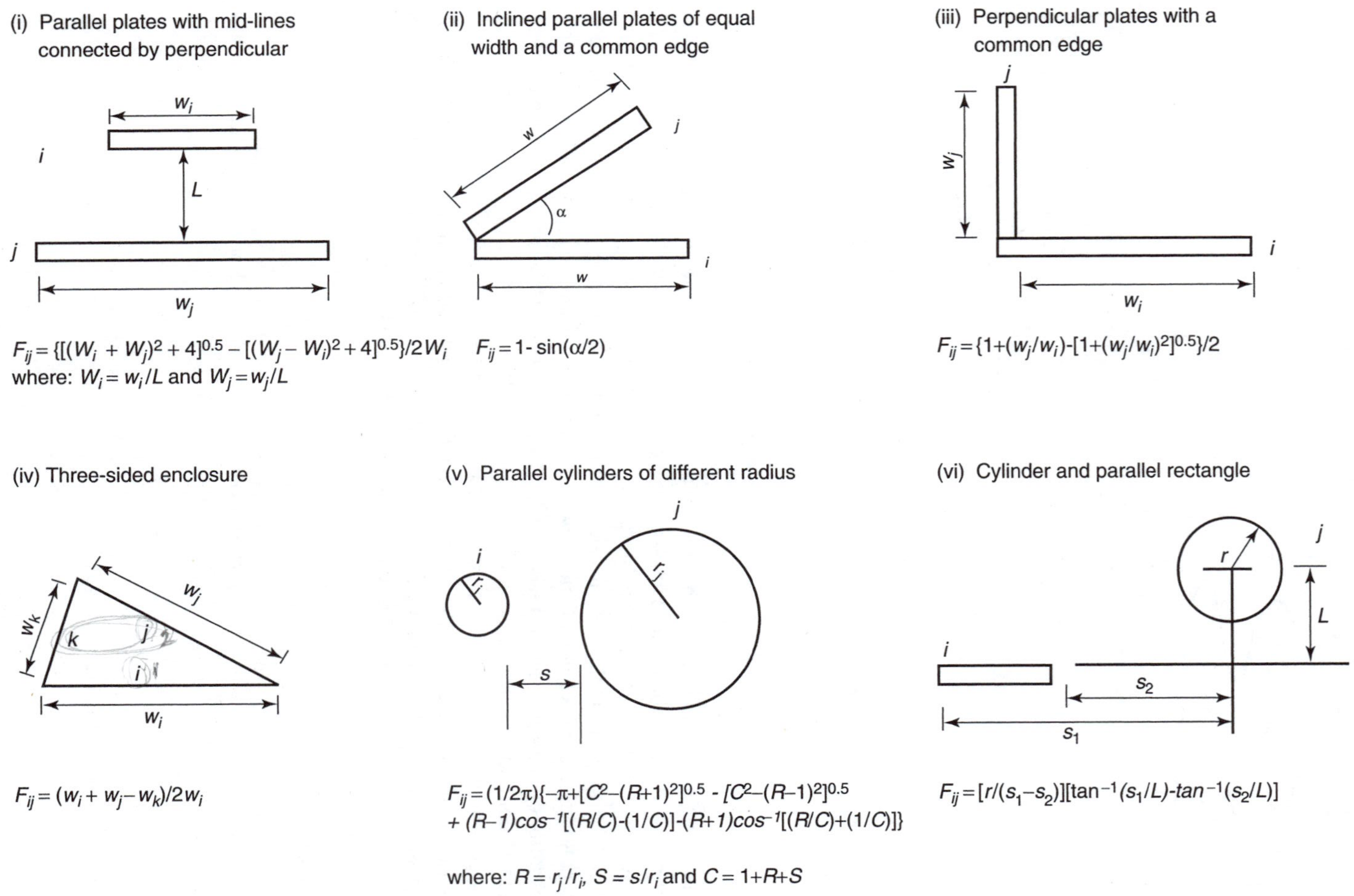

Figure 9.38. View factors for two-dimensional geometries.[45]

(i) Aligned parallel rectangles

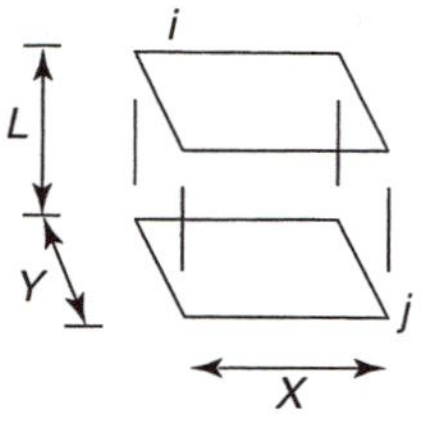

$$F_{ij} = [2/(\pi \bar{X}\bar{Y})]\{\ln[(1+\bar{X}^2)(1+\bar{Y}^2)/(1+\bar{X}^2+\bar{Y}^2)]^{0.5} + \bar{X}(1+\bar{Y}^2)^{0.5}\tan^{-1}[\bar{X}/(1+\bar{Y}^2)^{0.5}] + \bar{Y}(1+\bar{X}^2)^{0.5}\tan^{-1}[(\bar{Y}/(1+\bar{X}^2)^{0.5}) - \bar{X}\tan^{-1}\bar{X} - \bar{Y}\tan^{-1}\bar{Y}\}$$

where: $\bar{X} = X/L$ and $\bar{Y} = Y/L$

(ii) Coaxial parallel discs

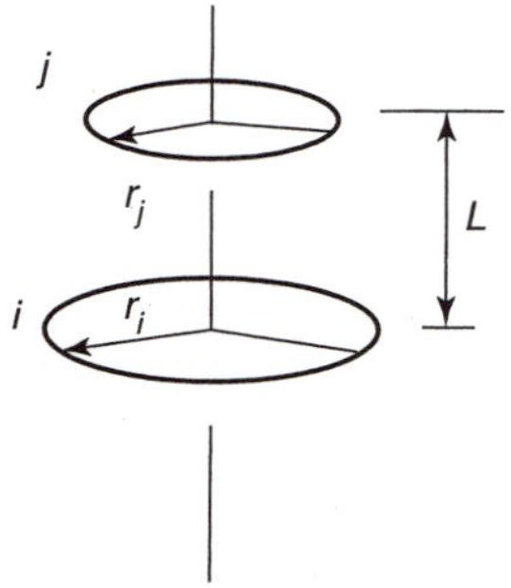

$$F_{ij} = 0.5\{S - [S^2 - 4(r_j/r_i)^2]^{0.5}\}$$

where: $R_i = r_i/L$, $R_j = r_j/L$ and $S = 1 + (1+R_j^2)/R_i^2$

(iii) Perpendicular rectangles with a common edge

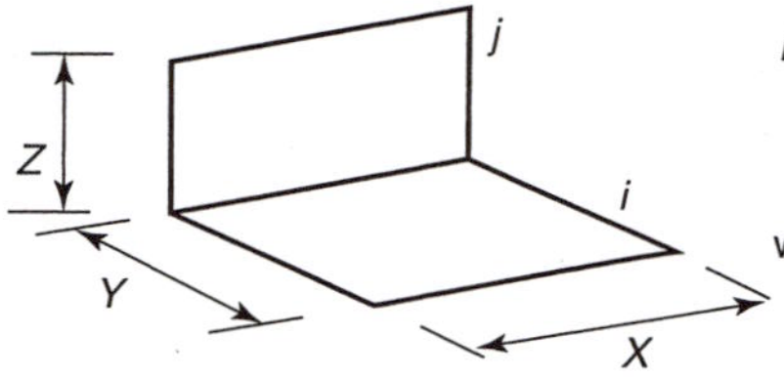

$$F_{ij} = (1/\pi W)\{W\tan^{-1}(1/W) + H\tan^{-1}(1/H) - (H^2+W^2)^{0.5}\tan^{-1}(H^2+W^2)^{-0.5} + 0.25\ln[(1+W^2)(1+H^2)/(1+W^2+H^2)][W^2(1+W^2+H^2)/(1+W^2)(W^2+H^2)]W^2 \times [H^2(1+H^2+W^2)/(1+H^2)(H^2+W^2)]H^2]\}$$

where: H=Z/X and W=Y/X

Figure 9.39. View factors for three-dimensional geometries[45]

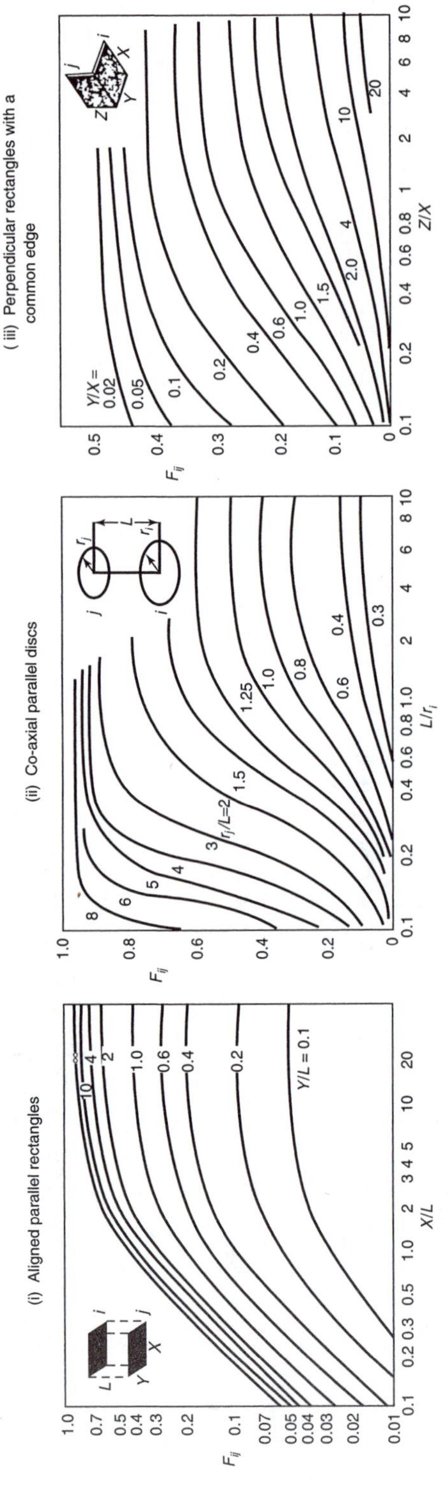

Figure 9.40. View factors for three-dimensional geometries[45]

(i) Two perpendicular rectangles
– between surfaces 1 and 6

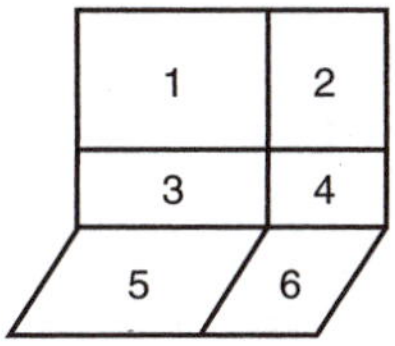

$$F_{16} = (A_6/A_1)[(1/2A_6)(A_{(1+2+3+4)}F_{(1+2+3+4)(5+6)}$$
$$+ A_6F_{6(2+4)} - A_5F_{5(1+3)} - (1/2A_6)(A_{(3+4)}F_{(3+4\times5+6)}$$
$$- A_6F_{6A} - A_5F_{53})]$$

(ii) Two parallel rectangles
– between surfaces 1 and 7

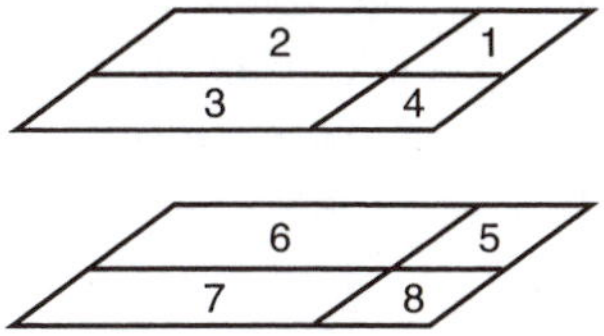

$$F_{17} = (1/4A_1)[A_{(1+2+3+4)}F_{(1+2+3+4)(5+6+7+8)}+A_1F_{15}+A_2F_{26}$$
$$+ A_3F_{37} + A_4F_{48}] - (1/4A_1)[A_{(1+2)}F_{(1+2)(5+6)}+A_{(1+4)}F_{(1+4)(5+8)}$$
$$+ A_{(3+4)}F_{(3+4)(7+8)} + A_{(2+3)}F_{(2+3)(6+7)}]$$

(iii) Two parallel circular rings
– between surfaces 2 and 3

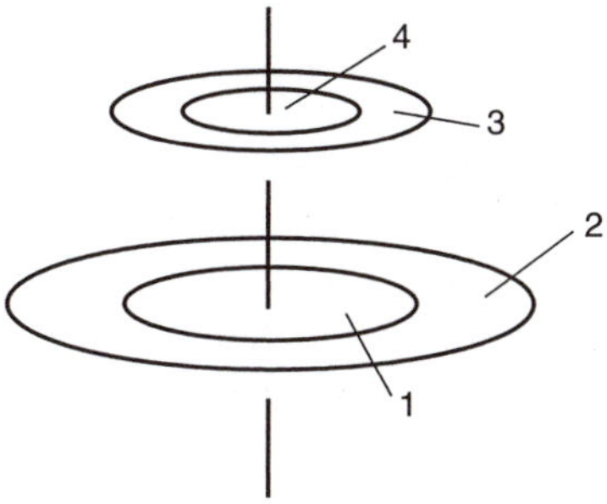

$$F_{23} = (A_{(1+2)}/A_2)[F_{(1+2)(3+4)} - F_{(1+2)4}]$$
$$- (A_1/A_2)[F_{1(3+4)} - F_{14}]$$

(iv) A circular tube and a disc between surface 3,
the inner wall of the tube of radius x_3 and surface 1,
the upper surface of the disc of radius x_1.

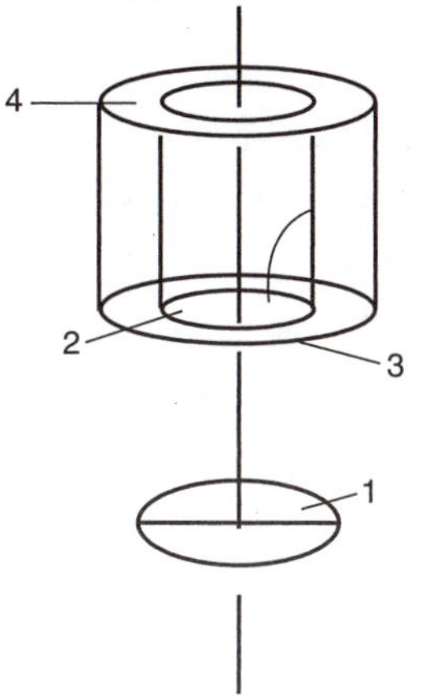

$$F_{13} = F_{12} - F_{14}$$
$$F_{31} = (x_3^2/x_1^2)(F_{12} + F_{14})$$

Figure 9.41. View factors obtained by using the summation rule[47]

From equation 9.126:

$$A_1F_{12} = A_2F_{21}$$

$$(3 \times 4)0.12 = (4 \times 6)F_{21}$$

and: $$F_{21} = \underline{\underline{0.06}}$$

(b) For the two spheres: $$F_{12} = 1 \text{ and } F_{21} = (r_1/r_2)^2 = (1/2)^2 = \underline{\underline{0.25}}$$

$$F_{22} = 1 - (r_1/r_2)^2 = 1 - 0.25 = \underline{\underline{0.75}}$$

For a given geometry, view factors are related to each other, one example being the reciprocity relationship given in equation 9.126. Another important relationship is the *summation rule* which may be applied to the surfaces of a complete enclosure. In this case, all the radiation leaving one surface, say i, must arrive at all other surfaces in the enclosure so that, for n surfaces:

$$F_{i1} + F_{i2} + F_{i3} + \cdots + F_{in} = 1$$

or: $$F_{ij} = 1 \tag{9.131}$$

from which: $$A_iF_{ij} = A_i \tag{9.132}$$

This means that the sum of the exchange areas associated with a surface in an enclosure must be same as the area of that surface. The principle of the summation rule may be extended to other geometries such as, for example, radiation from a vertical rectangle (area 1) to an adjacent horizontal rectangle (area 2), as shown in Figure 9.40iii, where they are joined to a second horizontal rectangle of the same width (area 3). In effect area 3 is an extension of area 2 but has a different view factor.
In this case:

$$A_1F_{1(2+3)} = A_1F_{12} + A_1F_{13}$$

or: $$A_1F_{13} = A_1F_{1(2+3)} - A_1F_{12} \tag{9.133}$$

Equation 9.133 allows F_{13} to be determined from the view factors F_{12} and $F_{1(2+3)}$ which can be obtained directly from Figure 9.40iii. Typical data obtained by using this technique are shown in Figure 9.41 which is based on the work of HOWELL[47].

Example 9.17

What is the view factor F_{23} for the two parallel rings shown in Figure 9.41iii if the inner and outer radii of the two rings are: upper = 0.2 m and 0.3 m; lower = 0.3 m and 0.4 m and the rings are 0.2 m apart?

Solution

From Figure 9.41iii:

$$F_{23} = (A_{(1+2)}/A_2)(F_{(1+2)(3+4)} - F_{(1+2)4}) - (A_1/A_2)(F_{1(3+4)} - F_{14})$$

Laying out the data in tabular form and obtaining F from Figure 9.40ii, then;

For:	r_i (m)	r_j (m)	L (m)	(r_j/L)	(L/r_i)	F
$F_{(1+2)(3+4)}$	0.4	0.3	0.2	1.5	0.5	0.40
$F_{(1+2)4}$	0.4	0.2	0.2	1.0	0.5	0.22
$F_{1(3+4)}$	0.3	0.3	0.2	1.5	0.67	0.55
F_{14}	0.3	0.2	0.2	1.0	0.67	0.30

$$A_{(1+2)}/A_2 = 0.4^2/(0.4^2 - 0.3^2) = 2.29$$

$$A_1/A_2 = 0.3^2/(0.4^2 - 0.3^2) = 1.29$$

and hence:

$$F_{23} = 2.29(0.40 - 0.22) + 1.29(0.55 - 0.30) = \underline{\underline{0.74}}$$

Equation 9.127 may be extended in order to determine the net rate of radiation heat transfer from a surface in an enclosure. If the enclosure contains n black surfaces, then the net heat transfer by radiation to surface i is given by:

$$Q_i = Q_{1i} + Q_{2i} + Q_{3i} + \cdots + Q_{ni}$$

or:

$$Q_i = \sum_{j=1}^{j=n} \sigma A_j F_{ji}(T_j^4 - T_i^4)$$

or, applying the reciprocity relationship:

$$Q_i = \sum_{j=1}^{j=n} \sigma A_i F_{ij}(T_j^4 - T_i^4) \qquad (9.134)$$

Example 9.18

A plate, 1 m in diameter at 750 K, is to be heated by placing it beneath a hemispherical dome of the same diameter at 1200 K; the distance between the plate and the bottom of the dome being 0.5 m, as shown in Figure 9.42. If the surroundings are maintained at 290 K, the surfaces may be regarded as black bodies and heat transfer from the underside of the plate is negligible, what is the net rate of heat transfer by radiation to the plate?

Solution

Taking area 1 as that of the plate, area 2 as the underside of the hemisphere, area 3 as an imaginary cylindrical surface linking the plate and the underside of the dome which represents the black surroundings and area 4 as an imaginary disc sealing the hemisphere and parallel to the plate, then, from equation 9.134, the net radiation to the surface of the plate 1 is given by:

$$Q_1 = \sigma A_2 F_{21}(T_2^4 - T_1^4) + \sigma A_3 F_{31}(T_3^4 - T_1^4)$$

or, using the reciprocity rule:

$$Q_1 = \sigma A_1 F_{12}(T_2^4 - T_1^4) + \sigma A_1 F_{13}(T_3^4 - T_1^4)$$

All radiation from the disc 1 to the dome 2 is intercepted by the imaginary disc 4 and hence $F_{12} = F_{14}$, which may be obtained from Figure 9.39ii, with i and j representing areas 1 and 4 respectively. Thus:

$$R_1 = r_1/L = (0.5/0.5) = 1; R_4 = r_4/L = (0.5/0.5) = 1$$

and:

$$S = 1 + (1 + R_4^2)/(R_1^2) = 1 + (1 + 1.0)/(1.0) = 3.0$$

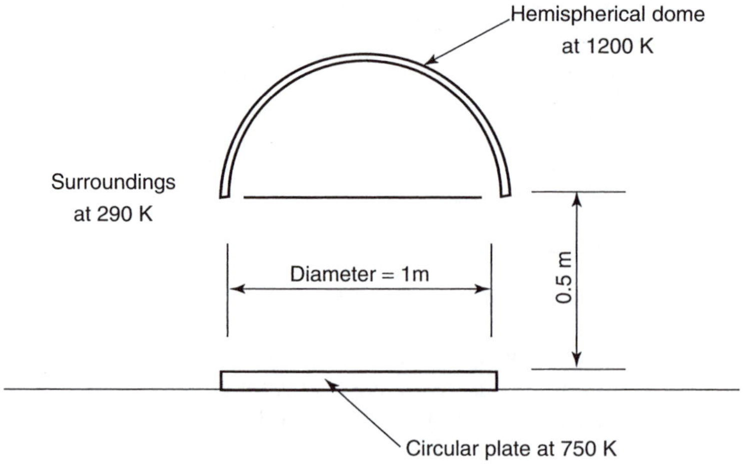

Figure 9.42. Data for Example 9.18

Thus: $F_{14} = 0.5\left[S - [(S^2 - 4(r_4/r_1)^2]^{0.5}\right] = 0.5\left[3 - [(3^2 - 4(0.5/0.5)^2]^{0.5}\right] = 0.38$

and: $F_{12} = F_{14} = 0.38$

The summation rule states that:

$$F_{11} + F_{12} + F_{13} = 1$$

and since, for a plane surface, $F_{11} = 0$, then:

$$F_{13} = (1 - 0.38) = 0.62$$

$$A_1 = (\pi 1.0^2)/4 = 0.785 \text{ m}^2$$

and hence:

$$\begin{aligned} Q_1 &= (5.67 \times 10^{-8} \times 0.785 \times 0.38)(1200^4 - 750^4) + (5.67 \times 10^{-8} \times 0.785 \times 0.62)(290^4 - 750^4) \\ &= (1.691 \times 10^{-8} \times 1.757 \times 10^{12}) - (2.760 \times 10^{-8} \times 3.093 \times 10^{11}) \\ &= 2.12 \times 10^4 \text{ W} = \underline{\underline{21.2 \text{ kW}}} \end{aligned}$$

Radiation between two black surfaces may be increased considerably by introducing a third surface which acts in effect as a re-radiator. For example, if a surface 1 of area A_1 at temperature T_1 is radiating to a second surface 2 of area A_2 at temperature T_2 joined to it as shown in Figure 9.39iii, then adding a further surface R consisting of insulating material so as to form a triangular enclosure will reduce the heat transfer to the surroundings considerably. Even though some heat will be conducted through the insulation, this will usually be small and most of the energy absorbed by the insulated surface will be re-radiated back into the enclosure.

The net rate of heat transfer to surface 2 is given by:

$$Q_2 = \sigma A_1 F_{12}(T_1^4 - T_2^4) + \sigma A_R F_{R2}(T_R^4 - T_2^4) \tag{9.135}$$

where T_R is the mean temperature of the insulation, though, in practice, there will be a temperature distribution across this surface. At steady-state, the net rate of radiation to surface R is equal to the heat loss from it to the surroundings, Q_{surr}, or:

$$Q_{surr} = \sigma A_1 F_{1R}(T_1^4 - T_R^4) + \sigma A_2 F_{2R}(T_2^4 - T_R^4) \tag{9.136}$$

If Q_{surr} is negligible, that is, the surface may be treated as adiabatic, then from equation 9.136:

$$\sigma A_1 F_{1R}(T_1^4 - T_R^4) + \sigma A_2 F_{2R}(T_2^4 - T_R^4) = 0$$

Rearranging:

$$T_R^4 = (A_1 F_{1R} T_1^4 + A_2 F_{2R} T_2^4)/(A_1 F_{1R} + A_2 F_{2R})$$

Substituting for T_R from this equation in equation 9.135 and noting, from the reciprocity relationship, that $A_2 F_{2R} = A_R F_{R2}$, then:

$$Q_2 = \sigma(T_1^4 - T_2^4)\{A_1 F_{12} + [(1/(A_1 F_{1R}) + (1/(A_R F_{R2})]^{-1}\} \tag{9.137}$$

Example 9.19

A flat-bottomed cylindrical vessel, 2 m in diameter, containing boiling water at 373 K, is mounted on a cylindrical section of insulating material, 1 m deep and 2 m ID at the base of which is a radiant heater, also 2 m in diameter, with a surface temperature of 1500 K. If the vessel base and the heater surfaces may be regarded as black bodies and conduction though the insulation is negligible, what is the rate of radiant heat transfer to the vessel? How would this be affected if the insulation were removed so that the system was open to the surroundings at 290 K?

Solution

If area 1 is the radiant heater surface and area 2 the under-surface of the vessel, with R the insulated cylinder, then:

$$A_1 = A_2 = (\pi \times 2^2/4) = 3.14 \text{ m}^2$$

and:

$$A_R = (\pi \times 2.0 \times 1.0) = 6.28 \text{ m}^2$$

From Figure 9.40ii, with $i = 1$, $j = 2$, $r_i = 1.0$ m, $r_j = 1.0$ m and $L = 1.0$ m,

$$(L/r_i) = (1.0/1.0) = 1.0; \quad \text{and } (r_j/L) = (1.0/1.0) = 1.0$$

and:

$$F_{12} = 0.40$$

[The view factor may also be obtained from Figure 9.39ii as follows:
Using the nomenclature of Figure 9.39:

$$R_1 = (r_1/L) = (1.0/1.0) = 1.0$$

$$R_2 = (r_2/L) = (1.0/1.0) = 1.0$$

$$S = 1 + [(1 + R_2^2)/R_1^2] = 1 + [(1 + 1^2)/1^2] = 3.0$$

and:

$$F_{12} = 0.5[S - [S^2 - 4(r_2/r_1)^2]^{0.5}] = 0.5[3 - [3^2 - (4 \times 1^2)]^{0.5}] = 0.382]$$

The summation rule states that:

$$F_{11} + F_{12} + F_{1R} = 1$$

and since, for a plane surface, $F_{11} = 0$, then: $F_{1R} = (1 - 0.382) = 0.618$

Since $A_1 = A_2$:

$$F_{21} = F_{12} \text{ and } F_{2R} = F_{1R} = 0.618$$

Also $A_R F_{R2} = A_2 F_{2R}$ and hence, from equation 9.137:

$$Q_2 = [A_1 F_{12} + ((1/A_1 F_{1R}) + (1/A_2 F_{2R}))^{-1}]\sigma(T_1^4 - T_2^4)$$

$$= \left[(3.14 \times 0.382) + [(1/(3.14 \times 0.618)] + [1/(3.14 \times 0.618)]^{-1}\right](5.67 \times 10^{-8})(1500^4 - 373^4)$$

$$= 6.205 \times 10^5 \text{ W or } \underline{\underline{620 \text{ kW}}}$$

If the surroundings without insulation are surface 3 at $T_3 = 290$ K, then $F_{23} = F_{2R} = 0.618$ and, from equation 9.135:

$$Q_2 = \sigma A_1 F_{12}(T_1^4 - T_2^4) + \sigma A_2 F_{23}(T_3^4 - T_2^4)$$

$$= (5.67 \times 10^{-8} \times 3.14 \times 0.382)(1500^4 - 373^4) + (5.67 \times 10^{-8} \times 3.14 \times 0.618)(290^4 - 373^4)$$

$$= 3.42 \times 10^5 \text{ W or } \underline{\underline{342 \text{ kW}}}\text{; a reduction of 45 per cent.}$$

9.5.5 Radiation transfer between grey surfaces

Since the absorptivity of a grey surface is less than unity, not all the incident radiation is absorbed and some is reflected diffusely causing multiple reflections to occur. This makes radiation between grey surfaces somewhat complex compared with black surfaces since, with grey surfaces, reflectivity as well as the geometrical configuration must be taken into account. With grey bodies, it is convenient to consider the total radiation leaving a surface Q_O, that is the emitted plus the reflected components. The equivalent flux, $Q_O/A = Q_O$ is termed *radiosity* and the total radiosity Q_{Oi}, which in the SI system has the units W/m^2, is the rate at which radiation leaves per unit area of surface i over the whole span of wavelengths. If the incident radiation arriving at a grey surface i in an enclosure is Q_{Ii}, corresponding to a flux $q_{Ii} = Q_{Ii}/A_i$, then the reflected flux, that is, energy per unit area, is $\mathbf{r}_i q_{Ii}$. The emitted flux is $\mathbf{e}_i E_{bi} = \mathbf{e}_i \sigma T_i^4$ and the radiosity is then given by:

$$q_{Oi} = \mathbf{e}_i E_{bi} + \mathbf{r}_i q_{Ii} \tag{9.138}$$

The net radiation from the surface is given by:

$$Q_i = \text{(rate at which energy leaves the surface)}$$
$$- \text{(rate at which energy arrives at the surface)}$$

or:

$$Q_i = Q_{Oi} - Q_{Ii} = A_i(q_{Oi} - q_{Ii}) \tag{9.139}$$

Substituting from equation 9.138 in equation 9.139 and noting that $(\mathbf{e}_i + \mathbf{r}_i) = 1$, then:

$$Q_i = A_i \mathbf{e}_i E_{bi}/\mathbf{r}_i + (A_i/\mathbf{r}_i)[q_{Oi}(1 - \mathbf{e}_i) - q_{Oi}]$$

$$= (A_i \mathbf{e}_i/\mathbf{r}_i)(E_{bi} - q_{Oi}) \tag{9.140}$$

If the temperature of a grey surface is known, then the net heat transfer to or from the surface may be determined from the value of the radiosity q_O. With regard to signs, the usual convention is that a positive value of Q_i indicates heat transfer from grey surfaces.

Example 9.20

Radiation arrives at a grey surface of emissivity 0.75 at a constant temperature of 400 K, at the rate of 3 kW/m^2. What is the radiosity and the net rate of radiation transfer to the surface? What coefficient of heat transfer is required to maintain the surface temperature at 300 K if the rear of the surface is perfectly insulated and the front surface is cooled by convective heat transfer to air at 295 K?

Solution

Since $\mathbf{e} + \mathbf{r} = 1$: $\quad \mathbf{r} = 0.25$

From equation 9.118: $\quad E_b = (5.67 \times 10^{-8} \times 400^4) = 1452\ \text{W/m}^2$

From equation 9.138:

$$q_O = \mathbf{e}E_b + \mathbf{r}q_I$$

$$= (0.75 \times 1452) + (0.25 \times 3000) = \underline{\underline{1839\ \text{W/m}^2}}$$

From equation 9.140:

$$Q/A = q = (1.0 \times 0.75/0.25)(1452 - 1839) = \underline{\underline{-1161\ \text{W/m}^2}}$$

where the negative value indicates heat transfer to the surface.

For convective heat transfer from the surface:

$$q_c = h(T_s - T_{\text{ambient}})$$

and: $\quad h_c = q_c/(T_s - T_{\text{ambient}}) = 1161/(400 - 295) = \underline{\underline{11.1\ \text{W/m}^2\ \text{K}}}$

For the simplest case of a *two-surface enclosure* in which surfaces 1 and 2 exchange radiation with each other only, then, assuming $T_1 > T_2$, Q_{12} is the net rate of transfer from 1, Q_1 or the rate of transfer to 2, Q_2.

Thus:

$$Q_{12} = Q_1 = -Q_2 \tag{9.141}$$

Substituting from equation 9.139:

$$Q_1 = A_1(q_{O1} - q_{I1}) \tag{9.142}$$

and:

$$q_{I1}A_1 = q_{O1}A_1F_{11} + q_{O2}A_2F_{21} \tag{9.143}$$

that is:

(rate of energy incident upon surface 1)

= (rate of energy arriving at surface 1 from itself)

+ (rate of energy arriving at surface 1 from surface 2)

From equations 9.142 and 9.143 and using $A_1F_{12} = A_2F_{21}$, then:

$$Q_1 = q_{O1}(A_1 - A_1F_{11}) - q_{O2}A_1F_{12} \tag{9.144}$$

Since, by the summation rule, $(A_1 - A_1F_{11}) = A_1F_{12}$, then:

$$Q_1 = (A_1F_{12})(q_{O1} - q_{O2}) \tag{9.145}$$

From equation 9.140:

$$Q_1 = (A_1\mathbf{e}_1/\mathbf{r}_1)(E_{b1} - q_{O1}) \text{ and } -Q_2 = (A_2\mathbf{e}_2/\mathbf{r}_2)(E_{b2} - q_{O2}) \tag{9.146}$$

Substituting from equation 9.146 into equation 9.145 and using the relationships in equation 9.141 gives:

$$Q_{12}[(1/A_1F_{12}) + (\mathbf{r}_1/A_1\mathbf{e}_1) + (\mathbf{r}_2/A_2\mathbf{e}_2)] = (E_{b1} - E_{b2})$$

and hence $$Q_{12} = (E_{b1} - E_{b2})/[(1/A_1F_{12}) + (\mathbf{r}_1/A_1\mathbf{e}_1) + (\mathbf{r}_2/A_2\mathbf{e}_2)] \quad (9.147)$$

Since $\mathbf{r} = 1 - \mathbf{e}$, then, writing $E_{b1} = \sigma T_1^4$ and $E_{b2} = \sigma T_2^4$:

$$Q_{12} = [\sigma(T_1^4 - T_2^4)]/[(1/A_1F_{12}) + (1 - \mathbf{e}_1)/(A_1\mathbf{e}_1) + (1 - \mathbf{e}_2)/(A_2\mathbf{e}_2)] \quad (9.148)$$

Equation 9.148 is the same as equation 9.127 for black body exchange with two additional terms $(1 - \mathbf{e}_1)/(A_1\mathbf{e}_1)$ and $(1 - \mathbf{e}_2)/(A_2\mathbf{e}_2)$ introduced in the denominator for surfaces 1 and 2.

Radiation between parallel plates

For two large parallel plates of equal areas, and separated by a small distance, it may be assumed that all of the radiation leaving plate 1 falls on plate 2, and similarly all of the radiation leaving plate 2 falls on plate 1.

Thus: $$F_{12} = F_{21} = 1$$

and: $$A_1 = A_2$$

Substituting in equation 9.148:

$$Q_{12} = \frac{A_1\sigma(T_1^4 - T_2^4)}{1 + (1 - \mathbf{e}_1)/\mathbf{e}_1 + (1 - \mathbf{e}_2)/\mathbf{e}_2}$$

and: $$\frac{Q_{12}}{A_1} = q_{12} = \frac{\sigma(T_1^4 - T_2^4)}{\dfrac{1}{\mathbf{e}_1} + \dfrac{1}{\mathbf{e}_2} - 1} \quad (9.149)$$

Other cases of interest include radiation between:

(i) two concentric spheres
(ii) two concentric cylinders where the length: diameter ratio is large.

In both of these cases, the inner surface 1 is convex, and all the radiation emitted by it falls on the outer surface 2.

Thus: $$F_{12} = 1$$

and from the reciprocal rule:

$$F_{21} = F_{12}\frac{A_1}{A_2} = \frac{A_1}{A_2} \quad \text{(equation 9.126)}$$

Substituting in equation 9.148:

$$Q_{12} = \frac{A_1\sigma(T_1^4 - T_2^4)}{1 + [(1 - \mathbf{e}_1)/\mathbf{e}_1] + [(1 - \mathbf{e}_2)/\mathbf{e}_2]\dfrac{A_1}{A_2}}$$

$$= \frac{A_1\sigma(T_1^4 - T_2^4)}{\dfrac{1}{\mathbf{e}_1} + \dfrac{1}{\mathbf{e}_2}(1 - \mathbf{e}_2)\dfrac{A_1}{A_2}}$$

and:

$$\frac{Q_{12}}{A_1} = q_{12} = \frac{\sigma(T_1^4 - T_2^4)}{\dfrac{1}{\mathbf{e}_1} + \left[\dfrac{(1 - \mathbf{e}_2)}{\mathbf{e}_2}\dfrac{A_1}{A_2}\right]} \tag{9.150}$$

For radiation from surface 1 to extensive surroundings ($A_1/A_2 \to 0$), then:

$$q_{12} = \mathbf{e}_1\sigma(T_1^4 - T_2^4) \tag{9.151}$$

Radiation shield

The rate of heat transfer by radiation between two surfaces may be reduced by inserting a shield, so that radiation from surface 1 does not fall directly on surface 2, but instead is intercepted by the shield at a temperature T_{sh} (where $T_1 > T_{sh} > T_2$) which then re-radiates to surface 2. An important application of this principle is in a furnace where it is necessary to protect the walls from high-temperature radiation.

The principle of the radiation shield may be illustrated by considering the simple geometric configuration in which surfaces 1 and 2 and the shield may be represented by large planes separated by a small distance as shown in Figure 9.43.

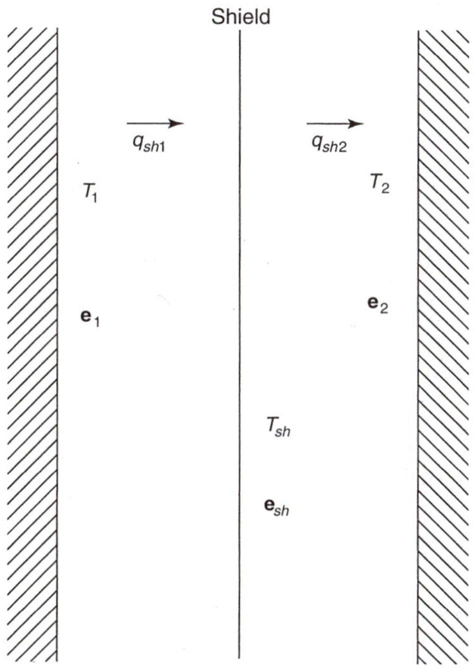

Figure 9.43. Radiation shield

Neglecting any temperature drop across the shield (which has a surface emissivity $\mathbf{e}_{sh}$), then in the steady state, the transfer rate of radiant heat to the shield from the surface 1 must equal the rate at which heat is radiated from the shield to surface 2.

Application of equation 9.149 then gives:

$$q_{sh} = q_{sh1} (= q_{sh2}) = \frac{\sigma(T_1^4 - T_{sh}^4)}{\left(\dfrac{1}{\mathbf{e}_1}\right) + \left(\dfrac{1}{\mathbf{e}_{sh}}\right) - 1} = \frac{\sigma(T_{sh}^4 - T_2^4)}{\left(\dfrac{1}{\mathbf{e}_{sh}}\right) + \left(\dfrac{1}{\mathbf{e}_2}\right) - 1} \tag{9.152}$$

Eliminating T_{sh}^4 in terms of T_1^4 and T_2^4 from equation 9.152:

$$T_1^4 - T_{sh}^4 = \frac{q_{sh}}{\sigma}\left(\frac{1}{\mathbf{e}_1} + \frac{1}{\mathbf{e}_{sh}} - 1\right)$$

$$T_{sh}^4 - T_2^4 = \frac{q_{sh}}{\sigma}\left(\frac{1}{\mathbf{e}_{sh}} + \frac{1}{\mathbf{e}_2} - 1\right)$$

Adding:
$$T_1^4 - T_2^4 = \frac{q_{sh}}{\sigma}\left(\frac{1}{\mathbf{e}_1} + \frac{2}{\mathbf{e}_{sh}} + \frac{1}{\mathbf{e}_2} - 2\right)$$

and:
$$q_{sh} = \frac{\sigma(T_1^4 - T_2^4)}{\left(\dfrac{1}{\mathbf{e}_1}\right) + \left(\dfrac{2}{\mathbf{e}_{sh}}\right) + \left(\dfrac{1}{\mathbf{e}_2}\right) - 2} \tag{9.153}$$

Then from equations 9.149 and 9.153:

$$\frac{q_{sh}}{q_{12}} = \frac{\left(\dfrac{1}{\mathbf{e}_1}\right) + \left(\dfrac{1}{\mathbf{e}_2}\right) - 1}{\dfrac{1}{\mathbf{e}_1} + \dfrac{2}{\mathbf{e}_{sh}} + \dfrac{1}{\mathbf{e}_2} - 2} \tag{9.154}$$

For the special case where all the emissivities are equal ($\mathbf{e}_1 = \mathbf{e}_{sh} = \mathbf{e}_2$):

$$\frac{q_{sh}}{q_{12}} = 1/2$$

Similarly, it can be shown that if n shields are arranged in series, then:

$$\frac{q_{sh}}{q_{12}} = \frac{1}{n+1}$$

In practice, as a result of introducing the radiation shield, the temperature T_2 will fall because a heat balance must hold for surface 2, and the heat transfer rate from it to the surroundings will have been reduced to q_{sh}. The extent to which T_2 is reduced depends on the heat transfer coefficient between surface 2 and the surroundings.

Multi-sided enclosures

For the more complex case of a *multi-sided enclosure* formed from n surfaces, the radiosities may be obtained from an energy balance for each surface in turn in the enclosure. Thus the energy falling on a typical surface i in an enclosure formed from

n surfaces is:

$$A_i q_{Ii} = q_{O1} A_1 F_{1i} + q_{O2} A_2 F_{2i} + q_{O3} A_3 F_{3i} + \cdots + q_{On} A_n F_{ni} \qquad (9.155)$$

where $A_i q_{Ii} = Q_i$ is the energy incident upon surface i, $q_{O1} A_1 F_{1i}$ is the energy leaving surface 1 which is intercepted by surface i and $q_{On} A_n F_{ni}$ is the energy leaving surface n which is intercepted by surface i. $q_{O1} A_1 = q_{O1}$ is the energy leaving surface 1 and F_{1i} is the fraction of this which is intercepted by surface i.

From equation 9.138:

$$q_{Ii} = (q_{Oi} - \mathbf{e}_i E_{bi})/\mathbf{r}_i$$

Substituting for q_{Ii} into equation 9.155 gives:

$$A_i(q_{Oi} - \mathbf{e}_i E_{bi})/\mathbf{r}_i = A_1 F_{1i} q_{O1} + A_2 F_{2i} q_{O2} + A_3 F_{3i} q_{O3} + \cdots + A_j F_{ji} q_{Oi} + \cdots + A_n F_{ni} q_{On} \qquad (9.156)$$

Noting, for example, that for surface 2, $i = 2$, then:

$$A_2(q_{O2} - \mathbf{e}_2 E_{b2})/\mathbf{r}_2 = A_1 F_{12} q_{O1} + A_2 F_{22} q_{O2} + A_3 F_{32} q_{O3} + \cdots + A_j F_{j2} q_{Oj} + \cdots + A_n F_{n2} q_{On} \qquad (9.157)$$

Rearranging:

$$A_1 F_{12} q_{O1} + [A_2 F_{22} - (A_2/\mathbf{r}_2)] q_{O2} + A_3 F_{32} q_{O3} + \cdots + A_j F_{j2} q_{Oi} + \cdots + A_n F_{n2} q_{On} = (A_2 \mathbf{e}_2/\mathbf{r}_2) E_{b2} \qquad (9.158)$$

Equations similar to equation 9.158 may be obtained for each of the surfaces in an enclosure, $i = 1$, $i = 2$, $i = 3$, $i = n$ and the resulting set of simultaneous equations may then be solved for the unknown radiosities, $q_{O1}, q_{O2}, \ldots q_{On}$. The radiation heat transfer is then obtained from equation 9.140. This approach requires data on the areas and view factors for all pairs of surfaces in the enclosure and the emissivity, reflectivity and the black body emissive power for each surface. Should any surface be well insulated, then, in this case, $Q_i = 0$ and:

$$A_i(\mathbf{e}_i/\mathbf{r}_i)(E_{bi} - q_{Oi}) = 0$$

Since, in general, $A_i(\mathbf{e}_i/\mathbf{r}_i) \neq 0$, then $E_{bi} = q_{Oi}$.

If a surface has a specified net thermal input flux, say q_{Ii}, then, from equation 9.140:

$$E_{bi} = (\mathbf{r}_i/(A_i \mathbf{e}_i)) q_{Ii} + q_{Oi}.$$

It may be noted that this approach assumes that the surfaces are grey and diffuse, that emissivity and reflectivity do not vary across a surface and that the temperature, irradiation and radiosity are constant over a surface. Since the technique uses average values over a surface, the subdivision of the enclosure into surfaces must be undertaken with care, noting that a number of surfaces may be regarded as a single surface, that it may be necessary to split one surface up into a number of smaller surfaces and also possibly to introduce an imaginary surface into the system, to represent the surroundings, for example. In a real situation, there may be both grey and black surfaces present and, for the latter, $\mathbf{r}_i$ tends to zero and $(A_i/\mathbf{r}_i)$ and $(A_i \mathbf{e}_i/\mathbf{r}_i)$ become very large.

Example 9.21

A horizontal circular plate, 1.0 m in diameter, is to be maintained at 500 K by placing it 0.20 m directly beneath a horizontal electrically heated plate, also 1.0 m in diameter, maintained at 1000 K. The assembly is exposed to black surroundings at 300 K, and convection heat transfer is negligible. Estimate the electrical input to the heater and the net rate of heat transfer to the plate if the emissivity of the heater is 0.75 and the emissivity of the plate 0.5.

Solution

Taking surface 1 as the heater, surface 2 as the heated plate and surface 3 as an imaginary enclosure consisting of a vertical cylindrical surface representing the surroundings, then, for each surface:

Surface	A (m^2)	**e**	**r**	$(A/\mathbf{r})$ (m^2)	$(A\mathbf{e}/\mathbf{r})$ (m^2)
1	1.07	0.75	0.25	4.28	3.21
2	1.07	0.50	0.50	2.14	1.07
3	0.628	1.0	0		

For surface 1:

For a plane surface: $F_{11} = 0$ and $A_1F_{11} = 0$
Using the nomenclature of Figure 9.39:
For co-axial parallel discs with $r_1 = r_2 = 0.5$ m and $L = 0.2$ m:

$$R_1 = r_1/L = (0.5/0.20) = 2.5$$

$$R_2 = r_2/L = (0.5/0.20) = 2.5$$

and:

$$S = 1 + (1 + R_2{}^2)/R_1{}^2 = 1 + (1 + 2.5^2)/2.5^2 = 2.16$$

From Figure 9.39ii:

$$F_{12} = 0.5\{S - [S^2 - 4(r_2/r_1)^2]^{0.5}\}$$

$$= 0.5 \times \{2.16 - [(2.16^2 - 4(0.5/0.5)^2]^{0.5}\} = 0.672$$

$$A_1F_{12} = (1.07 \times 0.672) = 0.719 \text{ m}^2$$

and, from the summation rule:

$$A_1F_{13} = A_1 - (A_1F_{11} + A_1F_{12}) = 1.07 - (0 + 0.719) = 0.350 \text{ m}^2$$

For surface 2:

For a plane surface: $A_2F_{22} = 0$

and by the reciprocity rule: $A_2F_{21} = A_1F_{12} = 0.719 \text{ m}^2$

By symmetry: $A_2F_{23} = A_1F_{13} = 0.350 \text{ m}^2$

For surface 3:

By the reciprocity rule: $A_3F_{31} = A_1F_{13} = 0.350 \text{ m}^2$

and: $A_3F_{32} = A_2F_{23} = 0.350 \text{ m}^2$

From the summation rule:

$$A_3F_{33} = A_3 - (A_3F_{31} + A_3F_{32})$$

$$= 0.785 - (0.350 + 0.350) = 0.085 \text{ m}^2$$

From equation 9.112:

$$E_{b1} = \sigma T_1^4 = (5.67 \times 10^{-8} \times 1000^4)$$

$$= 5.67 \times 10^4 \text{ W/m}^2 \text{ or } 56.7 \text{ kW/m}^2$$

$$E_{b2} = \sigma T_2^4 = (5.67 \times 10^{-8} \times 500^4)$$
$$= 3.54 \times 10^3 \text{ W/m}^2 \text{ or } 3.54 \text{ kW/m}^2$$
$$E_{b3} = \sigma T_3^4 = (5.67 \times 10^{-8} \times 300^4)$$
$$= 0.459 \times 10^3 \text{ W/m}^2 \text{ or } 0.459 \text{ kW/m}^2$$

Since surface 3 is a black body, $q_{o3} = E_{b3} = 0.459 \text{ kW/m}^2$

From equations 9.157 and 9.158:

$$(A_1F_{11} - A_1/\mathbf{r}_1)q_{o1} + A_2F_{21}q_{o2} + A_3F_{31}q_{o3} = -E_{b1}A_1\mathbf{e}_1/\mathbf{r}_1$$
$$\times (0 - 4.28)q_{o1} + 0.719q_{o2} + (0.350 \times 0.459)$$
$$= -(56.7 \times 1.07 \times 0.75)/0.25$$

or: $$0.719q_{o2} - 4.28q_{o1} = -182 \qquad (1)$$

and: $$(A_1F_{12}q_{o1}) + (A_2F_{22} - A_2/\mathbf{r}_2)q_{o2} = -E_{b2}A_2\mathbf{e}_2/\mathbf{r}_2$$
$$0.719q_{o1} + (0 - 1.07/0.5)q_{o2} = -(3.54 \times 1.07 \times 0.5)/0.5$$

or: $$0.719q_{o1} - 2.14q_{o2} = -3.79 \qquad (2)$$

Solving equations 1 and 2 simultaneously gives:

$$q_{o1} = 45.42 \text{ kW/m}^2 \text{ and } q_{o2} = 17.16 \text{ kW/m}^2$$

power input to the heater = rate of heat transfer from the heater

From equation 9.140:

$$Q_1 = (A_1\mathbf{e}_1/\mathbf{r}_1)(E_{b1} - q_{o1}) = (1.07 \times 0.75/0.25)(56.7 - 45.42) = \underline{\underline{36.2 \text{ kW}}}$$

Again, from equation 9.140, the rate of heat transfer to the plate is:

$$Q_2 = (A_2\mathbf{e}_2/\mathbf{r}_2)(E_{b2} - q_{o2}) = (1.07 \times 0.5/0.25)(3.54 - 17.16) = \underline{\underline{-14.57 \text{ kW}}}$$

where the negative sign indicates heat transfer to the plate.

9.5.6. Radiation from gases

In the previous discusion surfaces have been considered which are *isothermal, opaque* and *grey* which *emit* and *reflect diffusely* and are characterised by *uniform surface radiosity* and the medium separating the surfaces has been assumed to be *non-participating*, in that it neither absorbs nor scatters the surface radiation nor does it emit radiation itself. Whilst, in most cases, such assumptions are valid and permit reasonably accurate results to be calculated, there are occasions where such assumptions do not hold and more refined techniques are required such as those described in the specialist literature[(45,48–53)]. For *non-polar gases* such as N_2 and O_2, the foregoing assumptions are largely valid, since the gases do not emit radiation and they are essentially transparent to incident radiation. This is not the case with *polar molecules* such as CO_2 and H_2O vapour, NH_3 and hydrocarbon gases, however, since these not only emit and absorb over a wide temperature range, but also radiate in specific wavelength intervals called *bands*. Furthermore, gaseous radiation is a volumetric rather than a surface phenonemon.

Whilst the calculation of the radiant heat flux from a gas to an adjoining surface embraces inherent spectral and directional effects, a simplified approach has been developed by HOTTEL and MANGLESDORF[54], which involves the determination of radiation emission from a hemispherical mass of gas of radius L, at temperature T_g to a surface element, dA_1, near the centre of the base of the hemisphere. Emission from the gas per unit area of the surface is then:

$$E_g = \mathbf{e}_g \sigma T_g^4 \tag{9.159}$$

where the gas emissivity $\mathbf{e}_g$ is a function of T_g, the total pressure of the gas P, the partial pressure of the radiating gas P_g and the radius of the hemisphere L.

Data on the emissivity of water vapour at a total pressure of 101.3 kN/m^2 are plotted in Figure 9.44 for different values of the product of the vapour partial pressure P_w and the hemisphere radius L. For other values of the total pressure, the correction factor C_w also given in the figure must be used. Similar data for carbon dioxide are given in Figure 9.45. Although these data refer to water vapour or carbon dioxide alone in a mixture of non-radiating gases, they may be extended to situations where both are present in such a mixture by expressing the total emissivity as:

$$\mathbf{e}_g = \mathbf{e}_w + \mathbf{e}_c - \Delta\mathbf{e} \tag{9.160}$$

where $\Delta\mathbf{e}$ is a correction factor, shown in Figure 9.46, which allows for the reduction in emission associated with mutual absorption of radiation between the two species.

Although these data provide the emissivity of a hemispherical gas mass of radius L radiating to an element at the centre of the base, they may be extended to other geometries by using the concept of *mean beam length* L_e which correlates the dependence of gas emissivity with both the size and shape of the gas geometry in terms of a single parameter. Essentially the mean beam length is the radius of the hemisphere of gas whose *emissivity* is equivalent to that in the particular geometry considered, and typical values of L_e which are then used to replace L in Figures 9.38–40 are shown in Table 9.7. Using these data and Figures 9.38–9.40, the rate of transfer of radiant heat to a surface of area A_s due to emission from an adjoining gas is given by:

$$Q = \mathbf{e}_g A_s \sigma T_g^4 \tag{9.161}$$

A black surface will not only absorb all of this radiation but will also emit radiation, and the net rate at which radiation is exchanged between the gas and the surface at temperature T_s is given by:

$$Q_{\text{net}} = A_s \sigma(\mathbf{e}_g T_g^4 - \mathbf{a}_g T_s^4) \tag{9.162}$$

In this equation, the absorptivity $\mathbf{a}_g$ may be obtained from the emissivity using expressions of the form[54]:

water: $$\mathbf{a}_w = C_w \mathbf{e}_w (T_g/T_s)^{0.45} \tag{9.163}$$

carbon dioxide: $$\mathbf{a}_c = C_c \mathbf{e}_c (T_g/T_s)^{0.65} \tag{9.164}$$

where $\mathbf{e}_w$ and C_w and $\mathbf{e}_c$ and C_c are obtained from Figures 9.38 and 9.39 respectively, noting that T_g is replaced by T_s and $(P_w L_e)$ or $(P_c L_e)$ by $[P_w L_e(T_s/T_g)]$ or $[P_c L_e(T_s/T_g)]$

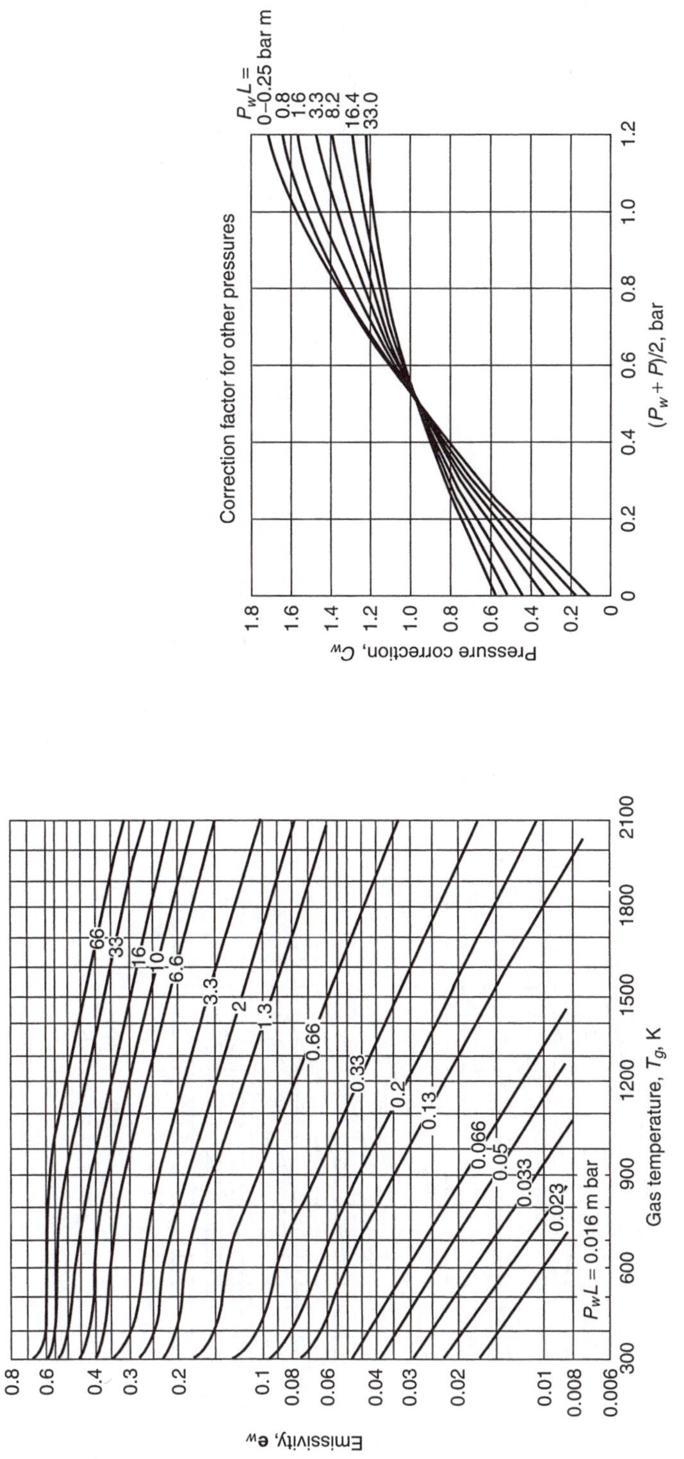

Figure 9.44. Emissivity of water vapour in a mixture of non-radiating gases at 101.3 kN/m²[(54)]

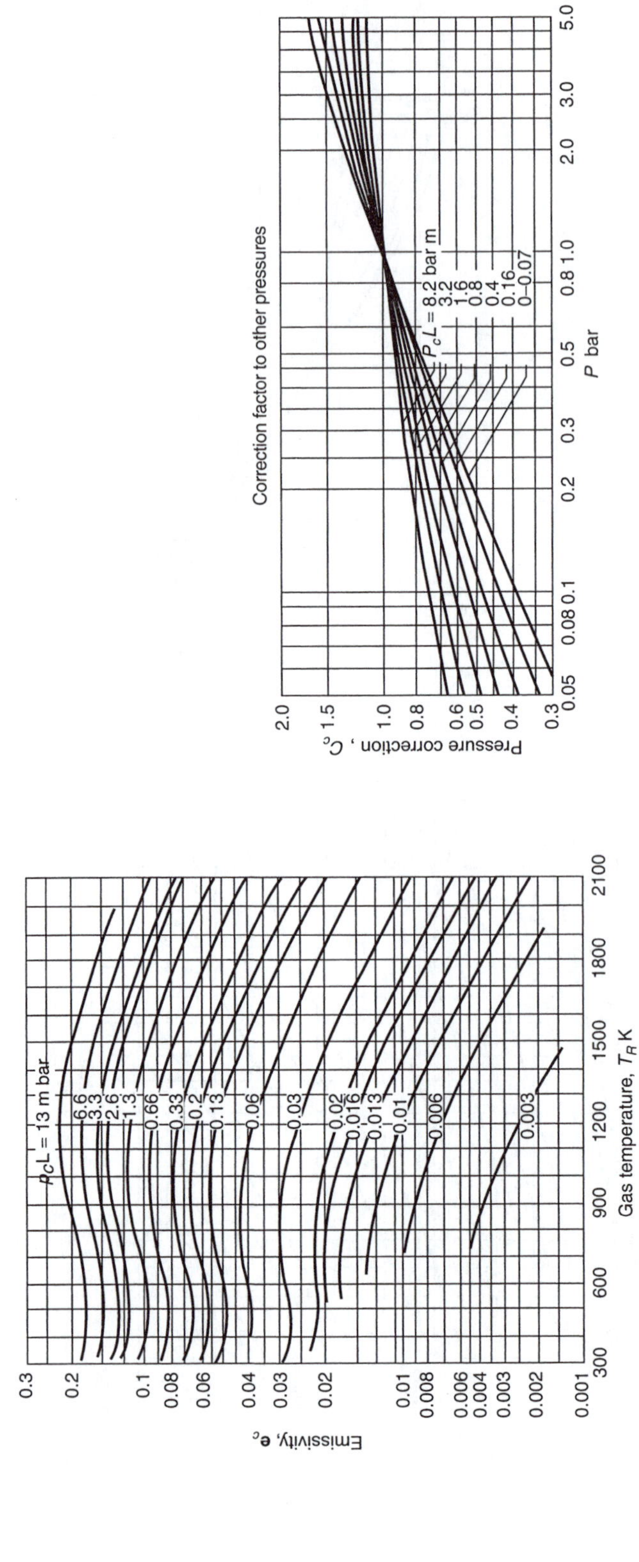

Figure 9.45. Emissivity of carbon dioxide in a mixture of non-radiating gases at 101.3 kN/m²[54]

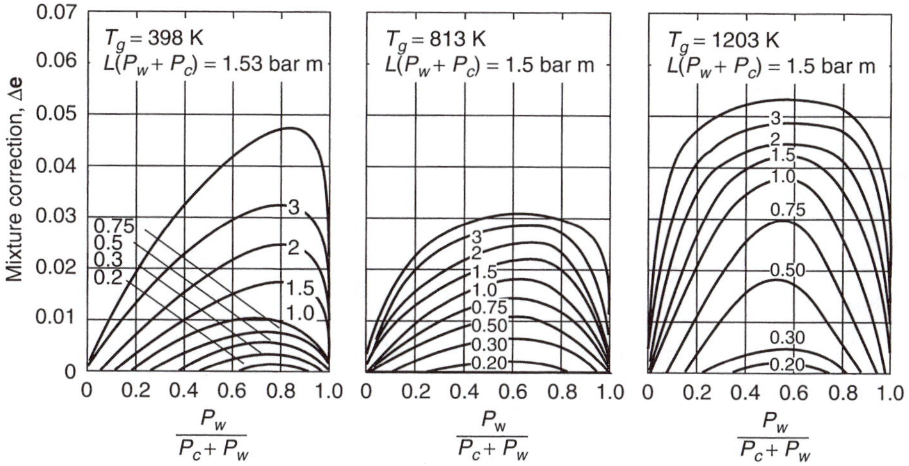

Figure 9.46. Correction factor for water vapour–carbon dioxide mixtures(54)

respectively. It may be noted that, in the presence of both water vapour and carbon dioxide, the total absorptivity is given by:

$$\mathbf{a}_g = \mathbf{a}_w + \mathbf{a}_c - \Delta\mathbf{a} \tag{9.165}$$

where $\Delta\mathbf{a} = \Delta\mathbf{e}$ is obtained from Figure 9.46. If the surrounding surface is grey, some of the radiation may be reflected and equation 9.162 may be modified by a factor, $\mathbf{e}_s/[1 - (1 - \mathbf{a}_g)(1 - \mathbf{e}_g)]$ to take this into account. This leads to the following equation for the heat transferred per unit time from the gas to the surface:

$$Q = \sigma \mathbf{e}_s A_s(\mathbf{e}_g T_g^4 - \mathbf{a}_g T_s^4)/[1 - (1 - \mathbf{a}_g)(1 - \mathbf{e}_g)] \tag{9.166}$$

Table 9.7. Mean beam lengths for various geometries(54)

Geometry	Characteristic length	Mean beam length, L_e
Sphere—radiation to surface	Diameter, D	$0.65D$
Infinite circular cylinder—radiation to curved surface	Diameter, D	$0.95D$
Semi-infinite cylinder—radiation to base	Diameter, D	$0.65D$
Cylinder of equal height and diameter—radiation to entire surface	Diameter, D	$0.60D$
Infinite parallel planes—radiation to planes	Spacing between planes, L	$1.80L$
Cube—radiation to any surface	Side, L	$0.66L$
Shape of volume, V—radiation to surface of area, A	Ratio: volume/area, (V/A)	$3.6(V/A)$

Example 9.22

The walls of a combustion chamber, 0.5 m in diameter and 2 m long, have an emissivity of 0.5 and are maintained at 750 K. If the combustion products containing 10 per cent carbon dioxide and 10 per cent water vapour are at 150 kN/m^2 and 1250 K, what is the net rate of radiation to the walls?

Solution

The partial pressures of carbon dioxide (P_c) and of water (P_w) are:

$$P_c = P_w = (10/100)150 = 15.0 \text{ kN/m}^2 \text{ or } (15.0/100) = 0.15 \text{ bar}$$

From Table 9.7:

$$L_e = 3.6V/A = 3.6(\pi/4 \times 0.5^2 \times 2)/[2\pi/4 \times 0.5^2) + (0.5\pi \times 2.0)] = 0.4 \text{ m}$$

For water vapour:

$$P_w L_e = (0.15 \times 0.4) = 0.06 \text{ bar m}$$

and from Figure 9.44, $\mathbf{e}_w = 0.075$

$$P = (150/100) = 1.5 \text{ bar}, P_w = 0.15 \text{ bar and:} \quad 0.5(P_w + P) = 0.825 \text{ bar}$$

Since $P_w L_e = 0.06$ bar m, then from Figure 9.44:

$$C_w = 1.4 \text{ and } \mathbf{e}_w = (1.4 \times 0.075) = 0.105$$

For carbon dioxide:

$$P_c L_e = (0.15 \times 0.4) = 0.06 \text{ bar m}$$

and from Figure 9.45, $\mathbf{e}_c = 0.037$
Since $P = 1.5$ bar, $P_c = 0.15$ bar and $P_c L_e = 0.06$ bar m, then, from Figure 9.38:

$$C_c = 1.2 \text{ and } \mathbf{e}_c = (1.2 \times 0.037) = 0.044$$

$$(P_w + P_c)L_e = (0.15 + 0.15)0.4 = 0.12 \text{ bar m}$$

and: $$P_c/(P_c + P_w) = 0.15/(0.15 + 0.15) = 0.5$$

Thus, from Figure 9.45 for $T_g > 1203$ K, $\Delta\mathbf{e} = 0.001$
and, from equation 9.160:

$$\mathbf{e}_g = \mathbf{e}_w - \Delta\mathbf{e} = (0.105 + 0.044 - 0.001) = 0.148$$

For water vapour:

$$P_w L_e(T_s/T_g) = 0.06(750/1250) = 0.036 \text{ bar m}$$

and, from Figure 9.44 at 750 K, $\mathbf{e}_w = 0.12$
Since $0.5(P_w + P) = 0.825$ bar and $P_w L_e(T_s/T_g) = P_c L_e(T_s/T_g) = 0.036$ bar m,
then, from Figure 9.44: $C_w = 1.40$ and $\mathbf{e}_w = (0.12 \times 1.40) = 0.168$
and the absorptivity, from equation 9.163 is:

$$\mathbf{a}_w = \mathbf{e}_w(T_g/T_s)^{0.45} = 0.168(1250/750)^{0.45} = 0.212$$

For carbon dioxide:
From Figure 9.45 at 750 K, $\mathbf{e}_c = 0.08$
From Figure 9.45 at $P = 1.5$ bar and $P_c L_e(T_s/T_g) = 0.036$ bar m:

and: $$C_c = 1.02 \text{ and } \mathbf{e}_c = (0.08 \times 1.02) = 0.082$$

and the absorptivity, from equation 9.164 is:

$$\mathbf{a}_c = \mathbf{e}_c(T_g/T_s)^{0.65} = 0.082(1250/750)^{0.65} = 0.114$$

$$P_w/(P_c + P_w) = 0.5 \text{ and } (P_c + P_w)L_e(T_s/T_g) = (0.036 + 0.036) = 0.072 \text{ bar m}$$

Thus, from Figure 9.46, for $T_g = 813$ K, $\Delta\mathbf{e} = \Delta\mathbf{a} < 0.01$ and this may be neglected.

Thus: $$\mathbf{a}_g = \mathbf{a}_w + \mathbf{a}_c - \Delta\mathbf{a} = (0.212 + 0.114 - 0) = 0.326$$

If the surrounding surface is black, then:

$$Q = \sigma A_s(\mathbf{e}_g T_g^4 - \alpha_g T_s^4) \quad \text{(equation 9.162)}$$

$$= (5.67 \times 10^{-8}[(2(\pi/4)0.5^2) + (0.5\pi \times 2.0)])[(0.148 \times 1250^4) - (0.326 \times 750^4)]$$

$$= 5.03 \times 10^4 \text{ W} = \underline{\underline{50.3 \text{ kW}}}$$

For grey walls, the correction factor allowing for multiple reflection of incident radiation is:

$$C_g = \mathbf{e}_s/[1 - (1 - \mathbf{a}_g)(1 - \mathbf{e}_g)] = 0.5/[1 - (1 - 0.326)(1 - 0.5)] = 0.754$$

and hence: net radiation to the walls, $Q_w = (50.3 \times 0.754) = \underline{\underline{37.9 \text{ kW}}}$

Radiation from gases containing suspended particles

The estimation of the radiation from pulverised-fuel flames, from dust particles in flames and from flames made luminous as a result of the thermal decomposition of hydrocarbons to soot, involves an evaluation of radiation from clouds of particles. In pulverised-fuel flames, the mean particle size is typically 25 μm and the composition varies from a very high carbon content to virtually pure ash. In contrast, the suspended matter in luminous flames, resulting from soot formation due to incomplete mixing of hydrocarbons with air before being heated, consists of carbon together with very heavy hydrocarbons with an initial particle size of some 0.3 μm. In general, pulverised-fuel particles are sufficiently large to be substantially opaque to incident radiation, whilst the particles in a luminous flame are so small that they act as semi-transparent bodies with respect to thermal or long wavelength radiation.

According to SCHACK(55), a single particle of soot transmits approximately 95 per cent of the incident radiation and a cloud must contain a very large number of particles before an appreciable emission can occur. If the concentration of particles is K', then the product of K' and the thickness of the layer L is equivalent to the product $P_g L_e$ in the radiation of gases. For a known or measured emissivity of the flame $\mathbf{e}_f$, the heat transfer rate per unit time to a wall is given by:

$$Q = \mathbf{e}_f \mathbf{e}_s \sigma(T_f^4 - T_w^4) \quad (9.167)$$

where $\mathbf{e}_s$ is the effective emissivity of the wall, and T_f and T_w are the temperatures of the flame and wall respectively. $\mathbf{e}_f$ varies, not only from point to point in a flame, but also depends on the type of fuel, the shape of the burner and combustion chamber, and on the air supply and the degree of preheating of the air and fuel.

9.6. HEAT TRANSFER IN THE CONDENSATION OF VAPOURS

9.6.1. Film coefficients for vertical and inclined surfaces

When a saturated vapour is brought into contact with a cool surface, heat is transferred from the vapour to the surface and a film of condensate is produced.

In considering the heat that is transferred, the method first put forward by NUSSELT[56] and later modified by subsequent workers is followed. If the vapour condenses on a vertical surface, the condensate film flows downwards under the influence of gravity, although it is retarded by the viscosity of the liquid. The flow will normally be streamline and the heat flows through the film by conduction. In Nusselt's work it is assumed that the temperature of the film at the cool surface is equal to that of the surface, and at the other side was at the temperature of the vapour. In practice, there must be some small difference in temperature between the vapour and the film, although this may generally be neglected except where non-condensable gas is present in the vapour.

It is shown in Chapter 3, that the mean velocity of a fluid flowing down a surface inclined at an angle ϕ to the horizontal is given by:

$$u = \frac{\rho g \sin\phi s^2}{3\mu} \qquad \text{(equation 3.87)}$$

For a vertical surface: $\sin\phi = 1$ and $u = \dfrac{\rho g s^2}{3\mu}$

The maximum velocity u_s which occurs at the free surface is:

$$u_s = \frac{\rho g \sin\phi\, s^2}{2\mu} \qquad \text{(equation 3.88)}$$

and this is 1.5 times the mean velocity of the liquid.

Since the liquid is produced by condensation, the thickness of the film will be zero at the top and will gradually increase towards the bottom. Under stable conditions the difference in the mass rates of flow at distances x and $x + \mathrm{d}x$ from the top of the surface will result from condensation over the small element of the surface of length $\mathrm{d}x$ and width w, as shown in Figure 9.47.

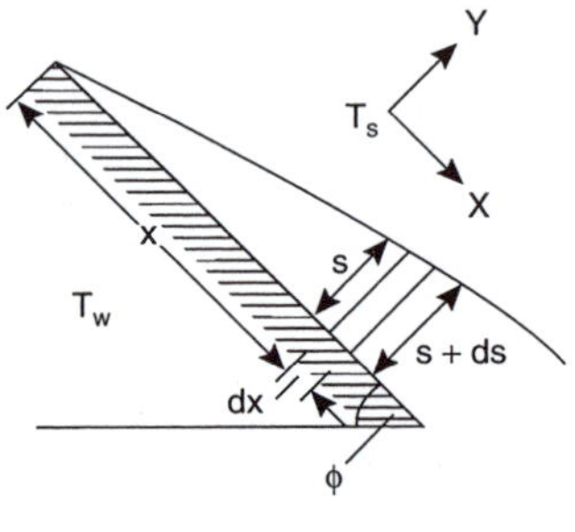

Figure 9.47. Condensation on an inclined surface

If the thickness of the liquid film increases from s to $s + \mathrm{d}s$ in that distance, the increase in the mass rate of flow of liquid $\mathrm{d}G$ is given by:

$$\frac{\mathrm{d}}{\mathrm{d}s}\left(\frac{\rho^2 g \sin\phi\, s^3 w}{3\mu}\right)\mathrm{d}s$$

$$= \frac{\rho^2 g \sin\phi}{\mu} w s^2\, \mathrm{d}s$$

If the vapour temperature is T_s and the wall temperature is T_w, the heat transferred by thermal conduction to an element of surface of length dx is:

$$\frac{k(T_s - T_w)}{s} w \, \mathrm{d}x$$

where k is the thermal conductivity of the condensate.

Thus the mass rate of condensation on this small area of surface is:

$$\frac{k(T_s - T_w)}{s\lambda} w \, \mathrm{d}x$$

where λ is the latent heat of vaporisation of the liquid.

Thus:
$$\frac{k(T_s - T_w)}{s\lambda} w \, \mathrm{d}x = \frac{\rho^2 g \sin\phi}{\mu} w s^2 \, \mathrm{d}s$$

On integration:
$$\mu k(T_s - T_w)x = \tfrac{1}{4}\rho^2 g \sin\phi \, s^4 \lambda$$

since $s = 0$ when $x = 0$.

Thus:
$$s = \sqrt[4]{\frac{4\mu k x(T_s - T_w)}{g \sin\phi \, \lambda \rho^2}} \tag{9.168}$$

Now the heat transfer coefficient h at $x = x$, $= k/s$, and hence:

Thus:
$$h = \sqrt[4]{\frac{\rho^2 g \sin\phi \, \lambda k^3}{4\mu x(T_s - T_w)}} \tag{9.169}$$

and:
$$Nu = \frac{hx}{k} = \sqrt[4]{\frac{\rho^2 g \sin\phi \, \lambda x^3}{4\mu k(T_s - T_w)}} \tag{9.170}$$

These expressions give point values of h and Nu_x at $x = x$. It is seen that the coefficient decreases from a theoretical value of infinity at the top as the condensate film thickens. The mean value of the heat transfer coefficient over the whole surface, between $x = 0$ and $x = x$ is given by:

$$h_m = \frac{1}{x}\int_0^x h \, \mathrm{d}x = \frac{1}{x}\int_0^x K x^{-1/4} \, \mathrm{d}x \quad \text{(where } K \text{ is independent of } x\text{)}$$

$$= \frac{1}{x} K \frac{x^{3/4}}{\frac{3}{4}} = \frac{4}{3} K x^{-1/4} = \frac{4}{3} h$$

$$= 0.943 \sqrt[4]{\frac{\rho^2 g \sin\phi \lambda k^3}{\mu x \Delta T_f}} \tag{9.171}$$

where ΔT_f is the temperature difference across the condensate film.

For a vertical surface, $\sin\phi = 1$ and:

$$h_m = 0.943 \sqrt[4]{\frac{\rho^2 g \lambda k^3}{\mu x \Delta T_f}} \tag{9.172}$$

9.6.2. Condensation on vertical and horizontal tubes

The Nusselt equation

If vapour condenses on the outside of a vertical tube of diameter d_o, then the hydraulic mean diameter for the film is:

$$\frac{4 \times \text{flow area}}{\text{wetted perimeter}} = \frac{4S}{b} \quad \text{(say)}$$

If G is the mass rate of flow of condensate, the mass rate of flow per unit area G' is G/S and the Reynolds number for the condensate film is then given by:

$$Re = \frac{(4S/b)(G/S)}{\mu} = \frac{4G}{\mu b} = \frac{4M}{\mu} \tag{9.173}$$

where M is the mass rate of flow of condensate per unit length of perimeter, or:

$$M = \frac{G}{\pi d_o}$$

For streamline conditions in the film, $4M/\mu \not> 2100$ and:

$$h_m = \frac{Q}{A\Delta T_f} = \frac{G\lambda}{bl\Delta T_f} = \frac{\lambda M}{l\Delta T_f}$$

From equation 9.172:

$$h_m = 0.943\left(\frac{k^3\rho^2 g}{\mu}\frac{\lambda}{l\Delta T_f}\right)^{1/4} = 0.943\left(\frac{k^3\rho^2 g}{\mu}\frac{h_m}{M}\right)^{1/4}$$

and hence:

$$h_m\left(\frac{\mu^2}{k^3\rho^2 g}\right)^{1/3} = 1.47\left(\frac{4M}{\mu}\right)^{-1/3} \tag{9.174}$$

For horizontal tubes, Nusselt proposes the equation:

$$h_m = 0.72\left(\frac{k^3\rho^2 g\lambda}{d_o\mu\Delta T_f}\right)^{1/4} \tag{9.175}$$

This may be rearranged to give:

$$h_m\left(\frac{\mu^2}{k^3\rho^2 g}\right)^{1/3} = 1.51\left(\frac{4M}{\mu}\right)^{-1/3} \tag{9.176}$$

where M is the mass rate of flow per unit length of tube.

This is approximately the same as equation 9.173 for vertical tubes and is a universal equation for condensation, noting that for vertical tubes $M = G/\pi d_o$ and for horizontal tubes $M = G/l$, where l is the length of the tube. Comparison of the two equations shows that, provided the length is more than three times the diameter, the horizontal tube will give a higher transfer coefficient for the same temperature conditions.

For j vertical rows of horizontal tubes, equation 9.175 may be modified to give:

$$h_m = 0.72\left(\frac{k^3\rho^2 g\lambda}{jd_o\mu\Delta T_f}\right)^{1/4} \tag{9.177}$$

KERN[28] suggests that, based on the performance of commercial exchangers, this equation is too conservative and that the exponent of j should be nearer $-\frac{1}{6}$ than $-\frac{1}{4}$. This topic is discussed in Volume 6, Chapter 12.

Experimental values

In testing Nusselt's equation it is important to ensure that the conditions comply with the requirements of the theory. In particular, it is necessary for the condensate to form a uniform film on the tubes, for the drainage of this film to be by gravity, and the flow streamline. Although some of these requirements have probably not been entirely fulfilled, results for pure vapours such as steam, benzene, toluene, diphenyl, ethanol, and so on, are sufficiently close to give support to the theory. Some data obtained by HASELDEN and PROSAD[57] for condensing oxygen and nitrogen vapours on a vertical surface, where precautions were taken to see that the conditions were met, are in very good agreement with Nusselt's theory. The results for most of the workers are within 15 per cent for horizontal tubes, although they tend to be substantially higher than the theoretical for vertical tubes. Typical values are given in Table 9.8 taken from MCADAMS[27] and elsewhere.

Table 9.8. Average values of film coefficients h_m for condensation of pure saturated vapours on horizontal tubes

Vapour	Value of h_m (W/m^2 K)	Value of h_m (Btu/h ft^2 °F)	Range of ΔT_f (deg K)
Steam	10,000–28,000	1700–5000	1–11
Steam	18,000–37,000	3200–6500	4–37
Benzene	1400–2200	240–380	23–37
Diphenyl	1300–2300	220–400	4–15
Toluene	1100–1400	190–240	31–40
Methanol	2800–3400	500–600	8–16
Ethanol	1800–2600	320–450	6–22
Propanol	1400–1700	250–300	13–20
Oxygen	3300–8000	570–1400	0.08–2.5
Nitrogen	2300–5700	400–1000	0.15–3.5
Ammonia	6000	1000	—
Freon-12	1100–2200	200–400	—

When considering commercial equipment, there are several factors which prevent the true conditions of Nusselt's theory being met. The temperature of the tube wall will not be constant, and for a vertical condenser with a ratio of ΔT at the bottom to ΔT at the top of five, the film coefficient should be increased by about 15 per cent.

Influence of vapour velocity

A high vapour velocity upwards tends to increase the thickness of the film and thus reduce h though the film may sometimes be disrupted mechanically as a result of the formation of small waves. For the downward flow of vapour, TEN BOSCH[58] has shown that h increases considerably at high vapour velocities and may increase to two or three times the value given by the Nusselt equation. It must be remembered that when a large fraction of the vapour is condensed, there may be a considerable change in velocity over the surface.

Under conditions of high vapour velocity CARPENTER and COLBURN(59) have shown that turbulence may occur with low values of the Reynolds number, in the range 200–400. When the vapour velocity is high, there will be an appreciable drag on the condensate film and the expression obtained for the heat transfer coefficient is difficult to manage.

Carpenter and Colburn have put forward a simple correlation of their results for condensation at varying vapour velocities on the inner surface of a vertical tube which takes the form:

$$h_m = 0.065 G'_m \sqrt{\frac{C_p \rho k (R'/\rho_v u^2)}{\mu \rho_v}} \tag{9.178}$$

where:

$$G'_m = \sqrt{\frac{(G_1'^2 + G'_1 G'_2 + G_2'^2)}{3}}$$

and u is the velocity calculated from G'_m. In this equation C_p, k, ρ, and μ are properties of the condensate and ρ_v refers to the vapour. G'_1 is the mass rate of flow per unit area at the top of the tube and G'_2 the corresponding value at the bottom. R' is the shear stress at the free surface of the condensate film.

As pointed out by COLBURN(60), the group $C_p \rho k/\mu \rho_v$ does not vary very much for a number of organic vapours so that a plot of h_m and G'_m will provide a simple approximate correlation with separate lines for steam and for organic vapours as shown in Figure 9.48(60,61). Whilst this must be regarded as an empirical approximation it is very useful for obtaining a good indication of the effect of vapour velocity.

Turbulence in the film

If *Re* is greater than 2100 during condensation on a vertical tube the mean coefficient h_m will increase as a result of turbulence. The data of KIRKBRIDE(62) and BADGER(63,64) for the condensation of diphenyl vapour and Dowtherm on nickel tubes are expressed in the form:

$$h_m \left(\frac{\mu^2}{k^3 \rho^2 g}\right)^{1/3} = 0.0077 \left(\frac{4M}{\mu}\right)^{0.4} \tag{9.179}$$

Comparing equation 9.176 for streamline flow of condensate and equation 9.179 for turbulent flow, it is seen that, with increasing Reynolds number, h_m decreases with streamline flow but increases with turbulent flow. These results are shown in Figure 9.49.

Design equations are given in Volume 6, Chapter 12, for condensation both inside and outside horizontal and vertical tubes, and the importance of avoiding flooding in vertical tubes is stressed.

9.6.3. Dropwise condensation

In the discussion so far, it is assumed that the condensing vapour, on coming into contact with the cold surface, wets the tube so that a continuous film of condensate is formed. If the droplets initially formed do not wet the surface, after growing slightly

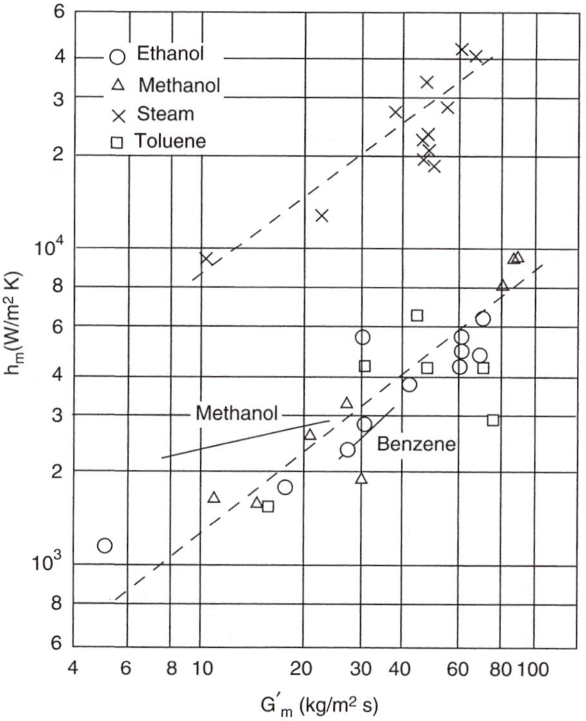

Figure 9.48. Average heat transfer data of CARPENTER and COLBURN[59] (shown as points) compared with those of TEPE and MUELLER[61] (shown as solid lines). Dashed lines represent equation 9.178

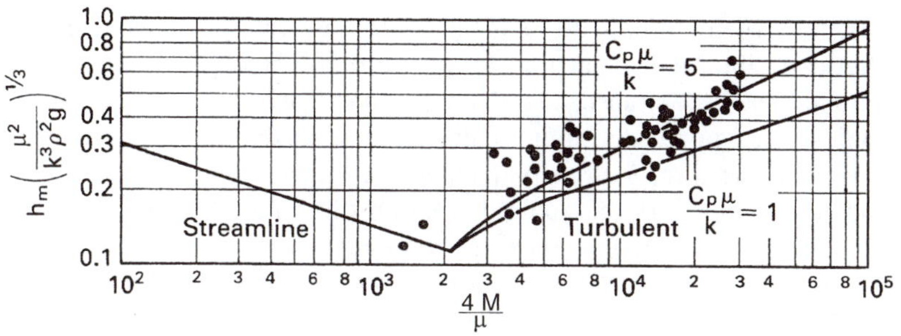

Figure 9.49. Effects of turbulence in condensate film

they will fall from the tube exposing fresh condensing surface. This is known as dropwise condensation and, since the heat does not have to flow through a film by conduction, much higher transfer coefficients are obtained. Steam is the only pure vapour for which definite dropwise condensation has been obtained, and values of h from 40 to 114 kW/m^2 K have been obtained, with much higher values on occasions. This question has been discussed by DREW, NAGLE and SMITH[65] who have shown that there are many materials which make the surface non-wettable although, of these, only those which are firmly held to the surface are of any practical use. Mercaptans and oleic acid have been used to promote dropwise condensation, but at present there is little practical application of this technique. Exceptionally high values of h will not give a corresponding increase in the overall coefficient, since for a condenser with steam, a value of about 11 kW/m^2 K can be obtained with film condensation. On the other hand, it may be helpful in experimental work to reduce the thermal resistance on one side of a surface to a negligible value.

9.6.4. Condensation of mixed vapours

In the previous discussion it has been assumed that the vapour is a pure material, such as steam or organic vapour. If it contains a proportion of non-condensable gas and is cooled below its dew point, a layer of condensate is formed on the surface with a mixture of non-condensable gas and vapour above it. The heat flow from the vapour to the surface then takes place in two ways. Firstly, sensible heat is passed to the surface because of the temperature difference. Secondly, since the concentration of vapour in the main stream is greater than that in the gas film at the condensate surface, vapour molecules diffuse to the surface and condense there, giving up their latent heat. The actual rate of condensation is then determined by the combination of these two effects, and its calculation requires a knowledge of mass transfer by diffusion, as discussed in Chapter 10.

In the design of a cooler–condenser for a mixture of vapour and a permanent gas, the method of COLBURN and HOUGEN[66] is considered. This requires a point-to-point calculation of the condensate–vapour interface conditions T_c and P_s. A trial and error solution is required of the equation:

$$q_v \quad + \quad q_\lambda \quad = \quad q_c \quad = U\Delta T \tag{9.180}$$

$$h_g(T_s - T_c) + k_G\lambda(P_g - P_s) = h_o(T_c - T_{cm}) = U\Delta T \tag{9.181}$$

where the first term q_v represents the sensible heat transferred to the condensing surface, the second term q_λ the latent heat transferred by the diffusing vapour molecules, and the third term q_c the heat transferred from the condensing surface through the pipe wall, dirt and scales, and water film to the cooling medium. h_g is the heat transfer coefficient over the gas film, h_o the conductance of the combined condensate film, tube wall, dirt and scale films, and the cooling medium film and U the overall heat transfer coefficient. T_s is the vapour temperature, T_c the temperature of the condensate, T_{cm} the cooling medium temperature, ΔT the overall temperature difference $= (T_s - T_{cm})$, P_g is the partial pressure of diffusing vapour, P_s the vapour pressure at T_c, λ the latent heat of vaporisation per unit mass, and k_G the mass transfer coefficient in mass per unit time, unit area and unit partial pressure difference.

To evaluate the required condenser area, point values of the group $U\Delta T$ as a function of q_c must be determined by a trial and error solution of equation 9.181. Integration of a plot of q_c against $1/U\Delta T$ will then give the required condenser area. This method takes into account point variations in temperature difference, overall coefficient and mass velocities and consequently produces a reasonably accurate value for the surface area required.

The individual terms in equation 9.181 are now examined to enable a trial solution to proceed. Values for h_g and k_G are most conveniently obtained from the CHILTON and

COLBURN[67] analogy discussed in Chapter 10.

Thus:
$$h_g = \frac{j_h G' C_p}{(C_p \mu / k)^{0.67}} \tag{9.182}$$

$$k_G = \frac{j_d G'}{P_{Bm}(\mu/\rho D)^{0.67}} \tag{9.183}$$

Values of j_h and j_d are obtained from a knowledge of the Reynolds number at a given point in the condenser. The combined conductance h_o is evaluated by determining the condensate film coefficient h_c from the Nusselt equation and combining this with the dirt and tube wall conductances and a cooling medium film conductance predicted from the Sieder–Tate relationships. Generally, h_o may be considered to be constant throughout the exchanger.

From a knowledge of h_g, k_G, and h_o and for a given T_s and T_{cm} values of the condensate surface temperature T_c are estimated until equation 9.181 is satisfied. The calculations are repeated, and in this manner several point values of the group $U\Delta T$ throughout the condenser may be obtained.

The design of a cooler condenser for the case of condensation of two vapours is more complicated than the preceding single vapour–permanent gas case[68], and an example has been given by JEFFREYS[69].

For the condensation of a vapour in the presence of a non-condensable gas, the following example is considered which is based on an the work of KERN[28].

Example 9.23

A mixture of 0.57 kg/s of steam and 0.20 kg/s of carbon dioxide at 308 kN/m^2 and its dew point enters a heat exchanger consisting of 246 tubes, 19 mm o.d., wall thickness 1.65 mm, 3.65 m long, arranged in four passes on 25 mm square pitch in a 0.54 m diameter shell and leaves at 322 K. Condensation is effected by cooling water entering and leaving the unit at 300 and 319 K respectively. If the diffusivity of steam–carbon dioxide mixtures is 0.000011 m^2/s and the group $(\mu/\rho D)^{0.67}$ may be taken to be constant at 0.62, estimate the overall coefficient of heat transfer and the dirt factor for the condenser.

Solution

In the steam entering the condenser, there is $\frac{0.57}{18} = 0.032$ kmol water
and $\frac{0.20}{44} = 0.0045$ kmol CO_2 } total = 0.0365 kmol.

Hence the partial pressure of water = (308 × 0.032/0.0365) = 270 kN/m^2 and from Table 11A in the Appendix, the dew point = 404 K.

Mean molecular weight of the mixture = (0.57 + 0.20)/0.0365 = 21.1 kg/kmol.

At the *inlet*: vapour pressure of water = 270 kN/m^2
inert pressure = (308 − 270) = 38 kN/m^2 } total = 308 kN/m^2

At the *outlet*: partial pressure of water at 322 K = 11.7 kN/m^2
inert pressure = (308 − 11.7) = 296.3 kN/m^2 } total = 308 kN/m^2

$\therefore$ steam at the outlet $= \frac{0.0045 \times 11.7}{296.3} = 0.000178$ kmol

and: steam condensed = (0.032 − 0.000178) = 0.03182 kmol.

The heat load is now estimated at each interval between the temperatures 404, 401, 397, 380, 339 and 322 K.

For the interval 404 to 401 K

From Table 11A in the Appendix, the partial pressure of steam at 401 K = 252.2 kN/m^2 and hence the partial pressure of CO_2 = (308 − 252.2) = 55.8 kN/m^2. Steam remaining = (0.0045 × 252.2/55.8) = 0.0203 kmol.

∴ Steam condensed = (0.032 − 0.0203) = 0.0117 kmol

Heat of condensation = (0.0117 × 18)(2180 + 1.93(404 − 401)) = 466 kW

Heat from uncondensed steam = (0.0203 × 18 × 1.93(404 − 401)) = 1.9 kW

Heat from carbon dioxide = (0.020 × 0.92(404 − 401)) = 0.5 kW

and the total for the interval = 468.4 kW

Repeating the calculation for the other intervals of temperature gives the following results:

Interval (K)	Heat load (kW)
404–401	468.4
401–397	323.5
397–380	343.5
380–339	220.1
339–322	57.9
Total	1407.3

and the flow of water = 1407.3/(4.187(319 − 300)) = 17.7 kg/s.

With this flow of water and a flow area per pass of 0.0120 m^2, the mass velocity of water is 1425 kg/m^{2}s, equivalent to a velocity of 1.44 m/s at which h_i = 6.36 kW/m^2 K. Basing this on the outside area, h_{io} = 5.25 kW/m^2 K.

Shell-side coefficient for entering gas mixture:

$$\text{The mean specific heat, } C_p = \frac{(0.20 \times 0.92) + (0.57 \times 1.93)}{0.77} = 1.704 \text{ kJ/kg K.}$$

Similarly, the mean thermal conductivity k = 0.025 kW/m K and the mean viscosity μ = 0.015 mN s/m^2

The area for flow through the shell = 0.0411 m^2 and the mass velocity on the shell side

$$= \frac{0.20 + 0.57}{0.0411} = 18.7 \text{ kg/m}^2\text{s}$$

Taking the equivalent diameter as 0.024 m, Re = 29,800

and: $$h_g = 0.107 \text{ kW/m}^2\text{ K or } 107 \text{ W/m}^2\text{ K.}$$

Now: $$\left(\frac{\mu}{\rho D}\right)^{0.67} = 0.62, \quad \left(\frac{C_p\mu}{k}\right)^{0.67} = 1.01$$

and: $$k_G = \frac{h_g(C_p\mu/k)^{0.67}}{C_p P_{sF}(\mu/\rho D)^{0.67}} = \frac{107 \times 1.01}{1704 P_{sF} \times 0.62}$$

$$= \frac{0.102}{P_{sF}}$$

At point 1

Temperature of the gas $T = 404$ K, partial pressure of steam $P_g = 270$ kN/m^2, partial pressure of the inert $P_s = 38$ kN/m^2, water temperature $T_w = 319$ K and $\Delta T = (404 - 319) = 85$ K. An estimate is now made for the temperature of the condensate film of $T_c = 391$ K. In this case $P_s = 185.4$ kN/m^2 and $P'_s = (308 - 185.4) = 122.6$ kN/m^2.

Thus: $$P_{sF} = \frac{122.6 - 38}{\ln(122.6/38)} = 72.2 \text{ kN/m}^2.$$

In equation 9.181:

$$h_g(T_s - T_c) + k_G\lambda(P_g - P_s) = h_{i0}(T_c - T_{cm})$$

$$0.107(404 - 391) + \left(\frac{0.102}{724}\right)2172(270 - 185.4) = 5.25(391 - 319)$$

$$259 = 378 \quad \text{i.e. there is no balance.}$$

Try $T_c = 378$ K, $P_s = 118.5$ kN/m^2, $P_g = (308 - 118.5) = 189.5$ kN/m^2

and: $$P_{sF} = \frac{189.5 - 38}{\ln(189.5/38)} = 94.2 \text{ kN/m}^2$$

$\therefore$ $$0.107(404 - 378) + \left(\frac{0.102}{94.2}\right)2172(270 - 118.5) = 5.25(378 - 319)$$

$$310 = 308 \quad \text{which agrees well}$$

$\therefore$ $$U\Delta T = 309 \text{ kW/m}^2 \quad \text{and} \quad U = \frac{309}{(404 - 319)}$$

$$= 3.64 \text{ kW/m}^2 \text{ K.}$$

Repeating this procedure at the various temperature points selected, the heat-exchanger area may then be obtained as the area under a plot of Σq vs. $1/U\Delta T$, or as $A = \Sigma q/U\Delta T$ according to the following tabulation:

Point	T_s (K)	T_c (K)	$U\Delta T$ (kW/m^2)	$(U\Delta T)_{ow}$ (kW/m^2)	Q (kW)	$A = Q/(U\Delta T)_{ow}$ (m^2)	ΔT (K)	ΔT_{ow} (K)	$Q/\Delta T_{ow}$ (kW/K)
1	404	378	309	—	—	—	84.4	—	—
2	401	356	228	268.5	468.4	1.75	88.1	86.3	5.42
3	397	336	145	186.5	323.5	1.74	88.6	88.4	3.66
4	380	312	40.6	88.1*	343.5	3.89	76.7	82.7	4.15
5	339	302	5.4	17.5*	220.1	12.58	38.1	55.2*	4.00
6	322	300	2.1	3.5*	51.9	14.83	22.2	29.6*	1.75
				Total:	1407.3	34.8			18.98

*based on LMTD.

If no condensation takes place, the logarithmic mean temperature difference is 46.6 K. In practice the value is $(1407.3/18.98) = 74.2$ K.

Assuming no scale resistance, the overall coefficient is $\dfrac{1407.3}{34.8 \times 74.2} = 0.545$ kW/m^2 K.

The available surface area on the outside of the tubes $= 0.060$ m^2/m

$$\text{or} \quad (246 \times 3.65 \times 0.060) = 53.9 \text{ m}^2$$

The actual coefficient is therefore $\dfrac{1407.3}{53.9 \times 74.2} = 0.352$ kW/m^2 K

and the dirt factor is $\dfrac{(0.545 - 0.352)}{(0.545 \times 0.352)} = \underline{\underline{1.01 \text{ m}^2\text{K/kW}}}$.

As shown in Figure 9.50, the clean coefficient varies from $\underline{\underline{3.64 \text{ kW/m}^2 \text{ K}}}$ at the inlet to $\underline{\underline{0.092 \text{ kW/m}^2 \text{ K}}}$ at the outlet.

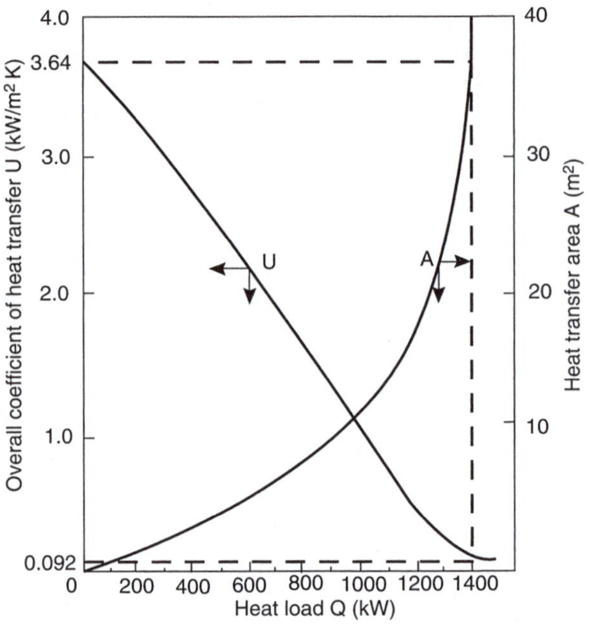

Figure 9.50. Results for Example 9.23

Condensation of mixed vapours is considered further in Volume 6, Chapter 12, where it is suggested that the local heat transfer coefficient may be expressed in terms of the local gas-film and condensate-film coefficients. For partial condensation where:

(i) *non-condensables < 0.5 per cent*; their effect can be ignored,
(ii) *non-condensables > 70 per cent*; the heat transfer can be taken as being by forced convection alone, and
(iii) *non-condensables 0.5–70 per cent*; both mechanisms are effective.

9.7. BOILING LIQUIDS

9.7.1 Conditions for boiling

In processing units, liquids are boiled either on submerged surfaces or on the inside of vertical tubes. Mechanical agitation may be applied in the first case and, in the second, the liquid may be driven through the tubes by means of an external pump. The boiling of liquids under either of these conditions normally leads to the formation of vapour first in the form of bubbles and later as a distinct vapour phase above a liquid interface. The conditions for boiling on the submerged surface are discussed here and the problems arising with boiling inside tubes are considered in Volume 2. Much of the fundamental work on the ideas of boiling has been presented by WESTWATER[(70)] and JAKOB[(71)], and subsequently by ROHSENOW and CLARK[(72)] and ROHSENOW[(73)] and by

FORSTER[74]. The boiling of solutions in which a solid phase is separated after evaporation has proceeded to a sufficient extent is considered in Volume 2.

For a bubble to be formed in a liquid, such as steam in water, for example, it is necessary for a surface of separation to be produced. Kelvin has shown that, as a result of the surface tension between the liquid and vapour, the vapour pressure on the inside of a concave surface will be less than that at a plane surface. As a result, the vapour pressure P_r inside the bubble is less than the saturation vapour pressure P_s at a plane surface. The relation between P_r and P_s is:

$$P_r = P_s - \left(\frac{2\sigma}{r}\right) \tag{9.184}$$

where r is the radius of curvature of the bubble, and σ is the surface tension.

Hence the liquid must be superheated near the surface of the bubble, the extent of the superheat increasing with decrease in the radius of the bubble. On this basis it follows that very small bubbles are difficult to form without excessive superheat. The formation of bubbles is made much easier by the fact that they will form on curved surfaces or on irregularities on the heating surface, so that only a small degree of superheat is normally required.

Nucleation at much lower values of superheat is believed to arise from the presence of existing nuclei such as non-condensing gas bubbles, or from the effect of the shape of the cavities in the surface. Of these, the current discussion on the influence of cavities is the most promising. In many cavities the angle θ will be greater than 90° and the effective contact angle, which includes the contact angle of the cavity β, will be considerably greater $[= \theta + (180 - \beta)/2]$, so that a much-reduced superheat is required to give nucleation. Thus the size of the mouth of the cavity and the shape of the cavity plays a significant part in nucleation[75].

It follows that for boiling to occur a small difference in temperature must exist between the liquid and the vapour. JAKOB and FRITZ[76] have measured the temperature distribution for water boiling above an electrically heated hot plate. The temperature dropped very steeply from about 383 K on the actual surface of the plate to 374 K about 0.1 mm from it. Beyond this point the temperature was reasonably constant until the water surface was reached. The mean superheat of the water above the temperature in the vapour space was about 0.5 deg K and this changed very little with the rate of evaporation. At higher pressures this superheating became smaller becoming 0.2 deg K at 5 MN/m^2 and 0.05 deg K at 101 MN/m^2. The temperature drop from the heating surface depends, however, very much on the rate of heat transfer and on the nature of the surface. Thus in order to maintain a heat flux of about 25.2 kW/m^2, a temperature difference of only 6 deg K was required with a rough surface as against 10.6 deg K with a smooth surface. The heat transfer coefficient on the boiling side is therefore dependent on the nature of the surface and on the difference in temperature available. For water boiling on copper plates JAKOB and FRITZ[76] give the following coefficients for a constant temperature difference of 5.6 deg K, with different surfaces:

(1) Surface after 8 h (28.8 ks) use and 48 h (172.8 ks) immersion in water	$h = 8000$ W/m^2 K
(2) Freshly sandblasted	$h = 3900$ W/m^2 K
(3) Sandblasted surface after long use	$h = 2600$ W/m^2 K

(4) Chromium plated $h = 2000\ \mathrm{W/m^2\,K}$

The initial surface, with freshly cut grooves, gave much higher figures than case (1).

The nature of the surface will have a marked effect on the physical form of the bubble and the area actually in contact with the surface, as shown in Figure 9.51.

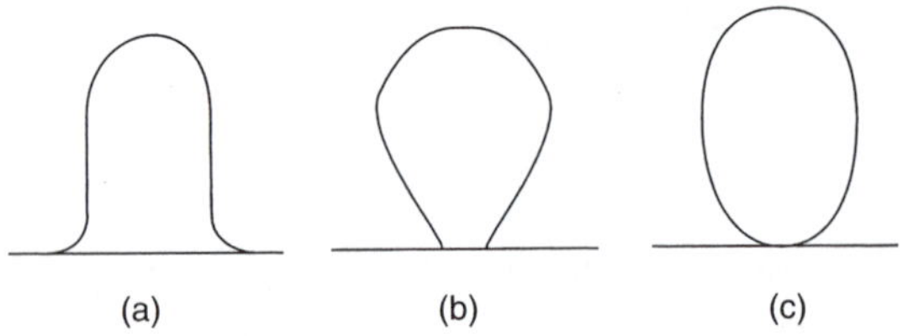

Figure 9.51. Shapes of bubbles (*a*) screen surface—thin oil layer (*b*) chromium plated and polished surface (*c*) screen surface—clean

The three cases are:

(a) *Non-wettable surface*, where the vapour bubbles spread out thus reducing the area available for heat transfer from the hot surface to the liquid.

(b) *Partially wettable surface*, which is the commonest form, where the bubbles rise from a larger number of sites and the rate of transfer is increased.

(c) *Entirely wetted surface*, such as that formed by a screen. This gives the minimum area of contact between vapour and surface and the bubbles leave the surface when still very small. It therefore follows that if the liquid has detergent properties this may give rise to much higher rates of heat transfer.

9.7.2. Types of boiling

Interface evaporation

In boiling liquids on a submerged surface it is found that the heat transfer coefficient depends very much on the temperature difference between the hot surface and the boiling liquid. The general relation between the temperature difference and heat transfer coefficient was first presented by NUKIYAMA(77) who boiled water on an electrically heated wire. The results obtained have been confirmed and extended by others, and Figure 9.52 shows the data of FARBER and SCORAH(78). The relationship here is complex and is best considered in stages.

In interface evaporation, the bubbles of vapour formed on the heated surface move to the vapour-liquid interface by natural convection and exert very little agitation on the liquid. The results are given by:

$$Nu = 0.61(Gr\ Pr)^{1/4} \tag{9.185}$$

which may be compared with the expression for natural convection:

$$Nu = C'(Gr\ Pr)^n \qquad \text{(equation 9.101)}$$

where $n = 0.25$ for streamline conditions and $n = 0.33$ for turbulent conditions.

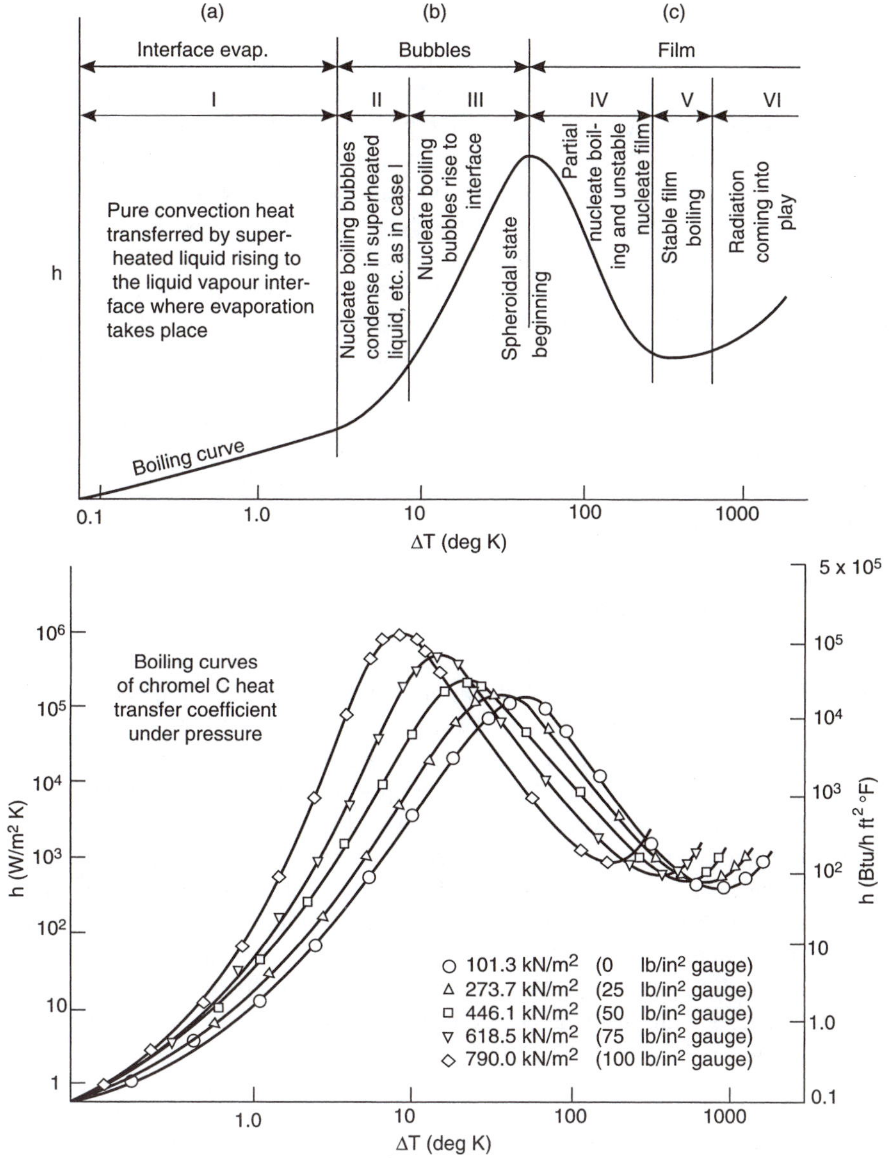

Figure 9.52. Heat transfer results of FARBER and SCORAH[78]

Nucleate boiling

At higher values of ΔT the bubbles form more rapidly and form more centres of nucleation. Under these conditions the bubbles exert an appreciable agitation on the liquid and the heat transfer coefficient rises rapidly. This is the most important region for boiling in industrial equipment.

Film boiling

With a sufficiently high value of ΔT, the bubbles are formed so rapidly that they cannot get away from the hot surface, and they therefore form a blanket over the surface. This means that the liquid is prevented from flowing on to the surface by the bubbles of vapour and the coefficient falls. The maximum coefficient occurs during nucleate boiling although this is an unstable region for operation. In passing from the *nucleate boiling* region to the *film boiling* region, two critical changes occur in the process. The first manifests itself in a decrease in the heat flux, the second is the prelude to stable film boiling. The intermediate region is generally known as the *transition* region. It may be noted that the first change in the process is an important hydrodynamic phenomenon which is common to other two-phase systems, such as flooding in countercurrent gas-liquid or vapour-liquid systems, for example.

With very high values of ΔT, the heat transfer coefficient rises again because of heat transfer by radiation. These very high values are rarely achieved in practice and usually the aim is to operate the plant at a temperature difference a little below the value giving the maximum heat transfer coefficient.

9.7.3. Heat transfer coefficients and heat flux

The values of the heat transfer coefficients for low values of temperature difference are given by equation 9.185. Figure 9.53 shows the values of h and for q for water boiling on a submerged surface. Whilst the actual values vary somewhat between investigations, they all give a maximum for a temperature difference of about 22 deg K. The maximum value of h is about 50 kW/m^2 K and the maximum flux is about 1100 kW/m^2.

Similar results have been obtained by BONILLA and PERRY(79), INSINGER and BLISS(80), and others for a number of organic liquids such as benzene, alcohols, acetone, and carbon tetrachloride. The data in Table 9.9 for liquids boiling at atmospheric pressure show that the maximum heat flux is much smaller with organic liquids than with water and the temperature difference at this condition is rather higher. In practice the critical value of ΔT may be exceeded. SAUER *et al.*(81) found that the overall transfer coefficient U for boiling ethyl acetate with steam at 377 kN/m^2 was only 14 per cent of that when the steam pressure was reduced to 115 kN/m^2.

Table 9.9. Maximum heat flux for various liquids boiling at atmospheric pressure

Liquid	Surface	Critical ΔT (deg K)	Maximum flux (kW/m^2)
Water	Chromium	25	910
50 mol% ethanol-water	Chromium	29	595
Ethanol	Chromium	33	455
n-Butanol	Chromium	44	455
iso-Butanol	Nickel	44	370
Acetone	Chromium	25	455
iso-Propanol	Chromium	33	340
Carbon tetrachloride	Copper	—	180
Benzene	Copper	—	170–230

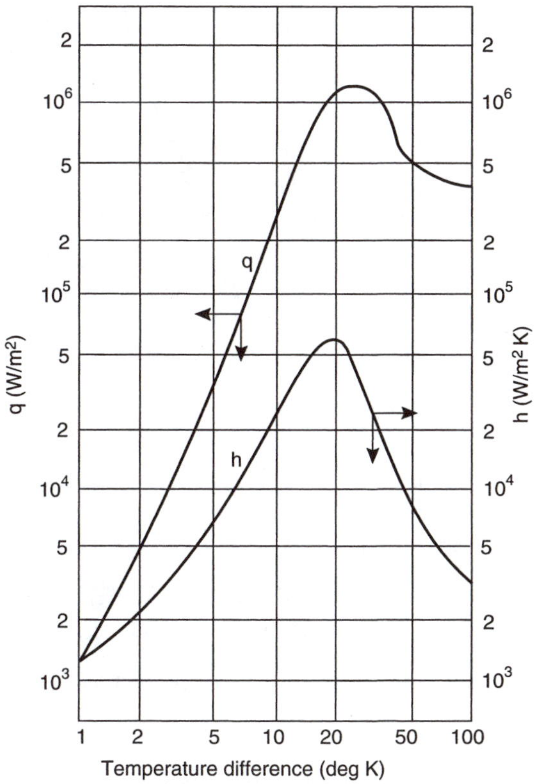

Figure 9.53. Effect of temperature difference on heat flux and heat transfer coefficient to water boiling at 373 K on a submerged surface

In considering the problem of nucleate boiling, the nature of the surface, the pressure, and the temperature difference must be taken into account as well as the actual physical properties of the liquid.

Apart from the question of scale, the nature of the clean surface has a pronounced influence on the rate of boiling. Thus BONILLA and PERRY[79] boiled ethanol at atmospheric pressure and a temperature difference of 23 deg K, and found that the heat flux at atmospheric pressure was 850 kW/m^2 for polished copper, 450 for gold plate, and 370 for fresh chromium plate, and only 140 for old chromium plate. This wide fluctuation means that care must be taken in anticipating the heat flux, since the high values that may be obtained initially may not persist in practice because of tarnishing of the surface.

Effect of temperature difference

CRYDER and FINALBORGO[82] boiled a number of liquids on a horizontal brass surface, both at atmospheric and at reduced pressure. Some of their results are shown in Figure 9.54, where the coefficient for the boiling liquid h is plotted against the temperature difference between the hot surface and the liquid. The points for the various liquids in Figure 9.54

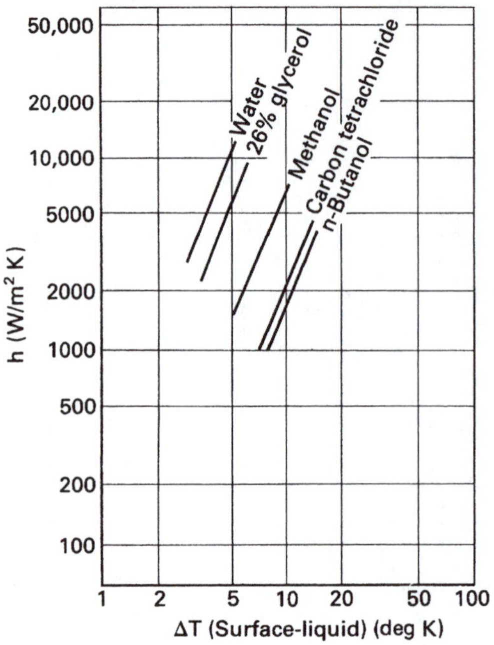

Figure 9.54. Effect of temperature difference on the heat transfer coefficient for boiling liquids (CRYDER and FINALBORGO[82])

lie on nearly parallel straight lines, which may be represented by:

$$h = \text{constant} \times \Delta T^{2.5} \tag{9.186}$$

This value for the index of ΔT has been found by other workers, although JAKOB and LINKE[83] found values as high as 4 for some of their work. It is important to note that this value of 2.5 is true only for temperature differences up to 19 deg K.

In some ways it is more convenient to show the results in the form of heat flux versus temperature difference, as shown in Figure 9.55, where some results from a number of workers are given.

Effect of pressure

CRYDER and FINALBORGO[82] found that h decreased uniformly as the pressure and hence the boiling point was reduced, according to the relation $h = \text{constant} \times B^{T''}$, where T'' is numerically equal to the temperature in K and B is a constant. Combining this with equation 9.186, their results for h were expressed in the empirical form:

$$h = \text{constant} \times \Delta T^{2.5} B^{T''}$$

or, using SI units:
$$\log\left(\frac{h}{5.67}\right) = a' + 2.5\log\Delta T + b'(T'' - 273) \tag{9.187}$$

where $(T'' - 273)$ is in °C.

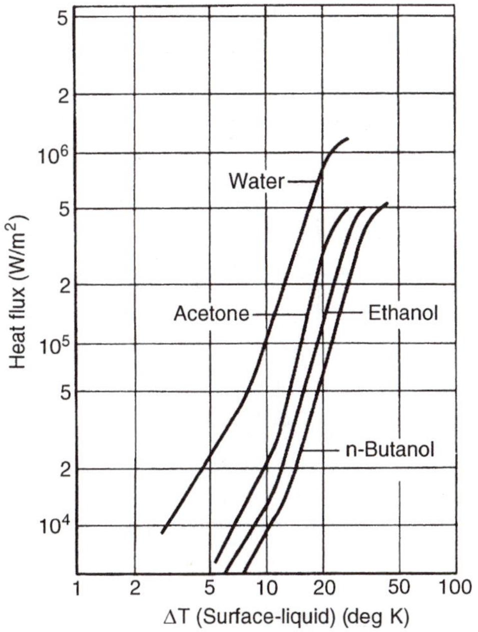

Figure 9.55. Effect of temperature difference on heat flux to boiling liquids (BONILLA and PERRY[79])

If a' and b' are given the following values, h is expressed in W/m^2 K:

	a'	b'		a'	b'
Water	−0.96	0.025	Kerosene	−4.13	0.022
Methanol	−1.11	0.027	10% Na_2SO_4	−1.47	0.029
CCl_4	−1.55	0.022	24% NaCl	−2.43	0.031

The values of a' will apply only to a particular apparatus although a value of b' of 0.025 is of more general application. *If* h_n is the coefficient at some standard boiling point T_n, and h at some other temperature T, equation 9.187 may be rearranged to give:

$$\log \frac{h}{h_n} = 0.025(T'' - T''_n) \tag{9.188}$$

for a given material and temperature difference.

As the pressure is raised above atmospheric pressure, the film coefficient increases for a constant temperature difference. CICHELLI and BONILLA[84] have examined this problem for pressures up to the critical value for the vapour, and have shown that ΔT for maximum rate of boiling decreases with the pressure. They obtained a single curve, shown in Figure 9.56, by plotting $q_{\max}/P_c$ against P_R, where P_c is the critical pressure and P_R the reduced pressure $= P/P_c$. This curve represents the data for water, ethanol, benzene, propane, n-heptane, and several mixtures with water. For water the results cover only a small range of P/P_c because of the high value of P_c. For the organic liquids

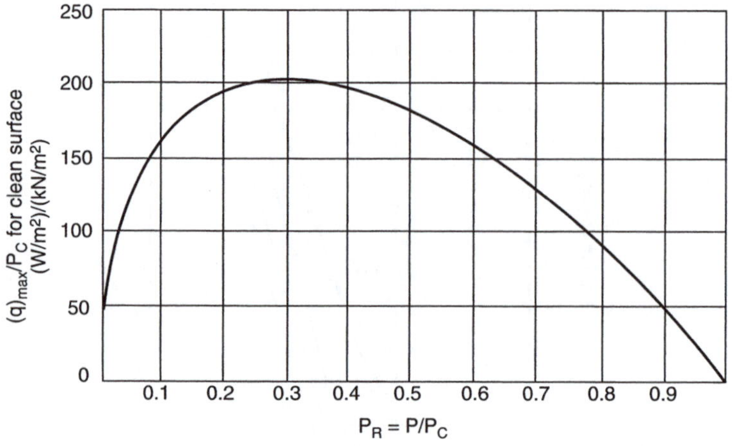

Figure 9.56. Effect of pressure on the maximum heat flux in nucleate boiling

investigated, it was shown that the maximum value of heat flux q occurs at a pressure P of about one-third of the critical pressure P_c. As shown in Table 9.10, the range of physical properties of the organic liquids is not wide and further data are required to substantiate the previous relation.

Table 9.10. Heat transfer coefficients for boiling liquids

Liquid	Boiling point (deg K)	ΔT (deg K)	h (W/m^2 K)	h (Btu/h ft^2 °F)
Water	372	4.7	9000	1600
	372	2.9	2700	500
	326	8.8	4700	850
	326	6.1	1300	250
Methanol	337	8.9	4800	850
	337	5.6	1500	250
	306	14.4	3000	500
	306	9.3	900	150
Carbon tetrachloride	349	12.6	3500	600
	349	7.2	1100	200
	315	20.1	2000	400
	315	11.8	700	100

9.7.4. Analysis based on bubble characteristics

It is a matter speculation as to why such high values of heat flux are obtained with the boiling process. It was once thought that the bubbles themselves were carriers of latent heat which was added to the liquid by their movement. It has now been shown, by determining the numbers of bubbles, that this mechanism would result in the transfer of only a moderate part of the heat that is actually transferred. The current views are that the high flux arises from the agitation produced by the bubbles, and two rather different explanations have been put forward. ROHSENOW and CLARK[72] and ROHSENOW[73] base their argument on the condition of the bubble on leaving the hot surface. By calculating the velocity and size of the bubble an expression may be derived for the heat transfer coefficient in the form of a Nusselt type equation, relating the Nusselt group to the Reynolds and Prandtl groups. FORSTER and ZUBER[85,86], however, argue that the important velocity is that of the growing bubble, and this is the term used to express the velocity. In either case the bubble movement is vital in obtaining a high flux. The liquid adjacent to the surface is agitated and exerts a mixing action by pushing hot liquid from the surface to the bulk of the stream.

Considering in more detail the argument proposed by ROHSENHOW and CLARK[72] and ROHSENHOW[73], the size of a bubble at the instant of breakaway from the surface has been determined by FRITZ[87] who has shown that d_b is given by:

$$d_b = C_1\phi\left(\frac{2\sigma}{g(\rho_l - \rho_v)}\right)^{1/2} \tag{9.189}$$

where σ is the surface tension, ρ_l and ρ_v the density of the liquid and vapour, ϕ is the contact angle, and C_1 is a constant depending on conditions.

The flowrate of vapour per unit area as bubbles u_b is given by:

$$u_b = \frac{fn\pi d_b^3}{6} \tag{9.190}$$

where f is the frequency of bubble formation at each bubble site and n is the number of sites of nucleation per unit area.

The heat transferred by the bubbles q_b is to a good approximation given by:

$$q_b = \tfrac{1}{6}\pi d_b^3 f n \rho_v \lambda \tag{9.191}$$

where λ is the latent heat of vaporisation.

It has been shown that for heat flux rates up to 3.2 kW/m^2 the product fd_b is constant and that the total heat flow per unit area q is proportional to n. From equation 9.191 it is seen that q_b is proportional to n at a given pressure, so that $q \propto q_b$.

Hence:

$$q = C_2\frac{\pi}{6}d_b^3 f n \rho_v \lambda \tag{9.192}$$

Substituting from equations 9.190 and 9.192, the mass flow per unit area:

$$\rho_v u_b = fn\frac{\pi}{6}d_b^3\rho_v = \frac{q}{C_2\lambda} \tag{9.193}$$

A Reynolds number for the bubble flow which represents the term for agitation may be defined as:

$$\begin{aligned} Re_b &= \frac{d_b\rho_v u_b}{\mu_l} \\ &= C_1\phi\left(\frac{2\sigma}{g(\rho_l - \rho_v)}\right)^{1/2}\left(\frac{q}{C_2\lambda}\right)\frac{1}{\mu_l} \end{aligned}$$

$$= C_3\phi\frac{q}{\lambda\mu_l}\left(\frac{\sigma}{g(\rho_l - \rho_v)}\right)^{1/2} \tag{9.194}$$

The Nusselt group for bubble flow, $Nu_b = h_b C_1 \dfrac{\phi}{k_l}\left(\dfrac{2\sigma}{g(\rho_l - \rho_v)}\right)^{1/2}$

$$= C_4 h_b \frac{\phi}{k_l}\left(\frac{\sigma}{g(\rho_l - \rho_v)}\right)^{1/2} \tag{9.195}$$

and hence a final correlation is obtained of the form:

$$Nu_b = \text{constant } Re_b^n Pr^m \tag{9.196}$$

or:

$$Nu_b = \text{constant}\left[\frac{C_3\phi q}{\mu_l\lambda}\left(\frac{\sigma}{g(\rho_l - \rho_v)}\right)^{1/2}\right]^n \left(\frac{C_l\mu_l}{k_l}\right)^m \tag{9.197}$$

where n and m have been found experimentally to be 0.67 and −0.7 respectively and the constant, which depends on the metal surface, ranges from 67–100 for polished chromium, 77 for platinum wire and 166 for brass.[73]

A comprehensive study of nucleate boiling of a wide range of liquids on thick plates of copper, aluminium, brass and stainless steel has been carried out by PIORO[88] who has evaluated the constants in equation 9.197 for different combinations of liquid and surface.

FORSTER and ZUBER[85,86] who employed a similar basic approach, although the radial rate of growth dr/dt was used for the bubble velocity in the Reynolds group, showed that:

$$\frac{dr}{dt} = \frac{\Delta T C_l \rho_l}{2\lambda\rho_v}\left(\frac{\pi D_{Hl}}{t}\right)^{1/2} \tag{9.198}$$

where D_{Hl} is the thermal diffusivity $(k_l/C_l\rho_l)$ of the liquid. Using this method, a final correlation in the form of equation 9.196 has been presented.

Although these two forms of analysis give rise to somewhat similar expressions, the basic terms are evaluated in quite different ways and the final expressions show many differences. Some data fit the Rohsenow equation reasonably well[88], and other data fit Forster's equation.

These expressions all indicate the importance of the bubbles on the rate of transfer, although as yet they have not been used for design purposes. INSINGER and BLISS[80] made the first approach by dimensional analysis and MCNELLY[89] has subsequently obtained a more satisfactory result. The influence of ΔT is taken into account by using the flux q, and the last term allows for the change in volume when the liquid vaporises. The following expression was obtained in which the numerical values of the indices were deduced from existing data:

$$\frac{hd}{k_l} = 0.225\left(\frac{C_l\mu_l}{k_l}\right)^{0.69}\left(\frac{qd}{\lambda\mu}\right)^{0.69}\left(\frac{Pd}{\sigma}\right)^{0.31}\left(\frac{\rho_l}{\rho_v} - 1\right)^{0.33} \tag{9.199}$$

9.7.5. Sub-cooled boiling

If bubbles are formed in a liquid which is much below its boiling point, then the bubbles will collapse in the bulk of the liquid. Thus if a liquid flows over a very hot surface then the bubbles formed are carried away from the surface by the liquid and sub-cooled boiling occurs. Under these conditions a very large number of small bubbles are formed and a very high heat flux is obtained. Some results for these conditions are given in Figure 9.57.

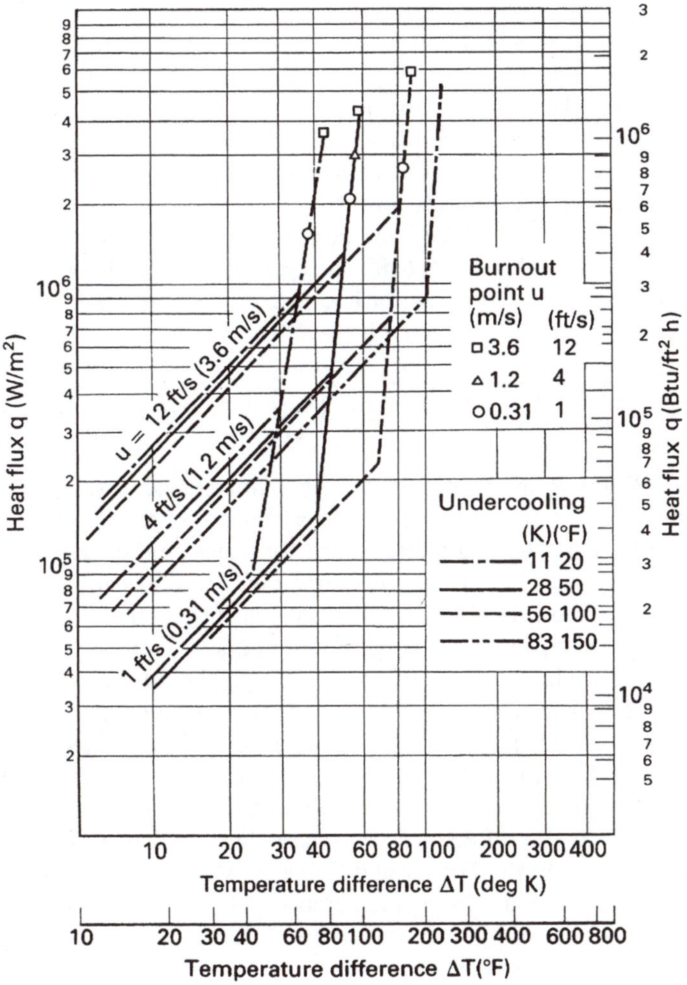

Figure 9.57. Heat flux in sub-cooled boiling

If a liquid flows through a tube heated on the outside then the heat flux q will increase with ΔT as shown in Figure 9.57. Beyond a certain value of ΔT the increase in q is very rapid. If the velocity through the tube is increased, then a similar plot is obtained with a higher value of q at low values of ΔT and then the points follow the first line. Over the first section, forced convection boiling exists where an increase in Reynolds number does not bring about a very great increase in q because the bubbles are themselves producing agitation in the boundary layer near the wall. Over the steep section, sub-cooled boiling exists where the velocity is not important provided it is sufficient to remove the bubbles rapidly from the surface. In the same way, mechanical agitation of a liquid boiling on a submerged surface will not markedly increase the heat flux.

9.7.6. Design considerations

In the design of vaporisers and reboilers, two types of boiling are important — nucleate boiling in a pool of liquid as in a kettle-type reboiler or a jacketed vessel, and convective boiling which occurs where the vaporising liquid flows over a heated surface and heat transfer is by both forced convection and nucleate boiling as, for example, in forced circulation or thermosyphon reboilers. The discussion here is a summary of that given in Volume 6 where a worked example is given.

In the absence of experimental data, the correlation given by FORSTER and ZUBER[86] may be used to estimate *pool boiling* coefficients, although the following reduced pressure correlation given by MOSTINSKI[90] is much simpler to use and gives reliable results for h (in W/m^2 K):

$$h = 0.104P_c^{0.69}q^{0.7}\left[1.8\left(\frac{P}{P_c}\right)^{0.17} + 4\left(\frac{P}{P_c}\right)^{1.2} + 10\left(\frac{P}{P_c}\right)^{10}\right] \tag{9.200}$$

In this equation, P_c and P are the critical and operating pressures (bar), respectively, and q is the heat flux (W/m^2). Both equations are for single component fluids, although they may also be used for close-boiling mixtures and for wider boiling ranges with a factor of safety. In reboiler and vaporiser design, it is important that the heat flux is well below the critical value. A correlation is given for the heat transfer coefficient for the case where film-boiling takes place on tubes submerged in the liquid.

Convective boiling, which occurs when the boiling liquid flows through a tube or over a tube bundle, depends on the state of the fluid at any point. The effective heat transfer coefficient can be considered to be made up of the convective and nucleate boiling components. The convective boiling coefficient is estimated using an equation for single-phase forced-convection heat transfer (equation 9.64, for example) modified by a factor to allow for the effects of two-phase flow. Similarly, the nucleate boiling coefficient is obtained from the Forster and Zuber or Mostinski correlation, modified by a factor dependent on the liquid Reynolds number and on the effects of two-phase flow. The estimation of convective boiling coefficients is illustrated by means of an example in Volume 6.

One of the most important areas of application of heat transfer to boiling liquids is in the use of evaporators to effect an increase in the concentration of a solution. This topic is considered in Volume 2.

For vaporising the liquid at the bottom of a distillation column a reboiler is used, as shown in Figure 9.58. The liquid from the still enters the boiler at the base, and, after flowing over the tubes, passes out over a weir. The vapour formed, together with any entrained liquid, passes from the top of the unit to the column. The liquid flow may be either by gravity or by forced circulation. In such equipment, provision is made for expansion of the tubes either by having a floating head as shown, or by arranging the tubes in the form of a hairpin bend (Figure 9.59). A vertical reboiler may also be used with steam condensing on the outside of the tube bundle. With all systems it is undesirable

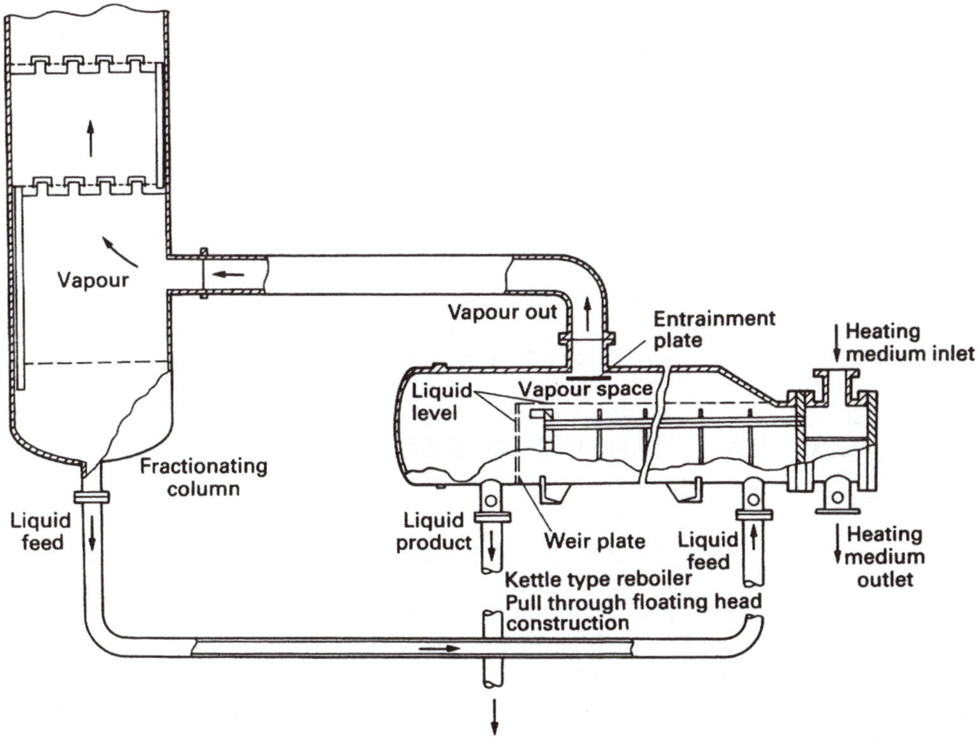

Figure 9.58. Reboiler installed on a distillation column

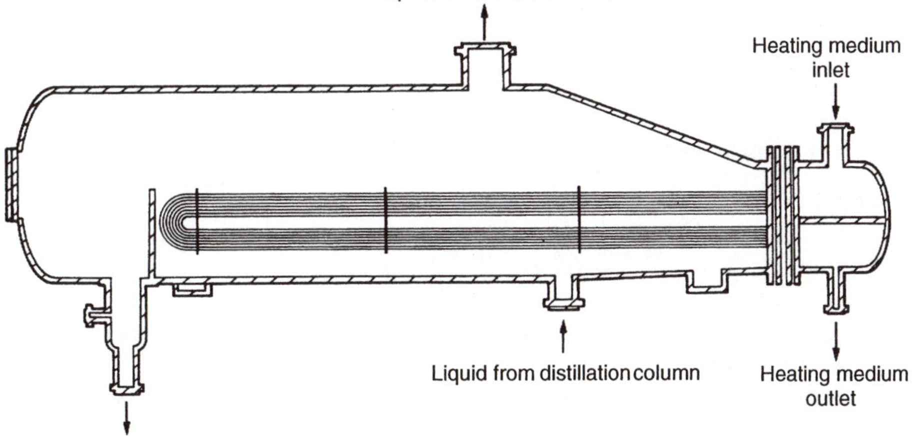

Figure 9.59. Kettle reboiler with hairpin tubes

to vaporise more than a small percentage of the feed since a good liquid flow over the tubes is necessary to avoid scale formation.

In the design of *forced convection* reboilers, the normal practice is to calculate the heat transfer coefficient on the assumption that heat is transferred by forced convection only, and this gives safe values. KERN(28) recommends that the heat flux should not exceed 60 kW/m^2 for organics and 90 kW/m^2 for dilute aqueous solutions. In *thermosyphon reboilers*, the fluid circulates at a rate at which the pressure losses in the system are just balanced by the hydrostatic head and the design involves an iterative procedure based on an assumed circulation rate through the exchanger. *Kettle reboilers*, such as that shown in Figure 9.59, are essentially pool boiling devices and their design, based on nucleate boiling data, uses the Zuber equation for single tubes, modified by a tube-density factor. This general approach is developed further in Volume 6.

9.8. HEAT TRANSFER IN REACTION VESSELS

9.8.1. Helical cooling coils

A simple jacketed pan or kettle is very commonly used in the processing industries as a reaction vessel. In many cases, such as in nitration or sulphonation reactions, heat has to be removed or added to the mixture in order either to control the rate of reaction or to bring it to completion. The addition or removal of heat is conveniently arranged by passing steam or water through a jacket fitted to the outside of the vessel or through a helical coil fitted inside the vessel. In either case some form of agitator is used to obtain even distribution in the vessel. This may be of the anchor type for very thick pastes or a propeller or turbine if the contents are not too viscous.

In such a vessel, the thermal resistances to heat transfer arise from the water film on the inside of the coil, the wall of the tube, the film on the outside of the coil, and any scale that may be present on either surface. The overall transfer coefficient may be expressed by:

$$\frac{1}{UA} = \frac{1}{h_i A_i} + \frac{x_w}{k_w A_w} + \frac{1}{h_o A_o} + \frac{R_o}{A_o} + \frac{R_i}{A_i} \tag{9.201}$$

where R_o and R_i are the scale resistances and the other terms have the usual definitions.

Inside film coefficient

The value of h_i may be obtained from a form of equation 9.64:

$$\frac{h_i d}{k} = 0.023 \left(\frac{du\rho}{\mu}\right)^{0.8} \left(\frac{C_p \mu}{k}\right)^{0.33} \tag{9.202}$$

if water is used in the coil, and the Sieder and Tate equation (equation 9.66) if a viscous brine is used for cooling.

These equations have been obtained for straight tubes; with a coil somewhat greater transfer is obtained for the same physical conditions. JESCHKE(91) cooled air in a 31 mm steel tube wound in the form of a helix and expressed his results in the form:

$$h_i(\text{coil}) = h_i(\text{straight pipe}) \left(1 + 3.5\frac{d}{d_c}\right) \tag{9.203}$$

where d is the inside diameter of the tube and d_c the diameter of the helix. PRATT[92] has examined this problem in greater detail for liquids and has given almost the same result. Combining equations 9.202 and 9.203, the inside film coefficient h_i for the coil may be calculated.

Outside film coefficient

The value of h_o is determined by the physical properties of the liquor and by the degree of agitation achieved. This latter quantity is difficult to express in a quantitative manner and the group $L^2N\rho/\mu$ has been used both for this problem and for the allied one of power used in agitation, as discussed in Chapter 7. In this group L is the length of the paddle and N the revolutions per unit time. CHILTON, DREW and JEBENS[93], working with a small tank only 0.3 m in diameter d_v, expressed their results by:

$$\frac{h_o d_v}{k}\left(\frac{\mu_s}{\mu}\right)^{0.14} = 0.87\left(\frac{C_p\mu}{k}\right)^{1/3}\left(\frac{L^2N\rho}{\mu}\right)^{0.62} \tag{9.204}$$

where the factor $(\mu_s/\mu)^{0.14}$ allows for the difference between the viscosity adjacent to the coil (μ_s) and that in the bulk of the liquor. A wide range of physical properties was achieved by using water, two oils, and glycerol.

PRATT[92] used both circular and square tanks up to 0.6 m in size and a series of different arrangements of a simple paddle as shown in Figure 9.60. The effect of altering the arrangement of the coil was investigated and the tube diameter d_o, the gap between the turns d_g, the diameter of the helix d_c, the height of the coil d_p, and the width of the stirrer W were all varied. The final equations tanks were:

For cylindrical tanks

$$\frac{h_o d_v}{k} = 34\left(\frac{L^2N\rho}{\mu}\right)^{0.5}\left(\frac{C_p\mu}{k}\right)^{0.3}\left(\frac{d_g}{d_p}\right)^{0.8}\left(\frac{W}{d_c}\right)^{0.25}\left(\frac{L^2 d_v}{d_o^3}\right)^{0.1} \tag{9.205}$$

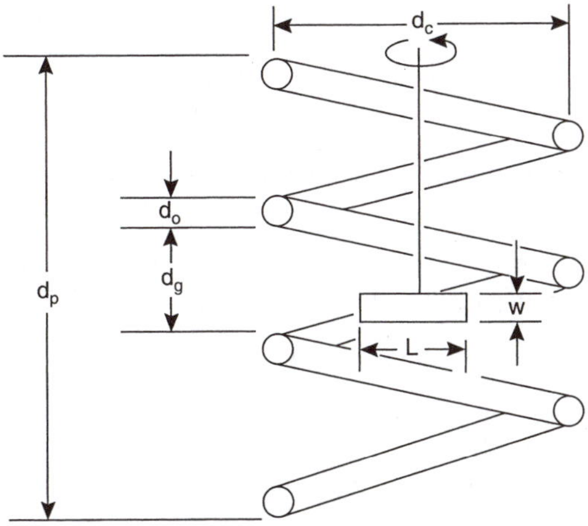

Figure 9.60. Arrangement of coil in Pratt's work[79]

For square tanks:

$$\frac{h_o l_v}{k} = 39\left(\frac{L^2 N\rho}{\mu}\right)^{0.5}\left(\frac{C_p\mu}{k}\right)^{0.3}\left(\frac{d_g}{d_p}\right)^{0.8}\left(\frac{W}{d_c}\right)^{0.25}\left(\frac{L^2 l_v}{d_o^3}\right)^{0.1} \tag{9.206}$$

where l_v is the length of the side of the vessel.

These give almost the same results as the earlier equations over a wide range of conditions. CUMMINGS and WEST(94) have tested these results with a much larger tank of 0.45 m^3 capacity and have given an expression similar to equation 9.204 but with a constant of 1.01 instead of 0.87. A retreating blade turbine impeller was used, and in many cases a second impeller was mounted above the first, giving an agitation which is probably more intense than that attained by the other workers. A constant of 0.9 seems a reasonable average from existing work.

Example 9.24

Toluene is continuously nitrated to mononitrotoluene in a cast-iron vessel, 1 m diameter, fitted with a propeller agitator 0.3 m diameter rotating at 2.5 Hz. The temperature is maintained at 310 K by circulating 0.5 kg/s cooling water through a stainless steel coil 25 mm o.d. and 22 mm i.d. wound in the form of a helix, 0.80 m in diameter. The conditions are such that the reacting material may be considered to have the same physical properties as 75 per cent sulphuric acid. If the mean water temperature is 290 K, what is the overall coefficient of heat transfer?

Solution

The overall coefficient U_o based on the outside area of the coil is given by equation 9.201:

$$\frac{1}{U_o} = \frac{1}{h_o} + \frac{x_w d_o}{k_w d_w} + \frac{d_o}{h_i d} + R_o + \frac{R_i d_o}{d}$$

where d_w is the mean diameter of the pipe.

From equations 9.202 and 9.203, the inside film coefficient for the water is given by:

$$h_i = \frac{k}{d}\left(1 + 3.5\frac{d}{d_c}\right)0.023\left(\frac{du\rho}{\mu}\right)^{0.8}\left(\frac{C_p\mu}{k}\right)^{0.4}$$

In this equation:

$$\rho u = \frac{0.5}{(\pi/4)\times 0.022^2} = 1315\ \text{kg/m}^2\text{s}$$

$d = 0.022$ m, $d_c = 0.80$ m, $k = 0.59$ W/m K, $\mu = 1.08$ mN s/m^2 or 1.08×10^{-3} N s/m^2, and $C_p = 4.18 \times 10^3$ J/kg K

Thus: $$h_i = \frac{0.59}{0.022}\left(1 + 3.5\times\frac{0.022}{0.80}\right)0.023\left(\frac{0.022\times 1315}{1.08\times 10^{-3}}\right)^{0.8}\left(\frac{4.18\times 10^3\times 1.08\times 10^{-3}}{0.59}\right)^{0.4}$$

$$= 0.680(26{,}780)^{0.8}(7.65)^{0.4} = 5490\ \text{W/m}^2\ \text{K}$$

The external film coefficient is given by equation 9.204:

$$\frac{h_o d_v}{k}\left(\frac{\mu_s}{\mu}\right)^{0.14} = 0.87\left(\frac{C_p\mu}{k}\right)^{0.33}\left(\frac{L^2 N\rho}{\mu}\right)^{0.62}$$

For 75 per cent sulphuric acid:

$k = 0.40$ W/m K, $\mu_s = 8.6 \times 10^{-3}$ N s/m^2 at 300 K, $\mu = 6.5 \times 10^{-3}$ N s/m^2 at 310 K, $C_p = 1.88 \times 10^3$ J/kg K, and $\rho = 1666$ kg/m^3

Thus: $$\frac{h_o \times 1.0}{0.40}\left(\frac{8.6}{6.5}\right)^{0.14} = 0.87\left(\frac{1.88 \times 10^3 \times 6.5 \times 10^{-3}}{0.40}\right)^{0.33}\left(\frac{0.3^2 \times 2.5 \times 1666}{6.5 \times 10^{-3}}\right)^{0.62}$$

$$2.5h_o \times 1.04 = 0.87 \times 3.09 \times 900$$

and: $$h_o = 930 \text{ W/m}^2\text{ K}$$

Taking $k_w = 15.9$ W/m K and R_o and R_i as 0.0004 and 0.0002 m^2 K/W, respectively:

$$\frac{1}{U_o} = \frac{1}{930} + \frac{0.0015 \times 0.025}{15.9 \times 0.0235} + \frac{0.025}{5490 \times 0.022} + 0.0004 + \frac{0.0002 \times 0.025}{0.022}$$

$$= 0.00107 + 0.00010 + 0.00021 + 0.00040 + 0.00023 = 0.00201$$

and: $$\underline{\underline{U_o = 498 \text{ W/m}^2\text{ K}}}$$

In this calculation a mean area of surface might have been used with sufficient accuracy. It is important to note the great importance of the scale terms which together form a major part of the thermal resistance.

9.8.2. Jacketed vessels

In many cases, heating or cooling of a reaction mixture is most satisfactorily achieved by condensing steam in a jacket or passing water through it — an arrangement which is often used for organic reactions where the mixture is too viscous for the use of coils and a high-speed agitator. CHILTON *et al.*[93] and CUMMINGS and WEST[94] have measured the transfer coefficients for this case by using an arrangement as shown in Figure 9.61, where heat is supplied to the jacket and simultaneously removed by passing water through the coil. Chilton measured the temperatures of the inside of the vessel wall, the bulk liquid, and the surface of the coil by means of thermocouples and thus obtained the film heat transfer coefficients directly. Cummings and West used an indirect method to give the film coefficient from measurements of the overall coefficients.

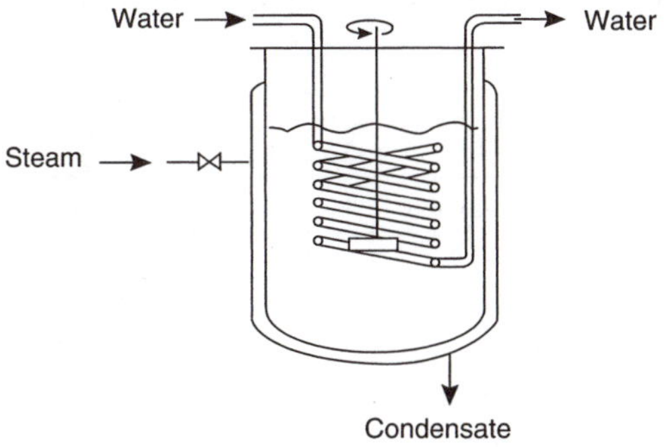

Figure 9.61. Reaction vessel with jacket and coil

CHILTON *et al.*[93] expressed their results by:

$$\frac{h_b d_v}{k}\left(\frac{\mu_s}{\mu}\right)^{0.14} = 0.36\left(\frac{L^2 N \rho}{\mu}\right)^{0.67}\left(\frac{C_p \mu}{k}\right)^{0.33} \tag{9.207}$$

where h_b is the film coefficient for the liquor adjacent to the wall of the vessel. CUMMINGS and WEST[94] used the same equation although the coefficient was 0.40. Considering that Chilton's vessel was only 0.3 m in diameter and fitted with a single paddle of 150 mm length, and that Cummings and West used a 0.45 m^3 vessel with two turbine impellers, agreement between their results is remarkably good. The group $(\mu_s/\mu)^{0.14}$ is again used to allow for the difference in the viscosities at the surface and in the bulk of the fluid.

BROWN *et al.*[95] have given data on the performance of 1.5 m diameter sulphonators and nitrators of 3.4 m^3 capacity as used in the dyestuffs industry. The sulphonators were of cast iron and had a wall thickness of 25.4 mm; the annular space in the jacket being also 25.4 mm. The agitator of the sulphonator was of the anchor type with a 127 mm clearance at the walls and was driven at 0.67 Hz. The nitrators were fitted with four-bladed propellers of 0.61 m diameter driven at 2 Hz. For cooling, the film coefficient h_b for the inside of the vessel was given by:

$$\frac{h_b d_v}{k}\left(\frac{\mu_s}{\mu}\right)^{0.14} = 0.55\left(\frac{L^2 N\rho}{\mu}\right)^{0.67}\left(\frac{C_p\mu}{k}\right)^{0.25} \tag{9.208}$$

which is very similar to that given by equation 9.207.

The film coefficients for the water jacket were in the range 635–1170 W/m^2 K for water rates of 1.44–9.23 l/s, respectively. It may be noted that 7.58 l/s corresponds to a vertical velocity of only 0.061 m/s and to a Reynolds number in the annulus of 5350. The thermal resistance of the wall of the pan was important, since with the sulphonator it accounted for 13 per cent of the total resistance at 323 K and 31 per cent at 403 K. The change in viscosity with temperature is important when considering these processes, since, for example, the viscosity of the sulphonation liquors ranged from 340 mN s/m^2 at 323 K to 22 mN s/m^2 at 403 K.

In discussing equations 9.207 and 9.208 FLETCHER[96] has summarised correlations obtained for a wide range of impeller and agitator designs in terms of the constant before the Reynolds number and the index on the Reynolds number as shown in Table 9.11.

Table 9.11. Data on common agitators for use in equations 9.207 and 9.208

Type of agitator	Constant	Index
Flat blade disc turbine		
unbaffled, or baffled vessel, $Re < 400$	0.54	0.67
baffled, $Re > 400$	0.74	0.67
Retreating-blade turbine with three blades, jacketed and baffled vessel, $Re = 2 \times 10^4$ to 2×10^6		
glassed steel impeller	0.33	0.67
alloy steel impeller	0.37	0.67
Propeller with three blades		
baffled vessel, $Re = 5500$ to 37,000	0.64	0.67
Flat blade paddle		
baffled or unbaffled vessel, $Re \geq 4000$	0.36	0.67
Anchor		
$Re = 30$ to 300	1.00	0.50
$Re = 300$ to 5000	0.38	0.67

9.8.3. Time required for heating or cooling

It is frequently necessary to heat or cool the contents of a large batch reactor or storage tank. In this case the physical constants of the liquor may alter and the overall transfer coefficient may change during the process. In practice, it is often possible to assume an average value of the transfer coefficient so as to simplify the calculation of the time required for heating or cooling. The heating of the contents of a storage tank is commonly effected by condensing steam, either in a coil or in some form of hairpin tube heater.

In the case of a storage tank with liquor of mass m and specific heat C_p, heated by steam condensing in a helical coil, it may be assumed that the overall transfer coefficient U is constant. If T_s is the temperature of the condensing steam, T_1 and T_2 the initial and final temperatures of the liquor, and A the area of heat transfer surface, and T is the temperature of the liquor at any time t, then the rate of transfer of heat is given by:

$$Q = mC_p \frac{dT}{dt} = UA(T_s - T)$$

$$\therefore \quad \frac{dT}{dt} = \frac{UA}{mC_p}(T_s - T)$$

$$\therefore \quad \int_{T_1}^{T_2} \frac{dT}{T_s - T} = \frac{UA}{mC_p} \int_0^t dt$$

$$\therefore \quad \ln \frac{T_s - T_1}{T_s - T_2} = \frac{UA}{mC_p} t \tag{9.209}$$

From this equation, the time t of heating from T_1 to T_2, may be calculated. The same analysis may be used if the steam condenses in a jacket of a reaction vessel.

This analysis does not allow for any heat losses during the heating, or, for that matter, cooling operation. Obviously the higher the temperature of the contents of the vessel, the greater are the heat losses and, in the limit, the heat supplied to the vessel is equal to the heat losses, at which stage no further rise in the temperature of the contents of the vessel is possible. This situation is illustrated in Example 9.25.

The heating-up time can be reduced by improving the rate of heat transfer to the fluid, by agitating of the fluid for example, and by reducing heat losses from the vessel by insulation. In the case of a large vessel there is a limit to the degree of agitation possible, and circulation of the fluid through an external heat exchanger is an attractive alternative.

Example 9.25

A vessel contains 1 tonne (1 Mg) of a liquid of specific heat capacity 4.0 kJ/kg K. The vessel is heated by steam at 393 K which is fed to a coil immersed in the agitated liquid and heat is lost to the surroundings at 293 K from the outside of the vessel. How long does it take to heat the liquid from 293 to 353 K and what is the maximum temperature to which the liquid can be heated? When the liquid temperature has reached 353 K, the steam supply is turned off for 2 hours (7.2 ks) and the vessel cools. How long will it take to reheat the material to 353 K? The surface area of the coil is 0.5 m^2 and the overall coefficient of heat transfer to the liquid may be taken as 600 W/m^2 K. The outside area of the vessel is 6 m^2 and the coefficient of heat transfer to the surroundings may be taken as 10 W/m^2 K.

Solution

If T K is the temperature of the liquid at time t s, then a heat balance on the vessel gives:

$$(1000 \times 4000)\frac{dT}{dt} = (600 \times 0.5)(393 - T) - (10 \times 6)(T - 293)$$

or:
$$4{,}000{,}000\frac{dT}{dt} = 135{,}480 - 360T$$

and:
$$11{,}111\frac{dT}{dt} = 376.3 - T.$$

The equilibrium temperature occurs when $dT/dt = 0$,

that is when: $\underline{\underline{T = 376.3 \text{ K.}}}$

In heating from 293 to 353 K, the time taken is:

$$t = 11{,}111 \int_{293}^{353} \frac{dT}{(376.3 - T)}$$

$$= 11{,}111 \ln\left(\frac{83.3}{23.3}\right)$$

$$= \underline{\underline{14{,}155 \text{ s}}} \text{ (or 3.93 h).}$$

The steam is turned off for 7200 s and during this time a heat balance gives:

$$(1000 \times 4000)\frac{dT}{dt} = -(10 \times 6)(T - 293)$$

$$66{,}700\frac{dT}{dt} = 293 - T$$

The change in temperature is then given by:

$$\int_{353}^{T} \frac{dT}{(293 - T)} = \frac{1}{66{,}700}\int_{0}^{7200} dt$$

$$\ln\frac{-60}{293 - T} = \frac{7200}{66{,}700} = 0.108$$

and:
$$T = 346.9 \text{ K.}$$

The time taken to reheat the liquid to 353 K is then given by:

$$t = 11{,}111 \int_{346.9}^{353} \frac{dT}{(376.3 - T)}$$

$$= 11{,}111 \ln\left(\frac{29.4}{23.3}\right)$$

$$= \underline{\underline{2584 \text{ s} \quad (0.72\text{h}).}}$$

9.9. SHELL AND TUBE HEAT EXCHANGERS

9.9.1. General description

Since shell and tube heat exchangers can be constructed with a very large heat transfer surface in a relatively small volume, fabricated from alloy steels to resist corrosion and be used for heating, cooling and for condensing a very wide range of fluids, they are the most widely used form of heat transfer equipment. Figures 9.62-9.64 show various forms of construction and a tube bundle is shown in Figure 9.65. The simplest type of unit, shown in Figure 9.62, has fixed tube plates at each end into which the tubes are expanded. The tubes are connected so that the internal fluid makes several passes up and

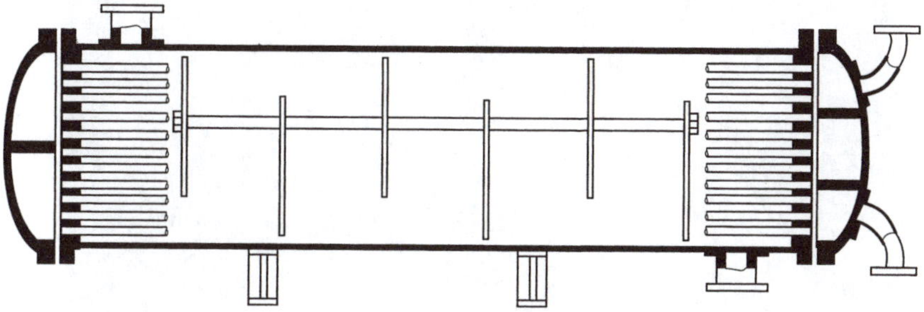

Figure 9.62. Heat exchanger with fixed tube plates (four tube, one shell-pass)

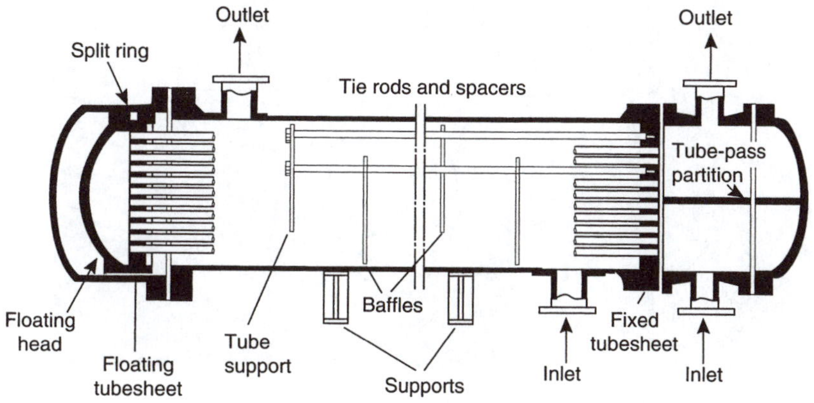

Figure 9.63. Heat exchanger with floating head (two tube-pass, one shell-pass)

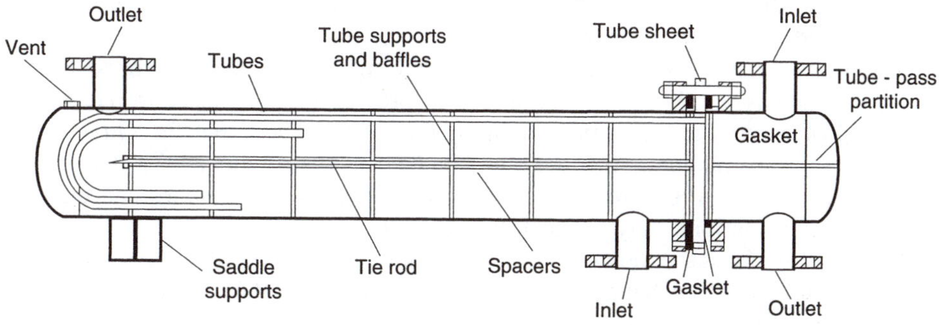

Figure 9.64. Heat exchanger with hairpin tubes

Figure 9.65. Expanding the ends of the tubes into the tube plate of a heat exchanger bundle

down the exchanger thus enabling a high velocity of flow to be obtained for a given heat transfer area and throughput of fluid. The fluid flowing in the shell is made to flow first in one sense and then in the opposite sense across the tube bundle by fitting a series of baffles along the length. These baffles are frequently of the segmental form with about 25 per cent cut away, as shown in Figure 9.29 to provide the free space to increase the velocity of flow across the tubes, thus giving higher rates of heat transfer. One problem with this type of construction is that the tube bundle cannot be removed for cleaning and no provision is made to allow for differential expansion between the tubes and the shell, although an expansion joint may be fitted to the shell.

In order to allow for the removal of the tube bundle and for considerable expansion of the tubes, a floating head exchanger is used, as shown in Figure 9.63. In this arrangement one tube plate is fixed as before, but the second is bolted to a floating head cover so that the tube bundle can move relative to the shell. This floating tube sheet is clamped between the floating head and a split backing flange in such a way that it is relatively easy to break the flanges at both ends and to draw out the tube bundle. It may be noted that the shell cover at the floating head end is larger than that at the other end. This enables the tubes to be placed as near as possible to the edge of the fixed tube plate, leaving very little unused space between the outer ring of tubes and the shell.

Another arrangement which provides for expansion involves the use of hairpin tubes, as shown in Figure 9.64. This design is very commonly used for the reboilers on large fractionating columns where steam is condensed inside the tubes.

In these designs there is one pass for the fluid on the shell-side and a number of passes on the tube-side. It is often an advantage to have two or more shell-side passes, although this considerably increases the difficulty of construction and, very often therefore, several smaller exchangers are connected together to obtain the same effect.

The essential requirements in the design of a heat exchanger are, firstly, the provision of a unit which is reliable and has the desired capacity, and secondly, the need to provide an exchanger at minimum overall cost. In general, this involves using standard components and fittings and making the design as simple as possible. In most cases, it is necessary to balance the capital cost in terms of the depreciation against the operating cost. Thus in a condenser, for example, a high heat transfer coefficient is obtained and hence a small exchanger is required if a higher water velocity is used in the tubes. Against this, the cost of pumping increases rapidly with increase in velocity and an economic balance must be struck. A typical graph showing the operating costs, depreciation and the total cost plotted as a function of the water velocity in the tubes is shown in Figure 9.66.

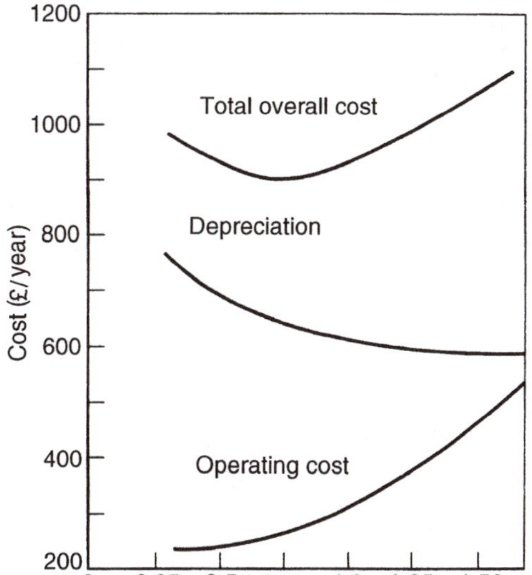

Figure 9.66. Effect of water velocity on annual operating cost of condenser

9.9.2. Basic components

The various components which make up a shell and tube heat exchanger are shown in Figures 9.63 and 9.64 and these are now considered. Many different mechanical arrangements are used and it is convenient to use a basis for classification. The standard published by the Tubular Exchanger Manufacturer's Association (TEMA[(97)]) is outlined here. It should be added that noting that SAUNDERS[(98)] has presented a detailed discussion of design codes and problems in fabrication.

Of the various *shell types* shown in Figure 9.67, the simplest, with entry and exit nozzles at opposite ends of a single pass exchanger, is the TEMA E-type on which most design methods are based, although these may be adapted for other shell types by allowing for the resulting velocity changes. The TEMA F-type has a longitudinal baffle giving two shell passes and this provides an alternative arrangement to the use of two shells required in order to cope with a close temperature approach or low shell-side flowrates. The pressure drop in two shells is some eight times greater than that encountered in the E-type design although any potential leakage between the longitudinal baffle and the shell in the F-type design may restrict the range of application. The so-called "split-flow" type of unit with a longitudinal baffle is classified as the TEMA G-type whose performance is superior although the pressure drop is similar to the E-type. This design is used mainly for reboilers and only occasionally for systems where there is no change of phase. The so-called "divided-flow" type, the TEMA J-type, has one inlet and two outlet nozzles and, with a pressure drop some one-eighth of the E-type, finds application in gas coolers and condensers operating at low pressures. The TEMA X-type shell has no cross baffles and hence the shell-side fluid is in pure counterflow giving extremely low pressure drops and again, this type of design is used for gas cooling and condensation at low pressures.

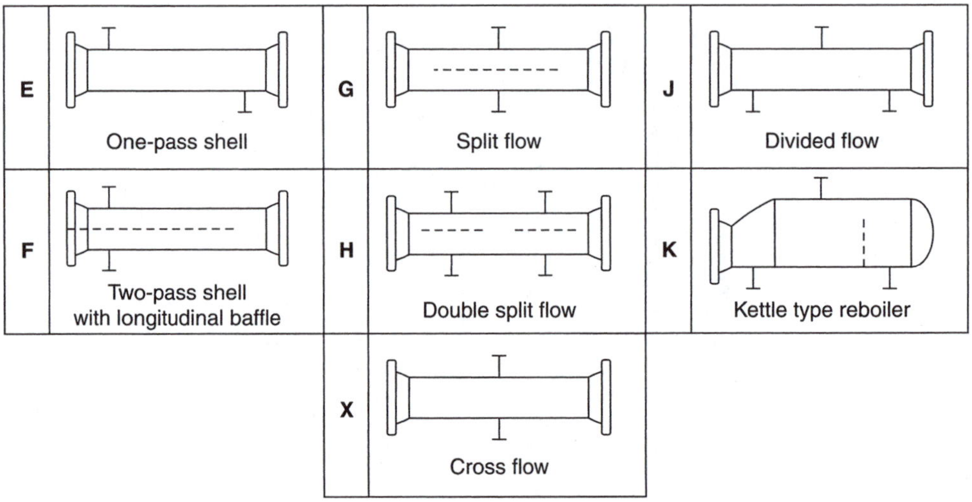

Figure 9.67. TEMA shell types

The *shell* of a heat exchanger is commonly made of carbon steel and standard pipes are used for the smaller sizes and rolled welded plate for the larger sizes (say 0.4–1.0 m).

The thickness of the shell may be calculated from the formula for thin-walled cylinders and a minimum thickness of 9.5 mm is used for shells over 0.33 m o.d. and 11.1 mm for shells over 0.9 m o.d. Unless the shell is designed to operate at very high pressures, the calculated wall thickness is usually less than these values although a corrosion allowance of 3.2 mm is commonly added to all carbon steel parts and thickness is determined more by rigidity requirements than simply internal pressure. The minimum shell thickness for various materials is given in BS3274[99]. A shell diameter should be such as to give as close a fit to the tube bundle as practical in order to reduce bypassing round the outside of the bundle. Typical values for the clearance between the outer tubes in the bundle and the inside diameter of the shell are given in Figure 9.68 for various types of exchanger.

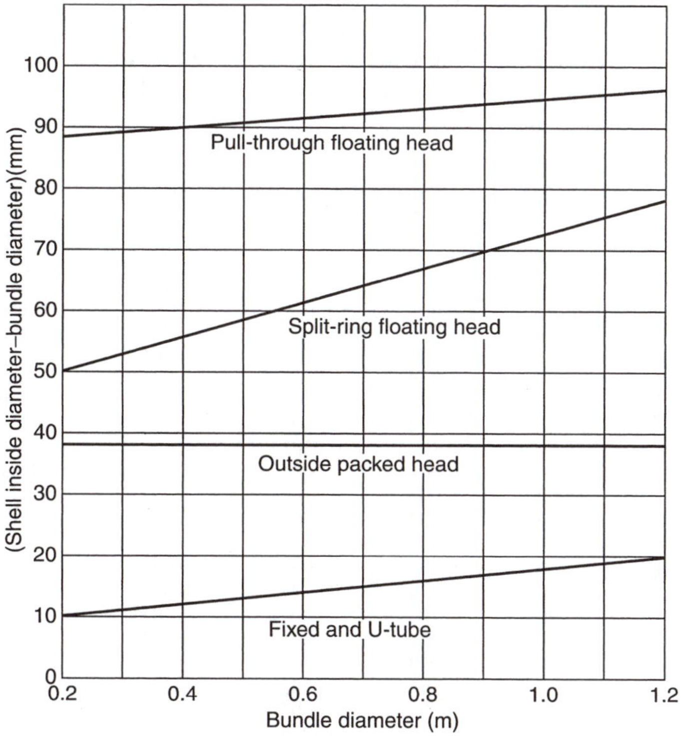

Figure 9.68. Shell-bundle clearance

The detailed design of the *tube bundle* must take into account both shell-side and tube-side pressures since these will both affect any potential leakage between the tube bundle and the shell which cannot be tolerated where high purity or uncontaminated materials are required. In general, tube bundles make use of a fixed tubesheet, a floating-head or U-tubes which are shown in Figures 9.62, 9.63 and 9.64 respectively. It may be noted here that the thickness of the fixed tubesheet may be obtained from a relationship of the form:

$$d_t = d_G\sqrt{0.25P/f)} \qquad (9.210)$$

where d_G is the diameter of the gasket (m), P the design pressure (MN/m^2), f the allowable working stress (MN/m^2) and d_t the thickness of the sheet measured at the bottom of the partition plate grooves. The thickness of the floating head tubesheet is very often calculated as $\sqrt{2d_t}$.

In selecting a *tube diameter*, it may be noted that smaller tubes give a larger heat transfer area for a given shell, although 19 mm o.d. tubes are normally the minimum size used in order to permit adequate cleaning. Although smaller diameters lead to shorter tubes, more holes have to be drilled in the tubesheet which adds to the cost of construction and increases the likelihood of tube vibration. Heat exchanger tubes are usually in the range 16 mm ($\frac{5}{8}$ in) to 50 mm (2 in) O.D.; the smaller diameter usually being preferred as these give more compact and therefore cheaper units. Against this, larger tubes are easier to clean especially by mechanical methods and are therefore widely used for heavily fouling fluids. The tube thickness or gauge must be such as to withstand the internal pressure and also to provide an adequate corrosion allowance. Details of steel tubes used in heat exchangers are given in BS3606[(100)] and summarised in Table 9.12, and standards for other materials are given in BS3274[(99)].

Table 9.12. Standard dimensions of steel tubes

Outside diameter d_o		Wall thickness		Cross sectional area for flow		Surface are per unit length	
(mm)	(in)	(mm)	(in)	(m^2)	(ft^2)	(m^2/m)	(ft^2/ft)
16	0.630	1.2	0.047	0.000145	0.00156	0.0503	0.165
		1.6	0.063	0.000129	0.00139		
		2.0	0.079	0.000113	0.00122		
20	0.787	1.6	0.063	0.000222	0.00239	0.0628	0.206
		2.0	0.079	0.000201	0.00216		
		2.6	0.102	0.000172	0.00185		
25	0.984	1.6	0.063	0.000373	0.00402	0.0785	0.258
		2.0	0.079	0.000346	0.00373		
		2.6	0.102	0.000308	0.00331		
		3.2	0.126	0.000272	0.00293		
30	1.181	1.6	0.063	0.000564	0.00607	0.0942	0.309
		2.0	0.079	0.000531	0.00572		
		2.6	0.102	0.000483	0.00512		
		3.2	0.126	0.000437	0.00470		
38	1.496	2.0	0.079	0.000908	0.00977	0.1194	0.392
		2.6	0.102	0.000845	0.00910		
		3.2	0.126	0.000784	0.00844		
50	1.969	2.0	0.079	0.001662	0.01789	0.1571	0.515
		2.6	0.102	0.001576	0.01697		
		3.2	0.126	0.001493	0.01607		

In general, the larger the *tube length*, the lower is the cost of an exchanger for a given surface area due to the smaller shell diameter, the thinner tube sheets and flanges and the smaller number of holes to be drilled, and the reduced complexity. Preferred tube lengths are 1.83 m (6 ft), 2.44 m (8 ft), 3.88 m (12 ft) and 4.88 m (16 ft); larger sizes are used where the total tube-side flow is low and fewer, longer tubes are required in order to obtain a required velocity. With the number of tubes per tube-side pass fixed in order to obtain a required velocity, the total length of tubes per tube-side pass is determined by the heat transfer surface required. It is then necessary to fit the tubes into a suitable shell to give the desired shell-side velocity. It may be noted that with long tube lengths and relatively few tubes in a shell, it may be difficult to arrange sufficient baffles for adequate support of the tubes. For good all-round performance, the ratio of tube length to shell diameter is usually in the range 5–10.

Tube layout and pitch, considered in Section 9.4.4 and shown in Figure 9.69, make use of equilateral triangular, square and staggered square arrays. The triangular layout gives a robust tube sheet although, because the vertical and horizontal distances between adjacent tubes is generally greater in a square layout compared with the equivalent triangular pitch design, the square array simplifies maintenance and particularly cleaning on the shell-side. Good practice requires a minimum pitch of 1.25 times the tube diameter and/or a minimum web thickness between tubes of about 3.2 mm to ensure adequate strength for tube rolling. In general, the smallest pitch in triangular 30° layout is used for clean fluids in both laminar and turbulent flow and a 90° or 45° layout with a 6.4 mm clearance where mechanical cleaning is required. The bundle diameter, d_b, may be estimated from the following empirical equation which is based on standard tube layouts:

$$\text{Number of tubes, } N_t = a(d_b/d_o)^b \tag{9.211}$$

where the values of the constants a and b are given in Table 9.13. Tables giving the number of tubes that can be accommodated in standard shells using various tube sizes, pitches and numbers of passes for different exchanger types are given, for example, in KERN[28] and LUDWIG[101].

Table 9.13. Constants for use with equation 9.211.

Number of passes		1	2	4	6	8
Triangular pitch*	a	0.319	0.249	0.175	0.0743	0.0365
	b	2.142	2.207	2.285	2.499	2.675
Square pitch*	a	0.215	0.156	0.158	0.0402	0.0331
	b	2.207	2.291	2.263	1.617	2.643

*Pitch = $1.25d_o$

Various *baffle designs* are shown in Figure 9.70. The cross-baffle is designed to direct the flow of the shell-side fluid across the tube bundle and to support the tubes against sagging and possible vibration, and the most common type is the segmental baffle which provides a baffle window. The ratio, baffle spacing/baffle cut, is very important in maximising the ratio of heat transfer rate to pressure drop. Where very low pressure drops are required, double segmental or "disc and doughnut" baffles are used to reduce the pressure drop by some 60 per cent. Triple segmental baffles and designs in which all the tubes are supported by all the baffles provide for low pressure drops and minimum tube vibration.

With regard to *baffle spacing*, TEMA[97] recommends that segmental baffles should not be spaced closer than 20 per cent of the shell inside diameter or 50 mm whichever is the greater and that the maximum spacing should be such that the unsupported tube lengths, given in Table 9.14, are not exceeded. It may be noted that the majority of failures due to vibration occur when the unsupported tube length is in excess of 80 per cent of the TEMA maximum; the best solution is to avoid having tubes in the baffle window.

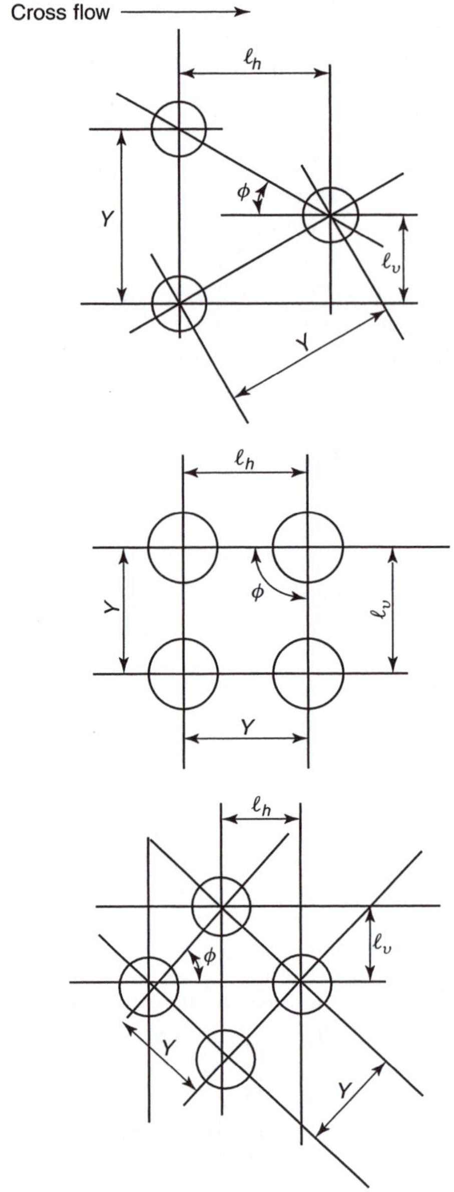

ϕ	ℓ_v	ℓ_h
30°	0.5Y	0.866Y
90°	Y	Y
45°	0.707Y	0.707Y

Figure 9.69. Examples of tube arrays(97)

9.9.3. Mean temperature difference in multipass exchangers

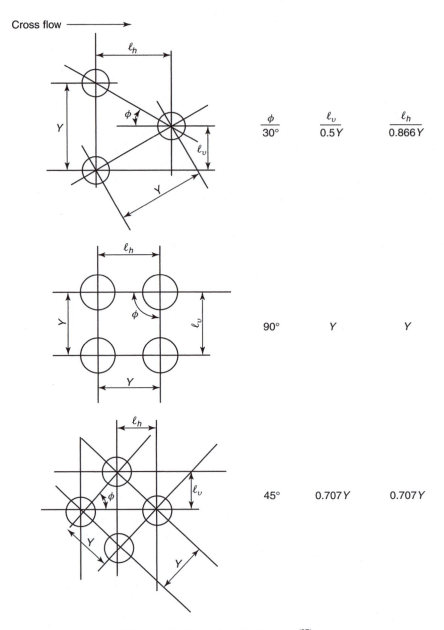

In an exchanger with one shell pass and several tube-side passes, the fluids in the tubes and shell will flow co-currently in some of the passes and countercurrently in the others. For given inlet and outlet temperatures, the mean temperature difference for countercurrent flow is greater than that for co-current or parallel flow, and there is no easy way of

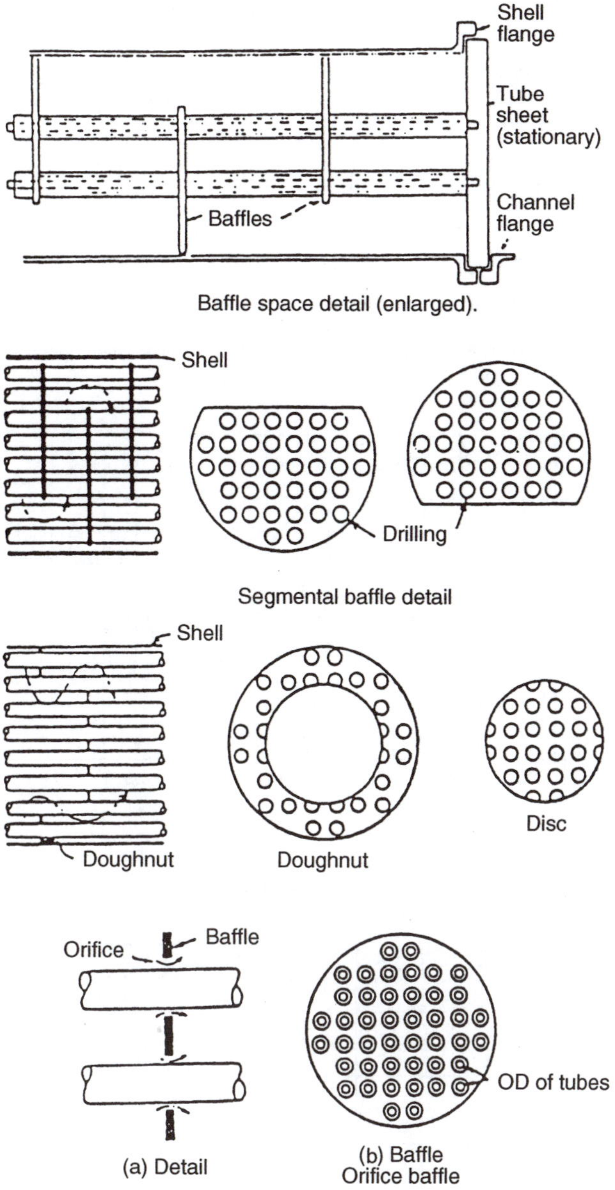

Figure 9.70. Baffle designs

finding the true temperature difference for the unit. The problem has been investigated by UNDERWOOD[102] and by BOWMAN *et al.*[103] who have presented graphical methods for calculating the true mean temperature difference in terms of the value of θ_m which would be obtained for countercurrent flow, and a correction factor F. Provided the following conditions are maintained or assumed, F can be found from the curves shown in Figures 9.71–9.74.

Table 9.14. Maximum unsupported spans for tubes

Aproximate tube OD. (mm)	Maximum unsupported span (mm)	
	Materials group A	Materials group B
19	1520	1321
25	1880	1626
32	2240	1930
38	2540	2210
50	3175	2794

Materials
Group A: Carbon and high alloy steel, low alloy steel, nickel-copper, nickel, nickel-chromium-iron.
Group B: Aluminium and aluminium alloys, copper and copper alloys, titanium and zirconium.

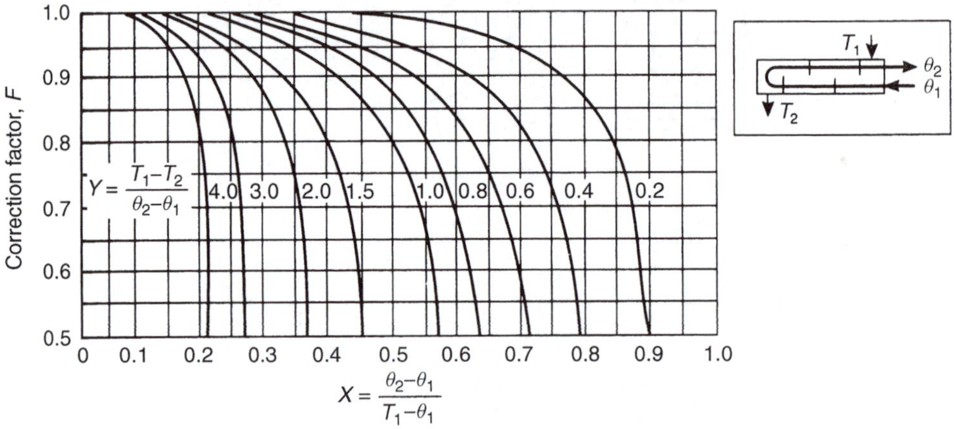

Figure 9.71. Correction for logarithmic mean temperature difference for single shell pass exchanger

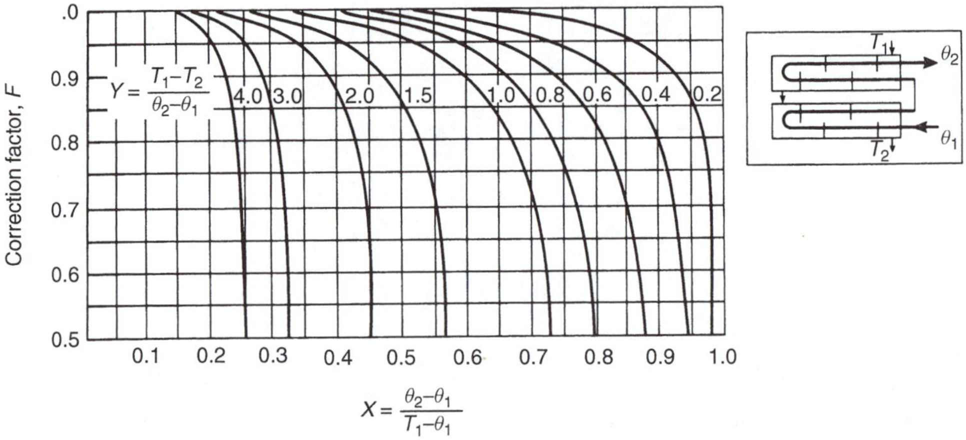

Figure 9.72. Correction for logarithmic mean temperature difference for double shell pass exchanger

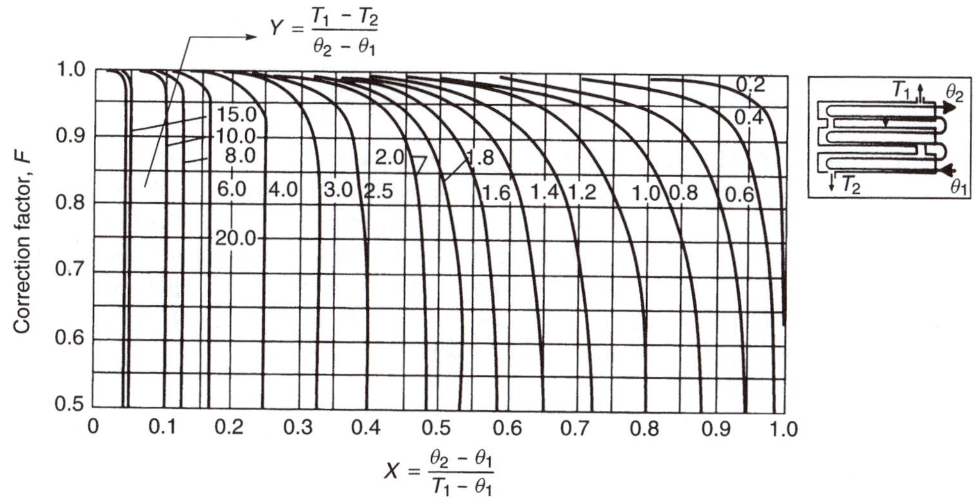

Figure 9.73. Correction for logarithmic mean temperature difference for three shell pass exchanger

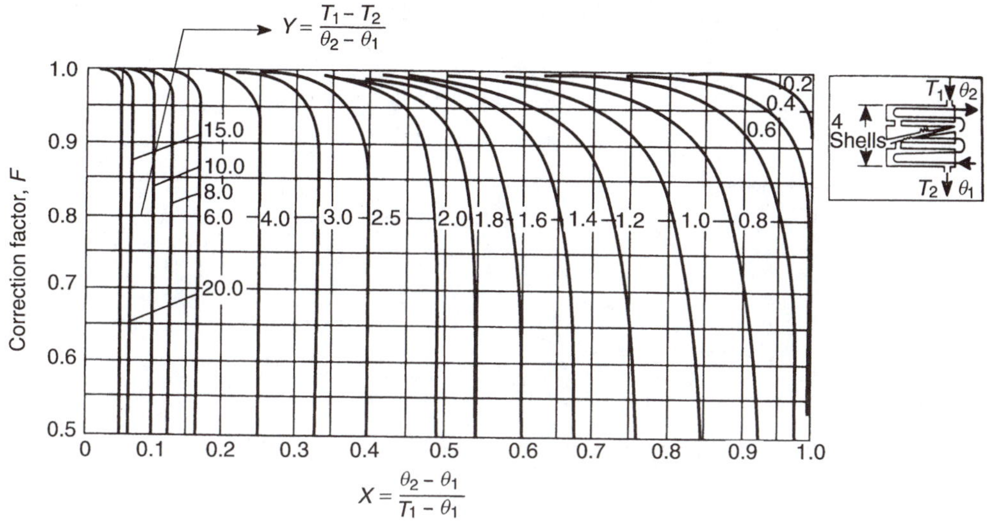

Figure 9.74. Correction for logarithmic mean temperature difference for four shell pass exchanger

(a) The shell fluid temperature is uniform over the cross-section considered as constituting a pass.
(b) There is equal heat transfer surface in each pass.
(c) The overall heat transfer coefficient U is constant throughout the exchanger.
(d) The heat capacities of the two fluids are constant over the temperature range.
(e) There is no change in phase of either fluid.
(f) Heat losses from the unit are negligible.

Then:
$$Q = UAF\theta_m \tag{9.212}$$

F is expressed as a function of two parameters:

$$X = \frac{\theta_2 - \theta_1}{T_1 - \theta_1} \quad \text{and} \quad Y = \frac{T_1 - T_2}{\theta_2 - \theta_1} \tag{9.213}$$

If a one shell-side system is used Figure 9.71 applies, for two shell-side passes Figure 9.72, for three shell-side passes Figure 9.73, and for four shell-passes Figure 9.74. For the case of a single shell-side pass and two tube-side passes illustrated in Figures 9.75*a* and 9.75*b* the temperature profile is as shown. Because one of the passes constitutes a parallel flow arrangement, the exit temperature of the cold fluid θ_2 cannot closely approach the hot fluid temperature T_1. This is true for the conditions shown in Figures 9.75*a* and 9.75*b* and UNDERWOOD[102] has shown that F is the same in both cases.

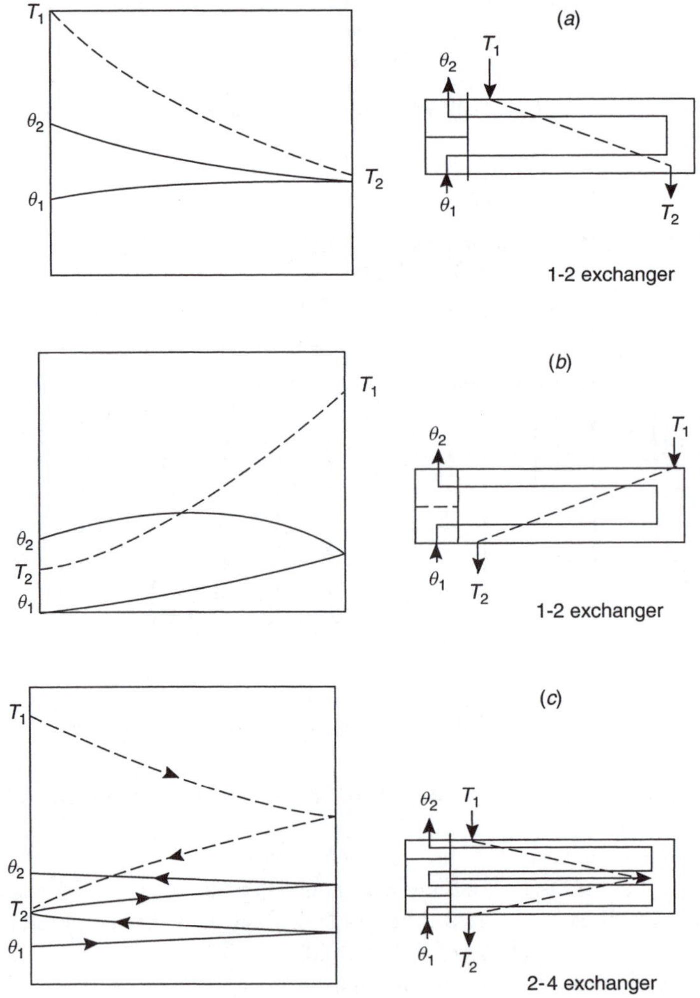

Figure 9.75. Temperature profiles in single and double shell pass exchangers

If, for example, an exchanger is required to operate over the following temperatures:

$$T_1 = 455 \text{ K}, \ T_2 = 372 \text{ K}$$

$$\theta_1 = 283 \text{ K}, \ \theta_2 = 388 \text{ K}$$

Then: $$X = \frac{\theta_2 - \theta_1}{T_1 - \theta_1} = \frac{388 - 283}{455 - 283} = 0.6$$

and: $$Y = \frac{T_1 - T_2}{\theta_2 - \theta_1} = \frac{455 - 372}{388 - 283} = 0.8$$

For a single shell pass arrangement, from Figure 9.71 F is 0.65 and, for a double shell pass arrangement, from Figure 9.72 F is 0.95. On this basis, a two shell-pass design would be used.

In order to obtain maximum heat recovery from the hot fluid, θ_2 should be as high as possible. The difference $(T_2 - \theta_2)$ is known as the *approach temperature* and, if $\theta_2 > T_2$, then a *temperature cross* is said to occur; a situation where the value of F decreases very rapidly when there is but a single pass on the shell-side. This implies that, in parts of the heat exchanger, heat is actually being transferred in the wrong direction. This may be illustrated by taking as an example the following data where equal ranges of temperature are considered:

Case	T_1	T_2	θ_1	θ_2	Approach $(T_2 - \theta_2)$	X	Y	F
1	613	513	363	463	50	0.4	1	0.92
2	573	473	373	473	0	0.5	1	0.80
3	543	443	363	463	cross of 20	0.55	1	0.66

If a temperature cross occurs with a single pass on the shell-side, a unit with two shell passes should be used. It is seen from Figure 9.75b that there may be some point where the temperature of the cold fluid is greater than θ_2 so that beyond this point the stream will be cooled rather than heated. This situation may be avoided by increasing the number of shell passes. The general form of the temperature profile for a two shell-side unit is as shown in Figure 9.75*c*. Longitudinal shell-side baffles are rather difficult to fit and there is a serious chance of leakage. For this reason, the use of two exchangers arranged in series, one below the other, is to be preferred. It is even more important to employ separate exchangers when three passes on the shell-side are required. On the very largest installations it may be necessary to link up a number of exchangers in parallel arranged as, say, sets of three in series as shown in Figure 9.76. This arrangement is preferable for any very large unit which would be unwieldy as a single system. When the total surface area is much greater than 250 m^2, consideration should be given to using multiple smaller units even though the initial cost may be higher.

In many processing operations there may be a large number of process streams, some of which need to be heated and others cooled. An overall heat balance will indicate whether, in total, there is a net surplus or deficit of heat available. It is of great economic importance to achieve the most effective match of the hot and cold streams in the heat exchanger network so as to reduce to a minimum both the heating and cooling duties placed on the works utilities, such as supplies of steam and cooling water. This necessitates making the best use of the temperature driving forces. In considering the overall requirements there will be some point where the temperature difference between the hot and cold streams is a

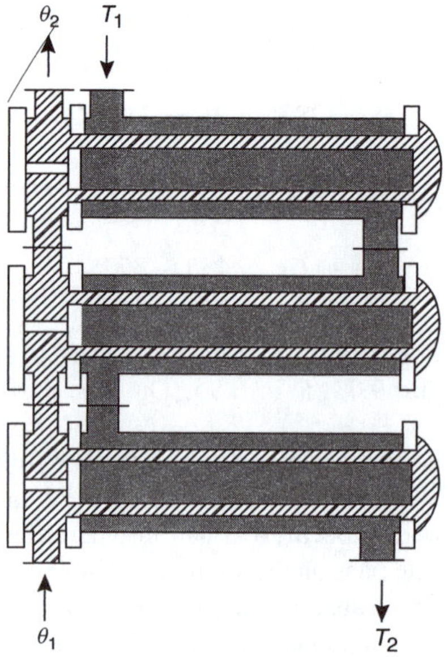

Figure 9.76. Set of three heat exchangers in series

minimum and this is referred to as the *pinch*. The lower the temperature difference at the pinch point, the lower will be the demand on the utilities, although it must be remembered that a greater area, (and hence cost) will be involved and an economic balance must therefore be struck. Heat exchanger networks are discussed in Volume 6 and, in detail, in the User Guide published by the Institution of Chemical Engineers[(104)]. Subsequently, LINNHOFF[(105)] has given an overview of the industrial application of pinch analysis to the design of networks in order to reduce both capital costs and energy requirements.

Example 9.26

Using the data of Example 9.1, calculate the surface area required to effect the given duty using a multipass heat exchanger in which the cold water makes two passes through the tubes and the hot water makes a single pass through the shell.

Solution

As in Example 9.1, the heat load = 1672 kW

With reference to Figure 9.71: $T_1 = 360$ K, $T_2 = 340$ K

and hence:
$$X = \frac{\theta_2 - \theta_1}{T_1 - \theta_1} = \frac{316 - 300}{360 - 300} = 0.267$$

and:
$$Y = \frac{T_1 - T_2}{\theta_2 - \theta_1} = \frac{360 - 340}{316 - 300} = 1.25$$

from Figure 9.58:
$$F = 0.97$$

and hence: $F\theta_m = (41.9 \times 0.97) = 40.6$ K

The heat transfer area is then: $A = \dfrac{1672}{2 \times 40.6} = \underline{\underline{20.6 \text{ m}^2}}$

9.9.4. Film coefficients

Practical values

In any item of heat transfer equipment, the required area of heat transfer surface for a given load is determined by the overall temperature difference θ_m, the overall heat transfer coefficient U and the correction factor F in equation 9.212. The determination of the individual film coefficients which determine the value of U has proved difficult even for simple cases, and it is quite common for equipment to be designed on the basis of practical values of U rather than from a series of film coefficients. For the important case of the transfer of heat from one fluid to another across a metal surface, two methods have been developed for measuring film coefficients. The first requires a knowledge of the temperature difference across each film and therefore involves measuring the temperatures of both fluids and the surface of separation. With a concentric tube system, it is very difficult to insert a thermocouple into the thin tube and to prevent the thermocouple connections from interfering with the flow of the fluid. Nevertheless, this method is commonly adopted, particularly when electrical heating is used. It must be noted that when the heat flux is very high, as with boiling liquids, there will be an appreciable temperature drop across the tube wall and the position of the thermocouple is then important. For this reason, working with stainless steel, which has a relatively low thermal conductivity, is difficult.

The second method uses a technique proposed by WILSON(106). If steam is condensing on the outside of a horizontal tube through which water is passed at various velocities, then the overall and film transfer coefficients one related by:

$$\frac{1}{U} = \frac{1}{h_o} + \frac{x_w}{k_w} + R_i + \frac{1}{h_i} \qquad \text{(from equation 9.201)}$$

provided that the transfer area on each side of the tube is approximately the same.

For conditions of turbulent flow the transfer coefficient for the water side, $h_i = \varepsilon u^{0.8}$, R_i the scale resistance is constant, and h_o the coefficient for the condensate film is almost independent of the water velocity. Thus, equation 9.201 reduces to:

$$\frac{1}{U} = (\text{constant}) + \frac{1}{\varepsilon u^{0.8}}$$

If $1/U$ is plotted against $1/u^{0.8}$ a straight line, known as a Wilson plot, is obtained with a slope of $1/\varepsilon$ and an intercept equal to the value of the constant. For a clean tube R_i should be nil, and hence h_o can be found from the value of the intercept, as x_w/k_w will generally be small for a metal tube. h_i may also be obtained at a given velocity from the difference between $1/U$ at that velocity and the intercept.

This technique has been applied by RHODES and YOUNGER(107) to obtain the values of h_o for condensation of a number of organic vapours, by PRATT(92) to obtain the inside coefficient for coiled tubes, and by COULSON and MEHTA(108) to obtain the coefficient for

Table 9.15. Thermal resistance of heat exchanger tubes

Gauge (BWG)	Thickness (mm)	Values of x_w/k_w (m^2K/kW)				
		Copper	Steel	Stainless steel	Admiralty metal	Aluminium
18	1.24	0.0031	0.019	0.083	0.011	0.0054
16	1.65	0.0042	0.025	0.109	0.015	0.0074
14	2.10	0.0055	0.032	0.141	0.019	0.0093
12	2.77	0.0072	0.042	0.176	0.046	0.0123
		Values of x_w/k_w ($ft^2h°F/Btu$)				
18	0.049	0.000018	0.00011	0.00047	0.000065	0.000031
16	0.065	0.000024	0.00014	0.00062	0.000086	0.000042
14	0.083	0.000031	0.00018	0.0008	0.00011	0.000053
12	0.109	0.000041	0.00024	0.001	0.00026	0.000070

Table 9.16. Thermal resistances of scale deposits from various fluids

	m^2K/kW	$ft^2h°F/Btu$		m^2K/kW	$ft^2h°F/Btu$
*Water**			*Steam*		
distilled	0.09	0.0005	good quality, oil-free	0.052	0.0003
sea	0.09	0.0005	poor quality, oil-free	0.09	0.0005
clear river	0.21	0.0012	exhaust from reciprocating engines	0.18	0.001
untreated cooling tower	0.58	0.0033			
treated cooling tower	0.26	0.0015	*Liquids*		
treated boiler feed	0.26	0.0015	treated brine	0.27	0.0015
hard well	0.58	0.0033	organics	0.18	0.001
Gases			fuel oils	1.0	0.006
air	0.25–0.50	0.0015–0.003	tars	2.0	0.01
solvent vapours	0.14	0.0008			

*For a velocity of 1 m/s (≈ 3 ft/s) and temperatures of less than 320 K (122°F)

an annulus. If the results are repeated over a period of time, the increase in the value of R_i can also be obtained by this method.

Typical values of thermal resistances and individual and overall heat transfer coefficients are given in Tables 9.15–9.18.

Correlated data

Heat transfer data for turbulent flow *inside* conduits of uniform cross-section are usually correlated by a form of equation 9.66:

$$Nu = CRe^{0.8}Pr^{0.33}(\mu/\mu_s)^{0.14} \tag{9.214}$$

where, based on the work of SIEDER and TATE[17], the index for the viscosity correction term is usually 0.14 although higher values have been reported. Using values of C of 0.021 for gases, 0.023 for non-viscous liquids and 0.027 for viscous liquids, equation 9.214 is sufficiently accurate for design purposes, and any errors are far outweighed by uncertainties in predicting shell-side coefficients. Rather more accurate tube-side data

Table 9.17. Approximate overall heat transfer coefficients U for shell and tube equipment

Hot side	Cold side	Overall U	
		W/m^{2}K	Btu/h ft^2 °F
Condensers			
Steam (pressure)	Water	2000–4000	350–750
Steam (vacuum)	Water	1700–3400	300–600
Saturated organic solvents (atmospheric)	Water	600–1200	100–200
Saturated organic solvents (vacuum some non-condensable)	Water–brine	300–700	50–120
Organic solvents (atmospheric and high non-condensable)	Water–brine	100–500	20–80
Organic solvents (vacuum and high non-condensable)	Water–brine	60–300	10–50
Low boiling hydrocarbons (atmospheric)	Water	400–1200	80–200
High boiling hydrocarbons (vacuum)	Water	60–200	10–30
Heaters			
Steam	Water	1500–4000	250–750
Steam	Light oils	300–900	50–150
Steam	Heavy oils	60–400	10–80
Steam	Organic solvents	600–1200	100–200
Steam	Gases	30–300	5–50
Dowtherm	Gases	20–200	4–40
Dowtherm	Heavy oils	50–400	8–60
Evaporators			
Steam	Water	2000–4000	350–750
Steam	Organic solvents	600–1200	100–200
Steam	Light oils	400–1000	80–180
Steam	Heavy oils (vacuum)	150–400	25–75
Water	Refrigerants	400–900	75–150
Organic solvents	Refrigerants	200–600	30–100
Heat exchangers (no change of state)			
Water	Water	900–1700	150–300
Organic solvents	Water	300–900	50–150
Gases	Water	20–300	3–50
Light oils	Water	400–900	60–160
Heavy oils	Water	60–300	10–50
Organic solvents	Light oil	100–400	20–70
Water	Brine	600–1200	100–200
Organic solvents	Brine	200–500	30–90
Gases	Brine	20–300	3–50
Organic solvents	Organic solvents	100–400	20–60
Heavy oils	Heavy oils	50–300	8–50

may be obtained by using correlations given by the Engineering Sciences Data Unit and, based on this work, BUTTERWORTH[109] offers the equation:

$$St = E\,\mathrm{Re}^{-0.205} Pr^{-0.505} \tag{9.215}$$

where:

the Stanton Number $St = NuRe^{-1}Pr^{-1}$

and $E = 0.22\exp[-0.0225(\ln Pr)^2]$

Equation 9.215 is valid for Reynolds Numbers in excess of 10,000. Where the Reynolds Number is less than 2000, the flow will be laminar and, provided natural convection effects

Table 9.18. Approximate film coefficients for heat transfer

	h_i or h_o	
	W/m^2 K	Btu/ft^{2}h °F
No change of state		
water	1700–11,000	300–2000
gases	20–300	3–50
organic solvents	350–3000	60–500
oils	60–700	10–120
Condensation		
steam	6000–17,000	1000–3000
organic solvents	900–2800	150–500
light oils	1200–2300	200–400
heavy oils (vacuum)	120–300	20–50
ammonia	3000–6000	500–1000
Evaporation		
water	2000–12,000	30–200
organic solvents	600–2000	100–300
ammonia	1100–2300	200–400
light oils	800–1700	150–300
heavy oils	60–300	10–50

are negligible, film coefficients may be estimated from a form of equation 9.85 modified to take account of the variation of viscosity over the cross-section:

$$Nu = 1.86(RePr)^{0.33}(d/l)^{0.33}(\mu/\mu_s)^{0.14} \tag{9.216}$$

The minimum value of the Nusselt Number for which equation 9.216 applies is 3.5. Reynolds Numbers in the range 2000–10,000 should be avoided in designing heat exchangers as the flow is then unstable and coefficients cannot be predicted with any degree of accuracy. If this cannot be avoided, the lesser of the values predicted by Equations 9.214 and 9.216 should be used.

As discussed in Section 9.4.3, heat transfer data are conveniently correlated in terms of a heat transfer factor j_h, again modified by the viscosity correction factor:

$$j_h = StPr^{0.67}(\mu/\mu_s)^{-0.14} \tag{9.217}$$

which enables data for laminar and turbulent flow to be included on the same plot, as shown in Figure 9.77. Data from Figure 9.77 may be used together with equation 9.217 to estimate coefficients with heat exchanger tubes and commercial pipes although, due to a higher roughness, the values for commercial pipes will be conservative. Equation 9.217 is rather more conveniently expressed as:

$$Nu = (hd/k) = j_h RePr^{0.33}(\mu/\mu_s)^{-0.14} \tag{9.218}$$

It may be noted that whilst Figure 9.77 is similar to Figure 9.24, the values of j_h differ due to the fact that KERN[28] and other workers define the heat transfer factor as:

$$j_H = NuPr^{-0.33}(\mu/\mu_s)^{-0.14} \tag{9.219}$$

Thus the relationship between j_h and j_H is:

$$j_h = j_H/Re \tag{9.220}$$

As discussed in Section 9.4.3, by incorporating physical properties into equations 9.214 and 9.216, correlations have been developed specifically for water and equation 9.221, based on data from EAGLE and FERGUSON(110) may be used:

$$h = 4280(0.00488T - 1)u^{0.8}/d^{0.2} \tag{9.221}$$

which is in SI units, with h (film coefficient) in W/m^{2}K, T in K, u in m/s and d in m.

Example 9.27

Estimate the heat transfer area required for the system considered in Examples 9.1 and 9.26, assuming that no data on the overall coefficient of heat transfer are available.

Solution

As in the previous examples,

$$\text{heat load} = 1672 \text{ kW}$$

and:

$$\text{corrected mean temperature difference, } F\theta_m = 40.6 \text{ deg K}$$

In the tubes;

$$\text{mean water temperature, } T = 0.5(360 + 340) = 350 \text{ K}$$

Assuming a tube diameter, $d = 19$ mm or 0.0019 m and a water velocity, $u = 1$ m/s, then, in equation 9.221:

$$h_i = 4280((0.00488 \times 350) - 1)1.0^{0.8}/0.0019^{0.2} = 10610 \text{ W/m}^2\text{K or } 10.6 \text{ kW/m}^2\text{K}$$

From Table 9.18, an estimate of the shell-side film coefficient is:

$$h_o = 0.5(1700 + 11000) = 6353 \text{ W/m}^2\text{K or } 6.35 \text{ kW/m}^2\text{K}$$

For steel tubes of a wall thickness of 1.6 mm, the thermal resistance of the wall, from Table 9.15 is:

$$x_w/k_w = 0.025 \text{ m}^2\text{K/kW}$$

and the thermal resistance for treated water, from Table 9.16, is 0.26 m^{2}K/kW for both layers of scale. Thus, in Equation 9.201:

$$(1/U) = (1/h_o) + (x_w/k_w) + R_i + R_o + (1/h_i)$$

$$= (1/6.35) + 0.025 + 0.52 + (1/10.6) = 0.797 \text{ m}^2\text{K/kW}$$

and:

$$U = 1.25 \text{ kW/m}^2\text{K}$$

The heat transfer area required is then:

$$A = Q/F\theta_m U = 1672/(40.6 \times 1.25) = \underline{\underline{32.9 \text{ m}^2}}$$

As discussed in Section 9.4.4, the complex flow pattern on the *shell-side* and the great number of variables involved make the prediction of coefficients and pressure drop very difficult, especially if leakage and bypass streams are taken into account. Until about 1960, empirical methods were used to account for the difference in the performance

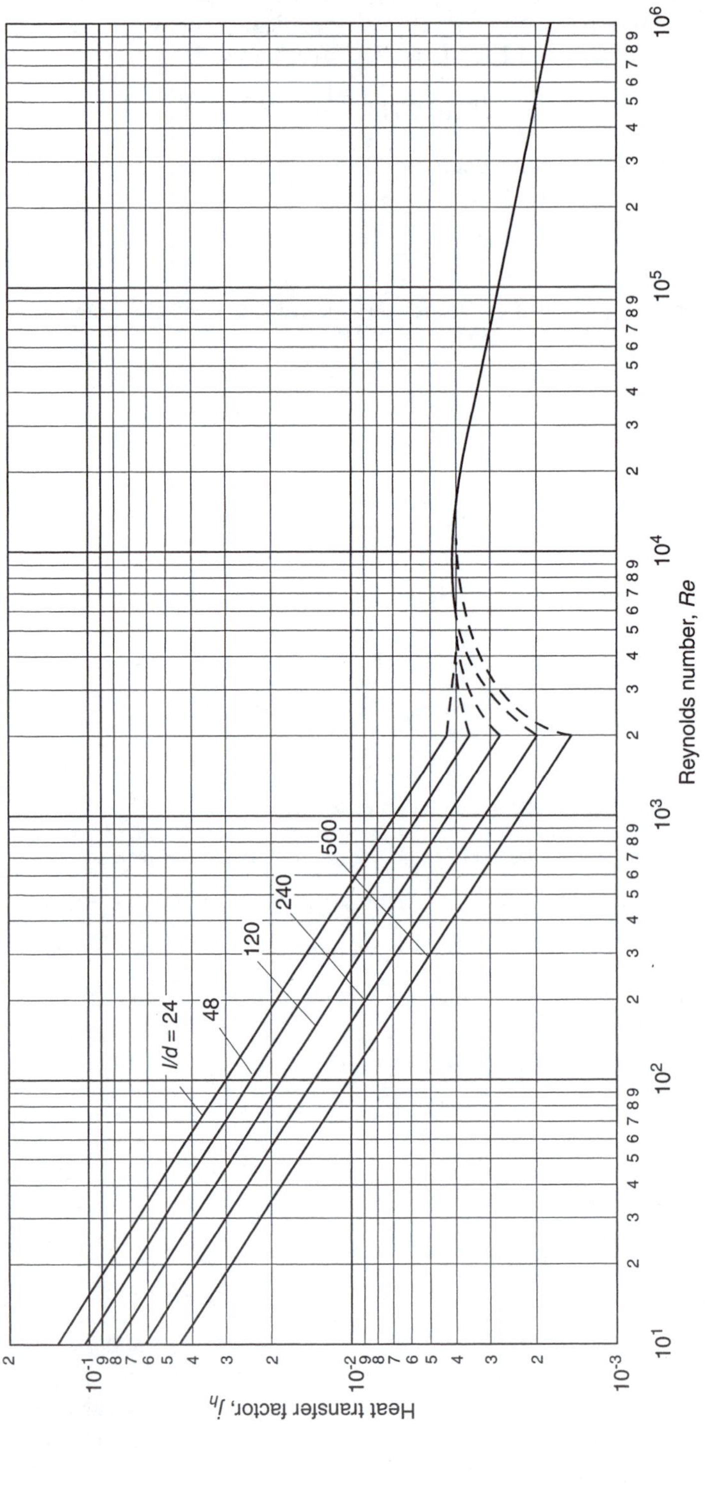

Figure 9.77. Heat transfer factor for flow inside tubes

of real exchangers as compared with that for cross-flow over ideal tube banks. The methods of KERN[28] and DONOHUE[111] are typical of these "bulk flow" methods and their approach, together with more recent methods involving an analysis of the contribution to heat transfer by individual streams in the shell, are discussed in Section 9.9.6.

Special correlations have also been developed for liquid metals, used in recent years in the nuclear industry with the aim of reducing the volume of fluid in the heat transfer circuits. Such fluids have high thermal conductivities, though in terms of heat capacity per unit volume, liquid sodium, for example, which finds relatively widespread application, has a value of $C_p\rho$ of only 1275 kJ/m^3 K.

Although water has a much greater value, it is unsuitable because of its high vapour pressure at the desired temperatures and the corresponding need to use high-pressure piping. Because of their high thermal conductivities, liquid metals have particularly low values of the Prandtl number (about 0.01) and they behave rather differently from normal fluids under conditions of forced convection. Some values for typical liquid metals are given in Table 9.19.

Table 9.19. Prandtl numbers of liquid metals

Metal	Temperature (K)	Prandtl number Pr
Potassium	975	0.003
Sodium	975	0.004
Na/K alloy (56:44)	975	0.06
Mercury	575	0.008
Lithium	475	0.065

The results of work on sodium, lithium, and mercury for forced convection in a pipe have been correlated by the expression:

$$Nu = 0.625(RePr)^{0.4} \tag{9.222}$$

although the accuracy of the correlation is not very good. With values of Reynolds number of about 18,000 it is quite possible to obtain a value of h of about 11 kW/m^2 K for flow in a pipe.

9.9.5. Pressure drop in heat exchangers

Tube-side

Pressure drop on the tube-side of a shell and tube exchanger is made up of the friction loss in the tubes and losses due to sudden contractions and expansions and flow reversals experienced by the tube-side fluid. The friction loss may be estimated by the methods outlined in Section 3.4.3 from which the basic equation for isothermal flow is given by equation 3.18 which can be written as:

$$-\Delta P_t = 4j_f(l/d_i)(\rho u^2) \tag{9.223}$$

where j_f is the dimensionless friction factor. Clearly the flow is not isothermal and it is usual to incorporate an empirical correction factor to allow for the change in physical properties, particularly viscosity, with temperature to give:

$$-\Delta P_t = 4j_f(l/d_i)(\rho u^2)(\mu/\mu_s)^m \tag{9.224}$$

where $m = -0.25$ for laminar flow ($Re < 2100$) and -0.14 for turbulent flow ($Re > 2100$). Values of j_f for heat exchanger tubes are given in Figure 9.78 which is based on Figure 3.7.

There is no entirely satisfactory method for estimating losses due to contraction at the tube inlets, expansion at the exits and flow reversals, although KERN[28] suggests adding four velocity heads per pass, FRANK[112] recommends 2.5 velocity heads and BUTTERWORTH[113] 1.8. LORD *et al.*[114] suggests that the loss per pass is equivalent to a tube length of 300 diameters for straight tubes and 200 for U-tubes, whilst EVANS[115] recommends the addition of 67 tube diameters per pass. Another approach is to estimate the number of velocity heads by using factors for pipe-fittings as discussed in Section 3.4.4 and given in Table 3.2. With four tube passes, for example, there will be four contractions equivalent to a loss of $(4 \times 0.5) = 2$ velocity heads, four expansions equivalent to a loss of $(4 \times 1.0) = 4$ velocity heads and three 180° bends equivalent to a loss of $(3 \times 1.5) = 4.5$ velocity heads. In this way, the total loss is 10.5 velocity heads, or 2.6 per pass, giving support to Frank's proposal of 2.5. Using this approach, equation 9.224 becomes:

$$-\Delta P_{\text{total}} = N_P[4j_f(l/d_i)(\mu/\mu_s)^m + 1.25](\rho u^2) \tag{9.225}$$

where N_P is the number of tube-side passes. Additionally, there will be expansion and contraction losses at the inlet and outlet nozzles respectively, and these losses may be estimated by adding one velocity head for the inlet, and 0.5 of a velocity head for the outlet, based on the nozzle velocities. Losses in the nozzles are only significant for gases at pressures below atmospheric.

Shell-side

As discussed in Section 9.4.4, the prediction of pressure drop, and indeed heat transfer coefficients, in the shell is very difficult due to the complex nature of the flow pattern in the segmentally baffled unit. Whilst the baffles are intended to direct fluid across the tubes, the actual flow is a combination of cross-flow between the baffles and axial or parallel flow in the baffle windows as shown in Figure 9.79, although even this does not represent the actual flow pattern because of leakage through the clearances necessary for the fabrication and assembly of the unit. This more realistic flow pattern is shown in Figure 9.80 which is based on the work of TINKER[116] who identifies the various streams in the shell as follows:

A–fluid flowing through the clearance between the tube and the hole in the baffle.
B–the actual cross-flow stream.
C–fluid flowing through the clearance between the outer tubes and the shell.
E–fluid flowing through the clearance between the baffle and the shell.
F–fluid flowing through the gap between the tubes because of any pass-partition plates. This is especially significant with a vertical gap.

Because stream A does not bypass the tubes, it is the pressure drop rather than the heat transfer which is affected. Streams C, E and F bypass the tubes, thus reducing the effective

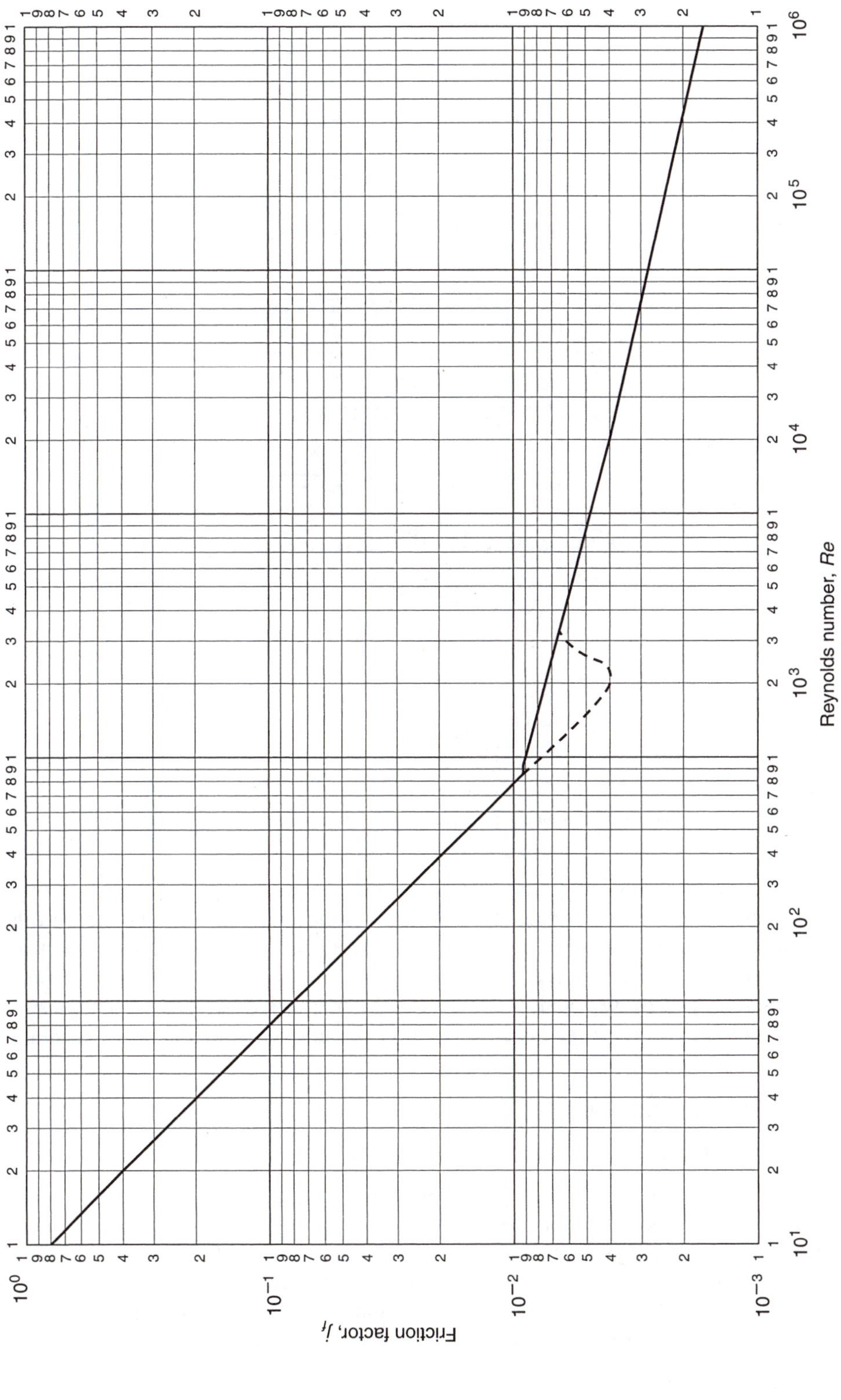

Figure 9.78. Tube-side friction factors[28].

Note: The friction factor j_f is the same as the friction factor for pipes $\phi(= (R/\rho u^2))$, defined in Chapter 3

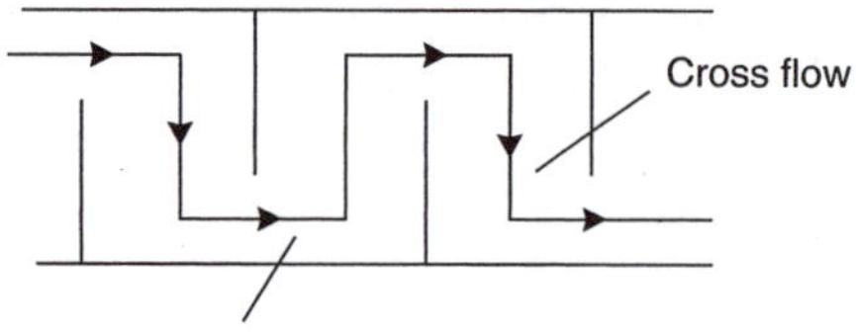

Figure 9.79. Idealised main stream flow

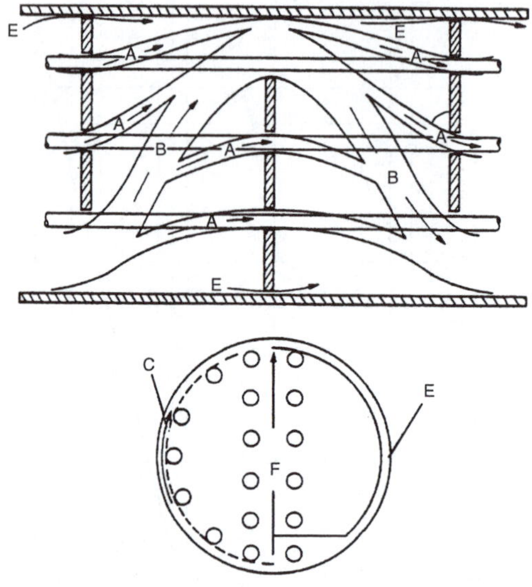

Figure 9.80. Shell-side leakage and by-pass paths[116]

heat transfer area. Stream C, the main bypass stream, is most significant in pull-through bundle units where there is of necessity a large clearance between the bundle and the shell, although this can be reduced by using horizontal sealing strips. In a similar way, the flow of stream F may be reduced by fitting dummy tubes. As an exchanger becomes fouled, clearances tend to plug and this increases the pressure drop. The whole question of shell-side pressure drop estimation in relation to design procedures is now discussed.

9.9.6. Heat exchanger design

Process conditions

A first-stage consideration in the design process is the allocation of fluids to either shell or tubes and, by and large, the more corrosive fluid is passed through the tubes to reduce the costs of expensive alloys and clad components. Similarly, the fluid with the greatest fouling tendency is also usually passed through the tubes where cleaning is easier. Furthermore, velocities through the tubes are generally higher and more readily controllable and can be adjusted to reduce fouling. Where special alloys are in contact with hot fluids, the fluids should be passed through the tubes to reduce costs. In addition the shell temperature is lowered, thus reducing lagging costs. Passing hazardous materials through the tubes leads to greater safety and, because high pressure tubes are cheaper than a high pressure shell, streams at high pressure are best handled on the tube-side. In a similar way, where a very low pressure drop is required as in vacuum operation for example, the fluids involved are best handled on the tube-side where higher film heat transfer coefficients are obtained for a given pressure drop. Provided the flow is turbulent, a higher heat transfer coefficient is usually obtained with a more viscous liquid in the shell because of the more complex flow patterns although, because the tube-side coefficient can be predicted with greater accuracy, it is better to place the fluid in the tubes if turbulent flow in the shell is not possible. Normally, the most economical design is achieved with the fluid with the lower flowrate in the shell.

In selecting a design velocity, it should be recognised that at high velocities high rates of heat transfer are achieved and fouling is reduced, but pressure drops are higher. Normally, the velocity must not be so high as to cause erosion which can be reduced at the tube inlet by fitting plastic inserts, and yet be such that any solids are kept in suspension. For process liquids, velocities are usually 0.3–1.0 m/s in the shell and 1.0–2.0 m/s in the tubes, with a maximum value of 4.0 m/s when fouling must be reduced. Typical water velocities are 1.5–2.5 m/s. For vapours, velocities lie in the range 5–10 m/s with high pressure fluids and 50–70 m/s with vacuum operation, the lower values being used for materials of high molecular weight.

In general, the higher of the temperature differences between the outlet temperature of one stream and the inlet temperature of the other should be 20 deg K and the lower temperature difference should be 5–7 deg K for water coolers and 3–5 deg K when using refrigerated brines, although optimum values can only be determined by an economic analysis of alternative designs.

Similar considerations apply to the selection of pressure drops where there is freedom of choice, although a full economic analysis is justified only in the case of very expensive units. For liquids, typical values in optimised units are 35 kN/m^2 where the viscosity is less than 1 mN s/m^2 and 50–70 kN/m^2 where the viscosity is 1–10 mN s/m^2; for gases, 0.4–0.8 kN/m^2 for high vacuum operation, 50 per cent of the system pressure at 100–200 kN/m^2, and 10 per cent of the system pressure above 1000 kN/m^2. Whatever pressure drop is used, it is important that erosion and flow-induced tube vibration caused by high velocity fluids are avoided.

Design methods

It is shown in Section 9.9.5 that, with the existence of various bypass and leakage streams in practical heat exchangers, the flow patterns of the shell-side fluid, as shown in Figure 9.79, are complex in the extreme and far removed from the idealised cross-flow situation discussed in Section 9.4.4. One simple way of using the equations for cross-flow presented in Section 9.4.4, however, is to multiply the shell-side coefficient obtained from these equations by the factor 0.6 in order to obtain at least an estimate of the shell-side coefficient in a practical situation. The pioneering work of KERN[(28)] and DONOHUE[(111)], who used correlations based on the total stream flow and empirical methods to allow for the performance of real exchangers compared with that for cross-flow over ideal tube banks, went much further and, although their early design method does not involve the calculation of bypass and leakage streams, it is simple to use and quite adequate for preliminary design calculations.

The method, which is based on experimental work with a great number of commercial exchangers with standard tolerances, gives a reasonably accurate prediction of heat transfer coefficients for standard designs, although predicted data on pressure drop is less satisfactory as it is more affected by leakage and bypassing. Using a similar approach to that for tube-side flow, shell-side heat transfer and friction factors are correlated using a hypothetical shell diameter and shell-side velocity where, because the cross-sectional area for flow varies across the shell diameter, linear and mass velocities are based on the maximum area for cross-flow; that is at the shell equator. The shell equivalent diameter is obtained from the flow area between the tubes taken parallel to the tubes, and the wetted perimeter, as outlined in Section 9.9.4 and illustrated in Figure 9.28. The shell-side factors, j_h and j_f, for various baffle cuts and tube arrangements based on the data given by KERN[28] and LUDWIG[101] are shown in Figures 9.81 and 9.82.

The general approach is to calculate the area for cross-flow for a hypothetical row of tubes at the shell equator from the equation given in Section 9.4.4:

$$A_s = d_s l_B C'/Y \tag{9.226}$$

where d_s is the shell diameter, l_B is the baffle length and (C'/Y) is the ratio of the clearance between the tubes and the distance between tube centres. The mass flow divided by the area A_s gives the mass velocity G'_s, and the linear velocity on the shell-side u_s is obtained by dividing the mass velocity by the mean density of the fluid. Again, using the equations in Section 9.4.4, the shell-side equivalent or hydraulic diameter is given by:

$$\text{For square pitch}: d_e = 4(Y^2 - \pi d_o^2)4/\pi d_0 = 1.27(Y^2 - 0.785d_o^2)/d_o \tag{9.227}$$

$$\text{and for triangular pitch}: d_e = 4[(0.87Y \times Y/2) - (0.5\pi d_o^2/4]/(\pi d_o/2)$$

$$= 1.10(Y^2 - 0.917d_o^2)/d_o \tag{9.228}$$

Using this equivalent diameter, the shell-side Reynolds number is then:

$$Re_s = G'_s d_e/\mu = u_s d_e \rho/\mu \tag{9.229}$$

where G'_s is the mass flowrate per unit area. Hence j_h may be obtained from Figure 9.81. The shell-side heat transfer coefficient is then obtained from a re-arrangement of equation 9.220:

$$Nu = (h_s d_e/k_f) = j_h Re Pr^{0.33}(\mu/\mu_s)^{0.14} \tag{9.230}$$

In a similar way, the factor j_f is obtained from Figure 9.82 and the pressure drop estimated from a modified form of equation 9.224:

$$-\Delta P_s = 4j_f(d_s/d_e)(l/l_B)(\rho u_s^2)(\mu/\mu_s)^{-0.14} \tag{9.231}$$

where (l/l_B) is the number of times the flow crosses the tube bundle $= (n + 1)$.

The pressure drop over the shell nozzles should be added to this value although this is usually only significant with gases. In general, the nozzle pressure loss is 1.5 velocity heads for the inlet and 0.5 velocity heads for the outlet, based on the nozzle area or the

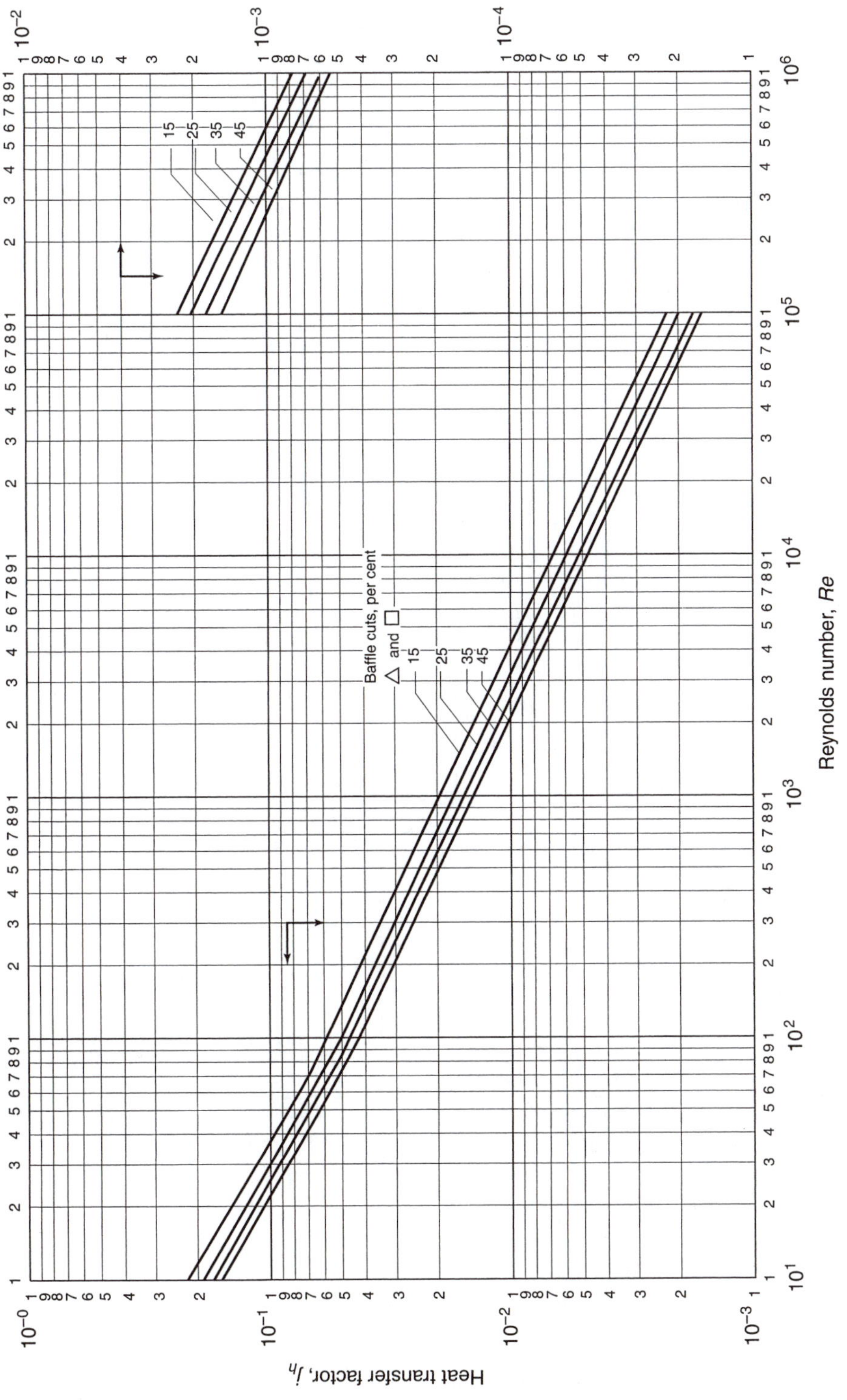

Figure 9.81. Shell-side heat-transfer factors with segmental baffles[28]

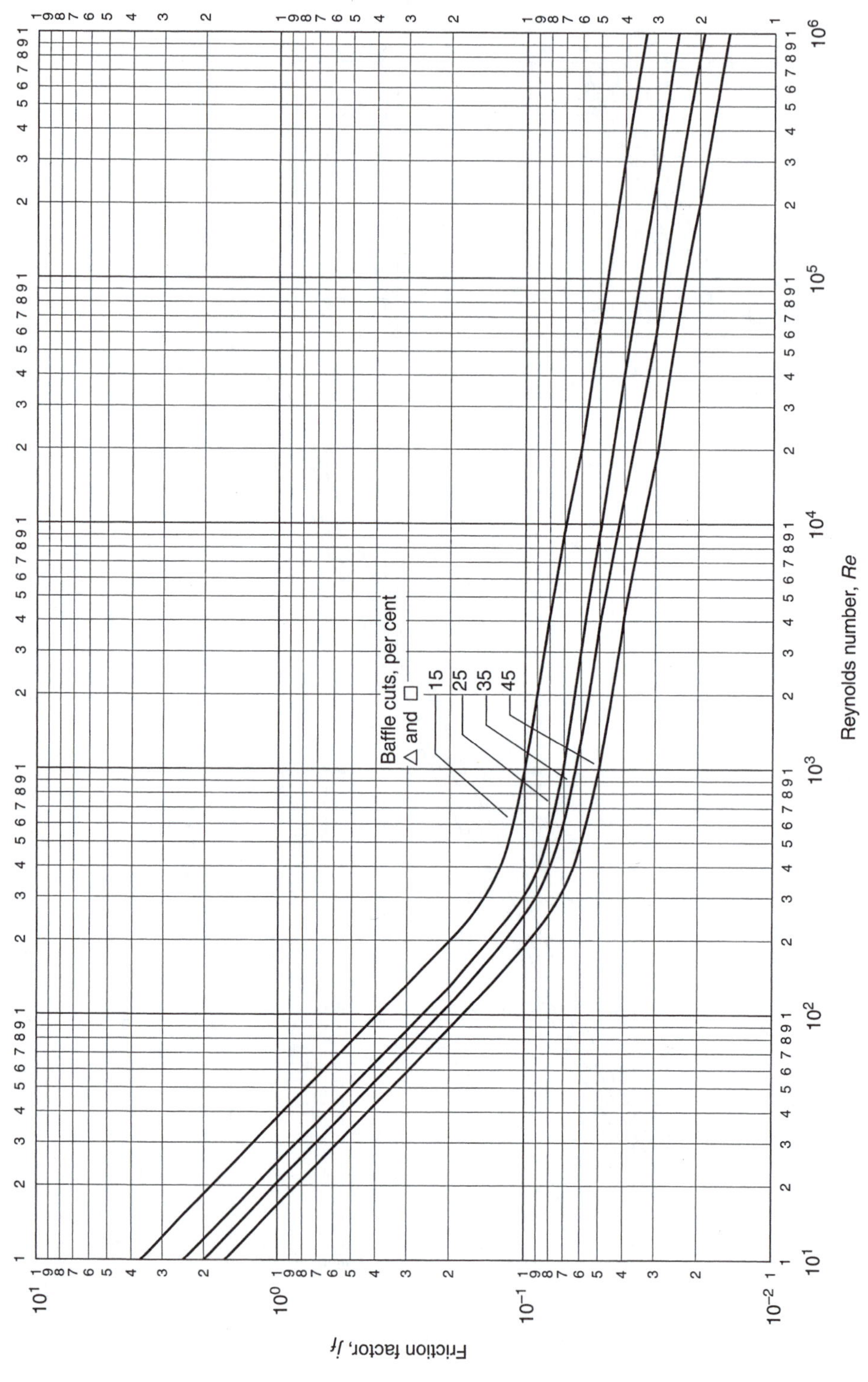

Figure 9.82. Shell-side friction factors with segmental baffles[28]

free area between the tubes in the row adjacent to the nozzle, whichever is the least. Kern's method is now illustrated in the following example.

Example 9.28

Using Kern's method, design a shell and tube heat exchanger to cool 30 kg/s of butyl alcohol from 370 to 315 K using treated water as the coolant. The water will enter at 300 K and leave at 315 K.

Solution

Since it is corrosive, the water will be passed through the tubes.

At a mean temperature of $0.5(370 + 315) = 343$ K, from Table 3, Appendix A1, the thermal capacity of butyl alcohol = 2.90 kJ/kg K and hence:

$$\text{Heat load} = (30 \times 2.90)(370 - 315) = 4785 \text{ kW}$$

If the heat capacity of water is 4.18 kJ/kg K, then:

$$\text{Flow of cooling water} = 4785/(4.18(315 - 300)) = 76.3 \text{ kg/s}$$

The logarithmic mean temperature difference,

$$\theta_m = [(370 - 315) - (315 - 300)]/\ln[(370 - 315)/(315 - 300)] = 30.7 \text{ deg K}$$

With one shell-side pass and two tube-side passes, then from equation 9.213:

$$X = (370 - 315)/(315 - 300) = 3.67 \text{ and } Y = (315 - 300)/(370 - 300) = 0.21$$

and from Figure 9.75:

$$F = 0.85 \text{ and } F\theta_m = (0.85 \times 30.7) = 26.1 \text{ deg K}$$

From Table 9.17, an estimated value of the overall coefficient is $U = 500$ W/m^{2}K and hence, the provisional area, from equation 9.212, is:

$$A = (4785 \times 10^3)/(26.1 \times 500) = 367 \text{ m}^2$$

It is convenient to use 20 mm OD, 16 mm ID tubes, 4.88 m long which, allowing for the tube-sheets, would provide an effective tube length of 4.83 m. Thus:

$$\text{Surface area of one tube} = \pi(20/1000) = 0.303 \text{ m}^2$$

and:

$$\text{Number of tubes required} = (367/0.303) = 1210$$

With a clean shell-side fluid, 1.25 triangular pitch may be used and, from equation 9.211:

$$1210 = 0.249(d_b/20)^{2.207}$$

from which:

$$d_b = 937 \text{ mm}$$

Using a split-ring floating head unit, then, from Figure 9.71, the diametrical clearance between the shell and the tubes = 68 mm and:

$$\text{Shell diameter, } d_s = (937 + 68) = 1005 \text{ mm}$$

which approximates to the nearest standard pipe size of 1016 mm.

Tube-side coefficient

The water-side coefficient may now be calculated using equation 9.218, although here, use will be made of the j_h factor.

$$\text{Cross-sectional area of one tube} = (\pi/40) \times 16^2 = 201 \text{ mm}^2$$

$$\text{Number of tubes/pass} = (1210/2) = 605$$

Thus: Tube-side flow area $= (605 \times 201 \times 10^{-6}) = 0.122\ \text{m}^2$

Mass velocity of the water $= (76.3/0.122) = 625\ \text{kg/m}^2\text{s}$

Thus, for a mean water density of 995 kg/m^3:

Water velocity, $u = (625/995) = 0.63$ m/s

At a mean water temperature of $0.5(315 + 300) = 308$ K, viscosity, $\mu = 0.8$ mN s/m^2 and thermal conductivity, $k = 0.59$ W/m K.
Thus:

$$Re = du\rho/\mu = (16 \times 10^{-3} \times 0.63 \times 995)/(0.8 \times 10^{-3}) = 12540$$

$$Pr = C_p\mu/k = (4.18 \times 10^3 \times 0.8 \times 10^{-3})/0.59 = 5.67$$

$$l/d_i = 4.83/(16 \times 10^{-3}) = 302$$

Thus, from Figure 9.77, $j_h = 3.7 \times 10^{-3}$, and, in equation 9.218, neglecting the viscosity term:

$$(h_i \times 16 \times 10^{-3})/0.59 = (3.7 \times 10^{-3} \times 12540 \times 5.67^{0.33})$$

and: $h_i = 3030\ \text{W/m}^2\text{K}$

Shell-side coefficient

The baffle spacing will be taken as 20 per cent of the shell diameter or $(1005 \times 20/100) = 201$ mm
The tube pitch $= (1.25 \times 20) = 25$ mm and, from equation 9.226:

$$\text{Cross-flow area, } A_s = [(25 - 20)/25](1005 \times 201 \times 10^{-6}) = 0.040\ \text{m}^2$$

Thus: Mass velocity in the shell, $G_s = (30/0.040) = 750\ \text{kg/m}^2\text{s}$

From equation 9.228:

$$\text{Equivalent diameter, } d_e = 1.10[25^2 - (0.917 \times 20^2)]/20 = 14.2\ \text{mm}$$

At a mean shell-side temperature of $0.5(370 + 315) = 343$ K, from Appendix A1:

density of butyl alcohol, $\rho = 780$ kg/m^3, viscosity, $\mu = 0.75$ mN s/m^2, heat capacity, $C_p = 3.1$ kJ/kg K

and thermal conductivity, $k = 0.16$ W/m K.
Thus, from equation 9.229:

$$Re = G_s d_e/\mu = (750 \times 14.2 \times 10^{-3})/(0.75 \times 10^{-3}) = 14200$$

$$Pr = C_p\mu/k = (3.1 \times 10^3 \times 0.75 \times 10^{-3})/0.16 = 14.5$$

Thus, with a 25 per cent segmental cut, from Figure 9.81: $j_h = 5.0 \times 10^{-3}$
Neglecting the viscosity correction term in equation 9.230:

$$(h_s \times 14.2 \times 10^{-3})/0.16 = 5.0 \times 10^{-3} \times 14200 \times 14.5^{0.33}$$

and: $h_s = 1933\ \text{W/m}^2\text{K}$

The mean butanol temperature = 343 K, the mean water temperature = 308 K and hence the mean wall temperature may be taken as $0.5(343 + 308) = 326$ K at which $\mu_s = 1.1$ mN s/m^2

Thus: $(\mu/\mu_s)^{0.14} = (0.75/1.1)^{0.14} = 0.95$

showing that the correction for a low viscosity fluid is negligible.

Overall coefficient

The thermal conductivity of cupro-nickel alloys = 50 W/m K and, from Table 9.16, scale resistances will be taken as 0.00020 m^{2}K/W for the water and 0.00018 m^{2}K/W for the organic.

Based on the outside area, the overall coefficient is given by:

$$1/U = 1/h_o + R_o + x_w/k_w + R_i/(d_o/d_i) + (1/h_i)(d_o/d_i)$$

$$= (1/1933) + 0.00020 + [0.5(20 - 16) \times 10^{-3}/50] + (0.00015 \times 20)/16 + 20/(3030 \times 16)$$

$$= 0.00052 + 0.00020 + 0.00004 + 0.000225 + 0.00041 = 0.00140 \text{ m}^2\text{K/W}$$

and: $U = 717$ W/m²K
which is well in excess of the assumed value of 500 W/m²K.

Pressure drop

On the *tube-side*, $Re = 12450$ and from Figure 9.78, $j_f = 4.5 \times 10^{-3}$
Neglecting the viscosity correction term, equation 9.225 becomes:

$$\Delta P_t = 2(4 \times 4.5 \times 10^{-3}(4830/16) + 1.25)(995 \times 0.63^2) = 5279 \text{ N/m}^2 \text{ or } 5.28 \text{ kN/m}^2$$

which is low, permitting a possible increase in the number of tube passes.
On the *shell-side*, the linear velocity, $(G_s/\rho) = (750/780) = 0.96$ m/s
From Figure 9.82, when $Re = 14200$, $j_f = 4.6 \times 10^{-2}$
Neglecting the viscosity correction term, in equation 9.231:

$$-\Delta P_s = (4 \times 4.6 \times 10^{-2})(1005/14.2)(4830/201)(780 \times 0.96^2)$$

$$= 224950 \text{ N/m}^2 \text{ or } 225 \text{ kN/m}^2$$

This value is very high and thought should be given to increasing the baffle spacing. If this is doubled, this will reduce the pressure drop by approximately $(1/2)^2 = 1/4$ and:

$$-\Delta P_s = (225/4) = 56.2 \text{ kN/m}^2 \text{ which is acceptable.}$$

Since $h_o \propto Re^{0.8} \propto u^{0.8}$,

$$h_o = 1933(1/2)^{0.8} = 1110 \text{ W/m}^2\text{K}$$

which gives an overall coefficient of 561 W/m²K which is still in excess of the assumed value of 500 W/m²K.

Further detailed discussion of Kern's method together with a worked example is presented in Volume 6.

Whilst Kern's method provides a simple approach and one which is quite adequate for preliminary design calculations, much more reliable predictions may be achieved by taking into account the contribution to heat transfer and pressure drop made by the various idealised flow streams shown in Figure 9.80. Such an approach was originally taken by TINKER[(116)] and many of the methods subsequently developed have been based on his model which unfortunately is difficult to follow and tedious to use. The approach has been simplified by DEVORE[(117)], however, who, in using standard tolerances for commercial exchangers and a limited number of baffle designs, gives nomographs which enable the method to be used with simple calculators. Devore's method has been further simplified by MUELLER[(118)] who gives an illustrative example. PALEN and TABOREK[(119)] and GRANT[(120)] have described how both Heat Transfer Inc. and Heat Transfer and Fluid Flow Services have used Tinker's method to develop proprietary computer-based methods.

Using Tinker's approach, BELL[(121,122)] has described a semi-analytical method, based on work at the University of Delaware, which allows for the effects of major bypass and leakage streams, and which is suitable for use with calculators. In this procedure, the heat transfer coefficient and the pressure drop are obtained from correlations for flow over ideal tube banks, applying correction factors to allow for the effects of leakage, bypassing and flow in the window zone. This approach gives more accurate predictions than Kern's method and can be used to determine the effects of constructional tolerances and the use of sealing tapes. This method is discussed in some detail in Volume 6, where an illustrative example is offered.

A more recent approach is that offered by WILLS and JOHNSTON[123] who have developed a simplified version of Tinker's method. This has been adopted by the Engineering Sciences Data Unit, ESDU[124], and it gives a useful calculation technique for providing realistic checks on 'black box' computer predictions. The basis of this approach is shown in Figure 9.83 which shows fluid flowing from A to B in two streams — over the tubes in cross-flow, and bypassing the tube bundle — which then combine to form a single stream. In addition, leakage occurs between the tubes and the baffle and between the baffle and the shell, as shown. For each of these streams, a coefficient is defined which permits the pressure drop for each stream to be expressed in terms of the square of the mass velocity for that stream. A knowledge of the total mass velocity and the sum of the pressure drops in each zone enables the coefficients for each stream to be estimated by an iterative procedure, and the flowrate of each stream to be obtained. The estimation of the heat transfer coefficient and the pressure drop is then a relatively simple operation. This method is of especial value in investigating the effect of various shell-to-baffle and baffle-to-tube tolerances on the performance of a heat exchanger, both in terms of heat transfer rates and the pressure losses incurred.

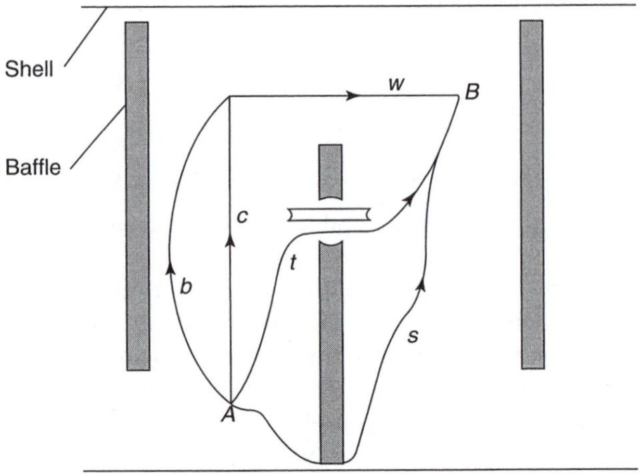

Figure 9.83. Flow streams in the Wills and Johnston method[123].

9.9.7. Heat exchanger performance

One of the most useful methods of evaluating the performance of an existing heat exchanger or to assess a proposed design is to determine its *effectiveness* η, which is defined as the ratio of the actual rate of heat transfer Q to the maximum rate Q_{max} that is thermodynamically possible or:

$$\eta = \frac{Q}{Q_{max}} \tag{9.232}$$

Q_{max} is the heat transfer rate which would be achieved if it were possible to bring the outlet temperature of the stream with the lower heat capacity to the inlet temperature of the other stream. Using the nomenclature in Figure 9.84, and taking stream 1 as having the lower value of GC_p, then:

$$Q_{max} = G_1 C_{p_1}(T_{11} - T_{21}) \tag{9.233}$$

An overall heat balance gives:

$$Q = G_1 C_{p_1}(T_{11} - T_{12}) = G_2 C_{p_2}(T_{22} - T_{21})$$

Thus, based on stream 1:

$$\eta = \frac{G_1 C_{p_1}(T_{11} - T_{12})}{G_1 C_{p_1}(T_{11} - T_{21})} = \frac{T_{11} - T_{12}}{T_{11} - T_{21}} \tag{9.234}$$

and, based on stream 2:

$$\eta = \frac{G_2 C_{p_2}(T_{22} - T_{21})}{G_1 C_{p_1}(T_{11} - T_{21})} \tag{9.235}$$

In calculating temperature differences, the positive value should always be taken.

Example 9.29

A flow of 1 kg/s of an organic liquid of heat capacity 2.0 kJ/kg K is cooled from 350 to 330 K by a stream of water flowing countercurrently through a double-pipe heat exchanger. Estimate the effectiveness of the unit if the water enters the exchanger at 290 K and leaves at 320 K.

Solution

Heat load, $Q = 1 \times 2.0(350 - 330) = 40$ kW

Flow of water, $G_{cool} = \dfrac{40}{4.187(320 - 290)} = 0.318$ kg/s

For organic: $(GC_p)_{hot} = (1 \times 2.0) = 2.0$ kW/K $(= G_2 C_{p_2})$

For water: $(GC_p)_{cold} = (0.318 \times 4.187) = 1.33$ kW/K$(= GC_p)_{min} = (G_1 C_{p_1})$

From equation 9.235:

$$\text{effectiveness } \eta = \frac{2.0(350 - 330)}{1.33(350 - 290)}$$

$$= \underline{\underline{0.50}}$$

9.9.8. Transfer units

The concept of a *transfer unit* is useful in the design of heat exchangers and in assessing their performance, since its magnitude is less dependent on the flowrate of the fluids than the heat transfer coefficient which has been used so far. The number of transfer units **N** is defined by:

$$\mathbf{N} = \frac{UA}{(GC_p)_{min}} \tag{9.236}$$

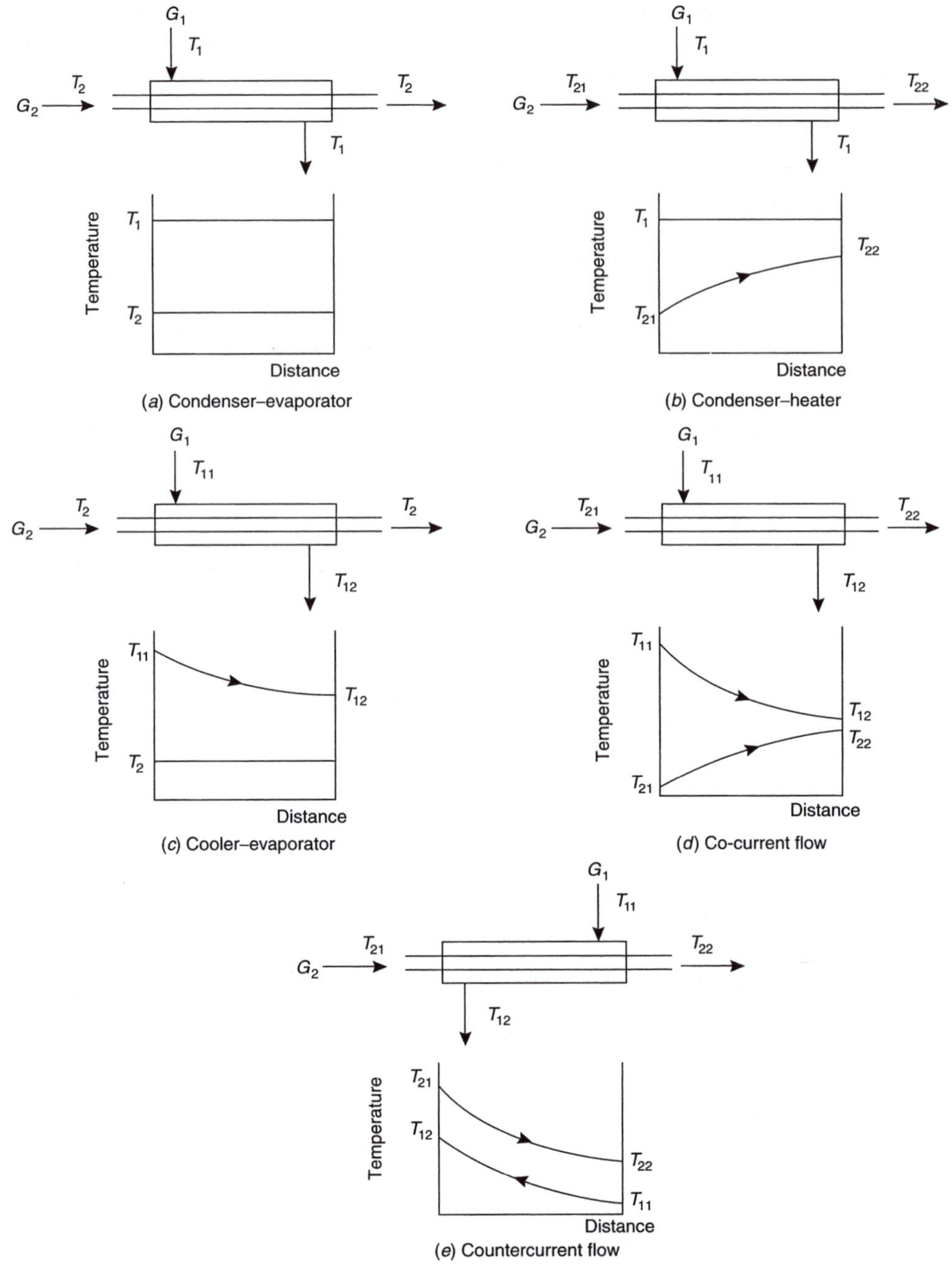

Figure 9.84. Nomenclature for effectiveness of heat exchangers

where $(GC_p)_{\min}$ is the lower of the two values $G_1C_{p_1}$ and $G_2C_{p_2}$. **N** is the ratio of the heat transferred for a unit temperature driving force to the heat absorbed by the fluid stream when its temperature is changed by 1 deg K. Thus, the number of transfer units gives a measure of the amount of heat which the heat exchanger can transfer. The relation for the effectiveness of the heat exchanger in terms of the heat capacities of the streams is now given for a number of flow conditions. The relevant nomenclature is given in Figure 9.84.

Transfer units are also used extensively in the calculation of mass transfer rates in countercurrent columns and reference should be made to Chapter 10.

Considering *co-current* flow as shown in Figure 9.84*d*, for an elemental area dA of a heat exchanger, the rate of transfer of heat dQ is given by:

$$dQ = U\,dA(T_1 - T_2) = U\,dA\,\theta \tag{9.237}$$

where T_1 and T_2 are the temperatures of the two streams and θ is the point value of the temperature difference between the streams.

In addition: $$dQ = G_2C_{p_2}\,dT_2 = -G_1C_{p_1}\,dT_1$$

Thus: $$dT_2 = \frac{dQ}{G_2C_{p_2}} \quad \text{and} \quad dT_1 = \frac{-dQ}{G_1C_{p_1}}$$

and: $$dT_1 - dT_2 = d(T_1 - T_2) = d\theta = -dQ\left(\frac{1}{G_1C_{p_1}} + \frac{1}{G_2C_{p_2}}\right)$$

Substituting from equation 9.237 for dQ:

$$\frac{d\theta}{\theta} = -U dA\left[\frac{1}{G_1C_{p_1}} + \frac{1}{G_2C_{p_2}}\right] \tag{9.238}$$

Integrating: $$\ln\frac{\theta_2}{\theta_1} = -UA\left[\frac{1}{G_1C_{p_1}} + \frac{1}{G_2C_{p_2}}\right]$$

or: $$\ln\frac{T_{12} - T_{22}}{T_{11} - T_{21}} = \frac{UA}{G_1C_{p_1}}\left[1 + \frac{G_1C_{p_1}}{G_2C_{p_2}}\right] \tag{9.239}$$

If $G_1C_{p_1} < G_2C_{p_2}$, $G_1C_{p_1} = (GC_p)_{\min}$

From equation 9.236: $$\mathbf{N} = \frac{UA}{G_1C_{p_1}}$$

Thus: $$\frac{T_{12} - T_{22}}{T_{11} - T_{21}} = \exp\left[-\mathbf{N}\left(1 + \frac{G_1C_{p_1}}{G_2C_{p_2}}\right)\right] \tag{9.240}$$

From equations 9.234 and 9.235:

$$T_{11} - T_{12} = \eta(T_{11} - T_{21})$$

$$T_{22} - T_{21} = \eta\frac{G_1C_{p_1}}{G_2C_{p_2}}(T_{11} - T_{21})$$

Adding: $$T_{11} - T_{12} + T_{22} - T_{21} = \eta \left(1 + \frac{G_1 C_{p1}}{G_2 C_{p2}}\right)(T_{11} - T_{21})$$

$$1 - \frac{T_{12} - T_{22}}{T_{11} - T_{21}} = \eta \left(1 + \frac{G_1 C_{p1}}{G_2 C_{p2}}\right)$$

Substituting in equation 9.240:

$$\eta = \frac{1 - \exp\left[-\mathbf{N}\left(1 + \frac{G_1 C_{p1}}{G_2 C_{p2}}\right)\right]}{1 + \frac{G_1 C_{p1}}{G_2 C_{p2}}} \qquad (9.241)$$

For the particular case where $G_1 C_{p_1} = G_2 C_{p_2}$:

$$\eta = 0.5[1 - \exp(-2\mathbf{N}] \qquad (9.242)$$

For a very large exchanger ($\mathbf{N} \rightarrow \infty$), $\eta \rightarrow 0.5$.

A similar procedure may be followed for *countercurrent flow* (Figure 9.84*e*), although it should be noted that, in this case, $\theta_1 = T_{11} - T_{22}$ and $\theta_2 = T_{12} - T_{21}$.

The corresponding equation for the effectiveness factor η is then:

$$\eta = \frac{1 - \exp\left[-\mathbf{N}\left(1 - \frac{G_1 C_{p1}}{G_2 C_{p2}}\right)\right]}{1 - \frac{G_1 C_{p1}}{G_2 C_{p2}} \exp\left[-\mathbf{N}\left(1 - \frac{G_1 C_{p1}}{G_2 C_{p2}}\right)\right]} \qquad (9.243)$$

For the case where $G_1 C_{p_1} = G_2 C_{p_2}$, it is necessary to expand the exponential terms to give:

$$\eta = \frac{\mathbf{N}}{1 + \mathbf{N}} \qquad (9.244)$$

In this case, for a very large exchanger ($\mathbf{N} \rightarrow \infty$), $\eta \rightarrow 1$.

If one component is merely undergoing *a phase change at constant temperature*, (Figure 9.84*b*, *c*) $G_1 C_{p_1}$ is effectively zero and both equations 9.241 and 9.243 reduce to:

$$\eta = 1 - \exp(-\mathbf{N}) \qquad (9.245)$$

Effectiveness factors η are plotted against number of transfer units **N** with $(G_1 C_{p_1}/G_2 C_{p_2})$ as parameter for a number of different configurations by KAYS and LONDON(125). Examples for countercurrent flow (based on equation 9.235) and an exchanger with one shell pass and two tube passes are plotted in Figures 9.85*a* and *b* respectively.

Example 9.30

A process requires a flow of 4 kg/s of purified water at 340 K to be heated from 320 K by 8 kg/s of untreated water which can be available at 380, 370, 360 or 350 K. Estimate the heat transfer surfaces of one shell pass, two tube pass heat exchangers suitable for these duties. In all cases, the mean heat capacity of the water streams is 4.18 kJ/kg K and the overall coefficient of heat transfer is 1.5 kW/m^2 K.

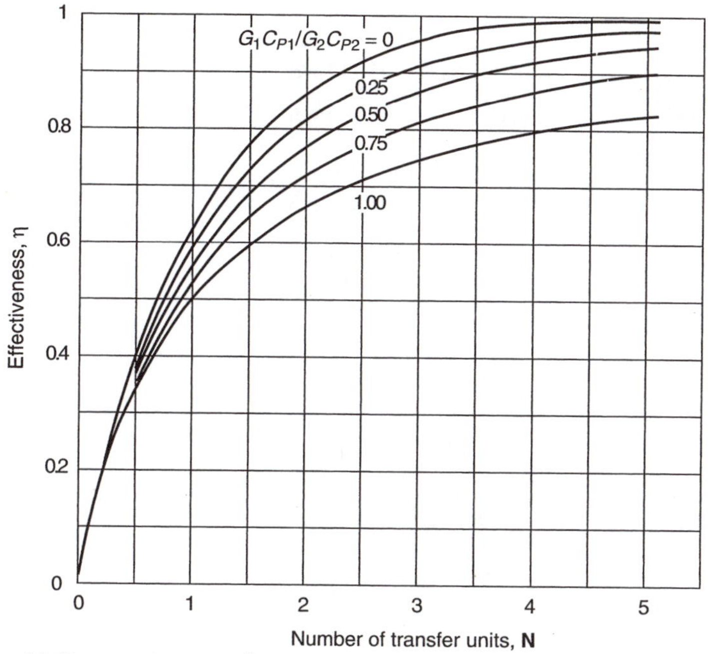

(*a*) True countercurrent flow

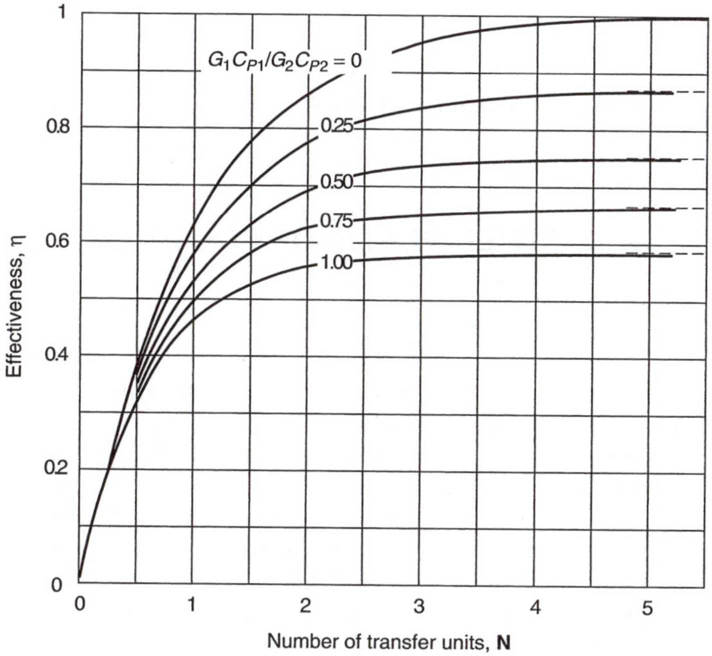

(*b*) One shell pass, two-tube pass exchanger

Figure 9.85. Effectiveness of heat exchangers as a function of number of transfer units[125]

Solution

For the untreated water: $GC_p = (8.0 \times 4.18) = 33.44$ kW/K

For the purified water: $GC_p = (4.0 \times 4.18) = 16.72$ kW/K

Thus:

$$(GC_p)_{\min} = 16.72 \text{ kW/K} = G_1 C_{p_1}$$

and:

$$\frac{G_1 C_{p_1}}{G_2 C_{p_2}} = \frac{16.72}{33.44} = 0.5$$

From equation 9.235:

$$\eta = \frac{4.0 \times 4.18(340 - 320)}{4.0 \times 4.18(T_{11} - 320)}$$

$$= \frac{20}{(T_{11} - 320)}.$$

Thus η may be calculated from this equation using values of $T_{11} = 380$, 370, 360 or 350 K and then **N** obtained from Figure 9.85*b*. The area required is then calculated from:

$$A = \frac{\mathbf{N}(GC_p)_{\min}}{U} \qquad \text{(equation 9.236)}$$

to give the following results:

T_{11} (K)	η (−)	**N** (−)	A (m^2)
380	0.33	0.45	5.0
370	0.4	0.6	6.6
360	0.5	0.9	10.0
350	0.67	1.7	18.9

Obviously, the use of a higher untreated water temperature is attractive in minimising the area required, although in practice any advantages would be offset by increased water costs, and an optimisation procedure would be necessary in obtaining the most effective design.

9.10. OTHER FORMS OF EQUIPMENT

9.10.1. Finned-tube units

Film coefficients

When viscous liquids are heated in a concentric tube or standard tubular exchanger by condensing steam or hot liquid of low viscosity, the film coefficient for the viscous liquid is much smaller than that on the hot side and it therefore controls the rate of heat transfer. This condition also arises with air or gas heaters where the coefficient on the gas side will be very low compared with that for the liquid or condensing vapour on the other side. It is often possible to obtain a much better performance by increasing the area of surface on the side with the limiting coefficient. This may be done conveniently by using a finned tube as in Figure 9.86 which shows one typical form of such equipment which may have either longitudinal or transverse fins.

The calculation of the film coefficients on the fin side is complex because each unit of surface on the fin is less effective than a unit of surface on the tube wall. This arises because there will be a temperature gradient along the fin so that the temperature difference

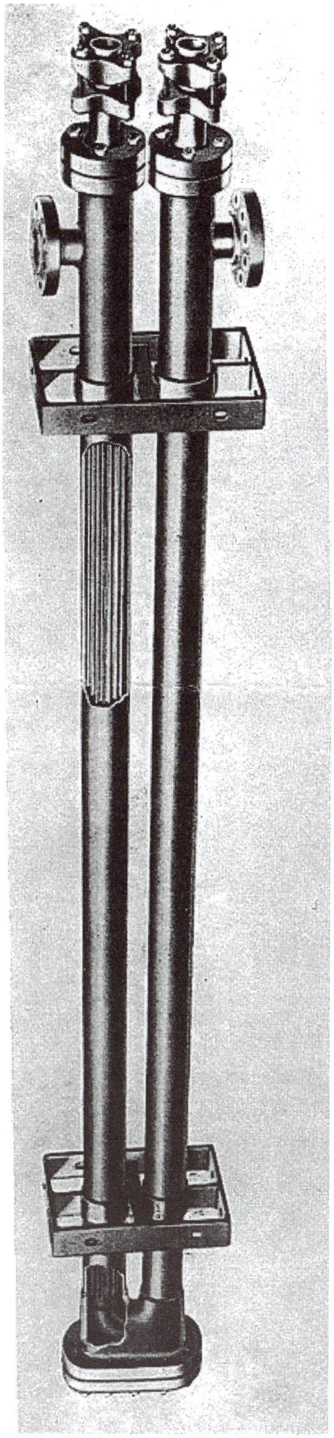

Figure 9.86*a*. Heat exchanger showing tubes with longitudinal fins

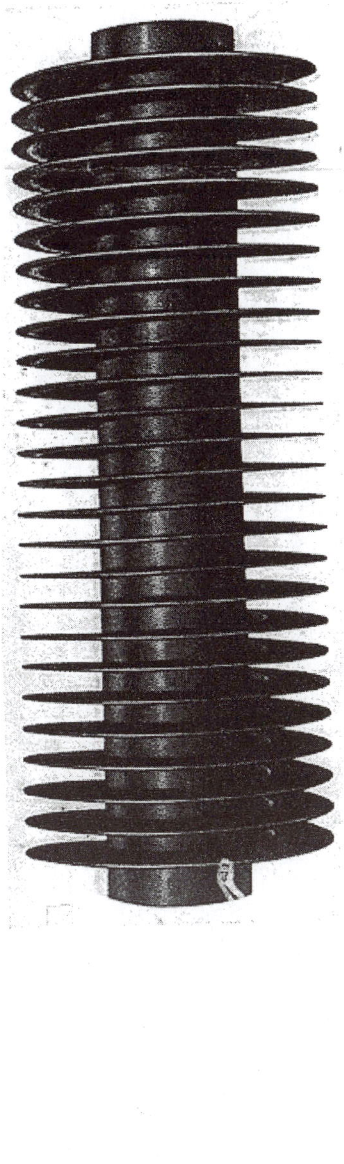

Figure 9.86*b*. Tube with radial fins

between the fin surface and the fluid will vary along the fin. To calculate the film coefficients it is convenient to consider firstly the extended surface as shown in Figure 9.87. A cylindrical rod of length L and cross-sectional area A and perimeter b is heated at one end by a surface at temperature T_1 and cooled throughout its length by a medium at temperature T_G so that the cold end is at a temperature T_2.

A heat balance over a length dx at distance x from the hot end gives:

$$\text{heat in} = \text{heat out along rod} + \text{heat lost to surroundings}$$

or:
$$-kA\frac{\mathrm{d}T}{\mathrm{d}x} = \left[-kA\frac{\mathrm{d}T}{\mathrm{d}x} + \frac{\mathrm{d}}{\mathrm{d}x}\left(-kA\frac{\mathrm{d}T}{\mathrm{d}x}\right)\mathrm{d}x\right] + hb\,\mathrm{d}x(T - T_G)$$

where h is the film coefficient from fin to surroundings.

Writing the temperature difference $T - T_G$ equal to θ:

$$kA\frac{\mathrm{d}^2T}{\mathrm{d}x^2}\mathrm{d}x = hb\,\mathrm{d}x\,\theta$$

Since T_G is constant, $\mathrm{d}^2T/\mathrm{d}x^2 = \mathrm{d}^2\theta/\mathrm{d}x^2$.

Thus:
$$\frac{\mathrm{d}^2\theta}{\mathrm{d}x^2} = \frac{hb}{kA}\theta = m^2\theta \quad \left(\text{where } m^2 = \frac{hb}{kA}\right) \tag{9.246}$$

and:
$$\theta = C_1\mathrm{e}^{mx} + C_2\mathrm{e}^{-mx} \tag{9.247}$$

In solving this equation, three important cases may be considered:

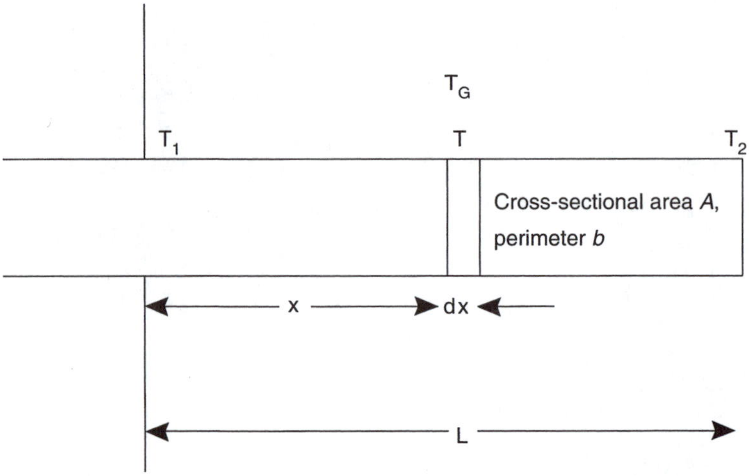

Figure 9.87. Heat flow in rod with heat loss to surroundings

(a) A long rod with temperature falling to that of surroundings, that is $\theta = 0$ when $x = \infty$.

In this case:
$$\theta = \theta_1\mathrm{e}^{-mx} \tag{9.248}$$

(b) A short rod from which heat loss from its end is neglected.

$$\text{At the hot end:} \quad x = 0, \quad \theta = \theta_1 = C_1 + C_2$$

$$\text{At the cold end:} \quad x = L, \quad \frac{d\theta}{dx} = 0$$

Thus: $$0 = C_1 m e^{mL} - C_2 m e^{-mL}$$

and: $$\theta_1 = C_1 + C_1 e^{2mL}$$

Thus: $$C_1 = \frac{\theta_1}{1 + e^{2mL}}, \quad C_2 = \frac{\theta_1}{1 + e^{-2mL}}$$

Hence: $$\theta = \frac{\theta_1 e^{mx}}{1 + e^{2mL}} + \frac{\theta_1 e^{2mL} e^{-mx}}{1 + e^{2mL}}$$

$$= \frac{\theta_1}{1 + e^{2mL}}[e^{mx} + e^{2mL}e^{-mx}] \tag{9.249}$$

or: $$\frac{\theta}{\theta_1} = \frac{e^{-mL}e^{mx} + e^{mL}e^{-mx}}{e^{-mL} + e^{mL}}$$

This may be written: $$\frac{\theta}{\theta_1} = \frac{\cosh m(L - x)}{\cosh mL} \tag{9.250}$$

(c) More accurately, allowing for heat loss from the end:

$$\text{At the hot end:} \quad x = 0, \quad \theta = \theta_1 = C_1 + C_2$$

$$\text{At the cold end:} \quad x = L, \quad Q = hA\theta_{x=L} = -kA\left(\frac{d\theta}{dx}\right)_{x=L}$$

The determination of C_1 and C_2 in equation 9.247 then gives:

$$\theta = \frac{\theta_1}{1 + Je^{-2mL}}(Je^{-2mL}e^{mx} + e^{-mx}) \tag{9.251}$$

where: $$J = \frac{km - h}{km + h}$$

or, again, noting that $\cosh x = \frac{1}{2}(e^x + e^{-x})$ and $\sinh x = \frac{1}{2}(e^x - e^{-x})$:

$$\frac{\theta}{\theta_1} = \frac{\cosh m(L - x) + (h/mk)\sinh m(L - x)}{\cosh mL + (h/mk)\sinh mL} \tag{9.252}$$

The *heat loss* from a finned tube is obtained initially by determining the heat flow into the base of the fin from the tube surface. Thus the heat flow to the root of the fin is:

$$Q_f = -kA\left(\frac{dT}{dx}\right)_{x=0} = -kA\left(\frac{d\theta}{dx}\right)_{x=0}$$

For case (a):

$$Q_f = -kA(-m\theta_1) = kA\sqrt{\frac{hb}{kA}}\theta_1 = \sqrt{hbkA}\,\theta_1 \tag{9.253}$$

For case (b):

$$Q_f = -kAm\theta_1\frac{1-e^{2mL}}{1+e^{2mL}} = \sqrt{hbkA}\,\theta_1 \tanh mL \tag{9.254}$$

For case (c):

$$Q_f = \sqrt{hbkA}\,\theta_1\left(\frac{1-Je^{-2mL}}{1+Je^{-2mL}}\right) \tag{9.255}$$

These expressions are valid provided that the cross-section for heat flow remains constant. When it is not constant, as with a radial or tapered fin, for example, the temperature distribution is in the form of a *Bessel function*(126).

If the fin were such that there was no drop in temperature along its length, then the maximum rate of heat loss from the fin would be:

$$Q_{f\,\max} = bLh\theta_1$$

The *fin effectiveness* is then given by $Q_f/Q_{f\,\max}$ and, for case (b), this becomes:

$$\frac{kAm\theta_1 \tanh mL}{bLh\theta_1} = \frac{\tanh mL}{mL} \tag{9.256}$$

Example 9.31

In order to measure the temperature of a gas flowing through a copper pipe, a thermometer pocket is fitted perpendicularly through the pipe wall, the open end making very good contact with the pipe wall. The pocket is made of copper tube, 10 mm o.d. and 0.9 mm wall, and it projects 75 mm into the pipe. A thermocouple is welded to the bottom of the tube and this gives a reading of 475 K when the wall temperature is at 365 K. If the coefficient of heat transfer between the gas and the copper tube is 140 W/m^2 K, calculate the gas temperature. The thermal conductivity of copper may be taken as 350 W/m K. This arrangement is shown in Figure 9.88.

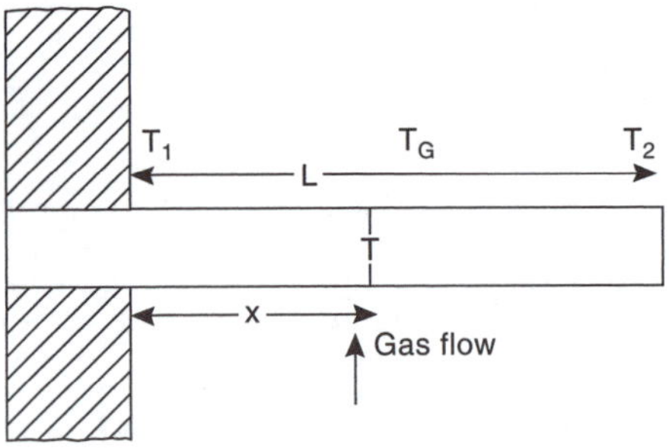

Figure 9.88. Heat transfer to thermometer pocket

Solution

If θ is the temperature difference $(T - T_G)$, then:

$$\theta = \theta_1\frac{\cosh m(L-x)}{\cosh mL}$$

At $x = L$: $$\theta = \frac{\theta_1}{\cosh mL}$$

$$m^2 = \frac{hb}{kA}$$

where the perimeter: $b = \pi \times 0.010$ m, tube i.d. = 8.2 mm or 0.0082 m

cross-sectional area of metal: $$A = \frac{\pi}{4}(10.0^2 - 8.2^2) = 8.19\ \pi\ \text{mm}^2 \quad \text{or} \quad 8.19\ \pi \times 10^{-6}\ \text{m}^2$$

$\therefore$ $$m^2 = \frac{(140 \times 0.010\pi)}{(350 \times 8.19\pi \times 10^{-6})} = 488\ \text{m}^{-2}$$

and: $$m = 22.1\ \text{m}^{-1}$$

$$\theta_1 = T_G - 365, \quad \theta_2 = T_G - 475$$

$$\frac{\theta_1}{\theta_2} = \cosh mL$$

$\therefore$ $$\frac{T_G - 365}{T_G - 475} = \cosh(22.1 \times 0.075) = 2.72$$

and: $$\underline{\underline{T_G = 539\ \text{K}}}$$

Example 9.32

A steel tube fitted with transverse circular steel fins of constant cross-section has the following specification:

tube o.d.: $d_2 = 54.0$ mm fin diameter $d_1 = 70.0$ mm

fin thickness: $w = 2.0$ mm number of fins/metre run = 230

Determine the heat loss per metre run of the tube when the surface temperature is 370 K and the temperature of the surroundings 280 K. The heat transfer coefficient between gas and fin is 30 W/m^2 K and the thermal conductivity of steel is 43 W/m K.

Solution

Assuming that the height of the fin is small compared with its circumference and that it may be treated as a straight fin of length $(\pi/2)(d_1 + d_2)$, then:

The perimeter: $$b = \frac{2\pi(d_1 + d_2)}{2} = \pi(d_1 + d_2)$$

The area: $$A = \frac{\pi(d_1 + d_2)w}{2},$$ i.e. the average area at right-angles to the heat flow

Then: $$m = \left(\frac{hb}{kA}\right)^{0.5} = \left\{\frac{h\pi(d_1 + d_2)}{[k\pi(d_1 + d_2)w/2]}\right\}^{0.5}$$

$$= \left(\frac{2h}{kw}\right)^{0.5}$$

$$= \left(\frac{2 \times 30}{43 \times 0.002}\right)^{0.5}$$

$$= 26.42\ \text{m}^{-1}$$

From equation 9.254, the heat flow is given for case (b) as:

$$Q_f = mkA\theta_1 \frac{e^{2mL} - 1}{1 + e^{2mL}}$$

In this equation: $$A = \frac{[\pi(70.0 + 54.0) \times 2.0]}{2} = 390 \text{ mm}^2 \quad \text{or} \quad 0.00039 \text{ m}^2$$

$$L = \frac{d_1 - d_2}{2} = 8.0 \text{ mm} \quad \text{or} \quad 0.008 \text{ m}$$

$$mL = 26.42 \times 0.008 = 0.211$$

$$\theta_1 = 370 - 280 = 90 \text{ deg K}$$

$$\therefore \quad Q_f = \frac{26.42 \times 43 \times 3.9 \times 10^{-4} \times 90(e^{0.422} - 1)}{1 + e^{0.422}}$$

$$= \frac{39.9 \times 0.525}{2.525} = 8.29 \text{ W per fin}$$

The heat loss per metre run of tube = 8.29 × 230 = 1907 W/m

or: 1.91 kW/m

In this case, the low value of mL indicates a fin efficiency of almost 1.0, though where mL tends to 1.0 the efficiency falls to about 0.8.

Practical data

A neat form of construction has been designed by the Brown Fintube Company of America. On both prongs of a hairpin tube are fitted horizontal fins which fit inside concentric tubes, joined at the base of the hairpin. Units of this form can be conveniently arranged in banks to give large heat transfer surfaces. It is usual for the extended surface to be at least five times greater than the inside surface, so that the low coefficient on the fin side is balanced by the increase in surface. An indication of the surface obtained is given in Table 9.20.

Table 9.20. Data on surface of finned tube units[(a)]

Pipe size outside diameter		Outside surface of pipe		Number	Surface of finned pipe (m^2/m) Height of fin		(ft^2/ft) Height of fin	
mm	(in)	(m^2/m)	(ft^2/ft length)	of fins	12.7 mm	25.4 mm	0.5 in	1 in
25.4	1	0.08	0.262	12	0.385	0.689	1.262	2.262
				16	0.486	0.893	1.595	2.929
				20	0.587	1.096	1.927	3.595
48.3	1.9	0.15	0.497	20	0.660	1.167	2.164	3.830
				24	0.761	1.371	2.497	4.497
				28	0.863	1.574	2.830	5.164
				36	1.066	1.980	3.498	6.497

(a) Brown Fintube Company.

A typical hairpin unit with an effective surface on the fin side of 9.4 m^2 has an overall length of 6.6 m, height of 0.34 m, and width of 0.2 m. The free area for flow on the fin side is 2645 mm^2 against 1320 mm^2 on the inside; the ratio of the transfer surface on the fin side to that inside the tubes is 5.93:1.

The fin side film coefficient h_f has been expressed by plotting:

$$\frac{h_f}{C_p G'}\left(\frac{C_p \mu}{k}\right)^{2/3}\left(\frac{\mu}{\mu_s}\right)^{-0.14} \quad \text{against} \quad \frac{d_e G'}{\mu}$$

where h_f is based on the total finside surface area (fin and tube), G' is the mass rate of flow per unit area, and d_e is the equivalent diameter, or:

$$d_e = \frac{4 \times \text{cross-sectional area for flow on fin side}}{\text{total wetted perimeter for flow (fin + outside of tube + inner surface of shell tube)}}$$

Experimental work has been carried out with exchangers in which the inside tube was 48 mm outside diameter and was fitted with 24, 28, or 36 fins (12.5 mm by 0.9 mm) in a 6.1 m length; the finned tubes were inserted inside tubes 90 mm inside diameter. With steam on the tube side, and tube oils and kerosene on the fin side, the experimental data were well correlated by plotting:

$$\frac{h_f}{C_p G'}\left(\frac{C_p \mu}{k}\right)^{2/3}\left(\frac{\mu}{\mu_s}\right)^{-0.14} \quad \text{against} \quad \frac{d_e G'}{\mu}$$

Typical values were:

$$\frac{h_f}{C_p G'}\left(\frac{C_p \mu}{k}\right)^{2/3}\left(\frac{\mu}{\mu_s}\right)^{-0.14} = 0.25 \quad 0.055 \quad 0.012 \quad 0.004$$

$$\frac{d_e G'}{\mu} = \quad 1 \quad 10 \quad 100 \quad 1000$$

Some indication of the performance obtained with *transverse finned tubes* is given in Table 9.21. The figures show the heat transferred per unit length of pipe when heating air on the fin side with steam or hot water on the tube side, using a temperature difference of 100 deg K. The results are given for three different spacings of the fins.

Table 9.21. Data on finned tubes

Inside diam. of tube Outside diam. of fin	19 mm 64 mm	25 mm 70 mm	38 mm 100 mm	50 mm 110 mm	75 mm 140 mm
No. of fins/m run	Heat transferred (kW/m)				
65	0.47	0.63	0.37	1.07	1.38
80	0.49	0.64	1.02	1.12	1.44
100	0.54	0.69	1.14	1.24	1.59
Inside diam. of tube	$\frac{3}{4}$ in	1 in	$1\frac{1}{2}$ in	2 in	3 in
Outside diam. of fin	$2\frac{1}{2}$ in	$2\frac{3}{4}$ in	$3\frac{7}{8}$ in	$4\frac{5}{16}$ in	$5\frac{1}{2}$ in
No. of fins/ft run	Heat transferred (Btu/h ft)				
20	485	650	1010	1115	1440
24	505	665	1060	1170	1495
30	565	720	1190	1295	1655

(Data taken from catalogue of G. A. Harvey and Co. Ltd. of London.)

9.10.2. Plate-type exchangers

A series of plate type heat exchangers which present some special features was first developed by the APV Company. The general construction is shown in Figure 9.89, which shows an Alfa-Laval exchanger and from which it is seen that the equipment consists of a series of parallel plates held firmly together between substantial head frames. The plates are one-piece pressings, frequently of stainless steel, and are spaced by rubber sealing gaskets cemented into a channel around the edge of each plate. Each plate has a number of troughs pressed out at right angles to the direction of flow and arranged so that they interlink with each other to form a channel of constantly changing direction and section. With normal construction the gap between the plates is 1.3–1.5 mm. Each liquid flows in alternate spaces and a large surface can be obtained in a small volume.

Because of the shape of the plates, the developed area of surface is appreciably greater than the projected area. This is shown in Table 9.22 for the four common sizes of plate.

Table 9.22. Plate areas

	Projected area		Developed area	
Plate type	m^2	ft^2	m^2	ft^2
HT	0.09	1.00	0.13	1.35
HX	0.13	1.45	0.17	1.81
HM	0.27	2.88	0.35	3.73
HF	0.36	3.85	0.43	4.60

A high degree of turbulence is obtained even at low flowrates and the high heat transfer coefficients obtained are illustrated by the data in Table 9.23. These refer to the heating of cold water by the equal flow of hot water in an HF type exchanger (aluminium or copper), at an average temperature of 310 K.

Table 9.23. Performance of plate-type exchanger type HF

Heat transferred per plate		Water flow		U based on developed area	
W/K	Btu/h °F	l/s	gal/h	kW/m^2 K	Btu/h ft^2 °F
1580	3000	0.700	550	3.70	650
2110	4000	1.075	850	4.94	870
2640	5000	1.580	1250	6.13	1080

(Courtesy of the APV Company.)

Using a stainless steel plate with a flow of 0.00114 m^3/s, the heat transferred is 1760 W/K for each plate.

The high transfer coefficient enables these exchangers to be operated with very small temperature differences, so that a high heat recovery is obtained. These units have been particularly successful in the dairy and brewing industries, where the low liquid capacity and the close control of temperature have been valuable features. A further advantage is that they are easily dismantled for inspection of the whole plate. The necessity for the long gasket is an inherent weakness, but the exchangers have been worked successfully

Figure 9.89. (*a*) Alfa-Laval plate heat exchanger (*b*) APV plate heat echanger

up to 423 K and at pressures of 930 kN/m^2. They are now being used in the processing and gas industries with solvents, sugar, acetic acid, ammoniacal liquor, and so on.

9.10.3. Spiral heat exchangers

A spiral plate exchanger is illustrated in Figure 9.90 in which two fluids flow through the channels formed between the spiral plates. With this form of construction the velocity may be as high as 2.1 m/s and overall transfer coefficients of 2.8 kW/m^2 K are frequently obtained. The size can therefore be kept relatively small and the cost becomes comparable or even less than that of shell and tube units, particularly when they are fabricated from alloy steels.

A further design of spiral heat exchanger, described by NEIL[(127)], is essentially a single pass counterflow heat exchanger with fixed tube plates distinguished by the spiral winding of the tubes, each consisting, typically of a 10 mm o.d. tube wound on to a 38 mm o.d. coil such that the inner heat transfer coefficient is 1.92 times greater than for a straight tube. The construction overcomes problems of differential expansion rates of the tubes and the shell and the characteristics of the design enable the unit to perform well with superheated steam where the combination of counterflow, high surface area per unit volume of shell and the high inside coefficient of heat transfer enables the superheat to be removed effectively in a unit of reasonable size and cost.

9.10.4. Compact heat exchangers

Advantages of compact units

In general, heat exchanger equipment accounts for some 10 per cent of the cost of a plant; a level at which there is no great incentive for innovation. Trends such as the growth in energy conservation schemes, the general move from bulk chemicals to value-added products and plant problems associated with drilling rigs have, however, all prompted the development and increased application of compact heat exchangers. Here compactness is a matter of degree, maximising the heat transfer area per unit volume of exchanger and this leads to narrow channels. Making shell and tube exchangers more compact presents construction problems and a more realistic approach is to use plate or plate and fin heat exchangers. The relation between various types of exchanger, in terms of the heat transfer area per unit volume of exchanger is shown in Figure 9.91, taken from REDMAN[(128)]. In order to obtain a thermal effectiveness in excess of 90 per cent, countercurrent flow is highly desirable and this is not easily achieved in shell and tube units which often have a number of tube-side passes in order to maintain reasonable velocities. Because of the baffle design on the shell-side, the flow at the best may be described as cross-flow, and the situation is only partly redeemed by having a train of exchangers which is an expensive solution. Again in dealing with high value added products, which could well be heat sensitive, a more controllable heat exchanger than a shell and tube unit, in which not all the fluid is heated to the same extent, might be desirable. One important application of compact heat exchangers is the cooling of natural gas on offshore rigs where space (costing as much as £120,000/m^2) is of paramount importance.

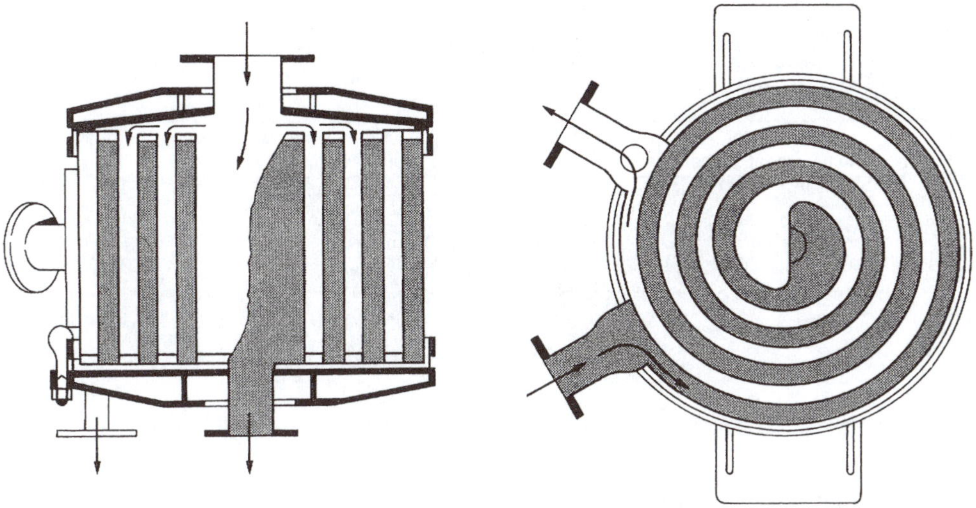

Figure 9.90. Spiral heat exchanger (*a*) Flow paths

Figure 9.90. Spiral plate exchanger (*b*) with cover removed

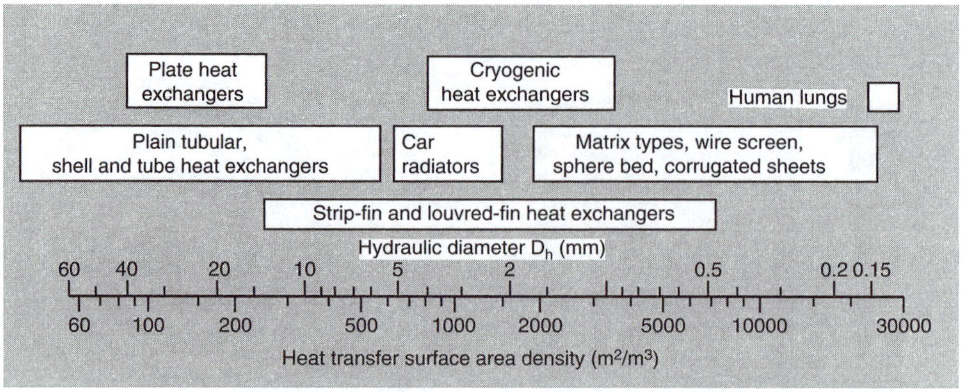

Figure 9.91. Surface area as a function of volume of exchanger for different types

Plate and fin exchangers

Plate and fin heat exchangers, used in the motor, aircraft and transport industries for many years, are finding increased application in the processing industries and in particular in natural gas liquefaction, cryogenic air separation, the production of olefins and in the separation of hydrogen and carbon monoxide. Potential applications include ammonia production, offshore processing, nuclear engineering and syngas production. As described by GREGORY(129), the concept is that of flat plates of metal, usually aluminium, with corrugated metal, which not only holds the plates together but also acts as a secondary surface for heat transfer. Bars at the edges of the plates retain each fluid between adjacent plates, and the space between each pair of plates, apportioned to each fluid according to the heat transfer and pressure drop requirements, is known as a layer. The heights of the bars and corrugations are standardised in the UK at 3.8, 6.35, and 8.9 mm and typical designs are shown in Figure 9.92.

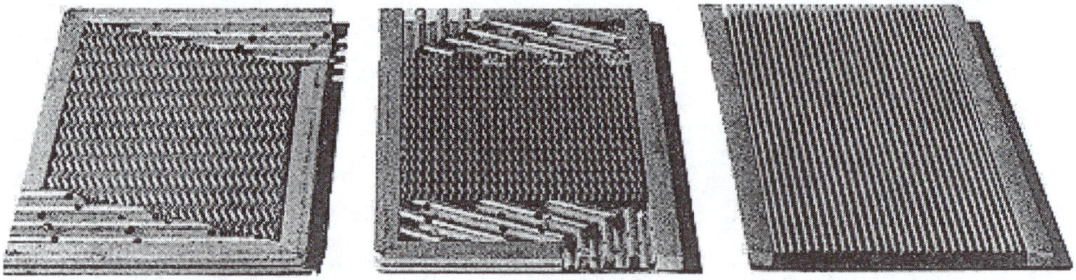

Figure 9.92. Plate and fin exchangers

There are four basic forms of corrugation: *plain* in which a sheet of metal is corrugated in the simplest way with fins at right angles to the plates; *serrated* where each cut is offset in relation to the preceding fin; *herringbone* corrugation made by displacing the fins sideways every 9.5 mm to give a zig-zag path; and *perforated* corrugation, a term used for a plain corrugation made from perforated material. Each stream to be heated or cooled can have different corrugation heights, different corrugation types, different numbers of layers and different entry and exit points including part length entry and exit. Because the surface in the hot fluid (or fluids) can vary widely from that in the cold fluid (or fluids) it is unrealistic to quote surface areas for either side, as in shell and tube units, though the overall surface area available to all the fluids is 1000 m^2/m^3 of exchanger.

In design, the general approach is to obtain the term (hA) for each stream, sum these for all the cold and all the hot streams and determine an overall value of (hA) given by:

$$\frac{1}{(hA)_{ov}} = \frac{1}{(hA)_w} + \frac{1}{(hA)_c} \tag{9.257}$$

where ov, w and c refer, respectively, to overall, hot-side and cold-side values.

Printed-circuit exchangers

Various devices such as small diameter tubes, fins, tube inserts and porous boiling surfaces may be used to improve the surface density and heat transfer coefficients in shell and tube heat exchangers and yet these are not universally applicable and, in general, such units remain essentially bulky. Plate and fin exchangers have either limited fluid compatibility or limited fin efficiency, problems which are overcome by using printed-circuit exchangers as described by JOHNSTON(130). These are constructed from flat metal plates which have fluid flow passages chemically milled into them by means of much the same techniques as are used to produce electrical printed circuits. These plates are diffusion-bonded together to form blocks riddled with precisely sized, shaped, routed and positioned passages, and these blocks are in turn welded together to form heat exchange cores of the required capacity. Fluid headers are attached to the core faces and sometimes the assembly is encapsulated. Passages are typically 0.3–1.5 mm deep giving surface areas of 1000–5000 m^2/m^3, an order of magnitude higher than surface densities in shell and tube designs; and in addition the fine passages tend to sustain relatively high heat transfer coefficients, undiminished by fin inefficiencies, and so less surface is required.

In designing a unit, each side of the exchanger is independently tailored to the duty required, and the exchanger effectiveness (discussed in Section 9.9.4) can range from 2–5 per cent to values in excess of 98 per cent without fundamental design or construction problems arising. Countercurrent, co-current and cross-flow contacting can be employed individually or in combination.

A note of caution on the use of photo-etched channels has been offered by RAMSHAW(131) who points out that the system is attractive in principle provided that severe practical problems such as fouling are not encountered. With laminar flow in matrices with a mean plate spacing of 0.3–1 mm, volumetric heat transfer coefficients of 7 MW/m^3 K have been obtained with modest pressure drops. Such values compare with 0.2 MW/m^3 K for shell and tube exchangers and 1.2 MW/m^3 K for plate heat exchangers.

9.10.5. Scraped-surface heat exchangers

In cases where a process fluid is likely to crystallise on cooling or the degree of fouling is very high or indeed the fluid is of very high viscosity, use is often made of scraped-surface heat exchangers in which a rotating element has spring-loaded scraper blades which wipe the inside surface of a tube which may typically be 0.15 m in diameter. Double-pipe construction is often employed with a jacket, say 0.20 m in diameter, and one common arrangement is to connect several sections in series or to install several pipes within a common shell. Scraped-surface units of this type are used in paraffin-wax plants and for evaporating viscous or heat-sensitive materials under high vacuum. This is an application to which the *thin-film device* is especially suited because of the very short residence times involved. In such a device the clearance between the agitator and the wall may be either fixed or variable since both rigid and hinged blades may be used. The process liquid is continuously spread in a thin layer over the vessel wall and it moves through the device either by the action of gravity or that of the agitator or of both. A tapered or helical agitator produces longitudinal forces on the liquid.

In describing chillers for the production of wax distillates, NELSON[132] points out that the rate of cooling depends very much on the effectiveness of the scrapers, and quotes overall coefficients of heat transfer ranging from 15 W/m^2 K with a poorly fitting scraper to 90 W/m^2 K where close fitting scrapers remove the wax effectively from the chilled surface.

The *Votator* design has two or more floating scraper-agitators which are forced against the cylinder wall by the hydrodynamic action of the fluid on the agitator and by centrifugal action; the blades are loosely attached to a central shaft called the mutator. The votator is used extensively in the food processing industries and also in the manufacture of greases and detergents. As the blades are free to move, the clearance between the blades and the wall varies with operating conditions and a typical installation may be 75–100 mm in diameter and 0.6–1.2 m long. In the *spring-loaded* type of scraped-surface heat exchanger, the scrapers are held against the wall by leaf springs, and again there is a variable clearance between the agitator and the cylinder wall since the spring force is balanced by the radial hydrodynamic force of the liquid on the scraper. Typical applications of this device are the processing of heavy waxes and oils and crystallising solutions. Generally the units are 0.15–0.3 m in diameter and up to 12 m long. Some of the more specialised heat exchangers and chemical reactors employ helical ribbons, augers or twisted tapes as agitators and, in general, these are fixed clearance devices used for high viscosity materials. There is no general rule as to maximum or minimum dimensions since each application is a special case.

One of the earliest investigations into the effectiveness of scrapers for improving heat transfer was that of HUGGINS[133] who found that although the improvement with water was negligible, cooling times for more viscous materials could be considerably reduced. This was confirmed by LAUGHLIN[134] who has presented operating data on a system where the process fluid changes from a thin liquid to a paste and finally to a powder. HOULTON[135], making tests on the votator, found that back-mixing was negligible and some useful data on a number of food products have been obtained by BOLANOWSKI and LINEBERRY[136] who, in addition to discussing the operation and uses of the votator, quote overall heat transfer coefficients for each food tested. Using a *liquid-full* system, SKELLAND *et al.*[137–139] have proposed the following general design correlation for the votator:

$$\frac{hd_v}{k} = c_1 \left(\frac{C_p\mu}{k}\right)^{c_2} \left(\frac{(d_v - d_r)u\rho}{\mu}\right) \left(\frac{d_vN}{u}\right)^{0.82} \left(\frac{d_r}{d_v}\right)^{0.55} (n_B)^{0.53} \tag{9.258}$$

where for cooling viscous liquids $c_1 = 0.014$ and $c_2 = 0.96$, and for thin mobile liquids $c_1 = 0.039$ and $c_2 = 0.70$. In this correlation d_v is the diameter of the vessel, d_r is the diameter of the rotor and u is the average axial velocity of the liquid. This correlation may only be applied to the range of experimental data upon which it is based, since h will not approach zero as n_B, d_r, $(d_v - d_r)$, N and u approach zero. Reference to the use of the votator for crystallisation is made in Volume 2, Chapter 15.

The majority of work on heat transfer in *thin-film* systems has been directed towards obtaining data on specific systems rather than developing general design methods, although BOTT *et al.*[(140–142)] have developed the following correlations for heating without change of phase:

$$\frac{hd_v}{k} = Nu = 0.018Re''^{0.6}Re^{0.46}Pr^{0.87}\left(\frac{d_v}{l}\right)^{0.48}(n_B)^{0.24} \tag{9.259}$$

and for evaporation:

$$\frac{hd_v}{k} = Nu = 0.65Re''^{0.43}Re^{0.25}Pr^{0.3}(n_B)^{0.33} \tag{9.260}$$

From both of these equations, it will be noted that the heat transfer coefficient is *not* a function of the temperature difference. Here $Re'' = (d_v^2 N\rho/\mu)$ and $Re = (ud_v\rho/\mu)$, where d_v is the tube diameter and u is the average velocity of the liquid in the film in the axial direction.

It is also of significance that the agitation suppresses nucleation in a fluid which might otherwise deposit crystals.

9.11. THERMAL INSULATION

9.11.1. Heat losses through lagging

A hot reaction or storage vessel or a steam pipe will lose heat to the atmosphere by radiation, conduction, and convection. The loss by radiation is a function of the fourth power of the absolute temperatures of the body and surroundings, and will be small for low temperature differences but will increase rapidly as the temperature difference increases. Air is a very poor conductor, and the heat loss by conduction will therefore be small except possibly through the supporting structure. On the other hand, since convection currents form very easily, the heat loss from an unlagged surface is considerable. The conservation of heat, and hence usually of total energy, is an economic necessity, and some form of lagging should normally be applied to hot surfaces. Lagging of plant operating at high temperatures is also necessary in order to achieve acceptable working conditions in the vicinity. In furnaces, as has already been seen, the surface temperature is reduced substantially by using a series of insulating bricks which are poor conductors.

The two main requirements of a good lagging material are that it should have a low thermal conductivity and that it should suppress convection currents. The materials that are frequently used are cork, 85 per cent magnesia, glass wool, and vermiculite. Cork is a very good insulator though it becomes charred at moderate temperatures and is used mainly in refrigerating plants. Eighty-five per cent magnesia is widely used for lagging steam pipes and may be applied either as a hot plastic material or in preformed sections. The preformed sections are quickly fitted and can frequently be dismantled and re-used whereas the plastic material must be applied to a hot surface and cannot be re-used. Thin metal sheeting is often used to protect the lagging.

The rate of heat loss per unit area is given by:

$$\frac{\text{total temperature difference}}{\text{total thermal resistance}}$$

For the case of heat loss to the atmosphere from a lagged steam pipe, the thermal resistance is due to that of the condensate film and dirt on the inside of the pipe, that of the pipe wall, that of the lagging, and that of the air film outside the lagging. Thus for unit length of a lagged pipe:

$$\frac{Q}{l} = \Sigma\Delta T \bigg/ \left[\frac{1}{h_i \pi d} + \frac{x_w}{k_w \pi d_w} + \frac{x_l}{k_r \pi d_m} + \frac{1}{(h_r + h_c)\pi d_s}\right] \tag{9.261}$$

where d is the inside diameter of pipe, d_w the mean diameter of pipe wall, d_m the logarithmic mean diameter of lagging, d_s the outside diameter of lagging, x_w, x_l are the pipe wall and lagging thickness respectively, k_w, k_r the thermal conductivity of the pipe wall and lagging, and h_i, h_r, h_c the inside film, radiation, and convection coefficients.

Example 9.33

A steam pipe, 150 mm i.d. and 168 mm o.d., is carrying steam at 444 K and is lagged with 50 mm of 85 per cent magnesia. What is the heat loss to air at 294 K?

Solution

In this case:

$$d = 150 \text{ mm} \quad \text{or} \quad 0.150 \text{ m}$$

$$d_o = 168 \text{ mm} \quad \text{or} \quad 0.168 \text{ m}$$

$$d_w = 159 \text{ mm} \quad \text{or} \quad 0.159 \text{ m}$$

$$d_s = 268 \text{ mm} \quad \text{or} \quad 0.268 \text{ m}$$

$$d_m, \text{ the log mean of } d_o \text{ and } d_s = 215 \text{ mm} \quad \text{or} \quad 0.215 \text{ m.}$$

The coefficient for condensing steam together with that for any scale will be taken as 8500 W/m^2 K, k_w as 45 W/m K, and k_l as 0.073 W/m K.

The temperature on the outside of the lagging is estimated at 314 K and $(h_r + h_c)$ will be taken as 10 W/m^2 K.

The thermal resistances are therefore:

$$\frac{1}{h_i \pi d} = \frac{1}{8500 \times \pi \times 0.150} = 0.00025$$

$$\frac{x_w}{k_w \pi d_w} = \frac{0.009}{45 \times \pi \times 0.159} = 0.00040$$

$$\frac{x_l}{k_l \pi d_m} = \frac{0.050}{0.073 \times \pi \times 0.215} = 1.013$$

$$\frac{1}{(h_r + h_c)\pi d_s} = \frac{1}{10 \times \pi \times 0.268} = 0.119$$

The first two terms may be neglected and hence the total thermal resistance is 1.132 m K/W.

The heat loss per metre length = (444 − 294)/1.132 = 132.5 W/m (from equation 9.261).

The temperature on the outside of the lagging may now be checked as follows:

$$\frac{\Delta T(\text{lagging})}{\Sigma \Delta T} = \frac{1.013}{1.132} = 0.895$$

$$\Delta T(\text{lagging}) = 0.895(444 - 294) = 134 \text{ deg K}$$

Thus the temperature on the outside of the lagging is (444 − 134) = 310 K, which approximates to the assumed value.

Taking an emissivity of 0.9, from equation 9.119:

$$h_r = \frac{[0.9 \times 5.67 \times 10^{-8}(310^4 - 294^4)]}{(310 - 294)} = 7.40 \text{ W/m}^2 \text{ K}$$

From Table 9.5 for air ($Gr\,Pr = 10^4 - 10^9$), $n = 0.25$ and $C'' = 1.32$.

Substituting in equation 9.105 (putting l = diameter = 0.268 m):

and:

$$h_c = C''(\Delta T)^n l^{3n-1} = 1.32\left[\frac{310 - 294}{0.268}\right]^{0.25} = 3.67 \text{ W/m}^2 \text{ K}$$

Thus $(h_r + h_c) = 11.1$ W/m^2 K, which is close to the assumed value. In practice it is rare for forced convection currents to be absent, and the heat loss is probably higher than this value.

If the pipe were unlagged, $(h_r + h_c)$ for $\Delta T = 150$ K would be about 20 W/m^2 K and the heat loss would then be:

$$\frac{Q}{l} = (h_r + h_c)\pi d_o \Delta T$$

$$= (20 \times \pi \times 0.168 \times 150) = 1584 \text{ W/m}$$

or: 1.58 kW/m

Under these conditions it is seen that the heat loss has been reduced by more than 90 per cent by the addition of a 50 mm thickness of lagging.

9.11.2. Economic thickness of lagging

Increasing the thickness of the lagging will reduce the loss of heat and thus give a saving in the operating costs. The cost of the lagging will increase with thickness and there will be an optimum thickness when further increase does not save sufficient heat to justify the cost. In general the smaller the pipe the smaller the thickness used, though it cannot be too strongly stressed that some lagging everywhere is better than excellent lagging in some places and none in others. For temperatures of 373–423 K, and for pipes up to 150 mm diameter, LYLE[143] recommends a 25 mm thickness of 85 per cent magnesia lagging and 50 mm for pipes over 230 mm diameter. With temperatures of 470–520 K 38 mm is suggested for pipes less than 75 mm diameter and 50 mm for pipes up to 230 mm diameter.

9.11.3. Critical thickness of lagging

As the thickness of the lagging is increased, resistance to heat transfer by thermal conduction increases. Simultaneously the outside area from which heat is lost to the surroundings also increases, giving rise to the possibility of increased heat loss. It is perhaps easiest to think of the lagging as acting as a fin of very low thermal conductivity. For a cylindrical pipe, there is the possibility of heat losses being increased by the application of lagging, only if $hr/k < 1$, where k is the thermal conductivity of the lagging, h is the outside film coefficient, and r is the outside diameter of the pipe. In practice, this situation is likely to arise only for pipes of small diameters.

The heat loss from a pipe at a temperature T_S to surroundings at temperature T_A is considered. Heat flows through lagging of thickness x across which the temperature falls from a constant value T_S at its inner radius r, to an outside temperature T_L which is a function of x, as shown in Figure 9.93. The rate of heat loss Q from a length l of pipe is given by equation 9.262, by considering the heat loss from the outside of the lagging; and by equations 9.263 and 9.264, which give the transfer rate by thermal conduction through the lagging of logarithmic mean radius r_m:

$$Q = 2\pi l(r + x)(T_L - T_A)h \tag{9.262}$$

$$Q = \frac{2\pi l r_m}{x} k(T_S - T_L) \tag{9.263}$$

$$= \frac{2\pi l k}{\ln \dfrac{r + x}{r}}(T_S - T_L) \tag{9.264}$$

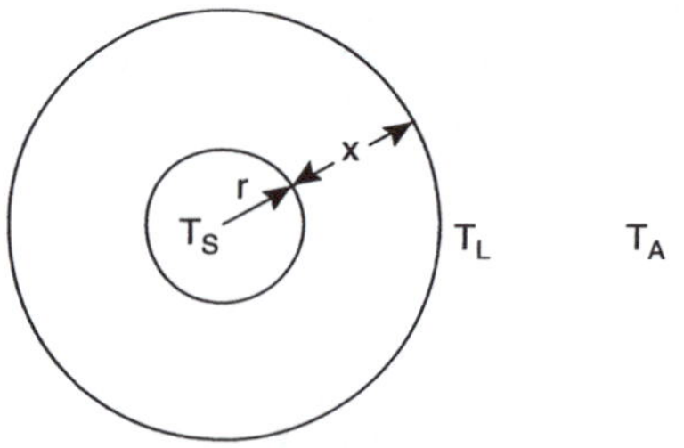

Figure 9.93. Critical lagging thickness

Equating the values given in equations 9.262 and 9.264:

$$(r + x)h(T_L - T_A) = \frac{k}{\ln \dfrac{r + x}{r}}(T_S - T_L)$$

Then:
$$a = (r + x)\frac{h}{k} \ln \frac{r + x}{r} = \frac{T_S - T_L}{T_L - T_A}$$

$$aT_L - aT_A = T_S - T_L$$

$$T_L = \frac{T_S + aT_A}{a + 1}$$

Substituting in equation 9.262:

$$Q = 2\pi l(r + x)\left\{\left(\frac{T_S + aT_A}{a + 1} - T_A\right)\right\}h$$

$$= \frac{2\pi lh(r+x)}{a+1}(T_S - T_A)$$

$$= 2\pi lh(T_S - T_A)\left\{(r+x)\frac{1}{1+(r+x)\frac{h}{k}\ln\left(\frac{r+x}{r}\right)}\right\} \tag{9.265}$$

Differentiating with respect to x:

$$\frac{1}{2\pi lh(T_S - T_A)}\frac{dQ}{dx} = \frac{\left\{1+(r+x)\frac{h}{k}\ln\frac{r+x}{r} - (r+x)\left[\frac{h}{k}\ln\frac{r+x}{r} + (r+x)\frac{h}{k}\frac{r}{(r+x)}\frac{1}{r}\right]\right\}}{\left[1+(r+x)\frac{h}{k}\ln\left(\frac{r+x}{r}\right)\right]^2}$$

The maximum value of $Q(Q_{\max})$ occurs when $dQ/dx = 0$.

that is, when: $$1 - (r+x)\frac{h}{k} = 0$$

or: $$x = \frac{k}{h} - r \tag{9.266}$$

When the relation between heat loss and lagging thickness exhibits a maximum for the unlagged pipe ($x = 0$), then:

$$\frac{hr}{k} = 1 \tag{9.267}$$

When $hr/k > 1$, the addition of lagging always reduces the heat loss.

When $hr/k < 1$, thin layers of lagging increase the heat loss and it is necessary to exceed the critical thickness given by equation 9.266 before any benefit is obtained from the lagging.

Substituting in equation 9.265 gives the maximum heat loss as:

$$Q_{\max} = 2\pi lh(T_S - T_A)\left\{\frac{k}{h}\frac{1}{1+\frac{k}{h}\frac{h}{k}\ln\frac{k}{hr}}\right\}$$

$$= 2\pi l(T_S - T_A)k\frac{1}{1+\ln\frac{k}{hr}} \tag{9.268}$$

For an unlagged pipe, $x = 0$ and $T_L = T_S$. Substitution in equation 9.262 gives the rate of heat loss Q_o as:

$$Q_o = 2\pi lr(T_S - T_A)h \tag{9.269}$$

Thus: $$\frac{Q_{\max}}{Q_o} = \frac{k}{rh}\bigg/\left(1+\ln\frac{k}{rh}\right) \tag{9.270}$$

The ratio Q/Q_o is plotted as a function of thickness of lagging (x) in Figure 9.94.

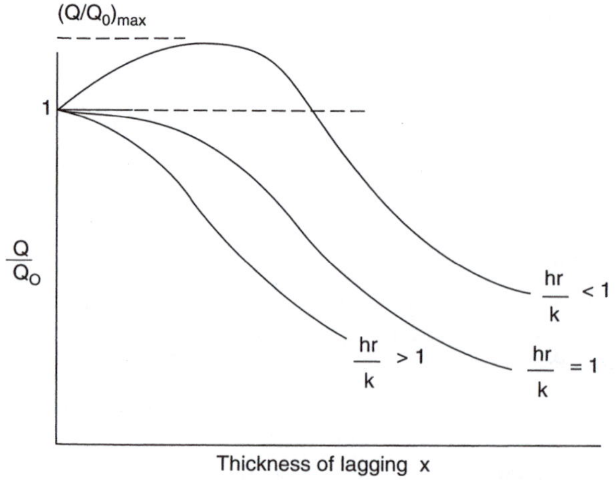

Figure 9.94. Critical thickness of lagging

Example 9.34

A pipeline of 100 mm outside diameter, carrying steam at 420 K, is to be insulated with a lagging material which costs £10/m^3 and which has a thermal conductivity of 0.1 W/m K. The ambient temperature may be taken as 285 K, and the coefficient of heat transfer from the outside of the lagging to the surroundings as 10 W/m^2 K. If the value of heat energy is 7.5×10^{-4} £/MJ and the capital cost of the lagging is to be depreciated over 5 years with an effective simple interest rate of 10 per cent per annum based on the initial investment, what is the economic thickness of the lagging?

Is there any possibility that the heat loss could actually be increased by the application of too thin a layer of lagging?

Solution

For a thick-walled cylinder, the rate of conduction of heat through lagging is given by equation 9.21:

$$Q = \frac{2\pi l k(T_i - T_o)}{\ln(d_o/d_i)} \quad \text{W}$$

where d_o and d_i are the external and internal diameters of the lagging and T_o and T_i the corresponding temperatures.

Substituting $k = 0.1$ W/mK, $T_o = 420$K (neglecting temperature drop across pipe wall), and $d_i = 0.1$ m, then:

$$\frac{Q}{l} = \frac{2\pi \times 0.1(420 - T_o)}{\ln(d_o/0.1)} \quad \text{W/m}$$

The term Q/l must also equal the heat loss from the outside of the lagging,

or: $$\frac{Q}{l} = h_o(T_o - 285)\pi d_o = 10(T_o - 285)\pi d_o \text{ W/m}$$

Thus: $$T_o = \left\{\frac{Q}{l}\frac{1}{10\pi d_o} + 285\right\} \quad \text{K}$$

Substituting: $$\frac{Q}{l} = \frac{2\pi \times 0.1\left[135 - \frac{Q}{l}\frac{1}{10\pi d_o}\right]}{\ln(d_o/0.1)}$$

or: $$\frac{Q}{l} = \frac{2\pi \times 0.1 \times 135}{\ln(d_o/0.1) + 2\pi \times 0.1 \times \dfrac{1}{10\pi d_o}} = \frac{84.82}{\ln(d_o/0.1) + (0.02/d_o)} \text{ W/m}$$

Value of heat lost = £7.5 × 10^{-4}/MJ

or: $$\frac{84.82}{\ln(d_o/0.1) + (0.02/d_o)} \times 7.5 \times 10^{-4} \times 10^{-6} = \frac{6.36 \times 10^{-8}}{\ln(d_o/2.1) + (0.02/d_o)} \text{ £/m s}$$

Volume of lagging per unit pipe length $= \dfrac{\pi}{4}[d_o^2 - (0.1)^2]$ m^3/m

Capital cost of lagging = £10/m^3 or $\dfrac{\pi}{4}[d_o^2 - 0.01]10 = 7.85(d_o^2 - 0.01)$ £/m

Noting that 1 year = 31.5 M s, then :

Depreciation $= 7.85[d_o^2 - 0.01]/(5 \times 31.5 \times 10^6) = 4.98 \times 10^{-8}(d_o^2 - 0.01)$ £/Ms

Interest charges $= (0.1 \times 7.85)(d_o^2 - 0.01)/(31.5 \times 10^6) = 2.49 \times 10^{-8}(d_o^2 - 0.01)$ £/Ms

Total capital charges $= 7.47 \times 10^{-8}(d_o^2 - 0.01)$ £/Ms

Total cost (capital charges + value of heat lost) is given by:

$$C = \left\{\frac{6.36}{\ln(d_o/0.1) + (0.02/d_o)} + 7.47(d_o^2 - 0.01)\right\} 10^{-8} \text{ £/Ms}$$

Differentiating with respect to d_o:

$$10^8 \frac{dC}{dd_o} = 6.36\left[\frac{-1}{[\ln(d_o/0.1) + (0.02/d_o)]^2}\right]\left[\frac{1}{d_o} - \frac{0.02}{d_o^2}\right] + 7.47(2d_o)$$

In order to obtain the minimum value of C, dC/dd_o must be put equal to zero.

Then: $$\frac{1}{[\ln(d_o/0.1) + (0.02/d_o)]^2} = \frac{(7.47 \times 2)}{6.36}\left[\frac{d_o}{(1/d_o) - (0.02/d_o^2)}\right]$$

that is: $$\frac{1}{[\ln(d_o/0.1) + (0.02/d_o)]^2} = 2.35\frac{d_o^3}{(d_o - 0.02)}$$

A trial and error solution gives $d_o = 0.426$m or 426 mm
Thus, the economic thickness of lagging = (426 − 100)/2 = 163 mm

For this pipeline: $$\frac{hr}{k} = \frac{10 \times (50 \times 10^{-3})}{0.1} = 5$$

From equation 9.267, the critical value of hr/k, below which the heat loss may be increased by a thin layer of lagging, is 1. For $hr/k > 1$, as in this problem, the situation will not arise.

9.12. FURTHER READING

ANDERSON, E. E.: *Solar Energy Fundamentals for Designers and Engineers* (Addison-Wesley, Reading, Mass., 1982)
AZBEL, D.: *Heat Transfer Applications in Process Engineering* (Noyes, New York, 1984).
BACKHURST, J. R. and HARKER, J. H.: *Process Plant Design* (Heinemann, London, 1973).
BIRD, R. B., STEWART, W. E., and LIGHTFOOT, E. N.: *Transport Phenomena* (Wiley, New York, 1960).
CHAPMAN, A. J.: *Heat Transfer*, 2nd edn (Macmillan, New York, 1967).
CHEREMISINOFF, N. P. (ed.): *Handbook of Heat and Mass Transfer*: Vol. 1, *Heat Transfer Operations*. (Gulf Publications, 1986).
COLLIER, J. G.: *Convective Boiling and Condensation* (McGraw-Hill, New York, 1972).
ECKERT, E. R. G. and DRAKE, R. M., Jr.: *Analysis of Heat and Mass Transfer* (McGraw-Hill, New York, 1972).

EDWARDS, D. K.: *Radiation Heat Transfer Notes* (Hemisphere Publishing, New York, 1981)
GEBHART, B.: *Heat Transfer*, 2nd edn (McGraw-Hill, New York, 1971).
GROBER, H., ERK, E., and GRIGULL, U.: *Fundamentals of Heat Transfer* (McGraw-Hill, New York, 1961).
HALLSTRÖM, B., SKJÖLDEBRAND, C. and TRÄCÄRDH, C.: *Heat Transfer and Food Products* (Elsevier Applied Science, London, 1988).
HEWITT, G. F. (Exec. ed.) *Heat Exchanger Design Handbook* (HEDH) (Begell House Publishers, 1998).
HEWITT, G. F., SHIRES, G. L., and BOTT, T. R.: *Process Heat Transfer* (CRC Press, 1994).
HOTTEL, H. C. and SAROFIM, A. F.: *Radiative Transfer* (McGraw-Hill, New York, 1967)
HOWELL, J. R.: *A Catalog of Radiation Configuration Factors* (McGraw-Hill, New York, 1982)
INCROPERA, F. P. and DE WITT, D. P.: *Introduction to Heat Transfer*, 4th edn (Wiley, New York, 1996).
INCROPERA, F. P. and DE WITT, D. P.: *Fundamentals of Heat and Mass Transfer* (Wiley, New York, 1985).
JAKOB, M.: *Heat Transfer*, Vol. 1 (Wiley, New York, 1949).
JAKOB, M.: *Heat Transfer*, Vol. II (Wiley, New York, 1957).
KAKAC, S., SHAH, R. K. and AUNG, W.: *Handbook of Single-phase Convective Heat Transfer* (Wiley, New York, 1987).
KAYS, W. M.: *Convective Heat and Mass Transfer* (McGraw-Hill, New York, 1966).
KAYS, W. M. and LONDON, A. L.: *Compact Heat Exchangers*, 2nd edn (McGraw-Hill, New York, 1964).
KERN, D. Q.: *Process Heat Transfer* (McGraw-Hill, New York, 1950).
KREITH, F.: *Principles of Heat Transfer* (Feffer & Simons, Scranton, Penn. 1965).
MCADAMS, W. H.: *Heat Transmission*, 3rd edn (McGraw-Hill, New York, 1954).
MINKOWYCZ, W.J.: *Handbook of Numerical Heat Transfer* (Wiley, New York, 1988).
MINTON, P.E.: *Handbook of Evaporation Technology* (Noyes, New York, 1986).
OZISIK, M.N.: *Boundary Value Problems of Heat Conduction* (Int. Textbook Co., Stanton, Penn., 1965).
PLANCK, M.: *The Theory of Heat Radiation* (Dover Publications, New York, 1959).
ROHSENHOW, W.M., HARTNETT, J.P. and CHO, Y. I., (eds): *Handbook of Heat Transfer Fundamentals* 3rd edn. (McGraw-Hill, New York, 1998).
SCHACK, A.: *Industrial Heat Transfer* (Chapman & Hall, London, 1965).
SMITH, R.A.: *Vaporisers: Selection, Design and Operation* (Longman, London, 1987).
SIEGEL, R. and HOWELL, J. R.: *Thermal Radiation Heat Transfer*, 2nd edn (McGraw-Hill, New York, 1981)
SPARROW, E. M. and CESS, R. D.: *Radiation Heat Transfer* (Hemisphere Publishing, New York, 1978)
TAYLOR, M. (ed.): *Plate-fin Heat Exchangers: Guide to their Specification and Use* (HTFS, Harwell, 1987).
TOULOUKIAN, Y. S.: *Thermophysical Properties of High Temperature Solid Materials* (Macmillan, New York, 1967)
WOOD, W.D., DEEM, H. W. and LUCKS, C. F.: *Thermal Radiation Properties* (Plenum Press, New York, 1964)

9.13. REFERENCES

1. UNDERWOOD, A.J.V.: *Industrial Chemist*, **9** (1933) 167. Graphical computation of logarithmic mean temperature difference.
2. SPALDING, D.B.: *Combustion and Mass Transfer* (Pergamon Press, Oxford, 1979).
3. LONG, V.D.: *J. Inst. Fuel.* **37** (1964) 522. A simple model of droplet combustion.
4. MONAGHAN, M.T., SIDDAL, R.G., and THRING, M.W.: *Comb and Flame* **17** (1968) 45. The influence of initial diameter on the combustion of single drops of liquid fuel.
5. CARSLAW, H.S. and JAEGER, J.C.: *Conduction of Heat in Solids.* 2nd edn (Oxford University Press, Oxford, 1959).
6. RICHARDSON, J.F.: *Fuel* **28** (1949) 265. Spread of fire by thermal conduction.
7. SCHMIDT, E.: in *Beitrage zur technischen Physik* (*Föppl Festschrift*) (Springer-Verlag, Berlin, 1924), 179–178;
8. GALLAGHER, R.H.: *Finite Element Analysis. Fundamentals* (Prentice Hall, Englewood Cliffs, NJ, 1975).
9. HEISLER, M. P.: *Trans ASME* **69** (1947) 227. Temperature charts for induction and constant temperature heating.
10. WALDIE, B.: *The Chemical Engineer (London)* No. 261 (1972) 188. Review of recent work on the processing of powders in high temperature plasmas: Pt. II. Particle dynamics, heat transfer and mass transfer.
11. TAYLOR, T.D.: *Physics of Fluids* **6** (1963) 987. Heat transfer from single spheres in a low Reynolds number slip flow.
12. FIELD, M.A., GILL, D.W., MORGAN, B.B., and HAWKSLEY, P.G.W.: *Combustion of Pulverised Coal* (BCURA, Leatherhead, 1967).

13. DITTUS, F.W. and BOELTER, L.M.K.: *Univ. of California, Berkeley, Pubns in Engineering* **2** (1930) 443. Heat transfer in automobile radiators of the tubular type. Reprinted: *Intl Comm. Heat Mass Transfer* **12** (1985) 3.
14. MCADAMS, W.H.: *Heat Transmission*, 2nd edn (McGraw-Hill, New York, 1942).
15. WINTERTON, R.H.S.: *Intl. Jl. Heat Mass Transfer* **41** (1998) 809. Where did the Dittus and Boelter equation come from?
16. COLBURN, A.P.: *Trans. Am. Inst. Chem. Eng.* **29** (1933) 174. A method of correlating forced convection heat transfer data and a comparison with fluid friction.
17. SIEDER, E.N. and TATE, G.E.: *Ind. Eng. Chem.* **28** (1936) 1429. Heat transfer and pressure drop of liquids in tubes.
18. FISHENDEN, M. and SAUNDERS, O.A.: *An Introduction to Heat Transfer* (Oxford University Press, Oxford, 1950).
19. COPE, W.F.: *Proc. Inst. Mech. Engrs.* **45** (1941) 99. The friction and heat transmission coefficients of rough pipes.
20. REIHER, M.: *Mitt. Forsch.* **269** (1925) 1. Wärmeübergang von strömender Luft an Röhren und Röhrenbündeln im Kreuzstrom.
21. HILPERT, R.: *Forsch. Geb. IngWes.* **4** (1933) 215. Wärmeabgabe von geheizten Drähten und Röhren.
22. GRIFFITHS, E. and AWBERY, J.H.: *Proc. Inst. Mech. Eng.* **125** (1933) 319. Heat transfer between metal pipes and a stream of air.
23. DAVIS, A.H.: *Phil. Mag.* **47** (1924) 1057. Convective cooling of wires in streams of viscous liquids.
24. GRIMISON, E.D.: *Trans. Am. Soc. Mech. Eng.* **59** (1937) 583: and *ibid.* **60** (1938) 381. Correlation and utilization of new data on flow resistance and heat transfer for cross flow of gases over tube banks.
25. HUGE, E.C.: *Trans. Am. Soc. Mech. Eng.* **59** (1937) 573. Experimental investigation of effects of equipment size on convection heat transfer and flow resistance in cross flow of gases over tube banks.
26. PIERSON, O.L.: *Trans. Am. Soc. Mech. Eng.* **59** (1937) 563. Experimental investigation of influence of tube arrangement on convection heat transfer and flow resistance in cross flow of gases over tube banks.
27. MCADAMS, W.H.: *Heat Transmission,* 3rd edn (McGraw-Hill, New York, 1954).
28. KERN, D.Q.: *Process Heat Transfer* (McGraw-Hill, New York, 1950).
29. SHORT, B.E.: *Univ. of Texas Pub.* No. 4324 (1943). Heat transfer and pressure drop in heat exchangers.
30. DONOHUE, D.A.: *Ind. Eng. Chem.* **41** (1949) 2499. Heat transfer and pressure drop in heat exchangers.
31. TINKER, T.: *Proceedings of the General Discussion on Heat Transfer, September,* 1951, p. 89. Analysis of the fluid flow pattern in shell and tube exchangers and the effect of flow distribution on the heat exchanger performance, (Inst. of Mech. Eng. and Am. Soc. Mech. Eng.).
32. COPE, W.F. and BAILEY, A.: *Aeronautical Research Comm. (Gt. Britain). Tech. Rept.* **43** (1933) 199. Heat transmission through circular, square, and rectangular tubes.
33. WASHINGTON, L. and MARKS, W.M.: *Ind. Eng. Chem.* **29** (1937) 337. Heat transfer and pressure drop in rectangular air passages.
34. DAVIS, E.S.: *Trans. Am. Soc. Mech. Eng.* **65** (1943) 755. Heat transfer and pressure drop in annuli.
35. CARPENTER, F.G., COLBURN, A.P., SCHOENBORN, E.M., and WURSTER, A.: *Trans. Am. Inst. Chem. Eng.* **42** (1946) 165. Heat transfer and friction of water in an annular space.
36. ROWE, P.N., CLAXTON, K.T., and LEWIS, J.B.: *Trans. Inst. Chem. Eng.* **41** (1965) T14. Heat and mass transfer from a single sphere in an extensive flowing fluid.
37. SCHMIDT, E.: *Z. ges. Kälte-Ind.* **35** (1928) 213. Versuche über den Wärmeübergang in ruhender Luft.
38. RAY, B.B.: *Proc. Indian Assoc. Cultivation of Science* **6** (1920) 95. Convection from heated cylinders in air.
39. DAVIS, A.H.: *Phil. Mag.* **44** (1922) 920. Natural convective cooling in fluids.
40. ACKERMANN, G.: *Forsch. Geb. IngWes.* **3** (1932) 42. Die Wärmeabgabe einer horizontal geheizten Röhre an kaltes Wasser bei natürlicher Konvektion.
41. SAUNDERS, O.A.: *Proc. Roy. Soc.* **A, 157** (1936) 278. Effect of pressure on natural convection to air.
42. PERRY, R. H. and GREEN D.W. (eds): *Perry's Chemical Engineers' Handbook*, 6th edn (McGraw-Hill, New York, 1984).
43. KATO H., NISHIWAKI, N. and HIRATA, M.: *Int. Jl. Heat Mass Transfer* **11** (1968) 1117. On the turbulent heat transfer by free convection from a vertical plate.
44. KRAUSSOLD, H.: *Forsch. Geb. IngWes.* **5** (1934) 186. Wärmeabgabe von zylindrischen Flüssigkeiten bei natürlicher Konvektion.
45. INCROPERA, F.P. and DE WITT, D.P.: *Introduction to Heat Transfer*, 3rd edn (Wiley, New York, 1996)
46. HOTTEL, H.C. and SAROFIM, A.F.: *Radiation Heat Transfer* (McGraw-Hill, New York, 1967)
47. HOWELL, J.R.: *A Catalog of Radiation Configuration Factors* (McGraw-Hill, New York, 1982)
48. SIEGEL, R. and HOWELL, J.R.: *Thermal Radiation Heat Transfer* (McGraw-Hill, New York, 1981)
49. TIEN, C.L.: Thermal Radiation Properties of Gases in Hartnett, J.P. and Irvine, T.F., eds: *Advances in Heat Transfer*, Volume 5 (Academic Press, New York, 1968)

50. SPARROW, E.M.: Radiant Interchange between Surfaces Separated by Non-absorbing and Non-emitting Media in Rosenhow, W.M. and Hartnett, J.P., eds: *Handbook of Heat Transfer* (McGraw-Hill, New York, 1973)
51. DUNKLE, R.V.: Radiation Exchange in an Enclosure with a Participating Gas in Rosenhow, W.M. and Hartnett, J.P., eds, *Handbook of Heat Transfer* (McGraw-Hill, New York, 1973)
52. SPARROW, E.M. and CESS, R.D.: *Radiation Heat Transfer* (Hemisphere Publishing, New York, 1978)
53. EDWARDS, D.K.: *Radiation Heat Transfer Notes* (Hemisphere Publishing, New York, 1981)
54. HOTTEL, H.C. and MANGELSDORF, H.G.: *Trans. Am. Inst. Chem. Eng.* **31** (1935) 517. Heat transmission by radiation from non-luminous gases. Experimental study of carbon dioxide and water vapour.
55. SCHACK, A.: *Z. tech. Phys.* **6** (1925) 530. Die Strahlung von leuchtenden Flammen.
56. NUSSELT, W.: *Z. Ver. deut. Ing.* **60** (1916) 541 and 569 Die Oberflächenkondensation des Wasserdampfes.
57. HASELDEN, G.G. and PROSAD, S.: *Trans. Inst. Chem. Eng.* **27** (1949) 195. Heat transfer from condensing oxygen and nitrogen vapours.
58. TEN BOSCH, M.: *Die Wärmeübertragung* (Springer, Berlin, 1936).
59. CARPENTER, E.F. and COLBURN, A.P.: *Proceedings of the General Discussion on Heat Transfer* (*September* 1951) 20. The effect of vapour velocity on condensation inside tubes (Inst. of Mech. Eng. and Am. Soc. Mech. Eng.).
60. COLBURN, A.P.: *Proceedings of the General Discussion on Heat Transfer, September,* 1951, p. 1. Problems in design and research on condensers of vapours and vapour mixtures. (Inst. of Mech. Eng. and Am. Soc. Mech. Eng.).
61. TEPE, J.B. and MUELLER, A.C.: *Chem. Eng. Prog.* **43** (1947) 267. Condensation and subcooling inside an inclined tube.
62. KIRKBRIDE, G.C.: *Ind. Eng. Chem.* **26** (1934) 425. Heat Transfer by condensing vapours on vertical tubes.
63. BADGER, W.L.: *Ind. Eng. Chem.* **22** (1930) 700. The evaporation of caustic soda to high concentrations by means of diphenyl vapour.
64. BADGER, W.L.: *Trans. Am. Inst. Chem. Eng.* **33** (1937) 441. Heat transfer coefficient for condensing Dowtherm films.
65. DREW, T.B., NAGLE, W.M., and SMITH, W.Q.: *Trans. Am. Inst. Chem. Eng.* **31** (1935) 605. The conditions for drop-wise condensation of steam.
66. COLBURN, A.P. and HOUGEN, O.A.: *Ind. Eng. Chem.* **26** (1934) 1178. Design of cooler condensers for mixtures of vapors with non-condensing gases.
67. CHILTON, T.H. and COLBURN, A.P.: *Ind. Eng. Chem.* **26** (1934) 1183. Mass transfer (absorption) coefficients.
68. REVILOCK, J.F., HURLBURT, H.Z., BRAKE, D.R., LANG, E.G., and KERN, D.Q.: *Chem. Eng. Prog.* Symposium Series No. 30, **56** (1960) 161. Heat and mass transfer analogy: An appraisal using plant scale data.
69. JEFFREYS, G.V.: *A Problem in Chemical Engineering Design* (Institution of Chemical Engineers, 1961). The manufacture of acetic anhydride.
70. WESTWATER, J.W.: in *Advances in Chemical Engineering* (Eds T.B. Drew and J.W. Hooper) (Academic Press, New York, 1956) Boiling of Liquids.
71. JAKOB, M.: *Mech. Eng.* **58** (1936) 643, 729. Heat transfer in evaporation and condensation.
72. ROHSENOW, W.M. and CLARK, J.A.: *Trans. Am. Soc. Mech. Eng.* **73** (1951) 609. A study of the mechanism of boiling heat transfer.
73. ROHSENOW, W.M.: *Trans. Am. Soc. Mech. Eng.* **74** (1952) 969. A method of correlating heat transfer data for surface boiling of liquids.
74. FORSTER, H.K.: *J. Appl. Phys.* **25** (1954) 1067. On the conduction of heat into a growing vapor bubble.
75. GRIFFITH, P. and WALLIS, J.D.: *Chem. Eng. Prog. Symposium Series* No. 30, **56** (1960) 49. The role of surface conditions in nuclear boiling.
76. JAKOB, M. and FRITZ, W.: *Forsch. Geb. IngWes.,* **2** (1931) 435. Versuche über den Verdampfungsvorgang.
77. NUKIYAMA, S.: *J. Soc. Mech. Eng. (Japan)* **37** (1934) 367. English abstract pp. S53–S54. The maximum and minimum values of the heat Q transmitted from metal to boiling water under atmospheric pressure.
78. FARBER, E.A. and SCORAH, R.L.: *Trans. Am. Soc. Mech. Eng.* **70** (1948) 369. Heat transfer to boiling water under pressure.
79. BONILLA, C.F. and PERRY, C.H.: *Trans. Am. Inst. Chem. Eng.* **37** (1941) 685. Heat transmission to boiling binary liquid mixtures.
80. INSINGER, T.H. and BLISS, H.: *Trans. Am. Inst. Chem. Eng.* **36** (1940) 491. Transmission of heat to boiling liquids.
81. SAUER, E.T., COOPER, H.B.H., AKIN, G.A. and MCADAMS, W.H.: *Mech. Eng.* **60** (1938) 669. Heat transfer to boiling liquids.
82. CRYDER, D.S. and FINALBORGO, A.C.: *Trans. Am. Inst. Chem. Eng.* **33** (1937) 346. Heat transmission from metal surfaces to boiling liquids.

83. JAKOB, M. and LINKE, W.: *Forsch. Geb. IngWes.* **4** (1933) 75. Der Wärmeübergang von einer waagerechten Platte an siedendes Wasser.
84. CICHELLI, M.T. and BONILLA, C.F.: *Trans. Am. Inst. Chem. Eng.* **41** (1945) 755. Heat transfer to liquids boiling under pressure.
85. FORSTER, H.K. and ZUBER, N.: *J. Appl. Phys.* **25** (1954) 474. Growth of a vapor bubble in a superheated liquid.
86. FORSTER, H.K. and ZUBER, N.: *A.I.Ch.E. Jl* **1** (1955) 531. Dynamics of vapor bubbles and boiling heat transfer.
87. FRITZ, W.: *Physik Z.* **36** (1935) 379. Berechnung des Maximalvolumens von Dampfblasen.
88. PIORO, I.L.: *Int. Jl. Heat Mass Transfer* **42** (1999) 2003. Experimental evaluation of constants for the Rohsenow pool boiling correlation.
89. MCNELLY, M.J.: *J. Imp. Coll. Chem. Eng. Soc.* **7** (1953) 18. A correlation of the rates of heat transfer to nucleate boiling liquids.
90. MOSTINSKI, I.L.: *Brit. Chem. Eng.* **8** (1963) 580. Calculation of boiling heat transfer coefficients, based on the law of corresponding states.
91. JESCHKE, D.: *Z. Ver. deut. Ing.* **69** (1925) 1526. Wärmeübergang und Druckverlust in Röhrschlangen.
92. PRATT, N.H.: *Trans. Inst. Chem. Eng.* **25** (1947), 163. The heat transfer in a reaction tank cooled by means of a coil.
93. CHILTON, T.H., DREW, T.B. and JEBENS, R.H.: *Ind. Eng. Chem.* **36** (1944), 570. Heat transfer coefficients in agitated vessels.
94. CUMMINGS, G.H. and WEST, A.S.: *Ind. Eng. Chem.* **42** (1950) 2303. Heat transfer data for kettles with jackets and coils.
95. BROWN, R.W., SCOTT, M.A. and TOYNE, C.: *Trans. Inst. Chem. Eng.* **25** (1947) 181. An investigation of heat transfer in agitated jacketed cast iron vessels.
96. FLETCHER, P.: *The Chemical Engineer, London* No 435 (1987) 33. Heat transfer coefficients for stirred batch reactor design.
97. *Standards of the Tubular Exchanger Manufacturers Association* (TEMA), 7th edn (New York, 1988).
98. SAUNDERS, E.A.D.: *Heat Exchangers Selection, Design and Construction* (Longman Scientific and Technical, Harlow 1988)
99. BS 3274: (British Standards Institution, London) *British Standard 3274* 1960: Tubular heat exchangers for general purposes.
100. BS 3606: (British Standards Institution, London) *British Standard 3606* 1978: Specification for steel tubes for heat exchangers.
101. LUDWIG, E.E.: *Applied Process Design for Chemical and Petroleum Plants* Volume 3 (Gulf, 1965)
102. UNDERWOOD, A.J.V.: *J. Inst. Petrol. Technol.* **20** (1934) 145. The calculation of the mean temperature difference in multipass heat exchangers.
103. BOWMAN, R.A., MUELLER, A.C. and NAGLE, W.M.: *Trans. Am. Soc. Mech. Eng.* **62** (1940) 283. Mean temperature difference in design.
104. LINNHOFF, B., TOWNSEND, D.W., BOLAND, D., HEWITT, G.F., THOMAS, B.E.A., GUY, A.R. and MARSLAND, R.H.: *A User Guide on Process Integration for the Efficient Use of Energy.* (I. Chem.E., Rugby, England, 1982).
105. LINNHOFF, B.: *Chem. Eng. Prog.* **90** (Aug. 1994) 32. Use pinch analysis to knock down capital costs and emissions.
106. WILSON, E.E.: *Trans. Am. Soc. Mech. Eng.* **37** (1915) 546. A basis for rational design of heat transfer apparatus.
107. RHODES, F.H. and YOUNGER, K.R.: *Ind. Eng. Chem.* **27** (1935) 957. Rate of heat transfer between condensing organic vapours and a metal tube.
108. COULSON, J.M. and MEHTA, R.R.: *Trans. Inst. Chem. Eng.* **31** (1953) 208. Heat transfer coefficients in a climbing film evaporator.
109. BUTTERWORTH, D.: *Conference on Advances in Thermal and Mechanical Design of Shell and Tube Heat Exchangers* NEL Report No. 590 (National Engineering Laboratory, East Kilbride, Glasgow 1973) A calculation method for shell and tube heat exchangers in which the overall coefficient varies along the length.
110. EAGLE, A. and FERGUSON, R.M.: *Proc. Roy. Soc.* **127** (1930) 540. On the coefficient of heat transfer from the internal surfaces of tube walls.
111. DONOHUE, D.A.: *Pet. Ref.* **34** (August, 1955) 94, (Oct) 128, (Dec) 175, **35** (1956) (Jan) 155. Heat exchanger design.
112. FRANK, O.: *Practical Aspects of Heat Transfer.* Chem. Eng. Prog. Tech. Manual (A.I.Ch.E. 1978) Simplified design procedure for tubular exchangers.
113. BUTTERWORTH, D.: *Introduction to Heat Transfer. Engineering Design Guide* 18 (Oxford University Press, Oxford, 1978)

114. LORD, R.C., MINTON, P.E. and SLUSSER, R.P.: *Chem. Eng. Albany* **77** (1, June 1970) 153. Guide to trouble-free heat exchangers.
115. EVANS, F.L.: *Equipment Design Handbook* Volume 2, 2nd edn (Gulf, 1980)
116. TINKER, T.: *Trans. Am. Soc. Mech. Eng.* **80** (1958) 36. Shell-side characteristics of shell and tube heat exchangers.
117. DEVORE, A.: *Hyd. Proc. & Pet. Ref.* **41** (December, 1962) 103. Use nomograms to speed exchanger design.
118. MUELLER, A.C.: *Heat Exchangers, Section 18* in ROHSENOW, W.M. HARTNETT, J.P. and CHO. Y.I., eds, *Handbook of Heat Transfer Fundamentals* 3rd edn. (McGraw-Hill, New York, 1998)
119. PALEN, J.W. and TABORAK, J.: *Chem. Eng. Prog. Sym. Ser.* No. 92, **65** (1969) 53. Solution of shell side flow pressure drop and heat transfer by stream analysis method.
120. GRANT, I.D.R.: *Conference on Advances in Thermal and Mechanical Design of Shell and Tube Heat Exchangers* NEL Report 590 (National Engineering Laboratory, East Kilbride 1973) Flow and pressure drop with single and two phase flow on the shell-side of segmentally baffled shell and tube exchangers.
121. BELL, K.J.: *Petro. Chem.* **32** (1960) C26. Exchanger design: based on the Delaware research report.
122. BELL, K.J.: *Final Report of the Co-operative Research Program on Shell and Tube Heat Exchangers*, University of Delaware Eng. Expt. Sta. Bull. 5 (University of Delaware, 1963)
123. WILLS, M.J.N. and JOHNSTON, D.: *22nd Nat. Heat Transfer Conf., HTD.*, **36**. (ASME, New York, 1984) A new and accurate hand calculation method for shell side pressure drop and flow distribution.
124. ESDU: *Engineering Sciences Data Unit Report 83038* Baffled shell and tube heat exchangers: flow distribution, pressure drop and heat transfer on the shell side. (ESDU International, London 1983)
125. KAYS, W.M. and LONDON, A.L.: *Compact Heat Exchangers* 3rd edn (Krieger, 1998).
126. MICKLEY, H.S., SHERWOOD, T.K. and REED, C.E.: *Applied Mathematics in Chemical Engineering*, 2nd edn, (McGraw-Hill, New York, 1957).
127. NEIL, D.S.: *Processing* **11** (1984) 12. The use of superheated steam in calorifiers.
128. REDMAN, J.: *The Chemical Engineer* (London) No. 452 (1988) 12. Compact future for heat exchangers.
129. GREGORY, E.: *The Chemical Engineer* (London) No. 440 (1987) 33. Plate and fin heat exchangers.
130. JOHNSTON, A.: *The Chemical Engineer* (London) No. 431 (1986) 36. Miniaturized heat exchangers for chemical processing.
131. RAMSHAW, C.: *The Chemical Engineer* (London) No. 415 (1985) 30. Process intensification–a game for *n* players.
132. NELSON, W.L.: *Petroleum Refinery Engineering*, 4th edn (McGraw-Hill, New York, 1958).
133. HUGGINS, F.E.: *Ind. Eng. Chem.* **23** (1931) 749. Effects of scrapers on heating, cooling and mixing.
134. LAUGHLIN, H.G.: *Trans. Am. Inst. Chem. Eng.* **36** (1940) 345. Data on evaporation and drying in a jacketed kettle.
135. HOULTON, H.G.: *Ind. Eng. Chem.* **36** (1944) 522. Heat transfer in the votator.
136. BOLANOWSKI, S.P. and LINEBERRY, D.D.: *Ind. Eng. Chem.* **44** (1952) 657. Special problems of the food industry.
137. SKELLAND, A.H.P.: *Chem. Eng. Sci.* **7** (1958) 166. Correlation of scraped-film heat transfer in the votator.
138. SKELLAND, A.H.P.: *Brit. Chem. Eng.* **3** (1958) 325. Scale-up relationships for heat transfer in the votator.
139. SKELLAND, A.H.P., OLIVER, D.R. and TOOKE, S.: *Brit. Chem. Eng.* **7** (1962) 346. Heat transfer in a water-cooled scraped-surface heat exchanger.
140. BOTT, T.R.: *Brit. Chem. Eng.* **11** (1966) 339. Design of scraped-surface heat exchangers.
141. BOTT, T.R. and ROMERO, J.J.B.: *Can. J. Chem. Eng.* **41** (1963) 213. Heat transfer across a scraped-surface.
142. BOTT, T.R. and SHEIKH, M.R.: *Brit. Chem. Eng.* **9** (1964) 229. Effects of blade design in scraped-surface heat transfer.
143. LYLE, O.: *Efficient Use of Steam* (HMSO, London, 1947).

9.14. NOMENCLATURE

		Units in SI System	Dimensions in **M**, **L**, **T**, θ (*or* **M**, **L**, **T**, θ, **H**)
A	Area available for heat transfer *or* area of radiating surface	m^2	$\mathbf{L}^2$
A_e	External area of body	m^2	$\mathbf{L}^2$
A_s	Maximum cross-flow area over tube bundle	m^2	$\mathbf{L}^2$
$\mathbf{a}$	Absorptivity	—	—
$\mathbf{a}_b$	Absorptivity of a black body	—	—

		Units in SI System	Dimensions in M, L, T, θ (*or* M, L, T, θ, H)
$\mathbf{a}_s$	Absorptivity of a gas	—	—
b	Wetted perimeter of condensation surface or perimeter of fin	m	**L**
C	Constant	—	—
C_1	First radiation constant	Wm^2	$\mathbf{ML^4T^{-3}}$ (*or* $\mathbf{HL^2T^{-1}}$)
C_2	Second radiation constant	mK	$\mathbf{L\theta}$
C_3	Third radiation constant	mK	$\mathbf{L\theta}$
C_4	Fourth radiation constant	W/m^3K^5	$\mathbf{ML^{-3}T^{-3}\theta^{-5}}$ (*or* $\mathbf{HL^{-3}T^{-1}\theta^{-5}}$)
C_f	Constant for friction in flow past a tube bundle	—	—
C_h	Constant for heat transfer in flow past a tube bundle	—	—
C'	Clearance between tubes in heat exchanger	m	**L**
C''	Coefficient in equation 9.95 (SI units only)	$J/m^{3n+1}sK^{n+1}$	$\mathbf{ML^{1-3n}\,T^{-3}\theta^{-n-1}}$ (*or* $\mathbf{HL^{-3n-1}T^{-1}\theta^{-n-1}}$)
C_p	Specific heat at constant pressure	J/kg K	$\mathbf{L^2\,T^{-2}\theta^{-1}}$ (*or* $\mathbf{HM^{-1}\theta^{-1}}$)
D	Diffusivity of vapour	m^2/s	$\mathbf{L^2\,T^{-1}}$
D	Diameter	m	**L**
D_H	Thermal diffusivity $(k/C_p\rho)$	m^2/s	$\mathbf{L^2\,T^{-1}}$
d	Diameter (internal or of sphere)	m	**L**
d_1, d_2	Inner and outer diameters of annulus	m	**L**
d_c	Diameter of helix	m	**L**
d_e	Hydraulic mean diameter	m	**L**
d_g	Gap between turns in coil	m	**L**
d_m	Logarithmic mean diameter of lagging	m	**L**
d_o	Outside diameter of tube	m	**L**
d_p	Height of coil	m	**L**
d_r	Diameter of shaft	m	**L**
d_s	Outside diameter of lagging *or* inside diameter of shell	m	**L**
d_t	Thickness of fixed tube sheet	m	**L**
d_v	Internal diameter of vessel	m	**L**
d_w	Mean diameter of pipe wall	m	**L**
E	Emissive power	W/m^2	$\mathbf{MT^{-3}}$ (*or* $\mathbf{HL^2\,T^{-1}}$)
E'_z	Energy emitted per unit area and unit time per unit wavelength	W/m^3	$\mathbf{ML^{-1}\,T^{-3}}$ (*or* $\mathbf{HL^{-3}\,T^{-1}}$)
E_λ	Energy emitted per unit area per unit wavelength	W/m^3	$\mathbf{MT^{-3}}$ (*or* $\mathbf{HL^{-3}T^{-1}}$)
$\mathbf{e}$	Emissivity	—	—
$\mathbf{e}'$	$\frac{1}{2}(e_s+1)$	—	—
$\mathbf{e}_F$	Emissivity of a flame	—	—
$\mathbf{e}_g$	Effective emissivity of a gas	—	—
$\mathbf{e}_{sh}$	Emissivity of shield	—	—
$\mathbf{e}_\lambda$	Spectral hemispherical emissivity	—	—
F	Geometric factor for radiation *or* correction factor for logarithmic mean temperature difference	—	—
f	Working stress	N/m^2	$\mathbf{ML^{-1}\,T^{-2}}$
f'	Shell-side friction factor	—	—
G	Mass rate of flow	kg/s	$\mathbf{MT^{-1}}$
G'	Mass rate of flow per unit area	kg/m^2s	$\mathbf{ML^{-2}\,T^{-1}}$
G'_s	Mass flow per unit area over tube bundle	kg/m^2s	$\mathbf{ML^{-2}\,T^{-1}}$
g	Acceleration due to gravity	m/s^2	$\mathbf{LT^{-2}}$
H	Ratio: Z/X	—	—
h	Heat transfer coefficient	W/m^2 K	$\mathbf{MT^{-3}\theta^{-1}}$ (*or* $\mathbf{HL^{-2}\,T^{-1}\theta^{-1}}$)
h_b	Film coefficient for liquid adjacent to vessel	W/m^2 K	$\mathbf{MT^{-3}\theta^{-1}}$ (*or* $\mathbf{HL^{-2}\,T^{-1}\theta^{-1}}$)
h_c	Heat transfer coefficient for convection	W/m^2 K	$\mathbf{MT^{-3}\theta^{-1}}$ (*or* $\mathbf{HL^{-2}\,T^{-1}\theta^{-1}}$)

		Units in SI System	Dimensions in **M, L, T, θ** (*or* **M, L, T, θ, H**)
h_f	Fin-side film coefficient	W/m² K	$\mathbf{MT^{-3}\theta^{-1}}$ (*or* $\mathbf{HL^{-2}\,T^{-1}\theta^{-1}}$)
h_m	Mean value of h over whole surface	W/m² K	$\mathbf{MT^{-3}\theta^{-1}}$ (*or* $\mathbf{HL^{-2}\,T^{-1}\theta^{-1}}$)
h_n	Heat transfer coefficient for liquid boiling at T_n	W/m² K	$\mathbf{MT^{-3}\theta^{-1}}$ (*or* $\mathbf{HL^{-2}\,T^{-1}\theta^{-1}}$)
h_r	Heat transfer coefficient for radiation	W/m² K	$\mathbf{MT^{-3}\theta^{-1}}$ (*or* $\mathbf{HL^{-2}\,T^{-1}\theta^{-1}}$)
I	Intensity of radiation	W/m²	$\mathbf{MT^{-3}}$ (*or* $\mathbf{HL^{-2}\,T^{-1}}$)
I'	Intensity of radiation falling on body	W/m²	$\mathbf{MT^{-3}}$ (*or* $\mathbf{HL^{-2}\,T^{-1}}$)
J	For fin: $(km - h)/(km + h)$	—	—
j	Number of vertical rows of tubes	—	—
j_d	j-factor for mass transfer	—	—
j_h	j-factor for heat transfer	—	—
K	Concentration of particles in a flame	m⁻³	$\mathbf{L^{-3}}$
K'	Factor describing particle concentration	N/m²	$\mathbf{ML^{-1}T^{-2}}$
k	Thermal conductivity	W/m K	$\mathbf{MLT^{-3}\theta^{-1}}$ (*or* $\mathbf{HL^{-1}\,T^{-1}\theta^{-1}}$)
k_a	Arithmetic mean thermal conductivity	W/m K	$\mathbf{MLT^{-3}\theta^{-1}}$ (*or* $\mathbf{HL^{-1}\,T^{-1}\theta^{-1}}$)
k_m	Mean thermal conductivity	W/m K	$\mathbf{MLT^{-3}\theta^{-1}}$ (*or* $\mathbf{HL^{-1}\,T^{-1}\theta^{-1}}$)
k_o	Thermal conductivity at zero temperature	W/m K	$\mathbf{MLT^{-3}\theta^{-1}}$ (*or* $\mathbf{HL^{-1}\,T^{-1}\theta^{-1}}$)
k'	Constant in equation 9.13	K⁻¹	$\mathbf{\theta^{-1}}$
k_G	Mass transfer coefficient (mass/unit area. unit time. unit partial pressure difference)	s/m	$\mathbf{L^{-1}\,T}$
L	Length of paddle, length of fin or characteristic dimension	m	$\mathbf{L}$
L	Separation of surfaces, length of a side or radius of hemispherical mass of gas	m	$\mathbf{L}$
L_e	Mean beam length	m	$\mathbf{L}$
l	Length of tube or plate, or distance apart of faces, *or* thickness of gas stream	m	$\mathbf{L}$
l_B	Distance between baffles	m	$\mathbf{L}$
l_v	Length of side of vessel	m	$\mathbf{L}$
M	Mass rate of flow of condensate per unit length of perimeter for vertical pipe and per unit length of pipe for horizontal pipe	kg/s m	$\mathbf{ML^{-1}\,T^{-1}}$
m	Mass of liquid	kg	$\mathbf{M}$
m	For fin: $(hb/kA)^{0.5}$	m⁻¹	$\mathbf{L^{-1}}$
n	Number of baffles	—	—
N	Number of revolutions in unit time	Hz	$\mathbf{T^{-1}}$
$\mathbf{N}$	Number of general term in series, *or* number of transfer units	—	—
n	An index	—	—
n_B	Number of blades on agitator	—	—
P	Pressure	N/m²	$\mathbf{ML^{-1}\,T^{-2}}$
P_{Bm}	Logarithmic mean partial pressure of inert gas **B**	N/m²	$\mathbf{ML^{-1}\,T^{-2}}$
P_c	Critical pressure	N/m²	$\mathbf{ML^{-1}\,T^{-2}}$
P_c	Partial pressure of carbon dioxide	N/m²	$\mathbf{ML^{-1}T^{-2}}$
P_g	Partial pressure of radiating gas	N/m²	$\mathbf{ML^{-1}T^{-2}}$
P_R	Reduced pressure (P/P_c)	—	—
P_r	Vapour pressure at surface of radius r	N/m²	$\mathbf{ML^{-1}\,T^{-2}}$
P_s	Saturation vapour pressure	N/m²	$\mathbf{ML^{-1}\,T^{-2}}$
P_w	Partial pressure of water vapour	N/m²	$\mathbf{ML^{-1}T^{-2}}$
p	Parameter in Laplace transform	s⁻¹	$\mathbf{T^{-1}}$
Q	Heat flow *or* generation per unit time	W	$\mathbf{ML^2\,T^{-3}}$ (*or* $\mathbf{HT^{-1}}$)
Q_e	Radiation emitted per unit time	W	$\mathbf{ML^2\,T^{-3}}$ (*or* $\mathbf{HT^{-1}}$)
Q_f	Heat flow to root of fin per unit time	W	$\mathbf{ML^2\,T^{-3}}$ (*or* $\mathbf{HT^{-1}}$)
Q_G	Rate of heat generation per unit volume	W/m³	$\mathbf{ML^{-1}\,T^{-3}}$ (*or* $\mathbf{HL^{-3}\,T^{-1}}$)
Q_I	Radiation incident on a surface	W	$\mathbf{ML^2T^{-3}}$ (*or* $\mathbf{HT^{-1}}$)
Q_i	Total incident radiation per unit time	W	$\mathbf{ML^2\,T^{-3}}$ (*or* $\mathbf{HT^{-1}}$)

		Units in SI System	Dimensions in **M, L, T, θ** (*or* **M, L, T, θ, H**)
Q_k	Heat flow per unit time by conduction in fluid	W	$\mathbf{ML^2\,T^{-3}}$ (*or* $\mathbf{HT^{-1}}$)
Q_o	Total radiation leaving surface per unit time	W	$\mathbf{ML^2\,T^{-3}}$ (*or* $\mathbf{HT^{-1}}$)
Q_r	Total heat reflected from surface per unit time	W	$\mathbf{ML^2\,T^{-3}}$ (*or* $\mathbf{HT^{-1}}$)
q	Heat flow per unit time and unit area	W/m^2	$\mathbf{MT^{-3}}$ (*or* $\mathbf{HL^{-2}\,T^{-1}}$)
q_I	Energy arriving at unit area of a grey surface	W/m^2	$\mathbf{MT^{-3}}$ (*or* $\mathbf{HL^{-2}T^{-1}}$)
q_{sh}	Value of q with shield	W/m^2	$\mathbf{MT^{-3}}$ (*or* $\mathbf{HL^{-2}T^{-1}}$)
q_0	Radiosity-energy leaving unit area of a grey surface	W/m^2	$\mathbf{MT^{-3}}$ (*or* $\mathbf{HL^{-2}T^{-1}}$)
R	Thermal resistance	m^2 K/W	$\mathbf{M^{-1}T^3\theta}$ (*or* $\mathbf{H^{-1}\,L^2T\theta}$)
R	Ratio: r_j/r_i or r/L	—	—
R_i, R_0	Thermal resistance of scale on inside, outside of tubes	m^2 K/W	$\mathbf{M^{-1}\,T^3\theta}$ (*or* $\mathbf{H^{-1}L^2T\theta}$)
R'	Shear stress at free surface of condensate film	N/m^2	$\mathbf{ML^{-1}\,T^{-2}}$
r	Reflectivity	—	—
r	Radius	m	**L**
r_1, r_2	Radius (inner, outer) of annulus or tube	m	**L**
r_a	Arithmetic mean radius	m	**L**
r_m	Logarithmic mean radius	m	**L**
S	Flow area for condensate film	m^2	$\mathbf{L^2}$
S	Ratio: s/r_i or $1 + (1 + R_j^2)/R_i^2$	—	—
s	Thickness of condensate film at a point	m	**L**
s	Distance between surfaces	m	**L**
T	Temperature	K	θ
T_c	Temperature of free surface of condensate	K	θ
T_{cm}	Temperature of cooling medium	K	θ
T_f	Mean temperature of film	K	θ
T_g	Temperature of gas	K	θ
T_G	Temperature of atmosphere surrounding fin	K	θ
T_m	Mean temperature of fluid	K	θ
T_n	Standard boiling point	K	θ
T_s	Temperature of condensing vapour	K	θ
T_{sh}	Temperature of shield	K	θ
T_w	Temperature of wall	K	θ
t	Transmissivity	—	—
t	Time	s	**T**
U	Overall heat transfer coefficient	W/m^2 K	$\mathbf{MT^{-3}\theta^{-1}}$ (*or* $\mathbf{HL^{-2}\,T^{-1}\theta^{-1}}$)
u	Velocity	m/s	$\mathbf{LT^{-1}}$
u_m	Maximum velocity in condensate film	m/s	$\mathbf{LT^{-1}}$
u_y	Velocity at distance y from surface	m/s	$\mathbf{LT^{-1}}$
V	Volume	m^3	$\mathbf{L^3}$
W	Ratio: w/L or Y/X	—	—
W	Width of stirrer	m	**L**
w	Width of fin or surface	m	**L**
$w_1, w_2, \ldots$	Indices in equation for heat transfer by convection	—	—
X	Length of surface	m	**L**
$\overline{X}$	Ratio: X/L	—	—
X	Distance between centres of tubes in direction of flow	m	**L**
X	Ratio of temperature differences used in calculation of mean temperature difference	—	—
x	Distance in direction of transfer or along surface	m	**L**
Y	Distance between centres of tubes at right angles to flow direction *or* width of surface	m	**L**

		Units in SI System	Dimensions in M, L, T, θ (*or* M, L, T, θ, H)
$\overline{Y}$	Ratio: Y/L	—	—
Y	Ratio of temperature differences used in calculation of mean temperature difference	—	—
y	Distance perpendicular to surface	m	**L**
Z	Width of vertical surface	m	**L**
Z	Wavelength	m	**L**
Z_m	Wavelength at which maximum energy is emitted	m	**L**
z	Distance in third principal direction	m	**L**
α	Angle between two surfaces *or*	m	**L**
	angle between normal and direction of radiation	—	—
β	Coefficient of cubical expansion	K^{-1}	θ^{-1}
η	Effectiveness of heat exchanger, defined by equation 9.153	—	—
ε	Coefficient relating h to $u^{0.8}$	J/s$^{0.2}$ m$^{2.8}$ K	$\mathbf{ML^{-0.8}\ T^{-2.2}}\theta^{-1}$ (*or* $\mathbf{HL^{-2.8}\ T^{-0.2}}\theta^{-1}$)
λ	Wavelength *or*	m	**L**
	latent heat of vaporisation per unit mass	J/kg	$\mathbf{L^2\ T^{-2}}$ (*or* $\mathbf{HM^{-1}}$)
μ	Viscosity	N s/m^2	$\mathbf{ML^{-1}\ T^{-1}}$
μ_s	Viscosity of fluid at surface	N s/m^2	$\mathbf{ML^{-1}\ T^{-1}}$
ρ	Density or density of liquid	kg/m^3	$\mathbf{ML^{-3}}$
ρ_v	Density of vapour	kg/m^3	$\mathbf{ML^{-3}}$
σ	Stefan—Boltzmann constant, *or*	W/m^2 K^4	$\mathbf{MT^{-3}}\theta^{-4}$ (*or* $\mathbf{HL^{-2}}\theta^{-4}$)
	Surface tension	J/m^2	$\mathbf{MT^{-2}}$
τ	Response time for heating or cooling	s	**T**
θ	Temperature or temperature difference	K	θ
θ_a	Temperature difference in Schmidt method	K	θ
θ_c	Temperature of centre of body	K	θ
θ_m	Logarithmic mean temperature difference	K	θ
θ_{xt}	Temperature at $t = t$, $x = x$	K	θ
θ_o	Initial uniform temperature of body	K	θ
θ'	Temperature of source or surroundings	K	θ
ψ	$\dfrac{G_1 C_{p1} + G_2 C_{p2}}{G_1 C_{p1} G_2 C_{p2}}$		
ϕ	Angle between surface and horizontal *or* angle of contact	—	—
$\overline{\theta}$	Laplace transform of temperature	s K	**T** θ
ω	Solid angle	—	—
Fo	Fourier number $D_H t/L^2$	—	—
Gr	Grashof number	—	—
Gz	Graetz number	—	—
Nu	Nusselt number	—	—
Nu'	Particle Nusselt number	—	—
Pr	Prandtl number	—	—
Re	Reynolds number	—	—
Re'	Particle Reynolds number	—	—
Re''	Rotational flow Reynolds number	—	—
Re_x	Reynolds number for flat plate	—	—
Δ	Finite difference in a property	—	—

Suffix, w refers to wall material
Suffixes i, o refer to inside, outside of wall or lagging; or inlet, outlet conditions
Suffixes b, l, v refer to bubble, liquid, vapour
Suffixes g, b, and s refer to gas, black body, and surface.

PART 3

Mass Transfer

CHAPTER 10

Mass Transfer

10.1. INTRODUCTION

The term mass transfer is used to denote the transference of a component in a mixture from a region where its concentration is high to a region where the concentration is lower. Mass transfer process can take place in a gas or vapour or in a liquid, and it can result from the random velocities of the molecules (*molecular diffusion*) or from the circulating or eddy currents present in a turbulent fluid (*eddy diffusion*).

In processing, it is frequently necessary to separate a mixture into its components and, in a physical process, differences in a particular property are exploited as the basis for the separation process. Thus, *fractional distillation* depends on differences in volatility, *gas absorption* on differences in solubility of the gases in a selective absorbent and, similarly, *liquid–liquid extraction* is based on on the selectivity of an immiscible liquid solvent for one of the constituents. The rate at which the process takes place is dependent both on the *driving force* (concentration difference) and on the mass transfer resistance. In most of these applications, mass transfer takes place across a phase boundary where the concentrations on either side of the interface are related by the phase equilibrium relationship. Where a chemical reaction takes place during the course of the mass transfer process, the overall transfer rate depends on both the chemical kinetics of the reaction and on the mass transfer resistance, and it is important to understand the relative significance of these two factors in any practical application.

In this chapter, consideration will be given to the basic principles underlying mass transfer both with and without chemical reaction, and to the models which have been proposed to enable the rates of transfer to be calculated. The applications of mass transfer to the design and operation of separation processes are discussed in Volume 2, and the design of reactors is dealt with in Volume 3.

A simple example of a mass transfer process is that occurring in a box consisting of two compartments, each containing a different gas, initially separated by an impermeable partition. When the partition is removed the gases start to mix and the mixing process continues at a constantly decreasing rate until eventually (theoretically after the elapse of an infinite time) the whole system acquires a uniform composition. The process is one of molecular diffusion in which the mixing is attributable solely to the random motion of the molecules. The rate of diffusion is governed by Fick's Law, first proposed by FICK[1] in 1855 which expresses the mass transfer rate as a linear function of the molar concentration gradient. In a mixture of two gases **A** and **B**, assumed ideal, Fick's Law for steady state diffusion may be written as:

$$N_A = -D_{AB}\frac{\mathrm{d}C_A}{\mathrm{d}y} \tag{10.1}$$

where N_A is the molar flux of **A** (moles per unit area per unit time),
C_A is the concentration of **A** (moles of **A** per unit volume),
D_{AB} is known as the diffusivity or diffusion coefficient for **A** in **B**, and
y is distance in the direction of transfer.

An equation of exactly the same form may be written for **B**:

$$N_B = -D_{BA}\frac{dC_B}{dy} \tag{10.2}$$

where D_{BA} is the diffusivity of **B** in **A**.

As indicated in the next section, for an ideal gas mixture, at constant pressure $(C_A + C_B)$, is constant (equation 10.9) and hence:

$$\frac{dC_A}{dy} = -\frac{dC_B}{dy} \tag{10.3}$$

The condition for the pressure or molar concentration to remain constant in such a system is that there should be no net transference of molecules. The process is then referred to as one of *equimolecular counterdiffusion*, and:

$$N_A + N_B = 0$$

This relation is satisfied only if $D_{BA} = D_{AB}$ and therefore the suffixes may be omitted and equation 10.1 becomes:

$$N_A = -D\frac{dC_A}{dy} \tag{10.4}$$

Equation 10.4, which describes the mass transfer rate arising solely from the random movement of molecules, is applicable to a stationary medium or a fluid in streamline flow. If circulating currents or eddies are present, then the molecular mechanism will be reinforced and the total mass transfer rate may be written as:

$$N_A = -(D + E_D)\frac{dC_A}{dy} \tag{10.5}$$

Whereas D is a physical property of the system and a function only of its composition, pressure and temperature, E_D, which is known as the *eddy diffusivity*, is dependent on the flow pattern and varies with position. The estimation of E_D presents some difficulty, and this problem is considered in Chapter 12.

The molecular diffusivity D may be expressed in terms of the molecular velocity u_m and the mean free path of the molecules λ_m. In Chapter 12 it is shown that for conditions where the kinetic theory of gases is applicable, the molecular diffusivity is proportional to the product $u_m\lambda_m$. Thus, the higher the velocity of the molecules, the greater is the distance they travel before colliding with other molecules, and the higher is the diffusivity D.

Because molecular velocities increase with rise of temperature T, so also does the diffusivity which, for a gas, is approximately proportional to T raised to the power of 1.5. As the pressure P increases, the molecules become closer together and the mean free path is shorter and consequently the diffusivity is reduced, with D for a gas becoming approximately inversely proportional to the pressure.

Thus:

$$D \propto T^{1.5}/P \tag{10.6}$$

A method of calculating D in a binary mixture of gases is given later (equation 10.43). For *liquids*, the molecular structure is far more complex and no such simple relationship exists, although various semi-empirical predictive methods, such as equation 10.96, are useful.

In the discussion so far, the fluid has been considered to be a continuum, and distances on the molecular scale have, in effect, been regarded as small compared with the dimensions of the containing vessel, and thus only a small proportion of the molecules collides directly with the walls. As the pressure of a gas is reduced, however, the mean free path may increase to such an extent that it becomes comparable with the dimensions of the vessel, and a significant proportion of the molecules may then collide directly with the walls rather than with other molecules. Similarly, if the linear dimensions of the system are reduced, as for instance when diffusion is occurring in the small pores of a catalyst particle (Section 10.7), the effects of collision with the walls of the pores may be important even at moderate pressures. Where the main resistance to diffusion arises from collisions of molecules with the walls, the process is referred to *Knudsen diffusion*, with a Knudsen diffusivity D_{Kn} which is proportional to the product $u_m l$, where l is a linear dimension of the containing vessel.

Since, from the *kinetic theory*[2], $u_m \propto (\mathbf{R}T/M)^{0.5}$:

$$D_{Kn} \propto l(\mathbf{R}T/M)^{0.5} \tag{10.7}$$

Each resistance to mass transfer is proportional to the reciprocal of the appropriate diffusivity and thus, when both molecular and Knudsen diffusion must be considered together, the effective diffusivity D_e is obtained by summing the resistances as:

$$1/D_e = 1/D + 1/D_{Kn} \tag{10.8}$$

In *liquids*, the effective mean path of the molecules is so small that the effects of Knudsen-type diffusion need not be considered.

10.2. DIFFUSION IN BINARY GAS MIXTURES

10.2.1. Properties of binary mixtures

If **A** and **B** are ideal gases in a mixture, the ideal gas law, equation 2.15, may be applied to each gas separately and to the mixture:

$$P_A V = n_A \mathbf{R}T \tag{10.9a}$$

$$P_B V = n_B \mathbf{R}T \tag{10.9b}$$

$$PV = n\mathbf{R}T \tag{10.9c}$$

where n_A and n_B are the number of moles of **A** and **B** and n is the total number of moles in a volume V, and P_A, P_B and P are the respective partial pressures and the total pressure.

Thus:

$$P_A = \frac{n_A}{V}\mathbf{R}T = C_A\mathbf{R}T = \frac{c_A}{M_A}\mathbf{R}T \tag{10.10a}$$

$$P_B = \frac{n_B}{V}\mathbf{R}T = C_B\mathbf{R}T = \frac{c_B}{M_B}\mathbf{R}T \tag{10.10b}$$

and:

$$P = \frac{n}{V}\mathbf{R}T = C_T\mathbf{R}T \tag{10.10c}$$

where c_A and c_B are mass concentrations and M_A and M_B molecular weights, and C_A, C_B, C_T are, the molar concentrations of **A** and **B** respectively, and the total molar concentration of the mixture.

From Dalton's Law of partial pressures:

$$P = P_A + P_B = \mathbf{R}T(C_A + C_B) = \mathbf{R}T\left(\frac{c_A}{M_A} + \frac{c_B}{M_B}\right) \tag{10.11}$$

Thus:

$$C_T = C_A + C_B \tag{10.12}$$

and:

$$1 = x_A + x_B \tag{10.13}$$

where x_A and x_B are the mole fractions of **A** and **B**.

Thus for a system at constant pressure P and constant molar concentration C_T:

$$\frac{dP_A}{dy} = -\frac{dP_B}{dy} \tag{10.14}$$

$$\frac{dC_A}{dy} = -\frac{dC_B}{dy} \tag{10.15}$$

$$\frac{dc_A}{dy} = -\frac{dc_B}{dy}\frac{M_A}{M_B} \tag{10.16}$$

and:

$$\frac{dx_A}{dy} = -\frac{dx_B}{dy} \tag{10.17}$$

By substituting from equations 10.7*a* and 10.7*b* into equation 10.4, the mass transfer rates N_A and N_B can be expressed in terms of partial pressure gradients rather than concentration gradients. Furthermore, N_A and N_B can be expressed in terms of gradients of mole fraction.

Thus:

$$N_A = -\frac{D}{\mathbf{R}T}\frac{dP_A}{dy} \tag{10.18}$$

or

$$N_A = -DC_T\frac{dx_A}{dy} \tag{10.19}$$

Similarly:

$$N_B = -\frac{D}{\mathbf{R}T}\frac{dP_B}{dy} = +\frac{D}{\mathbf{R}T}\frac{dP_A}{dy} \quad \text{(from equation 10.14)} \tag{10.20}$$

or

$$N_B = -DC_T\frac{dx_B}{dy} = +DC_T\frac{dx_A}{dy} \quad \text{(from equation 10.17)} \tag{10.21}$$

10.2.2. Equimolecular counterdiffusion

When the mass transfer rates of the two components are equal and opposite the process is said to be one of *equimolecular counterdiffusion*. Such a process occurs in the case of the box with a movable partition, referred to in Section 10.1. It occurs also in a distillation column when the molar latent heats of the two components are the same. At any point in the column a falling stream of liquid is brought into contact with a rising stream of vapour with which it is *not* in equilibrium. The less volatile component is transferred from the vapour to the liquid and the more volatile component is transferred in the opposite direction. If the molar latent heats of the components are equal, the condensation of a given amount of less volatile component releases exactly the amount of latent heat required to volatilise the same molar quantity of the more volatile component. Thus at the interface, and consequently throughout the liquid and vapour phases, equimolecular counterdiffusion is taking place.

Under these conditions, the differential forms of equation for N_A (10.4, 10.18 and 10.19) may be simply integrated, for constant temperature and pressure, to give respectively:

$$N_A = -D\frac{C_{A_2} - C_{A_1}}{y_2 - y_1} = \frac{D}{y_2 - y_1}(C_{A_1} - C_{A_2}) \tag{10.22}$$

$$N_A = -\frac{D}{\mathbf{R}T}\frac{P_{A_2} - P_{A_1}}{y_2 - y_1} = \frac{D}{\mathbf{R}T(y_2 - y_1)}(P_{A_1} - P_{A_2}) \tag{10.23}$$

$$N_A = -DC_T\frac{x_{A_2} - x_{A_1}}{y_2 - y_1} = \frac{DC_T}{y_2 - y_1}(x_{A_1} - x_{A_2}) \tag{10.24}$$

Similar equations apply to N_B which is equal to $-N_A$, and suffixes 1 and 2 represent the values of quantities at positions y_1 and y_2 respectively.

Equation 10.22 may be written as:

$$N_A = h_D(C_{A_1} - C_{A_2}) \tag{10.25}$$

where $h_D = D/(y_2 - y_1)$ is a *mass transfer coefficient* with the driving force expressed as a *difference in molar concentration*; its dimensions are those of velocity ($\mathbf{LT}^{-1}$).

Similarly, equation 10.23 may be written as:

$$N_A = k'_G(P_{A_1} - P_{A_2}) \tag{10.26}$$

where $k'_G = D/[\mathbf{R}T(y_2 - y_1)]$ is a *mass transfer coefficient* with the driving force expressed as a *difference in partial pressure*. It should be noted that its dimensions here, $\mathbf{NM}^{-1}\mathbf{L}^{-1}\mathbf{T}$, are different from those of h_D. It is always important to use the form of mass transfer coefficient corresponding to the appropriate driving force.

In a similar way, equation 10.24 may be written as:

$$N_A = k_x(x_{A_1} - x_{A_2}) \tag{10.27}$$

where $k_x = DC_T/(y_2 - y_1)$ is a *mass transfer coefficient* with the driving force in the form of a *difference in mole fraction*. The dimensions here are $\mathbf{NL}^{-2}\mathbf{T}^{-1}$.

10.2.3. Mass transfer through a stationary second component

In several important processes, one component in a gaseous mixture will be transported relative to a fixed plane, such as a liquid interface, for example, and the other will undergo no net movement. In gas absorption a soluble gas **A** is transferred to the liquid surface where it dissolves, whereas the insoluble gas **B** undergoes no net movement with respect to the interface. Similarly, in evaporation from a free surface, the vapour moves away from the surface but the air has no net movement. The mass transfer process therefore differs from that described in Section 10.2.2.

The concept of a stationary component may be envisaged by considering the effect of moving the box, discussed in Section 10.1, in the opposite direction to that in which **B** is diffusing, at a velocity equal to its diffusion velocity, so that to the external observer **B** appears to be stationary. The total velocity at which **A** is transferred will then be increased to its diffusion velocity plus the velocity of the box.

For the absorption of a soluble gas **A** from a mixture with an insoluble gas **B**, the respective diffusion rates are given by:

$$N_A = -D\frac{dC_A}{dy} \qquad \text{(equation 10.4)}$$

$$N_B = -D\frac{dC_B}{dy} = D\frac{dC_A}{dy} \qquad \text{(from equation 10.3)}$$

Since the total mass transfer rate of **B** is zero, there must be a "bulk flow" of the system towards the liquid surface exactly to counterbalance the diffusional flux away from the surface, as shown in Figure 10.1, where:

$$\text{Bulk flow of } \mathbf{B} = -D\frac{dC_A}{dy} \tag{10.28}$$

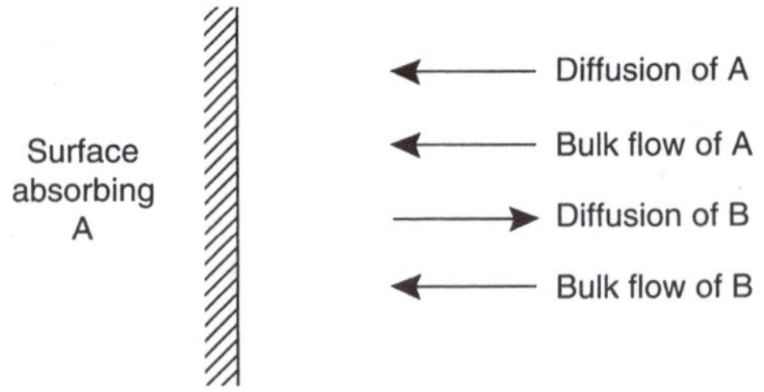

Figure 10.1. Mass transfer through a stationary gas **B**

The corresponding bulk flow of **A** must be C_A/C_B times that of **B**, since bulk flow implies that the gas moves *en masse*.

Thus:

$$\text{Bulk flow of } \mathbf{A} = -D\frac{dC_A}{dy}\frac{C_A}{C_B} \tag{10.29}$$

Therefore the total flux of **A**, N'_A, is given by:

$$N'_A = -D\frac{dC_A}{dy} - D\frac{dC_A}{dy}\frac{C_A}{C_B}$$

$$= -D\frac{dC_A}{dy}\frac{C_T}{C_B} \tag{10.30}$$

Equation 10.30 is known as Stefan's Law[3]. Thus the bulk flow enhances the mass transfer rate by a factor C_T/C_B, known as the *drift factor*. The fluxes of the components are given in Table 10.1.

Table 10.1. Fluxes of components of a gas mixture

	Component **A**	Component **B**	**A + B**
Diffusion	$-D\dfrac{dC_A}{dy}$	$+D\dfrac{dC_A}{dy}$	0
Bulk flow	$-D\dfrac{dC_A}{dy}\cdot\dfrac{C_A}{C_B}$	$-D\dfrac{dC_A}{dy}$	$-D\dfrac{dC_A}{dy}\cdot\dfrac{C_A+C_B}{C_B}$
Total	$-D\dfrac{dC_A}{dy}\cdot\dfrac{C_A+C_B}{C_B}$	0	$-D\dfrac{dC_A}{dy}\cdot\dfrac{C_A+C_B}{C_B}$
	$= -D\dfrac{dC_A}{dy}\cdot\dfrac{C_T}{C_B}$		$= -D\dfrac{dC_A}{dy}\cdot\dfrac{C_T}{C_B}$

Writing equation 10.30 as:

$$N'_A = D\frac{C_T}{C_B}\frac{dC_B}{dy} \qquad \text{(from equation 10.3)}$$

On integration:

$$N'_A = \frac{DC_T}{y_2 - y_1}\ln\frac{C_{B_2}}{C_{B_1}} \tag{10.31}$$

By definition, C_{Bm}, the logarithmic mean of C_{B_1} and C_{B_2}, is given by:

$$C_{Bm} = \frac{C_{B_2} - C_{B_1}}{\ln(C_{B_2}/C_{B_1})} \tag{10.32}$$

Thus, substituting for $\ln(C_{B_2}/C_{B_1})$ in equation 10.31:

$$N'_A = \left(\frac{DC_T}{y_2 - y_1}\right)\frac{C_{B_2} - C_{B_1}}{C_{Bm}}$$

$$= \left(\frac{D}{y_2 - y_1}\frac{C_T}{C_{Bm}}\right)(C_{A_1} - C_{A_2}) \tag{10.33}$$

or in terms of partial pressures:

$$N'_A = \left(\frac{D}{\mathbf{R}T(y_2 - y_1)}\frac{P}{P_{Bm}}\right)(P_{A_1} - P_{A_2}) \tag{10.34}$$

Similarly, in terms of mole fractions:

$$N'_A = \left(\frac{DC_T}{y_2 - y_1}\frac{1}{x_{Bm}}\right)(x_{A_1} - x_{A_2}) \tag{10.35}$$

Equation 10.31 can be simplified when the concentration of the diffusing component **A** is small. Under these conditions C_A is small compared with C_T, and equation 10.31 becomes:

$$N'_A = \frac{DC_T}{y_2 - y_1}\ln\left[1 - \left(\frac{C_{A_2} - C_{A_1}}{C_T - C_{A_1}}\right)\right]$$

$$= \frac{DC_T}{y_2 - y_1}\left[-\left(\frac{C_{A_2} - C_{A_1}}{C_T - C_{A_1}}\right) - \frac{1}{2}\left(\frac{C_{A_2} - C_{A_1}}{C_T - C_{A_1}}\right)^2 - \cdots\right]$$

For small values of C_A, $C_T - C_{A_1} \approx C_T$ and only the first term in the series is significant.

Thus:
$$N'_A \approx \frac{D}{y_2 - y_1}(C_{A_1} - C_{A_2}) \tag{10.36}$$

Equation 10.36 is identical to equation 10.22 for equimolecular counterdiffusion. Thus, the effects of bulk flow can be neglected at low concentrations.

Equation 10.33 can be written in terms of a mass transfer coefficient h_D to give:

$$N'_A = h_D(C_{A_1} - C_{A_2}) \tag{10.37}$$

where:
$$h_D = \frac{D}{y_2 - y_1}\frac{C_T}{C_{Bm}} \tag{10.38}$$

Similarly, working in terms of partial pressure difference as the driving force, equation 10.34 can be written:

$$N'_A = k_G(P_{A_1} - P_{A_2}) \tag{10.39}$$

where:
$$k_G = \frac{D}{\mathbf{R}T(y_2 - y_1)}\frac{P}{P_{Bm}} \tag{10.40}$$

Using mole fractions as the driving force, equation 10.35 becomes:

$$N'_A = k_x(x_{A_1} - x_{A_2}) \tag{10.41}$$

where:
$$k_x = \frac{DC_T}{y_2 - y_1}\frac{C_T}{C_{Bm}} = \frac{DC_T}{(y_2 - y_1)x_{Bm}} \tag{10.42}$$

It may be noted that all the transfer coefficients here are greater than those for equimolecular counterdiffusion by the factor $(C_T/C_{Bm})(= P/P_{Bm})$, which is an integrated form of the drift factor.

When the concentration C_A of the gas being transferred is low, C_T/C_{Bm} then approaches unity and the two sets of coefficients become identical.

Example 10.1

Ammonia gas is diffusing at a constant rate through a layer of stagnant air 1 mm thick. Conditions are such that the gas contains 50 per cent by volume ammonia at one boundary of the stagnant layer. The ammonia diffusing to the other boundary is quickly absorbed and the concentration is negligible at that plane. The temperature is 295 K and the pressure atmospheric, and under these conditions the diffusivity of ammonia in air is 1.8×10^{-5} m^2/s. Estimate the rate of diffusion of ammonia through the layer.

Solution

If the subscripts 1 and 2 refer to the two sides of the stagnant layer and the subscripts A and B refer to ammonia and air respectively, then the rate of diffusion through a stagnant layer is given by:

$$N_A = -\frac{D}{\mathbf{R}Tx}(P/P_{BM})(P_{A2} - P_{A1}) \qquad \text{(equation 10.31)}$$

In this case, $x = 0.001$ m, $D = 1.8 \times 10^{-5}$ m^2/s, $\mathbf{R} = 8314$ J/kmol K, $T = 295$ K and $P = 101.3$ kN/m^2 and hence:

$$P_{A1} = (0.50 \times 101.3) = 50.65 \text{ kN/m}^2$$

$P_{A2} = 0$

$P_{B1} = (101.3 - 50.65) = 50.65 \text{ kN/m}^2 = 5.065 \times 10^4 \text{ N/m}^2$

$P_{B2} = (101.3 - 0) = 101.3 \text{ kN/m}^2 = 1.013 \times 10^5 \text{ N/m}^2$

Thus: $P_{BM} = (101.3 - 50.65)/\ln(101.3/50.65) = 73.07 \text{ kN/m}^2 = 7.307 \times 10^4 \text{ N/m}^2$

and: $P/P_{BM} = (101.3/73.07) = 1.386.$

Thus, substituting in equation 10.31 gives:

$$N_A = -[1.8 \times 10^{-5}/(8314 \times 295 \times 0.001)]1.386(0 - 5.065 \times 10^4)$$

$$= \underline{\underline{5.15 \times 10^{-4} \text{ kmol/m}^2\text{s}}}$$

10.2.4. Diffusivities of gases and vapours

Experimental values of diffusivities are given in Table 10.2 for a number of gases and vapours in air at 298K and atmospheric pressure. The table also includes values of the Schmidt number *Sc*, the ratio of the kinematic viscosity (μ/ρ) to the diffusivity (D) for very low concentrations of the diffusing gas or vapour. The importance of the Schmidt number in problems involving mass transfer is discussed in Chapter 12.

Experimental determination of diffusivities

Diffusivities of vapours are most conveniently determined by the method developed by WINKELMANN[5] in which liquid is allowed to evaporate in a vertical glass tube over the top of which a stream of vapour-free gas is passed, at a rate such that the vapour

Table 10.2. Diffusivities (diffusion coefficients) of gases and vapours in air at 298 K and atmospheric pressure[4]

Substance	D ($\text{m}^2/\text{s} \times 10^6$)	$\mu/\rho D$	Substance	D ($\text{m}^2/\text{s} \times 10^6$)	$\mu/\rho D$
Ammonia	28.0	0.55	Valeric acid	6.7	2.31
Carbon dioxide	16.4	0.94	i-Caproic acid	6.0	2.58
Hydrogen	71.0	0.22	Diethyl amine	10.5	1.47
Oxygen	20.6	0.75	Butyl amine	10.1	1.53
Water	25.6	0.60	Aniline	7.2	2.14
Carbon disulphide	10.7	1.45	Chlorobenzene	7.3	2.12
Ethyl ether	9.3	1.66	Chlorotoluene	6.5	2.38
Methanol	15.9	0.97	Propyl bromide	10.5	1.47
Ethanol	11.9	1.30	Propyl iodide	9.6	1.61
Propanol	10.0	1.55	Benzene	8.8	1.76
Butanol	9.0	1.72	Toluene	8.4	1.84
Pentanol	7.0	2.21	Xylene	7.1	2.18
Hexanol	5.9	2.60	Ethyl benzene	7.7	2.01
Formic acid	15.9	0.97	Propyl benzene	5.9	2.62
Acetic acid	13.3	1.16	Diphenyl	6.8	2.28
Propionic acid	9.9	1.56	n-Octane	6.0	2.58
i-Butyric acid	8.1	1.91	Mesitylene	6.7	2.31

Note: the group $(\mu/\rho D)$ in the above table is evaluated for mixtures composed largely of air.

In this table, the figures taken from PERRY and GREEN[4] are based on data in *International Critical Tables* **5** (1928) and Landolt-Börnstein, *Physikalische-Chemische Tabellen* (1935).

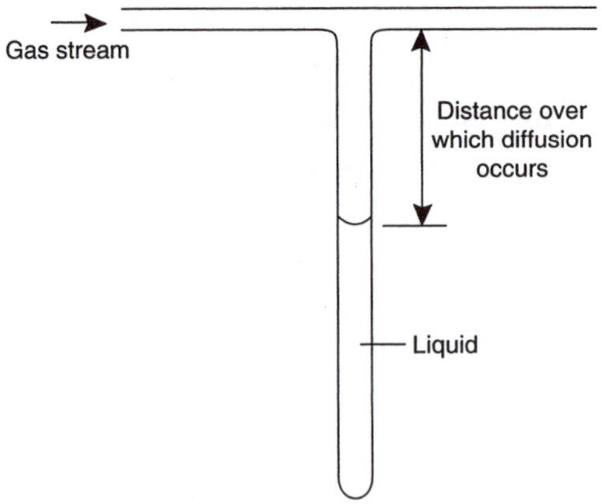

Figure 10.2. Determination of diffusivities of vapours

pressure is maintained almost at zero (Figure 10.2). If the apparatus is maintained at a steady temperature, there will be no eddy currents in the vertical tube and mass transfer will take place from the surface by molecular diffusion alone. The rate of evaporation can be followed by the rate of fall of the liquid surface, and since the concentration gradient is known, the diffusivity can then be calculated.

Example 10.2

The diffusivity of the vapour of a volatile liquid in air can be conveniently determined by Winkelmann's method in which liquid is contained in a narrow diameter vertical tube, maintained at a constant temperature, and an air stream is passed over the top of the tube sufficiently rapidly to ensure that the partial pressure of the vapour there remains approximately zero. On the assumption that the vapour is transferred from the surface of the liquid to the air stream by molecular diffusion alone, calculate the diffusivity of carbon tetrachloride vapour in air at 321 K and atmospheric pressure from the experimental data given in Table 10.3.

Table 10.3. Experimental data for diffusivity calculation

Time from commencement of experiment (h min)		(ks)	Liquid level (mm)	Time from commencement of experiment (h min)		(ks)	Liquid level (mm)
0	0	0.0	0.0	32	38	117.5	54.7
0	26	1.6	2.5	46	50	168.6	67.0
3	5	11.1	12.9	55	25	199.7	73.8
7	36	27.4	23.2	80	22	289.3	90.3
22	16	80.2	43.9	106	25	383.1	104.8

The vapour pressure of carbon tetrachloride at 321 K is 37.6 kN/m^2 and the density of the liquid is 1540 kg/m^3. The kilogram molecular volume may be taken as 22.4 m^3.

Solution

From equation 10.33 the rate of mass transfer is given by:

$$N'_A = D\frac{C_A}{L}\frac{C_T}{C_{Bm}}$$

where C_A is the saturation concentration at the interface and L is the effective distance through which mass transfer is taking place. Considering the evaporation of the liquid:

$$N'_A = \frac{\rho_L}{M}\frac{\mathrm{d}L}{\mathrm{d}t}$$

where ρ_L is the density of the liquid.

Hence:
$$\frac{\rho_L}{M}\frac{\mathrm{d}L}{\mathrm{d}t} = D\frac{C_A}{L}\frac{C_T}{C_{Bm}}$$

Integrating and putting $L = L_0$ at $t = 0$:

$$L^2 - L_0^2 = \frac{2MD}{\rho_L}\frac{C_A C_T}{C_{Bm}}t$$

L_0 will not be measured accurately nor is the effective distance for diffusion, L, at time t.

Accurate values of $(L - L_0)$ are available, however, and hence:

$$(L - L_0)(L - L_0 + 2L_0) = \frac{2MD}{\rho_L}\frac{C_A C_T}{C_{Bm}}t$$

or:
$$\frac{t}{L - L_0} = \frac{\rho_L}{2MD}\frac{C_{Bm}}{C_A C_T}(L - L_0) + \frac{\rho_L C_{Bm}}{MDC_A C_T}L_0$$

If s is the slope of a plot of $t/(L - L_0)$ against $(L - L_0)$, then:

$$s = \frac{\rho_L C_{Bm}}{2MDC_A C_T} \quad \text{or} \quad D = \frac{\rho_L C_{Bm}}{2MC_A C_T s}$$

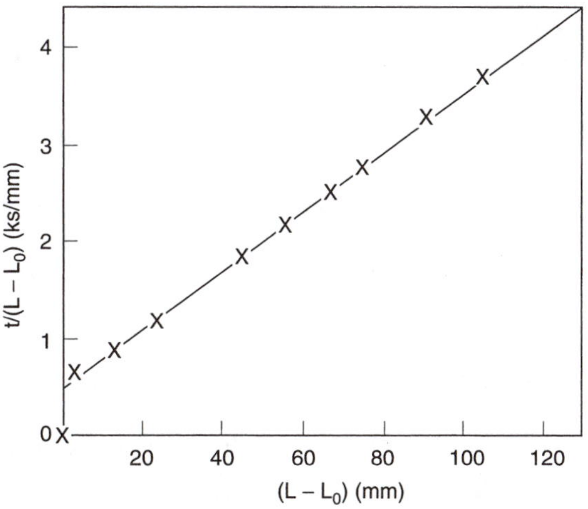

Figure 10.3. Plot of $t/(L - L_0)$ versus $(L - L_0)$ for Example 10.1

From a plot of $t/(L - L_0)$ against $(L - L_0)$ as shown in Figure 10.3:

$$s = 0.0310 \text{ ks/mm}^2 \quad \text{or} \quad 3.1 \times 10^7 \text{ s/m}^2$$

and:
$$C_T = \left(\frac{1}{22.4}\right)\left(\frac{273}{321}\right) = 0.0380 \text{ kmol/m}^3$$

$$M = 154 \text{ kg/kmol}$$

$$C_A = \left(\frac{37.6}{101.3}\right) 0.0380 = 0.0141 \text{ kmol/m}^3$$

$$\rho_L = 1540 \text{ kg/m}^3$$

$$C_{B1} = 0.0380 \text{ kmol/m}^3, \quad C_{B2} = \left(\frac{101.3 - 37.6}{101.3}\right) 0.0380 = 0.0238 \text{ kmol/m}^3$$

Thus:

$$C_{Bm} = \frac{(0.0380 - 0.0238)}{\ln(0.0380/0.0238)} = 0.0303 \text{ kmol/m}^3$$

and:

$$D = \frac{1540 \times 0.0303}{2 \times 154 \times 0.0141 \times 0.0380 \times 3.1 \times 10^7}$$

$$= \underline{\underline{9.12 \times 10^{-6} \text{ m}^2/\text{s}}}$$

Prediction of diffusivities

Where, the diffusivity D for the transfer of one gas in another is not known and experimental determination is not practicable, it is necessary to use one of the many predictive procedures. A commonly used method due to GILLILAND[6] is based on the "Stefan–Maxwell" hard sphere model and this takes the form:

$$D = \frac{4.3 \times 10^{-4} T^{1.5} \sqrt{(1/M_A) + (1/M_B)}}{P\left(\mathbf{V}_A^{1/3} + \mathbf{V}_B^{1/3}\right)^2} \tag{10.43}$$

where D is the diffusivity in m^2/s, T is the absolute temperature (K), M_A, M_B are the molecular masses of **A** and **B**, P is the total pressure in N/m^2, and $\mathbf{V}_A$, $\mathbf{V}_B$ are the molecular volumes of **A** and **B**. The molecular volume is the volume in m^3 of one kmol of the material in the form of liquid at its boiling point, and is a measure of the volume occupied by the molecules themselves. It may not always be known, although an approximate value can be obtained, for all but simple molecules, by application of *Kopp's law* of additive volumes. Kopp has presented a particular value for the equivalent atomic volume of each element[7], as given in Table 10.4, such that when the atomic volumes of the elements of the molecule in question are added in the appropriate proportions, an approximate value the equivalent molecular volume is obtained . There are certain exceptions to this rule, and corrections have to be made if the elements are combined in particular ways.

It will be noted from equation 10.43 that the diffusivity of a vapour is inversely proportional to the pressure and varies with the absolute temperature raised to the power of 1.5, although it has been suggested that this underestimates the temperature dependence.

A method, proposed more recently by FULLER, SCHETTLER and GIDDINGS[8], is claimed to give an improved correlation. In this approach the values of the "diffusion volume" have been "modified" to give a better correspondence with experimental values, and have then been adjusted arbitrarily to make the coefficient in the equation equal to unity. The method does contain some anomalies, however, particularly in relation to the values of $\mathbf{V}$ for nitrogen, oxygen and air. Details of this method are given in Volume 6.

Table 10.4. Atomic and structural diffusion volume increments (m^3/kmol)(7)

Antimony	0.0242	Oxygen, double-bonded	0.0074
Arsenic	0.0305	Coupled to two other elements:	
Bismuth	0.0480	in aldehydes and ketones	0.0074
Bromine	0.0270	in methyl esters	0.0091
Carbon	0.0148	in ethyl esters	0.0099
Chlorine, terminal, as in R—Cl	0.0216	in higher esters and ethers	0.0110
medial, as in R—CHCl—R′	0.0246	in acids	0.0120
Chromium	0.0274	in union with S, P, N	0.0083
Fluorine	0.0087	Phosphorus	0.0270
Germanium	0.0345	Silicon	0.0320
Hydrogen	0.0037	Sulphur	0.0256
Nitrogen, double-bonded	0.0156	Tin	0.0423
in primary amines	0.0105	Titanium	0.0357
in secondary amines	0.0120	Vanadium	0.0320
		Zinc	0.0204

For a three-membered ring, as in ethylene oxide,	deduct 0.0060.
For a four-membered ring, as in cyclobutane,	deduct 0.0085.
For a five-membered ring, as in furane,	deduct 0.0115.
For a six-membered ring, as in benzene, pyridine,	deduct 0.0150.
For an anthracene ring formation,	deduct 0.0475.
For naphthalene	deduct 0.0300.

Diffusion volumes of simple molecules (m^3/kmol)

H_2	0.0143	CO_2	0.0340	NH_3	0.0258
O_2	0.0256	H_2O	0.0189	H_2S	0.0329
N_2	0.0312	SO_2	0.0448	Cl_2	0.0484
Air	0.0299	NO	0.0236	Br_2	0.0532
CO	0.0307	N_2O	0.0364	I_2	0.0715

Example 10.3

Ammonia is absorbed in water from a mixture with air using a column operating at 1 bar and 295 K. The resistance to transfer may be regarded as lying entirely within the gas phase. At a point in the column, the partial pressure of the ammonia is 7.0 kN/m^2. The back pressure at the water interface is negligible and the resistance to transfer may be regarded as lying in a stationary gas film 1 mm thick. If the diffusivity of ammonia in air is 2.36×10^{-5} m^2/s, what is the transfer rate per unit area at that point in the column? How would the rate of transfer be affected if the ammonia air mixture were compressed to double the pressure?

Solution

Concentration of ammonia in the gas

$$= \left(\frac{1}{22.4}\right)\left(\frac{101.3}{101.3}\right)\left(\frac{273}{295}\right)\left(\frac{7.0}{101.3}\right) = 0.00285 \text{ kmol/m}^3$$

Thus:

$$\frac{C_T}{C_{Bm}} = \frac{101.3 \ln(101.3/94.3)}{101.3 - 94.3} = 1.036$$

From equation 10.33:

$$N'_A = \frac{D}{y_2 - y_1}\frac{C_T}{C_{Bm}}(C_{A1} - C_{A2})$$

$$= \left(\frac{2.36 \times 10^{-5}}{1 \times 10^{-3}}\right)(1.036 \times 0.00285)$$

$$= \underline{\underline{6.97 \times 10^{-5} \text{ kmol/m}^2\text{s}}}$$

If the pressure is doubled, the driving force is doubled, C_T/C_{Bm} is essentially unaltered, and the diffusivity, being inversely proportional to the pressure (equation 10.43) is halved. The mass transfer rate therefore remains the same.

10.2.5. Mass transfer velocities

It is convenient to express mass transfer rates in terms of velocities for the species under consideration where:

$$\text{Velocity} = \frac{\text{Flux}}{\text{Concentration}},$$

which, in the S.I system, has the units $(\text{kmol/m}^2\text{s})/(\text{kmol/m}^3) = \text{m/s}$.

For diffusion according to Fick's Law:

$$u_{DA} = \frac{N_A}{C_A} = -\frac{D}{C_A}\frac{\mathrm{d}C_A}{\mathrm{d}y} \tag{10.44a}$$

and:

$$u_{DB} = \frac{N_B}{C_B} = -\frac{D}{C_B}\frac{\mathrm{d}C_B}{\mathrm{d}y} = \frac{D}{C_B}\frac{\mathrm{d}C_A}{\mathrm{d}y} \tag{10.44b}$$

Since $N_B = -N_A$, then:

$$u_{DB} = -u_{DA}\frac{C_A}{C_B} = -u_{DA}\frac{x_A}{x_B} \tag{10.45}$$

As a result of the diffusional process, there is no net overall molecular flux arising from diffusion in a binary mixture, the two components being transferred at equal and opposite rates. In the process of equimolecular counterdiffusion which occurs, for example, in a distillation column when the two components have equal molar latent heats, the diffusional velocities are the same as the velocities of the molecular species relative to the walls of the equipment or the phase boundary.

If the physical constraints placed upon the system result in a bulk flow, the velocities of the molecular species relative to one another remain the same, but in order to obtain the velocity relative to a fixed point in the equipment, it is necessary to add the bulk flow velocity. An example of a system in which there is a bulk flow velocity is that in which one of the components is transferred through a second component which is undergoing no net transfer, as for example in the absorption of a soluble gas **A** from a mixture with an insoluble gas **B**. (See Section 10.2.3). In this case, because there is no set flow of **B**, the sum of its diffusional velocity and the bulk flow velocity must be zero.

In this case:

Component	**A**	**B**
Diffusional velocity	$u_{DA} = -\frac{D}{C_A}\frac{\mathrm{d}C_A}{\mathrm{d}y}$	$u_{DB} = +\frac{D}{C_B}\frac{\mathrm{d}C_A}{\mathrm{d}y}$
Bulk flow velocity	$u_F = -\frac{D}{C_B}\frac{\mathrm{d}C_A}{\mathrm{d}y}$	$u_F = -\frac{D}{C_B}\frac{\mathrm{d}C_A}{\mathrm{d}y}$
Total velocity	$u_A = -D\frac{C_T}{C_A C_B}\frac{\mathrm{d}C_A}{\mathrm{d}y}$	$u_B = 0$
Flux	$N'_A = u_A C_A = -D\frac{C_T}{C_B}\frac{\mathrm{d}C_A}{\mathrm{d}y}$	$N'_B = 0$

The flux of **A** has been given as Stefan's Law (equation 10.30).

10.2.6. General case for gas-phase mass transfer in a binary mixture

Whatever the physical constraints placed on the system, the diffusional process causes the two components to be transferred at equal and opposite rates and the values of the diffusional velocities u_{DA} and u_{DB} given in Section 10.2.5 are always applicable. It is the bulk flow velocity u_F which changes with imposed conditions and which gives rise to differences in overall mass transfer rates. In equimolecular counterdiffusion, u_F is zero. In the absorption of a soluble gas **A** from a mixture the bulk velocity must be equal and opposite to the diffusional velocity of **B** as this latter component undergoes no net transfer.

In general, for any component:

Total transfer = Transfer by diffusion + Transfer by bulk flow.

For component **A**:

Total transfer (moles/area time) $= N'_A$

Diffusional transfer according to Fick's Law $= N_A = -D\dfrac{dC_A}{dy}$

Transfer by bulk flow $= u_F C_A$

Thus for **A**: $$N'_A = N_A + u_F C_A \tag{10.46a}$$

and for **B**: $$N'_B = N_B + u_F C_B \tag{10.46b}$$

$$\text{The bulk flow velocity } u_F = \frac{\text{Total moles transferred/area time}}{\text{Total molar concentration}}$$

$$= \frac{(N'_A + N'_B)}{C_T} \tag{10.47}$$

Substituting:

$$N'_A = N_A + \frac{C_A}{C_T}(N'_A + N'_B)$$

$$N'_A = -D\frac{dC_A}{dy} + x_A(N'_A + N'_B)$$

$$N'_A = -DC_T\frac{dx_A}{dy} + x_A(N'_A + N'_B) \tag{10.48}$$

Similarly for **B**: $$N'_B = DC_T\frac{dx_A}{dy} + (1 - x_A)(N'_A + N'_B) \tag{10.49}$$

For equimolecular counterdiffusion $N'_A = -N'_B$ and equation 10.48 reduces to Fick's Law. For a system in which **B** undergoes no net transfer, $N'_B = 0$ and equation 10.48 is identical to Stefan's Law.

For the general case: $$fN'_A = -N'_B \tag{10.50}$$

If in a distillation column, for example the molar latent heat of **A** is f times that of **B**, the condensation of 1 mole of **A** (taken as the less volatile component) will result in the vaporisation of f moles of **B** and the mass transfer rate of **B** will be f times that of **A** in the opposite direction.

Substituting into equation 10.48:

$$N'_A = -DC_T\frac{dx_A}{dy} + x_A(N'_A - fN'_A) \tag{10.51}$$

Thus: $$[1 - x_A(1-f)]N'_A = -DC_T\frac{dx_A}{dy}$$

If x_A changes from x_{A_1} to x_{A_2} as y goes from y_1 to y_2, then:

$$N'_A\int_{y_1}^{y_2} dy = -DC_T\int_{x_{A_1}}^{x_{A_2}}\frac{dx_A}{1 - x_A(1-f)}$$

Thus: $$N'_A(y_2 - y_1) = -DC_T\frac{1}{1-f}\left[\ln\frac{1}{(1-f)^{-1} - x_A}\right]_{x_{A_1}}^{x_{A_2}}$$

or: $$N'_A = \frac{DC_T}{y_2 - y_1}\frac{1}{1-f}\ln\frac{1 - x_{A_2}(1-f)}{1 - x_{A_1}(1-f)} \tag{10.52}$$

10.2.7. Diffusion as a mass flux

Fick's Law of diffusion is normally expressed in *molar* units or:

$$N_A = -D\frac{dC_A}{dy} = -DC_T\frac{dx_A}{dy} \qquad \text{(equation 10.4)}$$

where x_A is the mole fraction of component **A.**

The corresponding equation for component **B** indicates that there is an equal and opposite molar flux of that component. If each side of equation 10.4 is multiplied by the molecular weight of **A**, M_A, then:

$$J_A = -D\frac{dc_A}{dy} = -DM_A\frac{dC_A}{dy} = -DC_TM_A\frac{dx_A}{dy} \tag{10.53}$$

where J_A is a flux in mass per unit area and unit time (kg/m^2 s in S.I units), and c_A is a concentration in mass terms, (kg/m^3 in S.I units).

Similarly, for component **B:**

$$J_B = -D\frac{dc_B}{dy} \tag{10.54}$$

Although the sum of the molar concentrations is constant in an ideal gas at constant pressure, the sum of the mass concentrations is not constant, and dc_A/dy and dc_B/dy are not equal and opposite,

Thus: $$C_A + C_B = C_T = \frac{c_A}{M_A} + \frac{c_B}{M_B} = \text{constant} \tag{10.55}$$

or: $$\frac{1}{M_A}\frac{dc_A}{dy} + \frac{1}{M_B}\frac{dc_B}{dy} = 0$$

and:
$$\frac{dc_B}{dy} = -\frac{M_B}{M_A}\frac{dc_A}{dy} \tag{10.56}$$

Thus, the diffusional process does not give rise to equal and opposite mass fluxes.

10.2.8. Thermal diffusion

If a temperature gradient is maintained in a binary gaseous mixture, a concentration gradient is established with the light component collecting preferentially at the hot end and the heavier one at the cold end. This phenomenon, known as the *Soret effect*, may be used as the basis of a separation technique of commercial significance in the separation of isotopes.

Conversely, when mass transfer is occurring as a result of a constant concentration gradient, a temperature gradient may be generated; this is known as the *Dufour effect*.

In a binary mixture consisting of two gaseous components **A** and **B** subject to a temperature gradient, the flux due to thermal diffusion is given by GREW and IBBS[9]:

$$(N_A)_{Th} = -D_{Th}\frac{1}{T}\frac{dT}{dy} \tag{10.57}$$

where $(N_A)_{Th}$ is the molar flux of **A** (kmol/m^2 s) in the Y-direction, and D_{Th} is the diffusion coefficient for thermal diffusion (kmol/m s).

Equation 10.57, with a positive value of D_{Th}, applies to the component which travels preferentially to the *low* temperature end of the system. For the component which moves to the high temperature end, D_{Th} is negative. In a binary mixture, the gas of higher molecular weight has the positive value of D_{Th} and this therefore tends towards the lower temperature end of the system.

If two vessels each containing completely mixed gas, one at temperature T_1 and the other at a temperature T_2, are connected by a lagged non-conducting pipe in which there are no turbulent eddies (such as a capillary tube), then under steady state conditions, the rate of transfer of **A** by thermal diffusion and molecular diffusion must be equal and opposite, or:

$$(N_A)_{Th} + N_A = 0 \tag{10.58}$$

N_A is given by Fick's Law as:

$$N_A = -D\frac{dC_A}{dy} = -DC_T\frac{dx_A}{dy} \qquad \text{(equation 10.53)}$$

where x_A is the mole fraction of **A**, and C_T is the total molar concentration at y and will not be quite constant because the temperature is varying.

Substituting equations 10.53 and 10.57 into equation 10.58 gives:

$$-D_{Th}\frac{1}{T}\frac{dT}{dy} - DC_T\frac{dx_A}{dy} = 0 \tag{10.59}$$

The relative magnitudes of the thermal diffusion and diffusion effects are represented by the dimensionless ratio:

$$\frac{D_{Th}}{DC_T} = K_{ABT}$$

where K_{ABT} is known as the *thermal diffusion ratio*.

Thus:
$$-K_{ABT}\frac{1}{T}\frac{dT}{dy} = \frac{dx_A}{dy} \tag{10.60}$$

If temperature gradients are small, C_T may be regarded as effectively constant. Furthermore, K_{ABT} is a function of composition, being approximately proportional to the product $x_A x_B$. It is therefore useful to work in terms of the *thermal diffusion factor* α, where:

$$\alpha = \frac{K_{ABT}}{x_A(1 - x_A)}$$

Substituting for α, assumed constant, in equation 10.60 and integrating, gives:

$$\alpha \ln \frac{T_1}{T_2} = \ln\left(\frac{x_{A_2}}{1 - x_{A_2}}\frac{1 - x_{A_1}}{x_{A_1}}\right)$$

Thus:
$$\frac{x_{A_2}}{x_{A_1}}\frac{x_{B_1}}{x_{B_2}} = \left(\frac{T_1}{T_2}\right)^{\alpha} \tag{10.61}$$

Equation 10.61 gives the mole fraction of the two components **A** and **B** as a function of the absolute temperatures and the thermal diffusion factor.

Values of α taken from data in GREW and IBBS[9] and HIRSCHFELDER, CURTISS and BIRD[10] are given in Table 10.5.

Table 10.5. Values of thermal diffusion factor (α) for binary gas mixtures (**A** is the heavier component, which moves towards the cooler end)

Systems A	Systems B	Temperature (K)	Mole fraction of **A** x_A	α
D_2	H_2	288–373	0.48	0.17
He	H_2	273–760	0.50	0.15
N_2	H_2	288–373	0.50	0.34
Ar	H_2	258	0.53	0.26
O_2	H_2	90–294	0.50	0.19
CO	H_2	288–373	0.50	0.33
CO_2	H_2	288–456	0.50	0.28
C_3H_8	H_2	230–520	0.50	0.30
N_2	He	287–373	0.50	0.36
		260	0.655	0.37
Ar	He	330	0.90	0.28
		330	0.70	0.31
		330	0.50	0.37
Ne	He	205	0.46	0.31
		330	0.46	0.315
		365	0.46	0.315
O_2	N_2	293	0.50	0.018

10.2.9. Unsteady-state mass transfer

In many practical mass transfer processes, unsteady state conditions prevail. Thus, in the example given in Section 10.1, a box is divided into two compartments each containing a different gas and the partition is removed. Molecular diffusion of the gases takes place and concentrations, and concentration gradients, change with time. If a bowl of liquid evaporates into an enclosed space, the partial pressure in the gas phase progressively increases, and the concentrations and the rate of evaporation are both time-dependent.

Considering an element of gas of cross-sectional area A and of thickness δy in the direction of mass transfer in which the concentrations C_A and C_B of the components **A** and **B** are a function of both position y and time t (Figure 10.4), then if the mass transfer flux is composed of two components, one attributable to diffusion according to Fick's Law and the other to a bulk flow velocity u_F, the fluxes of **A** and **B** at a distance y from the origin may be taken as N'_A and N'_B, respectively. These will increase to $N'_A + (\partial N'_A/\partial y)\delta y$ and $N'_B + (\partial N'_B/\partial y)\delta y$ at a distance $y + \delta y$ from the origin.

At position y, the fluxes N'_A and N'_B will be as given in Table 10.6. At a distance $y + \delta y$ from the origin, that is at the further boundary of the element, these fluxes will increase by the amounts shown in the lower part of Table 10.6.

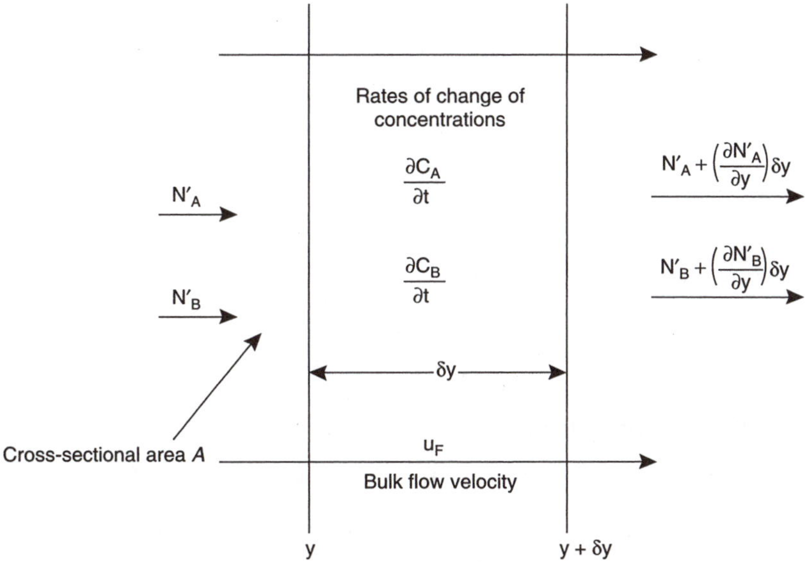

Figure 10.4. Unsteady state mass transfer

Thus, for **A**:

moles IN/unit time (at y)	−	moles OUT/unit time (at $y+\delta y$)	= rate of change of concn. × element volume

$$\left\{-D\frac{\partial C_A}{\partial y} + u_F C_A\right\}A - \left\{-D\frac{\partial C_A}{\partial y} + u_F C_A + \frac{\partial}{\partial y}\left[-D\frac{\partial C_A}{\partial y} + u_F C_A\right]\delta y\right\}A = \frac{\partial C_A}{\partial t}(\delta y.A)$$

Simplifying:
$$\frac{\partial C_A}{\partial t} = D\frac{\partial^2 C_A}{\partial y^2} - \frac{\partial(u_F C_A)}{\partial y} \tag{10.64a}$$

For component **B**:
$$\frac{\partial C_B}{\partial t} = D\frac{\partial^2 C_B}{\partial y^2} - \frac{\partial(u_F C_B)}{\partial y} \tag{10.64b}$$

and adding:
$$\frac{\partial(C_A + C_B)}{\partial t} = D\frac{\partial^2(C_A + C_B)}{\partial y^2} - \frac{\partial}{\partial y}[(C_A + C_B)u_F]$$

Table 10.6. Fluxes of a gas mixture

	Diffusional flux	Flux due to bulk flow	Total flux	
A	$-D\dfrac{\partial C_A}{\partial y}$	$u_F C_A$	$N'_A = -D\dfrac{\partial C_A}{\partial y} + u_F C_A$	(10.62)
B	$-D\dfrac{\partial C_B}{\partial y}$	$u_F C_B$	$N'_B = -D\dfrac{\partial C_B}{\partial y} + u_F C_B$	(10.63)
Changes in fluxes over distance δy:				
A	$-D\left(\dfrac{\partial^2 C_A}{\partial y^2}\right)\delta y$	$\dfrac{\partial(u_F C_A)}{\partial y}\delta y$	$\left\{-D\dfrac{\partial^2 C_A}{\partial y^2} + \dfrac{\partial(u_F C_A)}{\partial y}\right\}\delta y$	
B	$-D\left(\dfrac{\partial^2 C_B}{\partial y^2}\right)\delta y$	$\dfrac{\partial(u_F C_B)}{\partial y}\delta y$	$\left\{-D\dfrac{\partial^2 C_B}{\partial y^2} + \dfrac{\partial(u_F C_B)}{\partial y}\right\}\delta y$	

Since, for an ideal gas, $C_A + C_B = C_T = \text{constant}$ (equation 10.9):

$$0 = 0 - \frac{\partial}{\partial y}(u_F C_T)$$

and:

$$\frac{\partial u_F}{\partial y} = 0$$

where u_F is therefore independent of y.

Thus equation 10.64*a* can be written:

$$\frac{\partial C_A}{\partial t} = D\frac{\partial^2 C_A}{\partial y^2} - u_F\frac{\partial C_A}{\partial y} \tag{10.65}$$

Equimolecular counterdiffusion

For equimolecular counterdiffusion, $u_F = 0$ and equation 10.65 simplifies to:

$$\frac{\partial C_A}{\partial t} = D\frac{\partial^2 C_A}{\partial y^2} \tag{10.66}$$

Equation 10.66 is referred to as Fick's Second Law. This also applies when u_F is small, corresponding to conditions where C_A is always low. This equation can be solved for a number of important boundary conditions, and it should be compared with the corresponding equation for unsteady state heat transfer (equation 9.29).

For the more general three-dimensional case where concentration gradients are changing in the x, y and z directions, these changes must be added to give:

$$\frac{\partial C_A}{\partial t} = D\left[\frac{\partial^2 C_A}{\partial x^2} + \frac{\partial^2 C_A}{\partial y^2} + \frac{\partial^2 C_A}{\partial z^2}\right] \tag{10.67}$$

(c/f equation 9.28)

Gas absorption

In general, it is necessary to specify the physical constraints operating on the system in order to evaluate the bulk flow velocity u_F. In gas absorption, there will be no overall flux of the insoluble component **B** at the liquid interface ($y = 0$, say). In an unsteady state process, however, where, by definition, concentrations will be changing with time throughout the system, the flux of **B** will be zero only at $y = 0$. At the interface ($y = 0$), total flux of **B** (from equation 10.43*b*) is given by:

$$N'_B = -D\left(\frac{\partial C_B}{\partial y}\right)_{y=0} + u_F(C_B)_{y=0} = 0$$

or:
$$u_F = \frac{D\left(\dfrac{\partial C_B}{\partial y}\right)_{y=0}}{(C_B)_{y=0}} = \frac{-D}{(C_T - C_A)_{y=0}}\left(\frac{\partial C_A}{\partial y}\right)_{y=0} \tag{10.68}$$

Substituting in equation 10.65:

$$\frac{\partial C_A}{\partial t} = D\left[\frac{\partial^2 C_A}{\partial y^2} + \frac{1}{(C_T - C_A)_{y=0}}\left(\frac{\partial C_A}{\partial y}\right)_{y=0}\frac{\partial C_A}{\partial y}\right] \tag{10.69}$$

Thus at the interface ($y = 0$):

$$\left(\frac{\partial C_A}{\partial t}\right)_{y=0} = D\left[\left(\frac{\partial^2 C_A}{\partial y^2}\right)_{y=0} + \frac{1}{(C_T - C_A)_{y=0}}\left(\frac{\partial C_A}{\partial y}\right)^2_{y=0}\right] \tag{10.70}$$

This equation which is not capable of an exact analytical solution has been discussed by ARNOLD[11] in relation to evaporation from a free surface.

Substituting into equation 10.62 for N'_A:

$$N'_A = -D\frac{\partial C_A}{\partial y} - \frac{DC_A}{(C_T - C_A)_{y=0}}\left(\frac{\partial C_A}{\partial y}\right)_{y=0}$$

$$= -D\left[\frac{\partial C_A}{\partial y} + \frac{C_A}{(C_T - C_A)_{y=0}}\left(\frac{\partial C_A}{\partial y}\right)_{y=0}\right] \tag{10.71}$$

10.3. MULTICOMPONENT GAS-PHASE SYSTEMS

10.3.1. Molar flux in terms of effective diffusivity

For a multicomponent system, the bulk flow velocity u_F is given by:

$$u_F = \frac{1}{C_T}(N'_A + N'_B + N'_C + \cdots) \tag{10.72}$$

or:
$$u_F = x_A u_A + x_B u_B + x_C u_C + \cdots \tag{10.73}$$

or:
$$u_F = \sum x_A u_A = \frac{1}{C_T}\sum N'_A \tag{10.74}$$

Since $N'_A = N_A + u_F C_A$ (equation 10.46*a*), then:

$$N'_A = -D'\frac{dC_A}{dy} + \left[\frac{1}{C_T}\Sigma N'_A\right] C_A$$

$$= -D'C_T\frac{dx_A}{dy} + x_A\Sigma N'_A \tag{10.75}$$

where D' is the effective diffusivity for transfer of **A** in a mixture of **B**, **C**, **D**, and so on.

For the particular case, where N'_B, N'_C, and so on, are all zero:

$$N'_A = -D'C_T\frac{dx_A}{dy} + x_A N'_A$$

or:

$$N'_A = -D'\frac{C_T}{1 - x_A}\frac{dx_A}{dy} \tag{10.76}$$

A method of calculating the effective diffusivity D' in terms of each of the binary diffusivities is presented in Section 10.3.2.

10.3.2. Maxwell's law of diffusion

Maxwell's law for a binary system

MAXWELL[12] postulated that the partial pressure gradient in the direction of diffusion for a constituent of a two-component gaseous mixture was proportional to:

(a) the relative velocity of the molecules in the direction of diffusion, and
(b) the product of the molar concentrations of the components.

Thus:

$$-\frac{dP_A}{dy} = FC_A C_B(u_A - u_B) \tag{10.77}$$

where u_A and u_B are the mean molecular velocities of **A** and **B** respectively in the direction of mass transfer and F is a coefficient.

Noting that:

$$u_A = \frac{N'_A}{C_A} \tag{10.78}$$

and:

$$u_B = \frac{N'_B}{C_B} \tag{10.79}$$

and using: $P_A = C_A\mathbf{R}T$ (for an ideal gas) (equation 10.10*a*)

on substitution into equation 10.77 gives:

$$-\frac{dC_A}{dy} = \frac{F}{\mathbf{R}T}(N_A C_B - N_B C_A) \tag{10.80}$$

Equimolecular counterdiffusion

By definition: $$N'_A = -N'_B = N_A$$

Substituting in equation 10.80:

$$-\frac{dC_A}{dy} = \frac{FN_A}{\mathbf{R}T}(C_B + C_A) \tag{10.81}$$

or: $$N_A = -\frac{\mathbf{R}T}{FC_T}\frac{dC_A}{dy} \tag{10.82}$$

Then, by comparison with Fick's Law (equation 10.4):

$$D = \frac{\mathbf{R}T}{FC_T} \tag{10.83}$$

or: $$F = \frac{\mathbf{R}T}{DC_T} \tag{10.84}$$

Transfer of A *through stationary* B

By definition: $$N'_B = 0$$

Thus: $$-\frac{dC_A}{dy} = \frac{F}{\mathbf{R}T}N'_A C_B \quad \text{(from equation 10.80)}$$

or: $$N'_A = -\frac{\mathbf{R}T}{FC_T}\frac{C_T}{C_B}\frac{dC_A}{dy} \tag{10.85}$$

Substituting from equation 10.83:

$$N'_A = -D\frac{C_T}{C_B}\frac{dC_A}{dy} \tag{10.86}$$

It may be noted that equation 10.86 is identical to equation 10.30. (Stefan's Law) and, Stefan's law can therefore also be derived from Maxwell's Law of Diffusion.

Maxwell's Law for multicomponent mass transfer

This argument can be applied to the diffusion of a constituent of a multi-component gas. Considering the transfer of component **A** through a stationary gas consisting of components **B**, **C**, ... if the total partial pressure gradient can be regarded as being made up of a series of terms each representing the contribution of the individual component gases, then from equation 10.80:

$$-\frac{dC_A}{dy} = \frac{F_{AB}N'_A C_B}{\mathbf{R}T} + \frac{F_{AC}N'_A C_C}{\mathbf{R}T} + \cdots$$

or: $$-\frac{dC_A}{dy} = \frac{N'_A}{\mathbf{R}T}(F_{AB}C_B + F_{AC}C_C + \cdots) \tag{10.87}$$

From equation 10.84, writing:

$$F_{AB} = \frac{\mathbf{R}T}{D_{AB}C_T}, \quad \text{and so on.}$$

where D_{AB} is the diffusivity of **A** in **B**, and so on.:

$$-\frac{dC_A}{dy} = \frac{N'_A}{C_T}\left(\frac{C_B}{D_{AB}} + \frac{C_C}{D_{AC}} + \cdots\right) \tag{10.88}$$

$$\therefore \quad N'_A = -\frac{C_T}{\dfrac{C_B}{D_{AB}} + \dfrac{C_C}{D_{AC}} + \cdots}\frac{dC_A}{dy}$$

$$= -\frac{1}{\dfrac{C_B}{C_T - C_A}\dfrac{1}{D_{AB}} + \dfrac{C_C}{C_T - C_A}\dfrac{1}{D_{AC}} + \cdots}\frac{C_T}{C_T - C_A}\frac{dC_A}{dy}$$

$$= -\frac{1}{\dfrac{y'_B}{D_{AB}} + \dfrac{y'_C}{D_{AC}} + \cdots}\frac{C_T}{C_T - C_A}\frac{dC_A}{dy} \tag{10.89}$$

where y'_B is the mole fraction of **B** and so on in the stationary components of the gas.

By comparing equation 10.89 with Stefan's Law (equation 10.30) the effective diffusivity of **A** in the mixture (D') is given by:

$$\frac{1}{D'} = \frac{y'_B}{D_{AB}} + \frac{y'_C}{D_{AC}} + \cdots \tag{10.90}$$

Multicomponent mass transfer is discussed in more detail by TAYLOR and KRISHNA(13), CUSSLER(14) and ZIELINSKI and HANLEY(15)

10.4. DIFFUSION IN LIQUIDS

Whilst the diffusion of solution in a liquid is governed by the same equations as for the gas phase, the diffusion coefficient D is about two orders of magnitude smaller for a liquid than for a gas. Furthermore, the diffusion coefficient is a much more complex function of the molecular properties.

For an ideal gas, the total molar concentration C_T is constant at a given total pressure P and temperature T. This approximation holds quite well for real gases and vapours, except at high pressures. For a liquid however, C_T may show considerable variations as the concentrations of the components change and, in practice, the total mass concentration (density ρ of the mixture) is much more nearly constant. Thus for a mixture of ethanol and water for example, the mass density will range from about 790 to 1000 kg/m^3 whereas the molar density will range from about 17 to 56 kmol/m^3. For this reason the diffusion equations are frequently written in the form of a mass flux J_A (mass/area $\times$ time) and the concentration gradients in terms of mass concentrations, such as c_A.

Thus, for component **A**, the mass flux is given by:

$$J_A = -D\frac{dc_A}{dy} \tag{10.91}$$

$$= -D\rho \frac{d\omega_A}{dy} \tag{10.92}$$

where ρ is mass density (now taken as constant), and ω_A is the mass fraction of **A** in the liquid.

For component **B**:

$$J_B = -D\rho \frac{d\omega_B}{dy} \tag{10.93}$$

$$= D\rho \frac{d\omega_A}{dy} \quad \text{(since } \omega_A + \omega_B = 1\text{)}. \tag{10.94}$$

Thus, the diffusional process in a liquid gives rise to a situation where the components are being transferred at approximately equal and opposite mass (rather than molar) rates.

Liquid phase diffusivities are strongly dependent on the concentration of the diffusing component which is in strong contrast to gas phase diffusivities which are substantially independent of concentration. Values of liquid phase diffusivities which are normally quoted apply to very dilute concentrations of the diffusing component, the only condition under which analytical solutions can be produced for the diffusion equations. For this reason, only dilute solutions are considered here, and in these circumstances no serious error is involved in using Fick's first and second laws expressed in molar units.

The molar flux is given by:

$$N_A = -D\frac{dC_A}{dy} \qquad \text{(equation 10.4)}$$

and:

$$\frac{\partial C_A}{\partial t} = D\frac{\partial^2 C_A}{\partial y^2} \qquad \text{(equation 10.66)}$$

where D is now the liquid phase diffusivity and C_A is the molar concentration in the liquid phase.

On integration, equation 10.4 becomes:

$$N_A = -D\frac{C_{A_2} - C_{A_1}}{y_2 - y_1} = \frac{D}{y_2 - y_1}(C_{A_1} - C_{A_2}) \tag{10.95}$$

and $D/(y_2 - y_1)$ is the liquid phase mass transfer coefficient.

An example of the integration of equation 10.66 is given in Section 10.5.3.

10.4.1. Liquid phase diffusivities

Values of the diffusivities of various materials in water are given in Table 10.7. Where experimental values are not available, it is necessary to use one of the predictive methods which are available.

A useful equation for the calculation of liquid phase diffusivities of dilute solutions of non-electrolytes has been given by WILKE and CHANG[16]. This is not dimensionally consistent and therefore the value of the coefficient depends on the units employed. Using SI units:

$$D = \frac{1.173 \times 10^{-16}\phi_B^{1/2} M_B^{1/2} T}{\mu \mathbf{V}_A^{0.6}} \tag{10.96}$$

Table 10.7. Diffusivities (diffusion coefficients) and Schmidt numbers, in liquids at 293 K[4]

Solute	Solvent	D ($m^2/s \times 10^9$)	Sc ($\mu/\rho D$)*
O_2	Water	1.80	558
CO_2	Water	1.50	670
N_2O	Water	1.51	665
NH_3	Water	1.76	570
Cl_2	Water	1.22	824
Br_2	Water	1.2	840
H_2	Water	5.13	196
N_2	Water	1.64	613
HCl	Water	2.64	381
H_2S	Water	1.41	712
H_2SO_4	Water	1.73	580
HNO_3	Water	2.6	390
Acetylene	Water	1.56	645
Acetic acid	Water	0.88	1140
Methanol	Water	1.28	785
Ethanol	Water	1.00	1005
Propanol	Water	0.87	1150
Butanol	Water	0.77	1310
Allyl alcohol	Water	0.93	1080
Phenol	Water	0.84	1200
Glycerol	Water	0.72	1400
Pyrogallol	Water	0.70	1440
Hydroquinone	Water	0.77	1300
Urea	Water	1.06	946
Resorcinol	Water	0.80	1260
Urethane	Water	0.92	1090
Lactose	Water	0.43	2340
Maltose	Water	0.43	2340
Glucose	Water	0.60	1680
Mannitol	Water	0.58	1730
Raffinose	Water	0.37	2720
Sucrose	Water	0.45	2230
Sodium chloride	Water	1.35	745
Sodium hydroxide	Water	1.51	665
CO_2	Ethanol	3.4	445
Phenol	Ethanol	0.8	1900
Chloroform	Ethanol	1.23	1230
Phenol	Benzene	1.54	479
Chloroform	Benzene	2.11	350
Acetic acid	Benzene	1.92	384
Ethylene dichloride	Benzene	2.45	301

*Based on $\mu/\rho = 1.005 \times 10^{-6} m^2/s$ for water, 7.37×10^{-7} for benzene, and 1.511×10^{-6} for ethanol, all at 293 K. The data apply only for dilute solutions.

The values are based mainly on *International Critical Tables* **5** (1928).

where D is the diffusivity of solute **A** in solvent **B** (m^2/s),

ϕ_B is the association factor for the solvent (2.26 for water, 1.9 for methanol, 1.5 for ethanol and 1.0 for unassociated solvents such as hydrocarbons and ethers),

M_B is the molecular weight of the solvent,

μ is the viscosity of the solution ($N\,s/m^2$),

T is temperature (K), and

$\mathbf{V}_A$ is the molecular volume of the solute (m^3/kmol). Values for simple molecules are given in Table 10.4. For more complex molecules, $\mathbf{V}_A$ is calculated by summation of the atomic volume and other contributions given in Table 10.4.

It may be noted that for water a value of 0.0756 m^3/kmol should be used.

Equation 10.96 does not apply to either electrolytes or to concentrated solutions. REID, PRAUSNITZ and SHERWOOD(17) discuss diffusion in electrolytes. Little information is available on diffusivities in concentrated solutions although it appears that, for ideal mixtures, the product μD is a linear function of the molar concentration.

The calculation of liquid phase diffusivities is discussed further in Volume 6.

10.5. MASS TRANSFER ACROSS A PHASE BOUNDARY

The theoretical treatment which has been developed in Sections 10.2–10.4 relates to mass transfer within a single phase in which no discontinuities exist. In many important applications of mass transfer, however, material is transferred across a phase boundary. Thus, in distillation a vapour and liquid are brought into contact in the fractionating column and the more volatile material is transferred from the liquid to the vapour while the less volatile constituent is transferred in the opposite direction; this is an example of equimolecular counterdiffusion. In gas absorption, the soluble gas diffuses to the surface, dissolves in the liquid, and then passes into the bulk of the liquid, and the carrier gas is not transferred. In both of these examples, one phase is a liquid and the other a gas. In liquid–liquid extraction however, a solute is transferred from one liquid solvent to another across a phase boundary, and in the dissolution of a crystal the solute is transferred from a solid to a liquid.

Each of these processes is characterised by a transference of material across an interface. Because no material accumulates there, the rate of transfer on each side of the interface must be the same, and therefore the concentration gradients automatically adjust themselves so that they are proportional to the resistance to transfer in the particular phase. In addition, if there is no resistance to transfer at the interface, the concentrations on each side will be related to each other by the phase equilibrium relationship. Whilst the existence or otherwise of a resistance to transfer at the phase boundary is the subject of conflicting views(18), it appears likely that any resistance is not high, except in the case of crystallisation, and in the following discussion equilibrium between the phases will be assumed to exist at the interface. Interfacial resistance may occur, however, if a surfactant is present as it may accumulate at the interface (Section 10.5.5).

The mass transfer rate between two fluid phases will depend on the physical properties of the two phases, the concentration difference, the interfacial area, and the degree of turbulence. Mass transfer equipment is therefore designed to give a large area of contact between the phases and to promote turbulence in each of the fluids. In the majority of plants, the two phases flow continuously in a countercurrent manner. In a steady state process, therefore, although the composition of each element of fluid is changing as it passes through the equipment, conditions at any given point do not change with time. In most industrial equipment, the flow pattern is so complex that it is not capable of expression in mathematical terms, and the interfacial area is not known precisely.

A number of mechanisms have been suggested to represent conditions in the region of the phase boundary. The earliest of these is the *two-film* theory propounded by WHITMAN[19] in 1923 who suggested that the resistance to transfer in each phase could be regarded as lying in a thin film close to the interface. The transfer across these films is regarded as a steady state process of molecular diffusion following equations of the type of equation 10.22. The turbulence in the bulk fluid is considered to die out at the interface of the films. In 1935 HIGBIE[20] suggested that the transfer process was largely attributable to fresh material being brought by the eddies to the interface, where a process of unsteady state transfer took place for a fixed period at the freshly exposed surface. This theory is generally known as the *penetration theory*. DANCKWERTS[21] has since suggested a modification of this theory in which it is considered that the material brought to the surface will remain there for varying periods of time. Danckwerts also discusses the random age distribution of such elements from which the transfer is by an unsteady state process to the second phase. Subsequently, TOOR and MARCHELLO[22] have proposed a more general theory, the *film-penetration theory*, and have shown that each of the earlier theories is a particular limiting case of their own. A number of other theoretical treatments have also been proposed, including that of KISHINEVSKIJ[23]. The two-film theory and the penetration theory will now be considered, followed by an examination of the film-penetration theory.

10.5.1. The two-film theory

The two-film theory of WHITMAN[19] was the first serious attempt to represent conditions occurring when material is transferred from one fluid stream to another. Although it does not closely reproduce the conditions in most practical equipment, the theory gives expressions which can be applied to the experimental data which are generally available, and for that reason it is still extensively used.

In this approach, it is assumed that turbulence dies out at the interface and that a laminar layer exists in each of the two fluids. Outside the laminar layer, turbulent eddies supplement the action caused by the random movement of the molecules, and the resistance to transfer becomes progressively smaller. For equimolecular counterdiffusion the concentration gradient is therefore linear close to the interface, and gradually becomes less at greater distances as shown in Figure 10.5 by the full lines *ABC* and *DEF*. The basis of the theory is the assumption that the zones in which the resistance to transfer lies can be replaced by two hypothetical layers, one on each side of the interface, in which the transfer is entirely by molecular diffusion. The concentration gradient is therefore linear in each of these layers and zero outside. The broken lines *AGC* and *DHF* indicate the hypothetical concentration distributions, and the thicknesses of the two films are L_1 and L_2. Equilibrium is assumed to exist at the interface and therefore the relative positions of the points *C* and *D* are determined by the equilibrium relation between the phases. In Figure 10.5, the scales are not necessarily the same on the two sides of the interface.

The mass transfer is treated as a steady state process and therefore the theory can be applied only if the time taken for the concentration gradients to become established is very small compared with the time of transfer, or if the capacities of the films are negligible.

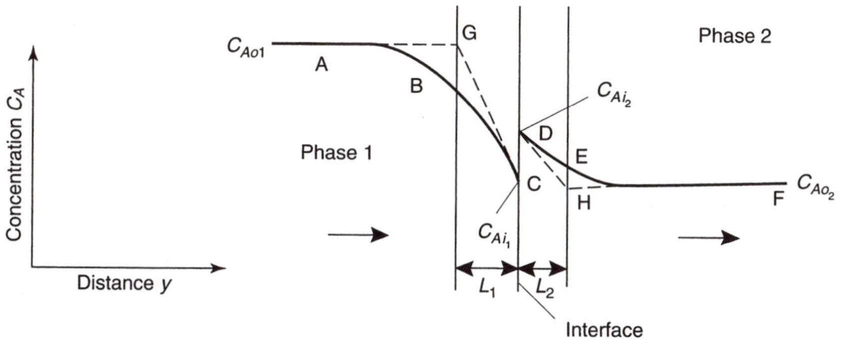

Figure 10.5. Two-film theory

From equation 10.22 the rate of transfer per unit area in terms of the two-film theory for equimolecular counterdiffusion is given for the first phase as:

$$N_A = \frac{D_1}{L_1}(C_{Ao_1} - C_{Ai_1}) = h_{D1}(C_{Ao_1} - C_{Ai_1}) \tag{10.97a}$$

where L_1 is the thickness of the film, C_{Ao_1} the molar concentration outside the film, and C_{Ai_1} the molar concentration at the interface.

For the second phase, with the same notation, the rate of transfer is:

$$N_A = \frac{D_2}{L_2}(C_{Ai_2} - C_{Ao_2}) = h_{D2}(C_{Ai_2} - C_{Ao_2}) \tag{10.97b}$$

Because material does not accumulate at the interface, the two rates of transfer must be the same and:

$$\frac{h_{D1}}{h_{D2}} = \frac{C_{Ai_2} - C_{Ao_2}}{C_{Ao_1} - C_{Ai_1}} \tag{10.98}$$

The relation between C_{Ai_1} and C_{Ai_2} is determined by the phase equilibrium relationship since the molecular layers on each side of the interface are assumed to be in equilibrium with one another. It may be noted that the ratio of the differences in concentrations is inversely proportional to the ratio of the mass transfer coefficients. If the bulk concentrations, C_{Ao_1} and C_{Ao_2} are fixed, the interface concentrations will adjust to values which satisfy equation 10.98. This means that, if the relative value of the coefficients changes, the interface concentrations will change too. In general, if the degree of turbulence of the fluid is increased, the effective film thicknesses will be reduced and the mass transfer coefficients will be correspondingly increased.

The theory is equally applicable when bulk flow occurs. In gas absorption, for example where may be expressed the mass transfer rate in terms of the concentration gradient in the gas phase:

$$N'_A = -D\frac{dC_A}{dy} \cdot \frac{C_T}{C_B} \qquad \text{(equation 10.30)}$$

In this case, for a steady state process, $(dC_A/dy)(C_T/C_B)$, as opposed to dC_A/dy, will be constant through the film and dC_A/dy will increase as C_A decreases. Thus lines GC and DH in Figure 10.5 will no longer be quite straight.

10.5.2. The Penetration Theory

The penetration theory was propounded in 1935 by HIGBIE[20] who was investigating whether or not a resistance to transfer existed at the interface when a pure gas was absorbed in a liquid. In his experiments, a slug-like bubble of carbon dioxide was allowed rise through a vertical column of water in a 3 mm diameter glass tube. As the bubble rose, the displaced liquid ran back as a thin film between the bubble and the tube. Higbie assumed that each element of surface in this liquid was exposed to the gas for the time taken for the gas bubble to pass it; that is for the time given by the quotient of the bubble length and its velocity. It was further supposed that during this short period, which varied between 0.01 and 0.1 s in the experiments, absorption took place as the result of unsteady state molecular diffusion into the liquid, and, for the purposes of calculation, the liquid was regarded as infinite in depth because the time of exposure was so short.

The way in which the concentration gradient builds up as a result of exposing a liquid — initially pure — to the action of a soluble gas is shown in Figure 10.6 which is based on Higbie's calculations. The percentage saturation of the liquid is plotted against the distance from the surface for a number of exposure times in arbitrary units. Initially only the surface layer contains solute and the concentration changes abruptly from 100 per cent to 0 per cent at the surface. For progressively longer exposure times the concentration profile develops as shown, until after an infinite time the whole of the liquid becomes saturated. The shape of the profiles is such that at any time the effective depth of liquid which contains an appreciable concentration of solute can be specified, and hence the theory is referred to as the *Penetration Theory*. If this depth of penetration is less than the total depth of liquid, no significant error is introduced by assuming that the total depth is infinite.

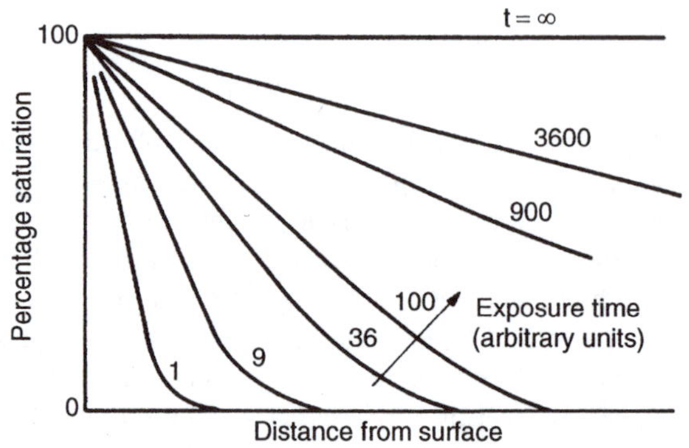

Figure 10.6. Penetration of solute into a solvent

The work of Higbie laid the basis of the penetration theory in which it is assumed that the eddies in the fluid bring an element of fluid to the interface where it is exposed to the second phase for a definite interval of time, after which the surface element is mixed with the bulk again. Thus, fluid whose initial composition corresponds with that of the bulk fluid remote from the interface is suddenly exposed to the second phase. It is assumed that equilibrium is immediately attained by the surface layers, that a process of unsteady state molecular diffusion then occurs and that the element is remixed after a fixed interval of time. In the calculation, the depth of the liquid element is assumed to be infinite and this is justifiable if the time of exposure is sufficiently short for penetration to be confined to the surface layers. Throughout, the existence of velocity gradients within the fluids is ignored and the fluid at all depths is assumed to be moving at the same rate as the interface.

The diffusion of solute **A** away from the interface (Y-direction) is thus given by equation 10.66:

$$\frac{\partial C_A}{\partial t} = D\frac{\partial^2 C_A}{\partial y^2} \qquad \text{(equation 10.64)}$$

for conditions of equimolecular counterdiffusion, or when the concentrations of diffusing materials are sufficiently low for the bulk flow velocity to be negligible. Because concentrations of **A** are low, there is no objection to using molar concentration for calculation of mass transfer rates in the liquid phase (see Section 10.4).

The following boundary conditions apply:

$$t = 0 \quad 0 < y < \infty \quad C_A = C_{Ao}$$

$$t > 0 \quad y = 0 \quad C_A = C_{Ai}$$

$$t > 0 \quad y = \infty \quad C_A = C_{Ao}$$

where C_{Ao} is the concentration in the bulk of the phase, and C_{Ai} the equilibrium value at the interface.

It is convenient to work in terms of a "deviation" variable C' as opposed to C_A, where C' is the amount by which the concentration of **A** exceeds the initial uniform concentration C_{Ao}. This change allows some simplification of the algebra.

With the substitution:

$$C' = C_A - C_{Ao} \qquad (10.99)$$

equation 10.66 becomes:

$$\frac{\partial C'}{\partial t} = D\frac{\partial^2 C'}{\partial y^2} \qquad (10.100)$$

because C_{Ao} is a constant with respect to both t and y; the boundary conditions are then:

$$t = 0 \quad 0 < y < \infty \quad C' = 0$$

$$t > 0 \quad y = 0 \quad C' = C'_i = C_{Ai} - C_{Ao}$$

$$t > 0 \quad y = \infty \quad C' = 0$$

These boundary conditions are necessary and sufficient for the solution of equation 10.100 which is first order with respect to t and second order with respect to y.

The equation is most conveniently solved by the method of Laplace transforms, used for the solution of the unsteady state thermal conduction problem in Chapter 9.

By definition, the Laplace transform $\overline{C'}$ of C' is given by:

$$\overline{C'} = \int_0^\infty e^{-pt} C' \, dt \qquad (10.101)$$

Then:
$$\overline{\frac{\partial C'}{\partial t}} = \int_0^\infty e^{-pt} \frac{\partial C'}{\partial t}\, dt \tag{10.102}$$
$$= [e^{-pt} C']_0^\infty + p \int_0^\infty e^{-pt} C'\, dt$$
$$= p\overline{C'} \tag{10.103}$$

Since the Laplace transform operation is independent of y:

$$\overline{\frac{\partial^2 C'}{\partial y^2}} = \frac{\partial^2 \overline{C'}}{\partial y^2} \tag{10.104}$$

Thus, taking Laplace transforms of both sides of equation 10.100:

$$p\overline{C'} = D \frac{\partial^2 \overline{C'}}{\partial y^2}$$

$$\therefore \quad \frac{\partial^2 \overline{C'}}{\partial y^2} - \frac{p}{D} \overline{C'} = 0$$

Equation 10.100 has therefore been converted from a partial differential equation in C' to an ordinary second order linear differential equation in $\overline{C'}$.

Thus:
$$\overline{C'} = B_1 e^{\sqrt{(p/D)}y} + B_2 e^{-\sqrt{(p/D)}y} \tag{10.105}$$

When: $y = 0,\ C_A = C_{Ai},\quad$ and $\quad C' = C_{Ai} - C_{Ao} = C'_i$

and when: $y = \infty,\ C_A = C_{Ao},\quad$ and $\quad C' = 0$

Hence: $B_1 = 0$

and:
$$\overline{C'} = B_2 e^{-\sqrt{(p/D)}y} = \overline{C'}_i e^{-\sqrt{(p/D)}y} \tag{10.106}$$

Now:
$$B_2 = \int_0^\infty (C_{Ai} - C_{Ao}) e^{-pt}\, dt$$
$$= \frac{1}{p}(C_{Ai} - C_{Ao})$$

Thus:
$$\overline{C'} = \frac{1}{p}(C_{Ai} - C_{Ao}) e^{-\sqrt{(p/D)}y} \tag{10.107}$$

Taking the inverse of the transform (Appendix A3 Table 12, No. 83), then:

$$C' = C_A - C_{Ao} = (C_{Ai} - C_{Ao}) \operatorname{erfc}\left(\frac{y}{2\sqrt{Dt}}\right)$$

or:
$$\frac{C_A - C_{Ao}}{C_{Ai} - C_{Ao}} = \operatorname{erfc}\left(\frac{y}{2\sqrt{Dt}}\right) = 1 - \operatorname{erf}\left(\frac{y}{2\sqrt{Dt}}\right) \tag{10.108}$$

This expression gives concentration C_A as a function of position y and of time t.

erf X is known as the error function and values are tabulated as for any other function of X; erfc X is the *complementary error function* $(1 - \text{erf } X)$.

By definition:

$$\text{erfc } X = \frac{2}{\sqrt{\pi}} \int_X^{\infty} e^{-x^2}\, dx \tag{10.109}$$

Since:

$$\int_0^{\infty} e^{-x^2}\, dx = \frac{\sqrt{\pi}}{2}, \tag{10.110}$$

erfc x goes from 1 to 0 as x goes from 0 to ∞.

The concentration gradient is then obtained by differentiation of equation 10.108 with respect to y.

Thus:

$$\frac{1}{C_{Ai} - C_{Ao}} \frac{\partial C_A}{\partial y} = \frac{\partial}{\partial y} \left[\frac{2}{\sqrt{\pi}} \int_{(y/2\sqrt{Dt})}^{\infty} e^{-y^2/4Dt}\, d\left(\frac{y}{2\sqrt{Dt}} \right) \right]$$

$$\therefore \quad \frac{\partial C_A}{\partial y} = -(C_{Ai} - C_{Ao}) \frac{2}{\sqrt{\pi}} \frac{1}{2\sqrt{Dt}} (e^{-y^2/4Dt})$$

$$= -(C_{Ai} - C_{Ao}) \frac{1}{\sqrt{\pi D t}} e^{-y^2/4Dt} \tag{10.111}$$

The mass transfer rate at any position y at time t is given by:

$$(N_A)_t = -D \frac{\partial C_A}{\partial y}$$

$$= (C_{Ai} - C_{Ao}) \sqrt{\frac{D}{\pi t}} e^{-y^2/4Dt} \tag{10.112}$$

The mass transfer rate per unit area of surface is then given by:

$$(N_A)_{t, y=0} = -D \left(\frac{\partial C_A}{\partial y} \right)_{y=0}$$

$$= (C_{Ai} - C_{Ao}) \sqrt{\frac{D}{\pi t}} \tag{10.113}$$

The point value of the mass transfer coefficient is therefore $\sqrt{D/\pi t}$.

Regular surface renewal

It is important to note that the mass transfer rate falls off progressively during the period of exposure, theoretically from infinity at $t = 0$ to zero at $t = \infty$.

Assuming that all the surface elements are exposed for the same time t_e (Higbie's assumption), from equation 10.113, the moles of A (n_A) transferred at an area A in time t_e is given by:

$$n_A = (C_{Ai} - C_{Ao}) \sqrt{\frac{D}{\pi}} A \int_0^{t_e} \frac{dt}{\sqrt{t}}$$

$$= 2(C_{Ai} - C_{Ao}) A \sqrt{\frac{D t_e}{\pi}} \tag{10.114}$$

and the average rate of transfer per unit area over the exposure time t_e is given by:

$$N_A = 2(C_{Ai} - C_{Ao})\sqrt{\frac{D}{\pi t_e}} \tag{10.115}$$

That is, the average rate over the time interval $t = 0$ to $t = t_e$ is twice the point value at $t = t_e$.

Thus, the shorter the time of exposure the greater is the rate of mass transfer. No precise value can be assigned to t_e in any industrial equipment, although its value will clearly become less as the degree of agitation of the fluid is increased.

If it is assumed that each element resides for the same time interval t_e in the surface, equation 10.115 gives the overall mean rate of transfer. It may be noted that the rate is a linear function of the driving force expressed as a concentration difference, as in the two-film theory, but that it is proportional to the diffusivity raised to the power of 0.5 instead of unity.

Equation 10.114 forms the basis of the *laminar-jet* method of determining the molecular diffusivity of a gas in a liquid. Liquid enters the the gas space from above through a sharp-edged circular hole formed in a thin horizontal plate, to give a vertical "rod" of liquid having a flat velocity profile, which is collected in a container of slightly larger diameter than the jet. The concentration of this outlet liquid is measured in order to determine the number of moles n_A of **A** transferred to the laminar jet during the exposure time t_e which can be varied by altering the velocity and the length of travel of the jet. A plot of n_A versus $t_e^{1/2}$ should give a straight line, the slope of which enables the molecular diffusivity D to be calculated, since C_{Ao} is zero and C_{Ai} is the saturation concentration. The assumptions and possible sources of error in this method are discussed by DANCKWERTS[24]; it is important that penetration depths must be small compared with the radius of the jet.

When mass transfer rates are very high, limitations may be placed on the rate at which a component may be transferred, by virtue of the limited frequency with which the molecules collide with the surface. For a gas, the collision rate can be calculated from the *kinetic theory* and allowance must then be made for the fact that only a fraction of these molecules may be absorbed, with the rest being reflected. Thus, when even a pure gas is brought suddenly into contact with a fresh solvent, the initial mass transfer rate may be controlled by the rate at which gas molecules can reach the surface, although the resistance to transfer rapidly builds up in the liquid phase to a level where this effect can be neglected. The point is well illustrated in Example 10.4.

Example 10.4

In an experimental wetted wall column, pure carbon dioxide is absorbed in water. The mass transfer rate is calculated using the penetration theory, application of which is limited by the fact that the concentration should not reach more than 1 per cent of the saturation value at a depth below the surface at which the velocity is 95 per cent of the surface velocity. What is the maximum length of column to which the theory can be applied if the flowrate of water is 3 cm^3/s per cm of perimeter?

Viscosity of water $= 10^{-3}$ N s/m^2. Diffusivity of carbon dioxide in water $= 1.5 \times 10^{-9}$ m^2/s

Solution

For the flow of a vertical film of fluid, the mean velocity of flow is governed by equation 3.87 in which $\sin\phi$ is put equal to unity for a vertical surface:

$$u_m = \frac{\rho g s^2}{3\mu}$$

where s is the thickness of the film.

The flowrate per unit perimeter $(\rho g s^3/3\mu) = 3 \times 10^{-4}\ \text{m}^2/\text{s}$

and:
$$s = \left(\frac{3 \times 10^{-4} \times 10^{-3} \times 3}{1000 \times 9.81}\right)^{1/3}$$
$$= 4.51 \times 10^{-4}\ \text{m}$$

The velocity u_x at a distance y' from the vertical column wall is given by equation 3.85 (using y' in place of y) as:

$$u_x = \frac{\rho g(sy' - \frac{1}{2}y'^2)}{\mu}$$

The free surface velocity u_s is given by substituting s for y' or:

$$u_s = \frac{\rho g s^2}{2\mu}$$

Thus:
$$\frac{u_x}{u_s} = 2\left(\frac{y'}{s}\right) - \left(\frac{y'}{s}\right)^2 = 1 - \left(1 - \frac{y'}{s}\right)^2$$

When $u_x/u_s = 0.95$, that is velocity is 95 per cent of surface velocity, then:

$$1 - \frac{y'}{s} = 0.224$$

and the distance below the free surface is $y = s - y' = 1.010 \times 10^{-4}$ m

The relationship between concentration C_A, time and depth is:

$$\frac{C_A - C_{Ao}}{C_{Ai} - C_{Ao}} = \text{erfc}\left(\frac{y}{2\sqrt{Dt}}\right) \qquad \text{(equation 10.108)}$$

The time at which concentration reaches 0.01 of saturation value at a depth of 1.010×10^{-4} m is given by:

$$0.01 = \text{erfc}\left(\frac{1.010 \times 10^{-4}}{2\sqrt{1.5 \times 10^{-9} t}}\right)$$

Thus:
$$\text{erf}\left(\frac{1.305}{\sqrt{t}}\right) = 0.99$$

Using Tables of error functions (Appendix A3, Table 12):

$$\frac{1.305}{\sqrt{t}} = 1.822$$

and:
$$t = 0.51 s$$

The surface velocity is then
$$u_s = \frac{\rho g s^2}{2\mu}$$
$$= \frac{1000 \times 9.81 \times (4.51 \times 10^{-4})^2}{2 \times 10^{-3}}$$
$$= 1\ \text{m/s}$$

and the maximum length of column is: $= (1 \times 0.51) = \underline{\underline{0.51\ \text{m}}}$

Example 10.5

In a gas–liquid contactor, a pure gas is absorbed in a solvent and the Penetration Theory provides a reasonable model by which to describe the transfer mechanism. As fresh solvent is exposed to the gas, the transfer rate is initially limited by the rate at which the gas molecules can reach the surface. If at 293 K and a pressure of 1 bar the maximum possible rate of transfer of gas is 50 m^3/m^2s, express this as an equivalent resistance, when the gas solubility is 0.04 $kmol/m^3$.

If the diffusivity in the liquid phase is 1.8×10^{-9} m^2/s, at what time after the initial exposure will the resistance attributable to access of gas be equal to about 10 per cent of the total resistance to transfer?

Solution

Bulk gas concentration $= \left(\frac{1}{22.4}\right)\left(\frac{273}{293}\right) = 0.0416$ kmol/m^3

Initial mass transfer rate $= 50$ m^3/m^2 s

$= (50 \times 0.0416) = 2.08$ kmol/m^2 s

Concentration driving force in liquid phase

$= (0.04 - 0) = 0.04$ kmol/m^3

Effective mass transfer coefficient initially

$= \frac{2.08}{0.04} = 52.0$ m/s

Equivalent resistance $= 1/52.0 = 0.0192$ s/m

When this constitutes 10 per cent of the total resistance,

liquid phase resistance $= 0.0192 \times 9 = 0.173$ s/m

liquid phase coefficient $= 5.78$ m/s

From equation 10.113, the point value of liquid phase mass transfer coefficient $= \sqrt{\frac{D}{\pi t}}$

$$= \sqrt{\frac{1.8 \times 10^{-9}}{\pi t}} = 2.394 \times 10^{-5} t^{-1/2} \text{ m/s}$$

$$\text{Resistance} = 4.18 \times 10^4 t^{1/2} \text{ s/m}$$

Thus: $4.18 \times 10^4 t^{1/2} = 0.173$

and: $t = \underline{\underline{1.72 \times 10^{-11} \text{ s}}}$

Thus the limited rate of access of gas molecules is not likely to be of any significance.

Example 10.6

A deep pool of ethanol is suddenly exposed to an atmosphere of pure carbon dioxide and unsteady state mass transfer, governed by Fick's Law, takes place for 100 s. What proportion of the absorbed carbon dioxide will have accumulated in the 1 mm layer closest to the surface in this period?

Diffusivity of carbon dioxide in ethanol $= 4 \times 10^{-9}$ m^2/s.

Solution

The accumulation in the 1 mm layer near the surface will be equal to the total amount of CO_2 entering the layer from the surface ($y = 0$) less that leaving ($y = 10^{-3}$ m) in the course of 100 s.

The mass rate of transfer at any position y and time t is given by equation 10.112:

$$(N_A)_t = C_{Ai}\sqrt{\frac{D}{\pi t}}e^{-y^2/4Dt}$$

where N_A is expressed in mols per unit area and unit time and C_{Ao} is zero because the solvent is pure ethanol.

Considering unit area of surface, the moles transferred in time t_e at depth y is given by:

$$= C_{Ai}\sqrt{\frac{D}{\pi}}\int_0^{t_e} t^{-1/2}e^{-y^2/4Dt}\,dt$$

Putting $y^2/4Dt = X^2$:
$$t^{-1/2} = \frac{2\sqrt{D}X}{y}$$

and:
$$dt = \frac{y^2}{4D}\frac{-2}{X^3}\,dX$$

Thus:
$$\text{Integral} = \int_\infty^{X_e}\frac{y^2}{4D}\frac{-2}{X^3}\frac{2\sqrt{D}X}{y}e^{-X^2}\,dX$$

$$= -\frac{y}{\sqrt{D}}\int_\infty^{X_e}X^{-2}e^{-X^2}\,dX$$

$$\text{Molar transfer per unit area} = C_{Ai}\sqrt{\frac{D}{\pi}}\left(\frac{-y}{\sqrt{D}}\right)\left\{\left[e^{-X^2}(-X^{-1})\right]_\infty^{X_e} - \int_\infty^{X_e}\left[-2Xe^{-X^2}(-X^{-1})\right]dX\right\}$$

$$= C_{Ai}\left(\frac{-y}{\sqrt{\pi}}\right)\left\{-X_e^{-1}e^{-X_e^2} + 2\int_{X_e}^{\infty}e^{-X^2}\,dX\right\}$$

$$= C_{Ai}\left(\frac{y}{\sqrt{\pi}}\right)\left\{\frac{2\sqrt{Dt_e}}{y}e^{-y^2/4Dt_e} - \sqrt{\pi}\,\text{erfc}\,\frac{y}{2\sqrt{Dt_e}}\right\}$$

$$= C_{Ai}\left\{2\sqrt{\frac{Dt_e}{\pi}}e^{-y^2/4Dt_e} - y\,\text{erfc}\,\frac{y}{2\sqrt{Dt_e}}\right\}$$

Putting $D = 4 \times 10^{-9}$ m^2/s and $t = 100$ s:

At $y = 0$:
$$\text{moles transferred} = 2C_{Ai}\sqrt{\frac{4 \times 10^{-9} \times 100}{\pi}} = 7.14 \times 10^{-4}C_{Ai}$$

At $y = 10^{-3}$ m:
$$\text{moles transferred} = C_{Ai}\{7.14 \times 10^{-4}e^{-0.626} - 10^{-3}\,\text{erfc}\,0.791\}$$

$$= C_{Ai}\{3.82 \times 10^{-4} - 2.63 \times 10^{-4}\}$$

$$= 1.19 \times 10^{-4}C_{Ai}$$

The proportion of material retained in layer = (7.14 − 1.19)/7.14 = 0.83 or 83 per cent

Random surface renewal

DANCKWERTS[21] suggested that each element of surface would not be exposed for the same time, but that a random distribution of ages would exist. It was assumed that the probability of any element of surface becoming destroyed and mixed with the bulk of the fluid was independent of the age of the element and, on this basis, the age distribution of the surface elements was calculated using the following approach.

Supposing that the rate of production of fresh surface per unit total area of surface is s, and that s is independent of the age of the element in question, the area of surface of age between t and $t + dt$ will be a function of t and may be written as $f(t)\,dt$. This will be equal to the area passing in time dt from the age range $[(t - dt)$ to $t]$ to the age range $[t$ to $(t + dt)]$. Further, this in turn will be equal to the area in the age group $[(t - dt)$ to $t]$, less that replaced by fresh surface in time dt, or:

$$f(t)dt = f(t - dt)dt - [f(t - dt)dt]s\,dt \qquad (10.116)$$

Thus:
$$\frac{f(t) - f(t - dt)}{dt} = -sf(t - dt)$$

$\therefore$
$$f'(t) + sf(t) = 0 \text{ (as } dt \to 0) \qquad (10.117)$$

$\therefore$
$$e^{st}f(t) = \text{constant}$$

$\therefore$
$$f(t) = \text{constant } e^{-st}$$

The total area of surface considered is unity, and hence:

$$\int_0^\infty f(t)dt = \text{constant} \int_0^\infty e^{-st}\,dt = 1 \qquad (10.118)$$

$\therefore$
$$\text{constant} \times \frac{1}{s} = 1$$

and:
$$f(t) = s\,e^{-st} \qquad (10.119)$$

Thus the age distribution of the surface is of an exponential form. From equation 10.113 the mass transfer rate at unit area of surface of age t is given by:

$$(N_A)_t = (C_{Ai} - C_{Ao})\sqrt{\frac{D}{\pi t}} \qquad \text{(equation 10.113)}$$

Thus, the overall rate of transfer per unit area when the surface is renewed in a random manner is:

$$N_A = (C_{Ai} - C_{Ao}) \int_{t=0}^{t=\infty} \sqrt{\frac{D}{\pi t}} s\,e^{-st}\,dt$$

or:
$$N_A = (C_{Ai} - C_{Ao})\,s\sqrt{\frac{D}{\pi}} \int_0^\infty t^{-1/2}e^{-st}\,dt \qquad (10.120)$$

Putting $st = \beta^2$, then $s\,dt = 2\beta d\beta$ and:

$$N_A = (C_{Ai} - C_{Ao})\,s\sqrt{\frac{D}{\pi}\frac{2}{s^{1/2}}} \int_0^\infty e^{-\beta^2}\,d\beta$$

Then, since the value of the integral is $\sqrt{\pi}/2$, then:

$$N_A = (C_{Ai} - C_{Ao})\sqrt{Ds} \qquad (10.121)$$

Equation 10.121 might be expected to underestimate the mass transfer rate because, in any practical equipment, there will be a finite upper limit to the age of any surface element. The proportion of the surface in the older age group is, however, very small and the overall rate is largely unaffected. It is be seen that the mass transfer rate is again proportional to the concentration difference and to the square root of the diffusivity. The numerical value of s is difficult to estimate, although this will clearly increase as the fluid becomes more turbulent. In a packed column, s will be of the same order as the ratio of the velocity of the liquid flowing over the packing to the length of packing.

Varying interface composition

The penetration theory has been used to calculate the rate of mass transfer across an interface for conditions where the concentration C_{Ai} of solute **A** in the interfacial layers ($y = 0$) remained constant throughout the process. When there is no resistance to mass transfer in the other phase, for instance when this consists of pure solute **A**, there will be no concentration gradient in that phase and the composition at the interface will therefore at all times be the same as the bulk composition. Since the composition of the interfacial layers of the *penetration* phase is determined by the phase equilibrium relationship, it, too, will remain constant and the conditions necessary for the penetration theory to apply will hold. If, however, the other phase offers a significant resistance to transfer this condition will not, in general, be fulfilled.

As an example, it may be supposed that in phase 1 there is a constant finite resistance to mass transfer which can in effect be represented as a resistance in a laminar film, and in phase 2 the penetration model is applicable. Immediately after surface renewal has taken place, the mass transfer resistance in phase 2 will be negligible and therefore the whole of the concentration driving force will lie across the film in phase 1. The interface compositions will therefore correspond to the bulk value in phase 2 (the penetration phase). As the time of exposure increases, the resistance to mass transfer in phase 2 will progressively increase and an increasing proportion of the total driving force will lie across this phase. Thus the interface composition, initially determined by the bulk composition in phase 2 (the penetration phase) will progressively approach the bulk composition in phase 1 as the time of exposure increases.

Because the boundary condition at $y = \infty$ ($C' = 0$) is unaltered by the fact that the concentration at the interface is a function of time, equation 10.106 is still applicable, although the evaluation of the constant B_2 is more complicated because $(C')_{y=0}$ is no longer constant.

$$\overline{C'} = B_2\,\mathrm{e}^{-\sqrt{(p/D)}y} \qquad \text{(equation 10.106)}$$

$$\frac{\mathrm{d}\overline{C'}}{\mathrm{d}y} = -\sqrt{\frac{p}{D}}B_2\,\mathrm{e}^{-\sqrt{(p/D)}y}$$

so that:

$$(\overline{C'})_{y=0} = B_2 \qquad (10.122)$$

and:

$$\left(\frac{\mathrm{d}\overline{C'}}{\mathrm{d}y}\right)_{y=0} = -\sqrt{\frac{p}{D}}B_2 \qquad (10.123)$$

In order to evaluate B_2 it is necessary to equate the mass transfer rates on each side of the interface.

The mass transfer rate per unit area across the film at any time t is given by:

$$(N_A)_t = -\frac{D_f}{L_f}(C''_{Ai} - C''_{Ao}) \tag{10.124}$$

where D_f is the diffusivity in the film of thickness L_f, and C''_{Ai} and C''_{Ao} are the concentrations of **A** at the interface and in the bulk.

The capacity of the film will be assumed to be small so that the hold-up of solute is negligible. If Henry's law is applicable, the interface concentration in the second (penetration) phase is given by:

$$C_{Ai} = \frac{1}{\mathscr{H}}C''_{Ai} \tag{10.125}$$

where C_A is used to denote concentration in the penetration phase. Thus, by substitution, the interface composition C_{Ai} is obtained in terms of the mass transfer rate $(N_A)_t$.

However, $(N_A)_t$ must also be given by applying Fick's law to the interfacial layers of phase 2 (the penetration phase).

Thus:
$$(N_A)_t = -D\left(\frac{\partial C_A}{\partial y}\right)_{y=0} \quad \text{(from equation 10.4)} \tag{10.126}$$

Combining equations 10.124, 10.125, and 10.126:

$$(C_A)_{y=0} = C_{Ai} = \frac{C''_{Ao}}{\mathscr{H}} + \frac{DL_f}{D_f\mathscr{H}}\left(\frac{\partial C_A}{\partial y}\right)_{y=0} \tag{10.127}$$

Replacing C_A by $C' + C_{Ao}$:

$$(C')_{y=0} = \left(\frac{C''_{Ao}}{\mathscr{H}} - C_{Ao}\right) + \frac{DL_f}{D_f\mathscr{H}}\left(\frac{\partial C'}{\partial y}\right)_{y=0}$$

Taking Laplace transforms of each side, noting that the first term on the right-hand side is constant:

$$(\overline{C'})_{y=0} = \frac{1}{p}\left(\frac{C''_{Ao}}{\mathscr{H}} - C_{Ao}\right) + \frac{DL_f}{D_f\mathscr{H}}\left(\frac{\mathrm{d}\overline{C'}}{\mathrm{d}y}\right)_{y=0} \tag{10.128}$$

Substituting into equation 10.128 from equations 10.122 and 10.123:

$$B_2 = \frac{1}{p}\left(\frac{C''_{Ao}}{\mathscr{H}} - C_{Ao}\right) + \frac{DL_f}{D_f\mathscr{H}}\left(-\sqrt{\frac{p}{D}}B_2\right)$$

or:
$$B_2 = \frac{D_f}{\sqrt{D}L_f}[C''_{Ao} - \mathscr{H}C_{Ao}] \bigg/ p\left\{\frac{D_f\mathscr{H}}{\sqrt{D}L_f} + \sqrt{p}\right\} \tag{10.129}$$

Substituting in equation 10.123:

$$\left(\frac{\mathrm{d}\overline{C'}}{\mathrm{d}y}\right)_{y=0} = \left(\frac{\mathrm{d}\overline{C'}}{\mathrm{d}y}\right)_{y=0} = -\frac{D_f}{DL_f}(C''_{Ao} - \mathscr{H}C_{Ao}) \bigg/ \sqrt{p}\left\{\frac{D_f\mathscr{H}}{\sqrt{D}L_f} + \sqrt{p}\right\} \tag{10.130}$$

On inversion (see Appendix, Table 12, No. 43):

$$\left(\frac{\mathrm{d}C'}{\mathrm{d}y}\right)_{y=0} = -\frac{D_f}{DL_f}[C''_{Ao} - \mathscr{H}C_{Ao}]\mathrm{e}^{[(D_f^2\mathscr{H}^2)/(DL_f^2)]t} \operatorname{erfc}\sqrt{\frac{D_f^2\mathscr{H}^2 t}{DL_f^2}} \tag{10.131}$$

The mass transfer rate at time t at the interface is then given by:

$$(N_A)_t = -D\left(\frac{\partial C_A}{\partial y}\right)_{y=0} = -D\left(\frac{\partial C'}{\partial y}\right)_{y=0}$$

or:

$$(N_A)_t = (C''_{Ao} - \mathscr{H}C_{Ao})\frac{D_f}{L_f}\left(\mathrm{e}^{[(D_f^2\mathscr{H}^2)/(DL_f^2)]t} \operatorname{erfc}\sqrt{\frac{D_f^2\mathscr{H}^2 t}{DL_f^2}}\right) \tag{10.132}$$

Average rates of mass transfer can be obtained, as previously, by using either the Higbie or the Danckwerts model for surface renewal.

Penetration model with laminar film at interface

HARRIOTT[25] suggested that, as a result of the effects of interfacial tension, the layers of fluid in the immediate vicinity of the interface would frequently be unaffected by the mixing process postulated in the penetration theory. There would then be a thin laminar layer unaffected by the mixing process and offering a constant resistance to mass transfer. The overall resistance may be calculated in a manner similar to that used in the previous section where the total resistance to transfer was made up of two components — a film resistance in one phase and a penetration model resistance in the other. It is necessary in equation 10.132 to put the Henry's law constant equal to unity and the diffusivity D_f in the film equal to that in the remainder of the fluid D. The driving force is then $C_{Ai} - C_{Ao}$ in place of $C'_{Ao} - \mathscr{H}C_{Ao}$, and the mass transfer rate at time t is given for a film thickness L by:

$$(N_A)_t = (C_{Ai} - C_{Ao})\frac{D}{L}\left(\mathrm{e}^{Dt/L^2} \operatorname{erfc}\sqrt{\frac{Dt}{L^2}}\right) \tag{10.133}$$

The average transfer rate according to the Higbie model for surface age distribution then becomes:

$$N_A = 2(C_{Ai} - C_{Ao})\sqrt{\frac{D}{\pi t_e}}\left[1 + \frac{1}{2}\sqrt{\frac{\pi L^2}{Dt_e}}\left(\mathrm{e}^{Dt_e/L^2} \operatorname{erfc}\sqrt{\frac{Dt_e}{L^2}} - 1\right)\right] \tag{10.134}$$

Using the Danckwerts model:

$$N_A = (C_{Ai} - C_{Ao})\sqrt{Ds}\left(1 + \sqrt{\frac{L^2 s}{D}}\right)^{-1} \tag{10.135}$$

10.5.3. The film–penetration theory

A theory which incorporates some of the principles of both the two-film theory and the penetration theory has been proposed by TOOR and MARCHELLO[(22)]. The whole of the resistance to transfer is regarded as lying within a laminar film at the interface, as in the two-film theory, but the mass transfer is regarded as an unsteady state process. It is assumed that fresh surface is formed at intervals from fluid which is brought from the bulk of the fluid to the interface by the action of the eddy currents. Mass transfer then takes place as in the penetration theory, except that the resistance is confined to the finite film, and material which traverses the film is immediately completely mixed with the bulk of the fluid. For short times of exposure, when none of the diffusing material has reached the far side of the layer, the process is identical to that postulated in the penetration theory. For prolonged periods of exposure when a steady concentration gradient has developed, conditions are similar to those considered in the two-film theory.

The mass transfer process is again governed by equation 10.66, but the third boundary condition is applied at $y = L$, the film thickness, and not at $y = \infty$. As before, the Laplace transform is then:

$$\overline{C'} = B_1 e^{\sqrt{(p/D)}y} + B_2 e^{-\sqrt{(p/D)}y} \qquad \text{(equation 10.105)}$$

$t > 0 \quad y = 0 \quad C_A = C_{Ai} \quad C' = C_{Ai} - C_{Ao} = C'_i \qquad \overline{C'} = (1/p)C'_i$

$t > 0 \quad y = L \quad C_A = C_{Ao} \quad C' = 0 \qquad \overline{C'} = 0$

Thus:
$$\frac{C'_i}{p} = B_1 + B_2$$

$$\therefore \quad 0 = B_1 e^{\sqrt{(p/D)}L} + B_2 e^{-\sqrt{(p/D)}L}$$

$$\therefore \quad B_1 = -B_2 e^{-2\sqrt{(p/D)}L} \quad \text{and} \quad B_2 = \frac{C'_i}{p}(1 - e^{-2\sqrt{(p/D)}L})^{-1}$$

$$B_1 = -\frac{C'_i}{p} e^{-2\sqrt{(p/D)}L}(1 - e^{-2\sqrt{(p/D)}L})^{-1}$$

$$\therefore \quad \overline{C'} = \frac{C'_i}{p}(1 - e^{-2\sqrt{(p/D)}L})^{-1}(e^{-\sqrt{(p/D)}y} - e^{-\sqrt{(p/D)}(2L-y)}) \qquad (10.136)$$

Since there is no inverse of equation 10.136 in its present form, it is necessary to expand using the binomial theorem. Noting that, since $2\sqrt{(p/D)}L$ is positive, $e^{-2\sqrt{(p/D)}L} < 1$ and from the binomial theorem:

$$(1 - e^{-2\sqrt{(p/D)}L})^{-1} = \{1 + e^{-2\sqrt{(p/D)}L} + \cdots + e^{-2n\sqrt{(p/D)}L} + \cdots \text{to } \infty\}$$

$$= \sum_{n=0}^{n=\infty} e^{-2n\sqrt{(p/D)}L}$$

Substituting in equation 10.136:

$$\overline{C'} = \frac{C'_i}{p}(e^{-\sqrt{(p/D)}y} - e^{-\sqrt{(p/D)}(2L-y)}) \sum_{n=0}^{n=\infty} e^{-2n\sqrt{(p/D)}L}$$

$$= C'_i \left[\sum_{n=0}^{n=\infty} \frac{1}{p} e^{-\sqrt{(p/D)}(2nL+y)} - \sum_{n=0}^{n=\infty} \frac{1}{p} e^{-\sqrt{(p/D)}\{2(n+1)L-y\}} \right] \quad (10.137)$$

On inversion of equation 10.137:

$$\frac{C_A - C_{Ao}}{C_{Ai} - C_{Ao}} = \frac{C'}{C'_i} = \sum_{n=0}^{n=\infty} \operatorname{erfc}\frac{(2nL + y)}{2\sqrt{Dt}} - \sum_{n=0}^{n=\infty} \operatorname{erfc}\frac{2(n+1)L - y}{2\sqrt{Dt}} \quad (10.138)$$

Differentiating with respect to y:

$$\frac{1}{C_{Ai} - C_{Ao}} \frac{\partial C_A}{\partial y} = \sum_{n=0}^{n=\infty} -\frac{2}{\sqrt{\pi}} \frac{1}{2\sqrt{Dt}} e^{-(2nL+y)^2/(4Dt)}$$

$$- \sum_{n=0}^{n=\infty} \frac{2}{\sqrt{\pi}} \frac{1}{2\sqrt{Dt}} e^{-[2(n+1)L-y]^2/4(Dt)}$$

At the free surface, $y = 0$ and:

$$\frac{1}{C_{Ai} - C_{Ao}} \left(\frac{\partial C_A}{\partial y}\right)_{y=0} = -\frac{1}{\sqrt{\pi Dt}} \left(\sum_{n=0}^{n=\infty} e^{-(n^2L^2)/(Dt)} + \sum_{n=0}^{n=\infty} e^{-[(n+1)^2L^2]/(Dt)} \right)$$

$$= -\frac{1}{\sqrt{\pi Dt}} \left(1 + 2\sum_{n=1}^{n=\infty} e^{-(n^2L^2)/(Dt)} \right)$$

Now: $$\sum_{n=0}^{n=\infty} e^{-(n^2L^2)/Dt} = 1 + \sum_{n=1}^{n=\infty} e^{-(n^2L^2)/Dt}$$

and $$\sum_{n=0}^{n=\infty} e^{-[(n+1)^2L^2]/Dt} = \sum_{n=1}^{n=\infty} e^{-(n^2L^2)/Dt}$$

The mass transfer rate across the interface per unit area is therefore given by:

$$(N_A)_t = -D\left(\frac{\partial C_A}{\partial y}\right)_{y=0}$$

$$= (C_{Ai} - C_{Ao})\sqrt{\frac{D}{\pi t}} \left(1 + 2\sum_{n=1}^{n=\infty} e^{-(n^2L^2)/(Dt)} \right) \quad (10.139)$$

Equation 10.139 converges rapidly for high values of L^2/Dt. For low values of L^2/Dt, it is convenient to employ an alternative form by using the identity[26], and:

$$\sqrt{\frac{\alpha}{\pi}} \left(1 + 2\sum_{n=1}^{n=\infty} e^{-n^2\alpha} \right) \equiv 1 + 2\sum_{n=1}^{n=\infty} e^{-(n^2\pi^2)/\alpha} \quad (10.140)$$

Taking $\alpha = L^2/Dt$:

$$(N_A)_t = (C_{Ai} - C_{Ao})\frac{D}{L}\left(1 + 2\sum_{n=1}^{n=\infty} e^{-(n^2\pi^2 Dt)/L^2}\right) \tag{10.141}$$

It will be noted that equations 10.139 and 10.141 become identical in form and in convergence when $L^2/Dt = \pi$.

$$\text{Then: } (N_A)_t = (C_{Ai} - C_{Ao})\frac{D}{L}\left(1 + 2\sum_{n=1}^{n=\infty} e^{-n^2\pi}\right)$$

$$= (C_{Ai} - C_{Ao})\frac{D}{L}[1 + 2(e^{-\pi} + e^{-4\pi} + e^{-9\pi} + \cdots)]$$

$$= (C_{Ai} - C_{Ao})\frac{D}{L}(1 + 0.0864 + 6.92 \times 10^{-6} + 1.03 \times 10^{-12} + \cdots)$$

Thus, provided the rate of convergence is not less than that for $L^2/Dt = \pi$, all terms other than the first in the series may be neglected. Equation 10.139 will converge more rapidly than this for $L^2/Dt > \pi$, and equation 10.141 will converge more rapidly for $L^2/Dt < \pi$.

Thus:

$$\pi \leqslant \frac{L^2}{Dt} < \infty \qquad (N_A)_t = (C_{Ai} - C_{Ao})\sqrt{\frac{D}{\pi t}}(1 + 2e^{-L^2/Dt}) \tag{10.142}$$

$$0 < \frac{L^2}{Dt} \leqslant \pi \qquad (N_A)_t = (C_{Ai} - C_{Ao})\frac{D}{L}(1 + 2e^{-(\pi^2 Dt)/L^2}) \tag{10.143}$$

It will be noted that the second terms in equations 10.142 and 10.143 never exceeds 8.64 per cent of the first term. Thus, with an error not exceeding 8.64 per cent,

$$\pi \leqslant \frac{L^2}{Dt} < \infty \qquad (N_A)_t = (C_{Ai} - C_{Ao})\sqrt{\frac{D}{\pi t}} \tag{10.144}$$

$$0 < \frac{L^2}{Dt} \leqslant \pi \qquad (N_A)_t = (C_{Ai} - C_{Ao})\frac{D}{L} \tag{10.145}$$

The concentration profiles near an interface on the basis of:

(a) the film theory (steady-state)
(b) the penetration-theory
(c) the film-penetration theory

are shown in Figure 10.7.

Thus either the penetration theory or the film theory (equation 10.144 or 10.145) respectively can be used to describe the mass transfer process. The error will not exceed some 9 per cent provided that the appropriate equation is used, equation 10.144 for $L^2/Dt > \pi$ and equation 10.145 for $L^2/Dt < \pi$. Equation 10.145 will frequently apply quite closely in a wetted-wall column or in a packed tower with large packings. Equation 10.144 will apply when one of the phases is dispersed in the form of droplets, as in a spray tower, or in a packed tower with small packing elements.

Equations 10.142 and 10.143 give the point value of N_A at time t. The average values N_A can then be obtained by applying the age distribution functions obtained by Higbie and by Danckwerts, respectively, as discussed Section 10.5.2.

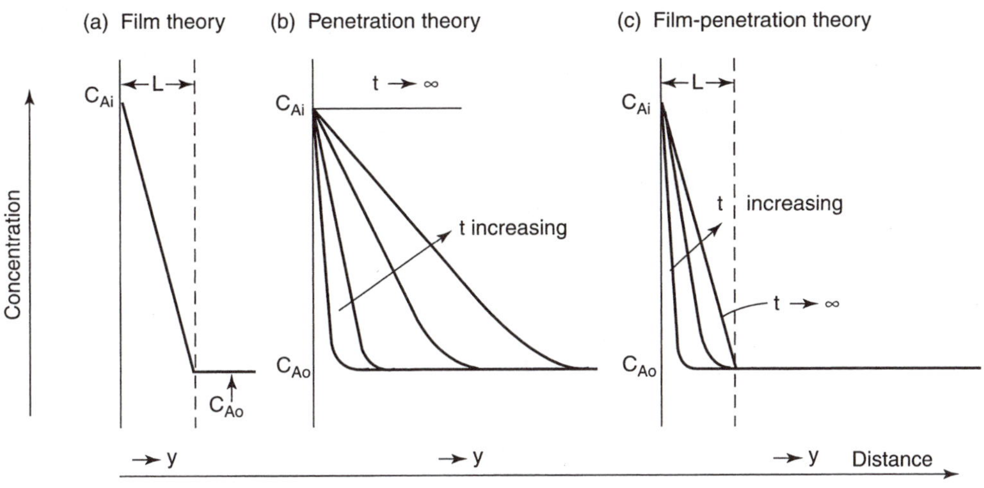

Figure 10.7. Concentration profiles near an interface

10.5.4 Mass transfer to a sphere in a homogenous fluid

So far in this chapter, consideration has been given to transfer taking place in a single direction of a rectangular coordinate system. In many applications of mass transfer, one of the fluids is injected as approximately spherical droplets into a second immiscible fluid, and transfer of the solute occurs as the droplet passes through the continuous medium.

The case of a sphere of pure liquid of radius r_0 being suddenly immersed in a gas, when the whole of the mass transfer resistance lying within the liquid, is now considered. It will be assumed that mass transfer is governed by Fick's law and that angular symmetry exists.

At radius r within sphere, mass transfer rate $= -(4\pi r^2)D\left(\frac{\partial C_A}{\partial r}\right)$

The change in mass transfer rate over a distance dr

$$= \frac{\partial}{\partial r}\left(-4\pi r^2 D\frac{\partial C_A}{\partial r}\right)\mathrm{d}r$$

Making a material balance over a shell gives:

$$-\frac{\partial}{\partial r}\left(-4\pi r^2 D\frac{\partial C_A}{\partial r}\right)\mathrm{d}r = \frac{\partial C_A}{\partial t}(4\pi r^2\mathrm{d}r)$$

$$\therefore \quad \frac{\partial C_A}{\partial t} = D\frac{1}{r^2}\frac{\partial}{\partial r}\left(r^2\frac{\partial C_A}{\partial r}\right) \tag{10.146}$$

Equation 10.146 may be solved by taking Laplace transforms for the boundary conditions:

$t = 0$	$0 < r < r_0$	$C_A = 0$
$t > 0$	$r = r_0$	$C_A = C_{Ai}$ (constant)
$t > 0$	$r = 0$	$\frac{\mathrm{d}C_A}{\mathrm{d}r} = 0$ (from symmetry)

The concentration gradient at the surface of the sphere ($r = r_0$) is then found to be given by:

$$\left(\frac{\partial C_A}{\partial r}\right)_{r=r_0} = \frac{-C_{Ai}}{r_0} + \frac{C_{Ai}}{\sqrt{\pi D t}}\left(1 + 2\sum_{n=1}^{n=\infty} e^{-(n^2 r_0^2)/Dt}\right) \tag{10.147}$$

The mass transfer rate at the surface of the sphere at time t

$$= 4\pi r_0^2(-D)\left(\frac{\partial C_A}{\partial r}\right)_{r=r_0}$$

$$= 4\pi r_0 D C_{Ai}\left[1 - \frac{r_0}{\sqrt{\pi D t}}\left(1 + 2\sum_{n=1}^{n=\infty} e^{-(n^2 r_0^2)/Dt}\right)\right] \tag{10.148}$$

The total mass transfer during the passage of a drop is therefore obtained by integration of equation 10.148 over the time of exposure.

10.5.5. Other theories of mass transfer

KISHINEVSKIJ[23] has developed a model for mass transfer across an interface in which molecular diffusion is assumed to play no part. In this, fresh material is continuously brought to the interface as a result of turbulence within the fluid and, after exposure to the second phase, the fluid element attains equilibrium with it and then becomes mixed again with the bulk of the phase. The model thus presupposes surface renewal without penetration by diffusion and therefore the effect of diffusivity should not be important. No reliable experimental results are available to test the theory adequately.

10.5.6. Interfacial turbulence

An important feature of the behaviour of the interface which has not been taken into account in the preceding treatment is the possibility of turbulence being generated by a means other than that associated with the fluid dynamics of the process. This *interfacial turbulence* may arise from local variations in the interfacial tension set up during the course of the mass transfer process. It can occur at both gas–liquid and liquid–liquid interfaces, although the latter case has commanded the greater attention. Interfacial turbulence may give rise to violent intermittent eruptions at localised regions in the interface, as a result of which rapid mixing occurs, and the mass transfer rate may be considerably enhanced.

The effect, which arises in cases where the interfacial tension is strongly dependent on the concentration of diffusing solute, will generally be dependent on the direction (sense) in which mass transfer is taking place.

If mass transfer causes interfacial tension to decrease, localised regions of low interfacial tension will form and, as a result, surface spreading will take place. If the gradient of interfacial tension is very high, the surface spreading will be rapid enough to give rise to intense ripples at the interface and a rapid increase in the mass transfer rate. On the other hand, if the mass transfer gives rise to increased interfacial tension, the surface shows no tendency to spread and tends to be stable.

This phenomenon, frequently referred to as the *Marangoni Effect*, explains some of the anomalously high mass transfer rates reported in the literature.

The effect may be reduced by the introduction of surfactants which tend to concentrate at the interface where they exert a stabilising influence, although they may introduce an interface resistance and substantially reduce the mass transfer rate. Thus, for instance, hexadecanol when added to open ponds of water will collect at the interface and substantially reduce the rate of evaporation.

Such effects are described in more detail by SHERWOOD, PIGFORD and WILKE[27].

10.5.7. Mass transfer coefficients

On the basis of each of the theories discussed, the rate of mass transfer in the absence of bulk flow is directly proportional to the driving force, expressed as a molar concentration difference, and, therefore:

$$N_A = h_D(C_{Ai} - C_{Ao}) \tag{10.149}$$

where h_D is a mass transfer coefficient (see equation 10.97). In the two-film theory, h_D is directly proportional to the diffusivity and inversely proportional to the film thickness. According to the penetration theory it is proportional to the square root of the diffusivity and, when all surface elements are exposed for an equal time, it is inversely proportional to the square root of time of exposure; when random surface renewal is assumed, it is proportional to the square root of the rate of renewal. In the film-penetration theory, the mass transfer coefficient is a complex function of the diffusivity, the film thickness, and either the time of exposure or the rate of renewal of surface.

In most cases, the value of the transfer coefficient cannot be calculated from first principles, although the way in which the coefficient will vary as operating conditions are altered can frequently be predicted by using the theory which is most closely applicable to the problem in question.

The penetration and film-penetration theories have been developed for conditions of equimolecular counterdiffusion only; the equations are too complex to solve explicitly for transfer through a stationary carrier gas. For gas absorption, therefore, they apply only when the concentration of the material under going mass transfer is low. On the other hand, in the two-film theory the additional contribution to the mass transfer which is caused by bulk flow is easily calculated and h_D (Section 10.23) is equal to $(D/L)(C_T/C_{Bm})$ instead of D/L.

In a process where mass transfer takes place across a phase boundary, the same theoretical approach can be applied to each of the phases, though it does not follow that the same theory is best applied to both phases. For example, the film model might be applicable to one phase and the penetration model to the other. This problem is discussed in the previous section.

When the film theory is applicable to each phase (the two-film theory), the process is steady state throughout and the interface composition does not then vary with time. For this case the two film coefficients can readily be combined. Because material does not accumulate at the interface, the mass transfer rate on each side of the phase boundary will be the same and for two phases it follows that:

$$N_A = h_{D1}(C_{Ao1} - C_{Ai1}) = h_{D2}(C_{Ai2} - C_{Ao2}) \tag{10.150}$$

If there is no resistance to transfer at the interface C_{Ai1} and C_{Ai2} will be corresponding values in the phase-equilibrium relationship.

Usually, the values of the concentration at the interface are not known and the mass transfer coefficient is considered for the overall process. Overall transfer coefficients are then defined by:

$$N_A = K_1(C_{Ao1} - C_{Ae1}) = K_2(C_{Ae2} - C_{Ao2}) \tag{10.151}$$

where C_{Ae1} is the concentration in phase 1 in equilibrium with C_{Ao2} in phase 2, and C_{Ae2} is the concentration in phase 2 in equilibrium with C_{Ao1} in phase 1. If the equilibrium relationship is linear:

$$\mathscr{H} = \frac{C_{Ai1}}{C_{Ai2}} = \frac{C_{Ae1}}{C_{Ao2}} = \frac{C_{Ao1}}{C_{Ae2}} \tag{10.152}$$

where $\mathscr{H}$ is a proportionality constant.

The relationships between the various transfer coefficients are obtained as follows. From equations 10.150 and 10.151

$$\frac{1}{K_1} = \frac{1}{h_{D1}}\frac{C_{Ao1} - C_{Ae1}}{C_{Ao1} - C_{Ai1}} = \frac{1}{h_{D1}}\frac{C_{Ai1} - C_{Ae1}}{C_{Ao1} - C_{Ai1}} + \frac{1}{h_{D1}}\frac{C_{Ao1} - C_{Ai1}}{C_{Ao1} - C_{Ai1}}$$

But: $$\frac{1}{h_{D1}} = \frac{1}{h_{D2}}\frac{C_{Ao1} - C_{Ai1}}{C_{Ai2} - C_{Ao2}} \quad \text{(from equation 10.150)}$$

and hence: $$\frac{1}{K_1} = \frac{1}{h_{D1}} + \frac{1}{h_{D2}}\left(\frac{C_{Ao1} - C_{Ai1}}{C_{Ai2} - C_{Ao2}}\right)\left(\frac{C_{Ai1} - C_{Ae1}}{C_{Ao1} - C_{Ai1}}\right)$$

From equation 10.152: $$\frac{C_{Ai1} - C_{Ae1}}{C_{Ai2} - C_{Ao2}} = \mathscr{H}$$

$$\therefore \quad \frac{1}{K_1} = \frac{1}{h_{D1}} + \frac{\mathscr{H}}{h_{D2}} \tag{10.153}$$

Similarly: $$\frac{1}{K_2} = \frac{1}{\mathscr{H}h_{D1}} + \frac{1}{h_{D2}} \tag{10.154}$$

and hence: $$\frac{1}{K_1} = \frac{\mathscr{H}}{K_2} \tag{10.155}$$

It follows, that when h_{D1} is large compared with h_{D2}, K_2 and h_{D2} are approximately equal, and, when h_{D2} is large compared with h_{D1}, K_1 and h_{D1} are almost equal.

These relations between the various coefficients are valid provided that the transfer rate is linearly related to the driving force and that the equilibrium relationship is a straight line. They are therefore applicable for the two-film theory, and for any instant of time for the penetration and film-penetration theories. In general, application to time-averaged coefficients obtained from the penetration and film-penetration theories is not permissible because the condition at the interface will be time-dependent unless all of the resistance lies in one of the phases.

Example 10.8

Ammonia is absorbed at 1 bar from an ammonia-air stream by passing it up a vertical tube, down which dilute sulphuric acid is flowing. The following laboratory data are available:

$$\text{Length of tube} = 825\ \text{mm}$$
$$\text{Diameter of tube} = 15\ \text{mm}$$
$$\text{Partial pressures of ammonia: at inlet} = 7.5\ \text{kN/m}^2; \quad \text{at outlet} = 2.0\ \text{kN/m}^2$$
$$\text{Air rate} = 2 \times 10^{-5}\ \text{kmol/s}$$

What is the overall transfer coefficient K_G based on the gas phase?

Solution

$$\text{Driving force at inlet} = 7500\ \text{N/m}^2$$
$$\text{Driving force at outlet} = 2000\ \text{N/m}^2$$
$$\text{Mean driving force} = \frac{(7500 - 2000)}{\ln(7.5/2.0)} = 4200\ \text{N/m}^2$$
$$\text{Ammonia absorbed} = 2 \times 10^{-5}\left(\frac{7.5}{93.8} - \frac{2.0}{99.3}\right) = 1.120 \times 10^{-6}\ \text{kmol/s}$$
$$\text{Wetted surface} = \pi \times 0.015 \times 0.825 = 0.0388\ \text{m}^2$$
$$\text{Hence } K_G = \frac{(1.120 \times 10^{-6})}{(0.0388 \times 4200)} = \underline{\underline{6.87 \times 10^{-9}\ \text{kmol/[m}^2\,\text{s(N/m}^2)]}}$$

10.5.8. Countercurrent mass transfer and transfer units

Mass transfer processes involving two fluid streams are frequently carried out in a column; countercurrent flow is usually employed although co-current flow may be advantageous in some circumstances. There are two principal ways in which the two streams may be brought into contact in a continuous process so as to permit mass transfer to take place between them, and these are termed *stagewise processes* and *continuous differential contact processes*.

Stagewise processes

In a stagewise process the fluid streams are mixed together for a period long enough for them to come close to thermodynamic equilibrium, following which they are separated, and each phase is then passed countercurrently to the next stage where the process is repeated. An example of this type of process is the plate-type of distillation column in which a liquid stream flows down the column, overflowing from each plate to the one below as shown in Figure 10.8. The vapour passes, up through the column, being dispersed into the liquid as it enters the plate, it rises through the liquid, disengaging from the liquid in the vapour space above the liquid surface, and then passes upwards to repeat the process on the next plate. Mass transfer takes place as the bubbles rise through the liquid, and as liquid droplets are thrown up into the vapour space above the liquid–vapour interface. On an *ideal plate*, the liquid and vapour streams leave in thermodynamic equilibrium with each another. In practice, equilibrium may not be achieved and a *plate efficiency*, based on the compositions in either the liquid or vapour stream, is defined as the ratio of the actual change in composition to that which would have been achieved in an ideal

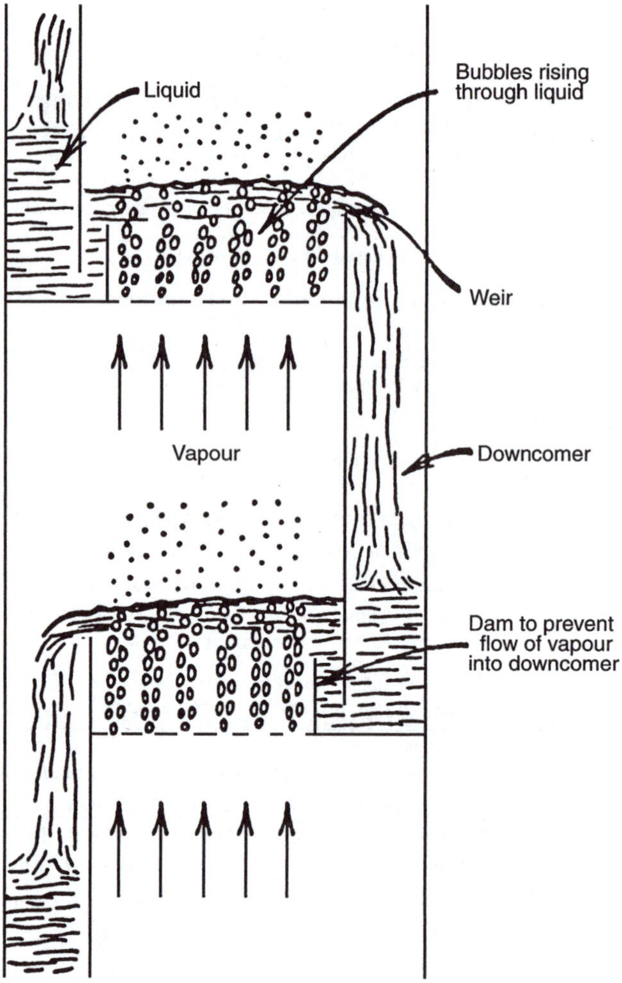

Figure 10.8. Vapour and liquid flow in a plate column

stage. The use of stagewise processes for distillation, gas absorption and liquid–liquid extraction is discussed in Volume 2.

Continuous differential contact processes

In this process, the two streams flow countercurrently through the column and undergo a continuous change in composition. At any location are in *dynamic* rather than *thermodynamic* equilibium. Such processes are frequently carried out in *packed columns*, in which the liquid (or one of the two liquids in the case of a liquid–liquid extraction process) wets the surface of the packing, thus increasing the interfacial area available for mass transfer and, in addition, promoting high film mass transfer coefficients within each phase.

In a packed distillation column, the vapour stream rises against the downward flow of a liquid reflux, and a state of dynamic equilibrium is set up in a steady state process.

The more volatile constituent is transferred under the action of a concentration gradient from the liquid to the interface where it evaporates and then is transferred into the vapour stream. The less volatile component is transferred in the opposite direction and, if the molar latent heats of the components are equal, equimolecular counterdiffusion takes place.

In a packed absorption column, the flow pattern is similar to that in a packed distillation column but the vapour stream is replaced by a mixture of carrier gas and solute gas. The solute diffuses through the gas phase to the liquid surface where it dissolves and is then transferred to the bulk of the liquid. In this case there is no mass transfer of the carrier fluid and the transfer rate of solute is supplemented by bulk flow.

In a liquid–liquid extraction column, the process is similar to that occurring in an absorption column except that both streams are liquids, and the lighter liquid rises through the denser one.

In distillation, equimolecular counterdiffusion takes place if the molar latent heats of the components are equal and the molar rate of flow of the two phases then remains approximately constant throughout the whole height of the column. In gas absorption, however, the mass transfer rate is increased as a result of bulk flow and, at high concentrations of soluble gas, the molar rate of flow at the top of the column will be less than that at the bottom. At low concentrations, however, bulk flow will contribute very little to mass transfer and, in addition, flowrates will be approximately constant over the whole column.

The conditions existing in a column during the steady state operation of a countercurrent process are shown in Figure 10.9. The molar rates of flow of the two streams are G_1 and G_2, which will be taken as constant over the whole column. Suffixes 1 and 2 denote the two phases, and suffixes t and b relate to the top and bottom of the column.

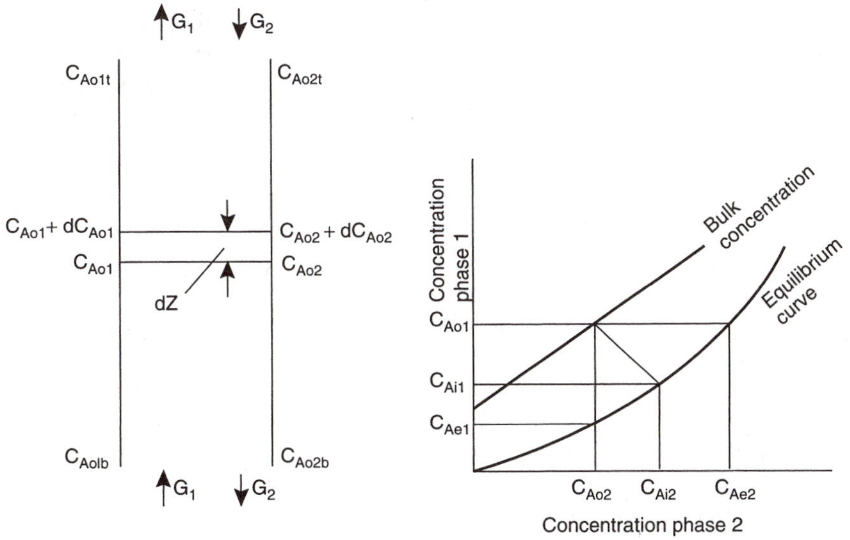

Figure 10.9. Countercurrent mass transfer in a column

If the height of the column is Z, its total cross-sectional area is S, and a is the interfacial area between the two phases per unit volume of column, then the rate of transfer of a component in a height dZ of column is given by:

$$G_1 \frac{1}{C_T} \mathrm{d}C_{Ao1} = h_{D1}(C_{Ai1} - C_{Ao1}) Sa \,\mathrm{d}Z \tag{10.156}$$

or:

$$\frac{\mathrm{d}C_{Ao1}/\mathrm{d}Z}{C_{Ai1} - C_{Ao1}} = \frac{h_{D1} a S C_T}{G_1} = \frac{h_{D1} a C_T}{G'_1} \tag{10.157}$$

where G'_1 is the molar rate of flow per unit cross-section of column. The exact interfacial area cannot normally be determined independently of the transfer coefficient, and therefore values of the product $h_{D1}a$ are usually quoted for any particular system.

The left-hand side of equation 10.157 is the rate of change of concentration with height for unit driving force, and is therefore a measure of the efficiency of the column and, in this way, a high value of $(h_{D1}aC_T)/G'_1$ is associated with a high efficiency. The reciprocal of this quantity is $G'_1/(h_{D1}aC_T)$ which has linear dimensions and is known as the *height of the transfer unit* $\mathbf{H}_1$ (HTU).

Rearranging equation 10.157 and integrating gives:

$$\int_{(C_{Ao1_b})}^{(C_{Ao1_t})} \frac{\mathrm{d}C_{Ao1}}{C_{Ai1} - C_{Ao1}} = \frac{h_{D1} a C_T}{G'_1} Z = \frac{Z}{G'_1/(h_{D1} a C_T)} = Z/\mathbf{H}_1 \tag{10.158}$$

The right-hand side of equation 10.158 is the height of the column divided by the HTU and this is known as the *number of transfer units* $\mathbf{N}_1$. It is obtained by evaluating the integral on the left-hand side of the equation.

Therefore:

$$\mathbf{N}_1 = \frac{Z}{\mathbf{H}_1} \tag{10.159}$$

In some cases, such as the evaporation of a liquid at an approximately constant temperature or the dissolving of a highly soluble gas in a liquid, the interface concentration may be either substantially constant or negligible in comparison with that of the bulk. In such cases, the integral on the left-hand side of equation 10.158 may be evaluated directly to give:

$$\int_{(C_{Ao1})_b}^{(C_{Ao1})_t} \frac{\mathrm{d}C_{Ao1}}{C_{Ai1} - C_{Ao1}} = \ln \frac{(C_{Ai1} - C_{Ao1})_b}{(C_{Ai1} - C_{Ao1})_t} = \frac{Z}{\mathbf{H}_1} \tag{10.160}$$

where b and t represent values at the bottom and the top, as shown in Figure 10.9.

Thus:

$$\frac{(C_{Ai1} - C_{Ao1})_b}{(C_{Ai1} - C_{Ao1})_t} = \mathrm{e}^{Z/\mathbf{H}_1} \tag{10.161}$$

Thus $\mathbf{H}_1$ is the height of column over which the *driving force changes by a factor of* e.

Equation 10.156 can be written in terms of the film coefficient for the second phase (h_{D2}) or either of the overall transfer coefficients (K_1, K_2). Transfer units based on either film coefficient or overall coefficient can therefore be defined, and the following equations are analogous to equation 10.159:

Number of transfer units based on phase 2, $\mathbf{N}_2 = Z/\mathbf{H}_2$ (10.162)

Number of overall transfer units based on phase 1, $\mathbf{N}_{o1} = Z/\mathbf{H}_{o1}$ (10.163)

Number of overall transfer units based on phase 2, $\mathbf{N}_{o2} = Z/\mathbf{H}_{o2}$ (10.164)

Using this notation the introduction of o into the suffix indicates an overall transfer unit.

The equations for $\mathbf{H}_1$, $\mathbf{H}_2$, $\mathbf{H}_{o1}$, $\mathbf{H}_{o2}$ are of the following form:

$$\mathbf{H}_1 = \frac{G'_1}{h_{D1} a C_T} \tag{10.165}$$

If one phase is a gas, as in gas absorption for example, it is often more convenient to express concentrations as partial pressures in which case:

$$\mathbf{H}_1 = \mathbf{H}_G = \frac{G'_1}{k_G a P} \tag{10.166}$$

where, in SI units, G'_1 is expressed in kmol/m^2 s, k_G in kmol/[m^2 s (N/m^2)], P in N/m^2, and a in m^2/m^3.

The overall values of the HTU may be expressed in terms of the film values by using equation 10.153 which gives the relation between the coefficients.

$$\frac{1}{K_1} = \frac{1}{h_{D1}} + \frac{\mathscr{H}}{h_{D2}} \qquad \text{(equation 10.153)}$$

Substituting for the coefficients in terms of the HTU:

$$\mathbf{H}_{o1}\frac{aC_T}{G'_1} = \mathbf{H}_1\frac{aC_T}{G'_1} + \frac{\mathscr{H} aC_T}{G'_2}\mathbf{H}_2 \tag{10.167}$$

or:

$$\mathbf{H}_{o1} = \mathbf{H}_1 + \mathscr{H}\frac{G'_1}{G'_2}\mathbf{H}_2 \tag{10.168}$$

Similarly:

$$\mathbf{H}_{o2} = \mathbf{H}_2 + \frac{1}{\mathscr{H}}\frac{G'_2}{G'_1}\mathbf{H}_1 \tag{10.169}$$

The advantage of using the transfer unit in preference to the transfer coefficient is that the former remains much more nearly constant as flow conditions are altered. This is particularly important in problems of gas absorption where the concentration of the solute gas is high and the flow pattern changes in the column because of the change in the total rate of flow of gas at different sections. In most cases the coefficient is proportional to the flowrate raised to a power slightly less than unity and therefore the HTU is substantially constant.

As noted previously, for equimolecular counterdiffusion, the film transfer coefficients, and hence the corresponding HTUs, may be expressed in terms of the physical properties of the system and the assumed film thickness or exposure time, using the two-film, the penetration, or the film-penetration theories. For conditions where bulk flow is important, however, the transfer rate of constituent **A** is increased by the factor C_T/C_{Bm} and the diffusion equations can be solved only on the basis of the two-film theory. In the design of equipment it is usual to work in terms of transfer coefficients or HTUs and not to endeavour to evaluate them in terms of properties of the system.

10.6. MASS TRANSFER AND CHEMICAL REACTION IN A CONTINUOUS PHASE

In many applications of mass transfer the solute reacts with the medium as in the case, for example, of the absorption of carbon dioxide in an alkaline solution. The mass transfer rate then decreases in the direction of diffusion as a result of the reaction. Considering the unidirectional molecular diffusion of a component **A** through a distance δy over area **A**. then, neglecting the effects of bulk flow, a material balance for an irreversible reaction of order n gives:

moles IN/unit time (at y)	− moles OUT/unit time (at $y + \delta y$)	= rate of change of concn. × element volume	+ reacted moles/unit volume × element volume

$$\left\{-D\frac{\partial C_A}{\partial y}\right\}A - \left\{-D\frac{\partial C_A}{\partial y} + \frac{\partial}{\partial y}\left[-D\frac{\partial C_A}{\partial y}\right]\delta y\right\}A = \frac{\partial C_A}{\partial t}(\partial y \cdot A) + (kC_A^n)(\delta y \cdot A)$$

or:

$$\frac{\partial C_A}{\partial t} = D\frac{\partial^2 C_A}{\partial y^2} - kC_A^n \tag{10.170}$$

where k is the reaction rate constant. This equation has no analytical solution for the general case.

10.6.1. Steady-state process

For a steady-state process, equation 10.170 becomes:

$$D\frac{d^2 C_A}{dy^2} - kC_A^n = 0 \tag{10.171}$$

Equation 10.171 may be integrated using the appropriate boundary conditions.

First-order reaction

For a first-order reaction, putting $n = 1$ in equation 10.171, then:

$$D\frac{d^2 C_A}{dy^2} - kC_A = 0 \tag{10.172}$$

The solution of equation 10.172 is:

$$C_A = B_1' e^{\sqrt{(k/D)}y} + B_2' e^{-\sqrt{(k/D)}y} \tag{10.173}$$

B_1' and B_2' must then be evaluated using the appropriate boundary conditions.

As an example, consideration is given to the case where the fluid into which mass transfer is taking place is initially free of solute and is semi-infinite in extent. The surface concentration C_{Ai} is taken as constant and the concentration at infinity as zero. The boundary conditions are therefore:

$$y = 0 \qquad C_A = C_{Ai}$$

$$y = \infty \qquad C_A = 0$$

Substituting these boundary conditions in equation 10.173 gives:

$$B_1' = 0 \qquad B_2' = C_{Ai}$$

$$C_A = C_{Ai} e^{-\sqrt{(k/D)}y} \tag{10.174}$$

$$\frac{dC_A}{dy} = -\sqrt{\frac{k}{D}} C_{Ai} e^{-\sqrt{(k/D)}y}$$

and: $$N_A = -D\frac{dC_A}{dy} = \sqrt{kD} C_{Ai} e^{-\sqrt{(k/D)}y} \qquad (10.175)$$

At the interface $y = 0$ and: $$N_A = \sqrt{kD} C_{Ai} \qquad (10.176)$$

*n*th-order reaction

The same boundary conditions will be used as for the first-order reaction. Equation 10.171 may be re-arranged to give:

$$\frac{d^2C_A}{dy^2} - \frac{k}{D} C_A^n = 0 \qquad (10.177)$$

Putting: $$\frac{dC_A}{dy} = q$$

$$\frac{d^2C_A}{dy^2} = \frac{dq}{dy} = \frac{dq}{dC_A}\frac{dC_A}{dy} = q\frac{dq}{dC_A}$$

and: $$q\frac{dq}{dC_A} - \frac{k}{D} C_A^n = 0$$

Multiplying through by dC_A and integrating:

$$\frac{q^2}{2} - \frac{1}{n+1}\frac{k}{D} C_A^{n+1} = B_3'$$

When $y = \infty$, $C_A = 0$ and $\dfrac{dC_A}{dy} = q = 0 \qquad \therefore B_3' = 0$

and: $$\left(\frac{dC_A}{dy}\right)^2 = \frac{2}{n+1}\frac{k}{D} C_A^{n+1}$$

Since $\dfrac{dC_A}{dy}$ is negative, the negative value of the square root will be taken to give:

$$\frac{dC_A}{dy} = -\sqrt{\frac{2}{n+1}}\sqrt{\frac{k}{D}} C_A^{\frac{n+1}{2}} \qquad (10.178)$$

$$N_A = -D\frac{dC_A}{dy} = \sqrt{\frac{2}{n+1}}\sqrt{kD} C_A^{\frac{n+1}{2}} \qquad (10.179)$$

At the free surface, $C_A = C_{Ai}$ and

$$N_A = \sqrt{\frac{2}{n+1}}\sqrt{kD} C_{Ai}^{\frac{n+1}{2}} \qquad (10.180)$$

which is identical to equation 10.176 for a first-order reaction when $n = 1$.

Integrating equation 10.178 gives:

$$C_A^{-\frac{n+1}{2}}\,\mathrm{d}C_A = -\sqrt{\frac{2}{n+1}}\sqrt{\frac{k}{D}}\mathrm{d}y$$

or:

$$\frac{2}{1-n}C_A^{\frac{1-n}{2}} = -\sqrt{\frac{2}{n+1}}\sqrt{\frac{k}{D}}y + B'_4$$

When $y = 0$, $C_A = C_{Ai}$ and $B'_4 = \dfrac{2}{1-n}C_{Ai}^{\frac{1-n}{2}}$

and:

$$C_A^{\frac{1-n}{2}} - C_{Ai}^{\frac{1-n}{2}} = (n-1)\sqrt{\frac{1}{2(n+1)}}\sqrt{\frac{k}{D}}y \qquad (10.181)$$

This solution cannot be used for a first-order reaction where $n = 1$ because it is then indeterminate.

Second-order reaction ($n = 2$)

In this case, equation 10.181 becomes:

$$C_A^{-\frac{1}{2}} - C_{Ai}^{-\frac{1}{2}} = \sqrt{\frac{k}{6D}}y \qquad (10.182)$$

and equation 10.180 becomes:

$$N_A = \sqrt{\frac{2}{3}}\sqrt{kD}C_A^{\frac{3}{2}} \qquad (10.183)$$

Example 10.9

In a gas absorption process, the solute gas **A** diffuses into a solvent liquid with which it reacts. The mass transfer is one of steady state unidirectional molecular diffusion and the concentration of **A** is always sufficiently small for bulk flow to be negligible. Under these conditions the reaction is first order with respect to the solute **A**.

At a depth l below the liquid surface, the concentration of **A** has fallen to one-half of the value at the surface. What is the ratio of the mass transfer rate at this depth l to the rate at the surface? Calculate the numerical value of the ratio when $l\sqrt{k/D} = 0.693$, where D is the molecular diffusivity and k the first-order rate constant.

Solution

This process is described by:

$$C_A = B'_1 e^{\sqrt{(k/D)}y} + B'_2 e^{-\sqrt{(k/D)}y} \qquad \text{(equation 10.173)}$$

If C_{Ai} is the surface concentration ($y = 0$):

$$C_{Ai} = B'_1 + B'_2$$

At $y = l$, $C_A = C_{Ai}/2$, and:

$$\therefore \quad \frac{C_{Ai}}{2} = B'_1 e^{\sqrt{(k/D)}l} + B'_2 e^{-\sqrt{(k/D)}l}$$

Solving for B'_1 and B'_2:

$$B'_1 = \frac{C_{Ai}}{2}(1 - 2e^{-\sqrt{(k/D)}l})(e^{\sqrt{(k/D)}l} - e^{-\sqrt{(k/D)}l})^{-1}$$

$$B'_2 = -\frac{C_{Ai}}{2}(1 - 2e^{\sqrt{(k/D)}l})(e^{\sqrt{(k/D)}l} - e^{-\sqrt{(k/D)}l})^{-1}$$

$$\frac{(N_A)_{y=l}}{(N_A)_{y=0}} = \frac{-D(dC_A/dy)_{y=l}}{-D(dC_A/dy)_{y=0}} = \frac{(dC_A/dy)_{y=l}}{(dC_A/dy)_{y=0}}$$

$$\frac{dC_A}{dy} = \sqrt{\frac{k}{D}}(B'_1 e^{\sqrt{(k/D)}y} - B'_2 e^{-\sqrt{(k/D)}y})$$

$$\frac{(N_A)_{y=l}}{(N_A)_{y=0}} = \frac{B'_1 e^{\sqrt{(k/D)}l} - B'_2 e^{-\sqrt{(k/D)}l}}{B'_1 - B'_2}$$

$$= \frac{(1 - 2e^{-\sqrt{(k/D)}l})e^{\sqrt{(k/D)}l} + (1 - 2e^{\sqrt{(k/D)}l})e^{-\sqrt{(k/D)}l}}{(1 - 2e^{-\sqrt{(k/D)}l}) + (1 - 2e^{\sqrt{(k/D)}l})}$$

$$= \frac{e^{\sqrt{(k/D)}l} + e^{-\sqrt{(k/D)}l} - 4}{2(1 - e^{-\sqrt{(k/D)}l} - e^{\sqrt{(k/D)}l})}$$

When: $$l\sqrt{\frac{k}{D}} = 0.693, \quad e^{\sqrt{(k/D)}l} = 2, \quad e^{-\sqrt{(k/D)}l} = 0.5$$

and: $$\frac{(N_A)_{y=l}}{(N_A)_{y=0}} = \frac{2 + \frac{1}{2} - 4}{2(1 - 2 - \frac{1}{2})} = \underline{\underline{0.5}}$$

Example 10.10

In a steady-state process, a gas is absorbed in a liquid with which it undergoes an irreversible reaction. The mass transfer process is governed by Fick's law, and the liquid is sufficiently deep for it to be regarded as effectively infinite in depth. On increasing the temperature, the concentration of reactant at the liquid surface C_{Ai} falls to 0.8 times its original value. The diffusivity is unchanged, but the reaction constant increases by a factor of 1.35. It is found that the mass transfer rate at the liquid surface falls to 0.83 times its original value. What is the order of the chemical reaction?

Solution

The mass transfer rate (moles/unit area and unit time) is given by equation 10.180, where denoting the original conditions by subscript 1 and the conditions at the higher temperature by subscript 2 gives:

$$N_{A1} = \sqrt{\frac{2}{n+1}}\sqrt{k_1 D}C_{Ai}^{\frac{n+1}{2}} \qquad \text{(equation 10.180)}$$

and: $$N_{A2} = 0.83N_{A1} = \sqrt{\frac{2}{n+1}}\sqrt{1.35k_1 D}(0.8C_{Ai})^{\frac{n+1}{2}}$$

Substituting the numerical values gives:

$$0.83 = \sqrt{1.35}(0.8)^{\frac{n+1}{2}}$$

or: $$0.8^{\frac{n+1}{2}} = 0.714$$

Thus: $$\frac{n+1}{2} = 1.506$$

and: $$n = 2.01$$

Thus, the reaction is of second-order.

Example 10.11

A pure gas is absorbed into a liquid with which it reacts. The concentration in the liquid is sufficiently low for the mass transfer to be covered by Fick's Law and the reaction is first-order with respect to the solute gas. It may be assumed that the film theory may be applied to the liquid and that the concentration of solute gas falls from the saturation value to zero across the film. The reaction is initially carried out at 293 K. By what factor will the mass transfer rate across the interface change, if the temperature is raised to 313 K?

The reaction rate constant at 293 K $= 2.5 \times 10^{-6}\ \text{s}^{-1}$

The energy of activation for reaction in the Arrhenuis equation = 26430 kJ/kmol = 2.643×10^7 J/kmol

Universal gas constant, $\mathbf{R} = 8314$ J/kmol K

Molecular diffusivity, $D = 10^{-9}\ \text{m}^2/\text{s}$

Film thickness, $L = 10$ mm

Solubility of gas at 313 K is 80 per cent of the solubility at 293 K.

Solution

For a first-order reaction:

$$D\frac{\partial^2 C_A}{\partial y^2} - kC_A = 0 \qquad \text{(equation 10.170)}$$

Solving:

$$C_A = B_1' e^{\sqrt{(k/D)}y} + B_2' e^{-\sqrt{(k/D)}y} \qquad \text{(equation 10.171)}$$

When $y = 0$, $C_A = C_{AS}$ and when $y = L$, $C_A = 0$.

Substituting:

$$B_1' = -C_{AS}e^{-2L\sqrt{(k/D)}}(1 - e^{-2L\sqrt{(k/D)}})^{-1}$$

and:

$$B_2' = C_{AS}(1 - e^{-2L\sqrt{(k/D)}})^{-1}$$

and thus:

$$\frac{C_A}{C_{AS}} = (e^{-\sqrt{(k/D)}y} - e^{-2L\sqrt{(k/D)}})(1 - e^{-2L\sqrt{(k/D)}})^{-1}$$

Differentiating and putting $y = 0$:

$$\frac{1}{C_{AS}}\left(\frac{dC_A}{dy}\right)_{y=0} = \frac{(-\sqrt{(k/D)} - \sqrt{(k/D)}e^{-2L\sqrt{(k/D)}})}{(1 - e^{-2L}\sqrt{(k/D)})}$$

The mass transfer rate at the interface is then:

$$-D\left(\frac{dC_A}{dy}\right)_{y=0} = C_{AS}\sqrt{(k/D)}(1 + e^{-2L\sqrt{(k/D)}})(1 - e^{-2L\sqrt{(k/D)}})^{-1}$$

When $D = 1 \times 10^{-9}\ \text{m}^2/\text{s}$, then at 293 K:

$$k = Ae^{-2.643\times10^7/(8314\times293)}$$

$$= 1.94 \times 10^{-5}A = 2.5 \times 10^{-6}\ \text{s}^{-1}$$

and: $$A = 0.129\ \text{s}^{-1}$$

Since $L = 0.01$ m, then:

$$2L\sqrt{(k/D)} = 2 \times 0.01(2.5 \times 10^{-6}/(1 \times 10^{-3}))^{0.5} = 1.0$$

$$e^{-2L\sqrt{(k/D)}} = e^{-1} = 0.368$$

and: $$C_{AS} = C_{AS1}$$

The mass transfer rate is then:

$$N_{A293} = C_{AS_1}\left(\frac{2.5 \times 10^{-6}}{1 \times 10^{-9}}\right)^{0.5}\frac{(1 + 0.368)}{(1 - 0.368)}$$

$$= 108.2C_{AS_1}$$

At 313 K:

$$k = Ae^{(-2.643 \times 10^7/(8314 \times 313))} = (0.129 \times 3.37 \times 10^{-5}) = 5.0 \times 10^{-6}\ \text{s}^{-1}$$

$$C_{AS} = 0.8C_{AS_1}$$

$$2L\sqrt{(k/D)} = (2 \times 0.01(5 \times 10^{-6}))/(1 \times 10^{-9})^{0.5} = 1.414$$

and: $$e^{-1.414} = 0.243$$

The mass transfer rate at 313 K is then:

$$N_{A313} = 0.8C_{AS_1}(5 \times 10^6 \times 1 \times 10^{-9})^{0.5}(1 + 0.243)/(1 - 0.243)$$

$$= 92.9C_{AS_1}$$

Hence the change in the mass transfer rate is given by the factor:

$$N_{A313}/N_{A293} = (92.9C_{AS_1}/108.2C_{AS_1})$$

$$= \underline{\underline{0.86}}$$

10.6.2. Unsteady-state process

For an unsteady-state process, equation 10.170 may be solved analytically only in the case of a first-order reaction ($n = 1$). In this case:

$$\frac{\partial C_A}{\partial t} = D\frac{\partial^2 C_A}{\partial y^2} - kC_A \tag{10.174}$$

The solution of this equation has been discussed by DANCKWERTS[28], and here a solution will be obtained using the Laplace transform method for a semi-infinite liquid initially free of solute. On the assumption that the liquid is in contact with pure solute gas, the concentration C_{Ai} at the liquid interface will be constant and equal to the saturation value. The boundary conditions will be those applicable to the penetration theory, that is:

$$t = 0 \quad 0 < y < \infty \quad C_A = 0$$

$$t > 0 \quad y = 0 \quad C_A = C_{Ai}$$

$$t > 0 \quad y = \infty \quad C_A = 0$$

From equations 10.103 and 10.104, taking Laplace transforms of both sides of equation 10.184 gives:

$$p\overline{C}_A = D\frac{d^2\overline{C}_A}{dy^2} - k\overline{C}_A$$

or:
$$\frac{d^2\overline{C}_A}{dy^2} - \frac{p+k}{D}\overline{C}_A = 0$$

Thus:
$$\overline{C}_A = B_1 e^{\sqrt{(p+k)/D}\,y} + B_2 e^{-\sqrt{(p+k)/D}\,y} \tag{10.185}$$

When $y = \infty$, $C_A = 0$ and therefore $\overline{C}_A = 0$, from which $B_1 = 0$.
When $y = 0$, $C_A = C_{Ai}$ and $\overline{C}_A = C_{Ai}/p$, from which $B_2 = C_{Ai}/p$

Equation 10.185 therefore becomes:

$$\overline{C}_A = \frac{C_{Ai}}{p} e^{-\sqrt{(p+k)/D}\,y} \tag{10.186}$$

The mass transfer rate N_A at the interface must be evaluated in order to obtain the rate at which gas is transferred to the liquid from the gas.

Differentiating equation 10.186 with respect to y gives:

$$\frac{d\overline{C}_A}{dy} = \frac{d\overline{C}_A}{dy} = -\sqrt{\frac{p+k}{D}}\frac{C_{Ai}}{p} e^{-\sqrt{(p+k)/D}\,y} \tag{10.187}$$

At the interface ($y = 0$):

$$\left(\frac{d\overline{C}_A}{dy}\right)_{y=0} = -\frac{C_{Ai}}{\sqrt{D}}\frac{\sqrt{p+k}}{p} \tag{10.188}$$

It is not possible to invert equation 10.188 directly using the transforms listed in Table 12 in Appendix A4. On putting $a = \sqrt{k}$, however, entry number 38 gives:

$$\text{Inverse of } \frac{\sqrt{p}}{p-k} = \frac{1}{\sqrt{\pi t}} + \sqrt{k}\, e^{kt} \operatorname{erf}(\sqrt{kt})$$

From the *shift theorem*, if $\overline{f(t)} = \overline{f}(p)$, and then:

$$\overline{f(t)e^{-kt}} = \int_0^\infty f(t)e^{-kt}e^{-pt}dt = \int_0^\infty f(t)e^{-(p+k)t}dt = \overline{f}(p+k)$$

Thus, inverse of $\dfrac{\sqrt{p+k}}{p}$ is $\dfrac{1}{\sqrt{\pi t}}e^{-kt} + \sqrt{k}\operatorname{erf}(\sqrt{kt})$

Hence, inverting equation 10.188 gives:

$$\left(\frac{dC_A}{dy}\right)_{y=0} = -\frac{C_{Ai}}{\sqrt{D}}\left\{\frac{1}{\sqrt{\pi t}}e^{-kt} + \sqrt{k}\operatorname{erf}\sqrt{kt}\right\}$$

and:
$$(N_A)_t = -D\left(\frac{dC_A}{dy}\right)_{y=0} = C_{Ai}\sqrt{\frac{D}{k}}\left\{\sqrt{\frac{k}{\pi t}}e^{-kt} + k\operatorname{erf}\sqrt{kt}\right\} \tag{10.189}$$

Equation 10.189 gives the instantaneous value of $(N_A)_t$ at time t.

The average value N_A for mass transfer over an exposure time t_e is given by:

$$N_A = \frac{1}{t_e}\int_0^{t_e} N_A\,\mathrm{d}t = \frac{1}{t_e}C_{Ai}\sqrt{\frac{D}{k}}\int_0^{t_e}\left\{\sqrt{\frac{k}{\pi t}}\mathrm{e}^{-kt} + k\ \mathrm{erf}\ \sqrt{kt}\right\}\mathrm{d}t \qquad (10.190)$$

The terms in the integral of equation 10.190 cannot be evaluated directly. Although, the integral can be re-arranged to give three terms [(i), (ii) and (iii)], each of which can be integrated, by both adding and subtracting $kt\sqrt{(k/\pi t)}\mathrm{e}^{-kt}$, and splitting the first term into two parts. This gives:

$$N_A = \frac{1}{t_e}C_{Ai}\sqrt{\frac{D}{k}}\int_0^{t_e}\left\{\underbrace{\left(k\ \mathrm{erf}\ \sqrt{kt} + kt\sqrt{\frac{k}{\pi t}}\mathrm{e}^{-kt}\right)}_{\text{(i)}}\right.$$

$$\left. + \underbrace{\left(-kt\sqrt{\frac{k}{\pi t}}\mathrm{e}^{-kt} + \frac{1}{2}\sqrt{\frac{k}{\pi t}}\mathrm{e}^{-kt}\right)}_{\text{(ii)}} + \underbrace{\frac{1}{2}\sqrt{\frac{k}{\pi t}}\mathrm{e}^{-kt}}_{\text{(iii)}}\right\}\mathrm{d}t$$

Considering each of the above terms (i), (ii), (iii) in turn:

(i) Noting that: $\dfrac{\mathrm{d}}{\mathrm{d}t}(k\mathrm{erf}\ \sqrt{kt}) = k\dfrac{\mathrm{d}}{\mathrm{d}t}\left\{\dfrac{2}{\sqrt{\pi}}\displaystyle\int_0^{\sqrt{kt}}\mathrm{e}^{-kt}\,\mathrm{d}(\sqrt{kt})\right\}$

$$= k\frac{\mathrm{d}}{\mathrm{d}t}\left\{\frac{2}{\sqrt{\pi}}\int_0^{t}\mathrm{e}^{-kt}\left(\frac{1}{2}\sqrt{\frac{k}{t}}\right)\mathrm{d}t\right\} = k\sqrt{\frac{k}{\pi t}}\mathrm{e}^{-kt}$$

$$\int_0^{t_e}\left(k\mathrm{erf}\ \sqrt{kt} + kt\sqrt{\frac{k}{\pi t}}\mathrm{e}^{-kt}\right)\mathrm{d}t = k\int_0^{t_e}\left(\frac{\mathrm{d}}{\mathrm{d}t}(t\ \mathrm{erf}\ \sqrt{kt})\right)\mathrm{d}t = kt_e\mathrm{erf}\ \sqrt{kt_e}$$

(ii) $\displaystyle\int_0^{t_e}\left(\sqrt{\frac{kt}{\pi}}(-k\mathrm{e}^{-kt}) + \frac{1}{2}\sqrt{\frac{k}{\pi t}}\mathrm{e}^{-kt}\right)\mathrm{d}t = \int_0^{t_e}\left(\frac{\mathrm{d}}{\mathrm{d}t}\left(\sqrt{\frac{kt}{\pi}}\mathrm{e}^{-kt}\right)\right)\mathrm{d}t = \sqrt{\frac{kt_e}{\pi}}\mathrm{e}^{-kt_e}$

(iii) $\displaystyle\int_0^{t_e}\left(\frac{1}{2}\sqrt{\frac{k}{\pi t}}\mathrm{e}^{-kt}\right)\mathrm{d}t = \int_0^{t_e}\left(\frac{1}{2}\sqrt{\frac{k}{\pi t}}\mathrm{e}^{-kt}\right)\left(2\sqrt{\frac{t}{k}}\mathrm{d}\sqrt{kt}\right)$

$$= \frac{1}{2}\cdot\frac{2}{\sqrt{\pi}}\int_0^{t_e}\mathrm{e}^{-kt}\mathrm{d}\sqrt{kt} = \frac{1}{2}\mathrm{erf}\ \sqrt{kt_e}$$

Thus: $\displaystyle N_A = \frac{1}{t_e}C_{Ai}\sqrt{\frac{D}{k}}\left\{kt_e\ \mathrm{erf}\ \sqrt{kt_e} + \sqrt{\frac{kt_e}{\pi}}\mathrm{e}^{-kt_e} + \frac{1}{2}\ \mathrm{erf}\ \sqrt{kt_e}\right\}$

$$= C_{Ai}\sqrt{\frac{D}{k}}\left\{\left(k + \frac{1}{2t_e}\right)\ \mathrm{erf}\ \sqrt{kt_e} + \sqrt{\frac{k}{\pi t_e}}\mathrm{e}^{-kt_e}\right\} \qquad (10.191)$$

Thus the mass transfer coefficient, enhanced by chemical reaction, h'_D is given by:

$$h'_D = \frac{N_A}{C_{Ai}} = \sqrt{\frac{D}{k}}\left\{\left(k + \frac{1}{2t_e}\right)\text{ erf }\sqrt{kt_e} + \sqrt{\frac{k}{\pi t_e}}\text{e}^{-kt_e}\right\} \qquad (10.192)$$

Two special cases are now considered:

(1) *When the reaction rate is very low* and $k \to 0$.

Using the Taylor series, for small values of $\sqrt{kt_e}$, and therefore neglecting higher powers of $\sqrt{kt_e}$:

$$\text{erf}\sqrt{kt_e} = (\text{erf}\sqrt{kt_e})_{\sqrt{kt_e}\to 0} + \left[\sqrt{kt_e}\left\{\frac{\text{d}}{\text{d}\sqrt{kt}}(\text{erf}\sqrt{kt})\right\}\right]_{\sqrt{kt}\to 0}$$

Then: $$\lim\left[\left(k + \frac{1}{2t_e}\right)\text{ erf}\sqrt{kt_e}\right]_{k\to 0} = \frac{1}{2t_e}\left[(\sqrt{kt_e})\frac{\text{d}}{\text{d}(\sqrt{kt})}\left\{\frac{2}{\sqrt{\pi}}\int_0^{t_e}\text{e}^{-kt}\text{d}(\sqrt{kt})\right\}\right]_{k\to 0}$$

$$= \frac{1}{2t_e}\left[\frac{2}{\sqrt{\pi}}\text{e}^{-kt}(\sqrt{kt_e})\right]_{k\to 0}$$

$$= \sqrt{\frac{k}{\pi t_e}}$$

Also: $$\lim\left[\sqrt{\frac{k}{\pi t_e}}\text{e}^{-kt_e}\right]_{k\to 0} = \sqrt{\frac{k}{\pi t_e}}$$

Substituting in equation 10.192 gives: $$h'_D \to \sqrt{\frac{D}{k}}\left(2\sqrt{\frac{k}{\pi t_e}}\right) = 2\sqrt{\frac{D}{\pi t_e}} \qquad (10.193)$$

as for mass transfer without chemical reaction (equation 10.115)

(2) *When the reaction rate is very high*, $\text{erf}\sqrt{kt_e} \to 1$, $k \gg (1/t_e)$, and from equation 10.192:

$$h'_D = \sqrt{\frac{D}{k}}\{[k \times 1] + 0\} = \sqrt{Dk} \qquad (10.194)$$

and is independent of exposure time.

10.7. MASS TRANSFER AND CHEMICAL REACTION IN A CATALYST PELLET

When an irreversible chemical reaction is carried out in a packed or fluidised bed composed of catalyst particles, the overall reaction rate is influenced by:

(i) the chemical kinetics
(ii) mass transfer resistance within the pores of the catalyst particles
and (iii) resistance to mass transfer of the reactant to the outer surface of the particles.

In general, the concentration of the reactant will decrease from C_{Ao} in the bulk of the fluid to C_{Ai} at the surface of the particle, to give a concentration driving force of $(C_{Ao} - C_{Ai})$. Thus, within the pellet, the concentration will fall progressively from C_{Ai} with distance from the surface. This presupposes that no distinct adsorbed phase is formed in the pores. In this section the combined effects of mass transfer and chemical reaction within the particle are considered, and the effects of external mass transfer are discussed in Section 10.8.4.

In the absence of a resistance to mass transfer, the concentration of reactant will be uniform throughout the the whole volume of the particle, and equal to that at its surface (C_{Ai}).

The reaction rate per unit volume of particle $\mathfrak{R}'_{vn}$ for an nth-order reaction is then given by:

$$\mathfrak{R}'_{vn} = kC_{Ai}^n \tag{10.195}$$

When the mass transfer resistance within the particle is significant, a concentration gradient of reactant is established within the particle, with the concentration, and hence the reaction rate, decreasing progressively with distance from the particle surface. The overall reaction rate is therefore less than that given by equation 10.195.

The internal structure of the catalyst particle is often of a complex labyrinth-like nature, with interconnected pores of a multiplicity of shapes and sizes. In some cases, the pore size may be less than the mean free path of the molecules, and both molecular and Knudsen diffusion may occur simultaneously. Furthermore, the average length of the diffusion path will be extended as a result of the tortuousity of the channels. In view of the difficulty of precisely defining the pore structure, the particle is assumed to be pseudo-homogeneous in composition, and the diffusion process is characterised by an effective diffusivity D_e (equation 10.8).

The ratio of the overall rate of reaction to that which would be achieved in the absence of a mass transfer resistance is referred to as the *effectiveness factor* η. SCOTT and DULLION[29] describe an apparatus incorporating a diffusion cell in which the effective diffusivity D_e of a gas in a porous medium may be measured. This approach allows for the combined effects of molecular and Knudsen diffusion, and takes into account the effect of the complex structure of the porous solid, and the influence of tortuosity which affects the path length to be traversed by the molecules.

The effectiveness factor depends, not only on the reaction rate constant and the effective diffusivity, but also on the size and shape of the catalyst pellets. In the following analysis detailed consideration is given to particles of two regular shapes:

(i) *Flat platelets* in which the mass transfer process can be regarded as one-dimensional, with mass transfer taking place perpendicular to the faces of the platelets. Furthermore, the platelets will be assumed to be sufficiently thin for deviations from unidirectional transfer due to end-effects to be negligible.
(ii) *Spherical particles* in which mass transfer takes place only in the radial direction and in which the area available for mass transfer decreases towards the centre of the particle.

Some consideration is also given to particles of other shapes (for example, long, thin cylinders), but the mathematics becomes complex and no detailed analysis will be given.

It will be shown, however, that the effectiveness factor does not critically depend on the shape of the particles, provided that their characteristic length is defined in an appropriate way. Some comparison is made be made between calculated results and experimental measurements with particles of frequently ill-defined shapes.

The treatment here is restricted to first-order irreversible reactions under steady-state conditions. Higher order reactions are considered by ARIS[30].

10.7.1. Flat platelets

In a thin flat platelet, the mass transfer process is symmetrical about the centre-plane, and it is necessary to consider only one half of the particle. Furthermore, again from considerations of symmetry, the concentration gradient, and mass transfer rate, at the centre-plane will be zero. The governing equation for the steady-state process involving a first-order reaction is obtained by substituting D_e for D in equation 10.172:

$$\frac{d^2C_A}{dy^2} - \frac{k}{D_e}C_A = 0 \qquad (10.196)$$

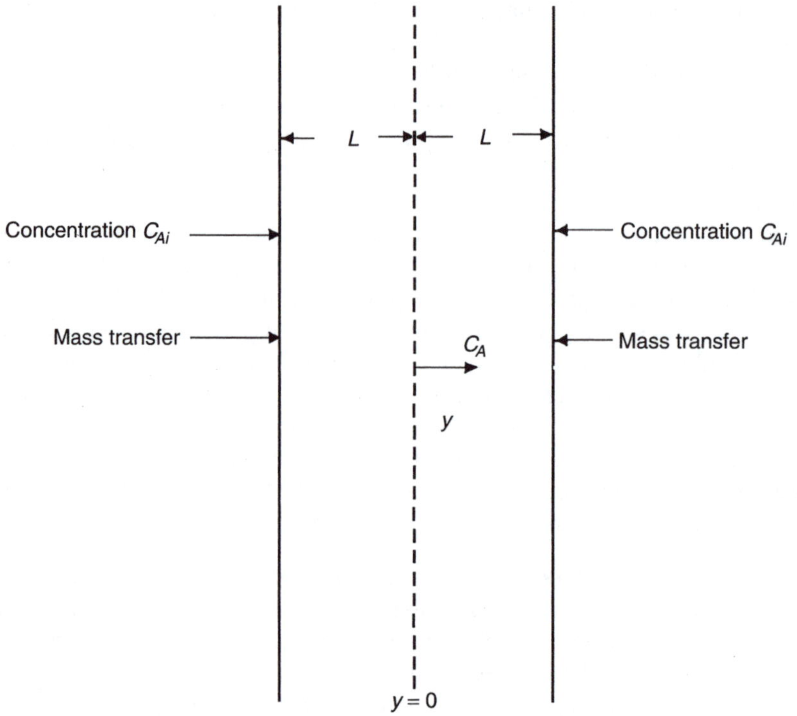

Figure 10.10. Mass transfer and reaction in a platelet

The process taking place only in the right-hand half of the pellet illustrated in Figure 10.10 is considered. If the plate has a total thickness $2L$, and the centre-plane is taken as the origin ($y = 0$), then the following boundary conditions will apply:

$$y = 0, \qquad \frac{dC_A}{dy} = 0 \qquad \text{(from symmetry)}$$

$$y = L, \qquad C_A = C_{Ai} \qquad \text{(the concentration at the surface of the particle)}$$

The general solution of equation 10.196 is:

$$C_A = B_1' e^{\sqrt{\frac{k}{D_e}}y} + B_2' e^{-\sqrt{\frac{k}{D_e}}y} \tag{10.197}$$

Putting $\dfrac{k}{D_e} = \lambda^2$:

$$C_A = B_1' e^{\lambda y} + B_2' e^{-\lambda y}$$

and:

$$\frac{dC_A}{dy} = \lambda(B_1' e^{\lambda y} - B_2' e^{-\lambda y})$$

When $y = 0$, $\dfrac{dC_A}{dy} = 0$, and: $B_1' = B_2'$

and: $C_A = B_2'(e^{\lambda y} + e^{-\lambda y})$

When $y = L$, $C_A = C_{Ai}$, and $B_2' = C_{Ai}(e^{\lambda L} + e^{-\lambda L})^{-1}$

and:

$$\frac{C_A}{C_{Ai}} = \frac{e^{\lambda y} + e^{-\lambda y}}{e^{\lambda L} + e^{-\lambda L}} \tag{10.198}$$

The concentration gradient at a distance y from the surface is therefore given by:

$$\frac{1}{C_{Ai}}\frac{dC_A}{dy} = \frac{\lambda(e^{\lambda y} - e^{-\lambda y})}{e^{\lambda L} + e^{-\lambda L}}$$

The mass transfer rate per unit area at the surface of the particle is then:

$$(N_A)_{y=L} = -D_e\left(\frac{dC_A}{dy}\right)_{y=L} = -C_{Ai}D_e\lambda\frac{e^{\lambda L} - e^{-\lambda L}}{e^{\lambda L} + e^{-\lambda L}}$$

or:

$$(N_A)_{y=L} = -C_{Ai}\sqrt{kD_e}\tanh \lambda L \tag{10.199}$$

The quantity $L\sqrt{\dfrac{k}{D_e}} = \lambda L$ is known as the *Thiele modulus*, ϕ[31]. The negative sign indicates that the transfer is in the direction of y negative, that is, towards the centre of the pellet. The rate of transfer of **A** (moles/unit area and unit time) from the external fluid to the surface of the half-pellet (and therefore, in a steady-state process, the rate at which it is reacting) is therefore:

$$(-N_A)_{y=L} = C_{Ai}\sqrt{kD_e}\tanh \lambda L \tag{10.200}$$

If there were no resistance to mass transfer, the concentration of **A** would be equal to C_{Ai} everywhere in the pellet and the reaction rate per unit area in the half-pellet (of volume per unit area equal to L) would be given by:

$$\mathfrak{R}_p = \boldsymbol{k} C_{Ai} L \tag{10.201}$$

The effectiveness factor η is equal to the ratio of the rates given by equations 10.200 and 10.201, or:

$$\eta = \frac{1}{\phi} \tanh \phi \tag{10.202}$$

A logarithmic plot of the dimensionless quantities ϕ and η is shown in Figure 10.11. It may be seen that η approaches unity at low values of ϕ, and becomes proportional to ϕ^{-1} at high values of ϕ.

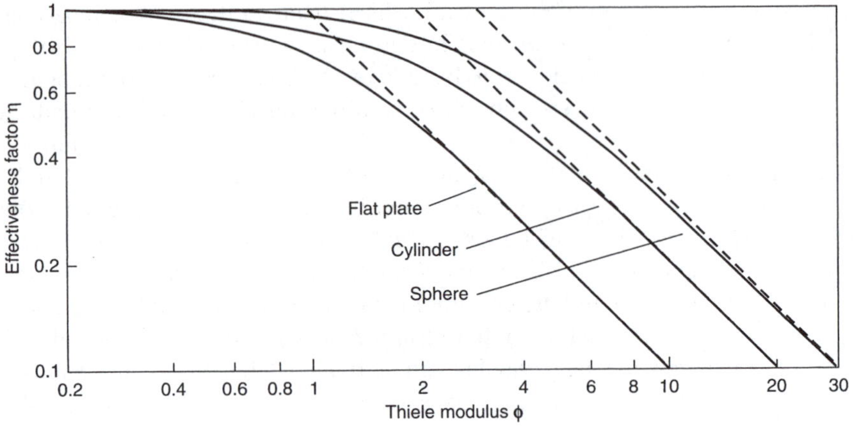

Figure 10.11. Effectiveness factor η as a function of Thiele modulus ϕ ($\phi = \lambda L$ for platelet, $\phi = \lambda r_c$ for cylinder, $\phi = \lambda r_o$ for sphere)

Three regions can be distinguished:

(i) $\phi <$ approximately 0.3 $\tanh \phi \to \phi, \quad \eta \to 1$
In this region, the mass transfer effects are small and the rate is determined almost entirely by the reaction kinetics.

(ii) $\phi >$ approximately 3 $\tanh \phi \to 1 \quad \eta \to \phi^{-1}$
Since $\phi = \sqrt{\frac{\boldsymbol{k}}{D_e}} L$, this corresponds to a region with high values of $\boldsymbol{k}$ (implying high reaction rates), coupled with low values of D_e and high values of L (implying a high mass transfer resistance). In this region mass transfer considerations are therefore of dominant importance and reaction tends to be confined to a thin region close to the particle surface.

(iii) $0.3 < \phi < 3$
This is a transitional region in which reaction kinetics and mass transfer resistance both affect the overall reaction rate.

10.7.2. Spherical pellets

The basic differential equation for mass transfer accompanied by an nth order chemical reaction in a spherical particle is obtained by taking a material balance over a spherical shell of inner radius r and outer radius $r + \delta r$, as shown in Figure 10.12.

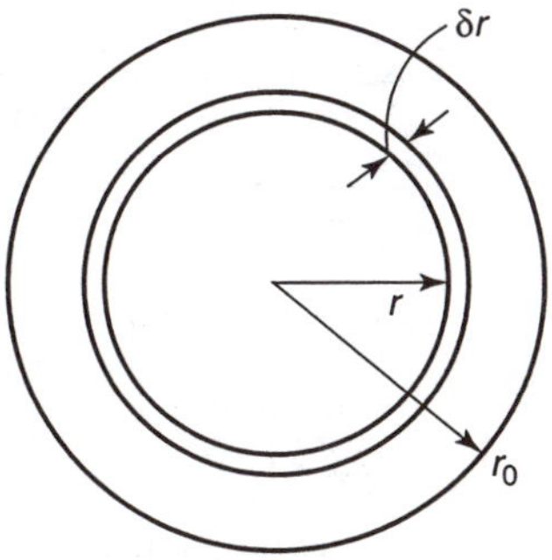

Figure 10.12. Mass transfer and chemical reaction in a spherical particle

At radius r, the mass transfer rate (moles per unit time) is given by:

$$-D_e \frac{\partial C_A}{\partial r} 4\pi r^2$$

The corresponding mass transfer rate at radius $r + \delta r$ is:

$$-D_e \frac{\partial C_A}{\partial r} 4\pi r^2 + \frac{\partial}{\partial r}\{-D_e \frac{\partial C_A}{\partial r} 4\pi r^2\}\delta r$$

The change in mass transfer rate over the distance δr is therefore:

$$4\pi D_e \frac{\partial}{\partial r}\left\{r^2 \frac{\partial C_A}{\partial r}\right\} \delta r$$

The rate of removal of **A** (moles/unit time) by chemical reaction is:

$$\boldsymbol{k} C_A^n (4\pi r^2 \delta r)$$

The rate of accumulation of **A** (moles/unit time) is given by:

$$\frac{\partial C_A}{\partial t}(4\pi r^2 \delta r)$$

A material balance gives:

$$4\pi D_e \frac{\partial}{\partial r}\left\{r^2 \frac{\partial C_A}{\partial r}\right\} \delta r = \boldsymbol{k} C_A^n (4\pi r^2 \delta r) + \frac{\partial C_A}{\partial t}(4\pi r^2 \delta r)$$

i.e.:

$$\frac{\partial C_A}{\partial t} = D_e \frac{1}{r^2} \frac{\partial}{\partial r}\left(r^2 \frac{\partial C_A}{\partial r}\right) - \boldsymbol{k} C_A^n \tag{10.203}$$

It may be noted that, in the absence of a chemical reaction, equation 10.203 reduces to equation 10.146. For a steady-state process $\partial C_A/\partial t = 0$, and for a first-order reaction $n = 1$. Thus:

$$\frac{d}{dr}\left(r^2 \frac{dC_A}{dr}\right) = \frac{\boldsymbol{k}}{D_e} r^2 C_A = \lambda^2 r^2 C_A$$

where $\lambda^2 = \boldsymbol{k}/D_e$.

Thus:
$$r^2\frac{d^2C_A}{dr^2} + 2r\frac{dC_A}{dr} = \frac{\boldsymbol{k}}{D_e}r^2C_A = \lambda^2r^2C_A \quad (10.204)$$

or:
$$\frac{d^2C_A}{dr^2} + \frac{2}{r}\frac{dC_A}{dr} - \lambda^2C_A = 0 \quad (10.205)$$

Equation 10.205, may be solved by putting $C_A = \frac{\psi}{r}$ to give:

$$r^2\frac{d^2\left(\frac{\psi}{r}\right)}{dr^2} + 2r\frac{d\left(\frac{\psi}{r}\right)}{dr} = \lambda^2r\psi \quad (10.206)$$

Since:
$$\frac{d\left(\frac{\psi}{r}\right)}{dr} = \frac{1}{r}\frac{d\psi}{dr} - \frac{1}{r^2}\psi$$

and:
$$\frac{d^2\left(\frac{\psi}{r}\right)}{dr^2} = \frac{1}{r}\frac{d^2\psi}{dr^2} - \frac{1}{r^2}\frac{d\psi}{dr} - \frac{1}{r^2}\frac{d\psi}{dr} + \frac{2}{r^3}\psi$$

Substituting in equation 10.206 gives:

$$r\frac{d^2\psi}{dr^2} - 2\frac{d\psi}{dr} + \frac{2}{r}\psi + 2\frac{d\psi}{dr} - \frac{2}{r}\psi = \lambda^2r$$

or:
$$\frac{d^2\psi}{dr^2} - \lambda^2\psi = 0$$

Solving:
$$\psi = B_1''e^{\lambda r} + B_2''e^{-\lambda r}$$

or:
$$C_A = B_1''\frac{1}{r}e^{\lambda r} + B_2''\frac{1}{r}e^{-\lambda r}$$

and:
$$\frac{dC_A}{dr} = B_1''\left(\frac{1}{r}\lambda e^{\lambda r} - \frac{1}{r^2}e^{\lambda r}\right) + B_2''\left(\frac{1}{r}(-\lambda)e^{-\lambda r} - \frac{1}{r^2}e^{-\lambda r}\right)$$

The boundary conditions are now considered. When $r = 0$, then from symmetry: $\frac{dC_A}{dr} = 0$

giving:
$$0 = B_1''\left(-\frac{1}{r^2}e^0\right) + B_2''\left(-\frac{1}{r^2}e^0\right) \quad \text{since} \quad \left(\frac{1}{r^2} \gg \frac{1}{r}\right)_{r\to 0}$$

Thus:
$$B_2'' = -B_1''$$

When $r = r_0$: $C_A = C_{Ai}$

and:
$$C_{Ai} = B_1''\frac{1}{r_0}e^{\lambda r_0} - B_1''\frac{1}{r_0}e^{-\lambda r_0}$$

and:
$$B_1'' = C_{Ai}r_0(e^{\lambda r_0} - e^{-\lambda r_0})^{-1}$$

Thus: $$\frac{C_A}{C_{Ai}} = \frac{r_0}{r}\frac{e^{\lambda r} - e^{-\lambda r}}{e^{\lambda r_0} - e^{-\lambda r_0}}$$

and: $$\frac{C_A}{C_{Ai}} = \frac{r_0}{r}\frac{\sinh \lambda r}{\sinh \lambda r_0} \tag{10.207}$$

Differentiating with respect to r to obtain the concentration gradient:

$$\frac{1}{C_{Ai}}\frac{dC_A}{dr} = \frac{r_0}{\sinh \lambda r_0}\left\{\frac{1}{r}\lambda \cosh \lambda r - \frac{1}{r^2}\sinh \lambda r\right\}$$

At the surface:

$$\left(\frac{dC_A}{dr}\right)_{r=r_0} = \frac{C_{Ai} r_0}{\sinh \lambda r_0}\left\{\frac{1}{r_0}\lambda \cosh \lambda r_0 - \frac{1}{r_0{}^2}\sinh \lambda r_0\right\}$$

$$= \frac{C_{Ai}}{r_0}\{\lambda r_0 \coth \lambda r_0 - 1\}$$

Thus the mass transfer at the outer surface of the pellet (moles/unit time) is:

$$-D_e\left(\frac{dC_A}{dr}\right)_{r=r_0}(4\pi r_0^2) = -C_{Ai}D_e 4\pi r_0\{\lambda r_0 \coth \lambda r_0 - 1\} \tag{10.208}$$

in the direction of r positive (away from the centre of the particle), and

$$C_{Ai}D_e 4\pi r_0\{\lambda r_0 \coth \lambda r_0 - 1\} \tag{10.209}$$

towards the centre of the pellet.

With no resistance to mass transfer, the concentration is C_{Ai} throughout the whole spherical pellet, and the reaction rate, which must be equal to the mass transfer rate in a steady-state process, is:

$$= \frac{4}{3}\pi r_0^3 \boldsymbol{k} C_{Ai} \tag{10.210}$$

The effectiveness factor η is obtained by dividing equation 10.209 by equation 10.210 to give:

$$\eta = \frac{C_{Ai}D_e 4\pi r_0\{\lambda r_0 \coth \lambda r_0 - 1\}}{\frac{4}{3}\pi r_0{}^3 \boldsymbol{k} C_{Ai}}$$

$$= 3\frac{D_e}{\boldsymbol{k}}\frac{1}{r_0{}^2}\{\lambda r_0 \coth \lambda r_0 - 1\}$$

$$= \frac{3}{r_0{}^2\lambda^2}\{\lambda r_0 \coth \lambda r_0 - 1\} \tag{10.211}$$

Thus: $$\eta = \frac{3}{\phi_0}\coth \phi_0 - \frac{3}{\phi_0{}^2} = \frac{3}{\phi_0}\left(\coth \phi_0 - \frac{1}{\phi_0}\right) \tag{10.212}$$

where $\phi_0 = \lambda r_0$ is the *Thiele modulus* for the spherical particle.

It is useful to redefine the characteristic linear dimension L of the spherical particle as its volume per unit surface area. This is, in effect, consistent with the definition of L adopted for the platelet where L is half its thickness. Then, for the sphere:

$$L = \frac{(4/3)\pi r_0^3}{4\pi r_0^2} = \frac{r_0}{3} \tag{10.213}$$

The Thiele modulus ϕ_L then becomes:

$$\phi_L = \lambda L = \sqrt{\frac{k}{D_e}}L \tag{10.214}$$

Substituting for r_0 in equation 10.215, the effectiveness factor η may be written as:

$$\eta = \frac{3}{9\lambda^2 L^2}\{3\lambda L \coth 3\lambda L - 1\}$$

$$= \frac{1}{\phi_L}\left(\coth 3\phi_L - \frac{1}{3\phi_L}\right) \tag{10.215}$$

A logarithmic plot of η versus ϕ_L in Figure 10.13 shows that, using this definition of L, the curves for the slab or platelet and the spherical particle come very close together.

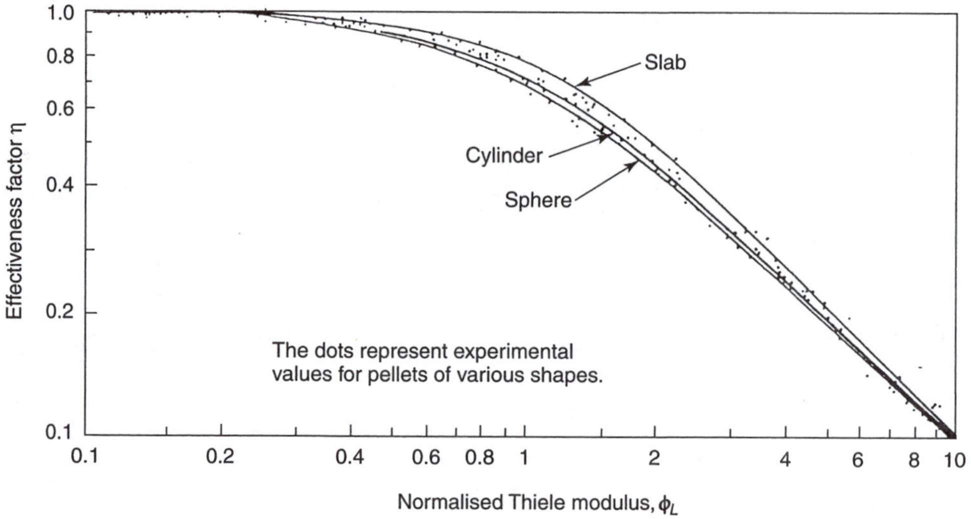

Figure 10.13. Effectiveness factors η as a function of normalised Thiele modulus $\phi_L = \lambda\frac{V_p}{A_p} = \lambda L$ for a first-order reaction

10.7.3. Other particle shapes

The relationship between effectiveness factor η and Thiele modulus ϕ_L may be calculated for several other regular shapes of particles, where again the characteristic dimension of the particle is defined as the ratio of its volume to its surface area. It is found that the results all fall quite closely together, irrespective of the particle shape. For a long cylindrical particle with a high length to diameter ratio, the effects of diffusion at the end faces may be neglected, and the mass transfer process may be regarded as taking place solely in the radial direction. The equation for mass transfer and chemical reaction in the cylinder may be derived in a manner analogous to that used for equation 10.205 for the sphere:

$$\frac{d^2C_A}{dr^2} + \frac{1}{r}\frac{dC_A}{dr} - \lambda^2 C_A = 0 \tag{10.216}$$

The solution of this equation is in the form of a Bessel function[32]. Again, the characteristic length of the cylinder may be defined as the ratio of its volume to its surface area; in this case, $L = r_c/2$. It may be seen in Figure 10.13 that, when the effectiveness factor η is plotted against the normalised Thiele modulus, the curve for the cylinder lies between the curves for the slab and the sphere. Furthermore, for these three particles, the effectiveness factor is not critically dependent on shape.

The results of investigations with particles of a variety of shapes, mainly irregular ones, have been reported by RESTER and ARIS[33], and the results are shown as data points in Figure 10.13. This provides additional evidence that the $\eta - \phi_L$ plot is not particularly sensitive to the shape of the particles, and a single curve can be used for most practical applications, particularly at high values of the Thiele modulus where reaction is confined to a thin region close to the surface whose curvature is then unimportant.

Example 10.12

Estimate the Thiele modulus and the effectiveness factor for a reactor in which the catalyst particles are:

(i) Thin rectangular platelets, the ends of which are sealed so that mass transfer is unidirectional and perpendicular to the surface of the particle. The total thickness of the particles is 8 mm.
(ii) Spherical particles, 10 mm in diameter.

The first-order rate constant is 5×10^{-4} s^{-1} and the effective diffusivity of the reactants in the pores of the particles is 2×10^{-9} m^2/s.

Solution

For the reacting system: $\lambda = \sqrt{\dfrac{k}{D_e}} = \sqrt{\dfrac{5 \times 10^{-4}}{2 \times 10^{-9}}} = 500 \text{ m}^{-1}$

(i) For the platelet of thickness 8 mm, $L = \left(\dfrac{1}{2} \times 8 \times 10^{-3}\right) = 0.004 \text{ m}^{-1}$

and the Thiele modulus $\phi = \lambda L = (500 \times 0.004) = 2.0$

From equation 10.202, the effectiveness factor η is given by:

$$\eta = \frac{1}{\phi}\tanh\phi = \frac{1}{2}\tanh 2 = \underline{\underline{0.482}}$$

(ii) For the sphere of diameter 10 mm, $r_0 = 0.005 \text{ m}^{-1}$.

and the Thiele modulus $\phi_0 = \lambda r_0 = (500 \times 0.005) = 2.5$

From equation 10.212, the effectiveness factor η is given by:

$$\eta = \frac{3}{\phi_0}\left(\coth\phi_0 - \frac{1}{\phi_0}\right) = \frac{3}{2.5}\left(\coth 2.5 - \frac{1}{2.5}\right) = \underline{\underline{0.736}}$$

Alternatively:

$$\phi_L = \lambda \frac{r_0}{3} = (500 \times 0.005)/3 = 0.833$$

and from equation 10.215:

$$\eta = \frac{1}{\phi_L}\left(\coth 3\phi_L - \frac{1}{3\phi_L}\right) = \frac{1}{0.833}\left(\coth(3 \times 0.833) - \frac{1}{3 \times 0.833}\right)$$

$$= \underline{\underline{0.736}}$$

Example 10.13

A first-order chemical reaction takes place in a reactor in which the catalyst pellets are platelets of thickness 5 mm. The effective diffusivity D_e for the reactants in the catalyst particle is 10^{-5} m^2/s and the first-order rate constant $\boldsymbol{k}$ is 14.4 s^{-1}.
Calculate:

(i) the effectiveness factor η.
(ii) the concentration of reactant at a position half-way between the centre and the outside of the pellet (i.e. at a position one quarter of the way across the particle from the outside).

Solution

For the particle, the Thiele modulus $\phi = \sqrt{\dfrac{k}{D_e}}L = \lambda L$

$$= \sqrt{\frac{14.4}{10^{-5}}} \times (2.5 \times 10^{-3}) = 3$$

From equation 10.202, the effectiveness factor, $\eta = \dfrac{1}{\phi}\tanh\phi$

$$= \frac{1}{3}\tanh 3 = 0.332$$

Thus $\eta\lambda \approx 1$, corresponding to the region where mass transfer effects dominate.
The concentration profile is given by equation 10.198 as:

$$\frac{C_A}{C_{Ai}} = \frac{\cosh \lambda y}{\cosh \lambda L}$$

The concentration C_A at $y = 1.25 \times 10^{-3}$ m

$$= 0.015\frac{\cosh 1.5}{\cosh 3} = \underline{\underline{0.035 \text{ kmol/m}^3}}$$

10.7.4. Mass transfer and chemical reaction with a mass transfer resistance external to the pellet

When the resistance to mass transfer to the external surface of the pellet is significant compared with that within the particle, part of the concentration driving force is required to overcome this external resistance, and the concentration of reacting material at the surface of the pellet C_{Ai} is less than that in the bulk of the fluid phase C_{Ao}. In Sections 10.7.1–10.7.3, the effect of mass transfer resistance within a porous particle is expressed as an effectiveness factor, by which the reaction rate within the particle is reduced as a result of this resistance. The reaction rate per unit volume of particle (moles/unit volume and unit time) for a first-order reaction is given by:

$$\mathfrak{R}_v = \eta \boldsymbol{k} C_{Ai} \tag{10.217}$$

For a particle of volume V_p, the reaction rate per unit time for the particle is given by:

$$\mathfrak{R}_p = \eta \boldsymbol{k} V_p C_{Ai} \tag{10.218}$$

When there is an external mass transfer resistance, the value of C_{Ai} (the concentration at the surface of the particle) is less than that in the bulk of the fluid (C_{Ao}) and will not be known. However, if the value of the external mass transfer coefficient is known, the mass transfer rate from the bulk of the fluid to the particle may be expressed as:

$$\mathfrak{R}_p = h_D A_p (C_{Ao} - C_{Ai}) \tag{10.219}$$

giving:

$$C_{Ai} = C_{Ao} - \frac{\mathfrak{R}_p}{h_D A_p} \tag{10.220}$$

Substituting in equation 10.219:

$$\mathfrak{R}_p = \frac{\boldsymbol{k} C_{Ao}}{\dfrac{1}{\eta V_p} + \dfrac{\boldsymbol{k}}{h_D A_p}} \tag{10.221}$$

Then, dividing by V_p gives the reaction rate $\mathfrak{R}_v$ (moles per unit volume of particle in unit time) as:

$$\mathfrak{R}_v = \frac{\boldsymbol{k} C_{Ao}}{\dfrac{1}{\eta} + \dfrac{\boldsymbol{k}}{h_D} L} \tag{10.222}$$

where $L = V_p/A_p$, the particle length term used in the normalised Thiele modulus. Information on external mass transfer coefficients for single spherical particle and particles in fixed beds is given in Section 10.8.4, and particles in fluidised beds are discussed in Volume 2.

Example 10.14

A hydrocarbon is cracked using a silica-alumina catalyst in the form of spherical pellets of mean diameter 2.0 mm. When the reactant concentration is 0.011 kmol/m^3, the reaction rate is 8.2×10^{-2} kmol/(m^3 catalyst) s. If the reaction is of first-order and the effective diffusivity D_e is 7.5×10^{-8} m^2/s, calculate the value of the effectiveness factor η. It may be assumed that the effect of mass transfer resistance in the fluid external to the particles may be neglected.

Solution

Since the value of the first-order rate constant is not given, λ and ϕ_L cannot be calculated directly. The reaction rate per unit volume of catalyst $\mathfrak{R}_v = \eta \boldsymbol{k} C_{Ai}$ (equation 10.217),

and the Thiele modulus

$$\phi_L = \sqrt{\frac{\boldsymbol{k}}{D_e}} L = \frac{\sqrt{\boldsymbol{k}}}{\sqrt{7.5 \times 10^{-8}}} \frac{1 \times 10^{-3}}{3}$$

$$= 1.217\sqrt{\boldsymbol{k}} \tag{i}$$

The effectiveness factor η is given by equation 10.215 as:

$$\eta = \frac{1}{\phi_L}\left(\coth 3\phi_L - \frac{1}{3\phi_L}\right)$$

If $\phi_L > 3$: $$\eta \approx \phi_L^{-1} \qquad \text{(ii)}$$

It is assumed that the reactor is operating in this regime and the assumption is then checked.

Substituting numerical values in equation 10.217:

$$8.2 \times 10^{-2} = \eta k(0.011) \qquad \text{(iii)}$$

From equations (i) and (ii):

$$\eta = \phi_L{}^{-1} = 0.822\frac{1}{\sqrt{k}}$$

From equation (iii):

$$8.2 \times 10^{-2} = 0.822\frac{1}{\sqrt{k}}k(0.011)$$

and: $$k = 82.2 \text{ s}^{-1}$$

From equation (i); $$\phi_L = 11.04 \text{ and } \underline{\underline{\eta = 0.0906}}$$

This result may be checked by using equation 10.215:

$$\eta = \frac{1}{11.04}\left(\coth 33.18 - \frac{1}{3 \times 11.04}\right)$$
$$= \underline{\underline{0.0878}}$$

This value is sufficiently close for practical purposes to the value of 0.0906, calculated previously. If necessary, a second iteration may be carried out.

10.8 PRACTICAL STUDIES OF MASS TRANSFER

The principal applications of mass transfer are in the fields of distillation, gas absorption and the other separation processes involving mass transfer which are discussed in Volume 2, In particular, mass transfer coefficients and heights of transfer units in distillation, and in gas absorption are discussed in Volume 2, . In this section an account is given of some of the experimental studies of mass transfer in equipment of simple geometry, in order to provide a historical perspective.

10.8.1. The *j*-factor of Chilton and Colburn for flow in tubes

Heat transfer

Because the mechanisms governing mass transfer are similar to those involved in both heat transfer by conduction and convection and in momentum transfer (fluid flow), quantitative relations exist between the three processes, and these are discussed in Chapter 12. There is generally more published information available on heat transfer than on mass transfer, and these relationships often therefore provide a useful means of estimating mass transfer coefficients.

Results of experimental studies of heat transfer may be conveniently represented by means of the j-factor method developed by COLBURN[34] and by CHILTON and COLBURN[35] for representing data on heat transfer between a turbulent fluid and the wall of a pipe. From equation 9.64:

$$Nu = 0.023Re^{0.8}Pr^{0.33} \qquad \text{(equation 9.64)}$$

where the viscosity is measured at the mean film temperature, and Nu, Re and Pr denote the Nusselt, Reynolds and Prandtl numbers, respectively.

If both sides of the equation are divided by the product $Re\ Pr$:

$$St = 0.023Re^{-0.2}Pr^{-0.67} \qquad (10.223)$$

where $St(= h/C_p\rho u)$ is the Stanton number.

Equation 10.223 may be rearranged to give:

$$j_h = St\ Pr^{0.67} = 0.023Re^{-0.2} \qquad (10.224)$$

The left-hand side of equation 10.224 is referred to as the j-factor for heat transfer (j_h). Chilton and Colburn found that a plot of j_h against Re gave approximately the same curve as the friction chart (ϕ versus Re) for turbulent flow of a fluid in a pipe.

The right-hand side of equation 10.224 gives numerical values which are very close to those obtained from the Blasius equation for the friction factor ϕ for the turbulent flow of a fluid through a smooth pipe at Reynolds numbers up to about 10^6.

$$\phi = 0.0396Re^{-0.25} \qquad \text{(equation 3.11)}$$

Re	3×10^3	10^4	3×10^4	10^5	3×10^5	10^6
ϕ (equation 3.11)	0.0054	0.0040	0.0030	0.0022	0.0017	0.0013
ϕ (equation 10.224)	0.0046	0.0037	0.0029	0.0023	0.0019	0.0015

Mass transfer

Several workers have measured the rate of transfer from a liquid flowing down the inside wall of a tube to a gas passing countercurrently upwards. GILLILAND and SHERWOOD[36] vaporised a number of liquids including water, toluene, aniline and propyl, amyl and butyl alcohols into an air stream flowing up the tube. A small tube, diameter $d = 25$ mm and length $= 450$ mm, was used fitted with calming sections at the top and bottom, and the pressure was varied from 14 to about 300 kN/m^2.

The experimental results were correlated using an equation of the form:

$$\frac{h_D}{u}\frac{C_{Bm}}{C_T}\left(\frac{\mu}{\rho D}\right)^{0.56} = 0.023Re^{-0.17} \qquad (10.225)$$

The index of the Schmidt group Sc is less than the value of 0.67 for the Prandtl group for heat transfer but the range of values of Sc used was very small.

There has for long been uncertainty concerning the appropriate value to be used for the exponent of the Schmidt number in equation 10.225. SHERWOOD, PIGFORD and WILKE[27] have analysed experimental results obtained by a number of workers for heat transfer to the walls of a tube for a wide range of gases and liquids (including water, organic liquids, oils and molten salts), and offered a logarithmic plot of the Stanton number ($St = h/C_p\rho u$) against Prandtl number (Pr) for a constant Reynolds number of 10,000. Superimposed on the graph are results for mass transfer obtained from experiments on the dissolution of solute from the walls of tubes composed of solid organics into liquids, and on the evaporation of liquid films from the walls of tubes to turbulent air streams using a wetted-wall column, again all at a Reynolds number of 10,000: these results were plotted as Stanton number for mass transfer (h_D/u) against Schmidt number (Sc). There is very close agreement between the results for heat transfer and for mass transfer, with a line slope of about −0.67 giving a satisfactory correlation of the results. The range of values of Prandtl and Schmidt numbers was from about 0.4 to 10,000. This established that the exponents for both the Prandtl and Schmidt numbers in the j-factors should be the same, namely 0.67. These conclusions are consistent with the experimental results of LINTON and SHERWOOD[37] who measured the rates of dissolution of cast tubes of benzoic acid, cinnamic acid and β-naphthol into water, giving Schmidt numbers in the range 1000 to 3000.

In defining a j-factor (j_d) for mass transfer there is therefore good experimental evidence for modifying the exponent of the Schmidt number in Gilliland and Sherwood's correlation (equation 10.225). Furthermore, there is no very strong case for maintaining the small differences in the exponent of Reynolds number. On this basis, the j-factor for mass transfer may be defined as follows:

$$j_d = \frac{h_D}{u}\frac{C_{Bm}}{C_T}Sc^{0.67} = 0.023Re^{-0.2} \qquad (10.226)$$

The term C_{Bm}/C_T (the ratio of the logarithmic mean concentration of the insoluble component to the total concentration) is introduced because $h_D(C_{Bm}/C_T)$ is less dependent than h_D on the concentrations of the components. This reflects the fact that the analogy between momentum, heat and mass transfer relates only to that part of the mass transfer which is *not* associated with the *bulk flow* mechanism; this is a fraction C_{Bm}/C_T of the total mass transfer. For equimolecular counterdiffusion, as in binary distillation when the molar latent heats of the components are equal, the term C_{Bm}/C_T is omitted as there is no bulk flow contributing to the mass transfer.

SHERWOOD and PIGFORD[7] have shown that if the data of GILLILAND and SHERWOOD[36] and others[35,38,39] are plotted with the Schmidt group raised to this power of 0.67, as shown in Figure 10.14, a reasonably good correlation is obtained. Although the points are rather more scattered than with heat transfer, it is reasonable to assume that both j_d and j_h are approximately equal to ϕ. Equations 10.224 and 10.226 apply in the absence of ripples which can be responsible for a very much increased rate of mass transfer. The constant of 0.023 in the equations will then have a higher value.

By equating j_h and j_d (equations 10.224 and 10.226), the mass transfer coefficient may be expressed in terms of the heat transfer coefficient, giving:

$$h_D = \left(\frac{h}{C_p\rho}\right)\left(\frac{C_T}{C_{Bm}}\right)\left(\frac{Pr}{Sc}\right)^{0.67} \qquad (10.227)$$

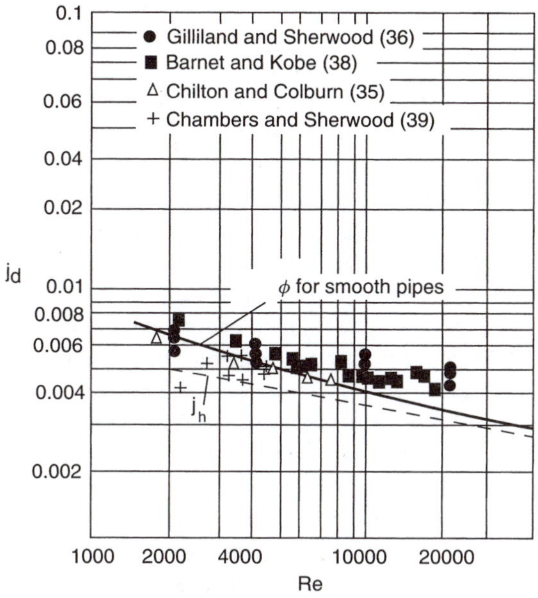

Figure 10.14. Mass transfer in wetted-wall columns

10.8.2. Mass transfer at plane surfaces

Many of the earlier studies of mass transfer involved measuring the rate of vaporisation of liquids by passing a turbulent air stream over a liquid surface. In addition, some investigations have been carried out in the absence of air flow, under what have been termed still air conditions. Most of these experiments have been carried out in some form of wind tunnel where the rate of flow of air and its temperature and humidity could be controlled and measured. In these experiments it was found to be important to keep the surface of the liquid level with the rim of the pan in order to avoid the generation of eddies at the leading edge.

HINCHLEY and HIMUS[40] measured the rate of evaporation from heated rectangular pans fitted flush with the floor of a wind tunnel (0.46m wide by 0.23m high), and showed that the rate of vaporisation was proportional to the difference between the saturation vapour pressure of the water P_s and the partial pressure of water in the air P_w. The results for the mass rate of evaporation W were represented by an empirical equation of the form:

$$W = \text{constant}\ (P_s - P_w) \tag{10.228}$$

where the constant varies with the geometry of the pan and the air velocity.

This early work showed that the driving force in the process was the pressure difference $(P_s - P_w)$. Systematic work in more elaborate equipment by POWELL and GRIFFITHS[41], WADE[42], and PASQUILL[43] then followed. Wade, who vaporised a variety of organic liquids, including acetone, benzene and tri-chloroethylene at atmospheric pressure, used a small pan 88 mm square in a wind tunnel. Powell and Griffiths stretched canvas sheeting over rectangular pans and, by keeping the canvas wet at all times, ensured that it behaved as a free water surface. In all of these experiments the rate of vaporisation showed a similar form of dependence on the partial pressure difference and the rate of flow of the air stream. Powell and Griffiths found that the vaporisation rate per unit area decreased in the downwind direction. For rectangular pans, of length L, the vaporisation rate was proportional to $L^{0.77}$. This can be explained in terms of the thickening of the boundary layer (see Chapter 11) and the increase in the partial pressure of vapour in the air stream arising from the evaporation at upstream positions.

In these experiments, it might be anticipated that, with high concentrations of vapour in the air, the rate of evaporation would no longer be linearly related to the partial pressure difference because of the contribution of bulk flow to the mass transfer process (Section 10.2.3), although there is no evidence of this even at mole fractions of vapour at the surface as high as 0.5. Possibly the experimental measurements were not sufficiently sensitive to detect this effect.

SHERWOOD and PIGFORD[7] have plotted the results of several workers[41,42,43,44,45] in terms of the Reynolds number Re_L, using the length L of the pan as the characteristic linear dimension. Figure 10.15, taken from this work, shows j_d plotted against Re_L for a number of liquids evaporating into an air stream. Although the individual points show some scatter, j_d is seen to follow the same general trend as j_h in this work. The Schmidt number was varied over such a small range that the correlation was not significantly poorer if it was omitted from the correlation.

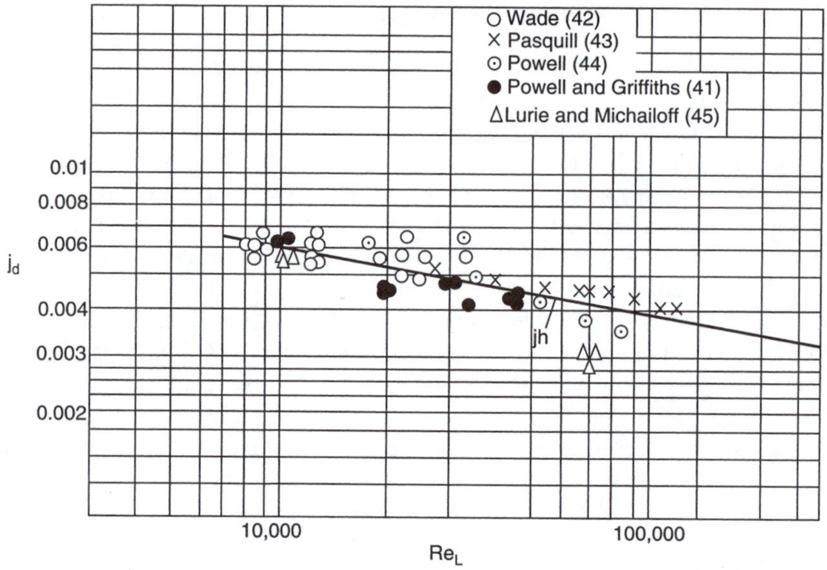

Figure 10.15. Evaporation from plane surfaces

MAISEL and SHERWOOD[46] also carried out experiments in a wind tunnel in which water was evaporated from a wet porous surface preceded by a dry surface of length L_0. Thus, a velocity boundary layer had become established in the air before it came into contact with the evaporating surface. The results were correlated by:

$$j_d = 0.0415 Re_L^{-0.2} \left[1 - \left(\frac{L_o}{L}\right)^{0.8}\right]^{-0.11} \tag{10.229}$$

where L is the total length of the surface (dry + wet).

10.8.3. Effect of surface roughness and form drag

The results discussed in Section 10.8.2 give reasonably good support to the treatment of heat, mass, and momentum transfer by the j-factor method, although it is important to remember that, in all the cases considered, the drag is almost entirely in the form of skin friction (that is, viscous drag at the surface). As soon as an attempt is made to apply the relation to cases where form drag (that is, additional drag caused by the eddies set up as a result of the fluid impinging on an obstruction) is important, such as beds of granular solids or evaporation from cylinders or spheres, the j-factor and the friction factor are found no longer to be equal. This problem receives further consideration in Volume 2. SHERWOOD[47,48] carried out experiments where the form drag was large compared with the skin friction, as calculated approximately by subtracting the form drag from the total drag force. In this way, reasonable agreement between the corresponding value of the friction factor ϕ and j_h and j_d was obtained.

GAMSON *et al.*[49] have successfully used the j-factor method to correlate their experimental results for heat and mass transfer between a bed of granular solids and a gas stream.

PRATT[50] has examined the effect of using artificially roughened surfaces and of introducing "turbulence promoters", which increase the amount of form drag. It was found that the values of ϕ and the heat and mass transfer coefficients were a minimum for smooth surfaces and all three quantities increased as the surface roughness was increased. ϕ increased far more rapidly than either of the other two quantities however, and the heat and mass transfer coefficients were found to reach a limiting value whereas ϕ could be increased almost indefinitely. Pratt has suggested that these limiting values are reached when the velocity gradient at the surface corresponds with that in the turbulent part of the fluid; that is, at a condition where the buffer layer ceases to exist (Chapter 11).

10.8.4 Mass transfer from a fluid to the surface of particles

It is necessary to calculate mass transfer coefficients between a fluid and the surface of a particle in a number of important cases, including:

(i) gas absorption in a spray tower,
(ii) evaporation of moisture from the surface of droplets in a spray tower,
(iii) reactions between a fluid and dispersed liquid droplets or solid particles as, for instance, in a combustion process where the oxygen in the air must gain access to the external surfaces, and
(iv) catalytic reactions involving porous particles where the reactant must be transferred to the outer surface of the particle before it can diffuse into the pores and make contact with the active sites on the catalyst.

There have been comparatively few experimental studies in this area and the results of different workers do not always show a high degree of consistency. Frequently, estimates of mass transfer coefficients have been made by applying the analogy between heat transfer and mass transfer, and thereby utilising the larger body of information which is available on heat transfer.

Interest extends from transfer to single particles to systems in which the particles are in the form of fixed or fluidised beds. The only case for which there is a rigorous analytical solution is that for heat by conduction and mass transfer by diffusion to a sphere.

Mass transfer to single particles

Mass transfer from a single spherical drop to still air is controlled by molecular diffusion and, at low concentrations when bulk flow is negligible, the problem is analogous to that of heat transfer by conduction from a sphere, which is considered in Chapter 9, Section 9.3.4. Thus, for steady-state radial diffusion into a large expanse of stationary fluid in which the partial pressure falls off to zero over an infinite distance, the equation for mass transfer will take the same form as that for heat transfer (equation 9.26):

$$Sh' = \frac{h_D d'}{D} = 2 \tag{10.230}$$

where Sh' is the Sherwood number which, for mass transfer, is the counterpart of the Nusselt number $Nu'(= hd'/k)$ for heat transfer to a sphere. This value of 2 for the Sherwood number is the theoretical minimum in any continuous medium and is increased if the concentration difference occurs over a finite, as opposed to an infinite, distance and if there is turbulence in the fluid.

For conditions of forced convection, FRÖSSLING[51] studied the evaporation of drops of nitrobenzene, aniline and water, and of spheres of naphthalene, into an air stream. The drops were mainly small and of the order of 1 mm diameter. POWELL[44] measured the evaporation of water from the surfaces of wet spheres up to 150 mm diameter and from spheres of ice.

The experimental results of Frössling may be represented by the equation:

$$Sh' = \frac{h_D d'}{D} = 2.0(1 + 0.276 Re'^{0.5} Sc^{0.33}) \tag{10.231}$$

SHERWOOD and PIGFORD[7] found that the effect of the Schmidt group was also influenced by the Reynolds group and that the available data were fairly well correlated as shown in Figure 10.16, in which $(h_D d')/D$ is plotted against $Re' Sc^{0.67}$.

GARNER and KEEY[52,53] dissolved pelleted spheres of organic acids in water in a low-speed water tunnel at particle Reynolds numbers between 2.3 and 255 and compared their results with other data available at Reynolds numbers up to 900. Natural convection was found to exert some influence at Reynolds numbers up to 750. At Reynolds numbers greater than 250, the results are correlated by equation 10.230:

$$Sh' = 0.94 Re'^{0.5} Sc^{0.33} \tag{10.232}$$

Mass transfer under conditions of natural convection was also investigated.

RANZ and MARSHALL[54] have carried out a comprehensive study of the evaporation of liquid drops and confirm that equation 10.231 correlates the results of a number of

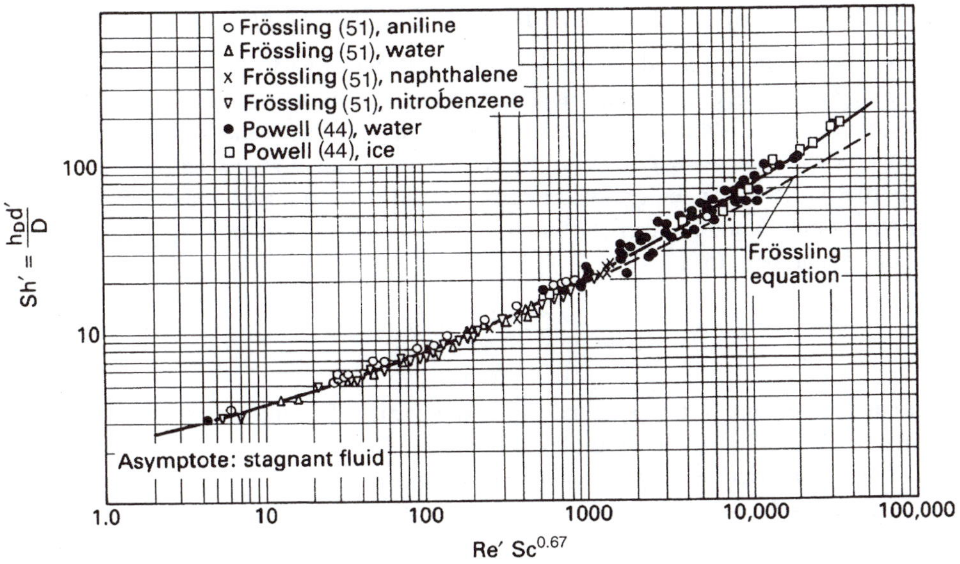

Figure 10.16. Mass transfer to single spheres

workers. A value of 0.3 in place of 0.276 for the coefficient is suggested, although the spread of the experimental results is such that the difference is not statistically significant.

ROWE *et al.*[55] have reviewed the literature on heat and mass transfer between spherical particles and a fluid. For heat transfer, their results which are discussed in Chapter 9, Section 9.4.6., are generally well represented by equation 9.100:

$$Nu' = 2 + \beta'' Re'^{0.5} Pr^{0.33} \quad (0.4 < \beta'' < 0.8) \tag{10.233}$$

For mass transfer:

$$Sh' = \alpha' + \beta' Re'^{0.5} Sc^{0.33} \quad (0.3 < \beta' < 1.0) \tag{10.234}$$

The constant α appears to be a function of the Grashof number, but approaches a value of about 2 as the Grashof number approaches zero.

In an experimental investigation[55] they confirmed that equations 10.233 and 10.234 can be used to represent the results obtained for transfer from both air and water to spheres. The constants β', β'' varied from 0.68 to 0.79.

There is therefore broad agreement between the results of FRÖSLING[51], RANZ and MARSHALL[54] and ROWE *et al.*[55]. The variations in the values of the coefficient are an indication of the degree of reproducibility of the experimental results However, BRIAN and HAYES[56] who carried a numerical solution of the equations for heat and mass transfer suggest that, at high values of $Re'\ Sc^{0.33}$, these equations tend to underestimate the value of the transfer coefficient, and an equation which can be expressed in the following form is proposed:

$$Sh' = [4.0 + 1.21(Re'Sc)^{0.67}]^{0.5} \tag{10.235}$$

Mass transfer to particles in a fixed or fluidised bed

Experimental results for fixed packed beds are very sensitive to the structure of the bed which may be strongly influenced by its method of formation. GUPTA and THODOS[57] have studied both heat transfer and mass transfer in fixed beds and have shown that the results for both processes may be correlated by similar equations based on j-factors (see Section 10.8.1). Re-arrangement of the terms in the mass transfer equation, permits the results for the Sherwood number (Sh') to be expressed as a function of the Reynolds (Re_c) and Schmidt numbers (Sc):

$$Sh' = 2.06\frac{1}{e}Re_c'^{0.425}Sc^{0.33} \tag{10.236}$$

where e is the voidage of the bed, and Re_c' is the particle Reynolds number incorporating the superficial velocity of the fluid (u_c).

KRAMERS[58] carried out experiments on heat transfer to particles in a fixed bed and has expressed his results in the form of a relation between the Nusselt, Prandtl and Reynolds numbers. This equation may be rewritten to apply to mass transfer, by using the analogy between the two processes, giving:

$$Sh' = 2.0 + 1.3Sc^{0.15} + 0.66Sc^{0.31}Re_c'^{0.5} \tag{10.198}$$

In selecting the most appropriate equation for any particular operation, it is recommended that the original references be checked to ascertain in which study the experimental conditions were closest.

Both heat transfer and mass transfer between a fluid and particles in a fluidised bed are discussed in Volume 2. The results are sensitive to the quality of fluidisation, and particularly to the uniformity of distribution of the particles in the fluid. In most cases, it is found that the same correlations for both heat transfer and mass transfer are applicable to fixed and fluidised beds.

10.9. FURTHER READING

BENNETT, C.O. and MYERS, J.E.: *Momentum, Heat, and Mass Transfer*, 3rd edn (McGraw-Hill, New York, 1983).

BIRD, R.B., STEWART, W.E. and LIGHTFOOT, E.N.: *Transport Phenomena* (Wiley, New York, 1960).

CUSSLER, E.L.: *Diffusion. Mass transfer in fluid systems*. 2nd edn (Cambridge University Press, Cambridge, 1997).

DANCKWERTS, P.V.: *Gas-liquid Reactions* (McGraw-Hill, New York, 1970).

ECKERT, E.R.G. and DRAKE, R.M.: *Analysis of Heat and Mass Transfer* (McGraw-Hill, New York, 1972).

EDWARDS, D. A., BRENNER, H. and WASAN, D. T. *Interface Transport Processes and Rheology* (Butterworth-Heinemann, Oxford, 1991).

HEWITT, G.F., SHIRES, G.L. and POLEZHAEV, Y.V.: *Encyclopedia of Heat and Mass Transfer* (CRC Press, Boca Raton, New York, 1997)

SHERWOOD, T.K.: *Chemical Engineering Education* (Fall, 1974), 204. A review of the development of mass transfer theory.

SHERWOOD, T.K., PIGFORD, R.L. and WILKE, C.R.: *Mass Transfer* (McGraw-Hill, New York, 1975).

TAYLOR, R., and KRISHNA, R.: *Multicomponent Mass Transfer* (Wiley, New York, 1993).

THOMAS, J.M. and THOMAS, W. J.: *Practice and Principles of Homogeneous Catalysis* (VCH, 1977).

TREYBAL, R.E.: *Mass-Transfer Operations*, 3rd edn (McGraw-Hill, New York, 1980).

10.10. REFERENCES

1. FICK, A.: *Ann. Phys.* **94** (1855) 59. Ueber Diffusion.
2. PRESENT, R.D.: *Kinetic Theory of Gases* (McGraw-Hill, New York, 1958).
3. STEFAN, J.: *Wiener Akad. Wissensch.* **68** (1873) 385; **79** (1879) 169; **98** (1889) 1418; *Ann. Physik* **41** (1890) 723. Versuche über die Verdampfung.
4. PERRY, R.H. and GREEN, D.W. (eds): *Perry's Chemical Engineers' Handbook.* 6th edn (McGraw-Hill, New York, 1984).
5. WINKELMANN, A.: *Ann. Physik.* **22** (1884) 1, 152. Ueber die Diffusion von Gasen und Dämpfen.
6. GILLILAND, E.R.: *Ind. Eng. Chem.* **26** (1934) 681. Diffusion coefficients in gaseous systems.
7. SHERWOOD, T.K. and PIGFORD, R.L.: *Absorption and Extraction* (McGraw-Hill, New York, 1952).
8. FULLER, E.N., SCHETTLER, P.D. and GIDDINGS, J.C.: *Ind. Eng. Chem.* **58** (1966) 19. A new method for prediction of binary gas-phase diffusion coefficients.
9. GREW, K.E. and IBBS, T.L.: *Thermal Diffusion in Gases* (Cambridge University Press, Cambridge, 1952).
10. HIRSCHFELDER, J.O., CURTISS, C.F. and BIRD, R.B.: *Molecular Theory of Gases and Liquids*, pp. 584–585 (Wiley, New York, 1954).
11. ARNOLD, J.H.: *Trans. Am. Inst. Chem. Eng.* **40** (1944) 361. Studies in diffusion: III, Steady-state vaporization and absorption.
12. MAXWELL, J.C.: *Phil. Trans. Roy. Soc.* **157** (1867) 49. The dynamical theory of gases.
13. TAYLOR, R. and KRISHNA, R.: *Multicomponent Mass Transfer* (Wiley, New York, 1993).
14. CUSSLER, E.L.: *Multicomponent Diffusion* (Elsevier, Amsterdam, 1976).
15. ZIELINSKI, J.M. and HANLEY, B.F.: *A.I.Ch.E.Jl.* **45** (1999) 1. Practical friction-based approach to modeling multicomponent diffusion.
16. WILKE C.R. and CHANG, P.: *A.I.Ch.E.Jl* **1** (1955) 264. Correlation of diffusion coefficients in dilute solutions.
17. REID, R.C., PRAUSNITZ, J.M. and SHERWOOD, T.K.: *The Properties of Gases and Liquids* (McGraw-Hill, New York, 1977).
18. GOODRIDGE, F. and BRICKNELL, D.J.: *Trans. Inst. Chem. Eng.* **40** (1962) 54. Interfacial resistance in the carbon dioxide-water system.
19. WHITMAN, W.G.: *Chem. and Met. Eng.* **29** (1923) 147. The two-film theory of absorption.
20. HIGBIE, R.: *Trans. Am. Inst. Chem. Eng.* **31** (1935) 365. The rate of absorption of pure gas into a still liquid during short periods of exposure.
21. DANCKWERTS, P.V.: *Ind. Eng. Chem.* **43** (1951) 1460. Significance of liquid film coefficients in gas absorption.
22. TOOR, H.L. and MARCHELLO, J.M.: *A.I.Ch.E.Jl* **4** (1958) 97. Film-penetration model for mass and heat transfer.
23. KISHINEVSKIJ, M.K.: *Zhur. Priklad. Khim.* **24** (1951) 542; *J. Appl. Chem. U.S.S.R.* **24** 593. The kinetics of absorption under intense mixing.
24. DANCKWERTS, P.V.: *Gas–Liquid Reactions* (McGraw-Hill, New York, 1970).
25. HARRIOTT, P.: *Chem. Eng. Sci.* **17** (1962) 149. A random eddy modification of the penetration theory.
26. DWIGHT, H.B.: *Tables of Integrals and Other Mathematical Data* (Macmillan, New York, 1957).
27. SHERWOOD, T.K., PIGFORD, R.L. and WILKE, C.R.: *Mass Transfer* (McGraw-Hill, New York, 1975).
28. DANCKWERTS, P. V.: *Trans. Faraday Soc.* **46** (1950) 300. Absorption by simultaneous diffusion and chemical reaction.
29. SCOTT, D.S. and DULLIEN, F.A.L.: *A.I.Ch.E.Jl.* **8** (1962) 113. Diffusion of ideal gases in capillaries and porous solids.
30. ARIS, R.: *The Mathematical Theory of Diffusion and Reaction in Permeable Catalysts.* (Oxford University Press, London, 1975).
31. THIELE, E.W.: *Ind. Eng. Chem.* **24** (1939) 916. Relation between catalyst activity and size of particle.
32. MICKLEY, H.S., SHERWOOD, T.K. and REED, C.E.: *Applied Mathematics in Chemical Engineering*, 2nd edn, (McGraw-Hill, New York, 1957).
33. RESTER, S. and ARIS, R.: *Chem. Eng. Sci.* **24** (1969) 793. Communications on the theory of diffusion and reaction–II The effect of shape on the effectiveness factor.
34. COLBURN, A.P.: *Trans. Am. Inst. Chem. Eng.* **29** (1933) 174. A method of correlating forced convection heat transfer data and a comparison with fluid friction.
35. CHILTON, T.H. and COLBURN, A.P.: *Ind. Eng. Chem.* **26** (1934) 1183. Mass transfer (absorption) coefficients —production from data on heat transfer and fluid friction.
36. GILLILAND, E.R. and SHERWOOD, T.K.: *Ind. Eng. Chem.* **26** (1934) 516. Diffusion of vapours into air streams.
37. LINTON, W.H. and SHERWOOD, T.K.: *Chem. Eng. Prog.* **46** (1950) 258. Mass transfer from solid shapes to water in streamline and turbulent flow.

38. BARNET, W.I. and KOBE, K.A.: *Ind. Eng. Chem.* **33** (1941) 436. Heat and vapour transfer in a wetted-wall column.
39. CHAMBERS, F.S. and SHERWOOD, T.K.: *Ind. Eng. Chem.* **29** (1937) 579. *Trans. Am. Inst. Chem. Eng.* **33** (1937) 579. Absorption of nitrogen dioxide by aqueous solutions.
40. HINCHLEY, J.W. and HIMUS, G.W.: *Trans. Inst. Chem. Eng.* **2** (1924) 57. Evaporation in currents of air.
41. POWELL, R.W. and GRIFFITHS, E.: *Trans. Inst. Chem. Eng.* **13** (1935) 175. The evaporation of water from plane and cylindrical surfaces.
42. WADE, S.H.: *Trans. Inst. Chem. Eng.* **20** (1942) 1. Evaporation of liquids in currents of air.
43. PASQUILL, F.: *Proc. Roy. Soc.* **A 182** (1943) 75. Evaporation from a plane free liquid surface into a turbulent air stream.
44. POWELL, R.W.: *Trans. Inst. Chem. Eng.* **18** (1940) 36. Further experiments on the evaporation of water from saturated surfaces.
45. LURIE, M. and MICHAILOFF, M.: *Ind. Eng. Chem.* **28** (1936) 345. Evaporation from free water surfaces.
46. MAISEL, D.S. and SHERWOOD, T.K.: *Chem. Eng. Prog.* **46** (1950) 131. Evaporation of liquids into turbulent gas streams.
47. SHERWOOD, T.K.: *Trans. Am. Inst. Chem. Eng.* **36** (1940) 817. Mass transfer and friction in turbulent flow.
48. SHERWOOD, T.K.: *Ind. Eng. Chem.* **42** (1950) 2077. Heat transfer, mass transfer, and fluid friction.
49. GAMSON, B.W., THODOS, G. and HOUGEN, O.A.: *Trans Am. Inst. Chem. Eng.* **39** (1943) 1. Heat mass and momentum transfer in the flow of gases through granular solids.
50. PRATT, H.R.C.: *Trans. Inst. Chem. Eng.* **28** (1950) 77. The application of turbulent flow theory to transfer processes in tubes containing turbulence promoters and packings.
51. FRÖSSLING, N.: *Gerlands Beitr. Geophys.* **52** (1938) 170. Über die Verdunstung fallender Tropfen.
52. GARNER, F.H. and KEEY, R.B.: *Chem. Eng. Sci.* **9** (1958) 119. Mass transfer from single solid spheres — I. Transfer at low Reynolds numbers.
53. GARNER, F.H. and KEEY, R.B.: *Chem. Eng. Sci.* **9** (1959) 218. Mass transfer from single solid spheres — II. Transfer in free convection.
54. RANZ, W.E and MARSHALL, W. R.: *Chem. Eng. Prog.* **48** (1952) 173. Evaporation from drops Part II.
55. ROWE, P.N., CLAXTON, K.T., and LEWIS, J.B.: *Trans. Inst. Chem. Eng.* **41** (1965) T14. Heat and mass transfer from a single sphere in an extensive flowing fluid.
56. BRIAN, P.L.T. and HAYES, H.B.: *A.I.Ch.E.Jl.* **15** (1969) 419. Effects of transpiration and changing diameter on heat and mass transfer.
57. GUPTA, A.S. and THODOS, G.: *A.I.Ch.E.Jl.* **9** (1963) 751. Direct analogy between heat and mass transfer in beds of spheres.
58. KRAMERS, H.: *Physica* **12** (1946) 61. Heat transfer from spheres to flowing media.

10.11. NOMENCLATURE

		Units in SI system	Dimensions in in **M, N, L, T, θ**
A_p	External surface area of particle	m^2	$\mathbf{L^2}$
a	Interfacial area per unit volume	m^2/m^3	$\mathbf{L^{-1}}$
B_1, B_2	Integration constants	kmol s/m^3	$\mathbf{NL^{-3}T}$
B'_1, B'_2	Integration constants	kmol/m^3	$\mathbf{NL^{-3}}$
C	Molar concentration	kmol/m^3	$\mathbf{NL^{-3}}$
C_A, C_B	Molar concentration of **A, B**	kmol/m^3	$\mathbf{NL^{-3}}$
C_p	Specific heat at constant pressure	J/kg K	$\mathbf{L^2T^{-2}}\boldsymbol{\theta}^{-1}$
C_T	Total molar concentration	kmol/m^3	$\mathbf{NL^{-3}}$
C_{Bm}	Logarithmic mean value of C_B	kmol/m^3	$\mathbf{NL^{-3}}$
C'	$C_A - C_{Ao}$	kmol/m^3	$\mathbf{NL^{-3}}$
C''_A, C''_B	Molar concentration of **A, B** in film	kmol/m^3	$\mathbf{NL^{-3}}$
$\overline{C'}$	Laplace transform of C'	kmol s/m^3	$\mathbf{NL^{-3}T}$
c	Mass concentration	kg/m^3	$\mathbf{ML^{-3}}$
D	Diffusivity	m^2/s	$\mathbf{L^2T^{-1}}$
D_e	Effective diffusivity within catalyst particle	m^2/s	$\mathbf{L^2T^{-1}}$
D_f	Diffusivity of fluid in film	m^2/s	$\mathbf{L^2T^{-1}}$
D_{Kn}	Knudsen diffusivity	m^2/s	$\mathbf{L^2T^{-1}}$

		Units in SI system	Dimensions in in **M**, **N**, **L**, **T**, θ
D_L	Liquid phase diffusivity	m^2/s	$\mathbf{L^2T^{-1}}$
D_{AB}	Diffusivity of **A** in **B**	m^2/s	$\mathbf{L^2T^{-1}}$
D_{Th}	Coefficient of thermal diffusion	kmol/ms	$\mathbf{NL^{-1}T^{-1}}$
D_{BA}	Diffusivity of **B** in **A**	m^2/s	$\mathbf{L^2T^{-1}}$
D'	Effective diffusivity in multicomponent system	m^2/s	$\mathbf{L^2T^{-1}}$
d	Pipe diameter	m	**L**
d'	Diameter of sphere	m	**L**
e	Bed voidage	—	—
E_D	Eddy diffusivity	m^2/s	$\mathbf{L^2T^{-1}}$
F	Coefficient in Maxwell's law of diffusion	m^3/kmols	$\mathbf{N^{-1}L^3T^{-1}}$
f	Ratio $-N'_B/N'_A$	—	—
G	Molar flow of stream	kmol/s	$\mathbf{NT^{-1}}$
G'	Molar flow per unit area	kmol/m^{2}s	$\mathbf{NL^{-2}T^{-1}}$
	Ratio of equilibrium values of concentrations in two phases	—	—
H	Height of transfer unit	m	**L**
h	Heat transfer coefficient	W/m^2 K	$\mathbf{MT^{-3}\theta^{-1}}$
h_D	Mass transfer coefficient	m/s	$\mathbf{LT^{-1}}$
h'_D	Mass transfer coefficient enhanced by chemical reaction	m/s	$\mathbf{LT^{-1}}$
$\mathscr{H}$	Henry's law constant, C''_{Ai}/C_{Ai} (equation 10.125)	—	—
j_d	j-factor for mass transfer	—	—
j_h	j-factor for heat transfer	—	—
K	Overall mass transfer coefficient	m/s	$\mathbf{LT^{-1}}$
J	Flux as mass per unit area and unit time	kg/m^{2}s	$\mathbf{ML^{-2}T^{-1}}$
K_{ABT}	Ratio of transport rate by thermal diffusion to that by Fick's law	—	—
k_G	Mass transfer coefficient for transfer through stationary fluid	kmol s/kg m	$\mathbf{NM^{-1}L^{-1}T}$
k'_G	Mass transfer coefficient for equimolecular counterdiffusion	kmol s/kg m	$\mathbf{NM^{-1}L^{-1}T}$
k_x	Mass transfer coefficient (mole fraction driving force)	kmol/m^{2}s	$\mathbf{NL^{-2}T^{-1}}$
$\boldsymbol{k}$	Reaction rate constant first-order reaction	s^{-1}	$\mathbf{T^{-1}}$
	nth-order reaction	kmol^{1-n} m^{3n-3} s^{-1}	$\mathbf{N^{1-n}\ L^{3n-3}\ T^{-1}}$
L	Length of surface, *or* film thickness, *or* half-thickness of platelet *or* V_p/A_p	m	L
L_f	Thickness of film	m	**L**
L_0	Length of surface, unheated	m	**L**
M	Molecular weight (Relative moleular mass)	kg/kmol	$\mathbf{MN^{-1}}$
N	Molar rate of diffusion per unit area (average value)	kmol/m^{2}s	$\mathbf{NL^{-2}T^{-1}}$
$(N)_t$	Molar rate of diffusion per unit area at time t	kmol/m^{2}s	$\mathbf{NL^{-2}T^{-1}}$
N'	Total molar rate of transfer per unit area	kmol/m^{2}s	$\mathbf{NL^{-2}T^{-1}}$
$(N)_{Th}$	Molar flux due to thermal diffusion	kmol/m^{2}s	$\mathbf{NL^{-2}T^{-1}}$
N	Number of transfer units	—	—
n	Number of moles of gas	kmol	**N**
n	Order of reaction, or number of term in series	—	—
P	Total pressure	N/m^2	$\mathbf{ML^{-1}T^{-2}}$
P_A, P_B	Partial pressure of **A, B**	N/m^2	$\mathbf{ML^{-1}T^{-2}}$
P_s	Vapour pressure of water	N/m^2	$\mathbf{ML^{-1}T^{-2}}$
P_w	Partial pressure of water in gas stream	N/m^2	$\mathbf{ML^{-1}T^{-2}}$
P_{Bm}	Logarithmic mean value of P_B	N/m^2	$\mathbf{ML^{-1}T^{-2}}$
p	Parameter in Laplace Transform	s^{-1}	$\mathbf{T^{-1}}$
q	Concentration gradient dC_A/dy	kmol/m^4	$\mathbf{NL^{-4}}$
R	Shear stress acting on surface	N/m^2	$\mathbf{ML^{-1}T^{-2}}$

		Units in SI system	Dimensions in in **M**, **N**, **L**, **T**, $\boldsymbol{\theta}$
$\mathbf{R}$	Universal gas constant	8314 J/kmol K	$\mathbf{MN^{-1}L^2T^{-2}\theta^{-1}}$
r	Radius within sphere or cylinder	m	$\mathbf{L}$
r_c	External radius of cylinder	m	$\mathbf{L}$
r_0	External radius of sphere	m	$\mathbf{L}$
$\Re_p$	Reaction rate per particle for first-order reaction	kmol/s	$\mathbf{N\ T^{-1}}$
$\Re_v$	Reaction rate per unit volume of particle for first-order reaction	kmol/m^{3}s	$\mathbf{N\ L^{-3}T^{-1}}$
$\Re'_{vn}$	Reaction rate per unit volume of particle for nth-order reaction (no mass transfer resistance)	kmol/m^{3}s	$\mathbf{N\ L^{-3}T^{-1}}$
S	Cross-sectional area of flow	m^2	$\mathbf{L^2}$
s	Rate of production of fresh surface per unit area	s^{-1}	$\mathbf{T^{-1}}$
T	Absolute temperature	K	$\boldsymbol{\theta}$
t	Time	s	$\mathbf{T}$
t_e	Time of exposure of surface element	s	$\mathbf{T}$
u	Mean velocity	m/s	$\mathbf{LT^{-1}}$
u_c	Superficial velocity (volumetric flowrate/ total area)	m/s	$\mathbf{LT^{-1}}$
u_A, u_B	Mean molecular velocity in direction of transfer	m/s	$\mathbf{LT^{-1}}$
u_{DA}, u_{DB}	Diffusional velocity of transfer	m/s	$\mathbf{LT^{-1}}$
u_F	Velocity due to bulk flow	m/s	$\mathbf{LT^{-1}}$
u_m	Molecular velocity	m/s	$\mathbf{LT^{-1}}$
u_s	Stream velocity	m/s	$\mathbf{LT^{-1}}$
V	Volume	m^3	$\mathbf{L^3}$
V_p	Volume of catalyst particle	m^3	L3
$\mathbf{V}$	Molecular volume	m^3/kmol	$\mathbf{N^{-1}L^3}$
$\mathbf{V}_o$	Correction term in equation 10.96	m^3/kmol	$\mathbf{N^{-1}L^3}$
W	Mass rate of evaporation	kg/s	$\mathbf{MT^{-1}}$
x	Distance from leading edge of surface or in X-direction	m	$\mathbf{L}$
	or mole fraction	—	—
y	Distance from surface or in direction of diffusion	m	$\mathbf{L}$
y'	Mol fraction in stationary gas	—	—
Z	Height of column	m	$\mathbf{L}$
z	Distance in Z-direction	m	$\mathbf{L}$
α	Thermal diffusion factor	—	—
α'	Term in equation 10.234	—	—
β', β''	coefficient in equations 10.234 and 10.233	—	—
η	Effectiveness factor	—	—
λ	$\sqrt{(k/D_e)}$	s^{-1}	$\mathbf{L^{-1}}$
λ_m	Mean free path of molecules	m	$\mathbf{L}$
μ	Viscosity of fluid	N s/m^2	$\mathbf{ML^{-1}T^{-1}}$
ρ	Density of fluid	kg/m^3	$\mathbf{ML^{-3}}$
ϕ	Friction factor ($R/\rho u^2$)	—	—
$\boldsymbol{\phi}$	Thiele modulus based on L, r_c or r_0	—	—
$\boldsymbol{\phi}_L$	Thiele modulus based on length term $L = V_p/A_p$	—	—
ψ	rC_A	kmol/m^2	$\mathbf{NL^{-2}}$
ω	Mass fraction	—	—
Nu	Nusselt number hd/k	—	—
Nu'	Nusselt number for sphere hd'/k	—	—
Pr	Prandtl number $C_p\mu/k$	—	—
Re	Reynolds number $ud\rho/\mu$	—	—
Re'	Reynolds number for sphere $ud'\rho/\mu$	—	—
Re'_c	Reynolds number for particle in fixed or fluidised bed $u_c d'\rho/\mu$	—	—
Re_L	Reynolds number for flat plate $u_s L\rho/\mu$	—	—
Sc	Schmidt number $\mu/\rho D$	—	—
Sh'	Sherwood number for sphere $h_D d'/D$	—	—

		Units in SI system	Dimensions in in **M**, **N**, **L**, **T**, θ
St	Stanton number $h/C_p\rho u$	—	—

* Dimensions depend on order of reaction.

Suffixes

0	Value in bulk of phase
1	Phase 1
2	Phase 2
A	Component **A**
B	Component **B**
AB	Of **A** in **B**
b	Bottom of column
e	Value in equilibrium with bulk of other phase
G	Gas phase
i	Interface value
L	Liquid phase
o	Overall value (for height and number of transfer units) *or* value in bulk of phase
t	Top of column

PART 4

Momentum, Heat and Mass Transfer

CHAPTER 11

The Boundary Layer

11.1. INTRODUCTION

When a fluid flows over a surface, that part of the stream which is close to the surface suffers a significant retardation, and a velocity profile develops in the fluid. The velocity gradients are steepest close to the surface and become progressively smaller with distance from the surface. Although theoretically there is no outer limit at which the velocity gradient becomes zero, it is convenient to divide the flow into two parts for practical purposes.

(1) A *boundary layer* close to the surface in which the velocity increases from zero at the surface itself to a near constant stream velocity at its outer boundary.
(2) A region outside the boundary layer in which the velocity gradient in a direction perpendicular to the surface is negligibly small and in which the velocity is everywhere equal to the stream velocity.

The thickness of the boundary layer may be arbitrarily defined as the distance from the surface at which the velocity reaches some proportion (such as 0.9, 0.99, 0.999) of the undisturbed stream velocity. Alternatively, it may be possible to approximate to the velocity profile by means of an equation which is then used to give the distance from the surface at which the velocity gradient is zero and the velocity is equal to the stream velocity. Difficulties arise in comparing the thicknesses obtained using these various definitions, because velocity is changing so slowly with distance that a small difference in the criterion used for the selection of velocity will account for a very large difference in the corresponding thickness of the boundary layer.

The flow conditions in the boundary layer are of considerable interest to chemical engineers because these influence, not only the drag effect of the fluid on the surface, but also the heat or mass transfer rates where a temperature or a concentration gradient exists.

It is convenient first to consider the flow over a thin plate inserted parallel to the flow of a fluid with a constant stream velocity u_s. It will be assumed that the plate is sufficiently wide for conditions to be constant across any finite width w of the plate which is being considered. Furthermore, the extent of the fluid in a direction perpendicular to the surface is considered as sufficiently large for the velocity of the fluid remote from the surface to be unaffected and to remain constant at the stream velocity u_s. Whilst part of the fluid flows on one side of the flat plate and part on the other, the flow on only one side is considered.

On the assumption that there is no slip at the surface (discussed in Section 11.3), the fluid velocity at all points on the surface, where $y = 0$, will be zero. At some position a distance x from the leading edge, the velocity will increase from zero at the surface to approach the stream velocity u_s asymptotically. At the leading edge, that is where $x = 0$, the fluid will have been influenced by the surface for only an infinitesimal time and therefore only the molecular layer of fluid at the surface will have been retarded. At progressively greater distances (x) along the surface, the fluid will have been retarded for a greater time and the effects will be felt to greater depths in the fluid. Thus the thickness (δ) of the boundary layer will increase, starting from a zero value at the leading edge. Furthermore, the velocity gradient at the surface $(du_x/dy)_{y=0}$ (where u_x is the velocity in the X-direction at a distance y from the surface) will become less because the velocity will change by the same amount (from $u_x = 0$ at $y = 0$ to $u_x = u_s$ at $y = \delta$) over a greater distance. The development of the boundary layer is illustrated in Figure 11.1.

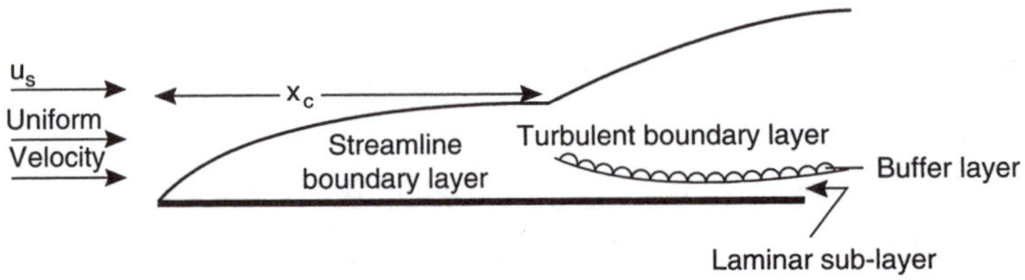

Figure 11.1. Development of the boundary layer

Near the leading edge of the surface where the boundary layer thickness is small, the flow will be streamline, or laminar, and the shear stresses will arise solely from viscous shear effects. When the boundary layer thickness exceeds a critical value, however, streamline flow ceases to be stable and turbulence sets in. The transition from streamline to turbulent flow is not as sharply defined as in pipe flow (discussed in Chapter 3) and may be appreciably influenced by small disturbances generated at the leading edge of the surface and by roughness and imperfections of the surface in itself, all of which lead to an early development of turbulence. However, for equivalent conditions, the important flow parameter is the Reynolds number $Re_\delta (= u_s \delta \rho / \mu)$. Because δ can be expressed as a function of x, the distance from the leading edge of the surface, the usual criterion is taken as the value of the Reynolds number $Re_x (= u_s x \rho / \mu)$. If the location of the transition point is at a distance x_c from the leading edge, then $Re_{x_c} = u_s x_c \rho / \mu$ is of the order of 10^5.

The transition from streamline to turbulent flow in the neighbourhood of a surface has been studied by BROWN[1] with the aid of a three-dimensional smoke tunnel with transparent faces. Photographs were taken using short-interval flash lamps giving effective exposure times of about 20 μs. Figure 11.2 illustrates the flow over an aerofoil and the manner in which vortices are created. Figure 11.3 shows the wake of a cascade of flat plates with sharpened leading edges and blunt trailing edges.

When the flow in the boundary layer is turbulent, streamline flow persists in a thin region close to the surface called the *laminar sub-layer*. This region is of particular importance because, in heat or mass transfer, it is where the greater part of the resistance to transfer lies. High heat and mass transfer rates therefore depend on the laminar sub-layer being thin. Separating the laminar sub-layer from the turbulent part of the boundary

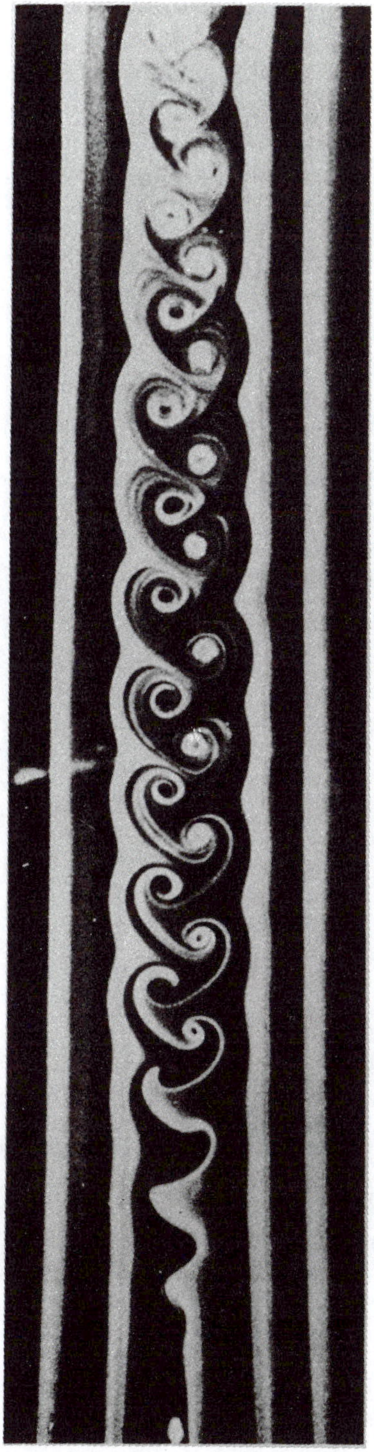

Figure 11.2. Formation of vortices in flow over an aerofoil

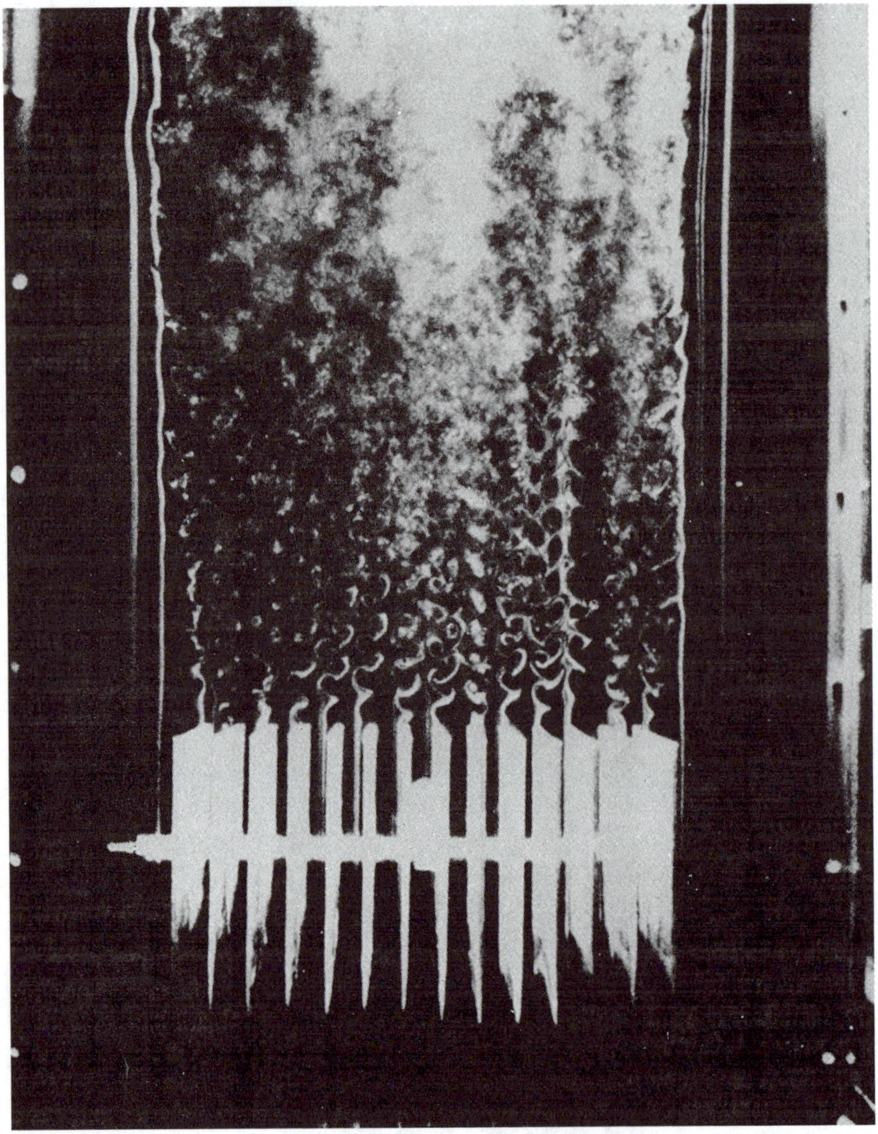

Figure 11.3. Wake produced by cascade of flat plates with sharpened leading edges and blunt trailing edges

layer is the *buffer layer* (Figure 11.1), in which the contributions of the viscous effects and of the turbulent eddies are of comparable magnitudes. This phenomenon is discussed in more detail in Chapter 12, although the general picture is as follows.

The average size of the eddies, or turbulent circulating currents, becomes progressively smaller as the surface is approached. The eddies are responsible for transference of momentum from the faster moving fluid remote from the surface to slower moving fluid near the surface. In this way momentum is transferred by a progression of eddies of diminishing dimensions down to about 1 mm, at which point they tend to die out and viscous forces (due to transfer by molecular-scale movement) predominate near the surface. Although the laminar sub-layer is taken to be the region where eddies are completely absent, there is evidence that occasionally eddies do penetrate right up to the surface. The situation is therefore highly complex and any attempt to model the process involves simplifying assumptions.

The following treatment, based on the simplified approach suggested by PRANDTL[(2)], involves the following three assumptions:

(1) That the flow may be considered essentially as unidirectional (X-direction) and that the effects of velocity components perpendicular to the surface *within* the boundary layer may be neglected (that is, $u_y \ll u_x$). This condition will not be met at very low Reynolds numbers where the boundary layer thickens rapidly.
(2) That the existence of the buffer layer may be neglected and that in turbulent flow the boundary layer may be considered as consisting of a turbulent region adjacent to a laminar sub-layer which separates it from the surface.
(3) That the stream velocity does not change in the direction of flow. On this basis, from Bernoulli's theorem, the pressure then does not change (that is, $\partial P/\partial x = 0$). In practice, $\partial P/\partial x$ may be positive or negative. If positive, a greater retardation of the fluid will result, and the boundary layer will thicken more rapidly. If $\partial P/\partial x$ is negative, the converse will be true.

For flow against a pressure gradient ($\partial P/\partial x$ positive in the direction of flow) the combined force due to pressure gradient and friction may be sufficient to bring the fluid

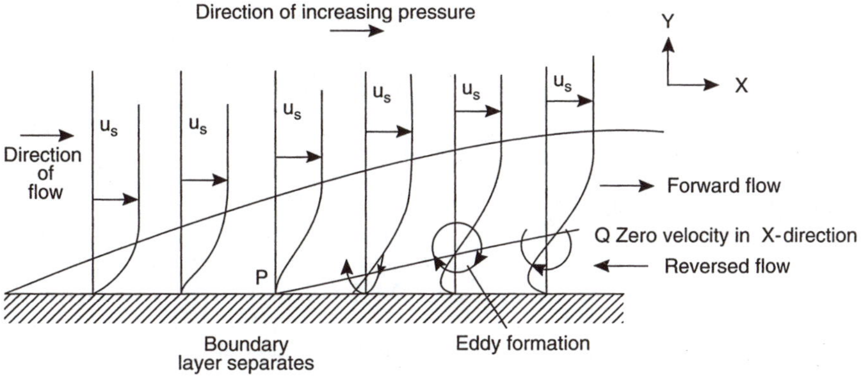

Figure 11.4. Boundary layer separation

completely to rest and to cause some backflow close to the surface. When this occurs the fluid velocity will be zero, not only at the surface, but also at a second position a small distance away. In these circumstances, the boundary layer is said to *separate* and circulating currents are set up as shown in Figure 11.4. An example of this phenomenon is described in Volume 2, Chapter 3, in which flow relative to a cylinder or sphere is considered.

The procedure adopted here consists of taking a momentum balance on an element of fluid. The resulting *Momentum Equation* involves no assumptions concerning the nature of the flow. However, it includes an integral, the evaluation of which requires a knowledge of the velocity profile $u_x = \mathrm{f}(y)$. At this stage assumptions must be made concerning the nature of the flow in order to obtain realistic expressions for the velocity profile.

11.2. THE MOMENTUM EQUATION

It will be assumed that a fluid of density ρ and viscosity μ flows over a plane surface and the velocity of flow outside the boundary layer is u_s. A boundary layer of thickness δ forms near the surface, and at a distance y from the surface the velocity of the fluid is reduced to a value u_x.

The equilibrium is considered of an element of fluid bounded by the planes 1–2 and 3–4 at distances x and $x + \mathrm{d}x$ respectively from the leading edge; the element is of length l in the direction of flow and is of depth w in the direction perpendicular to the plane 1-2-3-4. The distance l is greater than the boundary layer thickness δ (Figure 11.5), and conditions are constant over the width w. The velocities and forces in the X-direction are now considered.

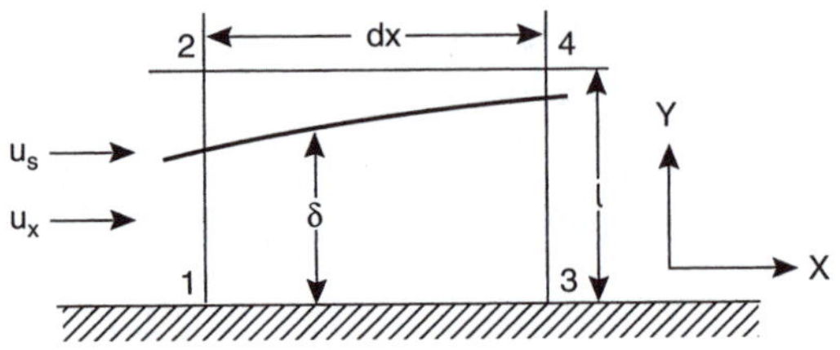

Figure 11.5. Element of boundary layer

At plane 1–2, mass rate of flow through a strip of thickness dy at distance y from the surface

$$= \rho u_x w\,\mathrm{d}y$$

The total flow through plane 1–2

$$= w \int_0^l \rho u_x\,\mathrm{d}y \qquad (11.1)$$

The rate of transfer of momentum through the elementary strip

$$= \rho u_x w\,\mathrm{d}y\,u_x = w\rho u_x^2\,\mathrm{d}y$$

The total rate of transfer of momentum through plane 1–2:

$$M_i = w \int_0^l \rho u_x^2 \, dy \tag{11.2}$$

In passing from plane 1–2 to plane 3–4, the mass flow changes by:

$$w \frac{\partial}{\partial x} \left(\int_0^l \rho u_x \, dy \right) dx \tag{11.3}$$

and the momentum flux changes by:

$$M_{ii} - M_i = w \frac{\partial}{\partial x} \left(\int_0^l \rho u_x^2 \, dy \right) dx \tag{11.4}$$

where M_{ii} is the momentum flux across the plane 3–4.

A mass flow of fluid equal to the difference between the flows at planes 3–4 and 1–2 (equation 11.3) must therefore occur through plane 2–4, as it is assumed that there is uniformity over the width of the element.

Since plane 2–4 lies outside the boundary layer, the fluid crossing this plane must have a velocity u_s in the X-direction. Because the fluid in the boundary layer is being retarded, there will be a smaller flow at plane 3–4 than at 1–2, and hence the flow through plane 2–4 is outwards, and fluid leaves the element of volume.

Thus the rate of transfer of momentum through plane 2–4 out of the element is:

$$M_{iii} = -w u_s \frac{\partial}{\partial x} \left(\int_0^l \rho u_x \, dy \right) dx \tag{11.5}$$

It will be noted that the derivative is negative, which indicates a positive outflow of momentum from the element.

Steady-state momentum balance over the element 1 - 2 - 3 - 4

The terms which must be considered in the momentum balance for the X-direction are:

(i) The momentum flux M_i through plane 1–2 *into* the element.
(ii) The momentum flux M_{ii} through plane 3–4 *out of* the element.
(iii) The momentum flux M_{iii} through plane 3–4 *out of* the element.

Thus, the net momentum flux out of the element M_{ex} is given by:

$$\underbrace{w \frac{\partial}{\partial x} \left(\int_0^l \rho u_x^2 \, dy \right) dx}_{M_{ii} - M_i} + \underbrace{\left\{ -w u_s \frac{\partial}{\partial x} \left(\int_0^l \rho u_x \, dy \right) dx \right\}}_{M_{iii}}$$

Then, since u_s is assumed not to vary with x:

$$M_{ex} = \left\{ -w \frac{\partial}{\partial x} \left(\int_0^l \rho u_x (u_s - u_x) \, dy \right) dx \right\} \tag{11.6}$$

The net rate of change of momentum in the X-direction on the element must be equal to the momentum added from outside, through plane 2–4, together with the net force acting on it.

The forces in the X-direction acting on the element of fluid are:

(1) A shear force resulting from the shear stress R_0 acting at the surface. This is a retarding force and therefore R_0 is negative.
(2) The force produced as a result of any difference in pressure dP between the planes 3–4 and 1–2. However, if the velocity u_s outside the boundary layer remains constant, from Bernoulli's theorem, there can be no pressure gradient in the X-direction and $\partial P/\partial x = 0$.

Thus, the net force acting F is just the retarding force attributable to the shear stress at the surface only and

Thus: $$F = R_0 w \, dx \tag{11.7}$$

Equating the net momentum flux out of the element to the net retarding force (equations 11.6 and 11.7) and simplifying gives:

$$\frac{\partial}{\partial x}\int_0^l \rho(u_s - u_x)u_x \, dy = -R_0 \tag{11.8}$$

This expression, known as the *momentum equation*, may be integrated provided that the relation between u_x and y is known.

If the velocity of the main stream remains constant at u_s and the density ρ may be taken as constant, equation 11.8 then becomes:

$$\rho\frac{\partial}{\partial x}\int_0^l (u_s - u_x)u_x \, dy = -R_0 \tag{11.9}$$

It may be noted that no assumptions have been made concerning the nature of the flow within the boundary layer and therefore this relation is applicable to both the streamline and the turbulent regions. The relation between u_x and y is derived for streamline and turbulent flow over a plane surface and the integral in equation 11.9 is evaluated.

11.3. THE STREAMLINE PORTION OF THE BOUNDARY LAYER

In the streamline boundary layer the only forces acting within the fluid are pure viscous forces and no transfer of momentum takes place by eddy motion.

Assuming that the relation between u_x and y can be expressed approximately by:

$$u_x = u_0 + ay + by^2 + cy^3 \tag{11.10}$$

The coefficients a, b, c and u_0 may be evaluated because the boundary conditions which the relation must satisfy are known, as shown in Figure 11.6.

In fluid dynamics it is generally assumed that the velocity of flow at a solid boundary, such as a pipe wall, is zero. This is referred to as the *no-slip condition*. If the fluid "wets" the surface, this assumption can be justified in physical terms since the molecules are

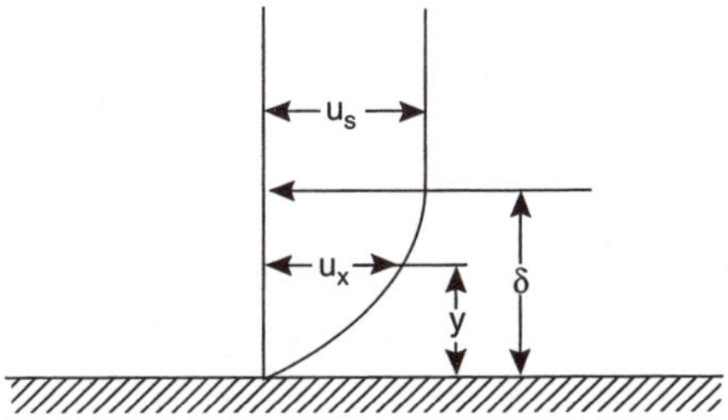

Figure 11.6. Velocity distribution in streamline boundary layer

small compared with the irregularities on the surface of even the smoothest pipe, and therefore a fluid layer is effectively held stationary at the wall. All gases and the majority of liquids have a sufficiently low contact angle for this condition to be met.

If the fluid does not wet the wall, the no-slip condition no longer applies and the pressure gradient at a given flowrate will be lower. This effect is particularly important with the flow of molten polymers, although it does not seem to be significant in other applications.

It is assumed here that the fluid in contact with the surface is at rest and therefore u_0 must be zero. Furthermore, all the fluid close to the surface is moving at very low velocity and therefore any changes in its momentum as it flows parallel to the surface must be extremely small. Consequently, the net shear force acting on any element of fluid near the surface is negligible, the retarding force at its lower boundary being balanced by the accelerating force at its upper boundary. Thus the shear stress R_0 in the fluid near the surface must approach a constant value.

Since $R_0 = -\mu(\partial u_x/\partial y)_{y=0}$, $\partial u_x/\partial y$ must also be constant at small values of y and:

$$\left(\frac{\partial^2 u_x}{\partial y^2}\right)_{y=0} = 0$$

At the distant edge of the boundary layer it is assumed that the velocity just equals the main stream velocity and that there is no discontinuity in the velocity profile.

Thus, when $y = \delta$: $\quad u_x = u_s \quad$ and $\quad \dfrac{\partial u_x}{\partial y} = 0$

Now with $u_0 = 0$, equation 11.10 becomes:

$$u_x = ay + by^2 + cy^3$$

$$\frac{\partial u_x}{\partial y} = a + 2by + 3cy^2$$

and: $$\frac{\partial^2 u_x}{\partial y^2} = 2b + 6cy$$

At $y = 0$: $$\frac{\partial^2 u_x}{\partial y^2} = 0$$

Thus: $$b = 0$$

At $y = \delta$: $$u_x = a\delta + c\delta^3 = u_s$$

and: $$\frac{\partial u_x}{\partial y} = a + 3c\delta^2 = 0$$

Thus: $$a = -3c\delta^2$$

Hence: $$c = -\frac{u_s}{2\delta^3} \quad \text{and} \quad a = \frac{3u_s}{2\delta}$$

The equation for the velocity profile is therefore:

$$u_x = \frac{3u_s}{2}\frac{y}{\delta} - \frac{u_s}{2}\left(\frac{y}{\delta}\right)^3 \tag{11.11}$$

$$\frac{u_x}{u_s} = \frac{3}{2}\left(\frac{y}{\delta}\right) - \frac{1}{2}\left(\frac{y}{\delta}\right)^3 \tag{11.12}$$

Equation 11.12 corresponds closely to experimentally determined velocity profiles in a laminar boundary layer.

This relation applies over the range $0 < y < \delta$.

When $y > \delta$, then: $$u_x = u_s \tag{11.13}$$

The integral in the momentum equation (11.9) can now be evaluated for the streamline boundary layer by considering the ranges $0 < y < \delta$ and $\delta < y < l$ separately.

Thus:

$$\int_0^l (u_s - u_x)u_x \, \mathrm{d}y = \int_0^\delta u_s^2 \left(1 - \frac{3}{2}\frac{y}{\delta} + \frac{y^3}{2\delta^3}\right)\left(\frac{3}{2}\frac{y}{\delta} - \frac{y^3}{2\delta^3}\right)\mathrm{d}y + \int_\delta^l (u_s - u_s)u_s \, \mathrm{d}y$$

$$= u_s^2 \int_0^\delta \left(\frac{3}{2}\frac{y}{\delta} - \frac{9}{4}\frac{y^2}{\delta^2} - \frac{1}{2}\frac{y^3}{\delta^3} + \frac{3}{2}\frac{y^4}{\delta^4} - \frac{1}{4}\frac{y^6}{\delta^6}\right)\mathrm{d}y$$

$$= u_s^2\delta\left(\frac{3}{4} - \frac{3}{4} - \frac{1}{8} + \frac{3}{10} - \frac{1}{28}\right)$$

$$= \frac{39}{280}\delta u_s^2 \tag{11.14}$$

In addition: $$R_0 = -\mu\left(\frac{\partial u_x}{\partial y}\right)_{y=0} = -\frac{3}{2}\mu\frac{u_s}{\delta} \tag{11.15}$$

Substitution from equations 11.14 and 11.15 in equation 11.9, gives:

$$\rho\frac{\partial}{\partial x}\left(\frac{39}{280}\delta u_s^2\right) = \frac{3}{2}\mu\frac{u_s}{\delta}$$

$$\therefore \quad \delta\,\mathrm{d}\delta = \left(\frac{140}{13}\right)\frac{\mu}{\rho}\frac{1}{u_s}\mathrm{d}x$$

$$\therefore \quad \frac{\delta^2}{2} = \left(\frac{140}{13}\right)\left(\frac{\mu x}{\rho u_s}\right) \quad \text{(since } \delta = 0 \text{ when } x = 0\text{)} \tag{11.16}$$

Thus:
$$\delta = 4.64\sqrt{\frac{\mu x}{\rho u_s}}$$

and:
$$\frac{\delta}{x} = 4.64\sqrt{\frac{\mu}{x\rho u_s}} = 4.64 Re_x^{-1/2} \tag{11.17}$$

The rate of thickening of the boundary layer is then obtained by differentiating equation 11.17:

$$\frac{d\delta}{dx} = 2.32 Re_x^{-1/2} \tag{11.18}$$

This relation for the thickness of the boundary layer has been obtained on the assumption that the velocity profile can be described by a polynomial of the form of equation 11.10 and that the main stream velocity is reached at a distance δ from the surface, whereas, in fact, the stream velocity is approached asymptotically. Although equation 11.11 gives the velocity u_x accurately as a function of y, it does not provide a means of calculating accurately the distance from the surface at which u_x has a particular value when u_x is near u_s, because $\partial u_x/\partial y$ is then small. The thickness of the boundary layer as calculated is therefore a function of the particular approximate relation which is taken to represent the velocity profile. This difficulty can be overcome by introducing a new concept, the *displacement thickness* δ^*.

When a viscous fluid flows over a surface it is retarded and the overall flowrate is therefore reduced. A non-viscous fluid, however, would not be retarded and therefore a boundary layer would not form. The displacement thickness δ^* is defined as the distance the surface would have to be moved in the Y-direction in order to obtain the same rate of flow with this non-viscous fluid as would be obtained for the viscous fluid with the surface retained at $x = 0$.

The mass rate of flow of a frictionless fluid between $y = \delta^*$ and $y = \infty$

$$= \rho \int_{\delta^*}^{\infty} u_s \, dy$$

The mass rate of flow of the real fluid between $y = 0$ and $y = \infty$

$$= \rho \int_{0}^{\infty} u_x \, dy$$

Then, by definition of the displacement thickness:

$$\rho \int_{\delta^*}^{\infty} u_s \, dy = \rho \int_{0}^{\infty} u_x \, dy$$

or:
$$\int_{0}^{\infty} u_s \, dy - \int_{0}^{\delta^*} u_s \, dy = \int_{0}^{\infty} u_x \, dy$$

and:
$$\delta^* = \int_{0}^{\infty} \left(1 - \frac{u_x}{u_s}\right) dy \tag{11.19}$$

Using equation 11.12 to give the velocity profile:

$$\delta^* = \int_{0}^{\delta} \left(1 - \frac{3}{2}\frac{y}{\delta} + \frac{1}{2}\frac{y^3}{\delta^3}\right) dy$$

since equation 11.12 applies only over the limits $0 < y < \delta$, and outside this region $u_x = u_s$ and the integral is zero.

Then: $$\delta^* = \delta(1 - \tfrac{3}{4} + \tfrac{1}{8})$$

and: $$\frac{\delta^*}{\delta} = 0.375 \qquad (11.20)$$

Shear stress at the surface

The shear stress in the fluid at the surface is given by:

$$R_0 = -\mu\left(\frac{\partial u_x}{\partial y}\right)_{y=0}$$

$$= -\frac{3}{2}\mu\frac{u_s}{\delta} \quad \text{(from equation 11.15)}$$

$$= -\frac{3}{2}\mu u_s \frac{1}{x}\frac{1}{4.64}\sqrt{\frac{x\rho u_s}{\mu}}$$

$$= -0.323\rho u_s^2\sqrt{\frac{\mu}{x\rho u_s}} = -0.323\rho u_s^2 Re_x^{-1/2}$$

The shear stress R acting on the surface itself is equal and opposite to the shear stress on the fluid at the surface; that is, $R = -R_0$.

Thus: $$\frac{R}{\rho u_s^2} = 0.323 Re_x^{-1/2} \qquad (11.21)$$

Equation 11.21 gives the point values of R and $R/\rho u_s^2$ at $x = x$. In order to calculate the total frictional force acting at the surface, it is necessary to multiply the average value of R between $x = 0$ and $x = x$ by the area of the surface.

The average value of $R/\rho u_s^2$ denoted by the symbol $(R/\rho u_s^2)_m$ is then given by:

$$\left(\frac{R}{\rho u_s^2}\right)_m x = \int_0^x \frac{R}{\rho u_s^2}\,\mathrm{d}x$$

$$= \int_0^x 0.323\sqrt{\frac{\mu}{x\rho u_s}}\,\mathrm{d}x \quad \text{(from equation 11.21)}$$

$$= 0.646x\sqrt{\frac{\mu}{x\rho u_s}}$$

$$\left(\frac{R}{\rho u_s^2}\right)_m = 0.646\sqrt{\frac{\mu}{x\rho u_s}}$$

$$= 0.646 Re_x^{-0.5} \approx 0.65 Re_x^{-0.5} \qquad (11.22)$$

11.4. THE TURBULENT BOUNDARY LAYER

11.4.1. The turbulent portion

Equation 11.12 does not fit velocity profiles measured in a turbulent boundary layer and an alternative approach must be used. In the simplified treatment of the flow conditions within the turbulent boundary layer the existence of the buffer layer, shown in Figure 11.1, is neglected and it is assumed that the boundary layer consists of a laminar sub-layer, in which momentum transfer is by molecular motion alone, outside which there is a turbulent region in which transfer is effected entirely by eddy motion (Figure 11.7). The approach is based on the assumption that the shear stress at a plane surface can be calculated from the simple power law developed by Blasius, already referred to in Chapter 3.

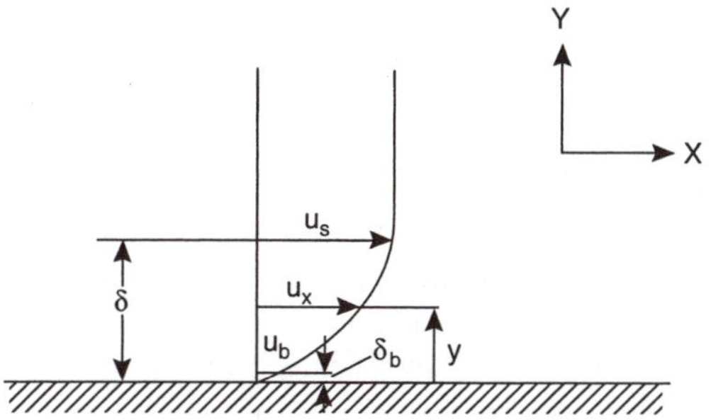

Figure 11.7. Turbulent boundary layer

BLASIUS[(3)] has given the following approximate expression for the shear stress at a plane smooth surface over which a fluid is flowing with a velocity u_s, for conditions where $Re_x < 10^7$:

$$\frac{R}{\rho u_s^2} = 0.0228\left(\frac{\mu}{u_s\delta\rho}\right)^{0.25} \tag{11.23}$$

Thus, the shear stress is expressed as a function of the boundary layer thickness δ and it is therefore implicitly assumed that a certain velocity profile exists in the fluid. As a first assumption, it may be assumed that a simple power relation exists between the velocity and the distance from the surface in the boundary layer, or:

$$\frac{u_x}{u_s} = \left(\frac{y}{\delta}\right)^f \tag{11.24}$$

Hence $\quad R = 0.0228\rho u_s^2\left(\dfrac{\mu}{u_s\delta\rho}\right)^{0.25}$

$$= 0.0228\rho^{0.75}\mu^{0.25}\delta^{-0.25}u_s^{1.75}$$

$$= 0.0228\rho^{0.75}\mu^{0.25}\delta^{-0.25}u_x^{1.75}\left(\frac{\delta}{y}\right)^{1.75f} \quad \text{(from equation 11.24)}$$

$$= 0.0228\rho^{0.75}\mu^{0.25}u_x^{1.75}y^{-1.75f}\delta^{1.75f-0.25} \tag{11.25}$$

If the velocity profile is the same for all stream velocities, the shear stress must be defined by specifying the velocity u_x at any distance y from the surface. The boundary layer thickness, determined by the velocity profile, is then no longer an independent variable so that the index of δ in equation 11.25 must be zero or:

$$1.75f - 0.25 = 0$$

and:

$$f = \tfrac{1}{7}$$

Thus:

$$\frac{u_x}{u_s} = \left(\frac{y}{\delta}\right)^{1/7} \tag{11.26}$$

Equation 11.26 is sometimes known as the *Prandtl seventh power law*.

Differentiating equation 11.26 with respect to y gives:

$$\frac{\partial u_x}{\partial y} = \frac{1}{7}u_s\delta^{-1/7}y^{-6/7} = \left(\frac{1}{7}\right)\left(\frac{u_s}{y}\right)\left(\frac{y}{\delta}\right)^{1/7} \tag{11.27}$$

This relation is not completely satisfactory in that it gives an infinite velocity gradient at the surface, where the laminar sub-layer exists, and a finite velocity gradient at the outer edge of the boundary layer. This is in contradiction to the conditions which must exist in the stream. However, little error is introduced by using this relation for the whole of the boundary layer in the momentum equation because, firstly both the velocities and hence the momentum fluxes near the surface are very low, and secondly it gives the correct value of the velocity at the edge of the boundary layer. Accepting equation 11.26 for the limits $0 < y < \delta$, the integral in equation 11.9 becomes:

$$\begin{aligned}
\int_0^l (u_s - u_x)u_x\,\mathrm{d}y &= u_s^2\left\{\int_0^\delta \left[1 - \left(\frac{y}{\delta}\right)^{1/7}\right]\left(\frac{y}{\delta}\right)^{1/7}\mathrm{d}y\right\} + \int_\delta^l (u_s - u_s)u_s\,\mathrm{d}y \\
&= u_s^2\int_0^\delta \left[\left(\frac{y}{\delta}\right)^{1/7} - \left(\frac{y}{\delta}\right)^{2/7}\right]\mathrm{d}y \\
&= u_s^2\delta\left(\frac{7}{8} - \frac{7}{9}\right) \\
&= \frac{7}{72}u_s^2\delta
\end{aligned} \tag{11.28}$$

From the Blasius equation:

$$-R_0 = R = 0.0228\rho u_s^2\left(\frac{\mu}{u_s\delta\rho}\right)^{1/4} \quad \text{(from equation 11.23)}$$

Substituting from equations 11.23 and 11.28 in equation 11.9 gives:

$$\rho\frac{\partial}{\partial x}\left[\frac{7}{72}u_s^2\delta\right] = 0.0228\rho u_s^2\left(\frac{\mu}{u_s\delta\rho}\right)^{1/4}$$

$$\delta^{1/4}\,\mathrm{d}\delta = 0.235\left(\frac{\mu}{u_s\rho}\right)^{1/4}\mathrm{d}x$$

and:
$$\frac{4}{5}\delta^{5/4} = 0.235x\left(\frac{\mu}{u_s\rho}\right)^{1/4} + \text{constant}$$

Putting the constant equal to zero, implies that $\delta = 0$ when $x = 0$, that is that the turbulent boundary layer extends to the leading edge of the surface. An error is introduced by this assumption, but it is found to be small except where the surface is only slightly longer than the critical distance x_c for the laminar-turbulent transition.

Thus:
$$\delta = 0.376x^{0.8}\left(\frac{\mu}{u_s\rho}\right)^{0.2} \tag{11.29}$$

$$= 0.376x\left(\frac{\mu}{u_s\rho x}\right)^{0.2}$$

or:
$$\frac{\delta}{x} = 0.376Re_x^{-0.2} \tag{11.30}$$

The rate of thickening of the boundary layer is obtained by differentiating equation 11.30 to give:

$$\frac{d\delta}{dx} = 0.301Re_x^{-0.2} \tag{11.31}$$

The displacement thickness δ^* is given by equation 11.19:

$$\delta^* = \int_0^\infty \left(1 - \frac{u_x}{u_s}\right) dy \qquad \text{(equation 11.19)}$$

$$= \int_0^\delta \left(1 - \left(\frac{y}{\delta}\right)^{1/7}\right) dy \quad \left(\text{since } 1 - \frac{u_x}{u_s} = 0 \text{ when } y > \delta\right)$$

$$= \tfrac{1}{8}\delta \tag{11.32}$$

As noted previously, δ^* is independent of the particular approximation used for the velocity profile.

It is of interest to compare the rates of thickening of the streamline and turbulent boundary layers at the transition point. Taking a typical value of $Re_{xc} = 10^5$, then:

For the streamline boundary layer, from equation 11.18, $\dfrac{d\delta}{dx} = 0.0073$

For the turbulent boundary layer, from equation 11.31, $\dfrac{d\delta}{dx} = 0.0301$

Thus the turbulent boundary layer is thickening at about four times the rate of the streamline boundary layer at the transition point.

11.4.2. The laminar sub-layer

If at a distance x from the leading edge the laminar sub-layer is of thickness δ_b and the total thickness of the boundary layer is δ, the properties of the laminar sub-layer can be found by equating the shear stress at the surface as given by the Blasius equation (11.23) to that obtained from the velocity gradient near the surface.

It has been noted that the shear stress and hence the velocity gradient are almost constant near the surface. Since the laminar sub-layer is very thin, the velocity gradient within it may therefore be taken as constant.

Thus the shear stress in the fluid at the surface,

$$R_0 = -\mu \left(\frac{\partial u_x}{\partial y}\right)_{y=0} = -\mu \frac{u_x}{y}, \quad \text{where } y < \delta_b$$

Equating this to the value obtained from equation 11.23 gives:

$$0.0228 \rho u_s^2 \left(\frac{\mu}{u_s \delta \rho}\right)^{1/4} = \mu \frac{u_x}{y}$$

and:
$$u_x = 0.0228 \rho u_s^2 \frac{1}{\mu} \left(\frac{\mu}{u_s \delta \rho}\right)^{1/4} y$$

If the velocity at the edge of the laminar sub-layer is u_b, that is, if $u_x = u_b$ when $y = \delta_b$:

$$u_b = 0.0228 \rho u_s^2 \frac{1}{\mu} \left(\frac{\mu}{u_s \delta \rho}\right)^{1/4} \delta_b$$

$$= 0.0228 \frac{\rho u_s^2}{\mu} \frac{\mu}{u_s \delta \rho} \delta_b \left(\frac{\mu}{u_s \delta \rho}\right)^{-3/4}$$

Thus:
$$\frac{\delta_b}{\delta} = \frac{1}{0.0228} \left(\frac{u_b}{u_s}\right) \left(\frac{\mu}{u_s \delta \rho}\right)^{3/4} \tag{11.33}$$

The velocity at the inner edge of the turbulent region must also be given by the equation for the velocity distribution in the turbulent region.

Hence:
$$\left(\frac{\delta_b}{\delta}\right)^{1/7} = \frac{u_b}{u_s} \quad \text{(from equation 11.26)}$$

Thus:
$$\left(\frac{u_b}{u_s}\right)^7 = \frac{1}{0.0228} \left(\frac{u_b}{u_s}\right) \left(\frac{\mu}{u_s \delta \rho}\right)^{3/4} \quad \text{(from equation 11.33)}$$

or:
$$\frac{u_b}{u_s} = 1.87 \left(\frac{\mu}{u_s \delta \rho}\right)^{1/8}$$

$$= 1.87 Re_\delta^{-1/8} \tag{11.34}$$

Since:
$$\delta = 0.376 x^{0.8} \left(\frac{\mu}{u_s \rho}\right)^{0.2} \quad \text{(equation 11.29)}$$

$$\frac{u_b}{u_s} = 1.87 \left(\frac{u_s \rho}{\mu} 0.376 \frac{x^{0.8} \mu^{0.2}}{u_s^{0.2} \rho^{0.2}}\right)^{-1/8}$$

$$= \frac{1.87}{0.376^{1/8}} \left(\frac{u_s^{0.8} x^{0.8} \rho^{0.8}}{\mu^{0.8}}\right)^{-1/8}$$

$$= 2.11 Re_x^{-0.1} \approx 2.1 Re_x^{-0.1} \tag{11.35}$$

The thickness of the laminar sub-layer is given by:

$$\frac{\delta_b}{\delta} = \left(\frac{u_b}{u_s}\right)^7 = \frac{190}{Re_x^{0.7}} \quad \text{(from equations 11.26 and 11.35)}$$

or:

$$\frac{\delta_b}{x} = \frac{190}{Re_x^{0.7}} \frac{0.376}{Re_x^{0.2}} \quad \text{(from equation 11.30)}$$

$$= 71.5 Re_x^{-0.9} \quad (11.36)$$

Thus $\delta_b \propto x^{0.1}$; that is, δ_b it increases very slowly as x increases. Further, $\delta_b \propto u_s^{-0.9}$ and therefore decreases rapidly as the velocity is increased, and heat and mass transfer coefficients are therefore considerably influenced by the velocity.

The shear stress at the surface, at a distance x from the leading edge, is given by:

$$R_0 = -\mu \frac{u_b}{\delta_b}$$

Since $R_0 = -R$, then:

$$R = \mu\, 2.11\, u_s\, Re_x^{-0.1} \frac{1}{x} \frac{1}{71.5} Re_x^{0.9} \quad \text{(from equations 11.35 and 11.36)}$$

$$= 0.0296 Re_x^{0.8} \frac{\mu u_s}{x}$$

$$= 0.0296 \rho u_s^2 Re_x^{-0.2} \approx 0.03 \rho u_s^2 Re_x^{-0.2} \quad (11.37)$$

or:

$$\frac{R}{\rho u_s^2} = 0.0296 Re_x^{-0.2} \quad (11.38)$$

or approximately:

$$\frac{R}{\rho u_s^2} = 0.03 Re_x^{-0.2} \quad (11.39)$$

The mean value of $R/\rho u_s^2$ over the range $x = 0$ to $x = x$ is given by:

$$\left(\frac{R}{\rho u_s^2}\right)_m x = \int_0^x \left(\frac{R}{\rho u_s^2}\right) \mathrm{d}x$$

$$= \int_0^x 0.0296 \left(\frac{\mu}{u_s x \rho}\right)^{0.2} \mathrm{d}x$$

$$= 0.0296 \left(\frac{\mu}{u_s x \rho}\right)^{0.2} \frac{x}{0.8}$$

or:

$$\left(\frac{R}{\rho u_s^2}\right)_m = 0.037 Re_x^{-0.2} \quad (11.40)$$

The total shear force acting on the surface is found by adding the forces acting in the streamline ($x < x_c$) and turbulent ($x > x_c$) regions. This can be done provided the critical value Re_{x_c}, is known.

In the streamline region:

$$\left(\frac{R}{\rho u_s^2}\right)_m = 0.646 Re_x^{-0.5} \quad \text{(equation 11.22)}$$

In the turbulent region: $$\left(\frac{R}{\rho u_s^2}\right)_m = 0.037Re_x^{-0.2} \quad \text{(equation 11.40)}$$

In calculating the mean value of $(R/\rho u_s^2)_m$ in the turbulent region, it was assumed that the turbulent boundary layer extended to the leading edge. A more accurate value for the mean value of $(R/\rho u_s^2)_m$ over the whole surface can be obtained by using the expression for streamline conditions over the range from $x = 0$ to $x = x_c$ (where x_c is the critical distance from the leading edge) and the expression for turbulent conditions in the range $x = x_c$ to $x = x$.

Thus: $$\left(\frac{R}{\rho u_s^2}\right)_m = \frac{1}{x}(0.646Re_{x_c}^{-0.5}x_c + 0.037Re_x^{-0.2}x - 0.037Re_{x_c}^{-0.2}x_c)$$

$$= 0.646Re_{x_c}^{-0.5}\frac{Re_{x_c}}{Re_x} + 0.037Re_x^{-0.2} - 0.037Re_{x_c}^{-0.2}\frac{Re_{x_c}}{Re_x}$$

$$= 0.037Re_x^{-0.2} + Re_x^{-1}(0.646Re_{x_c}^{0.5} - 0.037Re_{x_c}^{0.8}) \quad (11.41)$$

Example 11.1

Water flows at a velocity of 1 m/s over a plane surface 0.6 m wide and 1 m long. Calculate the total drag force acting on the surface if the transition from streamline to turbulent flow in the boundary layer occurs when the Reynolds group $Re_{x_c} = 10^5$.

Solution

Taking $\mu = 1$ mN s/m^2 $= 10^{-3}$ Ns/m^2, at the far end of the surface, $Re_x = (1 \times 1 \times 10^3)/10^{-3} = 10^6$
Mean value of $R/\rho u_s^2$ from equation 11.41

$$= 0.037(10^6)^{-0.2} + (10^6)^{-1}[0.646(10^5)^{0.5} - 0.037(10^5)^{0.8}]$$

$$= 0.00214$$

$$\text{Total drag force} = \frac{R}{\rho u_s^2}(\rho u_s^2) \times (\text{area of surface})$$

$$= (0.00214 \times 1000 \times 1^2 \times 1 \times 0.6)$$

$$= \underline{\underline{1.28 \text{ N}}}$$

Example 11.2

Calculate the thickness of the boundary layer at a distance of 150 mm from the leading edge of a surface over which oil, of viscosity 0.05 N s/m^2 and density 1000 kg/m^3 flows with a velocity of 0.3 m/s. What is the displacement thickness of the boundary layer?

Solution

$$Re_x = (0.150 \times 0.3 \times 1000/0.05) = 900$$

For streamline flow: $$\frac{\delta}{x} = \frac{4.64}{Re_x^{0.5}} \quad \text{(from equation 11.17)}$$

$$= \frac{4.64}{900^{0.5}} = 0.1545$$

Hence: $\delta = (0.1545 \times 0.150) = 0.0232 \text{ m} = \underline{\underline{23.2 \text{ mm}}}$

and from equation 11.20, the displacement thickness $\delta^* = (0.375 \times 23.2) = \underline{\underline{8.7 \text{ mm}}}$

11.5. BOUNDARY LAYER THEORY APPLIED TO PIPE FLOW

11.5.1. Entry conditions

When a fluid flowing with a uniform velocity enters a pipe, a boundary layer forms at the walls and gradually thickens with distance from the entry point. Since the fluid in the boundary layer is retarded and the total flow remains constant, the fluid in the central stream is accelerated. At a certain distance from the inlet, the boundary layers, which have formed in contact with the walls, join at the axis of the pipe, and, from that point onwards, occupy the whole cross-section and consequently remain of a constant thickness. *Fully developed flow* then exists. If the boundary layers are still streamline when fully developed flow commences, the flow in the pipe remains streamline. On the other hand, if the boundary layers are already turbulent, turbulent flow will persist, as shown in Figure 11.8.

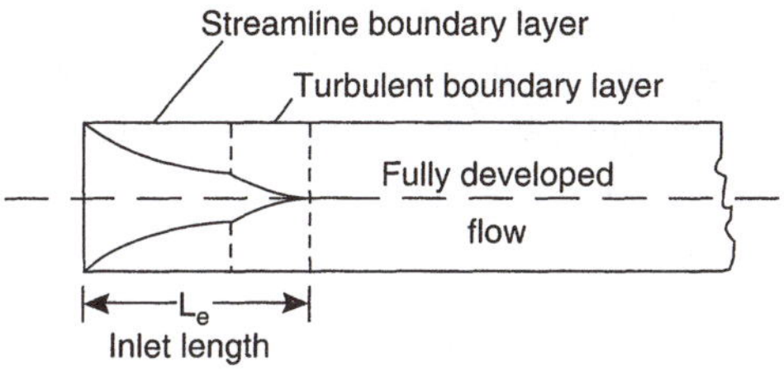

Figure 11.8. Conditions at entry to pipe

An approximate experimental expression for the inlet length L_e for laminar flow is:

$$\frac{L_e}{d} = 0.0575Re \qquad (11.42)$$

where d is the diameter of the pipe and Re is the Reynolds group with respect to pipe diameter, and based on the mean velocity of flow in the pipe. This expression is only approximate, and is inaccurate for Reynolds numbers in the region of 2500 because the boundary layer thickness increases very rapidly in this region. An average value of L_e at a Reynolds number of 2500 is about $100d$. The inlet length is somewhat arbitrary as steady conditions in the pipe are approached asymptotically, the boundary layer thickness being a function of the assumed velocity profile.

At the inlet to the pipe the velocity across the whole section is constant. The velocity at the pipe axis will progressively increase in the direction of flow and reach a maximum value when the boundary layers join. Beyond this point the velocity profile, and the velocity at the axis, will not change. Since the fluid at the axis has been accelerated, its kinetic energy per unit mass will increase and therefore there must be a corresponding fall in its pressure energy.

Under streamline conditions, the velocity at the axis u_s will increase from a value u at the inlet to a value $2u$ where fully-developed flow exists, as shown in Figure 11.9, because the mean velocity of flow u in the pipe is half of the axial velocity, from equation 3.36.

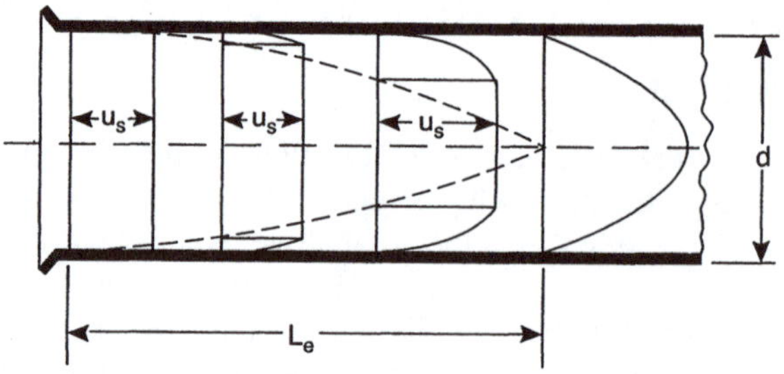

Figure 11.9. Development of the laminar velocity profile at the entry to a pipe

Thus the kinetic energy per unit mass of the fluid at the axis inlet $= \frac{1}{2}u^2$
The corresponding kinetic energy at the end of the inlet length $= \frac{1}{2}(2u)^2 = 2u^2$.
The increase in the kinetic energy per unit mass $= \frac{3}{2}u^2$.
Thus the fall in pressure due to the increase of velocity of the fluid $= \frac{3}{2}\rho u^2$.
If the flow in the pipe is turbulent, the velocity at the axis increases from u to only about $u/0.817$, as given by equation 3.63.
Under these conditions, the fall in pressure

$$= \frac{1}{2}\rho u^2\left(\frac{1}{0.817^2} - 1\right)$$

$$\approx \tfrac{1}{4}\rho u^2 \qquad (11.43)$$

If the fluid enters the pipe from a duct of larger cross-section, the existence of a radial velocity component gives rise to the formation of a *vena contracta* near the entry to the pipe but this has been neglected here.

11.5.2. Application of the boundary-layer theory

The velocity distribution and frictional resistance have been calculated from purely theoretical considerations for the streamline flow of a fluid in a pipe. The boundary layer theory can now be applied in order to calculate, approximately, the conditions when the fluid is turbulent. For this purpose it is assumed that the boundary layer expressions may be applied to flow over a cylindrical surface and that the flow conditions in the region of fully developed flow are the same as those when the boundary layers first join. The thickness of the boundary layer is thus taken to be equal to the radius of the pipe and the velocity at the outer edge of the boundary layer is assumed to be the velocity at the axis. Such assumptions are valid very close to the walls, although significant errors will arise near the centre of the pipe.

The velocity of the fluid may be assumed to obey the Prandtl one-seventh power law, given by equation 11.26. If the boundary layer thickness δ is replaced by the pipe radius r, this is then given by:

$$\frac{u_x}{u_s} = \left(\frac{y}{r}\right)^{1/7} \tag{11.44}$$

The relation between the mean velocity and the velocity at the axis is derived using this expression in Chapter 3. There, the mean velocity u is shown to be 0.82 times the velocity u_s at the axis, although in this calculation the thickness of the laminar sub-layer was neglected and the Prandtl velocity distribution assumed to apply over the whole cross-section. The result therefore is strictly applicable only at very high Reynolds numbers where the thickness of the laminar sub-layer is very small. At lower Reynolds numbers the mean velocity will be rather less than 0.82 times the velocity at the axis.

The expressions for the shear stress at the walls, the thickness of the laminar sub-layer, and the velocity at the outer edge of the laminar sub-layer may be applied to the turbulent flow of a fluid in a pipe. It is convenient to express these relations in terms of the mean velocity in the pipe, the pipe diameter, and the Reynolds group with respect to the mean velocity and diameter.

The shear stress at the walls is given by the Blasius equation (11.23) as:

$$\frac{R}{\rho u_s^2} = 0.0228\left(\frac{\mu}{u_s r \rho}\right)^{1/4}$$

Writing $u = 0.817u_s$ and $d = 2r$:

$$\frac{R}{\rho u^2} = 0.0386\left(\frac{\mu}{ud\rho}\right)^{1/4} = 0.0386Re^{-1/4} \tag{11.45}$$

This equation is more usually written:

$$\frac{R}{\rho u^2} = 0.0396Re^{-1/4} \quad \text{(See equation 3.11)} \tag{11.46}$$

The discrepancy between the coefficients in equations 11.45 and 11.46 is attributable to the fact that the effect of the curvature of the pipe wall has not been taken into account in applying the equation for flow over a plane surface to flow through a pipe. In addition, it takes no account of the existence of the laminar sub-layer at the walls.

Equation 11.46 is applicable for Reynolds numbers up to 10^5.

The velocity at the edge of the laminar sub-layer is given by:

$$\frac{u_b}{u_s} = 1.87\left(\frac{\mu}{u_s r \rho}\right)^{1/8} \quad \text{(equation 11.34)}$$

which becomes:

$$\frac{u_b}{u} = 2.49\left(\frac{\mu}{ud\rho}\right)^{1/8}$$

$$= 2.49Re^{-1/8} \tag{11.47}$$

and:

$$\frac{u_b}{u_s} = 2.0Re^{-1/8} \tag{11.48}$$

The thickness of the laminar sub-layer is given by:

$$\frac{\delta_b}{r} = \left(\frac{u_b}{u_s}\right)^7 \quad \text{(from equation 11.26)}$$

$$= (1.87)^7 \left(\frac{\mu}{u_s r \rho}\right)^{7/8} \quad \text{(from equation 11.34)}$$

Thus:

$$\frac{\delta_b}{d} = 62 \left(\frac{\mu}{u d \rho}\right)^{7/8}$$

$$= 62 Re^{-7/8} \tag{11.49}$$

The thickness of the laminar sub-layer is therefore almost inversely proportional to the Reynolds number, and hence to the velocity.

Example 11.3

Calculate the thickness of the laminar sub-layer when benzene flows through a pipe 50 mm in diameter at 2 l/s. What is the velocity of the benzene at the edge of the laminar sub-layer? Assume that fully developed flow exists within the pipe and that for benzene, $\rho = 870$ kg/m^3 and $\mu = 0.7$ mN s/m^2.

Solution

The mass flowrate of benzene $= (2 \times 10^{-3} \times 870)$

$= 1.74$ kg/s

Thus: Reynolds number $= \dfrac{4G}{\mu \pi D} = \dfrac{4 \times 1.74}{0.7 \times 10^{-3} \pi \times 0.050}$ (equation 4.52)

$= 63{,}290$

From equation 11.49: $\dfrac{\delta_b}{d} = 62 Re^{-7/8}$

$$\delta_b = \frac{(62 \times 0.050)}{63{,}290^{7/8}}$$

$= 1.95 \times 10^{-4}$ m

or: $\underline{\underline{0.195 \text{ mm}}}$

The mean velocity $= \dfrac{1.74}{(870 \times (\pi/4) 0.050^2)}$

$= 1.018$ m/s

From equation 11.47: $\dfrac{u_b}{u} = \dfrac{2.49}{Re^{1/8}}$

from which: $u_b = \dfrac{(2.49 \times 1.018)}{63{,}290^{1/8}}$

$= \underline{\underline{0.637 \text{ m/s}}}$

11.6. THE BOUNDARY LAYER FOR HEAT TRANSFER

11.6.1. Introduction

Where a fluid flows over a surface which is at a different temperature, heat transfer occurs and a temperature profile is established in the vicinity of the surface. A number of possible conditions may be considered. At the outset, the heat transfer rate may be sufficient to change the temperature of the fluid stream significantly or it may remain at a substantially constant temperature. Furthermore, a variety of conditions may apply at the surface. Thus the surface may be maintained at a constant temperature, particularly if it is in good thermal conduct with a heat source or sink of high thermal capacity. Alternatively, the heat flux at the surface may be maintained constant, or conditions may be intermediate between the constant temperature and the constant heat flux conditions. In general, there is likely to be a far greater variety of conditions as compared with those in the momentum transfer problem previously considered. Temperature gradients are likely to be highest in the vicinity of the surface and it is useful to develop the concept of a *thermal boundary layer*, analogous to the velocity boundary layer already considered, within which the whole of the temperature gradient may be regarded as existing.

Thus, a velocity boundary layer and a thermal boundary layer may develop simultaneously. If the physical properties of the fluid do not change significantly over the temperature range to which the fluid is subjected, the velocity boundary layer will not be affected by the heat transfer process. If physical properties are altered, there will be an interactive effect between the momentum and heat transfer processes, leading to a comparatively complex situation in which numerical methods of solution will be necessary.

In general, the thermal boundary layer will not correspond with the velocity boundary layer. In the following treatment, the simplest non-interacting case is considered with physical properties assumed to be constant. The stream temperature is taken as constant (θ_s). In the first case, the wall temperature is also taken as a constant, and then by choosing the temperature scale so that the wall temperature is zero, the boundary conditions are similar to those for momentum transfer.

It will be shown that the momentum and thermal boundary layers coincide only if the Prandtl number is unity, implying equal values for the kinematic viscosity (μ/ρ) and the thermal diffusivity ($D_H = k/C_p\rho$).

The condition of constant heat flux at the surface, as opposed to constant surface temperature, is then considered in a later section.

11.6.2. The heat balance

The procedure here is similar to that adopted previously. A heat balance, as opposed to a momentum balance, is taken over an element which extends beyond the limits of both the velocity and thermal boundary layers. In this way, any fluid entering or leaving the element through the face distant from the surface is at the stream velocity u_s and stream temperature θ_s. A heat balance is made therefore on the element shown in Figure 11.10 in which the length l is greater than the velocity boundary layer thickness δ and the thermal boundary layer thickness δ_t.

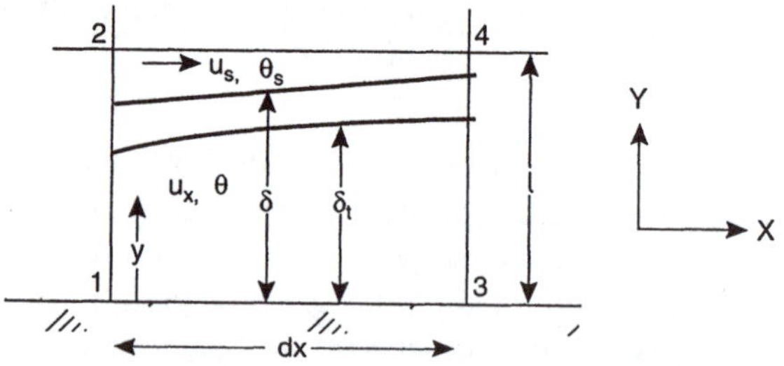

Figure 11.10. The thermal boundary layer

The rate of heat transfer through an element of width w of the plane 1–2, of thickness dy at a distance y from the surface is:

$$= C_p \rho \theta u_x w \, \mathrm{d}y \tag{11.50}$$

where C_p is the specific heat capacity of the fluid at constant pressure, ρ the density of the fluid, and θ and u_x are the temperature and velocity at a distance y from the surface.

The total rate of transfer of heat through the plane, 1–2 is then:

$$= C_p \rho w \int_0^l \theta u_x \, \mathrm{d}y \tag{11.51}$$

assuming that the physical properties of the fluid are independent of temperature. In the distance dx this heat flow changes by an amount given by:

$$C_p \rho w \frac{\partial}{\partial x}\left(\int_0^l \theta \, u_x \, \mathrm{d}y\right) \mathrm{d}x \tag{11.52}$$

It is shown in Section 11.2 that there is a mass rate of flow of fluid through plane, 2–4, out of the element equal to $\rho w(\partial/\partial x)(\int_0^l u_x \, \mathrm{d}y) \, \mathrm{d}x$.

Since the plane, 2–4, lies outside the boundary layers, the heat leaving the element through the plane as a result of this flow is:

$$C_p \rho \theta_s w \frac{\partial}{\partial x}\left(\int_0^l u_x \, \mathrm{d}y\right) \mathrm{d}x \tag{11.53}$$

where θ_s is the temperature outside the thermal boundary layer.

The heat transferred by thermal conduction into the element through plane, 1–3 is:

$$= -kw \, \mathrm{d}x \left(\frac{\partial \theta}{\partial y}\right)_{y=0} \tag{11.54}$$

If the temperature θ_s of the main stream is unchanged, a heat balance on the element gives:

$$C_p \rho w \left(\frac{\partial}{\partial x}\int_0^l \theta u_x \, \mathrm{d}y\right) \mathrm{d}x = C_p \rho \theta_s w \left(\frac{\partial}{\partial x}\int_0^l u_x \, \mathrm{d}y\right) \mathrm{d}x - k\left(\frac{\partial \theta}{\partial y}\right)_{y=0} w \, \mathrm{d}x$$

or:

$$\frac{\partial}{\partial x}\int_0^l u_x(\theta_s - \theta) \, \mathrm{d}y = D_H \left(\frac{\partial \theta}{\partial y}\right)_{y=0} \tag{11.55}$$

where D_H $(= k/C_p \rho)$ is the thermal diffusivity of the fluid.

The relations between u_x and y have already been obtained for both streamline and turbulent flow. A relation between θ and y for streamline conditions in the boundary layer is now derived, although it is not possible to define the conditions in the turbulent boundary layer sufficiently precisely to derive a similar expression for that case.

11.6.3. Heat transfer for streamline flow over a plane surface—constant surface temperature

The flow of fluid over a plane surface, heated at distances greater than x_0 from the leading edge, is now considered. As shown in Figure 11.11 the velocity boundary layer starts at the leading edge and the thermal boundary layer at a distance x_0 from it. If the temperature of the heated portion of the plate remains constant, this may be taken as the datum temperature. It is assumed that the temperature at a distance y from the surface may be represented by a polynomial of the form:

$$\theta = a_0 y + b_0 y^2 + c_0 y^3 \tag{11.56}$$

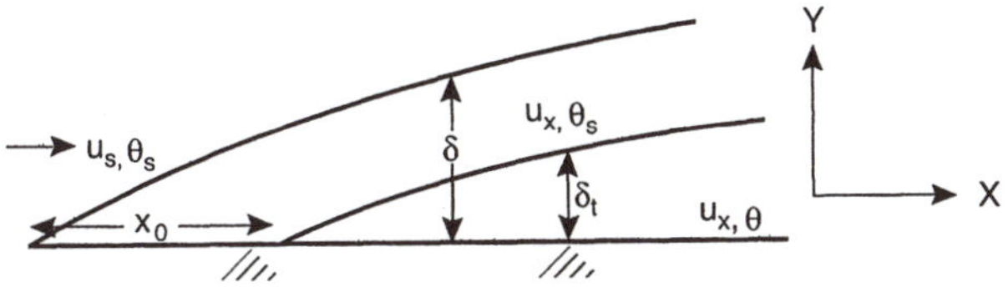

Figure 11.11. Thermal boundary layer—streamline flow

If the fluid layer in contact with the surface is assumed to be at rest, any heat flow in the vicinity of the surface must be by pure thermal conduction. Thus the heat transferred per unit area and unit time q_0 is given by:

$$q_0 = -k \left(\frac{\partial \theta}{\partial y}\right)_{y=0}$$

If the temperature of the fluid element in contact with the wall is to remain constant, the heat transfer rate into and out of the element must be the same, or:

$$\left(\frac{\partial \theta}{\partial y}\right)_{y=0} = \text{a constant} \quad \text{and} \quad \left(\frac{\partial^2 \theta}{\partial y^2}\right)_{y=0} = 0$$

At the outer edge of the thermal boundary layer, the temperature is θ_s and the temperature gradient $(\partial\theta/\partial y) = 0$ if there is to be no discontinuity in the temperature profile.

Thus the conditions for the thermal boundary layer, with respect to temperature, are the same as those for the velocity boundary layer with respect to velocity. Then, if the thickness of the thermal boundary layer is δ_t, the temperature distribution is given by:

$$\frac{\theta}{\theta_s} = \frac{3}{2}\left(\frac{y}{\delta_t}\right) - \frac{1}{2}\left(\frac{y}{\delta_t}\right)^3 \tag{11.57}$$

which may be compared with equation 11.12,

and:
$$\left(\frac{\partial\theta}{\partial y}\right)_{y=0} = \frac{3\theta_s}{2\delta_t} \tag{11.58}$$

It is assumed that the velocity boundary layer is everywhere thicker than the thermal boundary layer, so that $\delta > \delta_t$ (Figure 11.11). Thus the velocity distribution everywhere within the thermal boundary layer is given by equation 11.12. The implications of this assumption are discussed later.

The integral in equation 11.55 clearly has a finite value within the thermal boundary layer, although it is zero outside it. When the expression for the temperature distribution in the boundary layer is inserted, the upper limit of integration must be altered from l to δ_t.

Thus:
$$\int_0^l (\theta_s - \theta)u_x \,\mathrm{d}y = \theta_s u_s \int_0^{\delta_t} \left[1 - \frac{3}{2}\frac{y}{\delta_t} + \frac{1}{2}\left(\frac{y}{\delta_t}\right)^3\right]\left[\frac{3y}{2\delta} - \frac{1}{2}\left(\frac{y}{\delta}\right)^3\right]\mathrm{d}y$$

$$= \theta_s u_s \left[\frac{3}{4}\frac{\delta_t^2}{\delta} - \frac{3}{4}\frac{\delta_t^2}{\delta} - \frac{1}{8}\frac{\delta_t^4}{\delta^3} + \frac{3}{20}\left(\frac{\delta_t^2}{\delta} + \frac{\delta_t^4}{\delta^3}\right) - \frac{1}{28}\frac{\delta_t^4}{\delta^3}\right]$$

$$= \theta_s u_s \left(\frac{3}{20}\frac{\delta_t^2}{\delta} - \frac{3}{280}\frac{\delta_t^4}{\delta^3}\right)$$

$$= \theta_s u_s \delta \left(\frac{3}{20}\sigma^2 - \frac{3}{280}\sigma^4\right) \tag{11.59}$$

where $\sigma = \delta_t/\delta$.

Since $\delta_t < \delta$, the second term is small compared with the first, and:

$$\int_0^l (\theta_s - \theta)u_x \,\mathrm{d}y \approx \frac{3}{20}\theta_s u_s \delta\sigma^2 \tag{11.60}$$

Substituting from equations 11.58 and 11.60 in equation 11.55 gives:

$$\frac{\partial}{\partial x}\left(\frac{3}{20}\theta_s u_s \delta\sigma^2\right) = D_H \frac{3\theta_s}{2\delta_t} = D_H \frac{3\theta_s}{2\delta\sigma}$$

$$\therefore \quad \frac{1}{10}u_s\delta\sigma\frac{\partial}{\partial x}(\delta\sigma^2) = D_H$$

$$\therefore \quad \frac{1}{10}u_s\left(\delta\sigma^3\frac{\partial\delta}{\partial x} + 2\delta^2\sigma^2\frac{\partial\sigma}{\partial x}\right) = D_H \tag{11.61}$$

It has already been shown that:

$$\delta^2 = \frac{280\mu x}{13\rho u_s} = 21.5\frac{\mu x}{\rho u_s} \qquad \text{(equation 11.16)}$$

and hence:
$$\delta\frac{\partial\delta}{\partial x} = \frac{140}{13}\frac{\mu}{\rho u_s}$$

Substituting in equation 11.61:

$$\frac{u_s}{10}\frac{\mu}{\rho u_s}\left(\frac{140}{13}\sigma^3 + \frac{560}{13}x\sigma^2\frac{\partial\sigma}{\partial x}\right) = D_H$$

$\therefore$
$$\frac{14}{13}\frac{\mu}{\rho D_H}\left(\sigma^3 + 4x\sigma^2\frac{\partial\sigma}{\partial x}\right) = 1$$

and:
$$\sigma^3 + \frac{4x}{3}\frac{\partial\sigma^3}{\partial x} = \frac{13}{14}Pr^{-1}$$

where the Prandtl number $Pr = C_p\mu/k = \mu/\rho D_H$.

$\therefore$
$$\frac{3}{4}x^{-1}\sigma^3 + \frac{\partial\sigma^3}{\partial x} = \frac{13}{14}Pr^{-1}\frac{3}{4}x^{-1}$$

and:
$$\frac{3}{4}x^{-1/4}\sigma^3 + x^{3/4}\frac{\partial\sigma^3}{\partial x} = \frac{13}{14}Pr^{-1}\frac{3}{4}x^{-1/4}$$

Integrating:
$$x^{3/4}\sigma^3 = \frac{13}{14}Pr^{-1}x^{3/4} + \text{constant}$$

or:
$$\sigma^3 = \frac{13}{14}Pr^{-1} + \text{constant } x^{-3/4}$$

When $x = x_0$,
$$\sigma = 0$$

so that:
$$\text{constant} = -\frac{13}{14}Pr^{-1}x_0^{3/4}$$

Hence:
$$\sigma^3 = \frac{13}{14}Pr^{-1}\left[1 - \left(\frac{x_0}{x}\right)^{3/4}\right]$$

and:
$$\sigma = 0.976Pr^{-1/3}\left[1 - \left(\frac{x_0}{x}\right)^{3/4}\right]^{1/3} \qquad (11.62)$$

If the whole length of the plate is heated, $x_0 = 0$ and:

$$\sigma = 0.976Pr^{-1/3} \qquad (11.63)$$

In this derivation, it has been assumed that $\sigma < 1$.

For all liquids other than molten metals, $Pr > 1$ and hence, from equation 11.63, $\sigma < 1$.

For gases, $Pr \not< 0.6$, so that $\sigma \not> 1.18$.

Thus only a small error is introduced when this expression is applied to gases. The only serious deviations occur for molten metals, which have very low Prandtl numbers.

If h is the heat transfer coefficient, then:

$$q_0 = -h\theta_s$$

and:
$$-h\theta_s = -k\left(\frac{\partial\theta}{\partial y}\right)_{y=0}$$

or, from equation 11.58:
$$h = \frac{k}{\theta_s}\frac{3}{2}\frac{\theta_s}{\delta_t}$$

$$= \frac{3}{2}\frac{k}{\delta_t} = \frac{3}{2}\frac{k}{\delta\sigma} \qquad (11.64)$$

Substituting for δ from equation 11.17, and σ from equation 11.62 gives:

$$h = \frac{3k}{2}\frac{1}{4.64}\sqrt{\frac{\rho u_s}{\mu x}}\frac{Pr^{1/3}}{0.976[1-(x_0/x)^{3/4}]^{1/3}}$$

or:

$$\frac{hx}{k} = 0.332 Pr^{1/3} Re_x^{1/2} \frac{1}{[1-(x_0/x)^{3/4}]^{1/3}} \tag{11.65}$$

If the surface is heated over its entire length, so that $x_0 = 0$, then:

$$Nu_x = \frac{hx}{k} = 0.332 Pr^{1/3} Re_x^{1/2} \tag{11.66}$$

It is seen from equation 11.66 that the heat transfer coefficient theoretically has an infinite value at the leading edge, where the thickness of the thermal boundary layer is zero, and that it decreases progressively as the boundary layer thickens. Equation 11.66 gives the point value of the heat transfer coefficient at a distance x from the leading edge. The mean value between $x = 0$ and $x = x$ is given by:

$$h_m = x^{-1}\int_0^x h\,\mathrm{d}x \tag{11.67}$$

$$= x^{-1}\int_0^x \psi x^{-1/2}\,\mathrm{d}x$$

where ψ is not a function of x.

Thus:

$$h_m = x^{-1}\left[2\psi x^{1/2}\right]_0^x = 2h \tag{11.68}$$

The mean value of the heat transfer coefficient between $x = 0$ and $x = x$ is equal to twice the point value at $x = x$. The mean value of the Nusselt group is given by:

$$(Nu_x)_m = 0.664 Pr^{1/3} Re_x^{1/2} \tag{11.69}$$

11.6.4. Heat transfer for streamline flow over a plane surface—constant surface heat flux

Another important case is where the heat flux, as opposed to the temperature at the surface, is constant; this may occur where the surface is electrically heated. Then, the temperature difference $|\theta_s - \theta_0|$ will increase in the direction of flow (x-direction) as the value of the heat transfer coefficient decreases due to the thickening of the thermal boundary layer. The equation for the temperature profile in the boundary layer becomes:

$$\frac{\theta - \theta_0}{\theta_s - \theta_0} = \frac{3}{2}\left(\frac{y}{\delta_t}\right) - \frac{1}{2}\left(\frac{y}{\delta_t}\right)^3 \tag{11.70}$$

(from equation 11.57) and the temperature gradient at the walls is given by:

$$\left(\frac{\partial \theta}{\partial y}\right)_{y=0} = \frac{3(\theta_s - \theta_0)}{2\delta_t} \tag{11.71}$$

(from equation 11.58)

The value of the integral in the energy balance (equation 11.55) is again given by equation 11.60 [substituting $(\theta_s - \theta_0)$ for θ_s]. The heat flux q_0 at the surface is now constant, and the right-hand side of equation 11.55 may be expressed as $(-q_0/C_p\rho)$. Thus, for constant surface heat flux, equation 11.55 becomes:

$$\frac{\partial}{\partial x}\left(\frac{3}{20}(\theta_s - \theta_0)u_s\delta\sigma^2\right) = -\frac{q_0}{C_p\rho} \tag{11.72}$$

Equation 11.72 cannot be integrated directly, however, because the temperature driving force $(\theta_s - \theta_0)$ is not known as a function of location x on the plate. The solution of equation 11.72 involves a quite complex procedure which is given by KAYS and CRAWFORD(4) and takes the following form:

$$\frac{hx}{k} = Nu_x = 0.453Pr^{1/3}Re_x^{1/2} \tag{11.73}$$

By comparing equations 11.61 and 11.66, it is seen that the local Nusselt number and the heat transfer coefficient are both some 36 per cent higher for a constant surface heat flux as compared with a constant surface temperature.

The average value of the Nusselt group $(Nu_x)_m$ is obtained by integrating over the range $x = 0$ to $x = x$, giving:

$$(Nu_x)_m = 0.906Pr^{1/3}Re_x^{1/2} \tag{11.74}$$

11.7. THE BOUNDARY LAYER FOR MASS TRANSFER

If a concentration gradient exists within a fluid flowing over a surface, mass transfer will take place, and the whole of the resistance to transfer can be regarded as lying within a *diffusion boundary layer* in the vicinity of the surface. If the concentration gradients, and hence the mass transfer rates, are small, variations in physical properties may be neglected and it can be shown that the velocity and thermal boundary layers are unaffected(5). For low concentrations of the diffusing component, the effects of bulk flow will be small and the mass balance equation for component **A** is:

$$\frac{\partial}{\partial x}\int_0^l (C_{As} - C_A)u_x \, dy = D\left(\frac{\partial C_A}{\partial y}\right)_{y=0} \tag{11.75}$$

where C_A and C_{As} are the molar concentrations of **A** at a distance y from the surface and outside the boundary layer respectively, and l is a distance at right angles to the surface which is greater than the thickness of any of the boundary layers. Equation 11.70 is obtained in exactly the same manner as equation 11.55 for heat transfer.

Again, the form of the concentration profile in the diffusion boundary layer depends on the conditions which are assumed to exist at the surface and in the fluid stream. For the conditions corresponding to those used in consideration of the thermal boundary layer, that is constant concentrations both in the stream outside the boundary layer and at the surface, the concentration profile is of similar form to that given by equation 11.70:

$$\frac{C_A - C_{A0}}{C_{As} - C_{A0}} = \frac{3}{2}\left(\frac{y}{\delta_D}\right) - \frac{1}{2}\left(\frac{y}{\delta_D}\right)^3 \tag{11.76}$$

where δ_D is the thickness of the concentration boundary layer,
C_A is the concentration of **A** at $y = y$,
C_{A0} is the concentration of **A** at the surface ($y = 0$), and
C_{As} is the concentration of **A** outside the boundary layer

Substituting from equation 11.76 to evaluate the integral in equation 11.75, assuming that mass transfer takes place over the whole length of the surface ($x_0 = 0$), by analogy with equation 11.63 gives:

$$\frac{\delta_D}{\delta} \approx 0.976Sc^{-1/3} \tag{11.77}$$

where $Sc = \mu/\rho D$ is the Schmidt number. Equation 11.77 is applicable provided that $Sc > 1$. If Sc is only slightly less than 1, a negligible error is introduced and it is therefore applicable to most mixtures of gases, as seen in Table 10.2 (Chapter 10). The arguments are identical to those relating to the validity of equation 11.63 for heat transfer.

The point values of the Sherwood number Sh_x and mass transfer coefficient h_D are then given by:

$$Sh_x = \frac{h_D x}{D} = 0.331Sc^{1/3}Re_x^{1/2} \tag{11.78}$$

The mean value of the coefficient between $x = 0$ and $x = x$ is then given by:

$$(Sh_x)_m = 0.662Sc^{1/3}Re_x^{1/2} \tag{11.79}$$

In equations 11.78 and 11.79 Sh_x and $(Sh_x)_m$ represent the point and mean values respectively of the Sherwood numbers.

11.8. FURTHER READING

KAYS, W. M. and CRAWFORD, M. E.: *Convective Heat and Mass Transfer* 3rd edn. (McGraw-Hill, New York, 1998).

SCHLICHTING, H.: *Boundary Layer Theory* (trans. by KESTIN, J.) 6th edn (McGraw-Hill, New York, 1968).

WHITE, F. M.: *Viscous Fluid Flow* (McGraw-Hill, New York, 1974).

11.9. REFERENCES

1. BROWN, F. N. M.: *Proc. Midwest Conf. Fluid Mechanics* (Sept. 1959) 331. The organized boundary layer.
2. PRANDTL, L.: *The Essentials of Fluid Dynamics* (Hafner, New York, 1949).
3. BLASIUS, H.: *Forsch. Ver. deut. Ing.* **131** (1913). Das Ähnlichkeitsgesetz bei Reibungsvorgängen in Flüssigkeiten.
4. KAYS, W. M. and CRAWFORD, M. E.: *Convective Heat and Mass Transfer* 3rd edn. (McGraw-Hill, New York, 1998).
5. ECKERT, E. R. G. and DRAKE, R. M. Jr: *Analysis of Heat and Mass Transfer* (McGraw-Hill, New York, 1972).

11.10. NOMENCLATURE

		Units in SI system	Dimensions in **M, N, L, T, θ**
a	Coefficient of y	K/m	$\mathbf{L}^{-1}\boldsymbol{\theta}$
a_0	Coefficient of y	s^{-1}	$\mathbf{T}^{-1}$
b	Coefficient of y^2	$m^{-1}\ s^{-1}$	$\mathbf{L}^{-1}\mathbf{T}^{-1}$

		Units in SI system	Dimensions in M, N, L, T, θ
b_0	Coefficient of y^2	K/m^2	$\mathbf{L}^{-2}\theta$
C_A	Molar concentration of **A**	kmol/m^3	$\mathbf{NL}^{-3}$
C_{A0}	Molar concentration of **A** at surface ($y = 0$)	kmol/m^3	$\mathbf{NL}^{-3}$
C_{AS}	Molar concentration of **A** outside boundary layer	kmol/m^3	$\mathbf{NL}^{-3}$
C_p	Specific heat at constant pressure	J/kg K	$\mathbf{L}^2\mathbf{T}^{-2}\theta^{-1}$
c	Coefficient of y^3	m^{-2} s^{-1}	$\mathbf{L}^{-2}\mathbf{T}^{-1}$
c_0	Coefficient of y^3	K/m^3	$\mathbf{L}^{-3}\theta$
D	Molecular diffusivity	m^2/s	$\mathbf{L}^2\mathbf{T}^{-1}$
D_H	Thermal diffusivity	m^2/s	$\mathbf{L}^2\mathbf{T}^{-1}$
d	Pipe diameter	m	**L**
F	Retarding force	N	$\mathbf{MLT}^{-2}$
f	Index	—	—
h	Heat transfer coefficient	W/m^2 K	$\mathbf{MT}^{-3}\theta^{-1}$
h_D	Mass transfer coefficient	kmol/[(m^2)(s)(kmol/m^3)]	$\mathbf{LT}^{-1}$
h_m	Mean value of heat transfer coefficient	W/m^2 K	$\mathbf{MT}^{-3}\theta^{-1}$
k	Thermal conductivity	W/m K	$\mathbf{MLT}^{-3}\theta^{-1}$
L_e	Inlet length of pipe	m	**L**
l	Thickness of element of fluid	m	**L**
M	Momentun flux	N	$\mathbf{MLT}^{-2}$
P	Total pressure	N/m^2	$\mathbf{ML}^{-1}\mathbf{T}^{-2}$
q_0	Rate of transfer of heat per unit area at walls	W/m^2	$\mathbf{MT}^{-3}$
R	Shear stress acting on surface	N/m^2	$\mathbf{ML}^{-1}\mathbf{T}^{-2}$
R_0	Shear stress acting on fluid at surface	N/m^2	$\mathbf{ML}^{-1}\mathbf{T}^{-2}$
r	Radius of pipe	m	**L**
t	Time	s	**T**
u	Mean velocity	m/s	$\mathbf{LT}^{-1}$
u_b	Velocity at edge of laminar sub-layer	m/s	$\mathbf{LT}^{-1}$
u_0	Velocity of fluid at surface	m/s	$\mathbf{LT}^{-1}$
u_s	Velocity of fluid outside boundary layer, or at pipe axis	m/s	$\mathbf{LT}^{-1}$
u_x	Velocity in X-direction at $y = y$	m/s	$\mathbf{LT}^{-1}$
w	Width of surface	m	**L**
x	Distance from leading edge of surface in X-direction	m	**L**
x_c	Value of x at which flow becomes turbulent	m	**L**
x_0	Unheated length of surface	m	**L**
y	Distance from surface	m	**L**
δ	Thickness of boundary layer	m	**L**
δ_b	Thickness of laminar sub-layer	m	**L**
δ_D	Diffusion boundary layer thickness	m	**L**
δ_t	Thickness of thermal boundary layer	m	**L**
δ^*	Displacement thickness of boundary layer	m	**L**
θ	Temperature at $y = y$	K	θ
θ_s	Temperature outside boundary layer, or at pipe axis	K	θ
μ	Viscosity of fluid	N s/m^2	$\mathbf{ML}^{-1}\mathbf{T}^{-1}$
ρ	Density of fluid	kg/m^3	$\mathbf{ML}^{-3}$
σ	Ratio of δ_t to δ	—	—
Nu_x	Nusselt number hx/k	—	—
Re	Reynolds number $ud\rho/\mu$	—	—
Re_x	Reynolds number $u_s x\rho/\mu$	—	—
Re_{x_c}	Reynolds number $u_s x_c\rho/\mu$	—	—
Re_δ	Reynolds number $u_s\delta\rho/\mu$	—	—
Pr	Prandtl number $C_p\mu/k$	—	—
Sc	Schmidt number $\mu/\rho D$	—	—
Sh_x	Sherwood number $h_D x/D$	—	—

CHAPTER 12

Quantitative Relations between Transfer Processes

12.1. INTRODUCTION

In the previous chapters, the stresses arising from relative motion within a fluid, the transfer of heat by conduction and convection, and the mechanism of mass transfer are all discussed. These three major processes of momentum, heat, and mass transfer have, however, been regarded as independent problems.

In most of the unit operations encountered in the chemical and process industries, one or more of the processes of momentum, heat, and mass transfer is involved. Thus, in the flow of a fluid under adiabatic conditions through a bed of granular particles, a pressure gradient is set up in the direction of flow and a velocity gradient develops approximately perpendicularly to the direction of motion in each fluid stream; momentum transfer then takes place between the fluid elements which are moving at different velocities. If there is a temperature difference between the fluid and the pipe wall or the particles, heat transfer will take place as well, and the convective component of the heat transfer will be directly affected by the flow pattern of the fluid. Here, then, is an example of a process of simultaneous momentum and heat transfer in which the same fundamental mechanism is affecting both processes. Fractional distillation and gas absorption are frequently carried out in a packed column in which the gas or vapour stream rises countercurrently to a liquid. The function of the packing in this case is to provide a large interfacial area between the phases and to promote turbulence within the fluids. In a very turbulent fluid, the rates of transfer per unit area of both momentum and mass are high; and as the pressure drop rises the rates of transfer of both momentum and mass increase together. In some cases, momentum, heat, and mass transfer all occur simultaneously as, for example, in a water-cooling tower (see Chapter 13), where transfer of sensible heat and evaporation both take place from the surface of the water droplets. It will now be shown not only that the process of momentum, heat, and mass transfer are physically related, but also that quantitative relations between them can be developed.

Another form of interaction between the transfer processes is responsible for the phenomenon of *thermal diffusion* in which a component in a mixture moves under the action of a temperature gradient. Although there are important applications of thermal diffusion, the magnitude of the effect is usually small relative to that arising from concentration gradients.

When a fluid is flowing under streamline conditions over a surface, a forward component of velocity is superimposed on the random distribution of velocities of the molecules, and movement at right angles to the surface occurs solely as a result of the random motion of the molecules. Thus if two adjacent layers of fluid are moving at different velocities, there will be a tendency for the faster moving layer to be retarded and the slower moving layer to be accelerated by virtue of the continuous passage of molecules in each direction. There will therefore be a net transfer of momentum from the fast- to the slow-moving stream. Similarly, the molecular motion will tend to reduce any temperature gradient or any concentration gradient if the fluid consists of a mixture of two or more components. At the boundary the effects of the molecular transfer are balanced by the drag forces at the surface.

If the motion of the fluid is turbulent, the transfer of fluid by eddy motion is superimposed on the molecular transfer process. In this case, the rate of transfer to the surface will be a function of the degree of turbulence. When the fluid is highly turbulent, the rate of transfer by molecular motion will be negligible compared with that by eddy motion. For small degrees of turbulence the two may be of the same order.

It was shown in the previous chapter that when a fluid flows under turbulent conditions over a surface, the flow can conveniently be divided into three regions:

(1) At the surface, the laminar sub-layer, in which the only motion at right angles to the surface is due to molecular diffusion.
(2) Next, the buffer layer, in which molecular diffusion and eddy motion are of comparable magnitude.
(3) Finally, over the greater part of the fluid, the turbulent region in which eddy motion is large compared with molecular diffusion.

In addition to momentum, both heat and mass can be transferred either by molecular diffusion alone or by molecular diffusion combined with eddy diffusion. Because the effects of eddy diffusion are generally far greater than those of the molecular diffusion, the main resistance to transfer will lie in the regions where only molecular diffusion is occurring. Thus the main resistance to the flow of heat or mass to a surface lies within the laminar sub-layer. It is shown in Chapter 11 that the thickness of the laminar sub-layer is almost inversely proportional to the Reynolds number for fully developed turbulent flow in a pipe. Thus the heat and mass transfer coefficients are much higher at high Reynolds numbers.

There are strict limitations to the application of the analogy between momentum transfer on the one hand, and heat and mass transfer on the other. Firstly, it must be borne in mind that momentum is a vector quantity, whereas heat and mass are scalar quantities. Secondly, the quantitative relations apply only to that part of the momentum transfer which arises from *skin friction.* If *form drag* is increased there is little corresponding increase in the rates at which heat transfer and mass transfer will take place.

Skin friction is the drag force arising from shear stress attributable to the viscous force in the laminar region in the neighbourhood of a surface. *Form drag* is the inertial component arising from vortex formation arising from the presence of an obstruction of flow by, for instance, a baffle or a roughness element on the surface of the pipe. Thus, in the design of contacting devices such as column packings, it is important that they are so shaped that the greater part of the pressure drop is attributable to skin friction rather than to form drag.

12.2. TRANSFER BY MOLECULAR DIFFUSION

12.2.1. Momentum transfer

When the flow characteristics of the fluid are *Newtonian*, the shear stress R_y in a fluid is proportional to the velocity gradient and to the viscosity.

Thus, for constant density: $$R_y = -\mu\frac{du_x}{dy} = -\frac{\mu}{\rho}\frac{d(\rho u_x)}{dy} \quad \text{(cf. equation 3.3)} \qquad (12.1)$$

where u_x is the velocity of the fluid parallel to the surface at a distance y from it.

The shear stress R_y within the fluid, at a distance y from the boundary surface, is a measure of the rate of transfer of momentum per unit area at right angles to the surface.

Since (ρu_x) is the momentum per unit volume of the fluid, the rate of transfer of momentum per unit area is proportional to the gradient in the Y-direction of the momentum per unit volume. The negative sign indicates that momentum is transferred from the fast- to the slow-moving fluid and the shear stress acts in such a direction as to oppose the motion of the fluid.

12.2.2. Heat transfer

From the definition of thermal conductivity, the heat transferred per unit time through unit area at a distance y from the surface is given by:

$$q_y = -k\frac{d\theta}{dy} = -\left(\frac{k}{C_p\rho}\right)\frac{d(C_p\rho\theta)}{dy} \quad \text{(cf. equation 9.11)} \qquad (12.2)$$

where C_p is the specific heat of the fluid at constant pressure, θ the temperature, and k the thermal conductivity. C_p and ρ are both assumed to be constant.

The term $(C_p\rho\theta)$ represents the heat content per unit volume of fluid and therefore the flow of heat is proportional to the gradient in the Y-direction of the heat content per unit volume. The proportionality constant $k/C_p\rho$ is called the thermal diffusivity D_H.

12.2.3. Mass transfer

It is shown in Chapter 10, from Fick's Law of diffusion, that the rate of diffusion of a constituent **A** in a mixture is proportional to its concentration gradient.

Thus, from equation 10.4:

$$N_A = -D\frac{dC_A}{dy} \qquad (12.3)$$

where N_A is the molar rate of diffusion of constituent **A** per unit area, C_A the molar concentration of constituent **A** and D the diffusivity.

The essential similarity between the three processes is that the rates of transfer of momentum, heat, and mass are all proportional to the concentration gradients of these quantities. In the case of gases the proportionality constants μ/ρ, D_H, and D, all of which have the dimensions length^2/time, all have a physical significance. For liquids the constants cannot be interpreted in a similar manner. The viscosity, thermal conductivity, and diffusivity of a gas will now be considered.

12.2.4. Viscosity

Consider the flow of a gas parallel to a solid surface and the movement of molecules at right angles to this direction through a plane a–a of unit area, parallel to the surface and sufficiently close to it to be within the laminar sublayer (Figure 12.1). During an interval of time dt, molecules with an average velocity $i_1 u_m$ in the Y-direction will pass through the plane (where u_m is the root mean square velocity and i_1 is some fraction of it, depending on the actual distribution of velocities).

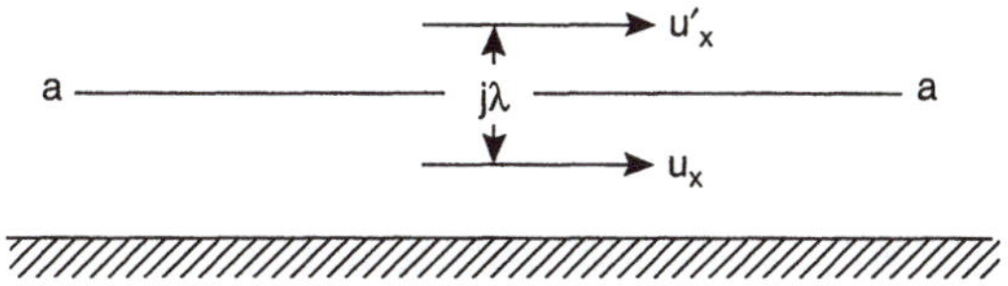

Figure 12.1. Transfer of momentum near a surface

If all these molecules can be considered as having the same component of velocity in the Y-direction, molecules from a volume $i_1 u_m \, dt$ will pass through the plane in time dt.

If **N** is the numerical concentration of molecules close to the surface, the number of molecules passing $= i_1 u_m \mathbf{N} \, dt$.

Thus the rate of passage of molecules $= i_1 u_m \mathbf{N}$.

These molecules have a mean velocity u_x (say) in the X-direction.

Thus the rate at which momentum is transferred across the plane away from the surface

$$= i_1 \mathbf{N} u_m m u_x$$

where m is the mass of each molecule.

By similar reasoning there must be an equivalent stream of molecules also passing through the plane in the opposite direction; otherwise there would be a resultant flow perpendicular to the surface.

If this other stream of molecules has originated at a distance $j\lambda$ from the previous ones, and the mean component of their velocities in the X-direction is u'_x (where λ is the mean free path of the molecules and j is some fraction of the order of unity) then:

The net rate of transfer of momentum away from the surface

$$= i_1 \mathbf{N} u_m m (u_x - u'_x)$$

The gradient of the velocity with respect to the Y-direction

$$= \frac{du_x}{dy} = \frac{(u'_x - u_x)}{j\lambda}$$

since λ is small.

Thus the rate of transfer of momentum per unit area which can be written as:

$$R_y = -i_1 \mathbf{N} u_m m j \lambda \frac{\mathrm{d}u_x}{\mathrm{d}y}$$

$$= -i_1 j \rho u_m \lambda \frac{\mathrm{d}u_x}{\mathrm{d}y} \tag{12.4}$$

(since $\mathbf{N}m = \rho$, the density of the fluid).

But: $$R_y = -\mu \frac{\mathrm{d}u_x}{\mathrm{d}y} \quad \text{(from equation 12.1)}$$

$\therefore$ $$\frac{\mu}{\rho} = i_1 j u_m \lambda \tag{12.5}$$

The value of the product $i_1 j$ has been variously given by different workers, from statistical treatment of the velocities of the molecules;[1] a value of 0.5 will be taken.

Thus: $$\frac{\mu}{\rho} = \frac{1}{2} u_m \lambda \tag{12.6}$$

It is now possible to give a physical interpretation to the Reynolds number:

$$Re = \frac{ud\rho}{\mu} = ud\frac{2}{u_m \lambda} = 2\frac{u}{u_m}\frac{d}{\lambda} \tag{12.7}$$

or *Re* is proportional to the product of the ratio of the flow velocity to the molecular velocity and the ratio of the characteristic linear dimension of the system to the mean free path of the molecules.

From the kinetic theory,[1] $u_m = \sqrt{(8\mathbf{R}T/\pi M)}$ and is independent of pressure, and $\rho\lambda$ is a constant.

Thus, the viscosity of a gas would be expected to be a function of temperature but not of pressure.

12.2.5. Thermal conductivity

Considering now the case where there is a temperature gradient in the Y-direction, the rate of passage of molecules through the unit plane a–a $= i_2 u_m \mathbf{N}$ (where i_2 is some fraction of the order of unity). If the temperature difference between two planes situated a distance $j\lambda$ apart is $(\theta - \theta')$, the net heat transferred as one molecule passes in one direction and another molecule passes in the opposite direction is $c_m(\theta - \theta')$, where c_m is the heat capacity per molecule.

The net rate of heat transfer per unit area $= i_2 u_m \mathbf{N} c_m (\theta - \theta')$.

The temperature gradient $\mathrm{d}\theta/\mathrm{d}y = (\theta' - \theta)/j\lambda$ since λ is small.

Thus the net rate of heat transfer per unit area

$$q = -i_2 j u_m \mathbf{N} c_m \lambda \frac{\mathrm{d}\theta}{\mathrm{d}y}$$

$$= -i_2 j u_m C_v \rho \lambda \frac{\mathrm{d}\theta}{\mathrm{d}y} \tag{12.8}$$

since $\mathbf{N}c_m = \rho C_v$, the specific heat per unit volume of fluid.

and: $$q = -k\frac{d\theta}{dy} \quad \text{(from equation 12.2)}$$

Thus, the thermal diffusivity:

$$\frac{k}{C_p\rho} = i_2 j u_m \lambda \frac{C_v}{C_p} \tag{12.9}$$

From statistical calculations[(1)] the value of $i_2 j$ has been given as $(9\gamma - 5)/8$ (where $\gamma = C_p/C_v$, the ratio of the specific heat at constant pressure to the specific heat at constant volume).

Thus: $$\frac{k}{C_p\rho} = u_m \lambda \frac{9\gamma - 5}{8\gamma} \tag{12.10}$$

The Prandtl number *Pr* is defined as the ratio of the kinematic viscosity to the thermal diffusivity.

Thus: $$Pr = \frac{\mu/\rho}{k/C_p\rho} = \frac{C_p\mu}{k} = \frac{\frac{1}{2}u_m\lambda}{u_m\lambda(9\gamma - 5)/8\gamma}$$

$$= \frac{4\gamma}{9\gamma - 5} \tag{12.11}$$

Values of *Pr* calculated from equation 12.11 are in close agreement with practical figures.

12.2.6. Diffusivity

Considering the diffusion, in the Y-direction, of one constituent **A** of a mixture across the plane a–a, if the numerical concentration is $\mathbf{N}_A$ on one side of the plane and $\mathbf{N}'_A$ on the other side at a distance of $j\lambda$, the net rate of passage of molecules per unit area

$$= i_3 u_m(\mathbf{N}_A - \mathbf{N}'_A)$$

where i_3 is an appropriate fraction of the order of unity.

The rate of mass transfer per unit area

$$= i_3 u_m(\mathbf{N}_A - \mathbf{N}'_A)m$$

The concentration gradient of **A** in the Y-direction

$$= \frac{dC_A}{dy} = \frac{(\mathbf{N}'_A - \mathbf{N}_A)m}{j\gamma}$$

Thus the rate of mass transfer per unit area

$$= -i_3 j\lambda u_m \frac{dC_A}{dy} \tag{12.12}$$

$$= -D\frac{dC_A}{dy} \quad \text{(from equation 12.3)}$$

Thus:

$$D = i_3 j u_m \lambda \tag{12.13}$$

There is, however, no satisfactory evaluation of the product $i_3 j$.

The ratio of the kinematic viscosity to the diffusivity is the Schmidt number, Sc, where:

$$Sc = \frac{(\mu/\rho)}{D} = \frac{\mu}{\rho D} \tag{12.14}$$

It is thus seen that the kinematic viscosity, the thermal diffusivity, and the diffusivity for mass transfer are all proportional to the product of the mean free path and the root mean square velocity of the molecules, and that the expressions for the transfer of momentum, heat, and mass are of the same form.

For liquids the same qualitative forms of relationships exist, but it is not possible to express the physical properties of the liquids in terms of molecular velocities and distances.

12.3. EDDY TRANSFER

In the previous section, the molecular basis for the processes of momentum transfer, heat transfer and mass transfer has been discussed. It has been shown that, in a fluid in which there is a momentum gradient, a temperature gradient or a concentration gradient, the consequential momentum, heat and mass transfer processes arise as a result of the random motion of the molecules. For an ideal gas, the kinetic theory of gases is applicable and the physical properties μ/ρ, $k/C_p\rho$ and D, which determine the transfer rates, are all seen to be proportional to the product of a molecular velocity and the mean free path of the molecules.

A fluid in turbulent flow is characterised by the presence of circulating or eddy currents, and these are responsible for fluid mixing which, in turn, gives rise to momentum, heat or mass transfer when there is an appropriate gradient of the "property" in question. The following simplified analysis of the transport processes in a turbulent fluid is based on the work and ideas of Prandtl. By analogy with the kinetic theory, it is suggested that the relationship between transfer rate and driving force should depend on quantities termed the *eddy kinematic viscosity* E, the *eddy thermal diffusivity* E_H and the *eddy diffusivity* E_D analogous to μ/ρ, $k/C_p\rho$ and D for molecular transport. Extending the analogy further, E, E_H and E_D might be expected to be proportional to the product of a velocity term and a length term, each of which is characteristic of the eddies in the fluid. Whereas μ/ρ, $k/C_p\rho$ and D are all physical properties of the fluid and, for a material of given composition at a specified temperature and pressure have unique values, the eddy terms E, E_H and E_D all depend on the intensity of the eddies. In general, therefore, they are a function of the flow pattern and vary from point to point within the fluid.

In Chapter 11 the concept of a *boundary layer* is discussed. It is suggested that, when a fluid is in turbulent flow over a surface, the eddy currents tend to die out in the region very close to the surface, giving rise to a *laminar sub-layer* in which E, E_H and E_D are all very small. With increasing distance from the surface these quantities become progressively greater, rising from zero in the laminar sub-layer to values considerably in excess of μ/ρ, $k/C_p\rho$ and D in regions remote from the surface. Immediately outside the laminar sub-layer is a buffer zone in which the molecular and eddy terms are of comparable magnitudes. At its outer edge, the eddy terms have become much larger than the molecular terms and the latter can then be neglected—in what can now be regarded as the fully turbulent region.

12.3.1. The nature of turbulent flow

In turbulent flow there is a complex interconnected series of circulating or eddy currents in the fluid, generally increasing in scale and intensity with increase of distance from any boundary surface. If, for steady-state turbulent flow, the velocity is measured at any fixed point in the fluid, both its magnitude and direction will be found to vary in a random manner with time. This is because a random velocity component, attributable to the circulation of the fluid in the eddies, is superimposed on the steady state mean velocity. No net motion arises from the eddies and therefore their time average in any direction must be zero. The instantaneous magnitude and direction of velocity at any point is therefore the vector sum of the steady and fluctuating components.

If the magnitude of the fluctuating velocity component is the same in each of the three principal directions, the flow is termed *isotropic*. If they are different the flow is said to be *anisotropic*. Thus, if the root mean square values of the random velocity components in the X, Y and Z directions are respectively $\sqrt{\overline{u_{Ex}^2}}$, $\sqrt{\overline{u_{Ey}^2}}$ and $\sqrt{\overline{u_{Ez}^2}}$, then for isotropic turbulence:

$$\sqrt{\overline{u_{Ex}^2}} = \sqrt{\overline{u_{Ey}^2}} = \sqrt{\overline{u_{Ez}^2}} \tag{12.15}$$

There are two principal characteristics of turbulence. One is the *scale* which is a measure of the mean size of the eddies, and the other is the *intensity* which is a function of the circulation velocity $(\sqrt{\overline{u_E^2}})$ within the eddies. Both the scale and the intensity increase as the distance from a solid boundary becomes greater. During turbulent flow in a pipe, momentum is transferred from large eddies in the central core through successively smaller eddies as the walls are approached. Eventually, when the laminar sub-layer is reached the eddies die out completely. However, the laminar sub-layer should not be regarded as a completely discrete region, because there is evidence that from time to time eddies do penetrate and occasionally completely disrupt it.

The intensity of turbulence I is defined as the ratio of the mean value of the fluctuating component of velocity to the steady state velocity. For flow in the X-direction parallel to a surface this may be written as:

$$I = \frac{\sqrt{\frac{1}{3}(\overline{u_{Ex}^2} + \overline{u_{Ey}^2} + \overline{u_{Ez}^2})}}{u_x} \tag{12.16}$$

For isotropic turbulence, from equation 12.15, this becomes:

$$I = \frac{\sqrt{\overline{u_E^2}}}{u_x} \tag{12.17}$$

The intensity of turbulence will vary with the geometry of the flow system. Typically, for a fluid flowing over a plane surface or through a pipe, it may have a value of between

0.005 and 0.02. In the presence of packings and turbulence promoting grids, very much higher values (0.05 to 0.1) are common.

The scale of turbulence is given approximately by the diameter of the eddy, or by the distance between the centres of successive eddies. The scale of turbulence is related to the dimensions of the system through which the fluid is flowing. The size of largest eddies is clearly limited by the diameter of the pipe or duct. As the wall is approached their average size becomes less and momentum transfer takes place by interchange through a succession of eddies of progressively smaller size (down to about 1 mm) until they finally die out as the laminar sub-layer is approached near the walls.

An idea of the scale of turbulence can be obtained by measuring instantaneous values of velocities at two different points within the fluid and examining how the correlation coefficient for the two sets of values changes as the distance between the points is increased.

When these are close together, most of the simultaneously measured velocities will relate to fluid in the same eddy and the correlation coefficient will be high. When the points are further apart the correlation coefficient will fall because in an appreciable number of the pairs of measurements the two velocities will relate to different eddies. Thus, the distance apart of the measuring stations at which the correlation coefficient becomes very poor is a measure of scale of turbulence. Frequently, different scales of turbulence can be present simultaneously. Thus, when a fluid in a tube flows past an obstacle or suspended particle, eddies may form in the wake of the particles and their size will be of the same order as the size of the particle; in addition, there will be larger eddies limited in size only by the diameter of the pipe.

12.3.2. Mixing length and eddy kinematic viscosity

PRANDTL[2,3] and TAYLOR[4] both developed the concept of a *mixing length* as a measure of the distance which an element of fluid must travel before it loses its original identity and becomes fully assimilated by the fluid in its new position. Its magnitude will be of the same order as the scale of turbulence or the eddy size. The mixing length is analogous in concept to the *mean free path* of gas molecules which, according to the kinetic theory is the mean distance a molecule travels before it collides with another molecule and loses its original identity.

In turbulent flow over a surface, a velocity gradient, and hence a momentum gradient, exists within the fluid. Any random movement perpendicular to the surface gives rise to a momentum transfer. Elements of fluid with high velocities are brought from remote regions towards the surface and change places with slower moving fluid elements. This mechanism is essentially similar to that involved in the random movement of molecules in a gas. It is therefore suggested that an *eddy kinematic viscosity* E for eddy transport may be defined which is analogous to the kinematic viscosity μ/ρ for molecular transport. Then for isotropic turbulence:

$$E \propto \lambda_E u_E \tag{12.18}$$

where λ_E is the mixing length, and

u_E is some measure of the linear velocity of the fluid in the eddies.

On this basis, the momentum transfer rate per unit area in a direction perpendicular to the surface at some position y is given by:

$$R_y = -E\frac{d(\rho u_x)}{dy} \tag{12.19}$$

For constant density:

$$R_y = -E\rho\frac{du_x}{dy} \tag{12.20}$$

Prandtl has suggested that u_E is likely to increase as both the mixing length λ_E and the modulus of the velocity gradient $|du_x/dy|$ increase. The simplest form of relation between the three quantities is that:

$$u_E \propto \lambda_E \left|\frac{du_x}{dy}\right| \tag{12.21}$$

This is tantamount to saying that the velocity change over a distance equal to the mixing length approximates to the eddy velocity. This cannot be established theoretically but is probably a reasonable assumption.

Combining equations 12.18 and 12.21 gives:

$$E \propto \lambda_E \left\{\lambda_E \left|\frac{du_x}{dy}\right|\right\} \tag{12.22}$$

Arbitrarily putting the proportionality constant equal to unity, then:

$$E = \lambda_E^2 \left|\frac{du_x}{dy}\right| \tag{12.23}$$

Equation 12.23 implies a small change in the definition of λ_E.

In the neighbourhood of a surface, the velocity gradient will be positive and the modulus sign in equation 12.23 may be dropped. On substitution into equation 12.20:

$$R_y = -\rho\lambda_E^2 \left(\frac{du_x}{dy}\right)^2 \tag{12.24}$$

or:

$$\sqrt{\frac{-R_y}{\rho}} = \lambda_E\frac{du_x}{dy} \tag{12.25}$$

In equations 12.19 and 12.20, R_y represents the momentum transferred per unit area and unit time. This momentum transfer *tends* to accelerate the slower moving fluid close to the surface and to retard the faster-moving fluid situated at a distance from the surface. It gives rise to a stress R_y at a distance y from the surface since, from Newton's Law of Motion, force equals rate of change of momentum. Such stresses, caused by the random motion in the eddies, are sometimes referred to as *Reynolds Stresses*.

The problem can also be approached in a slightly different manner. In Figure 12.2, the velocity profile is shown near a surface. At point 1, the velocity is u_x and at point 2, the velocity is u'_x. For an eddy velocity u_{Ey} in the direction perpendicular to the surface, the fluid is transported away from the surface at a mass rate per unit area equal to $u_{Ey}\rho$; this fluid must be replaced by an equal mass of fluid which is transferred in the opposite

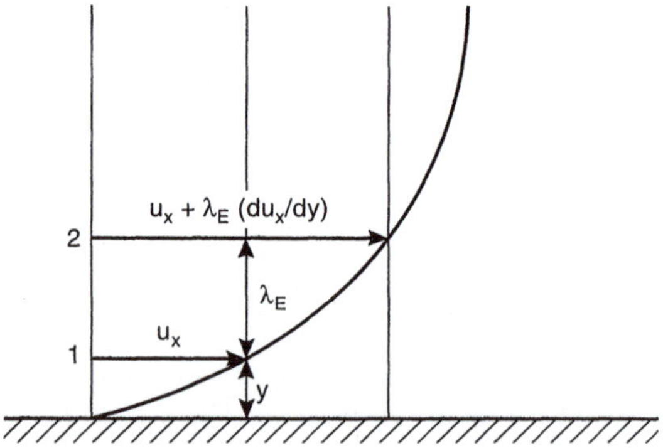

Figure 12.2. Prandtl mixing length

direction. The momentum transferred *away* from the surface per unit and unit time is given by:

$$R_y = \rho u_{Ey}(u_x - u'_x)$$

If the distance between the two locations is approximately equal to the mixing length λ_E, and if the velocity gradient is nearly constant over that distance:

$$\frac{u'_x - u_x}{\lambda_E} \approx \frac{du_x}{dy}$$

Again assuming that:

$$u'_x - u_x \approx u_{Ey}$$

then: $$R = -\rho\lambda_E^2\left(\frac{du_x}{dy}\right)^2 \qquad \text{(equation 12.24)}$$

It is assumed throughout that no mixing takes place with the intervening fluid when an eddy transports fluid elements over a distance equal to the mixing length.

Close to a surface $R_y \rightarrow R_0$, the value at the surface.

The shear stress R acting on the surface must be equal and opposite to that in the fluid in contact with the surface, that is $R = -R_0$, and:

$$\therefore \qquad \sqrt{\frac{R}{\rho}} = \lambda_E\frac{du_x}{dy} \tag{12.26}$$

$\sqrt{R/\rho}$ is known as the *shearing stress velocity* or *friction velocity* and is usually denoted by u^*.

In steady state flow over a plane surface, or close to the wall for flow in a pipe, u^* is constant and equation 12.26 can be integrated provided that the relation between λ_E and y is known. λ_E will increase with y and, if a linear relation is assumed, then:

$$\lambda_E = Ky \tag{12.27}$$

This is the simplest possible form of relation; its use is justified only if it leads to results which are in conformity with experimental results for velocity profiles.

Then: $$u^* = Ky\frac{du_x}{dy} \tag{12.28}$$

On integration: $$\frac{u_x}{u^*} = \frac{1}{K}\ln y + B \quad \text{where } B \text{ is a constant.}$$

or: $$\frac{u_x}{u^*} = \frac{1}{K}\ln\frac{yu^*\rho}{\mu} + B' \tag{12.29}$$

Since $(u^*\rho/\mu)$ is constant, B' will also be constant.
Writing the dimensionless velocity term $u_x/u^* = u^+$ and the dimensionless derivative of y $(yu^*\rho/\mu) = y^+$, then:

$$u^+ = \frac{1}{K}\ln y^+ + B' \tag{12.30}$$

If equation 12.29 is applied to the outer edge of the boundary layer when $y = \delta$ (boundary layer thickness) and $u_x = u_s$ (the stream velocity), then:

$$\frac{u_s}{u^*} = \frac{1}{K}\ln\frac{\delta u^*\rho}{\mu} + B' \tag{12.31}$$

Subtracting equation 12.29 from equation 12.31:

$$\frac{u_s - u_x}{u^*} = \frac{1}{K}\ln\frac{\delta}{y} \tag{12.32}$$

Using experimental results for flow of fluids over both smooth and rough surfaces, NIKURADSE[5,6] found K to have a value of 0.4.

Thus:

$$\frac{u_s - u_x}{u^*} = 2.5\ln\frac{\delta}{y} \tag{12.33}$$

For fully developed flow in a pipe, $\delta = r$ and u_s is the velocity at the axis, and then:

$$\frac{u_s - u_x}{u^*} = 2.5\ln\frac{r}{y} \tag{12.34}$$

Equation 12.34 is known as the *velocity defect law* (Figure 12.3).

The application to pipe flow is not strictly valid because u^* $(= \sqrt{R/\rho})$ is constant only in regions close to the wall. However, equation 12.34 appears to give a reasonable approximation to velocity profiles for turbulent flow, except near the pipe axis. The errors in this region can be seen from the fact that on differentiation of equation 12.34 and putting $y = r$, the velocity gradient on the centre line is $2.5u^*/r$ instead of zero.

Inserting $K = 0.4$ in equation 12.27 gives the relation between mixing length (λ_E) and distance (y) from the surface:

$$\frac{\lambda_E}{y} = 0.4 \tag{12.35}$$

Equation 12.35 applies only in those regions where eddy transfer dominates, i.e. outside both the laminar sub-layer and the buffer layer (see below).

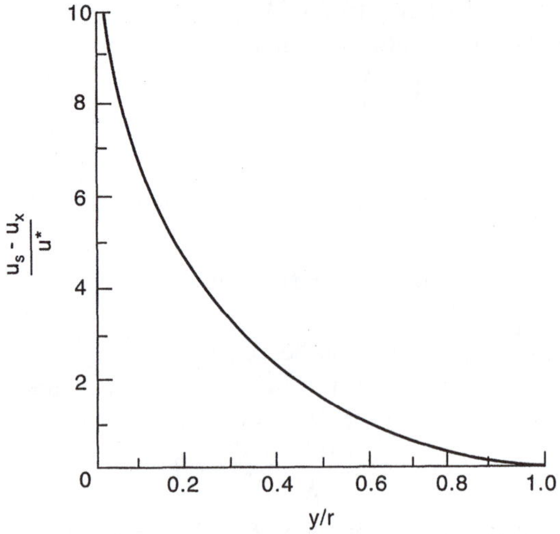

Figure 12.3. Velocity defect law

12.4. UNIVERSAL VELOCITY PROFILE

For fully developed turbulent flow in a pipe, the whole of the flow may be regarded as lying within the boundary layer. The cross-section can then conveniently be divided into three regions:

(a) The *turbulent core* in which the contribution of eddy transport is so much greater than that of molecular transport that the latter can be neglected.
(b) The *buffer layer* in which the two mechanisms are of comparable magnitude.
(c) The *laminar sub-layer* in which turbulent eddies have effectively died out so that only molecular transport need be considered.

It is now possible to consider each of these regions in turn and to develop a series of equations to represent the velocity over the whole cross section of a pipe. Together, they constitute the *Universal Velocity Profile.*

12.4.1. The turbulent core

Equation 12.30 applies in the turbulent core, except near the axis of the pipe where the shear stress is markedly different from that at the walls. Inserting the value of 0.4 for K:

$$u^+ = 2.5 \ln y^+ + B' \tag{12.36}$$

Plotting experimental data on velocity profiles as u^+ against $\log y^+$ (as in Figure 12.4) gives a series of parallel straight lines of slope 2.5 and with intercepts at $\ln y^+ = 0$ varying with the relative roughness of the surface (e/d). For smooth surfaces $(e/d = 0)$, $B' = 5.5$. B' becomes progressively smaller as the relative roughness increases.

Thus for a smooth pipe:

$$u^+ = 2.5 \ln y^+ + 5.5 \tag{12.37}$$

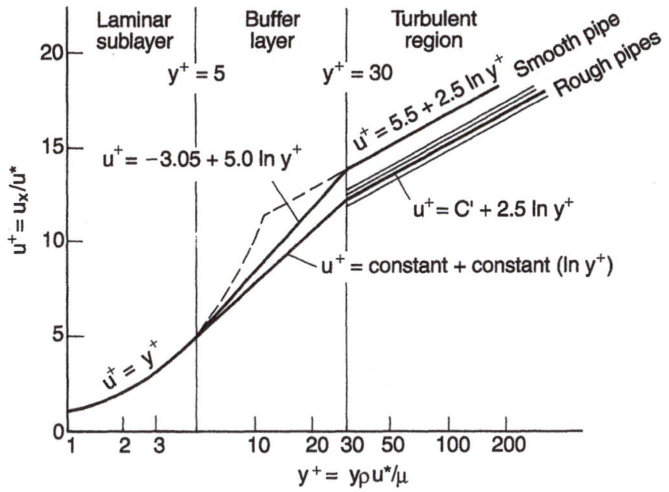

Figure 12.4. The universal velocity profile

and for a rough pipe:

$$u^+ = 2.5 \ln y^+ + B' \tag{12.38}$$

where B' is a function of e/d and is less than 5.5.

Equations 12.37 and 12.38 correlate experimental data well for values of y^+ exceeding 30.

12.4.2. The laminar sub-layer

In the laminar sub-layer, turbulence has died out and momentum transfer is attributable solely to viscous shear. Because the layer is thin, the velocity gradient is approximately linear and equal to u_b/δ_b where u_b is the velocity at the outer edge of a laminar sub-layer of thickness δ_b (see Chapter 11).

$$R = \mu \frac{u_b}{\delta_b} = \mu \frac{u_x}{y}$$

Then:

$$u^{*2} = \frac{R}{\rho} = \frac{\mu u_x}{\rho y}$$

$\therefore$

$$\frac{u_x}{u^*} = \frac{y u^* \rho}{\mu} \tag{12.39}$$

or:

$$u^+ = y^+ \tag{12.40}$$

This relationship holds reasonably well for values of y^+ up to about 5, and it applies to both rough and smooth surfaces.

12.4.3. The buffer layer

The buffer layer covers the intermediate range $5 < y^+ < 30$. A straight line may be drawn to connect the curve for the laminar sub-layer (equation 12.40) at $y^+ = 5$ with the line for the turbulent zone for flow over a smooth surface at $y^+ = 30$ (equation 12.37) (see Figure 12.4). The data for the intermediate region are well correlated by this line whose equation must be of the form:

$$u^+ = a \ln y^+ + a' \tag{12.41}$$

The line passes through the points ($y^+ = 5$, $u^+ = 5$) and ($y^+ = 30$, $u^+ = 2.5 \ln 30 + 5.5$) and therefore a and a' may be evaluated to give:

$$u^+ = 5.0 \ln y^+ - 3.05 \tag{12.42}$$

An equation, similar in form to equation 12.42, will be applicable for rough surfaces but the values of the two constants will be different.

12.4.4. Velocity profile for all regions

For a smooth pipe, therefore, the complete Universal Velocity Profile is given by:

$$0 < y^+ < 5 \qquad u^+ = y^+ \qquad \text{(equation 12.40)}$$

$$5 < y^+ < 30 \qquad u^+ = 5.0 \ln y^+ - 3.05 \qquad \text{(equation 12.42)}$$

$$y^+ > 30 \qquad u^+ = 2.5 \ln y^+ + 5.5 \qquad \text{(equation 12.37)}$$

A simplified form of velocity profile is obtained by neglecting the existence of the buffer layer and assuming that there is a sudden transition from the laminar sub-layer to an eddy-dominated turbulent regime. The transition will occur at the point of intersection of the curves (shown by broken lines in Figure 12.4) representing equations 12.40 and 12.37. Solving the equations simultaneously gives $y^+ = 11.6$ as the point of intersection.

For flow in a pipe, the dimensionless distance y^+ from the walls and the corresponding velocity u^+ may be expressed in terms of three dimensionless quantities; the pipe Reynolds number $ud\rho/\mu$, the pipe friction factor ϕ $(= R/\rho u^2)$ and the ratio y/d or y/r.

Since:

$$u^* = \sqrt{\frac{R}{\rho}} = \sqrt{\frac{R}{\rho u^2}}\,u = \phi^{1/2} u$$

$$u^+ = \frac{u_x}{u^*} = \frac{u_x}{u}\phi^{-1/2} \tag{12.43}$$

$$y^+ = \frac{y u^* \rho}{\mu} = \frac{y}{d}\frac{u^*}{u}\frac{u d \rho}{\mu} = \frac{y}{d}\phi^{1/2} Re \tag{12.44}$$

12.4.5. Velocity gradients

By differentiation of equations 12.40, 12.42 and 12.37 respectively, the corresponding values of the gradient du^+/dy^+ are obtained:

$$0 < y^+ < 5 \qquad \frac{du^+}{dy^+} = 1 \tag{12.45}$$

$$5 < y^+ < 30 \qquad \frac{du^+}{dy^+} = \frac{5.0}{y^+} \tag{12.46}$$

$$y^+ > 30 \qquad \frac{du^+}{dy^+} = \frac{2.5}{y^+} \tag{12.47}$$

Equations 12.45 and 12.47 are applicable to both rough and smooth surfaces; equation 12.46 is valid only for a smooth surface ($e/d \to 0$).

Velocity gradients are directly related to du^+/dy^+,

since:
$$\frac{du_x}{dy} = \frac{du^+}{dy^+}\frac{\rho u^{*2}}{\mu} \tag{12.48}$$

Thus:

$0 < y^+ < 5$
$$\frac{du_x}{dy} = \frac{\rho u^{*2}}{\mu} = \frac{R}{\mu} \tag{12.49}$$

$5 < y^+ < 30$ (smooth surfaces)
$$\frac{du_x}{dy} = \frac{5.0\rho u^{*2}}{y^+\mu} = \frac{5.0}{y^+}\frac{R}{\mu} \tag{12.50a}$$

$$\frac{du_x}{dy} = 5.0\frac{u^*}{y} = 5.0\phi^{1/2}\frac{u}{y} \tag{12.50b}$$

$y^+ > 30$
$$\frac{du_x}{dy} = \frac{2.5\rho u^{*2}}{y^+\mu} = \frac{2.5}{y^+}\frac{R}{\mu} \tag{12.51a}$$

$$\frac{du_x}{dy} = 2.5\frac{u^*}{y} = 2.5\phi^{1/2}\frac{u}{y} \tag{12.51b}$$

12.4.6. Laminar sub-layer and buffer layer thicknesses

On the basis of the Universal Velocity Profile, the laminar sub-layer extends from $y^+ = 0$ to $y^+ = 5$ and the buffer layer from $y^+ = 5$ to $y^+ = 30$.

From the definition of y^+:

$$y = y^+\frac{\mu}{u^*\rho}$$

$$= y^+\frac{\mu}{ud\rho}\frac{u}{u^*}d \tag{12.52}$$

Thus:
$$\frac{y}{d} = Re^{-1}\left(\frac{R}{\rho u^2}\right)^{-1/2} y^+ \tag{12.53}$$

$$= Re^{-1}\phi^{-1/2}y^+ \tag{12.53a}$$

Putting $y^+ = 5$, the laminar sub-layer thickness (δ_b) is given by:

$$\frac{\delta_b}{d} = 5Re^{-1}\phi^{-1/2} \tag{12.54}$$

The buffer layer extends to $y^+ = 30$, where:

$$\frac{y}{\delta} = 30Re^{-1}\phi^{-1/2} \tag{12.55}$$

If the buffer layer is neglected, it has been shown (Section 12.4.4) that the laminar sub-layer will extend to $y^+ = 11.6$ giving:

$$\frac{\delta_b}{d} = 11.6Re^{-1}\phi^{-1/2} \tag{12.56}$$

Using the Blasius equation (equation 11.46) to give an approximate value for $R/\rho u^2$ for a smooth pipe:

$$\phi = 0.0396Re^{-1/4} \quad \text{(from equation 11.46)}$$

$$\frac{\delta_b}{d} = 58Re^{-7/8} \tag{12.57}$$

The equation should be compared with equation 11.49 obtained using Prandtl's simplified approach to boundary layer theory which also disregards the existence of the buffer layer:

$$\frac{\delta_b}{d} = 62Re^{-7/8} \quad \text{(equation 11.49)}$$

Similarly, the velocity u_b at the edge of the laminar sub-layer is given by:

$$u^+ = \frac{u_b}{u^*} = 11.6 \tag{12.58}$$

(since $u^+ = y^+$ in the laminar sub-layer).

Thus using the Blasius equation:

$$\frac{u_b}{u} = 11.6\phi^{1/2} \tag{12.59}$$

Substituting for ϕ in terms of Re:

$$\frac{u_b}{u} = 2.32Re^{-1/8} \tag{12.60}$$

Again, equation 12.60 can be compared with equation 11.40, derived on the basis of the Prandtl approach:

$$\frac{u_b}{u} = 2.49Re^{-1/8} \quad \text{(equation 11.47)}$$

12.4.7. Variation of eddy kinematic viscosity

Since the buffer layer is very close to the wall, R_y can be replaced by R_0.

and:

$$R_0 = -(\mu + E\rho)\frac{du_x}{dy} \tag{12.61}$$

Thus:

$$\frac{-R_0}{\rho} = u^{*2} = \left(\frac{\mu}{\rho} + E\right)\frac{du_x}{dy} \tag{12.62}$$

For $5 < y^+ < 30$, substituting for du_x/dy from equation 12.50*a* for a smooth surface:

$$u^{*2} = \left(\frac{\mu}{\rho} + E\right)\frac{5\rho u^{*2}}{\mu y^+}$$

giving:

$$E = \frac{\mu}{\rho}\left(\frac{y^+}{5} - 1\right) \tag{12.63}$$

Thus E varies from zero at $y^+ = 5$, to $(5\mu/\rho)$ at $y^+ = 30$.

For values y^+ greater than 30, μ/ρ is usually neglected in comparison with E. The error is greatest at $y^+ = 30$ where $E/(\mu/\rho) = 5$, but it rapidly becomes smaller at larger distances from the surface.

Thus, for $y^+ > 30$, from equation 12.62:

$$u^{*2} = E\frac{du_x}{dy} \tag{12.64}$$

Substituting for du_x/dy from equation 12.51a:

$$u^{*2} = E\frac{2.5\rho u^{*2}}{\mu y^+}$$

Thus:

$$E = 0.4y^+\frac{\mu}{\rho} \tag{12.65a}$$

$$= 0.4u^* y. \tag{12.65b}$$

It should be noted that equation 12.65a gives $E = 12(\mu/\rho)$ at $y^+ = 30$, compared with $5(\mu/\rho)$ from equation 12.63. This arises because of the discontinuity in the Universal Velocity Profile at $y^+ = 30$.

12.4.8. Approximate form of velocity profile in turbulent region

A simple approximate form of the relation between u^+ and y^+ for the turbulent flow of a fluid in a pipe of circular cross-section may be obtained using the Prandtl one-seventh power law and the Blasius equation for a smooth surface. These two equations have been shown (Section 11.4) to be mutually consistent.

The Prandtl one-seventh power law gives:

$$\frac{u_x}{u_{CL}} = \left(\frac{y}{r}\right)^{1/7} \quad \text{(equation 3.59)}$$

Then:

$$u^+ = \frac{u_x}{u^*} = \frac{u_{CL}}{u^*}\left(\frac{y}{r}\right)^{1/7} = \frac{u_{CL}}{r^{1/7}}\frac{1}{u^*}\left(\frac{y^+\mu}{\rho u^*}\right)^{1/7} \tag{12.66}$$

The Blasius relation between friction factor and Reynolds number for turbulent flow is:

$$\phi = \frac{R}{\rho u^2} = 0.0396Re^{-1/4} \quad \text{(equation 3.11)}$$

Thus:

$$\frac{R}{\rho} = u^{*2} = 0.0396\rho^{-1/4}\mu^{1/4}(2r)^{-1/4}u^{1.75}$$

Again, from the Prandtl one-seventh power law:

$$\frac{u}{u_{CL}} = \frac{49}{60} \quad \text{(equation 3.63)}$$

Thus:

$$\frac{u_{CL}}{r^{1/7}} = (0.0396)^{-1/1.75}2^{1/7}\rho^{1/7}\mu^{-1/7}u^{*8/7}\frac{60}{49}$$

$$= 8.56\rho^{1/7}\mu^{-1/7}u^{*8/7}$$

Substituting in equation 12.66:

$$u^+ = (8.56\rho^{1/7}\mu^{-1/7}u^{*\,8/7})\frac{1}{u^*}\left(\frac{\mu}{\rho u^*}\right)^{1/7} y^{+\,1/7}$$

i.e.:
$$u^+ = 8.56y^{+\,1/7} \tag{12.67}$$

In Table 12.1, the values of u^+ calculated from equation 12.67 are compared with those given by the universal velocity profile (equations 12.37, 12.40 and 12.42). It will be seen that there is almost exact correspondence at $y^+ = 1000$ and differences are less than 6 per cent in the range $30 < y^+ < 3000$.

Table 12.1. Comparison of values of u^+ calculated from equation 12.67 with those given by the universal velocity profile

y^+	u^+ from equation 12.67	u^+ from UVP	% Difference (based on column 3)
10	11.89	8.46	+40.5
15	12.60	10.49	+20.1
20	13.13	11.93	+10.1
25	13.56	14.65	−7.4
30	13.92	14.00	−0.0
100	16.53	17.01	−2.9
300	19.33	19.76	−2.2
1,000	22.96	22.76	+0.0
2,000	25.35	24.50	+3.5
3,000	26.87	25.52	+5.3
4,000	27.99	26.23	+6.7
5,000	28.90	26.79	+7.9
10,000	31.91	28.52	+11.9
100,000	44.34	34.28	+29.3

12.4.9. Effect of curvature of pipe wall on shear stress

Close to the wall of a pipe, the effect of the curvature of the wall has been neglected and the shear stress in the fluid has been taken to be independent of the distance from the wall. However, this assumption is not justified near the centre of the pipe.

As shown in Chapter 3, the shear stress varies linearly over the cross-section rising from zero at the axis of the pipe to a maximum value at the walls.

A force balance taken over the whole cross section gives:

$$-\frac{\mathrm{d}P}{\mathrm{d}x}\pi r^2 = -R_0 2\pi r$$

Thus:
$$-R_0 = -\frac{\mathrm{d}P}{\mathrm{d}x}\frac{r}{2} \tag{12.68}$$

Taking a similar force balance over the central core of fluid lying at distances greater than y from the wall; that is for a plug of radius $(r - y)$:

$$-R_y = -\frac{\mathrm{d}P}{\mathrm{d}x}\frac{r-y}{2} \tag{12.69}$$

Thus:
$$\frac{R_y}{R_0} = 1 - \frac{y}{r} \tag{12.70}$$

At radius $(r - y)$, equation 12.61 becomes:

$$R_y = -(\mu + E\rho)\frac{du_x}{dy} \tag{12.71}$$

Neglecting μ/ρ compared with E, and substituting for R_y from equation 12.70, then:

$$\frac{-R_0}{\rho}\left(1 - \frac{y}{r}\right) = E\frac{du_x}{dy}$$

This leads to:
$$E = 0.4y^+\frac{\mu}{\rho}\left(1 - \frac{y}{r}\right) \tag{12.72a}$$

$$= 0.4u^* y\left(1 - \frac{y}{r}\right) \tag{12.72b}$$

$$= 0.4\phi^{1/2}uy\left(1 - \frac{y}{r}\right) \tag{12.72c}$$

12.5. FRICTION FACTOR FOR A SMOOTH PIPE

Equation 12.37 can be used in order to calculate the friction factor $\phi = R/\rho u^2$ for the turbulent flow of fluid in a pipe. It is first necessary to obtain an expression for the mean velocity u of the fluid from the relation:

$$u = \frac{\int_0^r [2\pi(r - y)\,dy\,u_x]}{\pi r^2}$$

$$= 2\int_0^1 u_x\left(1 - \frac{y}{r}\right) d\left(\frac{y}{r}\right) \tag{12.73}$$

The velocity at the pipe axis u_s is obtained by putting $y = r$ into equation 12.37.

Thus:
$$u_s = u^*\left(2.5 \ln\frac{r\rho u^*}{\mu} + 5.5\right) \tag{12.74}$$

This is not strictly justified because equation 12.74 gives a finite, instead of zero, velocity gradient when applied at the centre of the pipe.

Substituting for u_x in equation 12.73 from equation 12.34:

$$u = 2\int_0^1 \left(u_s - 2.5u^* \ln\frac{r}{y}\right)\left(1 - \frac{y}{r}\right) d\left(\frac{y}{r}\right)$$

$$\frac{u}{u_s} = 2\int_0^1 \left(1 + 2.5\frac{u^*}{u_s} \ln\frac{y}{r}\right)\left(1 - \frac{y}{r}\right) d\left(\frac{y}{r}\right)$$

$$= 2\left[\frac{y}{r} - \frac{1}{2}\left(\frac{y}{r}\right)^2\right]_0^1 + 5.0\frac{u^*}{u_s}\left\{\left[\left(\ln\frac{y}{r}\right)\left[\frac{y}{r} - \frac{1}{2}\left(\frac{y}{r}\right)^2\right]\right]_0^1 - \int_0^1 \left(\frac{y}{r}\right)^{-1}\left[\frac{y}{r} - \frac{1}{2}\left(\frac{y}{r}\right)^2\right] d\left(\frac{y}{r}\right)\right\}$$

$$= 1 + 5.0\frac{u^*}{u_s}\left\{0 - \left[\left(\frac{y}{r}\right) - \frac{1}{4}\left(\frac{y}{r}\right)^2\right]_0^1\right\}$$

$$= 1 + 5\frac{u^*}{u_s}\left(-\frac{3}{4}\right)$$

$$= 1 - 3.75\frac{u^*}{u_s} \tag{12.75}$$

Substituting into equation 12.74:

$$u + 3.75u^* = u^*\left\{2.5\ln\left[\left(\frac{d\rho u}{\mu}\right)\left(\frac{r}{d}\right)\left(\frac{u^*}{u}\right)\right] + 5.5\right\}$$

$$\frac{u}{u^*} = 2.5\ln\left\{(Re)\frac{u^*}{u}\right\}$$

Now: $$\phi = \frac{R}{\rho u^2} = \left(\frac{u^*}{u}\right)^2$$

$\therefore$ $$\phi^{-1/2} = 2.5\ln[(Re)\phi^{1/2}] \tag{12.76}$$

The experimental results for ϕ as a function of Re closely follow equation 12.76 modified by a correction term of 0.3 to give:

$$\phi^{-1/2} = 2.5\ln[(Re)\phi^{1/2}] + 0.3 \tag{12.77}$$

The correction is largely associated with the errors involved in using equation 12.74 at the pipe axis.

Equation 12.77 is identical to equation 3.12.

Example 12.1

Air flows through a smooth circular duct of internal diameter 250 mm at an average velocity of 15 m/s. Calculate the fluid velocity at points 50 mm and 5 mm from the wall. What will be the thickness of the laminar sub-layer if this extends to $u^+ = y^+ = 5$? The density and viscosity of air may be taken as 1.10 kg/m^3 and 20×10^{-6} N s/m^2 respectively.

Solution

Reynolds number: $Re = \dfrac{(0.250 \times 15 \times 1.10)}{(20 \times 10^{-6})} = 2.06 \times 10^5$

Hence, from Figure 3.7: $\dfrac{R}{\rho u^2} = 0.0018$

$$u_s = \frac{u}{0.817} = \left(\frac{15}{0.817}\right) = 18.4 \text{ m/s}$$

$$u^* = u\sqrt{\frac{R}{\rho u^2}} = 15\sqrt{0.0018} = 0.636 \text{ m/s}$$

At 50 mm from the wall: $\dfrac{y}{r} = \left(\dfrac{0.050}{0.125}\right) = 0.40$

Hence, from equation 12.34:

$$u_x = u_s + 2.5u^* \ln\left(\frac{y}{r}\right)$$
$$= 18.4 + (2.5 \times 0.636 \ln 0.4)$$
$$= \underline{\underline{16.9 \text{ m/s}}}$$

At 5 mm from the wall: $y/r = 0.005/0.125 = 0.04$

Hence: $$u_x = 18.4 + 2.5 \times 0.636 \ln 0.04$$
$$= \underline{\underline{13.3 \text{ m/s}}}$$

The thickness of the laminar sub-layer is given by equation 12.54:

$$\delta_b = \frac{5d}{Re\sqrt{(R/\rho u^2)}}$$
$$= \frac{(5 \times 0.250)}{(2.06 \times 10^5\sqrt{(0.0018)})}$$
$$= 1.43 \times 10^{-4} \text{ m}$$

or: $$\underline{\underline{0.143 \text{ mm}}}$$

12.6. EFFECT OF SURFACE ROUGHNESS ON SHEAR STRESS

Experiments have been carried out on artificially roughened surfaces in order to determine the effect of obstructions of various heights[5,6]. Experimentally, it has been shown that the shear force is not affected by the presence of an obstruction of height e unless:

$$\frac{u_e e \rho}{\mu} > 40 \tag{12.78}$$

where u_e is the velocity of the fluid at a distance e above the surface.

If the obstruction lies entirely within the laminar sub-layer the velocity u_e is given by:

$$R = \mu\left(\frac{\mathrm{d}u_x}{\mathrm{d}y}\right)_{y=0}$$
$$= \mu\left(\frac{u_e}{e}\right), \quad \text{approximately}$$

The shearing stress velocity:

$$u^* = \sqrt{\frac{R}{\rho}} = \sqrt{\frac{\mu u_e}{\rho e}}$$

so that: $$u_e = \frac{\rho e}{\mu}u^{*2}$$

Thus: $$\frac{u_e e \rho}{\mu} = \frac{e\rho}{\mu}u^{*2}\frac{e\rho}{\mu} = \left(\frac{e\rho u^*}{\mu}\right)^2 \tag{12.79}$$

$e\rho u^*/\mu$ is known as the *roughness Reynolds number*, Re_r.

For the flow of a fluid in a pipe:

$$u^* = \sqrt{\frac{R}{\rho}} = u\phi^{1/2}$$

where u is the mean velocity over the whole cross-section. Re_r will now be expressed in terms of the three dimensionless groups used in the friction chart.

Thus:

$$Re_r = \frac{e\rho u^*}{\mu}$$

$$= \left(\frac{d\rho u}{\mu}\right)\left(\frac{e}{d}\right)\phi^{1/2}$$

$$= Re\left(\frac{e}{d}\right)\phi^{1/2} \qquad (12.80)$$

The shear stress should then be unaffected by the obstruction if $Re_r < \sqrt{40}$, that is if $Re_r < 6.5$.

If the surface has a number of closely spaced obstructions, however, all of the same height e, the shear stress is affected when $Re_r >$ about 3. If the obstructions are of varying heights, with e as the arithmetic mean, the shear stress is increased if $Re_r >$ about 0.3, because the effect of one relatively large obstruction is greater than that of several small ones.

For hydrodynamically smooth pipes, through which fluid is flowing under turbulent conditions, the shear stress is given approximately by the Blasius equation:

$$\phi = \frac{R}{\rho u^2} \propto Re^{-1/4} \qquad \text{(from equation 11.46)}$$

so that:

$$R \propto u^{1.75}$$

and is independent of the roughness.

For smooth pipes, the frictional drag at the surface is known as *skin friction*. With rough pipes, however, an additional drag known as *form drag* results from the eddy currents caused by impact of the fluid on the obstructions and, when the surface is very rough, it becomes large compared with the skin friction. Since form drag involves dissipation of kinetic energy, the losses are proportional to the square of the velocity of the fluid, so that $R \propto u^2$. This applies when $Re_r > 50$.

Thus, when: $Re_r < 0.3, \quad R \propto u^{1.75}$

when: $Re_r > 50, \quad R \propto u^2$

and when: $0.3 < Re_r < 50, R \propto u^w$ where $1.75 < w < 2$.

When the thickness of the laminar sub-layer is large compared with the height of the obstructions, the pipe behaves as a smooth pipe (when $e < \delta_b/3$). Since the thickness of the laminar sub-layer decreases as the Reynolds number is increased, a surface which is hydrodynamically smooth at low Reynolds numbers may behave as a rough surface at higher values. This explains the shapes of the curves obtained for ϕ plotted against Reynolds number (Figure 3.7). The curves, for all but the roughest of pipes, follow the curve for the smooth pipe at low Reynolds numbers and then diverge at higher values. The greater the roughness of the surface, the lower is the Reynolds number at which the curve starts to diverge. At high Reynolds numbers, the curves for rough pipes become parallel to the Reynolds number axis, indicating that skin friction is negligible and $R \propto u^2$. Under these conditions, the shear stress can be calculated from equation 3.14.

Nikuradse's data[5,6] for rough pipes gives:

$$u^+ = 2.5 \ln\left(\frac{y}{e}\right) + 8.5 \tag{12.81}$$

12.7. SIMULTANEOUS MOMENTUM, HEAT AND MASS TRANSFER

It has been seen that when there is a velocity gradient in a fluid, the turbulent eddies are responsible for transferring momentum from the regions of high velocity to those of low velocity; this gives rise to shear stresses within the field. It has been suggested that the eddy kinematic viscosity E can be written as the product of an eddy velocity u_E and a mixing length λ_E giving:

$$E \propto \lambda_E u_E \qquad \text{(equation 12.18)}$$

If u_E is expressed as the product of the mixing length and the modulus of velocity gradient and if the proportionality constant is equal to unity:

$$E = \lambda_E^2 \left|\frac{\mathrm{d}u_x}{\mathrm{d}y}\right| \qquad \text{(equation 12.23)}$$

If there is a temperature gradient within the fluid, the eddies will be responsible for heat transfer and an eddy thermal diffusivity E_H may be defined in a similar way. It is suggested that, since the mechanism of transfer of heat by eddies is essentially the same as that for transfer of momentum, E_H is related to mixing length and velocity gradient in a similar manner.

Thus:

$$E_H = \lambda_E^2 \left|\frac{\mathrm{d}u_x}{\mathrm{d}y}\right| \tag{12.82}$$

On a similar basis an eddy diffusivity for mass transfer E_D can be defined for systems in which concentration gradients exist as:

$$E_D = \lambda_E^2 \left|\frac{\mathrm{d}u_x}{\mathrm{d}y}\right| \tag{12.83}$$

E/E_H is termed the *Turbulent Prandtl Number* and E/E_D the *Turbulent Schmidt Number*.

E, E_H and E_D are all nearly equal and therefore both the dimensionless numbers are approximately equal to unity.

Thus for the eddy transfer of heat:

$$q_y = -E_H \frac{\mathrm{d}(C_p \rho \theta)}{\mathrm{d}y} = -\lambda_E^2 \left|\frac{\mathrm{d}u_x}{\mathrm{d}y}\right| \frac{\mathrm{d}(C_p \rho \theta)}{\mathrm{d}y} \tag{12.84}$$

and similarly for mass transfer:

$$N_A = -E_D \frac{\mathrm{d}C_A}{\mathrm{d}y} = -\lambda_E^2 \left|\frac{\mathrm{d}u_x}{\mathrm{d}y}\right| \frac{\mathrm{d}C_A}{\mathrm{d}y} \tag{12.85}$$

In the neighbourhood of a surface du_x/dy will be positive and thus:

$$R_y = -\lambda_E^2 \frac{du_x}{dy} \frac{d(\rho u_x)}{dy} \tag{12.86}$$

$$q_y = -\lambda_E^2 \frac{du_x}{dy} \frac{d(C_p \rho \theta)}{dy} \tag{12.87}$$

and:

$$N_A = -\lambda_E^2 \frac{du_x}{dy} \frac{dC_A}{dy} \tag{12.88}$$

For conditions of constant density, equation 12.25 gives:

$$\sqrt{\frac{-R_y}{\rho}} = \lambda_E \frac{du_x}{dy} \quad \text{(equation 12.25)}$$

When molecular and eddy transport both contribute significantly, it may be assumed as a first approximation that their effects are additive. Then:

$$R_y = -\left(\frac{\mu}{\rho} + E\right) \frac{d(\rho u_x)}{dy} = -\mu \frac{du_x}{dy} - \lambda_E^2 \rho \left(\frac{du_x}{dy}\right)^2 \tag{12.89}$$

$$q_y = -\left(\frac{k}{C_p \rho} + E_H\right) \frac{d(C_p \rho \theta)}{dy} = -k \frac{d\theta}{dy} - \lambda_E^2 \rho C_p \left(\frac{du_x}{dy}\right) \left(\frac{d\theta}{dy}\right) \tag{12.90}$$

$$N_A = -(D + E_D) \frac{dC_A}{dy} = -D \frac{dC_A}{dy} - \lambda_E^2 \left(\frac{du_x}{dy}\right) \left(\frac{dC_A}{dy}\right) \tag{12.91}$$

and:

$$\frac{\lambda_E}{y} \approx 0.4 \quad \text{(equation 12.35)}$$

Whereas the kinematic viscosity μ/ρ, the thermal diffusivity $k/C_p\rho$, and the diffusivity D are physical properties of the system and can therefore be taken as constant provided that physical conditions do not vary appreciably, the eddy coefficients E, E_H, and E_D will be affected by the flow pattern and will vary throughout the fluid. Each of the eddy coefficients is proportional to the square of the mixing length. The mixing length will

Table 12.2. Relations between physical properties

	Molecular processes only	*Molecular and eddy transfer together*	*Eddy transfer predominating*
Momentum transfer	$R_y = -\frac{\mu}{\rho}\frac{d(\rho u_x)}{dy}$	$R_y = -\left(\frac{\mu}{\rho} + E\right)\frac{d(\rho u_x)}{dy}$	$R_y = -E\frac{d(\rho u_x)}{dy}$
Heat transfer	$q_y = -\frac{k}{C_p\rho}\frac{d(C_p\rho\theta)}{dy}$	$q_y = -\left(\frac{k}{C_p\rho} + E_H\right)\frac{d(C_p\rho\theta)}{dy}$	$q_y = -E_H\frac{d(C_p\rho\theta)}{dy}$
Mass transfer	$N_A = -D\frac{dC_A}{dy}$	$N_A = -(D + E_D)\frac{dC_A}{dy}$	$N_A = -E_D\frac{dC_A}{dy}$
		where $E \approx E_H \approx E_D \approx \lambda_E^2\left\lvert\frac{du_x}{dy}\right\rvert$ and $\lambda_E \approx 0.4y$	

normally increase with distance from a surface, and the eddy coefficients will therefore increase rapidly with position.

The relations are summarised in Table 12.2.

Before the equations given in columns 2 and 3 of Table 12.2 can be integrated, it is necessary to know how E, E_H and E_D vary with position. In Section 12.4 an estimate has been made of how E varies with the dimensionless distance $y^+ \left(= \dfrac{yu^*\rho}{\mu}\right)$ from the surface, by using the concept of the Universal Velocity Profile for smooth surfaces.

For molecular transport alone (the laminar sub-layer), $y^+ < 5$:

$$E \rightarrow 0$$

For combined molecular and eddy transport (the buffer zone), $5 < y < 30$:

$$E = \frac{\mu}{\rho}\left(\frac{y^+}{5} - 1\right) \qquad \text{(equation 12.63)}$$

For the region where the eddy mechanism predominates, $y^+ > 30$:

$$E = \frac{\mu}{\rho} \cdot \frac{y^+}{2.5} \qquad \text{(from equation 12.65}a\text{)}$$

Then, using the approximation $E = E_H = E_D$, these values may be inserted in the equations in Table 12.2. However, the limits of the buffer zone ($5 < y^+ < 30$) may be affected because of differences in the thicknesses of the boundary layers for momentum, heat and mass transfer (Chapter 11).

Mass Transfer

In the buffer zone: $5 < y^+ < 30$, and:

$$N_A = -(D + E_D)\frac{\mathrm{d}C_A}{\mathrm{d}y}$$

$$= -\left[D + \frac{\mu}{\rho}\left(\frac{y^+}{5} - 1\right)\right]\frac{\mathrm{d}C_A}{\mathrm{d}y^+} \cdot \frac{u^*\rho}{\mu}$$

$$= -u^*\left[Sc^{-1} + \frac{y^+}{5} - 1\right]\frac{\mathrm{d}C_A}{\mathrm{d}y^+}$$

Thus:

$$N_A \int_{y_1^+}^{y_2^+} \frac{\mathrm{d}y^+}{\dfrac{y^+}{5} + (Sc^{-1} - 1)} = -u^* \int_{C_{A1}}^{C_{A2}} \mathrm{d}C_A$$

giving:

$$N_A = \frac{u^*(C_{A1} - C_{A2})}{5 \ln\left[\dfrac{y_2^+ + 5(Sc^{-1} - 1)}{y_1^+ + 5(Sc^{-1} - 1)}\right]} \qquad (12.92)$$

In the fully turbulent region: $y^+ > 30$, and:

$$N_A = -E_D\frac{\mathrm{d}C_A}{\mathrm{d}y}$$

$$= -\frac{\mu}{\rho}\frac{y^+}{2.5}\frac{\mathrm{d}C_A}{\mathrm{d}y^+} \cdot \frac{u^*\rho}{\mu}$$

$$= -u^* \frac{y^+}{2.5} \frac{dC_A}{dy^+}$$

giving:
$$N_A = \frac{u^*(C_{A1} - C_{A2})}{2.5 \ln \dfrac{y_2^+}{y_1^+}} \qquad (12.93)$$

Heat transfer

The corresponding equations for heat transfer can be obtained in an exactly analogous manner.

In the buffer zone: $5 < y^+ < 30$, and:

$$q = \frac{C_p \rho u^*(T_1 - T_2)}{5 \ln \left[\dfrac{y_2^+ + 5(Pr^{-1} - 1)}{y_1^+ + 5(Pr^{-1} - 1)}\right]} \qquad (12.94)$$

In the fully turbulent region: $y^+ > 30$

$$q = \frac{C_p \rho u^*(T_1 - T_2)}{2.5 \ln \dfrac{y_2^+}{y_1^+}} \qquad (12.95)$$

12.8. REYNOLDS ANALOGY

12.8.1. Simple form of analogy between momentum, heat and mass transfer

The simple concept of the Reynolds Analogy was first suggested by REYNOLDS[7] to relate heat transfer rates to shear stress, but it is also applicable to mass transfer. It is assumed that elements of fluid are brought from remote regions to the surface by the action of the turbulent eddies; the elements do not undergo any mixing with the intermediate fluid through which they pass, and they instantaneously reach equilibrium on contact with the interfacial layers. An equal volume of fluid is, at the same time, displaced in the reverse direction. Thus in a flowing fluid there is a transference of momentum and a simultaneous transfer of heat if there is a temperature gradient, and of mass if there is a concentration gradient. The turbulent fluid is assumed to have direct access to the surface and the existence of a buffer layer and laminar sub-layer is neglected. Modification of the model has been made by TAYLOR[4] and PRANDTL[8,9] to take account of the laminar sub-layer. Subsequently, the effect of the buffer layer has been incorporated by applying the *universal velocity profile*.

Consider the equilibrium set up when an element of fluid moves from a region at high temperature, lying outside the boundary layer, to a solid surface at a lower temperature if no mixing with the intermediate fluid takes place. Turbulence is therefore assumed to persist right up to the surface. The relationship between the rates of transfer of momentum and heat can then be deduced as follows (Figure 12.5).

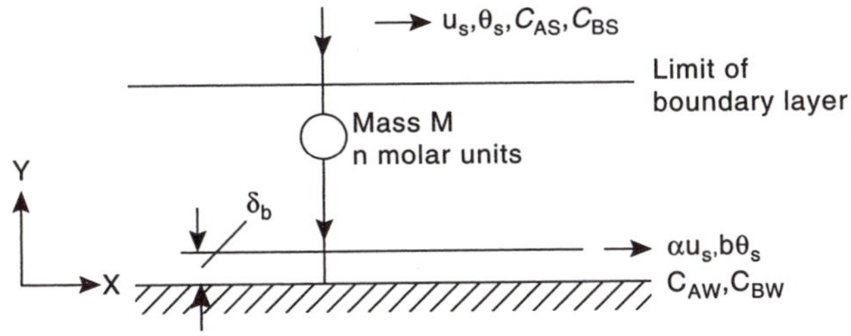

Figure 12.5. The Reynolds analogy — momentum, heat and mass transfer

Consider the fluid to be flowing in a direction parallel to the surface (X-direction) and for momentum and heat transfer to be taking place in a direction at right angles to the surface (Y-direction positive away from surface).

Suppose a mass **M** of fluid situated at a distance from the surface to be moving with a velocity u_s in the X-direction. If this element moves to the surface where the velocity is zero, it will give up its momentum $\mathbf{M}u_s$ in time t, say. If the temperature difference between the element and the surface is θ_s and C_p is the specific heat of the fluid, the heat transferred to the surface will be $\mathbf{M}\ C_p\theta_s$. If the surface is of area A, the rate of heat transfer is given by:

$$\frac{\mathbf{M}C_p\theta_s}{At} = -q_0 \tag{12.96}$$

where $-q_0$ is the heat transferred to the surface per unit area per unit time (NB — the negative sign has been introduced as the positive direction is away from the surface).

If the shear stress at the surface is R_0, it will equal the rate of change in momentum, per unit area.

Thus:
$$\frac{\mathbf{M}u_s}{At} = -R_0. \tag{12.97}$$

A similar argument can be applied to the mass transfer process when a concentration gradient exists. Thus, if the molar concentration of **A** remote from the surface is C_{As} and at the surface it is C_{Aw}, the moles of **A** transferred to the surface will be $(\mathbf{M}/\rho)(C_{As} - C_{Aw})$, if the density ($\rho$) can be assumed to be constant over the range of concentrations encountered. Thus the moles of **A** transferred to the surface per unit area and unit time $(-N_A)_{y=0}$ is given by:

$$\frac{1}{At}\frac{\mathbf{M}}{\rho}(C_{As} - C_{Aw}) = (-N_A)_{y=0} \tag{12.98}$$

Dividing equation 12.96 by 12.97:

$$\frac{C_p\theta_s}{u_s} = \frac{-q_0}{-R_0} \tag{12.99}$$

or:
$$\frac{-R_0}{u_s} = \frac{-q_0}{C_p\theta_s} \tag{12.100}$$

Again dividing equation 12.98 by equation 12.97:

$$\frac{C_{As} - C_{Aw}}{\rho u_s} = \frac{(-N_A)_{y=0}}{-R_0}$$

or:

$$\frac{-R_0}{\rho u_s} = \frac{(-N_A)_{y=0}}{C_{As} - C_{Aw}} \qquad (12.101)$$

Now R_0 (the shear stress in the fluid at the surface) is equal and opposite to R, the shear stress acting on the surface, $-q_0/\theta_s$ is by definition the heat transfer coefficient at the surface (h), and $(-N_A)_{y=0}/(C_{As} - C_{Aw})$ is the mass transfer coefficient (h_D). Then dividing both sides of equation 12.100 by ρu_s and of equation 12.101 by u_s to make them dimensionless:

$$\frac{R}{\rho u_s^2} = \frac{h}{C_p \rho u_s} = St \qquad (12.102)$$

where St denotes the Stanton Number $(h/C_p \rho u_s)$,

and:

$$\frac{R}{\rho u_s^2} = \frac{h_D}{u_s} \qquad (12.103)$$

h_D/u_s is sometimes referred to as the Stanton number for mass transfer.

Thus:

$$\frac{h_D}{u_s} = \frac{h}{C_p \rho u_s} \qquad (12.104)$$

or:

$$h_D = \frac{h}{C_p \rho} \qquad (12.105)$$

Equation 12.105 is often referred to as the *Lewis Relation*. It provides an approximate method for evaluating a mass transfer coefficient if the heat transfer coefficient is known. The assumption that the turbulent eddies can penetrate right up to the surface is justified however only in special circumstances and the problem is considered further in the next section.

For flow in a smooth pipe, the friction factor for turbulent flow is given approximately by the Blasius equation and is proportional to the Reynolds number (and hence the velocity) raised to a power of $-\frac{1}{4}$. From equations 12.102 and 12.103, therefore, the heat and mass transfer coefficients are both proportional to $u_s^{0.75}$.

The application of the analogies to the problems of heat and mass transfer to plane surfaces and to pipe walls for fully developed flow is discussed later.

Example 12.2

Air at 330 K, flowing at 10 m/s, enters a pipe of inner diameter 25 mm, maintained at 415 K. The drop of static pressure along the pipe is 80 N/m^2 per metre length. Using the Reynolds analogy between heat transfer and fluid friction, estimate the air temperature 0.6 m along the pipe.

Solution

From equations 12.102 and 3.18:

$$-\Delta P = 4\ (R/\rho u^2)(l/d)\rho u^2 = 4\ (h/C_p \rho u)(l/d)\rho u^2$$

(using the mean pipeline velocity u in the Reynolds analogy)

Then, in SI units.

$$-\Delta P/l = 80 = 4[h/(C_p \times 10)]\ (1/0.025) \times 10^2$$

and: $$h = 0.05C_p \quad \text{W/m}^2\ \text{K}$$

In passing through a length dL of pipe, the air temperature rises from T to $T + \text{d}T$.

The heat taken up per unit time by the air,

$$\text{d}Q = (\rho u C_p)(\pi d^2/4)\,\text{d}T$$

The density of air at 1 bar and 330 K $= (MP)/(\mathbf{R}T) = \dfrac{29 \times 10^5}{8314 \times 330} = 1.057\ \text{kg/m}^3$

Thus; $$\text{d}Q = (1.057 \times 10 \times C_p)[\pi(0.025)^2/4]\,\text{d}T$$

$$= 0.0052C_p\,\text{d}T \quad \text{W} \qquad \text{(i)}$$

The heat transferred through the pipe wall is also given by:

$$\text{d}Q = h(\pi d\,\text{d}L)(415 - T)$$

$$= (0.05C_p)(\pi \times 0.025\,\text{d}L)(415 - T) \qquad \text{(ii)}$$

$$= 0.039C_p(415 - T)\,\text{d}L \quad \text{W}$$

Equating (i) and (ii):

$$\int_{330}^{T_o} \frac{\text{d}T}{415 - T} = 0.75 \int_0^{0.6} \text{d}L$$

giving $$\ln[85/(415 - T_o)] = 0.45$$

and: $$85/(415 - T_o) = \text{e}^{0.45} = 1.57$$

and: $$\underline{\underline{T_o = 360\ \text{K}}}$$

12.8.2. Mass transfer with bulk flow

When the mass transfer process deviates significantly from equimolecular counter-diffusion, allowance must be made for the fact that there may be a very large difference in the molar rates of transfer of the two components. Thus, in a gas absorption process, there will be no transfer of the insoluble component **B** across the interface and only the soluble component **A** will be transferred. This problem will now be considered in relation to the Reynolds Analogy. However, it gives manageable results only if physical properties such as density are taken as constant and therefore results should be applied with care.

Consider the movement of an element of fluid consisting of n molar units of a mixture of two constituents **A** and **B** from a region outside the boundary layer, where the molecular concentrations are C_{As} and C_{Bs}, to the surface where the corresponding concentrations are C_{Aw} and C_{Bw}. The total molar concentration is everywhere C_T. The transfer is effected in a time t and takes place at an area A of surface.

There is no net transference of the component **B**. When n molar units of material are transferred from outside the boundary layer to the surface:

$$\text{Transfer of } \mathbf{A} \text{ towards surface} = n\frac{C_{As}}{C_T}$$

$$\text{Transfer of } \mathbf{B} \text{ towards surface} = n\frac{C_{Bs}}{C_T}$$

In this case the molar rate of transfer of **B** away from the surface is equal to the transfer towards the surface.

$$\text{Transfer of } \mathbf{B} \text{ away from surface} = n\frac{C_{Bs}}{C_T}$$

Associated transfer of **A** away from surface

$$= n\frac{C_{Bs}}{C_T}\frac{C_{Aw}}{C_{Bw}}$$

Thus the net transfer of **A** towards the surface

$$-N'_A At = n\left(\frac{C_{As}}{C_T} - \frac{C_{Bs}}{C_T}\frac{C_{Aw}}{C_{Bw}}\right)$$

$$= n\left(\frac{C_T C_{As} - C_{Aw}C_{As} - C_{Aw}C_T + C_{Aw}C_{As}}{C_T(C_T - C_{Aw})}\right)$$

$$= n\frac{(C_{As} - C_{Aw})}{C_{Bw}}$$

It is assumed that the total molar concentration is everywhere constant. Thus the rate of transfer per unit area and unit time is given by:

$$-N'_A = \frac{n(C_{As} - C_{Aw})}{C_{Bw}At} \tag{12.106}$$

The net transfer of momentum per unit time

$$= -R_0 A = \frac{n\rho u_s}{C_T t} \tag{12.107}$$

where ρ is taken as the mean mass density of the fluid.

$$\therefore \quad -R_0 = \frac{n\rho u_s}{C_T t A} \tag{12.108}$$

Dividing equations 12.106 and 12.108 gives:

$$\frac{N'_A}{R_0} = \frac{C_{As} - C_{Aw}}{\rho u_s}\frac{C_T}{C_{Bw}} \tag{12.109}$$

Writing $R_0 = -R$ and defining the mass transfer coefficient by the relation, then:

$$\frac{-N'_A}{C_{As} - C_{Aw}} = h_D \tag{12.110}$$

$$\frac{h_D}{u_s}\frac{C_{Bw}}{C_T} = \frac{R}{\rho u_s^2} \tag{12.111}$$

Thus, there is a direct proportionality between the momentum transfer and that portion of the mass transfer which is *not* attributable to bulk flow.

For heat transfer:

$$\frac{R}{\rho u_s^2} = \frac{h}{C_p \rho u_s} \qquad \text{(equation 12.102)}$$

Thus for simultaneous heat transfer and mass transfer with bulk flow giving rise to no net transfer of component **B**, combination of equations 12.111 and 12.102 gives:

$$\frac{h_D}{u_s}\frac{C_{Bw}}{C_T} = \frac{h}{C_p \rho u_s}$$

or:

$$h_D \frac{C_{Bw}}{C_T} = \frac{h}{C_p \rho} \tag{12.112}$$

Equation 12.112 is the form of the *Lewis Relation* which is applicable to mass transfer with bulk flow.

12.8.3. Taylor–Prandtl modification of Reynolds analogy for heat transfer and mass transfer

The original Reynolds analogy involves a number of simplifying assumptions which are justifiable only in a limited range of conditions. Thus it was assumed that fluid was transferred from outside the boundary layer to the surface without mixing with the intervening fluid, that it was brought to rest at the surface, and that thermal equilibrium was established. Various modifications have been made to this simple theory to take account of the existence of the laminar sub-layer and the buffer layer close to the surface.

TAYLOR[4] and PRANDTL[8,9] allowed for the existence of the laminar sub-layer but ignored the existence of the buffer layer in their treatment and assumed that the simple Reynolds analogy was applicable to the transfer of heat and momentum from the main stream to the edge of the laminar sub-layer of thickness δ_b. Transfer through the laminar sub-layer was then presumed to be attributable solely to molecular motion.

If αu_s and $b\theta_s$ are the velocity and temperature, respectively, at the edge of the laminar sub-layer (see Figure 12.5), applying the Reynolds analogy (equation 12.99) for transfer across the turbulent region:

$$\frac{-q_0}{-R_0} = \frac{C_p(\theta_s - b\theta_s)}{u_s - \alpha u_s} \tag{12.113}$$

The rate of transfer of heat by conduction through the laminar sub-layer from a surface of area A is given by:

$$-q_0 A = \frac{k b \theta_s A}{\delta_b} \tag{12.114}$$

The rate of transfer of momentum is equal to the shearing force and therefore:

$$-R_0 A = \frac{\mu \alpha u_s A}{\delta_b} = RA \tag{12.115}$$

Dividing equations 12.114 and 12.115 gives:

$$\frac{-q_0}{-R_0} = \frac{k b \theta_s}{\mu \alpha u_s} \tag{12.116}$$

Thus from equations 12.113 and 12.116:

$$\frac{k b \theta_s}{\mu \alpha u_s} = \frac{C_p(1-b)\theta_s}{(1-\alpha)u_s}$$

$$\therefore \quad \frac{Pr(1-b)}{b} = \frac{(1-\alpha)}{\alpha}$$

$$\therefore \quad \frac{b}{\alpha} = \frac{1}{\alpha + (1-\alpha)Pr^{-1}}$$

Substituting in equation 12.116:

$$\frac{-q_0}{-R_0} = \frac{C_p\theta_s}{u_s}\frac{1}{1+\alpha(Pr-1)} = \frac{h\theta_s}{R}$$

or:

$$St = \frac{h}{C_p\rho u_s} = \frac{R/\rho u_s^2}{1+\alpha(Pr-1)} \qquad (12.117)$$

The quantity α, which is the ratio of the velocity at the edge of the laminar sub-layer to the stream velocity, was evaluated in Chapter 11 in terms of the Reynolds number for flow over the surface. For flow over a plane surface, from Chapter 11:

$$\alpha = 2.1Re_x^{-0.1} \qquad \text{(equation 11.35)}$$

where Re_x is the Reynolds number $u_s x\rho/\mu$, x being the distance from the leading edge of the surface.

For flow through a pipe of diameter d (Chapter 11):

$$\alpha = 2.0Re^{-1/8} \qquad \text{(equation 11.48)}$$

where Re is the Reynolds number $ud\rho/\mu$.
Alternatively, from equation 12.59:

$$\alpha = \frac{u_b}{u} = 11.6\phi^{1/2} \approx 11.6\left(\frac{R}{\rho u_s^2}\right)^{1/2}$$

For mass transfer to a surface, a similar relation to equation 12.117 can be derived for equimolecular counterdiffusion except that the Prandtl number is replaced by the Schmidt number. It follows that:

$$\frac{h_D}{u_s} = \frac{R/\rho u_s^2}{1+\alpha(Sc-1)} \qquad (12.118)$$

where Sc is the Schmidt number $(\mu/\rho D)$. It is possible also to derive an expression to take account of bulk flow, but many simplifying assumptions must be made and the final result is not very useful.

It is thus seen that by taking account of the existence of the laminar sub-layer, correction factors are introduced into the simple Reynolds analogy.

For heat transfer, the factor is $[1+\alpha(Pr-1)]$ and for mass transfer it is $[1+\alpha(Sc-1)]$.

There are two sets of conditions under which the correction factor approaches unity:

(i) For gases both the Prandtl and Schmidt groups are approximately unity, and therefore the simple Reynolds analogy is closely followed. Furthermore, the Lewis relation which is based on the simple analogy would be expected to hold closely for gases. The Lewis relation will also hold for any system for which the Prandtl and Schmidt numbers are equal. In this case, the correction factors for heat and mass transfer will be approximately equal.

(ii) When the fluid is highly turbulent, the laminar sub-layer will become very thin and the velocity at the edge of the laminar sub-layer will be small. In these circumstances again, the correction factor will approach unity.

Equations 12.117 and 12.118 provide a means of expressing the mass transfer coefficient in terms of the heat transfer coefficient.

From equation 12.117:

$$\frac{R}{\rho u_s^2} = \frac{h}{C_p \rho u_s}[1 + \alpha(Pr - 1)] \tag{12.119}$$

From equation 12.118:

$$\frac{R}{\rho u_s^2} = \frac{h_D}{u_s}[1 + \alpha(Sc - 1)] \tag{12.120}$$

Thus:

$$h_D = \frac{h}{C_p \rho} \frac{1 + \alpha(Pr - 1)}{1 + \alpha(Sc - 1)} \tag{12.121}$$

Equation 12.121 is a modified form of the *Lewis Relation*, which takes into account the resistance to heat and mass transfer in the laminar sub-layer.

12.8.4. Use of universal velocity profile in Reynolds analogy

In the Taylor–Prandtl modification of the theory of heat transfer to a turbulent fluid, it was assumed that the heat passed directly from the turbulent fluid to the laminar sub-layer and the existence of the buffer layer was neglected. It was therefore possible to apply the simple theory for the boundary layer in order to calculate the heat transfer. In most cases, the results so obtained are sufficiently accurate, but errors become significant when the relations are used to calculate heat transfer to liquids of high viscosities. A more accurate expression can be obtained if the temperature difference across the buffer layer is taken into account. The exact conditions in the buffer layer are difficult to define and any mathematical treatment of the problem involves a number of assumptions. However, the conditions close to the surface over which fluid is flowing can be calculated approximately using the *universal velocity profile*.[10]

The method is based on the calculation of the total temperature difference between the fluid and the surface, by adding the components attributable to the laminar sub-layer, the buffer layer and the turbulent region. In the steady state, the heat flux (q_0) normal to the surface will be constant if the effects of curvature are neglected.

Laminar sub-layer ($0 < y^+ < 5$)

Since the laminar sub-layer is thin, the temperature gradient may be assumed to be approximately linear (see also Section 11.6.1).

Thus:

$$q_0 = -k\frac{\theta_5}{y_5} \tag{12.122}$$

where y_5 and θ_5 are respectively the values of y and θ at $y^+ = 5$. As before, the temperature scale is so chosen that the surface temperature is zero.

By definition:
$$y^+ = \frac{yu^*\rho}{\mu}$$

Thus:
$$y_5 = \frac{5\mu}{u^*\rho}$$

Substituting in equation 12.122:

$$\theta_5 = \frac{-q_0}{k}\frac{5\mu}{u^*\rho} \tag{12.123}$$

Buffer layer (5 < y^+ < 30)

It has been shown that it is reasonable to assume that the eddy kinematic viscosity E and the eddy thermal diffusivity E_H are equal. The variation of E through the buffer zone is given by:

$$E = \frac{\mu}{\rho}\left(\frac{y^+}{5} - 1\right) \quad \text{(equation 12.67)}$$

The heat transfer rate in the buffer zone is given by:

$$q_0 = -(k + E_H C_p\rho)\frac{d\theta}{dy} \quad \text{(from equation 12.94)}$$

Substituting from equation 12.63 and putting $E_H = E$:

$$-q_0 = \left[k + C_p\rho\frac{\mu}{\rho}\left(\frac{y^+}{5} - 1\right)\right]\frac{d\theta}{dy} \tag{12.124}$$

From the definition of y^+:

$$\frac{d\theta}{dy} = \frac{u^*\rho}{\mu}\frac{d\theta}{dy^+} \tag{12.125}$$

Substituting from equation 12.125 into equation 12.124:

$$-q_0 = \rho u^*\left[\frac{k}{\mu} + C_p\left(\frac{y^+}{5} - 1\right)\right]\frac{d\theta}{dy^+}$$

$$= \frac{C_p\rho u^*}{5}\left[\frac{5k}{C_p\mu} - 5 + y^+\right]\frac{d\theta}{dy^+}$$

$$\therefore \quad \frac{d\theta}{dy^+} = \frac{-5q_0}{C_p\rho u^*[5(Pr^{-1} - 1) + y^+]} \tag{12.126}$$

Integrating between the limits of $y^+ = 5$ and $y^+ = 30$:

$$\theta_{30} - \theta_5 = \frac{-5q_0}{C_p\rho u^*}\ln(5Pr + 1). \tag{12.127}$$

Turbulent zone ($y^+ > 30$)

In this region the Reynolds analogy can be applied. Equation 12.113 becomes:

$$\frac{-q_0}{-R_0} = \frac{C_p(\theta_s - \theta_{30})}{u_s - u_{30}}$$

The velocity at $y^+ = 30$ for a smooth surface is given by:

$$\frac{u_x}{u^*} = u^+ = 5.0 \ln y^+ - 3.05 \qquad \text{(equation 12.42)}$$

$$\therefore \qquad u_{30} = u^*(5.0 \ln 30 - 3.05)$$

and:

$$\theta_s - \theta_{30} = \frac{-q_0}{C_p \rho u^{*2}}(u_s - 5.0u^* \ln 30 + 3.05u^*) \qquad (12.128)$$

The overall temperature difference θ_s is obtained by addition of equations 12.123, 12.127, and 12.128:

$$\text{Thus} \quad \theta_s = \frac{-q_0}{C_p \rho u^{*2}}\left[\frac{5C_p\mu}{k}u^* + 5u^* \ln(5Pr + 1) + u_s - 5.0u^* \ln 30 + 3.05u^*\right]$$

$$= \frac{-q_0}{C_p \rho u^{*2}}\left[5Pr\, u^* + 5u^* \ln\left(\frac{Pr}{6} + \frac{1}{30}\right) + u_s + 3.05u^*\right] \qquad (12.129)$$

i.e.:

$$\frac{u^{*2}}{u_s^2} = \frac{-q_0}{C_p \rho u_s \theta_s}\left\{1 + 5\frac{u^*}{u_s}\left[Pr + \ln\left(\frac{Pr}{6} + \frac{1}{30}\right) + \ln 5 - 1\right]\right\} \qquad \text{(since } 5 \ln 5 = 8.05\text{)}$$

and:

$$\frac{R}{\rho u_s^2} = \frac{h}{C_p \rho u_s}\left\{1 + 5\sqrt{\frac{R}{\rho u_s^2}}\left[(Pr - 1) + \ln\left(\frac{5}{6}Pr + \frac{1}{6}\right)\right]\right\} \qquad (12.130)$$

where:

$$\frac{h}{C_p \rho u_s} = St \quad \text{(the Stanton number)}$$

A similar expression can also be derived for mass transfer in the absence of bulk flow:

$$\frac{R}{\rho u_s^2} = \frac{h_D}{u_s}\left\{1 + 5\sqrt{\frac{R}{\rho u_s^2}}\left[(Sc - 1) + \ln\left(\frac{5}{6}Sc + \frac{1}{6}\right)\right]\right\} \qquad (12.131)$$

12.8.5. Flow over a plane surface

The simple Reynolds analogy gives a relation between the friction factor $R/\rho u_s^2$ and the Stanton number for heat transfer:

$$\frac{R}{\rho u_s^2} = \frac{h}{C_p \rho u_s} \qquad \text{(equation 12.102)}$$

This equation can be used for calculating the point value of the heat transfer coefficient by substituting for $R/\rho u_s^2$ in terms of the Reynolds group Re_x using equation 11.39:

$$\frac{R}{\rho u_s^2} = 0.03 Re_x^{-0.2} \qquad \text{(equation 11.39)}$$

and:

$$St = \frac{h}{C_p \rho u_s} = 0.03 Re_x^{-0.2} \qquad (12.132)$$

Equation 12.132 gives the point value of the heat transfer coefficient. If the whole surface is effective for heat transfer, the mean value is given by:

$$St = \frac{h}{C_p \rho u_s} = \frac{1}{x}\int_0^x 0.03 Re_x^{-0.2}\,\mathrm{d}x$$

$$= 0.037 Re_x^{-0.2} \qquad (12.133)$$

These equations take no account of the existence of the laminar sub-layer and therefore give unduly high values for the transfer coefficient, especially with liquids. The effect of the laminar sub-layer is allowed for by using the Taylor–Prandtl modification:

$$St = \frac{h}{C_p \rho u_s} = \frac{R/\rho u_s^2}{1 + \alpha(Pr - 1)} \qquad \text{(equation 12.117)}$$

where:

$$\alpha = \frac{u_b}{u_s} = 2.1 Re_x^{-0.1} \qquad \text{(equation 11.35)}$$

Thus:

$$St = Nu_x Re_x^{-1} Pr^{-1} = \frac{0.03 Re_x^{-0.2}}{1 + 2.1 Re_x^{-0.1}(Pr - 1)} \qquad (12.134)$$

This expression will give the point value of the Stanton number and hence of the heat transfer coefficient. The mean value over the whole surface is obtained by integration. No general expression for the mean coefficient can be obtained and a graphical or numerical integration must be carried out after the insertion of the appropriate values of the constants.

Similarly, substitution may be made from equation 11.39 into equation 12.130 to give the point values of the Stanton number and the heat transfer coefficient, thus:

$$St = \frac{h}{C_p \rho u_s} = \frac{0.03 Re_x^{-0.2}}{1 + 0.87 Re_x^{-0.1}[(Pr - 1) + \ln(\frac{5}{6}Pr + \frac{1}{6})]} \qquad (12.135)$$

Mean values may be obtained by graphical or numerical integration.

The same procedure may be used for obtaining relationships for mass transfer coefficients, for equimolecular counterdiffusion or where the concentration of the non-diffusing constituent is small:

$$\frac{R}{\rho u_s^2} = \frac{h_D}{u_s} \qquad \text{(equation 12.103)}$$

For flow over a plane surface, substitution from equation 11.32 gives:

$$\frac{h_D}{u_s} = 0.03 Re_x^{-0.2} \qquad (12.136)$$

Equation 12.136 gives the point value of h_D. The mean value over the surface is obtained in the same manner as equation 12.133 as:

$$\frac{h_D}{u_s} = 0.037Re_x^{-0.2} \tag{12.137}$$

For mass transfer through a stationary second component:

$$\frac{R}{\rho u_s^2} = \frac{h_D}{u_s}\frac{C_{Bw}}{C_T} \qquad \text{(equation 12.111)}$$

The correction factor C_{Bw}/C_T must then be introduced into equations 12.136 and 12.137.

The above equations are applicable only when the Schmidt number Sc is very close to unity or where the velocity of flow is so high that the resistance of the laminar sub-layer is small. The resistance of the laminar sub-layer can be taken into account, however, for equimolecular counterdiffusion or for low concentration gradients by using equation 12.118.

$$\frac{h_D}{u_s} = \frac{R/\rho u_s^2}{1 + \alpha(Sc - 1)} \qquad \text{(equation 12.118)}$$

Substitution for $R/\rho u_s^2$ and α using equations 11.39 and 11.35 gives:

$$\frac{h_D}{u_s} = Sh_x Re_x^{-1} Sc^{-1} = \frac{0.03Re_x^{-0.2}}{1 + 2.1Re_x^{-0.1}(Sc - 1)} \tag{12.138}$$

12.8.6. Flow in a pipe

For the inlet length of a pipe in which the boundary layers are forming, the equations in the previous section will give an approximate value for the heat transfer coefficient. It should be remembered, however, that the flow in the boundary layer at the entrance to the pipe may be streamline and the point of transition to turbulent flow is not easily defined. The results therefore are, at best, approximate.

In fully developed flow, equations 12.102 and 12.117 can be used, but it is preferable to work in terms of the mean velocity of flow and the ordinary pipe Reynolds number Re. Furthermore, the heat transfer coefficient is generally expressed in terms of a driving force equal to the difference between the bulk fluid temperature and the wall temperature. If the fluid is highly turbulent, however, the bulk temperature will be quite close to the temperature θ_s at the axis.

Into equations 12.99 and 12.117, equation 3.63 may then be substituted:

$$u = 0.817u_s \qquad \text{(equation 3.63)}$$

$$\frac{u_b}{u} = 2.49Re^{-1/8} \qquad \text{(equation 11.47)}$$

$$\frac{R}{\rho u^2} = 0.0396Re^{-1/4} \qquad \text{(equation 11.46)}$$

Firstly, using the simple Reynolds analogy (equation 12.102):

$$\begin{aligned}\frac{h}{C_p\rho u} &= \left(\frac{h}{C_p\rho u_s}\right)\left(\frac{u_s}{u}\right)\\ &= \left(\frac{R}{\rho u_s^2}\right)\left(\frac{u_s}{u}\right)\\ &= \left(\frac{R}{\rho u^2}\right)\left(\frac{u}{u_s}\right)\\ &= 0.032Re^{-1/4}\end{aligned} \qquad (12.139)$$

Then, using the Taylor–Prandtl modification (equation 12.117):

$$\begin{aligned}\frac{h}{C_p\rho u} &= \frac{0.032Re^{-1/4}}{1+(u_b/u)(u/u_s)(Pr-1)}\\ &= \frac{0.032Re^{-1/4}}{1+2.0Re^{-1/8}(Pr-1)}\end{aligned} \qquad (12.140)$$

Finally, using equation 12.135:

$$\begin{aligned}St = \frac{h}{C_p\rho u} &= \frac{0.817(R/\rho u^2)}{1+0.817\sqrt{(R/\rho u^2)}\,5[(Pr-1)+\ln(\frac{5}{6}Pr+\frac{1}{6})]}\\ &= \frac{0.032Re^{-1/4}}{1+0.817Re^{-1/8}[(Pr-1)+\ln(\frac{5}{6}Pr+\frac{1}{6})]}\end{aligned} \qquad (12.141)$$

For mass transfer, the equations corresponding to equations 12.137 and 12.138 are obtained in the same way as the analogous heat transfer equations.

Thus, using the simple Reynolds analogy for equimolecular counterdiffusion:

$$\frac{h_D}{u} = 0.032Re^{-1/4} \qquad (12.142)$$

and for diffusion through a stationary gas:

$$\left(\frac{h_D}{u}\right)\left(\frac{C_{Bw}}{C_T}\right) = 0.032Re^{-1/4} \qquad (12.143)$$

Using the Taylor–Prandtl form for equimolecular counterdiffusion or low concentration gradients:

$$\frac{h_D}{u} = \frac{0.032Re^{-1/4}}{1+2.0Re^{-1/8}(Sc-1)} \qquad (12.144)$$

Example 12.2

Water flows at 0.50 m/s through a 20 mm tube lined with β-naphthol. What is the mass transfer coefficient if the Schmidt number is 2330?

Solution

Reynolds number: $Re = (0.020 \times 0.50 \times 1000/1 \times 10^{-3}) = 10{,}000$

From equation 12.144:

$$\frac{h_D}{u} = 0.032Re^{-1/4}[1 + 2.0(Sc - 1)Re^{-1/8}]^{-1}$$

$$\therefore \quad h_D = 0.032 \times 0.50 \times 0.1[1 + 2 \times 2329 \times 0.316]^{-1}$$

$$= \underline{\underline{1.085 \times 10^{-6}\ \text{kmol/m}^2\ \text{s}\ (\text{kmol/m}^3)}} = \underline{\underline{1.085 \times 10^{-6}\ \text{m/s}}}$$

Example 12.3

Calculate the rise in temperature of water which is passed at 3.5 m/s through a smooth 25 mm diameter pipe, 6 m long. The water enters at 300 K and the tube wall may be assumed constant at 330 K. The following methods may be used:

(a) the simple Reynolds analogy (equation 12.139);
(b) the Taylor–Prandtl modification (equation 12.140);
(c) the universal velocity profile (equation 12.141);
(d) $Nu = 0.023Re^{0.8}Pr^{0.33}$ (equation 9.64).

Solution

Taking the fluid properties at 310 K and assuming that fully developed flow exists, an approximate solution will be obtained neglecting the variation of properties with temperature.

$$Re = \frac{0.025 \times 3.5 \times 1000}{0.7 \times 10^{-3}} = 1.25 \times 10^5$$

$$Pr = \frac{4.18 \times 10^3 \times 0.7 \times 10^{-3}}{0.65} = 4.50$$

(a) *Reynolds analogy*

$$\frac{h}{C_p \rho u} = 0.032Re^{-0.25} \qquad \text{(equation 12.139)}$$

$$h = [4.18 \times 1000 \times 1000 \times 3.5 \times 0.032(1.25 \times 10^5)^{-0.25}]$$

$$= 24{,}902\ \text{W/m}^2\ \text{K} \quad \text{or} \quad 24.9\ \text{kW/m}^2\ \text{K}$$

Heat transferred per unit time in length dL of pipe $= h\pi\, 0.025\, dL\, (330 - \theta)$ kW, where θ is the temperature at a distance L m from the inlet.

Rate of increase of heat content of fluid $= \left(\frac{\pi}{4} 0.025^2 \times 3.5 \times 1000 \times 4.18\right) d\theta$ kW

The outlet temperature θ' is then given by:

$$\int_{300}^{\theta'} \frac{d\theta}{(330 - \theta)} = 0.0109h \int_0^6 dL$$

where h is in kW/m^2 K.

$$\therefore \quad \log_{10}(330 - \theta') = \log_{10} 30 - \left(\frac{0.0654h}{2.303}\right) = 1.477 - 0.0283h$$

In this case: $h = 24.9\ \text{kW/m}^2\ \text{K}$

$$\therefore \quad \log_{10}(330 - \theta') = (1.477 - 0.705) = 0.772$$

and: $\underline{\underline{\theta' = 324.1\ \text{K}}}$

(b) *Taylor–Prandtl equation*

$$\frac{h}{C_p \rho u} = 0.032Re^{-1/4}[1 + 2.0Re^{-1/8}(Pr - 1)]^{-1} \quad \text{(equation 12.140)}$$

$$\therefore \quad h = \frac{24.9}{(1 + 2.0 \times 3.5/4.34)}$$

$$= 9.53 \text{ kW/m}^2 \text{ K}$$

and: $$\log_{10}(330 - \theta') = 1.477 - (0.0283 \times 9.53) = 1.207$$

$$\underline{\underline{\theta' = 313.9 \text{ K}}}$$

(c) *Universal velocity profile equation*

$$\frac{h}{C_p \rho u} = 0.032Re^{-1/4}\{1 + 0.82Re^{-1/8}[(Pr - 1) + \ln(0.83Pr + 0.17)]\}^{-1} \quad \text{(equation 12.141)}$$

$$= \frac{24.9}{1 + (0.82/4.34)(3.5 + 2.303 \times 0.591)}$$

$$= 12.98 \text{ kW/m}^2 \text{ K}$$

$$\therefore \quad \log_{10}(330 - \theta') = 1.477 - (0.0283 \times 12.98) = 1.110$$

and: $$\underline{\underline{\theta' = 317.1 \text{ K}}}$$

(d) $Nu = 0.023Re^{0.8}Pr^{0.33}$

$$h = \frac{0.023 \times 0.65}{0.0250}(1.25 \times 10^5)^{0.8}(4.50)^{0.33} \quad \text{(equation 9.64)}$$

$$= 0.596 \times 1.195 \times 10^4 \times 1.64$$

$$= 1.168 \times 10^4 \text{ W/m}^2 \text{ K} \quad \text{or} \quad 11.68 \text{ kW/m}^2 \text{ K}$$

and: $$\log_{10}(330 - \theta') = 1.477 - (0.0283 \times 11.68) = 1.147$$

$$\underline{\underline{\theta' = 316.0 \text{ K}}}$$

Comparing the results:

Method	h(kW/m^2 K)	θ'(K)
(a)	24.9	324.1
(b)	9.5	313.9
(c)	13.0	317.1
(d)	11.7	316.0

It is seen that the simple Reynolds analogy is far from accurate in calculating heat transfer to a liquid.

Example 12.4

The tube in Example 12.3 is maintained at 350 K and air is passed through it at 3.5 m/s, the initial temperature of the air being 290 K. What is the outlet temperature of the air for the four cases used in Example 12.3?

Solution

Taking the physical properties of air at 310 K and assuming that fully developed flow exists in the pipe, then:

$$Re = 0.0250 \times 3.5 \times \frac{(29/22.4)(273/310)}{0.018 \times 10^{-3}} = 5535$$

$$Pr = \frac{1.003 \times 1000 \times 0.018 \times 10^{-3}}{0.024} = 0.75$$

The heat transfer coefficients and final temperatures are then calculated as in Example 12.3 to give:

Method	h(W/m^2 K)	θ(K)
(a)	15.5	348.1
(b)	18.3	349.0
(c)	17.9	348.9
(d)	21.2	349.4

In this case the result obtained using the Reynolds analogy agrees much more closely with the other three methods.

12.9. FURTHER READING

BENNETT, C. O. and MYERS, J. E.: *Momentum, Heat and Mass Transfer*, 3rd edn (McGraw-Hill, New York, 1983).
BRODKEY, R. S. and HERSHEY, H. C.: *Transport Phenomena* (McGraw-Hill, New York, 1988).
CUSSLER, E. L.: *Diffusion. Mass Transfer in Fluid Systems*, 2nd, edn. (Cambridge University Press, 1997).
HINZE, J. O.: *Turbulence*, 2nd edn (McGraw-Hill, New York, 1975).
MIDDLEMAN, S.: *An Introduction to Mass and Heat Transfer* (Wiley, 1997).

12.10. REFERENCES

1. JEANS, J. H.: *Kinetic Theory of Gases* (Cambridge University Press, 1940).
2. PRANDTL, L.: *Z. Angew. Math. u. Mech.* **5** (1925) 136. Untersuchungen zur ausgebildeten Turbulenz.
3. PRANDTL, L.: *Z. Ver. deut. Ing.* **77** (1933) 105. Neuere Ergebnisse der Turbulenzforschung.
4. TAYLOR, G. I.: *N.A.C.A. Rep. and Mem.* No. 272 (1916) 423. Conditions at the surface of a hot body exposed to the wind.
5. NIKURADSE, J.: *Forsch. Ver. deut. Ing.* **356** (1932). Gesetzmässigkeiten der turbulenten Strömung in glatten Röhren.
6. NIKURADSE, J.: *Forsch. Ver. deut. Ing.* **361** (1933). Strömungsgesetze in rauhen Röhren.
7. REYNOLDS, O.: *Proc. Manchester Lit. Phil. Soc.* **14** (1874) 7. On the extent and action of the heating surface for steam boilers.
8. PRANDTL, L.: *Physik. Z.* **11** (1910) 1072. Eine Beziehung zwischen Wärmeaustausch und Strömungswiderstand der Flüssigkeiten.
9. PRANDTL, L.: *Physik. Z.* **29** (1928) 487. Bemerkung über den Wärmeübergang im Röhr.
10. MARTINELLI, R. C.: *Trans. Am. Soc. Mech. Eng.* **69** (1947) 947. Heat transfer to molten metals.

12.11. NOMENCLATURE

		Units in SI system	Dimensions in **M, N, L, T, θ**
A	Area of surface	m^2	$\mathbf{L}^2$
a	Constant in equation 12.41	—	—
a'	Constant in equation 12.41	—	—
B	Integration constant	m/s	$\mathbf{LT}^{-1}$
B'	B/u^*	—	—

		Units in SI system	Dimensions in **M, N, L, T, θ**
b	Ratio of θ_b to θ_s	—	—
C	Molar concentration	kmol/m^3	$\mathbf{NL^{-3}}$
C_A, C_B	Molar concentration of **A**, **B**	kmol/m^3	$\mathbf{NL^{-3}}$
C_{As}, C_{Bs}	Molar concentration of **A**, **B** outside boundary layer	kmol/m^3	$\mathbf{NL^{-3}}$
C_{Aw}, C_{Bw}	Molar concentration of **A**, **B** at wall	kmol/m^3	$\mathbf{NL^{-3}}$
C_p	Specific heat at constant pressure	J/kg K	$\mathbf{L^2T^{-2}}\theta^{-1}$
C_T	Total molar concentration	kmol/m^3	$\mathbf{NL^{-3}}$
C_v	Specific heat at constant volume	J/kg K	$\mathbf{L^2T^{-2}}\theta^{-1}$
c_m	Heat capacity of one molecule	J/K	$\mathbf{ML^2T^{-2}}\theta^{-1}$
D	Diffusivity	m^2/s	$\mathbf{L^2T^{-1}}$
D_H	Thermal diffusivity	m^2/s	$\mathbf{L^2T^{-1}}$
d	Pipe or particle diameter	m	$\mathbf{L}$
E	Eddy kinematic viscosity	m^2/s	$\mathbf{L^2T^{-1}}$
E_D	Eddy diffusivity	m^2/s	$\mathbf{L^2T^{-1}}$
E_H	Eddy thermal diffusivity	m^2/s	$\mathbf{L^2T^{-1}}$
e	Surface roughness	m	$\mathbf{L}$
h	Heat transfer coefficient	W/m^2 K	$\mathbf{MT^{-3}}\theta^{-1}$
h_D	Mass transfer coefficient	m/s	$\mathbf{LT^{-1}}$
I	Intensity of turbulence	—	—
i	Fraction of root mean square velocity of molecules	—	—
j	Fraction of mean free path of molecules	—	—
K	Ratio of mixing length to distance from surface	—	—
k	Thermal conductivity	W/m K	$\mathbf{MLT^{-3}}\theta^{-1}$
M	Molecular weight	kg/kmol	$\mathbf{MN^{-1}}$
$\mathbf{M}$	Mass of fluid	kg	$\mathbf{M}$
m	Mass of gas molecule	kg	$\mathbf{M}$
N	Molar rate of diffusion per unit area	kmol/m^{2}s	$\mathbf{NL^{-2}T^{-1}}$
N'	Total molar rate of transfer per unit area	kmol/m^{2}s	$\mathbf{NL^{-2}T^{-1}}$
$\mathbf{N}$	Number of molecules per unit volume at $y = y$	m^{-3}	$\mathbf{L^{-3}}$
$\mathbf{N}'$	Number of molecules per unit volume $y = y + j\lambda$	m^{-3}	$\mathbf{L^{-3}}$
n	Number of molar units	kmol	$\mathbf{N}$
q_0	Rate of transfer of heat per unit area at walls	W/m^2	$\mathbf{MT^{-3}}$
q_y	Rate of transfer of heat per unit area at $y = y$	W/m^2	$\mathbf{MT^{-3}}$
R	Shear stress acting on surface	N/m^2	$\mathbf{ML^{-1}T^{-2}}$
R_0	Shear stress acting on fluid at surface	N/m^2	$\mathbf{ML^{-1}T^{-2}}$
R_y	Shear stress in fluid at $y = y$	N/m^2	$\mathbf{ML^{-1}T^{-2}}$
$\mathbf{R}$	Universal gas constant	8314 J/kmol K	$\mathbf{MN^{-1}L^2T^{-2}}\theta^{-1}$
r	Radius of pipe	m	$\mathbf{L}$
T	Absolute temperature	K	θ
t	Time	s	$\mathbf{T}$
u	Mean velocity	m/s	$\mathbf{LT^{-1}}$
u_b	Velocity at edge of laminar sub-layer	m/s	$\mathbf{LT^{-1}}$
u_E	Mean velocity in eddy	m/s	$\mathbf{LT^{-1}}$
u_{Ex}	Mean component of eddy velocity in X-direction	m/s	$\mathbf{LT^{-1}}$
u_{Ey}	Mean component of eddy velocity in Y-direction	m/s	$\mathbf{LT^{-1}}$
u_{Ez}	Mean component of eddy velocity in Z-direction	m/s	$\mathbf{LT^{-1}}$
u_e	Velocity at distance e from surface	m/s	$\mathbf{LT^{-1}}$
u_m	Root mean square velocity of molecules	m/s	$\mathbf{LT^{-1}}$
u_s	Velocity of fluid outside boundary layer, or at pipe axis	m/s	$\mathbf{LT^{-1}}$
u_x	Velocity in X-direction at $y = y$	m/s	$\mathbf{LT^{-1}}$
u'_x	Velocity in X-direction at $y = y + j\lambda$ or $y + \lambda_E$	m/s	$\mathbf{LT^{-1}}$
u^+	Ratio of u_y to u^*	—	—

		Units in SI system	Dimensions in **M, N, L, T, θ**
u^*	Shearing stress velocity, $\sqrt{R/\rho}$	m/s	$\mathbf{LT}^{-1}$
w	Exponent of velocity	—	—
y	Distance from surface	m	**L**
y^+	Ratio of y to $\mu/\rho u^*$	—	—
α	Ratio of u_b to u_s	—	—
γ	Ratio of C_p to C_v	—	—
δ	Thickness of boundary layer	m	**L**
δ_b	Thickness of laminar sub-layer	m	**L**
λ	Mean free path of molecules	m	**L**
λ_E	Mixing length	m	**L**
μ	Viscosity of fluid	N s/m^2	$\mathbf{ML}^{-1}\mathbf{T}^{-1}$
ϕ	Friction factor $R/\rho u^2$	—	—
ρ	Density of fluid	kg/m^3	$\mathbf{ML}^{-3}$
θ	Temperature at $y = y$	K	$\boldsymbol{\theta}$
θ_s	Temperature outside boundary layer, or at pipe axis	K	$\boldsymbol{\theta}$
θ'	Temperature at $y = y + j\lambda$	K	$\boldsymbol{\theta}$
Nu	Nusselt number hd/k	—	—
Nu_x	Nusselt number hx/k	—	—
Re	Reynolds number $ud\rho/\mu$	—	—
Re_r	Roughness Reynolds number $u^*e\rho/\mu$	—	—
Re_x	Reynolds number $u_s x\rho/\mu$	—	—
Pr	Prandtl number $C_p\mu/k$	—	—
Sc	Schmidt number $\mu/\rho D$	—	—
Sh_x	Sherwood number h_D/D	—	—
St	Stanton number $h/C_p\rho u$ or $h/C_p\rho u_s$	—	—

Subscripts

A, **B** For component **A**, **B**

CHAPTER 13

Applications in Humidification and Water Cooling

13.1. INTRODUCTION

In the processing of materials it is often necessary either to increase the amount of vapour present in a gas stream, an operation known as *humidification*; or to reduce the vapour present, a process referred to as *dehumidification*. In humidification, the vapour content may be increased by passing the gas over a liquid which then evaporates into the gas stream. This transfer into the main stream takes place by diffusion, and at the interface simultaneous heat and mass transfer take place according to the relations considered in previous chapters. In the reverse operation, that is dehumidification, partial condensation must be effected and the condensed vapour removed.

The most widespread application of humidification and dehumidification involves the air–water system, and a discussion of this system forms the greater part of the present chapter. Although the drying of wet solids is an example of a humidification operation, the reduction of the moisture content of the solids is the main objective, and the humidification of the air stream is a secondary effect. Much of the present chapter is, however, of vital significance in any drying operation. Air conditioning and gas drying also involve humidification and dehumidification operations. For example, moisture must be removed from wet chlorine so that the gas can be handled in steel equipment which otherwise would be severely corroded. Similarly, the gases used in the manufacture of sulphuric acid must be dried or dehumidified before entering the converters, and this is achieved by passing the gas through a dehydrating agent such as sulphuric acid, in essence an absorption operation, or by an alternative dehumidification process discussed later.

In order that hot condenser water may be re-used in a plant, it is normally cooled by contact with an air stream. The equipment usually takes the form of a tower in which the hot water is run in at the top and allowed to flow downwards over a packing against a countercurrent flow of air which enters at the bottom of the cooling tower. The design of such towers forms an important part of the present chapter, though at the outset it is necessary to consider basic definitions of the various quantities involved in humidification, in particular *wet-bulb* and *adiabatic saturation temperatures*, and the way in which humidity data are presented on charts and graphs. While the present discussion is devoted to the very important air-water system, which is in some ways unique, the same principles may be applied to other liquids and gases, and this topic is covered in a final section.

13.2. HUMIDIFICATION TERMS

13.2.1. Definitions

The more important terms used in relation to humidification are defined as follows:

Humidity ($\mathscr{H}$)	mass of vapour associated with unit mass of dry gas
Humidity of saturated gas ($\mathscr{H}_0$)	humidity of the gas when it is saturated with vapour at a given temperature
Percentage humidity	$100(\mathscr{H}/\mathscr{H}_0)$
Humid heat (s)	heat required to raise unit mass of dry gas and its associated vapour through unit temperature difference at constant pressure, or: $s = C_a + \mathscr{H}C_w$ where C_a and C_w are the specific heat capacities of the gas and the vapour, respectively. (For the air-water system, the humid heat is approximately: $s = 1.00 + 1.9\mathscr{H}$ kJ/kg K.)
Humid volume	volume occupied by unit mass of dry gas and its associated vapour
Saturated volume	humid volume of saturated gas
Dew point	temperature at which the gas is saturated with vapour. As a gas is cooled, the dew point is the temperature at which condensation will first occur.
Percentage relative humidity	$\left(\dfrac{\text{partial pressure of vapour in gas}}{\text{partial pressure of vapour in saturated gas}}\right) \times 100$

The above nomenclature conforms with the recommendations of BS1339[1], although there are some ambiguities in the standard.

The relationship between the partial pressure of the vapour and the humidity of a gas may be derived as follows. In unit volume of gas:

$$\text{mass of vapour} = \frac{P_w M_w}{\mathbf{R}T}$$

$$\text{and mass of non-condensable gas} = \frac{(P - P_w)M_A}{\mathbf{R}T}$$

The humidity is therefore given by:

$$\mathscr{H} = \frac{P_w}{P - P_w}\left(\frac{M_w}{M_A}\right) \tag{13.1}$$

and the humidity of the saturated gas is:

$$\mathscr{H}_0 = \frac{P_{w0}}{P - P_{w0}}\left(\frac{M_w}{M_A}\right) \tag{13.2}$$

where P_w is the partial pressure of vapour in the gas, P_{w0} the partial pressure of vapour in the saturated gas at the same temperature, M_A the mean molecular weight of the dry gas, M_w the molecular mass of the vapour, P the total pressure, $\mathbf{R}$ the gas constant (8314 J/kmol K in SI units), and T the absolute temperature.

For the air–water system, P_w is frequently small compared with P and hence, substituting for the molecular masses:

$$\mathscr{H} = \frac{18}{29}\left(\frac{P_w}{P}\right)$$

The relationship between the percentage humidity of a gas and the percentage relative humidity may be derived as follows:

The percentage humidity, by definition $= 100\mathscr{H}/\mathscr{H}_0$

Substituting from equations 13.1 and 13.2 and simplifying:

$$\text{Percentage humidity} = \left(\frac{P - P_{w0}}{P - P_w}\right)\cdot\left(\frac{P_w}{P_{w0}}\right) \times 100$$

$$= \frac{(P - P_{w0})}{(P - P_w)} \times (\text{percentage relative humidity}) \qquad (13.3)$$

When $(P - P_{w0})/(P - P_w) \approx 1$, the percentage relative humidity and the percentage humidity are equal. This condition is approached when the partial pressure of the vapour is only a small proportion of the total pressure or when the gas is almost saturated, that is as $P_w \to P_{w0}$.

Example 13.1

In a process in which it is used as a solvent, benzene is evaporated into dry nitrogen. At 297 K and 101.3 kN/m^2, the resulting mixture has a percentage relative humidity of 60. It is required to recover 80 per cent of the benzene present by cooling to 283 K and compressing to a suitable pressure. What should this pressure be? The vapour pressure of benzene is 12.2 kN/m^2 at 297 K and 6.0 kN/m^2 at 283 K.

Solution

From the definition of percentage relative humidity (RH):

$$P_w = P_{w0}\left(\frac{RH}{100}\right)$$

At 297 K:

$$P_w = (12.2 \times 1000) \times \left(\frac{60}{100}\right) = 7320 \text{ N/m}^2$$

In the benzene–nitrogen mixture:

$$\text{mass of benzene} = \frac{P_w M_w}{\mathbf{R}T} = \frac{(7320 \times 78)}{(8314 \times 297)} = 0.231 \text{ kg}$$

$$\text{mass of nitrogen} = \frac{(P - P_w)M_A}{\mathbf{R}T} = \frac{[(101.3 - 732) \times 1000 \times 28]}{(8314 \times 297)} = 1.066 \text{ kg}$$

Hence the humidity is:

$$\mathscr{H} = \left(\frac{0.231}{1.066}\right) = 0.217 \text{ kg/kg}$$

In order to recover 80 per cent of the benzene, the humidity must be reduced to 20 per cent of the initial value. As the vapour will be in contact with liquid benzene, the nitrogen will be saturated with benzene vapour and hence at 283 K:

$$\mathscr{H}_0 = \frac{(0.217 \times 20)}{100} = 0.0433 \text{ kg/kg}$$

Thus in equation 13.2:

$$0.0433 = \left(\frac{6000}{P - 6000}\right)\left(\frac{78}{28}\right)$$

from which:

$$\underline{\underline{P = 3.92 \times 10^5 \text{ N/m}^2 = 392 \text{ kN/m}^2}}$$

Example 13.2

In a vessel at 101.3 kN/m^2 and 300 K, the percentage relative humidity of the water vapour in the air is 25. If the partial pressure of water vapour when air is saturated with vapour at 300 K is 3.6 kN/m^2, calculate:

(a) the partial pressure of the water vapour in the vessel;
(b) the specific volumes of the air and water vapour;
(c) the humidity of the air and humid volume; and
(d) the percentage humidity.

Solution

(a) From the definition of percentage relative humidity:

$$P_w = P_{w0}\frac{RH}{100} = 3600 \times \left(\frac{25}{100}\right) = 900 \text{ N/m}^2 = \underline{\underline{0.9 \text{ kN/m}^2}}$$

(b) In 1 m^3 of air:

$$\text{mass of water vapour} = \frac{(900 \times 18)}{(8314 \times 300)} = 0.0065 \text{ kg}$$

$$\text{mass of air} = \frac{[(101.3 - 0.9) \times 1000 \times 29]}{(8314 \times 300)} = 1.167 \text{ kg}$$

Hence: specific volume of water vapour at 0.9 kN/m^2 $= \left(\frac{1}{0.0065}\right) = \underline{\underline{154 \text{ m}^3/\text{kg}}}$

specific volume of air at 100.4 kN/m^2 $= \left(\frac{1}{1.167}\right) = \underline{\underline{0.857 \text{ m}^3/\text{kg}}}$

(c) Humidity:

$$\mathscr{H} = \left(\frac{0.0065}{1.1673}\right) = \underline{\underline{0.0056 \text{ kg/kg}}}$$

(Using the approximate relationship:

$$\mathscr{H} = \frac{(18 \times 900)}{(29 \times 101.3 \times 1000)} = 0.0055 \text{ kg/kg.)}$$

∴ Humid volume = volume of 1 kg air + associated vapour = specific volume of air at 100.4 kN/m^2

$$= \underline{\underline{0.857 \text{ m}^3/\text{kg}}}$$

(d) From equation 13.3:

$$\text{percentage humidity} = \frac{[(101.3 - 3.6) \times 1000]}{[(101.3 - 0.9) \times 1000]} \times 25$$

$$= \underline{\underline{24.3 \text{ per cent}}}$$

13.2.2. Wet-bulb temperature

When a stream of unsaturated gas is passed over the surface of a liquid, the humidity of the gas is increased due to evaporation of the liquid. The temperature of the liquid falls below that of the gas and heat is transferred from the gas to the liquid. At equilibrium the rate of heat transfer from the gas just balances that required to vaporise the liquid and the liquid is said to be at the *wet-bulb temperature*. The rate at which this temperature is reached depends on the initial temperatures and the rate of flow of gas past the liquid surface. With a small area of contact between the gas and the liquid and a high gas flowrate, the temperature and the humidity of the gas stream remain virtually unchanged.

The rate of transfer of heat from the gas to the liquid can be written as:

$$Q = hA(\theta - \theta_w) \tag{13.4}$$

where Q is the heat flow, h the coefficient of heat transfer, A the area for transfer, and θ and θ_w are the temperatures of the gas and liquid phases.

The liquid evaporating into the gas is transferred by diffusion from the interface to the gas stream as a result of a concentration difference $(c_0 - c)$, where c_0 is the concentration of the vapour at the surface (mass per unit volume) and c is the concentration in the gas stream. The rate of evaporation is then given by:

$$W = h_D A(c_0 - c) = h_D A \frac{M_w}{\mathbf{R}T}(P_{w0} - P_w) \tag{13.5}$$

where h_D is the coefficient of mass transfer.

The partial pressures of the vapour, P_w and P_{w0}, may be expressed in terms of the corresponding humidities $\mathscr{H}$ and $\mathscr{H}_w$ by equations 13.1 and 13.2.

If P_w and P_{w0} are small compared with P, $(P - P_w)$ and $(P - P_{w0})$ may be replaced by a mean partial pressure of the gas P_A and:

$$\begin{aligned} W &= h_{DA} A \frac{(\mathscr{H}_w - \mathscr{H})M_w}{\mathbf{R}T} \cdot \left(P_A \frac{M_A}{M_w}\right) \\ &= h_D A \rho_A (\mathscr{H}_w - \mathscr{H}) \end{aligned} \tag{13.6}$$

where ρ_A is the density of the gas at the partial pressure P_A.

The heat transfer required to maintain this rate of evaporation is:

$$Q = h_D A \rho_A (\mathscr{H}_w - \mathscr{H})\lambda \tag{13.7}$$

where λ is the latent heat of vaporisation of the liquid.

Thus, equating equations 13.4 and 13.7:

$$(\mathscr{H} - \mathscr{H}_w) = -\frac{h}{h_D \rho_A \lambda}(\theta - \theta_w) \tag{13.8}$$

Both h and h_D are dependent on the equivalent gas film thickness, and thus any decrease in the thickness, as a result of increasing the gas velocity for example, increases both h and h_D. At normal temperatures, (h/h_D) is virtually independent of the gas velocity provided this is greater than about 5 m/s. Under these conditions, heat transfer by convection from the gas stream is large compared with that from the surroundings by radiation and conduction.

The wet-bulb temperature θ_w depends only on the temperature and the humidity of the gas and values normally quoted are determined for comparatively high gas velocities, such that the condition of the gas does not change appreciably as a result of being brought into contact with the liquid and the ratio (h/h_D) has reached a constant value. For the air–water system, the ratio $(h/h_D\rho_A)$ is about 1.0 kJ/kg K and varies from 1.5 to 2.0 kJ/kg K for organic liquids.

Example 13.3

Moist air at 310 K has a wet-bulb temperature of 300 K. If the latent heat of vaporisation of water at 300 K is 2440 kJ/kg, estimate the humidity of the air and the percentage relative humidity. The total pressure is 105 kN/m^2 and the vapour pressure of water vapour at 300 K is 3.60 kN/m^2 and 6.33 kN/m^2 at 310 K.

Solution

The humidity of air saturated at the wet-bulb temperature is given by:

$$\mathscr{H}_w = \frac{P_{w0}}{P - P_{w0}} \frac{M_w}{M_A} \qquad \text{(equation 13.2)}$$

$$= \left(\frac{3.6}{105.0 - 3.6}\right)\left(\frac{18}{29}\right) = 0.0220 \text{ kg/kg}$$

Therefore, taking $(h/h_D\rho_A)$ as 1.0 kJ/kg K, in equation 13.8:

$$(0.0220 - \mathscr{H}) = \left(\frac{1.0}{2440}\right)(310 - 300)$$

or: $$\underline{\underline{\mathscr{H} = 0.018 \text{ kg/kg}}}$$

At 310 K, $$P_{w0} = 6.33 \text{ kN/m}^2$$

In equation 13.2: $$0.0780 = \frac{18P_w}{(105.0 - P_w)29}$$

$\therefore$ $$P_w = 2.959 \text{ kN/m}^2$$

and the percentage relative humidity $$= \frac{(100 \times 2.959)}{6.33} = \underline{\underline{46.7 \text{ per cent}}}$$

13.2.3. Adiabatic saturation temperature

In the system just considered, neither the humidity nor the temperature of the gas is appreciably changed. If the gas is passed over the liquid at such a rate that the time of contact is sufficient for equilibrium to be established, the gas will become saturated and both phases will be brought to the same temperature. In a thermally insulated system, the total sensible heat falls by an amount equal to the latent heat of the liquid evaporated. As a result of continued passage of the gas, the temperature of the liquid gradually approaches an equilibrium value which is known as the *adiabatic saturation temperature*.

These conditions are achieved in an infinitely tall thermally insulated humidification column through which gas of a given initial temperature and humidity flows countercurrently to the liquid under conditions where the gas is completely saturated at the top of the column. If the liquid is continuously circulated round the column, and if any fresh liquid which is added is at the same temperature as the circulating liquid, the temperature of the liquid at the top and bottom of the column, and of the gas at the top, approach the adiabatic saturation temperature. Temperature and humidity differences are a maximum at the bottom and zero at the top, and therefore the rates of transfer of heat and mass decrease progressively from the bottom to the top of the tower. This is illustrated in Figure 13.1.

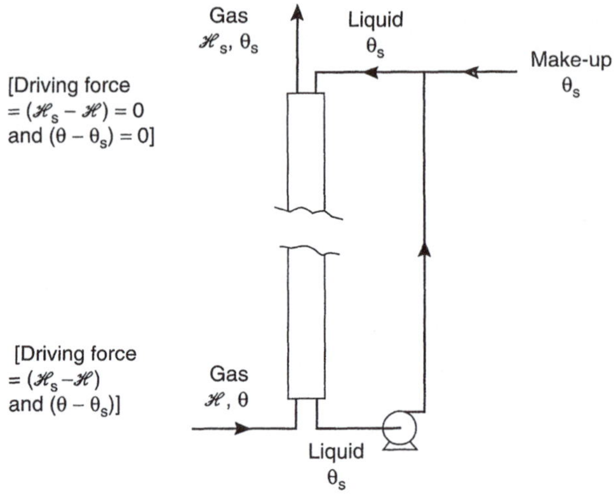

Figure 13.1. Adiabatic saturation temperature θ_s

Making a heat balance over the column, it is seen that the heat of vaporisation of the liquid must come from the sensible heat in the gas. The temperature of the gas falls from θ to the adiabatic saturation temperature θ_s, and its humidity increases from $\mathscr{H}$ to $\mathscr{H}_s$ (the saturation value at θ_s). Then working on the basis of unit mass of dry gas:

$$(\theta - \theta_s)s = (\mathscr{H}_s - \mathscr{H})\lambda$$

or:
$$(\mathscr{H} - \mathscr{H}_s) = -\frac{s}{\lambda}(\theta - \theta_s) \quad (13.9)$$

where s is the humid heat of the gas and λ the latent heat of vaporisation at θ_s. s is almost constant for small changes in $\mathscr{H}$.

Equation 13.9 indicates an approximately linear relationship between humidity and temperature for all mixtures of gas and vapour having the same adiabatic saturation temperature θ_s. A curve of humidity versus temperature for gases with a given adiabatic saturation temperature is known as an *adiabatic cooling line*. For a range of adiabatic saturation temperatures, a family of curves, approximating to straight lines of slopes equal to $-(s/\lambda)$, is obtained. These lines are not exactly straight and parallel because of variations in λ and s.

Comparing equations 13.8 and 13.9, it is seen that the adiabatic saturation temperature is equal to the wet-bulb temperature when $s = h/h_D\rho_A$. This is the case for most water vapour systems and accurately so when $\mathscr{H} = 0.047$. The ratio $(h/h_D\rho_A s) = b$ is sometimes known as the *psychrometric ratio* and, as indicated, b is approximately unity for the air–water system. For most systems involving air and an organic liquid, $b = 1.3 - 2.5$ and the wet-bulb temperature is higher than the adiabatic saturation temperature. This was confirmed in 1932 by SHERWOOD and COMINGS(2) who worked with water, ethanol, n-propanol, n-butanol, benzene, toluene, carbon tetrachloride, and n-propyl acetate, and found that the wet-bulb temperature was always higher than the adiabatic saturation temperature except in the case of water.

In Chapter 12 it is shown that when the Schmidt and Prandtl numbers for a mixture of gas and vapour are approximately equal to unity, the *Lewis relation* applies, or:

$$h_D = \frac{h}{C_p\rho} \qquad \text{(equation 12.105)}$$

where C_p and ρ are the mean specific heat and density of the vapour phase.

Therefore:

$$\frac{h}{h_D\rho_A} = \frac{C_p\rho}{\rho_A} \qquad (13.10)$$

Where the humidity is relatively low, $C_p \approx s$ and $\rho \approx \rho_A$ and hence:

$$s \approx \frac{h}{h_D\rho_A} \qquad (13.11)$$

For systems containing vapour other than that of water, s is only approximately equal to $h/h_D\rho_A$ and the difference between the two quantities may be as high as 50 per cent.

If an unsaturated gas is brought into contact with a liquid which is at the adiabatic saturation temperature of the gas, a simultaneous transfer of heat and mass takes place. The temperature of the gas falls and its humidity increases (Figure 13.2). The temperature of the liquid at any instant tends to change and approach the wet-bulb temperature corresponding to the particular condition of the gas at that moment. For a liquid other than water, the adiabatic saturation temperature is less than the wet-bulb temperature and therefore in the initial stages, the temperature of the liquid rises. As the gas becomes humidified, however, its wet-bulb temperature falls and consequently the temperature to

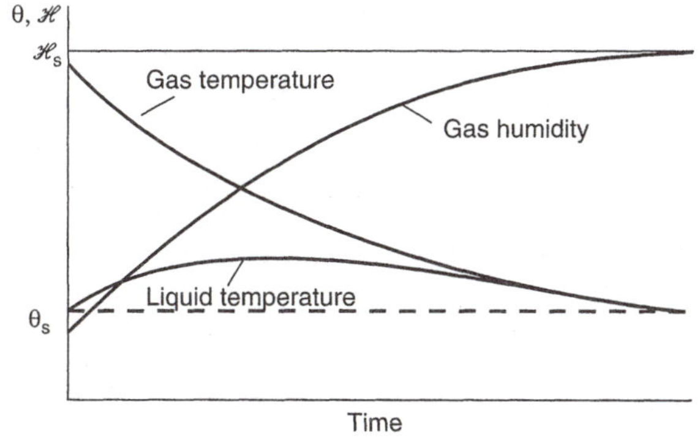

Figure 13.2. Saturation of gas with liquid other than water at the adiabatic saturation temperature

which the liquid is tending decreases as evaporation takes place. In due course, therefore, a point is reached where the liquid actually reaches the wet-bulb temperature of the gas in contact with it. It does not remain at this temperature, however, because the gas is not then completely saturated, and further humidification is accompanied by a continued lowering of the wet-bulb temperature. The temperature of the liquid therefore starts to fall and continues to fall until the gas is completely saturated. The liquid and gas are then both at the adiabatic saturation temperature.

The air–water system is unique, however, in that the Lewis relation holds quite accurately, so that the adiabatic saturation temperature is the same as the wet-bulb temperature. If, therefore, an unsaturated gas is brought into contact with water at the adiabatic saturation temperature of the gas, there is no tendency for the temperature of the water to change, and it remains in a condition of dynamic equilibrium through the whole of the humidification process (Figure 13.3). In this case, the adiabatic cooling line represents the conditions of gases of constant wet-bulb temperatures as well as constant adiabatic saturation temperatures. The change in the condition of a gas as it is humidified with water vapour is therefore represented by the adiabatic cooling line and the intermediate conditions of the gas during the process are readily obtained. This is particularly useful because only partial humidification is normally obtained in practice.

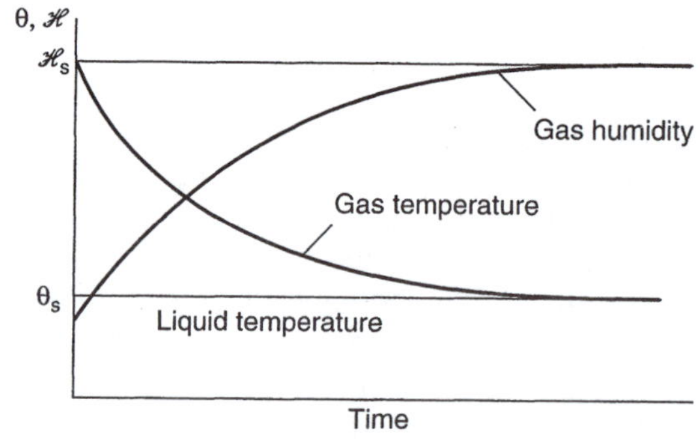

Figure 13.3. Saturation of air with water at adiabatic saturation temperature

13.3. HUMIDITY DATA FOR THE AIR–WATER SYSTEM

To facilitate calculations, various properties of the air–water system are plotted on a *psychrometric* or *humidity chart*. Such a chart is based on either the temperature or the enthalpy of the gas. The temperature–humidity chart is the more commonly used though the enthalpy–humidity chart is particularly useful for determining the effect of mixing two gases or of mixing a gas and a liquid. Each chart refers to a particular total pressure of the system. A humidity–temperature chart for the air–water system at atmospheric pressure, based on the original chart by GROSVENOR[3], is given in Figure 13.4 and the corresponding humidity–enthalpy chart is given in Figure 13.5.

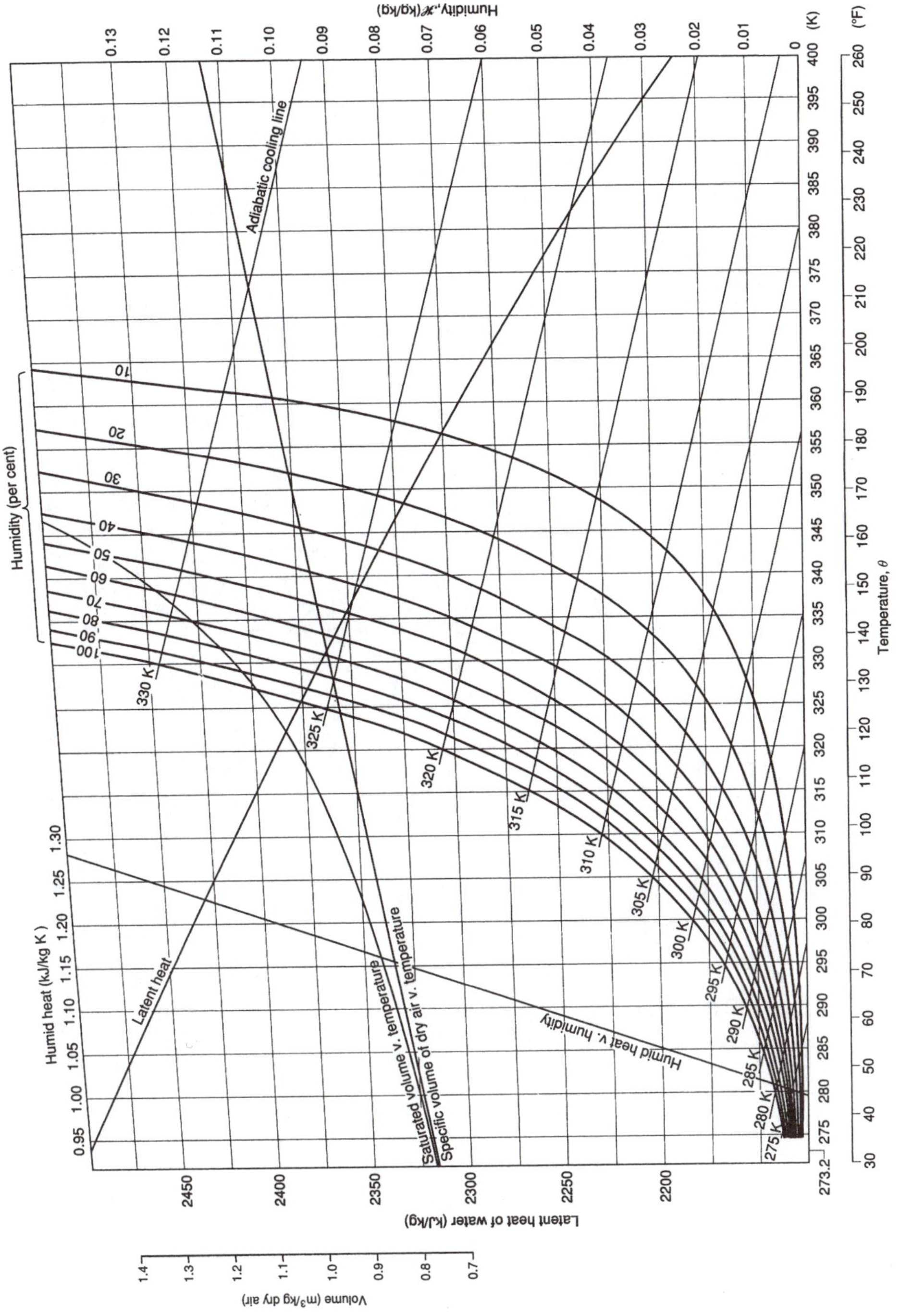

Figure 13.4. Humidity–temperature chart (See also the Appendix)

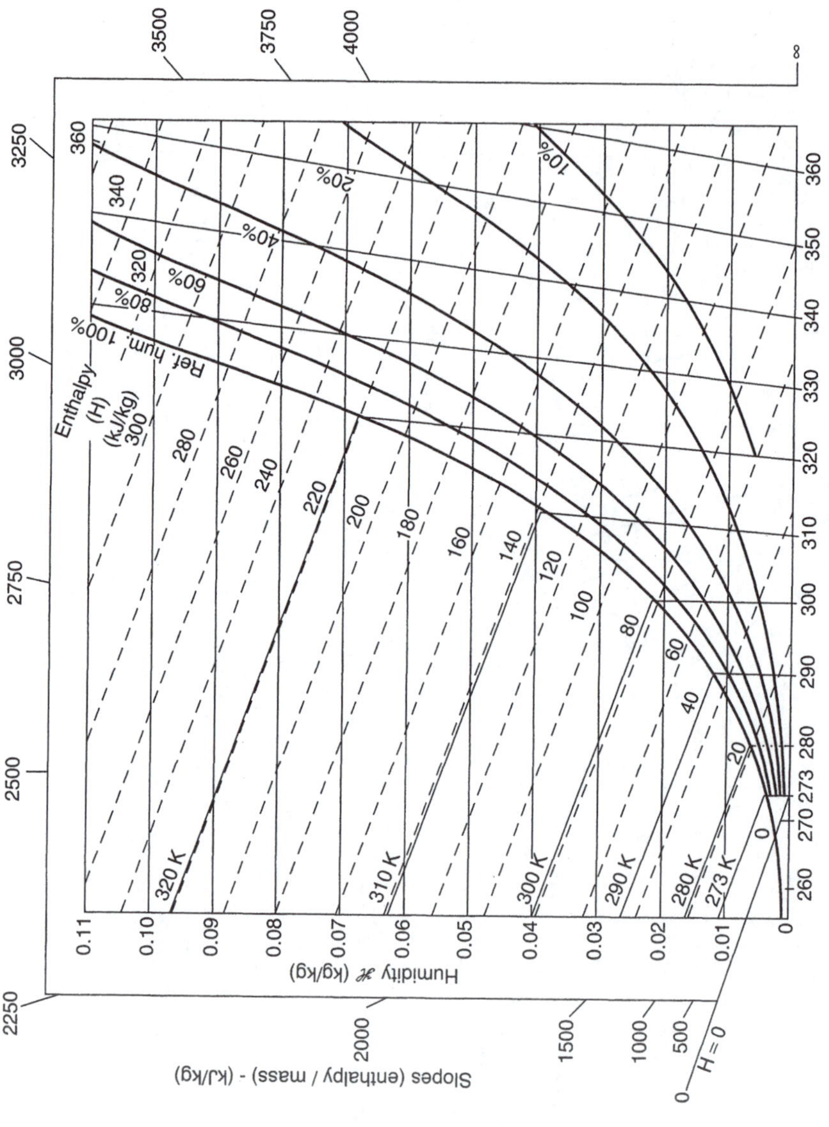

Figure 13.5. Humidity–enthalpy diagram for air-water vapour system at atmospheric pressure

13.3.1. Temperature–humidity chart

In Figure 13.4 it will be seen that the following quantities are plotted against temperature:

(i) The *humidity* $\mathscr{H}$ for various values of the percentage humidity.

For saturated gas: $$\mathscr{H}_0 = \frac{P_{w0}}{P - P_{w0}}\left(\frac{M_w}{M_A}\right) \quad \text{(equation 13.2)}$$

From equation 13.1 for a gas with a humidity less than the saturation value:

$$\mathscr{H} = \frac{P_w}{P - P_w}\left(\frac{M_w}{M_A}\right) = \mathscr{H}_0 \frac{P_w}{P_{w0}} \frac{P - P_{w0}}{P - P_w} \tag{13.12}$$

(ii) *The specific volume of dry gas*. This is a linear function of temperature.

(iii) *The saturated volume*. This increases more rapidly with temperature than the specific volume of dry gas because both the quantity and the specific volume of vapour increase with temperature. At a given temperature, the humid volume varies linearly with humidity and hence the humid volume of unsaturated gas can be found by interpolation.

(iv) *The latent heat of vaporisation*

In addition, the *humid heat* is plotted as the abscissa in Figure 13.4 with the humidity as the ordinate.

Adiabatic cooling lines are included in the diagram and, as already discussed, these have a slope of $-(s/\lambda)$ and they are slightly curved since s is a function of $\mathscr{H}$. On the chart they appear as straight lines, however, since the inclination of the axis has been correspondingly adjusted. Each adiabatic cooling line represents the composition of all gases whose adiabatic saturation temperature is given by its point of intersection with the 100 per cent humidity curve. For the air–water system, the adiabatic cooling lines represent conditions of constant wet-bulb temperature as well and, as previously mentioned, enable the change in composition of a gas to be followed as it is humidified by contact with water at the adiabatic saturation temperature of the gas.

Example 13.4

Air containing 0.005 kg water vapour per kg of dry air is heated to 325 K in a dryer and passed to the lower shelves. It leaves these shelves at 60 per cent humidity and is reheated to 325 K and passed over another set of shelves, again leaving at 60 per cent humidity. This is again repeated for the third and fourth sets of shelves, after which the air leaves the dryer. On the assumption that the material on each shelf has reached the wet-bulb temperature and that heat losses from the dryer may be neglected, determine:

(a) the temperature of the material on each tray;
(b) the amount of water removed in kg/s, if 5 m^3/s moist air leaves the dryer;
(c) the temperature to which the inlet air would have to be raised to carry out the drying in a single stage.

Solution

For each of the four sets of shelves, the condition of the air is changed to 60 per cent humidity along an adiabatic cooling line.

Initial condition of air: $\theta = 325$ K, $\mathscr{H} = 0.005$ kg/kg

On humidifying to 60 per cent humidity:

	$\theta = 301$ K,	$\mathscr{H} = 0.015$ kg/kg and $\theta_w = 296$ K
At the end of the second pass:	$\theta = 308$ K,	$\mathscr{H} = 0.022$ kg/kg and $\theta_w = 301$ K
At the end of the third pass:	$\theta = 312$ K,	$\mathscr{H} = 0.027$ kg/kg and $\theta_w = 305$ K
At the end of the fourth pass:	$\theta = 315$ K,	$\mathscr{H} = 0.032$ kg/kg and $\theta_w = 307$ K

Thus the temperatures of the material on each of the trays are:

296 K, 301 K, 305 K, and 307 K

Total increase in humidity $= (0.032 - 0.005) = 0.027$ kg/kg

The air leaving the system is at 315 K and 60 per cent humidity.

From Figure 13.4, specific volume of dry air $= 0.893$ m^3/kg

Specific volume of saturated air $(saturated volume) = 0.968$ m^3/kg

Therefore, by interpolation, the humid volume of air of 60 per cent humidity $= 0.937$ m^3/kg

$$\text{Mass of air passing through the dryer} = \left(\frac{5}{0.937}\right) = 5.34 \text{ kg/s}$$

Mass of water evaporated $= (5.34 \times 0.027) = 0.144$ kg/s

If the material is to be dried by air in a single pass, the air must be heated before entering the dryer such that its wet-bulb temperature is 307 K.

For air with a humidity of 0.005 kg/kg, this corresponds to a dry bulb temperature of 370 K.

The various steps in this calculation are shown in Figure 13.6.

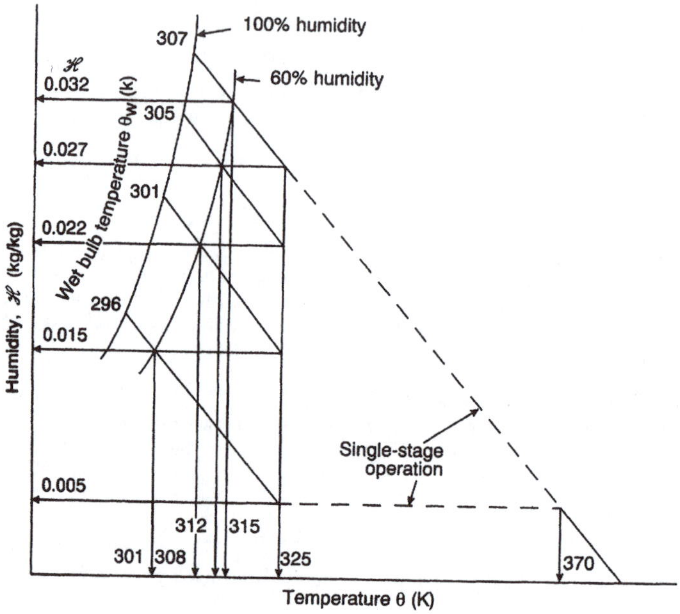

Figure 13.6. Humidification stages for Example 13.4 (schematic)

13.3.2. Enthalpy–humidity chart

In the calculation of enthalpies it is necessary to define some standard reference state at which the enthalpy is taken as zero. It is most convenient to take the melting point of the material constituting the vapour as the reference temperature, and the liquid state of the material as its standard state.

If H is the enthalpy of the humid gas per unit mass of dry gas, H_a the enthalpy of the dry gas per unit mass, H_w the enthalpy of the vapour per unit mass, C_a the specific heat of the gas at constant pressure, C_w the specific heat of the vapour at constant pressure, θ the temperature of the humid gas, θ_0 the reference temperature, λ the latent heat of vaporisation of the liquid at θ_0 and $\mathscr{H}$ the humidity of the gas,

then for an unsaturated gas: $H = H_a + H_w\mathscr{H}$ (13.13)

where: $$H_a = C_a(\theta - \theta_0) \tag{13.14}$$

and: $$H_w = C_w(\theta - \theta_0) + \lambda \tag{13.15}$$

Thus, in equation 13.13: $$H = (C_a + \mathscr{H}C_w)(\theta - \theta_0) + \mathscr{H}\lambda \tag{13.16}$$

$$= (\theta - \theta_0)(s + \mathscr{H}\lambda)$$

If the gas contains more liquid or vapour than is required to saturate it at the temperature in question, either the gas will be supersaturated or the excess material will be present in the form of liquid or solid according to whether the temperature θ is greater or less than the reference temperature θ_0. The supersaturated condition is unstable and will not be considered further.

If the temperature θ is greater than θ_0 and if the humidity $\mathscr{H}$ is greater than the humidity $\mathscr{H}_0$ of saturated gas, the enthalpy H per unit mass of dry gas is given by:

$$H = C_a(\theta - \theta_0) + \mathscr{H}_0[C_w(\theta - \theta_0) + \lambda] + C_L(\mathscr{H} - \mathscr{H}_0)(\theta - \theta_0) \tag{13.17}$$

where C_L is the specific heat of the liquid.

If the temperature θ is less than θ_0, the corresponding enthalpy H is given by:

$$H = C_a(\theta - \theta_0) + \mathscr{H}_0[C_w(\theta - \theta_0) + \lambda] + (\mathscr{H} - \mathscr{H}_0)[C_s(\theta - \theta_0) + \lambda_f] \tag{13.18}$$

where C_s is the specific heat of the solid and λ_f is the latent heat of freezing of the liquid, a negative quantity.

Equations 13.16 to 13.18 give the enthalpy in terms of the temperature and humidity of the humid gas for the three conditions: $\theta = \theta_0$, $\theta > \theta_0$, and $\theta < \theta_0$ respectively. Thus, given the percentage humidity and the temperature, the humidity may be obtained from Figure 13.4, the enthalpy calculated from equations 13.16, 13.17 or 13.18 and plotted against the humidity, usually with enthalpy as the abscissa. Such a plot is shown in Figure 13.7 for the air-water system, which includes the curves for 100 per cent humidity and for some lower value, say Z per cent.

Considering the nature of the isothermals for the three conditions dealt with previously, at constant temperature θ the relation between enthalpy and humidity for an unsaturated gas is:

$$H = \text{constant} + [C_w(\theta - \theta_0) + \lambda]\mathscr{H} \tag{13.19}$$

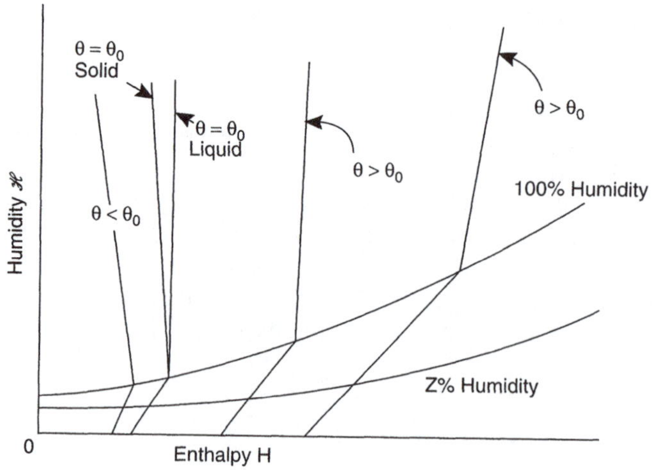

Figure 13.7. Humidity-enthalpy diagram for air-water system—rectangular axes

Thus, the isothermal is a straight line of slope $[C_w(\theta - \theta_0) + \lambda]$ with respect to the humidity axis. At the reference temperature θ_0, the slope is λ; at higher temperatures, the slope is greater than λ, and at lower temperatures it is less than λ. Because the latent heat is normally large compared with the sensible heat, the slope of the isothermals remains positive down to very low temperatures. Since the humidity is plotted as the ordinate, the slope of the isothermal relative to the X-axis decreases with increase in temperature. When $\theta > \theta_0$ and $\mathscr{H} > \mathscr{H}_0$, the saturation humidity, the vapour phase consists of a saturated gas with liquid droplets in suspension. The relation between enthalpy and humidity at constant temperature θ is:

$$H = \text{constant} + C_L(\theta - \theta_0)\mathscr{H} \tag{13.20}$$

The isothermal is therefore a straight line of slope $C_L(\theta - \theta_0)$. At the reference temperature θ_0, the slope is zero and the isothermal is parallel to the humidity axis. At higher temperatures, the slope has a small positive value. When $\theta < \theta_0$ and $\mathscr{H} > \mathscr{H}_0$, solid particles are formed and the equation of the isothermal is:

$$H = \text{constant} + [C_s(\theta - \theta_0) + \lambda_f]\mathscr{H} \tag{13.21}$$

This represents a straight line of slope $[C_s(\theta - \theta_0) + \lambda_f]$. Both $C_s(\theta - \theta_0)$ and λ_f are negative and therefore the slopes of all these isothermals are negative. When $\theta = \theta_0$, the slope is λ_f. In the supersaturated region therefore, there are two distinct isothermals at temperature θ_0; one corresponds to the condition where the excess vapour is present in the form of liquid droplets and the other to the condition where it is present as solid particles. The region between these isothermals represents conditions where a mixture of liquid and solid is present in the saturated gas at the temperature θ_0.

The shape of the humidity-enthalpy line for saturated air is such that the proportion of the total area of the diagram representing saturated, as opposed to supersaturated, air is small when rectangular axes are used. In order to enable greater accuracy to be obtained in the use of the diagram, oblique axes are normally used, as in Figure 13.5, so that the isothermal for unsaturated gas at the reference temperature θ_0 is parallel to the humidity axis.

It should be noted that the curves of humidity plotted against either temperature or enthalpy have a discontinuity at the point corresponding to the freezing point of the humidifying material. Above the temperature θ_0 the lines are determined by the vapour-liquid equilibrium and below it by the vapour–solid equilibrium.

Two cases may be considered to illustrate the use of enthalpy–humidity charts. These are the mixing of two streams of humid gas and the addition of liquid or vapour to a gas.

Mixing of two streams of humid gas

Consider the mixing of two gases of humidities $\mathscr{H}_1$ and $\mathscr{H}_2$, at temperatures θ_1 and θ_2, and with enthalpies H_1 and H_2 to give a mixed gas of temperature θ, enthalpy H, and humidity $\mathscr{H}$. If the masses of dry gas concerned are m_1, m_2, and m respectively, then taking a balance on the dry gas, vapour, and enthalpy:

$$m_1 + m_2 = m \tag{13.22}$$

$$m_1\mathscr{H}_1 + m_2\mathscr{H}_2 = m\mathscr{H} \tag{13.23}$$

and:

$$m_1H_1 + m_2H_2 = mH \tag{13.24}$$

Elimination of m gives:

$$m_1(\mathscr{H} - \mathscr{H}_1) = m_2(\mathscr{H}_2 - \mathscr{H}) \tag{13.25}$$

and:

$$m_1(H - H_1) = m_2(H_2 - H)$$

Dividing these two equations:

$$\frac{(\mathscr{H} - \mathscr{H}_1)}{(H - H_1)} = \frac{(\mathscr{H} - \mathscr{H}_2)}{(H - H_2)} \tag{13.26}$$

The condition of the resultant gas is therefore represented by a point on the straight line joining $(\mathscr{H}_1, H_1)$ and $(\mathscr{H}_2, H_2)$. The humidity $\mathscr{H}$ is given, from equation 13.25, by:

$$\frac{(\mathscr{H} - \mathscr{H}_1)}{(\mathscr{H}_2 - \mathscr{H})} = \frac{m_2}{m_1} \tag{13.27}$$

The gas formed by mixing two unsaturated gases may be either unsaturated, saturated, or supersaturated. The possibility of producing supersaturated gas arises because the 100 per cent humidity line on the humidity–enthalpy diagram is concave towards the humidity axis.

Example 13.5

In an air-conditioning system, 1 kg/s air at 350 K and 10 per cent humidity is mixed with 5 kg/s air at 300 K and 30 per cent humidity. What is the enthalpy, humidity, and temperature of the resultant stream?

Solution

From Figure 13.4:

at $\theta_1 = 350$ K and humidity = 10 per cent; $\mathscr{H}_1 = 0.043$ kg/kg

at $\theta_2 = 300$ K and humidity = 30 per cent; $\mathscr{H}_2 = 0.0065$ kg/kg

Thus, in equation 13.23:

$$(1 \times 0.043) + (5 \times 0.0065) = (1 + 5)\mathscr{H}$$

and:

$$\mathscr{H} = 0.0125 \text{ kg/kg}$$

From Figure 13.5:

$$\text{at } \theta_1 = 350 \text{ K and } \mathscr{H}_1 = 0.043 \text{ kg/kg}; \quad H_1 = 192 \text{ kJ/kg}$$

$$\text{at } \theta_2 = 300 \text{ K and } \mathscr{H}_2 = 0.0065 \text{ kg/kg}; \quad H_2 = 42 \text{ kJ/kg}$$

Thus, in equation 13.25:

$$1(H - 192) = 5(42 - H)$$

and:

$$H = 67 \text{ kJ/kg}$$

From Figure 13.5:

$$\text{at } H = 67 \text{ kJ/kg and } \mathscr{H} = 0.0125 \text{ kg/kg}$$

$$\theta = 309 \text{ K}$$

The data used in this example are shown in Figure 13.8.

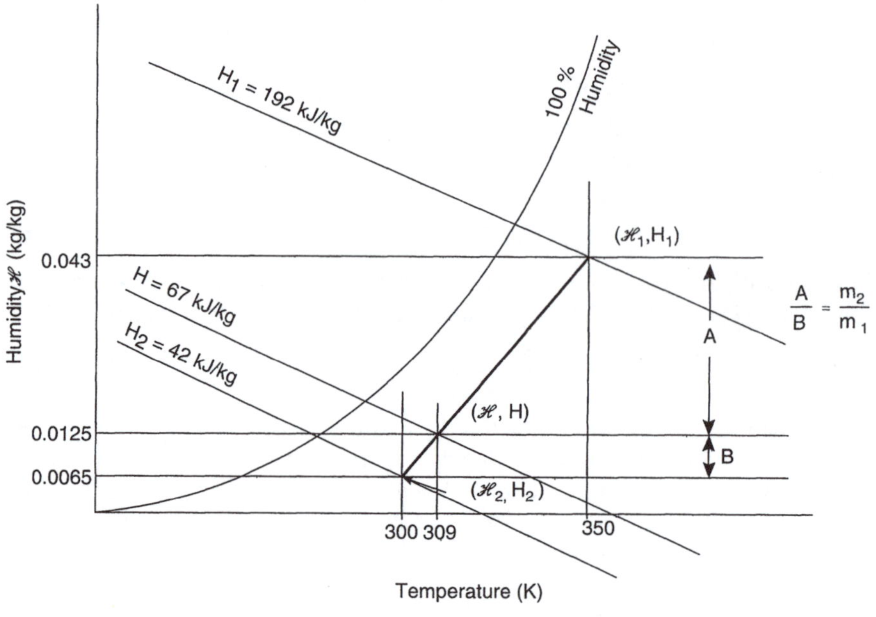

Figure 13.8. Data used in Example 13.5

Addition of liquid or vapour to a gas

If a mass m_3 of liquid or vapour of enthalpy H_3 is added to a gas of humidity $\mathscr{H}_1$ and enthalpy H_1 and containing a mass m_1 of dry gas, then:

$$m_1(\mathscr{H} - \mathscr{H}_1) = m_3 \tag{13.28}$$

$$m_1(H - H_1) = m_3 H_3 \tag{13.29}$$

Thus:
$$\frac{(H - H_1)}{(\mathscr{H} - \mathscr{H}_1)} = H_3 \tag{13.30}$$

where $\mathscr{H}$ and H are the humidity and enthalpy of the gas produced on mixing.

The composition and properties of the mixed stream are therefore represented by a point on the straight line of slope H_3, relative to the humidity axis, which passes through the point $(H_1, \mathscr{H}_1)$. In Figure 13.5 the edges of the plot are marked with points which, when joined to the origin, give a straight line of the slope indicated. Thus in using the chart, a line of slope H_3 is drawn through the origin and a parallel line drawn through the point $(H_1, \mathscr{H}_1)$. The point representing the final gas stream is then given from equation 13.28:

$$(\mathscr{H} - \mathscr{H}_1) = \frac{m_3}{m_1}$$

It can be seen from Figure 13.5 that for the air-water system a straight line, of slope equal to the enthalpy of dry saturated steam (2675 kJ/kg), is almost parallel to the isothermals, so that the addition of live steam has only a small effect on the temperature of the gas. The addition of water spray, even if the water is considerably above the temperature of the gas, results in a lowering of the temperature after the water has evaporated. This arises because the latent heat of vaporisation of the liquid constitutes the major part of the enthalpy of the vapour. Thus, when steam is added, it gives up a small amount of sensible heat to the gas, whereas when hot liquid is added a small amount of sensible heat is given up and a very much larger amount of latent heat is absorbed from the gas.

Example 13.6

0.15 kg/s steam at atmospheric pressure and superheated to 400 K is bled into an air stream at 320 K and 20 per cent relative humidity. What is the temperature, enthalpy, and relative humidity of the mixed stream if the air is flowing at 5 kg/s? How much steam would be required to provide an exit temperature of 330 K and what would be the humidity of this mixture?

Solution

Steam at atmospheric pressure is saturated at 373 K at which the latent heat

$$= 2258 \text{ kJ/kg}$$

Taking the specific heat of superheated steam as 2.0 kJ/kg K;

enthalpy of the steam: $H_3 = 4.18(373 - 273) + 2258 + 2.0(400 - 373)$

$$= 2730 \text{ kJ/kg}$$

From Figure 13.5:

at $\theta_1 = 320$ K and 20 per cent relative humidity; $\mathscr{H}_1 = 0.013$ kg/kg and $H_1 = 83$ kJ/kg

The line joining the axis and slope $H_3 = 2730$ kJ/kg at the edge of the chart is now drawn in and a parallel line is drawn through $(H_1, \mathscr{H}_1)$.

Thus:
$$(\mathscr{H} - \mathscr{H}_1) = \frac{m_3}{m_1} = \left(\frac{0.15}{5}\right) = 0.03 \text{ kg/kg}$$

and:
$$\mathscr{H} = (0.03 + 0.013) = 0.043 \text{ kg/kg}$$

At the intersection of $\mathscr{H} = 0.043$ kg/kg and the line through $(\mathscr{H}_1, H_1)$

$$\underline{\underline{H = 165 \text{ kJ/kg and } \theta = 324 \text{ K}}}$$

When $\theta = 330$ K the intersection of this isotherm and the line through $(\mathscr{H}_1, H_1)$ gives an outlet stream in which $\mathscr{H} = 0.094$ kg/kg (83 per cent relative humidity) and $H = 300$ kJ/kg.

Thus, in equation 13.28:

$$m_3 = 5(0.094 - 0.013) = \underline{\underline{0.41 \text{ kg/s}}}$$

The data used in this example are shown in Figure 13.9.

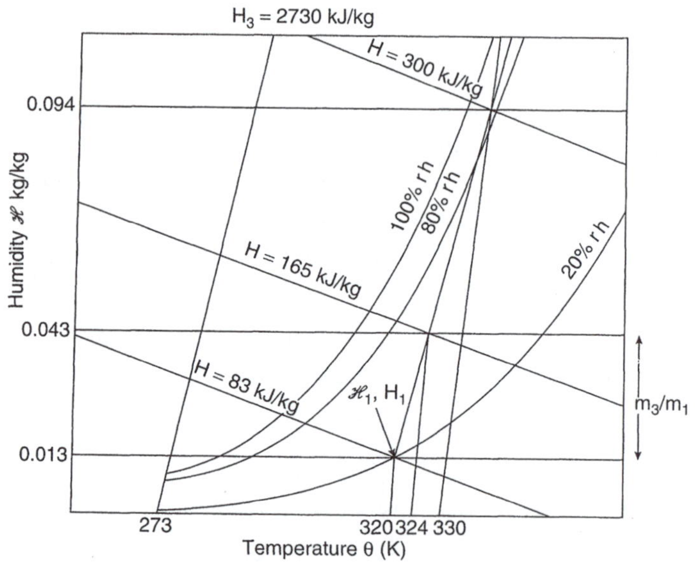

Figure 13.9. Data used in Example 13.6 (schematic)

13.4. DETERMINATION OF HUMIDITY

The most important methods for determining humidity are as follows:

(1) *Chemical methods.* A known volume of the gas is passed over a suitable absorbent, the increase in mass of which is measured. The efficiency of the process can be checked by arranging a number of vessels containing absorbent in series and ascertaining that the increase in mass in the last of these is negligible. The method is very accurate but is laborious. Satisfactory absorbents for water vapour are phosphorus pentoxide dispersed in pumice, and concentrated sulphuric acid.

(2) *Determination of the wet-bulb temperature.* Equation 13.8 gives the humidity of a gas in terms of its temperature, its wet-bulb temperature, and various physical properties of the gas and vapour. The wet-bulb temperature is normally determined as the temperature attained by the bulb of a thermometer which is covered with a piece of material which is maintained saturated with the liquid. The gas should be passed over the surface of the wet bulb at a high enough velocity (>5 m/s) (a) for the condition of the gas stream not to be affected appreciably by the evaporation of liquid, (b) for the heat transfer by convection to be large compared with that by radiation and conduction from the surroundings, and

(c) for the ratio of the coefficients of heat and mass transfer to have reached a constant value. The gas should be passed long enough for equilibrium to be attained and, for accurate work, the liquid should be cooled nearly to the wet-bulb temperature before it is applied to the material.

The stream of gas over the liquid surface may be produced by a small fan or other similar means (Figure 13.10*a*). The crude forms of wet-bulb thermometer which make

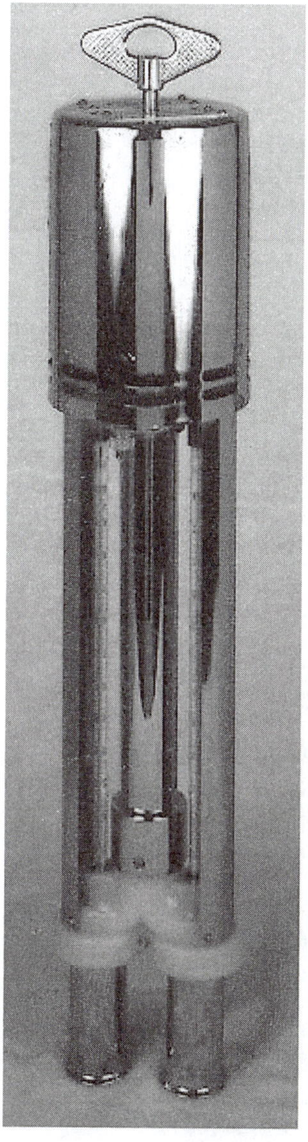

(a) Wet-bulb thermometer

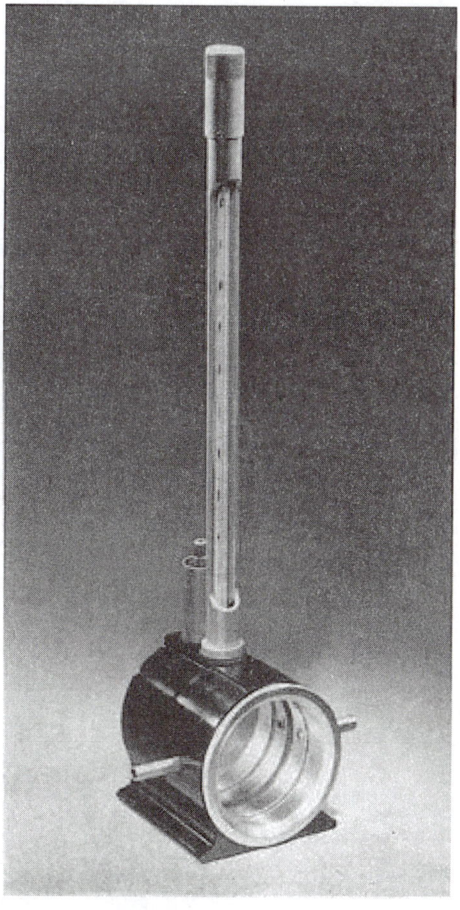

(b) Dew-point meter

Figure 13.10. Hygrometers

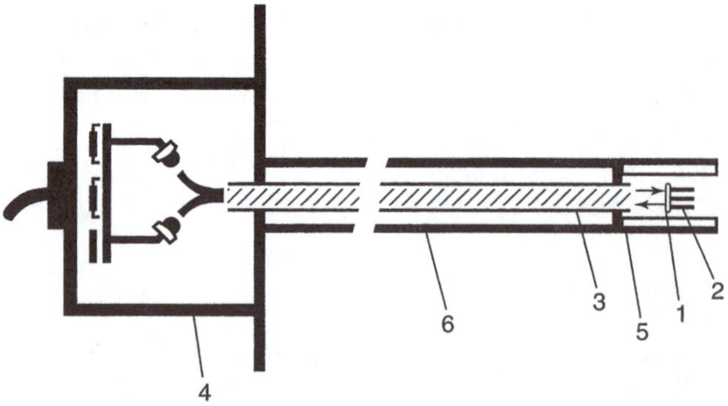

1. Solid gold mirror
2. Peltier cooling device
3. Optical fibres
4. Optical bridge
5. Dual stage filter
6. Carbon fibre

(*c*) Dew-point meter with cyclic chilled-mirror system

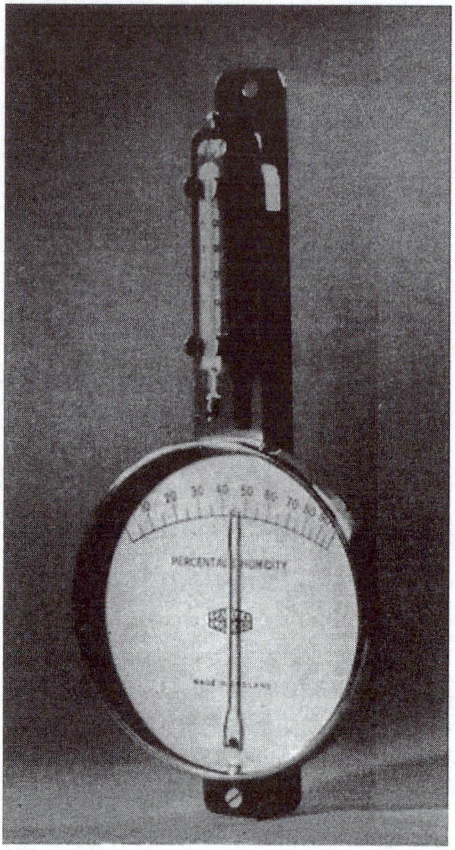

(*d*) Hair hygrometer

Figure 13.10. (*continued*)

no provision for the rapid passage of gas cannot be used for accurate determinations of humidity.

(3) *Determination of the dew point.* The dew point is determined by cooling a highly polished surface in the gas and observing the highest temperature at which condensation takes place (Figure 13.10*b*). The humidity of the gas is equal to the humidity of saturated gas at the dew-point. The instrument illustrated in Figure 13.10*c* incorporates a polished gold mirror which is cooled using a thermo-electric module which utilises the *Peltier effect.*

(4) *Measurement of the change in length of a hair or fibre.* The length of a hair or fibre is influenced by the humidity of the surrounding atmosphere. Many forms of apparatus for automatic recording of humidity depend on this property. The method has the disadvantage that the apparatus needs frequent calibration because the zero tends to shift. This difficulty is most serious when the instrument is used over a wide range of humidities. A typical hair hygrometer is shown in Figure 13.10*d*.

(5) *Measurement of conductivity of a fibre.* If a fibre is impregnated with an electrolyte, such as lithium chloride, its electrical resistance will be governed by its moisture content, which in turn depends on the humidity of the atmosphere in which it is situated. In a lithium chloride cell, a skein of very fine fibres is wound on a plastic frame carrying the electrodes and the current flowing at a constant applied voltage gives a direct measure of the relative humidity.

(6) *Measurement of heat of absorption on to a surface.*

(7) *Electrolytic hygrometry* in which the quantity of electricity required to electrolyse water absorbed from the atmosphere on to a thin film of desiccant is measured.

(8) *Piezo-electric hygrometry* employing a quartz crystal with a hygroscopic coating in which moisture is alternately absorbed from a wet-gas and desorbed in a dry-gas stream; the dynamics is a function of the gas humidity.

(9) *Capacitance meters* in which the electrical capacitance is a function of the degree of deposition of moisture from the atmosphere.

(10) *Observation of colour changes* in active ingredients, such as cobaltous chloride.

Further details of instruments for the measurement of humidity are given in Volume 3. Reference should also be made to standard works on psychrometry[4,5,6].

13.5. HUMIDIFICATION AND DEHUMIDIFICATION

13.5.1. Methods of increasing humidity

The following methods may be used for increasing the humidity of a gas:

(1) Live steam may be added directly in the required quantity. It has been shown that this produces only a slight increase in the temperature, but the method is not generally favoured because any impurities that are present in the steam may be added at the same time.

(2) Water may be sprayed into the gas at such a rate that, on complete vaporisation, it gives the required humidity. In this case, the temperature of the gas will fall as the latent heat of vaporisation must be supplied from the sensible heat of the gas and liquid.

(3) The gas may be mixed with a stream of gas of higher humidity. This method is frequently used in laboratory work when the humidity of a gas supplied to an apparatus is controlled by varying the proportions in which two gas streams are mixed.

(4) The gas may be brought into contact with water in such a way that only part of the liquid is evaporated. This is perhaps the most common method and will now be considered in more detail.

In order to obtain a high rate of humidification, the area of contact between the air and the water is made as large as possible by supplying the water in the form of a fine spray; alternatively, the interfacial area is increased by using a packed column. Evaporation occurs if the humidity at the surface is greater than that in the bulk of the air; that is, if the temperature of the water is above the dew point of the air.

When humidification is carried out in a packed column, the water which is not evaporated can be recirculated so as to reduce the requirements of fresh water. As a result of continued recirculation, the temperature of the water will approach the adiabatic saturation temperature of the air, and the air leaving the column will be cooled—in some cases to within 1 deg K of the temperature of the water. If the temperature of the air is to be maintained constant, or raised, the water must be heated.

Two methods of changing the humidity and temperature of a gas from $A(\theta_1, \mathscr{H}_1)$ to $B(\theta_2, \mathscr{H}_2)$ may be traced on the humidity chart as shown in Figure 13.11. The first method consists of saturating the air by water artificially maintained at the dew point of air of humidity $\mathscr{H}_2$ (line AC) and then heating at constant humidity to θ_2 (line CB). In the second method, the air is heated (line AD) so that its adiabatic saturation temperature corresponds with the dew point of air of humidity $\mathscr{H}_2$. It is then saturated by water at the adiabatic saturation temperature (line DC) and heated at constant humidity to θ_2 (line CB). In this second method, an additional operation—the preliminary heating—is carried out on the air, but the water temperature automatically adjusts itself to the required value.

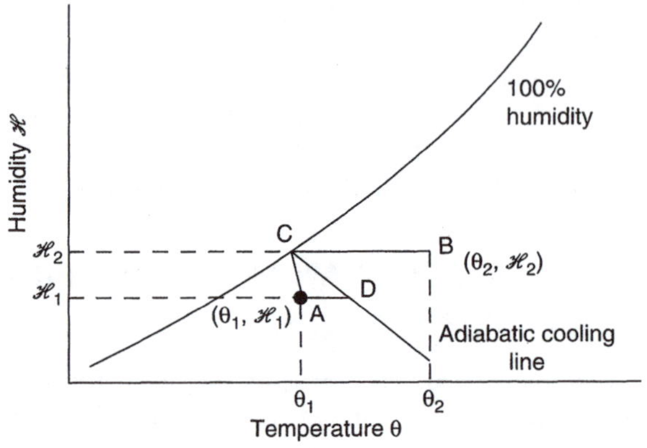

Figure 13.11. Two methods of changing conditions of gas from $(\theta_1, \mathscr{H}_1)$ to $(\theta_2, \mathscr{H}_2)$

Since complete humidification is not always attained, an allowance must be made when designing air humidification cycles. For example, if only 95 per cent saturation is attained the adiabatic cooling line should be followed only to the point corresponding to that degree of saturation, and therefore the gas must be heated to a slightly higher temperature before adiabatic cooling is commenced.

Example 13.7

Air at 300 K and 20 per cent humidity is to be heated in two stages with intermediate saturation with water to 90 per cent humidity so that the final stream is at 320 K and 20 per cent humidity. What is the humidity of the exit stream and the conditions at the end of each stage?

Solution

At $\theta_1 = 300$ K and 20 per cent humidity: $\mathscr{H}_1 = 0.0045$ kg/kg, from Figure 13.4, and
at $\theta_2 = 320$ K and 20 per cent humidity: $\mathscr{H}_2 = \underline{\underline{0.0140 \text{ kg/kg}}}$

When $\mathscr{H}_2 = 0.0140$ kg/kg, air is saturated at 292 K and has a humidity of 90 per cent at 293 K.

The adiabatic cooling line corresponding to 293 K intersects with $\mathscr{H} = 0.0045$ kg/kg at a temperature, $\theta = 318$ K.

Thus the stages are:

(i) Heat the air at $\mathscr{H} = 0.0045$ from 300 to 318 K.
(ii) Saturate with water at an adiabatic saturation temperature of 293 K until 90 per cent humidity is attained. At the end of this stage:

$$\mathscr{H} = 0.0140 \text{ kg/kg} \quad \text{and} \quad \theta = 294.5 \text{ K}$$

(iii) Heat the saturated air at $\underline{\underline{\mathscr{H} = 0.0140 \text{ kg/kg from 294.5 to 320 K}}}$

13.5.2. Dehumidification

Dehumidification of air can be effected by bringing it into contact with a cold surface, either liquid or solid. If the temperature of the surface is lower than the dew point of the gas, condensation takes place and the temperature of the gas falls. The temperature of the surface tends to rise because of the transfer of latent and sensible heat from the air. It would be expected that the air would cool at constant humidity until the dew point was reached, and that subsequent cooling would be accompanied by condensation. It is found, in practice, that this occurs only when the air is well mixed. Normally the temperature and humidity are reduced simultaneously throughout the whole of the process. The air in contact with the surface is cooled below its dew point, and condensation of vapour therefore occurs before the more distant air has time to cool. Where the gas stream is cooled by cold water, countercurrent flow should be employed because the temperature of the water and air are changing in opposite directions.

The humidity can be reduced by compressing air, allowing it to cool again to its original temperature, and draining off the water which has condensed. During compression, the partial pressure of the vapour is increased and condensation takes place as soon as it reaches the saturation value. Thus, if air is compressed to a high pressure, it becomes saturated with vapour, but the partial pressure is a small proportion of the total pressure. Compressed air from a cylinder therefore has a low humidity. Gas is frequently compressed before it is circulated so as to prevent condensation in the mains.

Many large air-conditioning plants incorporate automatic control of the humidity and temperature of the issuing air. Temperature control is effected with the aid of a thermocouple or resistance thermometer, and humidity control by means of a thermocouple recording the difference between the wet- and dry-bulb temperatures.

13.6. WATER COOLING

13.6.1. Cooling towers

Cooling of water can be carried out on a small scale either by allowing it to stand in an open pond or by the spray pond technique in which it is dispersed in spray form and then collected in a large, open pond. Cooling takes place both by the transference of sensible heat and by evaporative cooling as a result of which sensible heat in the water provides the latent heat of vaporisation.

On the large scale, air and water are brought into countercurrent contact in a cooling tower which may employ either natural draught or mechanical draught. The water flows down over a series of wooden slats which give a large interfacial area and promote turbulence in the liquid. The air is humidified and heated as it rises, while the water is cooled mainly by evaporation.

The natural draught cooling tower depends on the chimney effect produced by the presence in the tower of air and vapour of higher temperature and therefore of lower density than the surrounding atmosphere. Thus atmospheric conditions and the temperature and quantity of the water will exert a very important effect on the operation of the tower. Not only will these factors influence the quantity of air drawn through the tower, but they will also affect the velocities and flow patterns and hence the transfer coefficients between gas and liquid. One of the prime considerations in design therefore is to construct a tower in such a way that the resistance to air flow is low. Hence the packings and distributors must be arranged in open formation. The draught of a cooling tower at full load is usually only about 50 N/m^2,[7] and the air velocity in the region of 1.2–1.5 m/s, so that under the atmospheric conditions prevailing in the UK the air usually leaves the tower in a saturated condition. The density of the air stream at outlet is therefore determined by its temperature. Calculation of conditions within the tower is carried out in the manner described in the following pages. It is, however, necessary to work with a number of assumed air flowrates and to select the one which fits both the transfer conditions and the relationship between air rate and pressure difference in the tower.

The *natural draught cooling tower* consists of an empty shell, constructed either of timber or ferroconcrete, where the upper portion is empty and merely serves to increase the draught. The lower portion, amounting to about 10–12 per cent of the total height, is usually fitted with grids on to which the water is fed by means of distributors or sprays as shown in Figure 13.12. The shells of cooling towers are now generally constructed in ferroconcrete in a shape corresponding approximately to a hyperboloid of revolution. The shape is chosen mainly for constructional reasons, but it does take account of the fact that the entering air will have a radial velocity component; the increase in cross-section towards the top causes a reduction in the outlet velocity and there is a small recovery of kinetic energy into pressure energy.

The *mechanical draught cooling tower* may employ forced draught with the fan at the bottom, or induced draught with the fan driving the moist air out at the top. The air velocity can be increased appreciably above that in the natural draught tower, and a greater depth of packing can be used. The tower will extend only to the top of the packing unless atmospheric conditions are such that a chimney must be provided in order to prevent recirculation of the moist air. The danger of recirculation is considerably

Figure 13.12. Water-cooling tower. View of spray distribution system

less with the induced-draught type because the air is expelled with a higher velocity. Mechanical draught towers are generally confined to small installations and to conditions where the water must be cooled to as low a temperature as possible. In some cases it is possible to cool the water to within 1 deg K of the wet-bulb temperature of the air. Although the initial cost of the tower is less, maintenance and operating costs are of course higher than in natural draught towers which are now used for all large installations. A typical steel-framed mechanical draught cooling tower is shown in Figure 13.13.

The operation of the conventional water cooling tower is often characterised by the discharge of a plume consisting of a suspension of minute droplets of water in air. This is formed when the hot humid air issuing from the top of the tower mixes with the ambient atmosphere, and precipitation takes place as described earlier (Section 13.3.2). In the *hybrid* (or wet/dry) cooling tower[8], mist formation is avoided by cooling *part* of the water in a finned-tube exchanger bundle which thus generates a supply of warm dry air which is then blended with the air issuing from the evaporative section. By adjusting the proportion of the water fed to the heat exchanger, the plume can be completely eliminated.

In the cooling tower the temperature of the liquid falls and the temperature and humidity of the air rise, and its action is thus similar to that of an air humidifier. The limiting temperature to which the water can be cooled is the wet-bulb temperature corresponding to the condition of the air at inlet. The enthalpy of the air stream does not remain

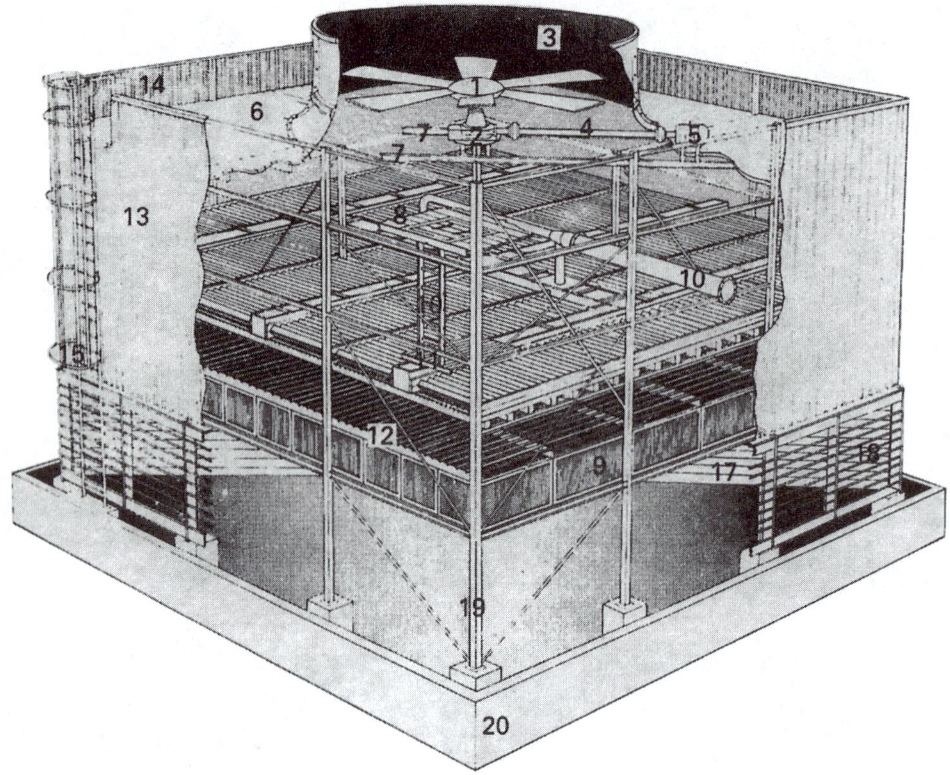

1 Fan assembly
2 Gearbox
3 Fan stack
4 Drive shaft assembly
5 Motor
6 Fan deck
7 Mechanical equipment supports
8 Drift eliminators (PVC or Timber - Timber shown)
9 Cooling tower packing (plastic plate or wooden lath)
10 Inlet water distribution pipe
11 Open type distribution system
12 Timber laths for even water distribution
13 Cladding
14 Cladding extended to form handrail
15 Access ladder
16 Internal access ladder to distribution system and drift eliminators
17 Diagonal wind baffles
18 Air inlet louvres
19 Steel structures with horizontal and diagonal ties
20 Cold water sump

Some structural members have been omitted for clarity

Figure 13.13. Visco 2000 series steel framed, mechanical draught, water cooling tower

constant since the temperature of the liquid changes rapidly in the upper portion of the tower. Towards the bottom, however, the temperature of the liquid changes less rapidly because the temperature differences are smaller. At the top of the tower, the temperature falls from the bulk of the liquid to the interface and then again from the interface to the bulk of the gas. Thus the liquid is cooled by transfer of sensible heat and by evaporation at the surface. At the bottom of a tall tower, however, the temperature gradient in the liquid is in the same direction, though smaller, but the temperature gradient in the gas is in the opposite direction. Transfer of sensible heat to the interface therefore takes place from the bulk of the liquid and from the bulk of the gas, and all the cooling is caused by the evaporation at the interface. In most cases, about 80 per cent of the heat loss from the water is accounted for by evaporative cooling.

13.6.2. Design of natural-draught towers

The air flow through a natural-draught or hyperbolic-type tower (Figure 13.14) is due largely to the difference in density between the warm air in the tower and the external ambient air; thus a draught is created in the stack by a chimney effect which eliminates the need for mechanical fans. It has been noted by McKelvey and Brooke[9] that natural-draught towers commonly operate at a pressure difference of some 50 N/m^2 under full load, and above the packing the mean air velocity is typically 1–2 m/s. The performance of a natural-draught tower differs from that of a mechanical-draught installation in that the cooling achieved depends upon the relative humidity as well as the wet-bulb temperature. It is important therefore, at the design stage, to determine correctly, and to specify, the density of the inlet and exit air streams in addition to the usual tower design conditions of water temperature range, how closely the water temperature should approach the wet bulb temperature of the air, and the quantity of water to be handled. Because the performance

Figure 13.14. Natural draught water-cooling towers

depends to a large extent on atmospheric humidity, the outlet water temperature is difficult to control with natural-draught towers.

In the design of natural-draught towers, a ratio of height to base diameter of 3:2 is normally used and a design method has been proposed by CHILTON(10). Chilton has shown that the duty coefficient D_t of a tower is approximately constant over the normal range of operation and is related to tower size by an efficiency factor or performance coefficient C_t given by:

$$D_t = \frac{19.50\,A_b\,z_t^{0.5}}{C_t^{1.5}} \tag{13.31}$$

where for water loadings in excess of 1 kg/m²s, C_t is usually about 5.2 though lower values are obtained with new packings which are being developed.

The duty coefficient is given by the following equation (in which SI units must be used as it is not dimensionally consistent):

$$\frac{W_L}{D_t} = 0.00369\frac{\Delta H'}{\Delta T}(\Delta T' + 0.0752\Delta H')^{0.5} \tag{13.32}$$

where W_L (kg/s) is the water load in the tower, $\Delta H'$ (kJ/kg) the change in enthalpy of the air passing through the tower, ΔT(deg K) the change in water temperature in passing through the tower and $\Delta T'$(deg K), the difference between the temperature of the air leaving the packing and the dry-bulb temperature of the inlet air. The air leaving the packing inside the tower is assumed to be saturated at the mean of the inlet and outlet water temperatures. Any divergence between theory and practice of a few degrees in this respect does not significantly affect the results as the draught component depends on the ratio of the change of density to change in enthalpy and not on change in temperature alone.(11) The use of equations 13.31 and 13.32 is illustrated in the following example.

Example 13.8

What are the diameter and height of a hyperbolic natural-draught cooling tower handling 4810 kg/s of water with the following temperature conditions:

water entering the tower = 301 K

water leaving the tower = 294 K.

air: dry bulb = 287 K

wet bulb = 284 K

Solution

Temperature range for the water, $\Delta T = (301 - 294) = 7$ deg K.
At a mean water temperature of $0.5(301 + 294) = 297.5$ K, the enthalpy = 92.6 kJ/kg
At a dry bulb temperature of 287 K, the enthalpy = 49.5 kJ/kg

$\therefore$ $$\Delta T' = (297.5 - 287) = 10.5 \text{ deg K}$$

and: $$\Delta H' = (92.6 - 49.5) = 43.1 \text{ kJ/kg}$$

In equation 13.32:

$$\frac{4810}{D_t} = 0.00369\left(\frac{43.1}{7}\right)[10.5 + (0.0752 \times 43.1)]^{0.5}$$

and: $$D_t = 57{,}110$$

Taking C_t as 5.0 and assuming as a first estimate a tower height of 100 m, then in equation 13.31:

$$57{,}110 = 19.50 A_b \frac{100^{0.5}}{5.0^{1.5}}$$

and:

$$A_b = 3274 \text{ m}^2$$

Thus the internal diameter of the column at sill level $= \left(\frac{3274 \times 4}{\pi}\right)^{0.5}$

$$= \underline{\underline{64.6 \text{ m}}}$$

Since this gives a height: diameter ratio of (100 : 64.6) $\approx$ 3 : 2, the design is acceptable.

13.6.3. Height of packing for both natural and mechanical draught towers

The height of a water-cooling tower can be determined[12] by setting up a material balance on the water, an enthalpy balance, and rate equations for the transfer of heat in the liquid and gas and for mass transfer in the gas phase. There is no concentration gradient in the liquid and therefore there is no resistance to mass transfer in the liquid phase.

Considering the countercurrent flow of water and air in a tower of height z (Figure 13.15), the mass rate of flow of air per unit cross-section G' is constant throughout the whole height of the tower and, because only a small proportion of the total supply of water is normally evaporated (1–5 per cent), the liquid rate per unit area L' can be taken as constant. The temperature, enthalpy, and humidity will be denoted by the symbols θ, H, and $\mathscr{H}$ respectively, suffixes G, L, 1, 2, and f being used to denote conditions in the gas and liquid, at the bottom and top of the column, and of the air in contact with the water.

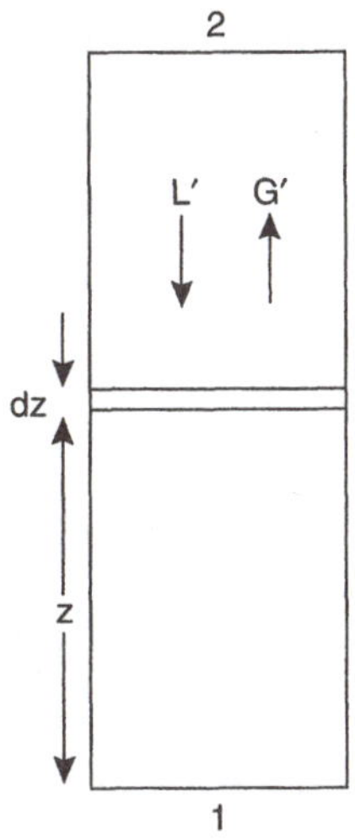

Figure 13.15. Flow in water-cooling tower

The five basic equations for an incremental height of column, dz, are:

(1) Water balance: $$\mathrm{d}L' = G'\,\mathrm{d}\mathscr{H} \tag{13.33}$$

(2) Enthalpy balance: $$G'\,\mathrm{d}H_G = L'\,\mathrm{d}H_L \tag{13.34}$$

since only a small proportion of the liquid is evaporated.

Now: $$H_G = s(\theta_G - \theta_0) + \lambda\mathscr{H} \tag{13.35}$$

and: $$H_L = C_L(\theta_L - \theta_0) \tag{13.36}$$

Thus: $$G'\,\mathrm{d}H_G = L'C_L\,\mathrm{d}\theta_L \tag{13.37}$$

and: $$\mathrm{d}H_G = s\mathrm{d}\theta_G + \lambda\,\mathrm{d}\mathscr{H} \tag{13.38}$$

Integration of this expression over the whole height of the column, on the assumption that the physical properties of the materials do not change appreciably, gives:

$$G'(H_{G2} - H_{G1}) = L'C_L(\theta_{L2} - \theta_{L1}) \tag{13.39}$$

(3) Heat transfer from the body of the liquid to the interface:

$$h_L a\,\mathrm{d}z(\theta_L - \theta_f) = L'C_L\,\mathrm{d}\theta_L \tag{13.40}$$

where h_L is the heat transfer coefficient in the liquid phase and a is the interfacial area per unit volume of column. It will be assumed that the area for heat transfer is equal to that available for mass transfer, though it may be somewhat greater if the packing is not completely wetted.

Rearranging equation 13.40:

$$\frac{\mathrm{d}\theta_L}{(\theta_L - \theta_f)} = \frac{h_L a}{L'C_L}\,\mathrm{d}z \tag{13.41}$$

(4) Heat transfer from the interface to the bulk of the gas:

$$h_G a\,\mathrm{d}z(\theta_f - \theta_G) = G's\,\mathrm{d}\theta_G \tag{13.42}$$

where h_G is the heat transfer coefficient in the gas phase.

Rearranging: $$\frac{\mathrm{d}\theta_G}{(\theta_f - \theta_G)} = \frac{h_G a}{G's}\,\mathrm{d}z \tag{13.43}$$

(5) Mass transfer from the interface to the gas:

$$h_D\rho a\,\mathrm{d}z(\mathscr{H}_f - \mathscr{H}) = G'\,\mathrm{d}\mathscr{H} \tag{13.44}$$

where h_D is the mass transfer coefficient for the gas and ρ is the mean density of the air (see equation 13.6).

Rearranging: $$\frac{\mathrm{d}\mathscr{H}}{(\mathscr{H}_f - \mathscr{H})} = \frac{h_D a\rho}{G'}\,\mathrm{d}z \tag{13.45}$$

These equations cannot be integrated directly since the conditions at the interface are not necessarily constant; nor can they be expressed directly in terms of the corresponding property in the bulk of the gas or liquid.

If the Lewis relation (equation 13.11) is applied, it is possible to obtain workable equations in terms of enthalpy instead of temperature and humidity. Thus, writing h_G as $h_D \rho s$, from equation 13.42:

$$G' s \, d\theta_G = h_D \rho a \, dz (s\theta_f - s\theta_G) \tag{13.46}$$

and from equation 13.44:

$$G' \lambda \, d\mathscr{H} = h_D \rho a \, dz (\lambda \mathscr{H}_f - \lambda \mathscr{H}) \tag{13.47}$$

Adding these two equations gives:

$$G'(s \, d\theta_G + \lambda \, d\mathscr{H}) = h_D \rho a \, dz[(s\theta_f + \lambda \mathscr{H}_f) - (s\theta_G + \lambda \mathscr{H})]$$

$$G' \, dH_G = h_D \rho a \, dz (H_f - H_G) \quad \text{(from equation 13.35)} \tag{13.48}$$

or:

$$\frac{dH_G}{(H_f - H_G)} = \frac{h_D a \rho}{G'} dz \tag{13.49}$$

The use of an enthalpy driving force, as in equation 13.48, was first suggested by MERKEL(13), and the following development of the treatment was proposed by MICKLEY(12).

Combining of equations 13.37, 13.40, and 13.48 gives:

$$\frac{(H_G - H_f)}{(\theta_L - \theta_f)} = -\frac{h_L}{h_D \rho} \tag{13.50}$$

From equations 13.46 and 13.48:

$$\frac{(H_G - H_f)}{(\theta_G - \theta_f)} = \frac{dH_G}{d\theta_G} \tag{13.51}$$

and from equations 13.46 and 13.44:

$$\frac{(\mathscr{H} - \mathscr{H}_f)}{(\theta_G - \theta_f)} = \frac{d\mathscr{H}}{d\theta_G} \tag{13.52}$$

These equations are now employed in the determination of the required height of a cooling tower for a given duty. The method consists of the graphical evaluation of the relation between the enthalpy of the body of gas and the enthalpy of the gas at the interface with the liquid. The required height of the tower is then obtained by integration of equation 13.49.

It is supposed that water is to be cooled at a mass rate L' per unit area from a temperature θ_{L2} to θ_{L1}. The air will be assumed to have a temperature θ_{G1}, a humidity $\mathscr{H}_1$, and an enthalpy H_{G1} (which can be calculated from the temperature and humidity), at the inlet point at the bottom of the tower, and its mass flow per unit area will be taken as G'. The change in the condition of the liquid and gas phases will now be followed on an enthalpy–temperature diagram (Figure 13.16). The enthalpy–temperature curve PQ for saturated air is plotted either using calculated data or from the humidity chart (Figure 13.4). The region below this line relates to unsaturated air and the region above it to supersaturated air. If it is assumed that the air in contact with the liquid surface

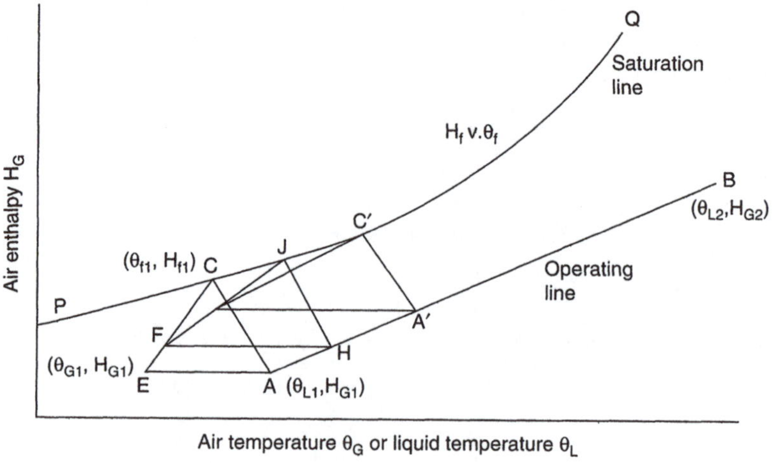

Figure 13.16. Construction for determining the height of water-cooling tower

is saturated with water vapour, this curve represents the relation between air enthalpy H_f and temperature θ_f at the interface.

The curve connecting air enthalpy and water temperature is now drawn using equation 13.39. This is known as the operating line and is a straight line of slope $(L'C_L/G')$, passing through the points $A(\theta_{L1}, H_{G1})$ and $B(\theta_{L2}, H_{G2})$. Since θ_{L1}, H_{G1} are specified, the procedure is to draw a line through (θ_{L1}, H_{G1}) of slope $L'C_L/G'$ and to produce it to a point whose abscissa is equal to θ_{L2}. This point B then corresponds to conditions at the top of the tower and the ordinate gives the enthalpy of the air leaving the column.

Equation 13.50 gives the relation between liquid temperature, air enthalpy, and conditions at the interface, for any position in the tower, and is represented by a family of straight lines of slope $-(h_L/h_D\rho)$. The line for the bottom of the column passes through the point $A(\theta_{L1}, H_{G1})$ and cuts the enthalpy-temperature curve for saturated air at the point C, representing conditions at the interface. The difference in ordinates of points A and C is the difference in the enthalpy of the air at the interface and that of the bulk air at the bottom of the column.

Similarly, line $A'C'$, parallel to AC, enables the difference in the enthalpies of the bulk air and the air at the interface to be determined at some other point in the column. The procedure can be repeated for a number of points and the value of $(H_f - H_G)$ obtained as a function of H_G for the whole tower.

Now:

$$\frac{dH_G}{(H_f - H_G)} = \frac{h_D a \rho}{G'} dz \qquad \text{(equation 13.49)}$$

On integration:

$$z = \int_1^2 dz = \frac{G'}{h_D a \rho} \int_1^2 \frac{dH_G}{(H_f - H_G)} \qquad (13.53)$$

assuming h_D to remain approximately constant.

Since $(H_f - H_G)$ is now known as a function of H_G, $1/(H_f - H_G)$ can be plotted against H_G and the integral evaluated between the required limits. The height of the tower is thus determined.

The integral in equation 13.53 cannot be evaluated by taking a logarithmic mean driving force because the saturation line PQ is far from linear. CAREY and WILLIAMSON[14] have given a useful approximate method of evaluating the integral. They assume that the enthalpy difference $(H_f - H_G) = \Delta H$ varies in a parabolic manner. The three fixed points taken to define the parabola are at the bottom and top of the column (ΔH_1 and ΔH_2 respectively) and ΔH_m, the value at the mean water temperature in the column. The effective mean driving force is $f\Delta H_m$, where f is a factor for converting the driving force at the mean water temperature to the effective value. In Figure 13.17, $(\Delta H_m/\Delta H_1)$ is plotted against $(\Delta H_m/\Delta H_2)$ and contours representing constant values of f are included.

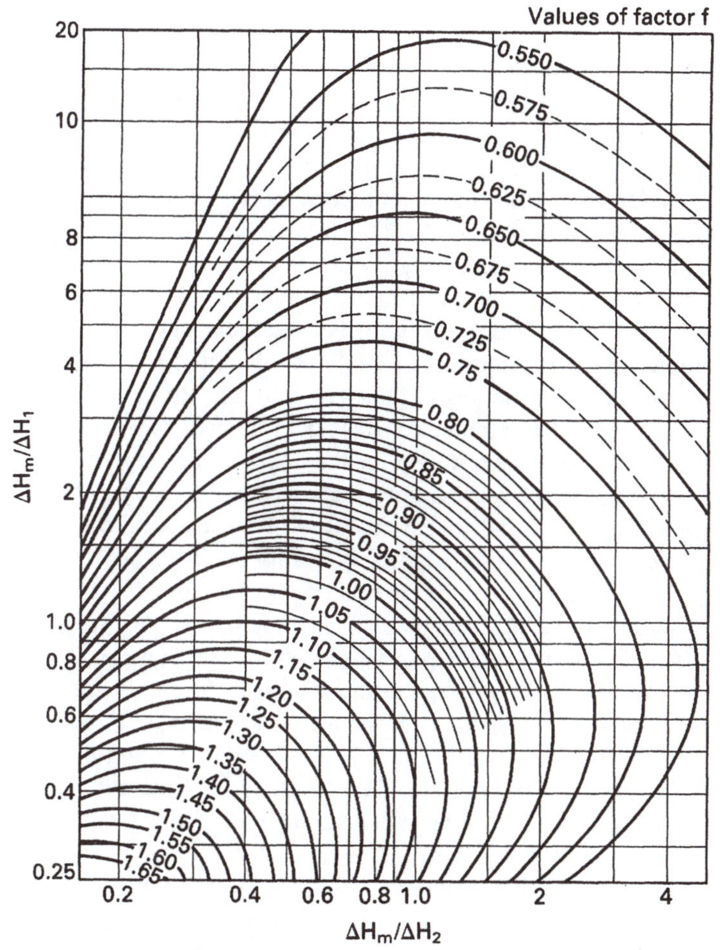

Figure 13.17. Correction factor f for obtaining the mean effective driving force in column

Using the mean driving force, integration of equation 13.53 gives:

$$\frac{(H_{G2} - H_{G1})}{f\Delta H_m} = \frac{h_D a \rho}{G'} z \tag{13.54}$$

or:

$$z = \frac{G'}{h_D a \rho} \frac{(H_{G2} - H_{G1})}{f\Delta H_m}$$

13.6.4. Change in air condition

The change in the humidity and temperature of the air is now obtained. The enthalpy and temperature of the air are known only at the bottom of the tower, where fresh air is admitted. Here the condition of the air may be represented by a point E with coordinates (H_{G1}, θ_{G1}). Thus the line AE (Figure 13.16) is parallel to the temperature axis.

Since:

$$\frac{H_G - H_f}{\theta_G - \theta_f} = \frac{dH_G}{d\theta_G} \tag{13.51}$$

the slope of the line EC is $(dH_G/d\theta_G)$ and represents the rate of change of air enthalpy with air temperature at the bottom of the column. If the gradient $(dH_G/d\theta_G)$ is taken as constant over a small section, the point F, on EC, will represent the condition of the gas at a small distance from the bottom. The corresponding liquid temperature is found by drawing through F a line parallel to the temperature axis. This cuts the operating line at some point H, which indicates the liquid temperature. The corresponding value of the temperature and enthalpy of the gas at the interface is then obtained by drawing a line through H, parallel to AC. This line then cuts the curve for saturated air at a point J, which represents the conditions of the gas at the interface. The rate of change of enthalpy with temperature for the gas is then given by the slope of the line FJ. Again, this slope can be considered to remain constant over a small height of the column, and the condition of the gas is thus determined for the next point in the tower. The procedure is then repeated until the curve representing the condition of the gas has been extended to a point whose ordinate is equal to the enthalpy of the gas at the top of the column. This point is obtained by drawing a straight line through B, parallel to the temperature axis. The final point on the line then represents the condition of the air which leaves the top of the water-cooling tower.

The size of the individual increments of height which are considered must be decided for the particular problem under consideration and will depend, primarily, on the rate of change of the gradient $(dH_G/d\theta_G)$. It should be noted that, for the gas to remain unsaturated throughout the whole of the tower, the line representing the condition of the gas must be below the curve for saturated gas. If at any point in the column, the air has become saturated, it is liable to become supersaturated as it passes further up the column and comes into contact with hotter liquid. It is difficult to define precisely what happens beyond this point as partial condensation may occur, giving rise to a mist. Under these conditions the preceding equations will no longer be applicable. However, an approximate solution is obtained by assuming that once the air stream becomes saturated it remains so during its subsequent contact with the water through the column.

13.6.5. Temperature and humidity gradients in a water cooling tower

In a water cooling tower, the temperature profiles depend on whether the air is cooler or hotter than the surface of the water. Near the top, hot water makes contact with the exit air which is at a lower temperature, and sensible heat is therefore transferred both from the water to the interface and from the interface to the air. The air in contact with the water is saturated at the interface temperature and humidity therefore falls from the interface to the air. Evaporation followed by mass transfer of water vapour therefore takes place and latent heat is carried away from the interface in the vapour. The sensible heat removed from the water is then equal to the sum of the latent and sensible heats transferred to the air. Temperature and humidity gradients are then as shown in Figure 13.18*a*.

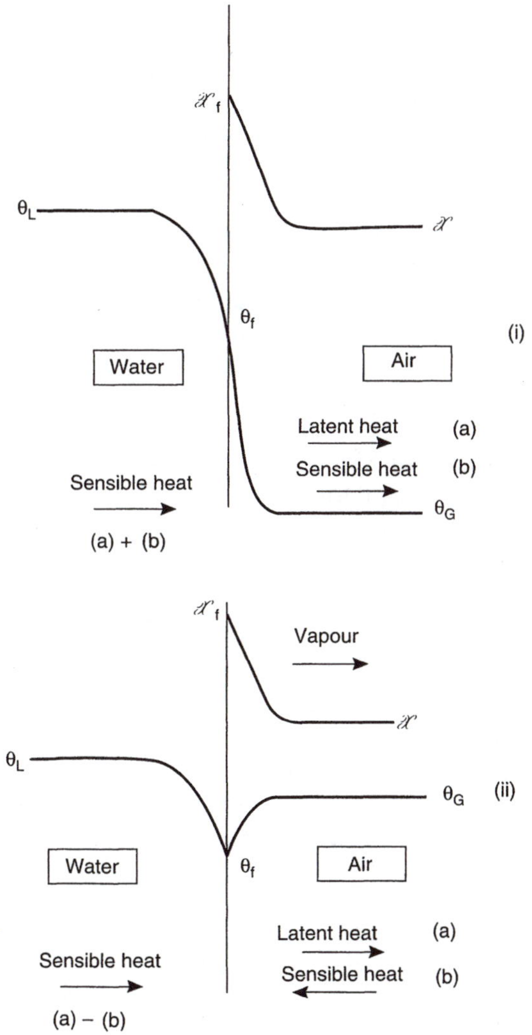

Figure 13.18. Temperature and humidity gradients in a water cooling tower (*i*) upper sections of tower (*ii*) bottom of tower

If the tower is sufficiently tall, the interface temperature can fall below the dry bulb temperature of the air (but not below its wet bulb temperature), and sensible heat will then be transferred from both the air and the water to the interface. The corresponding temperature and humidity profiles are given in Figure 13.18*b*. In this part of the tower, therefore, the sensible heat removed from the water will be that transferred as latent heat *less* the sensible heat transferred from the air.

13.6.6. Evaluation of heat and mass transfer coefficients

In general, coefficients of heat and mass transfer in the gas phase and the heat transfer coefficient for the liquid phase are not known. They may be determined, however, by carrying out tests in the laboratory or pilot scale using the same packing. If, for the air-water system, a small column is operated at steady water and air rates and the temperature of the water at the top and bottom and the initial and final temperatures and humidities of the air stream are noted, the operating line for the system is obtained. Assuming a value of the ratio $-(h_L/h_D\rho)$, for the slope of the tie-lines AC, etc., the graphical construction is carried out, starting with the conditions at the bottom of the tower. The condition of the gas at the top of the tower is thus calculated and compared with the measured value. If the difference is significant, another value of $-(h_L/h_D\rho)$ is assumed and the procedure repeated. Now that the slope of the tie line is known, the value of the integral of $\mathrm{d}H_G/(H_f - H_G)$ over the whole column can be calculated. Since the height of the column is known, the product h_Da is found by solution of equation 13.49. h_Ga may then be calculated using the Lewis relation. The values of the three transfer coefficients are therefore obtained at any given flow rates from a single experimental run. The effect of liquid and gas rate may be found by means of a series of similar experiments.

Several workers have measured heat and mass transfer coefficients in water-cooling towers and in humidifying towers. THOMAS and HOUSTON[15], using a tower 2 m high and 0.3 m square in cross-section, fitted with wooden slats, give the following equations for heat and mass transfer coefficients for packed heights greater than 75 mm:

$$h_Ga = 3.0L'^{0.26}G'^{0.72} \tag{13.55}$$

$$h_La = 1.04 \times 10^4 L'^{0.51}G'^{1.00} \tag{13.56}$$

$$h_Da = 2.95L'^{0.26}G'^{0.72} \tag{13.57}$$

In these equations, L' and G' are expressed in kg/m^{2}s, s in J/kg K, h_Ga and h_La in W/m^3 K, and h_Da in s^{-1}. A comparison of the gas and liquid film coefficients may then be made for a number of gas and liquid rates. Taking the humid heat s as 1.17×10^3 J/kg K:

	$L' = G' = 0.5$ kg/m^{2}s	$L' = G' = 1.0$ kg/m^{2}s	$L' = G' = 2.0$ kg/m^{2}s
h_Ga	1780	3510	6915
h_La	3650	10,400	29,600
h_La/h_Ga	2.05	2.96	4.28

CRIBB[16] quotes values of the ratio h_L/h_G ranging from 2.4 to 8.5.

It is seen that the liquid film coefficient is generally considerably higher than the gas film coefficient, but that it is not always safe to ignore the resistance to transfer in the liquid phase.

LOWE and CHRISTIE[17] used a 1.3 m square experimental column fitted with a number of different types of packing and measured heat and mass transfer coefficients and pressure drops. They showed that in most cases:

$$h_D a \propto L'^{1-n} G'^{n} \tag{13.58}$$

The index n was found to vary from about 0.4 to 0.8 according to the type of packing. It will be noted that when $n \approx 0.75$, there is close agreement with the results given by equation 13.57.

The heat-transfer coefficient for the liquid is often large compared with that for the gas phase. As a first approximation, therefore, it can be assumed that the whole of the resistance to heat transfer lies within the gas phase and that the temperature at the water-air interface is equal to the temperature of the bulk of the liquid. Thus, everywhere in the tower, $\theta_f = \theta_L$. This simplifies the calculations, since the lines AC, HJ, and so on, have a slope of $-\infty$, that is, they become parallel to the enthalpy axis.

Some workers have attempted to base the design of humidifiers on the overall heat transfer coefficient between the liquid and gas phases. This treatment is not satisfactory since the quantities of heat transferred through the liquid and through the gas are not the same, as some of the heat is utilised in effecting evaporation at the interface. In fact, at the bottom of a tall tower, the transfer of heat in both the liquid and the gas phases may be towards the interface, as already indicated. A further objection to the use of overall coefficients is that the Lewis relation may be applied only to the heat and mass transfer coefficients in the gas phase.

In the design of commercial units, nomographs[18,19] are available which give a performance characteristic (KaV/L'), where K is a mass transfer coefficient (kg water/m^{2}s) and V is the active cooling volume (m^3/m^2 plan area), as a function of θ, θ_w and (L'/G'). For a given duty (KaV/L') is calculated from:

$$\frac{KaV}{C_L L'} = \int_{\theta_1}^{\theta_2} \frac{d\theta}{(H_f - H_G)} \tag{13.59}$$

and then a suitable tower with this value of (KaV/L') is sought from performance curves[20,21]. In normal applications the performance characteristic varies between 0.5–2.5.

Example 13.9

Water is to be cooled from 328 to 293 K by means of a countercurrent air stream entering at 293 K with a relative humidity of 20 per cent. The flow of air is 0.68 m^3/m^{2}s and the water throughput is 0.26 kg/m^{2}s. The whole of the resistance to heat and mass transfer may be assumed to be in the gas phase and the product, $(h_D a)$, may be taken as 0.2 (m/s)(m^2/m^3), that is 0.2 s^{-1}.

What is the required height of packing and the condition of the exit air stream?

Solution

Assuming the latent heat of water at 273 K = 2495 kJ/kg
specific heat of air = 1.003 kJ/kg K
and specific heat of water vapour = 2.006 kJ/kg K

the enthalpy of the inlet air stream,

$$H_{G1} = 1.003(293 - 273) + \mathscr{H}[2495 + 2.006(293 - 273)]$$

From Figure 13.4:

at $\theta = 293$ K and 20 per cent RH, $\mathscr{H} = 0.003$ kg/kg, and hence

$$H_{G1} = (1.003 \times 20) + 0.003[2495 + (2.006 \times 20)]$$

$$= 27.67 \text{ kJ/kg}$$

In the inlet air, water vapour = 0.003 kg/kg dry air

or: $$\frac{(0.003/18)}{(1/29)} = 0.005 \text{ kmol/kmol dry air}$$

Thus flow of dry air $= (1 - 0.005)0.68 = 0.677 \text{ m}^3/\text{m}^2\text{s}$

Density of air at 293 K $= \left(\frac{29}{22.4}\right)\left(\frac{273}{293}\right) = 1.206 \text{ kg/m}^3$

and mass flow of dry air $= (1.206 \times 0.677) = 0.817 \text{ kg/m}^2\text{s}$

Slope of operating line: $(L'C_L/G') = \frac{(0.26 \times 4.18)}{0.817} = 1.33$

The coordinates of the bottom of the operating line are:

$$\theta_{L1} = 293 \text{ K}, \quad H_{G1} = 27.67 \text{ kJ/kg}$$

Hence on an enthalpy–temperature diagram, the operating line of slope 1.33 is drawn through the point $(293, 27.67) = (\theta_{L1}, H_{G1})$.

The top point of the operating line is given by $\theta_{L2} = 328$ K, and H_{G2} is found to be 76.5 kJ/kg (Figure 13.19).

From Figures 13.4 and 13.5 the curve representing the enthalpy of saturated air as a function of temperature is obtained and drawn in. Alternatively, this plot may be calculated from:

$$H_F = C_a(\theta_f - 273) + \mathscr{H}_0[C_w(\theta_f - 273) + \lambda] \text{ kJ/kg}$$

The curve represents the relation between enthalpy and temperature at the interface, that is H_f as a function of θ_f.

It now remains to evaluate the integral $\int dH_G/(H_f - H_G)$ between the limits, $H_{G1} = 27.7$ kJ/kg and $H_{G2} = 76.5$ kJ/kg. Various values of H_G between these limits are selected and the value of θ obtained from the operating line. At this value of θ_1, now θ_f, the corresponding value of H_f is obtained from the curve for saturated air. The working is as follows:

H_G	$\theta = \theta_f$	H_f	$(H_f - H_G)$	$1/(H_f - H_G)$
27.7	293	57.7	30	0.0330
30	294.5	65	35	0.0285
40	302	98	58	0.0172
50	309	137	87	0.0114
60	316	190	130	0.0076
70	323	265	195	0.0051
76.5	328	355	279	0.0035

A plot of $1/(H_f - H_G)$ and H_G is now made as shown in Figure 13.20 from which the area under the curve = 0.65. This value may be checked using the approximate solution of CAREY and WILLIAMSON[14].

At the bottom of the column:

$$H_{G1} = 27.7 \text{ kJ/kg}, \quad H_{f1} = 57.7 \text{ kJ/kg} \quad \therefore \Delta H_1 = 30 \text{ kJ/kg}$$

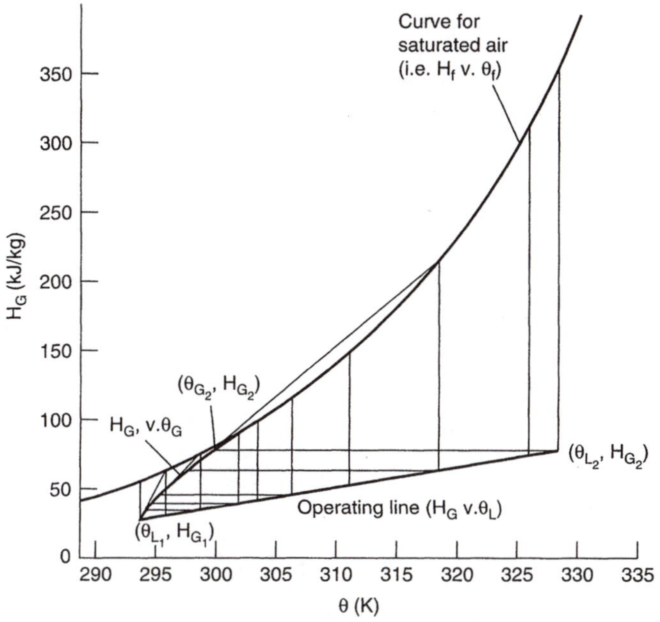

Figure 13.19. Calculation of the height of a water-cooling tower

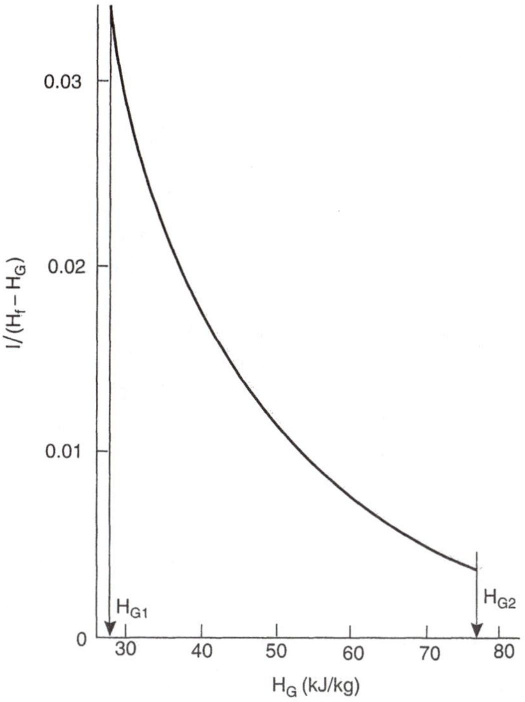

Figure 13.20. Evaluation of the integral of $dH_G/(H_f - H_G)$

At the top of the column:

$$H_{G2} = 76.5\ \text{kJ/kg}, \quad H_{f2} = 355\ \text{kJ/kg} \quad \therefore \Delta H_2 = 279\ \text{kJ/kg}$$

At the mean water temperature of $0.5(328 + 293) = 310.5$ K:

$$H_{Gm} = 52\ \text{kJ/kg}, \quad H_f = 145\ \text{kJ/kg} \quad \therefore \Delta H_m = 93\ \text{kJ/kg}$$

$$\frac{\Delta H_m}{\Delta H_1} = 3.10, \quad \frac{\Delta H_m}{\Delta H_2} = 0.333,$$

and from Figure 13.16: $f = 0.79$

Thus: $$\frac{(H_{G2} - H_{G1})}{f \Delta H_m} = \frac{(76.5 - 27.7)}{(0.79 \times 93)} = 0.66$$

which agrees well with the value (0.65) obtained by graphical integration.

Thus, in equation 13.53:

$$\text{height of packing, } z = \int_{H_{G1}}^{H_{G2}} \frac{dH_G}{(H_f - H_G)} \frac{G'}{h_D a \rho}$$

$$= \frac{(0.65 \times 0.817)}{(0.2 \times 1.206)}$$

$$= \underline{\underline{2.20\ \text{m}}}$$

Assuming that the resistance to mass transfer lies entirely within the gas phase, the lines connecting θ_L and θ_f are parallel with the enthalpy axis.

In Figure 13.18 a plot of H_G and θ_G is obtained using the construction given in Section 13.6.4 and shown in Figure 13.15. From this curve, the value of θ_{G2} corresponding to $H_{G2} = 76.5$ kJ/kg is 300 K. From Figure 13.5, under these conditions, the exit air has a humidity of 0.019 kg/kg which from Figure 13.4 corresponds to a relative humidity of 83 per cent.

13.6.7. Humidifying towers

If the main function of the tower is to produce a stream of humidified air, the final temperature of the liquid will not be specified, and the humidity of the gas leaving the top of the tower will be given instead. It is therefore not possible to fix any point on the operating line, though its slope can be calculated from the liquid and gas rates. In designing a humidifier, therefore, it is necessary to calculate the temperature and enthalpy, and hence the humidity, of the gas leaving the tower for a number of assumed water outlet temperatures and thereby determine the outlet water temperature resulting in the air leaving the tower with the required humidity. The operating line for this water-outlet temperature is then used in the calculation of the height of the tower required to effect this degree of humidification. The calculation of the dimensions of a humidifier is therefore rather more tedious than that for the water-cooling tower.

In a humidifier in which the make-up liquid is only a small proportion of the total liquid circulating, its temperature approaches the adiabatic saturation temperature θ_s, and remains constant, so that there is no temperature gradient in the liquid. The gas in contact with the liquid surface is approximately saturated and has a humidity $\mathscr{H}_s$.

Thus: $$d\theta_L = 0$$

and: $$\theta_{L1} = \theta_{L2} = \theta_L = \theta_f = \theta_s$$

Hence: $$-G's\, d\theta_G = h_G a\, dz(\theta_G - \theta_s) \quad \text{(from equation 13.42)}$$

and: $$-G'\, d\mathscr{H} = h_D \rho a\, dz(\mathscr{H} - \mathscr{H}_s) \quad \text{(from equation 13.44)}$$

Integration of these equations gives:

$$\ln \frac{(\theta_{G1} - \theta_s)}{(\theta_{G2} - \theta_s)} = \frac{h_G a}{G' s} z \tag{13.60}$$

and:

$$\ln \frac{(\mathscr{H}_s - \mathscr{H}_1)}{(\mathscr{H}_s - \mathscr{H}_2)} = \frac{h_D a \rho}{G'} z \tag{13.61}$$

assuming h_G, h_D, and s remain approximately constant.

From these equations the temperature θ_{G2} and the humidity $\mathscr{H}_2$ of the gas leaving the humidifier may be calculated in terms of the height of the tower. Rearrangement of equation 13.61 gives:

$$\ln \left(1 + \frac{\mathscr{H}_1 - \mathscr{H}_2}{\mathscr{H}_s - \mathscr{H}_1}\right) = -\frac{h_D a \rho}{G'} z$$

or:

$$\frac{(\mathscr{H}_2 - \mathscr{H}_1)}{(\mathscr{H}_s - \mathscr{H}_1)} = 1 - \mathrm{e}^{-h_D a \rho z/G'} \tag{13.62}$$

Thus the ratio of the actual increase in humidity produced in the saturator to the maximum possible increase in humidity (that is, the production of saturated gas) is equal to $(1 - \mathrm{e}^{-h_D a \rho z/G'})$, and complete saturation of the gas is reached exponentially. A similar relation exists for the change in the temperature of the gas stream:

$$\frac{(\theta_{G1} - \theta_{G2})}{(\theta_{G2} - \theta_s)} = 1 - \mathrm{e}^{-h_G a z/G' s} \tag{13.63}$$

Further, the relation between the temperature and the humidity of the gas at any stage in the adiabatic humidifier is given by:

$$\frac{\mathrm{d}\mathscr{H}}{\mathrm{d}\theta_G} = \frac{(\mathscr{H} - \mathscr{H}_s)}{(\theta_G - \theta_s)} \quad \text{(from equation 13.52)}$$

On integration:

$$\ln \frac{(\mathscr{H}_s - \mathscr{H}_2)}{(\mathscr{H}_s - \mathscr{H}_1)} = \ln \frac{(\theta_{G2} - \theta_s)}{(\theta_{G1} - \theta_s)} \tag{13.64}$$

or:

$$\frac{(\mathscr{H}_s - \mathscr{H}_2)}{(\mathscr{H}_s - \mathscr{H}_1)} = \frac{(\theta_{G2} - \theta_s)}{(\theta_{G1} - \theta_s)} \tag{13.65}$$

13.7. SYSTEMS OTHER THAN AIR–WATER

Calculations involving to systems where the Lewis relation is not applicable are very much more complicated because the adiabatic saturation temperature and the wet-bulb temperature do not coincide. Thus the significance of the adiabatic cooling lines on the psychrometric chart is very much restricted. They no longer represent the changes which take place in a gas as it is humidified by contact with liquid initially at the adiabatic saturation temperature of the gas, but simply give the compositions of all gases with the same adiabatic saturation temperature.

Calculation of the change in the condition of the liquid and the gas in a humidification tower is rendered more difficult since equation 13.49, which was derived for the air-water system, is no longer applicable. LEWIS and WHITE[22] have developed a method of calculation based on the use of a *modified enthalpy* in place of the true enthalpy of the system.

For the air–water system, from equation 13.11:

$$h_G = h_D \rho s \tag{13.66}$$

This relationship applies quite closely for the conditions normally encountered in practice. For other systems, the relation between the heat and mass transfer coefficients in the gas phase is given by:

$$h_G = b h_D \rho s \tag{13.67}$$

where b is approximately constant and generally has a value greater than unity.

For these systems, equation 13.46 becomes:

$$G' s\, \mathrm{d}\theta_G = b h_D \rho a\, \mathrm{d}z (s\theta_f - s\theta_G) \tag{13.68}$$

Adding equations 13.68 and 13.47 to obtain the relationship corresponding to equation 13.48 gives:

$$G'(s\, \mathrm{d}\theta_G + \lambda\, \mathrm{d}\mathscr{H}) = h_D \rho a\, \mathrm{d}z[(bs\theta_f + \lambda \mathscr{H}_f) - (bs\theta_G + \lambda \mathscr{H})] \tag{13.69}$$

Lewis and White use a *modified latent heat of vaporisation* λ' defined by:

$$b = \frac{\lambda}{\lambda'} \tag{13.70}$$

and a *modified enthalpy* per unit mass of dry gas defined by:

$$H'_G = s(\theta_G - \theta_0) + \lambda' \mathscr{H} \tag{13.71}$$

Substituting in equation 13.67; from equations 13.38, 13.70, and 13.71:

$$G'\, \mathrm{d}H_G = b h_D \rho a\, \mathrm{d}z (H'_f - H'_G) \tag{13.72}$$

and:

$$\frac{\mathrm{d}H_G}{(H'_f - H'_G)} = \frac{b h_D \rho a}{G'} \mathrm{d}z \tag{13.73}$$

Combining equations 13.37, 13.40, and 13.72:

$$\frac{(H'_G - H'_f)}{(\theta_L - \theta_f)} = -\frac{h_L}{h_D \rho b} \quad \text{(cf. equation 13.50)} \tag{13.74}$$

From equations 13.66 and 13.72:

$$\frac{(H'_G - H'_f)}{(\theta_G - \theta_f)} = \frac{\mathrm{d}H_G}{\mathrm{d}\theta_G} \quad \text{(cf. equation 13.51)} \tag{13.75}$$

From equations 13.44 and 13.67:

$$\frac{(\mathscr{H} - \mathscr{H}_f)}{(\theta_G - \theta_f)} = b \frac{\mathrm{d}\mathscr{H}}{\mathrm{d}\theta_G} \quad \text{(cf. equation 13.52)} \tag{13.76}$$

The calculation of conditions within a countercurrent column operating with a system other than air–water is carried out in a similar manner to that already described by applying equations 13.73, 13.74, and 13.75 in conjunction with equation 3.39:

$$G'(H_{G2} - H_{G1}) = L'C_L(\theta_{L2} - \theta_{L1}) \qquad \text{(equation 13.39)}$$

On an enthalpy–temperature diagram (Figure 13.20) the enthalpy of saturated gas is plotted against its temperature. If equilibrium between the liquid and gas exists at the interface, this curve *PQ* represents the relation between gas enthalpy and temperature at the interface (H_f v. θ_f). The modified enthalpy of saturated gas is then plotted against temperature (curve *RS*) to give the relation between H'_f and θ_f. Since b is greater than unity, *RS* will lie below *PQ*. By combining equations 13.35, 13.70, and 13.72, H'_G is obtained in terms of H_G.

$$H'_G = \frac{1}{b}[H_G + (b-1)s(\theta_G - \theta_0)] \qquad (13.77)$$

H'_G may be conveniently plotted against H_G for a number of constant temperatures. If b and s are constant, a series of straight lines is obtained. The operating line *AB* given by equation 13.39 is drawn in Figure 13.21. Point *A* has coordinates (θ_{L1}, H_{G1}) corresponding to the bottom of the column. Point *a* has coordinates (θ_{L1}, H'_{G1}), H'_{G1} being obtained from equation 13.77.

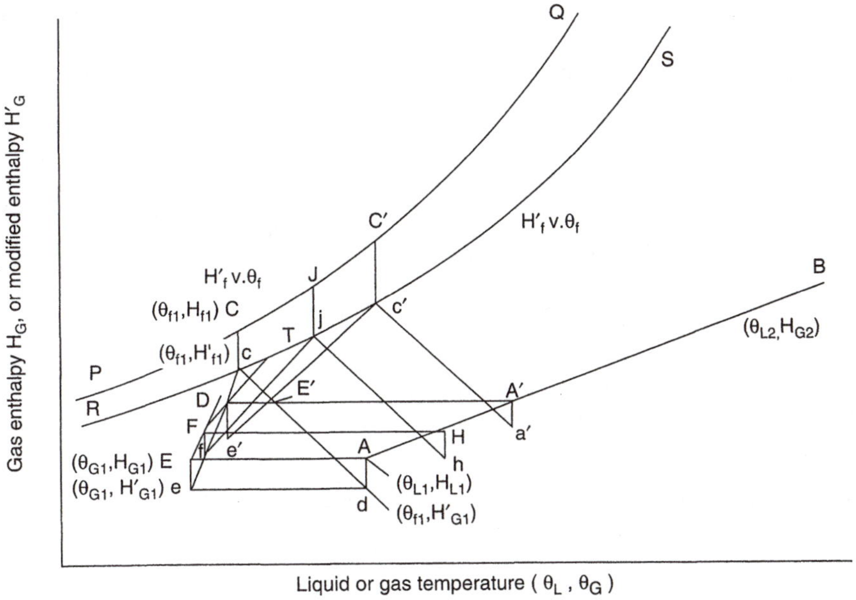

Figure 13.21. Construction for height of a column for vapour other than water

From equation 13.72, a line through a, of slope $-(h_L/h_D\rho b)$, will intersect curve *RS* at c, (θ_{f1}, H'_{f1}) to give the interface conditions at the bottom of the column. The corresponding air enthalpy is given by C, (θ_{f1}, H_{f1}). The difference between the ordinates of c and a then gives the driving force in terms of modified enthalpy at the bottom of the column ($H'_{f1} - H'_{G1}$). A similar construction at other points, such as A', enables the driving force to be calculated at any other point. Hence $(H'_f - H'_G)$ is obtained as a function of H_G throughout the column. The height of column corresponding to a given change in air enthalpy can be obtained from equation 13.71 since the left-hand side can now be evaluated.

Thus:

$$\int_{H_{G1}}^{H_{G2}} \frac{dH_G}{(H'_f - H'_G)} = \frac{bh_D a\rho}{G'} z \qquad (13.78)$$

The change in the condition of the gas stream is obtained as follows: E, with coordinates (θ_{G1}, H_{G1}), represents the condition of the inlet gas. The modified enthalpy of this gas is given by $e(\theta_{G1}, H'_{G1})$. From equation 13.75 it is seen that ec gives the rate of change of gas enthalpy with temperature $(dH_G/d\theta_G)$ at the bottom of the column. Thus ED, parallel to ec, describes the way in which gas enthalpy changes at the bottom of the column. At some arbitrary small distance from the bottom, F represents the condition of the gas and H gives the corresponding liquid temperature. In exactly the same way the next small change is obtained by drawing a line hj through h parallel to ac. The slope of fj gives the new value of $(dH_G/d\theta_G)$ and therefore the gas condition at a higher point in the column is obtained by drawing FT parallel to fj. In this way the change in the condition of the gas through the column can be followed by continuing the procedure until the gas enthalpy reaches the value H_{G2} corresponding to the top of the column.

A detailed description of the method of construction of psychrometric charts is given by SHALLCROSS and LOW[23], who illustrate their method by producing charts for three systems; air–water, air–benzene and air–toluene at pressures of 1 and 2 bar.

Example 13.10

In a countercurrent packed column, n-butanol flows down at a rate of 0.25 kg/m^2 s and is cooled from 330 to 295 K. Air at 290 K, initially free of n-butanol vapour, is passed up the column at the rate of 0.7 m^3/m^2 s. Calculate the required height of tower and the condition of the exit air.

Data:

Mass transfer coefficient per unit volume, $h_D a = 0.1$ s^{-1}
Psychrometric ratio, $h_G/(h_D \rho_A s) = b = 2.34$
Heat transfer coefficients, $h_L = 3h_G$
Latent heat of vaporisation of n-butanol, $\lambda = 590$ kJ/kg
Specific heat of liquid n-butanol, $C_L = 2.5$ kJ/kg K
Humid heat of gas, $s = 1.05$ kJ/kg K

Temperature (K)	Vapour pressure of butanol (kN/m^2)
295	0.59
300	0.86
305	1.27
310	1.75
315	2.48
320	3.32
325	4.49
330	5.99
335	7.89
340	10.36
345	14.97
350	17.50

Solution

The first stage is to calculate the enthalpy of the saturated gas by way of the saturated humidity, $\mathscr{H}_0$ given by:

$$\mathscr{H}_0 = \frac{P_{w0}}{P - P_{w0}}\frac{M_w}{M_A} = \frac{P_{w0}}{(101.3 - P_{w0})}\left(\frac{74}{29}\right)$$

The enthalpy is then:

$$H_f = \frac{1}{(1 + \mathscr{H}_0)} \times 1.001(\theta_f - 273) + \mathscr{H}_0[2.5(\theta_f - 273) + 590] \text{ kJ/kg}$$

where 1.001 kJ/kg K is the specific heat of dry air.

Thus:
$$H_f = \frac{1.001\theta_f - 273.27}{(1 + \mathscr{H}_0)} + \mathscr{H}_0(2.5\theta_f - 92.5) \text{ kJ/kg moist air}$$

The results of this calculation are presented in the following table and H_f is plotted against θ_f in Figure 13.21.

The modified enthalpy at saturation H'_f is given by:

$$H'_f = \frac{(1.001\theta_f - 273.27)}{(1 + \mathscr{H}_0)} + \mathscr{H}_0[2.5(\theta_f - 273) + \lambda']$$

where from equation 13.70: $\lambda' = \lambda/b = (590/2.34)$ or 252 kJ/kg

$$\therefore \quad H'_f = \frac{(1.001\theta_f - 273.27)}{(1 + \mathscr{H}_0)} + \mathscr{H}_0(2.5\theta_f - 430.5) \text{ kJ/kg moist air}$$

These results are also given in the following Table and plotted as H'_f against θ_f in Figure 13.21.

θ_f (K)	P_{w0} (kN/m^2)	$\mathscr{H}_0$ (kg/kg)	$(1.001\theta_f - 273.27)/(1 + \mathscr{H}_0)$ (kJ/kg)	$\mathscr{H}_0(2.5\theta_f - 92.5)$ (kJ/kg)	H_f (kJ/kg)	$\mathscr{H}_0(2.5\theta_f - 430.5)$ (kJ/kg)	H'_f (kJ/kg)
295	0.59	0.0149	21.70	9.61	31.31	4.57	26.28
300	0.86	0.0218	24.45	14.33	40.78	6.97	33.42
305	1.27	0.0324	31.03	21.71	52.74	10.76	41.79
310	1.75	0.0448	35.45	30.58	66.03	15.43	50.88
315	2.48	0.0640	39.52	44.48	84.00	22.85	62.37
320	3.32	0.0864	43.31	61.13	104.44	31.92	75.23
325	4.49	0.1183	46.55	85.18	131.73	45.19	91.74
330	5.99	0.1603	49.18	117.42	166.60	63.23	112.41
335	7.89	0.2154	51.07	160.47	211.54	87.67	138.73
340	10.36	0.2905	51.97	220.05	272.02	121.87	173.83
345	14.97	0.4422	49.98	340.49	390.47	191.03	241.01
350	17.50	0.5325	50.30	416.68	466.98	236.70	287.00

The bottom of the operating line (point a) has coordinates, $\theta_{L1} = 295$ K and H_{G1}, where:

$$H_{G1} = 1.05(290 - 273) = 17.9 \text{ kJ/kg.}$$

At a mean temperature of, say, 310 K, the density of air is:

$$\left(\frac{29}{22.4}\right)\left(\frac{273}{310}\right) = 1.140 \text{ kg/m}^3$$

and:
$$G' = (0.70 \times 1.140) = 0.798 \text{ kg/m}^2 \text{ s}$$

Thus, the slope of the operating line becomes:

$$\frac{L'C_L}{G'} = \frac{(0.25 \times 2.5)}{0.798} = 0.783 \text{ kJ/kg K}$$

and this is drawn in as AB in Figure 13.22 and at $\theta_{L2} = 330$ K, $H_{G2} = 46$ kJ/kg.

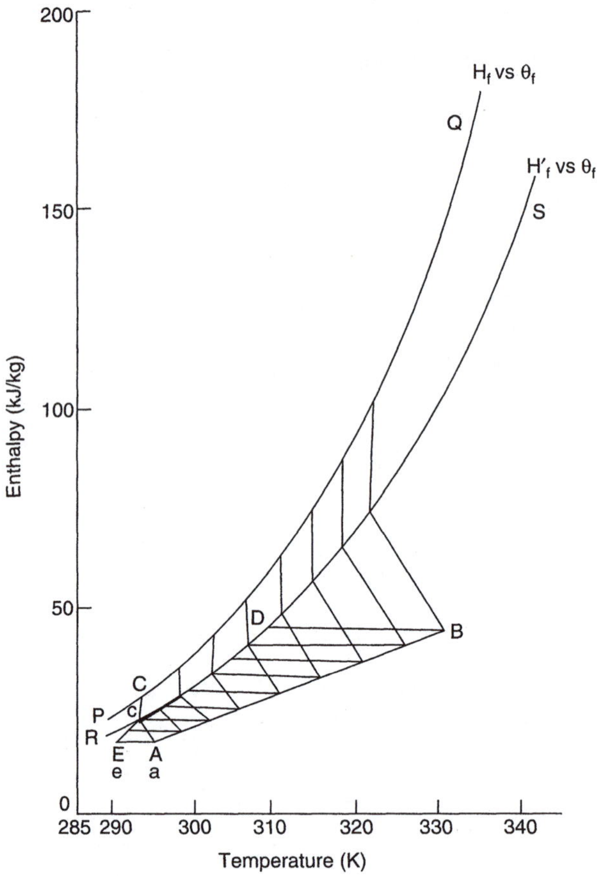

Figure 13.22. Graphical construction for Example 13.10

From equation 13.77: $H'_G = [H_G + (b - 1)s(\theta_G - \theta_0)]/b$

$$\therefore \qquad H'_{G1} = \frac{17.9 + (2.34 - 1)1.05(290 - 273)}{2.34} = 17.87 \text{ kJ/kg}$$

Point a coincides with the bottom of the column.

A line is drawn through a of slope $-\dfrac{h_L}{h_D \rho b} = -\left(\dfrac{3h_G}{h_D \rho b}\right) \cdot \left(\dfrac{h_D \rho s}{h_G}\right)$

$$= -3s \quad = -3.15 \text{ kJ/kg K}$$

This line meets curve RS at c (θ_{f1}, H'_{f1}) to give the interface conditions at the bottom of the column. The corresponding air enthalpy is given by point C whose co-ordinates are:

$$\theta_{f1} = 293 \text{ K} \qquad H_{f1} = 29.0 \text{ kJ/kg}$$

The difference between the ordinates of c and a gives the driving force in terms of the modified enthalpy at the bottom of the column, or:

$$(H'_{f1} - H'_{G1}) = (23.9 - 17.9) = 6.0 \text{ kJ/kg}$$

A similar construction is made at other points along the operating line with the results shown in the following table.

θ_f (K)	H_G (kJ/kg)	H'_G (kJ/kg)	H'_f (kJ/kg)	$(H'_f - H'_G)$ (kJ/kg)	$1/(H'_f - H'_G)$ (kg/kJ)	Mean value in interval	Interval	Value of integral over interval
295	17.9	17.9	23.9	6.0	0.167			
						0.155	4.1	0.636
300	22.0	22.0	29.0	7.0	0.143			
						0.126	4.0	0.504
305	26.0	26.0	35.3	9.3	0.108			
						0.096	4.0	0.384
310	30.0	30.0	42.1	12.1	0.083			
						0.073	4.0	0.292
315	34.0	34.0	50.0	16.0	0.063			
						0.057	4.1	0.234
320	38.1	38.1	57.9	19.8	0.051			
						0.046	3.9	0.179
325	42.0	42.0	66.7	24.7	0.041			
						0.0375	4.0	0.150
330	46.0	46.0	75.8	29.8	0.034			
						Value of integral = 2.379		

from which:

$$\int_{H_{G1}}^{H_{G2}} \frac{dH_G}{H'_f - H'_G} = 2.379$$

Substituting in equation 13.78:

$$\frac{bh_D \rho a z}{G'} = 2.379$$

and:

$$z = \frac{(2.379 \times 0.798)}{(2.34 \times 0.1)} = \underline{\underline{8.1 \text{ m}}}$$

It remains to evaluate the change in gas conditions.

Point e, ($\theta_{G1} = 290$ K, $H_{G1} = 17.9$ kJ/kg) represents the condition of the inlet gas. ec is now drawn in, and from equation 13.75, this represents $dH_G/d\theta_G$. As for the air-water system, this construction is continued until the gas enthalpy reaches H_{G2}. The final point is given by D at which $\underline{\underline{\theta_{G2} = 308 \text{ K}}}$.

It is fortuitous that, in this problem, $H'_G = H_G$. This is not always the case and reference should be made to Section 13.7 for elaboration of this point.

13.8. FURTHER READING

BACKHURST, J. R., HARKER, J. H., and PORTER, J. E.: *Problems in Heat and Mass Transfer* (Edward Arnold, London, 1974).

BURGER, R.: *Chem. Eng. Progr.*, **71**(7) (1975) 73. Cooling tower drift elimination.

DEMONBRUN, J. R.: *Chem. Eng.*, **75**(19) (1968) 106. Factors to consider in selecting a cooling tower.

DONOHUE, J. M. and NATHAN, C. C.: *Chem. Eng. Progr.*, **71**(7) (1975) 88. Unusual problems in cooling water treatment.

ECKERT, E. R. G. and DRAKE, R. M.: *Analysis of Heat and Mass Transfer* (McGraw-Hill, New York, 1972).

ELGAWHARY, A. W.: *Chem. Eng. Progr.*, **71**(7) (1975) 83. Spray cooling system design.

FRIAR, F.: *Chem. Eng.*, **81**(15) (1974) 122. Cooling-tower basin design.

GUTHRIE, K. M.: *Chem. Eng.*, **76**(6) (1969) 114. Capital cost estimating.

HALL, W. A.: *Chem. Eng. Progr.*, **67**(7) (1971) 52. Cooling tower plume abatement.

HANSEN, E. P., and PARKER, J. J.: *Power Eng.*, **71**(5) (1967) 38. Status of big cooling towers.

HOLZHAUER, R.: *Plant. Eng.*, **29**(15) (1975) 60. Industrial cooling towers.

INCROPERA, F. P. and DE WITT, D. P.: *Fundamentals of Heat and Mass Transfer* 4th edn (Wiley, New York, 1996).

INDUSTRIAL WATER SOCIETY: *Guide to Mechanical Draught Evaporative Cooling Towers: Selection, Operation and Maintenance* (London, 1987).

JACKSON, J.: *Cooling Towers* (Butterworths, London, 1951).

JORDAN, D. R., BEARDEN, M. D. and MCILHENNY, W. F.: *Chem. Eng. Progr.*, **71** (1975) 77. Blowdown concentration by electrodialysis.

JUONG, J. F.: *Hydrocarbon Process.*, **48**(7) (1969) 200. How to estimate cooling tower costs.

KELLY, G. M.: ASME Paper, 75-IPWR-9, (1975). Cooling tower design and evaluation parameters.
KOLFLAT, T. D.: *Power Eng.*, **78**(1) (1974) 32. Cooling tower practices.
MAZE, R. W.: *Hydrocarbon Process.*, **46**(2) (1967) 123. Practical tips on cooling tower sizing.
MAZE, R. W.: *Chem. Eng.*, **83**(1) (1975) 106. Air cooler or water tower — which for heat disposal?
MCCABE, W. L., SMITH, J. C. and HARRIOTT, P.: *Unit Operations of Chemical Engineering* (McGraw-Hill, New York, 1985).
MCKELVEY, K. K. and BROOKE, M.: *The Industrial Cooling Tower* (Elsevier, New York, 1959).
MEYTSAR, J.: *Hydrocarbon Process.*, **57**(11) (1978) 238. Estimate cooling tower requirements easily.
NELSON, W. L.: *Oil & Gas J.*, **65**(47) (1967) 182. What is cost of cooling towers?
NORMAN, W. S.: *Absorption, Distillation and Cooling Towers* (Longmans, London, 1961).
OLDS, F. C.: *Power Eng.*, **76**(12) (1972) 30. Cooling towers.
PAIGE, P. M.: *Chem. Eng.*, **74**(14) (1967) 93. Costlier cooling towers require a new approach to water-systems design.
PARK, J. E. and VANCE, J. M.: *Chem. Eng. Progr.*, **67** (1971) 55. Computer model of crossflow towers.
PICCIOTTI, M.: *Hydrocarbon Process.*, **56**(6) (1977) 163. Design quench water towers.
PICCIOTTI, M.: *Hydrocarbon Process.*, **56**(9) (1977) 179. Optimize quench water systems.
RABB, A.: *Hydrocarbon Process.*, **47**(2) (1968) 122. Are dry cooling towers economical?
UCHIYAMA, T.: *Hydrocarbon Process.*, **55**(12) (1976) 93. Cooling tower estimates made easy.
WALKER, R.: *Water Supply, Treatment and Distribution* (Prentice-Hall, New York, 1978).
WRINKLE, R. B.: *Chem. Eng. Progr.*, **67** (1971) 45. Performance of counterflow cooling tower cells.

13.9. REFERENCES

1. BS 1339: 1965(1981): British Standard 1339 (British Standards Institution, London) Definitions, formulae and constants relating to the humidity of the air.
2. SHERWOOD, T.K. and COMINGS, E.W.: *Trans. Am. Inst. Chem. Eng.* **28** (1932) 88. An experimental study of the wet bulb hygrometer.
3. GROSVENOR, M. M.: *Trans. Am. Inst. Chem. Eng.* **1** (1908) 184. Calculations for dryer design.
4. WEXLER, A.: In *Humidity and Moisture. Measurements and Control in Science and Industry.* Volume 1 Principles and Methods of Humidity Measurement in Gases, (Ed. R. E. Ruskin) (Reinhold, New York, 1965).
5. HICKMAN, M. J.: *Measurement of Humidity*, 4th edn (National Physical Laboratory. Notes on Applied Science No. 4). (HMSO, 1970).
6. MEADOWCROFT, D. B.: In *Instrumentation Reference Book*, (Ed. B. E. Noltingk), Chapter 6. Chemical analysis — moisture measurement (Butterworth, 1988).
7. WOOD, B. and BETTS, P.: *Proc. Inst. Mech. Eng. (Steam Group)* **163** (1950) 54. A contribution to the theory of natural draught cooling towers.
8. CLARK, R.: *The Chemical Engineer* (London) No. 529 (1992) 22. Cutting the fog.
9. MCKELVEY, K. K. and BROOKE, M.: *The Industrial Cooling Tower* (Elsevier, New York, 1959).
10. CHILTON, C. H.: *Proc. Inst. Elec. Engrs.* **99** (1952), 440. Performance of natural-draught cooling towers.
11. PERRY, R. H. and GREEN, D. W. (eds): *Perry's Chemical Engineers' Handbook.* 6th edn (McGraw-Hill, New York, 1984).
12. MICKLEY, H. S.: *Chem. Eng. Prog.* **45** (1949) 739. Design of forced draught air conditioning equipment.
13. MERKEL, F.: *Ver. deut. Ing. Forschungsarb.* No. 275 (1925). Verdunstungs-Kühlung.
14. CAREY, W. F. and WILLIAMSON, G. J.: *Proc. Inst. Mech. Eng. (Steam Group)* **163** (1950) 41. Gas cooling and humidification: design of packed towers from small scale tests.
15. THOMAS, W.J. and HOUSTON, P.: *Brit. Chem. Eng.* **4** (1959) 160, 217. Simultaneous heat and mass transfer in cooling towers.
16. CRIBB, G.: *Brit. Chem. Eng.* **4** (1959) 264. Liquid phase resistance in water cooling.
17. LOWE, H. J. and CHRISTIE, D. G.: *Inst. Mech. Eng. Symposium on Heat Transfer* (1962), Paper 113, 933. Heat transfer and pressure drop data on cooling tower packings, and model studies of the resistance of natural-draught towers to airflow.
18. WOOD, B. and BETTS, P.: *Engineer* **189** (1950) (4912) 337, (4913) 349. A total heat-temperature diagram for cooling tower calculations.
19. ZIVI, S. M. and BRAND, B. B.: *Refrig. Eng.* **64** (1956) 8, 31, 90. Analysis of cross-flow cooling towers.
20. *Counter-flow Cooling Tower Performance* (J. F. Pritchard & Co., Kansas City, 1957).
21. COOLING TOWER INSTITUTE: *Performance Curves* (Houston, 1967).
22. LEWIS, J. G. and WHITE, R. R.: *Ind. Eng. Chem.* **45** (1953), 486. Simplified humidification calculations.
23. SHALLCROSS, D. C. and LOW, S. L.: *Ch.E.R.D.* **72** (1994) 763. Construction of psychrometric charts for systems other than water vapour in air. Errata: *Ch.E.R.D.* **73** (1995) 865.

13.10. NOMENCLATURE

		Units in SI system	Dimensions in M, N, L, T, θ
A	Interfacial area	m^2	$\mathbf{L}^2$
A_b	Base area of hyperbolic tower	m^2	$\mathbf{L}^2$
a	Interfacial area per unit volume of column	m^2/m^3	$\mathbf{L}^{-1}$
b	Psychrometric ratio ($h/h_D \rho_A s$)	—	—
C_a	Specific heat of gas at constant pressure	J/kg K	$\mathbf{L}^2\ \mathbf{T}^{-2}\theta^{-1}$
C_L	Specific heat of liquid	J/kg K	$\mathbf{L}^2\ \mathbf{T}^{-2}\theta^{-1}$
C_p	Specific heat of gas and vapour mixture at constant pressure	J/kg K	$\mathbf{L}^2\ \mathbf{T}^{-2}\theta^{-1}$
C_s	Specific heat of solid	J/kg K	$\mathbf{L}^2\ \mathbf{T}^{-2}\theta^{-1}$
C_t	Performance coefficient or efficiency factor	—	—
C_w	Specific heat of vapour at constant pressure	J/kg K	$\mathbf{L}^2\ \mathbf{T}^{-2}\theta^{-1}$
c	Mass concentration of vapour	kg/m^3	$\mathbf{ML}^{-3}$
c_0	Mass concentration of vapour in saturated gas	kg/m^3	$\mathbf{ML}^{-3}$
D_t	Duty coefficient of tower (equation 13.31)	—	—
f	Correction factor for mean driving force	—	—
G'	Mass rate of flow of gas per unit area	kg/m^2 s	$\mathbf{ML}^{-2}\ \mathbf{T}^{-1}$
H	Enthalpy of humid gas per unit mass of dry gas	J/kg	$\mathbf{L}^2\ \mathbf{T}^{-2}$
H_a	Enthalpy per unit mass, of dry gas	J/kg	$\mathbf{L}^2\ \mathbf{T}^{-2}$
H_w	Enthalpy per unit mass, of vapour	J/kg	$\mathbf{L}^2\ \mathbf{T}^{-2}$
H_1	Enthalpy of stream of gas, per unit mass of dry gas	J/kg	$\mathbf{L}^2\ \mathbf{T}^{-2}$
H_2	Enthalpy of another stream of gas, per unit mass of dry gas	J/kg	$\mathbf{L}^2\ \mathbf{T}^{-2}$
H_3	Enthalpy per unit mass of liquid or vapour	J/kg	$\mathbf{L}^2\ \mathbf{T}^{-2}$
H'	Modified enthalpy of humid gas defined by (13.69)	J/kg	$\mathbf{L}^2\ \mathbf{T}^{-2}$
ΔH	Enthalpy driving force ($H_f - H_G$)	J/kg	$\mathbf{L}^2\ \mathbf{T}^{-2}$
$\Delta H'$	Change in air enthalpy on passing through tower	J/kg	$\mathbf{L}^2\ \mathbf{T}^{-2}$
h	Heat transfer coefficient	W/m^2 K	$\mathbf{MT}^{-3}\theta^{-1}$
h_D	Mass transfer coefficient	$kmol/(kmol/m^3)m^2$ s	$\mathbf{LT}^{-1}$
h_G	Heat transfer coefficient for gas phase	W/m^2 K	$\mathbf{MT}^{-3}\theta^{-1}$
h_L	Heat transfer coefficient for liquid phase	W/m^2 K	$\mathbf{MT}^{-3}\theta^{-1}$
$\mathscr{H}$	Humidity	kg/kg	—
$\mathscr{H}_s$	Humidity of gas saturated at the adiabatic saturation temperature	kg/kg	—
$\mathscr{H}_w$	Humidity of gas saturated at the wet-bulb temperature	kg/kg	—
$\mathscr{H}_0$	Humidity of saturated gas	kg/kg	—
$\mathscr{H}_1$	Humidity of a gas stream	kg/kg	—
$\mathscr{H}_2$	Humidity of second gas stream	kg/kg	—
L'	Mass rate of flow of liquid per unit area	kg/m^2 s	$\mathbf{ML}^{-2}\ \mathbf{T}^{-1}$
M_A	Molecular weight of gas	kg/kmol	$\mathbf{MN}^{-1}$
M_w	Molecular weight of vapour	kg/kmol	$\mathbf{MN}^{-1}$
m, m_1, m_2	Masses of dry gas	kg	$\mathbf{M}$
m_3	Mass of liquid or vapour	kg	$\mathbf{M}$
P	Total pressure	N/m^2	$\mathbf{ML}^{-1}\ \mathbf{T}^{-2}$
P_A	Mean partial pressure of gas	N/m^2	$\mathbf{ML}^{-1}\ \mathbf{T}^{-2}$
P_w	Partial pressure of vapour	N/m^2	$\mathbf{ML}^{-1}\ \mathbf{T}^{-2}$
P_{w0}	Partial pressure of vapour in saturated gas	N/m^2	$\mathbf{ML}^{-1}\ \mathbf{T}^{-2}$
Q	Rate of transfer of heat to liquid surface	W	$\mathbf{ML}^2\ \mathbf{T}^{-3}$
$\mathbf{R}$	Universal gas constant	8314 J/kmol K	$\mathbf{MN}^{-1}\ \mathbf{L}^2\ \mathbf{T}^{-2}\theta^{-1}$
s	Humid heat of gas	J/kg K	$\mathbf{L}^2\ \mathbf{T}^{-2}\theta^{-1}$
T	Absolute temperature	K	θ

		Units in SI system	Dimensions in **M, N, L, T, θ**
ΔT	Change in water temperature in passing through the tower	K	θ
$\Delta T'$	(Temperature of air leaving packing — ambient dry-bulb temperature)	K	θ
V	Active volume per plan area of column	m^3/m^2	**L**
W_L	Water loading on tower	kg/s	$\mathbf{MT^{-1}}$
w	Rate of evaporation	kg/s	$\mathbf{MT^{-1}}$
Z	Percentage humidity	—	—
z	Height from bottom of tower	m	**L**
z_t	Height of cooling tower	m	**L**
θ	Temperature of gas stream	K	θ
θ_0	Reference temperature, taken as the melting point of the material	K	θ
θ_s	Adiabatic saturation temperature	K	θ
θ_w	Wet bulb temperature	K	θ
λ	Latent heat of vaporisation per unit mass, at datum temperature	J/kg	$\mathbf{L^2\ T^{-2}}$
λ_f	Latent heat of freezing per unit mass, at datum temperature	J/kg	$\mathbf{L^2\ T^{-2}}$
λ'	Modified latent heat of vaporisation per unit mass defined by (13.68)	J/kg	$\mathbf{L^2\ T^{-2}}$
ρ	Mean density of gas and vapour	kg/m^3	$\mathbf{ML^{-3}}$
ρ_A	Mean density of gas at partial pressure P_A	kg/m^3	$\mathbf{ML^{-3}}$
Pr	Prandtl number	—	—
Sc	Schmidt number	—	—

Suffixes 1, 2, f, L, G denote conditions at the bottom of the tower, the top of the tower, the interface, the liquid, and the gas, respectively.
Suffix m refers to the mean water temperature.

Appendix

Table 1. *(continued)*

Liquid	k (W/m K)	(K)	k (Btu/h ft °F)
Cymene (*para*)	0.135	303	0.078
	0.137	333	0.079
Decane (n-)	0.147	303	0.085
	0.144	333	0.083
Dichlorodifluoromethane	0.099	266	0.057
	0.092	289	0.053
	0.083	311	0.048
	0.074	333	0.043
	0.066	355	0.038
Dichloroethane	0.142	323	0.082
Dichloromethane	0.192	258	0.111
	0.166	303	0.096
Ethyl acetate	0.175	293	0.101
Ethyl alcohol 100 per cent	0.182	293	0.105
80 per cent	0.237	293	0.137
60 per cent	0.305	293	0.176
40 per cent	0.388	293	0.224
20 per cent	0.486	293	0.281
100 per cent	0.151	323	0.087
Ethyl benzene	0.149	303	0.086
	0.142	333	0.082
Ethyl bromide	0.121	293	0.070
Ethyl ether	0.138	303	0.080
	0.135	348	0.078
Ethyl iodide	0.111	313	0.064
	0.109	348	0.063
Ethylene glycol	0.265	273	0.153
Gasoline	0.135	303	0.078
Glycerol 100 per cent	0.284	293	0.164
80 per cent	0.327	293	0.189
60 per cent	0.381	293	0.220
40 per cent	0.448	293	0.259
20 per cent	0.481	293	0.278
100 per cent	0.284	373	0.164
Heptane (n-)	0.140	303	0.081
	0.137	333	0.079

Liquid	k (W/m K)	(K)	k (Btu/h ft °F)
Oil, Olive	0.164	373	0.095
Paraldehyde	0.145	303	0.084
	0.135	373	0.078
Pentane (n-)	0.135	303	0.078
	0.128	348	0.074
Perchloroethylene	0.159	323	0.092
Petroleum ether	0.130	303	0.075
	0.126	348	0.073
Propyl alcohol (n-)	0.171	303	0.099
	0.164	348	0.095
Propyl alcohol (iso-)	0.157	303	0.091
	0.155	333	0.090
Sodium	0.85	373	49
	0.80	483	46
Sodium chloride brine 25.0 per cent	0.57	303	0.33
12.5 per cent	0.59	303	0.34
Sulphuric acid 90 per cent	0.36	303	0.21
60 per cent	0.43	303	0.25
30 per cent	0.52	303	0.30
Sulphur dioxide	0.22	258	0.128
	0.192	303	0.111
Toluene	0.149	303	0.086
	0.145	348	0.084
β-trichloroethane	0.133	323	0.077
Trichloroethylene	0.138	323	0.080
Turpentine	0.128	288	0.074
Vaseline	0.184	288	0.106
Water	0.57	273	0.330
	0.615	303	0.356
	0.658	333	0.381
	0.688	353	0.398
Xylene (ortho-)	0.155	293	0.090
(meta-)	0.155	293	0.090

*By permission from *Heat Transmission*, by W. H. McAdams, copyright 1942, McGraw-Hill.

A1. TABLES OF PHYSICAL PROPERTIES

Table 1. Thermal conductivities of liquids*

A linear variation with temperature may be assumed. The extreme values given constitute also the temperature limits over which the data are recommended.

Liquid	k (W/m K)	(K)	k (Btu/h ft °F)	Liquid	k (W/m K)	(K)	k (Btu/h ft °F)
Acetic acid 100 per cent	0.171	293	0.099	Hexane (n-)	0.138	303	0.080
50 per cent	0.35	293	0.20		0.135	333	0.078
Acetone	0.177	303	0.102	Heptyl alcohol (n-)	0.163	303	0.094
	0.164	348	0.095		0.157	348	0.091
Allyl alcohol	0.180	298 to 303	0.104	Hexyl alcohol (n-)	0.161	303	0.093
Ammonia	0.50	258 to 303	0.29		0.156	348	0.090
Ammonia, aqueous	0.45	293	0.261				
	0.50	333	0.29	Kerosene	0.149	293	0.086
Amyl acetate	0.144	283	0.083		0.140	348	0.081
Amyl alcohol (n-)	0.163	303	0.094				
	0.154	373	0.089	Mercury	8.36	301	4.83
Amyl alcohol (iso-)	0.152	303	0.088	Methyl alcohol 100 per cent	0.215	293	0.124
	0.151	348	0.087	80 per cent	0.267	293	0.154
Aniline	0.173	273 to 293	0.100	60 per cent	0.329	293	0.190
				40 per cent	0.405	293	0.234
Benzene	0.159	303	0.092	20 per cent	0.492	293	0.284
	0.151	333	0.087	100 per cent	0.197	323	0.114
Bromobenzene	0.128	303	0.074	Methyl chloride	0.192	258	0.111
	0.121	373	0.070		0.154	303	0.089
Butyl acetate (n-)	0.147	298 to 303	0.085				
Butyl alcohol (n-)	0.168	303	0.097	Nitrobenzene	0.164	303	0.095
	0.164	348	0.095		0.152	373	0.088
Butyl alcohol (iso-)	0.157	283	0.091	Nitromethane	0.216	303	0.125
					0.208	333	0.120
Calcium chloride brine							
30 per cent	0.55	303	0.32	Nonane (n-)	0.145	303	0.084
15 per cent	0.59	303	0.34		0.142	333	0.082
Carbon disulphide	0.161	303	0.093	Octane (n-)	0.144	303	0.083
	0.152	348	0.088		0.140	333	0.081
Carbon tetrachloride	0.185	273	0.107	Oils, Petroleum	0.138-0.156	273	0.08-0.09
	0.163	341	0.094	Oil, Castor	0.180	293	0.104
Chlorobenzene	0.144	283	0.083		0.173	373	0.100
Chloroform	0.138	303	0.080		0.168	293	0.097

Table 2. Latent heats of vaporisation*

Example: For water at 373 K, $\theta_c - \theta = (647 - 373) = 274$ K, and the latent heat of vaporisation is 2257 kJ/kg

No.	Compound	Range $\theta_c - \theta$ (°F)	θ_c (°F)	Range $\theta_c - \theta$ (K)	θ_c (K)	No.	Compound	Range $\theta_c - \theta$ (°F)	θ_c (°F)	Range $\theta_c - \theta$ (K)	θ_c (K)
18	Acetic acid	180–405	610	100–225	594	2	Freon-12 (CCl_2F_2)	72–360	232	40–200	384
22	Acetone	216–378	455	120–210	508	5	Freon-21 ($CHCl_2F$)	126–450	354	70–250	451
29	Ammonia	90–360	271	50–200	406	6	Freon-22 ($CHClF_2$)	90–306	205	50–170	369
13	Benzene	18–720	552	10–00	562	1	Freon-113 (CCl_2F-$CClF_2$)	162–450	417	90–250	487
16	Butane	162–360	307	90–200	426	10	Heptane	36–540	512	20–300	540
21	Carbon dioxide	18–180	88	10–100	304	11	Hexane	90–450	455	50–225	508
4	Carbon disulphide	252–495	523	140–275	546	15	Isobutane	144–360	273	80–200	407
2	Carbon tetrachloride	54–450	541	30–250	556	27	Methanol	72–450	464	40–250	513
7	Chloroform	252–495	505	140–275	536	20	Methyl chloride	126–450	289	70–250	416
8	Dichloromethane	270–450	421	150–250	489	19	Nitrous oxide	45–270	97	25–150	309
3	Diphenyl	315–720	981	175–400	800	9	Octane	54–540	565	30–300	569
25	Ethane	45–270	90	25–150	305	12	Pentane	36–360	387	20–200	470
26	Ethyl alcohol	36–252	469	20–140	516	23	Propane	72–360	205	40–200	369
28	Ethyl alcohol	252–540	469	140–300	516	24	Propyl alcohol	36–360	507	20–200	537
17	Ethyl chloride	180–450	369	100–250	460	14	Sulphur dioxide	162–288	314	90–160	430
13	Ethyl ether	18–720	381	10–00	467	30	Water	180–900	705	100–500	647
2	Freon-11 (CCl_3F)	126–450	389	70–250	471						

*By permission from *Heat Transmission*, by W. H. McAdams, Copyright 1942, McGraw-Hill.

Latent heats of vaporisation

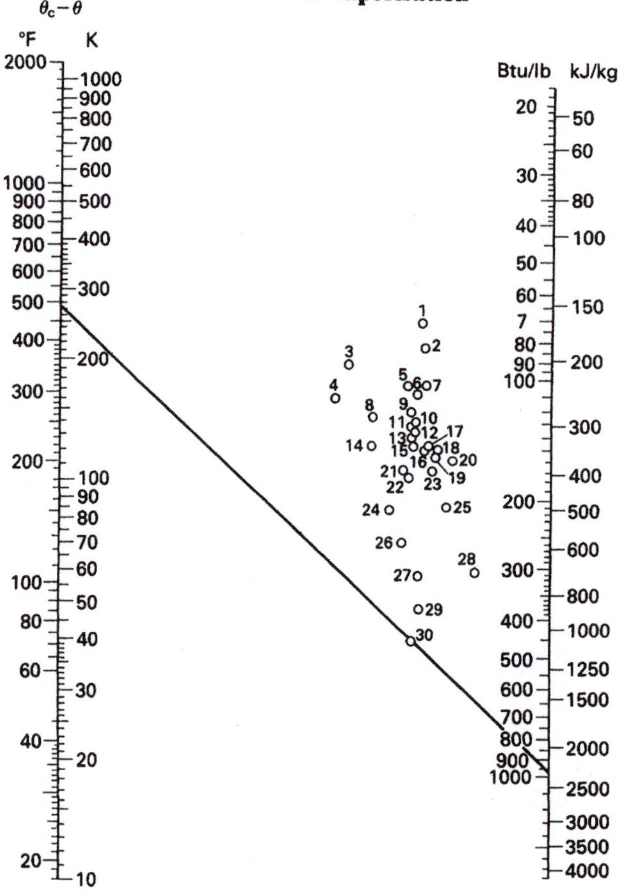

Table 3. Specific heats of liquids*

No.	Liquid	Range (K)
29	Acetic acid, 100 per cent	273–353
32	Acetone	293–323
52	Ammonia	203–323
37	Amyl alcohol	223–298
26	Amyl acetate	273–373
30	Aniline	273–403
23	Benzene	283–353
27	Benzyl alcohol	253–303
10	Benzyl chloride	243–303
49	Brine, 25 per cent $CaCl_2$	233–293
51	Brine, 25 per cent NaCl	233–293
44	Butyl alcohol	273–373
2	Carbon disulphide	173–298
3	Carbon tetrachloride	283–333
8	Chlorobenzene	273–373
4	Chloroform	273–323
21	Decane	193–298
6A	Dichloroethane	243–333
5	Dichloromethane	233–323
15	Diphenyl	353–393
22	Diphenylmethane	303–373
16	Diphenyl oxide	273–473
16	Dowtherm A	273–473
24	Ethyl acetate	223–298
42	Ethyl alcohol, 100 per cent	303–353
46	Ethyl alcohol, 95 per cent	293–353
50	Ethyl alcohol, 50 per cent	293–353
25	Ethyl benzene	273–373
1	Ethyl bromide	278–298
13	Ethyl chloride	243–313
36	Ethyl ether	173–298
7	Ethyl iodide	273–373
39	Ethylene glycol	233–473
2A	Freon-11 (CCl_3F)	253–343
6	Freon-12 (CCl_2F_2)	233–288
4A	Freon-21 ($CHCl_2F$)	253–343
7A	Freon-22 ($CHClF_2$)	253–333
3A	Freon-113 (CCl_2F-$CClF_2$)	253–343
38	Glycerol	233–293
28	Heptane	273–333
35	Hexane	193–293
48	Hydrochloric acid, 30 per cent	293–373
41	Isoamyl alcohol	283–373
43	Isobutyl alcohol	273–373
47	Isopropyl alcohol	253–323
31	Isopropyl ether	193–293
40	Methyl alcohol	233–293
13A	Methyl chloride	193–293
14	Naphthalene	363–473
12	Nitrobenzene	273–373
34	Nonane	223–298
33	Octane	223–298
3	Perchloroethylene	243–413
45	Propyl alcohol	253–373
20	Pyridine	223–298
9	Sulphuric acid, 98 per cent	283–318
11	Sulphur dioxide	253–373
23	Toluene	273–333
53	Water	283–473
19	Xylene (*ortho*)	273–373
18	Xylene (*meta*)	273–373
17	Xylene (*para*)	273–373

*By permission from *Heat Transmission*, by W. H. McAdams, copyright 1942, McGraw- Hill.

Specific heats of liquids

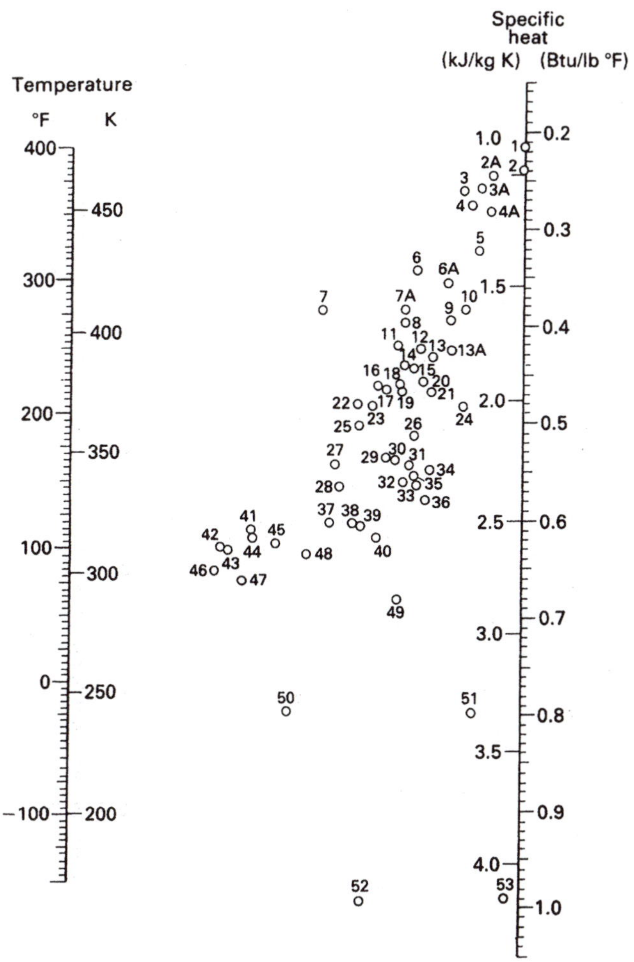

Table 4. Specific heats at constant pressure of gases and vapours at 101.3 kN/m^2*

No.	Gas	Range (K)	No.	Gas	Range (K)
10	Acetylene	273–473	1	Hydrogen	273–873
15	Acetylene	473–673	2	Hydrogen	873–1673
16	Acetylene	673–1673	35	Hydrogen bromide	273–1673
27	Air	273–1673	30	Hydrogen chloride	273–1673
12	Ammonia	273–873	20	Hydrogen fluoride	273–1673
14	Ammonia	873–1673	36	Hydrogen iodide	273–1673
18	Carbon dioxide	273–673	19	Hydrogen sulphide	273–973
24	Carbon dioxide	673–1673	21	Hydrogen sulphide	973–1673
26	Carbon monoxide	273–1673	5	Methane	273–573
32	Chlorine	273–473	6	Methane	573–973
34	Chlorine	473–1673	7	Methane	973–1673
3	Ethane	273–473	25	Nitric oxide	273–973
9	Ethane	473–873	28	Nitric oxide	973–1673
8	Ethane	873–1673	26	Nitrogen	273–1673
4	Ethylene	273–473	23	Oxygen	273–773
11	Ethylene	473–873	29	Oxygen	773–1673
13	Ethylene	873–1673	33	Sulphur	573–1673
17B	Freon-11 (CCl_3F)	273–423	22	Sulphur dioxide	273–673
17C	Freon-21 ($CHCl_2F$)	273–423	31	Sulphur dioxide	673–1673
17A	Freon-22 ($CHClF_2$)	273–423	17	Water	273–1673
17D	Freon-113 (CCl_2F-$CClF_2$)	273–423			

*By permission from *Heat Transmission*, by W. H. McAdams, copyright, 1942, McGraw-Hill.

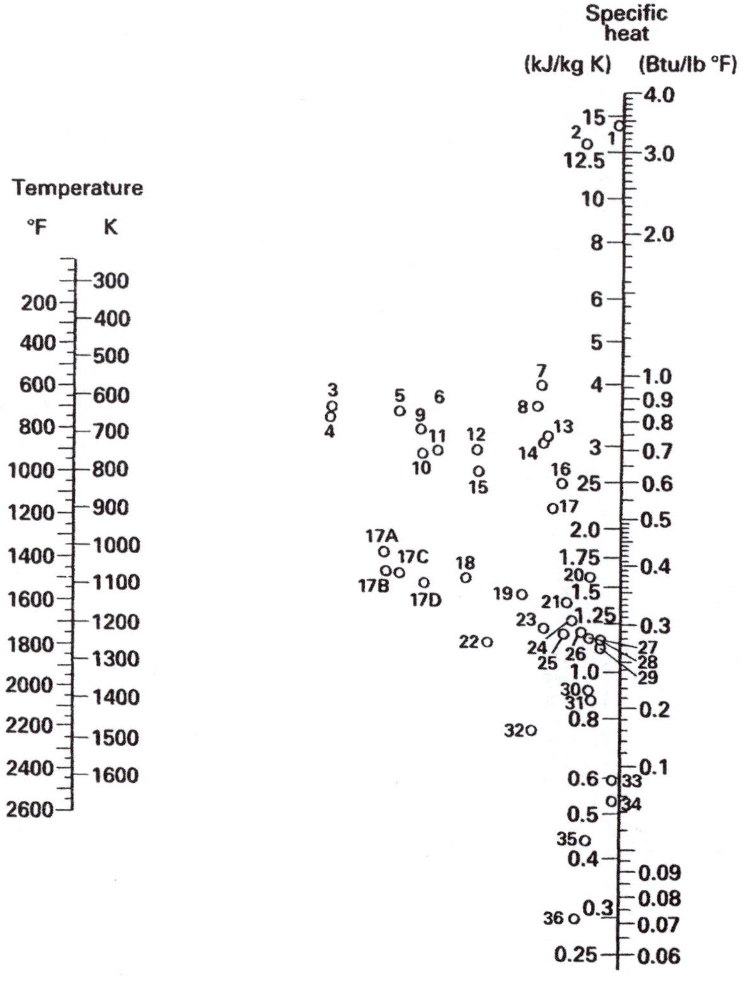

Table 5. Viscosity of water*

Temperature (θ) (K)	Viscosity (μ) (mN s/m^2)	Temperature (θ) (K)	Viscosity (μ) (mN s/m^2)	Temperature (θ) (K)	Viscosity (μ) (mN s/m^2)
273	1.7921	306	0.7523	340	0.4233
274	1.7313	307	0.7371	341	0.4174
275	1.6728	308	0.7225	342	0.4117
276	1.6191	309	0.7085	343	0.4061
277	1.5674	310	0.6947	344	0.4006
278	1.5188	311	0.6814	345	0.3952
279	1.4728	312	0.6685	346	0.3900
280	1.4284	313	0.6560	347	0.3849
281	1.3860	314	0.6439	348	0.3799
282	1.3462	315	0.6321	349	0.3750
283	1.3077	316	0.6207	350	0.3702
284	1.2713	317	0.6097	351	0.3655
285	1.2363	318	0.5988	352	0.3610
286	1.2028	319	0.5883	353	0.3565
287	1.1709	320	0.5782	354	0.3521
288	1.1404	321	0.5683	355	0.3478
289	1.1111	322	0.5588	356	0.3436
290	1.0828	323	0.5494	357	0.3395
291	1.0559	324	0.5404	358	0.3355
292	1.0299	325	0.5315	359	0.3315
293	1.0050	326	0.5229	360	0.3276
293.2	1.0000	327	0.5146	361	0.3239
294	0.9810	328	0.5064	362	0.3202
295	0.9579	329	0.4985	363	0.3165
296	0.9358	330	0.4907	364	0.3130
297	0.9142	331	0.4832	365	0.3095
298	0.8937	332	0.4759	366	0.3060
299	0.8737	333	0.4688	367	0.3027
300	0.8545	334	0.4618	368	0.2994
301	0.8360	335	0.4550	369	0.2962
302	0.8180	336	0.4483	370	0.2930
303	0.8007	337	0.4418	371	0.2899
304	0.7840	338	0.4355	372	0.2868
305	0.7679	339	0.4293	373	0.2838

*Calculated by the formula:

$$1/\mu = 21.482\left[(\theta - 281.435) + \sqrt{(8078.4 + (\theta - 281.435)^2}\right] - 1200 \quad (\mu \text{ in Ns/m}^2)$$

(By permission from *Fluidity and Plasticity*, by E.C. Bingham. Copyright 1922, McGraw-Hill Book Company Inc.)

Table 6. Thermal conductivities of gases and vapours

The extreme temperature values given constitute the experimental range. For extrapolation to other temperatures, it is suggested that the data given be plotted as log k *vs.* log T, or that use be made of the assumption that the ratio $C_p\mu/k$ is practically independent of temperature (and of pressure, within moderate limits).

Substance	k (W/m K)	(K)	k (Btu/h ft °F)
Acetone	0.0098	273	0.0057
	0.0128	319	0.0074
	0.0171	373	0.0099
	0.0254	457	0.0147
Acetylene	0.0118	198	0.0068
	0.0187	273	0.0108
	0.0242	323	0.0140
	0.0298	373	0.0172
Air	0.0164	173	0.0095
	0.0242	273	0.0140
	0.0317	373	0.0183
	0.0391	473	0.0226
	0.0459	573	0.0265
Ammonia	0.0164	213	0.0095
	0.0222	273	0.0128
	0.0272	323	0.0157
	0.0320	373	0.0185
Benzene	0.0090	273	0.0052
	0.0126	319	0.0073
	0.0178	373	0.0103
	0.0263	457	0.0152
	0.0305	485	0.0176
Butane (n-)	0.0135	273	0.0078
	0.0234	373	0.0135
(iso-)	0.0138	273	0.0080
	0.0241	373	0.0139
Carbon dioxide	0.0118	223	0.0068
	0.0147	273	0.0085
	0.0230	373	0.0133
	0.0313	473	0.0181
	0.0396	573	0.0228
disulphide	0.0069	273	0.0040
	0.0073	280	0.0042
monoxide	0.0071	84	0.0041
	0.0080	94	0.0046
	0.0234	213	0.0135
tetrachloride	0.0071	319	0.0041
	0.0090	373	0.0052
	0.0112	457	0.0065
Hydrogen	0.0113	173	0.065
	0.0144	223	0.083
	0.0173	273	0.100
	0.0199	323	0.115
	0.0223	373	0.129
	0.0308	573	0.178

Substance	k (W/m K)	(K)	k (Btu/h ft °F)
Chlorine	0.0074	273	0.0043
Chloroform	0.0066	273	0.0038
	0.0080	319	0.0046
	0.0100	373	0.0058
	0.0133	457	0.0077
Cyclohexane	0.0164	375	0.0095
Dichlorodifluoromethane	0.0083	273	0.0048
	0.0111	323	0.0064
	0.0139	373	0.0080
	0.0168	423	0.0097
Ethane	0.0114	203	0.0066
	0.0149	239	0.0086
	0.0183	273	0.0106
	0.0303	373	0.0175
Ethyl acetate	0.0125	319	0.0072
	0.0166	373	0.0096
	0.0244	457	0.0141
alcohol	0.0154	293	0.0089
	0.0215	373	0.0124
chloride	0.0095	273	0.0055
	0.0164	373	0.0095
	0.0234	457	0.0135
	0.0263	485	0.0152
ether	0.0133	273	0.0077
	0.0171	319	0.0099
	0.0227	373	0.0131
	0.0327	457	0.0189
	0.0362	485	0.0209
Ethylene	0.0111	202	0.0064
	0.0175	273	0.0101
	0.0267	323	0.0131
	0.0279	373	0.0161
Heptane (n-)	0.0194	473	0.0112
	0.0178	373	0.0103
Hexane (n-)	0.0125	273	0.0072
	0.0138	293	0.0080
Hexene	0.0106	273	0.0061
	0.0109	373	0.0189
	0.0225	457	0.0130
	0.0256	485	0.0148
Methylene chloride	0.0067	273	0.0039
	0.0085	319	0.0049
	0.0109	373	0.0063
	0.0164	485	0.0095

Table 6. *(continued)*

Substance	k (W/m K)	(K)	k (Btu/h ft °F)	Substance	k (W/m K)	(K)	k (Btu/h ft °F)
Hydrogen and carbon dioxide		273					
0 per cent H_2	0.0144		0.0083	Nitric oxide	0.0178	203	0.0103
20 per cent	0.0286		0.0165		0.0239	273	0.0138
40 per cent	0.0467		0.0270	Nitrogen	0.0164	173	0.0095
60 per cent	0.0709		0.0410		0.0242	273	0.0140
80 per cent	0.1070		0.0620		0.0277	323	0.0160
100 per cent	0.173		0.10		0.0312	373	0.0180
Hydrogen and nitrogen		273		Nitrous oxide	0.0116	201	0.0067
0 per cent H_2	0.0230		0.0133		0.0157	273	0.0087
20 per cent	0.0367		0.0212		0.0222	373	0.0128
40 per cent	0.0542		0.0313				
60 per cent	0.0758		0.0438	Oxygen	0.0164	173	0.0095
80 per cent	0.1098		0.0635		0.0206	223	0.0119
Hydrogen and nitrous oxide		273			0.0246	273	0.0142
0 per cent H_2	0.0159		0.0092		0.0284	323	0.0164
20 per cent	0.0294		0.0170		0.0321	373	0.0185
40 per cent	0.0467		0.0270				
60 per cent	0.0709		0.0410	Pentane (n-)	0.0128	273	0.0074
80 per cent	0.112		0.0650		0.0144	293	0.0083
Hydrogen sulphide	0.0132	273	0.0076	(iso-)	0.0125	273	0.0072
					0.0220	373	0.0127
Mercury	0.0341	473	0.0197	Propane	0.0151	273	0.0087
Methane	0.0173	173	0.0100		0.0261	373	0.0151
	0.0251	223	0.0145				
	0.0302	273	0.0175	Sulphur dioxide	0.0087	273	0.0050
	0.0372	323	0.0215		0.0119	373	0.0069
Methyl alcohol	0.0144	273	0.0083				
	0.0222	373	0.0128	Water vapour	0.0208	319	0.0120
acetate	0.0102	273	0.0059		0.0237	373	0.0137
	0.0118	293	0.0068		0.0324	473	0.0187
chloride	0.0092	273	0.0053		0.0429	573	0.0248
	0.0125	319	0.0072		0.0545	673	0.0315
	0.0163	373	0.0094		0.0763	773	0.0441

*By permission from *Heat Transmission*, by W. H. McAdams, copyright 1942, McGraw-Hill.

Table 7. Viscosities of gases*

Co-ordinates for use with graph on facing page

No.	Gas	*X*	*Y*
1	Acetic acid	7.7	14.3
2	Acetone	8.9	13.0
3	Acetylene	9.8	14.9
4	Air	11.0	20.0
5	Ammonia	8.4	16.0
6	Argon	10.5	22.4
7	Benzene	8.5	13.2
8	Bromine	8.9	19.2
9	Butene	9.2	13.7
10	Butylene	8.9	13.0
11	Carbon dioxide	9.5	18.7
12	Carbon disulphide	8.0	16.0
13	Carbon monoxide	11.0	20.0
14	Chlorine	9.0	18.4
15	Chloroform	8.9	15.7
16	Cyanogen	9.2	15.2
17	Cyclohexane	9.2	12.0
18	Ethane	9.1	14.5
19	Ethyl acetate	8.5	13.2
20	Ethyl alcohol	9.2	14.2
21	Ethyl chloride	8.5	15.6
22	Ethyl ether	8.9	13.0
23	Ethylene	9.5	15.1
24	Fluorine	7.3	23.8
25	Freon-11 (CCl_8F)	10.6	15.1
26	Freon-12 (CCl_2F_2)	11.1	16.0
27	Freon-21 ($CHCl_2F$)	10.8	15.3
28	Freon-22 ($CHClF_2$)	10.1	17.0
29	Freon-113 (CCl_2F-$CClF_3$)	11.3	14.0
30	Helium	10.9	20.5
31	Hexane	8.6	11.8
32	Hydrogen	11.2	12.4
33	$3H_2 + 1N_2$	11.2	17.2
34	Hydrogen bromide	8.8	20.9
35	Hydrogen chloride	8.8	18.7
36	Hydrogen cyanide	9.8	14.9
37	Hydrogen iodide	9.0	21.3
38	Hydrogen sulphide	8.6	18.0
39	Iodine	9.0	18.4
40	Mercury	5.3	22.9
41	Methane	9.9	15.5
42	Methyl alcohol	8.5	15.6
43	Nitric oxide	10.9	20.5
44	Nitrogen	10.6	20.0
45	Nitrosyl chloride	8.0	17.6
46	Nitrous oxide	8.8	19.0
47	Oxygen	11.0	21.3
48	Pentane	7.0	12.8
49	Propane	9.7	12.9
50	Propyl alcohol	8.4	13.4
51	Propylene	9.0	13.8
52	Sulphur dioxide	9.6	17.0
53	Toluene	8.6	12.4
54	2, 3, 3-trimethylbutane	9.5	10.5
55	Water	8.0	16.0
56	Xenon	9.3	23.0

(By permission from *Perry's Chemical Engineers' Handbook*, by Perry, R. H. and Green, D. W. (eds) 6th edn. Copyright 1984, McGraw-Hill Book Company Inc.)

[To convert to lb/ft h multiply by 2.42.]

Viscosities of gases

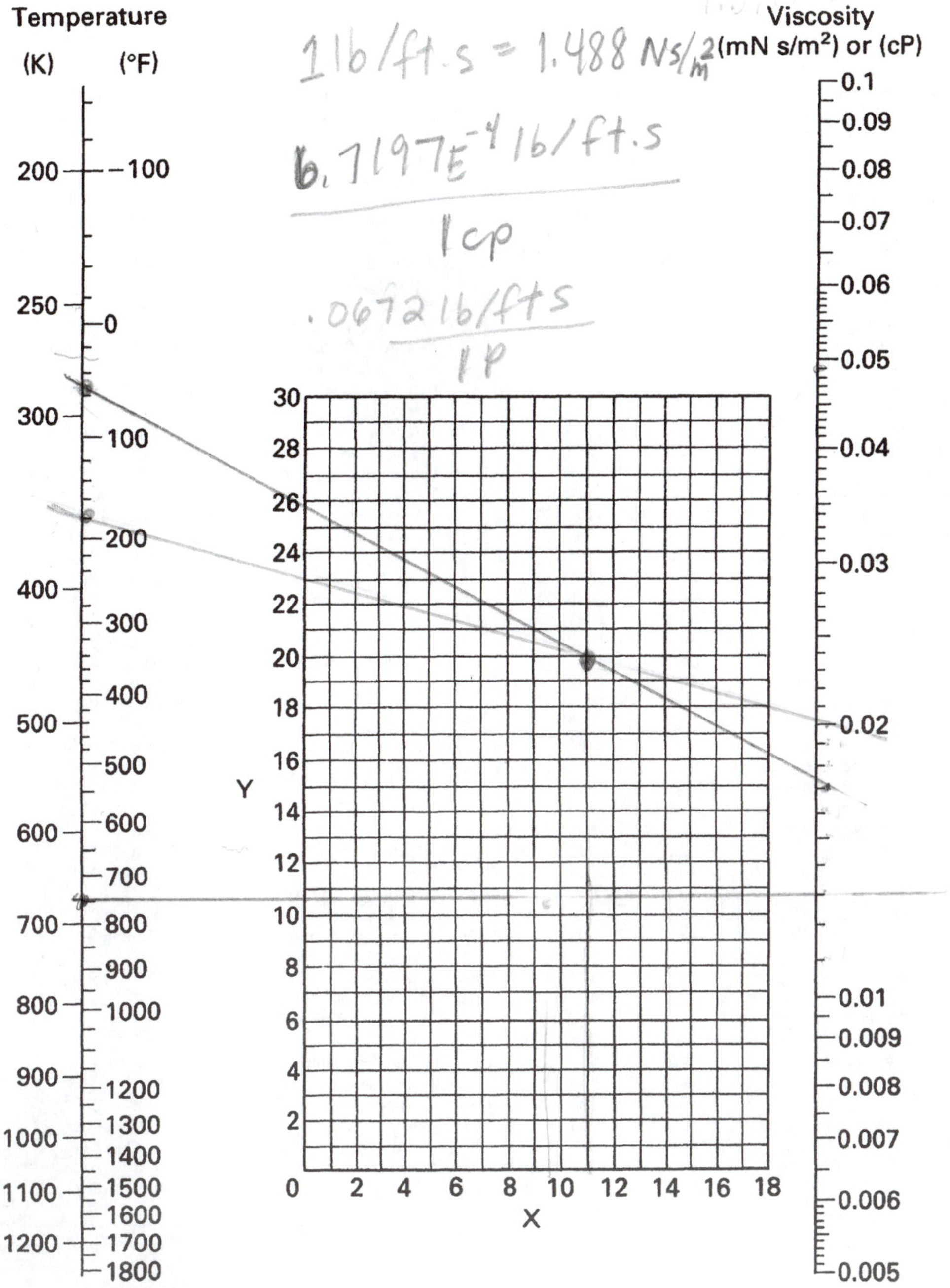

Table 8. Viscosities and densities of liquids*
Co-ordinates for graph on following page

No.	Liquid	X	Y	Density at 293 K (kg/m^3)
1	Acetaldehyde	15.2	4.8	783 (291 K)
2	Acetic acid, 100 per cent	12.1	14.2	1049
3	Acetic acid, 70 per cent	9.5	17.0	1069
4	Acetic anhydride	12.7	12.8	1083
5	Acetone, 100 per cent	14.5	7.2	792
6	Acetone, 35 per cent	7.9	15.0	948
7	Allyl alcohol	10.2	14.3	854
8	Ammonia, 100 per cent	12.6	2.0	817 (194 K)
9	Ammonia, 26 per cent	10.1	13.9	904
10	Amyl acetate	11.8	12.5	879
11	Amyl alcohol	7.5	18.4	817
12	Aniline	8.1	18.7	1022
13	Anisole	12.3	13.5	990
14	Arsenic trichloride	13.9	14.5	2163
15	Benzene	12.5	10.9	880
16	Brine, $CaCl_2$, 25 per cent	6.6	15.9	1228
17	Brine, NaCl, 25 per cent	10.2	16.6	1186 (298 K)
18	Bromine	14.2	13.2	3119
19	Bromotoluene	20.0	15.9	1410
20	Butyl acetate	12.3	11.0	882
21	Butyl alcohol	8.6	17.2	810
22	Butyric acid	12.1	15.3	964
23	Carbon dioxide	11.6	0.3	1101 (236 K)
24	Carbon disulphide	16.1	7.5	1263
25	Carbon tetrachloride	12.7	13.1	1595
26	Chlorobenzene	12.3	12.4	1107
27	Chloroform	14.4	10.2	1489
28	Chlorosulphonic acid	11.2	18.1	1787 (298 K)
29	Chlorotoluene, *ortho*	13.0	13.3	1082
30	Chlorotoluene, *meta*	13.3	12.5	1072
31	Chloroluene, *para*	13.3	12.5	1070
32	Cresol, *meta*	2.5	20.8	1034
33	Cyclohexanol	2.9	24.3	962
34	Dibromoethane	12.7	15.8	2495
35	Dichloroethane	13.2	12.2	1256
36	Dichloromethane	14.6	8.9	1336
37	Diethyl oxalate	11.0	16.4	1079
38	Dimethyl oxalate	12.3	15.8	1148 (327 K)
39	Diphenyl	12.0	18.3	992 (346 K)
40	Dipropyl oxalate	10.3	17.7	1038 (273 K)
41	Ethyl acetate	13.7	9.1	901
42	Ethyl alcohol, 100 per cent	10.5	13.8	789
43	Ethyl alcohol, 95 per cent	9.8	14.3	804
44	Ethyl alcohol, 40 per cent	6.5	16.6	935
45	Ethyl benzene	13.2	11.5	867
46	Ethyl bromide	14.5	8.1	1431
47	Ethyl chloride	14.8	6.0	917 (279 K)
48	Ethyl ether	14.5	5.3	708 (298 K)
49	Ethyl formate	14.2	8.4	923
50	Ethyl iodide	14.7	10.3	1933
51	Ethylene glycol	6.0	23.6	1113
52	Formic acid	10.7	15.8	1220
53	Freon-11 (CCl_3F)	14.4	9.0	1494 (290 K)
54	Freon-12 (CCl_2F_2)	16.8	5.6	1486 (293 K)
55	Freon-21 ($CHCl_2F$)	15.7	7.5	1426 (273 K)
56	Freon-22 ($CHClF_2$)	17.2	4.7	3870 (273 K)
57	Freon-113 (CCl_2F-$CClF_2$)	12.5	11.4	1576
58	Glycerol, 100 per cent	2.0	30.0	1261
59	Glycerol, 50 per cent	6.9	19.6	1126
60	Heptane	14.1	8.4	684
61	Hexane	14.7	7.0	659
62	Hydrochloric acid, 31.5 per cent	13.0	16.6	1157
63	Isobutyl alcohol	7.1	18.0	779 (299 K)
64	Isobutyric acid	12.2	14.4	949
65	Isopropyl alcohol	8.2	16.0	789
66	Kerosene	10.2	16.9	780–820
67	Linseed oil, raw	7.5	27.2	934 ± 4 (288 K)
68	Mercury	18.4	16.4	13546
69	Methanol, 100 per cent	12.4	10.5	792
70	Methanol, 90 per cent	12.3	11.8	820
71	Methanol, 40 per cent	7.8	15.5	935
72	Methyl acetate	14.2	8.2	924
73	Methyl chloride	15.0	3.8	952 (273 K)
74	Methyl ethyl ketone	13.9	8.6	805
75	Naphthalene	7.9	18.1	1145
76	Nitric acid, 95 per cent	12.8	13.8	1493
77	Nitric acid, 60 per cent	10.8	17.0	1367
78	Nitrobenzene	10.6	16.2	1205 (291 K)
79	Nitrotoluene	11.0	17.0	1160
80	Octane	13.7	10.0	703
81	Octyl alcohol	6.6	21.1	827
82	Pentachloroethane	10.9	17.3	1671 (298 K)
83	Pentane	14.9	5.2	630 (291 K)
84	Phenol	6.9	20.8	1071 (298 K)
85	Phosphorus tribromide	13.8	16.7	2852 (288 K)
86	Phosphorus trichloride	16.2	10.9	1574
87	Propionic acid	12.8	13.8	992
88	Propyl alcohol	9.1	16.5	804
89	Propyl bromide	14.5	9.6	1353
90	Propyl chloride	14.4	7.5	890
91	Propyl iodide	14.1	11.6	1749
92	Sodium	16.4	13.9	970
93	Sodium hydroxide, 50%	3.2	25.8	1525
94	Stannic chloride	13.5	12.8	2226
95	Sulphur dioxide	15.2	7.1	1434 (273 K)
96	Sulphuric acid, 110 per cent	7.2	27.4	1980
97	Sulphuric acid, 98 per cent	7.0	24.8	1836
98	Sulphuric acid, 60 per cent	10.2	21.3	1498
99	Sulphuryl chloride	15.2	12.4	1667
100	Tetrachloroethane	11.9	15.7	1600
101	Tetrachloroethylene	14.2	12.7	1624 (288 K)
102	Titanum tetrachloride	14.4	12.3	1726
103	Toluene	13.7	10.4	866
104	Trichloroethylene	14.8	10.5	1466
105	Turpentine	11.5	14.9	861–867
106	Vinyl acetate	14.0	8.8	932
107	Water	10.2	13.0	998
108	Xylene, *ortho*	13.5	12.1	881
109	Xylene, *meta*	13.9	10.6	867
110	Xylene, *para*	13.9	10.9	861

*By permission from *Perry's Chemical Engineers' Handbook*, by Perry, R. H. and Green, D. W. (eds), 6th edn. Copyright 1984, McGraw-Hill.

Viscosities of liquids

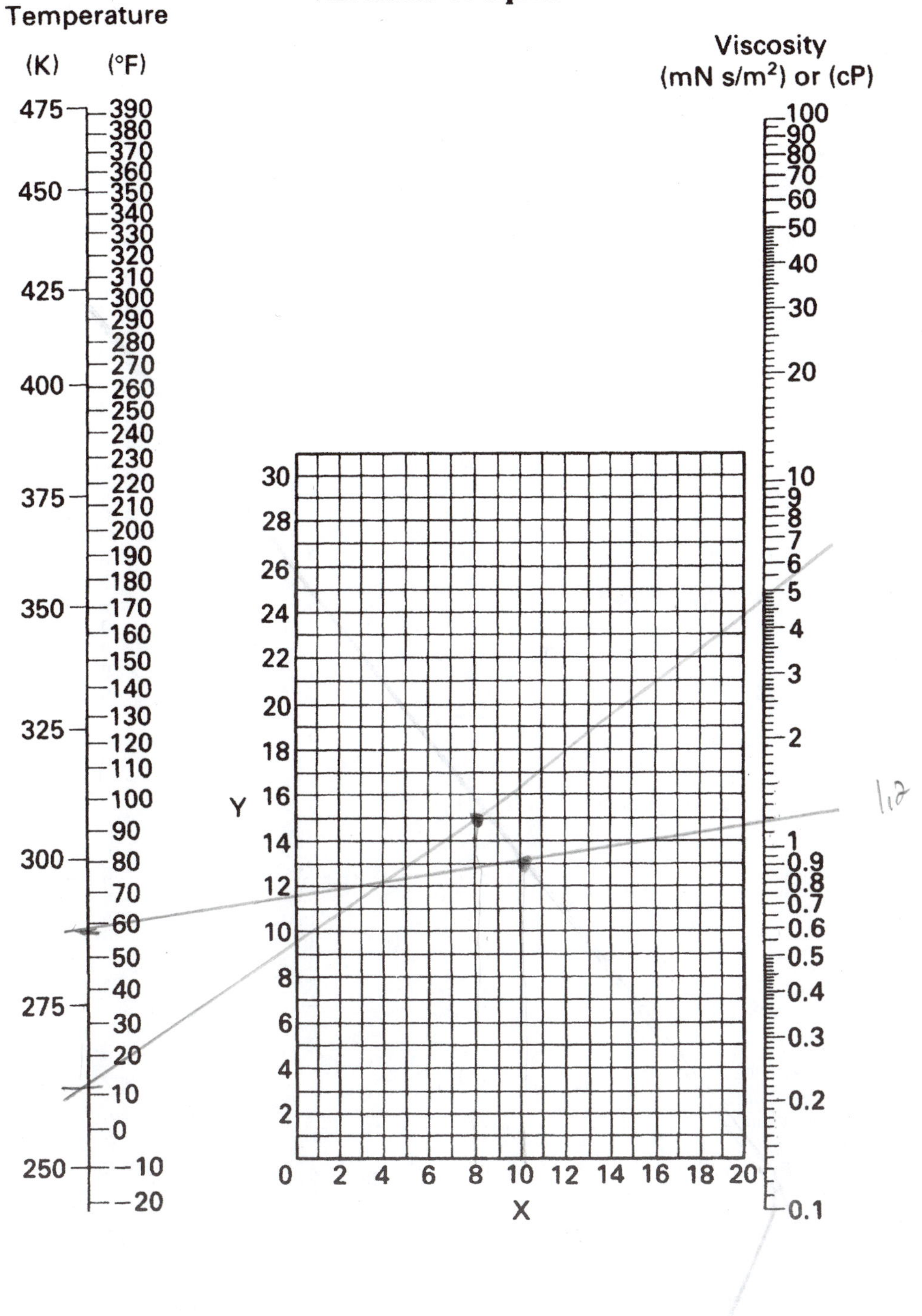

Table 9. Critical constants of gases*

	Critical temperature T_c(K)	Critical pressure P_c (MN/m^2)	Compressibility constant in critical state Z_c
Paraffins			
Methane	191	4.64	0.290
Ethane	306	4.88	0.284
Propane	370	4.25	0.276
n-Butane	425	3.80	0.274
Isobutane	408	3.65	0.282
n-Pentane	470	3.37	0.268
Isopentane	461	3.33	0.268
Neopentane	434	3.20	0.268
n-Hexane	508	3.03	0.264
n-Heptane	540	2.74	0.260
n-Octane	569	2.49	0.258
Mono-olefins			
Ethylene	282	5.07	0.268
Propylene	365	4.62	0.276
1-Butene	420	4.02	0.276
1-Pentene	474	4.05	
Miscellaneous organic compounds			
Acetic acid	595	5.78	0.200
Acetone	509	4.72	0.237
Acetylene	309	6.24	0.274
Benzene	562	4.92	0.274
1,3-Butadiene	425	4.33	0.270
Cyclohexane	553	4.05	0.271
Dichlorodifluoromethane (Freon-12)	385	4.01	0.273
Diethyl ether	467	3.61	0.261
Ethyl alcohol	516	6.38	0.249
Ethylene oxide	468	7.19	0.25
Methyl alcohol	513	7.95	0.220
Methyl chloride	416	6.68	0.276
Methyl ethyl ketone	533	4.00	0.26
Toluene	594	4.21	0.27
Trichlorofluoromethane (Freon-11)	471	4.38	0.277
Trichlorotrifluoroethane (Freon-113)	487	3.41	0.274
Elementary gases			
Bromine	584	10.33	0.307
Chlorine	417	7.71	0.276
Helium	5.3	0.23	0.300
Hydrogen	33.3	1.30	0.304
Neon	44.5	2.72	0.307
Nitrogen	126	3.39	0.291
Oxygen	155	5.08	0.29
Miscellaneous inorganic compounds			
Ammonia	406	11.24	0.242
Carbon dioxide	304	7.39	0.276
Carbon monoxide	133	3.50	0.294
Hydrogen chloride	325	8.26	0.266
Hydrogen sulphide	374	9.01	0.284
Nitric oxide (NO)	180	6.48	0.25
Nitrous oxide (N_2O)	310	7.26	0.271
Sulphur	1313	11.75	
Sulphur dioxide	431	7.88	0.268
Sulphur trioxide	491	8.49	0.262
Water	647	22.1	0.23

*Selected values from K. A. Kobe and R. E. Lynn, Jr, *Chem. Rev.*, **52**, 117 (1953). By permission.

Table 10. Emissivities of surfaces*

Surface	T(K)	Emissivity
A. Metals and metallic oxides		
Aluminium		
highly polished plate	500–850	0.039–0.057
polished plate	296	0.040
rough plate	299	0.055
plate oxidised at 872 K	472–872	0.11–0.19
aluminium-surfaced roofing	311	0.216
Brass		
hard-rolled, polished	294	0.038
polished	311–589	0.096
rolled-plate, natural surface	295	0.06
rubbed with coarse emery	295	0.20
dull plate	322–622	0.22
oxidised	472–872	0.61–0.59
Chromium-see Nickel alloys		
Copper		
polished electrolytic	353	0.018
commercial, emeried and polished	292	0.030
commercial, scraped shiny	295	0.072
polished	390	0.023
plate, covered with thick oxide	498	0.78
plate heated to 872 K	472–872	0.57–0.57
cuprous oxide	1072–1372	0.66–0.54
molten copper	1350–1550	0.16–0.13
Gold		
pure, highly polished	500–900	0.018–0.35
Iron and steel		
electrolytic iron, highly polished	450–500	0.052–0.064
polished iron	700–1300	0.144–0.377
freshly emeried iron	293	0.242
polished cast iron	473	0.21
wrought iron, highly polished	311–522	0.28
cast iron, newly turned	295	0.435
steel casting, polished	1044–1311	0.52–0.56
ground sheet steel	1211–1372	0.55–0.61
smooth sheet iron	1172–1311	0.55–0.60
cast iron, turned	1155–1261	0.60–0.70
oxidised surfaces		
iron plate, completely rusted	293	0.685
sheet steel, rolled and oxidised	295	0.657
iron	373	0.736
cast iron, oxidised at 872 K	472–872	0.64–0.78
steel, oxidised at 872 K	472–872	0.79–0.79
smooth electrolytic iron	500–800	0.78–0.82
iron oxide	772–1472	0.85–0.89
ingot iron, rough	1200–1390	0.87–0.95
sheet steel with rough oxide layer	297	0.80
cast iron, strongly oxidised	311–522	0.95
wrought iron, dull oxidised	294–633	0.94
steel plate, rough	311–644	0.94–0.97
molten metal		
cast iron	1572–1672	0.29–0.29
mild steel	1872–2070	0.28–0.28
Lead		
pure, unoxidised	400–500	0.057–0.075
grey, oxidised	297	0.281
oxidised at 472 K	472	0.03

(continued overleaf)

*From HOTTEL, H. C. and SAROFIM, A. F.: Radiation Heat Transfer (McGraw-Hill, New York, 1967)

Table 10. *(continued)*

Surface	T(K)	Emissivity
Mercury	273–373	0.09–0.12
Molybdenum		
filament	1000–2866	0.096–0.292
Monel		
metal oxidised at 872 K	472–872	0.41–0.46
Nickel		
electroplated on polished iron and polished	296	0.045
techically pure, polished	500–600	0.07–0.087
electroplated on pickled iron, unpolished	293	0.11
wire	460–1280	0.096–0.186
plate, oxidised by heating to 872 K	472–872	0.37–0.48
nickel oxide	922–1527	0.59–0.86
nickel alloys		
chromonickel	325–1308	0.64–0.76
nickelin, grey oxidised	294	0.262
KA–28 alloy, rough brown, after heating	489–763	0.44–0.36
KA–28 alloy, after heating at 800 K	489–800	0.62–0.73
NCT 3 alloy, oxidised from service	489–800	0.90–0.97
NCT 6 alloy, oxidised from service	544–836	0.89–0.82
Platinum		
pure, polished plate	500–900	0.054–0.104
strip	1200–1900	0.12–0.17
filament	300–1600	0.036–0.192
wire	500–1600	0.073–0.182
Silver		
polished, pure	500–900	0.0198–0.0324
polished	310–644	0.0221–0.0312
Steel–see Iron		
Tantalum		
filament	1600–3272	0.194–0.31
Tin		
bright tinned iron sheet	298	0.043 and 0.064
Tungsten		
filament, aged	300–3588	0.032–0.35
filament	3588	0.39
Zinc		
commercially pure, polished	500–600	0.045–0.053
oxidised by heating to 672 K	672	0.11
galvanized sheet iron, fairly bright	301	0.228
galvanized sheet iron, grey oxidised	297	0.276
B. Refractories, building materials, paints etc.		
Asbestos		
board	297	0.96
paper	311–644	0.93–0.945
Brick		
red, rough	294	0.93
silica, unglazed	1275	0.80
silica, glazed, rough	1475	0.85
grog, glazed	1475	0.75
Carbon		
T–carbon	400–900	0.81–0.79
filament	1311–1677	0.526
candle soot	372–544	0.952
lampblack–water–glass coating	372–456	0.957–0.952
thin layer on iron plate	294	0.927
thick coat	293	0.967
lampblack, 0.08 mm or thicker	311–644	0.945

Table 10. *(continued)*

Surface	T(K)	Emissivity
Enamel		
white fused on iron	292	0.897
Glass		
smooth	295	0.937
Gypsum		
0.5 mm thick on blackened plate	294	0.903
Marble		
light grey, polished	295	0.931
Oak		
planed	294	0.895
Oil layers		
on polished nickel		
polished surface alone		0.045
0.025 mm oil		0.27
0.050 mm oil		0.46
0.125 mm oil		0.72
thick oil layer		0.82
on aluminium foil		
aluminium foil alone	373	0.087
1 coat of oil	373	0.561
2 coats of oil	373	0.574
Paints, lacquers, varnishes		
snow white enamel on rough iron plate	296	0.906
black, shiny lacquer sprayed on iron	298	0.875
black, shiny shellac on tinned iron sheet	294	0.821
black matt shellac	350–420	0.91
black laquer	311–366	0.80–0.95
matt black lacquer	311–366	0.96–0.98
white lacquer	311–366	0.80–0.95
oil paints	373	0.92–0.96
aluminium paint	373	0.27–0.67
after heating to 600 K	422–622	0.35
aluminium lacquer	294	0.39
Paper, thin		
pasted on tinned iron plate	292	0.924
pasted on rough iron plate	292	0.929
pasted on black lacquered plate	292	0.944
roofing	294	0.91
Plaster, lime, rough	283–361	0.91
Porcellain, glazed	295	0.924
Quartz, rough, fused	294	0.932
Refractory materials		
poor radiators	872–1272	0.65–0.75
good radiators	872–1272	0.80–0.90
Rubber		
hard, glossy plate	296	0.945
soft, grey, rough	298	0.859
Serpentine, polished	296	0.900
Water	273–373	0.95–0.963

A2. STEAM TABLES

Tables 11A, 11B, 11C and 11D are adapted from the
Abridged Callendar Steam Tables
by permission of Messrs Edward Arnold (Publishers) Ltd.

Table 11A. Properties of saturated steam (S.I. units)

Absolute pressure (kN/m²)	Temperature (°C)	Temperature (K)	Enthalpy per unit mass (H_s) (kJ/kg)			Entropy per unit mass (S_s) (kJ/kg K)			Specific volume (ν) (m³/kg)	
	θ_s	T_s	water	latent	steam	water	latent	steam	water	steam
Datum: Triple point of water										
0.611	0.01	273.16	0.0	2501.6	2501.6	0	9.1575	9.1575	0.0010002	206.16
1.0	6.98	280.13	29.3	2485.0	2514.4	0.1060	8.8706	8.9767	0.001000	129.21
2.0	17.51	290.66	73.5	2460.2	2533.6	0.2606	8.4640	8.7246	0.001001	67.01
3.0	24.10	297.25	101.0	2444.6	2545.6	0.3543	8.2242	8.5785	0.001003	45.67
4.0	28.98	302.13	121.4	2433.1	2554.5	0.4225	8.0530	8.4755	0.001004	34.80
5.0	32.90	306.05	137.8	2423.8	2561.6	0.4763	7.9197	8.3960	0.001005	28.19
6.0	36.18	309.33	151.5	2416.0	2567.5	0.5209	7.8103	8.3312	0.001006	23.74
7.0	39.03	312.18	163.4	2409.2	2572.6	0.5591	7.7176	8.2767	0.001007	20.53
8.0	41.54	314.69	173.9	2403.2	2577.1	0.5926	7.6370	8.2295	0.001008	18.10
9.0	43.79	316.94	183.3	2397.9	2581.1	0.6224	7.5657	8.1881	0.001009	16.20
10.0	45.83	318.98	191.8	2392.9	2584.8	0.6493	7.5018	8.1511	0.001010	14.67
12.0	49.45	322.60	206.9	2384.2	2591.2	0.6964	7.3908	8.0872	0.001012	12.36
14.0	52.58	325.73	220.0	2376.7	2596.7	0.7367	7.2966	8.0333	0.001013	10.69
16.0	55.34	328.49	231.6	2370.0	2601.6	0.7721	7.2148	7.9868	0.001015	9.43
18.0	57.83	330.98	242.0	2363.9	2605.9	0.8036	7.1423	7.9459	0.001016	8.45
20.0	60.09	333.24	251.5	2358.4	2609.9	0.8321	7.0773	7.9094	0.001017	7.65
25.0	64.99	338.14	272.0	2346.4	2618.3	0.8933	6.9390	7.8323	0.001020	6.20
30.0	69.13	342.28	289.3	2336.1	2625.4	0.9441	6.8254	7.7695	0.001022	5.23
35.0	72.71	345.86	304.3	2327.2	2631.5	0.9878	6.7288	7.7166	0.001025	4.53
40.0	75.89	349.04	317.7	2319.2	2636.9	1.0261	6.6448	7.6709	0.001027	3.99
45.0	78.74	351.89	329.6	2312.0	2641.7	1.0603	6.5703	7.6306	0.001028	3.58
50.0	81.35	354.50	340.6	2305.4	2646.0	1.0912	6.5035	7.5947	0.001030	3.24
60.0	85.95	359.10	359.9	2293.6	2653.6	1.1455	6.3872	7.5327	0.001033	2.73
70.0	89.96	363.11	376.8	2283.3	2660.1	1.1921	6.2883	7.4804	0.001036	2.37
80.0	93.51	366.66	391.7	2274.0	2665.8	1.2330	6.2022	7.4352	0.001039	2.09
90.0	96.71	369.86	405.2	2265.6	2670.9	1.2696	6.1258	7.3954	0.001041	1.87
100.0	99.63	372.78	417.5	2257.9	2675.4	1.3027	6.0571	7.3598	0.001043	1.69
101.325	100.00	373.15	419.1	2256.9	2676.0	1.3069	6.0485	7.3554	0.0010437	1.6730
105	101.00	374.15	423.3	2254.3	2677.6	1.3182	6.0252	7.3434	0.001045	1.618
110	102.32	375.47	428.8	2250.8	2679.6	1.3330	5.9947	7.3277	0.001046	1.549
115	103.59	376.74	434.2	2247.4	2681.6	1.3472	5.9655	7.3127	0.001047	1.486
120	104.81	377.96	439.4	2244.1	2683.4	1.3609	5.9375	7.2984	0.001048	1.428
125	105.99	379.14	444.4	2240.9	2685.2	1.3741	5.9106	7.2846	0.001049	1.375
130	107.13	380.28	449.2	2237.8	2687.0	1.3868	5.8847	7.2715	0.001050	1.325
135	108.24	381.39	453.9	2234.8	2688.7	1.3991	5.8597	7.2588	0.001050	1.279
140	109.32	382.47	458.4	2231.9	2690.3	1.4109	5.8356	7.2465	0.001051	1.236
145	110.36	383.51	462.8	2229.0	2691.8	1.4225	5.8123	7.2347	0.001052	1.196
150	111.37	384.52	467.1	2226.2	2693.4	1.4336	5.7897	7.2234	0.001053	1.159
155	112.36	385.51	471.3	2223.5	2694.8	1.4445	5.7679	7.2123	0.001054	1.124
160	113.32	386.47	475.4	2220.9	2696.2	1.4550	5.7467	7.2017	0.001055	1.091
165	114.26	387.41	479.4	2218.3	2697.6	1.4652	5.7261	7.1913	0.001056	1.060
170	115.17	388.32	483.2	2215.7	2699.0	1.4752	5.7061	7.1813	0.001056	1.031
175	116.06	389.21	487.0	2213.3	2700.3	1.4849	5.6867	7.1716	0.001057	1.003
180	116.93	390.08	490.7	2210.8	2701.5	1.4944	5.6677	7.1622	0.001058	0.977
185	117.79	390.94	494.3	2208.5	2702.8	1.5036	5.6493	7.1530	0.001059	0.952
190	118.62	391.77	497.9	2206.1	2704.0	1.5127	5.6313	7.1440	0.001059	0.929
195	119.43	392.58	501.3	2203.8	2705.1	1.5215	5.6138	7.1353	0.001060	0.907
200	120.23	393.38	504.7	2201.6	2706.3	1.5301	5.5967	7.1268	0.001061	0.885

(continued overleaf)

Table 11A. *(continued)*

Absolute pressure (kN/m^2)	Temperature (°C)	Temperature (K)	Enthalpy per unit mass (H_s) (kJ/kg)			Entropy per unit mass (S_s) (kJ/kg K)			Specific volume (ν) (m^3/kg)	
	θ_s	T_s	water	latent	steam	water	latent	steam	water	steam
210	121.78	394.93	511.3	2197.2	2708.5	1.5468	5.5637	7.1105	0.001062	0.846
220	123.27	396.42	517.6	2193.0	2710.6	1.5628	5.5321	7.0949	0.001064	0.810
230	124.71	397.86	523.7	2188.9	2712.6	1.5781	5.5018	7.0800	0.001065	0.777
240	126.09	399.24	529.6	2184.9	2714.5	1.5929	5.4728	7.0657	0.001066	0.746
250	127.43	400.58	535.4	2181.0	2716.4	1.6072	5.4448	7.0520	0.001068	0.718
260	128.73	401.88	540.9	2177.3	2718.2	1.6209	5.4179	7.0389	0.001069	0.692
270	129.99	403.14	546.2	2173.6	2719.9	1.6342	5.3920	7.0262	0.001070	0.668
280	131.21	404.36	551.5	2170.1	2721.5	1.6471	5.3669	7.0140	0.001071	0.646
290	132.39	405.54	556.5	2166.6	2723.1	1.6596	5.3427	7.0022	0.001072	0.625
300	133.54	406.69	561.4	2163.2	2724.7	1.6717	5.3192	6.9909	0.001074	0.606
320	135.76	408.91	570.9	2156.7	2727.6	1.6948	5.2744	6.9692	0.001076	0.570
340	137.86	411.01	579.9	2150.4	2730.3	1.7168	5.2321	6.9489	0.001078	0.538
360	139.87	413.02	588.5	2144.4	2732.9	1.7376	5.1921	6.9297	0.001080	0.510
380	141.79	414.94	596.8	2138.6	2735.3	1.7575	5.1541	6.9115	0.001082	0.485
400	143.63	416.78	604.7	2132.9	2737.6	1.7764	5.1179	6.8943	0.001084	0.462
420	145.39	418.54	612.3	2127.5	2739.8	1.7946	5.0833	6.8779	0.001086	0.442
440	147.09	420.24	619.6	2122.3	2741.9	1.8120	5.0503	6.8622	0.001088	0.423
460	148.73	421.88	626.7	2117.2	2743.9	1.8287	5.0186	6.8473	0.001089	0.405
480	150.31	423.46	633.5	2112.2	2745.7	1.8448	4.9881	6.8329	0.001091	0.389
500	151.85	425.00	640.1	2107.4	2747.5	1.8604	4.9588	6.8192	0.001093	0.375
520	153.33	426.48	646.5	2102.7	2749.3	1.8754	4.9305	6.8059	0.001095	0.361
540	154.77	427.92	652.8	2098.1	2750.9	1.8899	4.9033	6.7932	0.001096	0.348
560	156.16	429.31	658.8	2093.7	2752.5	1.9040	4.8769	6.7809	0.001098	0.337
580	157.52	430.67	664.7	2089.3	2754.0	1.9176	4.8514	6.7690	0.001100	0.326
600	158.84	431.99	670.4	2085.0	2755.5	1.9308	4.8267	6.7575	0.001101	0.316
620	160.12	433.27	676.0	2080.8	2756.9	1.9437	4.8027	6.7464	0.001102	0.306
640	161.38	434.53	681.5	2076.7	2758.2	1.9562	4.7794	6.7356	0.001104	0.297
660	162.60	435.75	686.8	2072.7	2759.5	1.9684	4.7568	6.7252	0.001105	0.288
680	163.79	436.94	692.0	2068.8	2760.8	1.9803	4.7348	6.7150	0.001107	0.280
700	164.96	438.11	697.1	2064.9	2762.0	1.9918	4.7134	6.7052	0.001108	0.272
720	166.10	439.25	702.0	2061.1	2763.2	2.0031	4.6925	6.6956	0.001109	0.266
740	167.21	440.36	706.9	2057.4	2764.3	2.0141	4.6721	6.6862	0.001110	0.258
760	168.30	441.45	711.7	2053.7	2765.4	2.0249	4.6522	6.6771	0.001112	0.252
780	169.37	442.52	716.3	2050.1	2766.4	2.0354	4.6328	6.6683	0.001114	0.246
800	170.41	443.56	720.9	2046.5	2767.5	2.0457	4.6139	6.6596	0.001115	0.240
820	171.44	444.59	725.4	2043.0	2768.5	2.0558	4.5953	6.6511	0.001116	0.235
840	172.45	445.60	729.9	2039.6	2769.4	2.0657	4.5772	6.6429	0.001118	0.229
860	173.43	446.58	734.2	2036.2	2770.4	2.0753	4.5595	6.6348	0.001119	0.224
880	174.40	447.55	738.5	2032.8	2771.3	2.0848	4.5421	6.6269	0.001120	0.220
900	175.36	448.51	742.6	2029.5	2772.1	2.0941	4.5251	6.6192	0.001121	0.215
920	176.29	449.44	746.8	2026.2	2773.0	2.1033	4.5084	6.6116	0.001123	0.210
940	177.21	450.36	750.8	2023.0	2773.8	2.1122	4.4920	6.6042	0.001124	0.206
960	178.12	451.27	754.8	2019.8	2774.6	2.1210	4.4759	6.5969	0.001125	0.202
980	179.01	452.16	758.7	2016.7	2775.4	2.1297	4.4602	6.5898	0.001126	0.198
1000	179.88	453.03	762.6	2013.6	2776.2	2.1382	4.4447	6.5828	0.001127	0.194
1100	184.06	457.21	781.1	1998.6	2779.7	2.1786	4.3712	6.5498	0.001133	0.177
1200	187.96	461.11	798.4	1984.3	2782.7	2.2160	4.3034	6.5194	0.001139	0.163
1300	191.60	464.75	814.7	1970.7	2785.4	2.2509	4.2404	6.4913	0.001144	0.151
1400	195.04	468.19	830.1	1957.7	2787.8	2.2836	4.1815	6.4651	0.001149	0.141
1500	198.28	471.43	844.6	1945.3	2789.9	2.3144	4.1262	6.4406	0.001154	0.132
1600	201.37	474.52	858.5	1933.2	2791.7	2.3436	4.0740	6.4176	0.001159	0.124

Table 11A. *(continued)*

Absolute pressure (kN/m^2)	Temperature (°C) θ_s	Temperature (K) T_s	Enthalpy per unit mass (H_s) (kJ/kg) water	latent	steam	Entropy per unit mass (S_s) (kJ/kg K) water	latent	steam	Specific volume (v) (m^3/kg) water	steam
1700	204.30	477.45	871.8	1921.6	2793.4	2.3712	4.0246	6.3958	0.001163	0.117
1800	207.11	480.26	884.5	1910.3	2794.8	2.3976	3.9776	6.3751	0.001168	0.110
1900	209.79	482.94	896.8	1899.3	2796.1	2.4227	3.9327	6.3555	0.001172	0.105
2000	212.37	485.52	908.6	1888.7	2797.2	2.4468	3.8899	6.3367	0.001177	0.0996
2200	217.24	490.39	930.9	1868.1	2799.1	2.4921	3.8094	6.3015	0.001185	0.0907
2400	221.78	494.93	951.9	1848.5	2800.4	2.5342	3.7348	6.2690	0.001193	0.0832
2600	226.03	499.18	971.7	1829.7	2801.4	2.5736	3.6652	6.2388	0.001201	0.0769
3000	233.84	506.99	1008.3	1794.0	2802.3	2.6455	3.5383	6.1838	0.001216	0.0666
3500	242.54	515.69	1049.7	1752.2	2802.0	2.7252	3.3976	6.1229	0.001235	0.0570
4000	250.33	523.48	1087.4	1712.9	2800.3	2.7965	3.2720	6.0685	0.001252	0.0498
4500	257.41	530.56	1122.1	1675.6	2797.7	2.8612	3.1579	6.0191	0.001269	0.0440
5000	263.92	537.07	1154.5	1639.7	2794.2	2.9207	3.0528	5.9735	0.001286	0.0394
6000	275.56	548.71	1213.7	1571.3	2785.0	3.0274	2.8633	5.8907	0.001319	0.0324
7000	285.80	558.95	1267.5	1506.0	2773.4	3.1220	2.6541	5.8161	0.001351	0.0274
8000	294.98	568.13	1317.2	1442.7	2759.9	3.2077	2.5393	5.7470	0.001384	0.0235
9000	303.31	576.46	1363.8	1380.8	2744.6	3.2867	2.3952	5.6820	0.001418	0.0205
10000	310.96	584.11	1408.1	1319.7	2727.7	3.3606	2.2592	5.6198	0.001453	0.0180
11000	318.04	591.19	1450.6	1258.8	2709.3	3.4304	2.1292	5.5596	0.001489	0.0160
12000	324.64	597.79	1491.7	1197.5	2698.2	3.4971	2.0032	5.5003	0.001527	0.0143
14000	336.63	609.78	1571.5	1070.9	2642.4	3.6241	1.7564	5.3804	0.0016105	0.01150
16000	347.32	620.47	1650.4	934.5	2584.9	3.7470	1.5063	5.2533	0.0017102	0.00931
18000	356.96	630.11	1734.8	779.0	2513.9	3.8766	1.2362	5.1127	0.0018399	0.007497
20000	365.71	638.86	1826.6	591.6	2418.2	4.0151	0.9259	4.9410	0.0020374	0.005875
22000	373.68	646.83	2010.3	186.3	2196.6	4.2934	0.2881	4.5814	0.0026675	0.003735
22120	374.15	647.30	2107.4	0	2107.4	4.4429	0	4.4429	0.0031700	0.003170

Table 11B. Properties of saturated steam (Centigrade and Fahrenheit units)

Pressure		Temperature		Enthalpy per unit mass						Entropy (Btu/lb°F)		Specific volume (ft³/lb)
				Centigrade units (kcal/kg)			Fahrenheit units (Btu/lb)					
Absolute (lb/in.²)	Vacuum (in. Hg)	(°C)	(°F)	Water	Latent	Steam	Water	Latent	Steam	Water	Steam	Steam
0.5	28.99	26.42	79.6	**26.45**	582.50	**608.95**	**47.6**	1048.5	**1096.1**	0.0924	2.0367	643.0
0.6	28.79	29.57	85.3	**29.58**	580.76	**610.34**	**53.2**	1045.4	**1098.6**	0.1028	2.0214	540.6
0.7	28.58	32.28	90.1	**32.28**	579.27	**611.55**	**58.1**	1042.7	**1100.8**	0.1117	2.0082	466.6
0.8	28.38	34.67	94.4	**34.66**	577.95	**612.61**	**62.4**	1040.3	**1102.7**	0.1196	1.9970	411.7
0.9	28.17	36.80	98.2	**36.80**	576.74	**613.54**	**66.2**	1038.1	**1104.3**	0.1264	1.9871	368.7
1.0	27.97	38.74	101.7	**38.74**	575.60	**614.34**	**69.7**	1036.1	**1105.8**	0.1326	1.9783	334.0
1.1	27.76	40.52	104.9	**40.52**	574.57	**615.09**	**72.9**	1034.3	**1107.2**	0.1381	1.9702	305.2
1.2	27.56	42.17	107.9	**42.17**	573.63	**615.80**	**75.9**	1032.5	**1108.4**	0.1433	1.9630	281.1
1.3	27.35	43.70	110.7	**43.70**	572.75	**616.45**	**78.7**	1030.9	**1109.6**	0.1484	1.9563	260.5
1.4	27.15	45.14	113.3	**45.12**	571.94	**617.06**	**81.3**	1029.5	**1110.8**	0.1527	1.9501	243.0
1.5	26.95	46.49	115.7	**46.45**	571.16	**617.61**	**83.7**	1028.1	**1111.8**	0.1569	1.9442	228.0
1.6	26.74	47.77	118.0	**47.73**	570.41	**618.14**	**86.0**	1026.8	**1112.8**	0.1609	1.9387	214.3
1.7	26.54	48.98	120.2	**48.94**	569.71	**618.65**	**88.2**	1025.5	**1113.7**	0.1646	1.9336	202.5
1.8	26.33	50.13	122.2	**50.08**	569.06	**619.14**	**90.2**	1024.4	**1114.6**	0.1681	1.9288	191.8
1.9	26.13	51.22	124.2	**51.16**	568.47	**619.63**	**92.1**	1023.3	**1115.4**	0.1715	1.9243	182.3
2.0	25.92	52.27	126.1	**52.22**	567.89	**620.11**	**94.0**	1022.2	**1116.2**	0.1749	1.9200	173.7
3.0	23.88	60.83	141.5	**60.78**	562.89	**623.67**	**109.4**	1013.2	**1122.6**	0.2008	1.8869	118.7
4.0	21.84	67.23	153.0	**67.20**	559.29	**626.49**	**121.0**	1006.7	**1127.7**	0.2199	1.8632	90.63
5.0	19.80	72.38	162.3	**72.36**	556.24	**628.60**	**130.2**	1001.6	**1131.8**	0.2348	1.8449	73.52
6.0	17.76	76.72	170.1	**76.71**	553.62	**630.33**	**138.1**	996.6	**1134.7**	0.2473	1.8299	61.98
7.0	15.71	80.49	176.9	**80.52**	551.20	**631.72**	**144.9**	992.2	**1137.1**	0.2582	1.8176	53.64
8.0	13.67	83.84	182.9	**83.89**	549.16	**633.05**	**151.0**	988.5	**1139.5**	0.2676	1.8065	47.35
9.0	11.63	86.84	188.3	**86.88**	547.42	**634.30**	**156.5**	985.2	**1141.7**	0.2762	1.7968	42.40
10.0	9.59	89.58	193.2	**89.61**	545.82	**635.43**	**161.3**	982.5	**1143.8**	0.2836	1.7884	38.42
11.0	7.55	92.10	197.8	**92.15**	544.26	**636.41**	**165.9**	979.6	**1145.5**	0.2906	1.7807	35.14
12.0	5.50	94.44	202.0	**94.50**	542.75	**637.25**	**170.1**	976.9	**1147.0**	0.2970	1.7735	32.40
13.0	3.46	96.62	205.9	**96.69**	541.34	**638.03**	**173.9**	974.6	**1148.5**	0.3029	1.7672	30.05
14.0	1.42	98.65	209.6	**98.73**	540.06	**638.79**	**177.7**	972.2	**1149.9**	0.3086	1.7613	28.03
14.696	Gauge (lb/in.²)	100.00	212.0	**100.06**	539.22	**639.28**	**180.1**	970.6	**1150.7**	0.3122	1.7574	26.80
15	0.3	100.57	213.0	**100.65**	538.9	**639.5**	**181.2**	970.0	**1151.2**	0.3137	1.7556	26.28
16	1.3	102.40	216.3	**102.51**	537.7	**640.2**	**184.5**	967.9	**1152.4**	0.3187	1.7505	24.74
17	2.3	104.13	219.5	**104.27**	536.5	**640.8**	**187.6**	965.9	**1153.5**	0.3231	1.7456	23.38
18	3.3	105.78	222.4	**105.94**	535.5	**641.4**	**190.6**	964.0	**1154.6**	0.3276	1.7411	22.17
19	4.3	107.36	225.2	**107.53**	534.5	**642.0**	**193.5**	962.2	**1155.7**	0.3319	1.7368	21.07
20	5.3	108.87	228.0	**109.05**	533.6	**642.6**	**196.3**	960.4	**1156.7**	0.3358	1.7327	20.09
21	6.3	110.32	230.6	**110.53**	532.6	**643.1**	**198.9**	958.8	**1157.7**	0.3396	1.7287	19.19
22	7.3	111.71	233.1	**111.94**	531.7	**643.6**	**201.4**	957.2	**1158.6**	0.3433	1.7250	18.38
23	8.3	113.05	235.5	**113.30**	530.8	**644.1**	**203.9**	955.6	**1159.5**	0.3468	1.7215	17.63
24	9.3	114.34	237.8	**114.61**	530.0	**644.6**	**206.3**	954.0	**1160.3**	0.3502	1.7181	16.94
25	10.3	115.59	240.1	**115.87**	529.2	**645.1**	**208.6**	952.5	**1161.1**	0.3534	1.7148	16.30
26	11.3	116.80	242.2	**117.11**	528.4	**645.5**	**210.8**	951.1	**1161.9**	0.3565	1.7118	15.72
27	12.3	117.97	244.4	**118.31**	527.6	**645.9**	**212.9**	949.7	**1162.6**	0.3595	1.7089	15.17
28	13.3	119.11	246.4	**119.47**	526.8	**646.3**	**215.0**	948.3	**1163.3**	0.3625	1.7060	14.67
29	14.3	120.21	248.4	**120.58**	526.1	**646.7**	**217.0**	947.0	**1164.0**	0.3654	1.7032	14.19
30	15.3	121.3	250.3	**121.7**	525.4	**647.1**	**219.0**	945.6	**1164.6**	0.3682	1.7004	13.73
32	17.3	123.3	254.0	**123.8**	524.1	**647.9**	**222.7**	943.1	**1165.8**	0.3735	1.6952	12.93
34	19.3	125.3	257.6	**125.8**	522.8	**648.6**	**226.3**	940.7	**1167.0**	0.3785	1.6905	12.21

Table 11B. (*continued*)

Pressure		Temperature		Enthalpy per unit mass						Entropy (Btu/lb°F)		Specific volume (ft^3/lb)
				Centigrade units (kcal/kg)			Fahrenheit units (Btu/lb)					
Absolute (lb/in.^{2})	Vacuum (in. Hg)	(°C)	(°F)	Water	Latent	Steam	Water	Latent	Steam	Water	Steam	Steam
36	21.3	127.2	260.9	**127.7**	521.5	**649.2**	**229.7**	938.5	**1168.2**	0.3833	1.6860	11.58
38	23.3	128.9	264.1	**129.5**	520.3	**649.8**	**233.0**	936.4	**1169.4**	0.3879	1.6817	11.02
40	25.3	130.7	267.2	**131.2**	519.2	**650.4**	**236.1**	934.4	**1170.5**	0.3923	1.6776	10.50
42	27.3	132.3	270.3	**132.9**	518.0	**650.9**	**239.1**	932.3	**1171.4**	0.3964	1.6737	10.30
44	29.3	133.9	273.1	**134.5**	516.9	**651.4**	**242.0**	930.3	**1172.3**	0.4003	1.6700	9.600
46	31.3	135.4	275.8	**136.0**	515.9	**651.9**	**244.9**	928.3	**1173.2**	0.4041	1.6664	9.209
48	33.3	136.9	278.5	**137.5**	514.8	**652.3**	**247.6**	926.4	**1174.0**	0.4077	1.6630	8.848
50	35.3	138.3	281.0	**139.0**	513.8	**652.8**	**250.2**	924.6	**1174.8**	0.4112	1.6597	8.516
52	37.3	139.7	283.5	**140.4**	512.8	**653.2**	**252.7**	922.9	**1175.6**	0.4146	1.6566	8.208
54	39.3	141.0	285.9	**141.8**	511.8	**653.6**	**255.2**	921.1	**1176.3**	0.4179	1.6536	7.922
56	41.3	142.3	288.3	**143.1**	510.9	**654.0**	**257.6**	919.4	**1177.0**	0.4211	1.6507	7.656
58	43.3	143.6	290.5	**144.4**	510.0	**654.4**	**259.9**	917.8	**1177.7**	0.4242	1.6478	7.407
60	45.3	144.9	292.7	**145.6**	509.2	**654.8**	**262.2**	916.2	**1178.4**	0.4272	1.6450	7.175
62	47.3	146.1	294.9	**146.8**	508.4	**655.2**	**264.4**	914.6	**1179.0**	0.4302	1.6423	6.957
64	49.3	147.3	296.9	**148.0**	507.6	**655.6**	**266.5**	913.1	**1179.6**	0.4331	1.6398	6.752
66	51.3	148.4	299.0	**149.2**	506.7	**655.9**	**268.6**	911.6	**1180.2**	0.4359	1.6374	6.560
68	53.3	149.5	301.0	**150.3**	505.9	**656.2**	**270.7**	910.1	**1180.8**	0.4386	1.6350	6.378
70	55.3	150.6	302.9	**151.5**	505.0	**656.5**	**272.7**	908.7	**1181.4**	0.4412	1.6327	6.206
72	57.3	151.6	304.8	**152.6**	504.2	**656.8**	**274.6**	907.4	**1182.0**	0.4437	1.6304	6.044
74	59.3	152.6	306.7	**153.6**	503.4	**657.0**	**276.5**	906.0	**1182.5**	0.4462	1.6282	5.890
76	61.3	153.6	308.5	**154.7**	502.6	**657.3**	**278.4**	904.6	**1183.0**	0.4486	1.6261	5.743
78	63.3	154.6	310.3	**155.7**	501.8	**657.5**	**280.3**	903.2	**1183.5**	0.4510	1.6240	5.604
80	65.3	155.6	312.0	**156.7**	501.1	**657.8**	**282.1**	901.9	**1184.0**	0.4533	1.6219	5.472
82	67.3	156.5	313.7	**157.7**	500.3	**658.0**	**283.9**	900.6	**1184.5**	0.4556	1.6199	5.346
84	69.3	157.5	315.4	**158.6**	499.6	**658.2**	**285.6**	899.4	**1185.0**	0.4579	1.6180	5.226
86	71.3	158.4	317.1	**159.6**	498.9	**658.5**	**287.3**	898.1	**1185.4**	0.4601	1.6161	5.110
88	73.3	159.4	318.7	**160.5**	498.3	**658.8**	**289.0**	896.8	**1185.8**	0.4622	1.6142	5.000
90	75.3	160.3	320.3	**161.5**	497.6	**659.1**	**290.7**	895.5	**1186.2**	0.4643	1.6124	4.896
92	77.3	161.2	321.9	**162.4**	496.9	**659.3**	**292.3**	894.3	**1186.6**	0.4664	1.6106	4.796
94	79.3	162.0	323.3	**163.3**	496.3	**659.6**	**293.9**	893.1	**1187.0**	0.4684	1.6088	4.699
96	81.3	162.8	324.8	**164.1**	495.7	**659.8**	**295.5**	891.9	**1187.4**	0.4704	1.6071	4.607
98	83.3	163.6	326.6	**165.0**	495.0	**660.0**	**297.0**	890.8	**1187.8**	0.4723	1.6054	4.519
100	85.3	164.4	327.8	**165.8**	494.3	**660.1**	**298.5**	889.7	**1188.2**	0.4742	1.6038	4.434
105	90.3	166.4	331.3	**167.9**	492.7	**660.6**	**302.2**	886.9	**1189.1**	0.4789	1.6000	4.230
110	95.3	168.2	334.8	**169.8**	491.2	**661.0**	**305.7**	884.2	**1189.9**	0.4833	1.5963	4.046
115	100.3	170.0	338.1	**171.7**	489.8	**661.5**	**309.2**	881.5	**1190.7**	0.4876	1.5927	3.880
120	105.3	171.8	341.3	**173.6**	488.3	**661.9**	**312.5**	878.9	**1191.4**	0.4918	1.5891	3.729
125	110.3	173.5	344.4	**175.4**	486.9	**662.3**	**315.7**	876.4	**1192.1**	0.4958	1.5856	3.587
130	115.3	175.2	347.3	**177.1**	485.6	**662.7**	**318.8**	874.0	**1192.8**	0.4997	1.5823	3.456
135	120.3	176.8	350.2	**178.8**	484.2	**663.0**	**321.9**	871.5	**1193.4**	0.5035	1.5792	3.335
140	125.3	178.3	353.0	**180.5**	482.9	**663.4**	**324.9**	869.1	**1194.0**	0.5071	1.5763	3.222
145	130.3	179.8	355.8	**182.1**	481.6	**663.7**	**327.8**	866.8	**1194.6**	0.5106	1.5733	3.116
150	135.3	181.3	358.4	**183.7**	480.3	**664.0**	**330.6**	864.5	**1195.1**	0.5140	1.5705	3.015

Table 11C. Enthalpy of superheated steam, H (kJ/kg)

Pressure	Saturation		Temperature, θ (°C)	100	200	300	400	500	600	700	800
P (kN/m^2)	T_s (K)	S_s (kJ/kg)	Temperature, T(K)	373.15	473.15	573.15	673.15	773.15	873.15	973.15	1073.15
100	372.78	2675.4		2676.0	2875.4	3074.6	3278.0	3488.0	3705.0	3928.0	4159.0
200	393.38	2706.3			2870.4	3072.0	3276.4	3487.0	3704.0	3927.0	4158.0
300	406.69	2724.7			2866.0	3069.7	3275.0	3486.0	3703.1	3927.0	4158.0
400	416.78	2737.6			2861.3	3067.0	3273.5	3485.0	3702.0	3926.0	4157.0
500	425.00	2747.5			2856.0	3064.8	3272.1	3484.0	3701.2	3926.0	4156.8
600	431.99	2755.5			2850.7	3062.0	3270.0	3483.0	3701.0	3925.0	4156.2
700	438.11	2762.0			2845.5	3059.5	3269.0	3482.6	3700.2	3924.0	4156.0
800	443.56	2767.5			2839.7	3057.0	3266.8	3480.4	3699.0	3923.8	4155.0
900	448.56	2772.1			2834.0	3055.0	3266.2	3479.5	3698.6	3923.0	4155.0
1000	453.03	2776.2			2828.7	3051.7	3264.3	3478.0	3697.5	3922.8	4154.0
2000	485.59	2797.2				3024.8	3248.0	3467.0	3690.0	3916.0	4150.0
3000	506.98	2802.3				2994.8	3231.7	3456.0	3681.6	3910.4	4145.0
4000	523.49	2800.3				2962.0	3214.8	3445.0	3673.4	3904.0	4139.6
5000	537.09	2794.2				2926.0	3196.9	3433.8	3665.4	3898.0	4135.5
6000	548.71	2785.0				2886.0	3178.0	3421.7	3657.0	3891.8	4130.0
7000	558.95	2773.4				2840.0	3159.1	3410.0	3648.8	3886.0	4124.8
8000	568.13	2759.9				2785.0	3139.5	3398.0	3640.4	3880.8	4121.0
9000	576.46	2744.6					3119.0	3385.5	3632.0	3873.6	4116.0
10000	584.11	2727.7					3097.7	3373.6	3624.0	3867.2	4110.8
11000	591.19	2709.3					3075.6	3361.0	3615.5	3862.0	4106.0
12000	597.79	2698.2					3052.9	3349.0	3607.0	3855.3	4101.2

Table 11D. Entropy of superheated steam, S (kJ/kg K)

Pressure	Saturation		Temperature, θ (°C)	100	200	300	400	500	600	700	800
P (kN/m^2)	T_s (K)	H_s (kJ/kg)	Temperature, T(K)	373.15	473.15	573.15	673.15	773.15	873.15	973.15	1073.15
100	372.78	7.3598		7.362	7.834	8.216	8.544	8.834	9.100	9.344	9.565
200	393.38	7.1268			7.507	7.892	8.222	8.513	8.778	9.020	9.246
300	406.69	6.9909			7.312	7.702	8.033	8.325	8.591	8.833	9.057
400	416.78	6.8943			7.172	7.566	7.898	8.191	8.455	8.700	8.925
500	425.00	6.8192			7.060	7.460	7.794	8.087	8.352	8.596	8.820
600	431.99	6.7575			6.968	7.373	7.708	8.002	8.268	8.510	8.738
700	438.11	6.7052			6.888	7.298	7.635	7.930	8.195	8.438	8.665
800	443.56	6.6596			6.817	7.234	7.572	7.867	8.133	8.375	8.602
900	448.56	6.6192			6.753	7.176	7.515	7.812	8.077	8.321	8.550
1000	453.03	6.5828			6.695	7.124	7.465	7.762	8.028	8.272	8.502
2000	485.59	6.3367				6.768	7.128	7.431	7.702	7.950	8.176
3000	506.98	6.1838				6.541	6.922	7.233	7.508	7.756	7.985
4000	523.49	6.0685				6.364	6.770	7.090	7.368	7.620	7.850
5000	537.07	5.9735				6.211	6.647	6.977	7.258	7.510	7.744
6000	548.71	5.8907				6.060	6.542	6.880	7.166	7.422	7.655
7000	558.95	5.8161				5.933	6.450	6.798	7.088	7.345	7.581
8000	568.13	5.7470				5.792	6.365	6.724	7.020	7.280	7.515
9000	576.46	5.6820					6.288	6.659	6.958	7.220	7.457
10000	584.11	5.6198					6.215	6.598	6.902	7.166	7.405
11000	591.19	5.5596					6.145	6.540	6.850	7.117	7.357
12000	597.79	5.5003					6.077	6.488	6.802	7.072	7.312

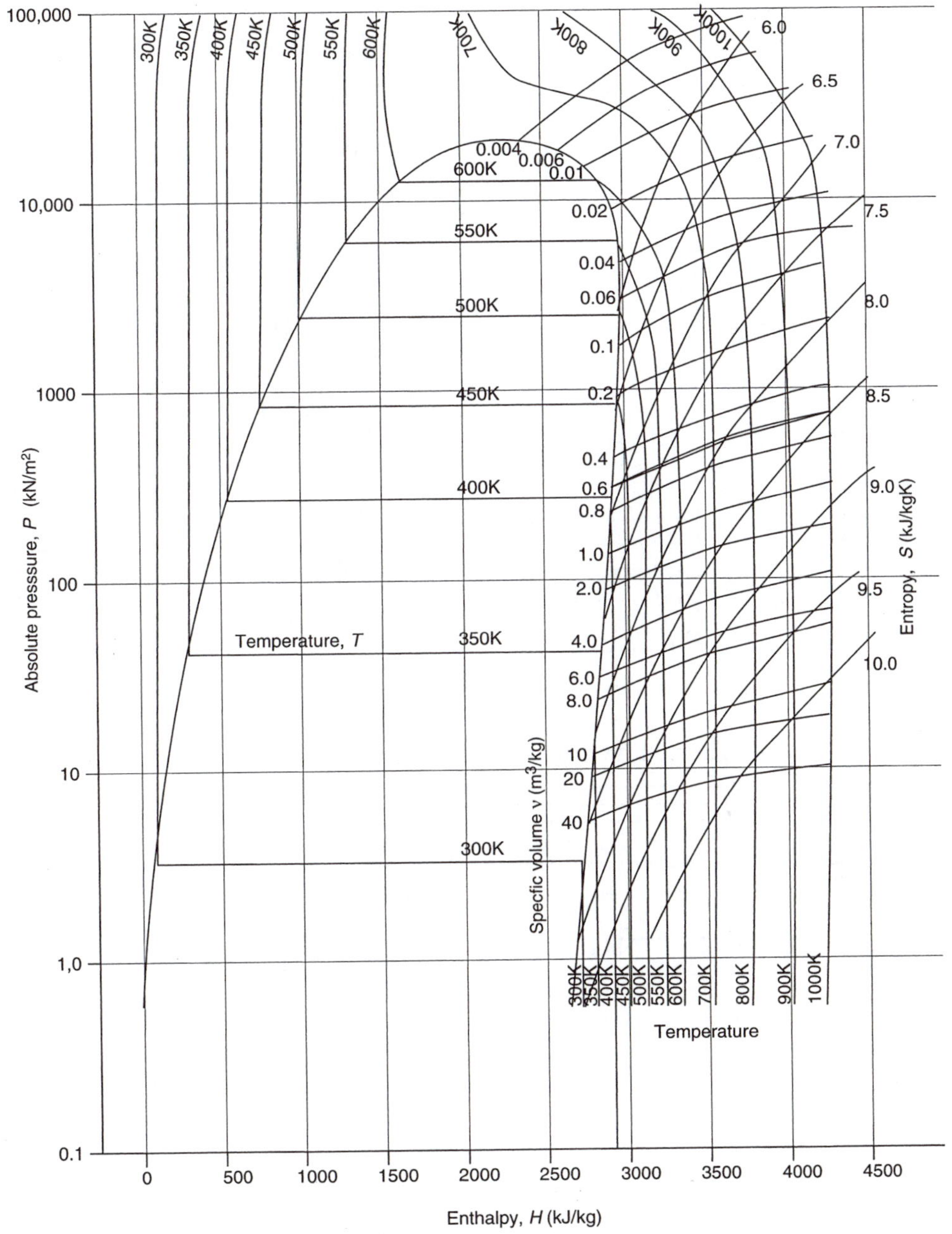

Figure 11A. Pressure–enthalpy diagram for water and steam

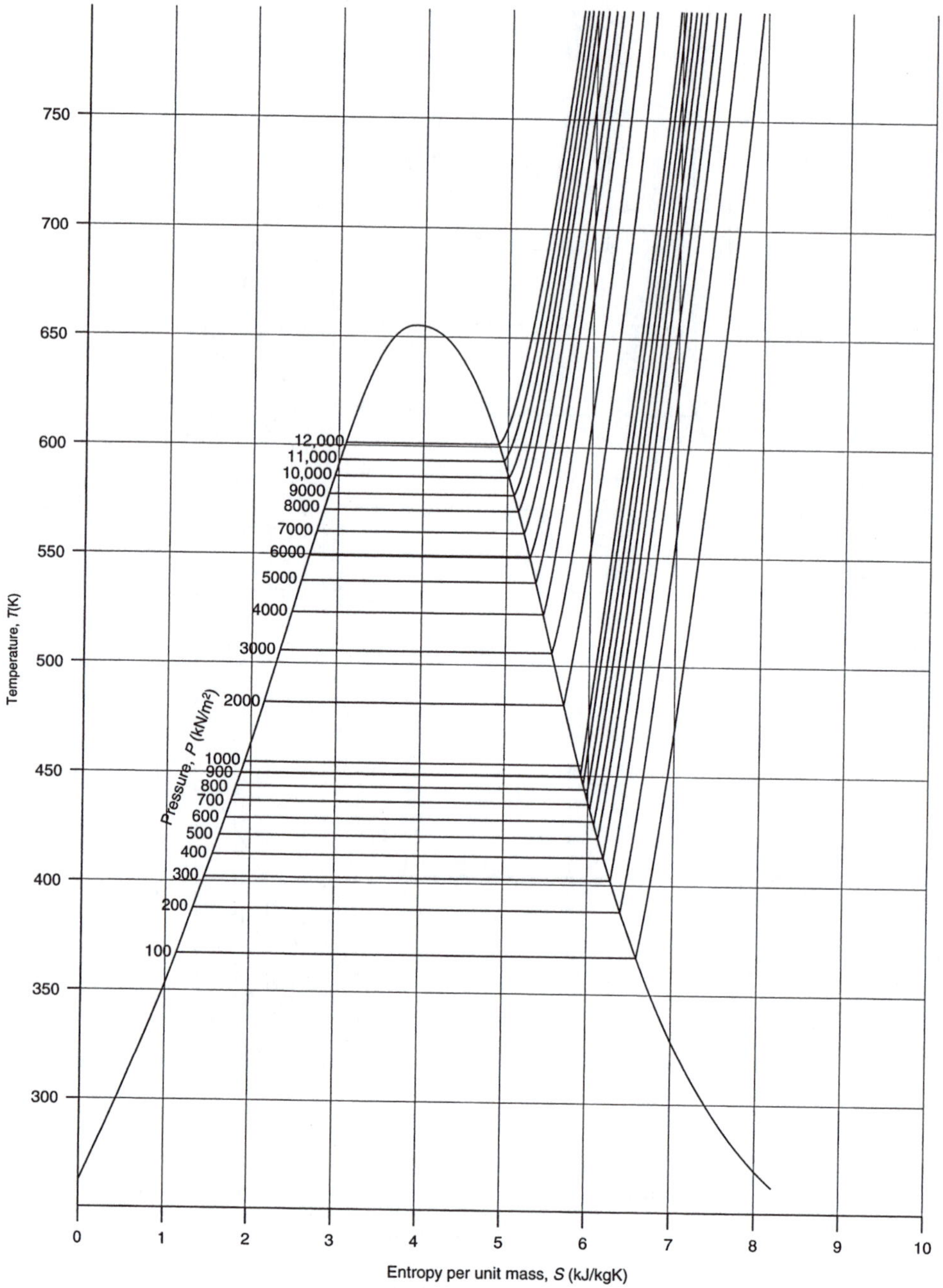

Figure 11B. Temperature–entropy diagram for water and steam

A3. MATHEMATICAL TABLES

Table 12. Laplace transforms*

No.	Transform $\bar{f}(p) = \int_0^\infty e^{-pt} f(t) dt$	Function f(t)
1	$\frac{1}{p}$	1
2	$\frac{1}{p^2}$	t
3	$\frac{1}{p^n}$ $n = 1, 2, 3, \ldots$	$\frac{t^{n-1}}{(n-1)!}$
4	$\frac{1}{\sqrt{p}}$	$\frac{1}{\sqrt{(\pi t)}}$
5	$\frac{1}{p^{3/2}}$	$2\sqrt{\frac{t}{\pi}}$
6	$\frac{1}{p^{n+\frac{1}{2}}}$ $n = 1, 2, 3, \ldots$	$\frac{2^n t^{n-\frac{1}{2}}}{[1.3.5 \ldots (2n-1)]\sqrt{\pi}}$
7	$\frac{\Gamma(k)}{p^k}$ $k > 0$	t^{k-1}
8	$\frac{1}{p-a}$	e^{at}
9	$\frac{1}{(p-a)^2}$	te^{at}
10	$\frac{1}{(p-a)^n}$ $n = 1, 2, 3, \ldots$	$\frac{1}{(n-1)!} t^{n-1} e^{at}$
11	$\frac{\Gamma(k)}{(p-a)^k}$ $k > 0$	$t^{k-1} e^{at}$
12	$\frac{1}{(p-a)(p-b)}$ $a \neq b$	$\frac{1}{a-b}(e^{at} - e^{bt})$
13	$\frac{p}{(p-a)(p-b)}$ $a \neq b$	$\frac{1}{a-b}(ae^{at} - be^{bt})$
14	$\frac{1}{(p-a)(p-b)(p-c)}$ $a \neq b \neq c$	$-\frac{(b-c)e^{at} + (c-a)e^{bt} + (a-b)e^{ct}}{(a-b)(b-c)(c-a)}$
15	$\frac{1}{p^2 + a^2}$	$\frac{1}{a} \sin at$
16	$\frac{p}{p^2 + a^2}$	$\cos at$
17	$\frac{1}{p^2 - a^2}$	$\frac{1}{a} \sinh at$

* By permission from *Operational Mathematics* by R.V. CHURCHILL, McGraw-Hill 1958.

Table 12. *(continued)*

No.	Transform $\bar{f}(p) = \int_0^\infty e^{-pt} f(t) dt$	Function f(t)
18	$\frac{p}{p^2 - a^2}$	$\cosh at$
19	$\frac{1}{p(p^2 + a^2)}$	$\frac{1}{a^2}(1 - \cos at)$
20	$\frac{1}{p^2(p^2 + a^2)}$	$\frac{1}{a^3}(at - \sin at)$
21	$\frac{1}{(p^2 + a^2)^2}$	$\frac{1}{2a^3}(\sin at - at \cos at)$
22	$\frac{p}{(p^2 + a^2)^2}$	$\frac{t}{2a} \sin at$
23	$\frac{p^2}{(p^2 + a^2)^2}$	$\frac{1}{2a}(\sin at + at \cos at)$
24	$\frac{p^2 - a^2}{(p^2 + a^2)^2}$	$t \cos at$
25	$\frac{p}{(p^2 + a^2)(p^2 + b^2)} \quad a^2 \neq b^2$	$\frac{\cos at - \cos bt}{b^2 - a^2}$
26	$\frac{1}{(p - a)^2 + b^2}$	$\frac{1}{b} e^{at} \sin bt$
27	$\frac{p - a}{(p - a)^2 + b^2}$	$e^{at} \cos bt$
28	$\frac{3a^2}{p^3 + a^3}$	$e^{-at} - e^{at/2} \left(\cos \frac{at\sqrt{3}}{2} - \sqrt{3} \sin \frac{at\sqrt{3}}{2} \right)$
29	$\frac{4a^3}{p^4 + 4a^4}$	$\sin at \cosh at - \cos at \sinh at$
30	$\frac{p}{p^4 + 4a^4}$	$\frac{1}{2a^2} \sin at \sinh at$
31	$\frac{1}{p^4 - a^4}$	$\frac{1}{2a^3}(\sinh at - \sin at)$
32	$\frac{p}{p^4 - a^4}$	$\frac{1}{2a^2}(\cosh at - \cos at)$
33	$\frac{8a^3 p^2}{(p^2 + a^2)^3}$	$(1 + a^2 t^2) \sin at - at \cos at$
34	$\frac{1}{p} \left(\frac{p - 1}{p} \right)^n$	$\frac{e^t}{n!} \frac{d^n}{dt^n}(t^n e^{-t})$ = Laguerre polynomial of degree n

Table 12. *(continued)*

No.	Transform $\bar{f}(p) = \int_0^\infty e^{-pt} f(t) dt$	Function f(t)
35	$\frac{p}{(p-a)^{3/2}}$	$\frac{1}{\sqrt{(\pi t)}} - e^{at}(1+2at)$
36	$\sqrt{(p-a)} - \sqrt{(p-b)}$	$\frac{1}{2\sqrt{(\pi t^3)}}(e^{bt} - e^{at})$
37	$\frac{1}{\sqrt{p+a}}$	$\frac{1}{\sqrt{(\pi t)}} - ae^{a^2 t}\ \text{erfc}\ (a\sqrt{t})$
38	$\frac{\sqrt{p}}{p-a^2}$	$\frac{1}{\sqrt{(\pi t)}} + ae^{a^2 t}\text{erf}(a\sqrt{t})$
39	$\frac{\sqrt{p}}{p+a^2}$	$\frac{1}{\sqrt{(\pi t)}} - \frac{2ae^{-a^2 t}}{\sqrt{\pi}} \int_0^{a\sqrt{t}} e^{\lambda} 2 d\lambda$
40	$\frac{1}{\sqrt{p(p-a^2)}}$	$\frac{1}{a} e^{a^2 t}\text{erf}(a\sqrt{t})$
41	$\frac{1}{\sqrt{p(p+a^2)}}$	$\frac{2e^{-a^2 t}}{a\sqrt{\pi}} \int_0^{a\sqrt{t}} e^{\lambda^2} d\lambda$
42	$\frac{b^2 - a^2}{(p-a^2)(b+\sqrt{p})}$	$e^{a^2 t}[b - a\text{erf}(a\sqrt{t})] - be^{b^2 t}\ \text{erfc}\ (b\sqrt{t})$
43	$\frac{1}{\sqrt{p}(\sqrt{p}+a)}$	$e^{a^2 t}\ \text{erfc}\ (a\sqrt{t})$
44	$\frac{1}{(p+a)\sqrt{(p+b)}}$	$\frac{1}{\sqrt{(b-a)}} e^{-at}\text{erf}[\sqrt{(b-a)}\sqrt{t}]$
45	$\frac{b^2-a^2}{\sqrt{p(p-a^2)}(\sqrt{p}+b)}$	$e^{a^2 t}\left[\frac{b}{a}\text{erf}(a\sqrt{t}) - 1\right] + e^{b^2 t}\ \text{erfc}\ (b\sqrt{t})$
46	$\frac{(1-p)^n}{p^{n+1/2}}$	$\frac{n!}{(2n)!\sqrt{(\pi t)}} H_{2n}(\sqrt{t})$ where $H_n(t) = e^{t^2} \frac{d^n}{dt^n} e^{-t^2}$ is the Hermite polynomial
47	$\frac{(1-p)^n}{p^{n+3/2}}$	$-\frac{n!}{\sqrt{\pi}(2n+1)!} H_{2n+1}(\sqrt{t})$
48	$\frac{\sqrt{(p+2a)}}{\sqrt{p}} - 1$	$ae^{-at}[I_1(at) + I_0(at)]$
49	$\frac{1}{\sqrt{(p+a)}\sqrt{(p+b)}}$	$e^{-\frac{1}{2}(a+b)t} I_0\left(\frac{a-b}{2}t\right)$

(continued overleaf)

Table 12. *(continued)*

No.	Transform $\bar{f}(p) = \int_0^\infty e^{-pt} f(t) dt$	Function f(t)
50	$\dfrac{\Gamma(k)}{(p+a)^k(p+b)^k} \quad k > 0$	$\sqrt{\pi}\left(\dfrac{t}{a-b}\right)^{k-\frac{1}{2}} e^{-\frac{1}{2}(a+b)t} I_{k-\frac{1}{2}}\left(\dfrac{a-b}{2}t\right)$
51	$\dfrac{1}{\sqrt{(p+a)}(p+b)^{3/2}}$	$te^{-\frac{1}{2}(a+b)t}\left[I_0\left(\dfrac{a-b}{2}t\right) + I_1\left(\dfrac{a-b}{2}t\right)\right]$
52	$\dfrac{\sqrt{(p+2a)} - \sqrt{p}}{\sqrt{(p+2a)} + \sqrt{p}}$	$\dfrac{1}{t} e^{-at} I_1(at)$
53	$\dfrac{(a-b)^k}{[\sqrt{(p+a)} + \sqrt{(p+b)}]^{2k}} \quad k > 0$	$\dfrac{k}{t} e^{-\frac{1}{2}(a+b)t} I_k\left(\dfrac{a-b}{2}t\right)$
54	$\dfrac{[\sqrt{(p+a)} + \sqrt{p}]^{-2j}}{\sqrt{p}\sqrt{(p+a)}} \quad j > -1$	$\dfrac{1}{a^j} e^{-\frac{1}{2}at} I_j\left(\dfrac{1}{2}at\right)$
55	$\dfrac{1}{\sqrt{(p^2+a^2)}}$	$J_0(at)$
56	$\dfrac{[\sqrt{(p^2+a^2)} - p]^j}{\sqrt{(p^2+a^2)}} \quad j > 1$	$a^j J_j(at)$
57	$\dfrac{1}{(p^2+a^2)^k} \quad k > 0$	$\dfrac{\sqrt{\pi}}{\Gamma(k)}\left(\dfrac{t}{2a}\right)^{k-\frac{1}{2}} J_{k-\frac{1}{2}}(at)$
58	$[\sqrt{(p^2+a^2)} - p]^k \quad k > 0$	$\dfrac{ka^k}{t} J_k(at)$
59	$\dfrac{[p - \sqrt{(p^2-a^2)}]^j}{\sqrt{(p^2-a^2)}} \quad j > -1$	$a^j I_j(at)$
60	$\dfrac{1}{(p^2-a^2)^k} \quad k > 0$	$\dfrac{\sqrt{\pi}}{\Gamma(k)}\left(\dfrac{t}{2a}\right)^{k-\frac{1}{2}} I_{k-\frac{1}{2}}(at)$
61	$\dfrac{e^{-kp}}{p}$	$S_k(t) = \begin{cases} 0 & \text{when } 0 < t < k \\ 1 & \text{when } t > k \end{cases}$
62	$\dfrac{e^{-kp}}{p^2}$	$\begin{cases} 0 & \text{when } 0 < t < k \\ t-k & \text{when } t > k \end{cases}$
63	$\dfrac{e^{-kp}}{p^j} \quad j > 0$	$\begin{cases} 0 & \text{when } 0 < t < k \\ \dfrac{(t-k)^{j-1}}{\Gamma(j)} & \text{when } t > k \end{cases}$
64	$\dfrac{1 - e^{-kp}}{p}$	$\begin{cases} 1 & \text{when } 0 < t < k \\ 0 & \text{when } t > k \end{cases}$

Table 12. *(continued)*

No.	Transform $\bar{f}(p) = \int_0^\infty e^{-pt} f(t) dt$	Function f(t)
65	$\dfrac{1}{p(1-e^{-kp})} = \dfrac{1+\coth\frac{1}{2}kp}{2p}$	$S(k, t) = n$ when $(n-1)k < t < nk \quad n = 1, 2, 3, \ldots$
66	$\dfrac{1}{p(e^{kp}-a)}$	$\begin{cases} 0 & \text{when } 0 < t < k \\ 1 + a + a^2 + \ldots + a^{n-1} & \\ \quad \text{when } nk < t < (n+1)k & n = 1, 2, 3, \ldots \end{cases}$
67	$\dfrac{1}{p}\tanh kp$	$M(2k, t) = (-1)^{n-1}$ when $2k(n-1) < t < 2kn \; n = 1, 2, 3, \ldots$
68	$\dfrac{1}{p(1+e^{-kp})}$	$\dfrac{1}{2}M(k, t) + \dfrac{1}{2} = \dfrac{1-(-1)^n}{2}$ when $(n-1)k < t < nk$
69	$\dfrac{1}{p^2}\tanh kp$	$H(2k, t) = \begin{cases} t & \text{when } 0 < t < 2k \\ 4k - t & \text{when } 2k < t < 4k \end{cases}$
70	$\dfrac{1}{p\sinh kp}$	$2S(2k, t+k) - 2 = 2(n-1)$ when $(2n-3)k < t < (2n-1)k \quad t > 0$
71	$\dfrac{1}{p\cosh kp}$	$M(2k, t+3k) + 1 = 1 + (-1)^n$ when $(2n-3)k < t < (2n-1)k \quad t > 0$
72	$\dfrac{1}{p}\coth kp$	$2S(2k, t) - 1 = 2n - 1$ when $2k(n-1) < t < 2kn$
73	$\dfrac{k}{p^2+k^2}\coth\dfrac{\pi p}{2k}$	$\lvert\sin kt\rvert$
74	$\dfrac{1}{(p^2+1)(1-e^{-\pi p})}$	$\begin{cases} \sin t & \text{when } (2n-2)\pi < t < (2n-1)\pi \\ 0 & \text{when } (2n-1)\pi < t < 2n\pi \end{cases}$
75	$\dfrac{1}{p}e^{-k/p}$	$J_0[2\sqrt{(kt)}]$
76	$\dfrac{1}{\sqrt{p}}e^{-k/p}$	$\dfrac{1}{\sqrt{(\pi t)}}\cos 2\sqrt{(kt)}$
77	$\dfrac{1}{\sqrt{p}}e^{k/p}$	$\dfrac{1}{\sqrt{(\pi t)}}\cosh 2\sqrt{(kt)}$
78	$\dfrac{1}{p^{3/2}}e^{-k/p}$	$\dfrac{1}{\sqrt{(\pi k)}}\sin 2\sqrt{(kt)}$
79	$\dfrac{1}{p^{3/2}}e^{k/p}$	$\dfrac{1}{\sqrt{(\pi k)}}\sinh 2\sqrt{(kt)}$
80	$\dfrac{1}{p^j}e^{k/p} \quad j > 0$	$\left(\dfrac{t}{k}\right)^{(j-1)/2} J_{j-1}[2\sqrt{(kt)}]$

(continued overleaf)

Table 12. (*continued*)

No.	Transform $\bar{f}(p) = \int_0^\infty e^{-pt} f(t) dt$	Function f(t)
81	$\frac{1}{p^j} e^{k/p} \quad j > 0$	$\left(\frac{t}{k}\right)^{(j-1)/2} I_{j-1}[2\sqrt{(kt)}]$
82	$e^{-k\sqrt{p}} \quad k > 0$	$\frac{k}{2\sqrt{(\pi t^3)}} \exp\left(-\frac{k^2}{4t}\right)$
83	$\frac{1}{p} e^{-k\sqrt{p}} \quad k \geq 0$	$\operatorname{erfc}\left(\frac{k}{2\sqrt{t}}\right)$
84	$\frac{1}{\sqrt{p}} e^{-k\sqrt{p}} \quad k \geq 0$	$\frac{1}{\sqrt{(\pi t)}} \exp\left(-\frac{k^2}{4t}\right)$
85	$p^{-3/2} e^{-k\sqrt{p}} \quad k \geq 0$	$2\sqrt{\frac{t}{\pi}} \left[\exp\left(-\frac{k^2}{4t}\right)\right] - k \operatorname{erfc}\left(\frac{k}{2\sqrt{t}}\right)$
86	$\frac{a e^{-k\sqrt{p}}}{p(a + \sqrt{p})} \quad k \geq 0$	$-\exp(ak)\exp(a^2 t) \operatorname{erfc}\left(a\sqrt{t} + \frac{k}{2\sqrt{t}}\right) + \operatorname{erfc}\left(\frac{k}{2\sqrt{t}}\right)$
87	$\frac{e^{-k\sqrt{p}}}{\sqrt{p(a + \sqrt{p})}} \quad k \geq 0$	$\exp(ak)\exp(a^2 t) \operatorname{erfc}\left(a\sqrt{t} + \frac{k}{2\sqrt{t}}\right)$
88	$\frac{e^{-k\sqrt{[p(p+a)]}}}{\sqrt{[p(p+a)]}}$	$\begin{cases} 0 & \text{when } 0 < t < k \\ \exp(-\frac{1}{2}at) I_0\left[\frac{1}{2}a\sqrt{(t^2 - k^2)}\right] & \text{when } t > k \end{cases}$
89	$\frac{e^{-k\sqrt{(p^2+a^2)}}}{\sqrt{(p^2 + a^2)}}$	$\begin{cases} 0 & \text{when } 0 < t < k \\ J_0[a\sqrt{(t^2 - k^2)}] & \text{when } t > k \end{cases}$
90	$\frac{e^{-k\sqrt{(p^2-a^2)}}}{\sqrt{(p^2 - a^2)}}$	$\begin{cases} 0 & \text{when } 0 < t < k \\ I_0[a\sqrt{(t^2 - k^2)}] & \text{when } t > k \end{cases}$
91	$\frac{e^{-k[\sqrt{(p^2+a^2)} - p]}}{\sqrt{(p^2 + a^2)}} \quad k \geq 0$	$J_0[a\sqrt{(t^2 + 2kt)}]$
92	$e^{-kp} - e^{-k\sqrt{(p^2+a^2)}}$	$\begin{cases} 0 \quad \text{when } 0 < t < k \\ \frac{ak}{\sqrt{(t^2 - k^2)}} J_1[a\sqrt{(t^2 - k^2)}] \\ \quad \text{when } t > k \end{cases}$
93	$e^{-k\sqrt{(p^2-a^2)}} - e^{-kp}$	$\begin{cases} 0 \quad \text{when } 0 < t < k \\ \frac{ak}{\sqrt{(t^2 - k^2)}} J_1[a\sqrt{(t^2 - k^2)}] \\ \quad \text{when } t > k \end{cases}$

Table 12. *(continued)*

No.	Transform $\bar{f}(p) = \int_0^\infty e^{-pt} f(t) dt$	Function f(t)
94	$\dfrac{a^j e^{-k\sqrt{(p^2+a^2)}}}{\sqrt{(p^2+a^2)}[\sqrt{(p^2+a^2)}+p]^j}$ $j > -1$	$\begin{cases} 0 & \text{when } 0 < t < k \\ \left(\dfrac{t-k}{t+k}\right)^{(1/2)j} J_j[a\sqrt{(t^2-k^2)}] & \text{when } t > k \end{cases}$
95	$\dfrac{1}{p}\ln p$	$\lambda - \ln t \quad \lambda = -0.5772\ldots$
96	$\dfrac{1}{p^k}\ln p \quad k > 0$	$t^{k-1}\left\{\dfrac{\lambda}{[\Gamma(k)]^2} - \dfrac{\ln t}{\Gamma(k)}\right\}$
97[b]	$\dfrac{\ln p}{p-a} \quad a > 0$	$(\exp at)[\ln a - Ei(-at)]$
98[c]	$\dfrac{\ln p}{p^2+1}$	$\cos t\,\mathrm{Si}(t) - \sin t\,Ci(t)$
99[c]	$\dfrac{p\ln p}{p^2+1}$	$-\sin t\,\mathrm{Si}(t) - \cos t\,\mathrm{Ci}(t)$
100[b]	$\dfrac{1}{p}\ln(1+kp) \quad k > 0$	$-\mathrm{Ei}\left(-\dfrac{t}{k}\right)$
101	$\ln\dfrac{p-a}{p-b}$	$\dfrac{1}{t}(e^{bt} - e^{at})$
102[c]	$\dfrac{1}{p}\ln(1+k^2p^2)$	$-2\mathrm{Ci}\left(\dfrac{t}{k}\right)$
103[c]	$\dfrac{1}{p}\ln(p^2+a^2) \quad a > 0$	$2\ln a - 2\mathrm{Ci}(at)$
104[c]	$\dfrac{1}{p^2}\ln(p^2+a^2) \quad a > 0$	$\dfrac{2}{a}[at\ln a + \sin at - at\mathrm{Ci}(at)]$
105	$\ln\dfrac{p^2+a^2}{p^2}$	$\dfrac{2}{t}(1-\cos at)$
106	$\ln\dfrac{p^2-a^2}{p^2}$	$\dfrac{2}{t}(1-\cosh at)$
107	$\tan^{-1}\dfrac{k}{p}$	$\dfrac{1}{t}\sin kt$

[b] $\mathrm{Ei}(-t) = -\int_t^\infty \dfrac{e^{-x}}{x} dx$ (for $t > 0$) = exponential integral function.

[c] $\mathrm{Si}(t) = \int_0^t \dfrac{\sin x}{x} dx$ = sine integral function.

$\mathrm{Ci}(t) = -\int_t^\infty \dfrac{\cos x}{x} dx$ = cosine integral function.

These functions are tabulated in M. Abramowitz and I.A. Stegun, *Handbook of Mathematical Functions*, Dover Publications, New York 1965.

Table 12. *(continued)*

No.	Transform $\bar{f}(p) = \int_0^\infty e^{-pt} f(t) dt$	Function f(t)
108[c]	$\frac{1}{p} \tan^{-1} \frac{k}{p}$	Si (kt)
109	$\exp(k^2 p^2)$ erfc $(kp) \quad k > 0$	$\frac{1}{k\sqrt{\pi}} \exp\left(-\frac{t^2}{4k^2}\right)$
110	$\frac{1}{p} \exp(k^2 p^2)$ erfc $(kp) \quad k > 0$	erf $\left(\frac{t}{2k}\right)$
111	$\exp(kp)$ erfc $[\sqrt{(kp)}] \quad k > 0$	$\frac{\sqrt{k}}{\pi\sqrt{t}(t+k)}$
112	$\frac{1}{\sqrt{p}}$ erfc $[\sqrt{(kp)}]$	$\begin{cases} 0 & \text{when } 0 < t < k \\ (\pi t)^{-\frac{1}{2}} & \text{when } t > k \end{cases}$
113	$\frac{1}{\sqrt{p}} \exp(kp)$ erfc $[\sqrt{(kp)}] \quad k > 0$	$\frac{1}{\sqrt{[\pi(t+k)]}}$
114	erf $\left(\frac{k}{\sqrt{p}}\right)$	$\frac{1}{\pi t} \sin(2k\sqrt{t})$
115	$\frac{1}{\sqrt{p}} \exp\left(\frac{k^2}{p}\right)$ erfc $\left(\frac{k}{\sqrt{p}}\right)$	$\frac{1}{\sqrt{(\pi t)}} \exp(-2k\sqrt{t})$
116[d]	$K_0(kp)$	$\begin{cases} 0 & \text{when } 0 < t < k \\ (t^2 - k^2)^{-\frac{1}{2}} & \text{when } t > k \end{cases}$
117[d]	$K_0(k\sqrt{p})$	$\frac{1}{2t} \exp\left(-\frac{k^2}{4t}\right)$
118[d]	$\frac{1}{p} \exp(kp) K_1(kp)$	$\frac{1}{k} \sqrt{[t(t+2k)]}$
119[d]	$\frac{1}{\sqrt{p}} K_1(k\sqrt{p})$	$\frac{1}{k} \exp\left(-\frac{k^2}{4t}\right)$
120[d]	$\frac{1}{\sqrt{p}} \exp\left(\frac{k}{p}\right) K_0\left(\frac{k}{p}\right)$	$\frac{2}{\sqrt{(\pi t)}} K_0[2\sqrt{(2kt)}]$
121[e]	$\pi \exp(-kp) I_0(kp)$	$\begin{cases} [t(2k-t)]^{-\frac{1}{2}} & \text{when } 0 < t < 2k \\ 0 & \text{when } t > 2k \end{cases}$
122[e]	$\exp(-kp) I_1(kp)$	$\begin{cases} \frac{k-t}{\pi k\sqrt{[t(2k-t)]}} & \text{when } 0 < t < 2k \\ 0 & \text{when } t > 2k \end{cases}$
123	unity	unit impulse

[d]$K_n(x)$ denotes the Bessel function of the second kind for the imaginary argument.
[e]$I_n(x)$ denotes the Bessel function of the first kind for the imaginary argument.

Table 13. Error function and its derivative

x	$\frac{2}{\sqrt{\pi}}e^{-x^2}$	erf$x$	x	$\frac{2}{\sqrt{\pi}}e^{-x^2}$	erf$x$
0.00	1.12837	0.00000	0.50	0.87878	0.52049
0.01	1.12826	0.01128	0.51	0.86995	0.52924
0.02	1.12792	0.02256	0.52	0.86103	0.53789
0.03	1.12736	0.03384	0.53	0.85204	0.54646
0.04	1.12657	0.04511	0.54	0.84297	0.55493
0.05	1.12556	0.05637	0.55	0.83383	0.56332
0.06	1.12432	0.06762	0.56	0.82463	0.57161
0.07	1.12286	0.07885	0.57	0.81536	0.57981
0.08	1.12118	0.09007	0.58	0.80604	0.58792
0.09	1.11927	0.10128	0.59	0.79666	0.59593
0.10	1.11715	0.11246	0.60	0.78724	0.60385
0.11	1.11480	0.12362	0.61	0.77777	0.61168
0.12	1.11224	0.13475	0.62	0.76826	0.61941
0.13	1.10946	0.14586	0.63	0.75872	0.62704
0.14	1.10647	0.15694	0.64	0.74914	0.63458
0.15	1.10327	0.16799	0.65	0.73954	0.64202
0.16	1.09985	0.17901	0.66	0.72992	0.64937
0.17	1.09623	0.18999	0.67	0.72027	0.65662
0.18	1.09240	0.20093	0.68	0.71061	0.66378
0.19	1.08837	0.21183	0.69	0.70095	0.67084
0.20	1.08413	0.22270	0.70	0.69127	0.67780
0.21	1.07969	0.23352	0.71	0.68159	0.68466
0.22	1.07506	0.24429	0.72	0.67191	0.69143
0.23	1.07023	0.25502	0.73	0.66224	0.69810
0.24	1.06522	0.26570	0.74	0.65258	0.70467
0.25	1.06001	0.27632	0.75	0.64293	0.71115
0.26	1.05462	0.28689	0.76	0.63329	0.71753
0.27	1.04904	0.29741	0.77	0.62368	0.72382
0.28	1.04329	0.30788	0.78	0.61408	0.73001
0.29	1.03736	0.31828	0.79	0.60452	0.73610
0.30	1.03126	0.32862	0.80	0.59498	0.74210
0.31	1.02498	0.33890	0.81	0.58548	0.74800
0.32	1.01855	0.34912	0.82	0.57601	0.75381
0.33	1.01195	0.35927	0.83	0.56659	0.75952
0.34	1.00519	0.36936	0.84	0.55720	0.76514
0.35	0.99828	0.37938	0.85	0.54786	0.77066
0.36	0.99122	0.38932	0.86	0.53858	0.77610
0.37	0.98401	0.39920	0.87	0.52934	0.78143
0.38	0.97665	0.40900	0.88	0.52016	0.78668
0.39	0.96916	0.41873	0.89	0.51103	0.79184
0.40	0.96154	0.42839	0.90	0.50196	0.79690
0.41	0.95378	0.43796	0.91	0.49296	0.80188
0.42	0.94590	0.44746	0.92	0.48402	0.80676
0.43	0.93789	0.45688	0.93	0.47515	0.81156
0.44	0.92977	0.46622	0.94	0.46635	0.81627
0.45	0.92153	0.47548	0.95	0.45761	0.82089
0.46	0.91318	0.48465	0.96	0.44896	0.82542
0.47	0.90473	0.49374	0.97	0.44037	0.82987
0.48	0.89617	0.50274	0.98	0.43187	0.83423
0.49	0.88752	0.51166	0.99	0.42345	0.83850

(*continued overleaf*)

Table 13. (*continued*)

x	$\frac{2}{\sqrt{\pi}}e^{-x^2}$	erf$x$	x	$\frac{2}{\sqrt{\pi}}e^{-x^2}$	erf$x$
1.00	0.41510	0.84270	1.50	0.11893	0.96610
1.01	0.40684	0.84681	1.51	0.11540	0.96727
1.02	0.39867	0.85083	1.52	0.11195	0.96841
1.03	0.39058	0.85478	1.53	0.10859	0.96951
1.04	0.38257	0.85864	1.54	0.10531	0.97058
1.05	0.37466	0.86243	1.55	0.10210	0.97162
1.06	0.36684	0.86614	1.56	0.09898	0.97262
1.07	0.35911	0.86977	1.57	0.09593	0.97360
1.08	0.35147	0.87332	1.58	0.09295	0.97454
1.09	0.34392	0.87680	1.59	0.09005	0.97546
1.10	0.33647	0.88020	1.60	0.08722	0.97634
1.11	0.32912	0.88353	1.61	0.08447	0.97720
1.12	0.32186	0.88678	1.62	0.08178	0.97803
1.13	0.31470	0.88997	1.63	0.07917	0.97884
1.14	0.30764	0.89308	1.64	0.07662	0.97962
1.15	0.30067	0.89612	1.65	0.07414	0.98037
1.16	0.29381	0.89909	1.66	0.07173	0.98110
1.17	0.28704	0.90200	1.67	0.06938	0.98181
1.18	0.28037	0.90483	1.68	0.06709	0.98249
1.19	0.27381	0.90760	1.69	0.06487	0.98315
1.20	0.26734	0.91031	1.70	0.06271	0.98379
1.21	0.26097	0.91295	1.71	0.06060	0.98440
1.22	0.25471	0.91553	1.72	0.05856	0.98500
1.23	0.24854	0.91805	1.73	0.05657	0.98557
1.24	0.24248	0.92050	1.74	0.05464	0.98613
1.25	0.23652	0.92290	1.75	0.05277	0.98667
1.26	0.23065	0.92523	1.76	0.05095	0.98719
1.27	0.22489	0.92751	1.77	0.04918	0.98769
1.28	0.21923	0.92973	1.78	0.04747	0.98817
1.29	0.21367	0.93189	1.79	0.04580	0.98864
1.30	0.20820	0.93400	1.80	0.04419	0.98909
1.31	0.20284	0.93606	1.81	0.04262	0.98952
1.32	0.19757	0.93806	1.82	0.04110	0.98994
1.33	0.19241	0.94001	1.83	0.03963	0.99034
1.34	0.18734	0.94191	1.84	0.03820	0.99073
1.35	0.18236	0.94376	1.85	0.03681	0.99111
1.36	0.17749	0.94556	1.86	0.03547	0.99147
1.37	0.17271	0.94731	1.87	0.03417	0.99182
1.38	0.16802	0.94901	1.88	0.03292	0.99215
1.39	0.16343	0.95067	1.89	0.03170	0.99247
1.40	0.15894	0.95228	1.90	0.03052	0.99279
1.41	0.15453	0.95385	1.91	0.02938	0.99308
1.42	0.15022	0.95537	1.92	0.02827	0.99337
1.43	0.14600	0.95685	1.93	0.02721	0.99365
1.44	0.14187	0.95829	1.94	0.02617	0.99392
1.45	0.13783	0.95969	1.95	0.02517	0.99417
1.46	0.13387	0.96105	1.96	0.02421	0.99442
1.47	0.13001	0.96237	1.97	0.02328	0.99466
1.48	0.12623	0.96365	1.98	0.02237	0.99489
1.49	0.12254	0.96489	1.99	0.02150	0.99511
			2.00	0.02066	0.99532

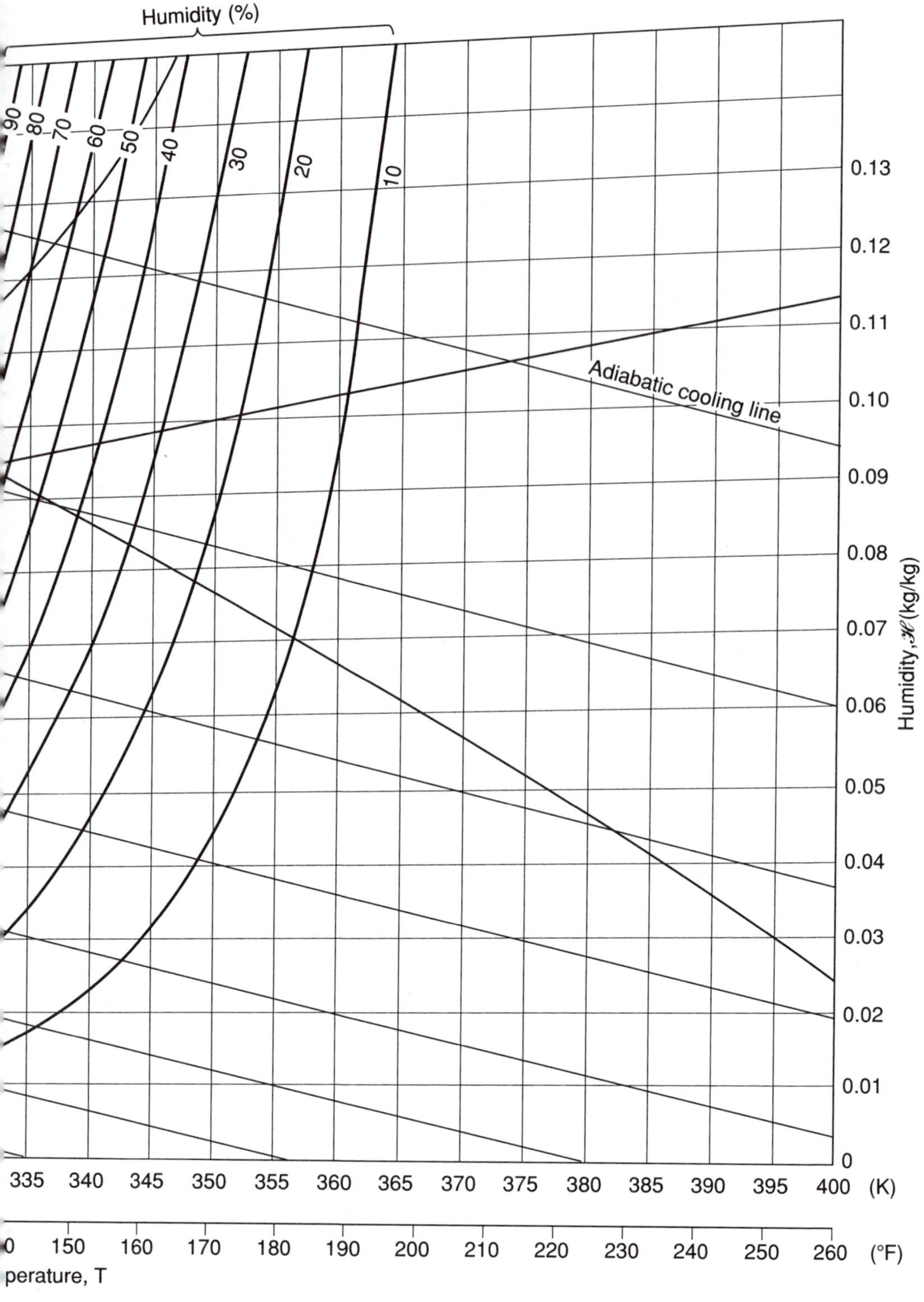

Humidity (%)
90
80
70
60
50
40
30
20
10
Adiabatic cooling line
0.13
0.12
0.11
0.10
0.09
0.08
0.07
0.06
0.05
0.04
0.03
0.02
0.01
0
Humidity, ℋ (kg/kg)
335 340 345 350 355 360 365 370 375 380 385 390 395 400 (K)
0 150 160 170 180 190 200 210 220 230 240 250 260 (°F)
perature, T

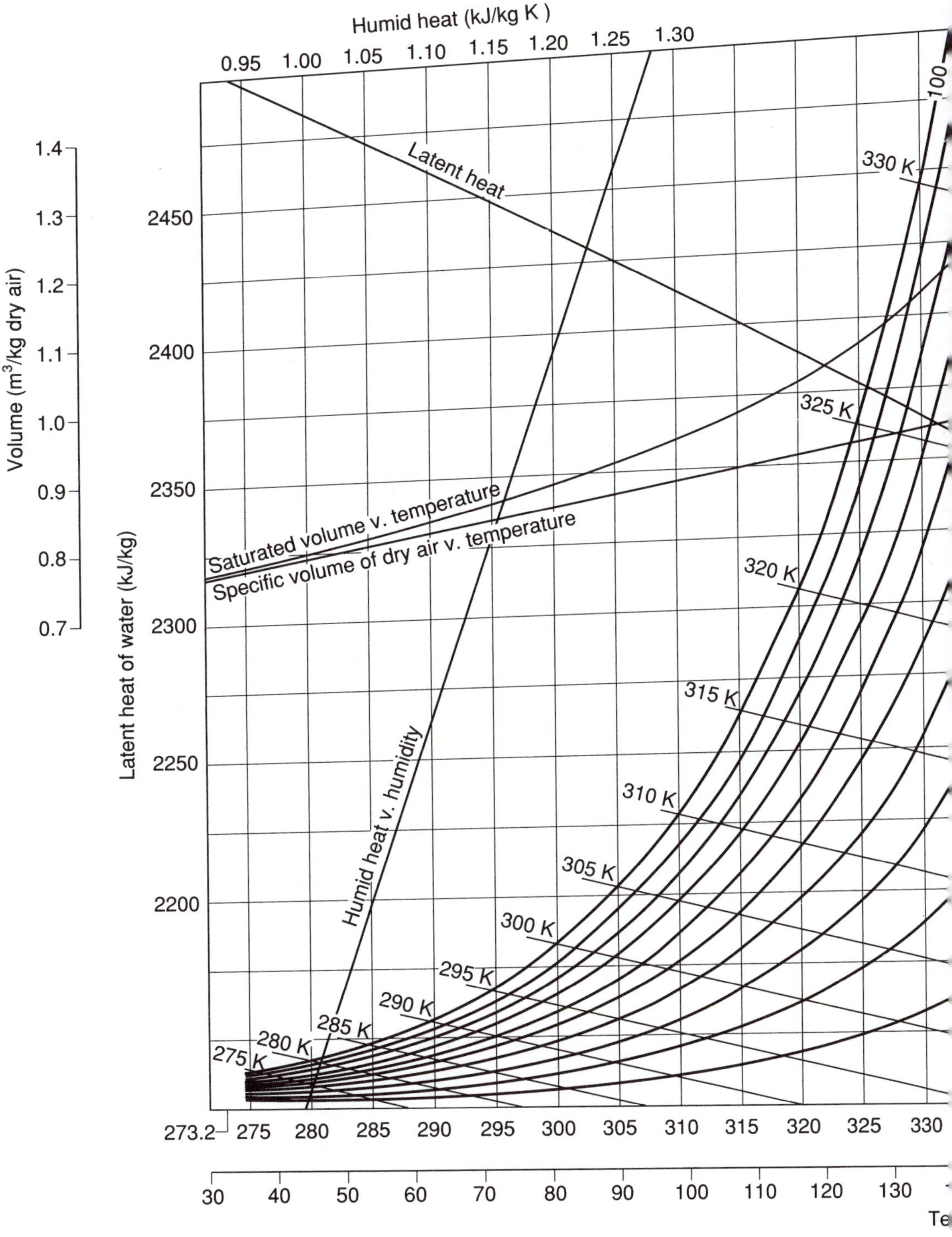

Figure 13.4. Humidity – temperature chart

Problems

A Solutions Manual is available for the Problems in Volume 1 of Chemical Engineering from booksellers and from:

Heinemann Customer Services
Halley Court
Jordan Hill
Oxford OX2 8YW
UK

Tel: 01865 888180
E-mail: bhuk.orders@repp.co.uk

1.1. 98% sulphuric acid of viscosity 0.025 Ns/m^2 and density 1840 kg/m^3 is pumped at 685 cm^3/s through a 25 mm line. Calculate the value of the Reynolds number.

1.2. Compare the costs of electricity at 1 p per kWh and gas at 15 p per therm.

1.3. A boiler plant raises 5.2 kg/s of steam at 1825 kN/m^2 pressure, using coal of calorific value 27.2 MJ/kg. If the boiler efficiency is 75%, how much coal is consumed per day? If the steam is used to generate electricity, what is the power generation in kilowatts, assuming a 20% conversion efficiency of the turbines and generators?

1.4. The power required by an agitator in a tank is a function of the following four variables:

(a) Diameter of impeller.
(b) Number of rotations of impeller per unit time.
(c) Viscosity of liquid.
(d) Density of liquid.

From a dimensional analysis, obtain a relation between the power and the four variables. The power consumption is found, experimentally, to be proportional to the square of the speed of rotation. By what factor would the power be expected to increase if the impeller diameter were doubled?

1.5. It is found experimentally that the terminal settling velocity u_0 of a spherical particle in a fluid is a function of the following quantities:

particle diameter d,
buoyant weight of particle (weight of particle-weight of displaced fluid) W,
fluid density ρ,
fluid viscosity μ.

Obtain a relationship for u_0 using dimensional analysis.

Stokes established, from theoretical considerations, that for small particles which settle at very low velocities, the settling velocity is independent of the density of the fluid except in so far as this affects the buoyancy. Show that the settling velocity *must* then be inversely proportional to the viscosity of the fluid.

1.6. A drop of liquid spreads over a horizontal surface. What are the factors which will influence:

(a) the rate at which the liquid spreads, and
(b) the final shape of the drop?

Obtain dimensionless groups involving the physical variables in the two cases.

1.7. Liquid is flowing at a volumetric flowrate Q per unit width down a vertical surface. Obtain from dimensional analysis the form of the relationship between flowrate and film thickness. If the flow is streamline, show that the volumetric flowrate is directly proportional to the density of the liquid.

1.8. Obtain, by dimensional analysis, a functional relationship for the heat transfer coefficient for forced convection at the inner wall of an annulus through which a cooling liquid is flowing.

1.9. Obtain by dimensional analysis a functional relationship for the wall heat transfer coefficient for a fluid flowing through a straight pipe of circular cross-section. Assume that the effects of natural convection can be neglected in comparison with those of forced convection.

It is found by experiment that, when the flow is turbulent, increasing the flowrate by a factor of 2 always results in a 50% increase in the coefficient. How would a 50% increase in density of the fluid be expected to affect the coefficient, all other variables remaining constant?

1.10. A stream of droplets of liquid is formed rapidly at an orifice submerged in a second, immiscible liquid. What physical properties would be expected to influence the mean size of droplet formed? Using dimensional analysis obtain a functional relation between the variables.

1.11. Liquid flows under steady-state conditions along an open channel of fixed inclination to the horizontal. On what factors will the depth of liquid in the channel depend? Obtain a relationship between the variables using dimensional analysis.

1.12. Liquid flows down an inclined surface as a film. On what variables will the thickness of the liquid film depend? Obtain the relevant dimensionless groups. It may be assumed that the surface is sufficiently wide for edge effects to be negligible.

1.13. A glass particle settles under the action of gravity in a liquid. Upon which variables would you expect the terminal velocity of the particle to depend? Obtain a relevant dimensionless grouping of the variables. The falling velocity is found to be proportional to the square of the particle diameter when other variables are kept constant. What will be the effect of doubling the viscosity of the liquid? What does this suggest about the nature of the flow?

1.14. Heat is transferred from condensing steam to a vertical surface and the resistance to heat transfer is attributable to the thermal resistance of the condensate layer on the surface.

What variables will be expected to affect the film thickness at a point?

Obtain the relevant dimensionless groups.

For streamline flow it is found that the film thickness is proportional to the one third power of the volumetric flowrate per unit width. Show that the heat transfer coefficient would be expected to be inversely proportional to the one third power of viscosity.

1.15. A spherical particle settles in a liquid contained in a narrow vessel. Upon what variables would you expect the falling velocity of the particle to depend? Obtain the relevant dimensionless groups.

For particles of a given density settling in a vessel of large diameter, the settling velocity is found to be inversely proportional to the viscosity of the liquid. How would you expect it to depend on particle size?

1.16. A liquid is in steady state flow in an open trough of rectangular cross-section inclined at an angle θ to the horizontal. On what variables would you expect the mass flow per unit time to depend? Obtain the dimensionless groups which are applicable to this problem.

1.17. The resistance force on a spherical particle settling in a fluid in given by Stokes' Law. Obtain an expression for the terminal falling velocity of the particle. It is convenient to express the results of experiments in the form of a dimensionless group which may be plotted against a Reynolds group with respect to the particle. Suggest a suitable form for this dimensionless group.

Force on particle from Stokes' Law $= 3\pi\mu du$; where μ is the fluid viscosity, d is the particle diameter and u is the velocity of the particle relative to the fluid.

What will be the terminal falling velocity of a particle of diameter 10 μm and of density 1600 kg/m^3 settling in a liquid of density 1000 kg/m^3 and of viscosity 0.001 Ns/m^2?

If Stokes' Law applies for particle Reynolds numbers up to 0.2, what is the diameter of the largest particle whose behaviour is governed by Stokes' Law for this solid and liquid?

1.18. A sphere, initially at a constant temperature, is immersed in a liquid whose temperature is maintained constant. The time t taken for the temperature of the centre of the sphere to reach a given temperature θ_c is a function of the following variables:

Diameter of sphere	d
Thermal conductivity of sphere	k
Density of sphere	ρ
Specific heat capacity of sphere	C_p
Temperature of fluid in which it is immersed	θ_s

Obtain relevant dimensionless groups for this problem.

1.19. Upon what variables would you expect the rate of filtration of a suspension of fine solid particles to depend? Consider the flow through unit area of filter medium and express the variables in the form of dimensionless groups.

It is found that the filtration rate is doubled if the pressure difference is doubled. What effect would you expect from raising the temperature of filtration from 293 to 313 K?

The viscosity of the liquid is given by:

$$\mu = \mu_0(1 - 0.015(T - 273))$$

where: μ is the viscosity at a temperature TK
and: μ_0 is the viscosity at 273 K.

2.1. Calculate the ideal available energy produced by the discharge to atmosphere through a nozzle of air stored in a cylinder of capacity 0.1 m^3 at a pressure of 5 MN/m^2. The initial temperature of the air is 290 K and the ratio of the specific heats is 1.4.

2.2. Obtain expressions for the variation of:

(a) internal energy with change of volume,
(b) internal energy with change of pressure, and
(c) enthalpy with change of pressure.

all at constant temperature, for a gas whose equation of state is given by van der Waals' law.

2.3. Calculate the energy stored in 1000 cm^3 of gas at 80 MN/m^2 and 290 K using a datum of STP.

2.4. Compressed gas is distributed from a works in cylinders which are filled to a pressure P by connecting them to a large reservoir of gas which remains at a steady pressure P and temperature T. If the small cylinders are initially at a temperature T and pressure P_0, what is the final temperature of the gas in the cylinders if heat losses can be neglected and if the compression can be regarded as reversible? Assume that the ideal gas laws are applicable.

3.1. Calculate the hydraulic mean diameter of the annular space between a 40 mm and a 50 mm tube.

3.2. 0.015 m^3/s of acetic acid is pumped through a 75 mm diameter horizontal pipe 70 m long. What is the pressure drop in the pipe?

Viscosity of acid $= 2.5$ mN s/m^2,

Density of acid $= 1060$ kg/m^3.

Roughness of pipe surface $= 6 \times 10^{-5}$ m.

3.3. A cylindrical tank, 5 m in diameter, discharges through a mild steel pipe 90 m long and 230 mm diameter connected to the base of the tank. Find the time taken for the water level in the tank to drop from 3 m to 1 m above the bottom. Take the viscosity of water as 1 mN s/m^2.

3.4. Two storage tanks A and B containing a petroleum product discharge through pipes each 0.3 m in diameter and 1.5 km long to a junction at D. From D the product is carried by a 0.5 m diameter pipe to a third storage tank C, 0.8 km away. The surface of the liquid in A is initially 10 m above that in C and the liquid level in B is 7 m higher than that in A. Calculate the initial rate of discharge of the liquid if the pipes are of mild steel. Take the density of the petroleum product as 870 kg/m^3 and the viscosity as 0.7 mN s/m^2.

3.5. Find the drop in pressure due to friction in a glazed porcelain pipe 300 m long and 150 mm diameter when water is flowing at the rate of 0.05 m^3/s.

3.6. Two tanks, the bases of which are at the same level, are connected with one another by a horizontal pipe 75 mm diameter and 300 m long. The pipe is bell-mouthed at each end so that losses on entry and exit are negligible. One tank is 7 m diameter and contains water to a depth of 7 m. The other tank is 5 m diameter and contains water to a depth of 3 m. If the tanks are connected to each other by means of the pipe, how long will it take before the water level in the larger tank has fallen to 6 m? Assume the pipe to be of aged mild steel.

3.7. Two immiscible fluids A and B, of viscosities μ_A and μ_B, flow under streamline conditions between two horizontal parallel planes of width b, situated a distance $2a$ apart (where a is much less than b), as two distinct parallel layers one above the other, each of depth a. Show that the volumetric rate of flow of A is:

$$\left(\frac{-\Delta P a^3 b}{12\mu_A l}\right) \times \left(\frac{7\mu_A + \mu_B}{\mu_A + \mu_B}\right),$$

where $-\Delta P$ is the pressure drop over a length l in the direction of flow.

3.8. Glycerol is pumped from storage tanks to rail cars through a single 50 mm diameter main 10 m long, which must be used for all grades of glycerol. After the line has been used for commercial material, how much pure glycerol must be pumped before the issuing liquid contains not more than 1% of the commercial material? The flow in the pipeline is streamline and the two grades of glycerol have identical densities and viscosities.

3.9. A viscous fluid flows through a pipe with slightly porous walls so that there is a leakage of kP m^3/m^2 s, where P is the local pressure measured above the discharge pressure and k is a constant. After a length L the liquid is discharged into a tank. If the internal diameter of the pipe is D m and the volumetric rate of flow at the inlet is Q m^3/s, show that the pressure drop in the pipe is given by:

$$-\Delta P = \left(\frac{Q}{\pi k D}\right) a \tanh aL,$$

where:

$$a = \left(\frac{128k\mu}{D^3}\right)^{0.5}$$

Assume a fully developed flow with $(R/\rho u^2) = 8Re^{-1}$.

3.10. A petroleum product of viscosity 0.5 mN s/m^2 and specific gravity 0.7 is pumped through a pipe of 0.15 m diameter to storage tanks situated 100 m away. The pressure drop along the pipe is 70 kN/m^2. The pipeline has to be repaired and it is necessary to pump the liquid by an alternative route consisting of 70 m of 20 cm pipe followed by 50 m of 10 cm pipe. If the existing pump is capable of developing a pressure of 300 kN/m^2, will it be suitable for use during the period required for the repairs? Take the roughness of the pipe surface as 0.00005 m.

3.11. Explain the phenomenon of hydraulic jump which occurs during the flow of a liquid in an open channel.

A liquid discharges from a tank into an open channel under a gate so that the liquid is initially travelling at a velocity of 1.5 m/s and a depth of 75 mm. Calculate, from first principles, the corresponding velocity and depth after the jump.

3.12. What is a non-Newtonian fluid? Describe the principal types of behaviour exhibited by these fluids. The viscosity of a non-Newtonian fluid changes with the rate of shear according to the approximate relationship:

$$\mu_a = k\left(-\frac{du_x}{dr}\right)^{-0.5}$$

where μ_a is the apparent viscosity, and du_x/dr is the velocity gradient normal to the direction of motion.

Show that the volumetric rate of streamline flow through a horizontal tube of radius a is:

$$\frac{\pi}{5}a^5\left(\frac{-\Delta P}{2kl}\right)^2$$

where $-\Delta P$ is the pressure drop over length l of the tube.

3.13. Calculate the pressure drop when 3 kg/s of sulphuric acid flows through 60 m of 25 mm pipe ($\rho =$ 1840 kg/m^3, $\mu = 0.025$ Ns/m^2).

3.14. The relation between cost per unit length C of a pipeline installation and its diameter d is given by:

$$C = a + bd,$$

where a and b are independent of pipe size. Annual charges are a fraction β of the capital cost. Obtain an expression for the optimum pipe diameter on a minimum cost basis for a fluid of density ρ and viscosity μ flowing at a mass rate of G. Assume that the fluid is in turbulent flow and that the Blasius equation is applicable, i.e. the friction factor is proportional to the Reynolds number to the power of minus one quarter. Indicate clearly how the optimum diameter depends on flowrate and fluid properties.

3.15. A heat exchanger is to consist of a number of tubes each 25 mm diameter and 5 m long arranged in parallel. The exchanger is to be used as a cooler with a rating of 4 MW and the temperature rise in the water feed to the tubes is to be 20 K.

If the pressure drop over the tubes is not to exceed 2 kN/m^2, calculate the minimum number of tubes that are required. Assume that the tube walls are smooth and that entrance and exit effects can be neglected.

$$\text{Viscosity of water} = 1 \text{ mN s/m}^2.$$

3.16. Sulphuric acid is pumped at 3 kg/s through a 60 m length of smooth 25 mm pipe. Calculate the drop in pressure. If the pressure drop falls by one half, what will be the new flowrate?

$$\text{Density of acid} = 1840 \text{ kg/m}^3.$$

$$\text{Viscosity of acid} = 25 \text{ mN s/m}^2.$$

3.17. A Bingham plastic material is flowing under streamline conditions in a pipe of circular cross-section. What are the conditions for one half of the total flow to be within the central core across which the velocity profile is flat? The shear stress acting within the fluid R_y varies with velocity gradient du_x/dy according to the relation:

$$R_y - R_c = -k\frac{du_x}{dy},$$

where R_c and k are constants for the material.

3.18. Oil of viscosity 10 mN s/m^2 and density 950 kg/m^3 is pumped 8 km from an oil refinery to a distribution depot through a 75 mm diameter pipeline and is then despatched to customers at a rate of 500 tonne/day. Allowance must be made for periods of maintenance which may interrupt the supply from the refinery for up to 72 hours. If the maximum permissible pressure drop over the pipeline is 3450 kN/m^2, what is the shortest time in which the storage tanks can be completely recharged after a 72 hour shutdown? Take the roughness of the pipe surface as 0.05 mm.

3.19. Water is pumped at 1.4 m^3/s from a tank at a treatment plant to a tank at a local works through two parallel pipes, 0.3 m and 0.6 m diameter respectively. What is the velocity in each pipe and, if a single pipe is used, what diameter will be needed if this flow of water is to be transported, the pressure drop being the same? Assume turbulent flow, with the friction factor inversely proportional to the one quarter power of the Reynolds number.

3.20. Oil of viscosity 10 mN s/m^2 and specific gravity 0.90, flows through 60 m of 100 mm diameter pipe and the pressure drop is 13.8 kN/m^2. What will be the pressure drop for a second oil of viscosity 30 mN s/m^2 and specific gravity 0.95 flowing at the same rate through the pipe? Assume the pipe wall to be smooth.

3.21. Crude oil is pumped from a terminal to a refinery through a foot diameter pipeline. As a result of frictional heating, the temperature of the oil is 20 deg K higher at the refinery end than at the terminal end of the pipe and the viscosity has fallen to one half its original value. What is the ratio of the pressure gradient in the pipeline at the refinery end to that at the terminal end?

$$\text{Viscosity of oil at terminal} = 90 \text{ mN s/m}^2$$

$$\text{Density of oil (approximately constant)} = 960 \text{ kg/m}^3$$

$$\text{Flowrate of oil} = 20{,}000 \text{ tonne/day}$$

Outline a method for calculating the temperature of the oil as a function of distance from the inlet for a given value of the heat transfer coefficient between the pipeline and the surroundings.

3.22. Oil with a viscosity of 10 mN s/m^2 and density 900 kg/m^3 is flowing through a 500 mm diameter pipe 10 km long. The pressure difference between the two ends of the pipe is 10^6N/m^2. What will the pressure drop be at the same flowrate if it is necessary to replace the pipe by one only 300 mm diameter? Assume the pipe surface to be smooth.

3.23. Oil of density 950 kg/m^3 and viscosity 10^{-2} Ns/m^2 is to be pumped 10 km through a pipeline and the pressure drop must not exceed 2×10^5N/m^2. What is the minimum diameter of pipe which will be suitable, if a flowrate of 50 tonne/h is to be maintained? Assume the pipe wall to be smooth. Use either the pipe friction chart *or* the Blasius equation ($R/\rho u^2 = 0.0396Re^{-1/4}$).

3.24. On the assumption that the velocity profile in a fluid in turbulent flow is given by the Prandtl one-seventh power law, calculate the radius at which the flow between it and the centre is equal to that between it and the wall, for a pipe 100 mm in diameter.

3.25. A pipeline 0.5 m diameter and 1200 m long is used for transporting an oil of density 950 kg/m^3 and of viscosity 0.01 Ns/m^2 at 0.4 m^3/s. If the roughness of the pipe surface is 0.5 mm, what is the pressure drop? With the same pressure drop, what will be the flowrate of a second oil of density 980 kg/m^3 and of viscosity 0.02 Ns/m^2?

3.26. Water (density 1000 kg/m^3, viscosity 1 mN s/m^2) is pumped through a 50 mm diameter pipeline at 4 kg/s and the pressure drop is 1 MN/m^2. What will be the pressure drop for a solution of glycerol in water (density 1050 kg/m^3, viscosity 10 mN s/m^2) when pumped at the same rate? Assume the pipe to be smooth.

3.27. A liquid is pumped in streamline flow through a pipe of diameter d. At what distance from the centre of the pipe will the fluid be flowing at the average velocity?

3.28. Cooling water is supplied to a heat exchanger and flows through 25 mm diameter tubes each 5 m long arranged in parallel. If the pressure drop over the heat exchanger is not to exceed 8000 N/m^2, how many tubes must be included for a total flowrate of water of 110 tonne/h?

Density of water 1000 kg/m^3

Viscosity of water 1 mN s/m^2

Assume pipes to be smooth-walled

If ten per cent of the tubes became blocked, what would the new pressure drop be?

3.29. The effective viscosity of a non-Newtonian fluid may be expressed by the relationship:

$$\mu_a = k''\left(-\frac{du_x}{dr}\right)$$

where k'' is constant.

Show that the volumetric flowrate of this fluid in a horizontal pipe of radius a under isothermal laminar flow conditions with a pressure gradient $-\Delta P/l$ per unit length is:

$$Q = \frac{2\pi}{7}a^{7/2}\left(\frac{-\Delta P}{2k''l}\right)^{1/2}$$

3.30. Determine the yield stress of a Bingham fluid of density 2000 kg/m^3 which will just flow out of an open-ended vertical tube of diameter 300 mm under the influence of its own weight.

3.31. A fluid of density 1.2×10^3 kg/m^3 flows down an inclined plane at 15° to the horizontal. If the viscous behaviour is described by the relationship:

$$R_{yx} = -k\left(\frac{du_x}{dy}\right)^n$$

where $k = 4.0$ Ns$^{0.4}$/m^2, and $n = 0.4$, calculate the volumetric flowrate per unit width if the fluid film is 10 mm thick.

3.32. A fluid with a finite yield stress is sheared between two concentric cylinders, 50 mm long. The inner cylinder is 30 mm diameter and the gap is 20 mm. The outer cylinder is held stationary while a torque is applied to the inner. The moment required just to produce motion was 0.01 N m. Calculate the torque needed to ensure all the fluid is flowing under shear if the plastic viscosity is 0.1 Ns/m^2.

3.33. Experiments, carried out with a capillary viscometer of length 100 mm and diameter 2 mm on a fluid, gave the following results:

Applied pressure difference $-\Delta P$ (N/m^2)	Volumetric flowrate Q (m^3/s)
1×10^3	1×10^{-7}
2×10^3	2.8×10^{-7}
5×10^3	1.1×10^{-7}
1×10^4	3×10^{-6}
2×10^4	9×10^{-6}
5×10^4	3.5×10^{-5}
1×10^5	1×10^{-4}

Suggest a suitable model to describe the fluid properties.

3.34. Data obtained with a cone and plate viscometer (cone half-angle 89° cone radius 50 mm) were:

cone speed (Hz)	measured torque (Nm)
0.1	4.6×10^{-1}
0.5	7×10^{-1}
1	1.0
5	3.4
10	6.4
50	3.0×10

Suggest a suitable model to describe the fluid properties.

3.35. Tomato purée of density 1300 kg/m^3 is pumped through a 50 mm diameter factory pipeline at a flowrate of 0.00028 m^3/s. It is suggested that in order to double production:

(a) a similar line with pump should be put in parallel to the existing one, or
(b) a large pump should force the material through the present line, or
(c) a large pump should supply the liquid through a line of twice the cross-sectional area.

Given that the flow properties of the purée can be described by the Casson equation:

$$(-R_y)^{1/2} = (-R_Y)^{1/2} + \left(-\mu_c \frac{du_x}{dy}\right)^{1/2}$$

where R_Y is a yield stress, here 20 N/m^2,

μ_c is a characteristic Casson plastic viscosity, 5 Ns/m^2, and

$\frac{du_x}{dy}$ is the velocity gradient.

evaluate the relative pressure drops of the three suggestions, assuming laminar flow throughout.

3.36. The rheological properties of a particular suspension can be approximated reasonably well by either a "power law" or a "Bingham plastic" model over the shear rate range of 10 to 50 s^{-1}. If the consistency k is 10 Ns^n/m^2 and the flow behaviour index n is 0.2 in the power law model, what will be the approximate values of the yield stress and of the plastic viscosity in the Bingham plastic model?

What will be the pressure drop, when the suspension is flowing under laminar conditions in a pipe 200 m long and 40 mm diameter, when the centre line velocity is 1 m/s, according to the power law model? Calculate the centre line velocity for this pressure drop for the Bingham plastic model and comment on the result.

3.37. Show how, by suitable selection of the index n, the power law may be used to describe the behaviour of both shear-thinning and shear-thickening non-Newtonian fluids over a limited range of shear rates. What are the main objections to the use of the power law? Give some examples of different types of shear-thinning fluids.

A power law fluid is flowing under laminar conditions through a pipe of circular cross-section. At what radial position is the fluid velocity equal to the mean velocity in the pipe? Where does this occur for a fluid with an n-value of 0.2?

3.38. A liquid whose rheology can be represented by the "power law" model is flowing under streamline conditions through a pipe of 5 mm diameter. If the mean velocity of flow in 1 m/s and the velocity at the pipe axis is 1.2 m/s, what is the value of the power law index n?

Water, of viscosity 1 mN s/m^2 flowing through the pipe at the same mean velocity gives rise to a pressure drop of 10^4 N/m^2 compared with 10^5 N/m^2 for the non-Newtonian fluid. What is the consistency ("k" value) of the non-Newtonian fluid?

3.39. Two liquids of equal densities, the one Newtonian and the other a non-Newtonian "power law" fluid, flow at equal volumetric rates down two wide vertical surfaces of the same widths. The non-Newtonian fluid has a power law index of 0.5 and has the same apparent viscosity in SI unit as the Newtonian fluid when its shear rate is 0.01 s^{-1}. Show that, for equal surface velocities of the two fluids, the film thickness for the Newtonian fluid is 1.125 times that of the non-Newtonian fluid.

3.40. A fluid which exhibits non-Newtonian behaviour is flowing in a pipe of diameter 70 mm and the pressure drop over a 2 m length of pipe is 4×10^4 N/m^2. (When the flowrate is doubled, the pressure drop increases by a factor of 1.5.) A pitot tube is used to measure the velocity profile over the cross-section. Confirm that the information given below is consistent with the laminar flow of a *power-law* fluid.

Any equations used must be derived from the basic relation between shear stress R and shear rate $\dot{\gamma}$:

$$R = k(\dot{\gamma})^n$$

Radial distance (s mm) from centre of pipe	Velocity (m/s)
0	0.80
10	0.77
20	0.62
30	0.27

3.41. A Bingham-plastic fluid (yield stress 14.35 N/m^2 and plastic viscosity 0.150 Ns/m^2) is flowing through a pipe of diameter 40 mm and length 200 m. Starting with the rheological equation, show that the relation between pressure gradient $-\Delta P/l$ and volumetric flowrate Q is:

$$Q = \frac{\pi(-\Delta P)r^4}{8l\mu_p}\left[1 - \frac{4}{3}X + \frac{1}{3}X^4\right]$$

where r is the pipe radius, μ_p is the plastic viscosity, and

X is the ratio of the yield stress to the shear stress at the pipe wall.

Calculate the flowrate for this pipeline when the pressure drop is 600 kN/m^2. It may be assumed that the flow is laminar.

4.1. A gas, having a molecular weight of 13 kg/kmol and a kinematic viscosity of 0.25 cm^2/s, is flowing through a pipe 0.25 m internal diameter and 5 km long at the rate of 0.4 m^3/s and is delivered at atmospheric pressure. Calculate the pressure required to maintain this rate of flow under isothermal conditions.

The volume occupied by 1 kmol at 273 K and 101.3 kN/m^2 is 22.4 m^3.

What would be the effect on the required pressure if the gas were to be delivered at a height of 150 m (i) above and (ii) below its point of entry into the pipe?

4.2. Nitrogen at 12 MN/m^2 is fed through a 25 mm diameter mild steel pipe to a synthetic ammonia plant at the rate of 1.25 kg/s. What will be the drop in pressure over a 30 m length of pipe for isothermal flow of the gas at 298 K?

Absolute roughness of the pipe surface = 0.005 mm.
Kilogram molecular volume = 22.4 m^3.
Viscosity of nitrogen = 0.02 mN s/m^2.

4.3. Hydrogen is pumped from a reservoir at 2 MN/m^2 pressure through a clean horizontal mild steel pipe 50 mm diameter and 500 m long. The downstream pressure is also 2 MN/m^2 and the pressure of this gas is raised to 2.6 MN/m^2 by a pump at the upstream end of the pipe. The conditions of flow are isothermal and the temperature of the gas is 293 K. What is the flowrate and what is the effective rate of working of the pump?

Viscosity of hydrogen = 0.009 mN s/m^2 at 293 K.

4.4. In a synthetic ammonia plant the hydrogen is fed through a 50 mm steel pipe to the converters. The pressure drop over the 30 m length of pipe is 500 kN/m^2, the pressure at the downstream end being 7.5 MN/m^2. What power is required in order to overcome friction losses in the pipe? Assume isothermal expansion of the gas at 298 K. What error is introduced by assuming the gas to be an incompressible fluid of density equal to that at the mean pressure in the pipe? $\mu = 0.02$ mN s/m^2.

4.5. A vacuum distillation plant operating at 7 kN/m^2 at the top has a boil-up rate of 0.125 kg/s of xylene. Calculate the pressure drop along a 150 mm bore vapour pipe used to connect the column to the condenser. The pipe length may be taken as equivalent to 6 m, $e/d = 0.002$ and $\mu = 0.01$ mN s/m^2.

4.6. Nitrogen at 12 MN/m^2 pressure is fed through a 25 mm diameter mild steel pipe to a synthetic ammonia plant at the rate of 0.4 kg/s. What will be the drop in pressure over a 30 m length of pipe assuming isothermal expansion of the gas at 300 K? What is the average quantity of heat per unit area of pipe surface that must pass through the walls in order to maintain isothermal conditions? What would be the pressure drop in the pipe if it were perfectly lagged? ($\mu = 0.02$ mN s/m^2)

4.7. Air, at a pressure of 10 MN/m^2 and a temperature of 290 K, flows from a reservoir through a mild steel pipe of 10 mm diameter and 30 m long into a second reservoir at a pressure P_2. Plot the mass rate of flow of the air as a function of the pressure P_2. Neglect any effects attributable to differences in level and assume an adiabatic expansion of the air. $\mu = 0.018$ mN s/m^2, $\gamma = 1.36$.

4.8. Over a 30 m length of 150 mm vacuum line carrying air at 293 K the pressure falls from 1 kN/m^2 to 0.1 kN/m^2. If the relative roughness e/d is 0.002, what is the approximate flowrate?

4.9. A vacuum system is required to handle 10 g/s of vapour (molecular weight 56 kg/kmol) so as to maintain a pressure of 1.5 kN/m^2 in a vessel situated 30 m from the vacuum pump. If the pump is able to maintain a pressure of 0.15 kN/m^2 at its suction point, what diameter pipe is required? The temperature is 290 K, and isothermal conditions may be assumed in the pipe, whose surface can be taken as smooth. The ideal gas law is followed.

Gas viscosity = 0.01 mN s/m^2.

4.10. In a vacuum system, air is flowing isothermally at 290 K through a 150 mm diameter pipeline 30 m long. If the relative roughness of the pipewall e/d is 0.002 and the downstream pressure is 130 N/m^2, what will the upstream pressure be if the flow rate of air is 0.025 kg/s?

Assume that the ideal gas law applies and that the viscosity of air is constant at 0.018 mN s/m^2.

What error would be introduced if the change in kinetic energy of the gas as a result of expansion were neglected?

4.11. Air is flowing at the rate of 30 kg/m^{2}s through a smooth pipe of 50 mm diameter and 300 m long. If the upstream pressure is 800 kN/m^2, what will the downstream pressure be if the flow is isothermal at 273 K? Take the viscosity of air as 0.015 mN s/m^2 and the kg molecular volume as 22.4 m^3. What is the significance of the change in kinetic energy of the fluid?

4.12. If temperature does not change with height, estimate the boiling point of water at a height of 3000 m above sea-level. The barometer reading at sea-level is 98.4 kN/m^2 and the temperature is 288.7 K. The vapour pressure of water at 288.7 K is 1.77 kN/m^2. The effective molecular weight of air is 29 kg/kmol.

4.13. A 150 mm gas main is used for transferring a gas (molecular weight 13 kg/kmol and kinematic viscosity 0.25 cm^2/s) at 295 K from a plant to a storage station 100 m away, at a rate of 1 m^3/s. Calculate the pressure drop, if the pipe can be considered to be smooth.

If the maximum permissible pressure drop is 10 kN/m^2, is it possible to increase the flowrate by 25%?

5.1. It is required to transport sand of particle size 1.25 mm and density 2600 kg/m^3 at the rate of 1 kg/s through a horizontal pipe, 200 m long. Estimate the air flowrate required, the pipe diameter and the pressure drop in the pipe-line.

5.2. Sand of mean diameter 0.2 mm is to be conveyed by water flowing at 0.5 kg/s in a 25 mm ID horizontal pipe, 100 m long. What is the maximum amount of sand which may be transported in this way if the head developed by the pump is limited to 300 kN/m^2? Assume fully suspended heterogeneous flow.

5.3. Explain the various mechanisms by which particles may be maintained in suspension during hydraulic transport in a horizontal pipeline and indicate when each is likely to be important.

A highly concentrated suspension of flocculated kaolin in water behaves as a pseudo-homogeneous fluid with shear-thinning characteristics which can be represented approximately by the Ostwald–de Waele power law, with an index of 0.15. It is found that, if air is injected into the suspension when in laminar flow, the pressure gradient may be reduced even though the flowrate of suspension is kept constant. Explain how this is possible in "slug" flow and estimate the possible reduction in pressure gradient for equal volumetric flowrates of suspension and air.

6.1. Sulphuric acid of density 1300 kg/m^3 is flowing through a pipe of 50 mm internal diameter. A thin-lipped orifice, 10 mm diameter, is fitted in the pipe and the differential pressure shown by a mercury manometer is 10 cm. Assuming that the leads to the manometer are filled with the acid, calculate (a) the mass of acid flowing per second, and (b) the approximate loss of pressure (in kN/m^2) caused by the orifice.

The coefficient of discharge of the orifice may be taken as 0.61, the density of mercury as 13,550 kg/m^3, and the density of water as 1000 kg/m^3.

6.2. The rate of discharge of water from a tank is measured by means of a notch for which the flowrate is directly proportional to the height of liquid above the bottom of the notch. Calculate and plot the profile of the notch if the flowrate is 0.1 m^3/s when the liquid level is 150 mm above the bottom of the notch.

6.3. Water flows at between 3000 and 4000 l/s through a 50 mm pipe and is metered by means of an orifice. Suggest a suitable size of orifice if the pressure difference is to be measured with a simple water manometer. What is the approximate pressure difference recorded at the maximum flowrate?

6.4. The rate of flow of water in a 150 mm diameter pipe is measured by means of a venturi meter with a 50 mm diameter throat. When the drop in head over the converging section is 100 mm of water, the flowrate is 2.7 kg/s. What is the coefficient for the converging cone of the meter at that flowrate and what is the head lost due to friction? If the total loss of head over the meter is 15 mm water, what is the coefficient for the diverging cone?

6.5. A venturi meter with a 50 mm throat is used to measure a flow of slightly salty water in a pipe of inside diameter 100 mm. The meter is checked by adding 20 cm^3/s of normal sodium chloride solution above the meter and analysing a sample of water downstream from the meter. Before addition of the salt, 1000 cm^3 of water requires 10 cm^3 of 0.1 M silver nitrate solution in a titration. 1000 cm^3 of the downstream sample required 23.5 cm^3 of 0.1 M silver nitrate. If a mercury-under-water manometer connected to the meter gives a reading of 221 mm, what is the discharge coefficient of the meter? Assume that the density of the liquid is not appreciably affected by the salt.

6.6. A gas cylinder containing 30 m^3 of air at 6 MN/m^2 pressure discharges to the atmosphere through a valve which may be taken as equivalent to a sharp-edged orifice of 6 mm diameter (coefficient of discharge = 0.6). Plot the rate of discharge against the pressure in the cylinder. How long will it take for the pressure in the cylinder to fall to (a) 1 MN/m^2, and (b) 150 kN/m^2?

Assume an adiabatic expansion of the gas through the valve and that the contents of the cylinder remain constant at 273 K.

6.7. Air at 1500 kN/m^2 and 370 K flows through an orifice of 30 mm^2 to atmospheric pressure. If the coefficient of discharge is 0.65, the critical pressure ratio is 0.527, and the ratio of the specific heats is 1.4, calculate the mass flowrate.

6.8. Water flows through an orifice of 25 mm diameter situated in a 75 mm pipe at the rate of 300 cm^3/s. What will be the difference in level on a water manometer connected across the meter? Take the viscosity of water as 1 mN s/m^2.

6.9. Water flowing at 1500 cm^3/s in a 50 mm diameter pipe is metered by means of a simple orifice of diameter 25 mm. If the coefficient of discharge of the meter is 0.62, what will be the reading on a mercury-under-water manometer connected to the meter?

What is the Reynolds number for the flow in the pipe?

$$\text{Density of water} = 1000 \text{ kg/m}^3,$$
$$\text{Viscosity of water} = 1 \text{ mN s/m}^2.$$

6.10. What size of orifice would give a pressure difference of 0.3 m water gauge for the flow of a petroleum product of specific gravity 0.9 at 0.05 m^3/s in a 150 mm diameter pipe?

6.11. The flow of water through a 50 mm pipe is measured by means of an orifice meter with a 40 mm aperture. The pressure drop recorded is 150 mm on a mercury-under-water manometer and the coefficient of discharge of the meter is 0.6. What is the Reynolds number in the pipe and what would you expect the pressure drop over a 30 m length of the pipe to be?

$$\text{Friction factor, } \phi = \frac{R}{\rho u^2} = 0.0025.$$
$$\text{Specific gravity of mercury} = 13.6$$
$$\text{Viscosity of water} = 1 \text{ mN s/m}^2.$$

What type of pump would you use, how would you drive it, and what material of construction would be suitable?

6.12. A rotameter has a tube 0.3 m long which has an internal diameter of 25 mm at the top and 20 mm at the bottom. The diameter of the float is 20 mm, its effective specific gravity is 4.80, and its volume 6.6 cm^3. If the coefficient of discharge is 0.72, at what height will the float be when metering water at 100 cm^3/s?

6.13. Explain why there is a critical pressure ratio across a nozzle at which, for a given upstream pressure, the flowrate is a maximum.

Obtain an expression for the maximum flow for a given upstream pressure for isentropic flow through a horizontal nozzle. Show that for air (ratio of specific heats $\gamma = 1.4$) the critical pressure ratio is 0.53 and calculate the maximum flow through an orifice of area 30 mm^2 and coefficient of discharge 0.65 when the upstream pressure is 1.5 MN/m^2 and the upstream temperature 293 K.

$$\text{Kilogram molecular volume} = 22.4 \text{ m}^3.$$

6.14. A gas cylinder containing air discharges to atmosphere through a valve whose characteristics may be considered similar to those of a sharp-edged orifice. If the pressure in the cylinder is initially 350 kN/m^2, by how much will the pressure have fallen when the flowrate has decreased to one-quarter of its initial value?

The flow through the valve may be taken as isentropic and the expansion in the cylinder as isothermal. The ratio of the specific heats at constant pressure and constant volume is 1.4.

6.15. Water discharges from the bottom outlet of an open tank 1.5 m by 1 m in cross-section. The outlet is equivalent to an orifice 40 mm diameter with a coefficient of discharge of 0.6. The water level in the tank is regulated by a float valve on the feed supply which shuts off completely when the height of water above the bottom of the tank is 1 m and which gives a flowrate which is directly proportional to the distance of the water surface below this maximum level. When the depth of water in the tank is 0.5 m the inflow and outflow are directly balanced.

As a result of a short interruption in the supply, the water level in the tank falls to 0.25 m above the bottom but is then restored again. How long will it take the level to rise to 0.45 m above the bottom?

6.16. The flowrate of air at 298 K in a 0.3 m diameter duct is measured with a pitot tube which is used to traverse the cross-section. Readings of the differential pressure recorded on a water manometer are taken with the pitot tube at ten different positions in the cross-section. These positions are so chosen as to be the midpoints of ten concentric annuli each of the same cross-sectional area. The readings are:

Position	1	2	3	4	5
Manometer reading (mm water)	18.5	18.0	17.5	16.8	15.7
Position	6	7	8	9	10
Manometer reading (mm water)	14.7	13.7	12.7	11.4	10.2

The flow is also metered using a 15 cm orifice plate across which the pressure differential is 50 mm on a mercury-under-water manometer. What is the coefficient of discharge of the orifice meter?

6.17. Explain the principle of operation of the pitot tube and indicate how it can be used in order to measure the total flowrate of fluid in a duct.

If a pitot tube is inserted in a circular cross-section pipe in which a fluid is in streamline flow, calculate at what point in the cross-section it should be situated so as to give a direct reading representative of the mean velocity of flow of the fluid.

6.18. The flowrate of a fluid in a pipe is measured using a pitot tube which gives a pressure differential equivalent to 40 mm of water when situated at the centre line of the pipe and 22.5 mm of water when midway between the axis and the wall. Show that these readings are consistent with streamline flow in the pipe.

6.19. Derive a relationship between the pressure difference recorded between the two orifices of a pitot tube and the velocity of flow of an incompressible fluid. A pitot tube is to be situated in a large circular duct in which fluid is in turbulent flow so that it gives a direct reading of the mean velocity in the duct. At what radius in the duct should it be located, if the radius of the duct is r?

The point velocity in the duct can be assumed to be proportional to the one-seventh power of the distance from the wall.

6.20. A gas of molecular weight 44 kg/kmol, temperature 373 K and pressure 202.6 kN/m^2 is flowing in a duct. A pitot tube is located at the centre of the duct and is connected to a differential manometer containing water. If the differential reading is 38.1 mm water, what is the velocity at the centre of the duct?

The volume occupied by 1 kmol at 273 K and 101.3 kN/m^2 is 22.4 m^3.

6.21. Glycerol, of density 1260 kg/m^3 and viscosity 50 mN s/m^2, is flowing through a 50 mm pipe and the flowrate is measured using an orifice meter with a 38 mm orifice. The pressure differential is 150 mm as indicated on a manometer filled with a liquid of the same density as the glycerol. There is reason to suppose that the orifice meter may have become partially blocked and that the meter is giving an erroneous reading. A check is therefore made by inserting a pitot tube at the centre of the pipe. It gives a reading of 100 mm on a water manometer. What does this suggest?

6.22. The flowrate of air in a 305 mm diameter duct is measured with a pitot tube which is used to traverse the cross-section. Readings of the differential pressure recorded on a water manometer are taken with the pitot tube at ten different positions in the cross-section. These positions are so chosen as to be the mid-points of ten concentric annuli each of the same cross-sectional area. The readings are as follows:

Position	1	2	3	4	5
Manometer reading (mm water)	18.5	18.0	17.5	16.8	15.8
Position	6	7	8	9	10
Manometer reading	14.7	13.7	12.7	11.4	10.2

The flow is also metered using a 50 mm orifice plate across which the pressure differential is 150 mm on a mercury-under-water manometer. What is the coefficient of discharge of the orifice meter?

6.23. The flow of liquid in a 25 mm diameter pipe is metered with an orifice meter in which the orifice has a diameter of 19 mm. The aperture becomes partially blocked with dirt from the liquid. What fraction of the area can become blocked before the error in flowrate at a given pressure differential exceeds 15 per cent? Assume that the coefficient of discharge of the meter remains constant when calculated on the basis of the actual free area of the orifice.

6.24. Water is flowing through a 100 mm diameter pipe and its flowrate is metered by means of a 50 mm diameter orifice across which the pressure drop is 13.8 kN/m^2. A second stream, flowing through a 75 mm diameter pipe, is also metered using a 50 mm diameter orifice across which the pressure differential is 150 mm measured on a mercury-under-water manometer. The two streams join and flow through a 150 mm diameter pipe. What would you expect the reading to be on a mercury-under-water manometer connected across a 75 mm diameter orifice plate inserted in this pipe?

The coefficients of discharge for all the orifice meters are equal.
(Density of mercury = 13600 kg/m^3)

6.25. Water is flowing through a 150 mm diameter pipe and its flowrate is measured by means of a 50 mm diameter orifice, across which the pressure differential is 2.27×10^4 N/m^2. The coefficient of discharge of the orifice meter is independently checked by means of a pitot tube which, when situated at the axis of the pipe, gave a reading of 15.6 mm on a mercury-under-water manometer. On the assumption that the flow in the pipe is turbulent and that the velocity distribution over the cross-section is given by the Prandtl one-seventh power law, calculate the coefficient of discharge of the orifice meter.

6.26. Air at 323 K and 152 kN/m^2 flows through a duct of circular cross-section, diameter 0.5 m. In order to measure the flow rate of air, the velocity profile across a diameter of the duct is measured using a Pitot-static tube connected to a water manometer inclined at an angle of $\cos^{-1}$ 0.1 to the vertical. The following results are obtained:

Distance from duct centre line (m)	Manometer Reading h_m(mm)
0	104
0.05	100
0.10	96
0.15	86
0.175	79
0.20	68
0.225	50

Calculate the mass flow rate of air through the duct, the average velocity, the ratio of the average to the maximum velocity and the Reynolds number. Comment on these results.

Discuss the application of this method of measuring gas flow rates with particular emphasis on the best distribution of experimental points across the duct and on the accuracy of the results.
(Take the viscosity of air as 1.9×10^{-2} mN s/m^2 and the molecular weight of air as 29 kg/kmol.)

7.1. A reaction is to be carried out in an agitated vessel. Pilot-plant experiments were performed under fully turbulent conditions in a tank 0.6 m in diameter, fitted with baffles and provided with a flat-bladed turbine. It was found that the satisfactory mixing was obtained at a rotor speed of 4 Hz, when the power consumption was 0.15 kW and the Reynolds number 160,000. What should be the rotor speed in order to retain the same mixing performance if the linear scale of the equipment is increased 6 times? What will be the power consumption and the Reynolds number?

7.2. A three-bladed propeller is used to mix a fluid in the laminar region. The stirrer is 0.3 m in diameter and is rotated at 1.5 Hz. Due to corrosion, the propeller has to be replaced by a flat two-bladed paddle, 0.75 m in diameter. If the same motor is used, at what speed should the paddle rotate?

7.3. Compare the capital and operating costs of a three-bladed propeller with those of a constant speed six-bladed turbine, both constructed from mild steel. The impeller diameters are 0.3 and 0.45 m respectively and both stirrers are driven by a 1 kW motor. What is the recommended speed of rotation in each case? Assume operation for 8000 hr/year, power at £0.01/kWh and interest and depreciation at 15%/year.

7.4. In a leaching operation, the rate at which solute goes into solution is given by an equation of the form:

$$\frac{dM}{dt} = k(c_s - c) \text{ kg/s}$$

where M kg is the amount of solute dissolving in t s, k m^3/s is a constant and c_s and c are the saturation and bulk concentrations of the solute respectively in kg/m^3. In a pilot test on a vessel 1 m^3 in volume, 75% saturation was attained in 10 s. If 300 kg of a solid containing 28% by mass of a water soluble solid is agitated with 100 m^3 of water, how long will it take for all the solute to dissolve assuming conditions are the same as in the pilot unit? Water is saturated with the solute at a concentration of 2.5 kg/m^3.

7.5. For producing an oil-water emulsion, two portable three-bladed propeller mixers are available; a 0.5 m diameter impeller rotating at 1 Hz and a 0.35 m impeller rotating at 2 Hz. Assuming turbulent conditions prevail, which unit will have the lower power consumption?

7.6. A reaction is to be carried out in an agitated vessel. Pilot-plant experiments were performed under fully turbulent conditions in a tank 0.6 m in diameter, fitted with baffles and provided with a flat-bladed turbine. It was found that satisfactory mixing was obtained at a rotor speed of 4 Hz, when the power consumption was 0.15 kW and the Reynolds number 160,000. What should be the rotor speed in order to retain the same mixing performance if the linear scale of the equipment is increased 6 times? What will be the power consumption and the Reynolds number?

7.7. Tests on a small scale tank 0.3 m diameter (Rushton impeller, diameter 0.1 m) have shown that a blending process between two miscible liquids (aqueous solutions, properties approximately the same as water, i.e. viscosity 1 mN s/m^2, density 1000 kg/m^3) is satisfactorily completed after 1 minute using an impeller speed of 250 rev/min. It is decided to scale up the process to a tank of 2.5 m diameter using the criterion of constant tip-speed.

(a) What speed should be chosen for the larger impeller?
(b) What power will be required?
(c) What will be the blend time in the large tank?

7.8. An agitated tank with a standard Rushton impeller is required to disperse gas in a solution of properties similar to those of water. The tank will be 3 m diameter (1 m diameter impeller). A power level of 0.8 kW/m^3 is chosen. Assuming fully turbulent conditions and that the presence of the gas does not significantly affect the relation between the Power and Reynolds numbers:

(a) What power will be required by the impeller?
(b) At what speed should the impeller be driven?
(c) If a small pilot scale tank 0.3 m diameter is to be constructed to test the process, at what speed should the impeller be driven?

8.1. A three-stage compressor is required to compress air from 140 kN/m^2 and 283 K to 4000 kN/m^2. Calculate the ideal intermediate pressures, the work required per kilogram of gas, and the isothermal efficiency of the process. Assume the compression to be adiabatic and the interstage cooling to cool the air to the initial temperature. Show qualitatively, by means of temperature-entropy diagrams, the effect of unequal work distribution and imperfect intercooling, on the performance of the compressor.

8.2. A twin-cylinder, single-acting compressor, working at 5 Hz, delivers air at 515 kN/m^2 pressure, at the rate of 0.2 m^3/s. If the diameter of the cylinder is 20 cm, the cylinder clearance ratio 5% and the temperature of the inlet air 283 K, calculate the length of stroke of the piston and the delivery temperature.

8.3. A single-stage double-acting compressor running at 3 Hz is used to compress air from 110 kN/m^2 and 282 K to 1150 kN/m^2. If the internal diameter of the cylinder is 20 cm, the length of stroke 25 cm and the piston clearance 5%, calculate (a) the maximum capacity of the machine, referred to air at the initial temperature and pressure, and (b) the theoretical power requirements under isentropic conditions.

8.4. Methane is to be compressed from atmospheric pressure to 30 MN/m^2 in four stages.

Calculate the ideal intermediate pressures and the work required per kilogram of gas. Assume compression to be isentropic and the gas to behave as an ideal gas. Indicate on a temperature-entropy diagram the effect of imperfect intercooling on the work done at each stage.

8.5. An air-lift raises 0.01 m^3/s of water from a well 100 m deep through a 100 mm diameter pipe. The level of the water is 40 m below the surface. The air consumed is 0.1 m^3/s of free air compressed to 800 kN/m^2.

Calculate the efficiency of the pump and the mean velocity of the mixture in the pipe.

8.6. In a single-stage compressor:

$$\begin{aligned} \text{Suction pressure} &= 101.3\ \text{kN/m}^2. \\ \text{Suction temperature} &= 283\ \text{K}. \\ \text{Final pressure} &= 380\ \text{kN/m}^2. \end{aligned}$$

If each new charge is heated 18 K by contact with the clearance gases, calculate the maximum temperature attained in the cylinder, assuming adiabatic compression.

8.7. A single-acting reciprocating pump has a cylinder diameter of 115 mm and a stroke of 230 mm. The suction line is 6 m long and 50 mm in diameter, and the level of the water in the suction tank is 3 m below the cylinder of the pump. What is the maximum speed at which the pump can run without an air vessel if separation is not to occur in the suction line? The piston undergoes approximately simple harmonic motion. Atmospheric pressure is equivalent to a head of 10.4 m of water and separation occurs at a pressure corresponding to a head of 1.22 m of water.

8.8. An air-lift pump is used for raising 0.8 l/s of a liquid of specific gravity 1.2 to a height of 20 m. Air is available at 450 kN/m^2. If the efficiency of the pump is 30%, calculate the power requirement, assuming isentropic compression of the air ($\gamma = 1.4$).

8.9. A single-acting air compressor supplies 0.1 m^3/s of air (at STP) compressed to 380 kN/m^2 from 101.3 kN/m^2 pressure. If the suction temperature is 288.5 K, the stroke is 250 mm, and the speed is 4 Hz, find the cylinder diameter. Assume the cylinder clearance is 4% and compression and re-expansion are isentropic ($\gamma = 1.4$). What is the theoretical power required for the compression?

8.10. Air at 290 K is compressed from 101.3 to 2000 kN/m^2 pressure in a two-stage compressor operating with a mechanical efficiency of 85%. The relation between pressure and volume during the compression stroke and expansion of the clearance gas is $PV^{1.25}$ = constant. The compression ratio in each of the two cylinders is the same and the interstage cooler may be taken as perfectly efficient. If the clearances in the two cylinders are 4% and 5% respectively, calculate:

(a) the work of compression per unit mass of gas compressed;
(b) the isothermal efficiency;
(c) the isentropic efficiency ($\gamma = 1.4$);
(d) the ratio of the swept volumes in the two cylinders.

8.11. Explain briefly the significance of the "specific speed" of a centrifugal or axial-flow pump.

A pump is designed to be driven at 10 Hz and to operate at a maximum efficiency when delivering 0.4 m^3/s of water against a head of 20 m. Calculate the specific speed. What type of pump does this value suggest?

A pump, built for these operating conditions, has a measured maximum overall efficiency of 70%. The same pump is now required to deliver water at 30 m head. At what speed should the pump be driven if it is to operate at maximum efficiency? What will be the new rate of delivery and the power required?

8.12. A centrifugal pump is to be used to extract water from a condenser in which the vacuum is 640 mm of mercury. At the rated discharge the net positive suction head must be at least 3 m above the cavitation vapour pressure of 710 mm mercury vacuum. If losses in the suction pipe account for a head of 1.5 m, what must be the least height of the liquid level in the condenser above the pump inlet?

8.13. What is meant by the Net Positive Suction Head (NPSH) required by a pump? Explain why it exists and how it can be made as low as possible. What happens if the necessary NPSH is not provided?

A centrifugal pump is to be used to circulate liquid, of density 800 kg/m^3 and viscosity 0.5 mN s/m^2, from the reboiler of a distillation column through a vaporiser at the rate of 400 cm^3/s, and to introduce the superheated liquid above the vapour space in the reboiler which contains liquid to a depth of 0.7 m. Suggest a suitable layout if a smooth bore 25 mm pipe is to be used. The pressure of the vapour in the reboiler is 1 kN/m^2 and the NPSH required by the pump is 2 m of liquid.

8.14. 1250 cm^3/s of water is to be pumped through a steel pipe, 25 mm diameter and 30 m long, to a tank 12 m higher than its reservoir. Calculate the approximate power required. What type of pump would you install for the purpose and what power motor (in kW) would you provide?

Viscosity of water = 1.30 mN s/m^2.

Density of water = 1000 kg/m^3.

8.15. Calculate the pressure drop in, and the power required to operate, a condenser consisting of 400 tubes 4.5 m long and 10 mm internal diameter. The coefficient of contraction at the entrance of the tubes is 0.6, and 0.04 m^3/s of water is to be pumped through the condenser.

8.16. 75% sulphuric acid, of density 1650 kg/m^3 and viscosity 8.6 mN s/m^2, is to be pumped for 0.8 km along a 50 mm internal diameter pipe at the rate of 3.0 kg/s, and then raised vertically 15 m by the pump. If the pump is electrically driven and has an efficiency of 50%, what power will be required? What type of pump would you use and of what material would you construct the pump and pipe?

8.17. 60% sulphuric acid is to be pumped at the rate of 4000 cm^3/s through a lead pipe 25 mm diameter and raised to a height of 25 m. The pipe is 30 m long and includes two right-angled bends. Calculate the theoretical power required.

The kinematic viscosity of the acid is 4.25×10^{-5} m^2/s and its density is 1531 kg/m^3. The density of water may be taken as 1000 kg/m^3.

8.18. 1.3 kg/s of 98% sulphuric acid is to be pumped through a 25 mm diameter pipe, 30 m long, to a tank 12 m higher than its reservoir. Calculate the power required and indicate the type of pump and material of construction of the line that you would choose.

Viscosity of acid = 0.025 N s/m^2.

Density = 1840 kg/m^3.

8.19. A petroleum fraction is pumped 2 km from a distillation plant to storage tanks through a mild steel pipeline, 150 mm in diameter, at the rate of 0.04 m^3/s. What is the pressure drop along the pipe and the power supplied to the pumping unit if it has an efficiency of 50%?

The pump impeller is eroded and the pressure at its delivery falls to one half. By how much is the flowrate reduced?

Density of the liquid = 705 kg/m^3.

Viscosity of the liquid = 0.5 mN s/m^2.

Roughness of pipe surface = 0.004 mm.

8.20. Calculate the power required to pump oil of density 850 kg/m^3 and viscosity 3 mN s/m^2 at 4000 cm^3/s through a 50 mm pipeline 100 m long, the outlet of which is 15 m higher than the inlet. The efficiency of the pump is 50%. What effect does the nature of the surface of the pipe have on the resistance?

8.21. 600 cm^3/s of water at 320 K is pumped in a 40 mm i.d. pipe through a length of 150 m in a horizontal direction and up through a vertical height of 10 m. In the pipe there is a control valve which may be taken as equivalent to 200 pipe diameters and other pipe fittings are equivalent to 60 pipe diameters. Also in the line there is a heat exchanger across which there is a loss in head of 1.5 m of water. If the main pipe has a roughness of 0.0002 m, what power must be delivered to the pump if the unit is 60% efficient?

8.22. A pump developing a pressure of 800 kN/m^2 is used to pump water through a 150 mm pipe 300 m long to a reservoir 60 m higher. With the valves fully open, the flowrate obtained is 0.05 m^3/s. As a result of corrosion and scaling the effective absolute roughness of the pipe surface increases by a factor of 10. By what percentage is the flowrate reduced?

$$\text{Viscosity of water} = 1\ \text{mN s/m}^2.$$

9.1. Calculate the time taken for the distant face of a brick wall, of thermal diffusivity, $D_H = 0.0042$ cm^2/s and thickness $l = 0.45$ m, initially at 290 K, to rise to 470 K if the near face is suddenly raised to a temperature of $\theta' = 870$ K and maintained at that temperature. Assume that all the heat flow is perpendicular to the faces of the wall and that the distant face is perfectly insulated.

9.2. Calculate the time for the distant face to reach 470 K under the same conditions as Problem 9.1, except that the distant face is not perfectly lagged but a very large thickness of material of the same thermal properties as the brickwork is stacked against it.

9.3. Benzene vapour, at atmospheric pressure, condenses on a plane surface 2 m long and 1 m wide, maintained at 300 K and inclined at an angle of 45° to the horizontal. Plot the thickness of the condensate film and the point heat transfer coefficient against distance from the top of the surface.

9.4. It is desired to warm 0.9 kg/s of air from 283 to 366 K by passing it through the pipes of a bank consisting of 20 rows with 20 pipes in each row. The arrangement is in-line with centre to centre spacing, in both directions, equal to twice the pipe diameter. Flue gas, entering at 700 K and leaving at 366 K with a free flow mass velocity of 10 kg/m^2s, is passed across the outside of the pipes.

Neglecting gas radiation, how long should the pipes be?

For simplicity, the outer and inner pipe diameters may be taken as 12 mm.

Values of k and μ, which may be used for both air and flue gases, are given below. The specific heat capacity of air and flue gases is 1.0 kJ/kg K.

Temperature (K)	Thermal conductivity k (W/m K)	Viscosity μ (mN s/m^2)
250	0.022	0.0165
500	0.040	0.0276
800	0.055	0.0367

9.5. A cooling coil, consisting of a single length of tubing through which water is circulated, is provided in a reaction vessel, the contents of which are kept uniformly at 360 K by means of a stirrer. The inlet and outlet temperatures of the cooling water are 280 and 320 K respectively. What would the outlet water temperature become if the length of the cooling coil were increased 5 times? Assume the overall heat transfer coefficient to be constant over the length of the tube and independent of the water temperature.

9.6. In an oil cooler 216 kg/h of hot oil enters a thin metal pipe of diameter 25 mm. An equal mass flow of cooling water passes through the annular space between the pipe and a larger concentric pipe with the oil and water moving in opposite directions. The oil enters at 420 K and is to be cooled to 320 K. If the water enters at 290 K, what length of pipe will be required? Take coefficients of 1.6 kW/m^2 K on the oil side and 3.6 kW/m^2 K on the water side and 2.0 kJ/kg K for the specific heat of the oil.

9.7. The walls of a furnace are built up to 150 mm thickness of a refractory of thermal conductivity 1.5 W/m K. The surface temperatures of the inner and outer faces of the refractory are 1400 and 540 K respectively.

If a layer of insulating material 25 mm thick, of thermal conductivity 0.3 W/m K, is added, what temperatures will its surfaces attain assuming the inner surface of the furnace to remain at 1400 K? The coefficient of heat transfer from the outer surface of the insulation to the surroundings, which are at 290 K, may be taken as 4.2, 5.0, 6.1, and 7.1 W/m K, for surface temperatures of 370, 420, 470, and 520 K respectively. What will be the reduction in heat loss?

9.8. A pipe of outer diameter 50 mm, maintained at 1100 K, is covered with 50 mm of insulation of thermal conductivity 0.17 W/m K.

Would it be feasible to use a magnesia insulation which will not stand temperatures above 615 K and has a thermal conductivity 0.09 W/m K for an additional layer thick enough to reduce the outer surface temperature to 370 K in surroundings at 280 K? Take the surface coefficient of heat transfer by radiation and convection as 10 W/m^2 K.

9.9. In order to warm 0.5 kg/s of a heavy oil from 311 to 327 K. it is passed through tubes of inside diameter 19 mm and length 1.5 m, forming a bank, on the outside of which steam is condensing at 373 K. How many tubes will be needed?

In calculating Nu, Pr, and Re, the thermal conductivity of the oil may be taken as 0.14 W/m K and the specific heat as 2.1 kJ/kg K, irrespective of temperature. The viscosity is to be taken at the mean oil temperature. Viscosity of the oil at 319 and 373 K is 154 and 19.2 mN s/m^2 respectively.

9.10. A metal pipe of 12 mm outer diameter is maintained at 420 K. Calculate the rate of heat loss per metre run in surroundings uniformly at 290 K, (a) when the pipe is covered with 12 mm thickness of a material of thermal conductivity 0.35 W/m K and surface emissivity 0.95, and (b) when the thickness of the covering material is reduced to 6 mm, but the outer surface is treated so as to reduce its emissivity to 0.10.

The coefficients of radiation from a perfectly black surface in surroundings at 290 K are 6.25, 8.18, and 10.68 W/m^2 K at 310, 370, and 420 K respectively.

The coefficients of convection may be taken as $1.22(\theta/d)^{0.25}$ W/m^2 K, where θ (K) is the temperature difference between the surface and the surrounding air, and d (m) is the outer diameter.

9.11. A condenser consists of 30 rows of parallel pipes of outer diameter 230 mm and thickness 1.3 mm, with 40 pipes, each 2 m long, per row. Water, inlet temperature 283 K, flows through the pipes at 1 m/s, and steam at 372 K condenses on the outside of the pipes. There is a layer of scale 0.25 mm thick, of thermal conductivity 2.1 W/m K, on the inside of the pipes.

Taking the coefficients of heat transfer on the water side as 4.0, and on the steam side as 8.5 kW/m^2 K, calculate the outlet water temperature and the total mass of steam condensed per second. The latent heat of steam at 372 K is 2250 kJ/kg. The density of water is 1000 kg/m^3.

9.12. In an oil cooler, water flows at the rate of 360 kg/h per tube through metal tubes of outer diameter 19 mm and thickness 1.3 mm, along the outside of which oil flows in the opposite direction at the rate of 75 g/s per tube.

If the tubes are 2 m long, and the inlet temperatures of the oil and water are respectively 370 and 280 K, what will be the outlet oil temperature? The coefficient of heat transfer on the oil side is 1.7 and on the water side 2.5 kW/m^2 K, and the specific heat of the oil is 1.9 kJ/kg K.

9.13. Waste gases flowing across the outside of a bank of pipes are being used to warm air which flows through the pipes. The bank consists of 12 rows of pipes with 20 pipes, each 0.7 m long, per row. They are arranged in-line, with centre-to-centre spacing equal in both directions to one-and-a-half times the pipe diameter. Both inner and outer diameter may be taken as 12 mm. Air, mass velocity 8 kg/m^{2}s, enters the pipes at 290 K. The initial gas temperature is 480 K and the total mass of the gases crossing the pipes per second is the same as the total mass of the air flowing through them.

Neglecting gas radiation, estimate the outlet temperature of the air. The physical constants for the waste gases may be assumed the same as for air, are:

Temperature (K)	Thermal conductivity (W/m K)	Viscosity (mN/ s/m^2)
250	0.022	0.0165
310	0.027	0.0189
370	0.030	0.0214
420	0.033	0.0239
480	0.037	0.0260

Specific heat = 1.00 kJ/kg K.

9.14. Oil is to be warmed from 300 to 344 K by passing it at 1 m/s through the pipes of a shell-and-tube heat exchanger. Steam at 377 K condenses on the outside of the pipes, which have outer and inner diameters of

48 and 41 mm respectively. Due to fouling, the inside diameter has been reduced to 38 mm, and the resistance to heat transfer of the pipe wall and dirt together, based on this diameter, is 0.0009 m^2 K/W.

It is known from previous measurements under similar conditions that the oil side coefficients of heat transfer for a velocity of 1 m/s. based on a diameter of 38 mm. vary with the temperature of the oil as follows:

Oil temperature (K)	300	311	322	333	344
Oil side coefficient of heat transfer (W/m^2 K)	74	80	97	136	244

The specific heat and density of the oil may be assumed constant at 1.9 kJ/kg K and 900 kg/m^3 respectively, and any resistance to heat transfer on the steam side neglected.

Find the length of tube bundle required.

9.15. It is proposed to construct a heat exchanger to condense 7.5 kg/s of n-hexane at a pressure of 150 kN/m^2, involving a heat load of 4.5 MW. The hexane is to reach the condenser from the top of a fractionating column at its condensing temperature of 356 K.

From experience it is anticipated that the overall heat transfer coefficient will be 450 W/m^2 K. Cooling water is available at 289 K.

Outline the proposals that you would make for the type and size of the exchanger and explain the details of the mechanical construction that you consider require special attention.

9.16. A heat exchanger is to be mounted at the top of a fractionating column about 15 m high to condense 4 kg/s of n-pentane at 205 kN/m^2, corresponding to a condensing temperature of 333 K. Give an outline of the calculations you would make to obtain an approximate idea of the size and construction of the exchanger required.

For purposes of standardisation, 19 mm outer diameter tubes of 1.65 mm wall thickness will be used, and these may be 2.5, 3.6, or 5 m in length. The film coefficient for condensing pentane on the outside of a horizontal tube bundle may be taken as 1.1 kW/m^2 K. The condensation is effected by pumping water through the tubes, the initial water temperature being 288 K.

The latent heat of condensation of pentane is 335 kJ/kg.

For these 19 mm tubes, a water velocity of 1 m/s corresponds to a flowrate of 200 g/s of water.

9.17. An organic liquid is boiling at 340 K on the inside of a metal surface of thermal conductivity 42 W/m K and thickness 3 mm. The outside of the surface is heated by condensing steam. Assuming that the heat transfer coefficient from steam to the outer metal surface is constant at 11 kW/m^2 K, irrespective of the steam temperature, find the value of the steam temperature to give a maximum rate of evaporation.

The coefficients of heat transfer from the inner metal surface to the boiling liquid which depend upon the temperature difference are:

Temperature difference metal surface to boiling liquid (K)	Heat transfer coefficient metal surface to boiling liquid (kW/m^2 K)
22.2	4.43
27.8	5.91
33.3	7.38
36.1	7.30
38.9	6.81
41.7	6.36
44.4	5.73
50.0	4.54

9.18. It is desired to warm an oil of specific heat 2.0 kJ/kg K from 300 to 325 K by passing it through a tubular heat exchanger with metal tubes of inner diameter 10 mm. Along the outside of the tubes flows water, inlet temperature 372 K and outlet temperature 361 K.

The overall heat transfer coefficient from water to oil, based on the inside area of the tubes, may be assumed constant at 230 W/m^2 K, and 75 g/s of oil is to be passed through each tube.

The oil is to make two passes through the heater. The water makes one pass along the outside of the tubes. Calculate the length of the tubes required.

9.19. A condenser consists of a number of metal pipes of outer diameter 25 mm and thickness 2.5 mm. Water, flowing at 0.6 m/s, enters the pipes at 290 K, and it is not permissible that it should be discharged at a temperature in excess of 310 K.

If 1.25 kg/s of a hydrocarbon vapour is to be condensed at 345 K on the outside of the pipes, how long should each pipe be and how many pipes should be needed?

Take the coefficient of heat transfer on the water side as 2.5 and on the vapour side as 0.8 kW/m^2 K and assume that the overall coefficient of heat transfer from vapour to water, based upon these figures, is reduced 20% by the effects of the pipe walls, dirt, and scale.

The latent heat of the hydrocarbon vapour at 345 K is 315 kJ/kg.

9.20. An organic vapour is being condensed at 350 K on the outside of a bundle of pipes through which water flows at 0.6 m/s, its inlet temperature being 290 K. The outer and inner diameters of the pipes are 19 mm and 15 mm respectively, but a layer of scale 0.25 mm thick and thermal conductivity 2.0 W/m K has formed on the inside of the pipes.

If the coefficients of heat transfer on the vapour and water sides respectively are 1.7 and 3.2 kW/m^2 K and it is required to condense 25 g/s of vapour on each of the pipes, how long should these be and what will be the outlet temperature of the water?

The latent heat of condensation is 330 kJ/kg.

Neglect any resistance to heat transfer in the pipe walls.

9.21. A heat exchanger is required to cool continuously 20 kg/s of warm water from 360 to 335 K by means of 25 kg/s of cold water, inlet temperature 300 K.

Assuming that the water velocities are such as to give an overall coefficient of heat transfer of 2 kW/m^2 K. assumed constant, calculate the total area of surface required (a) in a counterflow heat exchanger, i.e. one in which the hot and cold fluids flow in opposite directions, and (b) in a multipass heat exchanger, with the cold water making two passes through the tubes and the hot water making one pass along the outside of the tubes. In case (b) assume that the hot water flows in the same direction as the inlet cold water and that its temperature over any cross-section is uniform.

9.22. Find the heat loss per unit area of surface through a brick wall 0.5 m thick when the inner surface is at 400 K and the outside at 310 K. The thermal conductivity of the brick may be taken as 0.7 W/m K.

9.23. A furnace is constructed with 225 mm of firebrick, 120 mm of insulating brick, and 225 mm of building brick. The inside temperature is 1200 K and the outside temperature 330 K. If the thermal conductivities are 1.4, 0.2, and 0.7 W/m K, find the heat loss per unit area and the temperature at the junction of the firebrick and insulating brick.

9.24. Calculate the total heat loss by radiation and convection from an unlagged horizontal steam pipe of 50 mm outside diameter at 415 K to air at 290 K.

9.25. Toluene is continuously nitrated to mononitrotoluene in a cast-iron vessel of 1 m diameter fitted with a propeller agitator of 0.3 m diameter driven at 2 Hz. The temperature is maintained at 310 K by circulating cooling water at 0.5 kg/s through a stainless steel coil of 25 mm outside diameter and 22 mm inside diameter wound in the form of a helix of 0.81 m diameter. The conditions are such that the reacting material may be considered to have the same physical properties as 75% sulphuric acid. If the mean water temperature is 290 K, what is the overall heat transfer coefficient?

9.26. 7.5 kg/s of pure iso-butane is to be condensed at a temperature of 331.7 K in a horizontal tubular exchanger using a water inlet temperature of 301 K. It is proposed to use 19 mm outside diameter tubes of 1.6 mm wall arranged on a 25 mm triangular pitch. Under these conditions the resistance of the scale may be taken as 0.0005 m^2 K/W. Determine the number and arrangement of the tubes in the shell.

9.27. 37.5 kg/s of crude oil is to be heated from 295 to 330 K by heat exchange with the bottom product from a distillation column. The bottom product, flowing at 29.6 kg/s, is to be cooled from 420 to 380 K. There is available a tubular exchanger with an inside shell diameter of 0.60 m having one pass on the shell side and two passes on the tube side. It has 324 tubes, 19 mm outside diameter with 2.1 mm wall and 3.65 m long, arranged on a 25 mm square pitch and supported by baffles with a 25% cut, spaced at 230 mm intervals. Would this exchanger be suitable?

9.28. A 150 mm internal diameter steam pipe is carrying steam at 444 K and is lagged with 50 mm of 85% magnesia. What will be the heat loss to the air at 294 K?

9.29. A refractory material which has an emissivity of 0.40 at 1500 K and 0.43 at 1420 K is at a temperature of 1420 K and is exposed to black furnace walls at a temperature of 1500 K. What is the rate of gain of heat by radiation per unit area?

9.30. The total emissivity of clean chromium as a function of surface-temperature T K is given approximately by:

$$\mathbf{e} = 0.38\left(1 - \frac{263}{T}\right).$$

Obtain an expression for the absorptivity of solar radiation as a function of surface temperature and compare the absorptivity and emissivity at 300, 400, and 1000 K.

Assume that the sun behaves as a black body at 5500 K.

9.31. Repeat Problem 9.30 for the case of aluminium, assuming the emissivity to be 1.25 times that for chromium.

9.32. Calculate the heat transferred by solar radiation on the flat concrete roof of a building, 8 m by 9 m, if the surface temperature of the roof is 330 K. What would be the effect of covering the roof with a highly reflecting surface such as polished aluminium separated from the concrete by an efficient layer of insulation?

The total emissivity of concrete at 330 K is 0.89, whilst the total absorptivity of solar radiation (sun temperature = 5500 K) at this temperature is 0.60. Use the data from Problem 9.31 which should be solved first for aluminium.

9.33. A rectangular iron ingot 15 cm by 15 cm by 30 cm is supported at the centre of a reheating furnace. The furnace has walls of silica-brick at 1400 K, and the initial temperature of the ingot is 290 K. How long will it take to heat the ingot to 600 K?

It may be assumed that the furnace is large compared with the ingot-size, and that the ingot remains at uniform temperature throughout its volume. Convection effects are negligible.

The total emissivity of the oxidised iron surface is 0.78 and both emissivity and absorptivity are independent of the surface temperature.

Density of iron = 7.2 Mg/m^3.

Specific heat capacity of iron = 0.50 kJ/kg K.

9.34. A wall is made of brick, of thermal conductivity 1.0 W/m K, 230 mm thick, lined on the inner face with plaster of thermal conductivity 0.4 W/m K and of thickness 10 mm. If a temperature difference of 30 K is maintained between the two outer faces, what is the heat flow per unit area of wall?

9.35. A 50 mm diameter pipe of circular cross-section and with walls 3 mm thick is covered with two concentric layers of lagging, the inner layer having a thickness of 25 mm and a thermal conductivity of 0.08 W/m K, and the outer layer has a thickness of 40 mm and a thermal conductivity of 0.04 W/m K. What is the rate of heat loss per metre length of pipe if the temperature inside the pipe is 550 K and the outside surface temperature is 330 K?

9.36. The temperature of oil leaving a co-current flow cooler is to be reduced from 370 to 350 K by lengthening the cooler. The oil and water flow rates and inlet temperatures, and the other dimensions of the cooler, will remain constant. The water enters at 285 K and the oil at 420 K. The water leaves the original cooler at 310 K. If the original length is 1 m, what must be the new length?

9.37. In a countercurrent-flow heat exchanger, 1.25 kg/s of benzene (specific heat 1.9 kJ/kg K and density 880 kg/m^3) is to be cooled from 350 to 300 K with water at 290 K. In the heat exchanger, tubes of 25 mm external and 22 mm internal diameter are employed and the water passes through the tubes. If the film coefficients for the water and benzene are 0.85 and 1.70 kW/m^2 K respectively and the scale resistance can be neglected, what total length of tube will be required if the minimum quantity of water is to be used and its temperature is not to be allowed to rise above 320 K?

9.38. Calculate the rate of loss of heat from a 6 m long horizontal steam pipe of 50 mm internal diameter and 60 mm external diameter when carrying steam at 800 kN/m^2. The temperature of the surroundings is 290 K.

What would be the cost of steam saved by coating the pipe with a 50 mm thickness of 85% magnesia lagging of thermal conductivity 0.07 W/m K, if steam costs £0.5 per 100 kg? The emissivity of the surface of the bare pipe and of the lagging may be taken as 0.85, and the coefficient h for heat loss by natural convection is given by:

$$h = 1.65(\Delta T)^{0.25} \text{W/m}^2 \text{ K},$$

where ΔT is the temperature difference in deg K.

Take the Stefan-Boltzmann constant as 5.67×10^{-3} W/m^2 K^4.

9.39. A stirred reactor contains a batch of 700 kg reactants of specific heat 3.8 kJ/kg K initially at 290 K, which is heated by dry saturated steam at 170 kN/m^2 fed to a helical coil. During the heating period the steam supply rate is constant at 0.1 kg/s and condensate leaves at the temperature of the steam. If heat losses are neglected, calculate the true temperature of the reactants when a thermometer immersed in the material reads 360 K. The bulb of the thermometer is approximately cylindrical and is 100 mm long by 10 mm diameter with a water equivalent of 15 g, and the overall heat transfer coefficient to the thermometer is 300 W/m^2 K. What would a thermometer with a similar bulb of half the length and half the heat capacity indicate under these conditions?

9.40. How long will it take to heat 0.18 m^3 of liquid of density 900 kg/m^3 and specific heat 2.1 kJ/kg K from 293 to 377 K in a tank fitted with a coil of area 1 m^2? The coil is fed with steam at 383 K and the overall heat transfer coefficient can be taken as constant at 0.5 kW/m^2 K. The vessel has an external surface of 2.5 m^2, and the coefficient for heat transfer to the surroundings at 293 K is 5 W/m^2 K.

The batch system of heating is to be replaced by a continuous countercurrent heat exchanger in which the heating medium is a liquid entering at 388 K and leaving at 333 K. If the heat transfer coefficient is 250 W/m^2 K, what heat exchange area is required? Heat losses may be neglected.

9.41. The radiation received by the earth's surface on a clear day with the sun overhead is 1 kW/m^2 and an additional 0.3 kW/m^2 is absorbed by the earth's atmosphere. Calculate approximately the temperature of the sun, assuming its radius to be 700,000 km and the distance between the sun and the earth to be 150,000,000 km. The sun may be assumed to behave as a black body.

9.42. A thermometer is immersed in a liquid which is heated at the rate of 0.05 K/s. If the thermometer and the liquid are both initially at 290 K, what rate of passage of liquid over the bulb of the thermometer is required if the error in the thermometer reading after 600 s is to be no more than 1 deg K? Take the water equivalent of the thermometer as 30 g and the heat transfer coefficient to the bulb to be given by $U = 735\ u^{0.8}$ W/m^2 K. The area of the bulb is 0.01 m^2, where u is the velocity in m/s.

9.43. In a shell and tube heat exchanger with horizontal tubes 25 mm external diameter and 22 mm internal diameter, benzene is condensed on the outside by means of water flowing through the tubes at the rate of 0.03 m^3/s. If the water enters at 290 K and leaves at 300 K and the heat transfer coefficient on the water side is 850 W/m^2 K, what total length of tubing will be required?

9.44. In a contact sulphuric acid plant, the gases leaving the first converter are to be cooled from 845 to 675 K by means of the air required for the combustion of the sulphur. The air enters the heat exchanger at 495 K. If the flow of each of the streams is 2 m^3/s at NTP, suggest a suitable design for a shell and tube heat exchanger employing tubes of 25 mm internal diameter.

(a) Assume parallel co-current flow of the gas streams.
(b) Assume parallel countercurrent flow.
(c) Assume that the heat exchanger is fitted with baffles giving cross-flow outside the tubes.

9.45. A large block of material of thermal diffusivity $D_H = 0.0042$ cm^2/s is initially at a uniform temperature of 290 K and one face is raised suddenly to 875 K and maintained at that temperature. Calculate the time taken for the material at a depth of 0.45 m to reach a temperature of 475 K on the assumption of unidirectional heat transfer and that the material can be considered to be infinite in extent in the direction of transfer.

9.46. A 50% glycerol-water mixture is flowing at a Reynolds number of 1500 through a 25 mm diameter pipe. Plot the mean value of the heat transfer coefficient as a function of pipe length assuming that:

$$Nu = 1.62\left(RePr\frac{d}{l}\right)^{0.33}.$$

Indicate the conditions under which this is consistent with the predicted value $Nu = 4.1$ for fully developed flow.

9.47. A liquid is boiled at a temperature of 360 K using steam fed at a temperature of 380 K to a coil heater. Initially the heat transfer surfaces are clean and an evaporation rate of 0.08 kg/s is obtained from each square metre of heating surface. After a period, a layer of scale of resistance 0.0003 m^2 K/W is deposited by the boiling liquid on the heat transfer surface. On the assumption that the coefficient on the steam side remains unaltered and that the coefficient for the boiling liquid is proportional to its temperature difference raised to the power of 2.5, calculate the new rate of boiling.

9.48. A batch of reactants of specific heat 3.8 kJ/kg K and of mass 1000 kg is heated by means of a submerged steam coil of area 1 m^2 fed with steam at 390 K. If the overall heat transfer coefficient is 600 W/m^2 K, calculate the time taken to heat the material from 290 to 360 K, if heat losses to the surroundings are neglected.

If the external area of the vessel is 10 m^2 and the heat transfer coefficient to the surroundings at 290 K is 8.5 W/m^2 K, what will be the time taken to heat the reactants over the same temperature range and what is the maximum temperature to which the reactants can be raised?

What methods would you suggest for improving the rate of heat transfer?

9.49. What do you understand by the terms "black body" and "grey body" when applied to radiant heat transfer?

Two large, parallel plates with grey surfaces are situated 75 mm apart; one has an emissivity of 0.8 and is at a temperature of 350 K and the other has an emissivity of 0.4 and is at a temperature of 300 K. Calculate the net rate of heat exchange by radiation per unit area taking the Stefan-Boltzmann constant as 5.67×10^{-8} W/m^2 K^4. Any formula (other than Stefan's law) which you use must be proved.

9.50. A longitudinal fin on the outside of a circular pipe is 75 mm deep and 3 mm thick. If the pipe surface is at 400 K, calculate the heat dissipated per metre length from the fin to the atmosphere at 290 K if the coefficient of heat transfer from its surface may be assumed constant at 5 W/m^2 K. The thermal conductivity of the material of the fin is 50 W/m K and the heat loss from the extreme edge of the fin may be neglected. It should be assumed that the temperature is uniformly 400 K at the base of the fin.

9.51. Liquid oxygen is distributed by road in large spherical insulated vessels, 2 m internal diameter, well lagged on the outside. What thickness of magnesia lagging, of thermal conductivity 0.07 W/m K must be used so that not more than 1% of the liquid oxygen evaporates during a journey of 10 ks if the vessel is initially 80% full?

Latent heat of vaporisation of oxygen	= 215 kJ/kg.
Boiling point of oxygen	= 90 K.
Density of liquid oxygen	= 1140 kg/m^3
Atmospheric temperature	= 288 K.
Heat transfer coefficient from the outside surface of the lagging to atmosphere	= 4.5 W/m^2 K.

9.52. Benzene is to be condensed at the rate of 1.25 kg/s in a vertical shell and tube type of heat exchanger fitted with tubes of 25 mm outside diameter and 2.5 m long. The vapour condenses on the outside of the tubes and the cooling water enters at 295 K and passes through the tubes at 1.05 m/s. Calculate the number of tubes required if the heat exchanger is arranged for a single pass of the cooling water. The tube wall thickness is 1.6 mm.

9.53. One end of a metal bar 25 mm in diameter and 0.3 m long is maintained at 375 K and heat is dissipated from the whole length of the bar to surroundings at 295 K. If the coefficient of heat transfer from the surface is 10 W/m^2 K, what is the rate of loss of heat? Take the thermal conductivity of the metal as 85 W/m K.

9.54. A shell and tube heat exchanger consists of 120 tubes of internal diameter 22 mm and length 2.5 m. It is operated as a single pass condenser with benzene condensing at a temperature of 350 K on the outside of the tubes and water of inlet temperature 290 K passing through the tubes. Initially there is no scale on the walls and a rate of condensation of 4 kg/s is obtained with a water velocity of 0.7 m/s through the tubes. After prolonged operation, a scale of resistance 0.0002 m^2 K/W is formed on the inner surface of the tubes. To what value must the water velocity be increased in order to maintain the same rate of condensation on the assumption that the transfer coefficient on the water side is proportional to the velocity raised to the 0.8 power, and that the coefficient for the condensing vapour is 2.25 kW/m^2 K based on the inside area? The latent heat of vaporisation of benzene is 400 kJ/kg.

9.55. Derive an expression for the radiant heat transfer rate per unit area between two large parallel planes of emissivities $\mathbf{e}_1$ and $\mathbf{e}_2$ and at absolute temperatures T_1 and T_2 respectively.

Two such planes are situated 2.5 mm apart in air: one has an emissivity of 0.1 and is at 350 K and the other has an emissivity of 0.05 and is at 300 K. Calculate the percentage change in the total heat transfer rate by coating the first surface so as to reduce its emissivity to 0.025.

Stefan-Boltzmann constant $= 5.67 \times 10^{-8}$ W/m^2 K^4.
Thermal conductivity of air $= 0.026$ W/m K.

9.56. Water flows at 2 m/s through a 2.5 m length of a 25 mm diameter tube. If the tube is at 320 K and the water enters and leaves at 293 and 295 K respectively, what is the value of the heat transfer coefficient? How would the outlet temperature change if the velocity were increased by 50%?

9.57. A liquid hydrocarbon is fed at 295 K to a heat exchanger consisting of a 25 mm diameter tube heated on the outside by condensing steam at atmospheric pressure. The flow rate of the hydrocarbon is measured by means of a 19 mm orifice fitted to the 25 mm feed pipe. The reading on a differential manometer containing the hydrocarbon-over-water is 450 mm and the coefficient of discharge of the meter is 0.6.
Calculate the initial rate of rise of temperature (K/s) of the hydrocarbon as it enters the heat exchanger. The outside film coefficient = 6.0 kW/m^2 K. The inside film coefficient h is given by:

$$\frac{hd}{k} = 0.023\left(\frac{ud\rho}{\mu}\right)^{0.8}\left(\frac{C_p\mu}{k}\right)^{0.4}.$$

where:

u = linear velocity of hydrocarbon (m/s).

d = tube diameter (m).

ρ = liquid density (800 kg/m^3).

μ = liquid viscosity (9×10^{-4} Ns/m^2).

C_p = specific heat of liquid (1.7×10^3 J/kg K).

k = thermal conductivity of liquid (0.17 W/m K).

9.58. Water passes at 1.2 m/s through a series of 25 mm diameter tubes 5 m long maintained at 320 K. If the inlet temperature is 290 K, at what temperature would you expect it to leave?

9.59. Heat is transferred from one fluid stream to a second fluid across a heat transfer surface. If the film coefficients for the two fluids are, respectively, 1.0 and 1.5 kW/m^2 K, the metal is 6 mm thick (thermal conductivity 20 W/m K) and the scale coefficient is equivalent to 850 W/m^2 K, what is the overall heat transfer coefficient?

9.60. A pipe of outer diameter 50 mm carries hot fluid at 1100 K. It is covered with a 50 mm layer of insulation of thermal conductivity 0.17 W/m K. Would it be feasible to use magnesia insulation, which will not stand temperatures above 615 K and has a thermal conductivity of 0.09 W/m K for an additional layer thick enough to reduce the outer surface temperature to 370 K in surroundings at 280 K? Take the surface coefficient of transfer by radiation and convection as 10 W/m^2 K.

9.61. A jacketed reaction vessel containing 0.25 m^3 of liquid of specific gravity 0.9 and specific heat 3.3 kJ/kg K is heated by means of steam fed to a jacket on the walls. The contents of the tank are agitated by a stirrer rotating at 3 Hz. The heat transfer area is 2.5 m^2 and the steam temperature is 380 K. The outside film heat transfer coefficient is 1.7 kW/m^2 K and the 10 mm thick wall of the tank has a thermal conductivity of 6.0 W/m K.

The inside film coefficient was found to be 1.1 kW/m^2 K for a stirrer speed of 1.5 Hz and to be proportional to the two-thirds power of the speed of rotation.

Neglecting heat losses and the heat capacity of the tank, how long will it take to raise the temperature of the liquid from 295 to 375 K?

9.62. By dimensional analysis, derive a relationship for the heat transfer coefficient h for natural convection between a surface and a fluid on the assumption that the coefficient is a function of the following variables:

k = thermal conductivity of the fluid.
C_p = specific heat of the fluid.
ρ = density of the fluid.
μ = viscosity of the fluid.
βg = the product of the coefficient of cubical expansion of the fluid and the acceleration due to gravity.
l = a characteristic dimension of the surface.
ΔT = the temperature difference between the fluid and the surface.

Indicate why each of these quantities would be expected to influence the heat transfer coefficient and explain how the orientation of the surface affects the process.

Under what conditions is heat transfer by convection important in Chemical Engineering?

9.63. A shell and tube heat exchanger is used for preheating the feed to an evaporator. The liquid of specific heat 4.0 kJ/kg K and specific gravity 1.1 passes through the inside of the tubes and is heated by steam condensing at 395 K on the outside. The exchanger heats liquid at 295 K to an outlet temperature of 375 K when the flow rate is 175 cm^3/s and to 370 K when the flowrate is 325 cm^3/s. What is the heat transfer area and the value of the overall heat transfer coefficient when the flow is 175 cm^3/s?

Assume that the film heat transfer coefficient for the liquid in the tubes is proportional to the 0.8 power of the velocity, the transfer coefficient for the condensing steam remains constant at 3.4 kW/m^2 K and that the resistance of the tube wall and scale can be neglected.

9.64. 0.1 m^3 of liquid of specific heat capacity 3 kJ/kg K and density 950 kg/m^3 is heated in an agitated tank fitted with a coil, of heat transfer area 1 m^2, supplied with steam at 383 K. How long will it take to heat the liquid from 293 to 368 K, if the tank, of external area 20 m^2 is losing heat to surroundings at 293 K? To what temperature will the system fall in 1800 s if the steam is turned off?

Overall heat transfer coefficient in coil = 2000 W/m^2 K
Heat transfer coefficient to surroundings = 10 W/m^2 K

9.65. The contents of a reaction vessel are heated by means of steam at 393 K supplied to a heating coil which is totally immersed in the liquid. When the vessel has a layer of lagging 50 mm thick on its outer surfaces, it takes one hour to heat the liquid from 293 to 373 K. How long will it take if the thickness of lagging is doubled?

Outside temperature = 293 K
Thermal conductivity of lagging = 0.05 W/m K

Coefficient for heat loss by radiation and convection from outside surface of vessel = 10 W/m^2 K.

Outside area of vessel = 8 m^2
Coil area = 0.2 m^2

Overall heat transfer coefficient for steam coil = 300 W/m^2 K.

9.66. A smooth tube in a condenser which is 25 mm internal diameter and 10 m long is carrying cooling water and the pressure drop over the length of the tube is 2×10^4 N/m^2. If vapour at a temperature of 353 K is condensing on the outside of the tube and the temperature of the cooling water rises from 293 K at inlet to 333 K at outlet, what is the value of the overall heat transfer coefficient based on the inside area of the tube? If the coefficient for the condensing vapour is 15000 W/m^2 K, what is the film coefficient for the water? If the latent heat of vaporisation is 800 kJ/kg, what is the rate of condensation of vapour?

9.67. A chemical reactor, 1 m in diameter and 5 m long, operates at a temperature of 1073 K. It is covered with a 500 mm thickness of lagging of thermal conductivity 0.1 W/m K. The heat loss from the cylindrical surface to the surroundings is 3.5 kW. What is the heat transfer coefficient from the surface of the lagging to the surroundings at a temperature of 293 K? How would the heat loss be altered if the coefficient were halved?

9.68. An open cylindrical tank 500 mm diameter and 1 m deep is three quarters filled with a liquid of density 980 kg/m^3 and of specific heat capacity 3 kJ/kg K. If the heat transfer coefficient from the cylindrical walls and the base of the tank is 10 W/m^2 K and from the surface is 20 W/m^2 K, what area of heating coil, fed with steam at 383 K, is required to heat the contents from 288 K to 368 K in a half hour? The overall heat transfer coefficient for the coil may be taken as 100 W/m^2 K, the surroundings are at 288 K and the heat capacity of the tank itself may be neglected.

9.69. Liquid oxygen is distributed by road in large spherical vessels, 1.82 m in internal diameter. If the vessels were unlagged and the coefficient for heat transfer from the outside of the vessel to the atmosphere were 5 W/m^2 K, what proportion of the contents would evaporate during a journey lasting an hour? Initially the vessels are 80% full.

What thickness of lagging would be required to reduce the losses to one tenth?

Atmospheric temperature = 288 K
Boiling point of oxygen = 90 K,
Density of oxygen = 1140 kg/m^3,
Latent heat of vaporisation of oxygen = 214 kJ/kg,
Thermal conductivity of lagging = 0.07 W/m K.

9.70. Water at 293 K is heated by passing through a 6.1 m coil of 25 mm internal diameter pipe. The thermal conductivity of the pipe wall is 20 W/m K and the wall thickness is 3.2 mm. The coil is heated by condensing steam at 373 K for which the film coefficient is 8 kW/m^2 K. When the water velocity in the pipe is 1 m/s, its outlet temperature is 309 K. What will the outlet temperature be if the velocity is increased to 1.3 m/s, if the coefficient of heat transfer to the water in the tube is proportional to the velocity raised to the 0.8 power?

9.71. Liquid is heated in a vessel by means of steam which is supplied to an internal coil in the vessel. When the vessel contains 1000 kg of liquid it takes half an hour to heat the contents from 293 to 368 K if the coil is supplied with steam at 373 K. The process is modified so that liquid at 293 K is continuously fed to the vessel at the rate of 0.28 kg/s. The total contents of the vessel are always being maintained at 1000 kg. What is the equilibrium temperature which the contents of the vessel will reach, if heat losses to the surroundings are neglected and the overall heat transfer coefficient remains constant?

9.72. The heat loss through a firebrick furnace wall 0.2 m thick is to be reduced by addition of a layer of insulating brick to the outside. What is the thickness of insulating brick necessary to reduce the heat loss to 400 W/m^2? The inside furnace wall temperature is 1573 K, the ambient air adjacent to the furnace exterior is at 293 K and the natural convection heat transfer coefficient at the exterior surface is given by $h_o = 3.0\Delta T^{0.25}$ W/m^2 K, where ΔT is the temperature difference between the surface and the ambient air.

Thermal conductivity of firebrick = 1.5 W/m K
Thermal conductivity of insulating brick = 0.4 W/m K.

9.73. 2.8 kg/s of organic liquid of specific heat capacity 2.5 kJ/kg K is cooled in a heat exchanger from 363 to 313 K using water whose temperature rises from 293 to 318 K flowing countercurrently. After maintenance, the pipework is wrongly connected so that the two streams, flowing at the same rates as previously, are now in co-current flow. On the assumption that overall heat transfer coefficient is unaffected, show that the new outlet temperatures of the organic liquid and the water will be 320.6 K and 314.5 K, respectively.

9.74. An organic liquid is cooled from 353 to 328 K in a single-pass heat exchanger. When the cooling water of initial temperature 288 K flows countercurrently its outlet temperature is 333 K. With the water flowing co-currently, its feed rate has to be increased in order to give the same outlet temperature for the organic liquid, the new outlet temperature of the water is 313 K. When the cooling water is flowing countercurrently, the film heat transfer coefficient for the water is 600 W/m^2 K.

What is the coefficient when the exchanger is operating with co-current flow if its value is proportional to the 0.8 power of the water velocity?

Calculate the film coefficient from the organic liquid, on the assumptions that it remains unchanged, and that heat transfer resistances other than those attributable to the two liquids may be neglected.

9.75. A reaction vessel is heated by steam at 393 K supplied to a coil immersed in the liquid in the tank. It takes 1800 s to heat the contents from 293 K to 373 K when the outside temperature is 293 K. When the outside and initial temperatures are only 278 K, it takes 2700 s to heat the contents to 373 K. The area of the steam coil is 2.5 m^2 and of the external surface is 40 m^2. If the overall heat transfer coefficient from the coil to the liquid in the vessel is 400 W/m^2 K, show that the overall coefficient for transfer from the vessel to the surroundings is about 5 W/m^2 K.

9.76. Steam at 403 K is supplied through a pipe of 25 mm outside diameter. Calculate the heat loss per unit length to surroundings at 293 K, on the assumption that there is a negligible drop in temperature through the wall of the pipe. The heat transfer coefficient h from the outside of the pipe to the surroundings is given by:

$$h = 1.22\left(\frac{\Delta T}{d}\right)^{0.25} \quad \text{W/m}^2\ \text{K}$$

where d is the outside diameter of the pipe (m) and ΔT is the temperature difference (deg K) between the surface and surroundings.

The pipe is then lagged with a 50 mm thickness of lagging of thermal conductivity 0.1 W/m K. If the outside heat transfer coefficient is given by the same equation as for the bare pipe, by what factor is the heat loss reduced?

9.77. A vessel contains 1 tonne of liquid of specific heat capacity 4.0 kJ/kg K. It is heated by steam at 393 K which is fed to a coil immersed in the liquid and heat is lost to the surroundings at 293 K from the outside of the vessel. How long does it take to heat the liquid from 293 to 353 K and what is the maximum temperature to which the liquid can be heated? When the liquid temperature has reached 353 K, the steam supply is turned off for two hours and the vessel cools. How long will it take to reheat the material to 353 K?

Coil: Area 0.5 m^2. Overall heat transfer coefficient to liquid, 600 W/m^2 K
Outside of vessel area 6 m^2. Heat transfer coefficient to surroundings, 10 W/m^2 K.

9.78. A bare thermocouple is used to measure the temperature of a gas flowing through a hot pipe. The heat transfer coefficient between the gas and the thermocouple is proportional to the 0.8 power of the gas velocity and the heat transfer by radiation from the walls to the thermocouple is proportional to the temperature difference.

When the gas is flowing at 5 m/s the thermocouple reads 323 K. When it is flowing at 10 m/s it reads 313 K, and when it is flowing at 15.0 m/s it reads 309 K. Show that the gas temperature is about 298 K and calculate the approximate wall temperature. What temperature will the thermocouple indicate when the gas velocity is 20 m/s?

9.79. A hydrocarbon oil of density 950 kg/m^3 and specific heat capacity 2.5 kJ/kg K is cooled in a heat exchanger from 363 to 313 K by water flowing countercurrently. The temperature of the water rises from 293 to 323 K. If the flowrate of the hydrocarbon is 0.56 kg/s, what is the required flowrate of water?

After plant modifications, the heat exchanger is incorrectly connected so that the two streams are in co-current flow. What are the new outlet temperatures of hydrocarbon and water, if the overall heat transfer coefficient is unchanged?

9.80. A reaction mixture is heated in a vessel fitted with an agitator and a steam coil of area 10 m^2 fed with steam at 393 K. The heat capacity of the system is equal to that of 500 kg of water. The overall coefficient of heat transfer from the vessel of area 5 m^2 is 10 W/m^2 K. It takes 1800 s to heat the contents from ambient temperature of 293 to 333 K. How long will it take to heat the system to 363 K and what is the maximum temperature which can be reached?

Specific heat capacity of water = 4200 J/kg K.

9.81. A pipe, 50 mm outside diameter, is carrying steam at 413 K and the coefficient of heat transfer from its outer surface to the surroundings at 288 K is 10 W/m^2 K. What is the heat loss per unit length?

It is desired to add lagging of thermal conductivity 0.03 W/m K as a thick layer to the outside of the pipe in order to cut heat losses by 90%. If the heat transfer from the outside surface of the lagging is 5 W/m^2 K, what thickness of lagging is required?

9.82. It takes 1800 s (0.5 h) to heat a tank of liquid from 293 to 333 K using steam supplied to an immersed coil when the steam temperature is 383 K. How long will it take when the steam temperature is raised to 393 K? The overall heat transfer coefficient from the steam coil to the tank is 10 times the coefficient from the tank to surroundings at a temperature of 293 K, and the area of the steam coil is equal to the outside area of the tank.

9.83. A thermometer is situated in a duct in an air stream which is at a constant temperature. The reading varies with the gas flowrate as follows:

Air velocity (m/s)	Thermometer reading (K)
6.1	553
7.6	543
12.2	533

The wall of the duct and the gas stream are at somewhat different temperatures. If the heat transfer coefficient for radiant heat transfer from the wall to the thermometer remains constant, and the heat transfer coefficient between the gas stream and thermometer is proportional to the 0.8 power of the velocity, what is the true temperature of the air stream? Neglect any other forms of heat transfer.

10.1. Ammonia gas is diffusing at a constant rate through a layer of stagnant air 1 mm thick. Conditions are fixed so that the gas contains 50% by volume of ammonia at one boundary of the stagnant layer. The ammonia diffusing to the other boundary is quickly absorbed and the concentration is negligible at that plane. The temperature is 295 K and the pressure atmospheric, and under these conditions the diffusivity of ammonia in air is 0.18 cm^2/s. Calculate the rate of diffusion of ammonia through the layer.

10.2. A simple rectifying column consists of a tube arranged vertically and supplied at the bottom with a mixture of benzene and toluene as vapour. At the top a condenser returns some of the product as a reflux which flows in a thin film down the inner wall of the tube. The tube is insulated and heat losses can be neglected. At one point in the column the vapour contains 70 mol% benzene and the adjacent liquid reflux contains 59 mol% benzene. The temperature at this point is 365 K. Assuming the diffusional resistance to vapour transfer to be equivalent to the diffusional resistance of a stagnant vapour layer 0.2 mm thick, calculate the rate of interchange of benzene and toluene between vapour and liquid. The molar latent heats of the two materials can be taken as equal. The vapour pressure of toluene at 365 K is 54.0 kN/m^2 and the diffusivity of the vapours is 0.051 cm^2/s.

10.3. By what percentage would the rate of absorption be increased or decreased by increasing the total pressure from 100 to 200 kN/m^2 in the following cases?

(a) The absorption of ammonia from a mixture of ammonia and air containing 10% of ammonia by volume, using pure water as solvent. Assume that all the resistance to mass transfer lies within the gas phase.
(b) The same conditions as (a) but the absorbing solution exerts a partial vapour pressure of ammonia of 5 kN/m^2.

The diffusivity can be assumed to be inversely proportional to the absolute pressure.

10.4. In the Danckwerts model of mass transfer it is assumed that the fractional rate of surface renewal s is constant and independent of surface age. Under such conditions the expression for the surface age distribution function is se^{-st}. If the fractional rate of surface renewal were proportional to surface age (say $s = bt$, where b is a constant), show that the surface age distribution function would then assume the form:

$$\left(\frac{2b}{\pi}\right)^{1/2} e^{-bt^2/2}.$$

10.5. By considering of the appropriate element of a sphere show that the general equation for molecular diffusion in a stationary medium and in the absence of a chemical reaction is:

$$\frac{\partial C}{\partial t} = D\left(\frac{\partial^2 C}{\partial r^2} + \frac{1}{r^2}\frac{\partial^2 C}{\partial \beta^2} + \frac{1}{r^2 \sin^2\beta}\frac{\partial^2 C}{\partial \phi^2} + \frac{2}{r}\frac{\partial C}{\partial r} + \frac{\cot\beta}{r^2}\frac{\partial C}{\partial \beta}\right),$$

where C is the concentration of the diffusing substance, D the molecular diffusivity, t time, and r, β, and ϕ are spherical polar coordinates, β being the latitude angle.

10.6. Prove that for equimolecular counterdiffusion from a sphere to a surrounding stationary, infinite medium, the Sherwood number based on the diameter of the sphere is equal to 2.

10.7. Show that the concentration profile for unsteady-state diffusion into a bounded medium of thickness L, when the concentration at the interface is suddenly raised to a constant value C_i and kept constant at the initial value of C_o at the other boundary is:

$$\frac{C - C_o}{C_i - C_o} = 1 - \frac{z}{L} = \frac{2}{\pi}\left[\sum_{n=1}^{n=\infty} \frac{1}{n}\exp\left(-\frac{n^2\pi^2 Dt}{L^2}\right)\sin\frac{nz\pi}{L}\right].$$

Assume the solution to be the sum of the solution for infinite time (steady-state part) and the solution of a second unsteady-state part, which simplifies the boundary conditions for the second part.

10.8. Show that under the conditions specified in Problem 10.7 and assuming the Higbie model of surface renewal, the average mass flux at the interface is given by:

$$(N_A)_t = (C_i - C_o)\frac{D}{L}\left\{1 + \frac{2L^2}{\pi^2 Dt}\sum_{n=1}^{n=\infty}\left[\frac{\pi^2}{6} - \frac{1}{n^2}\exp\left(-\frac{n^2\pi^2 Dt}{L^2}\right)\right]\right\}.$$

Use the relation $\sum_{n=1}^{n=\infty} \frac{1}{n^2} = \frac{\pi^2}{6}$.

10.9. According to the simple penetration theory the instantaneous mass flux, $(N_A)_t$ is:

$$(N_A)_t = (C_i - C_o)\left(\frac{D}{\pi t}\right)^{0.5}$$

What is the equivalent expression for the instantaneous heat flux under analogous conditions?

Pure sulphur dioxide is absorbed at 295 K and atmospheric pressure into a laminar water jet. The solubility of SO_2, assumed constant over a small temperature range, is 1.54 kmol/m^3 under these conditions and the heat of solution is 28 kJ/kmol.

Calculate the resulting jet surface temperature if the Lewis number is 90. Neglect heat transfer between the water and the gas.

10.10. In a packed column, operating at approximately atmospheric pressure and 295 K, a 10% ammonia–air mixture is scrubbed with water and the concentration of ammonia is reduced to 0.1%. If the whole of the resistance to mass transfer may be regarded as lying within a thin laminar film on the gas side of the gas-liquid interface, derive from first principles an expression for the rate of absorption at any position in the column. At some intermediate point where the ammonia concentration in the gas phase has been reduced to 5%, the partial pressure of ammonia in equilibrium with the aqueous solution is 660 N/m^2 and the transfer rate is 10^{-3} kmol/m^{2}s. What is the thickness of the hypothetical gas film if the diffusivity of ammonia in air is 0.24 cm^2/s?

10.11. An open bowl, 0.3 m in diameter, contains water at 350 K evaporating into the atmosphere. If the air currents are sufficiently strong to remove the water vapour as it is formed and if the resistance to its mass transfer in air is equivalent to that of a 1 mm layer for conditions of molecular diffusion, what will be the rate of cooling due to evaporation? The water can be considered as well mixed and the water equivalent of the system is equal to 10 kg. The diffusivity of water vapour in air may be taken as 0.20 cm^2/s and the kilogram molecular volume at NTP as 22.4 m^3.

10.12. Show by substitution that when a gas of solubility C^+ is absorbed into a stagnant liquid of infinite depth, the concentration at time t and depth x is:

$$C^+ \operatorname{erfc} \frac{x}{2\sqrt{Dt}}.$$

Hence, on the basis of the simple penetration theory, show that the rate of absorption in a packed column will be proportional to the square root of the diffusivity.

10.13. Show that in steady-state diffusion through a film of liquid, accompanied by a first-order irreversible reaction, the concentration of solute in the film at depth z below the interface is given by:

$$\frac{C}{C_i} = \frac{\sinh \sqrt{(k/D)}(z_L - z)}{\sinh \sqrt{(k/D)} z_L}$$

if $C = 0$ at $z = z_L$ and $C = C_i$ at $z = 0$, corresponding to the interface. Hence show that according to the "film theory" of gas-absorption, the rate of absorption per unit area of interface N_A is given by:

$$N_A = K_L C_i \frac{\beta}{\tanh \beta},$$

where $\beta = \sqrt{(Dk)}/K_L$, D is the diffusivity of the solute, k the rate constant of the reaction, K_L the liquid film mass transfer coefficient for physical absorption, C_i the concentration of solute at the interface, z the distance normal to the interface, and z_L the liquid film thickness.

10.14. The diffusivity of the vapour of a volatile liquid in air can be conveniently determined by Winkelmann's method, in which liquid is contained in a narrow diameter vertical tube maintained at a constant temperature, and an air stream is passed over the top of the tube sufficiently rapidly to ensure that the partial pressure of the vapour there remains approximately zero. On the assumption that the vapour is transferred from the surface of the liquid to the air stream by molecular diffusion, calculate the diffusivity of carbon tetrachloride vapour in air at 321 K and atmospheric pressure from the following experimentally obtained data:

Time from commencement of experiment (ks)	Liquid level (cm)
0	0.00
1.6	0.25
11.1	1.29
27.4	2.32
80.2	4.39
117.5	5.47
168.6	6.70
199.7	7.38
289.3	9.03
383.1	10.48

The vapour pressure of carbon tetrachloride at 321 K is 37.6 kN/m^2 and the density of the liquid is 1540 kg/m^3. Take the kilogram molecular volume as 22.4 m^3.

10.15. Ammonia is absorbed in water from a mixture with air using a column operating at atmospheric pressure and 295 K. The resistance to transfer can be regarded as lying entirely within the gas phase. At a point in the column the partial pressure of the ammonia is 6.6 kN/m^2. The back pressure at the water interface is negligible and the resistance to transfer may be regarded as lying in a stationary gas film 1 mm thick. If the diffusivity of ammonia in air is 0.236 cm^2/s, what is the transfer rate per unit area at that point in the column? If the gas were compressed to 200 kN/m^2 pressure, how would the transfer rate be altered?

10.16. What are the general principles underlying the two-film, penetration and film-penetration theories for mass transfer across a phase boundary? Give the basic differential equations which have to be solved for these theories with the appropriate boundary conditions.

According to the penetration theory, the instantaneous rate of mass transfer per unit area (N_A) at some time t after the commencement of transfer is given by:

$$(N_A)_t = \Delta C\sqrt{\frac{D}{\pi t}},$$

where ΔC is the concentration force and D is the diffusivity.

Obtain expressions for the average rates of transfer on the basis of the Higbie and Danckwerts assumptions.

10.17. A solute diffuses from a liquid surface at which its molar concentration is C_i into a liquid with which it reacts. The mass transfer rate is given by Fick's law and the reaction is first order with respect to the solute. In a steady-state process the diffusion rate falls at a depth L to one half the value at the interface. Obtain an expression for the concentration C of solute at a depth z from the surface in terms of the molecular diffusivity D and the reaction rate constant $\boldsymbol{k}$. What is the molar flux at the surface?

10.18. 4 cm^3 of mixture formed by adding 2 cm^3 of acetone to 2 cm^3 of dibutyl phthalate is contained in a 6 mm diameter vertical glass tube immersed in a thermostat maintained at 315 K. A stream of air at 315 K and atmospheric pressure is passed over the open top of the tube to maintain a zero partial pressure of acetone vapour at that point. The liquid level is initially 11.5 mm below the top of the tube and the acetone vapour is transferred to the air stream by molecular diffusion alone. The dibutyl phthalate can be regarded as completely non-volatile and the partial pressure of acetone vapour may be calculated from Raoult's law on the assumption that the density of dibutyl phthalate is sufficiently greater than that of acetone for the liquid to be completely mixed.

Calculate the time taken for the liquid level to fall to 5 cm below the top of the tube, neglecting the effects of bulk flow in the vapour.

Kilogram molecular volume = 22.4 m^3
Molecular weights of acetone, dibutyl phthalate = 58 and 278 kg/kmol respectively.
Liquid densities of acetone, dibutyl phthalate = 764 and 1048 kg/m^3 respectively.
Vapour pressure of acetone at 315 K = 60.5 kN/m^2.
Diffusivity of acetone vapour in air at 315 K = 1.23×10^{-5} m^2/s.

10.19. A crystal is suspended in fresh solvent and 5% of the crystal dissolves in 300 s. How long will it take before 10% of the crystal has dissolved? Assume that the solvent can be regarded as infinite in extent, that the mass transfer in the solvent is governed by Fick's second law of diffusion and may be represented as a unidirectional process, and that changes in the surface area of the crystal may be neglected. Start your derivations using Fick's second law.

10.20. In a continuous steady state reactor, a slightly soluble gas is absorbed into a liquid in which it dissolves and reacts, the reaction being second order with respect to the dissolved gas. Calculate the reaction rate constant on the assumption that the liquid is semi-infinite in extent and that mass transfer resistance in the gas phase is negligible. The diffusivity of the gas in the liquid is 10^{-8} m^2/s, the gas concentration in the liquid falls to one half of its value in the liquid over a distance of 1 mm, and the rate of absorption at the interface is 4×10^{-6} kmol/m^2 s.

10.21. Experiments have been carried out on the mass transfer of acetone between air and a laminar water jet. Assuming that desorption produces random surface renewal with a constant fractional rate of surface renewal, s, but an upper limit on surface age equal to the life of the jet, τ, show that the surface age frequency distribution function, $\phi(t)$, for this case is given by:

$$\phi(t) = s\exp\frac{-st}{1-\exp(-st)} \quad \text{for} \quad 0 < t < \tau$$

$$\phi(t) = 0 \quad \text{for} \quad t > \tau.$$

Hence, show that the enhancement, E, for the increase in value of the liquid-phase mass transfer coefficient is:

$$E = \frac{(\pi s\tau)^{1/2}\ \text{erf}\ (s\tau)^{1/2}}{2[1-\exp(-s\tau)]}$$

where E is defined as the ratio of the mass transfer coefficient predicted by the conditions described above to the mass transfer coefficient obtained from the penetration theory for a jet with an undisturbed surface. Assume that the interfacial concentration of acetone is practically constant.

10.22. Solute gas is diffusing into a stationary liquid, virtually free of solvent, and of sufficient depth for it to be regarded as semi-infinite in extent. In what depth of fluid below the surface will 90% of the material which has been transferred across the interface have accumulated in the first minute?

Diffusivity of gas in liquid = 10^{-9} m^2/s.

10.23. A chamber, of volume 1 m^3, contains air at a temperature of 293 K and a pressure of 101.3 kN/m^2, with a partial pressure of water vapour of 0.8 kN/m^2. A bowl of liquid with a free surface of 0.01 m^2 and maintained at a temperature of 303 K is introduced into the chamber. How long will it take for the air to become 90% saturated at 293 K and how much water must be evaporated?

The diffusivity of water vapour in air is 2.4×10^{-5} m^2/s and the mass transfer resistance is equivalent to that of a stagnant gas film of thickness 0.25 mm. Neglect the effects of bulk flow.

Saturation vapour pressure of water = 4.3 kN/m^2 at 303 K and 2.3 kN/m^2 at 293 K.

10.24. A large deep bath contains molten steel, the surface of which is in contact with air. The oxygen concentration in the bulk of the molten steel is 0.03% by mass and the rate of transfer of oxygen from the air is sufficiently high to maintain the surface layers saturated at a concentration of 0.16% by weight. The surface of the liquid is disrupted by gas bubbles rising to the surface at a frequency of 120 bubbles per m^2 of surface per second, each bubble disrupts and mixes about 15 cm^2 of the surface layer into the bulk.

On the assumption that the oxygen transfer can be represented by a surface renewal model, obtain the appropriate equation for mass transfer by starting with Fick's second law of diffusion and calculate:

(a) The mass transfer coefficient
(b) The mean mass flux of oxygen at the surface
(c) The corresponding film thickness for a film model, giving the same mass transfer rate.

Diffusivity of oxygen in steel = 1.2×10^{-8} m^2/s
Density of molten steel = 7100 kg/m^3

10.25. Two large reservoirs of gas are connected by a pipe of length $2L$ with a full-bore valve at its mid-point. Initially a gas A fills one reservoir and the pipe up to the valve and gas B fills the other reservoir and the remainder of the pipe. The valve is opened rapidly and the gases in the pipe mix by molecular diffusion.

Obtain an expression for the concentration of gas A in that half of the pipe in which it is increasing, as a function of distance y from the valve and time t after opening. The whole system is at a constant pressure and the ideal gas law is applicable to both gases. It may be assumed that the rate of mixing in the vessels is high so that the gas concentration at the two ends of the pipe do not change.

10.26. A pure gas is absorbed into a liquid with which it reacts. The concentration in the liquid is sufficiently low for the mass transfer to be governed by Fick's law and the reaction is first order with respect to the solute gas. It may be assumed that the film theory may be applied to the liquid and that the concentration of solute gas falls from the saturation value to zero across the film. Obtain an expression for the mass transfer rate across the gas-liquid interface in terms of the molecular diffusivity, D, the first order reaction rate constant k, the film thickness L and the concentration C_{AS} of solute in a saturated solution. The reaction is initially carried out at 293 K. By what factor will the mass transfer rate across the interface change, if the temperature is raised to 313 K?

Reaction rate constant at 293 K = 2.5×10^{-6} s^{-1}
Energy of activation for reaction (in Arrhenius equation)
= 26,430 kJ/kmol
Universal gas constant **R** = 8314 J/kmol K
Molecular diffusivity D = 10^{-9} m^2/s
Film thickness, L = 10 mm
Solubility of gas at 313 K is 80% of solubility at 293 K.

10.27. Using Maxwell's law of diffusion obtain an expression for the effective diffusivity for a gas **A** in a binary mixture of **B** and **C**, in terms of the diffusivities of **A** in the two pure components and the molar concentrations of **A**, **B** and **C**.

Carbon dioxide is absorbed in water from a 25 per cent mixture in nitrogen. How will its absorption rate compare with that from a mixture containing 35 per cent carbon dioxide, 40 per cent hydrogen and 25 per cent nitrogen? It may be assumed that the gas-film resistance is controlling, that the partial pressure of carbon dioxide at the gas-liquid interface is negligible and that the two-film theory is applicable, with the gas film thickness the same in the two cases.

Diffusivity of CO_2: in hydrogen 3.5×10^{-5} m^2/s; in nitrogen 1.6×10^{-5} m^2/s.

10.28. Given that, from the penetration theory for mass transfer across an interface, the instantaneous rate of mass transfer is inversely proportional to the square root of the time of exposure, obtain a relationship between exposure time in the Higbie model and surface renewal rate in the Danckwerts model which will give the same average mass transfer rate. The age distribution function and average mass transfer rate from the Danckwerts theory must be derived from first principles.

10.29. Ammonia is absorbed in a falling film of water in an absorption apparatus and the film is disrupted and mixed at regular intervals as it flows down the column. The mass transfer rate is calculated from the penetration theory on the assumption that all the relevant conditions apply. It is found from measurements that the mass transfer rate immediately before mixing is only 16 per cent of that calculated from the theory and the difference has been attributed to the existence of a surface film which remains intact and unaffected by the mixing process. If the liquid mixing process takes place every second, what thickness of surface film would account for the discrepancy?

Diffusivity of ammonia in water $= 1.76 \times 10^{-9}$ m^2/s

10.30. A deep pool of ethanol is suddenly exposed to an atmosphere consisting of pure carbon dioxide and unsteady state mass transfer, governed by Fick's Law, takes place for 100 s. What proportion of the absorbed carbon dioxide will have accumulated in the 1 mm thick layer of ethanol closest to the surface?

Diffusivity of carbon dioxide in ethanol $= 4 \times 10^{-9}$ m^2/s.

10.31. A soluble gas is absorbed into a liquid with which it undergoes a second-order irreversible reaction. The process reaches a steady-state with the surface concentration of reacting material remaining constant at C_{As} and the depth of penetration of the reactant being small compared with the depth of liquid which can be regarded as infinite in extent. Derive the basic differential equation for the process and from this derive an expression for the concentration and mass transfer rate (moles per unit area and unit time) as a function of depth below the surface. Assume that mass transfer is by molecular diffusion.

If the surface concentration is maintained at 0.04 kmol/m^3, the second-order rate constant k_2 is 9.5×10^3 m^3/kmol.s and the liquid phase diffusivity D is 1.8×10^{-9} m^2/s, calculate:

(a) The concentration at a depth of 0.1 mm.
(b) The molar rate of transfer at the surface (kmol/m^{2}s).
(c) The molar rate of transfer at a depth of 0.1 mm.

It may be noted that if:

$$\frac{dC_A}{dy} = q, \text{ then } \frac{d^2C_A}{dy^2} = q\frac{dq}{dC_A}$$

10.32 In calculating the mass transfer rate from the *penetration theory*, two models for the age distribution of the surface elements are commonly used—those due to Higbie and to Danckwerts. Explain the difference between the two models and give examples of situations in which each of them would be appropriate.

In the Danckwerts model, it is assumed that elements of the surface have an age distribution ranging from zero to infinity. Obtain the age distribution function for this model and apply it to obtain the average mass transfer coefficient at the surface, given that from the penetration theory the mass transfer coefficient for surface of age t is $\sqrt{[D/(\pi t]}$, where D is the diffusivity.

If for unit area of surface the surface renewal rate is s, by how much will the mass transfer coefficient be changed if no surface has an age exceeding $2/s$?

If the probability of surface renewal is linearly related to age, as opposed to being constant, obtain the corresponding form of the age distribution function.

It may be noted that:

$$\int_0^{\infty} e^{-x^2}\,dx = \frac{\sqrt{\pi}}{2}$$

10.33. Explain the basis of the *penetration theory* for mass transfer across a phase boundary. What are the assumptions in the theory which lead to the result that the mass transfer rate is inversely proportional to the square root of the time for which a surface element has been expressed? (Do *not* present a solution of the differential equation.) Obtain the age distribution function for the surface:

(a) On the basis of the Danckwerts' assumption that the probability of surface renewal is independent of its age.
(b) On the assumption that the probability of surface renewal increases linearly with the age of the surface.

Using the Danckwerts surface renewal model, estimate:

(a) At what age of a surface element is the mass transfer rate equal to the mean value for the whole surface for a surface renewal rate (s) of 0.01 m^2/m^2s?
(b) For what proportion of the total mass transfer is surface of an age exceeding 10 seconds responsible?

10.34. At a particular location in a distillation column, where the temperature is 350 K and the pressure 500 m Hg, the mol fraction of the more volatile component in the vapour is 0.7 at the interface with the liquid and 0.5 in the bulk of the vapour. The molar latent heat of the more volatile component is 1.5 times that of the less volatile. Calculate the mass transfer rates (kmol $m^{-2}s^{-1}$) of the two components. The resistance to mass transfer in the vapour may be considered to lie in a stagnant film of thickness 0.5 mm at the interface. The diffusivity in the vapour mixture is 2×10^{-5} m^2s^{-1}.

Calculate the mol fractions and concentration gradients of the two components at the mid-point of the film. Assume that the ideal gas law is applicable and that the Universal Gas Constant $\mathbf{R} = 8314$ J/kmol K.

10.35. For the diffusion of carbon dioxide at atmospheric pressure and a temperature of 293 K, at what time will the concentration of solute 1 mm below the surface reach 1 per cent of the value at the surface? At that time, what will the mass transfer rate (kmol $m^{-2}s^{-1}$) be:

(a) At the free surface?
(b) At the depth of 1 mm?

The diffusivity of carbon dioxide in water may be taken as 1.5×10^{-9} m^2s^{-1}. In the literature, Henry's law constant K for carbon dioxide at 293 K is given as 1.08×10^6 where $K = P/X$, P being the partial pressure of carbon dioxide (mm Hg) and X the corresponding mol fraction in the water.

10.36. Experiments are carried out at atmospheric pressure on the absorption into water of ammonia from a mixture of hydrogen and nitrogen, both of which may be taken as insoluble in the water. For a constant mol fraction of 0.05 of ammonia, it is found that the absorption rate is 25 per cent higher when the molar ratio of hydrogen to nitrogen is changed from 1 : 1 to 4 : 1. Is this result consistent with the assumption of a steady-state gas-film controlled process and, if not, what suggestions have you to make to account for the discrepancy?

Neglect the partial pressure attributable to ammonia in the bulk solution.
Diffusivity of ammonia in hydrogen $= 52 \times 10^{-6}$ m^2s^{-1}
Diffusivity of ammonia in nitrogen $= 23 \times 10^{-6}$ m^2s^{-1}

10.37. Using a steady-state film model, obtain an expression for the mass transfer rate across a laminar film of thickness L in the vapour phase for the more volatile component in a binary distillation process:

(a) where the molar latent heats of two components are equal,
(b) where the molar latent heat of the less volatile component (LVC) is f times that of the more volatile component (MVC).

For the case where the ratio of the molar latent heats f is 1.5, what is the ratio of the mass transfer rate in case (b) to that in case (a) when the mole fraction of the MVC falls from 0.75 to 0.65 across the laminar film?

10.38. On the assumptions involved in the penetration theory of mass transfer across a phase boundary, the concentration C_A of a solute A at a depth y below the interface at a time t after the formation of the interface is given by:

$$\frac{C_A}{C_{Ai}} = \text{erfc}\left[\frac{y}{2\sqrt{(Dt)}}\right]$$

where C_{Ai} is the interface concentration, assumed constant and D is the molecular diffusivity of the solute in the solvent. The solvent initially contains no dissolved solute. Obtain an expression for the molar rate of transfer of A per unit area at time t and depth y, and at the free surface (at $y = 0$).

In a liquid–liquid extraction unit, spherical drops of solvent of uniform size are continuously fed to a continuous phase of lower density which is flowing vertically upwards, and hence countercurrently with respect to the droplets. The resistance to mass transfer may be regarded as lying wholly within the drops and the penetration theory may be applied. The upward velocity of the liquid, which may be taken as uniform over the cross-section of the vessel, is one-half of the terminal falling velocity of the droplets in the still liquid.

Occasionally, two droplets coalesce on formation giving rise to a single drop of twice the volume. What is the ratio of the mass transfer rate (kmol/s) to a coalesced drop to that of a single droplet when each has fallen the same distance, that is to the bottom of the equipment?

The fluid resistance force acting on the droplet should be taken as that given by Stokes' law, that is $3\pi\mu du$ where μ is the viscosity of the continuous phase, d the drop diameter and u its velocity relative to the continuous phase.

It may be noted that:

$$\text{erfc}(x) = \frac{2}{\sqrt{\pi}}\int_x^{\infty} e^{-x^2}\,dx.$$

10.39. In a drop extractor, a dense organic solvent is introduced in the form of spherical droplets of diameter d and extracts a solute from an aqueous stream which flows upwards at a velocity u_o equal to half the terminal falling velocity of the droplets. On increasing the flowrate of the aqueous stream by 50 per cent, whilst maintaining the solvent rate constant, it is found that the average concentration of solute in the outlet stream of organic phase is decreased by 10 per cent. By how much would the effective droplet size have had to change to account for this reduction in concentration? Assume that the penetration theory is applicable with the mass transfer coefficient inversely proportional to the square root of the contact time between the phases and the continuous phase resistance small compared with that in the droplets. The drag force F acting on the falling droplets may be calculated from Stokes' Law, $F = 3\pi\mu du_o$, where μ is the viscosity of the aqueous phase. Clearly state any assumptions made in your calculation.

10.40. According to the penetration theory for mass transfer across an interface, the ratio of the concentration C_A at a depth y and time t to the surface concentration C_{As} if the liquid is initially free of solute, is given by:

$$\frac{C_A}{C_{As}} = \text{erfc}\frac{y}{2\sqrt{Dt}}$$

where D is the diffusivity. Obtain a relation for the instantaneous rate of mass transfer at time t both at the surface ($y = 0$) and at a depth y.

What proportion of the total solute transferred into the liquid in the first 90 s of exposure will be retained in a 1 mm layer of liquid at the surface and what proportion will be retained in the next 0.5 mm? Take the diffusivity as 2×10^{-9} m^2/s.

10.41. Obtain an expression for the effective diffusivity of component A in a gaseous mixture of A, B and C, in terms of the binary diffusion coefficients D_{AB} for A in B, and D_{AC} for A in C.

The gas phase mass transfer coefficient for the absorption of ammonia into water from a mixture of composition NH_3 20%, N_2 73%, H_2 7% is found experimentally to be 0.030 m/s. What would you expect the transfer coefficient to be for a mixture of composition NH_3 5%, N_2 60%, H_2 35%? All compositions are given on a molar basis. The total pressure and temperature are the same in both cases. The transfer coefficients are based on a steady-state film model and the effective film thickness may be assumed constant. Neglect the solubility of N_2 and H_2 in water.

Diffusivity of NH_3 in $N_2 = 23 \times 10^{-6}$ m^2/s.
Diffusivity of NH_3 in $H_2 = 52 \times 10^{-6}$ m^2/s.

10.42. State the assumptions made in the penetration theory for the absorption of a pure gas into a liquid. The surface of an initially solute-free liquid is suddenly exposed to a soluble gas and the liquid is sufficiently deep for no solute to have time to reach the bottom of the liquid. Starting with Fick's second law of diffusion obtain an expression for (i) the concentration, and (ii) the mass transfer rate at a time t and a depth y below the surface.

After 50 s, at what depth y will the concentration have reached one tenth the value at the surface? What is the mass transfer rate (i) at the surface, and (ii) at the depth y, if the surface concentration has a constant value of 0.1 kmol/m^3?

10.43. In a drop extractor, liquid droplets of approximate uniform size and spherical shape are formed at a series of nozzles and rise countercurrently through the continuous phase which is flowing downwards at a velocity equal to one half of the terminal rising velocity of the droplets. The flowrates of both phases are then increased by 25 per cent. Because of the greater shear rate at the nozzles, the mean diameter of the droplets is however only 90 per cent of the original value. By what factor will the overall mass transfer rate change?

It may be assumed that the penetration model may be used to represent the mass transfer process. The depth of penetration is small compared with the radius of the droplets and the effects of surface curvature may be neglected. From the penetration theory, the concentration C_A at a depth y below the surface at time t is given by:

$$\frac{C_A}{C_{As}} = \text{erfc}\left[\frac{y}{2\sqrt{(Dt)}}\right] \quad \text{where} \quad \text{erfc } X = \frac{2}{\sqrt{\pi}}\int_X^\infty e^{-x^2}\text{d}x$$

where C_{As} is the surface concentration for the drops (assumed constant) and D is the diffusivity in the dispersed (droplet) phase. The droplets may be assumed to rise at their terminal velocities and the drag force F on the droplet may be calculated from Stokes' Law, $F = 3\pi\mu du$.

10.44. According to Maxwell's law, the partial pressure gradient in a gas which is diffusing in a two-component mixture is proportional to the product of the molar concentrations of the two components multiplied by its mass transfer velocity relative to that of the second component. Show how this relationship can be adapted to apply to the absorption of a soluble gas from a multicomponent mixture in which the other gases are insoluble and obtain an effective diffusivity for the multicomponent system in terms of the binary diffusion coefficients.

Carbon dioxide is absorbed in alkaline water from a mixture consisting of 30% CO_2 and 70% N_2 and the mass transfer rate is 0.1 kmol/s. The concentration of CO_2 in the gas in contact with the water is effectively zero. The gas is then mixed with an equal molar quantity of a second gas stream of molar composition 20% CO_2, 50%, N_2 and 30% H_2. What will be the new mass transfer rate, if the surface area, temperature and pressure remain unchanged? It may be assumed that a steady-state film model is applicable and that the film thickness is unchanged.

Diffusivity of CO_2 in $N_2 = 16 \times 10^{-6}$ m^2/s.
Diffusivity of CO_2 in $H_2 = 35 \times 10^{-6}$ m^2/s.

10.45. What is the penetration theory for mass transfer across a phase boundary? Give details of the underlying assumptions.

From the penetration theory, the mass transfer rate per unit area N_A is given in terms of the concentration difference ΔC_A between the interface and the bulk fluid, the molecular diffusivity D and the age t of the surface element by:

$$N_A = \sqrt{\frac{D}{\pi t}}\Delta C_A \qquad \text{kmol/m}^2\text{s (in SI units)}$$

What is the mean rate of transfer if all elements of the surface are exposed for the same time t_e before being remixed with the bulk?

Danckwerts assumed a random surface renewal process in which the probability of surface renewal is independent of its age. If s is the fraction of the total surface renewed per unit time, obtain the age distribution function for the surface and show that the mean mass transfer rate N_A over the whole surface is:

$$N_A = \sqrt{Ds}\Delta C_A \qquad \text{(kmol/m}^2\text{s, in SI units)}$$

In a particular application, it is found that the older surface is renewed more rapidly than the recently formed surface, and that after a time $1/s$, the surface renewal rate doubles, that is increases from s to $2s$. Obtain the new age distribution function.

10.46. Derive the partial differential equation for unsteady-state unidirectional diffusion accompanied by an nth-order chemical reaction (rate constant $\boldsymbol{k}$):

$$\frac{\partial C_A}{\partial t} = D\frac{\partial^2 C_A}{\partial y^2} - \boldsymbol{k}C_A{}^n$$

where C_A is the molar concentration of reactant at position y at time t.

Explain why, when applying the equation to reaction in a porous catalyst particle, it is necessary to replace the molecular diffusivity D by an effective diffusivity D_e.

Solve the above equation for a first-order reaction under steady-state conditions, and obtain an expression for the mass transfer rate per unit area at the surface of a catalyst particle which is in the form of a thin platelet of thickness $2L$.

Explain what is meant by the effectiveness factor η for a catalyst particle, and show that it is equal to $\frac{1}{\phi}\tanh\phi$ for the platelet referred to previously, where ϕ is the Thiele modulus, $L\sqrt{\frac{\boldsymbol{k}}{D_e}}$

For the case where there is a mass transfer resistance in the fluid external to the particle (mass transfer coefficient h_D), express the mass transfer rate in terms of the bulk concentration C_{Ao}, rather than the concentration C_{AS} at the surface of the particle.

For a bed of catalyst particles in the form of flat platelets it is found that the mass transfer rate is increased by a factor of 1.2 if the velocity of the external fluid is doubled. The mass transfer coefficient h_D is proportional to the velocity raised to the power of 0.6. What is the value of h_D at the original velocity?

$$\boldsymbol{k} = 1.6 \times 10^{-3}s^{-1}, D_e = 10^{-8}\text{m}^2\text{s}^{-1} \text{ catalyst thickness } (2L) = 10 \text{ mm}$$

10.47. Explain the basic concepts underlying the two-film theory for mass transfer across a phase boundary, and obtain an expression for film thickness.

Water evaporates from an open bowl at 349 K at the rate of 4.11×10^{-3} kg/m^{2}s. What is the effective gas-film thickness?

The water is replaced by ethanol at 343 K. What will be its rate of evaporation (in kg/m^{2}s) if the film thickness is unchanged?

What proportion of the total mass transfer will be attributable to bulk flow?

Data:

Vapour pressure of water at 349 K = 301 mm Hg
Vapour pressure of ethanol at 343 K = 541 mm Hg
Neglect the partial pressure of vapour in the surrounding atmosphere
Diffusivity of water vapour in air = 26×10^{-6} m^2/s
Diffusivity of ethanol in air = 12×10^{-6} m^2/s
Density of mercury = 13,600 kg/m^3
Universal gas constant $\mathbf{R}$ = 8314 J/kmol K

11.1. Calculate the thickness of the boundary layer at a distance of 75 mm from the leading edge of a plane surface over which water is flowing at the rate of 3 m/s. Assume that the flow in the boundary layer is streamline and that the velocity u of the fluid at a distance y from the surface may be represented by the relation $u = a + by + cy^2 + dy^3$, where the coefficients a, b, c, and d are independent of y. Take the viscosity of water as 1 mN s/m^2.

11.2. Water flows at a velocity of 1 m/s over a plane surface 0.6 m wide and 1 m long. Calculate the total drag force acting on the surface if the transition from streamline to turbulent flow in the boundary layer occurs when the Reynolds group $Re_x = 10^5$.

11.3. Calculate the thickness of the boundary layer at a distance of 150 mm from the leading edge of a surface over which oil, of viscosity 50 mN s/m^2 and density 990 kg/m^3, flows with a velocity of 0.3 m/s. What is the displacement thickness of the boundary layer?

11.4. Calculate the thickness of the laminar sub-layer when benzene flows through a pipe of 50 mm diameter at 0.003 m^3/s. What is the velocity of the benzene at the edge of the laminar sub-layer? Assume fully developed flow exists within the pipe.

11.5. Air is flowing at a velocity of 5 m/s over a plane surface. Derive an expression for the thickness of the laminar sub-layer and calculate its value at a distance of 1 m from the leading edge of the surface.

Assume that within the boundary layer outside the laminar sub-layer the velocity of flow is proportional to the one-seventh power of the distance from the surface and that the shear stress R at the surface is given by:

$$\frac{R}{\rho u_s^2} = 0.03\left(\frac{u_s \rho x}{\mu}\right)^{-0.2},$$

where ρ is the density of the fluid (1.3 kg/m^3 for air), μ the viscosity of the fluid (17×10^{-6} N s/m^2 for air), u_s the stream velocity (m/s), and x the distance from the leading edge (m).

11.6. Obtain the momentum equation for an element of boundary layer. If the velocity profile in the laminar region may be represented approximately by a sine function, calculate the boundary-layer thickness in terms of distance from the leading edge of the surface.

11.7. Explain the concepts of "momentum thickness" and "displacement thickness" for the boundary layer formed during flow over a plane surface. Develop a similar concept to displacement thickness in relation to heat flux across the surface for laminar flow and heat transfer by thermal conduction, for the case where the surface has a constant temperature and the thermal boundary layer is always thinner than the velocity boundary layer. Obtain an expression for this "thermal thickness" in terms of the thicknesses of the velocity and temperature boundary layers.

Similar forms of cubic equations may be used to express velocity and temperature variations with distance from the surface.

For a Prandtl number, Pr, less than unity, the ratio of the temperature to the velocity boundary layer thickness is equal to $Pr^{-1/3}$. Work out the "thermal thickness" in terms of the thickness of the velocity boundary layer for a value of $Pr = 0.7$.

11.8. Explain why it is necessary to use concepts, such as the displacement thickness and the momentum thickness, for a boundary layer in order to obtain a boundary layer thickness which is largely independent of the approximation used for the velocity profile in the neighbourhood of the surface.

It is found that the velocity u at a distance y from the surface may be expressed as a simple power function ($u \propto y^n$) for the turbulent boundary layer at a plane surface. What is the value of n if the ratio of the momentum thickness to the displacement thickness is 1.78?

11.9. Derive the momentum equation for the flow of a fluid over a plane surface for conditions where the pressure gradient along the surface is negligible. By assuming a sine function for the variation of velocity with distance from the surface (within the boundary layer) for streamline flow, obtain an expression for the boundary layer thickness as a function of distance from the leading edge of the surface.

11.10. Derive the momentum equation for the flow of a viscous fluid over a small plane surface.

Show that the velocity profile in the neighbourhood of the surface may be expressed as a sine function which satisfies the boundary conditions at the surface and at the outer edge of the boundary layer.

Obtain the boundary layer thickness and its displacement thickness as a function of the distance from the leading edge of the surface, when the velocity profile is expressed as a sine function.

11.11. Derive the momentum equation for the flow of a fluid over a plane surface for conditions where the pressure gradient along the surface is negligible. By assuming a sine function for the variation of velocity with distance from the surface (within the boundary layer) for streamline flow, obtain an expression for the boundary layer thickness as a function of distance from the leading edge of the surface.

11.12. Derive the momentum equation for the flow of a viscous fluid over a small plane surface. Show that the velocity profile in the neighbourhood of the surface may be expressed as a sine function which satisfies the boundary conditions at the surface and at the outer edge of the boundary layer.

Obtain the boundary layer thickness and its displacement thickness as a function of the distance from the leading edge of the surface, when the velocity profile is expressed as a sine function.

12.1. If the temperature rise per metre length along a pipe carrying air at 12.2 m/s is 66 deg K, what will be the corresponding pressure drop for a pipe temperature of 420 K and an air temperature of 310 K?

The density of air at 310 K is 1.14 kg/m^3.

12.2. It is required to warm a quantity of air from 289 to 313 K by passing it through a number of parallel metal tubes of inner diameter 50 mm maintained at 373 K. The pressure drop must not exceed 250 N/m^2. How long should the individual tubes be?

The density of air at 301 K is 1.19 kg/m^3 and the coefficients of heat transfer by convection from tube to air are 45, 62, and 77 W/m^{2}K for velocities of 20, 24, and 30 m/s at 301 K respectively.

12.3. Air at 330 K, flowing at 10 m/s, enters a pipe of inner diameter 25 mm, maintained at 415 K. The drop of static pressure along the pipe is 80 N/m^2 per metre length. Using the Reynolds analogy between heat transfer and friction, estimate the temperature of the air 0.6 m along the pipe.

12.4. Air flows at 12 m/s through a pipe of inside diameter 25 mm. The rate of heat transfer by convection between the pipe and the air is 60 W/m^{2}K. Neglecting the effects of temperature variation, estimate the pressure drop per metre length of pipe.

12.5. Air at 320 K and atmospheric pressure is flowing through a smooth pipe of 50 mm internal diameter and the pressure drop over a 4 m length is found to be 1.5 kN/m^2. Using Reynolds analogy, by how much would you expect the air temperature to fall over the first metre of pipe length if the wall temperature there is kept constant at 295 K?

$$\text{Viscosity of air} = 0.018 \text{ mN s/m}^2.$$
$$\text{Specific heat of air} = 1.05 \text{ kJ/kg K}.$$

12.6. Obtain an expression for the simple Reynolds analogy between heat transfer and friction. Indicate the assumptions which are made in the derivation and the conditions under which you would expect the relation to be applicable.

The Reynolds number of a gas flowing at 2.5 kg/m^{2}s through a smooth pipe is 20,000. If the specific heat of the gas at constant pressure is 1.67 kJ/kg K, what will the heat transfer coefficient be?

12.7. Explain Prandtl's concept of a "mixing length". What parallels can you draw between the mixing length and the mean free path of the molecules in a gas?

The ratio of the mixing length to the distance from the pipe wall has a constant value of 0.4 for the turbulent flow of a fluid in a pipe. What is the value of the pipe friction factor if the ratio of the mean velocity to the axial velocity is 0.8?

12.8. The velocity profile in the neighbourhood of a surface for a Newtonian fluid may be expressed in terms of a dimensionless velocity u^+ and a dimensionless distance y^+ from the surface. Obtain the relation between u^+ and y^+ in the laminar sub-layer. Outside the laminar sub-layer, the relation is:

$$u^+ = 2.5 \ln y^+ + 5.5$$

At what value of y^+ does the transition from the laminar sub-layer to the turbulent zone occur?

In the "Universal Velocity Profile", the laminar sub-layer extends to values of $y^+ = 5$ and the turbulent zone starts at $y^+ = 30$ and the range $5 < y^+ < 30$, the buffer layer, is covered by a second linear relation between u^+ and $\ln y^+$. What is the *maximum* difference between the values of u^+, in the range $5 < y^+ < 30$, using the two methods of representation of the velocity profile?

$$\text{Definitions: } u^+ = \frac{u_x}{u^*}$$

$$y^+ = \frac{yu^*\rho}{\mu}$$

$$u^{*2} = \frac{R}{\rho}$$

where u_x is the velocity at distance y from surface,
R is the wall shear stress, and
ρ, μ are the density, the viscosity of fluid respectively.

12.9. Calculate the rise in temperature of water passed at 4 m/s through a smooth 25 mm diameter pipe, 6 m long. The water enters at 300 K and the temperature of the wall of the tube can be taken as approximately constant at 330 K. Use:

(a) The simple Reynolds analogy.
(b) The Taylor-Prandtl modification.
(c) The buffer lays equation
(d) $Nu = 0.023Re^{0.8}Pr^{0.33}$.

Comment on the differences in the results so obtained.

12.10. Calculate the rise in temperature of a stream of air, entering at 290 K and passing at 4 m/s through the tube maintained at 350 K, other conditions remaining the same as detailed in Problem 12.9.

12.11. Air flows through a smooth circular duct of internal diameter 0.25 m at an average velocity of 15 m/s. Calculate the fluid velocity at points 50 mm and 5 mm from the wall. What will be the thickness of the laminar sub-layer if this extends to $u^+ = y^+ = 5$? The density of the air may be taken as 1.12 kg/m^3 and the viscosity as 0.02 mN s/m^2.

12.12. Obtain the Taylor-Prandtl modification of the Reynolds Analogy for momentum and heat transfer, and give the corresponding relation for mass transfer (no bulk flow).

An air stream at approximately atmospheric temperature and pressure and containing a low concentration of carbon disulphide vapour is flowing at 38 m/s through a series of 50 mm diameter tubes. The inside of the tubes is covered with a thin film of liquid and both heat and mass transfer are taking place between the gas stream and the liquid film. The film heat transfer coefficient is found to be 100 W/m^{2}K. Using a pipe friction chart and assuming the tubes to behave as smooth surfaces, calculate:

(a) the film mass transfer coefficient, and
(b) the gas velocity at the interface between the laminar sub-layer and the turbulent zone of the gas.

Specific heat of air = 1.0 kJ/kg K
Viscosity of air = 0.02 mN s/m^2
Diffusivity of carbon disulphide vapour in air = 1.1×10^{-5} m^2/s
Thermal conductivity of air 0.024 = W/m K.

12.13. Obtain the Taylor-Prandtl modification of the Reynolds' analogy between momentum and heat transfer and write down the corresponding analogy for mass transfer. For a particular system, a mass transfer coefficient of 8.71×10^{-8} m/s and a heat transfer coefficient of 2730 W/m^2 K were measured for similar flow conditions. Calculate the ratio of the velocity in the fluid where the laminar sub-layer terminates, to the stream velocity.

Molecular diffusivity = 1.5×10^{-9} m^2/s
Viscosity = 1 mN s/m^2
Density = 1000 kg/m^3
Thermal conductivity = 0.48 W/m K
Specific heat capacity = 4.0 kJ/kg K

12.14. Heat and mass transfer are taking place simultaneously to a surface under conditions where the Reynolds analogy between momentum, heat and mass transfer may be applied. The mass transfer is of a single component at a high concentration in a binary mixture and the other component of which undergoes no net transfer. Using the Reynolds analogy, obtain a relation between the coefficients for heat transfer and for mass transfer.

12.15. Derive the Taylor-Prandtl modification of the Reynolds Analogy between momentum and heat transfer.

In a shell and tube condenser, water flows through the tubes which are 10 m long and 40 mm diameter. The pressure drop across the tubes is 5.6 kN/m^2 and the effects of entry and exit losses may be neglected. The tube walls are smooth and flow may be taken as fully developed. The ratio of the velocity at the edge of the laminar sub-layer to the mean velocity of flow may be taken as $2\ Re^{-0.125}$, where Re is the Reynolds number in the pipeline.

If the tube walls are at an approximately constant temperature of 393 K and the inlet temperature of the water is 293 K, estimate the outlet temperature.

Physical properties of water: density 1000 kg/m^3
viscosity 1 mN s/m^2
thermal conductivity 0.6 W/m K
specific heat capacity 4.2 kJ/kg K

12.16. Explain the importance of the universal velocity profile and derive the relation between the dimensionless derivative of velocity u^+, and the dimensionless derivative of distance from the surface y^+, using the concept of Prandtl's mixing length λ_E.

It may be assumed that the fully turbulent portion of the boundary layer starts at $y^+ = 30$, that the ratio of the mixing length λ_E to the distance y from the surface, $\lambda_E/y = 0.4$, and that for a smooth surface $u^+ = 14$ at $y^+ = 30$.

If the laminar sub-layer extends from $y^+ = 0$ to $y^+ = 5$, obtain the equation for the relation between u^+ and y^+ in the buffer zone, and show that the ratio of the eddy viscosity to the molecular viscosity increases linearly from 0 to 5 through this buffer zone.

12.17. Derive the Taylor-Prandtl modification of the Reynolds analogy between heat and momentum transfer and express it in a form in which it is applicable to pipe flow.

If the relationship between the Nusselt number Nu, Reynolds number Re and Prandtl number Pr is:

$$Nu = 0.023Re^{0.8}Pr^{0.33}$$

calculate the ratio of the velocity at the edge of the laminar sub-layer to the velocity at the pipe axis for water ($Pr = 10$) flowing at a Reynolds number (Re) of 10,000 in a smooth pipe. Use the pipe friction chart.

12.18. Obtain a dimensionless relation for the velocity profile in the neighbourhood of a surface for the turbulent flow of a liquid, using Prandtl's concept of a "Mixing Length" (University Velocity Profile). Neglect the existence of the buffer layer and assume that, outside the laminar sub-layer, eddy transport mechanisms dominate. Assume that in the turbulent fluid the mixing length λ_E is equal to 0.4 times the distance y from the surface and that the dimensionless velocity u^+ is equal to 5.5 when the dimensionless distance y^+ is unity.

Show that, if the Blasius relation is used for the shear stress R at the surface, the thickness of the laminar sub-layer δ_b is approximately 1.07 times that calculated on the assumption that the velocity profile in the turbulent fluid is given by Prandtl's one seventh power law.

Blasius Equation:

$$\frac{R}{\rho u_s^2} = 0.0228\left(\frac{u_s\delta\rho}{\mu}\right)^{-0.25}$$

where ρ, μ are the density and viscosity of the fluid,
u_s is the stream velocity, and
δ is the total boundary layer thickness.

12.19. Obtain the Taylor-Prandtl modification of the Reynolds Analogy between momentum transfer and mass transfer (equimolecular counterdiffusion) for the turbulent flow of a fluid over a surface. Write down the corresponding analogy for heat transfer. State clearly the assumptions which are made. For turbulent flow over a surface, the film heat transfer coefficient for the fluid is found to be 4 kW/m^2 K. What would the corresponding value of the mass transfer coefficient be, given the following physical properties?

Diffusivity $D = 5 \times 10^{-9}$ m^2/s
Thermal conductivity $k = 0.6$ W/m K
Specific heat capacity $C_p = 4$ kJ/kg K
Density $\rho = 1000$ kg/m^3
Viscosity $\mu = 1$ mN s/m^2

Assume that the ratio of the velocity at the edge of the laminar sub-layer to the stream velocity is (a) 0.2, (b) 0.6.

Comment on the difference in the two results.

12.20. By using the simple Reynolds Analogy, obtain the relation between the heat transfer coefficient and the mass transfer coefficient for the gas phase for the absorption of a soluble component from a mixture of gases. If the heat transfer coefficient is 100 W/m^2 K, what will the mass transfer coefficient be for a gas of specific heat capacity C_p of 1.5 kJ/kg K and density 1.5 kg/m^3? The concentration of the gas is sufficiently low for bulk flow effects to be negligible.

12.21. The velocity profile in the neighbourhood of a surface for a Newtonian fluid may be expressed in terms of a dimensionless velocity u^+ and a dimensionless distance y^+ from the surface. Obtain the relation between u^+ and y^+ in the laminar sub-layer. Outside the laminar sub-layer, the relation takes the form:

$$u^+ = 2.5 \ln y^+ + 5.5$$

At what value of y^+ does the transition from the laminar sub-layer to the turbulent zone occur?

In the 'Universal Velocity Profile', the laminar sub-layer extends to values of $y^+ = 5$, the turbulent zone starts at $y^+ = 30$ and the range $5 < y^+ < 30$, the buffer layer, is covered by a second linear relation between u^+ and $\ln y^+$. What is the *maximum* difference between the values of u^+, in the range $5 < y^+ < 30$, using the two methods of representation of the velocity profile?

Definitions: $u^+ = \dfrac{u_x}{u^*}$

$$y^+ = \frac{yu^*\rho}{\mu}$$

$$u^{*2} = R/\rho$$

where u_x is velocity at distance y from surface
R is wall shear stress
ρ, μ are the density and viscosity of the fluid, respectively.

12.22. In the Universal Velocity Profile a "dimensionless" velocity u^+ is plotted against $\ln y^+$, where y^+ is a "dimensionless" distance from the surface. For the region where eddy transport dominates (eddy kinematic viscosity $\gg$ kinematic viscosity), the ratio of the mixing length (λ_E) to the distance (y) from the surface may be taken as approximately constant and equal to 0.4. Obtain an expression for du^+/dy^+ in terms of y^+.

In the buffer zone the ratio of du^+/dy^+ to y^+ is *twice* the value calculated above. Obtain an expression for the eddy kinematic viscosity E in terms of the kinematic viscosity (μ/ρ) and y^+. On the assumption that the eddy thermal diffusivity E_H and the eddy kinematic viscosity E are equal, calculate the value of the temperature gradient in a liquid flowing over the surface at $y^+ = 15$ (which lies within the buffer layer) for a surface heat flux of 1000 W/m^2 The liquid has a Prandtl number of 7 and a thermal conductivity of 0.62 W/m K.

12.23. Derive an expression relating the pressure drop for the turbulent flow of a fluid in a pipe to the heat transfer coefficient at the walls on the basis of the simple Reynolds analogy. Indicate the assumptions which are made and the conditions under which to apply closely. Air at 320 K and atmospheric pressure is flowing through a smooth pipe of 50 mm internal diameter, and the pressure drop over a 4 m length is found to be 150 mm water gauge. By how much would be expected the air temperature to fall over the first metre if the wall temperature there is 290 K?

Viscosity of air = 0.018 mN s/m^2.

Specific heat capacity (C_p) = 1.05 kJ/kg K.

Molecular volume = 22.4 m^3/kmol at 1 bar and 273 K.

13.1. In a process in which benzene is used as a solvent, it is evaporated into dry nitrogen. The resulting mixture at a temperature of 297 K and a pressure of 101.3 kN/m^2 has a relative humidity of 60%. It is desired to recover 80% of the benzene present by cooling to 283 K and compressing to a suitable pressure. What must this pressure be?

Vapour pressures of benzene: at 297 K = 12.2 kN/m^2; at 283 K = 6.0 kN/m^2.

13.2. 0.6 m^3/s of gas is to be dried from a dew point of 294 K to a dew point of 277.5 K. How much water must be removed and what will be the volume of the gas after drying?

$$\text{Vapour pressure of water at 294 K} = 2.5 \text{ kN/m}^2.$$
$$\text{Vapour pressure of water at 277.5 K} = 0.85 \text{ kN/m}^2.$$

13.3. Wet material, containing 70% moisture, is to be dried at the rate of 0.15 kg/s in a countercurrent dryer to give a product containing 5% moisture (both on a wet basis). The drying medium consists of air heated to 373 K and containing water vapour equivalent to a partial pressure of 1.0 kN/m^2. The air leaves the dryer at 313 K and 70% saturated. Calculate how much air will be required to remove the moisture. The vapour pressure of water at 313 K may be taken as 7.4 kN/m^2.

13.4. 30,000 m^3 of coal gas (measured at 289 K and 101.3 kN/m^2 saturated with water vapour) is compressed to 340 kN/m^2 pressure, cooled to 289 K and the condensed water is drained off. Subsequently the pressure is reduced to 170 kN/m^2 and the gas is distributed at this pressure and at 289 K. What is the percentage humidity of the gas after this treatment?

The vapour pressure of water at 289 K is 1.8 kN/m^2.

13.5. A rotary countercurrent dryer is fed with ammonium nitrate containing 5% moisture at the rate of 1.5 kg/s, and discharges the nitrate with 0.2% moisture. The air enters at 405 K and leaves at 355 K; the humidity of the entering air being 0.007 kg of moisture per kg of dry air. The nitrate enters at 294 K and leaves at 339 K.

Neglecting radiation losses, calculate the mass of dry air passing through the dryer and the humidity of the air leaving the dryer.

$$\begin{aligned} \text{Latent heat of water at 294 K} &= 2450 \text{ kJ/kg.} \\ \text{Specific heat of ammonium nitrate} &= 1.88 \text{ kJ/kg K.} \\ \text{Specific heat of dry air} &= 0.99 \text{ kJ/kg K.} \\ \text{Specific heat of water vapour} &= 2.01 \text{ kJ/kg K.} \end{aligned}$$

13.6. Material is fed to a dryer at the rate of 0.3 kg/s and the moisture removed is 35% of the wet charge. The stock enters and leaves the dryer at 324 K. The air temperature falls from 341 to 310 K, its humidity rising from 0.01 to 0.02 kg/kg.

Calculate the heat loss to the surroundings.

$$\begin{aligned} \text{Latent heat of water at 324 K} &= 2430 \text{ kJ/kg.} \\ \text{Specific heat of dry air} &= 0.99 \text{ kJ/kg K.} \\ \text{Specific heat of water vapour} &= 2.01 \text{ kJ/kg K.} \end{aligned}$$

13.7. A rotary dryer is fed with sand at the rate of 1 kg/s. The feed is 50% wet and the sand is discharged with 3% moisture. The air enters at 380 K with an solute humidity of 0.007 kg/kg. The wet sand enters at 294 K and leaves at 309 K and the air leaves at 310 K.

Calculate the mass of air passing through the dryer and the humidity of the air leaving the dryer. Allow a radiation loss of 25 kJ/kg of dry air.

$$\begin{aligned} \text{Latent heat of water at 294 K} &= 2450 \text{ kJ/kg.} \\ \text{Specific heat of sand} &= 0.88 \text{ kJ/kg K.} \\ \text{Specific heat of dry air} &= 0.99 \text{ kJ/kg K.} \\ \text{Specific heat of vapour} &= 2.01 \text{ kJ/kg K.} \end{aligned}$$

13.8. Water is to be cooled in a packed tower from 330 to 295 K by means of air flowing countercurrently. The liquid flows at the rate of 275 cm^3/m^2 s and the air at 0.7 m^3/m^2 s. The entering air has a temperature of 295 K and a humidity of 20%. Calculate the required height of tower and the condition of the air leaving at the top.

The whole of the resistance to heat and mass transfer may be considered as being within the gas phase and the product of the mass transfer coefficient and the transfer surface per unit volume of column ($h_D a$) may be taken as 0.2 s^{-1}.

13.9. Water is to be cooled in a small packed column from 330 to 285 K by means of air flowing countercurrently. The rate of flow of liquid is 1400 cm^3/m^2 s and the flow rate of the air, which enters at a temperature of 295 K and a humidity of 60%, is 3.0 m^3/m^2 s. Calculate the required height of tower if the whole of the resistance to heat and mass transfer may be considered as being in the gas phase and the product of the mass transfer coefficient and the transfer surface per unit volume of column is 2 s^{-1}.

What is the condition of the air which leaves at the top?

13.10. Air containing 0.005 kg of water vapour per kg of dry air is heated to 325 K in a dryer and passed to the lower shelves. It leaves these shelves at 60% humidity, reheated to 325 K and passed over another set of shelves, again leaving at 60% humidity. This is again reheated for the third and fourth sets of shelves, after which the air leaves the dryer. On the assumption that the material in each shelf has reached the wet bulb temperature and that heat losses from the dryer may be neglected, determine:

(a) the temperature of the material on each tray;
(b) the rate of water removal if 5 m^3/s of moist air leaves the dryer;
(c) the temperature to which the inlet air would have to be raised to carry out the drying in a single stage.

13.11. 0.08 m^3/s of air at 305 K and 60% humidity is to be cooled to 275 K. Calculate, by use of a psychrometric chart, the amount of heat to be removed for each 10 deg K interval of the cooling process. What total mass of moisture will be deposited? What is the humid heat of the air at the beginning and end of the process?

13.12. A hydrogen stream at 300 K and atmospheric pressure has a dew point of 275 K. It is to be further humidified by adding to it (through a nozzle) saturated stream at 240 kN/m^2 at the rate of 1 kg steam: 30 kg of hydrogen feed. What will be the temperature and humidity of the resultant stream?

13.13. In a countercurrent packed column n-butanol flows down at the rate of 0.25 kg/m^2 s and is cooled from 330 to 295 K. Air at 290 K, initially free of n-butanol vapour, is passed up the column at the rate of 0.7 m^3/m^2 s. Calculate the required height of tower and the condition of the exit air.
Data:

Mass transfer coefficient per unit volume: $h_D a = 0.1\ s^{-1}$

Psychrometric ratio: $\dfrac{h}{h_D \rho_A s} = 2.34$

Heat transfer coefficients: $h_L = 3h_G$

Latent heat of vaporisation of n-butanol, $\lambda = 590$ kJ/kg.

Specific heat of liquid n-butanol: $C_L = 2.5$ kJ/kg K.

Humid heat of gas: $s = 1.05$ kJ/kg K.

Temperature (K)	Vapour pressure of n-butanol (kN/m^2)
295	0.59
300	0.86
305	1.27
310	1.75
315	2.48
320	3.32
325	4.49
330	5.99
335	7.89
340	10.36
345	14.97
350	17.50

13.14. Estimate the height and base diameter of a natural draught hyperbolic cooling tower which will handle a flow of 5000 kg/s water entering at 300 K and leaving at 294 K. The dry-bulb air temperature is 287 K and the ambient wet bulb temperature is 284 K.

Index

LUDWIG. E. E. 509, 528, 565